Cable Television:

A Guide to Federal Regulations

Steven R. Rivkin

The issuance by the Federal Communications Commission of a comprehensive set of standards to govern cable television's future growth and development has prompted the need for an easy-to-understand explanation of the many broad and complex issues covered. This volume, based on research supported by a grant from the National Science Foundation, is designed to fill that need. It describes and analyzes laws and regulations applicable to cable television. It assembles the most relevant documents and clarifies their meanings. It serves as a reference tool for state and local decisionmakers, the cable industry, the legal community, and interested citizen groups. The author is a District of Columbia attorney and consultant to The Rand Corporation. The volume is in three parts: Part one discusses the status of Federal regulations and outlines FCC rules. Part two provides details of cable television policies and rules. Part three provides a compendium of relevant documents.

A Rand Cable Television Study

Rand Cable Television Series

Series Editor: Walter S. Baer
The Rand Corporation

Cable television, a little-known technology still in its infancy, may in time influence the way we live as radically as have the automobile and the telephone.

The Rand Cable Television *Series* represents the most comprehensive compilation of information on the subject yet published. This series will provide guidance to local government officials, educators, members of citizen groups, and all others concerned with the introduction, development, and utilization of cable television within their communities. *The Rand Cable Television Series* is also intended as an introduction to cable television for college and university classes in communications. The series is based on the results of a Rand Corporation study of cable television. The study was supported by a grant from the National Science Foundation to the Rand Communications Policy Program.

The first volume, intended as a handbook for local decisionmaking, provides basic information about cable television. It outlines the political, social, economic, legal, and technological issues a community will face in its decisionmaking.

The second volume provides detailed coverage of franchising considerations such as citizen participation in planning, the process of franchising, citizen participation after the franchise, and a guide to the technology.

The third volume is a guide to federal regulations.

The concluding volume discusses the development of community services and includes sections on public access, applications for municipal services, and uses in education.

Rand Cable Television Series
Walter S. Baer, *Series Editor*

•

Cable Television:
A Handbook for Decisionmaking
Walter S. Baer

Cable Television:
Franchising Considerations
Walter S. Baer, Michael Botein,
Leland L. Johnson, Carl Pilnick,
Monroe E. Price, Robert K. Yin

Cable Television:
A Guide to Federal Regulations
Steven R. Rivkin

Cable Television:
Developing Community Services
Polly Carpenter-Huffman,
Richard C. Kletter,
Robert K. Yin

CABLE TELEVISION:
A Guide to Federal Regulations

Steven R. Rivkin

C R

Crane, Russak & Company, Inc.

NEW YORK

Cable Television:
A Guide to Federal Regulations

Published in the United States by
Crane, Russak & Company, Inc.
347 Madison Avenue
New York, N.Y. 10007

ISBN 0-8448-0259-x

LC 73-90819

Printed in the United States of America

CONTENTS

v

PART III. COMPENDIUM OF RELEVANT DOCUMENTS

Preface

THIS book is the third of four volumes that present the results from a Rand Corporation study of cable television. The study was supported by a grant from the National Science Foundation to the Rand Communications Policy Program.

Rand began its research on cable television issues in 1969, under grants from The Ford Foundation and The John and Mary R. Markle Foundation. The central interest at that time was federal regulatory policy, still in its formative stages. Rand published more than a dozen reports related to that subject over the next three years. This phase of Rand's concern ended in February 1972 when the Federal Communications Commission issued its *Cable Television Report and Order.*

The *Report and Order* marked the end of a virtual freeze on cable development in the major metropolitan areas that had persisted since 1966. It asserted the FCC's authority to regulate cable development, laid down a number of firm requirements and restrictions, and at the same time permitted considerable latitude to communities in drawing up the terms of their franchises. It expressly encouraged communities to innovate, while reserving the authority to approve or disapprove many of their proposed actions.

The major decisions to be made next, and therefore the major focus of new cable research, will be on the local level. These decisions will be crucially important because cable television is no longer a modest technique for improving rural television reception. It is on the brink of turning into a genuine urban communication system, with profound implications for our entire society. Most important, cable systems in the major markets are yet to be built, and many cities feel great pressure to begin issuing franchises. The decisions shortly to be made will reverberate through the 1980s.

Aware of the importance of these events, the National Science Foundation asked Rand in December 1971 to compile a cable handbook for local decisionmaking. The handbook, Volume I in this series, presents basic information about cable television and outlines the political, social, economic, legal, and technological issues a community will face. This book (Volume III) discusses the cable television rules adopted by the Federal Communications Commission in 1972 and other federal regulations that apply to cable. Other volumes explore cable technology and franchising issues (Volume II) and the uses of a cable system for education, local government services, and public access to television (Volume IV).

The entire series is addressed to local government officials, educators, community group members, and other people concerned with the development of cable television in their communities. It also is intended as text and reference material for college and university classes in communications.

ix

The study director, Walter S. Baer, served as editor for the series. The author of this volume is Steven R. Rivkin, a partner in the law firm of Nicholson & Carter, Washington, D.C., and formerly counsel to the Sloan Commission on Cable Communications. Mr. Rivkin is a consultant to The Rand Corporation.

Special thanks are due to Philip R. Hochberg, Esq., of Washington, D.C. for assistance in developing the issues pertaining to copyright and program exclusivity; and to the several reviewers of earlier drafts who provided helpful criticisms. The faithful attention of the secretarial staffs of Nicholson & Carter and The Rand Corporation to details of textual materials that were often unfamiliar is also greatly appreciated.

The views expressed in this book are those of the author and do not necessarily reflect the opinions or policies of the National Science Foundation.

July, 1973

Walter S. Baer
Santa Monica, California

PART I.

SUMMARY AND OVERVIEW

Part I is a general overview of the materials included in this report. Its purpose is to inform the reader about the issues and serve as a point of departure for fuller inquiry in Parts II and III.

PRESENT STATUS OF FEDERAL REGULATION

The 1972 *Cable Television Report and Order* followed more than five years of federal regulatory restrictions on cable television growth. The FCC had imposed this restraint to protect local broadcast stations in major cities from competition with cable systems that picked up television programs from distant places and distributed them locally. The cable TV industry strenuously contested this restraint, considering "distant" signals a necessary inducement, in the form of wider program options, to justify the cost of viewer subscriptions. Pending the outcome of the controversy, cable growth virtually ground to a halt in major cities, while the FCC deliberated its long-range policy. Economic rivalries over cable's future role impeded the resolution of policy issues; and doubts about the scope of FCC's authority embroiled Congress, the Executive Branch, and the courts in the controversy.

It is important to keep in mind that the basic issues are still only partially resolved, since the FCC has made only its own preliminary determinations as to the terms on which it has found that cable growth will serve the public interest. Recently, the Supreme Court gave powerful momentum to this effort by affirming the FCC's power under the Communications Act of 1934 to order cable systems to engage in original programing. The Court's opinion in *U.S. v. Midwest Video Corporation*[1] goes far to dispel the line of legal and political uncertainty that had inhibited the development of the FCC's regulatory functions. Nonetheless, further court challenges are anticipated, and it is possible that legislative intervention will redefine

[1] 406 U.S. 649, 32 L Ed 2d 390 (1972). See Part III.

regulatory authority under the Communications Act—either by Congressional initiative or in response to an Administration bill.

On a second crucial issue—the rights of *program creators* to copyright protection where cable carriage extends the audience for television programs, and the rights of *local broadcasters* to preserve their audiences—another statutory anomaly poses barriers to cable growth. Following Supreme Court decisions in 1968 and 1974[2] that have eliminated the possibility under present law of copyright liability for cable carriage of television signals that have been broadcast, regardless of their point of origin, pressures have mounted to amend the Copyright Act of 1909 to impose statutory liability on cable systems for their unique activities with respect to copyright materials. At the same time, broadcasters have looked to the FCC to impose by regulation certain "blackout" provisions on cable systems to protect the contractual "exclusivity" of programs carried on local broadcast stations from competition with the same programs imported from other cities. A "Consensus Agreement" centering on these questions has been negotiated among the broadcast, copyright, and cable industries, and serves as a central assumption on which the *Report and Order* rests. It requires prompt enactment of new legislation, to whose key terms the parties to the agreement are said to be committed, but the prospects of early passage are doubtful. (See Part II, Sec. 2.) In this context, notwithstanding the precision of the FCC's design for the next phase of cable television growth, this pillar of the FCC's approach has yet to be moved into place.

Against this background of continuing uncertainty, the principal provisions of the FCC's rules are next briefly outlined, according to the structure of the fuller section-by-section analysis in Part II. Subsequently, this Part I addresses the overall scope for state and local initiative that the FCC's new regulatory program demarcates or implies, drawing a legal and practical perspective on this area whose transcendent importance pervades the subject matter of this study.

OUTLINE OF THE FCC RULES

The scope of the FCC's rules, as analyzed in Part II, encompasses six major categories:

- Carriage of Broadcast Television Signals (Sec. 1)
- Copyright and Program Exclusivity (Sec. 2)
- Cablecasting and Channel Capacity (Sec. 3)
- Role of State and Local Authorities (Sec. 4)
- Technical Standards (Sec. 5)
- Operating Requirements and Related Matters (Sec. 6)

While these categories reflect the FCC's terms of reference, they do not necessarily accord to state and local interests; consequently, some cross-referencing will be required. Here, the main import of each section is outlined.

---

[2] In 1968, in *Fortnightly Corp. v. United Artists Television, Inc.*, 392 U.S. 390, the Supreme Court held that cable systems were not liable to pay copyright royalties for carrying locally available television signals. Recently, in March 1974, the Court extended its *Fortnightly* rule in two related cases—*TelePrompTer v. Columbia Broadcasting System* (October Term 1972, No. 72-1628) and *Columbia Broadcasting System v. TelePrompTer* (October Term 1972, No. 72-1633), hereafter referred to jointly as *CBS-TelePrompTer*—to extinguish liability under the present copyright law for "distant signals" as well. The complete text of the *CBS-TelePrompTer* decision is set forth in Part III.

Section 1, "Carriage of Broadcast Television Signals," and Sec. 2, "Program Exclusivity and Copyright," deal with the steps taken by the FCC to mediate the long-standing conflict between cable interests and local broadcasters over the carriage of broadcast signals from one television market to another. One leg of the FCC's approach is to limit the number of additional signals that cable systems may import in major broadcast markets, according to quotas measured by the ability of markets of particular size to withstand competition from additional imported signals. Section 1 analyzes this approach. The FCC also imposes program-by-program controls on imported television programing to protect "exclusive" agreements between local broadcasters and copyright owners. As detailed in Sec. 2, the FCC rules require a cable system to "black out" the same program on an imported signal when it has been notified that "exclusivity" exists—the sole right to distribute a program in a particular market. In addition to this two-pronged regulatory approach, the FCC presumes that cable systems will pay copyright fees on imported programing as provided by statute—either under the present Copyright Act, as judicially construed, or under special provisions of a contemplated amendment to that Act, whose principal terms have been substantially negotiated among the broadcast, programing, and copyright industries.

The FCC's approach to carriage of broadcast signals from one market to another starts with a requirement for carriage of all "local" signals (upon request of the originating station), with the list of "local" signals specially defined according to market size. Markets are classified in descending size as "the first 50 major markets" (from New York City to Little Rock, Ark.), "the second 50 major markets" (from San Diego, California through Columbia, South Carolina), and "smaller television markets" (other communities to which television stations are licensed); there is also an added category of locations "outside of all television markets." (See Sec. 1.21 and the FCC's listing in § 76.51 of its rules.) For the 100 "major markets," local signals consist of the following:

- All market signals, i.e., those originating within 35 miles of the cable television system and those located in adjacent communities considered by the FCC as part of the same market;
- "Grade B" signals (signals capable of being viewed, according to the FCC formula, by viewers at 90 percent of the locations involved, at least 50 percent of the time);
- All commercial translator stations in the community with 100 watts or higher power, and all noncommercial translators in the community of greater than 5 watts power; and
- Television stations "significantly viewed" in the community according to county-by-county listings set forth in the FCC rules (see Sec. 1.212) or as separately measured.

In "smaller television markets," a fifth category of signals—Grade B signals from stations in other smaller markets—is added within the definition of "local" signals required to be carried. Cable systems located outside of all markets must carry on request: (1) all Grade B signals; (2) all 100-watt or higher translators; (3) all education stations from within 35 miles; and (4) television stations significantly viewed in the community.

After carriage of these local signals—principally intended by the FCC to ensure that local stations continue to reach cable subscribers—the FCC permits carriage of "distant" signals to bring the number and type of signals locally available up to quotas determined by the size of the local television market. These quotas are as follows:

- *The first 50 major markets:* three full *network* stations, three *independent* stations, and any number of additional education stations and non-English-language stations. Two additional signals from *independent* stations may be carried, to the extent such number has not been added under the above quota.[3]
- *The second 50 major markets:* three full *network* stations, two *independent* stations, and any number of additional education stations and non-English-language stations. Two additional signals from *independent* stations may be carried, to the extent such number has not been added under the above quota.[3]
- *Smaller markets:* three full *network* stations, one *independent* station, and any number of additional education stations and non-English-language stations.
- *Outside of any television market:* no quota (cable systems may carry any number of additional stations).

The foregoing quotas do not complete the FCC's program of restrictions on importation of "distant" signals, inasmuch as these signals must be selected in accordance with certain "leapfrogging" rules. These rules are applicable to cable systems in *all television markets,* i.e., they are inapplicable to systems *outside* of markets. They provide as follows:

- For *network* signal importation, the closest such station must be chosen, although the system has the option to select the closest such station within the same state.
- For *independent* signal importation, cable systems may choose the first *two* from any source, but if stations among those in the top 25 markets are selected, they must be taken from one or both of the two closest such markets. For any *third* station imported, cable systems must select from the following categories in order of their availability: a UHF independent within 200 miles; a VHF independent within 200 miles; and any UHF independent.

These rules governing "local" signals, "distant" signals, and "leapfrogging" are more fully treated in Secs. 1.0-1.24.

Section 2 deals with questions concerning copyright and program exclusivity. Resolution of the copyright issue is central to the practical effectiveness of the FCC's rules, but that issue remains subject to the uncertainties of pending litigation and an imminent legislative effort, prompted by doubts as to whether the holding in the

---

[3] In the first and second 50 major markets, the additional *independent* stations are charged against the basic quota, if signals are permitted to be carried under the quota; to the extent such *independent* signals are not under the applicable quota, they may be carried in addition to the quota.

1968 *Fortnightly* decision freeing cable carriage of local signals from copyright liability extends similar immunity for distant-signal carriage. The implications of this conjectural situation are discussed along with the FCC's specific regulations to protect program exclusivity. These related questions are underlying issues of the 1971 "Consensus Agreement," which foreshadowed the basic approach now taken by the FCC. The Commission's rules provide more protection for *independent* ("syndicated") programing in the first 50 major markets than in the second 50, eliminating other areas from exclusivity protection. Additional rules provide for exclusivity with respect to all *network* stations within the Grade B market area of each station or within the area of a translator carrying such a station.

With respect to *independent* programing in the first 50 markets, the rules bar cable showing of an exclusive program for one year after it is first shown anywhere in the country, and thereafter in accordance with specific contractual provisions applicable to each program. In the second 50 markets, prohibitions on carriage of varying duration are provided for specific categories of programs. For *network* programs, protection against simultaneous duplication in local showings is provided, and an ordering of priorities among stations entitled to such exclusivity protection is provided. Overall, the exclusivity provisions require that the station claiming protection give specific notification to each cable system. (See Sec. 2 for the details of the exclusivity rules and an appraisal of the implications of pending copyright issues.)

Section 3 deals with the new field of *cablecasting*—an activity which, to a greater or lesser extent (depending on size of market or of system), the FCC has mandated for many cable systems—and *channel capacity*—a concomitant of the cablecasting rules, defining the number of channels additional to broadcast channels that a cable system must make available for nonbroadcast purposes. Briefly, the activities embraced by the generic term "cablecasting" mandated by the FCC fall into the following categories:

- *"Origination cablecasting."* Each cable system with 3500 subscribers or more must operate "to a significant extent as a local outlet" by carrying "programing (exclusive of broadcast signals)" over "one or more channels and subject to the exclusive control of the cable operator . . ."
- *"Access cablecasting."* Each cable system, regardless of numbers of subscribers, *in the first 50 major markets and the second 50 major markets* must make available:
  1. At least one "public access" channel, "available without charge at all times on a first-come, first-served nondiscriminatory basis . . .";
  2. At least one "education access" channel, for use by "local educational authorities . . . for instructional programing and other educational purposes";
  3. At least one "local government access" channel; and
  4. "Lease access channels" for the remainder of channel capacity not used for other designated purposes, and for "any other available bandwidth."

An additional requirement, applicable to all cable systems in the *major* markets, is that all such systems shall "maintain a plant having technical capacity for nonvoice

return communications"; that is, the requirement prohibits technical capacities inconsistent with such developments but does not mandate actual service at this time.

The FCC's rule for minimum channel capacity in the major markets requires that at least 20 television channels be "available for immediate or potential use," and for every broadcast signal carried, an equivalent channel must be made available for nonbroadcast carriage. This is a companion requirement to the rules mandating availability of cablecasting channels. In fact, the equivalent-channel rule might in some circumstances result in a greater than 20-channel minimum capacity requirement. An additional rule sets procedures for making additional channels available whenever existing access channels are filled for a particular period of time, to ensure satisfaction of public demands for service. (See Sec. 3.22.) The rules are applicable to all systems put in operation after the date of their effectiveness (March 31, 1972), but existing systems which do not expand signal carriage under the new rules have up to five years to come into compliance. Existing systems that wish to add broadcast signals in advance of that future date must do so in accordance with a formula coupling added broadcast signals to access channels.

In addition to these rules, which mandate particular kinds of cablecasting service, Sec. 3 discusses other provisions that apply specifically to the nature and conduct of each form of cablecasting. For "origination cablecasting" by the cable operator, standards virtually identical to those applicable to broadcasting are provided for "equal time" for political candidates (Sec. 3.312), and lotteries (Sec. 3.316). Special rules for "pay-cablecasting" are set forth (Sec. 3.32), principally governing the types of programing that may be made available for a per-channel or per-program charge. With respect to "access cablecasting" the FCC provides standards for nondiscriminatory use of the "public access" channel, requiring cable systems to make facilities available to public groups at no cost for up to five minutes studio use, and mandating other reasonable charges (Sec. 3.3311). Similar standards are also set forth for "leased access" channels (Sec. 3.3314), and basic rules are stated governing the "education access" and "government access" channels (Sec. 3.3312-3.3313).

Section 4 discusses in detail the FCC's approach to "The Role of State and Local Authorities." Recognizing a conflict between its limited staff, budget, and statutory authority on the one hand, and on the other its desire to establish a stable framework for cable television growth, the FCC has articulated a concept of "Dual Jurisdiction" between itself and local franchising authorities. By using the term "local franchising authorities," the FCC recognizes that the determination of the locus of local regulation is a question of state rather than federal law, to be resolved by each state according to its own processes. In some instances regulation will be exercised by states, in others by municipalities or counties, and in still others by states and localities sharing responsibilities.

The concept of "Dual Jurisdiction" is in many respects unclear, and frankly experimental. It is expressed largely in procedural terms—that is, through a requirement that a cable system must seek federal authority (a "certificate of compliance" to carry broadcast signals) after March 31, 1972, when the rules became effective, and must make a showing that it has satisfied a precondition under locally applicable law. This precondition is that the cable system must have secured a "franchise or other appropriate authorization that contains recitations and provisions" that comply with certain enumerated federal guidelines, as follows:

- The franchise-holder's "legal, character, financial, technical and other qualifications, and the adequacy and feasibility of its construction arrangements" have been approved by the franchising authority after a "full public proceeding affording due process . . ." See below, Sec. 4.211.
- "Significant construction" will be accomplished within one year after federal certification and trunk cable will thereafter be energized "equitably and reasonably . . . to a substantial percentage" of its territory. Sec. 4.212.
- The initial franchise period must be no greater than 15 years and any renewal shall be of "reasonable duration." Sec. 4.213.
- Initial rates to subscribers have been specified or approved by the franchising authority and a procedure exists affording due process to the public with respect to rate changes. Sec. 4.214.
- Procedures have been locally specified for investigating and resolving service complaints, requiring a cable system to have a local office to resolve such issues. Sec. 4.215.
- Modifications in federal policy will be incorporated into local franchises within one year after adoption of such modifications. Sec. 4.216.

(For areas where no local franchising authority is identifiable, or where the proper authority is in controversy, the FCC states separate rules to achieve substantial compliance with the foregoing standards. See Sec. 4.21, footnote 26.) Separately, the FCC imposes a requirement restricting local franchise fees, levied upon cable operators, to "reasonable" percentages of gross subscriber revenues—"reasonable" being defined as "in the range of 3-5 percent" per year. The system and the franchising authority have the burden of satisfying the Commission, before federal certification will issue, that any fee over 3 percent "will not interfere with the effectuation of Federal regulatory goals" and is "appropriate in the light of the planned local regulatory program." (See Sec. 4.217.)

Additional federal standards variously prohibit local franchising authorities from imposing local restrictions on access channels that are "more restrictive" than federal regulations. (Sec. 4.22.) Although terminologically imprecise, the gist of these federal prohibitions is two-fold:

- In major markets (i.e., markets 1 through 100) franchising authorities may not "prescribe any other rules concerning the number or manner of operation of access channels" (for new systems and for existing systems after a five-year period). (See Sec. 4.221 below.)
- Outside of major markets, local authorities may not prescribe conditions that "exceed the provisions concerning the availability and administration of access channels" in federal regulations. (Sec. 4.222.)

The concluding section of this Part I summarizes the ramifications of these restrictions; they are detailed and evaluated in Sec. 4.3, "Evaluation of State and Local Responsibilities."

Section 5 deals with *Technical Standards*, established at this time by the FCC only with respect to the carriage and delivery to subscribers of broadcast signals. It leaves other aspects of technical performance to local regulation or to later federal determinations. The rules establish procedures and standards for measuring cable system performance, to inhibit technical degradation of broadcast signals.

Finally, Sec. 6 discusses "Operating Requirements and Related Matters." These are summarized in Sec. 6.1 from the perspective of local franchising authorities. Specifically discussed are provisions by which a cable system must petition the FCC for a federal "certificate of compliance" (Sec. 6.3), provisions for challenges to the rules and for oppositions to such petitions by interested parties (Sec. 6.2), and annual reports and fees due to the FCC from each cable system (Sec. 6.4). Sections 6.5 and 6.6 set forth additional standards governing diversification of ownership and discrimination in employment practices.

THE SCOPE FOR STATE AND LOCAL INITIATIVE

In the wake of the new federal standards for the development of cable television established by the FCC, the resulting scope for initiative by state and local governments is a question that holds utmost practical and legal importance. It is also a question to which some extraordinarily subtle answers can be given.

First of all, it is apparent from the face of the new federal rules and standards that an important role is envisaged for state and local regulation of cable television, in partnership with the FCC. Powerful reasons are stated for such a cooperative approach, centering on the limited resources of the federal regulatory agency and the familiarity of local governments with the particular social and economic circumstances prevailing in their localities. The FCC's concept of "Dual Jurisdiction" thus preserves a crucial role for state or local franchise authorities—the particular choice of the FCC's local regulatory partner having been carefully preserved as a question to be determined under local law.

The point of congruence between federal and state or local authority chosen by the FCC is the franchise process, wherein the FCC requires a local franchise as a precondition to federal "certification" of the carriage of broadcast signals on cable television systems. The mechanism by which dual jurisdiction is achieved is the grant or withholding of federal certification to cable systems franchised under local law, whose franchises fall within guidelines established by federal regulation. The FCC will observe what it calls a "go-no-go" approach, meaning that if all its requirements are met in an application for certification, it will issue a certificate expeditiously. If its requirements are not met, certification will be withheld or processed specially pursuant to a petition for waiver, wherein the party seeking to modify the Commission's rules has a burden of persuasion that its activities will serve "the public interest." Thus the FCC offers the reward of speed and sureness for literal compliance with its rules, and the risks of delay and uncertainty when the applicant's views of the public interest deviate from its own.

With wide variations of specificity, the FCC has acted to define the scope for local regulation. Within this scope, the FCC has mandated *procedures* local authorities must observe in the grant and administration of franchises. These require an investigation of the "legal, character, financial, technical and other qualifications and the adequacy and feasibility of its construction arrangements," and they require as well that approval follow a "full public proceeding affording due process." A further requirement that proceedings be competitive as an aspect of such a "full public proceeding" is suggested, and a requirement for a "public report" on the

reasons for selecting an applicant is specified in the explanatory *Report and Order*. While no similar provisions are set forth with respect to franchise renewals, such may properly be implied. Not further specified, these procedural rules are clearly pointed toward achieving essential fairness and broad public participation.

With respect to *substantive* aspects of local regulation, certain issues are dealt with by the FCC in more general terms also intended to be mandatory. In respect to several of these rules the FCC has indicated it will also entertain petitions for waiver or special relief premised on the "experimental" nature of locally sanctioned cable television service. In some matters, such as selection of franchise holders, the FCC has left determination solely in the hands of localities, subject only to the requirement that procedural fairness be observed. In other matters the FCC has required that localities take steps in franchises to impose their own local regulatory requirements on cable operators, while at the same time the FCC imposes limits on the ways this local discretion can be exercised. These questions include rate of construction, duration of initial and renewal terms, provision for responsiveness to service complaints, establishment of subscriber rates, and franchise fees. With respect to duration, subscriber rates, and franchise fees, for example, the FCC imposes standards of fairness or reasonableness, stated with varying specificity, which it views as appropriate.

In other areas, where the FCC sets up its own regulatory standards (which it imposes, directly, in the grant or withholding of certification), localities are expressly precluded from exercising their own discretion in ways deemed incompatible with federal regulation. The clearest instance, perhaps, concerns regulation of the various channels "designated" by the FCC for local access—public access, education access, government access, and lease access—and the "minimum channel capacity" rule with which access requirements are coupled. Here, local franchise authorities in major markets are prohibited from prescribing "any other rules concerning the number or manner of operation of access channels," and outside major markets are prohibited from "exceed[ing] the provisions concerning the availability and administration of access channels . . ." (See Secs. 3.21-22.) In much the same way, localities are squarely blocked from *prohibiting* program origination by cable operators who serve more than 3500 subscribers (as some may wish to do to attain common carrier regulation) because of the FCC's mandatory program origination rule recently upheld by the Supreme Court in *Midwest Video* and reissued with the *Report and Order*.

For many franchise authorities, the FCC's efforts to define a limited field for local discretion will represent a satisfactory basis for participation in launching cable television development, especially where local resources or interest in a more independent local regulatory approach may be lacking. On the other hand, other jurisdictions may prefer to investigate whether the regulatory approach specified by the FCC nationwide is appropriate to serve the needs of local citizens as locally determined. If certain local departures from federal norms are felt warranted, special concern will inevitably be given to how a locality might take the initiative to ensure that residents receive the improved cable television service local needs require. The question is particularly acute, for example, where localities view such fundamental questions as regulation of the rate of return on invested capital in ways diverging from the FCC's explicit or implicit standards.

The most straightforward approach, and the one obviously envisaged by the

FCC as preserving its primacy in controlling cable television growth, is to address areas of potential conflict as questions requiring petitions to the FCC for waiver or special relief. This is a conservative way of proceeding to limit controversy; it lets the FCC determine for itself whether it wishes to permit a particular innovative approach as an aspect of the "experimentation" it professedly favors. On the other hand, such a conservative approach can diminish the freedom of action of states and localities, beyond what is legally required, to find their own solutions to fundamental questions regarding the nature of cable service responsive to local needs; it would also ensure that the ultimate determination of any issue reflects the FCC's penchant for looking at cable issues from the *national* viewpoint.

By contrast, there may also be significant opportunities for state and local political entities to take a more aggressive course in serving the interests of their own citizens. The broad Constitutional framework should be kept in mind: Localities are legally preempted from regulatory programs that are incompatible with federal objectives (in areas appropriate for federal regulation), but preemption does not come about where local action is not inconsistent with federal regulation—both in fact and in the eyes of a court. Thus, where federal regulations are not precise, local powers may not be limited, and wherever federal regulation is beyond the scope of statutory authority, efforts at federal preemption may be unauthorized. Moreover, there would appear to be some scope for localities to encourage *voluntary* commitments by cable operators to deliver improved service in excess of explicit federal requirements.

Several areas for such local ingenuity are pointed out in this book, and have become subjects for continuing controversy between local and state authorities and citizens groups on the one hand, and the F.C.C. on the other. In an effort to resolve such controversies, the Commission appointed a Cable Television Advisory Committee on Federal/State-Local Regulatory Relationships, which rendered a bitterly contested and divided report in 1973. Subsequently, on April 22, 1974, the F.C.C. issued a *Proposed Clarification of Rules* (set forth in full in Part III as Item 19) which aims to plug a number of recognized loopholes and assert pre-emptive F.C.C. control over independent local viewpoints. The document is part "clarification" of present F.C.C. policy and part "notice of proposed rulemaking," on which public comment is sought (on or before June 7, 1974) following which rule changes can be anticipated. Because of the timeliness of such ongoing but still incomplete evolution in F.C.C. policy making, large asterisks (*) are prominently placed in the margin of paragraphs of this book likely to be affected, with marginal parentheses ( ) indicating the paragraphs of the F.C.C. *Clarification* to be noted and taken into consideration.

Moreover, the possibility must also be kept in mind that key franchise authorities could act together in dealing with cable television franchise issues and thereby accomplish more than any one authority acting individually. Since many potential applicants for franchises are applicants in more than one area, cities around the country have many common interests in the quality of cable services and could identify them as the basis for common bargaining positions. Thus, communication

between economically significant franchise authorities may give rise to effective and practical cooperation to ensure that local interests are fully protected in dealings with cable operators and the FCC.

In this light, a close analysis of FCC regulations suggests that highly motivated franchise authorities with clear views of their own local needs in the development of cable television may be able to accomplish much through independent action not inconsistent with the general approach of present federal rules. As indicated earlier, those rules themselves rest upon a legal and political foundation whose strength has yet to be finally tested. Hence, to consider FCC rules immutable or beyond challenge may be an unwarranted concession.

THE FUTURE OF CABLECASTING

In the months since the F.C.C. promulgated the *Cable Television Report and Order*, controversy over some of the key aspects of standards governing non-broadcast cable channels has deepened in ways not likely to be clarified in the immediate future. Over the short term, first of all, it is now clear that the "mandatory origination rule" requiring *origination cablecasting* by systems with 3500 or more subscribers (see page 5) is not now being enforced. The rule was stayed by the Commission pending the outcome of the *Midwest Video* case and—even though amended by the *Report and Order*—never put back in force. The economic burdens of building and operating programming studios proved onerous to many cable operators at a time of general recession and capital shortage, and the under-strength Federal Communications Commission has not yet determined its future policy directions.

While in the normal course of events, reissuance by the Commission could be forecast, the origination requirement goes to the heart of the basic national policy questions for the long-range future of cable services that were opened up in January 1974 by the issuance of the Report of the Cabinet Committee on Cable Communications. Calling for the ultimate separation of responsibility, control and ownership of cable's transmission and programming functions, the Whitehead Report would reorient cable television as a carrier of programming originated by independent programmers, a direction diametrically opposed to the integration of functions that the mandatory origination rule represents. In his State of the Union message, President Nixon undertook to submit implementing legislation to the Congress and encouraged a "widespread national debate on this subject which could play such a major role in all our lives during the future."

In this context, debate rather than decisive regulatory action can be reasonably foreseen during the months and years ahead. Pending the resolution of this debate, the F.C.C. has announced that it is considering the abandonment permanently of the mandatory origination rule or of modifications aimed at substituting stronger *access cablecasting* requirements for its present *origination cablecasting* rules. See 39 Fed. Reg. 12873 (April 9, 1974).

The Whitehead Report was fashioned by a Cabinet Committee under the direction of the President's Special Assistant for Telecommunications Policy and

was composed of various White House assistants and the Secretaries of Commerce, Housing and Urban Development, and Health, Education, and Welfare. Despite the lofty composition of the Committee (few of whom had remained in Government at the time the report was issued), the intense controversy over its findings which preceeded issuance can be expected to last long into the future. Even though the commitment has been made to propose implementing legislation, observers are virtually unanimous in discounting the likelihood that the Report's recommendations will ever become the precise model that Federal cable regulation will follow.

Nonetheless, a remarkable consensus also exists that the Whitehead Report has significantly elevated national consciousness of the public importance of cable television policies. Significantly, for the first time, the report gives legitimacy to cable's future as a broadband carrier, which many view as the appropriate organizing principle for cable development but which both the broadcasting and cable industries have long fought as a regulatory philosophy they were not ready to accept. That regulatory philosophy—separating the functions of cable transmission and programming—aims to curb the extent of private control over the medium (by confining cable's monopoly function to transmission and opening up programming to more or less unlimited competition) and thereby to limit the extent to which government intervention is required over the content of messages.

While the specific recommendations of the report are too numerous and, at this point, conjectural to discuss here—they are summarized in Item 18 of Part III—the broad outline of the report is useful to keep in mind. Its aim is to classify the "principal business of cable operators" as the lease of channels to other entities (although reservations of one or two such channels to the operator for his own use is to be permitted in a possibly significant deviation from the report's broad objectives). Limited cross-ownership provisions are proposed to protect that separation aim, and administrative regulation of content and pricing are precluded. By and large, a dualism in federal and local regulation is envisaged, and some publically funded demonstration programs are proposed.

Key to the vision of the report are provisions staying the full implementation of its radical separation philosophy until a significant portion of the country—50 percent—has been wired through the gradual growth of the cable television industry. Even though the notion of a "transition period" is intended to enhance the political and economic realism of the report's recommendations, the transition concept can also be viewed as a concession ultimately frustrating the achievement of the report's goals. Similar criticism can be leveled at a lack of emphasis given in the report to public programs stimulating other aspects of cable television's promise, such as service to minority, rural and other special communities. All of which, of course, will be among the sparks in the debate over the long-range future of cable communications, which the Whitehead Report has now, and much to its credit, brought into full public view.

Tom Hoy

STEVEN R. RIVKIN is a partner in the Washington, D.C., law firm of Nicholson & Carter. This book brings together the diverse strands of his professional background in antitrust, communications and municipal law, and the problems of technological innovation.

A graduate of Harvard College and Harvard Law School, Mr. Rivkin served as counsel to the White House Office of Science and Technology. As a staff assistant to the President's Science Advisory Committee and active participant in government-wide science planning, he developed a unique perspective on science and technology policy issues. In subsequent law practice in Boston and Washington, Mr. Rivkin has specialized in the legal and economic aspects of regulatory and antitrust questions that, in a practical sense, significantly influence the shape and timing of technological change.

Mr. Rivkin served as Counsel to the Sloan Commission on Cable Communications during the preparatory phases of its work. He has written extensively on cable television development and counseled municipalities, states, Federal government agencies, and voluntary organizations. The present focus of his interest in cable television issues is to develop public policies assuring that sufficient capital is available to bring mature broadband systems into being at an early date.

In addition to several articles in legal and general publications, Mr. Rivkin is the author of *Technology Unbound: Transferring Scientific and Engineering Resources from Defense to Civilian Purposes* (1968) and *The Building Code Burden* (1974, with Charles Field).

PART II.

EVISION POLICIES AND RULES

Section 1

CARRIAGE OF BROADCAST TELEVISION

As the FCC aptly indicates in its introduction to its broadcast television rules, "The carriage of distant television signals by cable television systems has been center stage in the continuing controversy before the Commission, the Congress, and the Courts." As noted earlier, cable operators regard distant-signal carriage as a widening of viewing options necessary to gain people's subscriptions, while broadcasters bitterly condemn it as the entering wedge for ruinous competition. Now the FCC has attempted to ease the conflict by according limited carriage authority to cable systems, conditioned on protection of local broadcasters' exclusive rights to programing in particular markets, and the development by cable of a range of nonbroadcast services. As a practical matter, the whole package will hang fire, to a large extent, until new copyright legislation is enacted. Given the complexities of the issues and the mutual suspicions of the parties, however, only time will tell whether the FCC's actions will succeed in placing cable development on the firm competitive foundation that the long-range growth of cable television will require.

By its recent actions, the FCC has legitimized some carriage by cable systems of broadcast signals in the nation's leading market areas. It has done so in the context of a broad effort to carve out a new role for the cable medium through a regulatory program whose basic objective is "to get cable moving so that the public may receive its benefits, and to do so without jeopardizing the basic structure of over-the-air television." *(Report and Order,* par. 55.) While many questioned whether regulatory action alone could stabilize the competitive relationship between cable systems and broadcasters, the FCC, with the assistance of the Executive Branch, succeeded in negotiating in the fall of 1971 a "Consensus Agreement" among broadcasters, cable interests, and copyright owners on which the present rules are based.[1] This Agreement sets forth the bare bones of an accommodation

---

[1] The text of the Consensus Agreement, Appendix D to the *Cable Television Report and Order,* is set forth in Part III. In the *Report and Order,* the FCC states that, "The Office of Telecommunications Policy [of the Executive Office of the President] provided valuable assistance in the negotiations that led to this agreement." (Par. 61.)

14

among the three affected industries.[2] It embraces the kinds of local and distant television signals cable systems may carry, the systems to which the rules apply, and the premises according to which the parties would agree to seek remedial copyright legislation. This perspective represents the core of the FCC's rules, whose signal-carriage aspects are treated in detail in this Sec. 1, after a brief background discussion. Section 2 considers related questions of program exclusivity and copyright.

1.1. BACKGROUND

Cable carriage of broadcast signals, while solely subject to federal control,[3] is critical to local authorities' planning for cable development. Without a broad array of viewer options—which the cable industry believes must include broadcast signals from other cities ("distant signals")—cable systems are by and large unable to offer viewers the inducement needed to justify cable subscriptions, except in areas where accidents of terrain make locally based signals otherwise inaccessible. Without distant carriage rights, investment in cable systems was seen as economically un-remunerative in consequence of the FCC's "freeze"[4] in 1966 on the further importation of distant signals. Subsequently, relaxation of this prohibition became a sine qua non of cable television growth.[5] But that relaxation prayed for by the cable industry promised to disadvantage both the broadcast industry (alleging irreparable harm to fledgling UHF stations from cable competition) and owners of copyrighted programing (exposed to disruption of exclusive market relationships with broadcast-

*

(8–10)

---

[2] The aim of the "television programing industry" as expressed by the FCC is "that suppliers of programing should receive compensation for use of their product by cable systems and that the exclusive sales of such programs in particular markets should be honored." *(Report and Order,* par. 51.)

[3] In 1969, the FCC ruled that local authorities had been preempted from interfering with federally authorized cable origination and advertising. *Clarification of CATV First Report,* 20 FCC 2d 741 (1969).

[4] In its *Second Report and Order in Docket Nos. 14895, 15233, and 15971,* 2 FCC 2d 725 (1966), the FCC adopted a rule (47 C.F.R. § 75.1107, now deleted) requiring its specific approval of new carriage by CATV systems extending the range of television signals beyond the "grade B contour" of a broadcast station, into the "grade A contour" of television stations in the 100 largest television markets. Such contours are concentric measures of the strength of a broadcast signal, defined in terms of intensity of an acceptable signal received by viewers at 90% of the locations embraced within each contour for the following percentages of time: Principal City Grade, 90%; Grade A, 70%; Grade B, 50%. 47 C.F.R. § 73.683. Practically speaking, the Commission's rule imposed a "freeze" on such authorizations by placing the burden on the CATV systems located in the top 100 markets of showing, in an evidentiary hearing, "that such extension would be consistent with the public interest, and specifically the establishment and healthy maintenance of television broadcast service in the area." 47 C.F.R. § 74.1107. In a celebrated test case filed immediately after the issuance of the *Second Report and Order,* the Commission held an extensive evidentiary hearing and prohibited several cable systems from extending their carriage of Los Angeles programs into San Diego. *Midwest Television, Inc., et al.,* 13 FCC 2d 478 (1968), affirmed *sub nom Midwest Television Inc. v. FCC,* 426 F2d 1222, (C.A.D.C., 1970). In December 1968, as part of a package of proposals reopening consideration of the Commission's distant signal policies, interim procedures were adopted suspending all evidentiary hearings under the 1966 rules, and action on requests for authorizations to carry signals was deferred pending completion of their rule-making proceeding—unless carriage was consistent with requirement of the proposed rules for "retransmission consent" of the originating station, an opportunity that proved nugatory for cable systems given the continuing impasse over long-range goals. See *Notice of Inquiry and Notice of Proposed Rule Making in Docket No. 18397,* 15 FCC 2d 417 (1968).

[5] The FCC's authority over certain aspects of cable television was initially validated by the Supreme Court in *U.S. v Southwest Cable Co.,* 392 U.S. 157 (1968), which found the FCC could take steps "reasonably ancillary to the effective performance of the Commission's various responsibilities for the regulation of television broadcasting." In that case, the FCC's power to prohibit carriage of distant signals was involved.

ers by the importation of distant signals without copyright liability for such importation having been established).[6] By reflecting concurrence in the broad outlines of a regulatory program in the consensus Agreement, the FCC presided over bargaining that led to terms on which it anticipates initiating a mutually acceptable phase of cable television growth.

But it is also essential to keep in mind that the FCC's rules are only the precondition—not the assurance—that such growth will now ensue. First of all, the achievement of legal copyright protections is central to the scheme of the Agreement, and they hinge on enactment of new copyright legislation that accords with the expressed views of the parties; it is by no means certain that legislation will be enacted promptly or that final action by Congress, if it comes about, will satisfy affected industries.[7] If the numerous hurdles remaining[8] are not promptly overcome, the "consensus" can come apart and the FCC can be subjected to pressures to modify or stay the applicability of its rules. Moreover, the economic effects of the FCC rules are in many key respects unknown and unknowable for the present, given their newness;[9] they will vary from market to market. Consequently, even if the FCC's rules were fully implemented, it is not certain rapid cable growth would ensue. Nonetheless, the FCC's commitment to the rules in their present form is, for the time being, a reality[10] and to that extent a palpable basis on which the near-term development of cable television can be anticipated.

[6] In *Fortnightly Corp. v. United Artists Television Inc.*, 392 U.S. 390 (1968), the Supreme Court held that mere relaying over cable locally available television signals was not a "performance" infringing a copyright holder's rights to "perform . . . in public for profit" under 17 U.S.C. §1(c), the cable system merely being a "passive beneficiary" of a broadcast signal analogous to the viewer's own TV set or antenna. Recently, the Supreme Court extended this holding in *CBS-TelePrompTer* to free cable systems from liability for carrying "distant signals" as well—i.e. signals broadcast in television markets geographically removed from the local area of the cable system. (The impact of these decisions is fully discussed in Section 2, Part II.) Thus, under present copyright law, cable systems are free from liability for any carriage of broadcast signals, although there is no question that they are, and have always been, liable for royalties for programming originated over cable. Without liability, cable competition may be harmful to local broadcasters in some circumstances. With liability, cable systems could be forced to engage in burdensome case-by-case negotiations with literally thousands of owners of copyright interests to obtain necessary clearances. The resulting inequities and burdens underline the need for new legislation to deal specifically with cable carriage of broadcast signals. See, illustratively, S.1361, 93d Cong., 1st. Sess., § 111 (1973), in Part III.

[7] Although the Chairman of the Senate Judiciary Committee's Subcommittee on the Patents and Copyrights has expressed his support of the terms of the Consensus Agreement, hearings have been delayed, opposition to its terms has arisen from participants and others, and Congressional delays until 1974 at the earliest are considered probable. As indicated more fully in Sec. 2, it is not clear that the parties to the Consensus Agreement are in accord on the much more specific terms of the legislation needed to implement the Consensus Agreement.

[8] The Supreme Court has recently upheld the authority of the FCC to order cable systems to originate programing in *United States v. Midwest Video Corp.*, 406 U.S. 649 (1972), and in the process accorded judicial sanction to an important key of the FCC's cable television regulatory program. See Sec. 3. Though further challenge is possible, FCC jurisdiction to regulate cable is now assured. However, further judicial challenges have been brought to the FCC's new rules, and now that reconsideration has been completed, such judicial reviews can be expected to multiply. Aside from questions of statutory jurisdiction, now largely resolved, plausible challenges can be anticipated on a number of grounds, including fairness to aggrieved persons, propriety of procedure and compliance with the anti-trust laws. Any of a number of possible deficiencies could lead a court in review to impose severe burdens on the fulfillment of the FCC's regulatory scheme. For background and analysis of pertinent issues see Rivkin, "The Changing Signals of Cable T.V.," 60 *Georgetown Law Journal,* 1470 (June 1972).

[9] See W. S. Baer and R. E. Park, "Financial Projections for the Dayton Metropolitan Area," in L. L. Johnson, et al., *Cable Communications in the Dayton Miami Valley: Basic Report,* The Rand Corporation, R-943-KF/FF, January 1972.

[10] In the process, the FCC has rescinded its prior advocacy of retransmission consent by distant broadcast stations for cable carriage in major markets and individual adjudications elsewhere. It has also

1.2. BROADCAST SIGNALS

The FCC's approach to broadcast signal carriage is two-pronged, summed up as follows:

> The approach we are adopting is to extend existing exclusivity rules so that they cover nonnetwork as well as network programing, and to restrict the number of distant signals that a system may carry based on the size of the market in which it is located and the estimated ability of that market to absorb additional competition. *(Report and Order,* par. 60)

Putting aside to later consideration in Sec. 2 the "exclusivity" and copyright issues, we first turn to the rules on permissible broadcast carriage. There, the FCC's standards classify broadcast signals according to three categories, summarized in the *Report and Order* (par. 74) in the following fashion:

> First, signals that a cable system, upon request of the appropriate station, must carry.
> Second, signals that, taking television market size into account, a cable system may carry.
> Third, signals that some systems may carry in addition to those required or permitted in the two above categories.

Each category reflects a different element of the balancing of interests in which the Commission is engaged. The first category—mandatory carriage on request of a local broadcast station—protects broadcasters against the possibility that a cable system, by shifting subscribers' reception to wired rather than over-the-air methods, might eliminate subscribers from a local station's market. The second category embodies a concept of "adequacy" in locally available programing; by defining an affirmative role for cable services, it encourages them to offer viewers a balanced package of programing, in many instances increasing the complement otherwise available from local television stations. Finally, the right to carry additional signals beyond those required or permitted—colloquially referred to as the "wild card"—can further enhance cable systems' ability to draw subscribers.

Conceptually, the first category embraces local broadcast signals, while the second and third are signals imported from distant broadcast stations. Thereunder, the FCC states a two-fold general aim:

> (1) To assure that "local" stations are carried on cable television systems and are not denied access to the audience they are licensed to serve; and (2) to gauge and, where appropriate, to ameliorate the competitive impact of "distant" signal carriage. *(Report and Order,* par. 78.)

Having so stated its aims, the agency goes on to preface its regulatory standards as to the signals permitted in various types of markets as follows:

> Because market patterns vary and there is only gradual deterioration in a station's receivability as the distance from its transmitter increases, there

---

put aside previous consideration of a "commercial substitution" rule (at one time paired with the concept of a "public dividend" levy to support public broadcasting). Also, in some respects it deviated from its August 5, 1971 "Letter of Intent" to Congress (31 FCC 2d 115, 1971). The Letter of Intent served, however, to establish some negotiating premises within which the Consensus Agreement was ultimately obtained.

is no necessarily clear dividing line between "distant" and "local" signals. Nevertheless, a line must be drawn somewhere. *(Report and Order,* par. 78.)

The next subsections discuss the FCC's signal definitions and their applications in particular market categories.

1.21: Local and Distant Signals

Starting from the premise that cable systems must first carry "local" signals in all markets before "distant" signals may be imported, the rules turn on an initial set of classifications specifically refined according to various categories of television markets. The FCC summarizes its concept of local signals, which cable systems must carry at the request of local braodcasters (and *may* carry in the absence of a request), as follows:

> Signals of stations within 35 miles of the cable system, signals meeting a significant viewing test, market signals in hyphenated markets, and in some cases Grade B signals. *(Report and Order,* par. 81.)

These particular types of television broadcast stations classified as local according to the *Report and Order* are the following:

1.211. **35-Mile and "Grade B" Signals.** Modifying a prior approach defining local signals in terms of the "Grade B" contour (see footnote 4), the FCC now draws separate rules for cable systems located within a 35-mile "specified zone"[11] of a community to which a broadcast station[12] is licensed and for systems wholly outside all markets. *(Report and Order,* par. 82.)

This "specified zone" distinction applies to all those localities that the FCC identifies as "television markets"—essentially, the nation's urbanized areas.[13] For systems wholly outside of any specified zone, the Grade B carriage rule applies to define the ambit of local signals.[14]

---

[11] *RULE: Specified Zone of a television broadcast station* is defined as "The area extending 35 air miles from the reference point in the community to which that station is licensed or authorized by the Commission." 47 C.F.R. § 76.5(f). A list of particular reference points for individual communities, identified by geographic coordinates (including communities for which construction permits have been issued) is specified. Planned stations are protected prior to going on the air by reserving to them a specified zone for 18 months after the issuance of a construction permit.

[12] The term "television broadcast stations" refers to commercial television stations, both U.H.F. and V.H.F. Cable systems in major and smaller markets (for definition see footnote 13) may, and on request must, carry noncommercial education broadcast stations within whose Grade B contours the community of the cable system is located, in whole or in part. (§ 76.59(a)(2), 76.61(a)(2), and 76.63(a).) If the cable system is outside of any market, carriage of noncommercial stations is restricted to the "specified zone," as next discussed in the text (§ 76.57(a)(3)).

[13] The FCC's rules differentiate between "major" and "smaller" markets. A major market is defined as the "specified zone of a commercial television station" assigned to the "first 50 major markets" and the "second 50 major markets," enumerated in § 76.51. A smaller television market is defined as the "specified zone of a commercial television station" licensed to another community. See §§ 76.5(g) and (i). Also, there are systems "operating in communities located outside of all major and smaller television markets" (§ 76.57).

[14] For all cable systems, local signals are defined as including "translator stations with 100 watts or higher power serving the the community of the system"; in addition, systems coming into operation after March 31, 1972 or adding broadcast signals after that date must carry "noncommercial educational translator stations with 5 watts or higher power serving the community of the system." §§ 76.57(a)(2), 76.59(a)(5), 76.61(a)(3), 76.63(a). Such noncommercial signals may also be carried voluntarily. *(Reconsideration,* par. 10.)

COMMENT: In adopting a "specified zone" concept for all urban television markets—as opposed to a measure of signal intensity, which varies with signal strength and terrain—the FCC aims to extend outward the carriage requirement for many limited-range UHF stations "into the area that we have determined is generally necessary for development of broadcast stations"(Report and Order, par. 82). Within that area UHF and VHF signals will now be carried on a equivalent basis. By retrenching from a Grade B rule for commercial stations, however, the FCC may have disadvantaged stations that are more than 35 miles distant from the cable system but whose Grade B signal covers the system's community when that community is within the specified zone of another broadcast station. A challenge on these grounds for the Rocky Mountain areas has been specifically rejected (Reconsideration, pars. 45-49).

1.212. Overlapping Markets: The "Significant Viewing" Standard. The FCC has long recognized the existence of a special situation in which television markets are closely adjacent to one another, most notably Baltimore and Washington, D.C. Called the "Footnote 69" situation after its designation in an earlier FCC report, it has been viewed as threatening considerable instability in the adjacent markets if their respective cable systems introduced nearby but nonlocal television stations as competitors for local broadcasters. This is what would actually happen, of course, if geography alone were used to define "local" signals, in which case FCC regulations would require such carriage. On the other hand, the FCC also recognizes how anomalous it would be if viewers in Baltimore, say, could pick up Washington signals over the air but not receive them on cable. This is a significant problem since acceptance of signals from an adjacent city as local signals would readily increase cable viewing options.

The FCC resolved the problem by adopting a "significant viewing standard"—subsequently defined—which operates as an alternative to the "specified zone concept." *(Report and Order,* par. 83). In effect, the FCC adds a basis for classifying signals as local, subject to an exception that in practice restricts its application to the core cities of Baltimore and Washington, D.C., as follows:

> (1) A *major market* cable system in a community to which a television station is licensed may carry as a local signal a station from *another major market* if that signal is "significantly viewed" (§§ 76.61(a)(5) and 76.63(a)), *provided that* the community of the cable system is located *wholly within* the specified zone of the other major market (§§ 76.61(a)(1) and 76.63(a)).

Thus, in the Washington-Baltimore situation for which the rule is fashioned, a cable system in Silver Spring, Maryland may carry a Baltimore television station as a local signal, but a cable system in Washington (to which television stations are licensed) may not carry such Baltimore station, since Washington is not wholly within Baltimore's specified (35-mile) zone.

> (2) A *smaller market* cable system may carry a major market station if that system is located in a community wholly within the specified zone of the originating station (§ 76.59(a)(1)) or if that signal is significantly viewed (§ 76.59(a)(6)).

A cable system in a *smaller market* may carry signals from *adjacent smaller markets* within any of three formulations:

(3) Such a signal may be carried as within the specified zone of the community of the cable system (§ 76.59(a)(1)); as a signal within another smaller market whose Grade B contours reach the community in whole or in part (§ 76.59(a)(3)); and as a station significantly viewed in the community of the system (§ 76.59(a)(6)).

For systems located in *communities outside all television markets,* the following standards apply:

A signal may be—or must be on request—carried as within the Grade B contour of the originating station (§ 76.57(a)(1)) or as significantly viewed (§ 76.57(a)(4)). (Note that § 76.57(b), which defines distant signals applicable to such systems, provides that "any additional television signals" may be carried.)

COMMENT: Though the complexities of the FCC's approach do not permit much simplification beyond indicating the various situations covered, the "overlapping market" concept is significant as defining the permissible limits of cable efforts to expand the basic complement of signals in many crowded metropolitan areas, where programing from other cities can occasionally be picked up off-the-air. These situations are the obvious target of the ingenuity with which the rules are drafted, and are also the source of their complexity and redundancy. The "significant viewing" standard mollifies the harsh result of adopting a "specified zone" approach, which the FCC felt was an essential key to compromise. In the process, the "significant viewing" standard applies on a case-by-case basis to widen the complement of local signals in areas other than the adjacent-major-market situation that originally prompted the FCC's concern. In some instances (see Reconsideration, *par. 57), the "significant viewing" standard exceeds the Grade B contour.*

In devising a significant viewing standard to apply to the foregoing situations, the FCC has looked to two measures of reception patterns: "net weekly circulation," measuring how many people watch a signal; and "share of viewing hours," measuring how much they watch it.[15] It applies these measures differently according to the nature of the originating station.

We have concluded that an out-of-market *network* affiliate should be considered to be significantly viewed if it obtains at least a 3-percent share of the viewing hours in television homes in the community and has a net weekly circulation of at least 25 percent. For *independent* stations, the test is a share of at least 2-percent viewing hours and a net weekly circulation of at least 5 percent. (Par. 84. Italics added.)[16]

To establish the extent to which particular signals satisfy its "significant viewing" test, the FCC turned to the findings of a private rating service, the American Re-

[15] "As used here the term *net weekly circulation* is a measure of the number of households that viewed a station for 5 minutes or more during an entire week, expressed as a percentage of the total television households in the community. *Share of viewing hours* is a measure of the total hours all television households in the community viewed a station during the week, expressed as a percentage of the total hours these households viewed all stations during the period surveyed." *(Report and Order,* par. 84, footnote 43; see § 76.5(n). Italics added.)

[16] The FCC defines a *network* station as a commercial television broadcast station that "generally carried, in weekly prime time hours, 85 percent of the hours of programing offered by one of the three major national television networks with which it has a prime affiliation," while an *independent* is defined as carrying "in prime time, not more than 10 hours" of such programing per week. (§§ 76.5(1) and (n).)

search Bureau (ARB). The ARB conducted two 1971 surveys, whose relevant findings the Commission adopted as its present determinations.[17] These determinations, set forth in Appendix B of the *Report and Order* as amended (see *Reconsideration*, pars. 55, 56, and 57) establish ratings by county, for every state. Recognizing that county-wide sampling "may not account for variations in viewing levels among communities within the county," and that there "may be other drawbacks in using these surveys, such as rounding and percentage errors," the FCC has nonetheless adopted the ARB standards as a "useable indication of viewing" and a needed source of "certainty, both from a cable and a broadcast point of view."[18] *(Report and Order,* par. 85.) To minimize initial controversy, the agency has indicated that it will not entertain special showings to modify its findings until March 31, 1973, and then only according to a specified probability test,[19] set "high" so as to minimize controversy.[20]

COMMENT: While cumbersome for general consumption, the FCC's rules are ultimately clear to particular communities in establishing initial guidelines and procedures for modification applicable to determining signals eligible to be considered "local." It should be pointed out that the FCC has indicated that broadcasters themselves may initiate surveys after the initial moratorium, either to gain or delete particular signal carriage. (Reconsideration, *par. 63.)*

1.213. Hyphenated Markets. There are a few special situations in which a single television market traditionally embraces more than one population center, and competing broadcast stations have been licensed to different cities within the market (such as Texarkana-Shreveport). In these situations the FCC "will permit and, on request of the station involved, require carriage of all stations licensed to designated communities in the market." *(Report and Order,* par. 87.) Since all homes in such a market may not receive Grade B signals from all stations in the market, the FCC is reserving to case-by-case review determination of what signals cable subscribers may receive within the category of local signals.

[17] The ARB surveys are "Television Circulation Share of Hours" and "Non-CATV Circulation and Share of Viewing Hours Study for ARB CATV-Controlled Counties." Its findings were challenged, and certain revisions were on reconsideration incorporated by the Commission in revised standards.

[18] On reconsideration, the FCC acknowledged imprecision but rejected a plea for ad hoc determinations, finding "The effort to more finely tune the information we have initially used to establish viewing levels through a process of survey and counter survey would, we believe, lead to continuing and pointless disputes about questions more subtle than the whole of our regulatory program is designed to deal with. *(Reconsideration,* par. 53.)

[19] *RULE:* "On or after March 31, 1973, significant viewing in a cable television community for signals not shown as significantly viewed under paragraph (a) of this section may be demonstrated by an independent professional audience survey of non-cable television homes that covers at least two weekly periods separated by at least thirty (30) days but no more than one of which shall be a week between the months of April and September. If two surveys are taken, they shall include samples sufficient to assure that the combined surveys result in an average figure at least one standard error above the required viewing level. If surveys are taken for more than two weekly periods in any 12 months, all such surveys must be submitted and the combined surveys must result in an average figure at least one standard error above the required viewing level." (47 C.F.R. § 76.54(b).)

[20] On reconsideration, the FCC adopted a notice and challenge procedure for such surveys. Under it, 30 days before a survey begins, notice must be served on all broadcast licensees and permittees within whose predicted Grade B contour the cable system is located, and on all cable systems, franchisees and applicants in the community. The notice must identify the survey organization and its procedures, to which objections may be made within 20 days. (§ 76.54(c).) In instances where the commission modified the ARB findings on reconsideration (shown by an asterisk in Appendix B), surveys may be submitted prior to March 31, 1973. *(Reconsideration,* par. 56(4), § 76.54(c) Note.)

1.22. "Additional Service": Distant Signals.

Upon a basis of the foregoing standards for complements of local signals that must be carried on request by television stations and may be carried in the absence of such request,the FCC has established a concept of "adequate service." "Adequacy" varies here according to market size, and on it hinges the determination of what "additional service" may be imported from other markets. The FCC formulated its concept according to its goal of "an equitable distribution of facilities" and its belief that "more television stations can be economically supported in areas of greatest population." *(Report and Order,* par. 88.) While its aim, in short, is not to afford everyone the variety of choice available in New York and Los Angeles, it is to endow all people with "the degree of choice that will afford them a substantial amount of diversity and the public services rendered by local stations."

The rules establish quotas of commercial signals that may be carried (including local stations) for four different categories of markets, whose principal features are next discussed. (Educational and foreign language carriage is discussed in Secs. 1.224 and 1.225.) There are further carriage provisions that apply across the board; these are considered subsequently. Because separate treatment of these issues prevents a concise overall view of the package of signals for any particular market size, the reader should refer to the rules themselves for listings of complete quotas of local and distant signals applicable to each market category. (See also Fig. 1.)

1.221. General Rules. After carriage of required local signals, i.e., stations within 35 miles, those from the same market, and those meeting the viewing test, the following are the complements of signals up to which additional service may be achieved if not thereby available:

(1) *In television markets 1-50* (as listed in § 76.51(a), New York City through Little Rock, Arkansas)
   (a) Three full network stations and
   (b) Three independent stations
(2) *In television markets 50-100* (as listed in § 76.51(b), San Diego through Columbia, S.C.)
   (a) Three full network stations and
   (b) Two independent stations
(3) *In "smaller television markets"* (i.e., in the "specified zone" of a commercial and television station in communities not listed in § 76.51) as defined in § 76.5(i)
   (a) Three full network stations and
   (b) One independent station
(4) *In communities outside of all major* and smaller television markets, "any additional television signals may be carried."[21]

---

[21] For the full text of the carriage rule for markets 1-50, see § 76.61; for markets 51-100, see § 76.63; and for smaller markets, see § 76.59.

For the full text of the carriage rule for systems operating in communities located outside of all major and smaller television markets, see § 76.57. In these areas, where distances between sparse and isolated populations are often great, the FCC indicates in the *Report and Order* a somewhat more stringent policy with respect to "local" signals than that expressed in the permissive language of § 76.51. In particular, the range of commercial stations required to be carried as "local" is extended from the "specified zone" (35 miles) applicable in other market situations, to the Grade B contour (§ 76.57(a)(1)). The FCC, noting petitions by smaller market broadcasters serving many of these areas, says that it is

These provisions represent basic quotas of signals, within which distant signals can be added (in accordance with stated FCC priorities), and hence represent a basis of economic opportunity on which cable systems must now approach the development of service. Note, however, that adding a minimum of two distant signals, next discussed, can somewhat expand these basic quotas for many markets.

1.222. Minimum Distant Signals: The "Wild Cards." The foregoing rules of limitation are subject to a qualification for major markets in the form of what has been called "wild cards," which permit certain increases over the basic quotas: "Cable systems in major markets are *in any case* permitted to carry two [additional independent] signals beyond those whose carriage is required under the mandatory carriage rules." *(Report and Order,* par. 90. Italics added.) For smaller markets (below 100), however, no signal-bonus is permitted.

The major market bonus (§§ 76.61(2)(c) and 76.63(a)) significantly applies only to *independent* signals and is reduced by the number of distant signals added pursuant to the above-mentioned quota. Thus, if a market is already served by its full complement of "local" signals, two additional independent signals may be added, but if two signals (network or independent) are needed to bring local service up to the applicable quota, the bonus has been discounted. (If one signal were added, one additional distant independent would remain available.)[22]

COMMENT: The obvious effect of the bonus rule is to ensure that in every market cable systems will be able to add at least two distant signals, so as to attract viewers by offering them more programing than is available locally over-the-air.

1.223. Manner of Selection: "Leapfrogging." Citing a "risk" that, in choosing distant stations for importation, "most cable systems would select stations from either Los Angeles, Chicago, New York, or one of the other larger markets," the FCC has adopted "leapfrogging" rules to ensure that regional programing "more likely to be of interest in the cable community" will be imported. *(Report and Order,* par. 92.)* To achieve this end, the FCC has adopted the following rules applicable to cable systems in all television markets:

(1) For importing *network* affiliates, the cable system "must afford priority of carriage to the closest such station or, at the option of the cable system, to the closest such station within the same State." *(Report and Order,* par. 93. See §§ 76.59(b)(1); 76.61(b)(1); and 76.63(a). Italics added.)

COMMENT: While most cable systems can be expected to import the closest station, some may have other preferences. Perhaps a cable system in Charlottesville,

---

... requiring that these smaller market signals, where significantly viewed, *must* be carried on all new cable systems and on all existing systems with sufficient channel capacity—even if the cable community is beyond Grade B contours—and, as to new systems must be afforded simultaneous nonduplication protection. (§§ 76.57(a)(4) and 76.91(c); *Report and Order,* par. 91. And see Sec. 2.3 below.)

The Commission also affords broadcasters the possibility of petitioning for additional relief, and assures continuing service to ensure that local signal nonduplication protections in these areas are adequate protections.

[22] On reconsideration, the FCC has also permitted cable systems in smaller and major markets to carry *network* programing that is not carried on network stations normally carried by the system. Thus, where a particular network program is not "cleared" by the network's local affiliate, the cable system may show that particular program. *(Reconsideration,* par. 19; §§ 76.59(d)(2); 76.61(e)(2); and 76.63(a).)*

Virginia may wish to import a network station from Norfolk, Virgina rather than Washington, D.C., in the interests of enhancing its regional identity. In notes to the aforementioned rules, the FCC has indicated that on petition under § 76.7, it may waive its priority rules.

> (2) With respect to *independent* stations, the first *two* such stations carried may be from any market at the choice of the operator, but if selected from the top 25 markets must be taken from one or both of the two closest markets. "Systems permitted to carry a *third* independent station are required to select a UHF station from within 200 miles," if available, or a VHF independent from within 200 miles, if available, or finally, any UHF independent. *(Report and Order,* par. 93. See §§ 76.59(b)(2); 76.61(b)(2); and 76.63(a). Italics added.)

Deletions of programing from distant independent stations are sometimes required because of exclusivity rules (see Secs 2.3-2.32 below), and are sometimes permitted for major market cable systems when such programing is of purely local interest in the community where the signal originates. Cable systems may replace such a deletion with "a program from any other television broadcast station" (not itself barred by program exclusivity protections.)[23] (§§ 76.61(b)(2)(iii) and 76.63(a)).

COMMENT: The leapfrogging rules aim to put at rest the long-standing fear that attractive New York and Los Angeles independent stations will balloon into cable networks and stunt the growth of regional television. The FCC's rules seem designed to strengthen regionalism, permitting economic considerations (the cost of microwaving signals from distant areas) to control cable interest in importation from far-distant markets. In addition, the FCC has opened up the limited possibility that particularly attractive programs from independent sources may be "brokered" to fill up limited portions of time blacked out on independent signals. Depending on the ingenuity of cable operators—and the real costs of making such programs available over great distances—these opportunities may become a significant increment to the attractiveness of cable services.

1.224. Educational Stations. The FCC recognizes "possible erosion of local support among cable television subscribers" as the "principal concern" of local educational stations with respect to the importation of distant educational stations. *(Report and Order,* par. 94.) Because local educational stations in large measure rely on voluntary contributions of viewers, this concern represents a justification for protectionism somewhat different from the situation of local commercial broadcasters.

The specific provisions of the FCC's rules require all cable systems to carry, on request, all educational stations within 35 miles and those placing a Grade B signal on all or part of the community of the cable system. (§§ 76.57(a)(13); 76.59(a)(2);

---

[23] § 76.61(b)(2)(ii) provides that such substituted programing "may be carried to its completion, and the cable system need not return to its regularly carried signal until it can do so without interrupting a program already in progress." On reconsideration, the FCC rejected proposals by cable systems that it broaden this proviso to a full "cherry picking" rule—permitting cable operators themselves to package all permitted distant programing in this fashion rather than use particular signals—and by broadcasters seeking to eliminate any flexibility in returning to normal programing schedules. *(Reconsideration,* pars. 13-18.) Recognizing there may be "unusual circumstances that do not fit the rule," the FCC preserves its flexibility to respond to waiver petitions under § 76.7.

76.61(a)(2); and 76.63(a).) In addition, for major and smaller markets the rules permit carriage of any noncommercial educational stations operated by the State in which the cable system is located, and also any other educational station, unless an objection is filed (pursuant to § 76.7) by any local noncommercial educational stations or State or local educational television authority. (§§ 76.59(d); 76.61(d); and 76.63(a).)

COMMENT: The FCC feels that, "In the absence of objection, . . . the widest possible dissemination of educational and public television programing is clearly of public benefit and should not be restricted." (Report and Order, *par. 95.) But the FCC also indicates that it does "not ordinarily anticipate precluding carriage of State-operated educational stations in the same State as the cable community." On reconsideration, the FCC recognized that it may be called upon to deal with waiver petitions where "carriage of both local and distant educational signals would cause an unwarranted profusion of educational signals."* (Reconsideration, *par. 12.)*

1.225: Foreign Language Stations. The FCC permits "cable systems to carry non-English language programing without limitation." *(Report and Order,* par. 96; §§ 76.59(d)(i); 76.61(e)(i); and 76.63(a).) The effect of these provisions, applicable in all markets, is to permit carriage of distant foreign language stations (originating most likely in the United States, the Caribbean, or across the Mexican and Canadian borders) without charging such carriage against signal quotas. While noting the opposition of some Spanish language stations in California, Texas, and Florida to suffering foreign competition, the FCC determined that such programing would in many instances be available over-the-air, that encouraging diversity among foreign language sources for non-English groups is desirable, and that special relief might be warranted if a "local station demonstrates that such importation will adversely affect its ability to serve the public." *(Report and Order,* par. 96.) This conclusion was reaffirmed on reconsideration; the FCC held that "it is in the public interest to make foreign language programing available without impediment." *(Reconsideration,* par. 23.)

COMMENT: In some areas of the country, carriage of foreign language stations could enhance the ability of cable systems to draw subscribers. Questions of copyright liability for carriage of signals from other countries will be among the questions requiring resolution by legislation as part of a package of copyright revisions.[24]

1.23. Manner of Carriage

The FCC established rules governing the carriage of local and distant signals. To ensure that cable carriage does not reduce the viewability of local broadcast signals—to the competitive disadvantage of local broadcasters—a rule prescribes the manner of carriage for such signals. Briefly, § 76.55(a) provides as follows for signals required to be carried:

(1) The signal shall be carried without material degradation in quality and subject to certain technical standards set forth in §§ 76.601 through 76.617.

---

[24] On reconsideration, the FCC declined to accord comparable treatment to "religious and other specially programed stations," finding the former category amply available throughout the country and the latter category incapable of reasonable differentiation. *(Reconsideration,* pars. 21 and 22.)

(2) On the request of the originating station, a cable system must carry the relevant signal on the channel number of the station, unless technically unfeasible.

(3) On request of the originating station, a cable system must carry the relevant signal on only one channel.

In addition, for all broadcast signals carried, a cable system must carry programing in full, "without deletion or alteration of any portion except as required by this part." (§ 76.55(b). See rules on exclusivity in Sec. 2.3.) Moreover, with respect to signals of translator stations required to be carried within the complement of local signals, such signals need not be carried if the cable system is also carrying the signal of the originating station or if the community of the cable system "is located, in whole or in part, within the Grade B contour of a station carried on the system whose programing is substantially duplicated by the translator station." (§ 76.55(c).)

COMMENT: Inasmuch as § 76.55 precludes alteration of broadcast signals required and permitted to be carried by a cable system, the cable system may also be protected from legal liability under State law for the content of programing so carried. This protection derives from the doctrine of Farmers Educational and Cooperative Union of America v. WDAY, Inc., 360 U.S. 525 (1959), which established that a broadcast station legally required to carry a political broadcast without power of censorship is sheltered from State laws against defamation. Since WDAY arose under a statutory requirement, however, its applicability to carriage mandated by agency regulation has been questioned. On the other hand, broadcast stations may, by virtue of FCC mandatory carriage rules, possess private rights of action against cable systems for failure to carry their signals in accordance with FCC rules, principally under common law tort theories, in addition to their rights of recourse to the agency itself. On reconsideration, the FCC declined to sanction payment of a consideration from a distant broadcast station to a cable system to ensure local carriage, and validated such practices with respect to "microwave costs to carry a signal that a cable system could not otherwise afford." (Reconsideration, par. 20.)

1.24. Applicability: "Grandfathering"

A final provision of the broadcast carriage rules governs their applicability. § 76.65 provides that carriage which was lawful as of the date the rules went into effect shall be permitted to continue, even if it conflicts with the provisions of the new rules. For systems whose carriage had been restricted by FCC order (for example, the situation in San Diego as the result of a celebrated 1966 challenge), expansion into hitherto restricted areas must in all respects be in accordance with the new rules. *(Report and Order, par. 107.)* Wherever "grandfather rights" exist, however, other systems operating or later commencing operations in the same community shall be entitled to the benefit of such authorizations, to avoid competive disadvantage. *(Report and Order, par. 105.)* For additional grandfathering provisions dealing with exclusivity protections, see Sec. 2.32 below.

Section 2

PROGRAM EXCLUSIVITY AND COPYRIGHT

The FCC has long recognized that resolving broadcast carriage issues is not enough to still the apprehensions of local broadcasters and program suppliers who are disturbed about the prospects of cable growth. They are disturbed because they view cable as threatening to make inroads on the significant amount of programming they distribute to individual broadcast markets.

Some programs are distributed through networks which make payments to individual affiliated stations in proportion to the audiences those stations furnish for network-provided advertising. Other distribution, typically to independent stations, is provided by "syndication" of programming direct to individual broadcasters, some of which is programming originally sold via network "primary markets" and later distributed to independent "secondary markets." These relationships are contracted for on an exclusive basis in each market, and the fees paid reflect the value to advertisers in each market of the size of the audience accumulated for each program. If cable systems are allowed to present the same programs that are sold in other markets, the audiences for local stations will drop. Local stations and copyright owners have contended that the destruction of secondary markets reduces the revenues that program producers can earn nationwide, and at the same time deprives many local stations of the protection of exclusivity for which they have bargained with program suppliers. Cable operators contend that their carriage of "local" signals to subscribers who might otherwise be unable to receive programs enhances revenues of local stations, while carriage of "distant" signals merely rearranges on more national or regional bases the points from which royalties should be calculated and copyright payments made.[1]

The potential disadvantage is especially great because the present state of copyright law does not require cable systems to make copyright payments for such programming. If it did, cable systems and broadcasters would be on a more equal footing. Both local broadcasters and program producers have therefore vociferously objected to cable importations as "unfair competition," a plaint that gave rise to the FCC's long-standing "freeze" on cable growth. In relaxing that freeze, the FCC has felt obliged to protect program exclusivity on signals allowed to be carried by mandating that major market cable systems delete certain programs that compete with local broadcasting. The industries involved have also, in the context of the 1971 "Consensus Agreement," committed themselves to sponsoring new copyright legislation intended to solve what problems of "unfair" competition may remain.

[1] See "CATV and Copyright Liability: On a Clear Day You Can see Forever," 52 *Virginia Law Review* 1505(1966) and "CATV and Copyright Liability," 80 *Harvard Law Review* 1544(1967).

2.1. THE COPYRIGHT BACKROUND

The 1968 decision of the Supreme Court in *Fortnightly v. United Artists Television, Inc.*[2] threw the communications industry into an uproar by ruling that cable systems were not liable to program owners for copyright payments for carriage of broadcast signals. Many commentators before *Fortnightly* had predicted liability, and foreseen a need for new copyright legislation to alleviate the resulting burden of program-by-program negotiations with diverse suppliers.[3] Without legislation to benefit cable interests, and to lessen retroactive liability, a bleak future had been anticipated for cable.

The *Fortnightly* case had been initiated by a group of large owners of copyrighted motion pictures to test their rights to acquire copyright royalties for CATV reception.[4] Systems owned by the Fortnightly Corporation in Clarksburg and Fairmont, West Virginia, carried television signals (some of which were "distant" but all off-the-air, without microwave) from Pittsburgh, Pennsylvania, and Steubenville and Wheeling, West Virginia. The Supreme Court's opinion was a stunning victory for the CATV industry. Viewing the functions of cable distribution in terms of a simple distinction between exhibition and viewing, the Court held that cable transmission would fall on the viewer's side of the line. Cable, said the Court, is essentially a cooperative function of a number of individual receiver owners—none of whom would be individually liable for copyright fees for merely turning on a TV set—cooperating through the instrumentality of the cable entrepreneur to acquire available programming. Cable's transmission role in making reception possible, the Court felt, is no more "active" than that of television sets themselves, since "broadcasters select the programs to be viewed; CATV systems simply carry, without editing, whatever programs they receive." On this basis the Supreme Court determined that merely picking up locally available television signals on a CATV antenna and distributing them by cable did not constitute a "performance" infringing a copyright holder's statutory rights to "perform . . . in public for profit" under 17 U.S.C. 1(c). The CATV system, according to the Court, was merely a "passive beneficiary" analogous to a viewer's own TV set or antenna. (392 U.S. at 399-400.)

By freeing cable systems from copyright liability, at least for the types of programs there involved, *Fortnightly* intensified pressure on the FCC to halt cable growth. That pressure arose predominantly with respect to the issue of carriage of "distant signals"—i.e. broadcast television not locally available by whatever technical definition might be applied—whose economic significance to cable operators, local broadcasters and copyright owners was vastly greater than that of local signals. Choosing to read *Fortnightly* as narrowly limited to its facts, broadcasters and copyright owners pressed additional litigation—*CBS-TelePrompTer*—to secure a favorable judicial distinction with respect to distant signals under existing copyright legislation. While that litigation ground its way at a slow pace through the Federal courts, cable access to additional distant signals was at first frozen by FCC regulation and only later relaxed—beginning in 1972 pursuant to the *Report and Order*, which limited use of distant signals (See Section 1, Part II), imposed further

---

[2] 392 U.S. 390(1968).
[3] See, e.g., Thomas G. Shack, "Copyright—The Need for Legislation," *CATV*, September 4, 1967, pp. 20-21.
[4] E. Stratford Smith, "The Emergence of CATV: A Look at the Evolution of a Revolution," *Proceedings of the IEEE*, 967, 973(1970).

protections to enforce exclusive market contracts, and reflected the commitment of all parties to a formula for early copyright law revision. With *CBS-TelePrompTer* resulting in March 1974 in another victory for cable interests and removing all ambiguities under existing legislation, the forces holding the parties together in an expedient compromise may have been significantly weakened.

The *CBS TelePrompTer* litigation was commenced in 1964 but delayed through a combination of chance and time-marking lethargy until final resolution by the Supreme Court in March 1974. In the case, five illustrative systems were sued by CBS and independent program production companies, in instances of signal transmission over great distances (in one case over 450 miles) via a combination of cable and point-to-point microwave. As the case reached the Supreme Court, the petitioners deferred to *Fortnightly's* line between broadcasters and viewers but claimed that two factors rendered distant signal carriage determinative of cable system functions as a broadcaster. Those factors were the making available of television not otherwise distributed in the general area of distribution and the circumstances that defendant cable systems also originated programming, shifted programming among themselves, and sold advertising on origination channels, thus rendering their service more analogous to a broadcast-type "performance" than the reception service freed from liability in *Fortnightly*. The District Court dismissed the suit on all grounds, 355 F. Supp. 618 (S.D.N.Y., 1972). The Court of Appeals, however, remanded the case to the District Court, for specific findings, 476 F. 2d 338 (2d Cir., 1973) construing the copyright law as compelling liability when signals could not be locally received over-the-air and imposing a burden of proving, where microwave is the method of transmission, that the signals were not "distant" signals.

The Supreme Court reversed the Court of Appeals, under *Fortnightly*, finding the acquisition of distant signals no more capable of transforming cable reception into a broadcast performance than where such signals are locally available. The Court found that in both instances, (where the reception is simultaneous with the delivery to cable subscribers[5]) reception services are essentially viewer functions, irrespective of the distance between the broadcasting station and the ultimate viewer. The Court further buttressed this holding by finding that cable systems exercised no control over the content or format of signals carried, as would broadcasters, notwithstanding the exercise of control over accompanying origination channels. The Court declined to view the changes in business practices reflected in cable competition in secondary markets as a concern of the Copyright act once copyright material has been originally released to the public. The Court viewed advertising markets dynamically, forecasting readjustment of rates paid by broadcasters in accordance with the totality of their viewers (over-the-air and via cable), rather than

---

[5] Since *Fortnightly*, additional copyright cases have established liability for cable systems where nonsimultaneous reproduction is involved. Liability has been upheld in one case in which a cable system taped programs off-the-air and replayed them a week later over the CATV system thousands of miles distant. *Walt Disney Production v. Alaska Television Network*, 310 F. Supp. 1073, (W.D. Wash., 1969). Moreover, CBS, a plaintiff in the *TelePrompTer* infringement suit, settled a copyright suit it had brought against a CATV system on the island of Guam, where a system had imported (on tape) a delayed signal from a mainland CBS affiliate. After suit was brought in the District Court for the Central District of California, a settlement reportedly required the cable system to pay $5000 plus $1.00 per subscriber per year for additional CBS programming not carried by CBS's Guam affiliate. See *Broadcasting*, March 27, 1972, p. 38.

permitting a system of double charges—upon both the broadcaster and the cable operator—to develop. Though a stern dissent by Justice Douglas criticized the cable industry as engaging in piracy and the Court as legislating changes in the Copyright Act, the six-member majority opinion (by Justice Stewart, who also wrote the Court's opinion in *Fortnightly*) viewed allegations of economic dislocation with equanimity—leaving necessary adjustments in the Copyright Act of 1909 to the Congress.[6]

2.2. THE REGULATORY RESPONSE

To stem the cries of anguish that followed *Fortnightly's* release of cable systems from much copyright liability, the FCC in 1968 adopted a provisional regulatory response attuned to the specific loophole that the Supreme Court apparently opened up with respect to distant signals. Terming importation of distant signals without payment of copyright "unfair competition" for local broadcasters, the Commission proposed to permit such importation where cable systems had secured "retransmission consent" from originating stations—a step that presumably would have led to contractual agreements removing any claims of inequity (Notice of Proposed Rule Making and Notice of Inquiry in Docket No. 18397, 15 FCC 2d 417 (1968). In the absence of such consent—which few cable interests (basking in the glow of *Fortnightly*) sought, and no broadcasters (conscious of their own vulnerabilities) gave—cable acquisition of distant signals ground to a halt. The resulting paralysis in cable growth merely shifted the source of pressure upon the FCC, broadcast interests seeking to halt cable now being replaced by cable interests, pressing for a new and more hospitable basis on which cable growth might resume.

In its *Letter of Intent* of August 5, 1971,[7] the Commission foreshadowed the approach it later took by indicating, in connection with its proposed limited relaxation of distant signal restraints, that program exclusivity "is a matter that has both copyright and regulatory implications," and that program exclusivity rules would be part and parcel of any relief. This declaration focused attention of affected industries on the possibility of reconciling adverse interests by a twofold approach under which some modicum of distant signal carriage would be permitted, while regulatory protections would curb particular forms of programing pending the passage of binding copyright legislation. The result was expressed in the "Consensus Agreement" negotiated in the fall of 1971 and later in the rules issued with the *Report and Order*, to which the text of the Consensus Agreement was appended.

> (i) First, the agreement will facilitate the passage of cable copyright legislation. It is essential that cable be brought within the television programing distribution market ...

> (ii) Passage of copyright legislation will in turn erase an uncertainty that now impairs cable's ability to attract the capital investment needed for substantial growth. The development of the industry, at least with respect to assessing copyright costs, would be settled by the new copyright legislation and its future no longer tied to the outcome of pending litigation.[8]

[6] The Slip Opinion of the Supreme Court, March 4, 1974, is set forth as Item 8, Part III.
[7] 31 FCC 2d 115 (1971).
[8] In its footnote to this text, the Commission expresses the view that passage of new legislation would make a decision in *CBS-TelePrompTer* "significant ... only with respect to past liability."

(iii) Finally, the enactment of cable copyright legislation by Congress—with the Commission's program before it—would in effect reaffirm the Commission's jurisdiction to carry out that program, including such important features as access to television facilities.[9]

Thus, the Commission's development of distant signal rules, as well as exclusivity protection and industry sponsorship of copyright revision, have been made parts of an interdependent package.[10] The agency says that, "Without Congressional validation . . . we would have to reexamine some aspects of the program." (*Report and Order*, par. 65.) Its steps with respect to distant signals having previously been considered, Secs. 2.3 and 2.4 deal with its exclusivity rules in the context of the framework agreed upon by the industry parties.

2.3. PROGRAM EXCLUSIVITY

The FCC's program exclusivity rules require cable systems to delete certain programs from distant stations, depending on the size of the cable system's market and the type of distant stations carried (network or independent). The rules also protect exclusive contracts of local stations from out-of-market competition.

The FCC's aim is "both to protect local broadcasters and to insure the continued supply of television programing"—that is, to protect both broadcast and cable program sources. (*Report and Order*, par. 73.) Its method imposes greater programing protections in the top 50 markets, "from which the bulk of program supplier revenue is derived and where these restrictions are consequently most needed to insure the continued health of the television programing industry." In markets 51-100, where network representation tends to be plentiful but the local availability of independent signals more restricted, greater opportunities for program carriage are created. Below the 100th market, the distant signal rules themselves are felt to restrict cable competition sufficiently so that no further protections of exclusivity are warranted.

COMMENT: The economic impact of the FCC's exclusivity rules on the prospects of cable growth are comprehensible only in terms of projected subscriber response for particular cable systems. Nonetheless, there is general agreement that the greatest opportunities for cable expansion are created in the 51st to 100th markets as a consequence of the limited availability of additional distant signals and the limited prohibition of the exclusivity provisions. Note, too, one compensating effect of the FCC's rules: major-market cable systems that black out programs to protect their exclusivity may fill in these time slots with signals from "any other television broadcast station." (§ § 76.61(b)(2)(ii); 76.63(a).)

[9] The Commission's reference to the issues in *Midwest Video* is obvious, and it believed copyright legislation would be curative of any adverse result.

[10] Confirming the adage that one man's meat is another's poison, Commissioner Nicholas Johnson (concurring in part and dissenting in part) attacked the Commission's rules as an expression of "government decision making at its worst" and an example of "Presidential interference" in the operation of an independent agency; he has forecast legal deficiencies under the Administrative Procedure Act, 5 U.S.C., § § 551 et seq., § § 701 et seq., and the Fifth Amendment to the U.S. Constitution, deriving from procedural aspects of the formulation of government policy and rules by delegation to an industry cabal. His preliminary and fuller opinions concurring in the *Report and Order* were not published in the *Federal Register*; they appear in Part III.

The complexities of the exclusivity rules are largely of specialized interest to cable television operators. The broad outlines of the rules, however—their mechanics and areas of applicability—can be set forth to permit appreciation of their role in the FCC's regulatory program.

2.31. Mechanics of Exclusivity Rules

The rules give rights to both parties to exclusive contracts—broadcast stations and copyright owners—to force cable systems carrying distant signals to observe the exclusive provisions of their contracts. § 76.153(a) accords such rights to copyright owners, and subparagraph (b) to local stations; such stations must have, however, "an exclusive right to broadcast [a particular] program against all other television stations licensed to the same designated community and against broadcast signal cable carriage of that program in the cable system community." (§ 76.153(c).)

§ 76.155 details the methods for notifying cable systems when they are to respect exclusivity. Cable systems in major markets are also required to keep and maintain for public inspection accurate logs of all distant programs carried. Once a protected party notifies a cable system that it wants exclusivity protection for a program, the FCC's rules come into play. They operate with significant differences for "network" and "syndicated" programing.

For *network* programing, there is an ordering of priorities among programs for which exclusivity is asserted, and the rules oblige cable systems to observe those priorities. Thus, § 76.91(a) provides that, on request of a local broadcast station, a cable system within the Grade B contour of the station or within the community of a 100-watt or higher translator must "maintain the station's exclusivity as an outlet for network programing against lower priority duplicating signals, but not against signals of equal priority;" exclusivity hence applies only to lower priority signals. § 76.91(b)'s order of priority is as follows:

First, all television broadcast stations within whose principal community contours the community of the system is located, in whole or in part;

Second, all television broadcast stations within whose Grade A contours the community of a system is located, in whole or in part;

Third, all television broadcast stations within whose Grade B contours the community of a system is located, in whole or in part;

Fourth, all television translator stations with 100 watts or higher power, licensed to the community of the system . . .

For *network* programing, in accordance with these priorities, protection against simultaneous duplication is provided, if proper notification has been received pursuant to § 76.93, with the possibility of special FCC relief providing for longer exclusive periods.[11] Exceptions involving priorities where cable system wish to transmit, in color, programs shown over the air in black and white, and allowing limited contractual agreements modifying exclusivity rules between broadcasters and operators, are permitted (§ 76.95) and the prospect of waivers is stated (§ 76.97).

[11] On reconsideration, the FCC amended § 76.93(b) to provide "same-day" exclusivity protection to network stations not in the top 50 markets in the Rocky Mountain time zone, essentially putting a burden of justification upon cable operators to cut back the exclusivity period. (*Reconsideration*, par. 29.)

Additional exclusivity rules in § 76.151 cover *independent, "syndicated"* television programing, but only against distant signals and varying according to market size. Thus, in markets 1 to 50, one-year protection is provided from the date a wholly new program is first sold or licensed anywhere in the country, during which cable systems may not import such programs. Thereafter, exclusive contracts are protected for their specific duration, if the contract specifically excludes showings by broadcast stations and cable distribution. In markets 50-100, the following standards apply for syndicated programs:

1. For series programs that have had a prior showing on network television, the programing is protected for one run or one year from the broadcast of the first episode of the series in the market, whichever is the shortest period.
2. For series that have had no previous network exposure, the protection will end two years from the broadcast of the first episode in the series in the market.
3. For programs that are nonseries but are neither feature films nor other special products, protection will end two years from the date that the program first becomes available in the market.
4. For feature films, protection will end two years from the date the feature film first becomes available in the market.
5. Other programs are protected for one day after first nonnetwork broadcast in the market or one year from date of purchase, whichever occurs first.
6. Where a distant station broadcasts a program in prime time and the same program is not broadcast in prime time by the local television station, the exclusivity protection rule does not apply.

Additional rules govern series previously shown on networks, and define applicable terminology—specifically, syndicated programs, off-network series, first-run series, first-run nonseries, and feature films. § 76.5(p)-(t).

COMMENT: The burden of enforcing these FCC rules will fall largely on the interested parties. Holders of exclusive rights must notify cable systems that they want exclusivity for specific programing, and both sides must then assure that exclusivity is in fact achieved. "To a real extent, the whole area of program exclusivity will work only if there is good faith."(Reconsideration, par. 42.) Given the diversity of situations covered, informal patterns of compliance are likely to evolve. Once copyright legislation is enacted, the prime justification for exclusivity protections may be eliminated, facilitating their repeal.

2.32. Applicability: "Grandfathering"

All carriage of signals prior to March 31, 1972, the effective date of the rule, is "grandfathered," pursuant to § 76.91(c) for network programing and § 76.159 for independent programing, but all new service is subject to the rules. Thus cable television service predating the effective date of the new rules will be subject to exclusivity standards differing from those for new service—particularly with respect to network programing which previously (under former § 74.1103) was entitled to "same-day," as opposed to "simultaneous," protection.

2.4. PENDING COPYRIGHT LEGISLATION

The terms of the Consensus Agreement obligate all signatories to support special cable copyright legislation and to seek its early passage.

Under the Agreement, liability for copyright infringement, including an obligation to respect valid exclusivity agreements, would be established for all radio and television broadcast signals except for independently owned cable systems now in existence with fewer than 3500 subscribers. On the other hand, no greater exclusivity against distant signals may be contracted for than the FCC may allow by rule. "Compulsory licenses" (i.e., rights to carry signals upon payment of reasonable royalties) would be granted for all *local* signals, those *distant* signals defined and authorized by the FCC, and grandfathered signals. While the FCC could authorize additional distant signals, no compulsory license would be granted for such signals, nor would the FCC limit the scope of exclusivity agreements for such signals beyond the limits applicable to over-the-air showings.

A schedule of fees would be adopted or, failing that, a commitment to binding arbitration. Broadcasters, as well as copyright owners, would have the right to enforce exclusivity rules through court actions for injunction and monetary relief.

The compromise was immediately subject to criticism, most notably in a letter from CBS President Dr. Frank Stanton. (See Part III.) Dr. Stanton believed the Consensus Agreement was arrived at too hurriedly, and criticized it for calling for compulsory licenses instead of negotiated payments, and for conceeding privileges without also erecting protections to broadcasters. On the other hand, the only pending legislation in the 93d Congress—S. 1361, whose pertinent part (§ 111) appears in Part III—has drawn a large measure of support from official representatives of the organizations that negotiated the Consensus Agreement. However, the cable operator's new freedom from all liability for carrying broadcast signals under *CBS-TelePrompTer* and *Fortnightly* now shifts the "burden of going forward" to broadcast interests and, incidentally, threatens their coalition with independent copyright interests (who might profit more from a policy of liberal cable growth than from a fee level that might curb that growth).

While hearings on S. 1361 have been held, a bill has not at this writing been reported out of the Senate Judiciary Committee and, in any event, early House passage is doubtful. Thus a status quo more congenial to cable interests than to broadcasters may be expected to continue, pending legislation or changes in regulation. In such a time of controversy, any consensus for policy change is likely to prove hard to come by. Significantly, renewed turmoil in the copyright field over cable has arisen at a time of growing "national debate"—to use the President's words—over the regulatory future of cable (prompted by the Whitehead Report), creating the likelihood of more and fresh scrutiny of national cable policies in the years ahead.

Section 3

CABLECASTING AND CHANNEL CAPACITY

In the *Report and Order*, the FCC established a set of comprehensive rules governing "cablecasting"—the cable presentation of video and nonvideo material not available to the public on over-the-air television. The Commission's aim is to realize some of the "technological and economic potential of an economy of abundance" that broadband communications can offer. *(Report and Order, par. 117.)* By so doing, the Commission has mandated the development of a major new factor "within a national communications structure," aiming to bring closer "the economy of abundance" and to move the cable medium well beyond the passive reception service it originally supplied.

In the process, extensive rules have been laid down, to which cable systems must adhere and which local government authorities must recognize in the grant and administration of franchises. The FCC has thereby preempted much of the regulatory responsibility for cablecasting from state and local authorities, the extent and implications of which are treated in Sec. 4. Federal regulations are doubly significant, in terms of both the local cable services they will permit and the opportunities they create for use of broadband facilities by local government entities. At the same time, however, the FCC's venture beyond broadcasting is a new departure in federal regulation, the lawfulness of which has recently been judicially upheld in large part and from which significant practical consequences will flow. The FCC has taken a bold step forward, not only in terms of cable regulation, but in enlarging the role that regulatory agencies can now play to protect the public's interest in its development. This section explores these considerations, concluding with an appraisal of longer-range implications for future federal regulation of cable television.

3.1. BACKGROUND

In 1969, the FCC broke with past precedent[1] and promulgated a rule requiring cable television systems with more than 3500 subscribers to originate special pro

---

[1] Prior to that time, the Commission had hewed to a line well within the boundaries of limited authority over cable systems recognized by the Supreme Court as "reasonably ancillary to the effective

graming as a condition for the carriage of television broadcast signals.[2] This rule—recently upheld by the Supreme Court (by a 5-to-4 vote) in *U.S. v. Midwest Video Corp.* (see text in Part III), against a challenge based on a literal view of the FCC's jurisdiction[3]—now serves in amended form as the core of the FCC's cablecasting rules issued in the *Report and Order* and codified at 47 C.F.R. §§ 76.201-76.331. The long-awaited outcome of *Midwest Video* has now permitted the FCC's rules to become effective in accordance with their terms, subject, of course, to the possibility of continuing challenges on other than jurisdictional grounds. In the process, the prospect of a legislative review of FCC jurisdiction, precipitated by an adverse court ruling, has now been averted.[4]

At the outset of this detailed consideration of the FCC's cablecasting regulations, some perspective on their overall role within the FCC's regulatory program may be helpful. Clearly, the cablecasting requirements are aimed at developing the public service capacities of cable television, thereby comporting with the FCC's obligations under the Communications Act "to make available . . . to all people of the United States a rapid, efficient nationwide, and world-wide wire and radio communications service with adequate facilities at reasonable charges . . ." (47 U.S.C. § 151 (1972).) On the other hand, the regulatory burdens imposed on cable systems in competition with other media will also limit the role of cable television, inasmuch as the costs of mandated services will inevitably affect the attractiveness of cable television both to potential subscribers and investors. While this is not the place to speculate regarding the precise aims or the wisdom of FCC policies in defining cable television as "a hybrid that requires identification and regulation as a separate force in communications,"[5] it should be appreciated that the impact of cablecasting regulations will probably be a significant competitive factor in defining cable's future role. Overall, of course, the FCC's cablecasting policies will contribute

---

performance of the Commission's various responsibilities for the regulation of television broadcasting." *U.S. v. Southwestern Cable Co.*, 392 U.S. 157, 178 (1968). In one notable instance, *Midwest Television, Inc.*, 13 FCC 2d 478 (1968), it utilized the sanction of withholding authority to carry broadcast signals to *prohibit* origination of advertising material, while otherwise taking care to *permit* program origination without restriction, subject to the applicability of any rules which may be adopted by the Commission in this area," 13 FCC 2d at 510, but it took no steps to *require* such origination.

[2] On October 27, 1969, in its *First Report and Order in Docket 18397*, 20 FCC 2d 201, the FCC issued the following "mandatory origination" rule:

> (a) Effective on and after April 1, 1971, no CATV system having more than 3500 or more subscribers shall carry the signal of any television broadcast station unless the system also operates to a significant extent as a local outlet by cablecasting and has available facilities for local production and presentation of programs other than automated services . . .

> (j) *Cablecasting.* The term "cablecasting" means programing distributed on a CATV system which has been originated by the CATV operator or by another entity, exclusive of broadcast signals carried on the system.

[3] Enacted in 1934, the Communications Act, 47 U.S.C. § 151 *et seq.*, long predated the development of local distribution of television signals by wire, a communications function that was not specifically embraced within the literal provisions of the Act. Numerous subsequent efforts to amend the Act with respect to cable television have failed. In *Southwestern Cable,* the Supreme Court held that the FCC could prohibit the carriage of broadcast signals over cable as an aspect of the agency's express authority over broadcasting; in *Midwest Video,* this authority was extended to permit the FCC to require large cable systems to originate programing because an overall effect of cablecasting upon broadcasting was recognized.

[4] However, the recent Report of the Cabinet Committee on Cable Communications has proposed a long-range policy largely inconsistent with a mandatory origination rule, which policy would permit but not require some limited operator origination, and White House legislative proposals are anticipated. Meanwhile, the applicability of 47 C.F.R. § 76.201(a) has been suspended pending the outcome of *Midwest Video*, 36 Fed. Reg. 10876 (1971) and the rule has not yet been put back in force.

[5] See, e.g., Rivkin, "The Changing Signals of Cable T.V.," 60 *Georgetown Law Journal,* 1470 (1972).

substantially to establishing both the nature of cable service to the public and its economic viability.

3.2. BASIC CABLECASTING POLICY

The FCC makes clear that the development of cablecasting services, in addition to the carriage of broadcast signals, is an inextricable part of its regulatory program over cable television. While its rules with respect to cablecasting do not uniformly apply to all cable systems (as is discussed subsequently), the FCC has made clear its determination that the pattern of growth its rules will now foster emphasizes making available extensive nonbroadcast services in addition to the carriage of broadcast television. Its determination rests on the following principle:

> We emphasize that the cable operator cannot accept the broadcast signals that will be made available without also accepting the obligation to provide the nonbroadcast bandwidth and the access services described below. The two are integrally linked in the public interest judgment we have made. *(Report and Order,* par. 118.)

At the same time, the agency also makes clear that its new regulations are not intended to constitute a "complete body of detailed regulations but a basic framework" within which it may "measure cable's technological promise, assess its role in our nationwide scheme of communications, and learn how to adapt its potential for energetic growth to serve the public." *(Report and Order,* par. 117.)

The cablecasting policies center on three specific problem areas: the designation of channels, and prescriptions with respect to minimum numbers and transmission capability.

3.21. Channel Designations

Federal standards for cable television systems embrace a four-fold classification of signals, within which specific standards are applied or intended to be applied. As defined in § 76.5, they are as follows:

Class I Channels:
Channels carrying broadcast video programs—"A signaling path provided by a cable television system to relay to subscriber terminals television broadcast programs that are received off-the-air or are obtained by microwave or by direct connection to a television broadcast station." (Subparagraph (z))

Class II Channels:
Channels carrying video programs originated to all subscribers—"A signaling path provided by a cable television system to deliver to subscriber terminals television signals that are intended for reception by a television broadcast receiver without the use of an auxiliary decoding device and which signals are not involved in a broadcast transmission path." (Subparagraph (aa))

Class III Channels:
Channels carrying nonvideo and limited-access video programs—"A signaling path provided by a cable television system to deliver to subscriber terminals signals that are intended for reception by equipment other than a television broadcast receiver or by a television broadcast receiver only when used with auxiliary decoding equipment." (Subparagraph (bb))

Class IV Channels:
Channels carrying signals originated by subscribers—"A signaling path provided by a cable television system to transmit signals of any type from a subscriber terminal to another point in the cable television system." (Subparagraph (cc))

In short, Class I channels carry broadcast television distributed over cable. Class II channels carry original video programing to *all* subscribers. Class III channels carry nonvideo programing, and programing (video and nonvideo) to be viewed on a subscriber's television set with auxiliary equipment. Class IV channels carry subscriber-originated messages (video and nonvideo). Pay-cablecasting—using filters or decoding devices—therefore fits within Class III along with other, more esoteric "narrowcasting," such as direct-retailing, which could also use auxiliary equipment to project images on the viewer's television screen.

Within this overall framework, the FCC has imposed requirements as to the "designated channels" that cable systems must make available depending upon the size of the television market[6] and number of subscribers. These are as follows:

3.211. Origination Cablecasting Channel(s). There must be at least one channel for "origination cablecasting" (Class II) defined in § 76.5 as "programing (exclusive of broadcast signals) carried on a cable television system over one or more channels and subject to the exclusive control of the cable operator."

The origination-channel requirement, set forth in § 76.201, is an amended version of the rule earlier established in the *First Report and Order in Docket 18397* in 1969, whose enforcement the Court of Appeals for the Eighth Circuit had enjoined in *Midwest Video,* now reversed by the Supreme Court on review.[7] Significantly, the requirement is limited only to systems with more than 3500 subscribers, a measure of a cable system's economic strength applying *without regard to the size of the particular television market,* where a nexus to the Commission's jurisdictional powers exists through cable carriage of broadcast signals. The Commission's rule is as follows:

RULE:
(a) No cable television system having 3500 or more subscribers shall carry the signal of any television broadcast station unless the system also operates to a significant extent as a local outlet by origination cablecasting

[6] The distinctions between "major television markets" (the first 100 markets), "smaller television markets," and markets elsewhere, were previously indicated in Sec. 1.211.

[7] The earlier rule, 47 C.F.R. § 74.1111, contained no specifications as to the number of channels for cablecasting, but differed, possibly significantly, in that origination was to be by the "CATV operator or by another entity," whereas the new § 76.5 requires that origination cablecasting be "subject to the exclusive control of the cable operator." The FCC suspended § 74.1111 pending final judgment in *Midwest Video,* 36 Fed. Reg. 10876 (1971), and deleted it when the new rules were issued. On reconsideration, the FCC has maintained that the "origination provisions in the *Report and Order* represent a recodification of our original requirements, with one or two minor changes designed only to eliminate any possible confusion with the new access rules." *(Reconsideration,* par. 68.)

and has available facilities for local production and presentation of programs other than automated services. Such origination cablecasting shall be limited to one or more designated channels which may be used for no other purpose. § 76.201.

Subparagraphs (b) and (c) establish requirements additional to paragraph (a), and cover situations where cable systems outside of all major markets—regardless of the numbers of subscribers—originate programing (either voluntarily or pursuant to subparagraph (a)). Paragraph (b) provides that no system *outside of major markets* shall:

RULE:
(b) ... enter into any contract, arrangement, or lease for use of its cablecasting facilities which prevents or inhibits the use of such facilities for a substantial period of time (including the time period 6-11 p.m.) for local programing designed to inform the public on controversial issues of public importance.

Paragraph (c) imposes on *all* cable systems carrying broadcast television—whatever the market or the number of their subscribers, and whether origination is "voluntary or pursuant to paragraph (a) of this section"—the requirement that origination conform to certain specific operating requirements (discussed in Sec. 3.31 below).

COMMENT: *The FCC's origination requirements, though mandatory only for large-sized systems (of which there are few at present), affect all cable systems that carry broadcast signals and also originate programing. Though the precise requirement for large systems is ill defined, it is clear that both operations ("to a significant extent as a local outlet") and facilities ("for local production and presentation of programs other than automated services") are intended to be covered. It is important to emphasize that no further determinations in this area have been made, and, indeed, with respect to cablecasting facilities, the Commission's rules on Technical Standards (see Sec. 5.0) state that "technical standards are not now being provided"; rather, the experimental nature of the Commission's regulatory program is stressed. Outside of the major markets, § 76.201(a) limits the prospect that major blocks of time may be dedicated to purposes other than the provision of local information programing, probably by the cable system itself. It is also pertinent to note that the Commission has now abandoned its earlier indefiniteness with respect to the number of channels on which operator origination would be permitted;[8] the operator may now cablecast on "one or more designated channels." It is not clear what practical purpose this warrant is intended to serve (given the likely marginal profitability of origination channels at the outset of service), although it is possible that*

---

[8] The Commission's earlier rule was based on its "tentative view that one entity should not control the content of the program material on all cable channels not used for carriage of broadcast signals." It is believed that diversity of program choices is "more apt to be achieved if others besides broadcasters and CATV operators have access to the cable channel capacity to reach the public with communications of their choice, free from control by the CATV operator as to content." *First Report and Order,* 20 FCC 2d 201. Of course, the Commission's rules with respect to leased channels offer some assurance that this aim of maximizing diversity will now be achieved, subject to the actual effectiveness of the Commission's rules in defining the economic rights of the programers who may have access to such channels (see Sec. 3.3313).

this loosening of the origination-channel requirement may become a basis for the cable operator's engaging in pay-cablecasting (on a Class III channel). Finally, another change from prior policy may be significant even though the FCC calls it "minor" (see footnote 7): By precluding "another entity" besides the operator from programing the origination channels, and placing origination under the "exclusive control of the cable operator," the FCC has moved to ensure that origination cablecasting is undertaken by operators responsible to the FCC. At the same time, the FCC has apparently eliminated the earlier possibility that a new class of cable programers other than the operators of cable systems might emerge as a high-priority goal of the Commission's policies.[9] Simultaneously, by § 76.201(b)'s prohibition of "any contract, arrangement or lease for use of cablecasting facilities," the FCC may also have eliminated such prospect where origination is undertaken voluntarily.

*
(12–35)

3.212. Access Cablecasting Channels.

Aside from origination channels, cable systems in the *major* markets must make the following additional channels available for "access cablecasting" (without a number-of-subscribers standard being applicable):

> *[A]t least one specially designated, noncommercial public access channel available on a first-come, nondiscriminatory basis. (§ 76.251(a)(4).)*

Local authorities in other areas are free, by local action, to make similar standards applicable.[10]

To satisfy an "increasing need for channels for community expression," the FCC requires cable systems to make available "a practical opportunity to participate in community dialogue through a mass medium." *(Report and Order,* par. 122.) The *Report and Order* provides, moreover, that to achieve its "goal of creating a low-cost, nondiscriminatory means of access," cable systems must make available "reasonable production facilities"; to this end, the FCC "expects" that cable systems will have capacity to facilitate public access programing in their own origination activities which could be made available, and that additional program production resources may be available from private institutions. *(Report and Order,* par. 142.) Thus, it

---

[9] To the same effect as the thrust of the above comment and the preceding footnote, the Commission's rules displace to the lesser-priority use of leased channels the possibility that vigorous programing entities—other than the cable operators themselves—might be encouraged to develop. The lesser priority to be accorded to the development of leased channels is clearly indicated by the FCC's ruling on reconsideration modifying the date at which its minimum channel and access channel rules are applicable to existing cable systems. As discussed also in Sec. 3.24, it now appears that existing systems seeking a federal certificate to add one or more broadcast signals before March 31, 1977 (when all major market systems must come into compliance) have to provide only one designated access channel for each broadcast signal added—in the following order of priority: "(1) public access, (2) education access, (3) local government access, and (4) leased access." (§ 76.251(c) and *Reconsideration,* par. 89.) The same order of shrinking priority is expressed in § 76.251(a)(7), which defines cable system responsibility for providing leased access channels.

[10] "While we encourage systems in markets below the top 100 to provide access channels, we are not at this time requiring them to do so. We will permit local franchising authorities in such areas to require systems to provide access service, but to no greater extent than we have specified for systems in the top markets. In that event, our access rules would be applicable." *(Report and Order,* par. 148. See §§ 76.201(b), (c), and 76.251.) Presumably, the FCC's reference in this context to its "access rules" relates only to the content of programing and terms of access to any particular type of access channel local authorities may require; otherwise, the FCC's aim to encourage such services could be frustrated by the requirement for packages of facilities and services more burdensome than federally required in major markets.

anticipates minimal inconvenience from the burden specifically imposed that "the cable operator maintain within the franchise area production facilities for use on the public access channel."[11] While use of the channel shall be "without charge" (*Report and Order,* par. 122), production costs of using the system's facilities may be charged to users at rates required to be "consistent with the goal of affording the public a low-cost means of television access," (§ 76.251(10)(ii)), "aside from live studio presentations not exceeding 5 minutes in length"—which is presumably free. *(For specific federal standards affecting the functions of public access channels, see Sec. 3.3311.)*

COMMENT: On reconsideration, the FCC addressed the question whether its requirements of "at least one specially designated" channel—applied to four categories of access channels in § 76.251(a)(4), (5), (6), and (7)—must be read as prohibiting local authorities from requiring more than one such channel. While the question of local authority is dealt with in detail elsewhere (see Secs. 4.22-4.222), it is pertinent here to note that the FCC's explanation of its own standards has not dispelled all doubt as to its meaning. The substance gleaned from the FCC's reflections is that more numerous channels are clearly envisaged for the future and that no barriers are being placed to local franchises' requiring "specified numbers of channels for educational purposes on a paid basis."(Reconsideration, par. 81.) Stating a penchant toward experimentation and a willingness to entertain petitions regarding experiments "with additional designated channels on a free basis or at reduced rates," the FCC thereby[12] avoided firmly closing the door on near-term local requirements—if we are to take the agency at its own more or less clear word. For the future, therefore, the agency retains the flexibility to respond to local requirements for several access channels of a particular type by treating them as petitions for experimental authorization.[13]

[11] RULE: "The system shall maintain and have available for public use at least the minimal equipment and facilities necessary for the production of programing for such a channel. See also § 76.201 [a reference to the mandatory origination requirement]." (§ 76.251(a)(4).)

[12] On reconsideration, the FCC said:

"81. It should be noted at the outset that, while one educational access channel is the minimum required, we specifically provide in Section 76.251(a)(8) for adding more access channels should the need for such channels be adequately demonstrated. Thus we envision an orderly growth of access channels, linked to demand. In addition, in the *Cable Telvision Report and Order* we stated that after a developmental period (to begin from the commencement of service until five years after completion of the basic trunk line), 'designed to encourage innovation in the educational uses of television'—we will be in a more informed position to determine in consultation with state and local authorities whether to expand or curtail the free use of channels for such purposes or to continue the developmental period. Clearly, as we have stated, this is an area which we will revisit. But without the further knowledge which can be gained only from allowing cable systems to experiment within our initial framework, we are not inclined to add extra burdens to the access requirements. Finally, we are in no way restricting arrangements between the local entity and the cable operator to provide specified numbers of channels for educational purposes on a paid basis. Such arrangements constitute the very type of new service which cable can and should provide. Further, we will entertain petitions from the franchising authority and the cable system when they wish to experiment with additional designated channels on a free basis or at reduced rates."

[13] In an apparent allusion to the susceptibility of the rules to ambiguous interpretation, the FCC's Chairman, Dean Burch, has thus sought to ward off potential challenges and heresies:

"Don't count on the Commission to bail you out. Indeed, I cannot urge too strongly and again the message is directed as much to state and local franchising authorities as it is to cable operators—don't *force* the Commission to bail you out. Game-playing, whether in the form of over-promise or under-performance, is not the mark of a responsible industry. And the simple solution is 'don't play games.'" Speech to National Cable Television Association, May 15, 1972, p. 3. (See Part III for full text.)

[A]t least one specially designated channel for use by local educational authorities. (§ 76.251(a)(5).)

The FCC foresees "varied" use of such a channel, an "important benefit" promising to be "greater community involvement in school affairs" and a "significant advance" through the linking of computers to cable systems with "two-way capability" to enhance the quality of educational programing. *(Report and Order,* par. 123.) The FCC's regulatory action envisages that such use will be free for five years after the completion of the system's basic trunk line (§ 76.251(10)(i)), subject to reevaluation at that time in consultation with state and local governments "to expand or curtail the free use of channels for such purposes or to continue the development period. The rule makes no provision for access to or use of production facilities. *(For specific federal standards affecting the functions of educational access channels, see Sec. 3.3312.)*

COMMENT: It would appear possible for local educational broadcast stations to qualify as a "local educational authority" able to make use of educational access channels, since such authority is not defined. No standards for mediating among uses by various educational institutions are provided. Nor does the FCC offer its views on the potentially thorny Constitutional and political question of use by parochial schools. On reconsideration, the FCC declined to act on urgings of the National Association of Educational Broadcasters "to amend our rules to enable leased channels to be used for educational purposes at lower rates and to provide that, at the termination of the free five-year developmental period for educational access channels, rates be kept at a minimum."(Reconsideration, par. 82.) Citing the virtues of an experimental approach, the FCC felt it desirable "to avoid any form of preferential policy with regard to whom may use and what must be paid for access channels."

[A]t least one specially designated channel for local government uses. (§ 76.251(a)(6).)

The *Report and Order* indicates that this channel is "designed to give maximum latitude for use by local governments," and that suggestions for its use "range across a broad spectrum and it is premature to establish precise requirements." *(Report and Order,* par. 124.) As with the educational access channel, terms are for free use for five years and no provision is made for production of programing. *(For specific federal standards affecting the functions of local government access channels, see Sec. 3.3314.)*

COMMENT: As with the educational channel, no federal provisions exist to identify the qualified users of the "local government access channel"—as between agencies of a city, or between a city or a state—and no provisions for use of production facilities are set forth. Determining a local approach to the utilization of this cable channel, and providing for local government programing over that channel, are among the significant areas for local initiatives discussed subsequently in Sec. 3.4.

"Leased access channels" for the remainder of channel capacity not used for other designated purposes, and for "any other available bandwidth."(See § 76.251(a)(7).)

Subject to specific conditions subsequently discussed in Sec. 3.3314, cable systems in major markets are required to make available, for leased use, channels within the limits of their required capacity (see Sec. 3.22) that are not occupied by required broadcast signals and designated cablecasting purposes. From the outset, therefore, it appears that the number of channels available for leased use is defined with reference to the FCC's minimum-channel capacity rule. However, additional opportunities are possible, beyond "the remainder of the required bandwidth" *(Report and Order,* par. 125). These would include the time periods when other access channels are not in use and the times at which broadcast channels are blacked out to protect local stations' exclusivity.[14] (These slots may also be used for fill-in broadcast programing, pursuant to § 76.61(b)(2)(ii) and § 76.63(a).) The rule makes clear that leased use of channels designated for other purposes shall be "undertaken with the express understanding that they are subject to displacement if there is a demand to use the channels for their specially designated purposes." Section 3.3314 below discusses the further requirement that "on at least one of the leased channels priority shall be given part-time users."

COMMENT: The FCC appears to have deliberately avoided being precise about its requirements for the number and manner of operation of leased channel operations. First of all, operation of the rules as to numbers of such channels is inseparable from the operation of the minimum channel rule, on the one hand, and from the broadcast carriage and exclusivity rules on the other. Secondly, standards governing use (discussed subsequently in Sec. 3.3314) are vague, with the exception of those applicable to the one channel for part-time users. Finally, the FCC has avoided coming to grips with the lament raised by the American Civil Liberties Union and others that the Commission has failed "to assure that channel capacity will be available over periods of sufficient duration to justify prudent investment by an entrepreneur wishing to supply broadband services." (Reconsideration, par. 83.) "We are somewhat puzzled by this view and can only state that we assume entrepreneurs will, in fact, lease channels over long periods for programing or other services." The Commission completes its view by projecting that "the problem is unlikely to arise in the next few years," declining to impose "artificial restraints," and—inconsistently—expressing the hope that "potential users of access services will . . . exercise a degree of restraint and will not, by over-enthusiastic reaction to cable television, force a determination of what, in fact, constitutes an 'unreasonable' demand for channels."

3.22. Minimum Channel Capacity

A central element of the FCC's program to achieve, through cable television, delivery of a novel communications service, is its aim of minimum channel capacity in major markets beyond that utilized for the delivery of broadcast signals. Were it not to do so, cable television would become a largely disappointing device that not only introduced economic instability into the broadcast media, but offered viewers no more than a wider choice of programs designed for mass tastes. With the availa-

*

(16)

---

[14] Though § 76.251(a)(7) refers to "other portions of its nonbroadcast bandwidth," the *Report and Order* makes it clear that "if a channel carrying *broadcast* programing is required to be blacked out because of our exclusivity rules or is otherwise not in use, that channel may be used for leased access purposes." *(Report and Order,* par. 125.)

bility of "constantly expanding channel capacity" devoted to new communication services, the FCC expects that new and attractive options for a full spectrum of subscribers can help sustain the emergence of new cable services at costs to cable systems that, over the long run, will be within their economic means.[15] *(Report and Order,* par. 126.) Despite opposition voiced by cable operators, the FCC plans, in making such channels available as a condition for cable growth, for a greater range of services to the public at an earlier date than could be anticipated from voluntary growth.[16]

Ultimately, according to the FCC, the development of cable television in this mold will create a framework by which "the principal services, channel uses, and potential sources of income will be from other than over-the-air signals." *(Report and Order,* par. 120.) Implementing this view of the future, the Commission counsels the planning of abundant capacity ("We urge cable operators and franchising authorities to consider that future demand may significantly exceed current projections, and we put them on notice that it is our intention to insist on the expansion of cable systems to accommodate all reasonable demands"). But it also wishes to practice restraint ("We wish to proceed conservatively, however, to avoid imposing unreasonable economic burdens on cable operators").

Its present step, therefore, is to establish a two-fold standard for cable systems *in the major markets:*

> In these markets, we believe that 20 channel capacity (actual or potential) is the minimum consistent with the public interest. We also require that for each broadcast signal carried, cable systems in these markets provide an additional channel 6 MHz in width [equivalent to one broadcast television channel] suitable for transmission of Class II or Class III signals.[17]

This initial approach, the FCC believes, is "a reasonable way to obtain necessary minimum channel capacity and yet gear it to particular community needs."

COMMENT: As with the companion leased-channel rule, "minimum channel capacity" is a concept of imprecise generality, ultimately defined by the FCC in terms of impalpable considerations of "reasonableness." Thus a burden rests on those seeking liberal definitions to show that the public interest requires a more generous interpretation, a circumstance that the FCC does not view as compelling for the immediate future. Moreover, by defining required channel capacity in terms of "minimum" federal obligations, the FCC may not have legally preempted local franchising authorities in the major markets from mandating larger capacity systems in their own franchising processes. This is possible despite the FCC's rule (§ 76.251(a)(11)(iv)) blocking local entities from prescribing "any other rules concerning the number or manner of operation of access channels" (emphasis added). Arguably, a local authority could still require a 30-channel system without expressly

---

[15] The FCC cites testimony from the CATV equipment industry to the effect that the "cost difference between building a 12 channel system and a 20 channel system would not appear to be substantial." *(Report and Order,* par. 120.)

[16] However, FCC modifications on reconsideration for existing cable systems (discussed in Sec. 3.24) have gone far toward slowing its original pace.

[17] The text of the FCC's rule, § 76.251(a)(1) and (2), relates to a cable system "operation in a community located *in whole or in part within a major television market"* and defines the 20-channel requirement in terms of "at least 120 MHz of bandwidth . . . available for immediate or potential use for the totality of cable services to be offered . . ." (italics added).

*providing rules to govern "the number or manner of operation of access channels."
More surely, channel capacity is not preempted from local authorities outside of
major markets (required not only "to exceed the provisions concerning the availability and administration of access channels . . ."). (§ 76.251(b).) Finally, it is apparent
in the largest major markets that the paired-capacity rule, dependent on the number
of broadcast signals the cable system chooses to carry, will actually determine
required capacity for operative channels, not the flat 20-channel requirement—
which is defined in terms of "actual or potential" service.*

Beyond the 20-channel capacity rule, the Commission states a "basic goal to
encourage cable television use that will lead to constantly expanding channel
capacity." *(Report and Order,* par. 126.) Hence, "Cable systems are therefore required to make additional bandwidth available as the demand rises." This "general
objective" raises questions about FCC standards for requiring system expansion
beyond 20-channel capacities—a significant barrier inasmuch as 20 channels may
represent an effective maximum capable of being carried on a single coaxial link.
The FCC states an initial "formula" approach[18] whose applicability is inferentially
limited to under a 20-channel maximum inasmuch as the agency acknowledges that,
"If it were necessary to rebuild or add an extensive new plant, this could not
reasonably be expected within a 6-month period [the period within which additional
channels, pursuant to the formula, must be made 'operational']." *(Report and Order,*
par. 127.) According to the formula (discussed in Sec. 3.332), once all channels are
in use for a particular percentage of the time during a particular period, the system
will have six months in which to make a new channel available.

* (22)

*COMMENT: As a practical matter, the FCC's approach will curtail tendencies on
the part of cable systems in major markets to restrict investment in installing cable
plant to less than what is reasonably compatible with the 20-channel maximum. In
and of itself, the requirement will not require the development of dual-cable systems. However, it does not appear likely that the FCC has totally foreclosed local
franchising authorities from requiring the development of larger capacity systems,
even though they may be precluded from developing regulatory programs concerning the use of such channels that are inconsistent with federal regulations.*

3.23. Two-Way Capacity

The final element of the FCC's overall formulation of nonbroadcast standards
is a requirement that cable systems must be engineered so as to "maintain a plant
having technical capacity for nonvoice return communications . . ." (§ 76.251(a)(3).)
The aim is to make available the potentiality, already existing in some cable systems, for a number of useful services: "surveys, marketing services, burglar alarm
devices, educational feedback, to name a few." *(Report and Order,* par. 128.)
Though opting for *potential* two-way service, the FCC has eschewed a requirement that cable systems "install necessary return communication devices at each

* (22)

[18] RULE: 47 C.F.R. § 76.251(a)(8) provides as follows: "Whenever all of the channels described in
subparagraphs (4) through (7) of this paragraph are in use during 80 percent of the time during any
consecutive 3-hour period for 6 consecutive weeks, such system shall have 6 months in which to make
a new channel available for any or all of the above-described purposes . . ."

subscriber terminal . . . ," holding such requirement "premature." *(Report and Order,* par. 129.) It views its requirement as being met either by installing the necessary auxiliary trunk equipment ("amplifiers and passive devices") or "equipment that could easily be altered to provide return services." Hence, nothing must be done in engineering that could impede the refitting of a cable system to provide return service.

In the *Report and Order,* and not by rule, the FCC indicates that, "When offered, activation of the return service must always be at the subscribers option." (Par. 129.)

*
(23–24)
COMMENT: The FCC's rule and explanation are generally self-explanatory.[19] *However, it is not clear whether the requirement stated in the* Report and Order, *that the subscriber must preserve the right to determine use of return service, now prohibits the widely appreciated possibility that cable operators can monitor viewer selections for the purpose of ascertaining the circulation of particular programing.*

3.24. Applicability

The time at which the FCC's cablecasting policies for major-market systems apply is determined by rules for the franchising process, discussed in Sec. 4. The effect of these rules is to prescribe the actual rate at which designated access channels and minimum channel capacity must be made available. For all new cable systems commencing operations after March 31, 1972, they apply immediately, while systems already in operation at that date are given five years in which to comply. By a significant modification to § 76.251(c) on reconsideration, the FCC has eased its earlier requirement that cable systems presently in operation seeking to add carriage of broadcast signals must comply with access and minimum channel capacity rules as soon as such signals are carried. Citing the prospect that this rule would force present systems seeking only one additional signal to add at least four new access channels, the FCC has substituted a one-for-one requirement to avoid "substantial or complete rebuilding, involving large investment." *(Reconsideration,* par. 87.) Accordingly, to avoid a flood of petitions for waivers in such cases in a "significant percentage of all presently operating cable television systems," the FCC now requires that "for each additional broadcast signal carried, such a system will have to provide one access channel." *(Reconsideration,* par. 89.) In order, these will be a public access channel, an educational access channel, a governmental access channel, and "any others by leased channels."

The application of the excess capacity rule to these systems will be deferred until March 31, 1977.

COMMENT: The FCC's postponement of its access and channel capacity requirements represents a significant retardation of the required rate of development of access channels—greater in proportion to the diminishing priority of access channels—and to that extent a boost to the economic vitality of existing cable television systems.

[19] But see also Baer and Feldman, "Review of Cable Systems Designs, Cable Equipment and Related Hardware Developments."

3.3. CABLECASTING REGULATIONS

To effectuate these policy determinations, the FCC has established specific regulatory provisions affecting nonbroadcast services applicable to three aspects of cable television operations: origination cablecasting, pay cablecasting, and access cablecasting.

3.31. Origination Cablecasting

The rules promulgated with the *Report and Order* for origination cablecasting very largely duplicate the rules promulgated in 1969 in the *First Report and Order,* recently affirmed in *Midwest Video* as to the FCC's jurisdictional authority. Their essential provisions are next set forth.

3.311. Applicability and Scope. Under § 76.201(a), a cable system with more than 3500 subscribers that carries broadcast signals must henceforth "operate ... to a significant extent as a local outlet by origination cablecasting and [have] available facilities for local production and presentation of programs other than automated services." One or more "designated channels" may be used for origination cablecasting and "no other purpose." Cable systems outside of all major television markets (without regard to number of subscribers) may not by "contract, arrangement or lease" prevent or inhibit the use of its facilities for local programing "designed to inform the public on controversial issues of public importance," pursuant to § 76.201(b). Finally, all cable systems that originate programing may do so only in accordance with substantive FCC rules next set forth. (§ 76.201(c).)

3.312. Equal Time for Political Candidates. In this category (§ 76.205) and in the related category of the "Fairness Doctrine" (§ 76.209), the Commission is applying to origination programing by cable systems its standards applicable to broadcast licensees under Section 315 of the Communications Act of 1934.[20] It establishes the following "general requirements":

RULE:
    If a cable television system shall permit any legally qualified candidate for public office to use its origination channel(s) and facilities therefor, it shall afford equal opportunities to all other such candidates for that office: *Provided, however,* that such system shall have no power of censorship over the material cablecast of any such candidate; *And provided, further,* That an appearance by a legally qualified candidate on any:

    (1) Bona fide newscast,

    (2) Bona fide interview,

    (3) Bona fide news documentary (if the appearance of the candidate is incidental to the presentation of the subject or subjects covered by the news documentary), or

    (4) On-the-spot coverage of bona fide news events (including, but not limited to, political conventions and activities incidental thereto),

---

[20] In its rules, the FCC's language is identical to pertinent parts of Sec. 315, with the exception of the substitution of language appropriate to cable origination for language in the Act referring to broadcast licensees.

shall not be deemed to be use of the facilities of the system within the meaning of this paragraph.

The rule concludes with a note that in these exempt categories, the fairness doctrine (§ 76.209) is applicable. Further regulations govern "rates and practices," which are required to be uniform (with respect to rates), not subject to rebates, not in excess of comparable rates for other purposes, and (with respect to practices) nondiscriminatory among competing candidates.

Under the heading of "Records," the FCC requires that cable systems keep track of all requests for origination time by candidates, and of the disposition of such requests, for a two-year period. As to the "Time of Request" for equal opportunities, the FCC requires it be made within one week of the first occurrence of the right, or within one week of the time a person who has become a candidate has become eligible for such opportunity by virtue of an occurrence subsequent to his candidacy. Finally, the "burden of proof" of eligibility for such opportunity rests on the claimant. (§§ 76.205(c), (d), and (e).)

COMMENT: The comparable treatment of cable systems and broadcast stations with respect to candidates for public office has been adopted by the Congress in the Federal Election Campaign Act of 1971, Amending Section 312 (dealing with revocation of broadcast licenses) and Section 315 of the Communications Act, effective April 7, 1972. The application of these standards to broadcasters, and now to cable systems, has been the subject of extensive controversy. The FCC has responded with the adoption of specific guidelines, some in question and answer form; reference to the statutes and to those guidelines is required. Accordingly, the complete text of these guidelines should be examined for answers to specific questions. See Use of Broadcast Facilities by Candidates for Public Office *("Political Broadcast Primer"), 35 Fed. Reg. 13048 (1970), and* Use of Broadcast and Cable Facilities by Candidates for Public Office, *37 Fed. Reg. 5795 (1972). The following problems with respect to cablecasts by candidates for political office merit specific comment:*

(a) Programs to which the rule applies. *Inasmuch as the terms of the rule embrace only appearances by candidates themselves, questions have been raised concerning speeches by others on candidates' behalf. The rule also covers any use by such candidates on origination channels, however brief and whatever the subject matter.* It is significant that the rule does not cover use of access channels, "because these channels are free of operator control and access is guaranteed." *Finally, while it is altogether possible that cable systems may avoid the imposition of the rule by barring* all *political candidates from its origination channels and facilities, it is implicit that such tactics would not qualify the system's origination cablecasting as operating "to a significant extent as a local outlet"; to the extent cable systems may seek to back away from the operation of the rule, its policies would be subject to complaint to the FCC—though such complaints, it should be noted, are often controverted and time-consuming.*

(b) Exemptions. *The four statutorily based exemptions to the rule—which may trigger the application of the companion "Fairness Doctrine"—are self-explanatory, centering on the good faith of the newscaster and the prerequisite that the news*

programing is entirely under the control of the cable system. Determination of these qualities requires subjective judgments, depending on the circumstances.

(c) Determination of a "legally qualified candidate." *This standard, which applies to both the first speaker and any potential rebuttalist, is a question to be determined under applicable state law, embracing both those whose names appear on the ballot and those who could be the subject of a write-in campaign. When in doubt, the cable system would be well advised to seek a ruling from a competent state or municipal official. No distinction being drawn between primary and general elections, the rule applies to both. The FCC now appears, in the "Zapple Ruling," to regard remarks by supporters of candidates as falling within the Equal Time rule.*

(d) Determination of who is an opposing candidate. *Although § 76.205(a) concerns candidates "for public office," subparagraph (b) relates to "the same office." Hence, a candidate for another office may not invoke the right to equal opportunities. Similarly, a candidate running in one party's primary may not invoke the rule to reply to a candidate in another party's primary campaign.*

(e) Determination of "equal opportunities." *The Commission takes pains in the* Report and Order *to emphasize that*

> Should a cablecast by a candidate for political office on the origination channel prompt the necessity for providing equal time to an opponent, it must be provided on the origination channel. In this situation, the opponent's appearance on an access channel will not suffice. (Report and Order, par. 145.)

The rule as a whole makes clear that "equal opportunities" means that the cable system may not discriminate in "charges, practices, regulations, facilities or services" rendered as to legally qualified candidates for the same office. While not subject to exact codification, the Commission has left exact determinations to candidates and cable systems in specific situations, keeping in mind the time of day concerned and the length of time that are, on balance, roughly comparable. Where system-controlled origination on more than one channel is involved, use of the same channel for reply would appear prudent.

(f) Censorship. *Though not explicit, the rule would appear to bar censorship of remarks made by any "legally qualified candidate," especially in the light of the requirement that it not "subject any such candidate to any prejudice of disadvantage," though it may request an advance script for planning purposes. (This restriction would not appear to apply to those speaking on behalf of a candidate.)*

(g) Rates and Practices. *Rates may not exceed normal commercial charges, and shall be comparable for all candidates.*

(h) Records and Inspections *The recordkeeping and inspection provisions (§ 76.205(c)) are self-explanatory. No form is prescribed.*[21]

---

[21] An "Agreement Form for Political Cablecasts," adopted from a form recommended by The National Association of Broadcasters, is set forth in Part III.

3.313. Fairness Doctrine; Personal Attacks; Political Editorials. The concomitant of the political candidates rule, also originally published with the *First Report and Order* but not reviewed by the Supreme Court in *Midwest Video,* are the Commission's Fairness Doctrine rules. These were also modeled after broadcasting standards that have stirred intense controversy over the years. Analytically, the rules (set forth in § 76.209) operate in three interrelated parts:

1. A requirement that origination cablecasting "shall afford reasonable opportunity for the discussion of conflicting views on issues of public importance." ("The Fairness Doctrine," subparagraph (a));

2. A requirement that when "an attack is made upon the honesty, character, integrity, or like personal qualities of an identified person or group," the cable system shall, within one week, notify the person or group so attacked, provide a tape or a script of the attack, and offer a reasonable opportunity to respond. (The "Personal Attack" rule, subparagraph (b));[22] and

3. A requirement that when, in a political editorial, the cable system endorses or opposes a legally qualified candidate, the system shall, within 24 hours, notify the other candidates, provide a tape or script of the editorial, and offer the other candidates or their spokesmen a reasonable opportunity to respond (The "Political Editorial" rule, subparagraph (d)).[23]

COMMENT: Despite its superficially self-evident thrust, the exact dimensions of the Fairness Doctrine *are another issue, which greatly exceeds the scope of this*

---

[22] The "Personal Attack" rule is as follows:

"(b)   When, during such origination cablecasting, an attack is made upon the honesty, character, integrity, or like personal qualities of an identified person or group, the cable television system shall, within a reasonable time and in no event later than one (1) week after the attack, transmit to the person or group attacked: (1) Notification of the date, time and identification of the cablecast; (2) a script or tape (or an accurate summary if a script or tape is not available) of the attack; and (3) an offer of a reasonable opportunity to respond over the system's facilities."

The rule is subject to the following exemptions:

"(c)   The provisions of paragraph (b) of this section shall not be applicable: (1) To attacks on foreign groups or foreign public figures; (2) to personal attacks which are made by legally qualified candidates, their authorized spokesmen, or persons associated with the candidates in the campaign; and (3) to bona fide newscasts, bona fide news interviews, and on-the-spot coverage of a bona fide news event (including commentary or analysis contained in the foregoing programs, but the provisions of paragraph [d]* of this section shall be applicable to editorials of the cable television system)."

*   The printed text of the rule reads "(b)," an obvious misprint for "(d)."

[23] The "Political Editorial" rule is as follows:

"(d)   Where a cable television system, in an editorial, (1) endorses or (2) opposes a legally qualified candidate or candidates, the system shall, within 24 hours after the editorial, transmit to respectively (i) the other qualified candidate or candidates for the same office, or (ii) the candidate opposed in the editorial, *(a)* notification of the date, time and channel of the editorial; *(b)* a script or tape of the editorial; and *(c)* an offer of a reasonable opportunity for a candidate or a spokesman of the candidate to respond over the system's facilities: *Provided, however,* That where such editorials are cablecast within 72 hours prior to the day of the election, the system shall comply with the provisions of this paragraph sufficiently far in advance of the broadcast to enable the candidate or candidates to have a reasonable opportunity to prepare a response and to present it in a timely fashion."

study. In specific situations, it will be necessary to refer to more detailed standards of the Commission. The FCC's principal policy guidelines are set forth in "Applicability of the Fairness Doctrine in Handling of Controversial Issues of Public Importance," 29 Fed. Reg. 10415 (1964), "Fairness Doctrine Primer."

In general, the FCC has indicated with respect to broadcasters that the rule is to be viewed with considerable flexibility:

> [T]he licensee, in applying the fairness doctrine is called upon to make reasonable judgments in good faith on the facts of each situation—as to whether a controversial issue of public importance is involved, as to what viewpoints have been or should be presented, as to the format and spokesmen to present the viewpoints, and all other facets of such programing... In passing on any complaint in this area, the Commission's rule is not to substitute its judgment for that of the licensee as to any of the above programing decisions, but rather to determine whether the licensee can be said to have acted reasonably and in good faith. There is thus room for considerably more discretion on the part of the licensee under the fairness doctrine than under the 'equal opportunities' requirement. Fairness Doctrine Primer, 29 Fed. Reg. 10415, 10416.

The FCC's policies represent an affirmative burden upon cable systems to seek out qualified spokesmen for relevant points of view on "issues of public importance," once one side has been expressed. What those issues may be defies exact definition, but they embrace advertising as well as regular informational programing, according to recent FCC and judicial interpretations. The operator's affirmative obligation includes a duty to announce the availability of time for reply, and to permit such reply even if commercial sponsorship is not forthcoming. While the rule is a federal responsibility imposed upon the cable operator, relevant local government entities must be aware of the implications of the "Fairness Doctrine" for the public issues with which they deal.

The more precise "Personal Attack" and "Political Editorial" rules set forth express obligations for cable systems. The former rule applies to attacks on identifiable individuals or groups, other than on exempted news programing (when the Fairness Doctrine applies), or on editorials (when the Political Editorial rule applies). Under both rules, some flexibility exists for the format and timing by which the right to reply must be exercised. Again, the duties of the cable operator in these regards are also the concerns of local government agencies through their implications for the making of local government policies.

Obscenity. The FCC has prescribed a rule, comparable to the rule long applicable to broadcasting, prohibiting transmission of "obscene or indecent" materials.[24] While the determination of what is obscene or indecent is a matter of subjective judgment, varying with cultural trends, cable systems engaged in origination cablecasting would be well advised to keep in mind the FCC's readiness to revoke

---

[24] § 76.215 provides as follows:

"No cable television system, when engaged in origination cablecasting shall transmit or permit to be transmitted on the origination cablecasting channel or channels material that is obscene or indecent."

broadcast licenses for violation of this rule.[25] The responsibility of the cable operator to establish rules to similar effect on access cablecasting channels (§ 76.251(a)(11)) should also be kept in mind.

Advertising and Sponsorship Identification. Twin rules, originally issued by the FCC in the *First Report and Order,* permit advertising on origination cablecasts at particular intervals and govern the relationship between cable operators and sponsors.

(1) *Advertising* on origination cablecasts is permitted, but unlike the situation in broadcasting may occur only at "the beginning and conclusion of each such program and at natural intermissions or breaks within a cablecast ..."[26] Interruptions to permit advertising are not permitted, and "natural intermissions" are defined as follows:

Any natural intermission in the program material which is beyond the control of the cable television operator, such as timeout in a sporting event, an intermission in a concert or dramatic performance, a recess in a city council meeting, an intermission in a long motion picture which was present at the time of theatre exhibition, etc. (§ 76.217, Note.)

COMMENT: Obviously designed to curb commercialization in cablecasting, the FCC's rule requires some special efforts by cable operators to make programing attractive to advertisers. The presentation of programs prepared for broadcast television, embodying frequent planned transitions to advertising materials, would not appear to violate the tenor of the rule.

(2) The rule with respect to *sponsorship identification* (§ 76.221) has four components. Broadly, whenever consideration is received, direct or indirect, for carrying a program, the cable operator must identify the source of the consideration, according to subparagraph (a). Subparagraph (b) embraces restrictions on relations with employees to prevent the practices known as "payola" or "plugola." According to subparagraph (c), attribution must be made when a source furnishes the material for the purpose of inducing carriage of political programing or programing involving discussions of controversial issues. All these requirements are waived with respect to feature films produced primarily for theater exhibition.

COMMENT: All of the foregoing restrictions, modeled after broadcasting restrictions, follow from the application to cable systems in origination cablecasting of the

[25] In other circumstances, the FCC invoked the sanction of refusing to renew the broadcast license of Channel 5 in Boston in Greater Boston Television Corp., 16 FCC 2d 1 (1969), aff'd. 444 F.2d 841 (C.A.D.C., 1970), *cert. denied* 403 U.S. 923 (1971).

[26] § 76.217 provides as follows:

"A cable television system engaged in origination cablecast programing may present advertising material at the beginning and conclusion of each such program and at natural intermissions or breaks within a cablecast: *Provided however,* that the system itself does not interrupt the presentation of program material in order to intersperse advertising: *And provided, further,* That advertising material is not presented on or in connection with origination cablecasting in any other manner."

On reconsideration, the FCC noted the argument of cable operators that advertisements should be permitted without limitation, as is the case for both broadcasting and leased access channels. It found the argument "not without attraction, especially in the light of the new relationship between origination cablecasting channels and access channels," but declined to act in this experimental area. *(Reconsideration,* par. 71.)

"fiduciary" principle long accorded to broadcasting; they must be contrasted with the limited applicability of this principle to access cablecasting channels, discussed subsequently in Sec. 3.3.

(f) Lotteries. In its advertising activities, the cable system must avoid activities that may be construed as "lotteries," pursuant to § 76.213. Specific reference to the rule is required, but the elements legally essential to constitute a lottery—chance, prize, and a consideration—must be present before the rule is breached.

* * * *

COMMENT: Overall, it is apparent that the FCC regulations endow cable operators with considerable responsibilities to the public as a whole in their obligation to "operate to a significant extent" as a local outlet. Inevitably, liability for defamatory utterances adheres to this programing role, as a consequence of this responsibility. Cable operators in their origination activities are subject to state laws in this respect, except to the extent that federal law has limited or preempted such liability. Two such limitations should be noted. First of all, it is apparent that Supreme Court decisions centering on New York Times Co. v. Sullivan, *376 U.S. 254 (1964) have significantly narrowed the play of state defamation laws where an area of public controversy or a "public figure" is concerned; plaintiffs must make showings of "actual malice" to sustain libel or slander actions. Secondly, the specific prohibition in § 76.205 against censoring cablecasts made by legally qualified candidates for public office—cable systems being denied the power to control the content of such cablecasts—may specifically preempt the operation of state laws, under the doctrine of* Farmers Educational and Cooperative Union of America v. WDAY, Inc., *360 U.S. 525 (1959). In other respects, cable operators, like broadcasters, are subject to applicable laws; one must keep in mind, however, the distinction between these defamation standards, applicable to channels where the operator has control over content, and those applicable to cable access channels (discussed in Sec. 3.3314 below).[27]*

3.32. Pay Cablecasting

Since 1968, the FCC has contemplated authorizing cable systems to make per-program charges[28] as distinguished from regular monthly subscription fees. In 1969, in companion rules to those challenged in *Midwest Video*, the FCC authorized imposition of per-program or per-channel charges; these rules have now been reissued as § 76.225. On September 8, 1971, the FCC took the position that "The Commis-

[27] In the *Report and Order*, the FCC notes the "possibility" that it may yet prohibit cable systems "from originating their own programing and be restricted entirely to a common carrier role," in response to suggestions that depriving a cable operator of an economic interest in cable programing may foster the widest development of cable services, through the programing of others. (Par. 146.) The Commission finds, however, that, "At this stage in the development of the cable industry, it is the system operator who has the greatest incentive to produce originated material attractive to existing and potential subscribers," so that such a restriction would be "premature." The Commission cautions, nonetheless, that "When cable penetration reaches high levels and demand increases for leased channel operations, we will revisit this matter."

COMMENT: One might well question the feasibility of making real such a "possibility" once cable systems have invested heavily in the origination operations and facilities the FCC has mandated.

[28] *Notice of Proposed Rule Making in Docket 18397* (1968).

sion has preempted the field of pay television cablecasting so that local franchise terms are inoperative . . . ," and that "pay television on cable may serve the public interest . . ."[29] As reissued by the FCC—expressly subject to future reevaluation *(Reconsideration,* par. 73)—the five-part rule embraces *both* origination cablecasting and access cablecasting (discussed further below), hence covering both Class II and Class III channels.

The pay-cablecasting rules closely integrate with the standards originally set forth by the FCC in 1968 with respect to subscription broadcasting. For a full appreciation of the background of the present rules, the reader must therefore refer to the FCC's broad policy statement, *Fourth Report and Order in Docket 11279,* 15 FCC 2d 466. The FCC's aim is to strike a balance between video and other methods of distribution—movie theaters and live exhibitions—and also between cable and broadcasting. Hence the rules operate principally to restrict the type of programing available over the cable medium. Since the FCC's powers to do so have been upheld with respect to broadcasting against a challenge based on the First Amendment, *National Association of Theater Owners v. FCC,* 420 F.2d 1 94, *cert. den.* 397 U.S. 922 (1970), similar restrictions on cable origination (FCC powers generally in this area having been upheld) would seem unassailable.

The rules deal principally with *films, sports events,* and *TV series* programming. Only the general points of these complex rules are set forth here.

With respect to *films,* the rule provides that films "which have had a general release in theaters anywhere in the United States more than two (2) years prior to their cablecast" may *not* be cablecast, subject to certain limited exceptions. These are, with respect to films *older than ten years,* that one film each month may be shown once or more during a single calendar week, and that films released *between two and ten years* previously may be shown only on a "convincing showing to the Commission" that an exhibition on broadcast television is impractical or not desired by the owner of broadcast rights for enumerated reasons. (§ 76.225(a)(1).) There are no restrictions on showing films that are *less than two years* old.

COMMENT: Two years being a film's normal circulation in theaters, older films are a staple of broadcasting. Hence, the FCC's rule permits only limited use of older films on cable. On the other hand, films less than two years old are valuable in the theater market; competing with such distribution on cable would be prohibitively costly. If film distributors can accelerate initial circulation of new films in movie theaters, however, they may find pay cable a profitable added market for such films. In areas where there are no movie theaters, distributors may even want to publicize cable distribution. As cable penetration increases in any market, moreover, the economic advantages of cable distribution may grow dramatically.

With respect to *sports events,* the rule prohibits cablecasting of events that have been "televised live on a nonsubscription, regular basis in the community during the two (2) years preceding their proposed cablecast . . ." (§ 76.225(a)(2).) If a "specific event" such as the Olympic Games was last televised more than two years previously, it also may not be cablecast.

---

[29] See Letter to Pierson, Ball & Dowd, 22 *Pike & Fischer R.R.* 2d 949, Part III.

COMMENT: Only a limited number of sports events are prohibited by the rule, which also aims to protect broadcast television. Permitted would appear to be sports events that were not telecast live on a regular basis (so that taped events, events that did not occur in both of the last two successive years, and events that were not "regular" may be seen on cable). This would mean, among other things, that if an entrepreneur wished to withhold an event from conventional television for one year, he could switch to cable, and, of course, that single events (such as a championship boxing match) could be shown. The FCC has pending a rule that would extend the two-year prohibition to five years. The matter is under study by Congress.

With respect to *series-type programing,* the FCC also protects broadcasting by barring cablecasts of any "series type of program with interconnected plot or substantially the same cast of principal characters . . ." (§ 76.225(a)(3).) An additional requirement prohibits excessive use of films and sports events (defined in § 76.225(a)(4) as more than 90 percent of total hours yearly, or 95 percent in any one month absent a showing of good cause), in order to foster the development of other types of programing, such as informational materials. Finally, all *advertising* on such channels is prohibited, except at the beginning or end of any pay cablecasts to promote other such programs.

3.33. Access Cablecasting

The FCC's broad policies regarding the availability of access channels, and the extent to which their regulation lies within federal or state and local control, are reviewed in Secs. 3.31-22 and 4.22-222. This section deals with the regulatory requirements imposed by the FCC on cable systems in the major television markets. At the outset, two basic principles must be noted. The first is that the FCC views its regulatory program as "intuitive" and "evolutionary," expressed in the following terms:

> We believe that the best course is to proceed with only minimal regulation in order to obtain experience. We emphasize, therefore, that the regulatory pattern is interim in nature—that we may alter the program as we gain the necessary insights. *(Report and Order,* par. 133.)

The FCC also indicates that its program consists of the imposition of "guidelines," in accordance with which, "We are requiring that cable systems promulgate rules to apply to access services, and that these rules be kept on public file at the system's local headquarters and with the commission." *(Report and Order,* par. 134.) Hence, the specific conditions for use of particular channels will be determined by *the cable system itself,* subject to review of the Commission—a prospect with respect to which the great number of individual cases and the small staff of the Commission must be kept in mind.

3.331. Rules for Designated Channels. Rules are specified by the FCC with respect to each of the categories of mandated access channels.

3.3311: Public Access Channels. Without further elaboration, the FCC requires that, for the public access channels, the system specify "nondiscriminatory access without charge on a first come, first-served basis." In addition,

*
(26–28)

RULE:
"The system shall maintain and have available for public use at least the minimal equipment and facilities necessary for the production of programing for such channel." (§ 76.251(a)(4).)

The Commission broadly hints that use of the cable system's origination studios and other production facilities will be adequate to satisfy its requirement. In another connection, to minimize cost burdens, it suggests that at this time it is not requiring that origination and access production facilities meet broadcast technical standards.[30] However, in another section of its rules, the FCC introduces major practical qualifications on its mandate of both free channel use and access to production facilities:

RULE:
"*One* of the public access channels described in subparagraph (4) of this paragraph shall always be made available without charge, except that production costs may be assessed for *live studio presentations exceeding 5 minutes*. Such production costs and any fees for use of other public access channels shall be consistent with the goal of affording the public a low-cost means of television access." (§ 76.251(a)(10)(ii); italics added.)

COMMENT: While the distinction between "the public access channel ... without charge" in the Report and Order *and the authorization in the above rule for other public access channels for which "fees for use" are permissible may be more apparent than intended, it is clear that the FCC rule envisages that some cost burden will be placed on public access users. Such charges would appear to be for "live studio presentations exceeding 5 minutes" as rental for production facilities, and production costs (without limit) and fees (without limit) for channel use for "other public access channels," which must be "consistent" with the ill-defined "goal of affording the public a low-cost means of television access." Use of system equipment for remote production is not covered, and presumably neither required nor regulated with respect to rental charges. Free use appears mandated only with respect to live studio presentations of less than 5 minutes duration, and to longer use not requiring any system facilities for production.*

Further operating rules must be established by the cable system with respect to the following particulars:

RULE:
"For the public access channel(s), such system shall establish rules requiring first-come nondiscriminatory access; prohibiting the presentation of: Any advertising material designed to promote the sale of commercial products or services (including advertising by or on behalf of candidates for public office); lottery information; and obscene or indecent matter (modeled after the prohibitions in §§ 76.213 and 76.215 respectively); and permitting public inspection of a complete record of the names and addresses of all

---

[30] "Thus, for the present, our technical standards will apply only to Class I channels ... We note specifically that the use of half-inch video tape is a growing and hopeful indication that low-cost recording equipment can and will be made available to the public. While such equipment does not now meet our technical standards for broadcasting, there is promise of its improvement and refinement. Further, since it provides an inexpensive means of program production, we see no reason why technical development of this nature should not be encouraged for use on cable systems." *(Report and Order,* par. 143.)

persons or groups requesting access time. Such a record shall be retained for a period of 2 years." (§ 76.251(a)(11)(i).)

In addition, § 76.251(a)(9) prohibits all control over content, although the FCC indicates that this prohibition shall not relieve cable operators from the requirements of subparagraph (11) with respect to prohibited materials. The net effect of these two requirements with respect to the exposure and obligations of operators and programers has not been clarified.

COMMENT: Other than reiterating the requirement for "rules requiring first-come nondiscriminatory access," the FCC established no guidance for establishing scheduling policies. This deficiency may prove particularly troublesome in the light of the relative desirability of particular segments of time, and the need of many groups desiring access to be free to engage in regularized, identifiable programing at particular times. (In this connection, while § 76.251(a)(11)(iv) precludes local franchising authority action in this area in the top 100 markets, for other communities where access service is made voluntary, subparagraph (b) restricts only local requirements that "exceed provisions concerning the availability and administration of access channels." In these areas, then, some local structuring of access scheduling would appear permissible.) The advertising rule is also imprecise, though it would appear to permit material not within the prohibited categories, such as notices of nonprofit activities and services and information directed toward public issues rather than candidates. In light of the prohibition of censorship in subparagraph (a)(9), a continuing conflict in the responsibilities of the cable system can be foreseen, given the companion requirement prohibiting certain forms of information; most likely, cable systems will wish to secure agreements from users that they will avoid prohibited formats, relying on self-restraint subject to the sanction that future access may be barred.

In this connection, the exposure of the cable system itself to liability for defamatory use of access channels has caused concern on the part of cable operators, which the FCC may have imperfectly appreciated. The agency feels that the New York Times and WDAY cases—the former setting a high standard of "actual malice" for issues of public concern and the latter suggesting that state laws frustrating important federal concerns may be preempted (Report and Order, par. 144)—adequately protect cable operators. But inasmuch as the WDAY case rested on legislation, rather than agency regulation, the FCC feels that, "if a problem should develop in this respect, it is readily remedied by Congress and, in this connection, we would welcome clarifying legislation." (Report and Order, par. 141.)

3.3312: *Educational Access Channels.* With respect to *educational access channels,* a similar prohibition of censorship and parallel requirements concerning advertising, lotteries, obscene and indecent matter, and public inspection (§ 76.251(a)(11)(ii)) are in force. Subparagraph (a)(10)(i) requires that the system offer use of educational channels "without charge" until a point five years after completion of the system's basic trunk line, with subsequent arrangements to be provided after consultation between the FCC and affected interests.

*

(29–31)

COMMENT: *As noted previously, the FCC's rules provide no standards for deter-mining qualified "local educational authorities" or allocating time among their use. No provisions are made to facilitate production of educational programing. (Note that the Commission's reference to "at least one" such channel raises the possibility that more than one channel may be devoted to such purposes—subject to the re-quirement for free use—by initiative of the cable system or the local franchise authority.)*

3.3313: Government Access Channels. The *government access channels* are also to be free of program content control, pursuant to § 76.251(a)(9), though no comparable general operating guidelines are prescribed; however, subparagraph (a)(11)(iv) makes clear that local authorities may prescribe "other rules concerning . . . the manner of operation" of the local government channel. Like the educational channel, this channel is to be free of charge for government use for a period of five years after trunk-line completion, and then subject to future rules.

COMMENT: *The absence of FCC restrictions on government channels makes regu-lation of this matter a clearly local responsibility, as to which stipulations may be made in franchises without conflict. Nothing in the FCC's rules permitting such regulation in connection with the "operation of the local government access chan-nel" would appear to prohibit clearly local franchise rules governing use of produc-tion facilities to effectuate such operation. The possibility that more than one such channel may be called for by local requirements may also not be precluded (see Sec. 3.3311). Overall, the problems of arbitrating among users and developing an appro-priate governmental program policy appear to be major responsibilities assigned to local government entities. In the light of the total lack of standards established by federal regulation, determination of rules of access will clearly be an area of poten-tial conflict between local regulatory authorities and cable operators and users.*

*
(32–35)

3.3314: Leased Access Channels. Leased access channels, which the FCC re-quires for the remainder of the required bandwidth and any other available band-width (i.e., unused time on broadcast channels or other access channels)[31] are hence required for whatever channels are unused by the cable system after it has "satisfied the origination cablecasting requirements . . . and the requirements . . . for specifi-cally designated access channels . . ." (§ 76.251(a)(7).) For leased channels, the FCC's rules require that cable systems establish rules requiring "first-come nondiscrimina-tory access" (§ 76.251(a)(11)(iii), and that "on at least one of the leased channels, priority shall be given to part-time users" (§ 76.251(a)(7)). Besides prohibiting system censorship (§ 76.251(a)(9)),[32] the FCC also mandates restrictions on advertising,

[31] Note, however, that § 76.251(a)(7) also requires, with respect to such "other available bandwidth" that:

> "However, these leased channel operations shall be undertaken with the ex-press understanding that they are subject to displacement if there is a demand to use the channels for their specially designated purposes."

COMMENT: *Thus it is clear that, although insertion of a leased program in a channel devoted to some clearly identifiable use may enhance that program's viewing (for example, if inserted into a blacked-out portion of a strongly viewed distant broadcast channel), the displacement requirement may also act to limit the regularity or stability of such a lease.*

[32] The FCC's view that its prohibition of censorship on access channels insulates cable system opera-tors from liability for defamation is equally applicable to leased as well as public access channels.

lotteries, and obscene or indecent matter similar to those applicable to other access channels. (§ 76.251(a)(11)(iii).) In addition to similar recordkeeping provisions (retention for two years of a list of all applicants and their addresses), cable systems are also required to develop "an appropriate rate schedule," not further specified.

COMMENT: Lacking a more precise definition of "first-come nondiscriminatory access" and of "an appropriate rate schedule," the FCC's rules introduce considerable uncertainty with respect to the exact terms and conditions to which leased channel activities may be subject. Although the FCC has indicated that, "On at least one of the leased channels part-time users must be given priority," no standards are suggested to moderate among and between such "part-time users" and others, presumably lessees of entire channels. Obviously, the FCC's view of the future development of leased channel operations hinges on its rule with respect to expansion of channels. Also, a significant aspect of its approach may be its "plan at a later date to institute a proceeding with a view to assuring that our requirement of capacity expansion is not frustrated through rate manipulation or by any other means." (Report and Order, par. 126.) This suggests that the FCC is already aware that some operators may try to curb the expansion of channel-leasing so as to reduce competition with their own origination channels or to cut down on expenses. It is evident that many hard questions about rates and terms of access have been left to the future—an uncertain state of affairs that may inhibit the desire of potential programers to lease channels.

3.332. Channel Capacity. A stated aim of the *Report and Order* is to encourage "cable television use that will lead to constantly expanding channel capacity." To that end it established basic principles with respect to "minimum channel capacity" (in the major markets, at least 20 channels, "available for immediate or potential use") and "equivalent bandwidth" (one access channel for each broadcast channel (§ 76.251(a)(1) and (2))). It supplements these system capacity rules with an initial formula "to determine when a new channel must be made operational":

*

(21)

> RULE:
> "Whenever all of the [access] channels . . . are in use during 80 percent of the weekdays (Monday-Friday) for 80 percent of the time during any consecutive 3-hour period for six consecutive weeks, such system shall have 6 months in which to make a new channel available for any or all of the above-described purposes . . ." (§ 76.251(a)(8).)

COMMENT: The FCC's channel capacity rules offer cable systems considerable leeway to operate in without facing the need for system expansion. It was noted in Sec. 3.22 that the FCC views expansion to dual cable facilities within the six-month period of the formula as "unreasonable," and hence subject to waiver. Moreover, there are significant possibilities for spacing out demand for access channels—in terms of timing, length of programs, use of unused channels, and rates—to permit the cable system wide latitude in staving off the point at which the formula becomes operative. For systems now in being, an amendment to § 76.251(c) now permits cable operators to postpone application of the expansion capacity rule for five years (see Sec. 3.24). Thus, the mathematical explicitness of the FCC's formulas actually works to minimize the likelihood that the channel expansion rule will significantly encour-

age expansion of access channel operations. This situation is likely to prevail until the FCC further clarifies and stabilizes potential users' right of access to leased channels.

3.4. EVALUATION AND PERSPECTIVE ON LOCAL INITIATIVES

The cablecasting rules are obviously experimental. This is their most significant feature. In the FCC's words, they are intended as

> ... a basic framework within which we may measure cable's technological promise, assess its role in our nationwide scheme of communications, and learn how to adapt its potential for energetic growth to serve the public. (*Report and Order*, par. 117.)

The FCC's approach is invaluable to cable systems as a long-sought public commitment to foster cable growth. It might also be called reasonable, and indeed ingenious, in view of the uncertainties that beset the cable medium and the market for its services, and the continuing hostility of its competitors. But the FCC's approach also entails numerous imprecisions and delaying actions that will generate uncertainty and controversy for years to come.

Now that the *Midwest Video* case has been decided—which questioned the FCC's basic authority to venture beyond broadcast regulation—much has been done to clear the air. Further essential realism will be injected into the environment in which regulatory issues are confronted if Congressional legislation ensues—either in the form of amendments to the Communications Act of 1934 or to the Copyright Act of 1909, itself only peripherally relevant to cablecasting but central to the economic stability of the cable industry. Sheer experience with this new medium, as the FCC appreciates, will engender further practical decisionmaking.

For the present, the new rules are a meaningful basis on which all concerned must adjust their planning—or initiate their lawsuits. The rules are not rigid. Being initial and experimental, they are not intended to stifle alternative approaches nor to prohibit efforts to initiate service that may depart from the patterns they presently set forth. Certainly, in areas of doubtful authority, a petition for special relief pursuant to 47 C.F.R. § 76.7 can be considered. Meanwhile, for those cable operators and programers requiring stability and certainty to get under way, the new rule represents a modicum of essential encouragement.

Even though the FCC has assumed substantial regulatory control over cable services, local authorities have significant responsibilities of their own. They must obviously adapt their governmental functions to the availability of access channels. In this respect, each government entity must now develop its policies, personnel, and facilities to take advantage of the new communications resource afforded by the growth of cable services. They will also be entitled to a powerful voice, before the FCC and Congress, in giving real meaning at local levels to the FCC's policies for cablecasting. It is to this crucial area that this analysis next turns.

Section 4

THE ROLE OF STATE AND LOCAL AUTHORITIES

The growth of cable television impinges on the concerns of each major echelon of American government. Municipalities, states, and the federal government must devise policies that incoporate cable television into older regulatory patterns, and find ways to use this promising new medium to achieve social and economic objectives. The emergence of cable also requires alignment of functions among governmental echelons that will ensure both its growth and its responsiveness to public controls.

These are sizable tasks, as the FCC pointed out in its *Report and Order.* The Commission repeated a previous finding that no "overall plan as to the Federal-local relationship" had yet emerged.

> This has resulted in a patchwork of disparate approaches affecting the development of cable television. While the Commission was pursuing a program to promote national cable policy, state and local governments were formulating policies to reflect local needs and desires. In many respects this dual approach worked well. To a growing extent, however, the rapid expansion of the cable television industry has led to overlapping and sometimes incompatible regulations. This resulted in confusion . . ." *(Report and Order,* par. 171.)

Since regulatory confusion can defeat the goal of controlled growth by imposing onerous burdens and inconsistent regulatory philosophies on cable companies, the Commission has adopted the broad outlines of a regulatory program centerd on what it has called a concept of "dual jurisdiction." By this approach, the FCC will manage certain aspects of cable regulation while appropriate authorities will handle others under the laws of their particular states.

After a brief historical and analytic review, this section presents the essential aspects of the dual regulatory program and indicates their practical implications for state and local government programs. The section concludes with an assessment of issues that can be expected to arise in intergovernmental relationships during the next phase of cable television growth.

61

4.1. THE HISTORICAL AND LEGAL CONTEXT

Since its inception in the late 1940's, regulation of cable television had been predominantly a local matter. Until 1962, the FCC took the view that cable television was neither a broadcaster nor an interstate common carrier, and hence was beyond its limited jurisdiction in the communications field.[1] In that year, the Commission first assumed *indirect* jurisdiction over cable systems receiving television broadcast signals via interstate microwave transmission services,[2] and in 1966 asserted limited *prohibitory* jurisdiction over cable systems carrying broadcast signals into major metropolitan areas from distant broadcast "markets."[3] It was not until 1969 that the FCC attempted to assert *affirmative* control over cable television systems through the jurisdictional nexus of their carriage of broadcast television.[4] It did so by ruling that cable television systems with more than 3500 subscribers must engage in original video programing as a condition for the carriage of any broadcast signals.

The FCC's gradual entry into cable regulation was due to its reluctance to expand the scope of its legislative mandate as demarcated in the Communication's Act of 1934. The Act of course did not mention cable television, since coaxial cable for local distribution of television programs first came into use in 1949, fifteen years later. The omission of cable from the Act raised the question of whether any form of federal regulation was permitted, and if so, which of the FCC's enumerated methods of regulations were required.[5] Federal regulatory policy reflected this uncertainty by initially disclaiming jurisdiction, later prohibiting cable use of broadcast signals, and finally imposing limited affirmative obligations on local distribution via cable, in accordance with a Supreme Court construction of the Communications Act permitting the FCC to impose requirements "reasonably ancillary" to the Commission's responsibilities for broadcasting.[6]

[1] The FCC took no action on the first request for cable regulation in 1954 by a broadcast station and again in 1959 explicitly declined to regulate cable television. *CATV and Repeater Services,* 26 FCC 403 (1959).

[2] *Carter Mountain Transmission Corp.,* 32 FCC 459, aff'd 321 F.2d 359 (C.A.D.C., 1962), cert. denied, 375 U.S. 951 (1963); *First Report and Order in Dockets 14895 and 15233,* 38 FCC 683 (1962).

[3] *Second Report and Order in Dockets 14895, 15233 and 15971,* 2 FCC 2d 725 (1966).

[4] *First Report and Order in Docket 18397,* 20 FCC 2d 201 (1969). Beginning in 1966, the FCC has also moved to bring the local activity of interstate telephone carriers under federal control pursuant to Title II of the Communications Act, by requiring the filing of federal tariffs for CATV services *(Common Carrier Tariffs for CATV systems,* 4 FCC 2d 257 (1966)); requiring federal certificates for construction of facilities to be used for CATV service *(General Tel. Co.,* 13 FCC 2d 448 (1968) *affirmed sub nom General Tel. Co. v. FCC,* 413 F.2d 448 (C.A.D.C., 1968), *cert. denied* 396 U.S. 888 (1969)); and by barring cross-ownership by such telephone companies of CATV companies in their service areas *(Final Report and Order* in Docket 18509, 21 FCC 2d 307 (1970)). Federal controls have thus also grown increasingly intrusive in this limited area.

[5] The Communications Act of 1934, 47 U.S.C. § 151 *et seq.* creates the FCC and endows it with comprehensive authority over interstate and foreign communications (§§ 151, 152(a), 153(a) and 153(b)), but its specific powers center on the licensing of radio and television broadcast stations (§§ 301-330) and certification of interstate service by "common carriers"—a terminology specifically defined to exclude broadcasters (§ 103(g)), but not further defined and in practice restricted to traditional telephone and telegraph service and the carriage of information for others. Similarly detailed and specific powers are not set forth for the use of broadband systems as local distributors of information.

[6] In *U.S. v. Southwestern Cable Company,* 392 U.S. 157 (1968), the Supreme Court upheld the FCC's restriction on carriage of distant broadcast signals (see footnote 4), on the grounds that the cable television systems are in interstate commerce and that the agency has shown a direct relationship between importation of distant signals and the economic health of local broadcast stations in San Diego. The court did not further define FCC powers, if any, under its broad mandate over interstate and foreign communications.

63

Thus, prior to the mandatory origination rule (1969), now upheld by the Supreme Court in *Midwest Video,* [7] the FCC had not attempted to exert direct regulatory authority over cable television. This contrasts with its extensive involvement in licensing broadcast stations (leaving such areas as construction and zoning for local concern), and its highly developed concern for rates and service of interstate common carriers (also establishing a line of demarcation between federal authority and intrastate regulation by state Public Utilities Commissions). As a result, state and local agencies have acted in a vacuum with respect to cable television, which, neither promoted nor forbidden by federal action, had been left to sink what local roots it could.

Against this background, state law has served as the principal basis for establishing the operating conditions of cable television systems.[8] In the absence of federal action, state authorities have chosen diverse courses. By action of state regulatory commissions or special legislation, a few states (presently nine) have brought cable television under regulation expressing utility principles to a greater or lesser extent comparable to those applied to the regulation of telephone suppliers.[9] Elsewhere, Public Utilities Commissions have been indifferent to cable television in many states[10] or dissuaded by litigation from assuming jurisdiction.[11] However, in the absence of consistent federal legislation or regulation, it has been authoritatively determined that states may regulate cable systems as public utilities.[12] Whether the federal mandatory origination rule subsequently upheld in *Midwest Video* will now significantly reduce state running-room remains to be judicially established.

Inasmuch as cities and counties are only political subdivisions of states, and have no independent legal basis, regulation of cable television other than under federal authority is entirely a question of applicable state law, for which two sepa-

[7] *United States v. Midwest Video Corp.* upheld federal authority to order cable systems to engage in origination cablecasting as within the scope of the FCC's broadcasting jurisdiction. The FCC has read the court's affirmation of its origination cablecasting jurisdiction as an affirmation of access cablecasting jurisdiction as well *(Reconsideration,* par. 71) both representing aspects of cable television service whose effects on broadcasting bring them within the *Midwest Video* rule.

[8] As a practical matter, federal prohibitions on the importation of distant signals, which began in 1966, have actually established crucial restraints on the economic viability and hence growth, of cable systems. Where cable systems have been established, it has been through loopholes in prohibitory federal regulation.

[9] States regulating cable television services under State-wide utility-type statutes are *Alaska,* Ala. Stats., Tit. 42, Chap. 05, Sec. 761 (see also Capital Cablevision, Inc., Dkt. V-71-77 (1971)); *Connecticut,* Conn. Gen. Stats., Chap. 289, Sec. 16-330 *et seq.; Hawaii,* Rev. Stats., Chap. 269 (see also the Opinion of the Attorney General of Hawaii No. 69-29, Dec. 2, 1969, CCH Utilities Law Reporter 21,206); *Illinois,* Ill. Rev. Stats. Chap. 111-2/3, Sec. 10-3(b); *Massachusetts,* Chap. 1103, Acts of 1971 creating a new Chap. 116A, Mass. Gen. Laws; *Nevada,* Nev. Rev. Stats., Sec. 711.010 et seq; *New York,* Chap. 28, McKinney's N.Y. Statutes, Executive Laws; *Rhode Island,* R.I. Gen. Laws, Sec. 39-19-1 et seq.; and *Vermont,* Vt. Stats. Ann., Title 30, Chap. 13.

[10] See, e.g., *Seneca Radio Corp.,* 57 P.U.R. 3d 67 (Ohio P.U.C., 1964) and *Southern Bell Tel. & Tel. Co.,* 65 P.U.R. 3d 117 (Fla. P.U.C. 1966).

[11] *Re Community Television Systems,* 23 P.U.R. 3d 444 (Wyo. P.S.C., 1958), reversed 17 R.R. 215 (Wyo. Dist. Ct., 1958) and *Cauch v. Television Transmission, Inc.,* 11 P.U.R. 3d 430 (Cal. P.U.C., 1956) reversed sub nom *Television Transmission, Inc. v. PUC,* 47 Cal. 2d 82, 301 P. 2d 862 (1956).

[12] In *TV Pix, Inc. v. Taylor,* 304 F. Supp. 459 (D.C. Nev., 1968) affirmed per curiam 396 U.S. 556 (1970), the District Court upheld the validity of a Nevada statute empowering the State Public Service Commission to impose utility regulation upon cable operators in the form of certification and supervision of rates and services. Denying a claim that such state regulation was an unconstitutional preemption of federal authority over interstate communications under the Communications Act of 1934, the court held that such state regulation is valid so long as any federal power remains "dormant and unexercised," citing *Head v. New Mexico Board,* 374 U.S. 424 (1963).

rate sources of power exist. On the one hand, state governments themselves may, directly or inferentially, exercise authority over cable television systems. Eight instances of explicit state public-utility-type regulation were cited above, under which principal local regulatory functions are exercised by a state regulatory commission "certificating" individual systems on a state-wide basis. Certain of these state functions may be delegated specifically to municipalities, as in Massachusetts, where the selection of license holders is to be a municipal function under state-imposed guidelines.[13] Also, state regulatory authority over telephone systems may implicitly preempt aspects of local cable television regulation from municipalities, typically in circumstances where courts have held a municipality may not require an additional franchise to make cable service available pursuant to a state-issued telephone franchise.[14]

On the other hand, to a greater or lesser extent depending on state constitutional and statutory delegations, municipalities usually possess powers of "home rule" that embrace the activities of cable distribution. These powers center on the regulation of the use of streets and alleys, under or over which cable must be strung. Many cities also exercise authorities as aspects of state "police powers" to protect the general health, safety, or economy. The scope accorded to municipalities to impose regulations on cable systems under state-delegated powers has been the subject of frequent litigation,[15] and varies widely in particular contexts. Municipal franchises have also varied widely, partly as a result of peculiar local authorities, but also in reflection of widely divergent regulatory objectives; the New York City franchise is widely cited as exemplifying an ambitious municipal concern for regulatory issues. When challenged, some elaborate franchises have been upheld as appropriate exercises of municipal authority, but some have been struck down.[16]

Against this background of legal authorities and philosophies has arisen the "patchwork of disparate approaches affecting the development of cable television" lamented by the FCC. Until it bestirred itself to identify affirmative goals for cable television growth, such a patchwork pattern was the inevitable consequence of the Commission's lethargy. The FCC having now established policies for the development of cable television nationally, a more systematic alignment corresponding to the FCC's goals can be expected to emerge.

The precise outlines of federal, state, and local relationships, however, will remain subject to much further refinement. As a matter of political and legal theory, the relationships among various governments are, or ought to be clear: Where federal powers over interstate commerce are "dormant and unexercised,"[17] state powers may be exercised to the extent they do not place an unreasonable burden on

[13] See Chap. 166A, Mass. Gen. Laws, Sec. 3, as added by Chap. 1103 of the Acts of 1971.

[14] See, e.g., *New York v. Comtel, Inc.* (1968), 57 Misc. 2d 585, 293 NYS2d 599, *aff'd* 30 App. Div. 2d 1049, 294 NYS2d 981 *aff'd* 25 NY2d 922, 304 NYS 2d 853, 252 NE 2d 285.

[15] See cases collected in "Municipal Regulation of CATV Systems," 41 A.L.R.3d 384 (1971).

[16] Contrast *DiBella v. Ontario,* 4 Ohio Misc. 120, 212 NE 2d 679 (1965), where the court upheld a cable franchise as within municipal powers, inasmuch as it did not conflict with any contrary state legislation and because municipalities should be free to respond to the changing nature of their responsibilities, with *Community Antenna Television of Wichita, Inc. v. Wichita,* 205 Kan. 537, 471 P2d 360 (1970), where a franchise regulating rates and imposing a franchise fee geared to gross receipts was held to exceed limited municipal authorities for regulating the streets.

[17] *TV Pix, Inc. v. Taylor,* 304 F. Supp. at 465 (see footnote 12).

interstate commerce;[18] but to the extent federal powers have been exercised, federal authority is supreme over inconsistent state regulation.[19] Prior to the issuance of the *Report and Order,* federal power was clearly "dormant and unexercised." FCC rules now purport to resolve key aspects of the future role of cable television, but they often do so inconclusively and with such vagueness, imprecision, and apparent flexibility that divergent state and local approaches may not therefore be lawfully "preempted."

It is clear that the FCC has moved to take a leading role in the development of cable television, but it is not clear that that role will effectively withstand purposeful assertions of state and local viewpoints. This is especially so because the FCC has obviously acknowledged a role for state and local authorities that it may be reluctant to stand against where clearly fashioned local interests are forcefully presented.[20] Key areas for local action and future accommodation of interests are treated in this section.

4.2. THE PATTERN OF DUAL JURISDICTION

The *Report and Order* projects its concept of dual jurisdiction in three principal contexts—"Federal-State-Local Relationships" (Sec. 4.21 below), wherein its overall regulatory pattern is set forth; "Access to and Use of Nonbroadcast Channels" (Sec. 4.22); and, inferentially, "Technical Standards" (Sec. 4.3). We consider these categories in turn. The latter two are also considered elsewhere in terms of direct federal responsibilities. The section concludes with a summary of the impact of these federal regulations on state and local responsibilities.

*

(41–44)

4.21. Federal-State-Local Relationships

The FCC has identified three options open to it with respect to intergovernmental relationships:

    (a)   Federal licensing of all cable television systems.

---

[19] For example, in *Greater Fremont, Inc. v. Fremont,* 302, F. Supp. 642 (D. Ohio, 1968), all'd *sub nom Wonderland Ventures, Inc. v. Sandusky,* 423 F.2d 548 (CA 6, 1970)—a case whose result has been questioned—it was held that a city's imposition of a franchise fee based on a percentage of gross receipts represented such an unreasonable burden on interstate commerce. There can be no doubt remaining, as the Supreme Court recently reiterated in *Midwest Video,* 406 U.S. at 662-663, that local cable service is within the reach of federal powers over interstate commerce.

[19] "This Constitution, and the Laws of the United States which shall be made in Pursuance thereof ... shall be the Supreme Law of the Land ..." U.S. Constitution, Article VI, Cl. 2.

[20] For example, as previously noted, a number of states regulate cable television under state regulatory statutes imposing public utility regulation on cable systems. In Connecticut, Conn. Gen. Stat. § 16-331 sets out an elaborate scheme of regulation by which certificates are issued of "perpetual" duration and certificate holders are subject to potential regulation to assure reasonable rates of return on invested capital; they are also subject to a separate tax on gross revenues, now eight percent. These circumstances have been raised by various broadcast interests in a petition to withhold certification from Valley Cable Vision, Inc., alleging failure to conform to federal requirements, based on a theory of preemption (FCC File Nos. CAC-115, 116, 119, petition filed May 12, 1972). The FCC has not acted on this petition and is apparently not likely to do so soon, given the head-on clash it threatens with a pervasive local regulatory scheme. Obviously, some practical accommodations will have to be reached to permit cable service to commence in Connecticut. Perhaps these will be among the issues to be mediated by the Advisory Committee created by the FCC and state and local interests pursuant to *Report and Order,* par. 188.

(b)   Maintenance of the current federal regulatory program enforced by Sec. 312(b) proceedings (cease-and-desist orders on an ad hoc basis).

(c)   Federal regulation of some aspects, with local regulation of others, under federal prescription of standards for local jurisdictions. *(Report and Order,* par. 171.)

The first alternative, which the FCC rejects, implies the establishment of a comprehensive regulatory regime resembling the FCC's control of broadcast stations. The Commission feels this would be "an unmanageable burden" beyond its administrative capabilities in the light of its limited budget and the vast number of proceedings that licensing for individual communities would entail. The second course, requiring the establishment of broad federal criteria enforced in individual cases by cease-and-desist orders, is implicitly rejected as an unwieldy method by which to promote (as opposed to retard) cable television growth. The final alternative, adopted by the FCC, expresses the outlines of a cooperative regulatory program, termed a kind of "creative federalism," intended to reconcile the interests of each government echelon within the context of overall federal promotion of the emergence of broadband services.[21]

Within the Commission's dual regulatory program, the role for local authority centers on its franchising process. The federal regulations do not purport to deal with the question of who has authority to grant a local "franchise, license, permit, or certificate" for any particular system; it remains primarily a question to be determined under applicable state law. But each cable system[22] must show, in its application for a certificate of compliance with the federal regulatory program,[23] that it has also complied with locally applicable law[24] in respects hereafter specified. The requirement for such a showing of local compliance, however, will not apply to a cable system in operation prior to the effective date of the federal rules until the

[21] The FCC bases its choice on the view that:

> [L]ocal governments are inescapably involved in the process because cable makes use of streets and ways and because local authorities are able to bring a special expertness to such matters, for example, as how best to parcel large urban areas into cable districts. Local authorities are also in a better position to follow up on service complaints. *(Report and Order,* par. 177.)

The FCC also views its "obligation to insure an efficient communications service with adequate facilities at reasonable charges" as requiring that it set "*at least minimum standards* for franchises issued by local authorities." (Italics added)

[22] To be subject to federal regulation, a cable system is defined as carrying broadcast signals for distribution to paying subscribers as follows:

> Any facilities that, in whole or in part, receives directly, or indirectly over the air, and amplifies or otherwise modifies the signals transmitting programs broadcast by one or more television or radio stations and distributes such signals by wire or cable to subscribing members of the public who pay for such service, but such term shall not include (1) any such facility that serves fewer than 50 subscribers, or (2) any such facility that serves only the residents of one or more apartment dwellings under common ownership, control, or management, and commercial establishments located on the premises of such an apartment house. (§ 76.5(a).)

[23] See specifically Sec. 6.3, dealing with FCC certification.

[24] Each system must separately demonstrate compliance with locally applicable law in each legal jurisdiction served:

> In general, each separate and distinct community or municipal entity (including single, discrete, unincorporated areas) served by cable television facilities constitutes a separate cable television system, even if there is a single headend and identical ownership of facilities extending into several communities . . ."

end of its current franchise period or March 31, 1977 (five years after the effective date of the rules), whichever comes first. (§§ 76.11(b), 76.31(a)(6) and (b).)[25]

An applicant for a federal certificate of compliance must show (§§ 76.13(a), (d)) a "franchise or other appropriate authorization that contains recitations and provisions ..." (§ 76.31(a))[26] that are consistent with the enumerated federal guidelines hereafter set forth.

4.211. Selection Process.

RULE:
"(1) The franchisee's legal, character, financial, technical, and other qualifications, and the adequacy and feasibility of its construction arrangements, have been approved by the franchising authority as part of a full public proceeding affording due process ..."

*
(50–64)

In the *Report and Order,* the rule is interpreted to include the specific requirement that a copy of the franchise be submitted, along with the required showings. The FCC also expresses in detail its "expectations" with respect to the nature of the "full public proceeding affording due process" which it requires:

> We expect that franchising authorities will publicly invite applications, that all applications will be placed on public file, that where appropriate a public hearing will be held to afford all interested persons an opportunity to testify on the qualifications of the applicants, and that the franchising authority will issue a public report setting forth the basis for its action. Such public participation in the franchising process is necessary to assure that the needs and desires of all segments of the community are carefully considered. *(Report and Order,* par. 178.)

COMMENT: While state courts have occasionally invalidated cable television franchises for failure to observe constitutional and statutory requirements,[27] the FCC's initiative represents a novel departure from past indifference to the methods by which franchises are issued. The FCC's requirement is couched in somewhat

[25] By its terms, the requirement to apply for a federal certificate showing compliance with local law applies to all new cable systems and systems adding a television broadcast signal after the date the rules become effective. (§ 76.11(a).) If an application by a system now in operation is filed 30 days prior to the date by which the rules require it to secure a certificate, § 76.11(c) authorizes continuing operations pending FCC review. § 76.251(c), however, permits existing cable systems to add and couple the addition to broadcast signals with the addition of individual access channels, and to delay the applicability of the channel capacity rule to 1977. See Sec. 3.24.

[26] On reconsideration, the FCC dealt with two situations not envisaged in its original rules: Where no local franchising authority exists—for example, in an unincorporated area where state authority is not specifically exercised or where a state court has curbed a municipality's franchising authority—the FCC will entertain waiver petitions proposing alternative assurance that the substance of § 76.31 has been achieved. *(Reconsideration,* par. 116.) Where there is doubt whether state or local authority is the appropriate franchise source, the FCC is prepared, upon a showing of service of the application of the cable system upon all authorities, to issue certificates on a conditional basis. *(Reconsideration,* par. 117.) Subsequently, the FCC has made such comprehensive service mandatory, along with a further requirement that where state or local franchising bodies do not make the contents of a cable system's federal application publicly available, the system do so itself. See "Procedures for Filing of Applications," 37 Fed. Reg. 23104-5 (October 28, 1972) in Part III. The Commission would also apparently extinguish the prospects that a cable operator could "avoid obtaining a franchise by obtaining channel service from a telephone company," by requiring a local franchise nonetheless. *(Reconsideration,* par. 116, footnote 39.)

[27] In terms of review of the substance of local and state regulatory determinations, however, state courts have traditionally held conservatively that legislative determinations are insulated from judicial intervention—e.g., *Monarch Cablevision, Inc. v. City Council of Pacific Grove,* 239 Cal. App. 2d 206 (1966) —while preserving a role in review for manifest abuse of discretion, *Palmeri v. Penn Hills,* 423 Pa. 454, 239 A2d 204 (1968).

tentative language, probably intended to permit some variations in local procedures with respect to form, especially apparent in the "where appropriate" provision with respect to a public hearing. Deviation from the FCC's overall requirement that "the needs and desires of all segments of the community [be] carefully considered" will occur at the risk of the franchising authority and the franchisee, inasmuch as the rules also provide that "any . . . interested person" can petition the FCC with respect to a ruling on "a complaint or disputed question." (47 C.F.R. § 76.7(a).) While the FCC is not further bound with respect to any such ruling, it clearly contemplates with-holding broadcast carriage rights as a sanction. The Administrative Procedure Act, 5 U.S.C. §§ 501 et seq. and 701 et seq., offers a significant potential that persons aggrieved by FCC determinations may have further resort to federal courts to test the grant or withholding of a certificate of compliance.

*
(67–71)

4.212. Construction Schedules. The Commission also offers further guidance in the *Report and Order* with respect to the "adequacy and feasibility of its construction arrangements." (Par. 181.) It finds the delineation of the franchise area is a responsibility "uniquely within the competence of local authorities," but stresses "that provision must be made for cable service to develop equitably and reasonably in all parts of the community." Thus, a "plan that would bring cable only to the more affluent parts of a city, ignoring the poorer areas, simply could not stand." Service must develop "reasonably" and "equitably" among the variety of ways deemed open to each community, though the choice is a matter "for local judgment."

COMMENT: Though imprecise, the Report and Order *envisages a showing of a good-faith plan to wire entire communities according to reasonable schedules. Where such a showing could not be made, the FCC warns that, "No broadcast signal would be authorized in such circumstances."*

Thus, the FCC requires in § 76.31 a showing that

> RULE:
> (2) The franchisee shall accomplish significant construction within one (1) year after receiving Commission certification, and shall thereafter equitably and reasonably extend energized trunk cable to a substantial percentage of its franchising authority . . .

The Commission points out that it is taking pains to establish "general timetables for construction and operation of systems to insure that franchises do not lie fallow or become the subject of trafficking." As a general proposition, not intended to become "an inflexible figure," an extension rate for energized trunk cable of at least 20 percent per year is regarded as appropriate, to be achieved within one year after certification, following construction of a "significant" amount (not further defined) in the first year. While the nature of the assurances needed to satisfy this requirement is not further defined by the FCC, such a timetable would presumably relate to the "adequacy and feasibility of . . . construction arrangements" with which the franchising authority must deal and which the franchise document must reflect.

COMMENT: It is clear, since § 76.31(6) provides that "consistency with these requirements shall not be expected of a cable television system that was in operation" at the time the rules go into effect until later certification, that the require-

ments regarding rate of expansion for such systems will not apply until such later date. Moreover, it should be noted that the FCC's rule applies only to the laying of energized trunk cable, not to the actual delivery of services to subscribers. The latter is an economic function that will be responsive to subscriber demand, reflecting rates charged and quality of services; hence the FCC rules do not purport to resolve all aspects of its "obligation to ensure an efficient communications service with adequate facilities at reasonable charges." Meanwhile, the FCC has been aware that New York City and various states have laid down requirements for more rapid construction. Acknowledging that "local circumstances may vary," it is important to note that the Commission, in this respect at least, has established a "minimum standard" that local franchising authorities may exceed without conflicting with federal regulation. (The implications of this and other such accommodations will be traced in Sec. 4.3.)

4.213. Franchise Periods. From an original position centering on a requirement of "reasonableness," the FCC on reconsideration has now specified the maximum duration of original franchise periods (par. 111), as follows:

*

(72–79)

> RULE:
> (3) The initial franchise period shall not exceed fifteen (15) years, and any renewal franchise period shall be of reasonable duration.

If good cause can be shown for some other franchise period, the FCC notes *(Reconsideration,* par. 111), that it will consider a petition for special relief. Its explanation for so limiting franchise periods was suggested originally in the *Report and Order:*

> Long terms have generally been found unsatisfactory by state and local regulatory authorities and are an invitation to obsolescence in the light of the momentum of cable technology. (Par. 182.)

Recognizing that local decisions may vary "in particular circumstances," the Commission acknowledged one basis for variance: that an applicant's proposal "to wire inner-city areas without charge or at reduced rates might call for a longer franchise." (Par. 182.)

COMMENT: By so curbing the prospect of ad hoc determinations, the FCC offers scope to local authorities to grant franchises of less than 15 years duration, but prohibits longer periods without special showings. It has thereby lessened the authority of localities to trade longer terms for special local purposes such as free or reduced-rate service. It has also created a special problem for public utility regulation, which does not traditionally impose limited terms as an aspect of a regulatory approach limiting overall profits. A challenge to such regulation in Connecticut has already been filed (see footnote 20 above), which might be conservatively forecast as a cause for future litigation testing FCC jurisdiction to so control the exercise of local regulatory powers.

4.214. Rates to Subscribers. Further, § 76.31(a) accords localities the power to specify subscriber rates, requiring the following showing of an applicant:

*

(84–85, 95)

> RULE:
> (4) The franchising authority has specified or approved the initial rates

which the franchisee charges subscribers for installation of equipment and regular subscriber services. No changes in rates charged to subscribers shall be made except as authorized by the franchising authority after an appropriate public proceeding affording due process . . .

The FCC requires that the franchising authority specify (presumably by franchise) or approve (presumably referring to subsequent ratification) rates for installation of equipment and *regular subscriber services,* omitting mention of subscriber services not embraced in a "regular" rate. The regulation also embraces rate changes, which must be made after "an *appropriate* public proceeding affording due process" (presumably to be distinguished from the "*full* public proceeding affording due process" required for initial choice of franchise-holder). In the *Report and Order* the FCC establishes an "appropriate standard" for rate-setting:

> . . . the maintenance of rates that are fair to the system and to the subscribing public—a matter that will turn on the facts of each particular case (after appropriate public proceeding affording due process) and the accumulated experience of other cable communities. *(Report and Order,* par. 183.)

COMMENT: Here, the FCC appears bent on encouraging the formation around the country of certain norms for cable fees, and urging local caution in authorizing rate increases subsequent to initial franchising decisions. While the sanction of withholding federal certification is presumably available, it is not clear by what standard it would be exercised. The federal guideline being merely a "minimum standard," it certainly offers no guidance as to which side—the public or the system—franchise authorities might lean in setting rates. This would suggest that what is "fair" should not deviate greatly from the "accumulated experience of other cable communities," which hitherto has come about through arbitrary and unsystematic bargaining. However, if lower than prevailing rates are embodied in a franchise agreement, the negotiating framework in which the issue would come to the FCC must be appreciated. That is, having accepted a curb on its rates as a condition for gaining a franchise, the franchise holder cannot practically turn against its contract and seek relief from the FCC. Finally, the FCC's authorization to local franchising authorities to regulate "regular subscriber services" is significant for what it omits: Since in an earlier proceeding the FCC has attempted to preempt "the field of pay television cablecasting so that local franchise terms are inoperative,"[28] it would appear that setting rates for pay-cablecasting is precluded. For reasons discussed subsequently in connection with regulation of access to nonbroadcast channels, it would appear that rate-setting for other nonbroadcast services is precluded as well. However, it should be noted that municipalities may take into consideration revenue that cable franchisees may earn from federally authorized pay-cablevision in determining what is "fair" to cable companies where rates for "regular subscriber services" are set.

---

[28] In a letter from the Secretary of the Commission to an attorney for two Manhattan cable companies, the Commission recently maintained that federal authorization of cablecasting on a per-program or per-channel basis, 47 C.F.R. § 76.225, preempted from local regulation a provision of Manhattan cable franchises requiring additional city authorization before such operations could commence. Letter to Pierson, Ball & Dowd, September 8, 1971, 22 *Pike & Fischer R.R.* 2d 949. It is certainly arguable that this declaration effectively blocks the *setting of rates* from local authority, especially in the light of possible procedural defects (notice and opportunity to be heard, etc.) which could be the subject of later litigation. Similar defects regarding federal preemption of other nonbroadcast rate-setting may support additional justiciable questions.

4.215. Service to Subscribers. The applicant must also show, pursuant to § 76.31(a), as follows:

*
(86–89)

RULE:
(5) The franchise shall specify procedures for the investigation and resolution of all complaints regarding the quality of service, equipment malfunction, and similar matters, and shall require that the franchisee maintain a local business office or agent for these purposes ...

Other than to note that "some local bodies are already considering detailed plans along these general lines," the FCC offers no elaboration of this requirement. *(Report and Order, par. 184.)*

COMMENT: By this requirement the FCC has established a local responsibility for attending to service complaints, for which many local authorities will wish to develop monitoring mechanisms. Establishment of a local regulatory office would be an appropriate method to relieve local public attorneys of much day-to-day supervision. Franchise terms establishing enforcement sanctions, including but not restricted to cancellation, would appear appropriate.

4.216. Modifications by FCC. Provision for subsequent FCC rule modifications must be shown:

*
(90)

RULE:
(6) Any modifications of the provisions of this section resulting from amendment by the Commission shall be incorporated into the franchise within one (1) year of adoption of the modification, or at the time of franchise renewal, whichever occurs first ...[29]

The implication of the guideline—the necessity for local amendments of franchises to accommodate subsequent FCC standards—is self-explanatory.

COMMENT: It is not clear what sanctions the FCC might bring to bear on local franchise authorities, other than withholding broadcast signals for cable systems, to attain local compliance with new federal standards, once initial federal certificates are issued.

4.217. Franchise Fees. In § 76.31(b) the FCC has also imposed a significant restriction on fees that may be charged by a local franchising authority:

*
(91–107)

RULE:
The franchise fee shall be reasonable (e.g., in the range of 3-5 percent of the franchisee's gross subscriber revenues per year from cable television operations in the community (including all forms of consideration, such as initial lump sum payments)). If the franchise fee exceeds 3 percent of such revenues, the cable television system shall not receive Commission certification until the reasonableness of the fee is approved by the Commission on showings, by the franchisee, that it will not interfere with the effectuation of federal regulatory goals in the field of cable television, and, by the fran-

[29] The subparagraph concludes with a provision with respect to "grandfathering," previously cited, which the FCC indicates is intended to "relieve both cable systems and local authorities of whatever minor dislocations our rules might otherwise cause."

chising authority, that is appropriate in light of the planned local regulatory program . . .[30]

In its explanation of the fee guideline, the *Report and Order* indicates that the prevailing franchise fee is about 5 or 6 percent of gross revenues, but some have been as high as 36 percent, exacted "more for revenue-raising than for regulatory purposes." *(Report and Order,* par. 185.) Since the ultimate effect of such an approach is "to levy an indirect and regressive tax on cable subscribers," the Commission aims to reduce such burdens on "our national communication policy," anticipating that the imposition of copyright payment requirements will create further substantial costs for cable systems. The Commission's aim is to "strike a balance that permits the achievement of federal goals and at the same time allows adequate revenues to defray the costs of local regulation."

In the light of its own regulatory levy of 0.5 percent of a cable system's gross revenues, the Commission deems a maximum franchise fee of between 3 and 5 percent appropriate "as a general standard, for specific local application" to achieve the goals of the particular regulatory program to be carried out in each jurisdiction. *(Report and Order,* par. 186.) Specifically, tracking the terms of the guideline, it envisages that where the fee exceeds 3 percent ("including all forms of consideration, such as initial lump sum payments"), the franchising authority must show that "the specified fee is appropriate in the light of the planned local regulatory program," and the franchisee must show that "the fee will not interfere with its ability to meet the obligations imposed by our rules."

COMMENT: The FCC has established a standard that seeks to stabilize franchise fees and put a burden of persuasion on franchises that exact fees above 3 percent. The FCC's standard incidentally validates gross-receipts fees in conformity with its rules, against the Constitutional defect based on interstate commerce considerations that was identified in the Wonderland Ventures *case. Again, it should be recognized that a franchise embodying such a fee would reflect an arms-length agreement of both parties, presumably supported by a local political consensus, and that the standards for a showing to the Commission are vague. It remains unclear whether the needs of the planned regulatory program are the only basis on which a franchise fee may exceed 3 percent, or whether there are other permissible reasons (such as funding access costs). Innovative approaches are surely not prohibited. It would appear, for instance, that if revenues exceeding 3 percent are earmarked for municipal cable service functions (perhaps to support programing on public-access channels), they are justified because they contribute to "the effectuation of Federal regulatory goals in the field of cable television." Moreover, a literal interpretation of the FCC's guidelines might not prohibit the imposition of special service requirements to which a cable system must devote a percentage of its gross receipts; such expenditure would not amount to consideration, in the strict legal sense of the term, since they would not move to the local franchising authority as or in lieu of a fee. In setting rates for service to subscribers, for example, the franchising authority would not be barred from requiring the cable system to apply some percentage of*

---

[30] The guideline concludes with a provision deferring application to cable systems in operation at the date the rules come into effect until the end of the present franchise period or March 31, 1977, whichever is sooner.

its revenues to delivery of services to certain subscribers at reduced or even zero rates, or to the capitalization and maintenance of studios and special programing facilities. This approach suggests that a pool or a trust fund might even be created from revenues that exceed the 5-percent maximum, with the cable operator acting as custodian. Such an arrangement would not seem to conflict with present federal regulation; arguably, the monies involved would not constitute consideration, and the arrangement would promote both federal and local regulatory goals. As indicated previously, for the FCC to deny certification in these circumstances would be to challenge a joint willingness by the cable operator and the franchise authority to commence service under mutually acceptable terms and conditions, and would be to risk precipitating difficult political conflicts.

Some controversy has arisen, and hence specific comment is required, with respect to the revenue base a local authority may use for computing its percentage franchise fee—"the franchisee's gross subscriber revenues per year from cable television operations," in the words of § 76.31(b). Subsequent to issuance of the Report and Order *and* Reconsideration, *the FCC's Cable Television Bureau has limited the permissible base to fees from cable subscribers for basic monthly service and, probably, installation charges for basic service. The limitation thereby excludes fees for special services ("pay television") and revenues to cable systems from channel leasing, rentals of access facilities and equipment, advertising on origination channels, and other sources (such as syndication of video materials to other systems).[31] The limitation has struck many representatives of franchising authorities as inequitable and inappropriate, and created controversy that is not likely to be resolved short of full review by the FCC. Until that time, the Bureau's limitation must be seen as only a statement of policy that could lead the bureau to withhold certifications pending final resolution. From the standpoint of local franchise authorities wishing to impress some or all of such revenues with a public purpose, the trust mechanism referred to above would appear an appropriate and permissible safeguard, at least on an interim basis.*

(119)

4.22. Access to and Use of Nonbroadcast Channels

The *Report and Order* and the rules promulgated therewith by the FCC contain extensive provisions dealing with access to and use of nonbroadcast channels. Because federal guidelines seek to limit the role of local franchising authorities, these provisions are relevant here. For a full appreciation of their effects, the reader should also refer to Sec. 3.33, which deals specifically with these rules.

The FCC's treatment of regulations for nonbroadcast channels begins by asserting that such concerns are appropriate for the Commission, inasmuch as programing on such channels will be indistinguishable from programing of broadcast stations carried on the cable, which is already subject to FCC regulation (see Sec. 3.1 above). The *Report and Order* states:

These channels fulfill Communications Act purposes and, in the context of our total program, are integrally bound up with the broadcast signals being

[31] See "Franchise Provisions at Variance with FCC Cable Television Rules," Public Notice 87959, August 22, 1972, p. 3, Answer to Question 8 (see Part III).

carried by cable. It is by no means clear that the viewing public will be able to distinguish between a broadcast program and an access program; rather, the subscriber will simply turn the dial from broadcast to access programing, much as he now selects television fare. Moreover, leased channels will undoubtedly carry interconnected programing via satellite or interstate terrestrial facilities, matters that are clearly within the Commission's jurisdiction. Finally, it is this Commission that must make the decisions as to conditions to be imposed on the operation of pay cable channels and we have already taken steps in that direction. (See § 76.225). Federal regulation is thus clearly called for. *(Report and Order,* par. 130.)

Against this background, the Commission proceeds to consider what role, if any, is appropriate for state and local regulation "where not inconsistent with federal purposes." It concludes negatively, as follows:

We think that in this area a dual form of regulation would be confusing and impracticable. Our objective in allowing a period for experimentation might be jeopardized if, for example, a local entity were to specify *more restrictive* regulations than we have prescribed. Thus, except for the Government channel, local regulation of access channels is precluded. If experience and further proceedings indicate its need or desirability, we can then delineate an appropriate local role. *(Report and Order,* par. 131; italics added.)

Thus the Commission states its purpose in establishing the following substantive rules in § 76.251:

RULE:
   (a)   No cable television system operating in a community located in whole or in part within a major television market, as defined in § 76.5, shall carry the signal of any television broadcast station unless the system also complies with the following requirements concerning the availability and administration of access channels: ...
   (11)(iv) ... Except on specific authorization, or with respect to the operation of the local government access channel, no local entity shall prescribe any other rules concerning the number or manner of operation of access channels; however, franchise specifications concerning the number of such channels for systems in operation prior to March 31, 1972, shall continue in effect.
   (b)   No cable television system operating in a community located wholly outside of all major television markets shall be required by a local entity to exceed the provisions concerning the availability and administration of access channels contained in paragraph (a) of this section. ...
   (c)   The provisions of this section shall apply to all cable television systems that commence operations on or after March 31, 1972, in a community located in whole or in part within a major television market. Systems that commenced operations prior to March 31, 1972, shall comply on or before March 31, 1977: *Provided, however,* that, if such systems begin to provide any of the access services described above at an earlier date, they shall comply with [the FCC's rules of access, see Sec. 3.33] at that time; *And provided, further,* that if such systems receive certificates of compliance to add television signals to their operations at an earlier date, pursuant to [the FCC's distant signal rules,] for each such signal added, such systems shall provide one (1) access channel in the following order of priority—(1) public access, (2) education access, (3) local government access, and (4) leased access —and shall comply with the appropriate requirements [of the FCC's rules of access] with respect thereto.

The following structure therefore emerges from the language of the FCC's rule and explanation:

4.221. Major Markets. Franchising authorities in the major markets may not "prescribe any other rules concerning the number or manner of operation of access channels" for systems commencing operation after the effective date of the rules. Preexisting standards for systems then in operation, however, may continue in effect no later than a date five years after the rules come into effect, or at such earlier date that an operating system receives a federal certificate of compliance to add television signals to its operations.

4.222. Outside Major Markets. Franchising authorities in areas outside the major markets may not prescribe conditions that "exceed the provisions concerning the availability and administration of access channels" set forth in federal standards, but may presumably prescribe conditions with respect to their "number."

4.223. Overview. The explanation of FCC policy in the *Report and Order* concludes with an amplification of the purposes and procedures of the rules. Without differentiating between major and nonmajor markets, the FCC states that "local entities will not be permitted, absent a special showing, to require that channels be assigned *for purposes other* than those specified" in the *Report and Order*. (Par. 132; italics added.) Stressing that "we are entering an experimental or developmental period," the FCC indicates that it will "entertain petitions and consider the appropriateness of authorizing . . . experiments" providing "additional channel capacity for such purposes as public, educational, and government access—on a free basis or at reduced charges . . ." In a footnote, the FCC cautions local franchising authority against "bidding contests . . . in awards" of franchises or "making selections based on the barter of extra channels." The *Report and Order* concludes that in areas outside the major markets, locally required access services may not be "*in excess of what we require for the major market*." (Italics added.)

COMMENT: Again, it is apparent that the FCC aims at stabilizing locally imposed requirements for access services. While there is some conceptual problem in harmonizing the FCC's declaration that its standards are "minimum standards" with its specific prohibition of local regulations ("any other" rules concerning "number or manner of operation" in major markets, and rules that "exceed the provisions concerning the availability and administration" in other areas), it is likely that the FCC's particular language takes precedence. Thus, while franchises may treat these questions, any divergence from FCC standards is a risk; the FCC envisages that such divergence must be justified by petition, presumably in the application for federal certification. Some room for local "experimentation" would appear to be possible, however. For example, the two parties to the franchise may establish principles for long-term leases of access channels, or the use of the cable system's studio facilities for program origination. They could then submit a mutually agreed-upon petition to the FCC for its approval. Outside of major markets, somewhat greater flexibility in FCC reviews can be anticipated, both with respect to the number of local-access channels and the construction of any rules for "availability and administration" that exceed federal standards.

4.23. Technical Standards

(108–118)

While the content of the FCC's technical standards rules are summarized in Sec. 5, it is apparent they bear inferentially on state and local powers.

The FCC's standards reveal no more than a limited concern for technical matters, and to this extent do not purport to exclude state or local regulation not inconsistent with federal standards. Indeed, on reconsideration, it specifically indicates it sees "no reason why franchising authorities may not now require more stringent technical standards" across the board. (*Reconsideration,* par. 91.) The FCC's rules deal principally with Class I (broadcast) channels, defined in § 76.5(z) as follows:

> RULE:
> A signaling path provided by a cable television system to relay to subscriber terminals television broadcast programs that are received off-the-air or are obtained by microwave or by direct connection to a television broadcast station.

In the *Report and Order,* the FCC states that only Class I standards are being promulgated, and that standards are not being promulgated or are only anticipated as an "ultimate intention" for channels in Class II (cablecasting), Class III (nontelevision), and Class IV (return signals). (*Report and Order,* par. 150.) Added rules now promulgated cover interference with broadcast stations (§ 76.613) and responsibility for received-generated interference (§ 76.617).

COMMENT: In the absence of present federal standards covering nonbroadcast programing, it is apparent that the field of regulation remains open for local authorities. Moreover, since as a practical matter system technical standards might not be easily segregatable among classes of programing, local franchise standards are generally permissible except to the extent that they might actually conflict with Federal Class I standards. As a matter of established Constitutional doctrine, actual conflicts of state and federal standards are not lightly assumed, and courts strive to construe particular regulatory programs to avoid finding such conflict. [32]

4.3. EVALUATION OF STATE AND LOCAL RESPONSIBILITIES

The foregoing analysis has treated the limited areas in which the FCC has established standards for the regulation of cable television systems, following the promulgation of its rules in the *Report and Order.* Overall, it is apparent that the FCC has moved in a limited fashion in establishing a basis for the development of cable television service to distribute broadcast television programs, but also to serve as a new channel for cablecasting and local public access services. It has also attempted to shelter cable systems from "inconsistent" local and state regulation, with a view to protecting cable against economic burdens that would jeopardize its ability to advance federal aims. Although it has established a principle of "dual jurisdic-

---

[32] See, e.g., *Florida Lime and Avocado Growers v. Paul,* 373 U.S. 132 (1963), *Head v. New Mexico Board,* 324 U.S. 430 (1964), and *TV Pix v. Taylor, supra.*

tion," the FCC has aimed at preserving for itself the determination of the overall nature of cable service, against conflicting policy determinations that other government entities might make.

The FCC's approach is significant vis-à-vis other government entities in three respects:

First of all, it is apparent that the FCC's view of cable television as a "hybrid" —neither a broadcaster nor a common carrier—makes accommodation to its regulatory philosophy conceptually difficult as an exercise in "creative federalism." In trying to determine whether a particular local regulatory approach has been preempted by federal authority, it is sometimes difficult to see clearly the purposes at which federal regulation aims. For example, in regulating conditions of access or in trying to demarcate a field of regulation of subscriber rates for local franchising authorities, the boundaries of local residual authority remain unclear. Given such lack of clarity, the FCC's regulations cannot be deemed, as a general matter, to wholly preclude the exercise of local ingenuity to attain locally significant regulatory purposes.

Secondly, it is apparent that the FCC's regulations have been drafted with some element of bluff with respect to conflicting local approaches, and with a considerable orientation toward pushing ahead problems that are of practical local significance in the franchising process. Overall, there is an aim to stabilize prevailing patterns of regulation, for example, with respect to subscriber rates and franchise fees, but these issues are not definitively resolved. Instead, the resolution of these issues is postponed to later case-by-case reviews, possibly in the hope they will recede and not require resolution by an agency notably preoccupied with its budgetary limitations and deeply embroiled in controversy. In this context, local franchising authorities have a significant bargaining ability to establish their preferred regulatory programs and get them accepted as a condition of granting a franchise to a particular applicant. Then, petitions to the FCC represent the product of arms-length negotiations and a local political consensus with respect to questions of public interest, which the FCC may not lightly disregard and may well respect under the guise of warmly supporting local initiative to "experiment." In any event, much scope in interpretation may remain for ad hoc determinations not conflicting with clear FCC policies. Firmer definition of these policies may also eventuate through the activities of the State-Local/Advisory Committee recently created by the Commission.

Finally, it is important to note that the question of FCC power to assume regulatory authority in nonbroadcast areas—while recently upheld in *Midwest Video* as a matter of broad federal authority—has by no means been finally clarified with respect to preemption of state and local roles. It is not at all certain that a court will perfunctorily construe FCC authority as blocking state powers to grant perpetual franchises or to impose rate-of-return regulation. Where such issues are clearly posed in a particular factual context, and the local viewpoint justified in terms of the full range of local considerations, judicial intervention may yet be available to stay or modify FCC rules—if conflict arises and federal statutory authority remains imprecise. It is surely pertinent to note that the FCC's jurisdiction over cable television under its present statutory mandate was judicially upheld by only one vote, and significant doubts were expressed by the four dissenters as to whether FCC initiative was within Congress's limited grant of statutory authority. Ultimately, Congress

remains the final arbiter of its intentions in relationships between local franchise authorities and the FCC.

Thus, it would seem premature to conclude that whole fields of regulatory concern have been definitively preempted from local franchising authorities. It is rarely prudent to provoke controversy, but some state and local authorities may choose to view it as their responsibility—while attempting to live within federal guidelines—to seek as well to devise independent policies in franchising and regulatory decisions that will promote the interests of their citizens. Such independent local determinations would be reflected in bargaining with cable operators, whose applications for federal certification could well be fashioned and substantiated to ensure that local purposes are served and confrontations over the essential meanings of the FCC's "minimum" standards are avoided. Of course, full deference should be given to the FCC's recognition that its present regulations are "interim." But the launching of independent initiatives—possibly coordinated with other local franchising authorities—would ensure that local desires are vigorously pressed in negotiating with the FCC and the cable industry. Such a coordinated response to the FCC's regulations would inevitably contribute to shaping further federal policies in these areas.

Section 5

TECHNICAL STANDARDS

In laying the regulatory groundwork for extensive cable television development, the FCC has seen a need to establish technical standards for the transmission of clear and reliable signals, to protect the interests of programers and subscribers. At the same time, it has recognized that cable systems and equipment manufacturers need measures of standardization to ensure that equipment is compatible and available in the widest possible market for technical innovation and capital investment.

Because the problems are complex and cable technology is developing rapidly, the FCC chose for the present to impose only limited standards, principally to prevent the degradation of broadcast signals through cable transmission, but also to provide minimum safeguards against interference with other electronic media. Noting its "ultimate intention to provide appropriate technical standards for the various kinds of signals that we expect cable systems will offer their subscribers," the FCC focuses at this time on the "need for standards governing the carriage of standard television signals that are picked up off the air." *(Report and Order,* par. 150.) This is so, doubtless, because such signals will probably make up the great volume of cable transmissions for the near future, and also because of competitive reasons; this approach will also allow other aspects of cable service "sufficient flexibility for further technical change." *(Report and Order,* par. 149.) The need for added technical standards—"in some measure possibly different for carriage of cablecast programs" (par. 150) is expected soon, as cable activities develop.

Instead of formulating comprehensive standards, the FCC merely adopts a four-part classification of cable channels according to their use. (See Sec. 3.21.) It indicates that, for now, only "Class I cable channels [for broadcast signals] are subject to the technical standards adopted herein." *(Report and Order,* par. 150.) Within this context, the FCC standards center on performance tests to ascertain compliance with particular standards of transmission to subscriber terminals, measured in terms of intensity according to specified procedures; indeed, suggestions that the FCC establish specific standards of performance for components and receiver hook ups have been specifically rejected. *(Reconsideration,* pars. 103, 105, 106.) In this

discussion, standards of transmission to terminals are next set forth with minimal explanation of their more technical details, the text of the FCC's rules being sufficiently self-explanatory for technical personnel.

COMMENT: The technical issues covered by the FCC's rules are limited essentially to carriage of broadcast signals; consequently, cablecasting, nonvideo, and return channels remain presently unregulated at the federal level. However, even with respect to broadcast signals, the FCC is primarily "interested in the end result," not in how it is achieved (Reconsideration, par. 103). Thus, key aspects of system design along with broad questions of safety, reliability, and standards of servicing remain open fields for state and local regulation. Even though the content of broadcast and cablecast signals would seem wholly or partially preempted from state and local regulation, the FCC sees "no reason why franchising authorities may not now require more stringent technical standards than those in Subpart K."(Reconsideration, *par. 91.)*

47 C.F.R. § 76.601 *("Performance Tests")* establishes a broad obligation for each cable television system (that carries broadcast signals, irrespective of market size) to be responsible "for insuring that each such system is designed, installed, and operated in a manner that fully complies with relevant FCC regulation, coupled with a broad requirement that compliance be demonstrated to the FCC upon authorized request. *(Paragraph (a).)*

To facilitate this requirement for showing compliance, the operator must maintain at his office a current listing of Class I channels and "shall specify for each subscriber the minimum visual signal level it maintains on each Class I cable television channel under normal operating conditions." (Paragraph (b).) "Visual signal level" is defined as, "The rms voltage produced by the visual signal during transmission of synchronizing pulses." (§ 76.5 (hh).)

Paragraph (c) requires that the system operator conduct "complete performance tests," subsequently defined (§ 76.609), at least *once each calendar year* (at intervals not to exceed 14 months) at *no less than three* widely separated points, at least one of which is representative of subscriber terminals most distant from the system's head-end. While the measurement points need not be subscriber terminals themselves, the "data shall be included to relate the measured performance to the system performance as would be viewed from a nearby subscriber terminal." Such records must describe the testing instruments and procedure used, and the qualifications of the person administering the tests; the cable system must retain the records for FCC inspection for at least five years.

Satisfactory tests alone, however, do not relieve cable systems of the requirement for complying "with all pertinent technical standards at all subscriber terminals," according to Paragraph (d), and the Commission may order additional tests to insure full compliance.

In § 76.605 *("Technical Standards")* the Commission establishes substantive measures of performance that tests must satisfy with respect to broadcast signals. Paragraph (a) deals with the following specific aspects of such carriage, for whose details the reader should refer to the rules themselves:

1. Frequency boundaries of cable channels;[1]
2. Frequencies of the visual carrier;
3. Frequencies of the aural carrier;
4. Visual signal levels;
5. Variations in the visual signal level;
6. Rms voltage of aural signals;
7. Peak-to-peak variation in visual signal level;
8. Channel frequency response;
9. Ratios of visual signal levels to system noise and to co-channel interference;
10. Ratios of visual signal levels to rms amplitude of coherent disturbances;
11. Terminal isolation provided each subscriber; and
12. External radiation from cable systems, generally.

In Paragraph (b), the possibility of exemptions from the application of specific rules is recognized for cable systems that distribute signals by multiple cables or through specialized receiving devices, but the Commission may also prescribe special technical requirements to ensure good-quality service.

Paragraph (c) makes the foregoing provisions governing external radiation applicable in all instances as of the date of the rules, but permits systems in operation at that time to delay implementation of other substantive standards for a five-year period.

Standards for measuring performance according to § 76.605 are set forth in § 76.609. Special safeguards are provided to assure testing for "conditions which reflect system performance during normal operations, including the effect of any microwave relay . . . intervening between pickup antenna and the cable distribution network."[2] These safeguards include operation of amplifiers, insertion of special measuring signals, and operation of pilot tones, auxiliary or substitute signals, and nontelevision signals normally carried on the system. Further standards govern removal of normal signals to facilitate performance measurements, additional or different tests prescribed by the Commission, alternative methods for determining frequency response, and methods of measuring system noise, amplitude of signal interferences, terminal isolation, and field strength.

COMMENT: The principal intent of the FCC's technical standards is to stabilize, and hence assure, carriage of broadcast signals. Specifically envisaged, however, is "the need for sufficient flexibility for further technical change" in the nonbroadcast area. Especially if economic conditions of cablecasting result in thin profit margins,

---

[1] § 76.605(a) (1) defines permissible boundaries as conforming to the preexisting rule § 73.603 (a) numerically designating channels to which television receivers manufactured in the United States are attuned; these are separately set forth in Part III. On special showings, alternative arrangements in the "public interest" may be justified to the Commission—a specific reference in the *Report and Order* (par. 155) to cable systems such as "Rediffusion," which utilizes unconventional transmission technology.

[2] The rule specifically includes the Community Antenna Relay Service ("CARS") operated by entities serving the cable industry, but omits service by common carriers. The *Report and Order* justifies this distinction on grounds that, "CAR facilities are under the direct control or supervision of the cable system; common carrier facilities are not. With respect to the CAR service, the cable operator is able to effect scheduling of the microwave facilities. When the microwave relay is operated by a common carrier, however, there is considerable difficulty in arranging measurement procedures." (Par. 164.) The FCC assumes, in the latter case, that cable operators will be able to contract to assure their ability to satisfy the rules.

the need for technical standards in other areas of cable transmission can be expected to become rapidly apparent.

A further brace of rules covers problems of interference with radio stations by cable transmissions. § 76.613 renders cable systems broadly responsible for remedying interference from cable transmission. § 76.617 provides that, although the receiver-operator is responsible, cable systems are required to suppress "receiver-generated interference that is distributed by the system when the interfering signals are introduced into the system at the receiver."

Section 6

OPERATING REQUIREMENTS AND RELATED MATTERS

The rules issued with the *Report and Order,* and subsequently amended on reconsideration, establish a comprehensive set of administrative requirements as the framework within which all cable television companies will hereafter operate. These have been collected from the various contexts in which they appear, and are discussed below. Omitted from coverage here are the very specialized regulations of Part 78, dealing with the operations of Community Antenna Relay Services a subject of limited interest. (See *Report and Order* in Part III for full text.)

6.1. OVERVIEW

In this Section 6, the key procedural provisions for future FCC disposition of cable television questions are set forth in the terms presented by the FCC in the *Report and Order* and the rules issued therewith. As such, these procedural rules will serve as the basis on which controverted questions will be resolved in specific cases—subject also to the agency's broad procedural standards as set forth in its regulations, the Communications Act, the Federal Administrative Procedure Act, and other federal sources—and they will be of interest to all affected persons. Nonetheless, some specific attention should be given here to the points at which these procedures will be most relevant to the concerns of state and local franchise authorities and community groups affected by particular FCC determinations.

The subject matter of this section can, from this perspective, be briefly summarized. The FCC's procedures for "special relief" or waiver of its rules are presented first. They apply to challenges to or requests for interpretation regarding any matter arising under the cable rules. The right to petition for special relief is open to a wide list of specified parties—indeed, to "any interested person." This general avenue of relief is to be distinguished from the FCC's specifically delineated certification process, which cable systems must traverse through an application procedure in order to gain certain specific rights and wherein "objections" may be filed—though not so

specified, presumably against the background of applicable Constitutional and administrative law—by "any interested person." Except for controversies over signal carriage and program exclusivity, these objections are treated as petitions for special relief by the FCC. Thus, the FCC's special relief provisions hold transcendent importance as the method by which local perspectives on important franchising questions are brought into focus before the FCC.[1]

As previously indicated in Part I and in Section 4.3 of this Part II, resort to a petition for special relief by a city is a course that is always available to clarify controversies over the meaning of the FCC's rules. Under some circumstances, as far as franchising authority is concerned—where there may be controversy over the scope of state or local vis-à-vis federal powers—it may also be a more conservative and possibly even overprudent course, in comparison with independent local action. From the point of view of citizen groups lacking independent legal powers, and the sanction of withholding a grant of local authority possessed by cities and states, the petition for special relief may be the logical avenue to bring grievances before the FCC for resolution. At times, of couse, the tendency of such a petition may effectively be to forestall other local or federal action to authorize cable system operations.

It should be kept in mind that there may be circumstances where federal administrative and constitutional law may require hearing, oral argument, and formal opinion by the FCC before a contested matter may be disposed of. The determination of when a controversy has reached this point involves administrative law questions running far beyond the scope of this handbook, and turns on the facts of particular situations. Suffice it to say that successful invocation of a requirement for a formal evidentiary hearing at the federal level can itself be an effective sanction available to franchise authorities and community groups to forestall precipitate or unwise action in specific cases, invoking as it does the further possibility of protracted judicial review.

The complexities of this question highlight the desirability of retaining competent counsel for advice and representation on questions of great significance to franchise authorities and community groups. It is of course true that full-scale legal representation is *not* a precondition to expressing local views to the FCC or even to being able to satisfy all formalities (assuming especially timely action in the filing of oppositions under the rules). It will not be necessary to retain a lawyer to be able to complain to the FCC that local franchising procedures did not satisfy FCC requirements of fairness or openness, for example. But, by contrast, it is also clear that the ability to take fullest advantage of procedural machinery to accomplish more intricate maneuvers, as well as to engage in perceptive interpretation of the FCC's substantive rules, may well depend upon such professional expertise being available to the challenger, as it assuredly is to both its adversaries and to the FCC.

---

[1] There is a range of apparent procedural distinctions between a petition for special relief and an objection to certification, which distinctions largely disappear when it is perceived that, once filed, an objection (except with respect to signal carriage and program exclusivity) "will be treated as a petition for special relief . . . " (§ 76.27.) It should be kept in mind that an objection is filed in opposition to an application from a cable operator for certification, and hence must be filed within a specified time—30 days after public notice of the filing of the operator's application—or else the FCC may (unless it determines to waive its rules, pursuant to a petition for special relief) consider the application unopposed. A petition for special relief, by contrast, is filed at the initiative of the petitioner. Inasmuch as § 76.7 indicates considerable informality in the mechanisms of petition, the rule's requirement that the petitioner serve a copy of its petition on any "interested person who may be directly affected if the relief requested in the petition should be granted" is the only assurance that community groups will be informed in time to submit comments or opposition.

6.2 SPECIAL RELIEF

Paralleling its general waiver provision (47 C.F.R. § 1.3) and the special waiver provisions it customarily includes in issuing new administrative regulations, the FCC establishes procedures in § 76.7 for special relief from the rules upon petition by a cable television system, an applicant, permittee, a licensee of a television broadcast translator, a microwave relay station, or any other interested person. (Subparagraph (a).) On such petition, the Commission may "waive any provision of the rules relating to cable television systems, impose additional or different requirements, or issue a ruling on a complaint or disputed question." Specific provisions of the rule defining the scope, content, and form of such provisions, of oppositions thereto and rulings thereon, are as follows and are largely self-explanatory.

RULE:
(b) The petition may be submitted informally, by letter, but shall be accompanied by an affidavit of service on any cable television system, franchising authority, station licensee, permittee, applicant, or other interested person who may be directly affected if the relief requested in the petition should be granted.

(c) (1) The petition shall state the relief requested and may contain alternative requests. It shall state fully and precisely all pertinent facts and considerations relied on to demonstrate the need for the relief requested and to support a determination that a grant of such relief would serve the public interest. Factual allegations shall be supported by affidavit of a person or persons with actual knowledge of the facts, and exhibits shall be verified by the person who prepares them.
(2) A petition for a ruling on a complaint or disputed question shall set forth all steps taken by the parties to resolve the problem, except where the only relief sought is a clarification or interpretation of the rules.
(3) If a petition involves more than one cable television community, three (3) copies of it should be filed pursuant to § 1.51 of this chapter [separately set forth in Part III].

(d) Interested persons may submit comments or opposition to the petition in thirty (30) days after it has been filed. For good cause shown in the petition, the Commission may, by letter or telegram to known interested persons, specify a shorter time for such submissions. Comments or oppositions shall be served on petitioner and on all persons listed in petitioner's affidavit of service, and shall contain a detailed full showing, supported by affidavit, of any facts or considerations relied on.

(e) The petitioner may file a reply to the comments or oppositions within twenty (20) days after their submission, which shall be served on all persons who have filed pleadings and shall also contain a detailed full showing, supported by affidavit of any additional facts or considerations relied on. For good cause shown, the Commission may specify a shorter time for the filing of reply comments.

(f) The Commission, after consideration of the pleadings, may determine whether the public interest would be served by the grant, in whole or in part, or denial of the request, or may issue a ruling on the complaint or dispute. The Commission may specify other procedures, such as oral argument, evidentiary hearing, or further written submissions directed to particular aspects, as it deems appropriate. In the event that an evidentiary hearing is

required, the Commission will determine, on the basis of the pleadings and such other procedures as it may specify, whether temporary relief should be afforded any party pending the hearings and the nature of any such temporary relief.

(g) Where a petition for waiver of the provisions of [local carriage rules], is filed within fifteen (15) days after a request for carriage, a cable television system need not carry the signal of the requesting station pending the Commission's ruling on the petition or on the question of temporary relief pending further proceedings.

[With any petition for special relief, § 1.1116 provides that a fee of $35.00 must be paid.]

A new provision added on reconsideration, § 76.7(h), provides for special ad hoc reviews upon petitions for waiver of requirements applicable to cable systems in states, territories, and possessions located outside of the contiguous 48 states, "since those areas are not likely to be strictly comparable to those areas for which the rules were designed." *(Reconsideration, par. 123.)*

COMMENT: With reference to this waiver provision, the agency has indicated, in the Report and Order *and the rules, a number of instances where the FCC contemplates that waivers will or will not be granted. See, e.g., §§ 76.59(b)(1) and (2), 76.61(b)(1) and (2). These provisions are obvious guides to predictable Commission policies in specific respects. Beyond these specific instances, future FCC policies cannot be confidently forecast, beyond noting that procedures exist for waiver of the rules, subject to showings of merit in each instance. However, the FCC has also indicated a general disposition—given extreme limitations of staff resources and the controversial nature of the issues it has attempted to compromise—to be freed from the ad hoc procedural morass that extensive waiver proceedings would create. Its rules will operate "on a 'go, no-go' basis—i.e., the carriage rules reflect our determination of what is, at this time, in the public interest with respect to cable carriage of local and distant signals." (Report and Order, par. 112.) The FCC's warning that "those seeking signal carriage restrictions on otherwise permitted signals have a substantial burden" comports with the Chairman's previously cited admonition, "Don't play games"—evincing an FCC penchant to limit special relief to instances where there is a clear showing of public interest. Delay and denial are the inevitable hazards of any petitions for special relief.*

6.3 FCC CERTIFICATION

The application of the rules centers on the requirement that each local cable system[2] obtain a federal certificate, to be issued in accordance with provisions laid

---

[2] A cable television system subject to the FCC's operating rules is defined as follows:

> Any facility that, in whole or in part, receives directly or indirectly over the air, and amplifies or otherwise modifies the signals transmitting programs broadcast by one or more television or radio stations and distributes such signals by wire or cable to subscribing members of the public who pay for such services, but such term shall not include (1) any facility that serves fewer than 50 subscribers, or (2) any such facility that serves only the residents of one or

down in § 76.11-.27, which the government has elsewhere described as a "quasi-license."[3]

§ 76.11 makes receipt of a certificate essential before a system may "commence operations or add a television broadcast signal to existing operations." No certificate is needed for continuing carriage of signals lawful before the rules enter into effect. Such carriage may continue until the later issuance of a federal certificate at the close of a system's current franchise period, but no later than five years after the rules have come into effect. Paragraph (c) provides that operations may continue on an interim basis pending processing of a timely application for a certificate.

While no "standard form is prescribed" for an application, according to § 76.13, the three copies of information must be provided for *cable systems not operational at the date the rules enter into effect* (§ 76.13 (a)):

RULE:
(1) The name and mailing address of the operator of the proposed system, community and area to be served, television signals to be carried (other than those permitted to be carried [as substitutes on blacked-out channels], ... proposed date on which cable operations will commence, and, if applicable, a statement that microwave radio facilities are to be used to relay one or more signals;

(2) A copy of the FCC Form 325, "Annual Report of Cable Television Systems" [see Part III], supplying the information requested as though the cable system were already in operation as proposed;

(3) A copy of the franchise, license, permit, or certificate granted to construct and operate a cable television system [by state or local authority];

(4) A statement that explains how the proposed system's franchise and its plans for availability and administration of access channels and other nonbroadcast cable services are consistent with the provisions of §§ 76.31, 76.201 and 76.251;[4]

(5) A statement that explains, in terms of the provisions of Subpart D of this part, how carriage of the proposed television signals is consistent with those provisions, including any special showings as to whether a signal is significantly viewed (see § 76.54(b));

(6) An affidavit of service of the information described in subparagraph (1) of this paragraph on the licensee or permittee of any television broadcast station within whose predicted Grade B contour or specified zone the community of the system is located, in whole or in part, the licensee or permittee of any 100-watt or higher power television translator station licensed to the community of the system, the superintendent of schools in the community of the system, and any local or state educational television authorities;

(7) A statement that a copy of the complete application has been served on the franchising authority, and that if such application is not made available for public inspection by the franchising authority, the applicant will provide for public inspection of the application at any accessible place (such

---

more apartment dwellings under common ownership, control or management, and commercial establishments located on the premises of such apartment house. (§ 76.5(a).)

A further note indicates that systems operating in more than one community are to be considered as separate systems for the purpose of these rules.

[3] See reply brief of the United States in *Midwest Video*, p. 4, footnote 4.

[4] A Note added to Subparagraph (a)(4) by a Correction to the Reconsideration, 37 Fed. Reg. 13990 (July 15, 1972), indicates that if the proposed system's franchise was issued prior to March 31, 1972, "only substantial consistency with the provisions of § 76.31 *the FCC's franchising standards* need be shown until the end of the current franchise period, or March 31, 1977, whichever occurs first.

as a public library, public registry for documents, or an attorney's office) in the community of the system at any time during regular business hours;[5]

(9)   A statement that the filing fee prescribed in § 1.1116 of this chapter is attached.[6]

For a cable system *proposing to add a television signal to existing operations or one authorized but not operational* prior to March 31, 1971, paragraph (b) requires the following information:

RULE:

(1)   The name and mailing address of the system, community and area served or to be served, television signals already being carried, television signals authorized to be carried but not carried prior to March 31, 1972, television signals not previously authorized and now proposed to be carried (other than ... [permitted fill-ins on blacked-out channels]), and, if applicable, a statement that microwave relay facilities are to be used to relay one or more signals;

(2)   If the system has not commenced operations but has been authorized to carry one or more television signals, a copy of FCC Form 325, "Annual Report of Cable Television Systems" [see Part III], supplying the information requested as though the cable system were already in operation as proposed;

(3)   If the system has not commenced operations but has been authorized to carry one or more television signals, a copy of the franchise, license, permit, or certificate granted to construct and operate a cable television system, and a statement that explains how the system's franchise is substantially consistent with the provisions of § 76.31;[7]

(4)   A statement that explains how the system's plans for availability and administration of access channels and other nonbroadcast cable services are consistent with the provisions of §§ 76.201 and 76.251;

(5)   A statement that explains ... how carriage of the television signals not previously authorized is consistent with [the FCC's distant signal] provisions, including any special showings as to whether a signal is significantly viewed (see § 76.54(b));

(6)   An affidavit of service of the information described in (b) (1) above on the parties named in paragraph (a) (6) of this section;

(7)   A statement that a copy of the complete application has been served on the franchising authority, and that if such application is not made available for public inspection by the franchising authority, the applicant will provide for public inspection of the application at any accessible place (such as a public library, public registry for documents, or an attorney's office) in the community of the system at any time during regular business hours;

(8)   A statement that the filing fee prescribed in § 1.1116 of this chapter [$35] is attached.

For a cable system seeking to certify its existing operations at a later date, the following is required in its application, pursuant to subparagraph (C).

---

[5] Subparagraph (8) appears to be inadvertently omitted.

[6] 57 C.F.R. § 1.1116(a) provides for a filing fee of $35, but if "multiple applications for a certificate of compliance are filed by cable television systems having a common headend and identical ownership but serving or proposing to serve more than one community, the full $35 fee will be required only for one of the communities, $10 will be required for each of the other communities." *(Report and Order,* par. 115.)

[7] A Note added to Subparagraph (b)(3) by a Correction to the Reconsideration, 37 Fed. Reg. 13990, 13991 (July 15, 1972) indicates that where "only substantial consistency with the provisions of § 76.31" [the FCC's franchising standards] is shown, a certificate granted hereunder shall be valid until the end of the system's current franchise period or March 31, 1977, whichever comes sooner.

RULE:

(1) The name and mailing address of the system, community and area served, television signals being carried (other than ... [permitted fill-ins on blacked-out channels]), television signals authorized or certified to be carried but not being carried, date on which operations commenced, and date on which its current franchise expires;

(2) A copy of the franchise, license, permit, or certificate under which the system will operate upon Commission certification (if such franchise has not previously been filed), and a statement that explains how the franchise is consistent with the provisions of § 76.31;

(3) A statement that explains how the system's plans for availability and administration of access channels and other nonbroadcast cable services are consistent with the provisions of §§ 76.201 and 76.251;

(4) An affidavit of service of the information described in (c)(1) above on the parties named in paragraph (a)(6) of this section;

(5) A statement that a copy of the complete application has been served on the franchising authority, and that if such application is not made available for public inspection by the franchising authority, the applicant will provide for public inspection of the application at any accessible place (such as a public library, public registry for documents, or an attorney's office) in the community of the system at any time during regular business hours;

(6) A statement that the filing fee prescribed in § 1.1116 of this chapter [$35] is attached.

Rule § 76.16 deals with methods and form of *signature* by or on behalf of applicants. Rule § 76.18 concerns the form and procedures for *amendments* to applications. The FCC's procedures for dismissal of applications are set forth in § 76.20.

Rule § 76.25 (renumbered from § 76.15) provides that the FCC must give *public notice* of applications and amendments in the *Federal Register* not later than thirty (30) days prior to the issuance of a certificate, and thus preserves opportunities for challenge.

§ 76.27 (renumbered from § 76.17) concerns *objections* to applications by interested persons, setting forth manner of filing any objections thereto, as follows:

RULE:

An objection to an application for certificate of compliance or an amendment thereto shall be filed within thirty (30) days of the public notice described in § 76.25. A reply may be filed within twenty (20) days after an objection is filed. Factual allegations shall be supported by affidavit of a person or persons with actual knowledge of the facts, and exhibits shall be verified by the person who prepares them. All pleadings shall be served on the persons specified in § 76.13, the cable television system, the franchising authority, and any other interested person. Controversies concerning carriage (Subpart D) and program exclusivity (§ 76.91) will be acted on in connection with the certificating process if raised within thirty (30) days of the public notice; any other objection will be treated as a petition for special relief filed pursuant to § 76.7.

COMMENT: The FCC's procedures for applications and challenges thereto are complex but self-explanatory for particular situations. Suffice it to note in this context that the existence of a valid local franchise is a precondition to any application, the the FCC must give public notice of applications filed, and that "any ... interested person" may file an objection to the issuance of a certificate, specifying reasons therefor, in accordance with the provisions of §§ 76.27 and 76.7.

6.4 ANNUAL REPORT AND FEE

To remedy a long-standing regulatory deficiency with respect to comprehensive information about the economics of cable operations, § 76.405 requires of each cable system the filing of a "Cable Television Annual Financial Report," on a form prescribed by the Commission ("Form 326," see Part III) before April 1 of each year. Systems in operation prior to December 1, 1971 may report within 90 days of the close of their fiscal years.

In addition, annual fees must be paid on or before April 1 of each year, according to § 76.406 under a schedule set forth in 57 C.F.R. § 1.1116.[8]

6.5 DIVERSIFICATION OF CONTROL

Preexisting rules, reissued with the *Report and Order* as Subpart J, § 76.501, prohibit "cross-ownership" of cable systems and national television networks, local broadcast stations (commercial and noncommercial), and local television translator stations. The rules do not come into effect until August 10, 1973 for interests in existence prior to July 1, 1970 (when the rules were originally issued), but prohibit all acquisitions of such interest after that 1970 date.

COMMENT: The cross-ownership rules have aroused much protest by broadcast interests. (Since 1966, similar rules have barred telephone companies from owning cable systems in their local service areas, subject to provision for waiver; see 47 C.F.R. §§ 64.601-602.) Noncommercial television stations have been especially anxious to secure modification of the cross-ownership rules to permit ownership of local cable systems, but the Report and Order *did not respond to this concern.*

6.6 NONDISCRIMINATION IN EMPLOYMENT PRACTICES

In an additional rulemaking, *Nondiscrimination in Employment Practices,* 37 Fed. Reg. 6586, issued March 29, 1972, the FCC has provided certain amendments to Parts 76 and 78, effective May 9, 1972, which are set forth verbatim in Part III.

---

[8] § 1.1116 provides that the annual fee shall be paid by April 1 of the next calendar year. It is equal to the number of subscribers, averaged on the last day of each calendar quarter, times 30 cents, bulk-rate subscriptions included individually.

PART III

COMPENDIUM OF RELEVANT DOCUMENTS

SATURDAY, FEBRUARY 12, 1972

WASHINGTON, D.C.

Volume 37 ■ Number 30

PART II

FEDERAL COMMUNICATIONS COMMISSION

■

Cable Television Service; Cable Television Relay Service

Title 17 TELECOMMUNICATION

Chapter I—Federal Communications Commission

[FCC 72-108; Dockets Nos. 18397, 18397-A, 18373, 18416, 18892, and 18894]

CABLE TELEVISION SERVICE; CABLE TELEVISION RELAY SERVICE

In the matter of Amendment of Part 74, Subpart K, of the Commission's rules and regulations relative to Community Antenna Television Systems; and inquiry into the development of Communications Technology and Services to formulate regulatory policy and rule making and/or legislative proposals, Docket No. 18397, Docket No. 18397-A. Amendment of § 74.1107 of the Commission's rules and regulations to avoid filing of repetitious requests, Docket No. 18373. Amendment of §§ 74.1031(c) and 74.1105 (a) and (b) of the Commission's rules and regulations as they relate to addition of new television signals, Docket No. 18416. Amendment of Part 74, Subpart K, of the Commission's rules and regulations relative to Federal-State or local relationships in the Community Antenna Television System Field; and/or formulation of legislative proposals in this respect, Docket No. 18892. Amendment of Subpart K of Part 74 of the Commission's rules and regulations with respect to technical standards for Community Antenna Television Systems, Docket No. 18894.

The Commission has the following before it for consideration:

(a) Notice of proposed rule making in Docket 18373,[1] notice of proposed rule making in Docket 18416,[2] notice of proposed rule making and notice of inquiry in Docket 18397,[3] further notice of proposed rule making in Docket 18397,[4] Public Notice Mimeo No. 35632 released July 23, 1969, second further notice of proposed rule making in Docket 18397-A,[5] all of which concern the carriage of television broadcast signals by CATV systems and/or the use of CATV channels for the distribution of nonbroadcast programing;

(b) Notice of proposed rule making in Docket 18894,[6] which concerns standards to govern the technical performance of CATV systems, minimum channel requirements, two-way transmission capability, and separate neighborhood program origination centers;

(c) Notice of proposed rule making in Docket 18892,[7] concerning the appropriate division of regulatory jurisdiction between the Federal and State and local levels of government and a limitation on

the local franchise fees paid by CATV systems;

(d) The comments and reply comments filed in each of the above;

(e) Transcript of oral argument in Docket 18397 held before the Commission en banc on February 3 and 4, 1969; and

(f) Transcript of panel discussions and oral presentations in Dockets 18397-A, 18891, 18892, and 18894 held with and before the Commission en banc on March 11, 12, 15, 18, 19, 22, 23, 25, and 26, 1971.[8]

I. INTRODUCTION

1. In our Notice of proposed rule making and notice of inquiry in Docket 18397, we launched an inquiry into the long-range development of cable television.[9] Our purpose was to explore:

* * * [H]ow best to obtain, consistent with the public interest standard of the Communications Act, the full benefits of developing communications technology for the public, with particular immediate reference to CATV technology and potential services * * *.

Though designed as a vehicle for eliciting comments and data, our notice recognized the variety of possible services that cable systems could offer. We did not attempt an all-inclusive listing of cable's potential uses, but took note of many.[10]

2. Our recognition of the importance and promise of cable development led to our proposing rules requiring program origination and a system of annual re-

---

[8] For orders establishing panel discussion procedure see 27 FCC 2d 303 (1971) and 27 FCC 2d 932 (1971).

[9] The Commission has heretofore generally referred to community-wide, broadband, co-axial cable, television broadcast signal distribution systems as Community Antenna Television or CATV systems. Because of the broader functions to be served by such facilities in the future, they are generally referred to herein by use of the more inclusive term cable television systems, although the older term is sometimes used.

[10] "[F]acsimile reproduction of newspapers, magazines, documents, etc.; electronic mail delivery; merchandising; business concern links to branch offices, primary customers or suppliers; access to computers; e.g., man to computer communications in the nature of inquiry and response (credit checks, airlines reservations, branch banking, etc.), information retrieval (library and other reference material, etc.), and computer to computer communications; the furtherance of various governmental programs on a Federal, State, and municipal level; e.g., employment services and manpower utilization, special communications systems to reach particular neighborhoods or ethnic groups within a community, and for municipal surveillance of public areas for protection against crime, fire detection, control of air pollution and traffic; various educational and training programs; e.g., job and literacy training, preschool programs in the nature of 'Project Headstart,' and to enable professional groups such as doctors to keep abreast of developments in their fields; and the provision of a low cost outlet for political candidates, advertisers, amateur expression (e.g., community or university drama groups) and for other moderately funded organizations or persons desiring access to the community or a particular segment of the community." 15 FCC 2d 417, 420.

ports. The Commission indicated further, that it intended to prescribe technical standards but that it would first issue a further notice proposing specific criteria. In addition, the Commission recognized, but did not propose rules to resolve, the problems of the proper relationship between local and Federal regulation. We noted that cable television service has tended to develop on a "noncompetitive, monopolistic basis in the areas served," thus denying cable subscribers "the normal protection afforded consumers by providing a choice between alternative suppliers." While we then declined to extend "our jurisdiction to the licensing of CATV systems," we expressed a belief that "local, State, and Federal governmental agencies must face up to providing some means of consumer protection in this area." And we emphasized that "[s]uch regulation, while called for in the case of present CATV operations, would be particularly appropriate in light of CATV operations with originations."

3. At the same time, the Commission undertook an inquiry into diversifying the ownership of cable in combination with other mass communications media. We made these specific proposals: To ban cross-ownership of cable with specified types of broadcast stations and to limit the number of commonly owned systems. The Commission also indicated its belief that encouraging cable systems to operate as common carriers on nonbroadcast channels would serve the public interest. Additionally, it was proposed that distant signal importation into television markets be conditioned on cable systems' obtaining "retransmission consent" from distant stations, and the Commission stated that it would authorize distant signal importation with retransmission consent in a limited number of cases in order to gain experience with the proposal. Finally, the Commission posed a number of related questions concerning the future development of CATV. 15 FCC 2d 417, 442.

4. The first report and order in Docket 18397[11] was the first significant action in the proceeding and established the ground rules for cable origination. Basically, the Commission decided that origination served the public interest, allowed cable systems to present commercials at natural breaks, encouraged the development of public access channels, approved interconnection of cable facilities: *Provided,* That cable systems with 3,500 or more subscribers would be required to originate, adopted antisiphoning rules for pay-cable operations, and adopted broadcast-type rules to deal with equal time, sponsorship identification, and fairness. Shortly thereafter, the Commission adopted rules permitting the use of private microwave facilities by cable systems for carrying locally originated programs.[12]

---

[11] 20 FCC 2d 201 (1969), stay denied, 20 FCC 2d 899 (1969), recon. denied, 23 FCC 2d 825 (1970).

[12] Report and order in Docket 18452, 20 FCC 2d 415 (1969); report and order in Docket 17999, 20 FCC 2d 422 (1969).

---

[1] FCC 68-1094, 33 F.R. 17855.
[2] FCC 69-9, 34 F.R. 872.
[3] FCC 68-1176, 15 FCC 2d 417 (1968), 33 F.R. 19028.
[4] FCC 69-516, 22 FCC 2d 603 (1969), 34 F.R. 7981.
[5] FCC 70-676, 24 FCC 2d 580 (1970), 35 F.R. 11045.
[6] FCC 70-679, 25 FCC 2d 38 (1970), 35 F.R. 11036.
[7] FCC 70-675, 22 FCC 2d 50 (1970), 35 F.R. 11044.

5. In June 1970 we issued further proposals on television broadcast signal carriage,[13] cross-ownership of cable systems and radio stations and cable and newspapers, multiple ownership,[14] technical performance standards, minimum channel capacity, two-way transmission capability, local origination centers,[15] and the division of jurisdiction between the Federal and State-local levels of government.[16] These were followed later by proposals concerning the logging of cablecast programing,[17] equal opportunities in employment practices,[18] and the use of call leters in connection with nonbroadcast channels.[19]

6. In Docket 18892, the Commission requested comments on the interrelationship of local and State regulation of cable with Federal regulation. It was also proposed that there be a limitation of 2 percent of revenues on local franchise fees. The Commission offered alternative models of Federal/local relationships, including Federal licensing and Federal standards for local application. Under the latter approach, the local entity would consider legal and financial questions and measure the character qualifications of franchise applicants. And local governments would, in turn, certify to the Commission that the various criteria had been considered.

7. In Docket 18397-A, the Commission proposed to permit cable systems in the top 100 markets to carry four distant independent signals if they deleted commercials on the distant signals and replaced them with commercials provided by local television stations. As a further condition to distant signal importation, systems would be required to pay 5 percent of their gross subscription revenues to support public broadcasting. Additionally, the Commission asked for comments on whether cable systems should be required to provide local government, public access, educational, and leased channels. Comments were also requested on a proposal that systems with 20 or more channels set aside half their capacity for such uses.

8. The preceding is illustrative of the range of regulatory controversy that has surrounded the cable television industry in recent years. Technological advances have multiplied the issues. At first, cable television systems served largely to provide subscribers with better quality reception and more channels of conventional broadcast television programing. While need for these services continues, increasingly sophisticated cable technology and cost reductions and improvements in the quality of program origi-

nation equipment have made possible increased channel capacity, low cost nonbroadcast programing, and a subscriber response capability. The confluence of these developments provides the basis for the next stage in cable television's evolution with which the rules now adopted are concerned. Additional services and further technological developments are under study as part of the industry's more distant future.

9. Our initial rule making proposals were issued in December 1968, and oral presentations with respect to those proposals were heard in February 1969. As discussed above, portions of that proceeding were resolved separately, additional rule making proposals were issued, and further comments received. In March 1971 further oral presentations were heard, part of which were in the form of panel discussions between the Commission and recognized authorities on specific issues. Following the public proceedings, the Commission formulated a cable program designed to allow for fulfillment of the technological promise of cable and, at the same time, to maintain the existing structure of broadcast television. The framework of the new program was described to the Congress in testimony before the Senate Communications Subcommittee on June 15, 1971, and before the House Communications and Power Subcommittee on July 22, 1971. In order to permit the committees and the Congress ample opportunity to consider its proposals prior to final adoption, the Commission on August 5, 1971, adopted a "Letter of Intent"[20] in which it described in detail the course it planned to adopt.

10. Over the years that the Commission has been evolving a cable program, it has had the benefit of a number of independent studies of the cable industry—of its possible impact on broadcast television, its potential for advancing national goals, and its appropriate role in a total communications structure. These have provided valuable input for the formulation of our regulatory policies. We have also witnessed over the last several years repeated attempts by the affected industries to resolve their differences. Following release of our letter of intent further negotiations were undertaken, and agreement was reached on a proposal that was supported by the National Cable Television Association, the National Association of Broadcasters, the Association of Maximum Service Telecasters, and a major group of program suppliers. This consensus agreement is fully discussed later in this report and it, too, has had significant impact on the direction of our settlement of the complex questions having to do with distant signals/copyright.

11. As indicated, the rules we are adopting are the result of a number of interwoven proceedings. The program is designed as a single package because each part has impact on all the others. Our

concerns may generally be divided into four main areas:

Television broadcast signal carriage;
Access to, and use of nonbroadcast cable channels, including minimum channel capacity;
Technical standards;
The appropriate division of regulatory jurisdiction between the Federal and State-local levels of government.

Each of these will be considered in order. Questions concerning patterns of ownership, including cross-ownership and multiple ownership, are under consideration in another proceeding and will be taken up separately.

II. TELEVISION BROADCAST SIGNAL CARRIAGE

PROPOSALS AND ALTERNATIVES

12. Within the frame described above, we turn to a consideration of the various proposals that have been advanced for settling the question of cable carriage of television broadcast signals.

1966 Rules

13. Under the rules adopted in March 1966, local broadcasters and the Commission had to be notified before any cable system could undertake to carry a television broadcast signal (§ 74.1105). A distant signal (that is, a signal carried beyond its Grade B contour) could not be carried into one of the 100 largest television markets without prior Commission authorization after evidentiary hearing (§ 74.1107). Carriage of local signals and carriage of distant signals in smaller markets could commence 30 days after notice, provided no objection had been filed (§ 74.1105(c)). If objected to, carriage could not be commenced until the Commission ruled on the merits of the objection (§§ 74.1105(c) and 74.1109). In every instance where the Commission was called on to judge whether a cable system should be permitted to carry distant or local signals, the test was the general public interest standard of the Communications Act, and more specifically the consistency of the carriage with "the establishment and healthy maintenance of television broadcast service in the area" (see § 74.1107). The 100 largest television markets were singled out for special attention because it was felt that the potential for independent television station growth, particularly for UHF stations, was most favorable in those areas. Additionally, all local stations on request had to be carried by cable systems within the stations' Grade B service areas and, again on request, systems generally were not to duplicate the programing of a higher priority station by carrying the same programing from a lower priority station during the same 24-hour period (§ 74.1103). The priority of a station for purposes of obtaining program exclusivity was based on the strength of its signal in the area, with stations of higher signal strength having higher priority (§ 74.1103(a)).

1968 Commission Proposal

14. By December 1968, the Commission concluded that its cable rules should be

---

[13] Second further notice of proposed rule making in Docket 18397-A, supra, note 5.

[14] Notice of proposed rule making in Docket 18891, 23 FCC 2d 833 (1971).

[15] Notice of proposed rule making in Docket 18894, supra, note 6.

[16] Notice of proposed rule making in Docket 18892, supra, note 7.

[17] Notice of proposed rule making in Docket 19128, 27 FCC 2d 18 (1971).

[18] Notice of proposed rule making in Docket 19246, 29 FCC 2d 18 (1971).

[19] Notice of proposed rule making in Docket 19334, FCC 71-1084 (1971).

[20] Cable Television Proposals, 31 FCC 2d 115 (1971), attached hereto as appendix C, filed as part of the original document.

revised to establish general guidelines and procedures governing television broadcast signal carriage so as to eliminate the necessity for the burdensome evidentiary hearings required by the 1966 rules. Adjudicatory proceedings had come to be viewed as unduly complex, and the types of issues involved did not

Retransmission consent	(1) All restrictions would be eliminated on the carriage of distant signal programing for which cable systems had obtained "retransmission consent" on a program-by-program basis from the originating station.
Top 100 markets	(2) Cable systems in communities within 35 miles of designated communities in the 100 largest television markets could carry no distant signal programing in the absence of retransmission consent.
Smaller markets	(3) Cable systems within 35 miles of commercial television station communities not in the top 100 markets could carry, without obtaining retransmission consent, sufficient distant signals to provide their subscribers with the signals of stations affiliated with each of the three national television networks and the signal of one commercial independent station.
Beyond all markets	(4) Cable systems in communities more than 35 miles from any commercial television station community could carry distant signals without restriction as to number.
Overlapping top 100 markets	(5) A cable system in a community within 35 miles of one top 100 market designated community could not, in the absence of retransmission consent, carry commercial programing from a station in another top 100 market designated community unless the cable community was also wholly within 35 miles of the second market.
Noncommercial educational stations	(6) No restrictions were placed on the carriage of noncommercial educational station signals. Prior to such carriage, however, notification to local noncommercial educational stations and educational authorities was to be required and those so notified would be afforded an opportunity to object to such carriage.
Leapfrogging	(7) In the absence of waiver for good cause, each distant signal carried had to be obtained from the closest station of the type sought or from the closest in-State station of that type.
Grandfathering	(8) Cable systems operating in compliance with existing rules on December 20, 1968, would be permitted to continue in operation even if inconsistent with the proposed rules.
Carriage and program exclusivity	(9) Existing rules concerning program exclusivity and mandatory carriage would remain essentially unchanged, except in overlapping top 100 market situations.

These rules were designed to achieve certain basic purposes: To insure at least a minimum of service in underserved areas, set limits to the impact of cable distant signal carriage on over-the-air broadcasting, and eliminate certain elements of competitive unfairness resulting from the fact that cable systems are not required under existing copyright laws to pay for the television broadcast programming they pick up and distribute. Carriage of the closest stations of particular types was required because they were more likely to be attuned to the needs and interests of the cable community.

16. At the time these rules were proposed, interim procedures were adopted. Under these procedures all hearings under the 1966 rules were suspended, and action on requests for authorizations to carry signals was deferred pending the completion of the rule making proceeding, unless carriage of the signals re-

appear capable of satisfactory resolution in individual proceedings. What was clearly indicated was the necessity for fixed standards that would lend certainty to the process of signal carriage.

15. The 1968 rules, proposed to replace the evidentiary hearing requirement, contained the following basic provisions:

quested was consistent with the proposed rules.

1970 Commission Proposal

17. In June 1970, another alternative to govern the carriage of television broadcast signals was proposed and released for comment. Under this proposal, cable systems within 35 miles of the designated communities in the 100 largest television markets would be permitted to carry four channels of distant nonnetwork television programming. Systems would be required to delete the advertising from these distant signals and insert advertising supplied by certain of the local stations. Preference in inserting commercials was to be based on a priority system, with those stations most threatened by cable competition receiving first priority. It was thought that by means of this proposal cable might be used affirmatively to promote the development of UHF stations.

10. Because of the commercial substitutions that would have been required in the distant signals carried, it was felt that the adoption of the proposal would have to dovetail with copyright legislation. While acknowledging that copyright was for Congress to resolve, a method of calculating the amount of compensation to which distant signal program owners would be entitled was included to show that the proposal could be designed to compensate program owners fully. As a further condition to carrying distant signals in this fashion, and affirmatively to support noncommercial broadcasting, cable systems would have been required to contribute 5 percent of their gross subscription revenues to public broadcasting.

19. Comments were also requested on other possible alternatives, such as an expansion of the existing program exclusivity rules to protect local independent stations from having their programing duplicated via cable-carried distant signals. Another alternative was a proposal for a system of direct payments by cable systems to local stations to make up revenues lost through the diversion of audience to distant stations. Other alternatives were received in comments filed and are discussed below along with the comments on the Commission's proposals.

Comments on Retransmission Consent Proposal

20. Section IV of our notice in Docket 18397 concerned the importation of television signals by cable systems and contained our retransmission consent proposal. Comments addressing this proposal focused on: The technical feasibility of the retransmission consent theory; the size of the specified zones around each market; the requirement that retransmission consent be obtained even for local (Grade B) signals when a cable system within the zone of one top 100 market carried local signals from an adjoining top 100 market; the makeup of the list of top 100 market designated communities; the requirement that, if a distant signal were to be carried, the closest in that class of stations be carried first (the leapfrogging rule); and the rules applying to the carriage of noncommercial educational stations.

21. *Comments by broadcast interests.* The National Association of Broadcasters (NAB), the Columbia Broadcasting System (CBS), the National Broadcasting Co. (NBC), and broadcast interests generally, supported the retransmission theory, although certain changes in the specifics of the proposal were recommended.[21] NAB, for example,

---

[21] These comments were filed during 1969, and it is recognized that the views of some of those commenting may have been changed by intervening events. During one panel discussion in 1971 the panelists, including cable, broadcast, and copyright owner representatives, were asked if they thought the retransmission consent concept was a valuable concept or had "any validity whatsoever as a practical matter." None of the panelists responded in support of the concept. Transcript vol. 4, p. 715, Mar. 18, 1971.

endorsed this approach "as a means to eliminate much of the unfair competition presently generated by distant signal importation." CBS supported the retransmission consent type of regulation only as an interim solution and indicated its belief that only Congress was capable of providing the comprehensive solution required. Some doubt was expressed as to the Commission's jurisdiction to create a regulatory framework of the type proposed. CBS stressed that its purpose was not to have stations insulated from competition but to ensure that competition be conducted fairly. Accordingly, it stated that there should be no restrictions on the carriage of distant signals into any market, but that no distant signal carriage should be permitted in the absence of retransmission consent or a congressionally enacted equivalent. Considerable doubt, however, was expressed as to how the proposal would operate in practice, because of the different market situations involved, the existing contractual and other relationships between program suppliers and broadcast stations, and the relative economic power of the cable, broadcast, and program supply interests. NBC felt the proposed regulations were well within the Commission's power, would eliminate the most undesirable elements of unfair competition, and should be the "keystone" of any regulatory provisions for cable systems. NBC would have had the requirements applied to all cable systems carrying distant signals, regardless of location, but suggested that cable systems, even without retransmission consent, should carry sufficient distant signals to provide their subscribers with at least one independent station's signal, one noncommercial educational signal, and one signal from a station affiliated with each of the major national networks. NBC visualized those stations that granted retransmission consent as acquiring rights from program suppliers to grant such retransmission consents and acting as small networks.[22] CBS, in contrast, thought that requiring distant stations to act as intermediaries between program suppliers and cable systems would be an "indirect and doubtfully effective" means of equalizing competition between cable systems and broadcast stations.

22. While the comments indicated general support for the retransmission consent proposal among the networks and broadcasters, some were opposed to it. The Association of Maximum Service

Telecasters (AMST), for example, found the retransmission consent requirement to be "* * * simply irrelevant to the critical problem of adverse impact on local broadcasting." AMST pointed out that the proposed rules would permit cable systems to carry an unlimited number of distant and overlapping market signals in any television market irrespective of impact on local broadcasting service to the public. AMST considered distant signal carriage pursuant to retransmission consents "* * * undesirable for all the reasons that CATV originations are undesirable, and more." The possibility was foreseen that cable systems might acquire a sufficiently large economic base to outbid local stations for the rights to carry certain programs, thus siphoning off exclusive rights to programs that are now broadcast over the air. As did NBC, AMST saw the possibility that a few strong stations would acquire from program suppliers the right to grant retransmission consents which would then be given freely to cable systems that would become in effect small networks, greatly expanding their markets, upsetting competitive patterns in their own markets, and destroying over-the-air broadcast service in distant markets.

23. In addition to endorsing the retransmission consent theory of regulation generally, there was broadcast support (including some who did not support the retransmission idea) for the proposal to establish a fixed list of designated top 100 market communities and to use fixed mileage zones. However, certain additions to the designated city list were suggested and the 35-mile zones proposed were generally considered to be too small. Zones of 45, 60, 75, or 100 miles were suggested, in addition to use of Grade A contours and a proposal that a sliding scale be used, with smaller markets having larger zones. Also, certain changes in the list of designated top 100 market communities were suggested. AMST, for example, provided a list of all allocations within the Grade A contours of the designated market list that were not "clearly" part of some other television market. It suggested that because stations operating on these allocations would be overshadowed by stations in the already-designated communities, they would not be network affiliates and would face all the difficulties of stations operating in the designated cities. It was claimed that the rating services do not yet consider operating stations in these communities as part of the designated markets because viewing of these stations has not yet reached the required level. AMST would have had us include all of these communities in the designated market list.

24. *Cable television interests.* The cable parties filing comments were without exception opposed to the retransmission consent proposal. The National Cable Television Association (NCTA) found the proposal completely unrealistic, arguing that consents could not be obtained on the required program-by-program basis and that by requiring pro-

gram-by-program consent the Commission would simply be turning over control of the cable industry to broadcasters and program suppliers. Many of the cable parties felt that the Commission was usurping the power of the Congress in the copyright area, because the consent requirement would have operated as though a change had been made in the copyright laws. It was argued that even if retransmission consents were theoretically available they would be impossible as a practical matter to obtain because they are not under control of one owner or entity but are bound up with exclusivity contracts, labor and residual rights agreements, music licensing agreements, and ownership disputes between stations having rights to broadcast the programs in specific areas and parties from whom such rights were obtained. It was argued that, because of these complications and the number of programs and channels involved, the paperwork required would in itself defeat all but the largest cable systems. Last minute changes in programs and failures in negotiations would mean that cable systems, if they overcame other problems, would be presenting a crazy quilt of programs interspersed with blacked-out channels. It was contended that lack of choice as to incoming distance signal programing would preclude meaningful price negotiations, and uncertainties as to future program availabilities would inhibit investment in system construction.

25. *Retransmission experiment.* In addition to the comments on retransmission consent, we have had a limited amount of experience with its operation. As part of the interim procedures of Docket 18397, we indicated that we would consider petitions for waiver of § 74.1107 of the rules for cable systems that would operate in accordance with the proposed retransmission consent requirement. Top Vision Cable Co., operator of a system in Owensboro, Ky., was granted authorization to carry programs from several distant stations for which it could obtain retransmission consent.[23] Top Vision has reported to the Commission every 60 days on its efforts to obtain retransmission consent.[24]

26. Top Vision's reports reveal a broad range of reactions to requests for retransmission consents. Some stations, two networks, a number of program owners, and music licensors refused consent, asserting that it would be inappropriate to give consents while the Commission was still considering whether carriage of distant signals was appropriate as a general policy matter; because others already had obtained exclusive rights to the programs for the Owensboro area; and, further, because uncertainties as to pending legislation, court decisions, and regulations made it

---

[22] It should also be noted that in an experiment with retransmission authorizations, consent was sought by a cable system to carry the local news program of a station. Because that program carried news films and other material supplied by the NBC News Program Service, the station referred the cable system to NBC to obtain consent. NBC refused, stating "* * * we have concluded that because of the nature of the material transmitted, as well as the manner of its transmission, we should not enter into arrangements to authorize other than affiliates to carry this service." Letter of Nov. 16, 1970, Ex. 6 to Top Vision's Sixth Report filed Dec. 28, 1970, in CATV 100-113.

[23] Initial authorization was granted in Top Vision Cable Co., 18 FCC 2d 1051 (1969).

[24] A second experiment was authorized, Tri-Cities Cable TV, Inc., 22 FCC 2d 533 (1970), but was terminated before useful results were obtained. Tri-Cities Cable TV, Inc., 27 FCC 2d 432 (1970).

inadvisable to grant consents. Some suppliers indicated that they were unable to grant consents to carriage of particular programs because the programs contained copyrighted musical compositions under the control of others. But consents were obtained to some programs without the payment of any fee or with the question of compensation deferred until the adoption of new legislation; other consents to some programing, including professional basketball games, were obtained in consideration for fees paid by Top Vision.

Comments on Commercial Substitution Proposals

27. Our proposal of June 1970, insofar as it required the deletion and insertion of advertising on distant signals was, without exception opposed by broadcast and copyright interests, and they were joined in this opposition by many cable parties. Objections went to the economic impact of distant signal carriage and the technical and economic feasibility of deletion and insertion procedures. Our proposal contemplated the possibility that distant stations might be required to insert electronic coding in their signal indicating the imminence and duraton of commercials.[25] It was expected that automatic switching equipment at the headend of each cable system could then be programed to perform the required advertising deletions and insertions. In the alternative, there was some thought that central switching centers in particular market areas might be created to perform the required switching operations simultaneously for all cable systems in the market. Many of the comments expressed the opinion that the complexity of performing these switching operations had been underestimated by the Commission.

28. *Comments by broadcast interests.* Storer Broadcasting Co.'s comments included an engineering statement discussing the substitution procedures that would be involved and the equipment, costs, and staff required. Storer posited a system where distant stations would transmit information, in coded form in the vertical blanking interval of their signal, as to upcoming advertising and its duration. This information would be decoded at the cable headend and relayed over telephone lines to the studio of the station inserting advertising. The local station would then transmit appropriate length advertisements via microwave to the cable headend for insertion on the channel of the distant signal in place of the advertising that would be deleted by the cable system. Equipment needed for this system, including a minicomputer, four video tape units, switching, decoding, and automatic logging equipment, and a one-hop, one-channel microwave system, was

estimated to cost $145,000. Wage payments to operate the system would total $36,000 per year. This arrangement would supply one cable system with the advertising to be inserted on one channel. The cost of the equipment required by the distant station was not included. As another alternative, some cost savings would have been achieved by having the distant signal delivered directly to the local station's studio, where all insertions would be made and the signal microwaved to all cable systems in the market.

29. NAB judged the commercial substitution proposal to be confiscatory, technically, and economically unworkable, and inconsistent with the realities of the marketplace. If each cable system performed the switching operations individually, each would need, it was alleged 12 video tape units (for 10-, 20-, and 30-second commercials on each of four imported channels), plus switching equipment, at a cost of $140,000. If switching for all cable systems in the market were performed at a single switching center, a cost of $800,000 for this facility was assumed. An estimated additional $175,000 annually would be required to maintain and staff the center. The carriage of sporting events and other live broadcasts during which commercial messages are not preprogramed would further complicate the operation of the substitution system and frustrate attempts to automate it. NAB concluded that, even if the system were technically workable, it would not provide sufficient additional advertising revenue to make it workable from an economic standpoint.

30. KOB-TV et al. regarded the proposal as "harebrained," "mind boggling," and a "Rube Goldberg device." In addition to the technical complexity and high cost of the required equipment, they rated the full cooperation of the distant stations, the local stations, and the cable system as essential, but saw little likelihood of its achievement. They reported that commercial time availabilities with audiences of only 3,000 to 5,000 or even more "cannot be sold at any price," and anticipated that the cost of selling the time would very likely exceed the revenue received. Other broadcast comments were pessimistic about the technical and economic validity of the proposal and opposed to it as a matter of principle.

31. *Comments by copyright interests.* The program suppliers were opposed to any proposal that involved compulsory licenses, at least in the larger markets, and were, therefore, opposed to the commercial substitution proposal.

32. *Comments by cable television interests.* In the cable television industry there was considerable diversity of opinion as to the proposal. The NCTA found it to be technically and economically feasible. The required deletions and insertions, it was believed, could be performed on three channels at a cost of $27,000 for equipment and $90,000 a year for operating expenses. These calculations were based on a system using closed circuit rather than standard broadcast equipment, no automatic signaling or switching, and no microwave expense.

The cost, it was suggested, would be prohibitively expensive for cable systems and should be borne by the broadcasters receiving the benefits of the commercial insertions. Athena Communications Corp. et al. stated that "commercial substitutions may be feasible and can probably work with the total cooperation of all parties concerned." Other cable operators—for example, Midwest Video Corp.—regarded the proposal as unworkable and an aid neither to cable nor to independent UHF stations. Those cable interests who regarded it as workable generally emphasized that the costs should be borne by the benefiting stations.

33. *Commercial substitution experiment.* Bucks County Cable TV, operator of a cable system in Falls Township, Pa., was authorized to carry three commercial signals and one noncommercial educational station from New York City on condition that it test and report on the technical feasibility of commercial substitution.[26] In this test the entire cost has been borne by the cable system with no assistance from either the distant stations carried or the local stations entitled to insert commercials. WPHL-TV, Philadelphia, authorized Bucks to record its commercials directly off the air and insert them in the distant signal channels. The commercials are inserted into the signals of the New York stations by means of a manual switching system that is dependent on the skill and efficiency of the persons operating the manual switches who are, in turn, dependent on visual and audio cues in the program material to indicate when the switch deleting the distant advertising and inserting local advertising should be thrown. Bucks judges that more than 90 percent of the insertions have been made perfectly and that "a manual switching system of commercial substitutions is feasible."

Comments on 5 Percent Payment to Public Broadcasting

34. As part of the commercial substitution plan, it was also proposed that cable systems be required to use a portion (5 percent) of their subscriber revenues to support public broadcasting. This payment was to be made by all cable systems without regard to whether they carried programing from noncommercial educational stations. The Corporation for Public Broadcasting, as a potential beneficiary of this proposal, supported it but suggested that the funds should be separately managed and used for purposes specifically benefiting subscribers to cable systems—such as the acquisition of hardware and the production of programing. An alternative to this proposal was suggested by the Suffolk County Organization for the Promotion of Education. Under this alternative, half of the 5 percent would be used to support cable-distributed instructional television programing and would be distributed through the U.S. Office of Edu-

---

[25] A filing by International Digisonics Corp. indicated that the nation's largest purchasers of television advertising are now placing monitoring "codes" in their commercials, and such "codes" might be suitable for commercial switching rather than signals inserted by the originating stations.

[26] Bucks County Cable TV, Inc., 27 FCC 2d 178 (1971); recon. denied, 28 FCC 2d 4 (1971).

cation to State education departments. School districts, institutions of higher learning, or nonpublic schools would then apply for the funds to produce, procure, and transmit educational, instructional, or school-community type programing. Many letters supporting this proposal were received from persons either engaged in instructional television activities or who believed that an expansion of instructional television would aid in the educational process. Cable parties generally questioned whether the Commission had the authority to enact regulations that would require cable systems to support public broadcasting. They pointed out that it was not the Commission's duty to provide financing for the public broadcasting system, that there were other methods of providing for financing, and that the requirement was discriminatory. Broadcasters generally were in agreement with the cable operators that the requirement would be beyond the jurisdiction of the Commission and should not be undertaken without legislation.

Comments on Leapfrogging Rule

35. In our notice of December 1968, we proposed generally that a cable system carrying a distant signal or signals would be required to carry the closest station of each type (ABC, NBC, or CBS network, partial network, independent, or noncommercial educational). In certain instances, the closest station of a type from within a State could be carried even if there were a closer station of the same type in another State. Provision was made for obtaining a waiver of this rule on a showing of good cause, such as a cable community's having a greater community of interest with the community of a more distant station. In our proposal of June 1970, some additional flexibility was permitted, with two of the four distant signals that we proposed to permit in the major markets having no restriction as to origin and two having to come from within the same State as the system.

36. The comments on this question were generally divided between broadcast and cable interests, with the former strongly supporting the rule and the latter either opposing it or supporting it with qualifications. Those in favor of a strict antileapfrogging rule stressed that such a rule would support our allocations policy, avoid undue concentrations of control in major market independent VHF stations, lead to carriage of stations more attuned to the needs and interests of the cable community, and result in the carriage of stations with less audience appeal, giving them the benefits of extra circulation and resulting in less audience diversion in the markets into which they were carried. Those opposing adoption of such a requirement felt that cable subscribers should be entitled to the best stations available without regard to place of origin, that concern over concentration of control could be discounted in light of the control of the existing networks from New York and Los Angeles, that community of interest

considerations might dictate carriage of more distant stations, and that frequently the choice is not between closer and more distant stations but between no additional stations and those available over existing microwave facilities. Western Microwave, Inc., argued that there were special problems that should be considered in areas of the sparsely populated West where additional service could only be obtained from a considerable distance. In these areas, without regard to which signals were carried, it was said to be necessary that all cable systems carry the same signals so that microwave costs might be shared. While many cable parties accepted the theory of such a rule, they pointed out the desirability of retaining sufficient flexibility to permit the carriage of more distant stations from communities with a greater community of interest, and of signals available on existing microwave routes when the construction cost of new microwave facilities to carry closer signals would not be economically feasible.

Comments on the Proposed Codification of Overlapping Market Rule

37. Our general policy has been to require that cable systems carry all Grade B signals (§ 74.1103). A possible exception to this rule was created in footnote 69 of the Second Report and Order in Docket 15971 [27] which suggested that in certain circumstances Grade B signals from one major market might be considered as distant signals for cable carriage in the vicinity of another major market. In Docket 18397 fixed rules were proposed to govern the carriage of overlapping major market signals. Under this proposal a cable system in a community within 35 miles of one major market could not, in the absence of retransmission consent, carry commercial television signals from another major market unless the cable community were also within 35 miles of the latter market.

38. Program suppliers and broadcasters, with a number of specific exceptions discussed below, supported the proposed rule. They urged that, because signals from one major market are generally not viewed in adjoining major markets, they should be treated as distant signals even in areas where predicted Grade B service is available. It was argued that cable carriage would alter existing viewing patterns and have the adverse consequences attributed to distant signal carriage. A chart filed by AMST indicated that these situations are in fact common, with the central cities of 72 of the top 100 television markets receiving some predicted Grade B service from other top 100 markets.

39. In contrast to this general support for the proposed rule, some specific cases were brought to our attention in which it was said that the zones proposed would cut broadcasters out of their normal markets. Camellia City Telecasters, Inc., licensee of KTXL-TV, Sacramento, Calif., and Kelly Broadcasting Co., licensee of KCRA-TV, Sacramento,

pointed out that portions of Contra Costa and Solano counties, which audience surveys show to be in the Sacramento-Stockton market and which are on "their" side of the mountains, are beyond their 35-mile zones and in the zone of San Francisco. They asked that any rules adopted be flexible with respect to their particular situations. Bay Broadcasting Co., licensee of television Station KUDO, San Francisco, similarly requested that the overlapping market rule not be used to bar carriage of its signal on cable systems in San Jose. It pointed out that San Jose, while 40 miles from the center of San Francisco, is within the Principal Community contours of KUDO and other San Francisco stations. A filing on behalf of eleven television stations (KCST et al.) included audience survey information indicating extensive viewing of out-of-market stations in several overlapping markets. New York City stations, for example, were shown as having a 43.7-percent share of audience in New Haven, and Boston stations were shown as having a 21.5-percent share in Providence.

40. In the oral presentations, a possible remedy for situations where there is actual viewing of out-of-market stations was suggested on behalf of U.S. Communications Corp., owner of several major market independent UHF television stations. In order that cable subscribers have available at least the signals that are actually viewed off the air, the Commission was urged to adopt a rule permitting cable carriage whenever there is "significant" off-the-air viewing of an overlapping market signal. Dr. Leland Johnson of the Rand Corp. suggested a similar approach and specifically that an overlapping market signal might be considered a local signal if 15 percent of the homes in the area watched the signal in question during an average day. AMST suggested there might be certain instances that lent themselves to ad hoc treatment, citing the Manchester, N.H., television market where more than 50 percent of viewing is of Boston stations.

41. Cable interests strongly oppose the adoption of any rule or the continuation of any policy barring the carriage of Grade B signals. It was their contention that any signal ordinarily receivable off the air should also be available to cable subscribers and that a contrary rule would restrict carriage of signals that are not only receivable but are in fact viewed. At a minimum, it was argued, cable subscribers should have access to what is available off the air. Prohibitions against carriage of out-of-market signals were said not only to discriminate against cable subscribers but to make it impossible to market cable service because the subscriber would be in the position of paying more to get less.

42. The proposed rule was also said to conflict with already established policy. Our decisions in Shen-Heights TV Association et al.[28] was cited as indicating the nature of this conflict. In that decision, it was held that a broadcast station "has

---

[27] 2 FCC 2d 725 (1966).

[28] 11 FCC 2d 814 (1968).

a responsibility of serving as an outlet for its entire service area" and that, as a counterpart to this obligation, cable systems within the predicted Grade B contour of a station "must observe the carriage and nonduplication requirements of our rules even though a viewable off-the-air picture is not available in any part of the CATV community." From this, and similar language in other decisions, it was argued that our proposed rule was in fundamental conflict with other policy decisions already made. Broadcast stations, for example, are intended and in fact required to serve the whole area within their Grade B contours.

Comments on the Carriage of Noncommercial Educational Television Stations

43. Our proposal to permit unrestricted carriage of noncommercial educational television stations in the absence of objection by local educational stations or local or state educational authorities was generally supported. The Joint Council on Educational Telecommunications, Nebraska Educational Television Commission et al., Eastern Educational Network, and the National Association of Educational Broadcasters objected to the proposal to the extent that it would force educational stations to become involved in hearings and other burdensome proceedings. They believed it discriminated against educational stations by forcing them to make specific objections to distant signal carriage, whereas commercial television station carriage would be regulated without such a requirement.

Additional Alternatives Proposed

44. Our notice of June 1970, in addition to containing the commercial substitution proposal, solicited comments on other alternatives and suggested as possibilities expanded program exclusivity and direct subsidization of UHF television by cable systems.

45. *Direct compensation for audience diversion.* Dr. Leland Johnson, in a report entitled "Cable Television and the Question of Protecting Local Broadcasting" prepared under a Markle Foundation grant, raised the possibility that UHF stations suffering audience diversion from cable should receive direct compensation from the cable systems in question. While Dr. Johnson did not believe broadcast stations in the larger markets would be harmed substantially by cable systems carrying four distant signals, it was his view that audience diversion resulting from such carriage could be ascertained through audience surveys and the revenue losses to UHF stations made up by direct payments. Payments could be made to all stations in the market, to those below a stated level of profitability, or to all stations on a sliding scale related to station profitability.

46. John J. McGowan, Roger G. Noll, and Merton J. Peck, as part of a Brookings Institution study financed by the Ford Foundation, also suggested the adoption of a form of direct compensation. They suggested that all cable systems be required to make payments into

a UHF development fund. The amount to be paid would be that amount sufficient to make up the deficit of all existing UHF independent stations. This fund would be distributed according to the following rules:

(i) Only unprofitable stations would be eligible and only to the extent of their deficit.

(ii) Payments would be related to the number of hours of programing devoted to first-run syndications or local live programs in order to encourage the development of new programing.

(iii) Eligibility would be limited to existing UHF independents or to one UHF independent in each market.

47. American Cable Television et al. suggested the elimination of the proposed 5-percent levy for public broadcasting and the establishment of 5 percent of gross revenues impact fund to compensate both commercial and educational stations for loss of revenue resulting from cable competition. Allen's TV Cable Service et al. suggested the creation of impact pools made up of 2 to 3 percent of the gross revenues of VHF stations and cable systems. UHF stations suffering economic impact from either cable or VHF would receive compensation from the respective impact pool. Others (both broadcasters and cable operators) were opposed to direct compensation on the ground that it is beyond the Commission's jurisdiction, would inhibit competition between the cable and broadcast industries, and would destroy the incentive of local stations to improve the quality of their programing.

48. Extended nonnetwork program exclusivity. Midwest Video Corp., a cable operator, suggested that the Commission adopt proposals that would implement the approach exemplified by section 111 of the copyright legislation pending in Congress.[29] The provisions of section 111, which are described here in simplified form, would subject all cable systems to copyright liability provided, however, that systems would have a compulsory license at a fixed fee to carry all local signals (including those from overlapping markets) and sufficient distant signals to provide a statutorily defined adequate service minimum. In the top 50 markets, adequate service is defined in the bill as carriage of signals from each of the three national networks, three independent signals, and one noncommercial educational signal. In markets 51 and below, adequate service is defined as signals from each of the national networks, two independents, and one noncommercial educational. If distant signals were carried to make up the adequate service complement, local stations would be entitled to protection for their nonnetwork programing. A provision was also included to restrict the carriage of live

---

[29] Section 111 is the CATV section of the Omnibus Copyright Revision Bill that was reported to the Senate Judiciary Committee by the Subcommittee on Patents, Trademarks, and Copyrights in December 1969. The bill was introduced in the 91st Congress, first session, as S. 543, and reintroduced in the 92d Congress, first session, as S. 644.

professional sports programs on distant signals.

49. Kaiser Broadcasting suggested a variation whereby cable systems in all markets could carry sufficient local and distant signals to provide their subscribers with at least three network signals, three independent signals, and one noncommercial educational signal. Full "run-of-contract" exclusivity would be afforded the nonnetwork programing of Grade A stations against Grade B stations and of Grade B stations against distant stations, carriage of all Grade B signals would be permitted, a compulsory copyright license would be provided for these signals, carriage of the closest distant signals of each type would be required, and professional sports blackouts would be protected. Westinghouse Broadcasting Co. suggested another variation. It would permit carriage of local and distant signals sufficient to provide at least service from the national networks, one noncommercial educational station, and three channels of nonnetwork programing. If this combination were already provided by signals from within the market, no distant signals or Grade B signals from other markets could be carried. If a local station were activated after the cable system had commenced operation, its signals would be substituted for a distant signal but only after a 7-year amortization period. The leapfrogging rules would be discarded and cable systems permitted to obtain authorized service from any station within 350 miles. No exclusivity of any type would be recognized beyond 35 miles of a station, but within that zone a station would be entitled to nonduplication protection for any programing it had contracted to broadcast. However, in order to avoid the shelving of programing, this exclusivity would apply only if the program were scheduled for broadcast within 2 years of purchase or, in the case of a series-type program, if telecast within 1 year. Compulsory copyright license legislation would be expected to accompany these regulations.

50. In connection with these proposals, a study by Harbridge House, Inc., attempted to calculate the loss of revenues to program producers resulting from the loss of first-run exclusivity. The study assumed, as the best estimate available, that loss of exclusivity reduces the value of a first-run feature film by 40 percent and even more in the case of series programs.

51. *Professional sports exclusivity.* A number of comments received from professional sports interests, including the Commissioner of Baseball, the National Football League, and the American Hockey League, requested that exclusivity rules be adopted to bar cable carriage of professional sports programing on distant signals. Their concern was primarily that carriage of such programing, especially into a blacked-out area, would decrease the gate attendance at professional sporting events. The Commissioner of Baseball first proposed a rule that would have precluded carriage of the signal of a distant station broadcasting a baseball game by a cable

system within 50 miles of a community to which a professional baseball club was franchised (when a home game was being played), unless consent had been obtained from the distant station and from the league of the baseball club being protected. Later, a broader rule was suggested that would preclude the carriage of any live organized professional team sporting event on a distant signal unless the cable system obtained the consent of the originating station and of the team that authorized the telecast. In addition, no live professional sporting event could be carried on a local signal if that signal were carried more than 35 miles from the community of the originating station to within 35 miles of a television station community in another market. It was suggested that the Commission use as its model section 111(a)(4)(c) of the proposed copyright revision bill. The American Hockey League and the National Football League supported this position. As an authority for the Commission's jurisdiction to adopt such restrictions Public Law 87-331 was cited.[30] This law provides some professional sports teams with a limited exemption from the provisions of the antitrust laws in order that they may make agreements designed to protect home game attendance. It confers no authority on the Commission but does indicate some public policy support for protecting the gate of professional sports teams.

52. *Allocations or market-tailored approach*. General Electric suggested that distant signal carriage might be regulated according to an allocations plan similar to that of FM radio and television facilities, but based on economic rather than interference considerations. Markets would be classified according to existing stations, allocations, populations, available revenues, etc., and judgments made as to the measure of distant signal carriage that should be permitted for each class of market. Under this proposal, it would be possible for areas not separately capable of supporting independent stations to be added together through cable carriage, so that these areas in the aggregate could support independent television service. General Electric was particularly concerned that, because of the rapidly changing nature of the industry, the Commission not attempt to define regulations that would permanently shape cable's evolution and represent the ultimate solution to all of its problems.

53. *Justice Department proposal*. The U.S. Department of Justice was critical of our cable regulations and proposals as being unnecessarily protective of the broadcast industry. It recommended that the Commission attempt to assure only a minimum of continued over-the-air service "consisting of one, two or perhaps even three stations." Beyond that minimum, there should be no restrictions on distant signal carriage, and copyright questions should be left entirely for congressional resolution. Cable systems, it

[30] 15 U.S.C. secs. 1291-1295.

was asserted, should be left to compete with broadcasters in the marketplace, and the marketplace should decide how many and what kind of facilities survive. Donald I. Baker, Deputy Director of Policy Planning for the Department, expressed the Justice position during the panel hearing as follows:

* * * our position is basically that the Commission is invited to embark on an elaborate scheme of social engineering, of handicapping here subsidizing there and so forth. We think that is an inappropriate role. This may have been a role that has been thrust on the Commission by the shortage of broadcasting spectrum in dealing with over-the-air broadcasting. It is not a necessary role with the abundance of cable.

* * * * *

* * * [B]asically the Commission should allow the people in the marketplace who want it or don't want it, who will pay or will not pay, to make the choice.

54. *Copyright approach*. The comments of the program suppliers (MCA, Inc., and Allied Artists Pictures Corp. et al.) indicated their belief in the crucial importance of full copyright liability for cable in the top 50 television markets. As a compromise solution for smaller markets, they suggested that cable systems be permitted to carry (on a compulsory license basis) sufficient local and distant signals to provide subscribers with signals from stations of the three national networks, one independent station, and one noncommercial educational station. In areas outside all markets and for local signals, a compulsory copyright license would be provided for all signals carried. In all markets where distant signals are imported, local stations would have their exclusive rights to feature films protected for 3 years, and series protection for 4 years. No distant signal programing could be imported unless the copyright owner had 2 years from the first nonnetwork showing to negotiate an exclusive sale in the market. Compulsory license fees would be established. Existing systems would be grandfathered to 150 percent of their subscribers as of a base date, and systems with fewer than 1,500 subscribers would be exempt unless affiliated with a multiple owner having more than 10,000 subscribers. Adoption of this proposal would require action by the Commission and the Congress.

55. *Elimination of exclusivity in the sale of television programing*. Leonard M. Ross of Harvard Law School made the point during our panel discussions that, while full copyright liability for distant signal programing "at first blush" appears to be fair, experience has shown that long-term exclusive sales of programing provide a substantial barrier to market entry—that is, a cable system attempting to purchase programing on a full copyright basis would find that the most desirable programing had already been sold on a long-term basis to broadcast stations in the area. Two approaches to limiting this barrier were suggested: The adoption of some type of compulsory license system or elimination or

limitation on exclusive program sales."
56. *NCTA proposal*. NCTA, during the panel discussions, proposed the following package:

Carriage of all Grade B signals;
Carriage of four distant independents, two from within State if possible;
Carriage of noncommercial educational stations in the absence of objection;
Preservation of professional sports team blackouts consistent with Public Law 87-331;
Nondiscriminatory first-come, first-served access to unreserved cable channels;
First run exclusivity provided to local independent UHF stations;
A failing-station doctrine under which special relief would be granted stations demonstrating inability to provide minimal service as the result of cable competition;
Existing systems grandfathered in their operating territories; and
Payment of reasonable copyright fees on a compulsory license basis to be decided by Congress.

RESOLUTION OF ISSUES CONCERNING TELEVISION BROADCAST SIGNAL CARRIAGE

57. The carriage of distant television broadcast signals by cable television systems has been center stage in the continuing controversy before the Commission, the Congress, and the Courts. The industries involved have variously argued—the cable industry, that cable technology will bring extra programing and other services to the public, both on distant signals and on locally originated channels; the broadcast industry, that distant signal importation will lead to smaller audiences and reduced revenues and thus threaten the existence of some broadcast stations or inhibit their ability to produce local public service programs; the television programing industry, that suppliers of programing should receive compensation for the use of their product by cable systems and that the exclusive sales of such programs in particular markets should be honored.

58. In resolving these issues, our basic objective is to get cable moving so that the public may receive its benefits, and to do so without jeopardizing the basic structure of over-the-air television.[32] We also desire to put to rest the problem of exclusivity protection for programs imported from distant cities by cable television systems and to open the way for

[31] These views were expressed in greater detail in Chazen and Ross, "Federal Regulation of Cable Television: The Visibile Hand," 83 Harv. L. Rev. 1820, 1839 (1969).
[32] We have previously set out the reasons why the public interest is served by preserving a healthy broadcast service. See Second Report and Order in Docket 15971 supra note 27; see also U.S. v. Southwestern Cable Co., 392 U.S. 157 (1968). It is sufficient to restate that we are guided by the standard of what will best serve the public interest and not by a desire to protect any industry from the impact of new technology.

resolution of the long-standing dispute over copyright payments. To achieve these goals, we have considered a number of alternative courses of action. Our existing rules, which require individual consideration of all distant signal carriage proposals for the top 100 television markets and of special relief requests in other markets, are unsuitable for reasons detailed in the notice in Docket 18397. The adjudications required in these cases have involved policy matters beyond the scope of the individual disputes. The procedures available for the settlement of these disputes have proved burdensome and have not furnished a dependable basis for regulation. The comments filed, almost without exception, support us in this decision.

59. We are also rejecting the retransmission consent proposal of Docket 18397. Experience has indicated that it simply will not achieve our basic objectives. Nor does the commercial substitution proposal of Docket 18397–A provide the answer. While the Bucks County Cable TV experiment (paragraph 33, supra) suggests that many of the technical objections to the proposal have been exaggerated, the prospect is not promising because of the necessity for close cooperation of all the parties—and such cooperation, as the comments indicate, is highly unlikely. We believe it imperative that our new approach above all be a pragmatic one, and have fashioned a program that melds techniques with which we have had experience—exclusivity and a limitation on the number of distant signals to be imported.

60. The approach we are adopting is to extend existing exclusivity rules so that they cover nonnetwork as well as network programing, and to restrict the number of distant signals that a system may carry based on the size of the market in which it is located and the estimated ability of that market to absorb additional competition. In so regulating distant signal carriage, we hope to give cable impetus to develop in the larger markets without creating an unacceptable risk of adverse impact on local television broadcast service. At the same time, these limits should serve to create an incentive for the development of those nonbroadcast services that represent the long term promise of cable television and are critical to the public interest judgment we have made.

The Consensus Agreement

61. In the course of developing a regulatory program, and because of Congressional concern over these important matters, the Commission in its letter of August 5, 1971 outlined to Congress the rules on which there was Commission agreement.[33] We noted there (p. 2) the recent efforts of the principal industries to reach agreement on the major issues at controversy and expressed the hope that these efforts would be successful. Following the letter's release, intensive efforts were made to achieve a consensus, and agreement has now

been reached. Because this consensus agreement is of particular significance to our deliberations, it is set out in full in Appendix D.[33a] The Office of Telecommunications Policy provided valuable assistance in the negotiations that led to this agreement.

62. The agreement does not alter in any respect the access, technical standards, or Federal-State/local aspects of the August 5 letter. It deals solely with Part I of the letter—television broadcast signal carriage. It proposes three modifications, as follows:

(i) *Exclusivity.* For syndicated programing, the agreement provides for extensive exclusivity in the top 50 markets, and more limited exclusivity in markets 51–100. For network programing, it substitutes simultaneous for same-day protection.

(ii) *Local signals.* The agreement changes the significant viewing standard applied to out-of-market independent stations in overlapping market situations from a 1 percent share of viewing hours to a 2 percent share; it does not alter the standard applied to network affiliates.

(iii) *Leapfrogging.* The agreement retains a UHF priority where a third distant signal is carried but changes the requirements for the first two signals. There is no restriction on these signals as to point of origin, except that if either is taken from any of the top 25 markets it must be from one of the two closest such markets. In the August 5 letter these signals were, in effect, channels of independent programing (conceivably a blend of several distant stations); now they are restricted to specified distant stations except during exclusivity protection periods.

63. The principal addition the agreement would make to the program we outlined in August is the provision of exclusivity for syndicated programing. In the August letter, we stated that "we intend to study whether present or future considerations call for altering our existing CATV program exclusivity rule (§ 74.1103), which in effect protects only the network programing of network affiliates." Clearly, even before the agreement was reached, the Commission recognized the need for considering action to protect syndicated programing.[34] Now a consensus has been hammered out by the principal industries themselves and they have agreed to support legislation that resolves the remaining aspect

of the copyright issue, that of copyright payments.

64. The provisions of the agreement would add exclusivity protection for syndicated programing—a matter that was in any event under study—and would work two changes in our earlier proposal. The changes in the viewing standard and in leapfrogging restrictions are consistent with our long-range goals for cable and represent merely variations on a theme. Adoption of the agreement does not mean that we would, absent agreement, have opted in its precise terms for the changes it contemplates. But their incorporation into our new rules for cable does not disturb the basic structure of our August 5 plan. And if, as we judge, the terms are within reasonable limits and the agreement is of public benefit, then it should be implemented in its entirety.

65. We believe that adoption of the consensus agreement will markedly serve the public interest:

(i) First, the agreement will facilitate the passage of cable copyright legislation. It is essential that cable be brought within the television programing distribution market. There have been several attempts to do so, but all have foundered on the opposition of one or more of the three industries involved. It is for this reason that Congress and the Commission have long urged the parties to compromise their differences.

(ii) Passage of copyright legislation will in turn erase an uncertainty that now impairs cable's ability to attract the capital investment needed for substantial growth. The development of the industry, at least with respect to assessing copyright costs, would be settled by the new copyright legislation and its future no longer tied to the outcome of pending litigation.[35]

(iii) Finally, the enactment of cable copyright legislation by Congress—with the Commission's program before it—would in effect reaffirm the Commission's jurisdiction to carry out that program, including such important features as access to television facilities.

It is important to emphasize that for full effectiveness the consensus agreement requires Congressional approval, not just that of the Commission. The rules will, of course, be put into effect promptly. Without Congressional validation, however, we would have to reexamine some aspects of the program. Congress, we believe, will share our conclusion—that implementation of the agreement clearly serves the public interest. (See exchange of letters between Chairman Burch and

---

[33] See full text, Appendix C filed as part of the original document.

[33a] Appendix D filed as part of the original document.

[34] The subject of exclusivity for syndicated programing was raised in our notices in Dockets 18397 and 18397–A, and numerous comments were received. Many of the suggestions received in the comments are now, in fact, being incorporated into the Commission's regulatory program. Exclusivity is a complex, dynamic subject that is most appropriately a matter for agency regulation. This is in accord with our view concerning S. 543 (Omnibus Copyright Revision Bill) where we urged that a revised copyright law leave detailed regulation of cable television signal carriage to administrative control. Letter of March 11, 1970, to Senator Warren G. Magnuson.

[35] Under the decision in Fortnightly Corporation v. United Artists, 392 U.S. 390 (1968), cable systems do not now make payment for broadcast programing. But the case of C.B.S. v. TelePrompTer, now pending in the Federal District Court of the Southern District of New York (64 Civil 3814) would test the limits of the Fortnightly decision. Assuming that Congress confirms the consensus agreement with the passage of copyright legislation, a decision in TelePrompTer could be significant, we believe, only with respect to past liability.

Senator McClellan attached as Appendix E.[35a])

66. There remains the question of the effect of the consensus agreement on the Commission's flexibility to shape cable's evolution. Our judgment is, to repeat, that the agreement serves the public and should thus be reflected in the rules here adopted. The legislation that we believe must follow will limit the number of distant signals to which compulsory copyright licenses apply to those specified in §§ 76.59, 76.61, and 76.63 of the rules. In all other respects—for example, the details of network and syndicated programing exclusivity protection, leapfrogging, the significant viewing standard, the definition of signals that must be carried—the Commission retains full freedom and, indeed, the responsibility to act as future developments warrant. We reiterate that we are affording cable the minimum number of distant signals necessary to promote its entry into some of the major television markets but that, ultimately, its success will depend on the provision of innovative nonbroadcast services. This is not to say that such matters as signal carriage, exclusivity, and leapfrogging are insignificant. These rules represent our best judgment as to broad policies that should govern cable's evolution. The Commission has no intention of setting out detailed regulations today, only to rewrite them tomorrow. But, as we gain experience and insight, we retain the flexibility to act accordingly—to make revisions, major or minor—and to keep pace with the future of this dynamic area of communications technology.

67. We have considered whether we should issue a further notice to solicit comment on the consensus agreement or turn to some other additional public proceeding. But we have concluded that it would serve little purpose to do so. It is not necessary to have further argument, for example, on which leapfrogging standard should be used. We are in position now to make that judgment. Indeed, all parties have had full opportunity to comment on this and all other matters covered by the agreement (e.g., exclusivity, significant viewing standard, overlapping markets). The decisive consideration is whether the public interest will be served by incorporating the consensus agreement in its entirety. And we have concluded that it clearly serves the public interest to do so. For more than 3 years we have been gathering data, soliciting views, hearing argument, evaluating studies, examining alternatives, authorizing experiments—turning finally to public panel discussions unique in communications rule making—and, in this effort, have necessarily delayed the substantial benefits of cable to the public. If it would serve some overriding national purpose, we would turn to further process even in the face of more delay. But it does not. It is time to act.

[35a] Appendix E filed as part of the original document.

Impact Considerations

68. Before proceeding to the specific provisions of the rules, some discussion would be useful on the judgments we have made as to: (a) The amount of distant signal competition that can be introduced into particular types of markets without having adverse impact on local television service, and (b) the effect of distant signal carriage on the supply of television programing. The answers rest in the complex economics of, and interrelationships between, the three industries involved as well as on expectations of future developments in the industries and in the economy generally.

69. With respect to the question of impact of distant signal carriage on local television broadcast service, a number of studies were undertaken to test our proposals in Docket 18397-A. These proposals would have permitted carriage of four distant independent signals in each of the top 100 markets. A study was undertaken by the Commission's staff, several studies were produced by the Rand Corp. under grants from the Ford and Markle Foundations, and studies and critical appraisals of the staff and Rand reports were submitted by various broadcast interests. In all of these it was assumed that four distant signals, including among them the strongest independents in the country, would be carried. There was no consensus as to the range of likely impact. The Rand studies concluded generally that carriage of four distant signals would not have significant adverse impact on local television broadcast service and that, in the short run at least, increased cable penetration would have a beneficial effect on local UHF stations because cable carriage eliminates the technical edge of VHF over UHF. The broadcast studies pointed out a number of alleged defects in the Commission staff and Rand studies and concluded that carriage of four distant signals as proposed would have a seriously detrimental impact on local broadcast service. The Commission staff study was somewhat less optimistic than the Rand studies but less pessimistic than those of the broadcasters.

70. The conflicting conclusions of these studies make abundantly clear the difficulties involved in attempting to predict the future where there are so many variables and unknowns. While the reports and studies have been useful in illuminating the various elements of our policy decision, we cannot rely on any particular report or study as a sure barometer of the future. We would simply point out there is no consensus, and we do not pretend that we can now forecast precisely how cable will evolve in major markets. There is inherent uncertainty. But this does not mean that we should stand still and block all possibility of new and diverse communications benefits. Rather, it means that we should act in a conservative, pragmatic fashion—in the sense of maintaining the present system and adding to it in a significant way, taking a sound and realistic first step and then evaluating our experience. That is the

approach we have taken. We have authorized not four distant signals, as proposed, but a more limited number (particularly in the smaller markets), and provided the added protection of nonnetwork program exclusivity (particularly in the larger markets where independent stations generally operate).

71. Based on our experience and on our study of the comments, we do not believe that this approach will have impact adverse to the public interest. On the contrary, it is our judgment that it would be wholly wrong to halt cable development on the basis of conjecture, for example, as to its impact on UHF stations. We believe the improvements that cable will make in clearer UHF pictures and wider UHF coverage will offset the inroads on UHF audiences made by the limited number of distant signals that our rules would permit. As to similar arguments concerning cable's impact on VHF in the smaller markets, it is our judgment—considering such factors as cable's rate of penetration and the growth of broadcast revenues—that our approach will not undermine these stations in their ability to serve the public. As with any general policy, there may well be exceptional cases—as to a particular market or, more likely, a particular station in that market. In such an event, we would be prepared to take appropriate action under the special relief provisions of the rules (§ 76.7).

72. The viewing patterns in cable and noncable homes will soon become apparent and serve as a measure of cable's possible impact on local broadcast service. We intend to obtain continuing reports from representative communities, and broadcasters will be free to submit such reports at any time. If these reports and the financial data from operating stations were to show the need for remedial action, we could and would act promptly. The range of possibilities here is broad. More extensive nonnetwork programing protection might be afforded to affected stations in markets below the top 50. Or, we might consider halting cable's growth with distant signals at discrete areas within the community—something we have done on occasion in the past.

73. The additional program exclusivity rules are designed both to protect local broadcasters and to insure the continued supply of television programing. The latter, of course, is fundamental to the continued functioning of broadcast and cable television alike. As with the basic signal carriage rules, the types of exclusivity incorporated into the rules vary according to market size: The most extensive protection is in the top 50 markets from which the bulk of program supplier revenue is derived and where these restrictions are consequently most needed to insure the continued health of the television programing industry.[36] This

---

[36] Our concern here with the continued supply of television programing has a counterpart in the prime time network access rules. See Network Television Broadcasting, 23 FCC 2d 382 (1970), aff'd, Mt. Mansfield Television, Inc. v. FCC, 492 F 2d 470 (2d Cir. 1971).

protection will also assist independent stations (including many UHF's) that are very largely concentrated in these markets. In markets 51–100 the rules afford additional, although limited, protection to local broadcasters. It has been necessary to find a middle ground: The stations are very largely network affiliated, and generally only two distant signals will be permitted; but these markets are mostly underserved, lacking independent stations, and thus there is a particular need for cable. No syndicated programing exclusivity is added in markets below 100 because the number of distant signals is very strictly limited under the rules. That limitation along with network programing protection, is, we believe, adequate to preserve local service, and no additional impediment should be placed on cable operations in these underserved markets.

SIGNAL CARRIAGE RULES

74. The following chart will give an overview of signals that will be permitted:

GENERAL OUTLINE OF THE RULES PERTAINING TO BROADCAST SIGNAL CARRIAGE

The television signal carriage rules divide all signals into three classifications:

First, signals that a cable system, upon request of the appropriate station, must carry.

Second, signals that, taking television market size into account, a cable system may carry.

Third, signals that some systems may carry in addition to those required or permitted in the two above categories.

These three classifications of signals are used in various market situations as outlined below:

CABLE SYSTEMS LOCATED OUTSIDE ALL TELEVISION MARKETS

PRIORITIES

First:

The following signals are required, upon request, to be carried:

(1) All Grade B signals.

(2) All translator stations in the cable community with 100 watts or higher power.

(3) All educational television stations within 35 miles.

(4) Television stations significantly viewed in the cable community.

Second:

The cable television system may carry any other additional signals.

CABLE SYSTEMS LOCATED IN SMALLER TELEVISION MARKETS

First:

The following signals are required, upon request, to be carried:

(1) All market signals (those within 35 miles and those located in other communities that are generally considered part of the same market.[1]

(2) Grade B signals of educational television stations.

(3) Grade B signals from stations in other smaller markets.

(4) All translator stations in the cable community with 100 watts or higher power.

(5) Television stations significantly viewed in the cable community.

---

[1] National audience rating services, e.g., ARB and Nielsen, recognize differing communities as being in the same market (hyphenated markets). These characterizations may be relied on for smaller markets; our new rules, however, designate specifically the hyphenated major markets.

Second:

A cable system may carry additional signals so that, including the signals required to be carried under the First priority, the following total may be provided.

(1) Three full network stations (subject to leapfrogging restrictions).

(2) One independent station (subject to leapfrogging restrictions).

Third:

Generally, the cable system may carry additional educational stations and one or more stations programed in non-English languages.

CABLE SYSTEMS LOCATED IN THE FIRST 50 MAJOR MARKETS

First:

The following signals are required, upon request, to be carried:

(1) All market signals (see smaller markets above).[2]

(2) Grade B signals of educational television stations.

(3) All translator stations in the cable community with 100 watts or higher power.

(4) Television stations significantly viewed in the cable community.

Second:

A cable system may carry additional signals so that, including the signals required to be carried under the First priority, the following total may be provided:

(1) Three full network stations (subject to leapfrogging restrictions).

(2) Three independent stations (subject to leapfrogging restrictions).

Third:

Generally, the cable system may carry educational and non-English language stations as described for smaller markets above.

The cable system may carry two additional independent stations (subject to leapfrogging restrictions): *Provided, however,* That the number of additional signals permitted under this priority is reduced by the number of signals added to the system under the second priority.

CABLE SYSTEMS LOCATED IN THE SECOND 50 MAJOR MARKETS

First:

The same requirements apply as for the First 50 Markets.

Second:

The cable system may carry additional signals so that, including the signals required to be carried under the First priority, the following total may be provided:

(1) Three full network stations (subject to leapfrogging restrictions).

(2) Two independent stations (subject to leapfrogging restrictions).

Third:

The same requirements apply as for the First 50 Markets.

NOTE: Cable systems located in overlapping markets where differing amounts of service are provided for under the rules, e.g., in the overlap of a smaller market and one of the first 50 markets, must operate in accordance with the rules for the larger market.

75. The signal carriage rules are tailored to markets of varying size in accordance with the estimated ability of these markets to withstand additional distant signal competition. The rules

---

[2] In the major markets, where a cable television system is located in the designated community of such a market, it shall not carry as a local signal the signal of a station licensed to a designated community in another major market, unless the designated community of the cable system is wholly within 35 miles of the reference point of the other community or unless the station meets the significant viewing standard.

---

vary according to whether the cable system is in the first 50 television markets, in markets 51–100, in a market below 100, or not in any television market. A list of the major markets (first 100) and their designated communities is made part of the rules (§ 76.51). The list is derived largely from the American Research Bureau's 1970 prime-time households ranking. The list will not be revised each time new rankings are issued; there must be stability in this area, so that plans and investment can go forward with confidence. A contrary approach would be disruptive to the viewing public. Previously, our rules (§ 74.1107) employed a market ranking system based on the net weekly circulation of the largest station in each market. We have now concluded that the prime-time households ranking will serve more appropriately because it more accurately reflects the audience and financial strength of each market.[37]

76. We have delineated the areas to which particular rules will be applicable. We define the basic area as a zone of 35-mile radius surrounding a specified reference point in each designated community in a market. A set of reference points fixing the center of the community to which each station is licensed is made part of the rules (§ 76.53). For new television stations where reference points have not been specified, the 35-mile zone will be drawn from the main post office in the television station community. The purpose of drawing these zones is to permit generally unrestricted cable operation in those outer areas where such operation would have insignificant effect on the revenues of local television stations.[38]

---

[37] Net weekly circulation is more an index of potential audience than actual audience. The latter can probably best be reflected by average prime-time rankings of all stations in the market. In employing these rankings we have changed some designations from those supplied by ARB where anomalous results would otherwise occur. There are also changes from the list attached to our Aug. 5, 1971 letter to the Congress: Little Rock, for example, is now ranked 50 and Wichita-Hutchinson 67 in order to reflect our earlier determination in the Prime Time Access Proceeding, Public Notice, 29 FCC 2d 212 (1971). Other markets have consequently also been renumbered.

[38] The 35-mile zone was first proposed in our proceeding in Docket 18397. It was based on experience and on analysis of a number of representative markets. In that proceeding the comments directed toward the size of the zone were predictably split: Cable interests desired smaller zones; broadcasters, larger ones. We are not convinced that our proposal for a 35-mile zone should be changed in either direction. The zone is particularly effective for UHF stations that generally have significantly smaller service areas than VHF stations. The comments filed by AMST indicated that it is the UHF stations—no matter where located—that have the substantial share of their audience within the 35-mile zone. In addition, as we stated in our proposal, a fixed mileage standard has the advantage of administrative ease and provides certainty to the affected industries.

77. Cable systems in communities partially within a 35-mile zone are treated as if they are entirely within the zone. There is, however, one exception to this rule: A cable system in a major market designated community is treated as within the zone of a station licensed to a designated community in another major market only if the 35-mile zone of the station covers the entire community of the cable system. In those instances where there is an overlapping of zones to which different carriage rules are applicable, the rules governing the larger market will be followed. Authorized stations with construction permits, but which have not yet commenced broadcasting, are treated as having a zone and as operational under the rules for a period of 18 months following initial grant of permit.[39] However, the emergence of new stations will not require displacement of existing signals because that would cause disruption of service to the public. Such new stations are likely only in the major markets where new systems will in any event have large channel capacity.

Signals Required to be Carried

78. Our objective in approaching the signal carriage issue has been generally twofold: (1) To assure that "local" stations are carried on cable television systems and are not denied access to the audience they are licensed to serve; and (2) to gage and, where appropriate, to ameliorate the competitive impact of "distant" signal carriage. Because market patterns vary and there is only gradual deterioration in a station's receivability as the distance from its transmitter increases, there is no necessarily clear dividing line between "distant" and "local" signals. Nevertheless, a line must be drawn somewhere.

79. Under prior rules, Grade B signals were generally considered to be local and, on request, cable systems were required to carry all Grade B signals covering their communities. Signals carried beyond their Grade B contours were considered to be distant. While the Grade B carriage rule has been a part of the Commission's cable television rules from the beginning, its operation has been complicated in practice as a result of footnote 69 to the Second Report and Order in Docket 15971. This footnote[40]

indicated that there might in rare instances be a question whether all local signals could be carried if the cable system were identified primarily with one market and some of the local signals came from an overlapping market.

80. Between March 1966, when the Second Report and Order was adopted, and our cable proposal of December 1968, many cable systems were precluded from carrying local stations because television broadcasters filed oppositions to, and petitions for special relief against, cable proposals seeking to carry signals that were in fact local but came from overlapping major markets. Under the Commission's rules, the filing of such oppositions resulted in a stay against carriage of the disputed signals until the Commission resolved the issue in each case.[41] In December 1968, we proposed to lend precision to the application of footnote 69 by providing that cable systems located in communities that were for all practical purposes part of two major markets (neither market could claim the community as its own because television viewers watched programs from both markets) could carry the signals of both markets but only in those cases where the community of the system lay wholly within the 35-mile zones of both overlapping markets.

81. We have now decided that the following classes of signals should be treated as local: Signals of stations within 35 miles of the cable system, signals meeting a significant viewing test, market signals in hyphenated markets, and in some cases Grade B signals.

82. *35-Mile and Grade B signals.* All cable systems must carry, on request, the signals of all stations licensed to communities within 35 miles of the cable system's community.[42] This requirement, based on policy considerations similar to those underlying existing carriage rules, is intended to aid stations—generally UHF—whose Grade B contours are limited. In this manner less powerful stations will be able to compete with more powerful stations in the same market more effectively than they could under our old carriage rules; they will be capable of extending their coverage into the area that we have determined is generally necessary for the development of broadcasting stations. With respect to

cable systems located wholly outside the specified zones of all stations, all Grade B signals must be carried. This, of course, maintains the earlier carriage rule and assures that all stations whose Grade B contours extend beyond 35-mile zones will be carried by systems located outside such zones.

83. *Overlapping market signals.* A more significant departure from our earlier carriage rules involves the overlapping market or footnote 69 situation. Audience measurements frequently show that stations from one market coming into another market do not receive audience shares of significant size in the latter even though they are of predicted Grade B strength. Such stations with no significant audience in a market may logically be treated as distant signals. The problem then is to draw a line between those stations that have sufficient audience to be considered local and those that do not. Cable development is not likely to be advanced if television choices on the cable are more limited than choices over the air, nor is it reasonable that signals significantly viewed over the air be excluded from carriage on cable systems. Thus, our rule permits and, on appropriate request, requires carriage of a signal from one major market into another if that signal—without regard to distance or contour—has a significant over-the-air audience in the cable system's community. Because the same rationale is applicable, the rule is also applicable to overlaps between major and smaller markets. In sum, cable systems in a smaller or major market may carry a signal from a major market as a local signal only if the system's community is wholly or partially within 35 miles of that market or if the signal in question is significantly viewed in the cable system community. However, where a cable system is located in the designated community of a major television market, it may carry the signal of a television station licensed to a designated community in another major television market only if the designated community in which the cable system is located is wholly within the specified 35-mile zone of the station. There will continue to be no restriction on carriage of Grade B signals or those significantly viewed from one smaller market into another, and network exclusivity will be applicable.

84. A significant viewing standard can reasonably be drawn at several points. We have concluded that an out-of-market network affiliate should be considered to be significantly viewed if it obtains at least a 3-percent share of the viewing hours in television homes in the community and has a net weekly circulation of at least 25 percent.[43] For

[39] A station that goes off the air will have no zone nor be treated as operational but can, under § 76.7, file appropriate pleadings to insure that the status quo not be altered during a reasonable period needed to put the station back on the air. However, the burden will be on the station to make a convincing showing that it will speedily return to broadcasting.

[40] The full text of footnote 69 is as follows: If two major markets each fall within one another's Grade B contour (e.g., Washington and Baltimore), this does not mean that there is no question as to the carriage by a Baltimore CATV system of the signals of Washington; for in doing so and thus equalizing the quality of the more distant Washington signals, it might be changing the viewing habits of the Baltimore population and thus affecting the development of the Baltimore independent UHF station or sta-

tions. Such instances rarely arise, and can, we think, be dealt with by appropriate petition or Commission consideration in the unusual case where a problem of this nature might arise.

[41] Under § 74.1105(c) an automatic stay against the carriage of signals objected to in a cable proposal became effective if the proposal was objected to within 30 days after notification was given to local broadcasters.

[42] All signals that systems must carry on request may also be carried in the absence of request. We also retain our present rule that all 100-watt or higher power translator stations licensed to the community of the system must be carried. We note, however, especially with respect to noncommercial educational stations, that translators may be operating with less than 100 watts of power. In many of these cases, it may be expected that the parent station will be carried. Should problems arise in this area, we will consider them either on an ad hoc or general basis.

[43] As used here the term net weekly circulation is a measure of the number of households that viewed a station for 5 minutes or more during an entire week, expressed as a percentage of the total television households in the community. Share of viewing hours is a measure of the total hours all television households in the community viewed a station during the week, expressed as a percentage of the total hours these households viewed all stations during the period surveyed.

independent stations, the test is a share of at least 2-percent viewing hours and a net weekly circulation of at least 5 percent. The two criteria reflect distinct concepts. Net weekly circulation reflects the extent to which signals are of any interest to television viewers but tends largely to reflect the availability or viewability of a signal as a technical matter. Audience share indicates the intensity of viewer interest. The combination of these two criteria provides greater assurance that the signal meeting the test is in fact significantly-viewed. The lower figures for independent stations are intended to reflect the smaller audiences that these stations generally attract even in their home markets.

85. For purposes of establishing that a station meets the significant viewing standard we are using the 1971 American Research Bureau "Television Circulation Share of Hours" survey information for those counties in which there is less than 10 percent cable television penetration. In those counties where there is 10 or more percent penetration we are using the ARB 1971 "Non-CATV Circulation and Share of Viewing Hours Study for ARB CATV-Controlled Counties." [44] The latter was prepared for the Commission by ARB so that in those counties with substantial existing cable penetration, over-the-air viewing in the absence of cable television can be measured. Because this data is provided on a countywide basis only, we recognize that it may not account for variations in viewing levels among communities within the county. There may be other drawbacks in using these surveys, such as rounding of percentages and sampling errors. We nevertheless propose to accept this countywide information to establish viewing levels for signals in all communities within these counties. In doing so, we note that survey information of this type is generally used by the television industry without differentiating among communities within counties, and that it gives a useable indication of viewing. But the most important consideration in our decision to accept these figures as conclusive is the strong desirability of certainty, both from a cable and a broadcast point of view. [45] Otherwise, rather than permitting cable to get moving, we believe there would be controversy in virtually every case. By proceeding in this fashion, we hope to reduce controversy, to provide a base of signals that cable systems will be assured they may carry, and to define areas in which stations will have rights to carriage. This approach strikes an appropriate balance—in 1966 we selected the Grade B contour, and in 1968 the 35-mile zone, neither of which

was specifically geared to actual viewing, while we now select a precise standard that is much more likely to reflect such viewing.

86. To minimize controversy at the outset of our new program, we are precluding special showings by cable systems or broadcasters until March 31, 1973. Thereafter, those wishing to make supplemental showings for the purpose of qualifying new signals under the significant viewing test may do so. Any survey data submitted must be based on the requirements specified in § 76.54 of the rules. This rule requires that surveys be made by disinterested professional organizations that are independent of the cable systems or television stations ordering the surveys. Two weekly periods separated by at least 30 days are required to offset any variations in viewing that may occur during a particular week, and one of the weeks must be outside the summer season when viewing patterns are unrepresentative of the entire year. We recognize that the results of sample surveys can only be determinative within a given probability. But because signals once permitted to be carried will not be deleted, we are setting our probability test high. We are providing that the sample result must exceed the significant viewing standard by at least one standard error. And although we will not require it, we believe it will reduce controversy if parties making studies were to inform other interested parties that the survey is to be made and of the methodology to be used. Objections, if any, to methodology should then be lodged so that corrections may be made before the survey is taken.

87. *Hyphenated markets.* In such markets, characterized by more than one major population center supporting all stations in the market but with competing stations licensed to different cities within the market area, we will permit and, on request of the station involved, require carriage of all stations licensed to designated communities in the market. [46] Because of the structure of these markets, including the terrain and the population distribution, portions of the market are occasionally located beyond the Grade B contours of some market stations. Consequently, we are adopting this rule in order to help equalize competition between stations in markets of this type, and to assure that stations will have access to cable subscribers in the market and that cable subscribers will have access to all stations in the market.

Additional Service

88. The Commission's television allocations policy to a large degree reflects population distribution: more channels

are allocated to densely populated areas than to those that are sparsely populated. This means more television service and more choices for those who live in the population centers of the country. This represents, however, not a judgment that inhabitants of the largest cities need the added service or have a public right to more diversity, but merely our decision as to an equitable distribution of facilities and that more television stations can be economically supported in areas of greatest population. Clearly, cable service can provide greater diversity—can, if permitted, provide the full television complement of a New York or a Los Angeles to all areas of the country. Although that would be a desirable achievement, it would pose a threat to broadcast television's ability to perform the obligations required in our system of television service. We believe, however, that those who are not accommodated as are New York or Los Angeles viewers should be entitled to the degree of choice that will afford them a substantial amount of diversity and the public services rendered by local stations.

89. Cable television can and should help in achieving the diversification sought by our allocations policies. It would, of course, be desirable to adopt one nationwide standard. However, because we seek to minimize possible impact on local broadcasting, we have decided to establish standards of television service that vary with market size. (Noncommercial educational and non-English language stations are not included in these standards and are discussed separately below.) It is our determination that the public interest will be served by allowing cable systems to make available the following complement of signals:

(1) In television markets 1–50:
Three full network stations. [47]
Three independent stations.
(2) In television markets 51–100:
Three full network stations.
Two independent stations.
(3) In smaller television markets (below 100):
Three full network stations.
One independent station.

If after carriage of stations within 35 miles, those from the same market, and those meeting the viewing test, the service authorized above is not available,

---

[44] For convenience, the Commission is herewith supplying as Appendix B the relevant information from these reports. In those instances, where ARB has divided counties for survey purposes, we have followed that pattern.

[45] To avoid disruption of viewing and to promote the needed certainty, we stress that the signals specified in the 1971 sweeps are not subject to deletion on the basis of some special showing or later survey.

[46] For example, a Mineral Springs, Ark., system would, on request, have to carry signals from stations licensed to Shreveport, even though they are of less than Grade B quality, because Mineral Springs is in the Texarkana-Shreveport market. Where smaller markets are involved, we will rely on industry practices as reflected by national audience rating services as to which markets are hyphenated. This is an area where decision will have to be made on the facts of each case.

[47] Some confusion existed under our former definitions of independent and network stations. For example, a fourth station in a market where the other three each had primary affiliations with a major network and where the fourth carried some network programing not otherwise available in the market, might have been construed to be a network station although essentially it was an independent. In order to clarify such ambiguities and to insure, particularly during prime time, that cable views will be provided with full network service, we have settled on the following definitions: (1) A full network station is one that generally carries during prime time 85 percent of the hours of programing offered by a single network with which it has a primary affiliation. (2) An independent station is one that generally carries during prime time not more than 10 hours of network programing per week.

distant signals are permitted to be carried to make up the defined level of service.[48]

90. Cable systems in major markets are in any case permitted to carry two signals beyond those whose carriage is required under the mandatory carriage rules. If the service standards set out in the preceding paragraph are met by the carriage of all stations required to be carried, two additional independent stations will be authorized. However, if the system adds distant signals—either network affiliates or independent stations—to meet the service standards, these will be counted against the two additional signals. If, for example, a system in a market ranked between 51 and 100 proposes to carry a distant network affiliate and a distant independent signal to reach the service standard, no further signals will be authorized. Cable systems in smaller markets (below 100) are not permitted to import network or independent television signals beyond the designated 3–1 service level. Noncommercial educational and non-English language stations may also be carried in accordance with the rules set out below. The rationale for permitting at least two additional signals in all major markets is simply this: It appears that two signals not available in the community is the minimum amount of new service needed to attract large amounts of investment capital for the construction of new systems and to open the way for the full development of cable's potential. We will, therefore, permit this complement of signals in the larger markets because it is necessary in terms of cable's requirements and because it is acceptable in terms of impact on broadcasting.

91. Cable systems in communities entirely outside the zone of any commercial television signal may carry television signals without restriction as to number and must carry all Grade B signals, all educational television stations within 35 miles, and all 100 watts or higher power translator stations licensed to the cable community. We have, however, given particular attention to the arguments of small market broadcasters that continuing cable penetration will adversely affect their ability to serve the public interest. Because these smaller stations serve sparsely populated areas, we agree that some relief is warranted. Accordingly, we are going beyond our August letter by requiring that these smaller market signals, where significantly viewed, must be carried on all new cable systems and on all existing systems with sufficient channel capacity—even if the cable community is beyond Grade B contours—and, as to new systems, must be afforded simultaneous nonduplication protection (§§ 76.57(a)(4) and 76.91

[48] In areas of overlap between markets in which different degrees of services are permitted, cable is required to operate in accordance with the rules governing the larger market. Generally, these overlapping areas, especially between major and smaller markets, comprise a small portion of both markets and do not encompass the populated centers of the markets.

(c)).[48] Smaller market broadcasters, particularly in the Rocky Mountain region, argue against 35-mile zones and contend that, in their case, an effective zone must be much greater (e.g., Grade B contour) to take into account audiences important to their operations. We recognize the validity of the contention that there is audience beyond the 35-mile zones. But our economic analysis—taking into account such factors as where cable can be feasibly constructed, the impact of existing cable penetration, and the revenues of such stations—simply does not bear out the need for any general rule that would have unpredictable consequences in other parts of the country. And we note that even the Rocky Mountain stations do not appear to fit into one mold: financially, some are doing better than the industry average, some the average, and some worse. In view of all these considerations, we have concluded that the appropriate way of proceeding at this time is to extend the special relief described above and to examine any showings filed by these stations in the certificating process. New cable systems must give notice before commencing operations, and broadcasters—with knowledge of their own situations—will thus have a full opportunity to make a case for additional relief. We will give these showings most careful scrutiny. Additionally, we will undertake our own in-depth analysis where the desirability of such study is indicated. The essential consideration is not the extent of cable penetration or audience fragmentation per se but rather a demonstration of the effect of cable operations on station revenues and profits and on their ability to serve the public interest. We intend to keep a close watch on future developments in the Rocky Mountain and other regions involving smaller station operations—in rural areas generally—and have directed our staff to prepare reports annually. We will be alert to any emerging trend and in position to adjust our program accordingly.

LEAPFROGGING

92. In establishing policy in this area we have had a number of conflicting considerations to reconcile. On the one hand, it is arguably desirable to allow cable systems the greatest possible choice, on the assumption that they will select those signals that will most appeal to their subscribers and are available at the least expense. But in that event there is a risk that most cable systems would select stations from either Los Angeles, Chicago, New York, or one of the other larger markets. There would then be no general participation by broadcast television stations in the benefits of cable carriage. There is the additional consideration that carriage of closer stations,

[49] New systems may wish to use microwave facilities in order to obtain a better quality picture. We recognize, however, that our requirements may impose undue burden on some systems and accordingly will give careful consideration to appropriate petitions for relief.

because they are usually in the same region and often in the state, supplies some programing that is more likely to be of interest in the cable community. We believe we have struck an appropriate balance.

93. The leapfrogging rules are applicable to cable systems in all television markets. With respect to network affiliates, a cable system must afford priority of carriage to the closest such station or, at the option of the cable system, to the closest such station within the same State. In selecting independent stations, cable systems have a choice as to the first two such stations carried, except that if stations from among those in the top 25 designated markets are selected, they must be taken from one or both of the two closest such markets. Systems permitted to carry a third independent station are required to select a UHF station from within 200 miles. In the absence of any UHF station in this area, a VHF independent from within the area may be carried or, at the option of the cable system, any UHF independent. During those periods when programing on a regularly carried independent station must be deleted by virtue of the program exclusivity rules, the system is free to insert unprotected programing from any other stations (including network affiliates) without regard to point of origin. Such substitute programing may be continued to its conclusion. The cable system may also substitute other programing when the material on the regularly carried independent is a program primarily of local interest to the distant community (e.g., local news or public affairs).

Educational Stations

94. The principal concern of noncommercial educational broadcasters with signal importation is not reduction in audience size but possible erosion of local support among cable television subscribers. The rule we are adopting will permit carriage of distant educational stations in the absence of objection from local educational stations or educational television authorities.

95. Educational television interests are concerned about such a rule only to the extent that it might involve them in difficult and expensive process. We recognize the difficulties that educational interests face if forced to spend time and money in protracted litigation before the Commission and will accordingly attempt to settle any questions that may arise through informal procedures. We will give their objections careful consideration, and will endeavor to work out accommodations that serve the public interest. In the absence of objection, however, the widest possible dissemination of educational and public television programing is clearly of public benefit and should not be restricted. The rules require cable systems to carry, on request, all educational stations within 35 miles and those placing a Grade B contour over the cable community. We are continuing to require that local educational stations and local and State edu-

cational authorities receive direct notification of proposals by cable television systems to carry educational stations. While all objections will be carefully considered, we do not ordinarily anticipate precluding carriage of State-operated educational stations in the same State as the cable community.

Foreign Language Stations

96. Except in a very few markets, all U.S. stations broadcast in the English language. Although there are areas of the country, especially along the Canadian and Mexican borders, with significant populations whose first or only language is French or Spanish, the economics of television broadcasting generally precludes providing these areas with other than English language programing. Cable systems, however, have the capability of overcoming this problem, and we believe this capability should be encouraged. We will, accordingly, permit cable systems to carry non-English language programing without limitation. Where there is a local station broadcasting predominantly in a foreign language the added diversity provided by the carriage of distant foreign language stations broadcasting in the same language will be permitted unless the local station demonstrates that such importation will adversely affect its ability to serve the public. In order to encourage this carriage, distant foreign language stations will not be counted as part of the additional signal quota discussed above and we will not impose any restriction as to which stations, either foreign or domestic, may be carried.[50] As with educational stations, foreign language stations fulfill an important need for what generally is an audience limited in number. As a consequence, we do not anticipate that their carriage will have significant impact on the totality of local television service.

PROGRAM EXCLUSIVITY

97. Our solution to the problem of distant signal carriage involves an extension of our existing program exclusivity rules to provide more effective protection to syndicated programing. Additionally, we believe a change is appropriate in the same-day exclusivity rule that applied as a practical matter only to network programing.

98. The previous exclusivity rule (§ 74.1103) was based on a system of priorities that generally protected a station of higher priority against having its programing duplicated on the same day

---

[50] Following our August letter to Congress, the licensees or permittees of Spanish-language stations in Los Angeles and Hanford, Calif., San Antonio, Tex., and Miami, Fla., wrote to the Commission requesting that importation from Mexico of Spanish language stations not be allowed where U.S. Spanish language programing is available either off the air or potentially available via microwave. We recognize the arguments in favor of supporting domestic stations. However, above all, we are attempting to encourage carriage of foreign language stations. Therefore, absent the unusual situation, we do not think any additional burden should be imposed on the cable systems involved.

by cable carriage of a lower priority station. From highest to lowest, the signal strength priorities are Principal Community, Grade A, and Grade B. With respect to network television programing, we are retaining this system of priorities but will only require cable systems, on request of a higher priority station, to refrain from simultaneous duplication of the higher priority station's network programing.[51] Except for this change from same-day to simultaneous protection, we retain the precedents and policies evolved under the prior rule.

99. The change, while serving effectively to protect an affiliate's all-important network programing (except in the time zone situation[52]), facilitates cable operation, particularly in the smaller markets. The new provision is also complementary to the changes in our signal carriage rules that permit new cable systems in both smaller and major markets to carry duplicate sets of network stations only if the signals are available under the significant viewing standard. Because these signals are generally available even without cable, it is appropriate that cable subscribers not be denied such time diversity as is available over the air.

100. Syndicated programing will now be effectively protected in the major markets.[53] In markets 1–50—cable systems, on receipt of appropriate notification, will be required to refrain from carrying syndicated programing on a distant signal as follows: (1) During a pre-clearance period of 1 year, syndicated programs sold for the first time anywhere in the United States for television broadcast exhibition; (2) during the run of the contract, programs under exclusive contract to a station licensed to a designated community in the market. In markets 51–100—cable systems, on receipt of appropriate notification, will be required to refrain from distant signal carriage of syndicated programs under exclusive contract to a station licensed to a designated community in the market, except in the following circumstances:

(1) *For off-network series programs:*

(A) Prior to the first nonnetwork broadcast in the market of an episode in the series;

---

[51] We do not afford exclusivity to foreign stations (§ 76.5(b)). We would, however, consider affording network exclusivity on petition filed under § 76.7 in the unusual situation where a U.S. network has obtained permission to have its programing transmitted into a U.S. market via a foreign station.

[52] We will, on appropriate petition, grant additional exclusivity relief in those situations where a signal is carried from one time zone into another.

[53] Syndicated programing is defined in the rules (§ 76.5(p)). Essentially, it encompasses nonnetwork programing sold in more than one market. This does not mean that if two stations (usually under common ownership) have a practice of saving on film costs by using a microwave interconnection for their syndicated presentations (e.g., an off-network series), the stations are not entitled to syndicated exclusivity protection. They are simply using a different means of presenting nonnetwork programing.

(B) After a first nonnetwork run of the series in the market or after 1 year from the date of the first nonnetwork broadcast in the market of an episode in the series, which ever occurs first;

(2) For first-run series programs:

(A) Prior to the first broadcast in the market of an episode in the series;

(B) After 2 years from the first broadcast in the market of an episode in the series;

(3) For first-run, nonseries programs:

(A) Prior to the date the program is available for broadcast in the market under the provisions of any contract or license of a television broadcast station in the market;

(B) After 2 years from the date of such first availability;

(4) For feature films:

(A) Prior to the date such film is available for nonnetwork broadcast in the market under the provisions of any contract or license of a television broadcast station in the market;

(B) Two years after the date of such first availability;

(5) For other programs: One day after the first nonnetwork broadcast in the market or 1 year from the date of purchase of the program for nonnetwork broadcast in the market, whichever occurs first.

Additionally, and with respect to each of these categories of programing, a cable system in markets 51–100 may carry any distant signal syndicated program during prime time unless the station asserting exclusivity has both an exclusive contract for that program and will broadcast that program during prime time hours.

101. The rules governing syndicated program exclusivity will be administered in the following manner. While contracts entered into before the effective date of these rules will be presumed to be exclusive, subsequent contracts must specifically provide for broadcast exclusivity (both over-the-air and by cable) before a program can be protected under the rules. At a minimum a television licensee seeking exclusivity protection must obtain (a) exclusivity against other television stations licensed to its designated community in the market[54] and (b) exclusivity against cable dissemination of

---

[54] We recognize that it may appear anomalous in some instances to require exclusivity only against stations licensed to the same designated community—e.g., a Minneapolis-St. Paul or a Dallas-Fort Worth situation. But the answer is that in such instances programs are not sold on an exclusive basis just in St. Paul or just in Fort Worth, but rather for both cities in each of the markets. As a practical matter the requirement for specific exclusivity for television broadcast in one of these designated cities insures that broadcast exclusivity has really been obtained in that market. Were we to specify that the contract should provide exclusivity for all the designated cities in the market, it might be requiring too much. In some markets, the designated communities are located so far apart that a sale in one does not and should not preclude a sale in the other. This matter of permissible geographical exclusivity is the subject of the proceeding in Docket 18179. We believe that by proceeding as above, we will largely avoid introducing 18179

the program within the 35-mile zone [55] via a distant signal. We think that this is a reasonable requirement. A broadcast station may now purchase the exclusive right to broadcast a television program in its market. Cable represents another way to distribute the program. The station may bargain for the exclusive right as to any cable television presentation (e.g., cable origination, pay-cable, or other leased channel presentation). But what it must obtain, in order to be entitled to protection, is the exclusive right with respect to broadcast exhibition— whether the broadcast exhibition stems from another station in the market or from a cable system in the market that is bringing in distant broadcast signals. This is reasonable market exclusivity which the broadcaster is entitled to seek and which he must obtain to claim exclusivity rights under § 76.151.

102. Because this is a complex subject, it may be helpful to give examples, using the Baltimore-Washington situation. A Washington station, even if significantly viewed in Baltimore, would have no right to preclude carriage of its syndicated programs on a distant signal (e.g., from Philadelphia) carried on a Baltimore cable system, because Baltimore is a designated major market community that does not fall wholly within 35 miles of Washington. A Washington station could preclude carriage of a protected program on a distant signal being carried on a Washington cable system and on other cable systems located within 35 miles of Washington (except on a cable system in Baltimore). In Laurel, Md, which lies between Washington and Baltimore, a cable system could carry both Washington and Baltimore signals, would protect the programing of neither against the other, and would protect the programing of both Baltimore and

problems in this area and yet will achieve our basic objective here. If there are abuses or the need for specific action because of some peculiar situation, we can handle those matters on complaint. The foregoing is the minimum requirement for exclusivity protection. If a broadcast station obtains in its contract exclusivity against stations located in other designated communities in its market, protection will also be afforded in the 35-mile zones of those communities. While this purchase of additional broadcast exclusivity (with explicit accompanying extension of cable protection) is clearly a permissible practice in many instances (e.g., the Dallas-Fort Worth situation) it could, as noted, raise policy questions under 18179 (e.g., Cleveland-Akron). These matters will be treated in Docket 18179, but may of course be raised in this area by any interested person, such as the cable system.

[55] A station located in a designated community of a major market is not entitled to exclusivity protection in a designated community located in another major market unless the latter community lies wholly within 35 miles of the station's community. This provision parallels § 76.61(a)(1) of the carriage rules. Further, stations from other markets carried by a cable system pursuant to the significant viewing test will not be entitled to syndicated program exclusivity on such systems. Nor will any of their programing have to be deleted to protect stations licensed to designated cities in the market in which the system is located.

Washington signals against distant signals. Assuming that a smaller television market community were located wholly or partially within the 35-mile zone of Washington, a Washington station would be entitled to top 50 market exclusivity protection in that community. If a community fell wholly or partially within 35 miles of both a top 50 station and a second 50 station, the 1 year preclearance period would be applicable, and the cable system could be called on to protect the programing of stations from both markets in accordance with the requirements respectively applicable to those markets.

103. In markets 1–50, preclearance protection is complementary to the way in which syndicated programs are sold— i.e., they are sold in the largest markets first and, without a preclearance period, cable carriage of signals from these larger markets into other markets in the first 50 could dilute exclusivity and the value of the product. We are also protecting exclusivity for the full term contracts in these markets, but we note that the duration of contracts is a matter that we have had under consideration in Docket 18179 where we stated:

> The issue is somewhat analogous to that in the motion picture field where the courts have held that clearances are reasonable only "when not unduly extended in area or duration" and are not reasonable if "in excess of what is reasonably necessary to protect the licensee in the run granted". U.S. v. Paramount Pictures, Inc., 66 F. Supp. 323, 70 F. Supp. 53 (S.D.N.Y., 1947), noted with approval by the Supreme Court, 334 U.S. 131, 145, 147 (1948).[56]

104. With respect to exclusivity in markets 51–100, a number of distinctions have been drawn among the types of programs involved and the length of protection each is afforded. In general, off-network programing (formerly on the network, now in syndication) is protected for a shorter period because it receives its initial protection under network exclusivity rules and because, with respect to series, a year is sufficient to establish viewer loyalty for the local station. We have also been attempting to encourage the production of first-run, nonnetwork syndicated programing through our prime time access rules, and the exclusivity afforded here will give additional encouragement to the production of that kind of programing.

105. With respect to series programs, all episodes are to be treated as a unit— i.e., for the period in which exclusivity protection is afforded, the whole series rather than individual episodes will be protected, and during that period a cable system will not only have to refrain from carrying on a distant signal the same episodes under contract in the market but all other episodes as well, regardless of whether any station in the market has an exclusive contract to broadcast the episodes against which exclusivity is sought. Similarly, a station's exclusivity rights expire as a unit so that, for example, protection ends for a first-run series 2 years after any station in the market first broadcasts an episode in

[56] 27 FCC 2d 13, 14 (1971).

the series. Thereafter, any episode of the series may be brought in by cable regardless of whether it has ever been shown by a station in the market or is under exclusive contract to a station in the market. Finally, in the first 50 markets preclearance applies only to series or packages of programs consisting wholly of newly created material.

106. The rules governing program protection specify that appropriate notification be given to cable systems when exclusivity rights are asserted. The preclearance rule for the first 50 markets is designed principally for the benefit of copyright holders. The burden is therefore placed on copyright holders or their designated agents to notify cable systems in these markets when a sale has been made and that the preclearance period is running. With respect to other requests for exclusivity, the burden is also placed on the party seeking protection, in these cases the broadcaster. But when program deletion on regularly carried distant signals is required, the burden of identifying substitute programing that may be carried shifts to the cable system. Section 76.155 specifies how proper notification is to be given and details the form of notification. Because the program protection obligations of cable systems turn on the terms specified in contracts between copyright holders and broadcast stations, the appropriate portions of such contracts are required to be included in the public files of broadcast stations where they will be available for examination.[57] Reciprocally, we are requiring cable systems to maintain a log of distant signals carried and the programs offered on those signals.

GRANDFATHERING

107. In light of the difficulty of withdrawing signals to which the public has become accustomed and in deference to the equities of existing system operators, we are not applying the new carriage rules to any signals that a cable system was authorized by the Commission to carry or was lawfully carrying prior to March 31, 1972.[58] If carriage of signals has been limited by Commission order to a discrete area of a community, any extension of service outside the discrete area will be subject to the new carriage rules. A cable television system currently operating with authorized signals, and not the subject of such an order, may freely expand in its community with such signals. Grandfathered cable systems may add signals of a class permitted by

[57] Arguably, full contracts should be in the file. We are not persuaded, however, that it is necessary to go that far, and are permitting the parties to withhold those terms of their contracts that do not relate to the exclusivity in question. But we expect to watch carefully how this arrangement works out in practice, and we will revisit the matter if abuse develops because all the terms of contracts are not revealed.

[58] Included among authorized signals are both those whose carriage has been permitted by specific decision of the Commission, and those authorized by operation of the provisions of former § 74.1105 of the rules and not inconsistent with former § 74.1107.

the rules (e.g., independent signal(s) if none is presently carried). The addition of such new signals where the system is located in one of the top 100 markets will also require compliance with the rules regarding access availability (§ 76.251 (a) (4) through (a) (11)). With respect to exclusivity, existing carriage is grandfathered so that an operating system need not comply with the syndicated exclusivity rules except for new signals added or if the system extends operations into a new community or beyond the discrete area to which it has been specifically limited by Commission order.

108. In addition, we have adopted the proposal in Docket 18373 [59] that permits signals authorized or grandfathered to one system in a community to be carried by other systems in the community. Systems availing themselves of this rule are governed by the syndicated programing exclusivity obligations applicable to the earlier system. This will eliminate competitive imbalances between systems operating in the same community and avoid the necessity for the filing of waiver requests.

PROCEDURAL MATTERS

109. With the adoption of our new program for cable television, we are also instituting new procedures. These have been designed to assure: That effective public notice of new proposals is given; that applications contain full information on the details of system operation; and that new cable proposals are, without exception, reviewed for consistency with our rules.

110. New service may not begin until a certificate of compliance is issued. An application for a certificate of compliance must contain the following information:

(1) The name, mailing address, and proposed starting date of service, the community to be served, a list of the broadcast stations to be carried (excluding those expected to be used for substitute programing under §§ 76.61(b) (2) (ii) and 76.63), and a statement of whether microwave service will be used to deliver any of the signals.

(2) A copy of FCC Form 325 "Annual Report of Cable Television Systems," supplying all applicable information.

(3) A copy of the franchise, license, permit, or certificate granted by the local authority.

(4) A statement demonstrating that the proposal complies with the cable television rules. This should indicate how the choice of signals to be regularly carried is consistent with the rules and should explain how the system's franchise and plans for availability and administration of access channels and other nonbroadcast services meets requirements.

After a cable system is certified, an application for a new certificate to add local or distant signals on a regular basis need not include the franchise or Form 325. A system in operation on March 31, 1972, does not have to file an application for certification if no new signals are added

[59] Supra, note 1.

to the system, but will have to apply for certification when its current franchise expires or by March 31, 1977, whichever comes first.

111. In issuing certificates, and for purposes of these new rules generally, we will continue the policy of treating cable operations, even if served by the same head end, as separate systems in each community served. Thus, when applications are filed for certificates of compliance, a separate application should be filed for each community in which the system will operate. Information pertaining to a number of communities need not be refiled separately for each community but may be incorporated by reference.

112. The Commission will issue public notices of all applications for certificates of compliance. Cable systems must give direct notice to local franchising authorities, local television stations, the superintendent of schools in the community, and local educational television authorities. Objections to proposed cable service may be made within 30 days after the Commission's public notice. Controversies concerning carriage (Subpart D) and network program exclusivity (§ 76.91) will be acted on in the certificating process if raised within 30 days of the public notice. Such matters may be raised at any time and will be considered under the special relief rules but outside the certificating process. The Commission will not certify new operations for 30 days after public notice and, whether or not objection is filed, a cable system may not commence new service before receipt of a certificate of compliance from the Commission. Absent special situations or showings, requests consistent with our rules will receive prompt certification. The rules will operate on a "go, no-go" basis—i.e., the carriage rules reflect our determination of what is, at this time, in the public interest with respect to cable carriage of local and distant signals. We will, of course, consider objections to signal carriage applications and have retained special relief rules, but those seeking signal carriage restrictions on otherwise permitted signals have a substantial burden. Before restrictions are imposed in such cases, there will have to be a clear showing that the proposed service is not consistent with the orderly integration of cable television service into the national communications structure and that the results would be inimical to the public interest. We have during the course of this proceeding fully considered the question of impact on local television service and we do not except to reevaluate that general question in individual cases. And, for the same reason, we have no intention of reevaluating on request of cable systems in individual proceedings the general questions settled in our carriage and exclusivity rules. Rather, we strongly believe that cable systems must generally operate under these rules and that, only after meaningful experience, will we be in position for a general reassessment.

113. In connection with our special relief provisions, we note that in our August letter we designated certain markets where it appeared that special treatment to restrict distant signal carriage might have to be considered. We are no longer singling out these cases because the inclusion of substantial exclusivity protection for syndicated programing limits the impact of cable on local television service and is a new factor that must be taken into account. We are leaving unusual situations to petition for special relief, but there must be substantial showing to warrant deviation from the "go, no-go" concept of the rules. Finally, our 1968 proposals contemplated waiver of the leapfrogging restrictions in several circumstances. We will continue to be flexible as to the leapfrogging provisions for network signals—the rules specify that waiver may be granted for good cause, e.g., to bring in a signal of greater interest or from the same State or to avoid excessive microwave costs. But waivers in the case of the leapfrogging provisions for independent signals are not contemplated.

114. *Pending cases, notices, and related matters.* Having described the contents and operation of the new signal carriage and exclusivity rules, it is appropriate to outline the Commission's intentions with respect to pleadings, notifications, and other documents filed pursuant to our earlier rules (Part 74, Subpart K) prior to the effective date of the rules adopted herein.[60]

A. Petitions relating to the carriage and program exclusivity provisions of § 74.1103:

(1) Petitions seeking waiver of the carriage rules will be dismissed as moot unless, within ninety (90) days of the effective date of the new rules, they are supplemented to demonstrate their relevance to the new regulatory program;

(2) Petitions seeking waiver of the program exclusivity rules will remain on file. Requests for same-day network program exclusivity will be presumed to have been modified to request only simultaneous network program exclusivity.

B. Notifications given pursuant to § 74.1105:

These notifications will remain on file and, where relevant, may be incorporated by reference into an application for certificate of compliance.

C. Petitions seeking waiver of § 74.1107 to import distant television signals into one of the 100 largest television markets:

These petitions will be dismissed as moot unless, within ninety (90) days of the effective date of the new rules, they are

[60] Although our discussion has been limited to the revised cable television rules, now to be found in a new Part 76 (Cable Television Service), we have also made conforming changes in Subpart I of Part 21 (Point-to-Point Microwave Radio Service), Subpart J of Part 74 (Community Antenna Relay Service), and Subpart L of Part 91 (Business Radio Service). We note that the Community Antenna Relay Service has been renamed the Cable Television Relay Service (still to be abbreviated CAR), and the rules have been rearranged in a new Part 78.

supplemented to convert them into applications for certificates of compliance filed pursuant to § 76.13 of the rules.

D. Petitions invoking § 74.1105(c) to stay the carriage of television signals:

These petitions will be dismissed as moot, without prejudice to the filing of new pleadings in response to related applications for certificates of compliance.

E. Petitions seeking authorization to import television signals into areas not within the top 100 markets:

These petitions will be dismissed as moot unless, within ninety (90) days of the effective date of the new rules, they are supplemented to convert them into applications for certificates of compliance filed pursuant to § 76.13 of the rules.

F. Petitions seeking interpretative rulings or the imposition of additional or different requirements, filed pursuant to § 74.109:

These petitions will be dismissed as moot unless, within ninety (90) days of the effective date of the new rules, they are supplemented to demonstrate their relevance to the new regulatory program.

G. Petitions seeking reconsideration or stay of prior Commission actions:

These petitions will remain on file. Their disposition will depend on the particulars of each case..

H. Hearing cases:

Cases in which hearings were ordered prior to December 13, 1968, and which were suspended pursuant to paragraph 51 of the December, 1968 notice in Docket 18397, will be disposed of by Hearing Examiners and other decision-making Commission personnel consistent with our action herein.

I. Microwave applications:

Applications for authorization in the Cable Television Relay (CAR) Service, Business Radio Service, or Domestic Public Point-to-Point Microwave Radio Service that are pending on March 31, 1972 will be dismissed unless, within ninety (90) days of the effective date of the new rules, they are supplemented to indicate that any necessary application for certificate of compliance, pursuant to §§ 76.11 and 76.13 of the rules, has been filed. The supplement shall identify the application for certificate of the cable television system by the name of the cable television system, the community and area served or to be served, the date on which the application was filed, and the file number (if available).

115. *Fees.* The revisions in these rules require corresponding changes in the Commission's schedule of fees for cable television (§ 1.1116 of the rules). In particular, the provisions concerning petitions for experimental operations pursuant to Docket 18397, and petitions for waiver of the top 100 market hearing requirement are now obsolete, and will be deleted. Rule making will soon be initiated concerning fees for the certificating process. In order to begin processing applications for certificates of compliance promptly, and because these applications are substantially equivalent to a combination of petitions for special relief

filed pursuant to § 74.1109 ($25 fee) and notifications pursuant to § 74.1105 ($10 fee), we have concluded that a filing fee of $35 per application can properly be assessed on an interim basis, pending the outcome of further fee rule making. We are amending the fee schedule accordingly. If multiple applications for certificates of compliance are filed by cable systems having a common headend and identical ownership but serving or proposing to serve more than one community, the full fee will be required only for one of the communities; $10 will be required for each of the other communities. This approach follows previous Commission practice with respect to multiple community filings.

116. Fees previously paid in connection with the filing of petitions that must now be supplemented to convert them into applications for certificates will be credited against the application fee, and, on request, refunds of previous fees in excess of the amount now required will be made. Fees paid in connection with the petitions dismissed as moot will also be refunded on request. Any objections to applications for certificates will be treated in the same manner as previous oppositions to petitions for waiver of the top 100 market hearing requirement: No fee will be due.

III. ACCESS TO AND USE OF NONBROADCAST CHANNELS

117. In its notice of proposed rule making in Docket 18894, the Commission stated that:

Cable television offers the technological and economic potential of an economy of abundance.[61]

On the basis of the record now assembled, we believe the time has come for cable television to realize some of that potential within a national communications structure. We recognize that in any matter involving future projections, there are necessarily certain imponderables. These access rules constitute not a complete body of detailed regulations but a basic framework within which we may measure cable's technological promise, assess its role in our nationwide scheme of communications, and learn how to adapt its potential for energetic growth to serve the public.

CHANNEL CAPACITY

118. Confronted with the need for more outlets for community expression on the one hand and, on the other, with cable television's capacity to provide an abundance of channels, we asserted in our second further notice of proposed rule making in Docket 18397-A the principle that the Commission "* * * must make an effort to ensure the development of sufficient channel availability on all new CATV systems to serve specific recognized functions."[62]

119. Most cable system operators and many others argue against the proposed establishment of a fixed minimum channel capacity. Some comments in Docket

18894 went further and suggested that the entire matter of channel capacity be left to experimentation.[63] While it is true that many existing cable systems have large channel capacities and seem at least technologically prepared to meet foreseeable demand, there are many systems apparently content to provide only broadcast signal carriage with no plans to expand service capabilities.

120. We envision a future for cable in which the principal services, channel uses, and potential sources of income will be from other than over-the-air signals. We note 40, 50, and 60 channel systems are currently being installed in some communities. The cost difference between building a 12 channel system and a 20 channel system would not appear to be substantial.[64] We urge cable operators and franchising authorities to consider that future demand may significantly exceed current projections, and we put them on notice that it is our intention to insist on the expansion of cable systems to accommodate all reasonable demands. We wish to proceed conservatively, however, to avoid imposing unreasonable economic burdens on cable operators. Accordingly, we will not require a minimum channel capacity in any except the top 100 markets. In these markets, we believe that 20 channel capacity (actual or potential) is the minimum consistent with the public interest. We also require that for each broadcast signal carried, cable systems in these markets provide an additional channel 6 MHz in width suitable for transmission of Class II or Class III signals. This seems a reasonable way to obtain necessary minimum channel capacity and yet gear it to particular community needs. We emphasize that the cable operator cannot accept the broadcast signals that will be made available without also accepting the obligation to provide the nonbroadcast bandwidth and the access services described below. The two are integrally linked in the public interest judgment we have made.

DESIGNATED CHANNELS

121. Broadcast signals are being used as a basic component in the establishment of cable systems, and it is therefore appropriate that the fundamental goals of a national communications structure be furthered by cable—the opening of new outlets for local expression, the promotion of diversity in television programing, the advancement of educational and instructional television, and increased informational services of local governments. Accordingly, cable television systems will have to provide one dedicated, noncommercial public access channel available without charge at all times on a first-come, first-served nondiscriminatory basis and, without charge during a developmental period, one channel for educational use and another channel for local government

---

[61] 25 FCC 2d 38, 39 (1970).
[62] 24 FCC 2d 580, 587 (1970).

[63] See, for instance, the comments of Storer Broadcasting Co.
[64] Testimony of Moses Shapiro on behalf of General Instrument Corp., Vol. 5, p. 982, transcript of hearings before the Commission in Docket 18397-A, Mar. 19, 1971.

use. We have already imposed an obligation on systems with 3,500 or more subscribers to originate programing and are now requiring that the origination channels be specifically designated.

122. *Public access channel*. It has long been a Commission objective to foster local service in broadcasting. To this end we have encouraged the growth of UHF television, and have looked to all broadcast stations to provide community-oriented programing. We expect no less of cable. In our July 1, 1970 notice we stated:

> The structure and operation of our system of radio and television broadcasting affects, among other things, the sense of "community" of those within the signal area of the station involved. Recently governmental programs have been directed toward increasing citizen involvement in community affairs. Cable television has the potential to be a vehicle to much needed community expression.[65]

We believe there is increasing need for channels for community expression, and the steps we are taking are designed to serve that need. The public access channel will offer a practical opportunity to participate in community dialogue through a mass medium. A system operator will be obliged to provide only use of the channel without charge, but production cost (aside from live studio presentations not exceeding 5 minutes in length) may be charged to users.

123. *Educational access channel*. It is our intention that local educational authorities have access to one designated channel for instructional programing and other educational purposes. Use of the educational channel will be without charge from the time subscriber service is inaugurated until 5 years after the completion of the cable system's basic trunk line. After this developmental period—designed to encourage innovation in the educational uses of television—we will be in a more informed position to determine in consultation with State and local authorities whether to expand or curtail the free use of channels for such purposes or to continue the developmental period. The potential uses of the educational channel are varied. An important benefit promises to be greater community involvement in school affairs. It is apparent, for instance, that combined with two-way capability, the quality of instructional programing can be greatly enhanced. Similarly, some envision significant advances in the educational field by the linking of computers to cable systems with two-way capability.[66] For the present, we are only requiring that systems provide an educational channel and, as noted below, some return communication capability, and will allow experiments in this field to proceed apace.

124. *Government access channel*. The Government access channel is designed to give maximum latitude for use by local governments. The suggestions for use

range across a broad spectrum and it is premature to establish precise requirements. As with the educational channel, use of the Government channel will be free from the time subscriber service is inaugurated until 5 years after the completion of the cable system's basic trunk line, at which time we will consider whether to expand or curtail such free use or to continue the developmental period.

LEASED ACCESS CHANNELS

125. In addition to the designated channels and broadcast channels, cable systems shall make available for leased use the remainder of the required bandwidth and any other available bandwidth (e.g., if a channel carrying broadcast programing is required to be blacked out because of our exclusivity rules or is otherwise not in use, that channel also may be used for leased access purposes). Additionally, to the extent that the public, education, and Government access channels are not being used, these channels may also be used for leased operation. But such operations may only be undertaken on the express condition that they are subject to immediate displacement if there is demand to use the channel for the dedicated purpose.

EXPANSION OF CAPACITY

126. Our basic goal is to encourage cable television use that will lead to constantly expanding channel capacity. Cable systems are therefore required to make additional bandwidth available as the demand arises. There are a number of ways to meet this general objective. Initially, we intend to use the following formula to determine when a new channel must be made operational: whenever all operational channels are in use during 80 percent of the weekdays (Monday-Friday), for 80 percent of the time during any consecutive 3-hour period for 6 weeks running, the system will then have 6 months in which to make a new channel available. This requirement should encourage use of the system with the knowledge that channel space will always be available, and also encourage the cable operator continually to expand and update his system. On at least one of the leased channels part-time users must be given priority. We plan at a later date to institute a proceeding with a view to assuring that our requirement of capacity expansion is not frustrated through rate manipulation or by any other means. This proceeding will also deal with such open questions as rates charged for leased channel operations.

127. We are aware of the possibility that the formula may impose undue burdens on system operations. If it were necessary to rebuild or add extensive new plant, this could not reasonably be expected within a 6-month period. The requirement for activating new capacity within 6 months is based on our understanding that only relatively modest effort is involved in converting existing potential to actual capacity. These considerations, however, point up the necessity for building now with a potential that takes the future into account. Be-

cause this part of our program is a relatively uncharted area, we will make it a matter for continuing regulatory concern.

TWO-WAY CAPACITY

128. On review of the comments received and our own engineering estimates, we have decided to require that there be built into cable systems the capacity for return communication on at least a non-voice basis. Such construction is now demonstrably feasible.[67] Two-way communication, even rudimentary in nature, can be useful in a number of ways—for surveys, marketing services, burglar alarm devices, educational feedback, to name a few.

129. We are not now requiring cable systems to install necessary return communication devices at each subscriber terminal. Such a requirement is premature in this early stage of cable's evolution. It will be sufficient for now that each cable system be constructed with the potential of eventually providing return communication without having to engage in time-consuming and costly system rebuilding. This requirement will be met if a new system is constructed either with the necessary auxiliary equipment (amplifiers and passive devices) or with equipment that could easily be altered to provide return service. When offered, activation of the return service must always be at the subscriber's option.

REGULATIONS APPLICABLE TO CHANNELS PRESENTING NONBROADCAST PROGRAMING

130. We now turn to the question of the regulation of access channels presenting nonbroadcast programing. We believe that such regulation is properly the concern of this Commission. These channels fulfill Communications Act purposes and, in the context of our total program, are integrally bound up with the broadcast signals being carried by cable. It is by no means clear that the viewing public will be able to distinguish between a broadcast program and an access program; rather, the subscriber will simply turn the dial from broadcast to access programing, much as he now selects television fare. Moreover, leased channels will undoubtedly carry interconnected programing via satellite or interstate terrestrial facilities, matters that are clearly within the Commission's jurisdiction. Finally, it is this Commission that must make the decisions as to conditions to be imposed on the operation of pay cable channels, and we have already taken steps in that direction. (See § 76. 225). Federal regulation is thus clearly called for.

131. There remains the issue of whether also to permit State or local regulation of these channels where not inconsistent with Federal purposes. We think that in this area a dual form of regulation would be confusing and impracticable. Our objective of allowing a period for experimentation might be

---

[65] Supra, note 66, at Pt. II, paragraph 6.

[66] Comments of the Stanford Cable Television Committee, Institute for Communications Research, in Docket 18397-A.

[67] We note the recent developments in this field by Sterling Manhattan Cable TV in New York City and Telecable Corp. in Overland Park, Kans.

jeopardized if, for example, a local entity were to specify more restrictive regulations than we have prescribed. Thus, except for the Government channel, local regulation of access channels is precluded. If experience and further proceedings indicate its need or desirability, we can then delineate an appropriate local role.[68]

132. Because of the Federal concern, local entities will not be permitted, absent a special showing to require that channels be assigned for purposes other than those specified above. We stress again that we are entering into an experimental or developmental period. Thus, where the cable operator and franchising authority wish to experiment by providing additional channel capacity for such purposes as public, educational, and Government access—on a free basis or at reduced charges—we will entertain petitions and consider the appropriateness of authorizing such experiments, to gain further insight and to guide future courses of action.[70] In communities outside the top 100 markets where access channels are not required by the Commission, we will permit local authorities to require access services, so long as they are not in excess of what we require for the major markets.

133. The question of what regulations we should impose at this time is most difficult. Our judgments on how these access services will evolve are at best intuitive. We believe that the best course is to proceed with only minimal regulation in order to obtain experience. We emphasize, therefore, that the regulatory pattern is interim in nature—that we may alter the program as we gain the necessary insights.

134. We are requiring that cable systems promulgate rules to apply to access services, and that these rules be kept on public file at the system's local headquarters and with the Commission. What matters during this experimental period is not form but substance, and we are specifying the guidelines that we believe are appropriate at this time. We believe we have full discretion to act in this fashion.[71]

135. With respect to the public access channel, the rules to be promulgated by the system must specify nondiscriminatory access without charge on a first-come, first-served basis. These rules shall also proscribe for all designated access channels (except the Government access

channel when it is being used for its designation purpose) the presentation of: Any advertising material designed to promote the sale of commercial products or services (including advertising by or on behalf of candidates for public office); lottery information and obscene or indecent matter (modeled after the prohibitions in §§ 76.213 and 76.215 respectively). The regulations shall also specify that persons or groups seeking access be identified, and their addresses obtained; this information should be publicly available and must be retained by the system for at least 2 years. The cable operator must not in any other way censor or exercise program content control of any kind over the material presented on the public access channel.

136. We recognize that open access carries with it certain risks. But some amount of risk is inherent in a democracy committed to fostering "uninhibited, robust, and wide-open" debate on public issues (New York Times Co. v. Sullivan, 376 U.S. 254, 270 (1964)). In any event, further regulation in this sensitive area should await experience. For example, we intend to explore whether it would be feasible or desirable to provide a locked switch to cut off the public access or leased channels, should subscribers wish to control channel selection.

137. In short, we recognize that the regulation of public access channels may result in many problems for the cable operator, especially during the break-in period. Effective operational procedures can evolve only from trial and error, and it is probable that systems will have different problems that do not now lend themselves to uniform regulation. We note, for example, the need to decide how applications for access time are to be made, what overall time limitations might be desirable, how copyrighted material will be protected, how production facilities will be provided, how the public can obtain advance notice of presentations, and so on. All these questions will probably be answered in a number of different ways. We will require that the rules adopted by cable systems in these respects be filed with us and made available to the public. But experimentation appears to be the best way to determine what will be workable for the long run.

138. The cable operator similarly must not censor or exercise program content control of any kind over the material presented on the leased access channels. Specifically, his rules shall provide for nondiscriminatory access on a first-come, first-served basis with the appropriate rate schedule specified. Again, he shall obtain the names and addresses of those leasing the channel, and shall adopt rules proscribing the presentation of: Lottery information; obscene or indecent matter; and advertising material not containing sponsorship identification.[72] Finally, in contrast with origination cablecasting rules (§ 76.217), we will not require commercials only at natural breaks on these channels. It is our expectation

that there will be experimentation, with some channels used entirely for advertising, some following the pattern of commercial broadcasts, and others that of § 76.217. We will continue to monitor developments in this area with a view to assuring that the public interest is served, particularly regarding such issues as false and misleading advertising.

139. The regulations we are imposing on systems engaging in cablecast origination are substantially the same as those first issued in the First Report and Order in Docket 18397. These regulations (§ 76.201 et seq.) include rules on lottery information, advertising, sponsorship identification, etc., and add a new specific proscription of obscenity.

LIABILITY

140. Many cable operators are concerned about potential civil and criminal liability resulting from use of these public and leased access channels. There is little likelihood of the possibility of a criminal suit in a situation where the system has no right of control and thus no specific intent to violate the law. See, e.g., Lambert v. California, 355 U.S. 225 (1957). The real fears of cable operators seem, in fact, to center on potential libel suits. The possible number and scope of such actions is, however, severely limited. In Rosenbloom v. Metromedia, Inc., 403 U.S. 29 (1971), the Court extended the "actual malice" rule of New York Times Co. v. Sullivan, supra,[73] to cover any situation where "the utterance involved concerns a matter of public or general interest." Since most users will presumably air opinion on matters that are of at least as much "public or general interest" as in the Rosenbloom case, it seems likely that their speech would come within the "actual malice" rule. It is doubtful that such malice could be imputed to a cable operator who has no control over the given program's content.

141. In the event that some material presented on these nonbroadcast channels were to fall outside the broad scope of the court's recent decisions such as Rosenbloom, this would not necessarily mean that the system is liable. (In this situation, recourse against the programer would be available.) We have adopted the no-censorship requirement in order to promote free discourse; this is, we believe, valid regulation having "the force of law." While the matter is of course one for resolution by the courts, State law imposing liability on a system that has no control over these channels may unconstitutionally frustrate Federal purposes. In any event, if a problem should develop in this respect, it is readily remedied by Congress and, in this connection, we would welcome clarifying legislation.[74]

PRODUCTION FACILITIES

142. It is apparent that our goal of creating a low-cost, nondiscriminatory means of access cannot be attained unless members of the public have reason-

---

[68] Franchise specifications concerning the number of dedicated channels for systems in operation prior to Mar. 31, 1972, will be permitted to continue in effect (§ 76.251(a) (11)(iv)).

[69] [Omitted]

[70] We are aware that bidding contests may result in awards that will unduly burden systems and possibly thwart achievement of our basic goals. We caution franchising authorities against encouraging such contests or making selections based on the barter of extra channels. If abuses arise in this respect, they will be examined in the course of the certificating process or on later petition.

[71] See Philadelphia Television Broadcasting Co. v. F.C.C., 123 U.S. App. D.C. 298, 359 F. 2d 282 (1966).

[72] Modeled after the prohibitions in §§ 76.-213, 76.215, and 76.221 of the Commission's rules, respectively.

[73] See also Curtis Publishing Co. v. Butts, 388 U.S. 130 (1967).

[74] Cf. Farmers Educational and Cooperative Union v. WDAY, 360 U.S. 525 (1959).

able production facilities available to them. We expect that many cable systems will have facilities with which to originate programing that will also be available to produce program material for public access. Hopefully, colleges and universities, high schools, recreation departments, churches, unions, and other community groups will have low-cost video-taping equipment for public use. In any event, we are requiring that the cable operator maintain within the franchise area production facilities for use on the public access channel.

143. In this experimental stage, it would be self-defeating to require cable systems to carry access programing and at the same time meet stringent technical standards. Thus, for the present, our technical standards will apply only to Class I channels (those used to distribute broadcast programing—see § 76.5(z) of the rules). We note specifically that the use of half-inch video tape is a growing and hopeful indication that low-cost recording equipment can and will be made available to the public. While such equipment does not now meet our technical standards for broadcasting, there is promise of its improvement and refinement. Further, since it provides and inexpensive means of program production, we see no reason why technical development of this nature should not be encouraged for use on cable systems.

144. Elaborate suggestions have been made for comprehensive community control plans such as neighborhood origination centers and neighborhood councils to oversee access channels. Here again the Commission will encourage experimentation rather than trying to impose a more formal structure at this time.

145. The access requirements we are imposing differ considerably in scope and purpose from our origination requirement of § 76.201. Because of the system operator's control over programing of originated material it was necessary to impose such obligations as are involved in the "equal time" and "fairness" doctrines. Such requirements are not being imposed on use of the access channels because these channels are free of operator control and access is guaranteed. But they do remain in effect for designated origination cablecasting channels. Should a cablecast by a candidate for political office on the origination channel prompt the necessity for providing equal time to an opponent, it must be provided on the origination channel. In this situation, the opponent's appearance on an access channel will not suffice. Similarly, should a controversial originated program raise a "fairness" issue, any countering views must also be presented on the origination channel.

146. The suggestion has been made that cable television systems be prohibited from originating their own programing and be restricted entirely to a common carrier role. We have considered these possibilities but feel that it would be premature to adopt either at this time. (See notice in Docket 18397, 15 FCC 2d 417 at paragraph 26 (1968).)

At this stage in the development of the cable industry, it is the system operator who has the greatest incentive to produce originated material attractive to existing and potential subscribers. We have tried to encourage this origination both through our origination rules (First Report and Order in Docket 18397, 20 FCC 2d 201 (1969)) and by structuring the broadcast signal carriage rules to stimulate the development of nonbroadcast services. At the same time, we have recognized that during this developmental stage we should not adopt rules that constrain experimentation and innovation in the services that cable systems provide but, rather, that we should seek to keep our future options open. When cable penetration reaches high levels and demand increases for leased channel operations, we will revisit this matter. For now, we remain of the view that the most appropriate mix for the orderly development of cable and for encouraging the maximization of its potential for public benefit is one that embraces "* * * a multipurpose CATV operation combining carriage of broadcast signals with program origination and common carrier service * * *" (First Report and Order in Docket 18397, supra, paragraph 3). The rules adopted here are designed to accomplish that.

APPLICABILITY

147. These access rules will be applicable to all new systems that become operational after March 31, 1972 in the top 100 television markets. Currently operating systems in those markets will have 5 years to comply fully with this section. We focus here on the top 100 markets because we have selected these markets as the recipients of certain benefits in order to stimulate cable growth. But, correspondingly, that growth should be accompanied by accesss obligations if the public is to receive the full benefits of this program. Further, cities in the top 100 markets have, as a general rule, more diverse minority groups (ethnic, racial, economic, or age) who are most greatly in need of both an opportunity to express their views and a more efficient method by which they can be apprised of governmental actions and educational opportunities. To the extent that the access requirements pose problems for systems operating in small communities in major markets, such systems are free to meet their obligations through joint building and related programs with cable operators in the larger core areas.

148. If these requirements should impose an undue burden on some isolated system, that is a matter to be dealt with in a waiver request, with an appropriate detailed showing. While we encourage systems in markets below the top 100 to provide access channels, we are not at this time requiring them to do so. We will permit local franchising authorities in such areas to require systems to provide access service, but to no greater extent than we have specified for systems in the top markets. In that event, our access rules would be applicable.

IV. TECHNICAL STANDARDS

149. In our June 24, 1970 notice of proposed rule making in Docket 18894, we proposed technical standards for the operation of cable television systems. Comments were received from diverse sources, including the National Cable Television Association, the consulting firm of Hammett and Edison, the Association of Maximum Service Telecasters, Inc., Archer S. Taylor, vice president of the engineering consulting firm, Malarkey, Taylor and Associates, Inc., and many others. All comments were reviewed and we are adopting a set of technical standards that we believe will provide much needed uniformity on a nationwide basis yet still allow sufficient flexibility for further technical change.

DEFINITIONS (§ 76.5)

150. It is our ultimate intention to provide appropriate technical standards for the various kinds of signals that we expect cable television systems will offer their subscribers. At this moment there is need for standards governing the carriage of standard television signals that are picked up off the air. We expect soon to need technical standards—in some measure possibly different—for carriage of cablecast programs. The burgeoning use of two-way or "return" communications will require the formulation of additional technical regulations in order to insure protection to channels used for television or other communications. Accordingly, at this time, we are adopting definitions for four categories of cable channels. These may be modified in the future, but at present we view them as a useful framework for administering the multifaceted development of cable distribution systems.

Class I cable channel. This definition is intended to designate those cable channels devoted to delivering standard broadcast television signals picked up off the air at the headend or delivered to the cable network by microwave or provided by direct connection to a local television broadcast station. Class I cable channels are subject to the technical standards adopted herein.

Class II cable channel. This is intended to designate those channels used for the delivery of cablecast programing. Technical standards are not now being provided for these channels. Class II cable channels are those used for television signals not obtained from television broadcast stations but that are intended to display pictures on subscriber television receivers without the use of decoding devices. Channels carrying television pictures purposely encoded or processed to permit reception by only selected subscribers are not included in this category.

Class III cable channel. In addition to television pictures, cable systems are likely to deliver to subscribers other forms of communication. We recognize the potential for a wide diversity of communications, some of which will require terminal equipment in subscriber homes. Some of these involve analog signals; others make use of digital signals. Not all

require a full 6 MHz of bandwidth. Class III cable channel uses might include: Encoded television signals which require special decoding equipment at the subscriber terminal, FM or AM broadcast signals, and facsimile and printed message material. Obviously, no single set of technical standards can embrace so many differing kinds of signals. We are not proposing standards for Class III cable channels at this time, but as the need becomes apparent, appropriate standards will be provided.

Class IV cable channel. This class will apply to "return" or "response" channels. At this time plans for use of those channels envision a relatively narrow band of frequencies that will be used to return limited amounts of information from subscriber to control point. Although it is too early to provide technical standards for such communications, it is expected that standards will be required.

Channel frequency response. This definition seeks to promote a common understanding that the frequency response requirements are those obtaining at subscriber terminals. We are not requiring frequency response standards for other points in the system or to other than Class I channels.

Subscriber terminal. This is defined as the point of interface between the facilities of the cable system and the receiving equipment normally the property of the subscriber. Thus, matching transformers, baluns, converters, or special amplifiers provided by the cable company are examples of facilities considered to be located on the system side of the subscriber terminal. Cable extensions that serve other premises and are not owned by the subscriber are considered the responsibility of the system.

PERFORMANCE TESTS (§ 76.601)

151. This section sets forth the responsibility of the cable system operator to make such tests and measurements as are necessary to offer reasonable assurance that the system performance is continuously satisfactory. The comments generally recognized the necessity of requiring some adequate measurement and monitoring schedule, although it was pointed out that the system subscribers are quick to report gross deficiencies in service. Our requirement is intended to reduce the incidence of malfunction by encouraging the system to institute procedures for regularly checking its operation. Many advised that requiring performance measurements at only three vaguely defined points would fall short of rigorously testing the system. Consideration has been given to requiring measurements at more than three points in order to insure "representative" sampling of system performance. But our view is that this requirement is not intended to establish that each subscriber will receive service in accordance with the standards—that can come only with a measurement at each subscriber terminal. The performance check is, rather, assurance to the operator and to the Commission (should the performance be questioned) that the signal path from head-end to check point is capable of conforming to the standards. We are therefore retaining the proposed requirement for three measurement points. Many systems, as a matter of good practice, will make routine observations at more than three points. The ultimate requirement, in any event, is that the technical standards must be met at each subscriber terminal.

152. Our aim is not to generate marginally useful measurement data for ourselves, but to encourage each cable operator to engage in systematic performance checking and preventive maintenance. Thus, we agree with those comments suggesting that the annual performance data not be filed with the Commission but be kept with the system where it will be available for inspection. The information required by our rule is minimal and should be readily available from every system. It will be useful in resolving service complaints and as reference data for identifying those cable television channels to which our technical standards apply and on which our required measurements are to be made.

153. It has been suggested that the cost of measuring equipment and the costs of hiring consultants to make the necessary measurements would be prohibitively high for small systems. While we recognize that compliance will involve some costs, we do not choose to sacrifice the public benefits derived from good technical performance. We have, therefore, carefully drawn our technical standards so that measuring equipment of reasonable cost can be used.

STANDARDS (§ 76.605)

154. Based on many persuasive comments, we are adopting certain revisions of our proposed rules. Some suggested that the phrase "pick up off-air" be eliminated from the opening sentence of subsection (a). Because we are adopting standards only for channels devoted to broadcast television programing, we are amending the disputed phrase. It is intended that the standards apply only to Class I channels—those carrying television broadcast signals picked up off the air, either at the cable system headend or relayed by CARS microwave from an off-air-pickup, or obtained by direct connection at a television station.

155. In (a)(1) we provide that the channels delivered to subscribers conform to the capability of the television broadcast receiver. This is not intended to limit the use of other channel arrangements within the system. We are also permitting, on adequate showing, the use of such arrangements as central switching systems similar to those identified with rediffusion.

156. We have relaxed the frequency tolerance standard originally proposed. Under the rule, systems which supply subscribers with an individual converter, tuner, or similar channel-selection device are required to meet a tolerance on the visual signal of only ±250 KHz. The proposed ±25 KHz tolerance is applicable to other systems. We are retaining the aural-visual separation tolerance as proposed.

157. With respect to visual signal level, we are requiring, in (a)(4), delivery of signals so that at no time is the signal on any cable television channel lower than the equivalent of 0 dBmV (across 75 ohms). We intended in our proposals to impose a limit on the difference in level permissible between any two adjacent cable channels (Channels 4 to 5 and 6 to 7 excepted), and also a limit on the maximum permissible difference between any other channels in the system. The ultimate purposes of such specifications are to insure an adequate signal on all channels, to prevent annoying visible differences of signal strength between channels, and to promote an optimum balance between signal level equality in distribution amplifiers and at subscriber terminals. These considerations and the comments on this subject have prompted us to adopt the signal level requirements set out in (a)(4) and (5). No specific maximum level is adopted. Instead, we are adopting a general rule which requires, in effect, that the signal level on any channel not exceed the level at which overload problems in the customer's receiver begin to occur.

158. After consideration of the comments directed to (a)(6) through (8), we have decided to retain these standards essentially as proposed, rejecting for the present the concern expressed by some that the 5-percent permissible amplitude for power frequency hum components is too high. In further proceedings we expect to reexamine this matter and may then decide from the information before us that a reduction on the maximum limit is necessary.

159. We have revised the signal-interference ratios proposed in (a)(9) and (10). Several parties pointed out that, as proposed, interference caused by undesired reception of a properly offset cochannel station would have to be reduced to a 46 dB ratio. However, off-air viewers are protected only to a ratio varying between 28 and 36 dB. These cochannel ratios involve certain assumptions about the percentage of audience that finds them acceptable and about the absolute value of the desired signals involved. Considering the superior quality of service every subscriber to cable should receive, we believe that some relaxation of the 46 dB cochannel ratio for offset signals may be appropriate, but not beyond 36 dB.

160. Most parties suggested that the value of terminal isolation we proposed in (a)(11) was too high and instead, recommended values in the range between 15 and 20 decibels. Further study suggests that a required terminal isolation of 18 decibels should be adequate to protect each subscriber from the effects of expected levels of spurious signals or impedance variations introduced at other subscriber terminals. We are amending our proposed rule accordingly.

161. We consider it appropriate to transfer (with some modification) from Part 15 to Part 76 our existing limits on radiation from cable systems. The

modifications we proposed—dropping the category of "sparsely inhabited areas" and tightening the radiation limit between 132 and 216 MHz—met general approval. They are now set forth in section 76.605(a)(12).

162. Special notice is taken of a letter dated February 24, 1971, addressed to our Acting Chief Engineer from the Office of Telecommunications Policy (OTP), expressing concern about possible interference to Air Traffic Control communications during periods of "CATV equipment malfunction." The frequency bands that OTP suggests might be excluded from use by cable systems include:

108 to 136 MHz.
162 to 174 MHz.
225 to 400 MHz.

Alternative suggestions include imposing power limitation on cable signals within the system, interleaving of channel assignments, and the installation of automatic shutoff devices that would remove power from amplifiers if malfunctions in the system might cause excessive radiation. While we recognize the desirability of eliminating the possibility of interference to air-ground communications, we are unable to share the OTP view of the hazard posed by possible cable malfunctions. After more than 20 years of cable operation, interference by cable radiation to aircraft communications has not been documented. We note also that OTP is not objecting to radiation from television receivers in the hands of the public. These, when tuned to Channels 4 through 13, may radiate signals within the bands OTP wishes to proscribe for cable systems. Television receiver oscillators may radiate fields as strong or stronger than those expected from cable systems. Spurious radiation from television transmitters also may occur in the aviation bands and, in our view, present a greater interference potential than cable systems. Because public benefits lie in encouraging full use of available radio spectrum within the cable, we are reluctant to hamper cable operations by so restricting the use of frequency space, particularly when that restriction would be based on a rather remote interference possibility. Accordingly, we are declining to adopt the frequency restrictions proposed by the Office of Telecommunications Policy.

MEASUREMENTS (§ 76.609)

163. Comments submitted with respect to this portion of our proposed rules reflected either a concern that the measuring techniques we proposed were inadequately detailed, or that other methods should be employed. Additionally, our proposal to include CARS microwave relays in the measurement caused concern. Our intention was to set forth a number of measurement procedures that we consider suitable for determining various aspects of system performance. Because cable systems operate under a variety of circumstances, we are adopting a flexible approach to determining system performance. As indicated in the rules, we are permitting the use of whatever alter-

native measuring methods can be fully justified. This should not be construed as permitting the use of rough-and-ready procedures that result in equivocal measurement data. We will insist upon a bona fide and authoritative attempt to measure system performance, and where the resulting data is inadequate, we may require remeasurement using specified equipment or procedures.

164. Some objected to our proposal to include CARS microwave relay circuits within the measurement loop and not to include similar facilities operated by common carriers. Others, noting this difference, suggested that we reserve the formulation of standards for microwave delivery for later rule making. We have made the distinction on the assumption that CAR facilities are under the direct control or supervision of the cable system; common carrier facilities are not. With respect to the CAR service, the cable operator is able to effect scheduling of the microwave facilities. When the microwave relay is operated by a common carrier, however, there is considerable difficulty in arranging measurement procedures. In the latter case, we are leaving it to the cable operator to insure contractually that the signals delivered to his system are adequate to permit him to conform to our technical requirements.

165. Comments directed to specific measurement procedures noted that we had failed to provide that, when antennas or other inputs are disconnected for system performance measurements, substitute carriers or pilot signals in some instances must be inserted in order to maintain proper operation of the rest of the system. In paragraphs (a) and (b) we now reflect this concern. We emphasize that measurement of a performance parameter in any cable television channel must be made under conditions that approximate those existing under normal operations. Signals should be present on all other channels on which signals normally are delivered. They should be of normal amplitude. Automatic gain controls or manually controlled gains should be normal.

166. With respect to measuring noise in a cable channel, it was suggested that the NCTA standard for noise measurement should be required. We agree. Thus, we are amending the language of § 76.609(e) to recognize that method. We are also taking note of the usual circumstances in which the variation of noise level over the width of a cable television channel is small, and are providing language to permit a "spot" measurement of noise. At the same time, we note a suggestion filed by Hammett and Edison to use an oscilloscope in a rapid sweep, single-trace mode to permit an acceptable visual observation of the noise voltage after demodulation. The peak-to-peak amplitude of the noise is directed compared to the peak-to-peak amplitude of the desired signal; to which ratio an appropriate peak-to-rms correction is added. The method is attractively simple and direct and, if performed with adequate precautions, appears to be acceptable.

167. Other comments questioned the appropriateness or necessity of various methods we had proposed. For example, the requirement that the measurement of noise in a channel must include the CARS microwave relay (if any) within the measurement loop poses the problem of measuring noise in the face of the channel carrier. This circumstance appears to dictate the use of a comparitively narrow band noise measuring technique. However, whatever the measurement procedure used, it will be subject to review as to accuracy and appropriateness.

168. As an exception to the approach used in § 76.609 (e), (f), and (g), we intend that the measurement procedures outlined in (h) of that section be followed strictly or, if special circumstances necessitate divergence from established procedures, the alternate procedures be thoroughly justified. The rule for measuring radiation from a cable system is essentially that which was established in Part 15 of our rules. The measuring procedure has been tested over a number of years. We see no indication that substantial change in the procedure is necessary.

RESPONSIBILITY FOR INTERFERENCE (§ 76.617)

169. We have noted the concern that a cable system would be held responsible for interfering signals radiated from television receivers connected to the system. We have long had a rule (§ 15.82) which places on the operator of a radio (television) receiver the responsibility for eliminating interference caused by that receiver. Section 76.617 is intended to place a similar restriction on the cable operator who must insure that his system does not distribute or reradiate an interfering signal generated in his customer's receiver, even if the latter generates a signal in excess of permissible limits. In our view, the obvious remedy when a receiver-generated interfering signal is found in the cable distribution system is to suspend service to the customer until the receiver is repaired.

ADDITIONAL TECHNICAL PROBLEMS

170. We are of the view that the technical standards we have adopted are minimal and should be augmented as soon as possible with standards covering other technical areas such as:

Standards for a cable television receiver (a television receiver specifically designed for use with a cable television system).
Frequency allocations within the cable network.
Standards for Class II, III, and IV channels.
Standards on envelope delay, differential gain, and phase.
Standards on permissible cross-modulation, "ghosting," hum.
Standards for cable carriage of aural broadcast programing.

We intend to intiate a new proceeding to deal with these matters. But we see the need for tapping a larger body of expertise in order to develop more technical and economic information than is ordinarily available through the rulemaking

process. Therefore, we will also establish a task force of experts to advise us in specifically designated areas.

V. FEDERAL-STATE/LOCAL RELATIONSHIPS

171. In our notice of proposed rule making in Docket 18892 [75] we observed that "actions have been taken in the cable field without any overall plan as to the Federal-local relationship." This has resulted in a patchwork of disparate approaches affecting the development of cable television. While the Commission was pursuing a program to promote national cable policy, State and local governments were formulating policies to reflect local needs and desires. In many respects this dual approach worked well. To a growing extent, however, the rapid expansion of the cable television industry has led to overlapping and sometimes incompatible regulations. This resulted in confusion, and we faced an obvious need to clarify the respective Federal, State, and local regulatory roles. Three possible approaches were outlined in Docket 18892:

(a) Federal licensing of all cable television systems.
(b) Maintenance of the current Federal regulatory program enforced by section 312 (b) proceedings.
(c) Federal regulation of some aspects, with local regulation of others under Federal prescription of standards for local jurisdictions.

As we noted in Docket 18892:

This last approach recognizes that although practical considerations argue in favor of leaving important aspects of cable regulation to State and local government, cable is nonetheless an integral part of the inter-State movement of electronic communications. United States v. Southwestern Cable Co., 392 U.S. 157 (1968). In these circumstances, it is appropriate for this agency to establish uniform or minimum standards to which local actions must conform.

We requested comments on the form such "uniform or minimum standards" might take. The filings differed in their specific proposals for resolution of the questions raised in our notice, thus indicating the wide diversity of opinion in this complex area of regulation.

ANALYSIS OF COMMENTS

172. *Broadcast interests.* To varying degrees, most broadcast interests favored a regulatory approach involving a distribution of authority between local government and the Commission. Views on the extent to which the Commission should impose guidelines for State or local action varied considerably, however. For instance, Storer Broadcasting Co. suggested that the Commission establish guidelines for character qualifications of franchise applicants. Others argued that the Commission should not establish guidelines for any aspect of the franchising process. Some favored Commission guidance for the regulation of subscriber rates, while most urged that this element of regulation might better be left to local authorities. The National Association of Broadcasters proposed

[75] 22 FCC 2d 50 (1970).

that the Commission impose minimum standards in most aspects of regulation, allowing local governments to impose additional requirements not inconsistent with the Federal standards. American Broadcasting Co., which atypically argued in favor of Federal licensing, still agreed that such matters as franchising and subscriber rates be left to local control. In general, broadcast interests did not favor the proposed 2-percent limitation on franchise fees, arguing that the Commission had provided no adequate basis for such a limitation. Westinghouse Broadcasting Co. thought the 2-percent figure acceptable as a starting point but would have permitted adjustment upward on appropriate showing. Those opposing our 2-percent proposal ventured no alternative figure, but most agreed that whatever the fee, it should be no more than is necessary to finance a local regulatory program.

173. *Cable television interests.* These parties uniformly were of the view that the present three-tiered regulatory approach is unsatisfactory. Pointing to the confusion and waste caused by such an approach, and arguing that on many issues local and State governments lack the expertise to oversee cable's development, they all favored some degree of Federal preemption. The National Cable Television Association urged that the Commission entirely preempt this field and limit local involvement to the selection among franchise applicants. Other groups such as Community Tele-Communications, Inc. agreed with this position and simply asserted that Federal licensing would be best. Recognizing, however, that such an approach might be burdensome, they supported a more flexible course whereby the Commission, as suggested in our third alternative, would preempt some areas, establish minimum standards for State and local authorities to follow in others, and leave purely local matters to the appropriate local entity. At the other end of the spectrum, Time-Life Broadcast, Inc., believed that Federal licensing would not be effective and suggested that the Commission adopt a dual-jurisdictional approach, establishing minimum guidelines for technical performance, legal and character qualifications and any other matters calling for regional or nationwide uniformity. Time-Life would leave the franchising function to local authorities who can best deal with such questions as local programming needs, compliance with local laws, rights of inspection, insurance, indemnity, performance bonds, grounds for revocation, property encumbrances, and the like.

174. Cable interests were clearly opposed to State regulation. They noted in particular that regulation by public utility commissions results in unconscionable delay and confusion. Their filings were also uniformly opposed to State rate regulation. Some acknowledged the need for some rate regulation but disagreed over whether this should more properly be a Federal or local function. The General Electric Co., for instance, maintained that rate regulation should be left to the local franchising entity be-

cause it can best gauge the requirements of its particular community. Comtel, on the other hand, argued for Federal standards for rate schedules. The National Cable Television Association said no rate regulations of any kind are needed. Most parties agreed that our proposed 2 percent franchise fee limitation was a reasonable point of departure. General Electric thought that the figure might be too low but that it was a matter best dealt with by the Commission. A joint filing by several multiple system operators called for the abolition of all franchise fees based on gross receipts, including those in existing franchise agreements. They argued that payment of anything more than reasonable regulatory costs would impede the growth of the industry. The National Cable Television Association favored a federally-established 2 percent maximum franchise fee.

175. *State and local governmental interests.* These interests unanimously opposed Federal preemption of cable regulation. It was maintained that the Commission, with its limited staff and uniform approach, cannot effectively regulate thousands of cable systems operating in communities across the country. Such regulation should be left to local governments which are responsible for the utilization of their physical facilities, familiar with local needs, and necessarily more responsive to community desires. Many local governments went further and argued that the Commission lacks the jurisdiction to regulate any local aspect of cable. Others, however, admitted that federally imposed technical standards would be desirable and some favored the establishment of minimum Federal guidelines, but only to the extent that local authority would not be diminished. The National Institute of Municipal Law Officers (NIMLO) urged a general approach similar to that suggested by the Commission: Federal regulation of some aspects of cable, plus local regulation of other aspects under prescribed Federal standards. State and local government interests uniformly opposed the 2 percent limitation on franchise fees. Although a few thought that some higher figure might be appropriate, most favored no Federal limitation at all.

176. *Other comments.* As in the case of cable and broadcast interests, most others recognized the need for some form of dual regulation, with the Commission issuing standards and guidelines for local franchising authorities to follow. They also acknowledged that, while Federal licensing might be the best solution, it is impractical and burdensome. The Ford Foundation stated its preference for nonprofit ownership, timely construction rules, reasonable duration of franchises, a requirement that construction extend to all areas within a franchise, the provision of local community program channels, and limitations on franchise fees. The Corporation for Public Broadcasting was concerned about the franchising process and urged that the Commission assure that adequate notice is given and all groups allowed to participate. Black Efforts for Soul in Television (BEST) supported the dual regulatory approach

and particularly noted the need for equal employment opportunities in this field. The American Civil Liberties Union urged that sufficient common carrier capacity for use at reasonable rates and terms be required but opposed Federal preemption, favoring regional and local experimentation instead. American Telephone and Telegraph Co. said that the Commission should regulate only the interstate aspects of cable. While this overview is not exhaustive, it does give a general picture of the diverse and helpful suggestions we have had available to us in this proceeding.

COMMISSION'S REGULATORY PROGRAM

177. *Dual jurisdiction.* The comments advance persuasive arguments against Federal licensing. We agree that conventional licensing would place an unmanageable burden on the Commission. Moreover, local governments are inescapably involved in the process because cable makes use of streets and ways and because local authorities are able to bring a special expertness to such matters, for example, as how best to parcel large urban areas into cable districts. Local authorities are also in better position to follow up on service complaints. Under the circumstances, a deliberately structured dualism is indicated; the industry seems uniquely suited to this kind of creative federalism. We are also persuaded that because of the limited resources of States and municipalities and our own obligation to insure an efficient communications service with adequate facilities at reasonable charges, we must set at least minimum standards for franchises issued by local authorities. These standards relate to such matters as the franchise selection process, construction deadlines, duration of the franchise, rates, and rate changes, the handling of service complaints, and the reasonableness of franchise fees. The standards will be administered in the certificating process.

178. *Franchising.* We are requiring that before a cable system commences operation with broadcast signals, it must obtain a certificate of compliance from the Commission. The application for such a certificate must contain (§ 76.31 (a)(1)) a copy of the franchise and a detailed statement showing that the franchising authority has considered in a public proceeding the system operator's legal, character, financial, technical, and other qualifications, and the adequacy and feasibility of construction arrangements. We expect that franchising authorities will publicly invite applications, that all applications will be placed on public file, that notice of such filings will be given, that where appropriate a public hearing will be held to afford all interested persons an opportunity to testify on the qualifications of the applicants, and that the franchising authority will issue a public report setting forth the basis for its action. Such public participation in the franchising process is necessary to assure that the needs and

desires of all segments of the community are carefully considered.

179. *Applicant qualifications.* We are authorizing the use of broadcast signals in order to obtain new benefits for the public. No such benefits will be forthcoming if the cable television applicant is not fully qualified to operate. The character of an applicant, for example, is of particular importance especially because he may be engaged in program origination. Some governmental body must insure that a franchise applicant's qualifications are consistent with the public interest, and we believe this matter is appropriate for local determination.

180. *Franchise area.* Another matter uniquely within the competence of local authorities is the delineation of franchise areas. We emphasize that provision must be made for cable service to develop equitably and reasonably in all parts of the community. A plan that would bring cable only to the more affluent parts of a city, ignoring the poorer areas, simply could not stand. No broadcast signals would be authorized under such circumstances. While it is obvious that a franchisee cannot build everywhere at once within a designated franchise area, provision must be made that he develop service reasonably and equitably. There are a variety of ways to divide up communities; the matter is one for local judgment.

181. *Construction.* We are establishing in § 76.31(a)(2) general timetables for construction and operation of systems to insure that franchises do not lie fallow or become the subject of trafficking. Specifically, we are providing that the franchise require the cable system to accomplish significant construction within 1 year after the certificate of compliance is issued, and that thereafter energized trunk cable be extended to a substantial percentage of the franchise area each year, the percentage to be determined by the franchising authority. As a general proposition, we believe that energized trunk cable should be extended to at least 20 percent of the franchise area per year, with the extension to begin within 1 year after the Commission issues its certificate of compliance. But we have not established 20 percent as an inflexible figure, recognizing that local circumstances may vary.[76]

182. *Franchise duration.* We are requiring in § 76.31(a)(3) that franchising authorities place reasonable limits on the duration of franchises. Long terms have generally been found unsatisfactory by

State and local regulatory authorities,[77] and are an invitation to obsolescence in light of the momentum of cable technology.[78] We believe that in most cases a franchise should not exceed 15 years and that renewal periods be of reasonable duration. We recognize that decisions of local franchising authorities may vary in particular circumstances. For instance, an applicant's proposal to wire inner-city areas without charge or at reduced rates might well call for a longer franchise. On the other hand, we note that there is some support for franchise periods of less than 15 years.[79]

183. *Subscriber rates.* In § 76.31(a)(4) we are permitting local authorities to regulate rates for services regularly furnished to all subscribers. The appropriate standard here is the maintenance of rates that are fair to the system and to the subscribing public—a matter that will turn on the facts of each particular case (after appropriate public proceedings affording due process) and the accumulated experience of other cable communities.

184. *Service complaints.* Section 76.31 (a)(5) requires that franchises provide for the investigation and resolution of local service complaints and also that the franchisee maintain a local business office or agent for these purposes. We note that some local bodies are already considering detailed plans along these general lines.

185. *Franchise fee.* While we have decided against adopting a 2 percent limitation on franchise fees, we believe some provision is necessary to insure reasonableness in this respect. First, many local authorities appear to have exacted high franchise fees more for revenue-raising than for regulatory purposes. Most fees are about 5 or 6 percent, but some have been known to run as high as 36 percent. The ultimate effect of any revenue-raising fee is to levy an indirect and regressive tax on cable subscribers. Second, and of great importance to the Commission, high local franchise fees may burden cable television to the extent that it will be unable to carry out its part in our na-

---

[76] Some municipalities may require expansion at a greater rate. The New York City contract, for example, requires that a cable television franchisee extend trunk cable to its whole franchise area within 4 years from the grant of the franchise. This 4-year period represents an increase from the 2- to 3-year period originally recommended by the Mayor's Advisory Task Force on CATV and Telecommunications, Report on Cable Television and Cable Communications in New York City (1968). Similar limitations appear to have been imposed throughout most of New York State. W. Jones, Regulation of Cable Television by the State of New York 134–35 (1970).

[77] E. Clemens, Economics and Public Utilities (N.Y.C.: Appleton-Century-Crofts, 1950) 75–76.

[78] R. Posner, Cable Television: The Problem of Local Monopoly 22–23 (1970), prepared for the Ford Foundation, Memorandum RM-6309-FF.

[79] At one extreme, two commentators have proposed 3-year franchise periods. R. Posner, id. at 26; Better Broadcasting Council, A Model Ordinance for Cable Television for the City of Chicago S2.16 (1970). An Illinois bill would have restricted franchises to 5 years. Illinois General Assembly, S. 169, § 6 (1971). And although the franchises ultimately granted by New York City were for 20 years, 10 years had been initially recommended, and the experimental initial grant was only 2 years. Mayor's Advisory Task Force on CATV and Telecommunications, Report on Cable Television and Cable Communications in New York City (1968); Bureau of Franchises, Report to the Board of Estimate Relating to Community Antenna Television and to the Petitions of Eight Applicants for the Consent of the City of New York to Install and Operate CATV Systems (1965).

tional communications policy.[80] Finally, cable systems are subject to substantial obligations under our new rules and may soon be subject to congressionally-imposed copyright payments. We are seeking to strike a balance that permits the achievement of Federal goals and at the same time allows adequate revenues to defray the costs of local regulation.

186. The Commission imposes an annual fee of 30 cents per subscriber to help finance its own cable regulatory program. Assuming average annual revenues to the cable system of $60 per subscriber, the Commission's fee amounts to one half of 1 percent of a system's gross receipts. The regulatory program to be carried out by local entities is different in scope and may vary from jurisdiction to jurisdiction. It is our judgment that maximum franchise fees should be between 3 and 5 percent of gross subscriber revenues. But we believe it more appropriate to specify this percentage range as a general standard, for specific local application. When the fee is in excess of 3 percent (including all forms of consideration, such as initial lump sum payments), the franchising authority is required to submit a showing that the specified fee is appropriate in light of the planned local regulatory program, and the franchisee must demonstrate that the fee will not interfere with its ability to meet the obligations imposed by our rules.

187. *Grandfathering.* The grandfathering provisions of our rules with respect to franchise standards seek to achieve a large measure of flexibility. An existing cable system will be required to certify within 5 years of the effective date of these rules or on renewal of its franchise, whichever comes first, that its franchise meets the requirements of the rules. This deferral should relieve both cable systems and local authorities of whatever minor dislocations our rules might otherwise cause.

188. *Advisory Committee.* We believe that we have provided a useful framework for the proper allocation of responsibility among the various levels of government. But much remains to be done as the industry evolves and experience accumulates. Recognizing that the rules are complex and break new ground, we are prepared to provide assistance, through our Cable Television Bureau, to all State and local governments requesting aid. We also intend to issue an explanatory handbook on cable television regulations. Further, because we expect significant development in cable television as a result of our action today, the Commission will seek the advice of a special committee composed of representatives of Federal, State, and local

governments, the cable industry, and public interest groups. This committee will aid the Commission as it attempts to define an appropriate allocation of responsibilities in cable regulation.

VI. CONCLUSION

189. Cable television is an emerging technology that promises a communications revolution. Inevitably, our regulatory pattern must evolve as cable evolves—and no one can say what the precise dimensions will be. This report and order represents the amount and the substance of regulation that we believe is essential, at this stage, for the orderly development of the industry. We have taken long overdue first steps after more than 3 years of exhaustive inquiry.

190. The rules will be effective March 31, 1972. Out of an abundance of caution, we are delaying the date beyond the 30 days ordinarily required so that we may have before us any petitions for reconsideration prior to the rules becoming operative. But for more than 3 years we have been gathering data, soliciting views, hearing argument, evaluating studies, examining alternatives, authorizing experiments—turning finally to public panel discussions unique in communications rule making—and, in this effort, have necessarily postponed the substantial public benefits that cable promises. In these circumstances, we do not foresee that there can be any case for further delay.

191. Authority for adoption of these rules is contained in sections 2, 3, 4 (i), and (j), 301, 303, 307, 308, and 309 of the Communications Act of 1934, as amended. We reaffirm our view that cable systems are neither broadcasters nor common carriers within the meaning of the Communications Act. Rather, cable is a hybrid that requires identification and regulation as a separate force in communications.

Accordingly, it is ordered, That effective March 31, 1972, Parts 1, 15, 21, 74, and 91 of the Commission's rules and regulations are amended as set forth in Appendix A below, and that new Parts 76 (Cable Television Service) and 78 (Cable Television Relay Service) of the Commission's rules and regulations are added as set forth in Appendix A.

It is further ordered, That the proceedings in Dockets 18397, 18397–A, 18373, 18892, and 18894 are terminated.

(Secs. 2, 3, 4, 301, 303, 307, 308, 309, 48 Stat., as amended, 1064, 1065, 1066, 1081, 1082, 1083, 1084, 1085; 47 U.S.C. 152, 153, 154, 301, 303, 307, 308, 309)

Adopted: February 2, 1972.

Released: February 3, 1972.

FEDERAL COMMUNICATIONS
COMMISSION,[81]
[SEAL]   BEN F. WAPLE,
Secretary.

APPENDIX A

Chapter 1 of Title 47 of the Code of Federal Regulations is amended as follows:

PART 1—PRACTICE AND PROCEDURE

1. In § 1.1116, the headnote and paragraphs (a) and (c) are revised to read as follows:

§ 1.1116 **Schedule of fees for Cable Television and Cable Television Relay Services.**

(a) Applications and petitions filed in the Cable Television and Cable Television Relay Services shall be accompanied by the fees prescribed below:

Applications in the Cable Television Relay (CAR) Service:	
For a construction permit	$50
For a license or renewal	15
For a modification of construction permit or license	15
Applications for certificates of compliance, pursuant to § 76.11	35

NOTE: If multiple applications for certificate of compliance are filed by cable television systems having a common headend and identical ownership but serving or proposing to serve more than one community, the full $35 fee will be required only for one of the communities; $10 will be required for each of the other communities.

Petitions for special relief, pursuant to § 76.7	25

*   *   *   *   *

(c) Fees are not required in the following instances:

(1) Petition for special relief filed pursuant to § 76.7 of this chapter by a noncommercial educational broadcast station.

PART 15—RADIO FREQUENCY DEVICES

§ 15.4 [Amended]

1. In § 15.4, paragraph (e) is deleted.

§§ 15.161–15.165 [Deleted]

2. Subpart D of Part 15, §§ 15.161–15.165, is deleted.

PART 21—DOMESTIC PUBLIC RADIO SERVICES (OTHER THAN MARITIME MOBILE)

1. In § 21.713, the headnote and text are revised to read as follows:

§ 21.713 **Applications for authorizations involving relay of television signals to cable television systems.**

An application in this service for authorization to establish new facilities or to modify existing facilities to be used to relay television signals to cable television systems shall contain a statement by the applicant that, to the best of his knowledge, each cable television system to be served has, on or before the filing date of the application, filed any necessary application for certificate of compliance, pursuant to §§ 76.11 and 76.13 of this chapter. Such statement by the applicant shall identify the application for certificate of compliance by the name of the cable television system for which

---

[80] We have from time to time stated our concern with the threat of other inhibiting factors. Cable television is also involved, for example, in a dispute over utility pole attachment rates and faces the burdening claims of the telephone and electric power industries that rental charges be increased. We are currently inquiring into pole rental practices (Docket 16928) and expect to address the question of what regulatory controls may appropriately be invoked.

[81] Commissioners Burch, chairman; Bartley, Reid and Wiley concurring and issuing statements to be released at a later date; Commissioner Robert E. Lee dissenting and issuing a statement to be released at a later date; Commissioner Johnson concurring in part and dissenting in part and issuing a statement filed as part of the original document; Commissioner H. Rex Lee absent.

the certificate is sought, the community and area served or to be served, the date on which the application was filed, and the file number (if available).

PART 74—EXPERIMENTAL, AUXIL-IARY, AND SPECIAL BROADCAST, AND OTHER PROGRAM DISTRIBU-TIONAL SERVICES

§§ 74.1001–74.1083 [Deleted]

1. Subpart J of Part 74, §§ 74.1001–74.1083, is deleted.

§§ 74.1101–74.1131 [Deleted]

2. Subpart K of Part 74, §§ 74.1101–74.1131, is deleted.

E. Part 76—Cable Television Service is added to read as follows:

PART 76—CABLE TELEVISION SERVICE

AUTHORITY: The provisions of this Part 76 issued under secs. 2, 3, 4, 301, 303, 307, 308, 309, 48 Stat., as amended, 1064, 1065, 1066, 1081, 1082, 1083, 1084, 1085; 47 U.S.C. 152, 153, 154, 301, 303, 307, 308, 309.

Subpart A—General

§ 76.1 Purpose.

The rules and regulations set forth in this part provide for the certification of cable television systems and for their operation in conformity with standards for carriage of television broadcast signals, program exclusivity, cablecasting, access channels, and related matters.

§ 76.3 Other pertinent rules.

Other pertinent provisions of the Commission's rules and regulations relating to the Cable Television Service are included in the following parts of this chapter:

Part 0—Commission Organization.
Part 1—Practice and Procedure.
Part 21—Domestic Public Radio Services (Other Than Maritime Mobile).
Part 63—Extension of Lines and Discontinuance of Service by Carriers.
Part 78—Cable Television Relay Service.
Part 91—Industrial Radio Services.

§ 76.5 Definitions.

(a) *Cable television system (or CATV system).* Any facility that, in whole or in part, receives directly, or indirectly over the air, and amplifies or otherwise modifies the signals transmitting programs broadcast by one or more television or radio stations and distributes such signals by wire or cable to subscribing members of the public who pay for such service, but such term shall not include (1) any such facility that serves fewer than 50 subscribers, or (2) any such facility that serves only the residents of one or more apartment dwellings under common ownership, control, or management, and commercial establishments located

on the premises of such an apartment house.

NOTE: In general, each separate and distinct community or municipal entity (including single, discrete, unincorporated areas) served by cable television facilities constitutes a separate cable television system, even if there is a single headend and identical ownership of facilities extending into several communities. See, e.g., Telerama, Inc., 3 FCC 2d 585 (1966); Mission Cable TV, Inc., 4 FCC 2d 236 (1966).

(b) *Television station; television broadcast station.* Any television broadcast station operating on a channel regularly assigned to its community by § 73.606 of this chapter, and any television broadcast station licensed by a foreign government: *Provided, however,* That a television broadcast station licensed by a foreign government shall not be entitled to assert a claim to carriage or program exclusivity, pursuant to Subpart D or F of this part, but may otherwise be carried if consistent with the rules.

(c) *Television translator station.* A television broadcast translator station as defined in § 74.701 of this chapter.

(d) *Principal community contour.* The signal contour that a television station is required to place over its entire principal community by § 73.685(a) of this chapter.

(e) *Grade A and Grade B contours.* The field intensity contours defined in § 73.683(a) of this chapter.

(f) *Specified zone of a television broadcast station.* The area extending 35 air miles from the reference point in the community to which that station is licensed or authorized by the Commission. A list of reference points is contained in § 76.53. A television broadcast station that is authorized but not operating has a specified zone that terminates eighteen (18) months after the initial grant of its construction permit.

(g) *Major television market.* The specified zone of a commercial television station licensed to a community listed in § 76.51, or a combination of such specified zones where more than one community is listed.

(h) *Designated community in a major television market.* A community listed in § 76.51.

(i) *Smaller television market.* The specified zone of a commercial television station licensed to a community that is not listed in § 76.51.

(j) *Substantially duplicated.* Regularly duplicated by the network programing of one or more stations in a week during the hours of 6 to 11 p.m., local time, for a total of 14 or more hours.

(k) *Significantly viewed.* Viewed in other than cable television households as follows: (1) For a full or partial network station—a share of viewing hours of at least 3 percent (total week hours), and a net weekly circulation of at least 25 percent; and (2) for an independent station—a share of viewing hours of at least 2 percent (total week hours), and a net weekly circulation of at least 5 percent. See § 76.54.

NOTE: As used in this paragraph, "share of viewing hours" means the total hours that noncable television households viewed the subject station during the week, expressed as a percentage of the total hours these households viewed all stations during the period, and "net weekly circulation" means the number of noncable television households that viewed the station for 5 minutes or more during the entire week, expressed as a percentage of the total noncable television households in the survey area.

(l) *Full network station.* A commercial television broadcast station that generally carries in weekly prime time hours 85 percent of the hours of programing offered by one of the three major national television networks with which it has a primary affiliation (i.e., right of first refusal or first call).

(m) *Partial network station.* A commercial television broadcast station that generally carries in prime time more than 10 hours of programing per week offered by the three major national television networks, but less than the amount specified in paragraph (l) of this section.

(n) *Independent station.* A commercial television broadcast station that generally carries in prime time not more than 10 hours of programing per week offered by the three major national television networks.

(o) *Network programing.* The programing supplied by a national or regional television network, commercial or noncommercial.

(p) *Syndicated program.* Any program sold, licensed, distributed, or offered to television station licensees in more than one market within the United States for noninterconnected (i.e., nonnetwork) television broadcast exhibition, but not including live presentations.

(q) *Series.* A group of two or more works which are centered around, and dominated by the same individual, or which have the same, or substantially the same, cast of principal characters or a continuous theme or plot.

(r) *Off-network series.* A series whose episodes have had a national network television exhibition in the United States or a regional network exhibition in the relevant market.

(s) *First-run series.* A series whose episodes have had no national network television exhibition in the United States and no regional network exhibition in the relevant market.

(t) *First-run nonseries programs.* Programs, other than series, that have had no national network television exhibition in the United States and no regional network exhibition in the relevant market.

(u) *Prime time.* The 5-hour period from 6 to 11 p.m., local time, except that in the central time zone the relevant period shall be between the hours of 5 and 10 p.m., and in the mountain time zone each station shall elect whether the period shall be 6 to 11 p.m. or 5 to 10 p.m.

NOTE: Unless the Commission is notified to the contrary, a station in the mountain time zone shall be presumed to have elected the 6 to 11 p.m. period.

(v) *Cablecasting.* Programing (exclusive of broadcast signals) carried on a cable television system. See paragraphs (aa), (bb), and (cc) (Classes II, III, and IV cable television channels) of this section.

(w) *Origination cablecasting.* Programing (exclusive of broadcast signals) carried on a cable television system over one or more channels and subject to the exclusive control of the cable operator.

(x) *Access cablecasting.* Services provided by a cable television system on its public, educational, local government, or leased channels.

(y) *Legally qualified candidate.* Any person who has publicly announced that he is a candidate for nomination by a convention of a political party or for nomination or election in a primary, special, or general election, municipal, county, State, or national, and who meets the qualifications prescribed by the applicable laws to hold the office for which he is a candidate, so that he may be voted for by the electorate directly or by means of delegates or electors, and who:

(1) Has qualified for a place on the ballot, or

(2) Is eligible under the applicable law to be voted for by sticker, by writing his name on the ballot, or other method, and (i) has been duly nominated by a political party which is commonly known and regarded as such, or (ii) makes a substantial showing that he is a bona fide candidate for nomination or office.

(z) *Class I cable television channel.* A signaling path provided by a cable television system to relay to subscriber terminals television broadcast programs that are received off-the-air or are obtained by microwave or by direct connection to a television broadcast station.

(aa) *Class II cable television channel.* A signaling path provided by a cable television system to deliver to subscriber terminals television signals that are intended for reception by a television broadcast receiver without the use of an auxiliary decoding device and which signals are not involved in a broadcast transmission path.

(bb) *Class III cable television channel.* A signaling path provided by a cable television system to deliver to subscriber terminals signals that are intended for reception by equipment other than a television broadcast receiver or by a television broadcast receiver only when used with auxiliary decoding equipment.

(cc) *Class IV cable television channel.* A signaling path provided by a cable television system to transmit signals of any type from a subscriber terminal to another point in the cable television system.

(dd) *Channel frequency response.* The relationship within a cable television channel between amplitude and frequency of a constant amplitude input signal as measured as a subscriber terminal.

(ee) *Subscriber terminal.* The cable television system terminal to which a subscriber's equipment is connected. Separate terminals may be provided for delivery of signals of various classes.

(ff) *System noise.* That combination of undesired and fluctuating disturbances within a cable television channel that degrades the transmission of the desired signal and that is due to modulation processes or thermal or other noise-producing effects, but does not include hum and other undesired signals of discrete frequency. System noise is specified in terms of its rms voltage or its mean power level as measured in the 4 MHz bandwidth between 1.25 and 5.25 MHz above the lower channel boundary of a cable television channel.

(gg) *Terminal isolation.* The attenuation, at any subscriber terminal, between that terminal and any other subscriber terminal in the cable television system.

(hh) *Visual signal level.* The rms voltage produced by the visual signal during the transmission of synchronizing pulses.

§ 76.7 Special relief.

(a) Upon petition by a cable television system, an applicant, permittee, or licensee of a television broadcast, translator, or microwave relay station, or by any other interested person, the Commission may waive any provision of the rules relating to cable television systems, impose additional or different requirements, or issue a ruling on a complaint or disputed question.

(b) The petition may be submitted informally, by letter, but shall be accompanied by an affidavit of service on any cable television systems, station licensee, permittee, applicant, or other interested person who may be directly affected if the relief requested in the petition should be granted.

(c) (1) The petition shall state the relief requested and may contain alternative requests. It shall state fully and precisely all pertinent facts and considerations relied on to demonstrate the need for the relief requested and to support a determination that a grant of such relief would serve the public interest. Factual allegations shall be supported by affidavit of a person or persons with actual knowledge of the facts, and exhibits shall be verified by the person who prepares them.

(2) A petition for a ruling on a complaint or disputed question shall set forth all steps taken by the parties to resolve the problem, except where the only relief sought is a clarification or interpretation of the rules.

(d) Interested persons may submit comments or opposition to the petition within thirty (30) days after it has been filed. For good cause shown in the petition, the Commission may, by letter or telegram to known interested persons, specify a shorter time for such submissions. Comments or oppositions shall be served on petitioner and on all persons listed in petitioner's affidavit of service, and shall contain a detailed full showing, supported by affidavit, of any facts or considerations relied on.

(e) The petitioner may file a reply to the comments or oppositions within twenty (20) days after their submission,

which shall be served on all persons who have filed pleadings and shall also contain a detailed full showing, supported by affidavit, of any additional facts or considerations relied on. For good cause shown, the Commission may specify a shorter time for the filing of reply comments.

(f) The Commission, after consideration of the pleadings, may determine whether the public interest would be served by the grant, in whole or in part, or denial of the request, or may issue a ruling on the complaint or dispute. The Commission may specify other procedures, such as oral argument, evidentiary hearing, or further written submissions directed to particular aspects, as it deems appropriate. In the event that an evidentiary hearing is required, the Commission will determine, on the basis of the pleadings and such other procedures as it may specify, whether temporary relief should be afforded any party pending the hearing and the nature of any such temporary relief.

(g) Where a petition for waiver of the provisions of §§ 76.57(a), 76.59(a), 76.61 (a), or 76.63(a), is filed within fifteen (15) days after a request for carriage, a cable television system need not carry the signal of the requesting station pending the Commission's ruling on the petition or on the question of temporary relief pending further proceedings.

Subpart B—Applications and Certificates of Compliance

§ 76.11 Certificate of compliance required.

(a) No cable television system shall commence operations or add a television broadcast signal to existing operations unless it receives a certificate of compliance from the Commission.

(b) No cable television system lawfully carrying television broadcast signals in a community prior to March 31, 1972, shall continue carriage of such signals beyond the end of its current franchise period, or March 31, 1977, whichever occurs first, unless it receives a certificate of compliance.

(c) A cable television system to which paragraph (b) of this section applies may continue to carry television broadcast signals after expiration of the period specified therein, if an application for certificate is filed at least thirty (30) days prior to the date on which a certificate would otherwise be required and the Commission has not acted on the application.

§ 76.13 Filing of applications.

No standard form is prescribed in connection with the filing of an application for a certificate of compliance; however, three (3) copies of the following information must be provided:

(a) For a cable television system not operational prior to March 31, 1972 (other than systems that were authorized to carry one or more television signals prior to March 31, 1972, but did not commence such carriage prior to that date), an application for certificate of compliance shall include:

(1) The name and mailing address of the operator of the proposed system, community and area to be served, television signals to be carried (other than those permitted to be carried pursuant to § 76.61(b)(2)(ii) or § 76.63(a) (as it related to § 76.61(b)(2)(ii)), proposed date on which cable operations will commence, and, if applicable, a statement that microwave radio facilities are to be used to relay one or more signals;

(2) A copy of FCC Form 325 "Annual Report of Cable Television Systems," supplying all applicable information;

(3) A copy of the franchise, license, permit, or certificate granted to construct and operate a cable television system;

(4) A statement that explains how the proposed system's franchise and its plans for availability and administration of access channels and other nonbroadcast cable services are consistent with the provisions of §§ 76.31 and 76.251;

(5) A statement that explains, in terms of the provisions of Subpart D of this part, how carriage of the proposed television signals is consistent with those provisions, including any special showings as to whether a signal is significantly viewed (see § 76.54(b));

(6) An affidavit of service of the information described in subparagraph (1) of this paragraph on the licensee or permittee of any television broadcast station within whose predicted Grade B contour or 35-mile zone the system will operate; the licensee or permittee of any 100-watt or higher power television translator station licensed to the community of the system, the franchising authority, the superintendent of schools in the community of the system, and any local or State educational television authorities;

(7) A statement that the filing fee prescribed in § 1.1116 of this chapter is attached.

(b) For a cable television system that was authorized to carry one or more television signals prior to March 31, 1972, but did not commence such carriage prior to that date, an application for certificate of compliance shall include:

(1) The name and mailing address of the system, community and area served or to be served, television signals authorized to be carried but not carried prior to March 31, 1972, and, if applicable, a statement that microwave relay facilities are to be used to relay one or more signals;

(2) A list of all television signals already being carried;

(3) A statement that explains how the system's plans for availability and administration of access channels and other nonbroadcast cable services are consistent with the provisions of § 76.251.

NOTE: The provisions of this subparagraph are applicable only to systems located in a community that is wholly or partially within a major television market.

(4) An affidavit of service of the information described in subparagraph (1) of this paragraph on the parties named in paragraph (a)(6) of this section;

(5) A statement that the filing fee prescribed in § 1.1116 of this chapter is attached.

(c) For a cable television system proposing to add a television signal to existing operations, an application for certificate of compliance shall include:

(1) The name and mailing address of the system, community and area served, television signals to be added (other than those permitted to be carried pursuant to § 76.61(b)(2)(ii) or § 76.63(a) (as it relates to § 76.61(b)(2)(ii)), and, if applicable, a statement that microwave relay facilities are to be used to relay one or more signals;

(2) A list of all television signals already being carried;

(3) A statement that explains, in terms of the provisions of Subpart D of this part, how carriage of the proposed television signals is consistent with those provisions, including any special showings on the question whether a signal is significantly viewed (see § 76.54(b));

(4) A statement that explains how the system's plans for availability and administration of access channels and other nonbroadcast cable services are consistent with the provisions of § 76.251;

NOTE: The provisions of this subparagraph are applicable only to systems operating in a community located in whole or in part within a major television market.

(5) An affidavit of service of the information described in subparagraph (1) of this paragraph on the parties named in paragraph (a)(6) of this section;

(6) A statement that the filing fee prescribed in § 1.1116 of this chapter is attached.

(d) For a cable television system seeking certification of existing operations in accordance with § 76.11(b), an application for certificate of compliance shall include:

(1) The name and mailing address of the system, community and area served, television signals being carried (other than those permitted to be carried pursuant to § 76.61(b)(2)(ii) or § 76.63(a) (as it relates to § 76.61(b)(2)(ii))), date on which operations commenced, and date on which its current franchise expires;

(2) A statement that explains how the franchise under which the system will operate upon Commission certification is consistent with the franchise standards specified in § 76.31;

(3) An affidavit of service of the information described in subparagraph (1) of this paragraph on the parties named in paragraph (a)(6) of this section;

(4) A statement that the filing fee prescribed by § 1.1116 of this chapter is attached.

NOTE: As used in § 76.13, the term "predicted Grade B contour" means the field intensity contour defined in § 73.683(a) of this chapter, the location of which is determined exclusively by means of the calculations prescribed in § 73.684 of this chapter.

§ 76.15 Public notice.

The Commission will give public notice of the filing of applications for certificates of compliance. A certificate will

not be issued sooner than thirty (30) days from the date of public notice.

§ 76.17 Objections to applications; related matters.

A petition challenging the service proposed in an application for certificate of compliance shall be filed within thirty (30) days of the public notice described in § 76.15. The procedures specified in § 76.7 shall be applicable to such petitions and to oppositions and replies. Controversies concerning carriage (Subpart D) and program exclusivity (§ 76.91) will be acted on in connection with the certificating process if raised within thirty (30) days of the public notice; any other objection will be treated as a petition for special relief filed pursuant to § 76.7.

Subpart C—Federal-State/Local Regulatory Relationships

§ 76.31 Franchise standards.

(a) In order to obtain a certificate of compliance, a proposed or existing cable television system shall have a franchise or other appropriate authorization that contains recitations and provisions consistent with the following requirements:

(1) The franchisee's legal, character, financial, technical, and other qualifications, and the adequacy and feasibility of its construction arrangements, have been approved by the franchising authority as part of a full public proceeding affording due process;

(2) The franchisee shall accomplish significant construction within one (1) year after receiving Commission certification, and shall thereafter equitably and reasonably extend energized trunk cable to a substantial percentage of its franchise area each year, such percentage to be determined by the franchising authority;

(3) The initial franchise period and any renewal franchise period shall be of reasonable duration;

(4) The franchising authority has specified or approved the initial rates which the franchisee charges subscribers for installation of equipment and regular subscriber services. No changes in rates charged to subscribers shall be made except as authorized by the franchising authority after an appropriate public proceeding affording due process;

(5) The franchise shall specify procedures for the investigation and resolution of all complaints regarding the quality of service, equipment malfunctions, and similar matters, and shall require that the franchisee maintain a local business office or agent for these purposes;

(6) Any modifications of the provisions of this section resulting from amendment by the Commission shall be incorporated into the franchise within one (1) year of adoption of the modification, or at the time of franchise renewal, whichever occurs first: *Provided, however,* That, in an application for certificate of compliance, consistency with these requirements shall not be expected

of a cable television system that was in operation prior to March 31, 1972, until the end of its current franchise period, or March 31, 1977, whichever occurs first.

(b) The franchise fee shall be reasonable (e.g., in the range of 3–5 percent of the franchisee's gross subscriber revenues per year from cable television operations in the community (including all forms of consideration, such as initial lump sum payments)). If the franchise fee exceeds 3 percent of such revenues, the cable television system shall not receive Commission certification until the reasonableness of the fee is approved by the Commission on showings, by the franchisee, that it will not interfere with the effectuation of Federal regulatory goals in the field of cable television, and, by the franchising authority, that it is appropriate in light of the planned local regulatory program. The provisions of this paragraph shall not be effective with respect to a cable television system that was in operation prior to March 31, 1972, until the end of its current franchise period, or March 31, 1977, whichever occurs first.

Subpart D—Carriage of Television Broadcast Signals

§ 76.51 Major television markets.

For purposes of the cable television rules, the following is a list of the major television markets and their designated communities:

(a) First 50 major television markets:

(1) New York, N.Y.—Linden-Paterson, N.J.
(2) Los Angeles-San Bernardino-Corona-Fontana, Calif.
(3) Chicago, Ill.
(4) Philadelphia, Pa.-Burlington, N.J.
(5) Detroit, Mich.
(6) Boston-Cambridge-Worcester, Mass.
(7) San Francisco-Oakland-San Jose, Calif.
(8) Cleveland-Lorain-Akron, Ohio.
(9) Washington, D.C.
(10) Pittsburgh, Pa.
(11) St. Louis, Mo.
(12) Dallas-Fort Worth, Tex.
(13) Minneapolis-St. Paul, Minn.
(14) Baltimore, Md.
(15) Houston, Tex.
(16) Indianapolis-Bloomington, Ind.
(17) Cincinnati, Ohio-Newport, Ky.
(18) Atlanta, Ga.
(19) Hartford-New Haven-New Britain-Waterbury, Conn.
(20) Seattle-Tacoma, Wash.
(21) Miami, Fla.
(22) Kansas City, Mo.
(23) Milwaukee, Wis.
(24) Buffalo, N.Y.
(25) Sacramento-Stockton-Modesto, Calif.
(26) Memphis, Tenn.
(27) Columbus, Ohio.
(28) Tampa-St. Petersburg, Fla.
(29) Portland, Oreg.
(30) Nashville, Tenn.
(31) New Orleans, La.
(32) Denver, Colo.
(33) Providence, R.I.-New Bedford, Mass.
(34) Albany-Schenectady-Troy, N.Y.
(35) Syracuse, N.Y.
(36) Charleston-Huntington, W. Va.
(37) Kalamazoo-Grand Rapids-Muskegon-Battle Creek, Mich.
(38) Louisville, Ky.
(39) Oklahoma City, Okla.

(40) Birmingham, Ala.
(41) Dayton-Kettering, Ohio.
(42) Charlotte, N.C.
(43) Phoenix-Mesa, Ariz.
(44) Norfolk-Newport News-Portsmouth-Hampton, Va.
(45) San Antonio, Tex.
(46) Greenville - Spartanburg - Anderson, S.C.-Asheville, N.C.
(47) Greensboro-High Point-Winston Salem, N.C.
(48) Salt Lake City, Utah.
(49) Wilkes Barre-Scranton, Pa.
(50) Little Rock, Ark.

(b) Second 50 major television markets:

(51) San Diego, Calif.
(52) Toledo, Ohio.
(53) Omaha, Nebr.
(54) Tulsa, Okla.
(55) Orlando-Daytona Beach, Fla.
(56) Rochester, N.Y.
(57) Harrisburg-Lebanon-Lancaster-York, Pa.
(58) Texarkana, Tex.-Shreveport, La.
(59) Mobile, Ala.-Pensacola, Fla.
(60) Davenport, Iowa-Rock Island-Moline, Ill.
(61) Flint-Bay City-Saginaw, Mich.
(62) Green Bay, Wis.
(63) Richmond-Petersburg, Va.
(64) Springfield - Decatur - Champaign-Jacksonville, Ill.
(65) Cedar Rapids-Waterloo, Iowa.
(66) Des Moines-Ames, Iowa.
(67) Wichita-Hutchinson, Kans.
(68) Jacksonville, Fla.
(69) Cape Girardeau, Mo.-Paducah, Ky.-Harrisburg, Ill.
(70) Roanoke-Lynchburg, Va.
(71) Knoxville, Tenn.
(72) Fresno, Calif.
(73) Raleigh-Durham, N.C.
(74) Johnstown-Altoona, Pa.
(75) Portland-Poland Spring, Maine.
(76) Spokane, Wash.
(77) Jackson, Miss.
(78) Chattanooga, Tenn.
(79) Youngstown, Ohio.
(80) South Bend-Elkhart, Ind.
(81) Albuquerque, N. Mex.
(82) Fort Wayne-Roanoke, Ind.
(83) Peoria, Ill.
(84) Greenville - Washington - New Bern, N.C.
(85) Sioux Falls-Mitchell, S. Dak.
(86) Evansville, Ind.
(87) Baton Rouge, La.
(88) Beaumont-Port Arthur, Tex.
(89) Duluth-Superior, Minn.
(90) Wheeling, W. Va.-Steubenville, Ohio.
(91) Lincoln-Hastings-Kearney, Nebr.
(92) Lansing-Onondaga, Mich.
(93) Madison, Wis.
(94) Columbus, Ga.
(95) Amarillo, Tex.
(96) Huntsville-Decatur, Ala.
(97) Rockford-Freeport, Ill.
(98) Fargo-Grand Forks-Valley City, N. Dak.
(99) Monroe, La.-El Dorado, Ark.
(100) Columbia, S.C.

§ 76.53 Reference points.

To determine the boundaries of the major and smaller television markets (defined in § 76.5), the following list of reference points for communities having licensed television broadcast stations and/or outstanding construction permits shall be used. Where a community's reference point is not given, the geographic coordinates of the main post office in the community shall be used.

State and community	Latitude ° ' "	Longitude ° ' "
Alabama:		
Anniston	33 30 40	87 40 47
Birmingham	33 31 01	86 48 36
Decatur	34 36 35	86 58 45
Demopolis	32 30 56	87 50 07
Dothan	31 13 27	85 23 35
Dozier	31 29 30	86 21 59
Florence	34 48 05	87 40 31
Huntsville	34 44 18	86 35 19
Louisville	31 47 00	85 33 09
Mobile	30 41 36	88 02 33
Montgomery	32 22 33	86 18 31
Mount Cheaha State Park	32 29 06	85 48 30
Selma	24 24 26	87 01 15
Tuscaloosa	33 12 05	87 33 44
Alaska:		
Anchorage	61 13 09	149 53 29
College	64 51 22	147 48 38
Fairbanks	64 50 35	147 41 31
Juneau	58 18 06	134 25 09
Sitka	57 02 58	135 20 12
Arizona:		
Flagstaff	35 11 54	111 39 02
Mesa	33 24 54	f11 49 41
Nogales	31 20 14	110 56 12
Phoenix	33 27 12	112 04 28
Tucson	32 13 15	110 58 08
Yuma	32 43 16	114 37 01
Arkansas:		
El Dorado	33 12 39	92 39 40
Fayetteville	36 03 41	94 09 38
Fort Smith	35 23 10	94 25 36
Jonesboro	35 50 14	90 42 11
Little Rock	34 44 42	92 16 37
California:		
Bakersfield	35 22 31	119 01 16
Chico	39 44 07	121 49 57
Concord	37 58 46	122 01 51
Corona	33 52 35	117 33 56
El Centro	32 47 25	115 32 45
Eureka	40 48 08	124 09 46
Fontana	34 05 45	117 26 29
Fresno	36 44 12	119 47 11
Guasti	34 03 48	117 35 10
Hanford	36 19 51	119 38 48
Los Angeles	34 03 15	118 14 28
Modesto	37 38 26	120 59 44
Monterey	36 35 44	121 53 39
Oakland	37 48 03	122 15 54
Palm Springs	33 49 22	116 32 46
Redding	40 34 57	122 23 34
Sacramento	38 34 57	121 29 41
Salinas	36 40 24	121 39 25
San Bernardino	34 06 30	117 17 28
San Diego	32 42 53	117 09 21
San Francisco	37 46 39	122 24 40
San Jose	37 20 16	121 53 24
San Luis Obispo	35 16 49	120 39 34
San Mateo	37 34 08	122 19 16
Santa Barbara	34 25 18	119 41 55
Santa Maria	34 57 02	120 26 10
Stockton	37 57 30	121 17 16
Tulare	36 12 31	119 20 35
Ventura	34 16 47	119 17 22
Visalia	36 19 46	119 17 30
Colorado:		
Colorado Springs	38 50 07	104 49 16
Denver	39 44 58	104 59 22
Durango	37 16 29	107 52 25
Grand Junction	39 04 06	108 33 54
Montrose	38 28 44	107 52 31
Pueblo	38 16 17	104 36 33
Sterling	40 37 29	103 12 25
Connecticut:		
Bridgeport	41 10 49	73 11 22
Hartford	41 46 12	72 40 49
New Britain	41 40 02	72 47 08
New Haven	41 18 25	72 55 30
Norwich	41 31 36	72 04 31
Waterbury	41 33 13	73 02 31
Delaware:		
Wilmington	39 44 46	75 32 51
District of Columbia:		
Washington	38 53 51	77 00 33
Florida:		
Clearwater	27 57 56	82 47 51
Daytona Beach	29 12 44	81 01 10
Fort Lauderdale	26 07 11	80 08 34
Fort Myers	26 38 42	81 52 06
Fort Pierce	27 26 48	80 19 38
Gainesville	29 38 56	82 19 19
Jacksonville	30 19 44	81 39 42
Largo	27 54 54	82 47 32
Leesburg	28 48 43	81 52 30
Melbourne	28 04 41	80 36 29
Miami	25 46 37	80 11 32
Ocala	29 11 34	82 08 14
Orlando	28 32 42	81 22 38
Panama City	30 09 24	85 39 46
Pensacola	30 24 51	87 12 56
St. Petersburg	27 46 18	82 38 19
Sarasota	27 20 05	82 32 20
Tallahassee	30 26 30	84 16 56
Tampa	27 56 58	82 27 25
West Palm Beach	26 42 36	80 03 07

State and community	Latitude ° ' "	Longitude ° ' "
Georgia:		
Albany	31 31 36	81 00 22
Athens	33 57 34	83 22 39
Atlanta	33 45 10	84 23 37
Augusta	33 28 20	81 58 00
Chatsworth	34 46 08	84 46 10
Cochran	32 23 18	83 21 18
Columbus	32 28 07	84 59 24
Dawson	31 46 33	84 26 20
Macon	32 50 12	83 37 36
Pelham	31 07 42	84 09 02
Savannah	32 04 42	81 05 37
Thomasville	30 50 25	83 58 59
Waycross	31 12 19	82 21 47
Wrens	33 12 21	82 23 23
Guam:		
Agana	13 28 23	144 45 00
Hawaii:		
Hilo	19 43 42	155 05 30
Honolulu	21 18 36	157 51 48
Wailuku	20 53 21	156 30 27
Idaho:		
Boise	43 37 07	116 11 58
Idaho Falls	43 29 39	112 02 28
Lewiston	46 25 05	117 01 10
Moscow	46 43 58	116 59 54
Pocatello	42 51 38	112 27 01
Twin Falls	42 33 25	114 28 21
Illinois:		
Aurora	41 45 22	88 18 56
Bloomington	40 45 58	88 59 32
Carbondale	37 43 38	89 13 00
Champaign	40 07 05	88 14 48
Chicago	41 52 28	87 38 22
Decatur	39 50 37	88 57 11
Elgin	42 02 14	88 16 53
Freeport	42 17 57	89 37 07
Harrisburg	37 44 20	88 32 25
Jacksonville	39 44 03	90 13 44
Joliet	41 31 37	88 04 52
La Salle	41 19 49	89 05 44
Moline	41 30 31	90 30 49
Mount Vernon	38 18 29	88 54 26
Olney	38 43 47	88 05 00
Peoria	40 41 42	89 35 33
Quincy	39 55 59	91 24 12
Rockford	42 16 07	89 05 48
Rock Island	41 30 40	90 34 24
Springfield	39 47 58	89 38 51
Urbana	40 06 41	88 13 13
Indiana:		
Bloomington	39 09 56	86 31 52
Elkhart	41 40 56	85 58 15
Evansville	37 58 20	87 34 21
Fort Wayne	41 04 21	85 08 26
Gary	41 35 59	87 20 07
Hammond	41 35 13	87 27 43
Indianapolis	39 46 07	86 09 46
Lafayette	40 25 11	86 53 39
Marion	40 33 17	85 39 49
Muncie	40 11 28	85 23 16
Richmond	39 49 49	86 53 26
Roanoke	40 57 50	85 22 30
St. John	41 27 00	87 28 13
South Bend	41 40 33	86 15 01
Terre Haute	39 28 03	87 24 26
Vincennes	38 40 52	87 31 12
Iowa:		
Ames	42 01 36	93 36 44
Cedar Rapids	41 58 48	91 39 48
Davenport	41 31 24	90 34 21
Des Moines	41 35 14	93 37 00
Dubuque	42 29 55	90 40 08
Fort Dodge	42 30 12	94 11 05
Iowa City	41 39 37	91 31 52
Mason City	43 09 15	93 12 00
Sioux City	42 29 46	96 24 30
Waterloo	42 29 40	92 20 20
Kansas:		
Ensign	37 38 48	100 14 00
Garden City	37 57 54	100 52 20
Goodland	39 20 53	101 42 35
Great Bend	38 22 04	98 45 58
Hays	38 52 16	99 19 57
Hutchinson	38 03 11	97 55 20
Pittsburg	37 24 50	97 42 11
Salina	38 50 36	97 36 46
Topeka	39 03 16	95 40 23
Wichita	37 41 30	97 20 16
Kentucky:		
Ashland	38 28 36	82 38 23
Bowling Green	36 59 41	86 26 33
Covington	39 05 00	84 30 29
Elizabethtown	38 41 38	85 51 35
Hazard	37 14 54	87 11 31
Lexington	38 02 50	84 29 46
Louisville	38 14 47	85 45 49
Madisonville	37 19 45	87 29 54
Morehead	38 10 53	83 26 08
Murray	36 36 35	88 18 39
Newport	39 05 28	84 29 20
Owensboro	37 46 27	87 06 46
Owenton	38 32 11	84 50 16
Paducah	37 05 13	88 35 56
Pikesville	37 28 49	82 31 09
Somerset	37 05 35	84 36 17

State and community	Latitude ° ' "	Longitude ° ' "
Louisiana:		
Alexandria	31 18 33	92 26 47
Baton Rouge	30 26 58	91 11 00
Houma	29 35 34	90 43 09
Lafayette	30 13 24	92 01 06
Lake Charles	30 13 45	93 12 52
Monroe	32 30 02	92 06 55
New Orleans	29 56 53	90 04 10
Shreveport	32 30 46	93 44 58
West Monroe	32 30 51	92 08 13
Maine:		
Augusta	44 18 53	69 46 29
Bangor	44 48 13	68 46 18
Calais	45 11 04	67 16 43
Orono	44 53 15	68 40 12
Poland Spring	44 01 42	70 21 40
Portland	43 39 33	70 15 19
Presque Isle	46 40 57	68 00 52
Maryland:		
Baltimore	39 17 26	76 36 45
Cumberland	39 39 01	78 45 45
Hagerstown	39 38 39	77 43 15
Salisbury	38 21 56	75 35 56
Massachusetts:		
Adams	42 37 30	73 07 05
Boston	42 21 24	71 03 25
Cambridge	42 21 58	71 06 24
Greenfield	42 35 15	72 35 54
New Bedford	41 38 13	70 55 41
Springfield	42 06 21	72 35 32
Worcester	42 15 37	71 48 17
Michigan:		
Allen Park	42 15 12	83 12 57
Battle Creek	42 18 58	85 10 48
Bay City	43 36 04	83 53 15
Cadillac	44 15 10	85 23 52
Cheboygan	45 38 38	84 28 38
Detroit	42 19 48	83 02 57
Escanaba	45 44 45	87 03 18
Flint	43 00 50	83 41 33
Grand Rapids	42 58 03	85 40 13
Jackson	42 14 43	84 24 22
Kalamazoo	42 17 29	85 35 14
Lansing	42 44 01	84 33 15
Marquette	46 32 37	87 23 43
Mount Pleasant	43 16 12	84 46 31
Muskegon	43 14 17	86 15 02
Onondaga	42 26 41	84 33 43
Saginaw	43 25 52	83 56 05
Sault Ste. Marie	46 29 58	84 20 37
Traverse City	44 45 47	85 37 25
University Center	43 33 31	83 59 09
Minnesota:		
Alexandria	45 53 06	95 22 49
Appleton	45 12 00	96 01 02
Austin	43 39 57	92 58 20
Duluth	46 46 56	92 06 24
Hibbing	47 25 43	92 56 21
Mankato	44 09 49	94 00 09
Minneapolis	44 58 57	93 15 43
Rochester	44 01 21	92 28 03
St. Cloud	45 33 35	94 09 38
St. Paul	44 56 50	93 05 11
Walker	47 05 57	94 35 12
Mississippi:		
Biloxi	30 23 43	88 53 08
Bude	31 27 46	90 50 34
Columbus	33 29 40	88 25 33
Greenwood	33 31 05	90 10 55
Gulfport	30 22 04	89 05 36
Jackson	32 17 56	90 11 06
Laurel	31 41 40	89 07 48
Meridian	32 21 57	88 42 02
Oxford	34 22 00	89 31 07
State College	33 27 18	88 47 13
Tupelo	34 15 26	88 42 30
Missouri:		
Cape Girardeau	37 18 29	89 31 29
Columbia	38 57 03	92 19 46
Hannibal	39 42 24	91 22 45
Jefferson City	38 34 40	92 10 24
Joplin	37 05 26	94 30 50
Kansas City	39 04 56	94 35 20
Kirksville	40 11 37	92 34 58
Poplar Bluff	36 45 20	90 23 38
St. Joseph	39 45 57	94 51 02
St. Louis	38 37 45	90 12 22
Sedalia	38 42 08	93 13 26
Springfield	37 13 03	93 17 32
Montana:		
Anaconda	46 07 40	112 57 12
Billings	45 47 00	108 30 04
Butte	46 01 06	112 32 11
Glendive	47 06 42	104 43 02
Great Falls	47 29 33	111 18 23
Helena	46 35 33	112 02 24
Kalispell	48 11 45	114 18 44
Miles City	46 24 34	105 50 30
Missoula	46 52 23	113 59 29
Nebraska:		
Albion	41 41 23	97 59 53
Alliance	42 00 04	102 52 08
Bassett	42 35 00	99 32 10
Grand Island	40 55 33	98 20 23
Hastings	40 35 21	98 23 20
Hayes Center	40 30 36	101 01 18
Hay Springs	42 41 03	102 41 22

State and community	Latitude ° ′ ″	Longitude ° ′ ″
Nebraska—Continued		
Kearney	40 41 58	99 04 53
Lexington	40 46 30	99 44 41
Lincoln	40 48 59	96 42 15
McCook	40 12 02	100 37 32
Merriman	42 55 07	101 42 02
Norfolk	42 01 56	97 24 42
North Platte	41 08 14	100 45 43
Omaha	41 15 42	95 56 14
Scottsbluff	41 51 40	103 39 00
Superior	40 01 12	98 04 00
Nevada:		
Elko	40 50 00	115 45 41
Henderson	36 02 00	114 58 57
Las Vegas	36 10 20	115 08 37
Reno	39 31 27	119 48 40
New Hampshire:		
Berlin	44 28 20	71 10 43
Durham	43 08 02	70 55 35
Hanover	43 42 03	72 17 24
Keene	42 56 02	72 16 44
Lebanon	43 38 34	72 15 12
Littleton	44 18 22	71 46 13
Manchester	42 59 28	71 27 41
New Jersey:		
Atlantic City	39 21 32	74 25 53
Burlington	40 04 21	74 51 47
Camden	39 56 45	75 07 20
Glen Ridge	40 48 16	74 12 14
Linden	40 37 57	74 15 22
Newark	40 44 14	74 10 19
New Brunswick	40 29 38	74 26 49
Paterson	40 54 51	74 09 51
Trenton	40 13 16	74 45 28
Vineland	39 29 13	75 01 17
Wildwood	38 59 18	74 48 43
New Mexico:		
Albuquerque	35 05 01	106 39 05
Carlsbad	32 25 09	104 13 47
Clovis	34 24 11	103 12 08
Portales	34 10 58	103 20 10
Roswell	33 23 47	104 31 26
New York:		
Albany	42 39 01	73 45 01
Binghamton	42 06 03	75 54 47
Buffalo	42 52 52	78 52 21
Carthage	43 58 50	75 36 26
Elmira	42 05 26	76 48 22
Garden City	40 43 26	73 38 03
Ithaca	42 26 33	76 29 42
Jamestown	42 05 45	79 14 40
New York	40 45 06	73 59 39
North Pole	44 23 59	73 51 00
Norwood	44 45 00	75 59 39
Oneonta	42 27 21	75 03 42
Patchogue	40 45 56	73 00 42
Plattsburgh	44 42 03	73 27 07
Riverhead	40 55 06	72 39 51
Rochester	43 09 41	77 36 21
Schenectady	42 48 52	73 56 24
Syracuse	43 03 04	76 09 14
Utica	43 06 12	75 13 33
Watertown	43 58 30	75 54 48
North Carolina:		
Asheville	35 35 42	82 33 26
Chapel Hill	35 54 51	79 03 11
Charlotte	35 13 44	80 50 45
Columbia	35 55 06	76 15 04
Concord	35 24 29	80 34 45
Durham	35 59 48	78 54 00
Fayetteville	35 03 12	78 52 54
Greensboro	36 04 17	79 47 25
Greenville	35 36 42	77 22 22
Hickory	35 43 54	81 20 20
High Point	35 57 14	80 00 15
Jacksonville	34 45 00	77 25 54
Linville	36 04 06	81 52 16
New Bern	35 06 33	77 02 03
Raleigh	35 46 38	78 38 21
Washington	35 32 35	77 03 16
Wilmington	34 14 14	77 56 58
Winston-Salem	36 05 52	80 14 42
North Dakota:		
Bismark	46 48 23	100 47 17
Devils Lake	48 06 42	98 51 29
Dickinson	46 52 55	102 47 06
Fargo	46 52 30	96 47 18
Minot	48 14 09	101 17 38
Pembina	48 58 00	97 14 37
Valley City	46 55 31	98 00 04
Williston	48 08 47	103 36 59
Ohio:		
Akron	41 05 00	81 30 44
Athens	39 19 38	82 06 09
Bowling Green	41 22 37	83 39 03
Canton	40 47 50	81 22 37
Cincinnati	39 06 07	84 05 06
Cleveland	41 30 51	81 41 50
Columbus	39 57 47	83 00 17
Dayton	39 45 52	84 11 43
Kettering	39 41 22	84 10 07
Lima	40 44 29	84 06 34
Lorain	41 27 48	82 10 23
Marion	40 35 14	83 07 36
Newark	40 03 35	82 24 15
Oxford	39 30 28	84 44 26

State and community	Latitude ° ′ ″	Longitude ° ′ ″
Ohio—Continued		
Portsmouth	38 44 06	82 59 39
Springfield	39 55 38	83 48 29
Steubenville	40 21 42	80 36 53
Toledo	41 39 14	83 32 39
Youngstown	41 05 57	80 39 02
Zanesville	39 56 59	82 00 56
Oklahoma:		
Ada	34 46 24	96 40 36
Ardmore	34 10 18	97 07 50
Lawton	34 36 20	98 23 11
Oklahoma City	35 28 26	97 31 04
Sayre	35 17 34	99 38 23
Tulsa	36 09 12	95 59 34
Oregon:		
Coos Bay	43 22 02	124 13 00
Corvallis	44 34 10	123 16 12
Eugene	44 03 16	123 05 30
Klamath Falls	42 13 32	121 46 32
La Grande	45 19 47	118 05 45
Medford	42 19 33	122 52 31
Portland	45 31 06	122 40 35
Roseburg	43 12 34	123 20 26
Salem	44 56 21	123 01 59
Pennsylvania:		
Allentown	40 36 11	75 28 06
Altoona	40 30 55	78 24 03
Bethlehem	40 37 57	75 21 36
Clearfield	41 01 20	78 26 10
Erie	42 07 15	80 04 57
Harrisburg	40 15 43	76 52 59
Hershey	40 17 04	76 39 01
Johnstown	40 19 35	78 55 03
Lancaster	40 02 25	76 18 29
Philadelphia	39 56 58	75 09 21
Pittsburgh	40 26 19	80 00 00
Reading	40 20 09	75 55 40
Scranton	41 24 32	75 39 46
Wilkes-Barre	41 14 32	75 53 17
York	39 57 35	76 43 36
Puerto Rico:		
Aguadilla	18 25 53	67 09 18
Arecibo	18 28 26	66 43 39
Caguas	18 13 59	66 02 06
Fajardo	18 19 35	65 39 21
Mayaguez	18 12 16	67 08 36
Ponce	18 00 51	66 36 58
San Juan	18 26 55	66 03 55
Rhode Island:		
Providence	41 49 32	71 24 41
South Carolina:		
Allendale	33 00 30	81 18 26
Anderson	34 30 06	82 38 54
Charleston	32 46 35	79 55 53
Columbia	34 00 02	81 02 00
Florence	34 11 49	79 46 06
Greenville	34 50 50	82 24 01
Spartanburg	34 57 03	81 56 06
South Dakota:		
Aberdeen	45 27 31	98 29 03
Brookings	44 18 38	96 47 53
Florence	45 03 14	97 19 35
Lead	44 21 07	103 46 03
Mitchell	43 42 48	98 01 36
Pierre	44 22 06	100 20 57
Rapid City	44 04 52	103 13 11
Reliance	43 52 45	99 36 18
Sioux Falls	43 32 36	96 43 35
Vermillion	42 46 52	96 55 35
Tennessee:		
Chattanooga	35 02 41	85 18 32
Jackson	35 36 48	88 49 15
Johnson City	36 19 04	82 20 56
Kingsport	36 32 57	82 33 44
Knoxville	35 57 39	83 55 07
Lexington	35 38 58	88 23 31
Memphis	35 08 46	90 03 13
Nashville	36 09 33	86 46 55
Sneedville	36 31 46	83 13 04
Texas:		
Abilene	32 27 05	99 43 51
Amarillo	35 12 27	101 50 04
Austin	30 16 09	97 44 37
Beaumont	30 05 20	94 06 09
Belton	31 03 31	97 27 39
Big Spring	32 15 03	101 28 38
Bryan	30 38 48	96 21 31
College Station	30 37 05	96 20 41
Corpus Christi	27 47 51	97 23 45
Dallas	32 47 09	96 47 37
El Paso	31 45 36	106 29 11
Fort Worth	32 44 55	97 19 44
Galveston	29 18 10	94 47 43
Harlingen	26 11 29	97 41 35
Houston	29 45 26	95 21 37
Laredo	27 30 22	99 30 30
Longview	32 29 24	94 42 43
Lubbock	33 35 05	101 50 33
Lufkin	31 20 14	94 43 21
Midland	31 59 54	102 04 31
Monahans	31 35 16	102 53 26
Nacogdoches	31 36 13	94 39 20
Odessa	31 50 49	102 22 01
Port Arthur	29 52 09	93 56 01
Richardson	32 57 06	96 44 05
Rosenberg	29 33 30	95 48 15

State and community	Latitude ° ′ ″	Longitude ° ′ ″
Texas—Continued		
San Angelo	31 27 39	100 26 03
San Antonio	29 25 37	98 29 06
Sweetwater	32 28 24	100 24 18
Temple	31 06 02	97 20 22
Texarkana	33 25 29	94 02 34
Tyler	32 21 21	95 17 52
Victoria	28 48 01	97 00 06
Waco	31 33 12	97 08 00
Weslaco	26 09 24	97 59 33
Wichita Falls	33 54 34	98 29 28
Utah:		
Logan	41 44 03	111 50 11
Ogden	41 12 71	111 58 21
Provo	40 14 07	111 39 34
Salt Lake City	40 45 23	111 53 26
Vermont:		
Burlington	44 28 34	73 12 46
Rutland	43 36 29	72 58 56
St. Johnsbury	44 25 16	72 01 13
Windsor	44 28 38	72 23 32
Virginia:		
Bristol	36 35 48	82 11 04
Charlottesville	38 01 52	78 28 50
Goldvein	38 26 54	77 39 19
Hampton	37 01 32	76 20 32
Harrisonburg	38 27 01	78 52 07
Lynchburg	37 24 51	79 08 37
Norfolk	36 51 10	76 17 21
Norton	36 56 05	82 37 31
Petersburg	37 13 40	77 24 15
Portsmouth	36 50 12	76 17 54
Richmond	37 32 15	77 26 09
Roanoke	37 16 13	79 56 44
Staunton	38 09 02	79 04 34
Virgin Islands:		
Charlotte Amalie	18 20 36	64 55 53
Christiansted	17 44 44	64 42 21
Washington:		
Bellingham	48 45 02	122 28 36
Kennewick	46 12 28	119 08 32
Lakewood Center	47 07 37	122 31 15
Pasco	46 13 50	119 05 27
Pullman	46 43 42	117 10 46
Richland	46 16 36	119 16 21
Seattle	47 36 32	122 20 12
Spokane	47 39 39	117 25 33
Tacoma	47 14 59	122 26 15
Yakima	46 36 09	120 30 39
West Virginia:		
Bluefield	37 15 29	81 13 20
Charleston	38 21 01	81 37 52
Clarksburg	39 16 50	80 20 38
Grandview	37 49 28	81 04 20
Huntington	38 25 13	82 26 33
Morgantown	39 37 41	79 57 28
Oak Hill	37 58 31	81 08 45
Parkersburg	39 15 57	81 33 46
Weston	39 02 19	80 28 05
Wheeling	40 04 03	80 43 20
Wisconsin:		
Eau Claire	44 48 31	91 29 49
Fond Du Lac	43 46 35	88 26 52
Green Bay	44 30 48	88 00 50
Janesville	42 40 52	89 01 39
Kenosha	42 35 04	87 49 14
La Crosse	43 48 48	91 15 02
Madison	43 04 23	89 22 55
Milwaukee	43 02 19	87 54 15
Rhinelander	45 38 09	89 24 50
Superior	46 43 14	92 06 07
Wausau	44 57 30	89 37 40
Wyoming:		
Casper	42 51 00	106 19 22
Cheyenne	41 08 09	104 49 07
Rawlins	41 47 23	107 14 37
Riverton	43 01 29	108 23 03

§ 76.54 **Significantly viewed signals; method to be followed for special showings.**

(a) Signals that are significantly viewed in a county (and thus are deemed to be significantly viewed within all communities within the county) are those that meet the test of significant viewing (see § 76.5(k)) according to the 1971 American Research Bureau "Television Circulation Share of Hours" survey, for counties in which there is less than 10 percent cable television penetration, and the 1971 American Research Bureau "Non-CATV Circulation and Share of Viewing Hours Study for ARB CATV-controlled Counties," for counties in which there is 10 percent or more cable television penetration.

NOTE: The relevant information from these surveys is available from the Commission.

(b) On or after March 31, 1973, significant viewing in a cable television community for signals not shown as significantly viewed under paragraph (a) of this section may be demonstrated by an independent professional audience survey of noncable television homes that covers at least two weekly periods separated by at least thirty (30) days but no more than one of which shall be a week between the months of April and September. If two surveys are taken, they shall include samples sufficient to assure that the combined surveys result in an average figure at least one standard error above the required viewing level. If surveys are taken for more than 2-weekly periods in any 12 months, all such surveys must be submitted and the combined surveys must result in an average figure at least one standard error above the required viewing level.

§ 76.55   Manner of carriage.

(a) Where a television broadcast signal is required to be carried by a cable television system, pursuant to the rules in this subpart:

(1) The signal shall be carried without material degradation in quality (within the limitations imposed by the technical state of the art), and, where applicable, in accordance with the technical standards of Subpart K of this part;

(2) The signal shall, on request of the station licensee or permittee, be carried on the system on the channel number on which the station is transmitting, except where technically infeasible;

(3) The signal shall, on request of the station licensee or permittee, be carried on the system on no more than one channel.

(b) Where a television broadcast signal is carried by a cable television system, pursuant to the rules in this subpart, the programs broadcast shall be carried in full, without deletion or alteration of any portion except as required by this part.

(c) A cable television system need not carry the signal of any television translator station if (1) the system is carrying the signal of the originating station, or (2) the community of the system is located, in whole or in part, within the Grade B contour of a station carried on the system whose programing is substantially duplicated by the translator station.

(d) If the community of a cable television system is located, in whole or in part, within the Grade B contour of both a satellite and its parent television station, and if the system would otherwise be required to carry both of them pursuant to the rules in this subpart, the system need carry only one of these signals, and may select between them.

§ 76.57   Provisions for systems operating in communities located outside of all major and smaller television markets.

A cable television system operating in a community located wholly outside all major and smaller television markets, as defined in § 76.5, shall carry television broadcast signals in accordance with the following provisions:

(a) Any such cable television system may carry or, on request of the relevant station licensee or permittee, shall carry the signals of:

(1) Television broadcast stations within whose Grade B contours the community of the system is located, in whole or in part;

(2) Television translator stations, with 100 watts or higher power, licensed to the communitiy of the system;

(3) Noncommercial educational television broadcast stations within whose specified zone the community of the system is located, in whole or in part;

(4) Commercial television broadcast stations that are significantly viewed in the community of the system. See § 76.54.

(b) In addition to the television broadcast signals carried pursuant to paragraph (a) of this section, any such cable television system may carry any additional television signals.

§ 76.59   Provisions for smaller television markets.

A cable television system operating in a community located in whole or in part within a smaller television market, as defined in § 76.5, shall carry television broadcast signals only in accordance with the following provisions:

(a) Any such cable television system may carry or, on request of the relevant station licensee or permittee, shall carry the signals of:

(1) Television broadcast stations within whose specified zone the community of the system is located, in whole or in part;

(2) Noncommercial educational television broadcast stations within whose Grade B contours the community of the system is located, in whole or in part;

(3) Commercial television broadcast stations licensed to communities in other smaller television markets, within whose Grade B contours the community of the system is located, in whole or in part;

(4) Television broadcast stations licensed to other communities which are generally considered to be part of the same smaller television market (Example: Burlington, Vt.—Plattsburgh, N.Y., television market);

(5) Television translator stations, with 100 watts or higher power, licensed to the community of the system;

(6) Commercial television broadcast stations that are significantly viewed in the community of the system. See § 76.54.

(b) Any such cable television system may carry sufficient additional signals so that, including the signals required to be carried pursuant to paragraph (a) of this section, it can provide the signals of a full network station of each of the major national television networks, and of one independent television station: Provided, however, That, in determining how many additional signals may be carried, any authorized but not operating television broadcast station that, if operational, would be required to be carried pursuant to paragraph (a)(1) of this section, shall be considered to be operational for a period terminating 18 months after grant of its initial construction permit. The following priorities are applicable to the additional television signals that may be carried:

(1) *Full network stations.* A cable television system may carry the nearest missing full network stations or the nearest in-state full network stations;

NOTE: The Commission may waive the requirements of this subparagraph for good cause shown in a petition filed pursuant to § 76.7.

(2) *Independent station.* A cable television system may carry any independent television station: *Provided, however,* That if a signal of a station in the first 25 major television markets (see § 76.51 (a)) is carried pursuant to this subparagraph, such signal shall be taken from one of the two closest such markets, where such signal is available.

NOTE: It is not contemplated that waiver of the provisions of this subparagraph will be granted.

(c) In addition to the noncommercial educational television broadcast signals carried pursuant to paragraph (a) of this section, any such cable television system may carry the signals of any noncommercial educational stations that are operated by an agency of the State within which the system is located. Such system may also carry any other noncommercial educational signals, in the absence of objection filed pursuant to § 76.7 by any local noncommercial educational station or State or local educational television authority.

(d) In addition to the television broadcast signals carried pursuant to paragraphs (a) through (c) of this section, any such cable television system may carry any television stations broadcasting predominantly in a non-English language.

(e) Where the community of a cable television system is wholly or partially within both one of the first 50 major television markets and a smaller television market, the carriage provisions for the first 50 major markets shall apply. Where the community of a system is wholly or partially within both one of the second 50 major television markets and a smaller television market, the carriage provisions for the second 50 major markets shall apply.

§ 76.61   Provisions for first 50 major television markets.

A cable television system operating in a community located in whole or in part within one of the first 50 major television markets listed in § 76.51(a) shall carry television broadcast signals only in accordance with the following provisions:

(a) Any such cable television system may carry, or on request of the relevant station licensee or permittee, shall carry the signals of:

(1) Television broadcast stations within whose specified zone the community of the system is located, in whole or in part: *Provided, however,* That where

a cable television system is located in the designated community of a major television market, it shall not carry the signal of a television station licensed to a designated community in another major television market, unless the designated community in which the cable system is located is wholly within the specified zone (see § 76.5(f)) of the station, except as otherwise provided in this section:

(2) Noncommercial educational television broadcast stations within whose Grade B contours the community of the system is located, in whole or in part;

(3) Television translator stations, with 100 watts or high power, licensed to the community of the system;

(4) Television broadcast stations licensed to other designated communities of the same major television market (Example: Cincinnati, Ohio-Newport, Ky., television market);

(5) Commercial television broadcast stations that are significantly viewed in the community of the system. See § 76.54.

(b) Any such cable television system may carry sufficient additional signals so that, including the signals required to be carried pursuant to paragraph (a) of this section, it can provide the signals of a full network station of each of the major national television networks, and of three independent television stations: *Provided, however,* That in determining how many additional signals may be carried, any authorized but not operating television broadcast station that, if operational, would be required to be carried pursuant to paragraph (a)(1) of this section, shall be considered to be operational for a period terminating 18 months after grant of its initial construction permit. The following priorities are applicable to the additional television signals that may be carried:

(1) *Full network stations.* A cable television system may carry the nearest missing full network stations, or the nearest in-State full network stations;

NOTE: The Commission may waive the requirements of this subparagraph for good cause shown in a petition filed pursuant to § 76.7.

(2) *Independent stations.* (i) For the first and second additional signals, if any, a cable television system may carry the signals of any independent television station: *Provided, however,* That if signals of stations in the first 25 major television markets (see § 76.51(a)) are carried pursuant to this subparagraph, such signals shall be taken from one or both of the two closest such markets, where such signals are available. If a third additional signal may be carried, a system shall carry the signal of any independent UHF television station located within 200 air miles of the reference point for the community of the system (see § 76.53), or, if there is no such station, either the signal of any independent VHF television station located within 200 air miles of the reference point for the community of the system, or the signal of any independent UHF television station.

NOTE: It is not contemplated that waiver of the provisions of this subparagraph will be granted.

(ii) Whenever, pursuant to Subpart F of this part, a cable television system is required to delete a television program on a signal carried pursuant to subdivision (i) of this subparagraph or paragraph (c) of this section, or a program on such a signal is primarily of local interest to the distant community (e.g., a local news or public affairs program), such system may, consistent with the program exclusivity rules of Subpart F of this part, substitute a program from any other television broadcast station. A program substituted may be carried to its completion, and the cable system need not return to its regularly carried signal until it can do so without interrupting a program already in progress.

(c) After the service standards specified in paragraph (b) of this section have been satisfied, a cable television system may carry two additional independent television broadcast signals, chosen in accordance with the priorities specified in paragraph (b)(2) of this section: *Provided, however,* That the number of additional signals permitted under this paragraph shall be reduced by the number of signals added to the system pursuant to paragraph (b) of this section.

(d) In addition to the noncommercial educational television broadcast signals carried pursuant to paragraph (a) of this section, any such cable television system may carry the signals of any noncommericial educational stations that are operated by an agency of the State within which the system is located. Such system may also carry any other noncommercial educational signals, in the absence of objection filed pursuant to § 76.7 by any local noncommercial educational station or State or local educational television authority.

(e) In addition to the television broadcast signals carried pursuant to paragraphs (a) through (d) of this section, any such cable television system may carry any television stations broadcasting predominantly in a non-English language.

(f) Where the community of a cable television system is wholly or partially within both one of the first 50 major television markets and another television market, the provisions of this section shall apply.

§ 76.63 Provisions for second 50 major television markets.

(a) A cable television system operating in a community located in whole or in part within one of the second 50 major television markets listed in § 76.51(b) shall carry television broadcast signals only in accordance with the provisions of § 76.61, except that in paragraph (b) of § 76.61, the number of additional independent television signals that may be carried is two (2).

(b) Where the community of a cable television system is wholly or partially within both one of the second 50 major television markets and one of the first 50 major television markets, the carriage provisions for the first 50 major markets shall apply. Where the community of a system is wholly or partially within both

one of the second 50 major television markets and a smaller television market, the provisions of this section shall apply.

§ 76.65 Grandfathering provisions.

The provisions of §§ 76.57, 76.59, 76.61, and 76.63 shall not be deemed to require the deletion of any television broadcast or translator signals which a cable television system was authorized to carry or was lawfully carrying prior to March 31, 1972: *Provided, however,* That if carriage of a signal has been limited by Commission order to discrete areas of a community, any expansion of service will be subject to the appropriate provisions of this subpart. If a cable television system in a community is authorized to carry signals, either by virtue of specific Commission authorization or otherwise, any other cable television system already operating or subsequently commencing operations in the same community may carry the same signals. (Any such new system shall, before instituting service, obtain a certificate of compliance, pursuant to § 76.11.)

Subpart E—[Reserved]

Subpart F—Program Exclusivity

§ 76.91 Stations entitled to network program exclusivity.

(a) Any cable television system operating in a community, in whole or in part, within the Grade B contour of any television broadcast station, or within the community of a 100-watt or higher power television translator station, and that carries the signal of such station shall, on request of the station licensee or permittee, maintain the station's exclusivity as an outlet for network programing against lower priority duplicating signals, but not against signals of equal priority, in the manner and to the extent specified in §§ 76.93 and 76.95.

(b) For purposes of this section, the order of priority of television signals carried by a cable television system is as follows:

(1) First, all television broadcast stations within whose principal community contours the community of the system is located, in whole or in part;

(2) Second, all television broadcast stations within whose Grade A contours the community of the system is located, in whole or in part;

(3) Third, all television broadcast stations within whose Grade B contours the community of the system is located, in whole or in part;

(4) Fourth, all television translator stations with 100 watts or higher power, licensed to the community of the system.

(c) If the signal of a television broadcast station licensed to a community in a smaller television market is carried by a cable television system, pursuant to § 76.57(a)(4), such signal shall, on request, be afforded network program exclusivity. This provision shall not be applicable to any signal authorized or lawfully carried by a cable television system prior to March 31, 1972.

§ 76.93 Extent of protection.

(a) Where the network programing of a television station is entitled to pro-

gram exclusivity, the cable television system, shall, on request of the station licensee or permittee, refrain from simultaneously duplicating any network program broadcast by such station, if the cable operator has received notification from the requesting station of the date and time of its broadcast of the program and the date and time of any broadcast to be deleted, as soon as possible and in any event no later than 48 hours prior to the broadcast to be deleted. On request of the cable system, such notice shall be given no later than the Monday preceding the calendar week (Sunday–Saturday) during which exclusivity is sought.

(b) On petition filed pursuant to § 76.7, the Commission will afford additional, limited program exclusivity to a network-affiliated station where, because of the time-zone situation, the affording of simultaneous program exclusivity would result in duplication of a substantial amount of such station's network programing. Where a station is currently receiving same-day program exclusivity and files for such relief within fifteen (15) days of the effective date of this rule, it shall continue to receive same-day program exclusivity pending the Commission's ruling on the petition. During such period, and if same-day program exclusivity is required thereafter, the following provisions shall be applicable:

(1) A cable television system need not delete reception of a network program if, in so doing, it would leave available for reception by subscribers, at any time, less than the programs of two networks (including those broadcast by any stations whose signals are being carried and whose program exclusivity is being protected pursuant to the requirements of this section);

(2) A system need not delete reception of a network program which is scheduled by the network between the hours of 6 and 11 p.m., eastern time, but is broadcast by the station requesting deletion, in whole or in part, outside of the period which would normally be considered prime time for network programing in the time zone involved.

§ 76.95 Exceptions.

Notwithstanding the requirements of § 76.93:

(a) A cable television system need not delete reception of any program which would be carried on the system in color but will be broadcast in black and white by the station requesting deletion.

(b) The Commission will give full effect to private agreements between operators of cable television systems and local television stations which provide for a type or degree of network exclusivity which differs from the requirements of §§ 76.91 and 76.93.

§ 76.97 Waiver petitions.

Where a petition for waiver of the provisions of §§ 76.91 and 76.93 is filed within fifteen (15) days after a request for program exclusivity is received by the operator of a cable television system,

such system need not provide program exclusivity pending the Commission's ruling on the petition or on the question of temporary relief pending further proceedings.

§ 76.151 Syndicated program exclusivity; extent of protection.

Upon receiving notification pursuant to § 76.155:

(a) No cable television system, operating in a community in whole or in part within one of the first 50 major television markets, shall carry a syndicated program, pursuant to § 76.61 (b), (c), (d), or (e), for a period of 1 year from the date that program is first licensed or sold as a syndicated program to a television station in the United States for television broadcast exhibition;

(b) No cable television system, operating in a community in whole or in part within a major television market, shall carry a syndicated program, pursuant to §§ 76.61 (b), (c), (d), or (e), or 76.63(a) (as it refers to § 76.61 (b), (c), (d), or (e)), while a commercial television station licensed to a designated community in that market has exclusive broadcast exhibition rights (both over-the-air and by cable) to that program: *Provided, however,* That if a commercial station licensed to a designated community in one of the second 50 major television markets has such exclusive rights, a cable television system located in whole or in part within the market of such station may carry such syndicated programs in the following circumstances:

(1) If the program is carried by the cable television system in prime time and will not also be broadcast by a commercial market station in prime time during the period for which there is exclusivity for the program;

(2) For off-network series programs:
(i) Prior to the first nonnetwork broadcast in the market of an episode in the series;
(ii) After a nonnetwork first-run of the series in the market or after 1 year from the date of the first nonnetwork broadcast in the market of an episode in the series, whichever occurs first;

(3) For first-run series programs:
(i) Prior to the first broadcast in the market of an episode in the series;
(ii) After two (2) years from the first broadcast in the market of an episode in the series;

(4) For first-run, nonseries programs:
(i) Prior to the date the program is available for broadcast in the market under the provision of any contract or license of a television broadcast station in the market;
(ii) After two (2) years from the date of such first availability;

(5) For feature films:
(i) Prior to the date such film is available for nonnetwork broadcast in the market under the provisions of any contract or license of a television broadcast station in the market;
(ii) Two (2) years after the date of such first availability;

(6) For other programs: 1 day after the first nonnetwork broadcast in the market or 1 year from the date of pur-

chase of the program for nonnetwork broadcast in the market, whichever occurs first.

NOTE 1: For purposes of § 76.151, a series will be treated as a unit, that is:

(i) No episode of a series (including an episode in a different package of programs in the same series) may be carried by a cable television system, pursuant to §§ 76.61 (b), (c), (d), or (e) or 76.63(a) (as it refers to § 76.61 (b), (c), (d), or (e)) while any episodes of the series are subject to exclusivity protection.

(ii) In the second 50 major television markets, no exclusivity will be afforded a different package of programs in the same series after the initial exclusivity period has terminated.

NOTE 2: As used in this section, the phrase "broadcast in the market" or "broadcast by a market station" refers to a broadcast by a television station licensed to a designated community in the market.

§ 76.153 Parties entitled to exclusivity.

(a) Copyright holders of syndicated programs shall be entitled to the exclusivity provided by § 76.151(a). In order to receive such exclusivity, the copyright holder shall notify each cable system of the exclusivity sought in accordance with the requirements of § 76.155.

(b) Television broadcast stations licensed to designated communities in the major television markets shall be entitled to the exclusivity provided by § 76.151(b). In order to receive such exclusivity, such television stations shall notify each cable system of the exclusivity sought in accordance with the requirements of § 76.155.

(c) In order to be entitled to exclusivity for a program under § 76.151(b), a television station must have an exclusive right to broadcast that program against all other television stations licensed to the same designated community and against broadcast signal cable carriage of that program in the cable system community: *Provided, however,* That such exclusivity will not be recognized in a designated community of another major television market unless such community is wholly within the television market of the station seeking exclusivity. In hyphenated markets, exclusivity will be recognized beyond the specified zone of a station only to the extent the station has exclusivity against other stations in the designated communities of the market. In such instances, exclusivity to the extent a station has obtained it will be recognized within the specified zones of such other stations. It shall be presumed that broadcast rights acquired prior to March 31, 1972, are exclusive for the specified zones of all stations in the market in which the station is located.

§ 76.155 Notification.

(a) Syndicated program exclusivity notifications shall include the following information:

(1) For purposes of § 76.151(a):
(i) The name and address of the copyright holder requesting exclusivity;
(ii) The name of the program or series for which exclusivity is sought;
(iii) The date of first sale or license of the program for television broadcast

as a syndicated program in the United States.

(2) For purposes of § 76.151(b):

(i) The name and address of the television broadcast station requesting exclusivity;

(ii) The name of the program or series for which exclusivity is sought;

(iii) The dates on which exclusivity is to commence and terminate;

(iv) As to programs to be deleted from signals regularly carried by the system pursuant to §§ 76.61 (b), (c), (d), or (e) and 76.63(a) (as it refers to § 76.61 (b), (c), (d), or (e)): the name of the program; the call letters of the station from which the deletion is to be made; and the date, time, and duration of the deletion. Information, once supplied pursuant to subparagraphs (2) (i), (ii), (iii), or (3) of this paragraph, need not be repeated in any notification supplying the information required by this subparagraph.

(3) For purposes of § 76.151(b) (as it relates to television stations licensed to designated communities in the second 50 major television markets), the following information shall be supplied in addition to that required by subparagraph (2) of this paragraph:

(i) Whether the program will be broadcast in prime time by the station requesting exclusivity during the period of protection provided in § 76.151(b);

(ii) The specific rule pursuant to which exclusivity is requested (e.g., § 76.151(b)(2)—off-network series, § 76.151(b)(3)—first-run series);

(iii) For off-network series programs, the number of showings contracted for, including the number of repeat presentations, if any, and the date when the first run is to end.

(b) Subject to the provisions of paragraph (c) of this section, notifications given pursuant to § 76.151 must be received no later than the Monday preceding the calendar week (Sunday-Saturday) during which exclusivity is sought.

(c) Direct notice of a change in the schedule of a television station against which exclusivity is sought, given to a cable television system by a television station seeking exclusivity, shall, if given more than 36 hours prior to the time a deletion is to be made, supersede prior notifications containing the information required by paragraph (a) of this section and any information otherwise relied on pursuant to paragraph (d) of this section.

(d) In determining which programs must be deleted from a television signal when such information is not required to be provided pursuant to paragraph (a) of this section, a cable television system may rely on information from any of the following sources published or made available during the week the deletion is to be made or during the prior week:

(i) Newspapers or journals of general circulation in the service area of a television station whose programs may be subject to deletion;

(ii) A television station whose programs may be subject to deletion;

(iii) Any television station requesting exclusivity.

§ 76.157 Exclusivity contracts.

With respect to each program as to which a television broadcast station licensee or permittee requests exclusivity pursuant to § 76.151, such licensee or permittee shall maintain in its public file an exact copy of those portions of the exclusivity contract, such portions to be signed by both the copyright holder and the licensee or permittee, setting forth in full the provisions pertinent to the duration, nature, and extent of the exclusivity terms concerning broadcast signal exhibition (whether over-the-air or by cable) to which the parties have agreed.

§ 76.159 Grandfathering.

The provisions of § 76.151 shall not be deemed to require a cable television system to delete programing from any signal that was carried prior to March 31, 1972, or that any other cable television system in the same community was carrying prior to March 31, 1972: *Provided, however,* That if carriage of a signal has been limited by Commission order to discrete areas of a community, any expansion of service will be subject to the appropriate provisions of the subpart.

Subpart G—Cablecasting

§ 76.201 Origination cablecasting in conjunction with carriage of broadcast signals.

(a) No cable television system having 3,500 or more subscribers shall carry the signal of any television broadcast station unless the system also operates to a significant extent as a local outlet by origination cablecasting and has available facilities for local production and presentation of programs other than automated services. Such origination cablecasting shall be limited to one or more designated channels which may be used for no other purpose.

(b) No cable television system located outside of all major television markets shall enter into any contract, arrangement, or lease for use of its cablecasting facilities which prevents or inhibits the use of such facilities for a substantial portion of time (including the time period 6–11 p.m.) for local programing designed to inform the public on controversial issues of public importance.

(c) No cable television system shall carry the signal of any television broadcast station if the system engages in origination cablecasting, either voluntarily or pursuant to paragraph (a) of this section, unless such cablecasting is conducted in accordance with the provisions of §§ 76.205, 76.209, 76.213, 76.215, 76.217, 76.221, and 76.225.

§ 76.205 Origination cablecasts by candidates for public office.

(a) *General requirements.* If a cable television system shall permit any legally qualified candidate for public office to use its origination channel(s) and facilities therefor, it shall afford equal opportunities to all other such candidates for

that office: *Provided, however,* That such system shall have no power of censorship over the material cablecast of any such candidate; *And provided, further,* That an appearance by a legally qualified candidate on any:

(1) Bona fide newscast,

(2) Bona fide news interview,

(3) Bona fide news documentary (if the appearance of the candidate is incidental to the presentation of the subject or subjects covered by the news documentary), or

(4) On the spot coverage of bona fide news events (including but not limited to political conventions and activities incidental thereto),

shall not be deemed to be use of the facilities of the system within the meaning of this paragraph.

NOTE: The fairness doctrine is applicable to these exempt categories. See § 76.209.

(b) *Rates and practices.* (1) The rates, if any, charged all such candidates for the same office shall be uniform, shall not be rebated by any means direct or indirect, and shall not exceed the charges made for comparable origination use of such facilities for other purposes.

(2) In making facilities available to candidates for public office no cable television system shall make any discrimination between candidates in charges, practices, regulations, facilities, or services for or in connection with the service rendered, or make or give any preference to any candidate for public office or subject any such candidate to any prejudice or disadvantage; nor shall any cable television system make any contract or other agreement which shall have the effect of permitting any legally qualified candidate for any public office to cablecast to the exclusion of other legally qualified candidates for the same public office.

(c) *Records, inspections.* Every cable television system shall keep and permit public inspection of a complete record of all requests for origination cablecasting time made by or on behalf of candidates for public office, together with an appropriate notation showing the disposition made by the system of such requests, the charges made, if any, and the length and time of cablecast, if the request is granted. Such records shall be retained for a period of 2 years.

(d) *Time of request.* A request for equal opportunities for use of the origination channel(s) must be submitted to the cable television system within one (1) week of the day on which the first prior use, giving rise to the right of equal opportunities, occurred: *Provided, however,* That where a person was not a candidate at the time of such first prior use, he shall submit his request within one (1) week of the first subsequent use after he has become a legally qualified candidate for the office in question.

(e) *Burden of proof.* A candidate requesting such equal opportunities of the cable television system, or complaining of noncompliance to the Commission, shall have the burden of proving that he and his opponent are legally qualified candidates for the same public office.

§ 76.209 Fairness doctrine; personal attacks; political editorials.

(a) A cable television system engaging in origination cablecasting shall afford reasonable opportunity for the discussion of conflicting views on issues of public importance.

NOTE: See public notice, "Applicability of the Fairness Doctrine in the Handling of Controversial Issues of Public Importance," 29 F.R. 10415.

(b) When, during such origination cablecasting, an attack is made upon the honesty, character, integrity, or like personal qualities of an identified person or group, the cable television system shall, within a reasonable time and in no event later than one (1) week after the attack, transmit to the person or group attacked: (1) Notification of the date, time, and identification of the cablecast; (2) a script or tape (or an accurate summary if a script or tape is not available) of the attack; and (3) an offer of a reasonable opportunity to respond over the system's facilities.

(c) The provisions of paragraph (b) of this section shall not be applicable: (1) To attacks on foreign groups or foreign public figures; (2) to personal attacks which are made by legally qualified candidates, their authorized spokesmen, or those associated with them in the campaign, on other such candidates, their authorized spokesmen, or persons associated with the candidates in the campaign; and (3) to bona fide newscasts, bonafide news interviews, and on-the-spot coverage of a bona fide news event (including commentary or analysis contained in the foregoing programs, but the provisions of paragraph (b) of this section shall be applicable to editorials of the cable television system).

(d) Where a cable television system, in an editorial, (1) endorses or (2) opposes a legally qualified candidate or candidates, the system shall, within 24 hours after the editorial, transmit to respectively (i) the other qualified candidate or candidates for the same office, or (ii) the candidate opposed in the editorial, (a) notification of the date, time, and channel of the editorial; (b) a script or tape of the editorial; and (c) an offer of a reasonable opportunity for a candidate or a spokesman of the candidate to respond over the system's facilities: *Provided, however,* That where such editorials are cablecast within 72 hours prior to the day of the election, the system shall comply with the provisions of this paragraph sufficiently far in advance of the broadcast to enable the candidate or candidates to have a reasonable opportunity to prepare a response and to present it in a timely fashion.

§ 76.213 Lotteries.

(a) No cable television system when engaged in origination cablecasting shall transmit or permit to be transmitted on the origination cablecasting channel or channels any advertisement of or information concerning any lottery, gift enterprise, or similar scheme, offering prizes dependent in whole or in part upon lot or chance, or any list of the prizes drawn or awarded by means of any such lottery, gift enterprise, or scheme, whether said list contains any part or all of such prizes.

(b) The determination whether a particular program comes within the provisions of paragraph (a) of this section depends on the facts of each case. However, the Commission will in any event consider that a program comes within the provisions of paragraph (a) of this section if in connection with such program a prize consisting of money or thing of value is awarded to any person whose selection is dependent in whole or in part upon lot or chance, if as a condition of winning or competing for such prize, such winner or winners are required to furnish any money or thing of value or are required to have in their possession any product sold, manufactured, furnished, or distributed by a sponsor of a program cablecast on the system in question.

§ 76.215 Obscenity.

No cable television system when engaged in origination cablecasting shall transmit or permit to be transmitted on the origination cablecasting channel or channels material that is obscene or indecent.

§ 76.217 Advertising.

A cable television system engaged in origination cablecast programing may present advertising material at the beginning and conclusion of each such program and at natural intermissions or breaks within a cablecast: *Provided, however,* That the system itself does not interrupt the presentation of program material in order to intersperse advertising: *And provided, further,* That advertising material is not presented on or in connection with origination cablecasting in any other manner.

NOTE: The term "natural intermissions or breaks within a cablecast" means any natural intermission in the program material which is beyond the control of the cable television operator, such as time-out in a sporting event, an intermission in a concert or dramatic performance, a recess in a city council meeting, an intermission in a long motion picture which was present at the time of theatre exhibition, etc.

§ 76.221 Sponsorship identification.

(a) When a cable television system engaged in origination cablecasting presents any matter for which money, services, or other valuable consideration is either directly or indirectly paid or promised to, or charged or received by, such system, the system shall make an announcement that such matter is sponsored, paid for, or furnished, either in whole or in part, and by whom or on whose behalf such consideration was supplied: *Provided, however,* That "service or other valuable consideration" shall not include any service or property furnished without charge or at a nominal charge for use on, or in connection with, such cablecasting unless it is so furnished as consideration for an identification in a cablecast of any person, product, service, trademark, or brand name beyond an identification which is reasonably related to the use of such service or property on the cablecast.

(b) Each system engaged in origination cablecasting shall exercise reasonable diligence to obtain from its employees, and from other persons with whom it deals directly in connection with any program matter for origination cablecasting, information to enable it to make the announcement required by this section.

(c) In the case of any political program or any program involving the discussion of public controversial issues for which any films, records, transcriptions, talent, script, or other material or services of any kind are furnished, either directly or indirectly, to a cable television system as an inducement to the origination cablecasting of such program, an announcement to this effect shall be made at the beginning and conclusion of such program: *Provided, however,* That only one such announcement need be made in the case of any such program of five (5) minutes' duration or less, either at the beginning or conclusion of the program.

(d) The announcements required by this section are waived with respect to feature motion picture films produced initially and primarily for theater exhibition.

§ 76.225 Per-program or per-channel charges for reception of cablecasts.

(a) Origination or access cablecasting operations for which a per-program or per-channel charge is made shall comply with the following requirements:

(1) Feature films shall not be cablecast which have had general release in theaters anywhere in the United States more than two (2) years prior to their cablecast: *Provided, however,* That during 1 week of each calendar month one feature film the general release of which occurred more than ten (10) years previously may be cablecast, and more than a single showing of such film may be made during that week: *Provided, further,* That feature films the general release of which occurred between two (2) and ten (10) years before proposed cablecast may be cablecast upon a convincing showing to the Commission that bona fide attempt has been made to sell the films for conventional television broadcasting and that they have been refused, or that the owner of the broadcast rights to the films will not permit them to be televised on conventional television because he has been unable to work out satisfactory arrangements concerning editing for presentation thereon, or perhaps because he intends never to show them on conventional television since to do so might impair their repetitive box office potential in the future.

NOTE: As used in this subparagraph, "general release" means the first-run showing of a feature film in a theatre or theatres in an area, on a nonreserved-seat basis, with continuous performances. For first-run showing of feature films on a nonreserved-seat basis which are not considered to be "general release" for purposes of this subparagraph, see note 56 in Fourth Report and Order in Docket No. 11279, 15 FCC 2d 466.

(2) Sports events shall not be cablecast which have been televised live on a nonsubscription, regular basis in the community during the two (2) years preceding their proposed cablecast: *Provided, however,* That if the last regular occurrence of a specific event (e.g., summer Olympic games) was more than two (2) years before proposed showing on cable television in a community and the event was at that time televised on conventional television in that community, it shall not be cablecast.

NOTE 1. In determining whether a sports event has been televised in a community on a nonsubscription basis, only commercial television broadcast stations which place a Grade A contour over the entire community will be considered. Such stations need not necessarily be licensed to serve that community.

NOTE 2: The manner in which this subparagraph will be administered and in which "sports," "sports events," and "televised live on a nonsubscription regular basis" will be construed is explained in paragraphs 288–305 in Fourth Report and Order in Docket No. 11279, 15 FCC 2d 466.

(3) No series type of program with interconnected plot or substantially the same cast of principal characters shall be cablecast.

(4) Not more than 90 percent of the total cablecast programing hours shall consist of feature films and sports events combined. The percentage calculations may be made on a yearly basis, but, absent a showing of good cause, the percentage of such programing hours may not exceed 95 percent of the total cablecast programing hours in any calendar month.

(5) No commercial advertising announcements shall be carried on such channels during such operations except, before and after such programs, for promotion of other programs for which a per-program or per-channel charge is made.

§ 76.251 Minimum channel capacity; access channels.

(a) No cable television system operating in a community located in whole or in part within a major television market, as defined in § 76.5, shall carry the signal of any television broadcast station unless the system also complies with the following requirements concerning the availability and administration of access channels:

(1) *Minimum channel capacity.* Each such system shall have at least 120 MHz of bandwidth (the equivalent of 20 television broadcast channels) available for immediate or potential use for the totality of cable services to be offered;

(2) *Equivalent amount of bandwidth.* For each Class I cable channel that is utilized, such system shall provide an additional channel, 6 MHz in width, suitable for transmission of Class II or Class III signals (see § 76.5 for cable channel definitions);

(3) *Two-way communications.* Each such system shall maintain a plant having technical capacity for nonvoice return communications;

(4) *Public access channel.* Each such system shall maintain at least one specially designated, noncommercial public access channel available on a first-come, nondiscriminatory basis. The system shall maintain and have available for public use at least the minimal equipment and facilities necessary for the production of programing for such a channel. See also § 76.201;

(5) *Education access channel.* Each such system shall maintain at least one specially designated channel for use by local educational authorities;

(6) *Local government access channel.* Each such system shall maintain at least one specially designated channel for local government uses;

(7) *Leased access channels.* Having satisfied the origination cablecasting requirements of § 76.201, and the requirements of subparagraphs (4), (5), and (6) of this paragraph for specially designated access channels, such system shall offer other portions of its nonbroadcast bandwidth, including unused portions of the specially designated channels, for leased access services. However, these leased channel operations shall be undertaken with the express understanding that they are subject to displacement if there is a demand to use the channels for their specially designated purposes. On at least one of the leased channels, priority shall be given part-time users;

(8) *Expansion of access channel capacity.* Whenever all of the channels described in subparagraphs (4) through (7) of this paragraph are in use during 80 percent of the weekdays (Monday–Friday) for 80 percent of the time during any consecutive 3-hour period for 6 consecutive weeks, such system shall have 6 months in which to make a new channel available for any or all of the above-described purposes;

(9) *Program content control.* Each such system shall exercise no control over program content on any of the channels described in subparagraphs (4) through (7) of this paragraph; however, this limitation shall not prevent it from taking appropriate steps to insure compliance with the operating rules described in subparagraph (11) of this paragraph;

(10) *Assessment of costs.* (i) From the commencement of cable television service in the community of such system until five (5) years after completion of the system's basic trunk line, the channels described in subparagraphs (5) and (6) of this paragraph shall be made available without charge.

(ii) One of the public access channels described in subparagraph (4) of this paragraph shall always be made available without charge, except that production costs may be assessed for live studio presentations exceeding 5 minutes. Such production costs and any fees for use of other public access channels shall be consistent with the goal of affording the public a low-cost means of television access;

(11) *Operating rules.* (i) For the public access channel(s), such system shall establish rules requiring first-come non-discriminatory access; prohibiting the presentation of: Any advertising material designed to promote the sale of commercial products or services (including advertising by or on behalf of candidates for public office); lottery information; and obscene or indecent matter (modeled after the prohibitions in §§ 76.213 and 76.215, respectively); and permitting public inspection of a complete record of the names and addresses of all persons or groups requesting access time. Such a record shall be retained for a period of 2 years.

(ii) For the educational access channel(s), such system shall establish rules prohibiting the presentation of: Any advertising material designed to promote the sale of commercial products or services (including advertising by or on behalf of candidates for public office); lottery information; and obscene or indecent matter (modeled after the prohibitions in §§ 76.213 and 76.215, respectively) and permitting public inspection of a complete record of the names and addresses of all persons or groups requesting access time. Such a record shall be retained for a period of 2 years.

(iii) For the leased channel(s), such system shall establish rules requiring first-come, nondiscriminatory access; prohibiting the presentation of lottery information and obscene or indecent matter (modeled after the prohibitions in §§ 76.213 and 76.215, respectively); requiring sponsorship identification (see § 76.221); specifying an appropriate rate schedule and permitting public inspection of a complete record of the names and addresses of all persons or groups requesting time. Such a record shall be retained for a period of 2 years.

(iv) The operating rules governing public access, educational, and leased channels shall be filed with the Commission within 90 days after a system first activates any such channels, and shall be available for public inspection at the system's offices. Except on specific authorization, or with respect to the operation of the local government access channel, no local entity shall prescribe any other rules concerning the number or manner of operation of access channels; however, franchise specifications concerning the number of such channels for systems in operation prior to March 31, 1972, shall continue in effect.

(b) No cable television system operating in a community located wholly outside of all major television markets shall be required by a local entity to exceed the provisions concerning the availability and administration of access channels contained in paragraph (a) of this section. If a system provides any access programing, it shall comply with paragraphs (a) (9), (10), and (11) of this section.

(c) The provisions of this section shall apply to all cable television systems that commence operations on or after March 31, 1972, in a community located in whole or in part within a major television market. Systems that commenced operations prior to March 31, 1972, shall comply on or before March 31, 1977: *Provided, however,* That, if such systems begin to provide any of the access services described

above at an earlier date, they shall comply with paragraph (a) (9), (10), and (11) of this section at that time *And provided, further,* That if such systems receive certificates of compliance to add television signals to their operations at an earlier date, they shall comply with paragraph (a) (4) through (11) of this section at the time of such addition.

Subpart H—General Operating Requirements

§ 76.301   Copies of rules.

The operator of a cable television system shall have a current copy of Part 76, and is expected to be familiar with the rules governing cable television systems. Copies of the Commission's rules may be obtained from the Superintendent of Documents, Government Printing Office, Washington, D.C. 20402, at nominal cost.

§ 76.305   Logging and recordkeeping requirements.

(a) *Carriage of certain television signals.* (1) A cable television system operating in a community located in whole or in part within a major television market shall keep and permit public inspection of a record of all television signals carried pursuant to §§ 76.61 (b), (c), (d), or (e) or 76.63(a) (as it refers to § 76.61 (b), (c), (d), or (e)). Such record shall include the call letters and location of each such station whose signals are carried, the date and specific starting and ending time of such carriage, and the names of the programs scheduled to be shown. This record shall be retained for a period of 2 years.

(2) This paragraph shall be applicable only to television signals whose carriage commenced on or after March 31, 1972.

(b) *Origination cablecasts by candidates for public office.* See § 76.205(c).

(c) *Public access channels.* See § 76.251 (a) (11).

(d) *Educational access channels.* See § 76.251(a) (11).

(e) *Leased access channels.* See § 76.251(a) (11).

Subpart I—Forms and Reports

§ 76.401   Annual report of cable television systems.

An "Annual Report of Cable Television Systems" (FCC Form 325) shall be filed with the Commission for each cable television system, as defined in § 76.5, on or before March 1 of each year, for the preceding calendar year.

§ 76.405   Cable television annual financial report.

A "Cable Television Annual Financial Report" (FCC Form 326) shall be filed with the Commission for each cable television system, as defined in § 76.5, on or before April 1 of each year, for the preceding calendar year: *Provided, however,* That a cable television system which commences operations prior to December 1, 1971, may report on a fiscal year basis, in which case Form 326 shall be filed annually no more than ninety (90) days after the close of the system's fiscal year.

§ 76.406   Computation of cable television annual fee.

A "Computation of Cable Television Annual Fee" (FCC Form 326–A) shall be filed with the Commission for each cable television system, as defined in § 76.5, on or before April 1 of each year, for the preceding calendar year, to accompany payment of the cable television annual fee. See §§ 1.1101 and 1.1116.

Subpart J—Diversification of Control

§ 76.501   Cross-ownership.

(a) No cable television system (including all parties under common control) shall carry the signal of any television broadcast station if such system directly or indirectly owns, operates, controls, or has an interest in:

(1) A national television network (such as ABC, CBS, or NBC); or

(2) A television broadcast station whose predicted Grade B contour, computed in accordance with § 73.684 of this chapter, overlaps in whole or in part the service area of such system (i.e., the area within which the system is serving subscribers); or

(3) A television translator station licensed to the community of such system.

NOTE 1: The word "control" as used herein is not limited to majority stock ownership, but includes actual working control in whatever manner exercised.

NOTE 2: The word "interest" as used herein includes, in the case of corporations, common officers or directors, and partial (as well as total) ownership interests represented by ownership of voting stock.

NOTE 3: In applying the provisions of paragraph (a) of this section to the stockholders of a corporation which has more than 50 stockholders:

(a) Only those stockholders need be considered who are officers or directors or who directly or indirectly own 1 percent or more of the outstanding voting stock.

(b) Stock ownership by an investment company as defined in 15 U.S.C. section 80a–3 (commonly called a mutual fund) need be considered only if it directly or indirectly owns 3 percent or more of the outstanding voting stock or if officers or directors of the corporation are representatives of the investment company. Holdings by investment companies under common management shall be aggregated. If an investment company directly or indirectly owns voting stock in an intermediate company which in turn directly or indirectly owns 50 percent or more of the voting stock of the corporation, the investment company shall be considered to own the same percentage of outstanding shares of such corporation as it owns of the intermediate company: *Provided, however,* That the holding of the investment company need not be considered where the intermediate company owns less than 50 percent of the voting stock, but officers or directors of the corporation who are representatives of the intermediate company shall be deemed to be representatives of the investment company.

(c) In cases where record and beneficial ownership of voting stock is not identical (e.g., bank nominees holding stock as record owners for the benefit of mutual funds, brokerage houses holding stock in street name for the benefit of customers, trusts holding stock as record owners for the benefit of designated parties), the party having the right to determine how the stock will be voted will be considered to own it for the purposes of this section.

(b) The provisions of paragraph (a) of this section are not effective until August 10, 1972, as to ownership interests proscribed herein if such interests were in existence on or before July 1, 1970 (e.g., if a franchise were in existence on or before July 1, 1970): *Provided, however,* That the provisions of paragraph (a) of this section are effective on August 10, 1970, as to such interests acquired after July 1, 1970.

Subpart K—Technical Standards

§ 76.601   Performance tests.

(a) The operator of each cable television system shall be responsible for insuring that each such system is designed, installed, and operated in a manner that fully complies with the provisions of this subpart. Each system operator shall be prepared to show, on request by an authorized representative of the Commission, that the system does, in fact, comply with the rules.

(b) The operator of each cable television system shall maintain at its local office a current listing of the cable television channels which that system delivers to its subscribers and the station or stations whose signals are delivered on each Class I cable television channel, and shall specify for each subscriber the minimum visual signal level it maintains on each Class I cable television channel under normal operating conditions.

(c) The operator of each cable television system shall conduct complete performance tests of that system at least once each calendar year (at intervals not to exceed 14 months) and shall maintain the resulting test data on file at the system's local office for at least five (5) years. It shall be made available for inspection by the Commission on request. The performance tests shall be directed at determining the extent to which the system complies with all the technical standards set forth in § 76.605. The tests shall be made on each Class I cable television channel specified pursuant to paragraph (b) of this section, and shall include measurements made at no less than three widely separated points in the system, at least one of which is representative of terminals most distant from the system input in terms of cable distance. The measurements may be taken at convenient monitoring points in the cable network: *Provided,* That data shall be included to relate the measured performance to the system performance as would be viewed from a nearby subscriber terminal. A description of intruments and procedure and a statement of the qualifications of the person performing the tests shall be included.

(d) Successful completion of the performance tests required by paragraph (c) of this section does not relieve the system of the obligation to comply with all pertinent technical standards at all subscriber terminals. Additional tests, repeat tests, or tests involving specified subscriber terminals may be required by the Commission in order to secure compliance with the technical standards.

(c) All of the provisions of this section shall become effective March 31, 1972.

§ 76.605 Technical standards.

(a) The following requirements apply to the performance of a cable television system as measured at any subscriber terminal with a matched termination, and to each of the Class I cable television channels in the system:

(1) The frequency boundaries of cable television channels delivered to subscriber terminals shall conform to those set forth in § 73.603(a) of this chapter: *Provided, however,* That on special application including an adequate showing of public interest, other channel arrangements may be approved.

(2) The frequency of the visual carrier shall be maintained 1.25 MHz±25 kHz above the lower boundary of the cable television channel, except that, in those systems that supply subscribers with a converter in order to facilitate delivery of cable television channels, the frequency of the visual carrier at the output of each such converter shall be maintained 1.25 MHz±250 kHz above the lower frequency boundary of the cable television channel.

(3) The frequency of the aural carrier shall be 4.5 MHz±1 kHz above the frequency of the visual carrier.

(4) The visual signal level, across a terminating impedance which correctly matches the internal impedance of the cable system as viewed from the subscriber terminals, shall be not less than the following appropriate value:

Internal impedance:
75 ohms.
300 ohms.
Visual signal level:
1 millivolt.
2 millivolts.

(At other impedance values, the minimum visual signal level shall be $\sqrt{0.0133}$ Z millivolts, where Z is the appropriate impedance value.)

(5) The visual signal level on each channel shall not vary more than 12 decibels overall, and shall be maintained within

(i) 3 decibels of the visual signal level of any visual carrier within 6 MHz nominal frequency separation, and

(ii) 12 decibels of the visual signal level on any other channel, and

(iii) A maximum level such that signal degradation due to overload in the subscriber's receiver does not occur.

(6) The rms voltage of the aural signal shall be maintained between 13 and 17 decibels below the associated visual signal level.

(7) The peak-to-peak variation in visual signal level caused undesired low frequency disturbances (hum or repetitive transients) generated within the system, or by inadequate low frequency response, shall not exceed 5 percent of the visual signal level.

(8) The channel frequency response shall be within a range of ±2 decibels for all frequencies within −1 MHz and +4 MHz of the visual carrier frequency.

(9) The ratio of visual signal level to system noise, and of visual signal level to

any undesired cochannel television signal operating on proper offset assignment, shall be not less than 36 decibels. This requirement is applicable to:

(i) Each signal which is delivered by a cable television system to subscribers within the predicted Grade B contour for that signal, or

(ii) Each signal which is first picked up within its predicted Grade B contour.

(10) The ratio of visual signal level to the rms amplitude of any coherent disturbances such as intermodulation products or discrete-frequency interfering signals not operating on proper offset assignments shall not be less than 46 decibels.

(11) The terminal isolation provided each subscriber shall be not less than 18 decibels, but in any event, shall be sufficient to prevent reflections caused by open-circuited or short-circuited subscriber terminals from producing visible picture impairments at any other subscriber terminal.

(12) Radiation from a cable television system shall be limited as follows:

Frequencies	Radiation limit (microvolts/meter)	Distance (feet)
Up to and including 54 MHz	15	100
Over 54 up to and including 216 MHz.	20	10
Over 216 MHz	15	100

(b) Cable television systems distributing signals by using multiple cable techniques or specialized receiving devices, and which, because of their basic design, cannot comply with one or more of the technical standards set forth in paragraph (a) of this section, may be permitted to operate provided that an adequate showing is made which establishes that the public interest is benefited. In such instances the Commission may prescribe special technical requirements to ensure that subscribers to such systems are provided with a good quality of service.

(c) Paragraph (a)(12) of this section shall become effective March 31, 1972. All other provisions of this section shall become effective in accordance with the following schedule:

	Effective date
Cable television systems in operation prior to March 31, 1972	Mar. 31, 1977
Cable television systems commencing operations on or after March 31, 1972	Mar. 31, 1972

§ 76.609 Measurements.

(a) Measurements made to demonstrate conformity with the performance requirements set forth in §§ 76.701 and 76.605 shall be made under conditions which reflect system performance during normal operations, including the effect of any microwave relay operated in the Cable Television Relay (CAR) Service intervening between pickup antenna and the cable distribution network. Amplifiers shall be operated at normal gains, either by the insertion of appro-

priate signals or by manual adjustment. Special signals inserted in a cable television channel for measurement purposes should be operated at levels approximating those used for normal operation. Pilot tones, auxiliary or substitute signals, and nontelevision signals normally carried on the cable television system should be operated at normal levels to the extent possible. Some exemplary, but not mandatory, measurement procedures are set forth in this section.

(b) When it may be necessary to remove the television signal normally carried on a cable television channel in order to facilitate a performance measurement, it will be permissible to disconnect the antenna which serves the channel under measurement and to substitute therefor a matching resistance termination. Other antennas and inputs should remain connected and normal signal levels should be maintained on other channels.

(c) As may be necessary to ensure satisfactory service to a subscriber, the Commission may require additional tests to demonstrate system performance or may specify the use of different test procedures.

(d) The frequency response of a cable television channel may be determined by one of the following methods, as appropriate:

(1) By using a swept frequency or a manually variable signal generator at the sending end and a calibrated attenuator and frequency-selective voltmeter at the subscriber terminal; or

(2) By using a multiburst generator and modulator at the sending end and a demodulator and oscilloscope display at the subscriber terminal.

(e) System noise may be measured using a frequency-selective voltmeter (field strength meter) which has been suitably calibrated to indicate rms noise or average power level and which has a known bandwidth. With the system operating at normal level and with a properly matched resistive termination substituted for the antenna, noise power indications at the subscriber terminal are taken in successive increments of frequency equal to the bandwidth of the frequency-selective voltmeter, summing the power indications to obtain the total noise power present over a 4 MHz band centered within the cable television channel. If it is established that the noise level is constant within this bandwidth, a single measurement may be taken which is corrected by an appropriate factor representing the ratio of 4 MHz to the noise bandwidth of the frequency-selective voltmeter. If an amplifier is inserted between the frequency-selective voltmeter and the subscriber terminal in order to facilitate this measurement, it should have a bandwidth of at least 4 MHz and appropriate corrections must be made to account for its gain and noise figure. Alternatively, measurements made in accordance with the NCTA standard on noise measurement (NCTA Standard 005-0669) may be employed.

(f) The amplitude of discrete frequency interfering signals within a cable television channel may be determined with either a spectrum analyzer or with a frequency-selective voltmeter (field strength meter), which instruments have been calibrated for adequate accuracy. If calibration accuracy is in doubt, measurements may be referenced to a calibrated signal generator, or a calibrated variable attenuator, substituted at the point of measurement. If an amplifier is used between the subscriber terminal and the measuring instrument, appropriate corrections must be made to account for its gain.

(g) The terminal isolation between any two terminals in the system may be measured by applying a signal of known amplitude to one and measuring the amplitude of that signal at the other terminal. The frequency of the signal should be close to the midfrequency of the channel being tested.

(h) Measurements to determine the field strength of radio frequency energy radiated by cable television systems shall be made in accordance with standard engineering procedures. Measurements made on frequencies above 25 MHz shall include the following:

(1) A field strength meter of adequate accuracy using a horizontal dipole antenna shall be employed.

(2) Field strength shall be expressed in terms of the rms value of synchronizing peak for each cable television channel for which radiation can be measured.

(3) The dipole antenna shall be placed 10 feet above the ground and positioned directly below the system components. Where such placement results in a separation of less than 10 feet between the center of the dipole antenna and the system components, the dipole shall be repositioned to provide a separation of 10 feet.

(4) The horizontal dipole antenna shall be rotated about a vertical axis and the maximum meter reading shall be used.

(5) Measurements shall be made where other conductors are 10 or more feet away from the measuring antenna.

§ 76.613 Interference from a cable television system.

In the event that the operation of a cable television system causes harmful interference to reception of authorized radio stations, the operation of the system shall immediately take whatever steps are necessary to remedy the interference.

§ 76.617 Responsibility for receiver-generated interference.

Interference generated by a radio or television receiver shall be the responsibility of the receiver operator in accordance with the provisions of Part 15, Subpart C, of this chapter: Provided, however, That the operator of a cable television system to which the receiver is connected shall be responsible for the suppression of receiver-generated interference that is distributed by the system when the interfering signals are introduced into the system at the receiver.

PART 78—CABLE TELEVISION RELAY SERVICE

F. Part 78—Cable Television Relay Service—is added to read as follows:

Subpart A—General

Sec.
78.1 Purpose.
78.3 Other pertinent rules.
78.5 Definitions.

Subpart B—Applications and Licenses

78.11 Permissible service.
78.13 Eligibility for license.
78.15 Contents of applications.
78.17 Frequency assignments.
78.19 Interference.
78.21 Notification of filing of applications.
78.23 Equipment tests.
78.25 Service or program tests.
78.27 License conditions.
78.29 License period.
78.31 Temporary extension of license.

Subpart C—General Operating Requirements

78.51 Remote control operation.
78.53 Unattended operation.
78.55 Time of operation.
78.57 Station inspection.
78.59 Posting of station and operator licenses.
78.61 Operator requirements.
78.63 Painting and lighting of antenna structures.
78.65 Additional orders.
78.67 Copies of rules.
78.69 Operating log.

Subpart D—Technical Regulations

78.101 Power limitations.
78.103 Emissions and bandwidth.
78.105 Antennas.
78.107 Equipment and installation.
78.109 Equipment changes.
78.111 Frequency tolerance.
78.113 Frequency monitors and measurements.
78.115 Modulation limits.

AUTHORITY: The provisions of this Part 78 issued under secs. 2, 3, 4, 301, 303, 307, 308, 309, 48 Stat., as amended, 1064, 1065, 1066, 1081, 1082, 1083, 1084, 1085; 47 U.S.C. 152, 153, 154, 301, 303, 307, 308, 309.

Subpart A—General

§ 78.1 Purpose.

The rules and regulations set forth in this part provide for the licensing and operation of fixed or mobile cable television relay stations used for the transmission of television and related audio signals, signals of standard and FM broadcast stations, signals of instructional television fixed stations, and cablecasting from the point of reception to a terminal point from which the signals are distributed to the public by cable.

§ 78.3 Other pertinent rules.

Other pertinent provisions of the Commission's rules and regulations relating to the Cable Television Relay Service are included in the following parts of this chapter:

Part 0—Commission Organization.
Part 1—Practice and Procedure.
Part 76—Cable Television Service.

§ 78.5 Definitions.

For purposes of this part, the following definitions are applicable. For other definitions, see Part 76 (Cable Television Service) of this chapter.

(a) Cable television relay (CAR) station. A fixed or mobile station used for the transmission of television and related audio signals, signals of standard and FM broadcast stations, signals of instructional television fixed stations, and cablecasting from the point of reception to a terminal point from which the signals are distributed to the public by cable.

NOTE: Except where the rules contained in this part make separate provision, the term "cable television relay" or "CAR" includes the term "local distribution service" or "LDS", the term "cable television relay studio to headend link" or "SHL," and the term "cable television relay pickup," as defined in paragraphs (b), (c), and (d) of this section.

(b) Local distribution service (LDS) station. A fixed CAR station used within a cable television system or systems for the transmission of television signals and related audio signals, signals of standard and FM broadcast stations, signals of instructional television fixed stations, and cablecasting from a local transmission point to one or more receiving points, from which the communications are distributed to the public by cable. LDS stations may also engage in repeatered operation.

(c) Cable television relay studio to headend link (SHL) station. A fixed CAR station used for the transmission of television program material and related communications from a cable television studio to the headend of a cable television system.

(d) Cable television relay pickup station. A land mobile CAR station used for the transmission of television signals and related communications from the scenes of events occurring at points removed from cable television studios to cable television studios or headends.

(e) Remote control operation. Operation of a station by a qualified operator on duty at a control position from which the transmitter is not visible but which control position is equipped with suitable control and telemetering circuits so that the essential functions that could be performed at the transmitter can also be performed from the control point.

(f) Attended operation. Operation of a station by a qualified operator on duty at the place where the transmitting apparatus is located with the transmitter in plain view of the operator.

(g) Unattended operation. Operation of a station by automatic means whereby the transmitter is turned on and off and performs its functions without attention by a qualified operator.

Subpart B—Applications and Licenses

§ 78.11 Permissible service.

(a) Cable television relay stations are authorized to relay television broadcast and related audio signals, the signals of standard and FM broadcast stations, signals of instructional television fixed stations, and cablecasting intended for use solely by one or more cable television systems. LDS stations are authorized to relay television broadcast and related audio signals, the signals of standard

and FM broadcast stations, signals of instructional television fixed stations, cablecasting, and such other communications as may be authorized by the Commission, Relaying includes retransmission of signals by intermediate relay stations in the system. CAR licensees may interconnect their facilities with those of other CAR or common carrier licensees, and may also retransmit the signals of such CAR or common carrier stations, provided that the program material retransmitted meets the requirements of this paragraph.

(b) The transmitter of a cable television relay station using FM transmission may be multiplexed to provide additional communication channels for the transmission of standard and FM broadcast station programs and operational communications directly related to the technical operation of the relay system (including voice communications, telemetry signals, alerting signals, fault reporting signals, and control signals). A cable television relay station will be authorized only where the principal use is the transmission of television broadcast program material or cablecasting: *Provided, however,* That this requirement shall not apply to LDS stations.

(c) Cable television relay station licenses may be issued to cable television owners or operators and to cooperative enterprises wholly owned by cable television owners or operators.

(d) Cable television relay systems shall supply program material to cable television systems only in the following circumstances:

(1) Where the licensee of the CAR station or system is owner or operator of the cable television systems supplied with program material; or

(2) Where the licensee of the CAR station or system supplies program material to cable television systems either without charge or on a nonprofit, cost-sharing basis pursuant to a written contract between the parties involved which provides that the CAR licensee shall have exclusive control over the operation of the cable television relay stations licensed to him and that contributions to capital and operating expenses are accepted only on a cost-sharing, nonprofit basis, prorated on an equitable basis among all cable television systems being supplied with program material in whole or in part. Records showing the cost of the service and its nonprofit, cost-sharing nature shall be maintained by the CAR licensee and held available for inspection by the Commission.

(e) A CAR licensee shall file a notification with the Commission thirty (30) days prior to supplying program material to any cable television system that has not been specified in its license application or in a prior notification to the Commission containing the following information:

(1) A copy of the contract between the parties pursuant to which the program material will be supplied;

(2) Network and station origin of the signals to be transmitted or, if cablecasting, the intended source and general nature of the programing;

(3) Location of the point at which reception will be made;

(4) Location of intermediate relay stations in the system through which the signal will be transmitted;

(5) Location of the relay station that will supply the program material to the cable television system;

(6) Name of each community to be served by the cable television system;

(7) Current number of subscribers of the cable television system; and

(8) Identity of the owner or owners of the cable television system.

The CAR licensee may institute the service described in such notification 30 days after filing unless the Commission during that period notifies the licensee that the information supplied is inadequate or that the proposed service is not authorized under these rules, and the licensee shall then have the right to amend or file another notification to remedy the inadequacy or defect and to institute the service 30 days thereafter, or at such earlier date as the Commission may set upon finding that the inadequacy or defect has been remedied.

(f) Each CAR licensee providing program material to a cable television system pursuant to paragraph (d)(2) of this section shall file an annual report with the Commission within 90 days of the close of its fiscal year containing:

(1) A financial statement of such operations in sufficient detail to show compliance with the requirements of this section;

(2) The names of those who have shared the use of the licensed facilities;

(3) A brief statement as to the use of the facilities made by each person sharing the use and an estimate of the approximate percentage of use by each participant; and

(4) Any change in the items previously reported to the Commission in the application for the license or in a notification under this section.

(g) The provisions of paragraph (d) of this section and § 78.13 shall not apply to a licensee who has been licensed in the CAR service pursuant to § 21.709 of this chapter, except that paragraph (d) of this section shall apply with respect to facilities added or cable television systems first served after February 1, 1966.

(h) Except during momentary circuit failure and brief transition periods, a cable television relay station shall not be permitted to radiate unless it is supplying programs to one or more users.

(i) The license of a CAR pickup station authorizes the transmission of program material, and related communications necessary to the accomplishment of such transmission, from the scenes of events occurring in places other than a cable television studio, to the studio or headend of its associated cable television system, or to such other cable television systems as are carrying the same program material. CAR pickup stations may be used to provide temporary CAR studio to headend links or CAR circuits consistent with this part without further authority of the Commission: *Provided, however,* That prior Commission author-

ity shall be obtained if the transmitting antenna to be installed will increase the height of any natural formation or manmade structure by more than 20 feet and will be in existence for a period of more than 2 consecutive days.

§ 78.13 Eligibility for license.

A license for a cable television relay station will be issued only to the owner of a cable television system or to a cooperative enterprise wholly owned by cable television owners or operators upon a showing that applicant is qualified under the Communications Act of 1934, that frequencies are available for the proposed operation, and that the public interest, convenience, and necessity will be served by a grant thereof.

§ 78.15 Contents of applications.

(a) An application for a new cable television relay station or for changes in the facilities of an existing station shall specify the call sign and location of any television, standard, or FM broadcast stations or instructional television fixed stations to be received and the intended source and general nature of any cablecasting to be relayed, the location of the point at which reception will be made, the number and location of any intermediate relay stations in the system, the location of the terminal receiving point(s) in the system, the name or names of the communities to be served by the cable television system or systems to which the programs will be delivered, the current number of subscribers of each such cable television system, and the name of any other licensee to whom the same program will be delivered through interconnection facilities. An application for a new LDS station or for changes in the facilities of an existing station shall specify in detail the precise nature and technical operation of any service other than the relay of television broadcast signals proposed to be provided on the LDS facilities, including any sections of this part for which waiver is sought.

(b) An application for any authorization subject to § 78.27 for a station used or to be used for the transmission of television broadcast signals shall contain a statement that the applicant has notified the licensee or permittee of any television broadcast station within whose predicted Grade B contour the system operates or will operate, the licensee or permittee of any 100-watt or higher power television translator station licensed to the community of the system, the franchising authority, the superintendent of schools in the community of the system, and any local or state educational television authorities, of the filing of the application. Such statement of the applicant shall be supported by copies of the letters of notification. The notice shall include the fact of intended filing by the applicant, the name and mailing address of each cable television system served or to be served under the authorization sought, the community and area served by each cable television system, and the television, standard broadcast, FM, and instructional television fixed

stations whose signals will be carried by each cable television system.

(c) An application for a construction permit for a new CAR pickup station or for renewal of license of an existing station shall designate the cable television system with which it is to be operated and specify the area in which the proposed operation is intended.

(d) An application for a CAR studio to headend link or LDS station construction permit shall contain a statement that the applicant has investigated the possibility of using cable rather than microwave and the reasons why it was decided to use microwave rather than cable.

NOTE: As used in this § 78.15 the term "predicted Grade B contour" means the field intensity contour defined in § 73.683(a) of this chapter, the location of which is determined exclusively by means of the calculations prescribed in § 73.684 of this chapter.

§ 78.17 Frequency assignments.

(a) The following channels may be assigned to cable television relay stations:

(1) For cable television relay stations using FM transmission:

Group A MHz	Group B MHz
12,700–12,725	12,712.5–12,737.5
12,725–12,750	12,737.5–12,762.5
12,750–12,775	12,762.5–12,787.5
12,775–12,800	12,787.5–12,812.5
12,800–12,825	12,812.5–12,837.5
12,825–12,850	12,837.5–12,862.5
12,850–12,875	12,862.5–12,887.5
12,875–12,900	12,887.5–12,912.5
12,900–12,925	12,912.5–12,937.5
12,925–12,950	

(2) Cable television relay stations using vestigial sideband AM transmission:

Group C MHz	Group D MHz
12,700–12,706.5	12,759.7–12,765.7
12,706.5–12,712.5	12,765.7–12,771.7
12,712.5–12,718.5	12,771.7–12,777.7
12,718.5–12,722.5 [1]	12,777.7–12,781.7 [1]
12,722.5–12,728.5	12,781.7–12,787.7
12,728.5–12,734.5	12,787.7–12,793.7
12,734.5–12,740.5	12,793.7–12,799.7
12,740.5–12,746.5	12,799.7–12,805.7
12,746.5–12,752.5	12,805.7–12,811.7
12,752.5–12,758.5	12,811.1–12,817.7
12,820.5–12,826.5	12,879.7–12,885.7
12,826.5–12,832.5	12,885.7–12,891.7
12,832.5–12,838.5	12,891.7–12,897.7
12,838.5–12,844.5	12,897.7–12,903.7
12,844.5–12,850.5	12,903.1–12,909.7
12,850.5–12,856.5	12,909.7–12,915.7
12,856.5–12,862.5	12,915.7–12,921.7
12,862.5–12,868.5	12,921.7–12,927.7
12,868.5–12,874.5	12,927.7–12,933.7

[1] For transmission of pilot subcarriers, or other authorized narrow band signals.

Auxiliary Channels MHz

12,939.7–12,945.7	12,933.7–12,939.7

(3) For cable television relay stations using frequency modulation to transmit a baseband of frequency-division multiplexed standard television signals:

(i) When the baseband comprises three or four standard television signals:

Group E MHz	Group F MHz
12,700–12,775	12,725–12,800
12,775–12,850	12,800–12,875
12,850–12,925	12,875–12,950

(ii) When the baseband comprises five to eight standard television signals:

Group G MHz	
12,825–12,950	12,700–12,825

(iii) When the baseband comprises nine or more standard television signals:

Group H MHz

12,700–12,950

(b) Television pickup, STL, and intercity relay stations may be assigned channels in the band 12,700–12,950 MHz subject to the conditions that no harmful interference is caused to cable television relay stations authorized at the time of such grants. Similarly, new cable television relay stations shall not cause harmful interference to television STL and intercity relay stations authorized at the time of such grants. Television pickup stations and CAR pickup stations will be assigned channels in the band on a coequal basis subject to the condition that they accept interference from and cause no interference to existing or subsequently authorized television STL, television intercity relay, fixed CAR, CAR SHL, or LDS stations. A cable television system operator will normally be limited in any one area to the assignment of not more than three channels for CAR pickup use: *Provided, however,* That additional channels may be assigned upon a satisfactory showing that additional channels are necessary and are available.

(c) An application for a cable television relay station shall be specific with regard to the channel or channels requested. Channels shall be identified by the channel-edge frequencies listed in paragraph (a) of this section.

(d) For cable television relay stations using frequency modulation to transmit a single television signal, channels normally shall be selected from Group A. Channels in Group B will be assigned only on a case-by-case basis upon an adequate showing that Group A channels cannot be used and that such use will not degrade the technical quality of service provided in Group A channels to the extent that the Group A channels could not be used. On-the-air tests may be required before channels in Group B are permitted to be placed in regular use.

(e) For cable television relay stations using vestigial sideband AM transmission, channels from only Group C or Group D normally will be assigned a station, although upon adequate showing variations in the use of channels in Groups C and D may be authorized on a case-by-case basis in order to avoid potential interference or to permit a more efficient use. The use of channels in both Groups C and D may be authorized for repeated operation, or where the channels in one group are not sufficient to accommodate the services proposed to be provided on the cable television system, if the Commission finds that such use of channels in both groups would serve the public interest.

(f) For vestigial sideband AM transmission, the assigned visual carrier frequency for each channel listed in Group C or Group D shall be 1.25 MHz above the lower channel-edge frequency. The center frequency for the accompanying FM aural carrier in each channel shall be 4.5 MHz above the corresponding visual carrier frequency.

(g) For cable television relay stations using frequency modulation to transmit a baseband of frequency-division multiplexed standard television signals, channels will be assigned from Groups E, F, G, and H according to the number of standard television signals which comprise the baseband, as set forth in paragraph (a)(3) of this section. The station license will indicate the number of standard television signals authorized to be multiplexed for transmission in the assigned channel. The transmission of additional standard television signals may be authorized upon a showing that such can be provided without degradation of the technical quality of the service, and that interference will not be caused to existing operations.

(h) Should any conflict arise among applications for stations in this band, priority will be based on the filing date of an application completed in accordance with the instructions thereon.

§ 78.19 Interference.

(a) Applicants for cable television relay stations shall endeavor to select an assignable frequency or frequencies which will be least likely to result in interference to other licensees in the same area.

(b) Applicants for cable television relay stations shall take full advantage of all known techniques, such as the geometric arrangement of transmitters and receivers, the use of minimum power required to provide the needed service, and the use of highly directive transmitting and receiving antenna systems, to prevent interference to the reception of television STL, television intercity relay, and other CAR stations.

§ 78.21 Notification of filing of applications.

(a) *Radio astronomy and radio research installations.* In order to minimize harmful interference at the National Radio Astronomy Observatory site located at Green Bank, Pocahontas County, W. Va., and at the Naval Radio Research Observatory at Sugar Grove, Pendleton County, W. Va., an applicant for authority to construct a cable television relay station, except a CAR pickup station, or for authority to make changes in the frequency, power, antenna height, or antenna directivity of an existing station within the area bounded by 39°15′ N. on the north, 78°30′ W., on the east, 37°30′ N. on the south and 80°30′ W. on the west shall, at the time of filing such application with the Commission, simultaneously notify the Director, National Radio Astronomy Observatory, Post Office Box No. 2, Green Bank, WV 24944, in writing, of the technical particulars of the proposed station. Such notification shall include the geographical coordinates of the antenna, antenna height, antenna directivity if any, proposed frequency, type of emission, and power. In

addition, the applicant shall indicate in his application to the Commission the date notification was made to the Observatory. After receipt of such application, the Commission will allow a period of 20 days for comments or objections in response to the notifications indicated. If an objection to the proposed operation is received during the 20-day period from the National Radio Astronomy Observatory for itself or on behalf of the Naval Radio Research Observatory, the Commission will consider all aspects of the problem and take whatever action is deemed appropriate.

(b) *Location on Government land.* Applicants proposing to construct a cable television relay station on a site located under the jurisdiction of the U.S. Forest Service, U.S. Department of Agriculture, or the Bureau of Land Management, U.S. Department of the Interior, must supply the information and must follow the procedure prescribed by § 1.70 of this chapter.

§ 78.23 Equipment tests.

(a) During the process of construction of a cable television relay station, the permittee, after notifying the Commission and Engineer in Charge of the district in which the station is located, may, without further authority of the Commission, conduct equipment tests for the purpose of such adjustments and measurements as may be necessary to assure compliance with the terms of the construction permit, the technical provisions of the application therefor, the rules and regulations, and the applicable engineering standards.

(b) The Commission may notify the permittee to conduct no tests or may cancel, suspend, or change the date for the beginning of equipment tests as and when such action may appear to be in the public interest, convenience, and necessity.

(c) Equipment tests may be continued so long as the construction permit shall remain valid.

(d) The authorization for tests contained in this section shall not be construed as constituting a license to operate but as a necessary part of construction.

§ 78.25 Service or program tests.

(a) Upon completion of construction of a cable television relay station in accordance with the terms of the construction permit, the technical provisions of the application therefor, and the rules and regulations and applicable engineering standards, and when an application for station license has been filed showing the station to be in satisfactory operating condition, the permittee of such station may, without further authority of the Commission, conduct service or program tests: *Provided, however,* That the Engineer in Charge of the district in which the station is located and the Commission are notified at least two (2) days (not including Sundays and Saturdays and legal holidays when the offices of the Commission are not open) in advance of the beginning of such operation.

(b) The Commission may notify the permittee to conduct no tests or may

cancel, suspend, or change the date for the beginning of such tests as and when such action may appear to be in the public interest, convenience, and necessity.

(c) Unless sooner suspended or revoked, program test authority will continue valid during Commission consideration of the application for license, and during this period further extension of the construction permit is not required. Program test authority shall be automatically terminated by final determination upon the application for station license.

(d) The authorization for tests contained in this section shall not be construed as approval by the Commission of the application for station license.

§ 78.27 License conditions.

Authorizations (including initial grants, modifications, assignments or transfers of control, and renewals) in the Cable Television Relay Service to construct or operate fixed or mobile stations to relay television and related audio signals, signals of standard and FM broadcast stations, signals of instructional television fixed stations, and cablecasting to cable television systems, either directly or indirectly, shall contain the condition that such cable television systems shall operate in compliance with the provisions of Part 76 (Cable Television Service) of this chapter.

§ 78.29 License period.

Licenses for cable television relay stations will be issued for a period not to exceed five (5) years. On and after February 1, 1966, licenses for CAR stations ordinarily will be issued for a period expiring on February 1, 1971, and, when regularly renewed, at 5-year intervals thereafter. When a license is granted subsequent to the last renewal date for CAR stations, the license will be issued only for the unexpired period of the current license term of such stations. The license renewal date applicable to CAR stations may be varied as necessary to permit the orderly processing of renewal applications, and individual station licenses may be granted or renewed for a shorter period of time than that generally prescribed for CAR stations, if the Commission finds that the public interest, convenience, and necessity would be served by such action.

§ 78.31 Temporary extension of license.

Where there is pending before the Commission any application, investigation, or proceeding which, after hearing, might lead to or make necessary the modification of, revocation of or the refusal to renew an existing cable television relay station license, the Commission will grant a temporary extension of such license: *Provided, however,* That no such temporary extension shall be construed as a finding by the Commission that the operation of any CAR station thereunder will serve the public interest, convenience, and necessity beyond the express terms of such temporary extension of license: *And provided, further,* That such temporary extension of license will in nowise affect or limit the action

of the Commission with respect to any pending application or proceeding.

Subpart C—General Operating Requirements

§ 78.51 Remote control operation.

(a) A cable television relay station may be operated by remote control provided the following conditions are met:

(1) The transmitter and associated control system shall be installed and protected in a manner designed to prevent tampering or operation by unauthorized persons.

(2) An operator meeting the requirements of § 78.61 shall be on duty at the remote control position and in actual charge thereof at all times when the station is in operation.

(3) Facilities shall be provided at the control position which will permit the operator to turn the transmitter on and off at will. The control position shall also be equipped with suitable devices for observing the overall characteristics of the transmissions and a carrier operated device which will give a continuous visual indication whenever the transmitting antenna is radiating a signal. The transmitting apparatus shall be inspected as often as may be necessary to insure proper operation.

(4) The control circuits shall be so designed and installed that short circuits, open circuits, other line faults, or any other cause which would result in loss of control of the transmitter will automatically cause the transmitter to cease radiating.

(b) An application for authority to construct a new station or to make changes in the facilities of an existing station and which proposes operation by remote control shall include an adequate showing of the manner of compliance with the requirements of this section.

§ 78.53 Unattended operation.

(a) A cable television relay station (other than a CAR pickup station) may be operated unattended provided that the following requirements are met:

(1) The transmitter and associated control circuits shall be installed and protected in a manner designed to prevent tampering or operation by unauthorized persons.

(2) The transmitter shall be equipped with an automatic control which will permit it to radiate only when it is relaying an incoming signal. The automatic control may be either a time clock or a signal sensing device. Allowances may be made for momentary circuit failures and brief transition periods when no incoming signal is available for retransmission.

(3) If the transmitting apparatus is located at a site which is not readily accessible at all hours and in all seasons, means shall be provided for turning the transmitter on and off at will from a location which can be reached promptly at all hours and in all seasons.

(4) Licensed radio personnel responsible for the maintenance of the station shall be available on call at a location which will assure expeditious performance of such technical servicing and

maintenance as may be necessary whenever the station is operating. In lieu thereof, arrangements may be made to have an unlicensed person or persons available at all times when the transmitter is operating, to turn the transmitter off in the event that it is operating improperly. The transmitter may not be restored to operation until the malfunction has been corrected by a technically qualified person.

(5) The station licensee shall be responsible for the proper operation of the station at all times and is expected to provide for observations, servicing, and maintenance as often as may be necessary to insure proper operation. All adjustments or tests during or coincident with the installation, servicing, or maintenance of the station which may affect its operation shall be performed by or under the immediate supervision of a licensed radio operator as provided in § 78.61.

(b) An application for authority to construct a new station or make changes in the facilities of an existing station and which proposes unattended operation shall include an adequate showing as to the manner of compliance with the requirements of this section.

§ 78.55   **Time of operation.**

(a) A cable television relay station is not expected to adhere to any prescribed schedule of operation. However, it is limited to operation only when the originating station, or stations, is transmitting the programs which it relays except as provided in paragraph (b) of this section.

(b) The transmitter may be operated for short periods of time to permit necessary tests and adjustments. The radiation of an unmodulated carrier for extended periods of time or other unnecessary transmissions are forbidden.

§ 78.57   **Station inspection.**

The station and all records required to be kept by the licensee shall be made available for inspection upon request by any authorized representative of the Commission.

§ 78.59   **Posting of station and operator licenses.**

(a) The station license and any other instrument of authorization or individual order concerning the construction or the equipment or manner of operation shall be posted at the place where the transmitter is located, so that all terms thereof are visible except as otherwise provided in paragraphs (b) and (c) of this section.

(b) In cases where the transmitter is operated by remote control, the documents referred to in paragraph (a) of this section shall be posted in the manner described at the control point of the transmitter.

(c) In cases where the transmitter is operated unattended, the name of the licensee and the call sign of the unattended station shall be displayed at the transmitter site on the structure supporting the transmitting antenna, so as

to be visible to a person standing on the ground at the transmitter site. The display shall be prepared so as to withstand normal weathering for a reasonable period of time and shall be maintained in a legible condition at all times by the licensee. The station license and other documents referred to in paragraph (a) of this section shall be kept at the nearest attended station or, in cases where the licensee of the unattended station does not operate attended stations, at the point of destination of the signals relayed by the unattended station.

(d) The original of each station operator license shall be posted at the place where the operator is on duty: *Provided, however,* That if the original license of a station operator is posted at another radio transmitting station in accordance with the rules governing the class of station and is there available for inspection a representative of the Commission, a verification card (FCC Form 758–F) is acceptable in lieu of the posting of such license: *And provided, further,* That if the operator on duty holds an operator permit of the card form (as distinguished from the diploma form), he shall not post that permit but shall keep it in his personal possession.

§ 78.61   **Operator requirements.**

(a) Except in cases where a cable television relay station is operated unattended in accordance with § 78.53, an operator holding a valid radiotelephone first- or second-class operator license shall be on duty at the place where the transmitting apparatus is located, in plain view and in actual charge of its operation or at a remote control point established pursuant to the provisions of § 78.51, at all times when the station is in operation. Control and monitoring equipment at a remote control point shall be readily accessible and clearly visible to the operator at that position.

(b) In cases where the cable television relay station is operated unattended pursuant to the provisions of § 78.53, the licensed personnel referred to in paragraph (a)(4) of that section shall hold a valid radiotelephone first- or second-class operator license.

(c) Any transmitter tests, adjustments, or repairs during or coincident with the installation, servicing, operation, or maintenance of a cable television relay station which may affect the proper operation of such station shall be made by or under the immediate supervision and responsibility of a person holding a valid first- or second-class radiotelephone operator license, who shall be fully responsible for proper functioning of the station equipment.

(d) The licensed operator on duty and in charge of a cable television relay station may, at the discretion of the licensee, be employed for other duties or for the operation of another station or stations in accordance with the class of operator license which he holds and the rules governing such stations. However, such duties shall in no way impair or impede the required supervision of the cable television relay station.

§ 78.62   **Painting and lighting of antenna structures.**

The painting and lighting of antenna structures employed by the stations licensed under this part, where required, will be specified in the authorization issued by the Commission. Part 17 of this chapter sets forth the conditions under which painting and lighting will be required and the responsibility of the licensee with regard thereto.

§ 78.65   **Additional orders.**

In case the rules of this part do not cover all phases of operation with respect to external effects, the Commission may make supplemental or additional orders in each case as may be deemed necessary.

§ 78.67   **Copies of rules.**

The licensee of a cable television relay station shall have a current copy of this Part 78, and, in cases where aeronautical obstruction marking of antennas is required, Part 17 of this chapter shall be available for use by the operator in charge. Both the licensee and the operator or operators responsible for the proper operation of the station are expected to be familiar with the rules governing cable television relay stations. Copies of the Commission's rules may be obtained from the Superintendent of Documents, Government Printing Office, Washington, D.C. 20402, at nominal cost.

§ 78.69   **Operating log.**

(a) The licensee of a cable television relay station shall maintain an operating log showing the following:

(1) The date and time of the beginning and end of each period of operation of each transmitter;

(2) The date and time of any unscheduled interruptions to the transmissions of the station, the duration of such interruptions, and the causes thereof;

(3) A record of repairs, adjustments, tests, maintenance, and equipment changes;

(4) Entries required by § 17.49 of this chapter concerning daily observations of tower lights and quarterly inspections of the condition of the tower lights and associated control equipment and an entry when towers are cleaned or repainted as required by § 17.50 of this chapter.

(b) Log entries shall be made in an orderly and legible manner by the person or persons competent to do so, having actual knowledge of the facts required, who shall sign the log when starting duty and again when going off duty.

(c) No log or portion thereof shall be erased, obliterated, or willfully destroyed within the period of retention required by rule. Any necessary correction may be made only by the person who made the original entry who shall strike out the erroneous portion, initial the correction made, and show the date the correction was made.

(d) Operating logs shall be retained for a period of not less than 2 years. The Commission reserves the right to order retention of logs for a longer period of time. In cases where the licensee has no-

tice of any claim or complaint, the log shall be retained until such claim or complaint has been fully satisfied or until the same has been barred by statute limiting the time for filing of suits upon such claims.

Subpart D—Technical Regulations

§ 78.101 Power limitations.

(a) Transmitter peak output power shall not be greater than necessary, and in any event, shall not exceed 5 watts on any channel; except that, stations using frequency modulation to transmit a baseband of frequency-division multiplexed standard television signals may be authorized to use peak power of 15 watts on frequency assignments in Groups E and F, 30 watts on frequency assignments in Group G, and 60 watts on assignments in Group H.

(b) LDS stations shall use for the visual signal either vestigial sideband AM transmission or frequency-division multiplexed FM transmission. When vestigial sideband AM transmission is used, the peak power of the visual signal on all channels shall be maintained within 2 decibels of equality. The mean power of the aural signals on each channel shall not exceed a level 7 decibels below the peak power of the visual signal.

§ 78.103 Emissions and bandwidth.

(a) A cable television relay station may be authorized to employ any type of emission suitable for the simultaneous transmission of visual and aural television signals.

(b) Any emission appearing on a frequency outside of the channel authorized for a transmitter shall be attenuated below the peak power of emission in accordance with the following schedule:

(1) For CAR stations using FM transmission (including those modulated by a frequency-division baseband of standard television signals):

(i) On any frequency above the upper channel limit and below the lower channel limit by between zero and 50 percent of the assigned channel width: At least 25 decibels;

(ii) On any frequency above the upper channel limit or below the lower channel limit by more than 50 percent and up to 150 percent of the assigned channel width: At least 35 decibels; and

(iii) On any frequency above the upper channel limit or below the lower channel limit by more than 150 percent of the assigned channel width: At least $43+10 \log_{10}$ (power in watts) decibels.

(2) For CAR stations using vestigial sideband AM transmission: At least 50 decibels.

(c) In the event that interference to other stations is caused by emissions outside the authorized channel, the Commission may require greater attenuation than that specified in paragraph (b) of this section.

§ 78.105 Antennas.

(a) Cable television relay stations shall use directive transmitting antennas. The maximum beamwidth in the horizontal plane between half power points

of the major lobe shall not exceed 3°: *Provided, however,* That, upon adequate showing of need to serve a larger sector, or more than a single sector, greater beamwidth or multiple antennas may be authorized for LDS stations. Either vertical, horizontal, or elliptical polarization may be employed. The Commission reserves the right to specify the polarization of the transmitted signal.

(b) The choice of receiving antennas is left to the discretion of the licensee. However, licensees will not be protected from interference which results from the lack of adequate antenna discrimination against unwanted signals.

§ 78.107 Equipment and installation.

(a) From time to time the Commission publishes a revised list of type approved and type accepted equipment entitled "Radio Equipment List." Copies of this list are available for inspection at the Commission's offices in Washington, D.C., and at each of its field offices.

(b) Each transmitter authorized for use in the Cable Television Relay Service (other than a CAR pickup station) must be of a type which has been type accepted pursuant to Part 2 (Subpart F) of this chapter, as capable of meeting the requirements of §§ 78.17, 78.101, 78.111, and 78.115.

(c) The installation of a cable television relay station shall be made by or under the immediate supervision of a qualified engineer. Any tests or adjustments requiring the radiation of signals and which could result in improper operation shall be conducted by or under the immediate supervision of an operator holding a valid first- or second-class radiotelephone operator license.

(d) Simple repairs such as the replacement of tubes, fuses, or other plug-in components which require no particular skill may be made by an unskilled person. Repairs requiring replacement of attached components or the adjustment of critical circuits or corroborative measurements shall be made only by a person with required knowledge and skill to perform such tasks.

§ 78.109 Equipment changes.

(a) Formal application is required for any of the following changes:

(1) Replacement of the transmitter as a whole, except replacement with an identical transmitter, or any change in equipment which could result in a change in the electrical characteristics or performance of the station;

(2) Any change in the transmitting antenna system of a station (other than a CAR pickup station), including the direction of the main radiation lobe, directive pattern, antenna gain or transmission line;

(3) Any change in the height of the antenna of a station (other than a CAR pickup station) above ground, or any horizontal change in the location of the antenna;

(4) Any change in the transmitter control system;

(5) Any change in the location of a station transmitter (other than a CAR

pickup station transmitter), except a move within the same building or upon the tower or mast or a change in the area of operation of a CAR pickup station;

(6) Any change in frequency assignment;

(7) Any change of authorized operation power.

(b) Other equipment changes not specifically referred to in paragraph (a) of this section may be made at the discretion of the licensee, provided that the Engineer in Charge of the radio district in which the station is located and the Commission in Washington, D.C., are notified in writing upon the completion of such changes and provided further, that the changes are appropriately reflected in the next application for renewal of licenses of the station.

§ 78.111 Frequency tolerance.

(a) The frequency of the unmodulated carrier as radiated by a cable television relay station using FM transmission (including those modulated by a frequency-division baseband of standard television signals) shall be maintained within 0.02 percent of the center of the assigned channel.

(b) The frequency of the visual carrier of a CAR station using vestigial sideband AM transmission shall be maintained within 0.0005 percent of the assigned frequency, and the center frequency of the accompanying aural signal shall be maintained 4.5 MHz±1 kHz above the visual frequency.

§ 78.113 Frequency monitors and measurements.

(a) Suitable means shall be provided to insure that the operating frequency is within the prescribed tolerance at all times. The operating frequency shall be checked as often as is necessary to insure compliance with § 78.111 and in any case at intervals of no more than 1 month.

(b) The choice of apparatus to measure the operating frequency is left to the discretion of the licensee. However, failure of the apparatus to detect departures of the operating frequency in excess of the prescribed tolerance will not be deemed an acceptable excuse for the violation.

§ 78.115 Modulation limits.

(a) If amplitude modulation is employed, negative modulation peaks shall not exceed 100 percent modulation.

(b) If frequency modulation is employed, carrier excursions shall be limited to the extent necessary to comply with the requirements of § 78.103 and shall in no event extend beyond the channel limits.

PART 91—INDUSTRIAL RADIO SERVICES

§ 91.557 [Amended]

1. In § 91.557, the text of paragraph (a) is deleted and the word "Reserved" is substituted therefor.

2. In § 91.559, the headnote and text are revised to read as follows:

§ 91.559 Authorizations for operational fixed stations to relay television signals to cable television systems.

Authorizations (including initial grants, modifications, assignments or transfers of control, and renewals) in the Business Radio Service to construct or operate point-to-point operational fixed stations to relay television signals to cable television systems shall contain the condition that such cable television systems shall operate in compliance with the provisions of Part 76 (Cable Television Service) of this chapter.

3. Section 91.561 is amended to read as follows:

§ 91.561 Notification by applicant.

An application for any authorization subject to § 91.559 shall contain a statement that the applicant has notified the licensee or permittee of any television broadcast station within whose predicted Grade B contour the cable television system served or to be served operates or will operate, the licensee or permittee of any 100-watt or higher power television translator station licensed to the community of the system, the franchising authority, the superintendent of schools in the community of the system, and any local or State educational television authorities, of the filing of the application. Such statement of the applicant shall be supported by copies of the letters of notification. The notice shall include the fact of intended filing by the applicant, the name and mailing address of each cable television system served or to be served under the authorization sought, the community and area served or to be served by each cable television system, and the television signals to be carried by each cable television system.

NOTE: As used in § 91.561, the term "predicted Grade B contour" means the field intensity contour defined in § 73.683(a) of this chapter, the location of which is determined exclusively by means of the calculations prescribed in § 73.684 of this chapter.

APPENDIX B

SIGNIFICANTLY VIEWED TELEVISION STATIONS

This table lists the television stations significantly for purposes of cable television carriage, in accordance with § 76.54(a) of the Commission's rules. All stations meeting the significant viewing test are listed, including market and other stations that might be subject to required or permissible carriage under other provisions of the rules.

Information in the table is derived from the American Research Bureau's 1971 Share of Hours Study for all counties with less than 10 percent cable television penetration and ARB's special study of Non-CATV Circulation and Share of Viewing Hours for those counties with between 10 percent and 90 percent cable television penetration. No data is shown for the following seven counties that are reported to have more than 90 percent cable penetration: El Dorado, Calif.; Northumberland, Montour, Pa.; Concho, Terrell, Tex.; Mineral, W. Va.; and Sweetwater, Wyo.

Cities not politically part of any county, are listed with the county in which they were included for survey purposes. A description of how split counties have been divided is included after each State listing that includes split counties. The description indicates which Census County Divisions (1960) are included in each division of the county. Maps of Census County Divisions may be found in U.S. Census of Population: 1960, Vol. I, Characteristics of the Population, Part A, Number of Inhabitants.

FEDERAL COMMUNICATIONS COMMISSION,
Washington, D.C. 20554.

JANUARY 1972.

The table of "Significantly Viewed Signals," occupying pp. 3299–3340, was reworked and reissued as part of the *Reconsideration* (see Item 2, following).

(1) For off-network series, commencing with first showing until first run completed, but no longer than 1 year.

(2) For first-run syndicated series, commencing with first showing and for 2 years thereafter.

(3) For feature films and first-run, non-series syndicated programs, commencing with availability date and for 2 years thereafter.

(4) For other programing, commencing with purchase and until day after first run, but no longer than 1 year.

Provided, however, That no exclusivity protection would be afforded against a program imported by a cable system during prime time unless the local station is running or will run that program during prime time.

Existing contracts will be presumed to be exclusive. No preclearance in these markets.

C. *Smaller Markets.*

No change in the FCC proposals.

Exclusivity for Network Programing:

The same-day exclusivity now provided for network programing would be reduced to simultaneous exclusivity (with special relief for time-zone problems) to be provided in all markets.

Leapfrogging:

A. For each of the first two signals imported, no restriction on point of origin, except that if it is taken from the top-25 markets it must be from one or the two closest such markets. Whenever a CATV system must black out programing from a distant top-25 market station signals it normally carries, it may substitute any distant signals without restriction.

B. For the third signal, the UHF priority, as set forth in the FCC's letter of August 5, 1971, p. 16.

Copyright Legislation:

A. All parties would agree to support separate CATV copyright legislation as described below, and to seek its early passage.

B. Liability to copyright, including the obligation to respect valid exclusivity agreements, will be established for all CATV carriage of all radio and television broadcast signals except carriage by independently owned systems now in existence with fewer than 3,500 subscribers. As against distant signals importable under the FCC's initial package, no greater exclusivity may be contracted for than the Commission may allow.

C. Compulsory licenses would be granted for all local signals as defined by the FCC, and additionally for those distant signals defined and authorized under the FCC's initial package and those signals grandfathered when the initial package goes into effect. The FCC would retain the power to authorize additional distant signals for CATV carriage; there would, however, be no compulsory license granted with respect to such signals, nor would the FCC be able to limit the scope of exclusivity agreements as applied to such signals beyond the limits applicable to over-the-air showings.

D. Unless a schedule of fees covering the compulsory licenses or some other payment mechanism can be agreed upon between the copyright owners and the CATV owners in time for inclusion in the new copyright statute, the legislation would simply provide for compulsory arbitration failing private agreement on copyright fees.

E. Broadcasters, as well as copyright owners, would have the right to enforce exclusivity rules through court actions for injunction and monetary relief.

Radio Carriage:

When a CATV system carries a signal from an AM or FM radio station licensed to a community beyond a 35-mile radius of the system, it must, on request, carry the signals of all local AM or FM stations, respectively.

Grandfathering:

The new requirements as to signals which may be carried are applicable only to new systems. Existing CATV systems are "grandfathered." They can thus freely expand currently offered service throughout their presently franchised areas with one exception: In the top 100 markets, if the system expands beyond discrete areas specified in FCC order (e.g., the San Diego situation), operations in the new portions must comply with the new requirements.

Grandfathering exempts from future obligation to respect copyright exclusivity agreements, but does not exempt from future liability for copyright payments.

[FR Doc.72-1826 Filed 2-11-72;8:45 am]

Local Signals:

Local signals defined as proposed by the FCC, except that the significant viewing standard to be applied to "out-of-market" independent stations in overlapping market situations would be a viewing hour share of at least 2 percent and a net weekly circulation of at least 5 percent.

Distant Signals:

No change from what the FCC has proposed.

Exclusivity for Nonnetwork Programing (*against distant signals only*): A series shall be treated as a unit for all exclusivity purposes.

The burden will be upon the copyright owner or upon the broadcaster to notify cable systems of the right to protection in these circumstances.

A. *Markets 1-50.*

A 12-month presale period running from the date when a program in syndication is first sold any place in the United States, plus run-of-contract exclusivity where exclusivity is written into the contract between the station and the program supplier (existing contracts will be presumed to be exclusive).

B. *Markets 51-100.*

For syndicated programing which has had no previous nonnetwork broadcast showing in the market, the following contractual exclusivity will be allowed:

FRIDAY, JULY 14, 1972

WASHINGTON, D.C.

Volume 37 ■ Number 136

PART II

FEDERAL
COMMUNICATIONS
COMMISSION

CABLE TELEVISION SERVICE

Reconsideration of Report
and Order

Title 47—TELECOMMUNICATION

Chapter I—Federal Communications Commission

[Docket No. 18397 etc.; FCC 72-530]

PART 1—PRACTICE AND PROCEDURE
PART 76—CABLE TELEVISION SERVICE

Reconsideration of Report and Order

Memorandum Opinion and Order. In the matter of amendment of Part 74, Subpart K, of the Commission's rules and regulations relative to Community Antenna Television Systems; and inquiry into the development of communications technology and services to formulate regulatory policy and rulemaking and/or legislative proposals (Dockets Nos. 18397, 18397-A); amendment of § 74.1107 of the Commission's rules and regulations to avoid filing of repetitious requests (Docket No. 18373); amendment of §§ 74.1031(c) and 74.1105 (a) and (b) of the Commission's rules and regulations as they relate to addition of new television signals (Docket No. 18416); amendment of Part 74, Subpart K, of the Commission's rules and regulations relative to Federal-State or local relationships in the Community Antenna Television System field; and/or formulation of legislative proposals in this respect (Docket No. 18892); amendment of Subpart K of Part 74 of the Commission's rules and regulations with respect to technical standards for Community Antenna Television Systems (Docket No. 18894).

1. On February 2, 1972, we adopted the Cable Television Report and Order.[1] Petitions for reconsideration, oppositions thereto and the reply comments have been filed, and we now address the objections. At the outset—and before taking up exceptions to specific rules—it is appropriate to deal with two matters of overriding and more general concern, i.e., whether, in adopting the rules, the Commission followed the requirements of the Administrative Procedure Act and whether the effective date of the rules should be delayed pending the enactment of copyright legislation.

2. *Compliance with Administrative Procedure Act.* Dispute continues over whether the Commission observed the requirements of the Administrative Procedure Act in adopting the Cable Television Report and Order. That report recites in detail the more than 3-year history of the cable proceeding and spells out step by step how the Commission probed the issues raised by its proposals for rulemaking. In the course of finalizing rules, the Commission considered whether or not to extend the proceeding for further comment or more oral argument. We decided that additional process was neither required nor likely to serve a useful purpose. A summary of the reasons may be found in paragraph 67 of the report. We continue to hold the view that after years of gathering data, soliciting

[1] FCC 72-108, 37 F.R. 3252 (1972).

views, hearing argument, evaluating studies, examining alternatives, authorizing experiments, and holding final discussions—during which every principal aspect of cable was examined—the requirements governing the adoption of new rules have been satisfied.

3. More recently, the Commission, on March 23, 1972, denied a "Motion to Stay Pending Appeal," filed by the Nevada Independent Broadcasting Corp. The motion raised the same issue that is now before us on reconsideration, and we rely on the language of the memorandum opinion and order supporting the refusal to stay the rules, 34 FCC 2d 165 (1972). (See Appendix C for text.) In that decision, we stated that notice was given of the subject matters and issues in each area. We received extensive comments and proposals and held hearings and thus had discretion to fashion rules applicable to the issues raised. This was true when we outlined the regulatory program in our August 5 letter. This situation did not suddenly change when the Commission took into account the November consensus agreement. For, in all matters involved in the agreement—exclusivity, leapfrogging, overlapping market signals—the Commission gave full notice of the "subject matter and issues" and had permitted and received extensive comments. The Commission studied the industry agreement on exclusivity, leapfrogging, and overlapping markets, and it made its judgment on whether to adopt the agreement against the background of the extensive comments received. It concluded that adoption of exclusivity factors clearly would serve the public interest—that this detailed, complex facet of regulation was appropriate for agency rather than legislative process. Overall, the Commission concluded that adoption of the features of the agreement—found to be reasonable—would markedly serve the public interest in promoting the development of cable television for the reasons set out in paragraph 67 of the Cable Television Report and Order—most important of which was the agreement's effect on the crucial underlying controversy, cable's standing vis-à-vis the TV programing distribution market. Significantly, all parties have had an opportunity to address themselves to the fundamental judgment on reconsideration, and none has shown it to be in error. Indeed, the great majority of parties in opposition have largely ignored this crucial point to make technical APA arguments. But cases such as NLRB v. Wyman-Gordon Co., 394 U.S. 759 (1969), are not in point here. That case focused on a requirement adopted in an adjudicatory proceeding that was to have prospective, general application. Here, rules were adopted pursuant to statutory rule making procedure. We have held proceeding upon proceeding and given ample opportunity for all interested parties to address themselves to this subject matter and all pertinent issues. Indeed, the criticism is made with some force that with all this process we have held back cable's development too long—that formulation of governmental policy has much delayed a new and vital technology. At some point,

action is called for. That point is reached here.

4. *Copyright.* Several petitioners, including the National Association of Broadcasters (NAB), the Rocky Mountain Broadcasters Association (RMBA), Columbia Broadcasting System, Inc. (CBS), American Broadcasting Co., Inc. (ABC), and KMSO-TV, Inc., argue that the effective date of our new rules should be delayed until settlement has been reached on the copyright issue. The RMBA argues that the rules should be delayed until congressional enactment of copyright legislation or at least until a draft for legislation is agreed on by the parties to the consensus agreement. The NAB, making the same argument, states: "It is our understanding that such a draft was to have been proposed to Congress by the time the Commission's rules were released." ABC contends that "* * * should copyright legislation not be forthcoming within a reasonable time, it will be necessary, in ABC's view, for the Commission to halt authorization of CATV operation."

5. Over the years, the ultimate integration of cable television into the Nation's communications structure has been deadlocked on the copyright question—how to weave the cable industry into the market for distributing television programs, a process that distributes the costs of programing among those who use it. The tying of cable's development to the settlement of copyright has in the past served to harden the impasse, not unblock it. We now expect agreement of the industries and that legislation will be forthcoming. We are convinced that putting our program into effect only after legislation is enacted will effectively diminish the prospect for settlement and will not promote our goal of fostering the orderly development of cable television.

6. This view finds additional support in the exchange of letters (reprinted in full in Appendix E to the Cable Television Report and Order) between Chairman Burch and Senator McClellan, Chairman of the Subcommittee on Patents, Trademarks, and Copyrights. Chairman Burch wrote, "* * * a primary factor in our judgment as to the course of action that would best serve the public interest is the probability that Commission implementation of the consensus agreement will, in fact, facilitate the passage of cable copyright legislation. The parties themselves pledge to work for this result." In his reply, the Senator stated, "* * * I concur in the judgment set forth in your letter that implementation of the agreement will markedly facilitate passage of such legislation." The Senator went on to say that all parties have been notified in a letter of December 15, 1971, that the subcommittee intended immediately to "* * * resume active consideration of the copyright legislation upon the implementation of the Commission's new cable rules".

7. Finally, we reach ABC's contention that the Commission will have to take action if copyright legislation is not forthcoming within a reasonable period

and have so stated in paragraph 65 of the report. It would be premature to speculate now what action would be necessary in that event. We hope never to have to reach that point since it is our expectation that the parties will expeditiously reach an accord and that copyright legislation will be enacted once these rules become effective.[2] We have decided after much study and debate to take the first step. We will revisit the matter if our estimate proves wrong that adoption of our program will facilitate copyright legislation.

RECONSIDERATION OF SPECIFIC RULES

8. Most of the requests to reconsider have been principally directed to specific rules. To simplify reconsideration, we have taken the objections by subject matter and grouped them to conform to the arrangement of the Cable Television Report and Order, under the following principal headings—signal carriage, access, technical standards, Federal-State/local relationships, procedure.

SIGNAL CARRIAGE RULES

9. *Translators.* The Montana Network, licensee of KOOK–TV in Billings, Mont., and of 100-watt VHF Translator K131Y at Lewistown, Mont., which rebroadcasts the signal of KOOK–TV, seeks reconsideration of our definition of a television station contained in § 76.5(b). Montana Network contends that the definition should include any 100-watt or higher translator operating on a channel regularly assigned to its community. It is alleged that this would provide a 35-mile zone around such translators and, by protecting them from the impact of unlimited distant signal importation, would conceivably enable them to develop into regular television stations. The Rocky Mountain Broadcasters Association seeks reconsideration of our carriage rules so that translators of less than 100 watts will be required to be carried. The National Association of Educational Broadcasters notes that the rules require carriage of translator stations only where the station is licensed to serve the cable community but that some translators, in fact, serve areas beyond the community of license. NAEB seeks modification of the carirage rule so that translators will be carried in such areas.

10. Under the rules adopted in 1966, second report and order in Docket No. 14895,[3] we require the carriage of translator stations with 100 watts or higher power. It was our view that carriage was desirable because it afforded access to cable subscribers without which the incentive to establish new translator service might be diminished. We are not persuaded that there is any reason to require cable systems to carry commercial translators of less than 100 watts transmitter

output power. As to noncommercial educational translators, however, the need for carriage of translators of power of less than 100 watts has become apparent because such translators often represent the only means of bringing educational programs to remote communities and schools. Further, translators are frequently licensed to serve areas rather than identifiable communities. The factor that will determine whether carriage of a translator is required, therefore, will be whether the translator serves the cable community. We recognize that, in specific situations, questions may arise as to whether a translator "serves" a cable community and we will deal with such problems on an ad hoc basis. We will, of course, expect the translator licensee to do what is necessary to make a quality signal available to the cable system if it wishes to be carried on the system. For the reasons discussed, we have decided to retain the requirement that cable systems carry commercial translators of 100 watts or more serving the cable community, where the system is not carrying the primary station, but we are revising the rules to require that a cable system that was not operational before March 31, 1972, or that expands its channel capacity must carry, on request, any noncommercial educational translator station with 5 watts or higher power serving the community of the system where the system is not carrying the primary station. A noncommercial educational translator is defined as one that carries the programing of a noncommercial educational television station, irrespective of the identity of the licensee of the translator.

11. *Educational stations.* NAEB argues that the compulsory carriage of educational stations throughout their Grade B contour "will generate an intolerable profusion of educational signals, in many communities." NAEB believes that carriage should only be required within specified mileage zones.

12. We have required carriage of educational stations throughout the Grade B contours of such stations because of the public interest in wide dissemination of their programing and the difficulty in devising a significant viewing standard for educational stations. It would be a disservice to the public to deprive them of educational television service from local stations, and our market analysis does not indicate that large numbers of educational signals are available locally even in overlapping market situations. As a safeguard, however, local educational stations, which must receive notification of cable signal carriage proposals, are free to object to carriage of educational stations where the carriage of both local and distant educational stations would cause an unwarranted profusion of educational signals.

13. *Carriage of independent distant signals.* Buckeye Cablevision, Inc., et al., NewChannels Corp., Athena Communications, and Jerrold Electronics Corp., seek reconsideration of our rules that require cable systems importing distant signals to specifically designate the dis-

tant stations to be carried. Under the rules, a cable system may "program" distant signals—i.e., "cherry pick" or switch distant signals to obtain desired programing—when the system blacks out the designated independent signal under the syndicated program exclusivity rules. See §§ 76.61 and 76.151. These parties argue that unless a "programed independent service" is permitted, the Commission's desire to "get cable moving" and to encourage UHF development will be frustrated. It is contended that in many areas of the country, independent stations may only be imported via long haul microwave, and that the expense of such carriage can, in many instances, only be justified if the cable system is able thereby to obtain an attractive package of programing from a distant source. It is submitted that the making up of channels of programing by selecting the best from available distant sources would result in greater carriage of UHF programing, that there is not likely to be greater impact on local stations if cable operators are permitted to program such channels, and that cable subscribers would be afforded more diversity of programing. The same parties also request reconsideration of §§ 76.59 (b) and 76.61(b) so that network programs not available in the community may be imported on distant signals. It is also requested that stations other than independents, e.g., full or partial network stations or cherry pickers, be permitted to be carried. It is argued that in those areas of the country where it is not economically feasible to import independent stations there should be an alternative, such as importation of nonnetwork programs from television stations that do not meet the definition of independent stations.

14. Kaiser Broadcasting, the Association of Maximum Service Telecasters (MST), NAB, and ABC request reconsideration of the rule that allows for substitution of programs of greater length than the program that is blacked out and of the rule permitting deletion of programs of local interest to the distant communities. With respect to the first, it is argued that the removal of a distant signal from a cable system for any period longer than necessary to provide exclusivity lessens the opportunity for the distant station to obtain identity and viewer acceptance in the cable community and is inconsistent with the industry consensus agreement. With respect to substituting for programs that are primarily of local interest, it is argued that the rule (§ 76.61(b)(2)(ii)) is internally inconsistent with other cable television policy, e.g., the leapfrogging provisions of the rules and § 76.55 (b) which concerns carriage of programs in full without deletion or alteration except as otherwise required. It is submitted that the rule encourages distant stations not to broadcast local news and public affairs programs and that the cable operator will have discretion to make judgments about programing that he has not seen—perhaps deleting programing of relevance or of interest to subscribers in the cable community. As a further matter, NAB and Kaiser state

---

[2] In its opposition to petitions for reconsideration, the NCTA states that meetings between copyright owners and cable industry representatives are presently taking place in order that draft legislation may be proposed to the Congress.

[3] 2 FCC 2d 725 (1966).

that because many stations may wish to participate in the benefits of carriage, a cable system is in a position to "auction" its channels to the distant stations that are the highest bidders—resulting in profit to the cable system but not necessarily furthering the Commission's public interest goals. It is requested that the Commission amend its rules to prohibit a cable system from extracting payment from a distant station as consideration for carriage of the station's signal

15. In resolving these questions, it is useful to focus on our statement at paragraph 62 of the Cable Television Report and Order.

In the August 5 letter these [distant] signals were, in effect, channels of independent programing (conceivably a blend of several distant stations); now they are restricted to specific distant station's except during exclusivity protection periods.

The change referred to above was one that grew out of the consensus agreement. See Cable Television Report and Order at paragraphs 61–67 and Appendix D. Although the rule adopted is different from the August 5 formulation, we have determined that it is in the public interest to adopt the rule in order, inter alia, to implement the consensus agreement. While less diversity may result than under the cherry picking concept, cable systems will be able to select programing when blacking out protected programs. The rule also offers broadcasters carriage on a more uniform basis than if cherry picking were allowed and thus a more saleable commodity to advertisers—particularly when a station is carried near its home market. Admittedly, the benefit of carriage on a distant cable system is diminished by the rules requiring that syndicated programs be protected. Because distant signals will be blacked out from time to time, carriage of a signal is probably of less value than it would have been in the absence of exclusivity protection. On balance, exclusivity protection is probably of greater advantage to broadcasters than is carriage in distant markets, but carriage on a fairly regular basis should in some measure be of benefit to broadcasters.

16. We have carefully considered the need to give cable systems the full benefit of carriage of distant signals, the promotion of broadcast stations, particularly UHF, through fairly regular carriage, and the provision of new diversity to the viewing public. We have balanced these considerations with the factor that we have limited cable in all television market zones to one or two and at the most three distant signals and the requirement in the major markets for considerable exclusivity protection. Consequently, we have adopted two rules that do afford some flexibility to program signals. First, where a distant station is televising a program that is primarily of interest to its own community, e.g., local news, public affairs, or other locally produced programs, the cable system may delete such programs from the distant signal and insert others. Second, when substituting for blacked out programs—either to pro-

vide protection for syndicated programs or because they are primarily of local interest—the cable system may insert any nonprotected program and carry it to its completion. We believe it necessary that cable systems have flexibility in finding substitute programs that are not protected when exclusivity protection is given to a local station. It will often be difficult to find substitute programs, and the inability to carry a program that is no longer than the one being blacked out will mean, in many instances, that no programs will be available or that subscribers will be given the opportunity to watch only portions of programs.

17. With regard to substitutions on distant signals in place of programs that are primarily of local interest in the distant community, we believe this too will provide greater diversity to the public. Our rule does not, however, give cable systems wide discretion not to carry programing from regularly carried distant stations. The programing in question involves such fare as a local newscast or a locally produced program dealing with a local issue. It makes little sense to require that a cable system bring to its subscribers news and public affairs coverage of matters of interest to distant communities. Nor are broadcasters likely to curtail local news and public service programs merely because such programs might not be carried in distant markets. This kind of programing is a critical part of a commitment to serve local viewers, not distant ones. Finally, our rule permitting substitution for local interest programs is not inconsistent with § 76.55(b) which prohibits deletion or alteration "except as otherwise required" and, by implication, whenever permitted. In operating under this provision of the rules we will expect cable operators to exercise care so that the intended purpose of the rules is not subverted. Should abuses develop we will be prepared to take another look at this provision.

18. As with all rules designed to balance competing interests, there will be unusual circumstances that do not fit the rule. Section 76.7 preserves the Commission's flexibility to deal with these situations. So, for example, in certain areas of the country, carriage of syndicated programing from full or partial network stations instead of from independents might be indicated because of inordinate costs involved in obtaining independent signals. In the event such a system later obtains independent distant signals, it could only do so in accordance with the rules and may have to delete carriage of syndicated programs from network stations.

19. Cable interests have urged that we permit carriage of network programs from distant stations when those programs are not broadcast by local network affiliates. We find merit in this suggestion and will amend the rules accordingly. One of our goals in this proceeding, with which there has been little basic disagreement, has been to assure that all cable subscribers have full network service available. To the extent that network affiliates of the national networks are not

available locally, we have permitted carriage of distant affiliates. (See §§ 76.59 (b), 76.61(b), and 76.63.) In line with this policy of assuring the availability of full network service, it appears appropriate to permit carriage of those programs offered by the networks but not cleared by local affiliates.[4] This is of particular importance in those cases where the programs not otherwise available include network news or other public affairs programing. In any event, our analysis reveals that primary network affiliates generally carry a high percentage of the programs offered by the networks so that the impact of this rule revision should be limited .

20. Finally, with respect to the passing of consideration from a broadcaster to a cable system in order to be carried as a distant signal, we do not believe that a rule prohibiting such arrangements is now necessary. There was no restriction in our previous rules against entering into such agreements. Cf. paragraph 56 of the second report and order in Docket 14895, 2 FCC 2d 725 (1966). And there has been no indication that there have been abuses in this area. In many cases, consideration may properly take the form of payment of microwave costs to carry a signal that a cable system could not otherwise afford. In these circumstances, we find no compelling reason to adopt a rule. We will take any necessary action if abuses develop in this area.

21. *Religious and other specially programed stations.* Jerrold Electronics Corp., Athena Communications, Buckeye Cablevision, Inc., et al., and New Channels Corp. request that religious stations and other specially programed stations be permitted to be carried as distant signals without counting such signals against the applicable distant signal quota.[5] It is argued that religious stations, like non-English language stations, generally attract select small audiences, and will not be carried by cable systems unless an exemption from the distant signal quota is provided. While petitioner's assertions may be true, there is a fundamental difference between the considerations that prompted us to adopt a rule for non-English language stations and those pertaining to religious programing. In the case of the first, local service is available in very few places in the country. But religious programing is generally available both on radio and television broadcast stations throughout the county, and the resulting impact of unlimited carriage is likely to be more pervasive.

---

[4] Because there may not be much advance notice of the nonclearance of network programs by local stations and the availability from a distant station and due to the limited number of programs likely to be involved we will not require specific reference to the possibility of such carriage in certificate of compliance applications nor will the leapfrogging rules be applied to such carriage. (Compare § 76.61(b)(2)(ii) as it refers to carriage of programs substituted for programs deleted under the program exclusivity rules.)

[5] Oppositions were filed by Connecticut Television, Inc., and MST.

22. As to specially programed stations,[6] petitioners allege that such stations should also be treated outside the confines of distant signal quotas. But the lack of standards by which to measure "specially programed stations" and the failure of petitioners to demonstrate how the public interest would be served by assuming the risks of greater impact on local stations from widened distribution of the programing of such stations compel rejection of the proposal.

23. *Foreign language stations.* The Spanish International Communications Corp. has filed for reconsideration of the rules regarding the importation of foreign language stations.[7] As we noted in the Cable Television Report and Order at footnote 50, petitioner requested, following the issuance of our letter of intent, that importation from Mexico of Spanish-language stations not be allowed where U.S. Spanish-language programing is available either off the air or potentially available via microwave. The petition for reconsideration restates that request. But we considered the request in finalizing the rules and see no reason to alter our view. We are attempting to encourage the carriage of foreign language programing. Where there is a local, Spanish-language station, it will of course get carriage priority. But outside its own market, where there is no "right" of carriage and no special need for protection against other stations programed in the same language, it is in the public interest to make foreign language programing available without impediment. In unusual situations where a domestic Spanish-language station makes a compelling demonstration for relief with respect to a particular application, we can afford such relief under § 76.7. This should serve to maintain the vitality of local foreign language services without general restrictions on the right of cable systems to distribute the programing of foreign stations.

24. *Leapfrogging.* MST contends that neither the letter of intent nor the consensus agreement addressed leapfrogging in areas beyond the 35-mile zones of television stations and that the rationale for leapfrogging supports the imposition of restrictions in areas outside such zones. KFIZ Broadcasting Co. seeks modification of the leapfrogging rules so that independent UHF stations would be required to be carried as a first priority by cable systems within those portions of the Grade B contours of such stations where carriage is otherwise not now required. NewChannels Corp., Athena Communication, Jerrold Electronics Corp., and Buckeye Cablevision, et al. request reconsideration of the leapfrogging rules in two respects: (1) It is submitted that the leapfrogging rules should not apply to smaller markets because cable systems in those markets are limited to only one independent distant signal; (2) it is also urged that the Commission's rule that contemplates no waiving of the leapfrogging rule for independent stations is inequitable. Nevada Independent Broadcasting Corp. (NIB), in a motion for stay, states that our leapfrogging rule will permit vast expansion of four Los Angeles independent television stations throughout the West and will thus engender the birth of superstations.[8] NIB asserts that all pending microwave requests will be granted because of the procedural rules adopted by the Commission. This, it is asserted, will be to the detriment of NIB because spot advertising will now go to the Los Angeles independents to reach areas via cable that small market broadcasters could serve instead.

25. Our treatment of the leapfrogging question is based on the following factors: First, we thought it desirable to move away from the limits of our 1968 proposal because it did not provide enough flexibility to cable operators, with the result that the Commission was inundated with requests for waiver filed pursuant to our interim processing procedures. Second, we were concerned that permitting the greatest possible choice could lead to the selection of stations from only a few of the largest markets, thereby foreclosing any benefit of cable carriage to many stations. We believe that the consensus agreement provides a sound resolution of these two considerations. The implementation of the leapfrogging restriction in all markets is necessary to insure that the benefits of carriage are more evenly distributed. In doing so, there is no need to require the restriction in areas outside television markets where it would just be an unnecessary restriction because the risk of impact on local broadcast service from carriage of distant signals is diminished. The rule adopted strikes the appropriate balance, and we reassert that we do not contemplate its waiver. We do not intend to return to the process whereby waiver is requested in case after case because of microwave savings; to do so would undermine the leapfrogging rule. But we are not unmindful of the need for relief in unusual circumstances, Sun Cable T-V, 27 FCC 2d 261 (1971), and will respond accordingly. See United States v. Storer Broadcasting Co., 351 U.S. 192.

26. With respect to the petition for reconsideration filed by KFIZ, it should be noted that for reasons discussed above and in the Cable Television Report and Order, we have changed our leapfrogging rule from the formulation in our letter of intent. The UHF priority is now third rather than first. We believe that in most situations the provision of syndicated programing protection more than offsets this change. And we expect that there will be significant carriage of UHF stations under the first two priorities. It appears that petitioner's circumstance in the Fond du Lac market may be an unusual one more appropriately to be dealt with in individual proceedings involving that market rather than in this rulemaking proceeding.

27. Finally, we believe that the contentions of NIB are also without merit. NIB is the licensee of Television Station KVVU, Henderson, Nev. Because Henderson is in a smaller market area that already has available locally three network stations and an independent station, the rules do not permit the importation of any additional English language commercial television stations. See § 76.59 of the rules and paragraph 48 infra. At least with respect to this market, NIB is therefore incorrect that national spot advertising dollars will be drawn away from it to the Los Angeles independent stations. Additionally, the rules are designed to place as few impediments as possible on the carriage of stations such as KVVU so that it may compete for cable carriage. It is entitled, on request, to carriage on cable systems within its Grade B contour and in those areas where it is significantly viewed and its carriage is not proscribed to any degree by the leapfrogging rules. Finally, it is not correct that microwave applications involving carriage of Los Angeles signals are automatically granted under our new rules, nor are oppositions to these proposals rendered moot. See our decision denying NIB's motion for stay, 34 FCC 2d 165 (1972).

28. *Network program exclusivity.* The central issue relating to network program exclusivity in the reconsideration petitions is whether such exclusivity should be simultaneous only, or same-day, as in former § 74.1103. A number of smaller market television stations, including KBOI–TV, Boise, Idaho, KOAI, Flagstaff, Ariz., and KID–TV, Idaho Falls, Idaho, argue that same-day protection is the minimum necessary to maintain their audiences from serious fractionalization and advertiser by-pass in favor of imported distant signals that will not be blacked out because their programs are not simultaneously duplicated by the smaller market stations. The inadequacies of simultaneous exclusivity are allegedly most severe in the mountain standard time zone where, the Rocky Mountain Broadcasting Association argues, a combination of lack of direct network feeds and the common practices of "bicycling" network programs or taping a network feed and replaying it on a delayed basis works the result that a significant amount of network programing is not simultaneously duplicated. Petitioners maintain that network located within the mountain standard zone do not even have uniform or near-uniform schedules among themselves, and that simultaneous-only protection will force these stations into identical programing schedules, contrary to the public interest. Springfield Television, Inc. and Mid-Continent Telecasting, Inc. assert that simultaneous exclusivity provides insufficient protection in the central standard

---

[6] Petitioners state, for example, that KWHY–TV "provides highly specialized financial programing including stock market ticker service over substantial portions of its broadcast day."

[7] Opposition petitions were filed by Trans Video Corp. and Sammons Communications, Inc.

[8] Pursuant to memorandum opinion and order, 34 FCC 2d 165 (1972), we are treating the merits of petitioner's motion for stay as a petition for reconsideration.

zone as well, and urge that § 76.93(b) be amended to provide automatic same-day exclusivity whenever any smaller market network station and a network station licensed to a community in a different time zone are carried by a cable system. They further argue that the special relief provisions of § 76.7 will be too cumbersome, costly, and time-consuming as an ad hoc alternative to amending the rules.

29. The Commission recognized in the second report and order in Docket 15971, that "Simultaneous nonduplication protects the bulk of the popular network programing of most network affiliates * * *," and indicated that although it was adopting a same-day exclusivity rule, it would continue to give full effect to private agreements between cable operators and local television stations that provided for a different degree of protection for local stations, such as simultaneous-only exclusivity. In adjudicatory proceedings the Commission also concluded that simultaneous exclusivity could provide adequate protection to local stations. E.g., Black Hills Video Corp., FCC 65–989, 1 FCC 2d 1458; Hardin Cable TV, Inc., FCC 69–1098, 20 FCC 2d 56. We have set forth in paragraph 99 of the Cable Television Report and Order the reasons for now adopting simultaneous exclusivity rules. Except with respect to that situation peculiar to stations operating in the mountain standard time zone we reaffirm our view that simultaneous exclusivity affords adequate protection to network stations and appropriately balances the interest of local stations in not having their programing duplicated by lower priority stations and that of cable subscribers in such time diversity as may be available from different network stations. However, with respect to the concerns expressed regarding operation of the rule in situations where there are time zone problems, we believe some change is in order. In the rules (§ 76.93(b)), we recognized that simultaneous network exclusivity might not afford adequate protection to stations involved in certain time zone situations. The rules provided for attention to such problems on petition for special relief by stations involved. On further consideration, it appears that the problem involves stations in the mountain standard time zone almost entirely and that there is sufficient similarity in situations throughout that zone to permit the adoption of a general rule for this area. Briefly, it appears that stations in this zone follow no uniform network program distribution pattern because prime time viewing hours in the zone do not coincide with the network feed of prime time programs. To correct this situation some stations in the area tape and replay network programs out of sequence. The result is that the simultaneous exclusivity rule is not effective to protect a station's network programing. Accordingly, we believe it appropriate to modify the rule as it applies to stations in the mountain standard time zone. The action we take will essentially shift the burden of seeking relief from

the general rule in time zone situations from stations in the zone to cable operators. We will amend the rules to provide that stations licensed to communities in the mountain standard time zone, if they are not licensed to communities in the first 50 major television markets, will be entitled as a general rule to same day network program exclusivity. Stations licensed to the first 50 market cities in the zone, other stations outside the zone, and cable operators providing exclusivity to stations within the zone will then have the burden under § 76.7 of the rules of seeking waiver of the general requirement if it is thought either to provide insufficient protection or to be unduly restrictive.[9]

30. In the event the Commission is not disposed to restore same-day exclusivity for all programing, MST and ABC argue that it should at least apply to network news, especially where a local station is broadcasting the network feed "live." In a similar vein, Duhamel Broadcasting Enterprises argues that stations that carry any network feed "live" should be entitled to same day exclusivity for the "live" programing. But, other petitioners point out that the Commission encourages scheduling flexibility, particularly in a program area such as news where maximizing the choice of viewing hours helps insure that the public will be able conveniently to view programs of key interest. Cable may be able to contribute to "time diversity" in the news area. Hence, we decline to adopt special exclusivity for programing merely because it is taken directly from a network feed, or to provide special protection for network news.

31. The National Association of Educational Broadcasters argues that, regardless of the degree of exclusivity given to commercial stations, noncommercial educational network programing should be accorded same-day exclusivity in view of its deemphasis of simultaneous broadcasts. Although the need for additional exclusivity is a matter that might appropriately be raised in connection with certificate of compliance applications proposing carriage of distant educational stations, we do not see the need for a general rule revision. To the extent feasible, we think it desirable to permit the time diversity of programing that carriage of more than one educational station makes possible under the simultaneous exclusivity rule.

32. WBEN, Inc., Taft Broadcasting Co., and Capital Cities Broadcasting Corp. propose that the exclusivity rules be amended to prohibit a cable system from carrying network programs broadcast by foreign stations at any time prior to their first domestic broadcast. A similar argument was made in petitions filed in connection with the reconsideration of the Second Report and Order in Docket 14895. At that time, the Commission rejected the suggestion and indicated that special treatment would be accorded

petitions seeking relief from prereleased programs. Memorandum opinion and order in Docket 14895, 6 FCC 2d 309, 315–316 (1967). Subsequently, in Colorcable, Inc., 25 FCC 2d 195 (1970), the Commission determined that the prerelease problem was not especially significant and that whatever "problem" existed appeared to be on the verge of elimination. The latest petitions concerning this matter contain no new matter on the extent of the problem, such as how many foreign stations and domestic programs are involved and how widely these stations are carried by cable systems. Lacking such information, we find no basis for amending the rules. Special relief remains available, pursuant to § 76.7 on appropriate showing of need.

33. Duhamel Broadcasting Enterprises has asked for clarification of the extent of simultaneity that is necessary to qualify for simultaneous exclusivity protection. It envisions instances in which certain programs, such as sports events, may run beyond their schedule, or where stations may delay the start of their taped network programing to read news bulletins or provide special election results. Duhamel suggests that an overlap of 50 percent of the same programing should be sufficient. Although we agree that some allowance should be made for the absence of exact overlap where simultaneous exclusivity is concerned, we believe that a 50 percent overlap is too much, because the television viewer will probably not be able to see the missing 50 percent at any other time. To qualify for simultaneous exclusivity protection, no more than 5 or 10 minutes of a program may be overlooked. If significant omissions of a station's network programing occur frequently, the Commission may grant special relief from the exclusivity requirement to affected cable systems.

34. Duhamel, KID Broadcasting Corp., Mid-Continent Telecasting, Inc., and Springfield Television, Inc. urge that, as in syndicated exclusivity, network exclusivity should treat all episodes of a series as a single unit rather than separately. Although it might be simpler administratively for cable operators to be required to delete all episodes of a series instead of only those that are simultaneously duplicated, throughout the history of the program exclusivity rules the Commission has taken the position that application of the rules should not result in the loss of any program content, e.g., Black Hills Video Corp., 1 FCC 2d 1458 (1965). Since the possibility exists that indiscriminate deletion of all episodes of a series might permanently deprive viewers of the opportunity of seeing some (for, with preemptions and other schedule changes, there is no guarantee that every network station will show every episode of a series), we are not making the suggested change in the rules.[10]

---

[9] Petitions filed pursuant to § 76.93(b) will be dismissed as moot, unless, within 60 days of the publication of this document in the FEDERAL REGISTER, they are supplemented to demonstrate their continued relevance.

[10] Commission experience with exclusivity notification schedules indicates that many stations and cable systems have apparently agreed to treat series as single units for exclusivity purposes; we will not override these arrangements.

35. The Association of Maximum Service Telecasters favors the deletion of § 76.97 which stays a cable system from having to provide network exclusivity until the Commission rules on any timely filed waiver request. It argues that a cable operator should be required to seek waiver of the exclusivity rules at the time that it files an application for certificate of compliance and that, if it does not, it should be required to provide exclusivity even before the Commission rules on any waiver request. We cannot accept this approach, for two reasons. First, the waiver provision of § 76.97 is grounded on the established policy of maintaining the status quo while the Commission considers the application of a rule that would require the expenditure of a substantial sum of money to achieve compliance, and petitioner does not explain why we should depart from this policy here. Second, MST assumes that requests for exclusivity protection will only be received when a proposed cable system is about to go into operation, or that every existing cable operator will soon be applying for a certificate of compliance. The fact is that existing systems receive exclusivity requests even years after operations have commenced (particularly where newly licensed stations are involved), that often a new system does not receive any requests for exclusivity until well after it has commenced operations (and, hence, has no need to contemplate exclusivity waiver petitions at the time that it files an application for certificate of compliance), and that many cable operators will not have to obtain certificates of compliance until March 31, 1977 (see § 76.11(b)). MST fails to explain why an existing system that seeks waiver of a rule should be placed in a less favorable position with respect to the maintenance of the status quo during an adjudicatory proceeding than an emergent one. Further, under § 76.17 broadcasters may raise carriage and exclusivity matters in connection with the certificating process, if they so desire.

36. WBRE–TV, Inc. asks that the rules be amended to indicate that, regardless of the outcome of Dockets 16004 and 18052 (proposed amendments of part 73 of the rules concerning field strength measurements and curves for FM and television broadcast stations), in determining the obligations of a cable system to carry or provide program exclusivity to stations, the field strength curves in effect at the time the cable system commenced operations should be utilized. This kind of grandfathering provision is more properly within the scope of Dockets 16004 and 18052, and will not be considered here.

37. Finally, clarification is needed concerning the meaning of § 76.91(c), which outlines the exclusivity rights of certain significantly viewed television signals. It is important to note that this provision applies only to smaller market signals carried by cable systems located outside of all major and smaller television markets. Secondly, the provision must be viewed in the context of § 76.91 as a whole. Thus, subsection (c) means only

that a significantly viewed smaller market signal that does not place a Grade B contour over the community of a system located outside of all major and smaller television markets has priority over a nonsignificantly viewed signal that likewise does not place a Grade B contour over the community of the system—it does not have priority over a nonsignificantly viewed Grade B signal. Similarly, a significantly viewed smaller market Grade B signal has only equal priority (and, hence, no right to exclusivity) with a nonsignificantly viewed Grade B signal.

38. *Syndicated program exclusivity.* Rust Craft Broadcasting Co., Mid-Continent Telecasting, Inc., Duhammel Broadcasting Enterprises, and RMBA assert that syndicated programing exclusivity should be extended to smaller television markets. It is argued that smaller market stations pay for exclusive rights in their markets and that the rules should protect those rights. Similarly, NAEB states that syndicated program exclusivity should be extended to educational stations. It is alleged that educational stations deserve the same protection as commercial stations for syndicated programs that they purchase. MST states, as its understanding of the consensus agreement, that broadcast stations need obtain exclusivity only against other broadcast stations in order to obtain exclusivity against cable systems, but that the rules require exclusivity to be obtained against cable systems as well as against broadcast stations. With respect to procedural matters, MST contends that cable systems should be required to notify broadcasters of their intention to comply or not comply with requests for exclusivity. New Channels, Athena, Jerrold, Buckeye et al., MST, and Kaiser suggest that the notification process would be made easier if the Commission were to encourage or require television stations to make available to any broadcast station or cable system requesting it, information concerning their program schedules as is regularly made available to advertisers, sales representatives, etc. Kaiser alleges that this information, especially with respect to feature films, is generally available 2 weeks or more before the scheduled broadcast. MST objects to the requirement that relevant excerpts from program contracts be kept on public file by stations requesting protection for those programs. It is argued that, because stations are Commission licensees, they will not give inappropriate notices, and that stations do not want competitors to obtain information concerning their syndicated program libraries.

39. We have not provided syndicated program exclusivity for smaller market stations and, on reconsideration, are not persuaded to now do so. Distant signal importation in these markets is severely limited—only one distant independent signal may be imported. It may well be that this limitation will impede significant new cable construction in smaller markets. But we have determined that smaller markets can least withstand additional signal importation, and have

fashioned our rules accordingly. To add syndicated exclusivity protection would make these markets even less desirable for new cable construction. The primary consideration, however, is whether syndicated program exclusivity is needed in smaller markets. We think it is not. Certainly, it is of only marginal benefit to copyright holders who derive the substantial bulk of their revenues from the top markets. And we believe that network exclusivity will afford sufficient protection to stations in smaller markets. In unusual circumstances, our special relief provisions allow us to provide other relief where appropriate. See, for example, El Paso Cablevision, Inc., 27 FCC 2d 835 (1971).

40. As to educational stations, it does not appear that the absence of additional exclusivity protection will have a significant adverse impact on their operations. And it does not appear to be desirable to curtail the amount of programing available to cable subscribers from educational stations. Furthermore, we note that the pleadings and comments of educational broadcasters in all our rule making proceedings uniformly asked for simplified procedures for educational stations. Compliance with syndicated program exclusivity notification requirements would involve educational broadcasters with cable systems on a day-to-day basis at considerable expense in time and money.

41. With respect to the type of exclusivity required to be purchased before a broadcast station may claim protection under our rules, we believe there is good reason to require that exclusivity be obtained both against other broadcasts and against cable carriage. The rules in this area are designed to permit copyright holders to distribute programing in particular markets either by broadcast alone or, if they wish, by both broadcast and through cable distant signal carriage. In fact, broadcasters do not now obtain exclusivity against other local stations by FCC fiat; they obtain it by contracting with the copyright owner. The same pattern should obtain with respect to exclusivity against cable distribution of programs. Consequently, our rules also provide exclusivity based upon contractual relationships. Many broadcasters will not desire blanket exclusivity against all systems in their market but only against particular systems. In such cases, broadcasters may be able to obtain programing at less cost than if exclusivity is presumed in the bargained-for price of programing.

42. As to procedural matters, we agree that television stations should endeavor where possible to make their program schedules available to both broadcasters and cable operators at the earliest possible date. We believe that our notification system will work without requiring broadcasters to do more than we have required. We will monitor this situation carefully to see if other rules are required. In any event, we expect that cooperative arrangements will be made between broadcasters, copyright owners, and cable operators to insure the effec-

tiveness of the rules. We are not inclined to accede to MST's request for another round of notifications—this time from cable operators to broadcasters as to compliance with exclusivity requests. It it sufficient that cable operators will have to keep records of programs carried on distant signals. We did not require counter-notifications under our former exclusivity rules. We expect and are assuming that there will be good faith on the part of broadcasters and cable operators. To a real extent, the whole area of program exclusivity will work only if there is good faith. We will be alert to the development of abuse on either side, and are prepared to take action where necessary. Finally, we do not question that licensees of this Commission will obey our rules. But we are retaining the requirement that broadcast stations maintain for inspection pertinent excerpts from their contracts covering programs for which they seek protection. We believe that in order for our syndicated exclusivity rules to work effectively, cable operators, applicants for franchises, and others who desire to know what programming will be available in a community over a period of time be able to find out. Otherwise, investments in cable, program planning, resolution of disputes concerning exclusivity and, most importantly, the rights and obligations under exclusivity contracts could not be readily determined.

43. *Logging.* Comments on our logging requirements, § 76.305, were filed by MST, RMBA, NewChannels Corp., Jerrold Electronics Corp., Buckeye Cablevision, Inc., et al., and Athena Communications Corp. The broadcasting associations argued that the logging requirements should be extended to all markets and for all signals. The cable operators asked for clarification of the rules.

44. The purpose of the logging rules is to assure that our new syndicated program exclusivity rules, which depend on many complex factors, are properly carried out. We stated in paragraph 106 of the Cable Television Report and Order that logging would be required of distant signals carried and the programs offered on those signals. Those are the only signals that are affected by the new syndicated exclusivity rules. Because signals carried prior to March 31, 1972, are not subject to the syndicated program exclusivity rules, they do not fall into the group of signals for which logs are needed. As to the argument that logging should be required of systems that are not located in major markets, this seems to be an unnecessary burden since the syndicated program exclusivity rules do not extend to those markets. MST states that a general logging requirement would assist in assuring compliance with the network exclusivity rules. The network exclusivity rules have been in force in some form for at least 5 years, and compliance has been secured without the added burden of logging. We see no reason for adding that burden now.

45. *Markets.* RMBA, Rust Craft, KID Broadcasting, KOAI (TV), Mid-Continent Broadcasting, Bi-States Co.,

Springfield Television, Inc., KMSO-TV, Duhammel Broadcasting, Boise Valley Broadcasting, and others seek reconsideration of our rule that limits smaller television markets to a zone of 35-mile radius. Athena Communications, Buckeye Cablevision, Colony Communications, Cox Cable Communications, Jerrold Electronics, NewChannels Corp. and Sammons Communications filed oppositions to the requests for expanding the size of the zone.

46. The broadcasters restate the position that they have maintained throughout these proceedings—that a 35-mile zone is inadequate for smaller market stations located in the Rocky Mountain area. Generally, the broadcasters desire a zone coterminous with a station's Grade B contours. The effect of such a rule would be to limit distant signal importation to one independent signal throughout the Grade B contour instead of within a zone of 35 miles. Other suggestions were that the zone encompass Grade A contours or a station's Area of Dominant Influence (ADI). It is alleged that the Rocky Mountain stations place Grade B contours. The effect of such a miles, that transmitter locations are often at a considerable distance from station locations, and that the 35-mile zone will not include a large percentage of the area within 35 miles of the station transmitter. The argument is made that the Rocky Mountain stations depend on audience and revenues from areas beyond the 35-mile zone, that in some cases nearly half the homes reached are beyond the zone, that substantial portions of a station's local advertising revenues come from areas outside the zone, and that stations must look to the entire Grade B contour for homes served on which to base network and advertiser support, both national and local, because advertisers purchase total audience.

47. The question of size of zones was examined at great length and perhaps in more depth than any other issue in this rulemaking proceeding. In determining that a zone of 35 miles would be appropriate in the Rocky Mountain area, we did our own independent analysis of this area of the country. We considered:

1. Station revenues.
2. Station rate cards.
3. Cable penetration within 35-mile zones.
4. Cable penetration outside 35-mile zones.
5. The number and size of cities where additional cable penetration is likely to occur.
6. The difference between the Rocky Mountain area and other areas of the country where there are smaller market stations.
7. The possibility of local advertising being directed away from local stations because of distant signal competition.
8. The practices of national advertisers with respect to the Rocky Mountain stations.
9. The interrelationships of all the above with our new rules concerning signal carriage, program exclusivity, leapfrogging, and grandfathering.

In considering these matters, there are obviously no definitive answers. Necessarily, we are left to judgements—with estimates as to future effects. But based on the above considerations and the experience of years of cable development in these areas, we concluded that the 35-mile zone was appropriate. The petitions for reconsideration add no new information to that which we have previously considered. And in our deliberations based, in part, on information received from the Rocky Mountain stations, we could not find deleterious effect from cable operations on the ability of Rocky Mountain stations to obtain local or national advertising. The Commission is concerned that the Rocky Mountain stations not be harmed in their ability to serve the public by virtue of the adoption of the new rules. However, a case for changing the size of the zone has not been made. We emphasize again our high interest in this matter and our intention to keep abreast of developments as cable expands. As stated in the Cable Television Report and Order at paragraph 91:

New cable systems must give notice before commencing operations, and broadcasters—with knowledge of their own situations—will thus have a full opportunity to make a case for additional relief. We will give these showings most careful scrutiny. Additionally, we will undertake our own in-depth analysis where the desirability of such study is indicated. The essential consideration is not the extent of cable penetration or audience fragmentation per se but rather a demonstration of the effect of cable operation on station revenues and profits and on their ability to serve the public interest. We intend to keep a close watch on future developments in the Rocky Mountain and other regions involving smaller station operations—in rural areas generally—and have directed our staff to prepare reports annually. We will be alert to any emerging trend and in position to adjust our program accordingly.

48. In some markets, the rules may foreclose cable entirely—e.g., a smaller market where one independent station already exists, as in Las Vegas, Nev.[11] Because smaller markets with independent service are likely to be the most vulnerable to distant signal impact, we do not believe that we can make a determination that as a general matter distant signal carriage should be permitted in such circumstances.

49. Several petitions for reconsideration were directed toward the applicability of the rules to specific markets and the need for special relief in those markets. See, for example, petitions of WHYN-TV, KID-TV, KFIZ, WKNX, and KNOI-TV. Although the claims made in these petitions may be meritorious, we do not believe that it would be appropriate to deal with them in this rulemaking proceeding. We have established procedures for obtaining special relief. See § 76.7 of the rules. In connection with the certificating procedure or upon appropriate petition, we will examine all such claims and the responses to them.

---

[11] See petitions of Diversified Communication Investors, Inc. and Community Cable TV.

It would be unfair to make, at this time, ex parte determinations of whether special relief will be given with respect to each or any of these markets. Any petitions for reconsideration that referred to particular markets may subsequently be incorporated by reference into pleadings filed in connection with cable certificating or special relief proceedings.

50. *Significant viewing.* The rules contain a number of sections permitting or requiring carriage of signals meeting a defined viewing level. Viewing at the required level (3 percent share of audience and 25 percent net weekly circulation for network stations, and 2 percent share and 5 percent NWC for independent stations) may be established either by reference to a county-by-county list (published as attachment B to the Cable Television Report and Order (§ 76.54(a)) or by the use of individual surveys in accordance with specified requirements (§ 76.54(b)). Showings of the latter type may only be submitted after March 31, 1973 for the purpose of showing that signals not included in Appendix B are significantly viewed.

51. Reconsideration of these rules is sought by a number of parties, including MST, NAB, Hubbard Broadcasting, Poole Broadcasting, Mahoning Valley Cablevision, Capital Cities Broadcasting, NewChannels, Jerrold Electronics, and Buckeye Cablevision et al. Comments of the National Cable Television Association are included in an opposition petition. In general, the petitions are directed not to the viewing levels adopted as "significant" but to the procedures adopted for demonstrating that individual signals meet the test.

52. A number of petitions filed on behalf of broadcast interests question the use of county-by-county survey data. It is said that the use of countywide data fails to sufficiently account for differences in viewing within counties and that the American Research Bureau (ARB) data may not provide a sufficiently reliable indication of actual viewing. Thus, it is argued, the data from ARB should be used only as indicative of viewing and should be subject to challenge based on further surveys.

53. In the Cable Television Report and Order, paragraph 95, we acknowledged that countywide data might "not account for viewing levels among communities within the county" and that the survey data might have other drawbacks "such as rounding of percentages and sampling errors." We nevertheless determined that these disadvantages were outweighed by the desirability of certainty and were not of sufficient magnitude to preclude use of the data to cure a signal carriage problem where an uncertain standard and the possibility of protracted hearings had created years of uncertainty for both broadcasters and cable operators. The course petitioners ask us to take would completely defeat our goal of providing certainty, with no significant public benefits. In addition to the desirability of certainty there are a number of other factors that should be noted in considering the desirabiilty and equity of the

rule adopted. Initially, as we noted in the report and order, paragraph 85, data of the type used here has been commonly used by advertisers and broadcasters without the fine distinctions between communities within counties which it is here suggested that we make. In the course of filing comments and economic studies in this proceeding, countywide data obtained from the American Research Bureau was frequently used. And, when consideration was given by the Commission to a rule of this type, and viewing levels were selected, countywide data was again used and consideration was given to patterns of carriage that would develop from the various tests of viewing under consideration.[12] Thus, in developing the rule, judgments as to what level of viewing is appropriate under the rule, what proof will be accepted as showing compliance with the rule, and as to impact on broadcast service have become intertwined. It would not, therefore, be appropriate to reconsider the standard for showing compliance without also reconsidering the levels that have been established. We see no reason to do that. The effort to more finely tune the information we have initially used to establish viewing levels through a process of survey and counter survey would, we believe, lead to continuing and pointless disputes about questions more subtle than the whole of our regulatory program is designed to deal with. As noted in the report and order, paragraph 84, the significant viewing levels adopted could reasonably have been drawn at several points. Recognizing that the selection was at best a choice among reasonable alternatives, we do not believe that there would be any point now in encouraging quibbles over fractions of percentages if a method is available for establishing a clear dividing line. The rule adopted establishes such a line and we see no public interest reason for altering the rule and adopting a procedure that would result in extended controversy and would not produce results of any greater decisional significance.

54. Some petitioners have questioned the reliability of the significant viewing list attached as Appendix B to the Cable Television Report and Order on the grounds that it reflected, to some extent, viewing by cable subscribers rather than just off-the-air viewers. Since publishing that list we have further refined it to more accurately reflect off-the-air viewing patterns throughout the country and that revised list is attached hereto as Appendix B.

---

[12] The final levels selected were conservative. As our deliberations in this proceeding progressed a number of alternative resolutions to the overlapping market (or footnote 69) problem were considered. Thus, use of 60-mile zones, Grade B and Grade A contours were considered. Later the significant viewing concept was developed and was successively altered from a test involving share or net weekly circulation to share and net weekly circulation and finally, after the consensus agreement, the test for independent stations was changed from a 1 percent to a 2 percent share. At each tightening of the standard we looked at the resulting carriage patterns in the major markets.

55. The original survey from which Appendix B was developed included cable viewing in counties that ARB estimated to have less than 10 percent cable penetration.[13] In order to improve the accuracy of the list and eliminate all effects of cable viewing, we ordered a second study from ARB that eliminated all cable viewing so that only off-the-air viewing is reflected in our new data. The new study obtained included two (November 1970 and February/March 1971) of the three survey periods used in the original study.[14] We have retained the data from the original ARB study for those counties that were cable controlled and for those counties where there was, in fact, no cable. In determining the presence of cable we compared ARB's controls with that of trade publications and information derived from our own reporting forms. With respect to signals subject to required carriage, we have retained the original list.

56. Where cable is present but was not controlled by ARB in its original study and where there is a discrepancy between the two ARB studies, we have taken the following steps:

(1) Where a signal, present in the first study, did not appear in the second study, we have deleted it from the list of significantly viewed signals.

(2) Where a signal not present in the first study appears in the second study we did not add the signal to the list of those significantly viewed.

(3) In those few counties where cable penetration was so great (90 percent or more) or where adequate data was not available, we have excluded such counties from our list.

(4) Whenever the procedures discussed above were implemented we have removed the 1-year moratorium on surveys for particular communities. Such counties are denoted by an asterisk on the revised Appendix B.

57. In following this procedure we have eliminated all distortions in the original list that might have resulted from cable rather than over-the-air viewing and, we believe, have significantly improved the accuracy of the original list. No signals were added to the list even if the second survey showed that they had met the significant viewing test. We have proceeded cautiously in this area and will permit carriage of these signals under the significant viewing rules only upon individual showings. Although some additional audience surveying may be required by private parties in those counties where there have been deletions, we do not expect that there will be many disputes raised by this requirement. Many of the counties involved are not within the zone of any station and cable systems there will be in a position to carry the signals

---

[13] The original data obtained from ARB was used because it appeared to be the only form in which the viewing information was available and with the expectation that cable penetration of less than 10 percent would not significantly alter the results.

[14] Information for the third survey period (May 1970) was not available from ARB because the computer tapes for that survey period had been erased.

RULES AND REGULATIONS

13856

in question under other provisions of the rules without regard to whether or not they are significantly viewed.

58. The use of a significant viewing test beyond the predicted Grade B contour of a station has also been objected to, but we think without good cause. It is clearly not uncommon for stations to have audience beyond their Grade B contour, and if this is the case, the rationale for using the viewing test is applicable regardless of the location of the station's contour. Mahoning Valley Cablevision requests that three signals from the Cleveland-Lorain-Akron market be included as significantly viewed in Trumbull County, Ohio which is in the Youngstown market. The only rationale urged for doing so is the fact that the three signals are of Grade B quality in the cable communities involved, are UHF, and are needed to "get cable moving" in Trumbull County. Without any showing as to the actual audience of these stations it would be inconsistent with the regulatory program to take the requested action.

59. Several cable television parties, as well as Hubbard Broadcasting, licensee of independent UHF television station WTOG, St. Petersburg, Fla., urge changes in § 76.54(b) of the rules. This section establishes a procedure for taking individual community surveys to show what signals meet the significant viewing test. The cable parties request that we permit such individual surveys prior to March 31, 1973, so that investments may be made during the coming year with certainty and stations that have come on the air following the 1971 ARB survey are not deprived of carriage during the year. Hubbard contends that the 1-year moratorium on filings in conjunction with the prescribed survey methodology cuts it off from areas which it has considered to be part of its market (specifically Charlotte and Highland Counties). It is requested that we waive the 1-year moratorium where ARB's initial survey fails to comply with the survey standards set forth in § 76.54(b), permit the use of nondiary type surveys, and permit the use of countywide rather than community by community surveys.

60. The moratorium on surveys to demonstrate additional signals significantly viewed was generated by a desire to lend certainty to the certificating process during the early stage of our new program. We are adhering to it because we are persuaded that to do otherwise would result in a clogging of processing lines over the disputes certain to arise from the taking of special surveys. After March 31, 1973, we will undertake the task of receiving such surveys and the countering evidence likely to be offered. We see no reason, however, to permit surveys of this type to be made on a countywide basis. There is a basic difference between this kind of survey and that which formed the basis for Appendix B. The purpose of the list in Appendix B is to establish with certainty a base of signals meeting the test, based on information commonly used by the

television industry. Community by community viewing data is simply not now available. Additional signals may also be shown after March 31, 1972, to meet the test, but we have established certain standards that have to be followed in taking individual surveys, so that survey methods, survey times, etc. are not keyed to produce only the desired results and so that numerous surveys are not taken with the hope that through random variations a favorable sample and result are finally achieved. We see no reason not also to require that the required viewing level is attained in the cable community in question where the showing is in support of a specific application.

61. Hubbard's final point concerns the type of study that may be presented under § 76.54(b). The point is made that if we rule out telephone-type surveys we will deprive stations of the opportunity to use a survey mechanism that is far less expensive than the meter or diary survey that would otherwise be required. Our concern in this area is that we have some reasonable assurance that the survey information presented to us has not been manipulated to produce the desired results. To assure this, we have specified a desirable degree of accuracy (one standard error above the required viewing level), when surveys may be taken (during two weekly periods separated by at least 30 days, but no more than one of which shall be a week between the months of April and September), and that the survey be taken by an independent professional survey organization. Within these limits any reliable survey method may be used. While we do not think it appropriate at this time to amend the rules to accommodate particular survey methods, we do not exclude the possibility that, with proper foundation, the telephone survey method proposed by Hubbard can be used.

62. MST raises a further question concerning how we will administer § 76.54(b) of the rules so that parties objecting to carriage will have an opportunity to complete and submit their own survey information. In the Cable Television Report and Order, paragraph 86, we suggested that parties taking individual surveys under this provision of the rules inform other interested parties that a survey was to be made and of the methodology to be used so that questions about methodology could be raised and possibly resolved prior to the survey taking place. We now think it appropriate, based on the concerns expressed by MST, to adopt this suggestion as a rule. Accordingly, we will amend § 76.54(b) to require: (1) Notice at least 30 days prior to the initial survey period to all television station licensees and permittees placing a predicted Grade B contour over the cable community and to all cable television systems, franchisees and franchise applicants, that a survey is being undertaken, the identity of the survey organization taking the survey, and the procedure to be used in the survey, and (2) that objections to survey organiza-

tions or procedures be made within 20 days after receipt of such notice to the party undertaking the survey. By following this procedure, it should be possible to resolve questions concerning surveys at a point when there is still time to correct problems that are found to exist. Additionally, this procedure will provide an opportunity for counter surveys, where appropriate, to be undertaken.

63. Finally, concern has been expressed that, as viewing patterns change, systems may make individual surveys and add additional signals without deleting signals that no longer meet the significant viewing test. This is a matter which we think may warrant our further attention in the certificating process if it appears that signals are being added to systems simply through random fluctuations in survey information. But the issue is not one we are prepared to settle simply by requiring substitutions in every instance. First, we have set our sample probability test high to avoid problems of this type (see paragraph 86 of the Cable Television Report and Order) and it is therefore not likely that there will be many instances of this occurring. Second, the rules not only permit surveys and carriage by cable systems but it also entitles broadcasters to take surveys and request carriage. In some reconsideration petitions concern has been expressed by broadcasters with the difficulty they may have in obtaining carriage under these rules, especially if they are new stations or have recently improved their facilities or programing.[15] In these circumstances it would not be appropriate to set conditions automatically discouraging carriage or act to penalize cable systems seeking carriage of such stations by forcing a choice between deletion of stations to which subscribers have become accustomed and the addition of stations whose off-the-air audience has improved. Thus, while we do not anticipate problems in this area, if problems do arise they can best be considered in individual proceedings.

64. *Grandfathering.* Jerrold Electronics Corp., Athena Communications Corp., New Channels Corp., and Buckeye Cablevision et al. request reconsideration of the grandfathering rule which exempts cable systems operational as of March 31, 1972, from compliance with the syndicated exclusivity rules, but does not do so for nonoperational systems authorized to carry signals prior to that date. Petitioners state that there is no reason why this distinction should be made. KID Broadcasting Corp., MST, The Rocky Mountain Broadcasters Association, Duhamel Broadcasting Enterprises, Grand Canyon Television Co., WGAL Television, Inc., Stainless, Inc., Bi-States Co., the NAB, and WBRE-TV filed comments concerning the grandfathering date and the applicability of grandfathering to program exclusivity. The broadcasters argue that the grandfathering date should not be March 31, 1972,

[15] See for example petitions filed by Hubbard Broadcasting and Rust Craft Broadcasting.

but some earlier date.[16] It is stated that cable systems built in recent years were constructed during a period when the Commission was studying cable television and with the awareness that the Commission might adopt limitations on cable operations in smaller markets. It is alleged that the grandfathering provisions are inconsistent with the consensus agreement and with previous Commission statements concerning grandfathering. And it is submitted that there has been a large number of § 74.1105 notifications mailed to broadcasters since the adoption of the new rules and that these should confer no grandfathering rights. MST states that an authorization pursuant to § 74.1105 "is not really an authorization at all." As to the grandfathering provisions for program exclusivity, the Rocky Mountain Broadcasters Association urges that all smaller market stations that were receiving same-day exclusivity on August 5, 1971, should continue to receive it. WBRE-TV, Inc. and WGAL Television, Inc. argue that the new rules eliminate the same-day, non-network exclusivity rights that stations could receive under former § 74.1103, as well as preclude any station from receiving syndicated exclusivity pursuant to § 76.151 if it is carried by new systems located outside of all major markets.

65. *Grandfathering is essentially a balancing process.* A line must be drawn somewhere. And wherever it is drawn there will be parties affected by the decision that would prefer the line to be drawn somewhere else. In establishing the cut-off date, we selected March 31, so that all rights of parties affected by the rule would vest or divest on the same day. This is not an inappropriate date because our former rules were in force until the effective date of the new rules.

66. With respect to the consensus agreement and the § 74.1105 notifications that have recently been filed, we believe that our decision in El Paso Cablevision, 27 FCC 2d 835 (1971), concerning § 74.1105 authorizations, is in point, and we have framed the grandfathering provisions concerning signal carriage accordingly. We have been monitoring the § 74.1105 notifications recently filed. We note that broadcasters have it in their power to object to any notifications and thereby stay their effect, and that they have generally done so. Most significantly, we have discovered no recently filed notifications for designated cities of major markets that are unopposed, and have found that other notifications have also been opposed. In any event, any notification filed after the end of February 1972, conferred no rights on cable systems because the effective date of the rules preceded the time for filing objection to the notifications. We have also

provided that all such systems must obtain certificates of compliance before commencing operation. Furthermore, any system that may be authorized to carry signals but was not operating on March 31, 1972, will not be grandfathered with respect to syndicated exclusivity protection. We do not believe that such systems should be grandfathered with respect to exclusivity just because the signals are grandfathered. There are systems with approved signals that have not commenced operation for a variety of reasons. We will not disturb signals where rights have vested, even where the system has not gone into operation. There is no upsetting of viewing patterns in insisting on compliance with exclusivity requirements. Nor do we believe that any proposals concerning grandfathering made in Docket 18397 are fundamentally inconsistent with the rule adopted. In any case we are not required to adopt the exact terms of our original proposal.[17]

67. As explained in paragraph 29, we adhere to our previous determination that simultaneous program exclusivity effectively protects a network affiliate's network programing; hence, we see no reason to perpetuate, via grandfathering, the extra burdens imposed on cable operators and subscribers by same-day protection. On the other hand, we find merit in the proposal that we restore to stations whose signals were carried by a cable system prior to March 31, 1972, the nonnetwork exclusivity rights that they enjoyed under former § 74.1103, but on a simultaneous-only basis.

ACCESS TO AND USE OF NONBROADCAST CHANNELS

68. *Program origination.* The American Civil Liberties Union questions our authority and our decision to require cable systems serving 3,500 or more subscribers to originate their own programing and urges common carrier regulations for cable systems.[18] The origination provisions in the report and order represent a recodification of our original requirements, with one or two minor changes designed only to eliminate any possible confusion with the new access rules. It seems unnecessary, therefore, to engage in any lengthy reconsideration of the jurisdictional issue again here.

69. With respect to our judgment in requiring origination programing and the question of common carrier regulation, at paragraph 146 of the report and order we dealt with these issues stating:

We have considered these possibilities but feel that it would be premature to adopt either at this time. (See notice in Docket 18397, 15 FCC 2d 417 at paragraph 26 (1968).) At this stage in the development of the cable industry it is the system operator who has

the greatest incentive to produce originated material attractive to existing and potential subscribers. We have tried to encourage this origination both through our origination rules (first report and order in Docket 18397, 20 FCC 2d 201 (1969)) and by structuring the broadcast signal carriage rules to stimulate the development of nonbroadcast services. At the same time, we have recognized that during this developmental stage we should not adopt rules that constrain experimentation and innovation in the services that cable systems provide but, rather, that we should seek to keep our future options open. When cable penetration reaches high levels and demand increases for leased channel operations, we will revisit this matter. For now, we remain of the view that the most appropriate mix for the orderly development of cable and for encouraging the maximization of its potential for public benefit is one that embraces "* * * a multipurpose CATV operation combining carriage of broadcast signals with program origination and common carrier service * * *" (first report and order in Docket 18397, supra, paragraph 3). The rules adopted here are designed to accomplish that.

We have fully considered the positions urged on us by the ACLU and have explained, as above, why we have elected to proceed as we are. The Union has supplied useful insights into cable's potential. But we remain of the view that it is unwise at this stage to fasten unnecessarily restrictive formulas on the evolution of the new technology. Cable's success is by no means assured in all these large markets with a plethora of broadcast service. The cable entrepreneur should be given appropriate leeway during this critical period of development. The ACLU's approach, which may prove sound eventually, at the present time does not afford the industry the flexibility that we desire to encourage experimentation and innovation. Further, we doubt very much if, in new systems in major markets, a scarcity of access channels will arise from a cable operator's excessive use of bandwidth for his own origination purposes; but if a problem should arise, we shall be alert to take action to maintain our emphasis on the provision of access channels.

70. *Natural breaks.* Buckeye Cablevision et al. and Jerrold Electronics Corp. have requested reconsideration of § 76.217 which permits advertising on origination cablecasting channels only at the beginning and conclusion of each program and at natural intermissions or breaks within a cablecast.[19] Petitioners argue that advertising on such channels should be permitted to the same extent as for broadcasters. They maintain that more advertising revenue will tend to alleviate the financial burden of providing free access channels. Further, they suggest that since we have placed no advertising restrictions on leased access channels, it no longer seems reasonable to maintain the restrictions on the origination channel.

71. The argument is not without attraction, especially in light of the new relationship between origination cablecasting channels and access channels.

---

[16] Suggested dates range from Aug. 5, 1971 (date of the letter of intent), to Feb. 12, 1972 (date of publication of the rules in the FEDERAL REGISTER), with several suggestions in between those two dates. KID Broadcasting suggests that distant signals on existing cable systems in smaller markets be deleted, one per year, until the system conforms to the three network-one independent formulation of § 76.39 of the rules.

[17] The statement concerning grandfathering that appeared in our letter of intent was admittedly in error. Among other things it did not purport to grandfather systems commencing operations subsequent to the second report and order and prior to our notice of proposed rule making in Docket 18397.

[18] Our authority to require cable origination has been confirmed in U.S. v. Midwest Video Corporation, —— U.S. —— (Case No. 71–506) June 7, 1972.

[19] Section 76.217 is the recodification of former § 74.1117 of our rules.

At this stage, however, we have not received enough information in this experimental area to enable us to ascertain the likely source and extent of a cable operator's revenues. It may be, for instance, that the revenues derived from leased operations will more than suffice to offset whatever losses are incurred as a result of our advertising limitations on the origination cablecasting channel. It is too early to determine. We expect to be watching developments in the nonbroadcast area closely and, should it become necessary or desirable, we will revisit this problem.

72. *Pay-Cable.* ABC, the Motion Picture Association of America, et al., (MPAA), and the National Association of Theatre Owners, Inc. (NATO), filed comments on the questions of pay-cable and the siphoning of broadcast programing. All three parties request a thorough review of the Commission's policy toward pay-cable, ABC and NATO with the perspective of prohibiting it, and MPAA, representing major program producers and distributors, urging that restrictions be eliminated. ABC argues further that antisiphoning rules should be considered for any originated cable programing—not just pay-cable—on the ground that any siphoning of programing from broadcasters, especially considering the potential of interconnected cable origination, would be harmful. ABC recommends that the Commission " * * * through appropriate further rulemaking proceedings, undertake to inform itself and take appropriate action * * *" respecting pay-cable and siphoning. The program producers and distributors, on the other hand, contend that the market place should be free as to program availability and not hampered by the restrictions imposed by our pay-cable rules.

73. It should be noted here that we intend to act separately on previously received petitions to reconsider the pay-cable rules. These rules were adopted by Commission action of June 24, 1970, FCC 70-677, 23 FCC 2d 825, 35 F.R. 1090. Our new cable rules have carried over the pay-cable regulations (old § 74.1121) simply to provide continuity in codification. All the rights of the parties requesting reconsideration of § 74.1121 remain intact. Further, on request of the parties, any petitions for reconsideration of the Cable Television Report and Order dealing with pay-cable will be included in our reconsideration of the pay-cable rules which we intend to act on shortly.[29]

74. *Jurisdiction to compel access.* The Columbia Broadcasting System has suggested that the Commission lacks sufficient jurisdiction to impose access obligations on cable television systems. We disagree. Cable television, as it grows, must be integrated into a nationwide communications structure. Were we to permit an uncontrolled development of cable we would be breaching our obli-

gations under the Communications Act of 1934, as amended. This Commission was created, amid the chaotic developments in the field of radio, " * * * to make available, so far as possible, to all the people of the United States a rapid, efficient, nationwide, and worldwide wire and radio communications service * * *"[21] As "an integral part of interstate broadcast transmission," cable operators "cannot have the economic benefits of such carriage as they perform and be free of the necessarily pervasive jurisdiction of the Commission."[22] Thus, we conceive it to be our obligation to consider the actual and potential services of cable television and create a Federal policy which insures that these services can be distributed equitably, on a nationwide basis as merely one link in our communications systems. Much as we impose standards of public responsibility on broadcasters, so too must we fashion a role for cable television. We have attempted to construct only an initial framework within which cable may develop its potential for public service. We believe that cable's integral relationship to broadcast transmission, recognized in United States v. Southwestern Cable Co. and United States v. Midwest Video Corporation[23] and the duties imposed on us by the Communications Act of 1934 make it only reasonable and necessary for us to do so.

75. *Smaller market minimum channel capacity.* Publi-Cable, Inc. suggests that we complement the minimum channel capacity rules with a requirement that new systems in smaller markets have a minimum of 12 channels and that existing systems in these markets have 5 years (or until the renewal of their franchises, whichever occure first) to attain a 12-channel capacity.

76. Our reason for limiting our channel capacity requirements to systems in major markets at this stage was "to avoid imposing unreasonable economic burdens on cable operators."[24] In any event there are few, if any, cable systems being built anywhere today with less than a 12-channel capacity. We will give careful scrutiny to any application for certification proposing less than a 12-channel capacity. With respect to older systems, we envision that rebuilding, whether because of general obsolescence or because of the necessity for compliance with our technical standards, will eventually result in all systems in small markets having at least 12-channel capacity.

77. *Equal bandwidth.* Section 76.251 (a) (2) of the new access rules provides, in effect, that cable systems will have to provide as much bandwidth for nonbroadcast services as they use for the carriage of broadcast signals. Thus, for each broadcast signal carried, an equal

amount of bandwidth will have to be available for nonbroadcast use.

78. NewChannels Corp. and others have advanced the argument that educational, religious, and foreign language broadcast stations not be counted when making their determination of channel capacity. Petitioners suggest that in some cases, where the requirement will work hardship, systems may decide not to carry all the broadcast signals legally available. They argue that the channel capacity expansion formula of § 76.251(a) (8) provides an adequate assurance of bandwidth for nonbroadcast purposes.

79. We do not find these arguments persuasive. In our rules dealing with channel capacity, our goal was to insure that cable systems in major markets would not underbuild. "We urge[d] cable operators and franchising authorities to consider that future demand may significantly exceed current projections, and we put them on notice that it is our intention to insist on the expansion of cable systems to accommodate all reasonable demands."[25] We believe this consideration to be controlling and find it difficult to believe that cable operators will not carry all the broadcast signals available to them.

80. *Number of designated access channels.* Publi-Cable, Inc., the National Association of Educational Broadcasters (NAEB), and the National Education Association (NEA) have questioned what they regard as an unduly severe limitation on the number of designated access channels to be provided by cable systems pursuant to § 76.251(a) (4), (5), and (6) of the rules. They argue, particularly with respect to educational channels, that the potential for use far exceeds the limit of one channel. NEA has sug-

---

[29] We note that AMST, among others, has filed a response to petitions for reconsideration which includes argument on the pay-cable antisiphoning issues. These too, upon request, will be included in our separate pay-cable proceeding.

[21] Section 1, 47 U.S.C. 151.
[22] General Telephone Co. of California v. Federal Communications Commission, 413 F. 2d 390, 401 (C.A.D.C.) (1969), cert. denied 396 U.S. 888.
[23] 392 U.S. 157 and 40 USLW 4626.
[24] Cable Television Report and Order, paragraph 120.

[25] Cable Television Report and Order, paragraph 120. The question has arisen whether we have preempted the area of channel capacity so that local governmental entities could not require more than twenty channel capacity or more than required under the equal bandwidth rule, § 76.251(a)(2). We believe that our requirement for expansion of channel capacity will insure that cable systems will be constructed with sufficient capacity. However, if a local governmental entity considers that greater channel capacity is needed than is required under the rules, we would not foreclose a system from meeting local requirements upon a demonstration of need for such channel capacity and the system's ability to provide it. A similar question has been raised with respect to two-way capability. We find no reason why a cable operator wishing to experiment with a more sophisticated two-way capability than that which we have required should be precluded from doing so. However, we do not believe that franchising authorities should require more than we have provided for in our rule because it is possible that any such requirement will exceed the state of the art or place undue burdens on cable operators in this stage of cable development in the major markets. Where a franchising authority has a plan for actual use of a more sophisticated two-way capability and the cable operator can demonstrate its feasibility both practically and economically we will consider, in the certificating process, allowing such a requirement.

gested once more, that a minimum of 20 percent of system capacity be set aside for educational use.

81. It should be noted at the outset that, while one educational access channel is the minimum required, we specifically provide in § 76.251(a)(8) for adding more access channels should the need for such channels be adequately demonstrated. Thus we envision an orderly growth of access channels, linked to demand.[26] In addition, in the Cable Television Report and Order we stated that after a developmental period (to begin from the commencement of service until 5 years after completion of the basic trunk line) "designed to encourage innovation in the educational uses of television—we will be in a more informed position to determine in consultation with State and local authorities whether to expand or curtail the free use of channels for such purposes or to continue the developmental period."[27] Clearly, as we have stated, this is an area which we will revisit. But without the further knowledge which can be gained only from allowing cable systems to experiment within our initial framework, we are not inclined to add extra burdens to the access requirements. Finally, we are in no way restricting arrangements between the local entity and the cable operator to provide specified numbers of channels for educational purposes on a paid basis. Such arrangements constitute the very type of new service which cable can and should provide. Further, we will entertain petitions from the franchising authority and the cable system when they wish to experiment with additional designated channels on a free basis or at reduced rates.

82. *Rates for educational users of leased channels.* The NAEB urges us to amend our rules to enable leased channels to be used for educational purposes at lower rates and to provide that, at the termination of the free 5-year developmental period for educational access channels, rates be kept at a minimum. As stated, we are entering into a period of experiment. The access rules will, without question, require further study and future deliberations. The question of access channel rates is but one of the matters which we will have to confront again. Our initial feeling in this matter is to avoid any form of preferential policy with regard to who may be use and what must be paid for access channels. For the present, we deem it desirable to allow the experiment to proceed apace.

83. *Leased channel availability.* The ACLU claims that § 76.251(a)(11)(iii), requiring that the cable operator establish rules for first-come, nondiscrimina-

tory access to leased access channels, fails to assure that channel capacity will be available over periods of sufficient duration to justify prudent investment by an entrepreneur wishing to supply broadband services.[28] We are somewhat puzzled by this view and can only state that we assume entrepreneurs will, in fact, lease channels over long periods for programming or other services. It is for this reason that we specifically require in § 76.251(a)(7) that "* * * on at least one of the leased channels, priority shall be given part-time users."

84. We do, however, feel constrained to inject a note of caution at this point. Our view that cable systems be required to accept "all reasonable demand" for access use is predicated on the knowledge that cable technology embraces very large amounts of bandwidth. In the report and order we noted the existence of a few 40-, 50-, and even 60-channel systems. We are confident that, as technological developments proceed, the majority of cable systems will be able to offer similar and even greater channel capacities. It would be unrealistic, however, to assume that we are dealing with an infinite entity. Any channel capacity, no matter how large, can in theory be completely consumed. We have proceeded in what we believe is a reasonable fashion. We have not required all systems to offer 60-channel capacities. For many if not most systems, to do so at the present time might be impossible and/or economically unwise. It is clear, therefore, that until such time as channel capacity can practically approach the huge numbers we envision, any single person or group claiming access to large numbers of channels will create problems for both the system and for the Commission. But our present judgment is that the problem is unlikely to arise in the next few years. We do not, therefore, believe it necessary now to place artificial restrictions on the number of channels any one person or group can use. Indeed, we may not be confronted with the issue because technological advances may outstrip even huge channel demands. If we are wrong in our present estimates, potential users of access services will, hopefully, exercise a degree of restraint and will not, by overenthusiastic reaction to cable television, force a determination of what, in fact, constitutes an "unreasonable" demand for channels. Finally, while the cable operator remains fully responsible for compliance with our rules, local groups providing assistance to the cable operator can be most helpful and should therefore receive the co-

operation of the cable operator in their appropriate activities.

85. *Access channel liability.* Various parties have questioned our judgment that there seems little likelihood of civil or criminal liability against cable operators from the use of access channels. The parties contend, understandably, that our feeling in this matter, however persuasive, is hardly a guarantee. They note, further, that although the cable operator will have no control over program content on access channels, he is charged with proscribing the presentation of obscene material. It is suggested that to this extent, at least, the operator will, in effect, be required to exercise control. To clarify this area, we are requested to seek legislation to grant immunity to a system operating under our access rule. We, of course, appreciate petitioners' concern over the liability issue. We still believe, however, that existing case law solves most problems in this area.[29]

86. *Provision of access services for operating systems.* A number of parties[30] have requested reconsideration of § 76.251(c) of the rules. This section makes the minimum channel capacity and access channel requirements applicable to all cable television systems which commence operations in a major television market after March 30, 1972. Systems already in operation prior to March 31, 1972, are given until March 31, 1977, to comply. Finally, the section provides that if a major market system in operation prior to March 31, 1972, receives a certificate of compliance to add television signals to its operations before March 31, 1977, it shall comply with various elements of the access rules at the time of such addition of signals.[31] It is this provision of § 76.251(c) which has prompted petitioners' concern.

87. The access provisions made applicable by § 76.251(c) are the requirements for designated public, educational and local government access channels, and leased access channels, plus the rules applicable to the operation of such channels and the requirement of "expansion of access channel capacity" (§ 76.251(a)(8)). We did not require the immediate compliance with the new minimum channel capacity rules of § 76.251(a)(1) and (2). It is clear, however, that a currently operating system without the bandwidth required by these new rules, may well not have the channel capacity to add whatever additional channels might be required by the expansion-of-capacity formula. Even without this provision, a system would still be faced with the prospect of adding at least four channels to what in many cases will be a channel capacity capable

---

[26] Paragraph 123.

[27] In paragraph 132 of the report and order we specifically note that in instances where the system operator and franchising authority may wish to experiment by providing additional channel capacity for educational, as well as other access channels—on a free basis or at reduced charges—we will consider appropriate showings.

---

[28] The ACLU also suggests that our rules do not provide for the leasing of channels for "data grade" or "audio grade" transmissions, even though such transmissions do not require a large amount of bandwidth. The rules and the report and order in fact do contemplate such channel uses. Section 76.251(a)(2) requires that nonbroadcast capacity be suitable for carriage of "Class III" signals, which are clearly described in the report and order as used for many nonvideo activities requiring less than a full 6 MHz of bandwidth.

---

[29] Cable Television Report and Order, at paragraph 141.

[30] New Channels Corp., Athena Communications Corp., Jerrold Electronics Corp., Buckeye Cablevision, et al.

[31] Sec. 76.251(c) would require compliance with paragraphs (a)(4) through (11) of that section.

of absorbing only one or two.[32] Petitioners contend that the addition of such a large number of channels to most existing systems will require substantial or complete rebuilding, involving large investment. They note, further, that additional broadcast signals might well provide some revenue base to underwrite such a rebuilding program when it becomes necessary at some later date. Petitioners would prefer to add additional signals where they can, and then have some reasonable period in which to provide the access services. Such a changeover period would be decided on, apparently, by a series of rulings by the Commission. While we do not find such a request unreasonable, we are reluctant to submit ourselves to a flood of petitions, the inevitable effect of which will be to delay the certificating process. We stated in the Cable Television Report and Order that should some isolated system be unduly burdened by the access requirements, such a matter could be dealt with in a waiver request.[33] The number of systems potentially affected here, however, constitutes a significant percentage of all presently operating cable television systems. We do not choose to burden both ourselves and the industry with the necessity of making hundreds of individual determinations if it is possible to preserve our "go, no-go" concept with the application of a general rule.

88. We have stated that our focus with respect to the access requirements is "* * * on the top 100 markets because we have selected these markets as the recipients of special benefits in order to stimulated cable growth."[34] It was our intention that "growth should be accompanied by access obligations if the public is to receive the full benefit of this program."[35] It seems appropriate that a presently operating system in a major market, which receives the benefits of additional signal carriage as a result of our new rules, should be required to provide some of the access services which would not otherwise be required until 1977. The question raised is the degree of such compliance and the burdens it would place on cable systems.

89. We have decided to modify our original approach to lessen its immediate impact on the affected systems, while at the same time preserving much of its underlying philosophy. Thus, § 76.251(c) is being amended to require that, for each additional broadcast signal carried, such a system will have to provide one access channel. The first additional signal will be complemented by a public access channel, the second by an educational access channel, the third by a governmental access channel, and any others by leased channels. The expansion of channel capacity rule will not be required, until March 31, 1977, at which time, of course, all the access requirements become applicable to existing systems in major markets.

90. Our new access cablecasting rules apply to each system in major market areas. However, we are not unmindful of the existence of multiple systems served by a single headend. In most of these situations, each system has the same channel capacity, and carries the same broadcast programing. The "system" as a whole is not designed to carry program material selectively to each component system. The ability of an existing conglomerate of systems to comply with the access channel requirements will necessarily vary with the proximity of the component systems, the basic design of the system, and, of course, the channel capacity.[36] Clearly, we cannot establish a rule of general applicability in this area. To the extent possible, however, within the technical and geographic parameters of the systems involved we intend to safeguard the integrity of our access requirements. This can best be done if, during the certificating process we are provided with sufficient detailed information concerning the systems' ability to comply. Again, we will require compliance to the greatest possible extent. In some cases it may be possible for individual systems to share channel time. If this is the case we may be persuaded for instance that, at least two shared public access channels will suffice for some conglomerate systems. Where boards of education are under the same jurisdiction, the problems may be alleviated. Local governments may agree to share time on one or two channels. We must, however, be given as much information in these respects, as possible, together with specific proposals on the part of the systems. Until we receive such material certificates will not issue.

TECHNICAL STANDARDS

91. The general question of Federal preemption of technical standards has been informally raised by a number of parties. Our technical standards provide only a start. They will be expanded and refined to meet changes in the state of the art. We see no reason why franchising authorities may not now require more stringent technical standards than those in Subpart K.

92. *Definition of "subscriber terminal"*. Zenith Radio Corp. requests clarification of the definition for "subscriber terminal" in § 76.5(ee). They suggest modification of the definition so that, if a converter is used, the subscriber terminal will be considered to be at the output of the converter. In the majority of cases, there would appear to be no practical difference between the two definitions. In most cases, the converter output terminals are connected directly to the subscriber's receiver input terminals. But this may not always be the case, and we must anticipate variations on this practice. We also must anticipate that the subscriber may connect his own converter between his television receiver and the cable system. The cable system should not be burdened with the responsibility for the performance of a privately owned converter. We are of the view that our definition of "subscriber terminal" is to be preferred—it is most appropriate to define it as the point at which the facilities supplied by the cable system connect to the equipment supplied by the subscriber.

93. *Tolerance.* Zenith protests that the frequency tolerance applicable at subscriber terminals where converters are used (§ 76.605(a)(2)) is inadequate to prevent adjacent channel interference. Zenith points out that when converters are adjusted to deliver signals to subscriber receivers on channel 12 with strong ambient field present on channels 11 or 13—although the strong local fields induce signals directly into the subscriber's receiver input—these signals are adequately rejected by traps within the receiver when it is tuned properly. However, when the receiver is detuned toward the maximum departure (250 KHz) of the desired channel 12 signal coming from the cable, the internal traps are no longer able to reject the undesired adjacent channel signals which are picked up directly within the receiver. Zenith proposes that, although a ± KHz tolerance must be permitted for the present, the Commission should provide now for a scheduled reduction of permissible frequency tolerance within 2 or 3 years.

94. Other information which we have considered in this matter (for example, data filed by NCTA which report a substantial number of receiver measurements; comments of Sterling Information Systems regarding tuner performance) indicate that difficulties with converter drift may not be as critical as the Zenith comments suggest. NCTA contended that because there are a number of other practical factors which also affect susceptibility to adjacent channel interference, the benefits theoretically obtainable by tight control of converter output frequencies may be obviated. A review of measurement data on representative receivers indicates that although some of the attenuation of undesired adjacent channel signals is lost by a tuning shift of 250 KHz, it is reasonable to expect that most receivers will still provide between 25 and 40 decibels of discrimination against the undesired signals.

95. Zenith's position also is directly challenged in the TPT Sterling joint comments. In discussing the matter of adjacent channel rejection by trap circuits within home receivers it is stated

This is simply not the case when the signals on the CATV system proper (excluding the converter) are maintained within the

---

[32] A strict interpretation of § 6.251(c) reveals that it does not distinguish between the addition of signals which must be carried on request and the addition of signals which the cable system may choose to carry, but which are not mandatory. In some instances, our new carriage rules will require a system to carry, on request, a signal not previously required or carried on the system. It was not our intention that carriage of such a signal would trigger the access requirements of § 76.251(c). We are amending that section accordingly.

[33] Paragraph 148.

[34] Cable Television Report and Order, Paragraph 147.

[35] Ibid.

[36] While we do not require existing systems to undergo radical redesigning, we expect that newly built "conglomerate systems" will be designed to comply with the access requirements.

±25 KHz required in § 76.605(a)(2) of the rules. Zenith evidently concedes that there will be no interference problem if tolerances are held to the ±25 KHz standard. The subscriber's converter, will translate all carriers in the pass band of the converter, maintaining the original spacing and tolerances of the carriers on the system proper.

96. Nevertheless, we are persuaded that Zenith's pleading for a more strict frequency tolerance has merit. At present, converters meeting the stability requirements Zenith suggests are, to our best knowledge, not available in quantities or at costs which would permit us to impose a tighter tolerance. We are reluctant, therefore, to adopt a schedule for a reduced tolerance at this time, but we look to revised rules which we expect to consider after a reasonable period of experience. Within the next several years we shall have the benefit of practical experience with the effects of the ±25 KHz tolerance, and will also have advice from the Cable Television Advisory Committee which was established on February 2, 1972.

97. In their joint opposition to the Zenith petition, TPT-Sterling also request clarification of § 76.605(a)(2) in a manner which would require a visual carrier frequency stability of no more than ±25 KHz at subscriber terminals which are served through a converter. The intent of the suggested clarification and our rule appear to be the same. We recognize that, in a converter subject to manual adjustment by the subscribed, the frequency of the visual carrier normally can be adjusted into or out of the desired channel by the subscriber. The practical effect of our rule is to require that, once the visual carrier frequency is adjusted properly within channel, it be maintained between 1 and 1.5 MHz above the lower boundary of the channel. The TPT-Sterling recommendation would apply the same latitude for frequency drift (±250 KHz) to the visual carrier as delivered to the subscriber, but would not require that the signal be kept within the desired channel. We are not persuaded that the rule should be "clarified" in the manner requested by TPT and Sterling.

98. *Maximum visual signal level.* Both Zenith and MST request the establishment of a specific maximum limit for the visual signal level which may be delivered to subscriber terminals. Zenith proposes a maximum of 5 millivolts across 75 ohms; MST apparently requests a specified limit which will prevent overloading subscriber receivers. Section 76.605(a)(5) now requires that the visual signal level at any subscriber terminal not exceed a value that would produce signal degradation due to overloading the customer's receiver. Because the level at which overload effects become noticeable varies widely from receiver to receiver, we preferred not to set a fixed specific limit and elected to leave it to the discretion of the cable system to deliver whatever maximum level it found advisable, so long as it does not cause signal degradation in the customer's receiver. Zenith's concern seems to be centered around the possibility that radia-

tion from the receiver input circuits might cause interference to nearby non-subscribers. Zenith notes that, in a different action, the Commission is proposing to limit the permissible output level from Class I television devices (Docket 19281) on the grounds of potential interference, and considers that there is a similar interference potential from cable installations.

99. Our view is that there are important differences between cable and Class I television devices. Although both types of signals usually would be delivered to home television receivers by direct connection of coaxial cables, cable television connections would be made by service technicians whose objective is to provide a proper connection and adjustment of levels to insure good picture quality. Class I television devices, on the other hand, are expected to be consumer items and are expected to be tampered and experimented with by the owner. Further, they are subject to operation with an outside antenna connected to the television receiver when the device is in use. Measurements we have made on several such devices show a very serious interference potential. On the other hand, although thousands of cable systems have operated with unregulated signal levels, our Field Engineering Bureau is unable to report any cases of interference due to radiation from subscriber receivers, as suggested by Zenith. Thus, we are not persuaded that a limit on maximum visual signal level is required at this time. In this respect, we note also the comments supportive of our position which were filed by TPT and Sterling.

100. There is a point with respect to the MST petition for modification of § 76.605(a)(5) with which we concur. MST objects to the reference in the rule to the term "12 decibels overall" without defining the requirement more clearly. The aim of this requirement is to accomodate a reasonable variation in levels throughout the system. Present techniques appear to be adequate to hold such variations within the 12 dB latitude, and we wish to be certain that this capability is applied for the benefit of subscribers. In response to MST's suggestion, we are revising the rule to indicate that, over any 24-hour period, levels on individual channels must be maintained within a 12 dB range.

101. *Channel response.* Oak Electro/netics Corp. requests amendment of the channel response requirements when converters are used. They maintain that converters, in order to provide substantially greater rejection of adjacent channel signals, must be permitted to "roll off" the desired channel response closer to the visual carrier frequency than § 76.605(a)(8) would permit. In support of this position, Oak submits two studies representing the response characteristics of converters which they now manufacture or plan to manufacture. Both bear a notation which indicates that they fail to meet the requirements of § 76.605(a)(8). However, we observe that by a slight retuning to relocate the visual carrier closer to the bandpass center, the requirements of § 76.605(a)(8) are met

by both response curves. Oak's comments are, in our view, based on a desire to provide a substantial increase in adjacent channel selectivity. We recognize that many existing receivers are deficient with respect to rejecting adjacent channel signals, but we are reluctant to sanction remedies of the deficiency which would use converters that substantially modify response characteristics near the visual carrier frequency in order to provide added selectivity. The intention of § 76.605(a)(8) is to insure that the cable system does not modify unduly the spectrum of frequencies presented to the subscriber's receiver, particularly near the picture carrier frequency. Accordingly, we will not adopt the amendment proposed by Oak.

102. *Radiation from cable systems.* Zenith points out an inadvertent inconsistency in the technical standards that we now correct. Section 76.605 embraces a series of performance requirements that are applicable to system performance as measured at each subscriber terminal. However, the radiation limitation set forth in subparagraph (12) obviously is not applicable to measurements at a subscriber terminal. It is intended that measurements of radiation from the system be made in accordance with the provisions of § 76.609(h). We are adopting an appropriate amendment to clarify this matter.

103. *Component standards.* There have been requests that the Commission adopt standards for various components of cable systems. For instance, standards were suggested for television antenna and preamp design, cable headend equipment, and cable receiving antennas. The thrust of our technical standards has been to refrain from specifying either equipment type or characteristics. Rather, we are concerned primarily with performance standards as measured at subscriber terminals. We are interested in the end result. Many alternative approaches have been considered, but on balance we have endeavored to take into account the considerable technical diversity found in the new cable technology and have thus adopted "* * * a set of standards that we believe will provide much needed uniformity on a nationwide basis' yet still allow sufficient flexibility for further technical change." We are not inclined, therefore, to specify types of equipment or equipment design.

104. It should be apparent, of course, that some cable systems may well have to invest in new or different equipment in order to comply with our standards. For instance, where antennas are located adjacent to power sources capable of generating interference so that our standards cannot be met, the antennas will have to be moved. We do not believe, however, that it is advisable to require this and similar measures in our rules. Conditions vary, and it will be up to each cable system to comply with our technical standards by whatever means become necessary.

105. *Receiver modification.* Some have suggested that we require the cable system to modify its subscribers' television

receivers in a number of ways. Trans-Tel, for instance, would have us require cable systems to modify tuner knobs to specify the particular UHF channels available in the market and being carried on the system. Others, including Mid-Continent, urge us to require that the operator install a shielded lead between the television tuner and its antenna terminals to reduce the possibility of ghosting.

106. We are reluctant to require any cable television system to engage in these or other television receiver modifications. We note in this regard the opposition to petition for reconsideration filed by TransVideo Corp., indicating that to require receiver modification may involve financial burdens, loss of customer good will and in some instances excessive technical difficulty. We choose to leave this problem in the hands of the cable system. Should the system and subscriber agree to make certain receiver modifications, we will have no objection.

107. *"Local" and "distant" signals.* Several parties are still particularly concerned that our technical standards could result in a lack of comparable signal quality between "distant" or microwaved stations and "off-air" local transmissions. We simply do not believe that this will be the case. The standards we have adopted, when supplemented by appropriate color and ghosting standards,[37] should result in substantial parity of all signals received at a subscriber's terminal.

108. *"On-channel" carriage.* We have been urged to require that broadcast signals on cable systems be carried on the same channel number with which the station is identified. Section 76.55(a)(2) of our rules does require this, on request, but not where technically infeasible. It will be up to the cable system to comply with § 76.55(a)(2) to the extent permitted by our technical standards. The system would be offering no benefit at all were it to sacrifice technical quality for on-channel carriage.

109. *Applicability.* MST has requested that we make the cable television technical standards applicable immediately to expansions of existing systems, applicable in 1 year to existing systems serving more than 3,500 subscribers, and applicable in 2 years to smaller existing systems.[38] Clearly, an effective date for rules requiring large expenditures should be applied with reason. We have had to balance the obvious burden we are imposing on many existing cable systems against the resulting benefits to the public. Our original proposal in Docket 18894 was that existing cable television systems should have 3 years within

which to comply with the technical standards. Comments received in that docket and our own reevaluation persuaded us that a 3-year period would in most cases not be sufficient time. Instead, we have tried what we believe to be the most realistic approach, and have made different sections of our technical requirements applicable to existing systems at different times. Thus, existing cable systems will be required to comply with our performance test requirement immediately. Similarly, the radiation limitations of § 76.601(a)(12) will have to be met now. The bulk of our technical standards (§ 76.605(a)(1) through (11)) will have to be complied with within 5 years. We are not persuaded to change this schedule.

FEDERAL-STATE/LOCAL RELATIONSHIPS

110. *Multiple franchising.* Publi-Cable, Inc., urges the Commission to adopt more comprehensive rules encouraging multiple franchise arrangements for large cities and promoting more citizen participation. As we noted in the Cable Television Report and Order, we are looking forward to a period of experimentation in the development of cable television. While Publi-Cable's comments on the desirability of multiple franchising and citizen participation are valuable and hopefully will be implemented in various localities, it would be premature at this time to institute specific comprehensive rules of this nature. We are attempting to give great latitude to local entities to experiment with the various regulatory and franchising modes for cable television. We do not wish to hamper that flexibility any more than is necessary

111. *Franchise duration.* Publi-Cable also argues that franchises should be limited to 10 years, with renewal periods not to exceed 3 years. In § 76.31(a)(3) of our rules, we required only that initial franchise periods and renewals be of "reasonable duration." We noted in the report and order, however, our general belief that a franchise period should not exceed 15 years. While there may be situations where a 15-year franchise period is inappropriate, it appears to be a reasonable point of departure. Because our requirement of "reasonable duration" seems to have confused some parties, we have decided that our rules should more directly reflect the statements made in the report and have therefore now set 15 years as the standard to be followed (See revised § 76.31(a)(3)). If good cause can be shown in a particular instance for some other franchise period, we will of course entertain such a documented showing in a petition for special relief.

112. Also by way of clarification, while it was apparently clear that if, after the initial franchise expires, a new applicant receives a franchise he would have to obtain a certificate of compliance, some question was raised about simple renewals. It is our intent that whenever a franchise expires, whether it is subsequently renewed or a new franchise is granted, a new certificate of compliance will be required.

113. Questions have been also received by the Commission regarding our power

to require a cable system to remain operational during a period when the operator's local franchise has expired and a new applicant has been selected by the locality. The problem arises in cases where the operator holds the potential threat of stopping service if he does not get a franchise renewal and refuses to sell or lease the existing plant to the new franchise holder, be it another private party or the city. We do not at this time intend to extend our requirements for a certificate of compliance to cover this potential problem, but would strongly recommend that local officials include specific "buy-back" or continuation of service provisions in their franchises. If we find at a later date that this is still a recurring problem we may well then include such requirements in our rules in order to protect the public's right to continuity of service.

114. *Interconnection of franchise areas.* The National Association of Educational Broadcasters is concerned with how cable is to develop to assure the interconnection of franchise areas (regionally or statewide) and the adequate planning of equitable service expansion from urban to rural areas. Petitioner argues that local officials may not be able to meet such a challenge for compatible development and interconnection across political boundaries. Again, we feel that it would be premature to codify such rules as the petitioner suggests. However, we do agree with the NAEB that such guidelines should be identified as a priority problem for the Cable Television Advisory Committee on Federal-State/local relationships.

115. *Franchise grandfathering.* Many parties have raised questions on what procedure the Commission will follow regarding franchises that were granted prior to March 31, 1972, where there are franchise provisions inconsistent with the new rules. While all franchises are required to comply with our rules by March 31, 1977, we have indicated that some renegotiation of franchises may be required immediately in localities where the franchise was granted but the system was not built and operational. In some cases, this requirement appears to be creating unreasonable hardships and delays. Therefore, we are modifying the rule so that franchises granted prior to March 31, 1972, will be processed even though they do not meet all the requirements of our new rules so long as there is substantial compliance. For instance, the delay attendant to renegotiation of a franchise requiring a 6 percent franchise fee would do more of a disservice to the public we are trying to protect than would the fee itself, which will have, in any case, to be modified within 5 years. Further, any system that, in reliance on the existing franchise, has made a significant financial investment or entered into binding contractual agreements prior to the effective date of the rules but was not operational by that date, may request that its inconsistent franchise be grandfathered until March 31, 1977, upon such a showing in a petition for special relief. We would of course welcome the participation of the affected franchising au-

---

[37] We expect shortly to request comments on such standards. In addition it is expected that they will be a subject of study by the Technical Standards Committee established on Feb. 2, 1972.

[38] Although we have not adopted this proposal, it should be clear to operating systems that any substantial system expansion before the date required by the Commission should be accomplished with a view toward the necessity of compliance with our technical standards by that date.

thority in any such proceeding. As the rules already make clear, the franchising authority must be given notice by the applicant whenever such relief is sought. We are making this point even clearer by amending § 76.7 to indicate that the franchising authority should always be considered an interested party in any filing to the Commission affecting a cable system to which he has issued a franchise.

116. *Franchising authority.* Community Antenna Television of Wichita, Inc. has noted a need for clarification of § 76.31(a)(1-6) in the situation where there is no "franchise or other appropriate authorization" available for the cable operator to submit in his application for a certificate of compliance. It appears that the best way to deal with this situation is on a case-by-case basis through the special relief provisions of § 76.7 of the new rules. Such a petition, seeking a certificate of compliance under these unusual circumstances would have to include an acceptable alternative proposal for assuring that the substance of our rules, and specifically § 76.31, is complied with.[30]

117. A related matter concerns proposed cable operation in areas where there is doubt as to whether the appropriate franchising authority is on the State or local level. At a minimum, when there is such a dispute we believe that notice of the filing of an application for a certificate of compliance should be served on all authorities that are asserting a claim to jurisdiction. To the extent feasible, we will attempt to administer our rules so that, if otherwise permitted by local laws, cable operations need not be indefinitely held up while local jurisdictional disputes are settled. It may, for example, be possible to issue certificates on a conditional basis, subject to review when the local issues have been finally resolved. Our ability to do this will, of course, depend on the facts of particular cases.

118. *Enforcement.* ABC, MST, and RMBA note their concern that the rules do not provide the Commission with adequate enforcement tools to assure that cable systems which have obtained certificates of compliance continue to abide by the rules and operate consistently with the public interest. ABC in particular, in its initial and reply petitions, urges that the Commission license cable television systems following the existing pattern in the television broadcast station area.

119. These suggestions were before us and were considered in connection with the issuance of the Cable Television Report and Order. While these matters are appropriate for continuing consideration, we do not think that action is required at this time. The Commission now

exercises its authority to issue cease and desist orders for violations of its rules. Penalties for violation of the rules may be imposed under section 502 of the Communications Act. Legislation has been sought to permit the Commission to assess forfeitures for rule violations by cable television systems. Under the rules, the Commission also shares responsibility with local authorities for assuring that cable operations are consistent with the public interest, and local authorities will be in a position to review the performance of cable systems at franchise renewal times and as otherwise permitted under local laws. The desirability of traditional Federal licensing is a point which ABC has commented on a number of times in the past (see, for example, paragraph 172 of the Cable Television Report and Order) and which was raised as a possible alternative in Docket 18892. We find in the reconsideration petitions nothing to convince us that traditional licensing is a burden which we should undertake or which dissuades us from proceeding with the dual jurisdictional approach we have adopted.

PROCEDURE, SPECIAL RELIEF, ETC.

120. *Processing procedures.* Buckeye Cablevision, Inc., et al., NewChannels Corp., Athena Communications, and Jerrold Electronics Corp. have commented on the processing procedures that govern the certification of cable television systems. In order to effectuate the "go, no-go" concept of the new rules, these parties suggest that:

(1) The Cable Television Bureau be given delegated authority to grant automatically all unopposed applications at the end of 60 days after public notice expires.

(2) The Commission, either on reconsideration or in the first few decisions in the certificating process, set forth pleading standards so that dilatory, unmeritorious objections will not curtail the processing procedure beyond 90 days after final pleadings are filed.

121. We believe that delegated authority in certain areas will be given to the Cable Television Bureau in order that processing procedures may be streamlined and applications acted on within a reasonable time. However, we believe it would be desirable to obtain experience with the new processing procedures before delegations are conferred. As to the processing of applications, we only need restate our intentions outlined in the Cable Television Report and Order that:

Absent special situations or showings, requests consistent with our rules will receive prompt certification. The rules will operate on a "go, no-go" basis—i.e., the carriage rules reflect our determination of what is, at this time, in the public interest with respect to cable carriage of local and distant signals. We will, of course, consider objections to signal carriage applications and have retained special relief rules, but those seeking signal carriage restrictions on otherwise permitted signals have a substantial burden. Before restrictions are imposed in such cases, there will have to be a clear showing that the proposed service is not consistent with the

orderly integration of cable television service into the national communications structure and that the results would be inimical to the public interest. We have during the course of this proceeding fully considered the question of impact on local television service and we do not expect to reevaluate that general question in individual cases. And, for the same reason, we have no intention of reevaluating on request of cable systems in individual proceedings the general questions settled in our carriage and exclusivity rules. Rather, we strongly believe that cable systems must generally operate under these rules and that, only after meaningful experience, will we be in position for a general reassessment. [Paragraph 112.]

In connection with our special relief provisions, we note that in our August letter we designated certain markets where it appeared that special treatment to restrict distant signal carriage might have to be considered. We are no longer singling out these cases because the inclusion of substantial exclusivity protection for syndicated programing limits the impact of cable on local television service and is a new factor that must be taken into account. We are leaving unusual situations to petition for special relief, but there must be substantial showing to warrant deviation from the "go, no-go" concept of the rules. [Paragraph 113.][40]

We do not believe that it would be appropriate now to set time limits on the processing procedures. As with delegated authority, experience with the new procedures is necessary before any evaluations may be made as to the time required to process applications and other petitions.

122. *Service of applications.* MST suggests that full copies of applications for certificates of compliance should be served on all stations placing a Grade B or better signal over the cable community. We see no need for such service. It would be burdensome and unnecessary for cable systems to have to supply voluminous documents when the key information needed by the station can be easily provided by the notification procedure we have specified. Of course, the full application will always be available for public inspection both at the cable system's office and at the Commission. This is sufficient. However, it appears that it would be useful if a copy of the application for certification were available for public inspection in the community of the system. Consequently, we are requiring applicants to serve the franchising authority with the complete application for certification. We strongly urge franchising authorities to make the application available for public inspection. However, if the application is not, in this manner, available for inspection, the cable operator must, in some other way, make it available in the community of the system.

123. *Regulation in areas outside of the 48 contiguous States.* Although no reconsideration petitions were directed to the

---

[30] Because we have made local franchising an integral part of our program, we do not believe it appropriate for a cable system to avoid obtaining a franchise by obtaining channel service from a telephone company. Consequently, all applications for certificates from channel service customers must conform to § 76.31 of the rules.

[40] The quotation from paragraph 113 reflects our policy concerning markets that were designated with an asterisk in our letter of intent. We do not believe it necessary to modify its language as requested by Connecticut Television, Inc., to reflect that we intend to specially examine "asterisk" markets.

point, we believe it appropriate on our own motion that some additional consideration be given to the applicability of the rules to cable systems operating in Alaska, Puerto Rico, Hawaii, and other areas not included within the 48 contiguous States. Because of the unique situation with respect to broadcasting and cable television in these areas we believe some special consideration may be called for. Thus, for example, it is clear that § 76.59(d) of the rules regulating the carriage of non-English language stations could not be applied literally in Puerto Rico where most of the stations regularly broadcast in the Spanish language. Alaska, as was recently noted in the proposed second report and order in the domestic satellite proceeding (FCC 72–220, paragraph 144), is characterized by geographical remoteness from the contiguous States, has vast area and small population. There are only a few existing television stations and cable television systems, and in some instances both the television station and the cable system in the same community receive their programing on tape. Neither Hawaii nor Puerto Rico has distant signal programing readily available, and both have major cities where it could be argued our access rules should apply.[41] It is likely that other areas such as the Virgin Islands are likewise dissimilar from otherwise comparable areas within the 48 States. Because of the peculiar circumstances with respect to cable in these areas we believe it appropriate to treat certificate of compliance applications from these areas on an ad hoc basis, measuring the applications filed against the policies and standards contained in the new rules and specifically with regard to the rules concerning carriage priorities, program exclusivity, origination restrictions, and the applicability of the access requirements. We believe this is an appropriate method of proceeding, since these areas are not likely to be strictly comparable to those areas for which the rules were designed.

124. *Clarification of certification sections.* On our own motion we have reworded parts of § 76.13 to clarify the elements of applications for certificates of compliance in different situations, and we have added to this section the requirement that the applicant explain how he plans to comply with the origination cablecasting requirements of § 76.201. In this connection, we have also added new § 76.16 *Who may sign applications.*, § 76.18 *Amendment of applications.*, and § 76.20 *Dismissal of applications.*, and have amended the public notice provision (renumbered § 76.25) and the section concerning objections to applications (renumbered § 76.27) to indicate that signal carriage amendments to applications will be placed on

---

[41] San Juan has a population of 452,749, Ponce 128,233. The Honolulu television market has a population of 630,528, with 324,871 in the city itself.

public notice and may be the subject of objections. A number of other minor editorial changes have also been made.[41]

125. *Concluding matters.* In a number of places in this document, we have described modifications in the rules made either on our own motion or in response to arguments by petitioners. Since these amendments are essential elements of our overall cable television regulatory program, delay in their implementation would confuse the public and would be contrary to the public interest; hence the amendments will be effective immediately on publication in the FEDERAL REGISTER (7–14–72).

Authority for the rule amendments adopted herein is contained in sections 2, 3, 4 (i), and (j), 301, 303, 307, 308, and 309 of the Communications Act of 1934, as amended.

Accordingly, *it is ordered,* That effective July 14, 1972, the modifications in Parts 1 and 76 of the Commission's rules and regulations that are set out below are adopted.

It is further ordered, That the petitions for reconsideration or declaratory ruling are denied in all other respects.

It is further ordered, That, in order to consider pending petitions for reconsideration of the second report and order in Docket 18397, 23 FCC 2d 816, 35 F.R. 10903 (1970), and the memorandum opinion and order in Docket 18397, 23 FCC 2d 825 (1970), at a different time, the proceedings in Docket 18397, previously terminated in the Cable Television Report and Order, FCC 72–108, 37 F.R. 3252, are reopened.

It is further ordered, That the proceedings in Docket 18416, the subject matter of which was examined in the Cable Television Report and Order, 37 F.R. 3252, are terminated.

(Secs. 2, 3, 4, 301, 303, 307, 308, 309, 48 Stat., as amended, 1064, 1065, 1066, 1081, 1082, 1084, 1085; 47 U.S.C. 152, 153, 154, 301, 303, 307, 308, 309)

Adopted: June 16, 1972.

Released: June 26, 1972.

FEDERAL COMMUNICATIONS
COMMISSION,[42]

[SEAL]   BEN F. WAPLE,
Secretary.

---

[41] The major market list has been revised, deleting three cities that have no television stations licensed to them. The reference point of Pittsburg, Kans. has been changed to correct a typographical error. Sec. 76.17 has been amended to clear up an ambiguity in the pleading schedule with respect to certificate of compliance applications. Sec. 76.31 has been amended to make it clear that reductions in subscription charges make be made without a public proceeding.

[42] Commissioner Bartley concurring in the result; Commissioner Robert E. Lee dissenting and issuing a statement; Commissioners Johnson, H. Rex Lee and Reid concurring and issuing statements; Commissioner Wiley concurring in the result. Statements of Commissioners Robert E. Lee, Johnson, and H. Rex Lee are filed as part of the original document. Statement of Commissioner Reid to be released at a later date.

Chapter I of Title 47 of the Code of Federal Regulations is amended as follows:

A. Part 1—Practice and Procedure:
1. In § 1.1116(a), a note is added to read as follows:

§ 1.1116 Schedule of fees for Cable Television and Cable Television Relay Services.

(a) * * *

Petitions for special relief, pursuant to § 76.7 ------------------------------- $25.

NOTE: If a petition for special relief involves more than one cable television community, and the communities are served by cable facilities having a common headend and identical ownership, only a single $25 fee is required.

B. Part 76—Cable Television Service:
1. Section 76.3 is revised to read as follows:

§ 76.3 Other pertinent rules.

Other pertinent provisions of the Commission's rules and regulations relating to the Cable Television Service are included in the following parts of this chapter:

Part 0 —Commission Organization.
Part 1 —Practice and Procedure.
Part 21—Domestic Public Radio Services (Other Than Maritime Mobile).
Part 63—Extension of Lines and Discontinuance of Service by Carriers.
Part 64—Miscellaneous Rules Relating to Common Carriers.
Part 78—Cable Television Relay Service.
Part 91—Industrial Radio Services.

2. In § 76.5(a), the note is revised to read as follows:

§ 76.5 Definitions.

(a) * * *

NOTE: In general, each separate and distinct community or municipal entity (including unincorporated communities within unincorporated areas and single, discrete unincorporated areas) served by cable television facilities constitutes a separate cable television system, even if there is a single headend and identical ownership of facilities extending into several communities. See, e.g., Telerama, Inc., 3 FCC 2d 585 (1966); Mission Cable TV, Inc., 4 FCC 2d 236 (1966).

* * * * *

3. In § 76.7, paragraphs (a) and (b) are amended, and paragraphs (c)(3) and (h) are added, as follows:

§ 76.7 Special relief.

(a) On petition by a cable television system, a franchising authority, an applicant, permittee, or licensee of a television broadcast, translator, or microwave relay station, or by any other interested person, the Commission may waive any provision of the rules relating to cable television systems, impose additional or different requirements, or issue a ruling on a complaint or disputed question.

(b) The petition may be submitted informally, by letter, but shall be accompanied by an affidavit of service on any cable television system, franchising authority, station licensee, permittee, or

applicant, or other interested person who may be directly affected if the relief requested in the petition should be granted.

(c) * * *

(3) If a petition involves more than one cable television community, three (3) copies of it should be filed for each such community, in addition to the number of copies otherwise required to be filed pursuant to § 1.51 of this chapter.

* * * * *

(h) On a finding that the public interest so requires, the Commission may determine that a cable television system operating or proposing to operate in a community located outside of the 48 contiguous states shall comply with provisions of Subparts D, F, and G of this part in addition to the provisions thereof otherwise applicable. In such instances, any additional signal carriage that is authorized shall be deemed to be pursuant to the appropriate provision of §§ 76.61 (b) or 76.63(a) (as it relates to § 76.61 (b)).

4. In § 76.11, a new paragraph (d) is added, as follows:

§ 76.11 Certificate of compliance required.

* * * * *

(d) A certificate of compliance that is granted pursuant to this section shall be valid until the unamended expiration date of the franchise under which the certificated cable television system is operating or will operate, unless the Commission otherwise orders. A cable system may continue to carry television broadcast signals after the expiration of its certificate, if an application for a new certificate is filed at least thirty (30) days prior to the expiration date of the existing certificate and the Commission has not acted on the application.

5. In § 76.13, paragraphs (a), (2), (4), (6), (7) and (9) and (b) are amended, paragraph (c) is revised and paragraph (d) is deleted, as follows:

§ 76.13 Filing of applications.

* * * * *

(a) * * *

(2) A copy of FCC Form 325, "Annual Report of Cable Television Systems," supplying the information requested as though the cable system were already in operation as proposed;

* * * * *

(4) A statement that explains how the proposed system's franchise and its plans for availability and administration of access channels and other nonbroadcast cable services are consistent with the provisions of §§ 76.31, 76.201, and 76.251;

* * * * *

(6) An affidavit of service of the information described in subparagraph (1) of this paragraph on the licensee or permittee of any television broadcast station within whose predicted Grade B contour or specified zone the community of the system is located, in whole or in part, the licensee or permittee of any 100-watt or higher power television translator station licensed to the community of the system, the superintendent of schools in the community of the

system and any local or state educational television authorities;

(7) A statement that a copy of the complete application has been served on the franchising authority, and that if such application is not made available for public inspection by the franchising authority, the applicant will provide for public inspection of the application at any accessible place (such as a public library, public registry for documents, or an attorney's office) in the community of the system at any time during regular business hours;

* * * * *

(9) A statement that the filing fee prescribed in § 1.1116 of this chapter is attached.

(b) For a cable television system that proposes to add a television signal to existing operations, or that was authorized to carry one or more television signals prior to March 31, 1972, but did not commence such carriage prior to that date, an application for certificate of compliance shall include:

(1) The name and mailing address of the system, community and area served or to be served, television signals already being carried, television signals authorized to be carried but not carried prior to March 31, 1972, television signals not previously authorized and now proposed to be carried (other than those permitted to be carried pursuant to § 76.61(b)(2) (ii) or § 76.63(a) (as it relates to § 76.61 (b)(2)(ii)), and, if applicable, a statement that microwave relay facilities are to be used to relay one or more signals;

(2) If the system has not commenced operations but has been authorized to carry one or more television signals, a copy of FCC Form 325, "Annual Report of Cable Television Systems," supplying the information requested as though the cable system were already in operation as proposed;

(3) If the system has not commenced operations but has been authorized to carry one or more television signals, a copy of the franchise, license, permit, or certificate granted to construct and operate a cable television system, and a statement that explains how the system's franchise is substantially consistent with the provisions of § 76.31;

(4) A statement that explains how the system's plans for availability and administration of access channels and other nonbroadcast cable services are consistent with the provisions of §§ 76.201 and 76.251;

(5) A statement that explains, in terms of the provisions of Subpart D of this part, how carriage of the television signals not previously authorized is consistent with those provisions, including any special showings as to whether a signal is significantly viewed (see § 76.54(b));

(6) An affidavit of service of the information described in subparagraph (1) of this paragraph on the parties named in paragraph (a)(6) of this section;

(7) A statement that a copy of the complete application has been served on the franchising authority, and that if such application is not made available

for public inspection by the franchising authority, the applicant will provide for public inspection of the application at any accessible place (such as a public library, public registry for documents, or an attorney's office) in the community of the system at any time during regular business hours;

(8) A statement that the filing fee prescribed in § 1.1116 of this chapter is attached.

(c) For a cable television system seeking certification of existing operations in accordance with § 76.11(b), an application for certificate of compliance shall include:

(1) The name and mailing address of the system, community and area served, television signals being carried (other than those permitted to be carried pursuant to § 76.61(b)(2)(ii) or § 76.63(a) (as it relates to § 76.61(b)(2)(ii)), television signals authorized or certified to be carried but not being carried, date on which operations commenced, and date on which its current franchise expires;

(2) A copy of the franchise, license, permit, or certificate under which the system will operate upon Commission certification (if such franchise has not previously been filed), and a statement that explains how the franchise is consistent with the provisions of § 76.31;

(3) A statement that explains how the system's plans for availability and administration of access channels and other nonbroadcast cable services are consistent with the provisions of §§ 76.201 and 76.251;

(4) An affidavit of service of the information described in subparagraph (1) of this paragraph on the parties named in paragraph (a)(6) of this section;

(5) A statement that a copy of the complete application has been served on the franchising authority, and that if such application is not made available for public inspection by the franchising authority, the applicant will provide for public inspection of the application at any accessible place (such as a public library, public registry for documents, or an attorney's office) in the community of the system at any time during regular business hours;

(6) A statement that the filing fee prescribed in § 1.1116 of this chapter is attached.

(d) [Deleted]

NOTE: As used in § 76.13, the term "predicted Grade B contour" means the field intensity contour defined in § 73.683(a) of this chapter, the location of which is determined exclusively by means of the calculations prescribed in § 73.684 of this chapter.

6. A new § 76.16 is added, as follows:

§ 76.16 Who may sign applications.

(a) Applications for certificates of compliance, amendments thereto, and related statements of fact required by the Commission shall be personally signed by the applicant, if the applicant is an individual; by one of the partners, if the applicant is a partnership; by an officer, if the applicant is a corporation; or by a member who is an officer, if the applicant is an unincorporated association. Applications, amendments, and related statements of fact filed on behalf

of Government entities shall be signed by such duly elected or appointed officials as may be competent to do so under the laws of the applicable jurisdiction.

(b) Applications, amendments thereto, and related statements of fact required by the Commission may be signed by the applicant's attorney in case of the applicant's physical disability or of his absence from the United States. The attorney shall in that event separately set forth the reasons why the application is not signed by the applicant. In addition, if any matter is stated on the basis of the attorney's belief only (rather than his knowedge), he shall separately set forth his reasons for believing that such statements are true.

(c) Only the original of applications, amendments, or related statements of fact need be signed; copies may be conformed.

7. A new § 76.18 is added, as follows:

§ 76.18 Amendment of applications.

An application for a certificate of compliance may be amended as a matter of right prior to the adoption date of any final action taken by the Commission with respect to the application, merely by filing three (3) copies of the amendment in question duly executed in accordance with § 76.16. All amendments shall be served on the franchising authority, on all parties that have filed pleadings responsive to the application, and, if the addition or deletion of a television broadcast signal is involved, on all parties served pursuant to § 76.13. Amendments shall be made available for public inspection in the same manner as the application.

80. A new § 76.20 is added, as follows:

§ 76.20 Dismissal of applications.

(a) An application for a certificate of compliance may, upon request of the applicant, be dismissed without prejudice as a matter of right prior to the adoption date of any final action taken by the Commission with respect to the application. An applicant's request for the return of an application will be regarded as a request for dismissal.

(b) Failure to prosecute an application, or failure to respond to official correspondence or request for additional information, will be cause for dismissal. Such dismissal will be without prejudice if it occurs prior to the adoption date of any final action taken by the Commission with respect to the application.

9. Section 76.15 is renumbered as § 76.25 and is amended, as follows:

§ 76.25 Public notice.

The Commission will give public notice of the filing of applications for certificates of compliance and of amendments thereto that add or delete television signals. A certificate will not be issued sooner than thirty (30) days from the date of public notice.

10. Section 76.17 is renumbered as § 76.27 and is amended, as follows:

§ 76.27 Objections to applications; related matters.

An objection to an application for certificate of compliance or an amendment thereto shall be filed within thirty (30) days of the public notice described in § 76.25. A reply may be filed within twenty (20) days after an objection is filed. Factual allegations shall be supported by affidavit of a person or persons with actual knowledge of the facts, and exhibits shall be verified by the person who prepares them. All pleadings shall be served on the persons specified in § 76.13, the cable television system, the franchising authority, and any other interested person. Controversies concerning carriage (Subpart D) and program exclusivity (§ 76.91) will be acted on in connection with the certificating process if raised within thirty (30) days of the public notice; any other objection will be treated as a petition for special relief filed pursuant to § 76.7.

11. In § 76.31, paragraph (a)(3) and (4) and the proviso after (a)(6) are amended, as follows:

§ 76.31 Franchise standards.

(a) * * *

(3) The initial franchise period shall not exceed fifteen (15) years, and any renewal franchise period shall be of reasonable duration;

(4) The franchising authority has specified or approved the initial rates that the franchisee charges subscribers for installation of equipment and regular subscriber services. No increases in rates charged to subscribers shall be made except as authorized by the franchising authority after an appropriate public proceeding affording due process;

*   *   *   *   *

Provided, however, That, in an application for certificate of compliance, consistency with these requirements shall not be expected of a cable television system that was in operation prior to March 31, 1972, until the end of its current franchise period, or March 31, 1977, whichever occurs first; *And provided, further,* That on a petition filed pursuant to § 76.7, in connection with an application for certificate of compliance, the Commission may waive consistency with these requirements for a cable system that was not in operation prior to March 31, 1972, and that, relying on an existing franchise, made a significant financial investment or entered into binding contractual agreements prior to March 31, 1972, until the end of its current franchise period, or March 31, 1977, whichever comes first.

*   *   *   *   *

12. In § 76.51, paragraph (b) is amended, as follows:

§ 76.51 Major television markets.

(b) Second 50 major television markets:

*   *   *   *   *

(57) Harrisburg-Lancaster-York, Pa.

*   *   *   *   *

(89) Duluth, Minn.-Superior, Wis.

*   *   *   *   *

(98) Fargo-Valley City, N.D.

*   *   *   *   *

13. In § 76.53, the geographic coordinates of Pittsburg, Kans. are corrected, as follows:

§ 76.53 Reference points.

*   *   *   *   *

State and community	Latitude	Longitude
KANSAS * * *	* * *	* * *
Pittsburg	37°24'50"	94°42'11"
* * *	* * *	* * *

14. In § 76.54, paragraph (a) is revised, and a new paragraph (c) and a note are added, as follows:

§ 76.54 Significantly viewed signals; method to be followed for special showings.

(a) Signals that are significantly viewed in a county (and thus are deemed to be significantly viewed within all communities within the county) are those that are listed in Appendix B of the memorandum opinion and order on reconsideration of the Cable Television Report and Order (Docket 18397 et al.), FCC 72-530.

*   *   *   *   *

(c) Notice of a survey to be made pursuant to paragraph (b) of this section shall be served on all licensees or permittees of television broadcast stations within whose predicted Grade B contour the cable community is located, in whole or in part, and on all cable systems, franchisees, and franchise applicants in the cable community at least thirty (30) days prior to the initial survey period. Such notice shall include the name of the survey organization and a description of the procedures to be used. Objections to survey organizations or procedures shall be served on the party sponsoring the survey within twenty (20) days after receipt of such notice.

NOTE: With respect to those counties designated by an asterisk in Appendix B of the memorandum opinion and order on reconsideration of the Cable Television Report and Order (Docket 18397 et al.), FCC 72-530, surveys of significant viewing made pursuant to § 76.54(b) may be submitted prior to March 31, 1973.

15. In § 76.57, paragraph (a)(2) is revised, as follows:

§ 76.57 Provisions for systems operating in communities located outside of all major and smaller television markets.

*   *   *   *   *

(a) * * *

(2) Television translator stations with 100 watts or higher power serving the community of the system and, as to cable systems that commence operations or expand channel capacity after March 30, 1972, noncommercial educational translator stations with 5 watts or higher power serving the community of the sys-

tem. In addition, any cable system may elect to carry the signal of any noncommercial educational translator station,

* * * * *

16. In § 76.59, paragraphs (a)(5), (b)(1), and (d) are amended, as follows:

§ 76.59 Provisions for smaller television markets.

* * * * *

(a) * * *

(5) Television translator stations with 100 watts or higher power serving the community of the system and, as to cable systems that commence operations or expand channel capacity after March 30, 1972, noncommercial educational translator stations with 5 watts or higher power serving the community of the system. In addition, any cable system may elect to carry the signal of any noncommercial educational translator station;

* * * * *

(b) * * *

(1) Full network stations. A cable television system may carry the nearest full network stations or the nearest in-state full network stations;

* * * * *

(d) In addition to the television broadcast signals carried pursuant to paragraphs (a) through (c) of this section, any such cable television system may carry:

(1) Any television stations broadcasting predominantly in a non-English language; and

(2) Any television station broadcasting a network program that will not be carried by a station normally carried on the system. Carriage of such additional stations shall be only for the duration of the network programs not otherwise available, and shall not require prior Commission notification or approval in the certificating process.

* * * * *

17. In § 76.61, paragraphs (a)(3), (b)(1), and (e) are amended, as follows:

§ 76.61 Provisions for first 50 major television markets.

* * * * *

(a) * * *

(3) Television translator stations with 100 watts or higher power serving the community of the system and, as to cable systems that commence operations or expand channel capacity after March 30, 1972, noncommercial educational translator stations with 5 watts or higher power serving the community of the system. In addition, any cable system may elect to carry the signal of any noncommercial educational translator station;

* * * * *

(b) * * *

(1) Full network stations. A cable television system may carry the nearest full network stations, or the nearest in-State full network stations;

* * * * *

(e) In addition to the television broadcast signals carried pursuant to

paragraphs (a) through (d) of this section, any such cable television system may carry:

(1) Any television stations broadcasting predominantly in a non-English language; and

(2) Any television station broadcasting a network program that will not be carried by a station normally carried on the system. Carriage of such additional stations shall be only for the duration of the network programs not otherwise available, and shall not require prior Commission notification or approval in the certificating process.

* * * * *

18. In § 76.93, paragraph (b) is amended, as follows:

§ 76.93 Extent of protection.

* * * * *

(b) Notwithstanding the provisions of paragraph (a) of this section, on request of a television station licensed to a community in the Mountain Standard Time Zone that is not one of the designated communities in the first 50 major television markets, a cable television system shall refrain from duplicating any network program broadcast by such station on the same day as its broadcast by the station. Where a cable system is required to provide same-day program exclusivity, the following provisions shall be applicable:

* * * * *

19. A new § 76.99 is added, as follows:

§ 76.99 Grandfathering.

The provisions of §§ 76.91, 76.93, 76.151, and 76.153 shall not be deemed to deprive a television station whose signal was carried by a cable television system prior to March 31, 1972, of the nonnetwork program exclusivity rights that such station had on March 30, 1972: Provided, however, That such exclusivity rights shall extend only to simultaneous duplication of programing by lower priority television stations, unless the station whose exclusivity rights are at issue is entitled to same-day network program exclusivity pursuant to § 76.93(b), in which case that station shall also be entitled to continued same-day nonnetwork program exclusivity.

20. In § 76.201, paragraph (a) is amended, as follows:

§ 76.201 Origination cablecasting in conjunction with carriage of broadcast signals.

(a) No cable television system having 3,500 or more subscribers shall carry the signal of any television broadcast station unless the system also operates to a significant extent as a local outlet by origination cablecasting and has available facilities for local production and presentation of programs other than automated services. Such origination cablecasting shall be limited to one or more designated channels which may be used for no other cablecasting purpose.

* * * * *

21. In § 76.251, paragraphs (a)(2) and (c) are amended, as follows:

§ 76.251 Minimum channel capacity; access channels.

(a) * * *

(2) Equivalent amount of bandwidth. For each Class I cable channel that is utilized, such system shall be capable of providing an additional channel, 6 MHz in width, suitable for transmission of Class II or Class III signals (see § 76.5 for cable channel definitions);

* * * * *

(c) The provisions of this section shall apply to all cable television systems that commence operations on or after March 31, 1972, in a community located in whole or in part within a major television market. Systems that commenced operations prior to March 31, 1972, shall comply on or before March 31, 1977: Provided, however, That if such systems begin to provide any of the access services described above at an earlier date, they shall comply with paragraph (a)(9), (10), and (11) of this section at that time: And provided, further, That if such systems receive certificates of compliance to add television signals to their operations at an earlier date, pursuant to § 76.61(b) or (c), or § 76.63(a) (as it relates to § 76.61(b) or (c)), for each such signal added, such systems shall provide one (1) access channel in the following order of priority—(1) public access, (2) education access, (3) local government access, and (4) leased access—and shall comply with the appropriate requirements of paragraphs (a)(4)–(7) and (a)(9)–(11) of this section with respect thereto.

22. In § 76.605(a), subparagraph (5) and the introductory text of subparagraph (12) are amended, as follows:

§ 76.605 Technical standards.

(a) * * *

(5) The visual signal level on each channel shall not vary more than 12 decibels within any 24-hour period, and shall be maintained within:

* * * * *

(12) As an exception to the general provision requiring measurements to be made at subscriber terminals, and without regard to the class of cable television channel involved, radiation from a cable television system shall be measured in accordance with procedures outlined in § 76.609(h), and shall be limited as follows:

* * * * *

§ 76.609 [Amended]

23. In § 76.609, the reference in the first sentence of paragraph (a) to "§§ 76.701 and 76.605" should read "§§ 76.601 and 76.605."

APPENDIX B—REVISED MAY 1972

SIGNIFICANTLY VIEWED TELEVISION STATIONS

This table lists the television stations significantly viewed for purposes of cable television carriage, in accordance with § 76.54(a)

RULES AND REGULATIONS

of the Commission's rules. All stations meeting the significant viewing test are listed, including market and other stations that might be subject to required or permissible carriage under other provisions of the rules.

Cities, not politically part of any county, are listed with the county in which they were included for survey purposes. A description of how split counties have been divided is included after each State listing that includes split counties. The description indicates which Census County Divisions (1960) are included in each division of the county. Maps of Census County Divisions may be found in U.S. Census of Population: 1960, Vol. 1, Characteristics of the Population, Part A, Number of Inhabitants.

A plus sign (+) following the station call letters indicates that the viewing is due to either the parent station or a satellite. In this situation either the parent or the satellite station may be carried.

Information in the table is derived from 1971 American Research Bureau surveys. See memorandum opinion and order on reconsideration of the Cable Television Report and Order, FCC 72–530. In counties marked by an asterisk, individual survey may be submitted under § 76.54(b) of the rules prior to March 31, 1973. No stations are listed for those counties which had over 90 percent cable television penetration. For counties in which less than 5 non-CATV diaries were tabulated, no stations are listed (shown as NA) if the results showed significant signals from more than one market.

SIGNIFICANTLY VIEWED SIGNALS

County	Call letters, channel number, and market name		
ALABAMA			
Autauga	WSFA	12	Montgomery.
	WCOV	20	Do.
	WKAB	32	Do.
Baldwin	WEAR	3	Mobile-Pensacola.
	WKRG	5	Do.
	WALA	10	Do.
Barbour	WRBL	3	Columbus, Ga.
	WTVM	9	Do.
	WTVY	4	Dothan.
	WSFA	12	Montgomery.
Bibb	WBRC	6	Birmingham.
	WAPI	13	Do.
	WBMG	42	Do.
Blount	WBRC	6	Do.
	WAPI	13	Do.
	WHNT	19	Huntsville-Decatur-Florence.
Bullock	WSFA	12	Montgomery.
	WRBL	3	Columbus, Ga.
	WTVM	9	Do.
Butler *	WSFA	12	Montgomery.
Calhoun	WBRC	6	Birmingham.
	WAPI	13	Do.
	WHMA	40	Anniston, Ala.
Chambers *	WRBL	3	Columbus, Ga.
	WTVM	9	Do.
Cherokee	WSB	2	Atlanta.
	WAGA	5	Do.
	WQXI	11	Do.
	WBRC	6	Birmingham.
Chilton	WBRC	6	Do.
	WAPI	13	Do.
	WBMG	42	Do.
	WSFA	12	Montgomery.
Choctaw	WTOK	11	Meridian.
Clarke	WEAR	3	Mobile-Pensacola.
	WKRG	5	Do.
	WALA	10	Do.
Clay	WBRC	6	Birmingham.
	WAPI	13	Do.
	WRBL	3	Columbus, Ga.
Cleburne	WSB	2	Atlanta.
	WAGA	5	Do.
	WQXI	11	Do.
	WBRC	6	Birmingham.
	WAPI	13	Do.

SIGNIFICANTLY VIEWED SIGNALS—Continued

County	Call letters, channel number, and market name		
ALABAMA—continued			
Coffee*	WTVY	4	Dothan.
	WSFA	12	Montgomery.
Colbert	WOWL	15	Huntsville-Decatur-Florence.
	WHNT	19	Do.
	WAAY	31	Do.
	WMSL	48	Do.
Conecuh	WEAR	3	Mobile-Pensacola.
	WKRG	5	Do.
	WALA	10	Do.
	WSFA	12	Montgomery.
Coosa	WBRC	6	Birmingham.
	WAPI	13	Do.
	WSFA	12	Montgomery.
Covington	WSFA	12	Montgomery.
	WTVY	4	Dothan.
Crenshaw	WSFA	12	Montgomery.
	WCOV	20	Do.
	WTVY	4	Dothan.
Cullman	WBRC	6	Birmingham.
	WAPI	13	Do.
	WHNT	19	Huntsville-Decatur-Florence.
	WAAY	31	Do.
Dale	WTVY	4	Dothan.
	WTVM	9	Columbus, Ga.
	WRBL	3	Do.
	WSFA	12	Montgomery.
Dallas	WSFA	12	Do.
	WCOV	20	Do.
	WKAB	32	Do.
	WBRC	6	Birmingham.
De Kalb*	WRCB	3	Chattanooga.
	WTVC	9	Do.
	WDEF	12	Do.
Elmore	WSFA	12	Montgomery.
	WCOV	20	Do.
	WKAB	32	Do.
Escambia	WEAR	3	Mobile-Pensacola.
	WKRG	5	Do.
	WALA	10	Do.
Etowah	WBRC	6	Birmingham.
	WAPI	13	Do.
	WHMA	40	Anniston, Ala.
Fayette	WBRC	6	Birmingham.
	WAPI	13	Do.
	WCBI	4	Columbus, Miss.
Franklin	WHNT	19	Huntsville-Decatur-Florence.
	WBRC	6	Birmingham.
	WCBI	4	Columbus, Miss.
	WTWV	9	Tupelo.
Geneva	WTVY	4	Dothan.
	WDHN	18	Do.
	WJHG	7	Panama City.
Greene	WBRC	6	Birmingham.
	WAPI	13	Do.
	WTOK	11	Meridian.
	WCFT	33	Tuscaloosa.
Hale	WBRC	6	Birmingham.
	WAPI	13	Do.
	WTOK	11	Meridian.
	WCFT	33	Tuscaloosa.
Henry	WTVY	4	Dothan.
	WRBL	3	Columbus, Ga.
	WTVM	9	Do.
Houston	WTVY	4	Dothan.
	WDHN	18	Do.
	WJHG	7	Panama City.
Jackson	WRCB	3	Chattanooga.
	WTVC	9	Do.
	WDEF	12	Do.
Jefferson	WBRC	6	Birmingham.
	WAPI	13	Do.
	WBMG	42	Do.
Lamar *	WCBI	4	Columbus, Miss.
	WBRC	6	Birmingham.
	WAPI	13	Do.
Lauderdale	WOWL	15	Huntsville-Decatur-Florence.
	WHNT	19	Do.
	WAAY	31	Do.
	WMSL	48	Do.
Lawrence	WHNT	19	Do.
	WAAY	31	Do.
	WMSL	48	Do.
	WBRC	6	Birmingham.
Lee	WRBL	3	Columbus, Ga.
	WTVM	9	Do.
	WSFA	12	Montgomery.
Limestone	WHNT	19	Huntsville-Decatur-Florence.
	WAAY	31	Do.
	WMSL	48	Do.
Lowndes	WSFA	12	Montgomery.
	WCOV	20	Do.
	WKAB	32	Do.
Macon	WSFA	12	Do.
	WCOV	20	Do.
	WRBL	3	Columbus, Ga.
	WTVM	9	Do.

SIGNIFICANTLY VIEWED SIGNALS—Continued

County	Call letters, channel number, and market name		
ALABAMA—continued			
Madison	WHNT	19	Huntsville-Decatur-Florence.
	WAAY	31	Do.
	WMSL	48	Do.
Marengo	WTOK	11	Meridian.
Marion	WBRC	6	Birmingham.
	WAPI	13	Do.
	WCBI	4	Columbus, Miss.
Marshall	WHNT	19	Huntsville-Decatur-Florence.
	WAAY	31	Do.
	WMSL	48	Do.
	WBRC	6	Birmingham.
	WAPI	13	Do.
Mobile	WEAR	3	Mobile-Pensacola.
	WKRG	5	Do.
	WALA	10	Do.
Monroe	WEAR	3	Do.
	WRRG	5	Do.
	WALA	10	Do.
Montgomery	WSFA	12	Montgomery.
	WCOV	20	Do.
	WKAB	32	Do.
Morgan	WHNT	19	Huntsville-Decatur-Florence.
	WAAY	31	Do.
	WMSL	48	Do.
	WBRC	6	Birmingham.
Perry	WBRC	6	Do.
	WAPI	13	Do.
	WSFA	12	Montgomery.
Pickens	WBRC	6	Birmingham.
	WAPI	13	Do.
	WCBI	4	Columbus, Miss.
Pike	WSFA	12	Montgomery.
	WRBL	3	Columbus, Ga.
	WTVM	9	Do.
	WTVY	4	Dothan.
Randolph	WRBL	3	Columbus, Ga.
	WTVM	9	Do.
	WSB	2	Atlanta.
	WAGA	5	Do.
	WQXI	11	Do.
Russell	WRBL	3	Columbus, Ga.
	WTVM	9	Do.
	WYEA	38	Do.
St. Clair	WBRC	6	Birmingham.
	WAPI	13	Do.
Shelby	WBRC	6	Do.
	WAPI	13	Do.
	WBMG	42	Do.
Sumter	WTOK	11	Meridian.
Talladega	WBRC	6	Birmingham.
	WAPI	13	Do.
	WBMG	42	Do.
Tallapoosa	WBRC	6	Do.
	WAPI	13	Do.
	WRBL	3	Columbus, Ga.
	WTVM	9	Do.
	WSFA	12	Montgomery.
Tuscaloosa	WBRC	6	Birmingham.
	WAPI	13	Do.
	WCFT	33	Tuscaloosa.
Walker	WBRC	6	Birmingham.
	WAPI	13	Do.
	WBMG	42	Do.
Washington	WEAR	3	Mobile-Pensacola.
	WKRG	5	Do.
	WALA	10	Do.
Wilcox	WSFA	12	Montgomery.
	WEAR	3	Mobile-Pensacola.
	WKRG	5	Do.
Winston	WBRC	6	Birmingham.
	WAPI	13	Do.
	WBMG	42	Do.
	WHNT	19	Huntsville-Decatur-Florence.

County	Call letters, channel number, and market name		
ARIZONA			
Apache	KVOA	4	Tucson.
	KGUN	9	Do.
	KOLD	13	Do.
	KOB	4	Albuquerque.
	KOAT	7	Do.
	KGGM	13	Do.
Cochise	KVOA	4	Tucson.
	KGUN	9	Do.
	KOLD	13	Do.
Coconino	KOAI	2	Phoenix.
	KTVK	3	Do.
	KPHO	5	Do.
	KOOL	10	Do.
	KTAR	12	Do.
Gila	KTVK	3	Do.
	KPHO	5	Do.
	KOOL	10	Do.
	KTAR	12	Do.

SIGNIFICANTLY VIEWED SIGNALS—Continued

County	Call letters, channel number, and market name		

ARIZONA—continued

County	Call letters	Channel	Market
Graham	KTAR	12	Do.
	KVOA	4	Tucson.
	KGUN	9	Do.
	KOLD	13	Do.
Greenlee	KVOA	4	Do.
	KGUN	9	Do.
	KOLD	13	Do.
Maricopa	KTVK	3	Phoenix.
	KPHO	5	Do.
	KOOL	10	Do.
	KTAR	12	Do.
Mohave	KTVK	3	Do.
	KPHO	5	Do.
	KOOL	10	Do.
	KTAR	12	Do.
	KORK	3	Las Vegas.
Navajo	KVOA	4	Tucson.
	KGUN	9	Do.
	KOLD	13	Do.
	KOAI	2	Phoenix.
	KOOL	10	Do.
Pima East	KVOA	4	Tucson.
	KGUN	9	Do.
	KZAZ	11	Do.
	KOLD	13	Do.
Pima West	KVOA	4	Do.
	KGUN	9	Do.
	KOLD	13	Do.
	KPHO	5	Phoenix.
Pinal	KTVK	3	Do.
	KPHO	5	Do.
	KOOL	10	Do.
	KTAR	12	Do.
Santa Cruz	KVOA	4	Tucson.
	KGUN	9	Do.
	KZAZ	11	Do.
	KOLD	13	Do.
	KPHO	5	Phoenix.
	XHFA	2	Mexico.
Yavapai	KTVK	3	Phoenix.
	KPHO	5	Do.
	KOOL	10	Do.
	KTAR	12	Do.
Yuma	KPHO	5	Do.
	KBLU	13	Yuma.
	KECC	9	El Centro.

Census County Divisions in Split Counties

Pima West: Ajo, Papago.
Pima East: All other.

ARKANSAS

County	Call letters	Channel	Market
Arkansas	KARK	4	Little Rock.
	KATV	7	Do.
	KTHV	11	Do.
Ashley*	KNOE	8	Monroe-El Dorado.
	KTVE	10	Do.
Baxter	KYTV	3	Springfield, Mo.
	KTTS	10	Do.
	KMTC	27	Do.
Benton	KOAM	7	Joplin-Pittsburg.
	KODE	12	Do.
	KUHI	16	Do.
	KFSA	5	Fort Smith.
	KOTV	6	Tulsa.
	KTUL	8	Do.
Boone	KYTV	3	Springfield, Mo.
	KTTS	10	Do.
Bradley	KARK	4	Little Rock.
	KATV	7	Do.
	KTHV	11	Do.
	KTVE	10	Monroe-El Dorado.
Calhoun	KARK	4	Little Rock.
	KATV	7	Do.
	KNOE	8	Monroe-El Dorado.
	KTVE	10	Do.
Carroll	KYTV	3	Springfield, Mo.
	KTTS	10	Do.
Chicot*	KNOE	8	Monroe-El Dorado.
	KTVE	10	Do.
	WABC	6	Greenwood-Greenville.
Clark	KARK	4	Little Rock.
	KATV	7	Do.
	KTHV	11	Do.
Clay	WREC	3	Memphis.
	WMC	5	Do.
	WHBQ	13	Do.
	KAIT	8	Jonesboro.
Cleburne	KARK	4	Little Rock.
	KATV	7	Do.
	KTHV	11	Do.
Cleveland	KARK	4	Do.
	KATV	7	Do.
	KTHV	11	Do.

SIGNIFICANTLY VIEWED SIGNALS—Continued

ARKANSAS—continued

County	Call letters	Channel	Market
Columbia	KTBS	3	Shreveport-Texarkana.
	KTAL	6	Do.
	KSLA	12	Do.
Conway	KARK	4	Little Rock.
	KATV	7	Do.
	KTHV	11	Do.
Craighead	KAIT	8	Jonesboro.
	WREC	3	Memphis.
	WMC	5	Do.
	WHBQ	13	Do.
Crawford*	KFSA	5	Fort Smith.
	KTUL	8	Tulsa.
Crittenden	WREC	3	Memphis.
	WMC	5	Do.
	WHBQ	13	Do.
Cross	WREC	3	Do.
	WMC	5	Do.
	WHBQ	13	Do.
Dallas	KARK	4	Little Rock.
	KATV	7	Do.
	KTHV	11	Do.
Desha	KARK	4	Do.
	KATV	7	Do.
	KTHV	11	Do.
	KTVE	10	Monroe-El Dorado.
Drew	KARK	4	Little Rock.
	KATV	7	Do.
	KTHV	11	Do.
	KTVE	10	Monroe-El Dorado.
Faulkner	KARK	4	Little Rock.
	KATV	7	Do.
	KTHV	11	Do.
Franklin*	KFSA	5	Fort Smith.
	KARK	4	Little Rock.
	KTHV	11	Do.
Fulton*	KYTV	3	Springfield, Mo.
	KAIT	8	Jonesboro.
Grant	KARK	4	Little Rock.
	KATV	7	Do.
	KTHV	11	Do.
Garland	KARK	4	Do.
	KATV	7	Do.
	KTHV	11	Do.
Greene	WREC	3	Memphis.
	WMC	5	Do.
	WHBQ	13	Do.
	KAIT	8	Jonesboro.
Hempstead	KTBS	3	Shreveport-Texarkana.
	KTAL	6	Do.
	KSLA	12	Do.
Hot Spring	KARK	4	Little Rock.
	KATV	7	Do.
	KTHV	11	Do.
Howard	KTBS	3	Shreveport-Texarkana.
	KTAL	6	Do.
Independence	KARK	4	Little Rock.
	KATV	7	Do.
	KTHV	11	Do.
Izard	KARK	4	Do.
	KTHV	11	Do.
	KYTV	3	Springfiled, Mo.
Jackson	KARK	4	Little Rock.
	KTHV	11	Do.
	KAIT	8	Jonesboro.
	WREC	3	Memphis.
	WMC	5	Do.
Jefferson	KARK	4	Little Rock.
	KATV	7	Do.
	KTHV	11	Do.
Johnson	KARK	4	Do.
	KTHV	11	Do.
	KFSA	5	Fort Smith.
	KATV	7	Little Rock.
Lafayette	KTBS	3	Shreveport-Texarkana.
	KTAL	6	Do.
	KSLA	12	Do.
Lawrence	KAIT	8	Jonesboro.
	WREC	3	Memphis.
	WMC	5	Do.
Lee	WREC	3	Do.
	WMC	5	Do.
	WHBQ	13	Do.
	KATV	7	Little Rock.
Lincoln	KARK	4	Do.
	KATV	7	Do.
	KTHV	11	Do.
Little River	KTBS	3	Shreveport-Texarkana.
	KTAL	6	Do.
	KSLA	12	Do.
Logan	KFSA	5	Fort Smith.
	KARK	4	Little Rock.
	KTHV	11	Do.
Lonoke	KARK	4	Do.
	KATV	7	Do.
	KTHV	11	Do.
Madison*	KYTV	3	Springfield, Mo.
	KFSA	5	Fort Smith.

SIGNIFICANTLY VIEWED SIGNALS—Continued

ARKANSAS—continued

County	Call letters	Channel	Market
Marion	KYTV	3	Springfield, Mo.
	KTTS	10	Do.
	KMTC	27	Do.
	KARK	4	Little Rock.
	KTHV	11	Do.
Miller	KTBS	3	Shreveport-Texarkana.
	KTAL	6	Do.
	KSLA	12	Do.
Mississippi	WREC	3	Memphis.
	WMC	5	Do.
	WHBQ	13	Do.
Monroe	KARK	4	Little Rock.
	KATV	7	Do.
	KTHV	11	Do.
Montgomery	KARK	4	Do.
	KATV	7	Do.
	KTHV	11	Do.
Nevada	KARK	4	Do.
	KATV	7	Do.
	KTHV	11	Do.
	KTBS	3	Shreveport-Texarkana.
	KTAL	6	Do.
	KSLA	12	Do.
Newton	KYTV	3	Springfield, Mo.
	KARK	4	Little Rock.
	KTHV	11	Do.
Ouachita	KARK	4	Do.
	KATV	7	Do.
	KTHV	11	Do.
	KTVE	10	Monroe-El Dorado.
Perry	KARK	4	Little Rock.
	KATV	7	Do.
	KTHV	11	Do.
Phillips	WREC	3	Memphis.
	WMC	5	Do.
	WHBQ	13	Do.
	KATV	7	Little Rock.
Pike	KARK	4	Do.
	KATV	7	Do.
	KTHV	11	Do.
	KTBS	3	Shreveport-Texarkana.
	KTAL	6	Do.
Poinsett	WREC	3	Memphis.
	WMC	5	Do.
	WHBQ	13	Do.
	KAIT	8	Jonesboro.
Polk	KARK	4	Little Rock.
	KTHV	11	Do.
	KFSA	5	Fort Smith.
	KTAL	6	Shreveport-Texarkana.
	KATV	7	Little Rock.
Pope	KARK	4	Do.
	KATV	7	Do.
	KTHV	11	Do.
Prairie	KARK	4	Do.
	KATV	7	Do.
	KTHV	11	Do.
Pulaski	KARK	4	Do.
	KATV	7	Do.
	KTHV	11	Do.
Randolph	KAIT	8	Jonesboro.
	WREC	3	Memphis.
	WMC	5	Do.
St. Francis	WREC	3	Do.
	WMC	5	Do.
	WHBQ	13	Do.
	KATV	7	Little Rock.
Saline	KARK	4	Do.
	KATV	7	Do.
	KTHV	11	Do.
Scott	KFSA	5	Fort Smith.
	KARK	4	Little Rock.
	KTUL	8	Tulsa.
Searcy	KARK	4	Little Rock.
	KATV	7	Do.
	KTHV	11	Do.
	KYTV	3	Springfield, Mo.
Sebastian	KFSA	5	Fort Smith.
	KTUL	8	Tulsa.
Sevier	KTBS	3	Shreveport-Texarkana.
	KTAL	6	Do.
	KSLA	12	Do.
Sharp*	KAIT	8	Jonesboro.
	KARK	4	Little Rock.
	WMC	5	Memphis.
Stone	KARK	4	Little Rock.
	KATV	7	Do.
	KTHV	11	Do.
Union	KNOE	8	Monroe-El Dorado.
	KTVE	10	Do.
	KATV	7	Little Rock.
	KTBS	3	Shreveport-Texarkana.
	KTAL	6	Do.
Van Buren	KARK	4	Little Rock.
	KATV	7	Do.
	KTHV	11	Do.
Washington	KOTV	6	Tulsa.
	KTUL	8	Do.
	KFSA	5	Fort Smith.
	KODE	12	Joplin-Pittsburg.

SIGNIFICANTLY VIEWED SIGNALS—Continued

County	Call letters, channel number, and market name		
ARKANSAS—continued			
White	KARK	4	Little Rock.
	KATV	7	Do.
	KTHV	11	Do.
Woodruff	KARK	4	Do.
	KATV	7	Do
	KTHV	11	Do.
Yell	KARK	4	Do.
	KATV	7	Do.
	KTHV	11	Do.
	KFSA	5	Fort Smith.
CALIFORNIA			
Alameda East	KTVU	2	San Francisco.
	KRON	4	Do.
	KPIX	5	Do.
	KGO	7	Do.
	KGSC	36	Do.
Alameda West	KTVU	2	Do.
	KRON	4	Do.
	KPIX	5	Do.
	KGO	7	Do.
	KEMO	20	Do.
	KBHK	44	Do.
Alpine	KTVN	2	Reno.
	KCRL	4	Do.
	KOLO	8	Do.
Amador	KCRA	3	Sacramento-Stockton.
	KXTV	10	Do.
	KOVR	13	Do.
Butte	KRCR	7	Chico-Redding.
	KHSL	12	Do.
	KCRA	3	Sacramento-Stockton.
	KXTV	10	Do.
	KOVR	13	Do.
Calaveras	KCRA	3	Do.
	KXTV	10	Do.
	KOVR	13	Do.
Colusa	KCRA	3	Do.
	KXTV	10	Do.
	KOVR	13	Do.
	KRCR	7	Chico-Redding.
	KHSL	12	Do.
Contra Costa East.	KCRA	3	Sacramento-Stockton.
	KXTV	10	Do.
	KOVR	13	Do.
	KTXL	40	Do.
	KTVU	2	San Francisco.
	KPIX	5	Do.
Contra Costa West.	KTVU	2	Do.
	KRON	4	Do.
	KPIX	5	Do.
	KGO	7	Do.
	KEMO	20	Do.
	KBHK	44	Do.
Del Norte	KIEM	3	Eureka.
	KVIQ	6	Do.
El Dorado east over 90 percent cable penetration.			
El Dorado West	KCRA	3	Sacramento-Stockton.
	KXTV	10	Do.
	KOVR	13	Do.
Fresno	KMJ	24	Fresno.
	KFRE	30	Do.
	KJEO	47	Do.
Glenn	KRCR	7	Chico-Redding.
	KHSL	12	Do.
Humboldt	KIEM	3	Eureka.
	KVIQ	6	Do.
Imperial	KECC	9	El Centro.
	KCOP	13	Los Angeles.
	KBLU	13	Yuma.
	XHBC	3	Mexico.
Inyo	KNXT	2	Los Angeles.
	KNBC	4	Do.
	KTLA	5	Do.
	KABC	7	Do.
	KOLO	8	Reno.
Kern East	KNXT	2	Los Angeles.
	KNBC	4	Do.
	KTLA	5	Do.
	KABC	7	Do.
	KHJ	9	Do.
	KTTV	11	Do.
	KCOP	13	Do.
Kern West	KJTV	17	Bakersfield.
	KERO	23	Do.
	KBAK	29	Do.
Kings	KMJ	24	Fresno.
	KFRE	30	Do.
	KJEO	47	Do.
	KERO	23	Bakersfield.
	KBAK	29	Do.
Lake	KCRA	3	Sacramento-Stockton.
	KOVR	13	Do.
	KTVU	2	San Francisco.
Lassen	KTVN	2	Reno.
	KCRL	4	Do.
	KOLO	8	Do.

SIGNIFICANTLY VIEWED SIGNALS—Continued

County	Call letters, channel number, and market name		
CALIFORNIA—continued			
Los Angeles	KNXT	2	Los Angeles.
	KNBC	4	Do.
	KTLA	5	Do.
	KABC	7	Do.
	KHJ	9	Do.
	KTTV	11	Do.
	KCOP	13	Do.
Madera	KMJ	24	Fresno.
	KFRE	30	Do.
	KJEO	47	Do.
Marin	KTVU	2	San Francisco.
	KRON	4	Do.
	KPIX	5	Do.
	KGO	7	Do.
Mariposa	KCRA	3	Sacramento-Stockton.
	KXTV	10	Do.
	KMJ	24	Fresno.
	KFRE	30	Do.
	KJEO	47	Do.
	KOVR	13	Sacramento-Stockton.
Mendocino	KTVU	2	San Francisco.
	KRON	4	Do.
	KPIX	5	Do.
	KGO	7	Do.
	KIEM	3	Eureka.
Merced	KMJ	24	Fresno.
	KFRE	30	Do.
	KJEO	47	Do.
Modoc	KRCR	7	Chico-Redding.
	KOTI	2	Klamath Falls.
	KMED	10	Medford.
	KOLO	8	Reno.
Mono	KOLO	8	Do.
	KCRA	3	Sacramento-Stockton.
	KPIX	5	San Francisco.
	KGO	7	Do.
	KTVU	2	Do.
Monterey East	KSBW	8	Salinas-Monterey.
	KNTV	10	Do.
	KMST	46	Do.
	KTVU	2	San Francisco.
	KNTV	11	Do.
Monterey West	KSBW	8	Salinas-Monterey.
	KNTV	11	Do.
	KMST	46	Do.
	KTVU	2	San Francisco.
Napa North	KTVU	2	Do.
	KRON	4	Do.
	KPIX	5	Do.
	KGO	7	Do.
Napa South	KTVU	2	Do.
	KRON	4	Do.
	KPIX	5	Do.
	KGO	7	Do.
	KEMO	20	Do.
Nevada East	KCRA	3	Sacramento-Stockton.
	KXTV	10	Do.
	KOVR	13	Do.
	KOLO	8	Reno.
	KTVU	2	San Francisco.
Nevada West	KCRA	3	Sacramento-Stockton.
	KXTV	10	Do.
	KOVR	13	Do.
Orange North	KNXT	2	Los Angeles.
	KNBC	4	Do.
	KTLA	5	Do.
	KABC	7	Do.
	KHJ	9	Do.
	KTTV	11	Do.
	KCOP	13	Do.
Orange South	KNXT	2	Do.
	KNBC	4	Do.
	KTLA	5	Do.
	KABC	7	Do.
	KHJ	9	Do.
	KTTV	11	Do.
	KCOP	13	Do.
Placer East	KOLO	8	Reno.
Placer West	KCRA	3	Sacramento-Stockton.
	KXTV	10	Do.
	KOVR	13	Do.
	KTXL	40	Do.
Plumas	KCRA	3	Do.
	KTXL	40	Do.
	KHSL	12	Chico-Redding.
Riverside East	KTVK	3	Phoenix.
	KPHO	5	Do.
	KOOL	10	Do.
	KTAR	12	Do.
Riverside West	KNXT	2	Los Angeles.
	KNBC	4	Do.
	KTLA	5	Do.
	KABC	7	Do.
	KHJ	9	Do.
	KTTV	11	Do.
	KCOP	13	Do.
Riverside Central.	KNXT	2	Do.
	KNBC	4	Do.
	KTLA	5	Do.
	KABC	7	Do.
	KHJ	9	Do.
	KTTV	11	Do.
	KCOP	13	Do.

SIGNIFICANTLY VIEWED SIGNALS—Continued

County	Call letters, channel number, and market name		
CALIFORNIA continued			
Sacramento	KCRA	3	Sacramento-Stockton.
	KXTV	10	Do.
	KOVR	13	Do.
	KTXL	40	Do.
San Benito	KTVU	2	San Francisco.
	KRON	4	Do.
	KPIX	5	Do.
	KSBW	8	Salinas-Monterey.
	KNTV	11	Do.
San Bernardino East.	KTVK	3	Phoenix.
	KPHO	5	Do.
	KOOL	10	Do.
	KTAR	12	Do.
San Bernardino West.	KNXT	2	Los Angeles.
	KNBC	4	Do.
	KTLA	5	Do.
	KABC	7	Do.
	KHJ	9	Do.
	KTTV	11	Do.
	KCOP	13	Do.
San Diego	XETV	6	San Diego.
	KFBM	8	Do.
	KOGO	10	Do.
	KCST	39	Do.
	KCOP	13	Los Angeles.
	KNBC	4	Do.
San Francisco	KTVU	2	San Francisco.
	KRON	4	Do.
	KPIX	5	Do.
	KGO	7	Do.
	KEMO	20	Do.
San Joaquin	KCRA	3	Sacramento-Stockton.
	KXTV	10	Do.
	KOVR	13	Do.
	KTXL	40	Do.
San Luis Obispo	KSBY	6	Salinas-Monterey.
	KEYT	3	Santa Barbara-Santa Maria.
	KCOY	12	Do.
San Mateo	KTVU	2	San Francisco.
	KRON	4	Do.
	KPIX	5	Do.
	KGO	7	Do.
	KEMO	20	Do.
	KBHK	44	Do.
Santa Barbara North.	KEYT	3	Santa Barbara-Santa Maria.
	KCOY	12	Do.
	KSBY	6	Salinas-Monterey.
Santa Barbara South.	KEYT	3	Santa Barbara-Santa Maria.
	KNXT	2	Los Angeles.
	KNBC	4	Do.
	KTLA	5	Do.
	KABC	7	Do.
	KHJ	9	Do.
	KTTV	11	Do.
	KCOP	13	Do.
Santa Clara East.*	KTVU	2	San Francisco.
	KRON	4	Do.
	KPIX	5	Do.
	KGO	7	Do.
	KEMO	20	Do.
	KBHK	44	Do.
	KSBW	8	Salinas-Monterey.
	KNTV	11	Do.
Santa Clara West.	KTVU	2	San Francisco.
	KRON	4	Do.
	KPIX	5	Do.
	KGO	7	Do.
	KEMO	20	Do.
	KBHK	44	Do.
	KNTV	11	Salinas-Monterey.
Santa Cruz	KSBW	8	Do.
	KNTV	11	Do.
	KMST	46	Do.
	KTVU	2	San Francisco.
Shasta	KRCR	7	Chico-Redding
	KHSL	12	Do.
Sierra	KCRL	4	Reno.
	KCRA	3	Sacramento-Stockton.
	KXTV	10	Do.
	KTVU	2	San Francisco.
	KRON	4	Do.
Siskiyou	KRCR	7	Chico-Redding.
	KHSL	12	Do.
	KMED	10	Medford.
Solano East	KCRA	3	Sacramento-Stockton.
	KXTV	10	Do
	KOVR	13	Do
	KTXL	40	Do.
	KTVU	2	San Francisco.
	KPIX	5	Do.
	KGO	7	Do.
	KRON	4	Do.
Solano West	KTVU	2	Do.
	KRON	4	Do.
	KPIX	5	Do.
	KGO	7	Do.

SIGNIFICANTLY VIEWED SIGNALS—Continued

County	Call letters, channel number, and market name		

CALIFORNIA—continued

County	Call letters	Ch.	Market name
Sonoma North	KTVU	2	Do.
	KRON	4	Do.
	KPIX	5	Do.
	KGO	7	Do.
Sonoma South	KTVU	2	Do.
	KRON	4	Do.
	KPIX	5	Do.
	KGO	7	Do.
Stanislaus	KCRA	3	Sacramento-Stockton.
	KXTV	10	Do.
	KOVR	13	Do.
	KTXL	40	Do.
Sutter*	KCRA	3	Do.
	KXTV	10	Do.
	KOVR	13	Do.
	KHSL	12	Chico-Redding.
	KTVU	2	San Francisco.
Tehama	KRCR	7	Chico-Redding.
	KHSL	12	Do.
Trinity	KRCR	7	Do.
	KHSL	12	Do.
Tulare	KMJ	24	Fresno.
	KFRE	30	Do.
	KJEO	47	Do.
	KJTV	17	Bakersfield.
	KERO	23	Do.
	KBAK	29	Do.
Tuolumne	KCRA	3	Sacramento-Stockton.
	KXTV	10	Do.
	KOVR	13	Do.
	KSBW	8	Salinas-Monterey.
	KTVU	2	San Francisco.
	KRON	4	Do.
Ventura	KNXT	2	Los Angeles.
	KNBC	4	Do.
	KTLA	5	Do.
	KABC	7	Do.
	KHJ	9	Do.
	KTTV	11	Do.
	KCOP	13	Do.
Yolo	KCRA	3	Sacramento-Stockton.
	KXTV	10	Do.
	KOVR	13	Do.
	KTXL	40	Do.
Yuba	KCRA	3	Do.
	KXTV	10	Do.
	KOVR	13	Do.
	KTXL	40	Do.
	KHSL	12	Chico-Redding.

Census county divisions in split counties

Alameda East: Livermore, Pleasanton.
Alameda West: All other.
Contra Costa East: Ambrose, Antioch, Brentwood-Bryon, Clayton-Tassajara, Martinez, Oakley-Bethel, Pittsburg, Pleasant Hill, Port Chicago.
Contra Costa West: All other.
El Dorado East: Lake Valley.
El Dorado West: All other.
Kern East: East Kern, Tehachapi.
Kern West: All other.
Monterey West: Carmel, Carmel Valley, Ford Ord, Monterey, Monterey Penin., Pacific Grove, Seaside.
Monterey East: All other.
Napa North: Angwin, Berryessa, Calistoga, St. Helena.
Napa South: All other.
Nevada East: Donner.
Nevada West: All other.
Orange North: Anaheim-Garden Grove, Buena Park-Cypress, Fullerton-La Habra, Santa Ana Canyon, Santa Ana-Orange.
Orange South: All other.
Placer East: Lake Tahoe.
Placer West: All other.
Riverside East: Palo Verde.
Riverside Central: Cathedral City-Palm Desert, Chuck-walla, Coachella Valley, Desert Hot Springs, Idyllwild, Palm Springs, San Gorgonio Pass.
Riverside West: All other.
San Bernardino East: Needles.
San Bernardino West: All other.
Santa Barbara North: Cuyama, Guadalupe, Lompoc Valley, Santa Maria, Santa Maria Valley, Santa Ynez Valley.
Santa Barbara South: All other.
Santa Clara East: Diablo Range, Gilroy, Llagas-Uvas, Morgan Hill, San Martin.
Santa Clara West: All other.
Solano East: Dixon, Fairfield-Suisun, Rio Vista, Vacaville.
Solano West: All other.
Sonoma South: Petaluma, Petaluma Rural, Sonoma.
Sonoma North: All other.

SIGNIFICANTLY VIEWED SIGNALS—Continued

County	Call letters, channel number, and market name		

COLORADO

County	Call letters	Ch.	Market name
Adams	KWGN	2	Denver.
	KOA	4	Do.
	KLZ	7	Do.
	KBTV	9	Do.
Alamosa	KOB	4	Albuquerque.
	KOAT	7	Do.
	KGGM	13	Do.
	KRDO	13	Colorado Springs-Pueblo.
Arapahoe	KWGN	2	Denver.
	KOA	4	Do.
	KLZ	7	Do.
	KBTV	9	Do.
Archuleta	KOB	4	Albuquerque.
	KOAT	7	Do.
	KGGM	13	Do.
Baca	KOAA	5	Colorado Springs-Pueblo.
	KKTV	11	Do.
	KRDO	13	Do.
Bent	KOAA	5	Do.
	KKTV	11	Do.
	KRDO	13	Do.
Boulder	KWGN	2	Denver.
	KOA	4	Do.
	KLZ	7	Do.
	KBTV	9	Do.
Chaffee	KOA	4	Do.
	KLZ	7	Do.
	KBTV	9	Do.
	KOAA	5	Colorado Springs-Pueblo.
Cheyenne	KAYS+	7	Wichita-Hutchinson.
	KKTV	11	Colorado Springs-Pueblo.
	KRDO	13	Do.
Clear Creek	KWGN	2	Denver.
	KOA	4	Do.
	KLZ	7	Do.
	KBTV	9	Do.
Conejos	KOB	4	Albuquerque.
	KOAT	7	Do.
	KGGM	13	Do.
Costilla	KOB	4	Do.
	KOAT	7	Do.
	KGGM	13	Do.
Crowley	KOAA	5	Colorado Springs-Pueblo.
	KKTV	11	Do.
	KRDO	13	Do.
Custer	KOAA	5	Do.
	KKTV	11	Do.
	KRDO	13	Do.
Delta	KREX	5	Grand Junction.
	KREY	10	Do.
	KBTV	9	Denver.
Denver	KWGN	2	Do.
	KOA	4	Do.
	KLZ	7	Do.
	KBTV	9	Do.
Dolores	KOB	4	Albuquerque.
	KOAT	7	Do.
	KGGM	13	Do.
Douglas	KWGN	2	Denver.
	KOA	4	Do.
	KLZ	7	Do.
	KBTV	9	Do.
	KRDO	13	Colorado Springs-Pueblo.
Eagle	KOA	4	Denver.
	KLZ	7	Do.
	KBTV	9	Do.
Elbert	KWGN	2	Do.
	KOA	4	Do.
	KLZ	7	Do.
	KBTV	9	Do.
	KKTV	11	Colorado Springs-Pueblo.
	KRDO	13	Do.
El Paso	KOAA	5	Do.
	KKTV	11	Do.
	KRDO	13	Do.
Fremont	KOAA	5	Do.
	KKTV	11	Do.
	KRDO	13	Do.
Garfield	KREX	5	Grand Junction.
	KOA	4	Denver.
Gilpin	KWGN	2	Do.
	KOA	4	Do.
	KLZ	7	Do.
	KBTV	9	Do.
Grand	KWGN	2	Do.
	KOA	4	Do.
	KLZ	7	Do.
	KBTV	9	Do.

SIGNIFICANTLY VIEWED SIGNALS—Continued

County	Call letters, channel number, and market name		

COLORADO—continued

County	Call letters	Ch.	Market name
Gunnison	KOAA	5	Colorado Springs-Pueblo.
	KBTV	9	Denver.
	KREX	5	Grand Junction.
	KREY	10	Do.
Hinsdale	KREX	5	Do.
	KOAA	5	Colorado Springs-Pueblo.
Huerfano	KOAA	5	Do.
	KKTV	11	Do.
	KRDO	13	Do.
Jackson	KOA	4	Denver.
	KFBC	5	Cheyenne.
Jefferson	KWGN	2	Denver.
	KOA	4	Do.
	KLZ	7	Do.
	KBTV	9	Do.
Kiowa	KOAA	5	Colorado Springs-Pueblo.
	KKTV	11	Do.
	KRDO	13	Do.
Kit Carson	KAYS+	7	Wichita-Hutchinson.
Lake	KAYS+	7	Do.
	KWGN	2	Denver.
	KOA	4	Do.
	KLZ	7	Do.
	KBTV	9	Do.
La Plata	KOB	4	Albuquerque.
	KOAT	7	Do.
	KGGM	13	Do.
	KREZ	6	Grand Junction.
Larimer	KWGN	2	Denver.
	KOA	4	Do.
	KLZ	7	Do.
	KBTV	9	Do.
	KFBC	5	Cheyenne.
Las Animas	KOAA	5	Colorado Springs-Pueblo.
	KKTV	11	Do.
	KRDO	13	Do.
Lincoln	KOAA	5	Do.
	KKTV	11	Do.
	KRDO	13	Do.
	KWGN	2	Denver.
	KOA	4	Do.
Logan	KTVS	3	Cheyenne.
Mesa	KREX	5	Grand Junction.
Mineral	KOAA	5	Colorado Springs-Pueblo.
	KOAT	7	Albuquerque.
	KGGM	13	Do.
Moffat	KOA	4	Denver.
	KLZ	7	Do.
	KBTV	9	Do.
Montezuma	KOB	4	Albuquerque.
	KOAT	7	Do.
	KGGM	13	Do.
Montrose	KREY	10	Grand Junction.
	KOAA	5	Colorado Springs-Pueblo.
	KBTV	9	Denver.
	KUTV	2	Salt Lake City.
Morgan	KWGN	2	Denver.
	KOA	4	Do.
	KLZ	7	Do.
	KBTV	9	Do.
	KTVS	3	Cheyenne.
Otero	KOAA	5	Colorado Springs-Pueblo.
	KKTV	11	Do.
	KRDO	13	Do.
Ouray	KREX	5	Grand Junction.
Park	KWGN	2	Denver.
	KOA	4	Do.
	KLZ	7	Do.
	KBTV	9	Do.
Phillips	KTVS	3	Cheyenne.
	KHOL+	13	Lincoln-Hastings-Kearney.
Pitkin*	NA		
Prowers	KOAA	5	Colorado Springs-Pueblo.
	KKTV	11	Do.
	KRDO	13	Do.
	KGLD	11	Wichita-Hutchinson.
Pueblo	KOAA	5	Colorado Springs-Pueblo.
	KKTV	11	Do.
	KRDO	13	Do.
Rio Blanco	KUTV	2	Salt Lake City.
	KCPX	4	Do.
	KSL	5	Do.
Rio Grande*	KOB	4	Albuquerque.
	KOAT	7	Do.
	KGGM	13	Do.

RULES AND REGULATIONS

County	Call letters, channel number, and market name		

COLORADO—continued

County	Call letters	Ch.	Market name
Routt	KOA	4	Denver.
	KLZ	7	Do.
	KBTV	9	Do.
Saguache	KOB	4	Albuquerque.
	KOAT	7	Do.
	KGGM	13	Do.
	KOAA	5	Colorado Springs-Pueblo.
San Juan	KREX	5	Grand Junction.
San Miguel	KREX	5	Do.
Sedgwick	KTVS	3	Cheyenne.
	KHOL+	13	Lincoln-Hastings-Kearney.
	KNOP	2	North Platte.
Summit	KWGN	2	Denver.
	KOA	4	Do.
	KLZ	7	Do.
	KBTV	9	Do.
Teller	KOA	4	Do.
	KLZ	7	Do.
	KBTV	9	Do.
	KKTV	11	Colorado Springs-Pueblo.
	KRDO	13	Do.
Washington	KWGN	2	Denver.
	KOA	4	Do.
	KLZ	7	Do.
	KBTV	9	Do.
	KTVS	3	Cheyenne.
Weld	KWGN	2	Denver.
	KOA	4	Do.
	KLZ	7	Do.
	KBTV	9	Do.
Yuma	KAYS+	7	Wichita-Hutchinson.
	KOMC	8	Do.
	KTVS	3	Cheyenne.
	KHOL+	13	Lincoln-Hastings-Kearney.

CONNECTICUT

County	Call letters	Ch.	Market name
Fairfield	WCBS	2	New York.
	WNBC	4	Do.
	WNEW	5	Do.
	WABC	7	Do.
	WOR	9	Do.
	WPIX	11	Do.
	WNHC	8	Hartford-New Haven.
Hartford	WTIC	3	Do.
	WNHC	8	Do.
	WHCT	18	Do.
	WHNB	30	Do.
Litchfield	WTIC	3	Do.
	WNHC	8	Do.
	WHNB	30	Do.
	WCBS	2	New York.
	WNBC	4	Do.
	WNEW	5	Do.
	WPIX	11	Do.
Middlesex	WTIC	3	Hartford-New Haven.
	WNHC	8	Do.
	WHNB	30	Do.
	WNEW	5	New York.
New Haven	WTIC	3	Hartford-New Haven.
	WNHC	8	Do.
	WCBS	2	New York.
	WNBC	4	Do.
	WNEW	5	Do.
	WADO	7	Do.
	WOR	9	Do.
	WPIX	11	Do.
New London	WTEV	6	Providence.
	WJAR	10	Do.
	WPRI	12	Do.
	WHDH	5	Boston.
	WTIC	3	Hartford-New Haven.
	WNHC	8	Do.
Tolland	WTIC	3	Do.
	WNHC	8	Do.
	WHNB	30	Do.
	WBZ	4	Boston.
	WHYN	40	Springfield, Mass.
Windham	WTEV	6	Providence.
	WJAR	10	Do.
	WPRI	12	Do.
	WBZ	4	Boston.
	WHDH	5	Do.
	WNAC	7	Do.
	WTIC	3	Hartford-New Haven.
	WNHC	8	Do.

DELAWARE

County	Call letters	Ch.	Market name
Kent	KYW	3	Philadelphia.
	WFIL	6	Do.
	WCAU	10	Do.
	WPHL	17	Do.

DELAWARE—continued

County	Call letters	Ch.	Market name
	WMAR	2	Baltimore.
	WBAL	11	Do.
New Castle	KYW	3	Philadelphia.
	WFIL	6	Do.
	WCAU	10	Do.
	WPHL	17	Do.
	WTAF	29	Do.
	WKBS	48	Do.
Sussex	WBOC	16	Salisbury.
	WMAR	2	Baltimore.
	WBAL	11	Do.
	WJZ	13	Do.
	WTTG	5	Washington, D.C.

DISTRICT OF COLUMBIA

County	Call letters	Ch.	Market name
District of Columbia.	WRC	4	Washington, D.C.
	WTTG	5	Do.
	WMAL	7	Do.
	WTOP	9	Do.
	WDCA	20	Do.

FLORIDA

County	Call letters	Ch.	Market name
Alachua	WJXT	4	Jacksonville, Fla.
	WFGA	12	Do.
	WESH	2	Orlando-Daytona Beach.
Baker	WJXT	4	Jacksonville, Fla.
	WFGA	12	Do.
	WJKS	17	Do.
Bay	WJHG	7	Panama City.
	WTVY	4	Dothan.
Bradford	WJXT	4	Jacksonville, Fla.
	WFGA	12	Do.
	WJKS	17	Do.
Brevard	WESH	2	Orlando-Daytona Beach.
	WDBO	6	Do.
	WFTV	9	Do.
Broward	WTVJ	4	Miami.
	WCKT	7	Do.
	WPLG	10	Do.
	WAJA	23	Do.
	WPTV	5	West Palm Beach.
	WEAT	12	Do.
Calhoun	WJHG	7	Panama City.
	WTVY	4	Dothan.
	WCTV	6	Tallahassee.
Charlotte	WFLA	8	Tampa-St. Petersburg.
	WTVT	13	Do.
	WINK	11	Fort Myers.
Citrus	WFLA	8	Tampa-St. Petersburg.
	WLCY	10	Do.
	WTVT	13	Do.
	WESH	2	Orlando-Daytona Beach.
	WDBO	6	Do.
	WFTV	9	Do.
Clay	WJXT	4	Jacksonville, Fla.
	WFGA	12	Do.
	WJKS	17	Do.
Collier	90 percent cable penetration.		
Columbia	WJXT	4	Jacksonville, Fla.
	WFGA	12	Do.
Dade	WTVJ	4	Miami.
	WCIX	6	Do.
	WCKT	7	Do.
	WPLG	10	Do.
	WAJA	23	Do.
De Soto	WFLA	8	Tampa-St. Petersburg.
	WTVT	13	Do.
	WTOG	44	Do.
	WINK	11	Fort Myers.
Dixie	WJXT	4	Jacksonville, Fla.
	WFGA	12	Do.
	WESH	2	Orlando-Daytona Beach.
	WCTV	6	Tallahassee.
	WLCY	10	Tampa-St. Petersburg.
Duval	WJXT	4	Jacksonville, Fla.
	WFGA	12	Do.
	WJKS	17	Do.
Escambia	WEAR	3	Mobile-Pensacola.
	WKRG	5	Do.
	WALA	10	Do.
Flagler	WESH	2	Orlando-Daytona Beach.
	WDBO	6	Do.
	WFTV	9	Do.
	WJXT	4	Jacksonville, Fla.
Franklin	WCTV	6	Tallahassee.
	WJHG	7	Panama City.
Gadsden	WCTV	6	Tallahassee.
	WTVY	4	Dothan.
	WJHG	7	Panama City.

FLORIDA—continued

County	Call letters	Ch.	Market name
Gilchrist	WJXT	4	Jacksonville, Fla.
	WFGA	12	Do.
	WESH	2	Orlando-Daytona Beach.
Glades	WPTV	5	West Palm Beach.
	WEAT	12	Do.
	WINK	11	Fort Myers.
	WTVJ	4	Miami.
Gulf	WJHG	7	Panama City.
	WTVY	4	Dothan.
	WCTV	6	Tallahassee.
Hamilton *	WJXT	4	Jacksonville, Fla.
	WCTV	6	Tallahassee.
Hardee	WFLA	8	Tampa-St. Petersburg.
	WTVT	13	Do.
	WTOG	44	Do.
Hendry *	WINK	11	Fort Myers.
	WBBH	20	Do.
	WPTV	5	West Palm Beach.
	WEAT	12	Do.
Hernando	WFLA	8	Tampa-St. Petersburg.
	WLCY	10	Do.
	WTVT	13	Do.
	WTOG	44	Do.
Highlands	WFLA	8	Tampa-St. Petersburg.
	WTVT	13	Do.
	WINK	11	Fort Myers.
Hillsborough	WFLA	8	Tampa-St. Petersburg.
	WLCY	10	Do.
	WTVT	13	Do.
	WTOG	44	Do.
Holmes	WTVY	4	Dothan.
	WJHG	7	Panama City.
Indiana River	WPTV	5	West Palm Beach.
	WEAT	12	Do.
	WTVX	34	Fort Pierce-Vero Beach.
Jackson	WTVY	4	Dothan.
	WJHG	7	Panama City.
	WCTV	6	Tallahassee.
Jefferson	WCTV	6	Tallahassee.
	WALB	10	Albany, Ga.
Lafayette	WCTV	6	Tallahassee.
Lake*	WESH	2	Orlando-Daytona Beach.
	WDBO	6	Do.
	WFTV	9	Do.
Lee	WINK	11	Fort Myers.
	WBBH	20	Do.
Leon	WCTV	6	Tallahassee.
	WALB	10	Albany, Ga.
	WJHG	7	Panama City.
Levy*	WESH	2	Orlando-Daytona Beach.
	WJXT	4	Jacksonville, Fla.
	WLCY	10	Tampa-St. Petersburg.
	WTVT	13	Do.
Liberty	WCTV	6	Tallahassee.
	WJHG	7	Panama City.
Madison	WCTV	6	Tallahassee.
	WALB	10	Albany, Ga.
Manatee	WFLA	8	Tampa-St. Petersburg.
	WLCY	10	Do.
	WTVT	13	Do.
	WTOG	44	Do.
Marion	WESH	2	Orlando-Daytona Beach.
	WDBO	6	Do.
	WFTV	9	Do.
Martin	WPTV	5	West Palm Beach.
	WEAT	12	Do.
	WTVJ	4	Miami.
Monroe	WTVJ	4	Do.
	WCIX	6	Do.
	WCKT	7	Do.
	WPLG	10	Do.
Nassau	WJXT	4	Jacksonville, Fla.
	WFGA	12	Do.
	WJKS	17	Do.
Okaloosa	WEAR	3	Mobile-Pensacola.
	WKRG	5	Do.
	WALA	10	Do.
	WJHG	7	Panama City.
Okeechobee	WPTV	5	West Palm Beach.
	WEAT	12	Do.
Orange	WESH	2	Orlando-Daytona Beach.
	WDBO	6	Do.
	WFTV	9	Do.
Osceola	WESH	2	Do.
	WDBO	6	Do.
	WFTV	9	Do.
Palm Beach	WPTV	5	West Palm Beach.
	WEAT	12	Do.
	WTVJ	4	Miami.
	WCKT	7	Do.
	WPLG	10	Do.
Pasco	WFLA	8	Tampa-St. Petersburg.
	WLCY	10	Do.
	WTVT	13	Do.
	WTOG	44	Do.

SIGNIFICANTLY VIEWED SIGNALS—Continued   SIGNIFICANTLY VIEWED SIGNALS—Continued   SIGNIFICANTLY VIEWED SIGNALS—Continued

County	Call letters, channel number, and market name		

FLORIDA—continued

County	Call letters	Ch.	Market name
Pinellas	WFLA	8	Do.
	WLCY	10	Do.
	WTVT	13	Do.
	WTOG	44	Do.
Polk	WFLA	8	Do.
	WLCY	10	Do.
	WTVT	13	Do.
	WTOG	44	Do.
	WDBO	6	Orlando-Daytona Beach.
	WFTV	9	Do.
Putnam	WJXT	4	Jacksonville, Fla.
	WFGA	12	Do.
	WJKS	17	Do.
	WESH	2	Orlando-Daytona Beach.
	WDBO	6	Do.
St. Johns	WJXT	4	Jacksonville, Fla.
	WFGA	12	Do.
	WJKS	17	Do.
St. Lucie	WPTV	5	West Palm Beach.
	WEAT	12	Do.
	WTVX	34	Fort Pierce-Vero Beach.
Santa Rosa	WEAR	3	Mobile-Pensacola.
	WKRG	5	Do.
	WALA	10	Do.
Sarasota	WFLA	8	Tampa-St. Petersburg.
	WLCY	10	Do.
	WTVT	13	Do.
	WTOG	44	Do.
Seminole	WESH	2	Orlando-Daytona Beach.
	WDBO	6	Do.
	WFTV	9	Do.
Sumter	WESH	2	Do.
	WDBO	6	Do.
	WFTV	9	Do.
	WFLA	8	Tampa-St. Petersburg.
	WTVT	13	Do.
Suwannee	WJXT	4	Jacksonville, Fla.
	WFGA	12	Do.
	WCTV	6	Tallahassee.
Taylor	WCTV	6	Do.
Union	WJXT	4	Jacksonville, Fla.
	WFGA	12	Do.
	WJKS	17	Do.
Volusia	WESH	2	Orlando-Daytona Beach.
	WDBO	6	Do.
	WFTV	9	Do.
Wakulla	WCTV	6	Tallahassee.
	WJHG	7	Panama City.
Walton	WJHG	7	Do.
	WTVY	4	Dothan.
	WEAR	3	Mobile-Pensacola.
Washington	WTVY	4	Dothan.
	WJHG	7	Panama City.

GEORGIA

County	Call letters	Ch.	Market name
Appling	WSAV	3	Savannah.
	WTOC	11	Do.
	WJCL	22	Do.
	WJBF	6	Augusta.
	WJXT	4	Jacksonville, Fla.
	WCSC	5	Charleston, S.C.
Atkinson	WALB	10	Albany, Ga.
	WCTV	6	Tallahassee.
Bacon	WALB	10	Albany, Ga.
	WJXT	4	Jacksonville, Fla.
	WSAV	3	Savannah.
Baker	WALB	10	Albany, Ga.
	WRBL	3	Columbus, Ga.
	WTVM	9	Do.
	WTVY	4	Dothan.
	WCTV	6	Tallahassee.
Baldwin	WMAZ	13	Macon.
	WSB	2	Atlanta.
	WAGA	5	Do.
Banks	WSB	2	Do.
	WAGA	5	Do.
	WQXI	11	Do.
	WFBC	4	Greenville-Spartanburg-Asheville.
	WSPA	7	Do.
Barrow	WSB	2	Atlanta.
	WAGA	5	Do.
	WQXI	11	Do.
	WTCG	17	Do.
Bartow	WSB	2	Do.
	WAGA	5	Do.
	WQXI	11	Do.
	WATL	36	Do.
Ben Hill	WALB	10	Albany, Ga.
	WMAZ	13	Macon.
Berrien	WALB	10	Albany, Ga.
	WCTV	6	Tallahassee.

GEORGIA—continued

County	Call letters	Ch.	Market name
Bibb	WMAZ	13	Macon.
	WCWB	41	Do.
	WSB	2	Atlanta.
	WTVM	9	Columbus, Ga.
Bleckley	WMAZ	13	Macon.
Brantley	WJXT	4	Jacksonville, Fla.
	WFGA	12	Do.
Brooks	WCTV	6	Tallahassee.
	WALB	10	Albany, Ga.
Bryan	WSAV	3	Savannah.
	WTOC	11	Do.
	WJCL	22	Do.
Bulloch	WSAV	3	Do.
	WTOC	11	Do.
	WJBF	6	Augusta.
	WRDW	12	Do.
Burke	WJBF	6	Do.
	WRDW	12	Do.
Butts	WSB	2	Atlanta.
	WAGA	5	Do.
	WQXI	11	Do.
	WTCG	17	Do.
	WATL	36	Do.
Calhoun	WALB	10	Albany, Ga.
	WRBL	3	Columbus, Ga.
	WTVM	9	Do.
	WTVY	4	Dothan.
	WCTV	6	Tallahassee.
Camden	WJXT	4	Jacksonville, Fla.
	WFGA	12	Do.
	WJKS	17	Do.
Candler	WJBF	6	Augusta.
	WRDW	12	Do.
	WSAV	3	Savannah.
	WTOC	11	Do.
Carroll	WSB	2	Atlanta.
	WAGA	5	Do.
	WQXI	11	Do.
Catoosa	WRCB	3	Chattanooga.
	WTVC	9	Do.
	WDEF	12	Do.
Charlton	WJXT	4	Jacksonville, Fla.
	WFGA	12	Do.
	WJKS	17	Do.
Chatham	WSAV	3	Savannah.
	WTOC	11	Do.
	WJCL	22	Do.
Chattahoochee	WRBL	3	Columbus, Ga.
	WTVM	9	Do.
Chattooga	WRCB	3	Chattanooga.
	WTVC	9	Do.
	WDEF	12	Do.
	WSB	2	Atlanta.
	WAGA	5	Do.
	WQXI	11	Do.
Cherokee	WSB	2	Do.
	WAGA	5	Do.
	WQXI	11	Do.
	WTCG	17	Do.
	WATL	36	Do.
Clarke	WSB	2	Do.
	WAGA	5	Do.
	WQXI	11	Do.
	WFBC	4	Greenville-Spartanburg-Asheville.
Clay	WRBL	3	Columbus, Ga.
	WTVM	9	Do.
	WTVY	4	Dothan.
Clayton	WSB	2	Atlanta.
	WAGA	5	Do.
	WQXI	11	Do.
	WTCG	17	Do.
	WATL	36	Do.
Clinch	WJXT	4	Jacksonville, Fla.
	WFGA	12	Do.
	WALB	10	Albany, Ga.
	WCTV	6	Tallahassee.
Cobb	WSB	2	Atlanta.
	WAGA	5	Do.
	WQXI	11	Do.
	WTCG	17	Do.
	WATL	36	Do.
Coffee	WALB	10	Albany, Ga.
Colquitt	WALB	10	Do.
	WCTV	6	Tallahassee.
Columbia	WJBF	6	Augusta.
	WRDW	12	Do.
	WATU	26	Do.
Cook	WALB	10	Albany, Ga.
	WCTV	6	Tallahassee.
Coweta	WSB	2	Atlanta.
	WAGA	5	Do.
	WQXI	11	Do.
	WATL	36	Do.
Crawford	WMAZ	13	Macon.
	WRBL	3	Columbus, Ga.
	WTVM	9	Do.
	WSB	2	Atlanta.
Crisp	WRBL	3	Columbus, Ga.
	WTVM	9	Do.
	WALB	10	Albany, Ga.
	WMAZ	13	Macon.

GEORGIA—continued

County	Call letters	Ch.	Market name
Dade	WRCB	3	Chattanooga.
	WTVC	9	Do.
	WDEF	12	Do.
Dawson	WSB	2	Atlanta.
	WAGA	5	Do.
	WQXI	11	Do.
Decatur	WCTV	6	Tallahassee.
	WALB	10	Albany, Ga.
	WTVY	4	Dothan.
De Kalb	WSB	2	Atlanta.
	WAGA	5	Do.
	WQXI	11	Do.
	WTCG	17	Do.
	WATL	36	Do.
Dodge	WMAZ	13	Macon.
	WALB	10	Albany, Ga.
Dooly	WMAZ	13	Macon.
	WALB	10	Albany, Ga.
	WRBL	3	Columbus, Ga.
	WTVM	9	Do.
Dougherty	WALB	10	Albany, Ga.
	WRBL	3	Columbus, Ga.
	WTVM	9	Do.
	WCTV	6	Tallahassee.
Douglas	WSB	2	Atlanta.
	WAGA	5	Do.
	WQXI	11	Do.
	WTCG	17	Do.
	WATL	36	Do.
Early	WTVY	4	Dothan.
	WTVM	9	Columbus, Ga.
	WCTV	6	Tallahassee.
	WALB	10	Albany, Ga.
Echols	WCTV	6	Tallahassee.
	WALB	10	Albany, Ga.
Effingham	WSAV	3	Savannah.
	WTOC	11	Do.
	WJCL	22	Do.
Elbert	WFBC	4	Greenville-Spartanburg-Asheville.
	WSPA	7	Do.
	WLOS	13	Do.
	WJBF	6	Augusta.
Emanuel	WJBF	6	Do.
	WRDW	12	Do.
Evans	WSAV	3	Savannah.
	WTOC	11	Do.
	WJBF	6	Augusta.
Fannin	WRCB	3	Chattanooga.
	WTVC	9	Do.
	WDEF	12	Do.
	WSB	2	Atlanta.
	WAGA	5	Do.
	WATL	36	Do.
Fayette	WSB	2	Do.
	WAGA	5	Do.
	WQXI	11	Do.
	WATL	36	Do.
Floyd	WSB	2	Do.
	WAGA	5	Do.
	WQXI	11	Do.
	WRCB	3	Chattanooga.
	WTVC	9	Do.
	WDEF	12	Do.
Forsyth	WSB	2	Atlanta.
	WAGA	5	Do.
	WQXI	11	Do.
	WTCG	17	Do.
	WATL	36	Do.
Franklin	WFBC	4	Greenville-Spartanburg-Asheville.
	WSPA	7	Do.
	WLOS	13	Do.
Fulton	WSB	2	Atlanta.
	WAGA	5	Do.
	WQXI	11	Do.
	WTCG	17	Do.
	WATL	36	Do.
Gilmer*	WSB	2	Do.
	WAGA	5	Do.
	WRCB	3	Chattanooga.
	WTVC	9	Do.
	WDEF	12	Do.
Glascock	WJBF	6	Augusta.
	WRDW	12	Do.
Glynn	WJXT	4	Jacksonville, Fla.
	WFGA	12	Do.
Gordon	WRCB	3	Chattanooga.
	WTVC	9	Do.
	WDEF	12	Do.
	WSB	2	Atlanta.
	WAGA	5	Do.
	WQXI	11	Do.
	WTCG	17	Do.
Grady	WCTV	6	Tallahassee.
	WALB	10	Albany, Ga.
Greene	WSB	2	Atlanta.
	WAGA	5	Do.
	WQXI	11	Do.
	WJBF	6	Augusta.

SIGNIFICANTLY VIEWED SIGNALS—Continued

GEORGIA—continued

County	Call letters, channel number, and market name		
Gwinnett	WSB	2	Atlanta.
	WAGA	5	Do.
	WQXI	11	Do.
	WTCG	17	Do.
	WATL	36	Do.
Habersham*	WSB	2	Do.
	WAGA	5	Do.
	WQXI	11	Do.
	WFBC	4	Greenville-Spartanburg-Asheville.
	WSPA	7	Do.
Hall	WSB	2	Atlanta.
	WAGA	5	Do.
	WQXI	11	Do.
Hancock	WJBF	6	Augusta
	WRDW	12	Do.
	WMAZ	13	Macon.
Haralson	WSB	2	Atlanta.
	WAGA	5	Do.
	WQXI	11	Do.
	WTCG	17	Do.
	WATL	36	Do.
Harris	WRBL	3	Columbus, Ga.
	WTVM	9	Do.
	WSB	2	Atlanta.
Hart	WFBC	4	Greenville-Spartanburg-Asheville.
	WSPA	7	Do.
	WLOS	13	Do.
Heard	WSB	2	Atlanta.
	WAGA	5	Do.
	WQXI	11	Do.
Henry	WSB	2	Do.
	WAGA	5	Do.
	WQXI	11	Do.
	WATL	36	Do.
Houston	WMAZ	13	Macon.
	WCWB	41	Do.
	WRBL	3	Columbus, Ga.
	WTVM	9	Do.
Irwin*	WALB	10	Albany, Ga.
	WCTV	6	Tallahassee.
Jackson	WSB	2	Atlanta.
	WAGA	5	Do.
	WQXI	11	Do.
	WFBC	4	Greenville-Spartanburg-Asheville.
Jasper	WSB	2	Atlanta.
	WAGA	5	Do.
	WQXI	11	Do.
	WTCG	17	Do.
	WATL	36	Do.
	WMAZ	13	Macon.
Jeff Davis	WALB	10	Albany, Ga.
	WJXT	4	Jacksonville, Fla.
	WSAV	3	Savannah.
	WTOC	11	Do.
Jefferson	WJBF	6	Augusta.
	WRDW	12	Do.
Jenkins	WJBF	6	Do.
	WRDW	12	Do.
Johnson	WMAZ	13	Macon.
	WJBF	6	Augusta.
	WRDW	12	Do.
Jones	WMAZ	13	Macon.
	WCWB	41	Do.
	WSB	2	Atlanta.
	WAGA	5	Do.
	WQXI	11	Do.
Lamar	WSB	2	Do.
	WAGA	5	Do.
	WQXI	11	Do.
	WMAZ	13	Macon.
Lanier	WALB	10	Albany, Ga.
	WCTV	6	Tallahassee.
Laurens	WMAZ	13	Macon.
	WCWB	41	Do.
Lee	WRBL	3	Columbus, Ga.
	WTVM	9	Do.
	WALB	10	Albany, Ga.
Liberty	WSAV	3	Savannah.
	WTOC	11	Do.
	WJCL	22	Do.
Lincoln	WJBF	6	Augusta.
	WRDW	12	Do.
	WFBC	4	Greenville-Spartanburg-Asheville.
Long	WSAV	3	Savannah.
	WTOC	11	Do.
	WJCL	22	Do.
Lowndes	WCTV	6	Tallahassee.
	WALB	10	Albany, Ga.
Lumpkin	WSB	2	Atlanta.
	WAGA	5	Do.
	WQXI	11	Do.
McDuffie	WJBF	6	Augusta.
	WRDW	12	Do.
McIntosh	WSAV	3	Savannah.
	WTOC	11	Do.
	WJXT	4	Jacksonville, Fla.
Macon	WRBL	3	Columbus, Ga.
	WTVM	9	Do.
	WALB	10	Albany, Ga.
	WMAZ	13	Macon.
Madison	WFBC	4	Greenville-Spartanburg-Asheville.
	WSPA	7	Do.
	WLOS	13	Do.
	WSB	2	Atlanta.
	WAGA	5	Do.
	WQXI	11	Do.
Marion	WRBL	3	Columbus, Ga.
	WTVM	9	Do.
Meriwether	WSB	2	Atlanta.
	WAGA	5	Do.
	WQXI	11	Do..
	WRBL	3	Columbus, Ga.
	WTVM	9	Do.
Miller	WTVY	4	Dothan.
	WALB	10	Albany, Ga.
	WCTV	6	Tallahassee.
Mitchell	WALB	10	Albany, Ga.
	WCTV	6	Tallahassee.
Monroe	WSB	2	Atlanta.
	WAGA	5	Do.
	WQXI	11	Do.
	WMAZ	13	Macon.
Montgomery	WMAZ	13	Do.
	WJBF	6	Augusta.
	WSAV	3	Savannah.
	WTOC	11	Do.
Morgan	WSB	2	Atlanta.
	WAGA	5	Do.
	WQXI	11	Do.
Murray	WRCB	3	Chattanooga.
	WTVC	9	Do.
	WDEF	12	Do.
	WSB	2	Atlanta.
Muscogee	WRBL	3	Columbus, Ga.
	WTVM	9	Do.
	WYEA	38	Do.
Newton	WSB	2	Atlanta.
	WAGA	5	Do.
	WQXI	11	Do.
	WATL	36	Do.
Oconee	WSB	2	Do.
	WAGA	5	Do.
	WQXI	11	Do.
Oglethorpe	WSB	2	Do.
	WAGA	5	Do.
	WQXI	11	Do.
	WJBF	6	Augusta.
	WFBC	4	Greenville-Spartanburg-Asheville.
	WSPA	7	Do.
	WLOS	13	Do.
Paulding	WSB	2	Atlanta.
	WAGA	5	Do.
	WQXI	11	Do.
	WATL	36	Do.
Peach	WMAZ	13	Macon.
	WCWB	41	Do.
	WRBL	3	Columbus, Ga.
	WTVM	9	Do.
Pickens	WSB	2	Atlanta.
	WAGA	5	Do.
	WQXI	11	Do.
	WTCG	17	Do.
	WATL	36	Do.
Pierce	WJXT	4	Jacksonville, Fla.
	WFGA	12	Do.
Pike	WSB	2	Atlanta.
	WAGA	5	Do.
	WQXI	11	Do.
	WTCG	17	Do.
	WATL	36	Do.
Polk	WSB	2	Do.
	WAGA	5	Do.
	WQXI	11	Do.
Pulaski	WMAZ	13	Macon.
	WRBL	3	Columbus, Ga.
Putnam	WSB	2	Atlanta.
	WAGA	5	Do.
	WQXI	11	Do.
	WMAZ	13	Macon.
Quitman	WRBL	3	Columbus, Ga.
	WTVM	9	Do.
	WSFA	12	Montgomery.
Rabun	WFBC	4	Greenville-Spartanburg-Asheville.
	WFPA	7	Do.
	WSB	2	Atlanta.
	WAGA	5	Do.
	WQXI	11	Do.
Randolph	WRBL	3	Columbus, Ga.
	WTVM	9	Do.
	WALB	10	Albany, Ga.
	WTVY	4	Dothan.
Richmond	WJBF	6	Augusta.
	WRDW	12	Do.
	WATU	26	Do.
Rockdale	WSB	2	Atlanta.
	WAGA	5	Do.
	WQXI	11	Do.
	WTCG	17	Do.
	WATL	36	Do.
Schley	WRBL	3	Columbus, Ga.
	WTVM	9	Do.
	WALB	10	Albany, Ga.
Screven	WJBF	6	Augusta.
	WRDW	12	Do.
	WSAV	3	Savannah.
	WTOC	11	Do.
Seminole	WTVY	4	Dothan.
	WALB	10	Albany, Ga.
	WCTV	6	Tallahassee.
Spalding	WSB	2	Atlanta.
	WAGA	5	Do.
	WQXI	11	Do.
	WATL	36	Do.
Stephens	WFBC	4	Greenville-Spartanburg-Asheville.
	WSPA	7	Do.
	WLOS	13	Do.
Stewart	WRBL	3	Columbus, Ga.
	WTVM	9	Do.
Sumter	WRBL	3	Do.
	WTVM	9	Do.
	WALB	10	Albany, Ga.
Talbot	WRBL	3	Columbus, Ga.
	WTVM	9	Do.
	WSB	2	Atlanta.
	WAGA	5	Do.
	WQXI	11	Do.
Taliaferro	WJBF	6	Augusta.
	WRDW	12	Do.
	WAGA	5	Atlanta.
Tattnall	WSAV	3	Savannah.
	WTOC	11	Do.
Taylor	WRBL	3	Columbus, Ga.
	WTVM	9	Do.
	WMAZ	13	Macon.
Telfair	WALB	10	Albany, Ga.
	WMAZ	13	Macon.
Terrell	WRLB	3	Columbus, Ga.
	WTVM	9	Do.
	WALB	10	Albany, Ga.
Thomas	WCTV	6	Tallahassee.
	WALB	10	Albany, Ga.
Tift	WALB	10	Do.
	WCTV	6	Tallahassee.
Toombs	WSAV	3	Savannah.
	WTOC	11	Do.
	WJCL	22	Do.
	WJBF	6	Augusta.
	WRDW	12	Do.
Towns	WSB	2	Atlanta.
	WAGA	5	Do.
	WQXI	11	Do.
	WRCB	3	Chattanooga.
Treutlen	WMAZ	13	Macon.
	WJBF	6	Augusta.
	WRDW	12	Do.
Troup	WSB	2	Atlanta.
	WAGA	5	Do.
	WQXI	11	Do.
	WRBL	3	Columbus, Ga.
	WTVM	9	Do.
Turner	WALB	10	Albany, Ga.
	WRBL	3	Columbus, Ga.
	WTVM	9	Do.
	WCTV	6	Tallahassee.
Twiggs	WMAZ	13	Macon.
	WCWB	41	Do.
Union	WSB	2	Atlanta.
	WAGA	5	Do.
	WQXI	11	Do.
	WRCB	3	Chattanooga.
Upson	WSB	2	Atlanta.
	WAGA	5	Do.
	WQXI	11	Do.
	WRBL	3	Columbus, Ga.
	WTVM	9	Do.
	WMAZ	13	Macon.
Walker	WRCB	3	Chattanooga.
	WTVC	9	Do.
	WDEF	12	Do.
Walton	WSB	2	Atlanta.
	WAGA	5	Do.
	WQXI	11	Do.
	WTCG	17	Do.
	WATL	36	Do.
Ware	WJXT	4	Jacksonville, Fla.
	WFGA	12	Do.
Warren	WJBF	6	Augusta.
	WRDW	12	Do.
	WATU	26	Do.
Washington	WJBF	6	Do.
	WRDW	12	Do.
	WMAZ	13	Macon.

County	Call letters, channel number, and market name		

GEORGIA—continued

County	Call letters	Ch.	Market name
Wayne	WSAV	3	Savannah.
	WTOC	11	Do.
	WJCL	22	Do.
	WJXT	4	Jacksonville, Fla.
Webster	WRBL	3	Columbus, Ga.
	WTVM	9	Do.
Wheeler	WMAZ	13	Macon.
	WJBF	6	Augusta.
White	WSB	2	Atlanta.
	WAGA	5	Do.
	WQXI	11	Do.
Whitfield	WRCB	3	Chattanooga.
	WTVC	9	Do.
	WDEF	12	Do.
	WSB	2	Atlanta.
	WAGA	5	Do.
Wilcox	WMAZ	13	Macon.
	WALB	10	Albany, Ga.
	WRBL	3	Columbus, Ga.
	WTVM	9	Do.
Wilkes	WJBF	6	Augusta.
	WRDW	12	Do.
	WFBC	4	Greenville-Spartan-burg-Asheville.
Wilkinson	WMAZ	13	Macon.
	WCWB	41	Do.
Worth	WALB	10	Albany, Ga.
	WTVM	9	Columbus, Ga.
	WCTV	6	Tallahassee.
	WRBL	3	Columbus.

HAWAII

County	Call letters	Ch.	Market name
Hawaii 1	KHON+	2	Honolulu.
	KHVH+	4	Do.
	KGMB+	9	Do.
Hawaii 2	KHON+	2	Do.
	KHVH+	4	Do.
	KGMB+	9	Do.
Hawaii 3	KHON+	2	Do.
	KHVH+	4	Do.
	KGMB+	9	Do.
Hawaii 4	KHON+	2	Do.
	KHVH+	4	Do.
	KGMB+	9	Do.
Hawaii 5	KHVH+	4	Fo.
Honolulu 1	KHON+	2	Do.
	KHVH+	4	Do.
	KGMB+	9	Do.
Honolulu 2	KHON+	2	Do.
	KHVH+	4	Do.
	KGMB+	9	Do.
	KIKU+	13	Do.
Honolulu 3	KHON+	2	Do.
	KHVH+	4	Do.
	KGMB+	9	Do.
Honolulu 4	KHON+	2	Do.
	KHVH+	4	Do.
	KGMB+	9	Do.
	KIKU+	13	Do.
Kauai	KHON+	2	Do.
	KHVH+	4	Do.
	KGMB+	9	Do.
Maui 1	KHON+	2	Do.
	KHVH+	4	Do.
	KGMB+	9	Do.
	KIKU+	13	Do.
Maui 2	KHON+	2	Do.
	KHVN+	4	Do.
	KGMB+	9	Do.
Maui 3	KHON+	2	Do.
	KHVH+	4	Do.
	KGMB+	9	Do.
Maui 4	KHON+	2	Do.
	KHVH+	4	Do.
	KGMB+	9	Do.

Census County Divisions in Split Counties

Hawaii 1: North Kona, South Kona.
Hawaii 2: Keaau-Mountain View, Pahoa-Kalapana.
Hawaii 3: Hilo North Hilo, Papaikow-Wailea.
Hawaii 4: Honokaa-Kukuihaela, North Hohala, Paau-hau-Paauilo, South Kohala.
Hawaii 5: Kau.
Honolulu 1: Koolaupoko.
Honolulu 2: Koolauloa, Waialua, Wahiawa.
Honolulu 3: Waianae.
Honolulu 4: Ewa, Honolulu.
Maui 1: Lahaina, Hanai City.
Maui 2: Kahaului, Kihei, Puunene, Sprecklesville, Waihee-Waikapu, Wailuku.
Maui 3: Hoiku-Pauwela, Hana, Kula, Makawao-Paia.
Maui 4: East Molokai, West Molokai, Kalawao.

IDAHO

County	Call letters	Ch.	Market name
Ada	KBOI	2	Boise.
	KTVB+	7	Do.
Adams	KBOI	2	Do.
	KTVB+	7	Do.
Bannock	KID	3	Idaho Falls-Pocatello.
	KTLE	6	Do.
	KIFI	8	Do.
Bear Lake	KUTV	2	Salt Lake City.
	KCPX	4	Do.
	KSL	5	Do.
Benewah	KREM	2	Spokane.
	KXLY	4	Do.
	KHQ	6	Do.
Bingham	KID	3	Idaho Falls-Pocatello.
	KIFI	8	Do.
Blaine	KMVT	11	Twin Falls.
	KID	3	Idaho Falls-Pocatello.
Boise	KBOI	2	Boise.
	KTVB+	7	Do.
Bonner	KREM	2	Spokane.
	KXLY	4	Do.
	KHQ	6	Do.
Bonneville	KID	3	Idaho Falls-Pocatello.
	KIFI	8	Do.
Boundary	KREM	2	Spokane.
	KXLY	4	Do.
	KHQ	6	Do.
Butte	KID	3	Idaho Falls-Pocatello.
	KIFI	8	Do.
Camas	KMVT	11	Twin Falls.
Canyon	KBOI	2	Boise.
	KTVB+	7	Do.
Caribou	KUTV	2	Salt Lake City.
	KCPX	4	Do.
	KSL	5	Do.
	KID	3	Idaho Falls-Pocatello.
	KIFI	8	Do.
Cassia	KMVT	11	Twin Falls.
	KID	3	Idaho Falls-Pocatello.
	KTLE	6	Do.
	KIFI	8	Do.
Clark	KID	3	Do.
	KIFI	8	Do.
Clearwater	KREM	2	Spokane.
	KXLY	4	Do.
	KHQ	6	Do.
	KLEW	3	Yakima.
Custer	KID	3	Idaho Falls-Pocatello.
	KIFI	8	Do.
Elmore	KBOI	2	Boise.
	KTVB+	7	Do.
Franklin	KUTV	2	Salt Lake City.
	KCPX	4	Do.
	KSL	5	Do.
Fremont	KID	3	Idaho Falls-Pocatello.
	KIFI	8	Do.
Gem	KBOI	2	Boise.
	KTVB+	7	Do.
Gooding*	KMVT	11	Twin Falls.
Idaho	KREM	2	Spokane.
	KXLY	4	Do.
	KHQ	6	Do.
	KLEW	3	Yakima.
Jefferson	KID	3	Idaho Falls-Pocatello.
	KIFI	8	Do.
Jerome	KMVT	11	Twin Falls.
Kootenai	KREM	2	Spokane.
	KXLY	4	Do.
	KHQ	6	Do.
Latah	KREM	2	Do.
	KXLY	4	Do.
	KHQ	6	Do.
	KLEW	3	Yakima.
Lemhi	KID	3	Idaho Falls-Pocatello.
	KGVO+	13	Missoula.
Lewis	KREM	2	Spokane.
	KXLY	4	Do.
	KHQ	6	Do.
Lincoln	KMVT	11	Twin Falls.
Madison	KID	3	Idaho Falls-Pocatello.
	KIFI	8	Do.
Minidoka	KMVT	11	Twin Falls.
	KID	3	Idaho Falls-Pocatello.
	KIFI	8	Do.
Nez Perce	KLEW	3	Yakima.
	KREM	2	Spokane.
	KXLY	4	Do.
	KHQ	6	Do.
Oneida	KUTV	2	Salt Lake City.
	KCPX	4	Do.
	KSL	5	Do.
Owyhee	KBOI	2	Boise.
	KTVB+	7	Do.
Payette	KBOI	2	Boise.
	KTVB+	7	Do.
Power	KID	3	Idaho Falls-Pocatello.
	KTLE	6	Do.
	KIFI	8	Do.

IDAHO—continued

County	Call letters	Ch.	Market name
Shoshone	KREM	2	Spokane.
	KXLY	4	Do.
	KHQ	6	Do.
Teton	KID	3	Idaho Falls-Pocatello.
	KIFI	8	Do.
Twin Falls	KMVT	11	Twin Falls.
	KTVB+	7	Boise.
Valley	KBOI	2	Do.
	KTVB+	7	Do.
Washington	KBOI	2	Do.
	KTVB+	7	Do.

ILLINOIS

County	Call letters	Ch.	Market name
Adams	KHQA	7	Quincy-Hannibal.
	WGEM	10	Do.
	WJJY	14	Jacksonville, Ill.
Alexander	WSIL+	3	Paducah-Cape Girardeau-Harrisburg.
	WPSD	6	Do.
	KFVS	12	Do.
Bond	KTVI	2	St. Louis.
	KMOX	4	Do.
	KSD	5	Do.
	KPLR	11	Do.
Boone	WREX	13	Rockford.
	WTVO	17	Do.
	WCEE	23	Do.
	WGN	9	Chicago.
Brown	KHQA	7	Quincy-Hannibal.
	WGEM	10	Do.
	WJJY	14	Jacksonville, Ill.
Bureau*	WHBF	4	Davenport-Rock Island (Quad City).
	WOC	6	Do.
	WQAD	8	Do.
Calhoun	KTVI	2	St. Louis.
	KMOX	4	Do.
	KSD	5	Do.
	KPLR	11	Do.
Carroll	WHBF	4	Davenport-Rock Island (Quad City).
	WOC	6	Do.
	WQAD	8	Do.
	WREX	13	Rockford.
Cass	KHQA	7	Quincy-Hannibal.
	WGEM	10	Do.
	WJJY	14	Jacksonville, Ill.
	WMBD	31	Peoria.
	WICS	20	Springfield-Decatur-Champaign.
	WIRL	19	Peoria.
Champaign	WCIA	3	Springfield-Decatur-Champaign.
	WICD	15	Do.
	WAND	17	Do.
Christian	WCIA	3	Do.
	WAND	17	Do.
	WICS	20	Do.
Clark	WTWO	2	Terre Haute.
	WTHI	10	Do.
	WTTV	4	Indianapolis.
Clay	WTWO	2	Terre Haute.
	WTHI	10	Do.
	WTVW	7	Evansville.
	KMOX	4	St. Louis.
Clinton	KTVI	2	Do.
	KMOX	4	Do.
	KSD	5	Do.
	KPLR	11	Do.
Coles	WCIA	3	Springfield-Decatur-Champaign.
	WICD	15	Do.
	WAND	17	Do.
	WTWO	2	Terre Haute.
	WTHI	10	Do.
Cook	WBBM	2	Chicago.
	WMAQ	5	Do.
	WLS	7	Do.
	WGN	9	Do.
	WFLD	32	Do.
Crawford	WTWO	2	Terre Haute.
	WTHI	10	Do.
Cumberland	WTWO	2	Do.
	WTHI	10	Do.
	WCIA	3	Springfield-Decatur-Champaign.
	WICS	20	Do.
	WAND	17	Do.
De Kalb	WBBM	2	Chicago.
	WMAQ	5	Do.
	WLS	7	Do.
	WGN	9	Do.
	WREX	13	Rockford.
	WTVO	17	Do.
	WCEE	23	Do.

SIGNIFICANTLY VIEWED SIGNALS—Continued

ILLINOIS—continued

County	Call letters	Channel	Market name
De Witt	WCIA	3	Springfield-Decatur-Champaign.
	WAND	17	Do.
	WICS	20	Do.
	WEEK	25	Peoria.
Douglas	WCIA	3	Springfield-Decatur-Champaign.
	WICD	15	Do.
	WAND	17	Do.
Du Page	WBBM	2	Chicago.
	WMAQ	5	Do.
	WLS	7	Do.
	WGN	9	Do.
	WFLD	32	Do.
Edgar	WTWO	2	Terre Haute.
	WTHI	10	Do.
	WTTV	4	Indianapolis.
	WCIA	3	Springfield-Decatur-Champaign.
Edwards	WTVW	7	Evansville.
	WFIE	14	Do.
	WEHT	25	Do.
Effingham	WCIA	3	Springfield-Decatur-Champaign.
	WTWO	2	Terre Haute.
	WTHI	10	Do.
Fayette	KTVI	2	St. Louis.
	KMOX	4	Do.
	KSD	5	Do.
	KPLR	11	Do.
Ford	WCIA	3	Springfield-Decatur-Champaign.
	WICD	15	Do.
	WAND	17	Do.
Franklin	WSIL+	3	Paducah-Cape Girardeau-Harrisburg.
	WPSD	6	Do.
	KFVS	12	Do.
Fulton	WIRL	19	Peoria.
	WEEK+	25	Do.
	WMBD	31	Do.
Gallatin	WSIL+	3	Paducah-Cape Girardeau-Harrisburg.
	WPSD	6	Do.
	KFVS	12	Do.
	WTVW	7	Evansville.
	WFIE	14	Do.
	WEHT	25	Do.
Greene	KTVI	2	St. Louis.
	KMOX	4	Do.
	KSD	5	Do.
	KPLR	11	Do.
Grundy	WBBM	2	Chicago.
	WMAQ	5	Do.
	WLS	7	Do.
	WGN	9	Do.
	WFLD	32	Do.
Hamilton	WSIL+	3	Paducah-Cape Girardeau-Harrisburg.
	WPSD	6	Do.
	KFVS	12	Do.
	WTVW	7	Evansville.
Hancock	KHQA	7	Quincy-Hannibal.
	WGEM	10	Do.
	WJJY	14	Jacksonville, Ill.
Hardin	WSIL+	3	Paducah-Cape Girardeau-Harrisburg.
	WPSD	6	Do.
	KFVS	12	Do.
Henderson	WHBF	4	Davenport-Rock Island (Quad City).
	WOC	6	Do.
	WQAD	8	Do.
Henry	WHBF	4	Do.
	WOC	6	Do.
	WQAD	8	Do.
Iroquois	WCIA	3	Springfield-Decatur-Champaign.
	WICD	15	Do.
	WBBM	2	Chicago.
	WMAQ	5	Do.
	WLS	7	Do.
	WGN	9	Do.
	WFLD	32	Do.
Jackson	WSIL+	3	Paducah-Cape Girardeau-Harrisburg.
	WPSD	6	Do.
	KFVS	12	Do.
	KPLR	11	St. Louis.
Jasper	WTWO	2	Terre Haute.
	WTHI	10	Do.
Jefferson	KTVI	2	St. Louis.
	KMOX	4	Do.
	KSD	5	Do.
	KPLR	11	Do.

County	Call letters	Channel	Market name
	WSIL+	3	Paducah-Cape Girardeau-Harrisburg.
	WPSD	6	Do.
	KFVS	12	Do.
Jersey	KTVI	2	St. Louis.
	KMOX	4	Do.
	KSD	5	Do.
	KPLR	11	Do.
	KDNL	30	Do.
Jo Daviess	WHBF	4	Davenport-Rock Island (Quad City).
	WOC	6	Do.
	WQAD	8	Do.
	WISC	3	Madison.
	WREX	13	Rockford.
	WTVO	17	Do.
Johnson	WSIL+	3	Paducah-Cape Girardeau-Harrisburg.
	WPSD	6	Do.
	KFVS	12	Do.
Kane	WBBM	2	Chicago.
	WMAQ	5	Do.
	WLS	7	Do.
	WGN	9	Do.
	WFLD	32	Do.
Kankakee	WBBM	2	Chicago.
	WMAQ	5	Do.
	WLS	7	Do.
	WGN	9	Do.
	WFLD	32	Do.
Kendall	WBBM	2	Chicago.
	WMAQ	5	Do.
	WLS	7	Do.
	WGN	9	Do.
	WFLD	32	Do.
Knox	WHBF	4	Davenport-Rock Island (Quad City).
	WOC	6	Do.
	WQAD	8	Do.
	WIRL	19	Peoria.
	WEEK+	25	Do.
	WMBD	31	Do.
Lake	WBBM	2	Chicago.
	WMAQ	5	Do.
	WLS	7	Do.
	WGN	9	Do.
	WFLD	32	Do.
La Salle	WBBM	2	Chicago.
	WMAQ	5	Do.
	WLS	7	Do.
	WGN	9	Do.
	WHBF	4	Davenport-Rock Island (Quad City).
	WEEK+	25	Peoria.
Lawrence	WTWO	2	Terre Haute.
	WTHI	10	Do.
	WTVW	7	Evansville.
Lee	WHBF	4	Davenport-Rock Island (Quad City).
	WOC	6	Do.
	WQAD	8	Do.
	WGN	9	Chicago.
	WREX	13	Rockford.
	WTVO	17	Do.
	WCEE	23	Do.
Livingston	WBBM	2	Chicago.
	WMAQ	5	Do.
	WLS	7	Do.
	WGN	9	Do.
	WIRL	19	Peoria.
	WEEK+	25	Do.
	WMBD	31	Do.
	WCIA	3	Springfield-Decatur-Champaign.
Logan	WIRL	19	Peoria.
	WEEK+	25	Do.
	WMBD	31	Do.
	WCIA	3	Springfield-Decatur-Champaign.
	WAND	17	Do.
	WICS	20	Do.
McDonough	KHQA	7	Quincy-Hannibal.
	WGEM	10	Do.
	WHBF	4	Davenport-Rock Island (Quad City).
	WOC	6	Do.
	WQAD	8	Do.
McHenry	WBBM	2	Chicago.
	WMAQ	5	Do.
	WLS	7	Do.
	WGN	9	Do.
	WFLD	32	Do.
McLean	WIRL	19	Peoria.
	WEEK+	25	Do.
	WMBD	31	Do.
	WCIA	3	Springfield-Decatur-Champaign.
	WAND	17	Do.
Macon	WCIA	3	Do.
	WAND	17	Do.
	WICS	20	Do.

County	Call letters	Channel	Market name
Macoupin	KTVI	2	St. Louis.
	KMOX	4	Do.
	KSD	5	Do.
	KPLR	11	Do.
	KDNL	30	Do.
Madison	KTVI	2	Do.
	KMOX	4	Do.
	KSD	5	Do.
	KPLR	11	Do.
	KDNL	30	Do.
Marion	KTVI	2	Do.
	KMOX	4	Do.
	KSD	5	Do.
	KPLR	11	Do.
Marshall	WIRL	19	Peoria.
	WEEK	25	Do.
	WMBD	31	Do.
Mason	WIRL	19	Peoria.
	WEEK+	25	Do.
	WMBD	31	Do.
Massac	WSIL+	3	Paducah-Cape Girardeau-Harrisburg.
	WPSD	6	Do.
	KFVS	12	Do.
Menard	WIRL	19	Peoria.
	WEEK+	25	Do.
	WMBD	31	Do.
	WAND	17	Springfield-Decatur-Champaign.
	WICS	20	Do.
Mercer	WHBF	4	Davenport-Rock Island (Quad City).
	WOC	6	Do.
	WQAD	8	Do.
Monroe	KTVI	2	St. Louis.
	KMOX	4	Do.
	KSD	5	Do.
	KPLR	11	Do.
	KDNL	30	Do.
Montgomery	KTVI	2	Do.
	KMOX	4	Do.
	KSD	5	Do.
	KPLR	11	Do.
	WICS	20	Springfield-Decatur-Champaign.
Morgan	KHQA	7	Quincy-Hannibal.
	WGEM	10	Do.
	WJJY	14	Jacksonville, Ill.
	KTVI	2	St. Louis.
	KPLR	11	Do.
	WICS	20	Springfield-Decatur-Champaign.
Moultrie	WCIA	3	Do.
	WICD	15	Do.
	WAND	17	Do.
	WICS	20	Do.
Ogle	WREX	13	Rockford.
	WTVO	17	Do.
	WCEE	23	Do.
Peoria	WIRL	19	Peoria.
	WEEK+	25	Do.
	WMDD	31	Do.
Perry	KTVI	2	St. Louis.
	KMOX	4	Do.
	KSD	5	Do.
	KPLR	11	Do.
	WSIL+	3	Paducah-Cape Girardeau-Harrisburg.
	KFVS	11	Do.
Piatt	WCIA	3	Springfield-Decatur-Champaign.
	WICD	15	Do.
	WAND	17	Do.
	WICS	20	Do.
Pike	KHQA	7	Quincy-Hannibal.
	WGEM	10	Do.
	WJJY	14	Jacksonville, Ill.
	KTVI	2	St. Louis.
	KSD	5	Do.
	KPLR	11	Do.
Pope	WSIL+	3	Paducah-Cape Girardeau-Harrisburg.
	WPSD	6	Do.
	KFVS	12	Do.
Pulaski	WSIL+	3	Do.
	WPSD	6	Do.
	KFVS	12	Do.
Putnam	WIRL	19	Peoria.
	WEEK+	25	Do.
	WMBD	31	Do.
	WHBF	4	Davenport-Rock Island (Quad City).
	WOC	6	Do.
	WQAD	8	Do.
Randolph	KTVI	2	St. Louis.
	KMOX	4	Do.
	KSD	5	Do.
	KPLR	11	Do.
	KDNL	30	Do.

County	Call letters	Channel	Market name
ILLINOIS—continued			
Richland	WTWO	2	Terre Haute.
	WTHI	10	Do.
	WTVW	7	Evansville.
Rock Island	WHBF	4	Davenport-Rock Island (Quad City).
	WOC	6	Do.
	WQAD	8	Do.
St. Clair	KTVI	2	St. Louis.
	KMOX	4	Do.
	KSD	5	Do.
	KPLR	11	Do.
	KDNL	30	Do.
Saline	WSIL+	3	Paducah-Cape Girardeau-Harrisburg.
	WPSD	6	Do.
	KFVS	12	Do.
Sangamon	WCIA	3	Springfield-Decatur-Champaign.
	WAND	17	Do.
	WICS	20	Do.
Schuyler	KHQA	7	Quincy-Hannibal.
	WGEM	10	Do.
	WJJY	14	Jacksonville, Ill.
Scott	KHQA	7	Quincy-Hannibal.
	WGEM	10	Do.
	WJJY	14	Jacksonville, Ill.
	KTVI	2	St. Louis.
	KSD	5	Do.
	KPLR	11	Do.
Shelby	WCIA	3	Springfield-Decatur-Champaign.
	WAND	17	Do.
	WICS	20	Do.
Stark	WHBF	4	Davenport-Rock Island (Quad City).
	WOC	6	Do.
	WQAD	8	Do.
	WIRL	19	Peoria.
	WEEK+	25	Do.
	WMBD	31	Do.
Stephenson	WREX	13	Rockford.
	WTVO	17	Do.
	WCEE	23	Do.
	WISC	3	Madison.
Tazewell	WIRL	19	Peoria.
	WEEK+	25	Do.
	WMBD	31	Do.
Union	SWIL+	3	Paducah-Cape Girardeau-Harrisburg.
	WPSD	6	Do.
	KFVS	12	Do.
Vermilion	WCIA	3	Springfield-Decatur-Champaign.
	WICD	15	Do.
	WAND	17	Do.
	WTWO	2	Terre Haute.
Wabash	WTVW	7	Evansville.
	WFIE	14	Do.
	WEHT	25	Do.
Warren	WHBF	4	Davenport-Rock Island (Quad City).
	WOC	6	Do.
	WQAD	8	Do.
Washington	KTVI	2	St. Louis.
	KMOX	4	Do.
	KSD	5	Do.
	KPLR	11	Do.
	KDNL	30	Do.
Wayne	WTVW	7	Evansville.
	WFIE	14	Do.
	WEHT	25	Do.
	WSIL+	3	Paducah-Cape Girardeau-Harrisburg.
	WPSD	6	Do.
	KFVS	12	Do.
White	WTVW	7	Evansville.
	WFIE	14	Do.
	WEHT	25	Do.
	WSIL+	3	Paducah-Cape Girardeau-Harrisburg.
	WPSD	6	Do.
Whiteside	WHBF	4	Davenport-Rock Island (Quad City).
	WOC	6	Do.
	WQAD	8	Do.
Will	WBBM	2	Chicago.
	WMAQ	5	Do.
	WLS	7	Do.
	WGN	9	Do.
	WFLD	32	Do.
Williamson	WSIL+	3	Paducah-Cape Girardeau-Harrisburg.
	WPSD	6	Do.
	KFVS	12	Do.

County	Call letters	Channel	Market name
ILLINOIS—continued			
Winnebago	WREX	13	Rockford.
	WTVO	17	Do.
	WCEE	23	Do.
Woodford	WIRL	19	Peoria.
	WEEK+	25	Do.
	WMBD	31	Do.
INDIANA			
Adams	WANE	15	Fort Wayne.
	WPTA	21	Do.
	WKJG	33	Do.
Allen	WANE	15	Do.
	WPTA	21	Do.
	WKJG	33	Do.
Bartholomew	WTTV	4	Indianapolis.
	WFBM	6	Do.
	WISH	8	Do.
	WLWI	13	Do.
Benton	WTTV	4	Do.
	WFBM	6	Do.
	WLWI	13	Do.
	WGN	9	Chicago.
	WLFI	18	Lafayette, Ind.
	WCIA	3	Springfield-Decatur-Champaign.
	WICD	15	Do.
Blackford	WTTV	4	Indianapolis.
	WFBM	6	Do.
	WISH	8	Do.
	WLWI	13	Do.
	WANE	15	Fort Wayne.
	WPTA	21	Do.
	WKJG	33	Do.
Boone	WTTV	4	Indianapolis.
	WFBM	6	Do.
	WISH	8	Do.
	WLWI	13	Do.
Brown	WTTV	4	Do.
	WFBM	6	Do.
	WISH	8	Do.
	WLWI	13	Do.
Carroll	WTTV	4	Do.
	WFBM	6	Do.
	WISH	8	Do.
	WLWI	13	Do.
	WLFI	18	Lafayette, Ind.
Cass	WFBM	6	Indianapolis.
	WISH	8	Do.
	WLWI	13	Do.
	WLFI	18	Lafayette, Ind.
Clark	WAVE	3	Louisville.
	WHAS	11	Do.
	WLKY	32	Do.
Clay	WTWO	2	Terre Haute.
	WTHI	10	Do.
	WTTV	4	Indianapolis.
	WFBM	6	Do.
	WISH	8	Do.
	WLWI	13	Do.
Clinton	WTTV	4	Do.
	WFBM	6	Do.
	WISH	8	Do.
	WLWI	13	Do.
Crawford	WAVE	3	Louisville.
	WHAS	11	Do.
	WLKY	32	Do.
	WTVW	7	Evansville.
Daviess	WTWO	2	Terre Haute.
	WTHI	10	Do.
	WTVW	7	Evansville.
	WTTV	4	Indianapolis.
Dearborn	WLWT	5	Cincinnati.
	WCPO	9	Do.
	WKRC	12	Do.
	WXIX	19	Do.
Decatur	WTTV	4	Indianapolis.
	WFBM	6	Do.
	WISH	8	Do.
	WLWI	13	Do.
De Kalb	WANE	15	Fort Wayne.
	WPTA	21	Do.
	WKJG	33	Do.
Delaware	WTTV	4	Indianapolis.
	WFBM	6	Do.
	WISH	8	Do.
	WLWI	13	Do.
Dubois	WTVW	7	Evansville.
	WFIE	14	Do.
	WEHT	25	Do.
	WTTV	4	Indianapolis.
	WAVE	3	Louisville.
	WHAS	11	Do.
	WTWO	2	Terre Haute.
	WTHI	10	Do.

County	Call letters	Channel	Market name
INDIANA—continued			
Elkhart	WNDU	16	South Bend-Elkhart.
	WSBT	22	Do.
	WSJV	28	Do.
Fayette	WLWT	5	Cincinnati.
	WCPO	9	Do.
	WKRC	12	Do.
	WHIO	7	Dayton.
	WTTV	4	Indianapolis.
	WFBM	6	Do.
	WISH	8	Do.
Floyd	WAVE	3	Louisville.
	WHAS	11	Do.
	WLKY	32	Do.
Fountain	WTTV	4	Indianapolis.
	WFBM	6	Do.
	WLWI	13	Do.
	WCIA	3	Springfield-Decatur-Champaign.
	WTWO	2	Terre Haute.
	WTHI	10	Do.
Franklin	WLWT	5	Cincinnati.
	WCPO	9	Do.
	WKRC	12	Do.
	WTTV	4	Indianapolis.
Fulton	WNDU	16	South Bend-Elkhart.
	WSBT	22	Do.
	WSJV	28	Do.
Gibson	WTVW	7	Evansville.
	WFIE	14	Do.
	WEHT	25	Do.
Grant	WTTV	4	Indianapolis.
	WFBM	6	Do.
	WISH	8	Do.
	WLWI	13	Do.
Greene	WTWO	2	Terre Haute.
	WTHI	10	Do.
	WTVW	7	Evansville.
	WTTV	4	Indianapolis.
	WFBM	6	Do.
	WLWI	13	Do.
Hamilton	WTTV –	4	Do.
	WFBM	6	Do.
	WISH	8	Do.
	WLWI	13	Do.
Hancock	WTTV	4	Do.
	WFBM	6	Do.
	WISH	8	Do.
	WLWI	13	Do.
Harrison	WAVE	3	Louisville.
	WHAS	11	Do.
	WLKY	32	Do.
Hendricks	WTTV	4	Indianpolis.
	WFBM	6	Do.
	WISH	8	Do.
	WLWI	13	Do.
Henry	WTTV	4	Do.
	WFBM	6	Do.
	WISH	8	Do.
	WLWI	13	Do.
Howard	WTTV	4	Do.
	WFBM	6	Do.
	WISH	8	Do.
	WLWI	13	Do.
Huntington	WANE	15	Fort Wayne.
	WPTA	21	Do.
	WKJG	33	Do.
Jackson	WAVE	3	Louisville.
	WHAS	11	Do.
	WLKY	32	Do.
	WTTV	4	Indianapolis.
	WFBM	6	Do.
	WISH	8	Do.
	WLWI	13	Do.
Jasper	WBBM	2	Chicago.
	WMAQ	5	Do.
	WLS	7	Do.
	WGN	9	Do.
Jay	WTTV	4	Indianapolis.
	WFBM	6	Do.
	WISH	8	Do.
	WLWI	13	Do.
	WANE	15	Fort Wayne.
	WPTA	21	Do.
	WKJG	33	Do.
Jefferson	WAVE	3	Louisville.
	WHAS	11	Do.
	WLKY	32	Do.
	WLWT	5	Cincinnati.
	WCPO	9	Do.
	WKRC	12	Do.
	WTTV	4	Indianapolis.
Jennings	WTTV	4	Do.
	WFBM	6	Do.
	WISH	8	Do.
	WLWI	13	Do.
	WAVE	3	Louisville.
	WHAS	11	Do.
	WLKY	32	Do.
Johnson	WTTV	4	Indianapolis.
	WFBM	6	Do.
	WISH	8	Do.
	WLWI	13	Do.

County	Call letters, channel number and market name		
INDIANA—continued			
Knox	WTWO	2	Terre Haute.
	WTHI	10	Do.
	WTVW	7	Evansville.
Kosciusko	WNDU	16	South Bend-Elkhart.
	WSBT	22	Do.
	WSJV	28	Do.
La Grange	WNDU	16	Do.
	WSBT	22	Do.
	WSJV	28	Do.
	WANE	15	Fort Wayne.
	WPTA	21	Do.
	WKJG	33	Do.
Lake	WBBM	2	Chicago.
	WMAQ	5	Do.
	WLS	7	Do.
	WGN	9	Do.
	WFLD	32	Do.
La Porte	WBBM	2	Do.
	WMAQ	5	Do.
	WLS	7	Do.
	WGN	9	Do.
	WFLD	32	Do.
	WNDU	16	South Bend-Elkhart.
	WSBT	22	Do.
Lawrence	WTTV	4	Indianapolis.
	WFBM	6	Do.
	WISH	8	Do.
	WAVE	3	Louisville.
	WHAS	11	Do.
	WTWO	2	Terre Haute.
	WTHI	10	Do.
Madison	WTTV	4	Indianapolis.
	WFBM	6	Do.
	WISH	8	Do.
	WLWI	13	Do.
Marion	WTTV	4	Indianapolis.
	WFBM	6	Do.
	WISH	8	Do.
	WLWI	13	Do.
Marshall	WNDU	16	South Bend-Elkhart.
	WSBT	22	Do.
	WSJV	28	Do.
	WGN	9	Chicago.
Martin	WTWO	2	Terre Haute.
	WTHI	10	Do.
	WTVW	7	Evansville.
	WTTV	4	Indianapolis.
	WAVE	3	Louisville.
Miami *	WTTV	4	Indianapolis.
	WFBM	6	Do.
	WISH	8	Do.
	WLWI	13	Do.
	WNDU	16	South Bend-Elkhart.
Monroe	WTTV	4	Indianapolis.
	WFBM	6	Do.
	WISH	8	Do.
	WLWI	13	Do.
	WTWO	2	Terre Haute.
	WTHI	10	Do.
Montgomery	WTTV	4	Indianapolis.
	WFBM	6	Do.
	WISH	8	Do.
	WLWI	13	Do.
Morgan	WTTV	4	Indianapolis.
	WFBM	6	Do.
	WISH	8	Do.
	WLWI	13	Do.
Newton*	WBBM	2	Chicago.
	WMAQ	5	Do.
	WLS	7	Do.
	WGN	9	Do.
Noble	WANE	15	Fort Wayne.
	WPTA	21	Do.
	WKJG	33	Do.
	WNDU	16	South Bend-Elkhart.
	WSBT	22	Do.
	WSJV	28	Do.
Ohio	WLWT	5	Cincinnati.
	WCPO	9	Do.
	WKRC	12	Do.
	WXIX	19	Do.
Orange	WAVE	3	Louisville.
	WHAS	11	Do.
	WLKY	32	Do.
	WTTV	4	Indianapolis.
Owen	WTTV	4	Do.
	WFBM	6	Do.
	WISH	8	Do.
	WLWI	13	Do.
	WTWO	2	Terre Haute.
	WTHI	10	Do.
Parke	WTWO	2	Do.
	WTHI	10	Do.
	WTTV	4	Indianapolis.
	WFBM	6	Do.
	WLWI	13	Do.
Perry	WTVW	7	Evansville.
	WFIE	14	Do.
	WEHT	25	Do.
	WAVE	3	Louisville.
	WHAS	11	Do.

County	Call letters, channel number and market name		
INDIANA—continued			
Pike	WTVW	7	Evansville.
	WFIE	14	Do.
	WEHT	25	Do.
	WTTV	4	Indianapolis.
	WTWO	2	Terre Haute.
	WTHI	10	Do.
Porter	WBBM	2	Chicago.
	WMAQ	5	Do.
	WLS	7	Do.
	WGN	9	Do.
	WFLD	32	Do.
Posey	WTVW	7	Evansville.
	WFIE	14	Do.
	WEHT	25	Do.
Pulaski	WNDU	16	South Bend-Elkhart.
	WSBT	22	Do.
	WSJV	28	Do.
	WBBM	2	Chicago.
	WMAQ	5	Do.
	WLS	7	Do.
	WGN	9	Do.
Putnam	WTTV	4	Indianapolis.
	WFBM	6	Do.
	WISH	8	Do.
	WLWI	13	Do.
	WTWO	2	Terre Haute.
	WTHI	10	Do.
Randolph	WTTV	4	Indianapolis.
	WFBM	6	Do.
	WISH	8	Do.
	WLWI	13	Do.
	WLWD	2	Dayton.
	WHIO	7	Do.
Ripley	WLWT	5	Cincinnati.
	WCPO	9	Do.
	WKRC	12	Do.
	WXIX	19	Do.
	WTTV	4	Indianapolis.
Rush	WTTV	4	Do.
	WFBM	6	Do.
	WISH	8	Do.
	WLWI	13	Do.
St. Joseph	WNDU	16	South Bend-Elkhart.
	WSBT	22	Do.
	WSJV	28	Do.
	WGN	9	Chicago.
Scott	WAVE	3	Louisville.
	WHAS	11	Do.
	WLKY	32	Do.
	WTTV	4	Indianapolis.
Shelby	WTTV	4	Do.
	WFBM	6	Do.
	WISH	8	Do.
	WLWI	13	Do.
Spencer	WTVW	7	Evansville.
	WFIE	14	Do.
	WEHT	25	Do.
Starke	WNDU	16	South Bend-Elkhart.
	WSBT	22	Do.
	WSJV	28	Do.
	WBBM	2	Chicago.
	WMAQ	5	Do.
	WLS	7	Do.
	WGN	9	Do.
Steuben	WANE	15	Fort Wayne.
	WPTA	21	Do.
	WKJG	33	Do.
	WKZO	3	Grand Rapids-Kalamazoo.
Sullivan	WTWO	2	Terre Haute.
	WTHI	10	Do.
	WTTV	4	Indianapolis.
Switzerland	WLWT	5	Cincinnati.
	WCPO	9	Do.
	WKRC	12	Do.
	WXIX	19	Do.
Tippecanoe	WLFI	18	Lafayette, Ind.
	WTTV	4	Indianapolis.
	WFBM	6	Do.
	WISH	8	Do.
	WLWI	13	Do.
Tipton	WTTV	4	Do.
	WFBM	6	Do.
	WISH	8	Do.
	WLWI	13	Do.
Union	WLWT	5	Cincinnati.
	WCPO	9	Do.
	WKRC	12	Do.
	WXIX	19	Do.
	WHIO	7	Dayton.
	WKTR	10	Do.
	WTTV	4	Indianapolis.
Vanderburgh	WTVW	7	Evansville.
	WFIE	14	Do.
	WEHT	25	Do.
Vermillion	WTWO	2	Terre Haute.
	WTHI	10	Do.
	WTTV	4	Indianapolis.
	WFBM	6	Do.
	WLWI	13	Do.
	WCIA	3	Springfield-Decatur-Champaign.

County	Call letters, channel number and market name		
INDIANA—continued			
Vigo	WTWO	2	Terre Haute.
	WTHI	10	Do.
	WTTV	4	Indianapolis.
Wabash	WFBM	6	Do.
	WISH	8	Do.
	WLWI	13	Do.
	WANE	15	Fort Wayne.
	WPTA	21	Do.
	WKJG	33	Do.
Warren	WTTV	4	Indianapolis.
	WFBM	6	Do.
	WLWI	13	Do.
	WCIA	3	Springfield-Decatur-Champaign.
	WICD	15	Do.
	WTHI	10	Terre Haute.
Warrick	WTVW	7	Evansville.
	WFIE	14	Do.
	WEHT	25	Do.
Washington	WAVE	3	Louisville.
	WHAS	11	Do.
	WLKY	32	Do.
	WTTV	4	Indianapolis.
Wayne	WLWD	2	Dayton.
	WHIO	7	Do.
	WLWT	5	Cincinnati.
	WCPO	9	Do.
	WKRC	12	Do.
	WTTV	4	Indianapolis.
	WFBM	6	Do.
	WISH	8	Do.
	WLWI	13	Do.
Wells	WANE	15	Fort Wayne.
	WPTA	21	Do.
	WKJG	33	Do.
White	WTTV	4	Indianapolis.
	WFBM	6	Do.
	WISH	8	Do.
	WLWI	13	Do.
	WGN	9	Chicago.
	WLFI	18	Lafayette, Ind.
Whitley	WANE	15	Fort Wayne.
	WPTA	21	Do.
	WKJG	33	Do.
IOWA			
Adair	WOI	5	Des Moines.
	KRNT	8	Do.
	WHO	13	Do.
Adams	KMTV	3	Omaha.
	WOW	6	Do.
	KETV	7	Do.
Allamakee	WMT	2	Cedar Rapids-Waterloo.
	KWWL	7	Do.
	KCRG	9	Do.
	WKBT	8	La Crosse-Eau Claire.
	KROC	10	Rochester-Mason City-Austin.
Appanoose	KTVO	3	Ottumwa-Kirksville.
	KRNT	8	Des Moines.
	WHO	13	Do.
Audubon	KMTV	3	Omaha.
	WOW	6	Do.
	KETV	7	Do.
Benton	WMT	2	Cedar Rapids-Waterloo.
	KWWL	7	Do.
	KCRG	9	Do.
Black Hawk	WMT	2	Do.
	KWWL	7	Do.
	KCRG	9	Do.
Boone	WOI	5	Des Moines.
	KRNT	8	Do.
	WHO	13	Do.
Bremer	WMT	2	Cedar Rapids-Waterloo.
	KWWL	7	Do.
	KCRG	9	Do.
Buchanan	WMT	2	Do.
	KWWL	7	Do.
	KCRG	9	Do.
Buena Vista	KTIV	4	Sioux City.
	KCAU	9	Do.
	KMEG	14	Do.
Butler	WMT	2	Cedar Rapids-Waterloo.
	KWWL	7	Do.
	KCRG	9	Do.
	KGLO	3	Rochester-Mason City-Austin.
Calhoun	WOI	5	Des Moines.
	KRNT	8	Do.
	KVFD	21	Fort Dodge.
	KTIV	4	Sioux City.
	KCAU	9	Do.

SIGNIFICANTLY VIEWED SIGNALS—Continued

IOWA—continued

County	Call letters	Channel	Market name
Carroll	WOI	5	Des Moines.
	KRNT	8	Do.
	WHO	13	Do.
	WOW	6	Omaha.
Cass	KMTV	3	Do.
	WOW	6	Do.
	KETV	7	Do.
Cedar	WMT	2	Cedar Rapids-Waterloo.
	KWWL	7	Do.
	KCRG	9	Do.
	WHBF	4	Davenport-Rock Island (Quad City).
	WOC	6	Do.
	WQAD	8	Do.
Cerro Gordo	KGLO	3	Rochester-Mason City-Austin
	KAUS	6	Do.
	KROC	10	Do.
Cherokee	KTIV	4	Sioux City.
	KCAU	9	Do.
	KMEG	14	Do.
	KELO+	11	Sioux Falls-Mitchell.
	KSOO	13	Do.
Chickasaw	WMT	2	Cedar Rapids-Waterloo.
	KWWL	7	Do.
	KCRG	9	Do.
	KGLO	3	Rochester-Mason City-Austin.
	KROC	10	Do.
Clarke	WOI	5	Des Moines.
	KRNT	8	Do.
	WHO	13	Do.
Clay	KTIV	4	Sioux City.
	KCAU	9	Do.
	KELO+	11	Sioux Falls-Mitchell.
	KSOO+	13	Do.
Clayton	WMT	2	Cedar Rapids-Waterloo.
	KWWL	7	Do.
	KCRG	9	Do.
Clinton	WHBF	4	Davenport-Rock Island (Quad City).
	WOC	6	Do.
	WQAD	8	Do.
Crawford	KMTV	3	Omaha.
	WOW	6	Do.
	KETV	7	Do.
	KTIV	4	Sioux City.
	KCAU	9	Do.
Dallas	WOI	5	Des Moines.
	KRNT	8	Do.
	WHO	13	Do.
Davis	KTVO	3	Ottumwa-Kirksville.
	KRNT	8	Des Moines.
	WHO	13	Do.
	KHQA	7	Quincy-Hannibal.
	WGEM	10	Do.
Decatur	KRNT	8	Des Moines.
	WHO	13	Do.
	KTVO	3	Ottumwa-Kirksville.
Delaware	WMT	2	Cedar Rapids-Waterloo.
	KWWL	7	Do.
	KCRG	9	Do.
Des Moines	WHBF	4	Davenport-Rock Island (Quad City).
	WOC	6	Do.
	WQAD	8	Do.
Dickinson	KTIV	4	Sioux City.
	KCAU	9	Do.
	KEYC	12	Mankato.
	KELO+	11	Sioux Falls-Mitchell.
	KSOO+	13	Do.
Dubuque	WMT	2	Cedar Rapids-Waterloo.
	KWWL	7	Do.
	KCRG	9	Do.
	KDUB	40	Dubuque, Ia.
Emmet	KEYC	12	Mankato.
	KAUS	6	Rochester-Mason City-Austin.
	KCAU	9	Sioux City.
Fayette	WMT	2	Cedar Rapids-Waterloo.
	KWWL	7	Do.
	KCRG	9	Do.
Floyd	KGLO	3	Rochester-Mason City-Austin.
	KAUS	6	Do.
	KROC	10	Do.
	WMT	2	Cedar Rapids-Waterloo.
	KWWL	7	Do.
	KCRG	9	Do.
Franklin	KGLO	3	Rochester-Mason City-Austin.
	KROC	10	Do.
	WMT	2	Cedar Rapids-Waterloo.
	KWWL	7	Do.
	KCRG	9	Do.
	WOI	5	Des Moines.
Fremont	KMTV	3	Omaha.
	WOW	6	Do.
	KETV	7	Do.
Greene	WOI	5	Des Moines.
	KRNT	8	Do.
	WHO	13	Do.
Grundy	WMT	2	Cedar Rapids-Waterloo.
	KWWL	7	Do.
	KCRG	9	Do.
Guthrie	WOI	5	Des Moines.
	KRNT	8	Do.
	WHO	13	Do.
Hamilton	WOI	5	Des Moines.
	KRNT	8	Do.
	WHO	13	Do.
Hancock	KGLO	3	Rochester-Mason City-Austin.
	KAUS	6	Do.
	KROC	10	Do.
Hardin	WMT	2	Cedar Rapids-Waterloo.
	KWWL	7	Do.
	KCRG	9	Do.
	WOI	5	Des Moines.
	KRNT	8	Do.
	WHO	13	Do.
Harrison	KMTV	3	Omaha.
	WOW	6	Do.
	KETV	7	Do.
Henry	WHBF	4	Davenport-Rock Island (Quad City).
	WOC	6	Do.
	WQAD	8	Do.
	KTVO	3	Ottumwa-Kirksville.
Howard	KGLO	3	Rochester-Mason City-Austin.
	KAUS	6	Do.
	KROC	10	Do.
	KWWL	7	Cedar Rapids-Waterloo.
	KCRG	9	Do.
Humboldt	WOI	5	Des Moines.
	KRNT	8	Do.
	KVFD	21	Fort Dodge.
Ida	KTIV	4	Sioux City.
	KCAU	9	Do.
	KMEG	14	Do.
Iowa	WMT	2	Cedar Rapids-Waterloo.
	KWWL	7	Do.
	KCRG	9	Do.
Jackson	WHBF	4	Davenport-Rock Island (Quad City).
	WOC	6	Do.
	WQAD	8	Do.
	WMT	2	Cedar Rapids-Waterloo.
	KWWL	7	Do.
	KCRG	9	Do.
Jasper	WOI	5	Des Moines.
	KRNT	8	Do.
	WHO	13	Do.
Jefferson	KTVO	3	Ottumwa-Kirksville.
	WMT	2	Cedar Rapids-Waterloo.
	KCRG	9	Do.
	WOC	6	Davenport-Rock Island (Quad City).
	KHQA	7	Quincy-Hannibal.
Johnson	WMT	2	Cedar Rapids-Waterloo.
	KWWL	7	Do.
	KCRG	9	Do.
	WHBF	4	Davenport-Rock Island (Quad City).
	WOC	6	Do.
Jones	WMT	2	Cedar Rapids-Waterloo.
	KWWL	7	Do.
	KCRG	9	Do.
Keokuk	WMT	2	Do.
	KWWL	7	Do.
	KCRG	9	Do.
	WHO	13	Des Moines.
	KTVO	3	Ottumwa-Kirksville.
Kossuth	KGLO	3	Rochester-Mason City-Austin.
	KAUS	6	Do.
	KVFD	21	Fort Dodge.
	KEYC	12	Mankato.
Lee	KHQA	7	Quincy-Hannibal.
	WGEM	10	Do.
	KTVO	3	Ottumwa-Kirksville.
Linn	WMT	2	Cedar Rapids-Waterloo.
	KWWL	7	Do.
	KCRG	9	Do.
Louisa	WHBF	4	Davenport-Rock Island (Quad City).
	WOC	6	Do.
	WQAD	8	Do.
Lucas	WOI	5	Des Moines.
	KRNT	8	Do.
	WHO	13	Do.
Lyon	KELO+	11	Sioux Falls-Mitchell.
	KSOO+	13	Do.
	KTIV	4	Sioux City.
	KCAU	9	Do.
Madison	WOI	5	Des Moines.
	KRNT	8	Do.
	WHO	13	Do.
Mahaska	KRNT	8	Do.
	WHO	13	Do.
	WMT	2	Cedar Rapids-Waterloo.
	KWWL	7	Do.
	KCRG	9	Do.
	KTVO	3	Ottumwa-Kirksville.
Marion	WOI	5	Des Moines.
	KRNT	8	Do.
	WHO	13	Do.
Marshall	WMT	2	Cedar Rapids-Waterloo.
	KWWL	7	Do.
	KCRG	9	Do.
	WOI	5	Des Moines.
	KRNT	8	Do.
	WHO	13	Do.
Mills	KMTV	3	Omaha.
	WOW	6	Do.
	KETV	7	Do.
Mitchell	KGLO	3	Rochester-Mason City-Austin
	KAUS	6	Do.
	KROC	10	Do.
Monona	KMTV	3	Omaha.
	WOW	6	Do.
	KETV	7	Do.
	KTIV	4	Sioux City.
	KCAU	9	Do.
Monroe	KRNT	8	Des Moines.
	WHO	13	Do.
	KTVO	3	Ottumwa-Kirksville.
Montgomery	KMTV	3	Omaha.
	WOW	6	Do.
	KETV	7	Do.
Muscatine	WHBF	4	Davenport-Rock Island (Quad City).
	WOC	6	Do.
	WQAD	8	Do.
O'Brien	KTIV	4	Sioux City.
	KCAU	9	Do.
	KELO+	11	Sioux Falls-Mitchell.
	KSOO+	13	Do.
Osceola	KELO+	11	Do.
	KSOO+	13	Do.
	KTIV	4	Sioux City.
	KCAU	9	Do.
Page	KMTV	3	Omaha.
	WOW	6	Do.
	KETV	7	Do.
Palo Alto	KTIV	4	Sioux City.
	KCAU	9	Do.
	KVFD	21	Fort Dodge.
	KEYC	12	Mankato.
Plymouth	KTIV	4	Sioux City.
	KCAU	9	Do.
	KMEG	14	Do.
	KELO+	11	Sioux Falls-Mitchell.
Pocahontas	KTIV	4	Sioux City.
	KCAU	9	Do.
	KRNT	8	Des Moines.
	KVFD	21	Fort Dodge.
Polk	WOI	5	Des Moines.
	KRNT	8	Do.
	WHO	13	Do.
Pottawattamie	KMTV	3	Omaha.
	WOW	6	Do.
	KETV	7	Do.
Poweshiek	WMT	2	Cedar Rapids-Waterloo.
	KWWL	7	Do.
	KCRG	9	Do.
	WOI	5	Des Moines.
	KRNT	8	Do.
	WHO	13	Do.

SIGNIFICANTLY VIEWED SIGNALS—Continued

County	Call letters, channel number and market name		

IOWA—continued

County	Call letters	Channel	Market name
Ringgold	WOI	5	Do.
	KRNT	8	Do.
	WHO	13	Do.
	WOW	6	Omaha.
	KQTV	2	St. Joseph.
Sac	KTIV	4	Sioux City.
	KCAU	9	Do.
	WOW	6	Omaha.
Scott	WHBF	4	Davenport-Rock Island (Quad City).
	WOC	6	Do.
	WQAD	8	Do.
Shelby	KMTV	3	Omaha.
	WOW	6	Do.
	KETV	7	Do.
Sioux	KTIV	4	Sioux City.
	KCAU	9	Do.
	KMEG	14	Do.
	KELO+	11	Sioux Falls-Mitchell.
	KSOO+	13	Do.
Story	WOI	5	Des Moines.
	KRNT	8	Do.
	WHO	13	Do.
Tama	WMT	2	Cedar Rapids-Waterloo.
	KWWL	7	Do.
	KCRG	9	Do.
	WHO	13	Des Moines.
Taylor	KMTV	3	Omaha.
	WOW	6	Do.
	KETV	7	Do.
Union	WOI	5	Des Moines.
	KRNT	8	Do.
	WHO	13	Do.
	KETV	7	Omaha.
Van Buren	KTVO	3	Ottumwa-Kirksville.
	KHQA	7	Quincy-Hannibal.
	WGEM	10	Do.
Wapello	KRNT	8	Des Moines.
	WHO	13	Do.
	KTVO	3	Ottumwa-Kirksville.
Warren	WOI	5	Des Moines.
	KRNT	8	Do.
	WHO	13	Do.
Washington	WMT	2	Cedar Rapids-Waterloo.
	KWWL	7	Do.
	KCRG	9	Do.
	WHBF	4	Davenport-Rock Island (Quad City).
	WOC	6	Do.
Wayne	KRNT	8	Des Moines.
	WHO	13	Do.
	KTVO	3	Ottumwa-Kirksville.
Webster	WOI	5	Des Moines.
	KRNT	8	Do.
	KVFD	21	Fort Dodge.
Winnebago	KGLO	3	Rochester-Mason City-Austin.
	KAUS	6	Do.
	KROC	10	Do.
	KEYC	12	Mankato.
Winneshiek	WMT	2	Cedar Rapids-Waterloo.
	KWWL	7	Do.
	KCRG	9	Do.
	WKBT	8	La Crosse-Eau Claire.
	KGLO	3	Rochester-Mason City-Austin.
	KROC	10	Do.
Woodbury	KTIV	4	Sioux City.
	KCAU	9	Do.
	KMEG	14	Do.
Worth	KGLO	3	Rochester-Mason City-Austin.
	KAUS	6	Do.
	KROC	10	Do.
Wright*	WOI	5	Des Moines.
	KVFD	21	Fort Dodge.
	KGLO	3	Rochester-Mason City-Austin.

KANSAS

County	Call letters	Channel	Market name
Allen	KOAM	7	Joplin-Pittsburg.
	KODE	12	Do.
Anderson	WDAF	4	Kansas City.
	KCMO	5	Do.
	KMBC	9	Do.
	WIBW	13	Topeka.
Atchison	WDAF	4	Kansas City.
	KCMO	5	Do.
	KMBC	9	Do.
	KQTV	2	St. Joseph.
	KTSB	27	Topeka.
Barber	KARD	3	Wichita-Hutchinson.
	KAKE	10	Do.
	KTVH	12	Do.
	KTEN	10	Ardmore-Ada.

SIGNIFICANTLY VIEWED SIGNALS—Continued

KANSAS—continued

County	Call letters	Channel	Market name
Barton	KCKT	2	Wichita-Hutchinson.
	KAKE	10	Do.
	KTVH	12	Do.
Bourbon	KOAM	7	Joplin-Pittsburg.
	KODE	12	Do.
	KCMO	5	Kansas City.
Brown	WDAF	4	Do.
	KCMO	5	Do.
	KMBC	9	Do.
	KQTV	2	St. Joseph.
	WIBW	13	Topeka.
Butler	KARD	3	Wichita-Hutchinson.
	KAKE	10	Do.
	KTVH	12	Do.
Chase	KARD	3	Do.
	KAKE	10	Do.
	KTVH	12	Do.
	WIBW	13	Topeka.
Chautauqua	KTEW	2	Tulsa.
	KOTV	6	Do.
	KTUL	8	Do.
Cherokee	KOAM	7	Joplin-Pittsburg.
	KODE	12	Do.
	KUHI	16	Do.
Cheyenne	KAYS+	7	Wichita-Hutchinson.
	KOMC	8	Do.
	KHOL+	13	Lincoln-Hastings-Kearney.
Clark	KTVC	6	Wichita-Hutchinson.
	KGLD	11	Do.
	KUPK	13	Do.
Clay	WIBW	13	Topeka.
	KHTL	4	Lincoln-Hastings-Kearney.
Cloud	KHTL	4	Do.
	KOLN +10		Lincoln-Hastings-Kearney.
	WIBW	13	Topeka.
	KTVH	12	Wichita-Hutchinson.
Coffey	WIBW	13	Topeka.
	KOAM	7	Joplin-Pittsburg.
	WDAF	4	Kansas City.
	KCMO	5	Do.
	KMBC	9	Do.
Comanche	KTVC	6	Wichita-Hutchinson.
	KUPK	13	Do.
Cowley	KARD	3	Do.
	KAKE	10	Do.
	KTVH	12	Do.
Crawford	KOAM	7	Joplin-Pittsburg.
	KODE	12	Do.
	KUHI	16	Do.
Decatur	KOMC	8	Wichita-Hutchinson.
	KHOL+	13	Lincoln-Hastings-Kearney.
Dickinson	KARD	3	Wichita-Hutchinson.
	KAKE	10	Do.
	KTVH	12	Do.
	WIBW	13	Topeka.
Doniphan	WDAF	4	Kansas City.
	KCMO	5	Do.
	KMBC	9	Do.
	KQTV	2	St. Joseph.
Douglas	WDAF	4	Kansas City.
	KCMO	5	Do.
	KMBC	9	Do.
	WIBW	13	Topeka.
Edwards	KCKT	2	Wichita-Hutchinson.
	KTVC	6	Do.
	KAYS+	7	Do.
	KTVH	12	Do.
Elk	KTEW	2	Tulsa.
	KOTV	6	Do.
	KTUL	8	Do.
	KOAM	7	Joplin-Pittsburg.
	KARD	3	Wichita-Hutchinson.
	KAKE	10	Do.
Ellis	KCKT	2	Do.
	KAYS+	7	Do.
Ellsworth	KCKT	2	Do.
	KAKE	10	Do.
	KTVH	12	Do.
Finney	KTVC	6	Do.
	KGLD	11	Do.
	KUPK	13	Do.
Ford	KTVC	6	Do.
	KGLD	11	Wichita-Hutchinson.
	KUPK	13	Do.
Franklin	WDAF	4	Kansas City.
	KCMO	5	Do.
	KMBC	9	Do.
	WIBW	13	Topeka.
Geary	WIBW	13	Do.
	KAKE	10	Wichita-Hutchinson.
Gove	KAYS+	7	Do.
	KOMC	8	Do.
Graham	KAYS+	7	Do.
	KOMC	8	Do.
Grant	KTVC	6	Do.
	KGLD	11	Do.
	KUPK	13	Do.

SIGNIFICANTLY VIEWED SIGNALS—Continued

KANSAS—continued

County	Call letters	Channel	Market name
Gray	KTVC	6	Do.
	KGLD	11	Do.
	KUPK	13	Do.
Greeley	KAYS+	7	Do.
	KGLD	11	Do.
Greenwood	KARD	3	Do.
	KAKE	10	Do.
	WIBW	13	Topeka.
Hamilton	KGLD	11	Wichita-Hutchinson.
	KUPK	13	Do.
Harper	KARD	3	Do.
	KAKE	10	Do.
	KTVH	12	Do.
Harvey	KARD	3	Do.
	KAKE	10	Do.
	KTVH	12	Do.
Haskell	KTVC	6	Do.
	KGLD	11	Do.
	KUPK	13	Do.
Hodgeman	KCKT	2	Do.
	KTVC	6	Do.
	KAYS+	7	Do.
	KUPK	13	Do.
Jackson	WIBW	13	Topeka.
	KTSB	27	Do.
	WDAF	4	Kansas City.
	KCMO	5	Do.
	KMBC	9	Do.
	KQTV	2	St. Joseph.
Jefferson	WIBW	13	Topeka.
	KTSB	27	Do.
	WDAF	4	Kansas City.
	KCMO	5	Do.
	KMBC	9	Do.
	KQTV	2	St. Joseph.
Jewell	KHTL	4	Lincoln-Hastings-Kearney.
	KHAS	5	Do.
	KOLN+	10	Do.
Johnson	WDAF	4	Kansas City.
	KCMO	5	Do.
	KMBC	9	Do.
	KCIT	50	Do.
	KBMA	41	Do.
Kearney	KTVC	6	Wichita-Hutchinson.
	KGLD	11	Do.
	KUPK	13	Do.
Kingman	KARD	3	Do.
	KAKE	10	Do.
	KTVH	12	Do.
Kiowa	KCKT	2	Do.
	KTVC	6	Do.
	KAKE	10	Do.
	KTVH	12	Do.
Labette	KOAM	7	Joplin-Pittsburg.
	KODE	12	Do.
	KUHI	16	Do.
Lane	KTVC	6	Wichita-Hutchinson.
	KGLD	11	Do.
	KUPK	13	Do.
Leavenworth	WDAF	4	Kansas City.
	KCMO	5	Do.
	KMBC	9	Do.
	KBMA	41	Do.
	KCIT	50	Do.
Lincoln	KCKT	2	Wichita-Hutchinson.
	KAYS+	7	Do.
	KTVH	12	Do.
Linn	WDAF	4	Kansas City.
	KCMO	5	Do.
	KMBC	9	Do.
	KOAM	7	Joplin-Pittsburg.
Logan	KAYS+	7	Wichita-Hutchinson.
	KOMC	8	Do.
Lyon	WIBW	13	Topeka.
	KTSB	27	Do.
McPherson	KARD	3	Wichita-Hutchinson.
	KAKE	10	Do.
	KTVH	12	Do.
Marion	KARD	3	Do.
	KAKE	10	Do.
	KTVH	12	Do.
Marshall	WIBW	13	Topeka.
	KHTL	4	Lincoln-Hastings-Kearney.
	KOLN+	10	Do.
Meade	KTVC	6	Wichita-Hutchinson.
	KGLD	11	Do.
	KUPK	13	Do.
Miami	WDAF	4	Kansas City.
	KCMO	5	Do.
	KMBC	9	Do.
	KBMA	41	Do.
Mitchell	KCKT	2	Wichita-Hutchinson.
	KAYS+	7	Do.
	KHTL	4	Lincoln-Hastings-Kearney.
Montgomery	KTEW	2	Tulsa.
	KOTV	6	Do.
	KTUL	8	Do.
	KOAM	7	Joplin-Pittsburg.
	KODE	12	Do.

SIGNIFICANTLY VIEWED SIGNALS—Continued

KANSAS—continued

County	Call letters, channel number and market name		
Morris	WIBW	13	Topeka.
	KTSB	27	Do.
	KARD	3	Wichita-Hutchinson.
	KAKE	10	Do.
Morton	KGLD	11	Do.
	KUPK	13	Do.
Nemaha	WIBW	13	Topeka.
	WDAF	4	Kansas City.
	KQTV	2	St. Joseph.
Neosho	KOAM	7	Joplin-Pittsburg.
	KODE	12	Do.
	KUHI	16	Do.
Ness	KCKT	2	Wichita-Hutchinson.
	KAYS+	7	Do.
Norton	KAYS+	7	Do.
	KOMC	8	Do.
	KOLN+	10	Lincoln-Hastings-Kearney.
	KHOL+	13	Do.
Osage	WIBW	13	Topeka.
	KTSB	27	Do.
	WDAF	4	Kansas City.
	KCMO	5	Do.
	KMBC	9	Do.
Osborne	KCKT	2	Wichita-Hutchinson.
	KAYS+	7	Do.
	KHTL	4	Lincoln-Hastings-Kearney.
Ottawa	KCKT	2	Wichita-Hutchinson.
	KAKE	10	Do.
	KTVH	12	Do.
	KHTL	4	Lincoln-Hastings-Kearney.
Pawnee	KCKT	2	Wichita-Hutchinson.
	KAYS+	7	Do.
	KAKE	10	Do.
	KTVH	12	Do.
Phillips	KOLN+	10	Lincoln-Hastings-Kearney.
	KHOL+	13	Do.
	KHAS	5	Do.
	KAYS+	7	Wichita-Hutchinson.
Pottawatomie	WIBW	13	Topeka.
	KTSB	27	Do.
Pratt	KARD	3	Wichita-Hutchinson.
	KAKE	10	Do.
	KTVH	12	Do.
Rawlins	KAYS+	7	Do.
	KOMC	8	Do.
	KHOL+	13	Lincoln-Hastings-Kearney.
Reno	KARD	3	Wichita-Hutchinson.
	KAKE	10	Do.
	KTVH	12	Do.
Republic	KHTL	4	Lincoln-Hastings-Kearney.
	KHAS	5	Do.
	KOLN+	10	Do.
Rice	KCKT	2	Wichita-Hutchinson.
	KARD	3	Do.
	KAKE	10	Do.
	KTVH	12	Do.
Riley	WIBW	13	Topeka.
	KTSB	27	Do.
Rooks*	KCKT	2	Wichita-Hutchinson.
	KAYS+	7	Do.
Rush	KCKT	2	Do.
	KAYS+	7	Do.
Russell	KCKT	2	Do.
	KAYS+	7	Do.
Saline	KARD	3	Do.
	KAKE	10	Do.
	KTVH	12	Do.
Scott	KTVC	6	Do.
	KGLD	11	Do.
	KUPK	13	Do.
Sedgwick	KARD	3	Do.
	KAKE	10	Do.
	KTVH	12	Do.
Seward	KTVC	6	Do.
	KGLD	11	Do.
	KUPK	13	Do.
Shawnee	WIBW	13	Topeka.
	KTSB	27	Do.
	WDAF	4	Kansas City.
	KCMO	5	Do.
	KMBC	9	Do.
Sheridan	KAYS+	7	Wichita-Hutchinson.
	KOMC	8	Do.
Sherman	KAYS+	7	Do.
	KOMC	8	Do.
Smith	KHTL	4	Lincoln-Hastings-Kearney.
	KHAS	5	Do.
	KOLN+	10	Do.
	KHOL+	13	Do.
Stafford	KCKT	2	Wichita-Hutchinson.
	KAKE	10	Do.
	KTVH	12	Do.
Stanton	KTVC	6	Do.
	KGLD	11	Do.
	KUPK	13	Do.
Stevens	KTVC	6	Do.
	KGLD	11	Do.
	KUPK	13	Do.
Sumner	KARD	3	Do.
	KAKE	10	Do.
	KTVH	12	Do.
Thomas	KAYS+	7	Do.
	KOMC	8	Do.
Trego	KCKT	2	Do.
	KAYS+	7	Do.
Wabaunsee	WIBW	13	Topeka.
	KTSB	27	Do.
	KCMO	5	Kansas City.
	KMBC	9	Do.
Wallace	KAYS+	7	Wichita-Hutchinson.
Washington	KHTL	4	Lincoln-Hastings-Kearney.
	KOLN+	10	Do.
	WIBW	13	Topeka.
Wichita	KAYS+	7	Wichita-Hutchinson.
	KGLD	11	Do.
	KUPK	13	Do.
Wilson	KOAM	7	Joplin-Pittsburg.
	KODE	12	Do.
	KOTV	6	Tulsa.
Woodson	KOAM	7	Joplin-Pittsburg.
	KODE	12	Do.
	WIBW	13	Topeka.
Wyandotte	WDAF	4	Kansas City.
	KCMO	5	Do.
	KMBC	9	Do.
	KBMA	41	Do.
	KCIT	50	Do.

KENTUCKY

County	Call letters, channel number and market name		
Adair	WAVE	3	Louisville.
	WHAS	11	Do.
	WLAC	5	Nashville.
Allen	WSM	4	Do.
	WLAC	5	Do.
	WSIX	8	Do.
Anderson	WAVE	3	Louisville.
	WHAS	11	Do.
	WLKY	32	Do.
	WLEX	18	Lexington.
	WKYT	27	Do.
	WBLG	62	Do.
Ballard	WSIL+	3	Paducah-Cape Girardeau-Harrisburg.
	WPSD	6	Do.
	KFVS	12	Do.
Barren	WSM	4	Nashville.
	WLAC	5	Do.
	WSIX	8	Do.
Bath	WLEX	18	Lexington.
	WKYT	27	Do.
	WBLG	62	Do.
	WLWT	5	Cincinnati.
	WCPO	9	Do.
	WKRC	12	Do.
Bell	WATE	6	Knoxville.
	WBIR	10	Do.
Boone	WLWT	5	Cincinnati.
	WCPO	9	Do.
	WKRC	12	Do.
	WXIX	19	Do.
Bourbon	WLEX	18	Lexington.
	WKYT	27	Do.
	WBLG	62	Do.
	WLWT	5	Cincinnati.
	WCPO	9	Do.
	WKRC	12	Do.
Boyd	WSAZ	3	Charleston-Huntington.
	WCHS	8	Do.
	WHTN	13	Do.
Boyle	WLEX	18	Lexington.
	WKYT	27	Do.
	WBLG	62	Do.
	WAVE	3	Louisville.
	WHAS	11	Do.
Bracken	WLWT	5	Cincinnati.
	WCPO	9	Do.
	WKRC	12	Do.
	WXIX	19	Do.
Breathitt	WSAZ	3	Charleston-Huntington.
	WLEX	18	Lexington.
	WKYT	27	Do.
	WBLG	62	Do.
Breckinridge	WAVE	3	Louisville.
	WHAS	11	Do.
	WLKY	32	Do.
	WTVW	7	Evansville.
Bullitt	WAVE	3	Louisville.
	WHAS	11	Do.
	WLKY	32	Do.

KENTUCKY—continued

County	Call letters, channel number and market name		
Butler	WSM	4	Nashville.
	WLAC	5	Do.
	WBKO	13	Bowling Green.
	WTVW	7	Evansville.
Caldwell	WSIL+	3	Paducah-Cape Girardeau-Harrisburg.
	WPSD	6	Do.
	KFVS	12	Do.
	WSM	4	Nashville.
	WLAC	5	Do.
Calloway	WSIL+	3	Paducah-Cape Girardeau-Harrisburg.
	WPSD	6	Do.
	KFVS	12	Do.
	WSM	4	Nashville.
	WLAC	5	Do.
	WSIX	8	Do.
Campbell	WLWT	5	Cincinnati.
	WCPO	9	Do.
	WKRC	12	Do.
	WXIX	19	Do.
Carlisle	WSIL+	3	Paducah-Cape Girardeau-Harrisburg.
	WPSD	6	Do.
	KFVS	12	Do.
Carroll	WLWT	5	Cincinnati.
	WCPO	9	Do.
	WKRC	12	Do.
	WXIX	19	Do.
	WTTV	4	Indianapolis.
	WAVE	3	Louisville.
	WHAS	11	Do.
	WLKY	32	Do.
Carter	WSAZ	3	Charleston-Huntington.
	WCHS	8	Do.
	WHTN	13	Do.
Casey	WLEX	18	Lexington.
	WKYT	27	Do.
	WAVE	3	Louisville.
	WHAS	11	Do.
Christian	WSM	4	Nashville.
	WLAC	5	Do.
	WSIX	8	Do.
Clark	WLEX	18	Lexington.
	WKYT	27	Do.
	WBLG	62	Do.
Clay	WATE	6	Knoxville.
	WBIR	10	Do.
Clinton	WSM	4	Nashville.
	WLAC	5	Do.
	WSIX	8	Do.
Crittenden	WSIL+	3	Paducah-Cape Girardeau-Harrisburg.
	WPSD	6	Do.
	KFVS	12	Do.
	WTVW	7	Evansville.
Cumberland	WSM	4	Nashville.
	WLAC	5	Do.
	WSIX	8	Do.
Daviess	WTVW	7	Evansville.
	WFIE	14	Do.
	WEHT	25	Do.
Edmonson	WSM	4	Nashville.
	WLAC	5	Do.
	WSIX	8	Do.
	WBKO	13	Bowling Green.
Elliott	WSAZ	3	Charleston-Huntington.
	WCHS	8	Do.
	WHTN	13	Do.
Estill	WLEX	18	Lexington.
	WKYT	27	Do.
	WBLG	62	Do.
Fayette	WLEX	18	Do.
	WKYT	27	Do.
	WBLG	62	Do.
Fleming	WLWT	5	Cincinnati.
	WCPO	9	Do.
	WKRC	12	Do.
	WLEX	18	Lexington.
Floyd	WSAZ	3	Charleston-Huntington.
	WCHS	8	Do.
	WHTN	13	Do.
Franklin*	WAVE	3	Louisville.
	WHAS	11	Do.
	WLKY	32	Do.
	WKRC	12	Cincinnati.
	WLEX	18	Lexington.
	WKYT	27	Do.
	WBLG	62	Do.
Fulton	WSIL+	3	Paducah-Cape Girardeau-Harrisburg.
	WPSD	6	Do.
	KFVS	12	Do.
Gallatin	WLWT	5	Cincinnati.
	WCPO	9	Do.
	WKRC	12	Do.
	WXIX	19	Do.

RULES AND REGULATIONS

SIGNIFICANTLY VIEWED SIGNALS—Continued

County	Call letters, channel number and market name		
KENTUCKY—continued			
Garrard*	WLEX	18	Lexington.
	WKYT	27	Do.
	WBLG	62	Do.
Grant	WLWT	5	Cincinnati.
	WCPO	9	Do.
	WKRC	12	Do.
	WXIX	19	Do.
Graves	WSIL+	3	Paducah-Cape Girardeau-Harrisburg
	WPSD	6	Do.
	KFVS	12	Do.
Grayson	WAVE	3	Louisville.
	WHAS	11	Do.
	WBKO	13	Bowling Green.
	WTVW	7	Evansville.
Green	WAVE	3	Louisville.
	WHAS	11	Do.
	WLKY	32	Do.
Greenup	WSAZ	3	Charleston-Huntington.
	WCHS	8	Do.
	WHTN	13	Do.
Hancock	WTVW	7	Evansville.
	WFIE	14	Do.
	WEHT	25	Do.
	WAVE	3	Louisville.
	WHAS	11	Do.
Hardin	WAVE	3	Louisville.
	WHAS	11	Do.
	WLKY	32	Do.
Harlan	WATE	6	Knoxville.
	WBIR	10	Do.
	WLOS	13	Greenville-Spartanburg-Asheville.
Harrison	WLWT	5	Cincinnati.
	WCPO	9	Do.
	WKRC	12	Do.
	WLEX	18	Lexington.
Hart	WSM	4	Nashville.
	WLAC	5	Do.
	WBKO	13	Bowling Green.
	WAVE	3	Louisville.
	WHAS	11	Do.
Henderson	WTVW	7	Evansville.
	WFIE	14	Do.
	WEHT	25	Do.
Henry	WAVE	3	Louisville.
	WHAS	11	Do.
	WLKY	32	Do.
	WLWT	5	Cincinnati.
	WCPO	9	Do.
	WKRC	12	Do.
Hickman	WSIL+	3	Paducah-Cape Girardeau-Harrisburg.
	WPSD	6	Do.
	KFVS	12	Do.
Hopkins	WTVW	7	Evansville.
	WEHT	25	Do.
	WSM	4	Nashville.
	WLAC	5	Do.
	WPSD	6	Paducah-Cape Girardeau-Harrisburg.
Jackson	WLEX	18	Lexington.
	WKYT	27	Do.
	WBLG	62	Do.
	WATE	6	Knoxville.
	WBIR	10	Do.
Jefferson	WAVE	3	Louisville.
	WHAS	11	Do.
	WLKY	32	Do.
Jessamine	WLEX	18	Lexington.
	WKYT	27	Do.
	WBLG	62	Do.
Johnson	WSAZ	3	Charleston-Huntington.
	WCHS	8	Do.
	WHTN	13	Do.
Kenton	WLWT	5	Cincinnati.
	WCPO	9	Do.
	WKRC	12	Do.
	WXIX	19	Do.
Knott	WCYB	5	Bristol-Kingsport-Johnson City.
	WJHL	11	Do.
	WSAZ	3	Charleston-Huntington.
	WLOS	13	Greenville-Spartanburg-Asheville.
	WLEX	18	Lexington.
Knox	WATE	6	Knoxville.
	WBIR	10	Do.
Larue	WAVE	3	Louisville.
	WHAS	11	Do.
	WLKY	32	Do.
Laurel	WATE	6	Knoxville.
	WBIR	10	Do.

SIGNIFICANTLY VIEWED SIGNALS—Continued

County	Call letters, channel number and market name		
KENTUCKY—continued			
Lawrence	WSAZ	3	Charleston-Huntington.
	WCHS	8	Do.
	WHTN	13	Do.
Lee	WLEX	18	Lexington.
	WKYT	27	Do.
	WCYB	5	Bristol-Kingsport-Johnson City.
Leslie	WCYB	5	Do.
	WJHL	11	Do.
	WBIR	10	Knoxville.
Letcher	WHTN	13	Charleston-Huntington.
	WCYB	5	Bristol-Kingsport-Johnson City.
Lewis *	WSAZ	3	Charleston-Huntington.
	WLWT	5	Cincinnati.
	WCPO	9	Do.
	WKRC	12	Do.
Lincoln	WLEX	18	Lexington.
	WKYT	27	Do.
	WBLG	62	Do.
Livingston	WSIL+	3	Paducah-Cape Girardeau-Harrisburg.
	WPSD	6	Do.
	KFVS	12	Do.
Logan	WSM	4	Nashville.
	WLAC	5	Do.
	WSIX	8	Do.
Lyon	WSIL+	3	Paducah-Cape Girardeau-Harrisburg.
	WPSD	6	Do.
	KFVS	12	Do.
McCracken	WSIL+	3	Do.
	WPSD	6	Do.
	KFVS	12	Do.
McCreary	WATE	6	Knoxville.
	WBIR	10	Do.
McLean	WTVW	7	Evansville.
	WFIE	14	Do.
	WEHT	25	Do.
	WLAC	5	Nashville.
Madison	WLEX	18	Lexington.
	WKYT	27	Do.
	WBLG	62	Do.
Magoffin	WSAZ	3	Charleston-Huntington.
Marion	WAVE	3	Louisville.
	WHAS	11	Do.
	WLKY	32	Do.
Marshall	WSIL+	3	Paducah-Cape Girardeau-Harrisburg.
	WPSD	6	Do.
	KFVS	12	Do.
Martin	WSAZ	3	Charleston-Huntington.
	WCHS	8	Do.
	WHTN	13	Do.
Mason	WLWT	5	Cincinnati.
	WCPO	9	Do.
	WKRC	12	Do.
Meade	WAVE	3	Louisville.
	WHAS	11	Do.
	WLKY	32	Do.
Menifee	WLEX	18	Lexington.
	WKYT	27	Do.
	WBLG	62	Do.
Mercer	WLEX	18	Lexington.
	WKYT	27	Do.
	WBLG	62	Do.
	WAVE	3	Louisville.
	WHAS	11	Do.
Metcalfe	WSM	4	Nashville.
	WLAC	5	Do.
	WSIX	8	Do.
Monroe	WSM	4	Nashville.
	WLAC	5	Do.
	WSIX	8	Do.
Montgomery *	WLEX	18	Lexington.
	WKYT	27	Do.
	WBLG	62	Do.
Morgan	WSAZ	3	Charleston-Huntington.
	WCHS	8	Do.
	WHTN	13	Do.
Muhlenberg	WSM	4	Nashville.
	WLAC	5	Do.
	WSIX	8	Do.
	WBKO	13	Bowling Green.
	WTVW	7	Evansville.
	WEHT	25	Do.
Nelson	WAVE	3	Louisville.
	WHAS	11	Do.
	WLKY	32	Do.
Nicholas	WLEX	18	Lexington.
	WKYT	27	Do.
	WBLG	62	Do.
	WLWT	5	Cincinnati.
	WCPO	9	Do.
	WKRC	12	Do.

SIGNIFICANTLY VIEWED SIGNALS—Continued

County	Call letters, channel number and market name		
KENTUCKY—continued			
Ohio	WTVW	7	Evansville.
	WEHT	25	Do.
	WBKO	13	Bowling Green.
	WSM	4	Nashville.
	WLAC	5	Do.
Oldham	WAVE	3	Louisville.
	WHAS	11	Do.
	WLKY	32	Do.
Owen	WLWT	5	Cincinnati.
	WCPO	9	Do.
	WKRC	12	Do.
	WAVE	3	Louisville.
	WHAS	11	Do.
Owsley	WLEX	18	Lexington.
	WKYT	27	Do.
	WATE	6	Knoxville.
Pendleton	WLWT	5	Cincinnati.
	WCPO	9	Do.
	WKRC	12	Do.
	WXIX	19	Do.
Perry	WCYB	5	Bristol-Kingsport-Johnson City.
	WJHL	11	Do.
Pike	WSAZ	3	Charleston-Huntington.
	WCHS	8	Do.
	WHTN	13	Do.
	WHIS	6	Bluefield-Beckley-Oak Hill.
Powell	WLEX	18	Lexington.
	WKYT	27	Do.
	WBLG	62	Do.
Pulaski	WATE	6	Knoxville.
	WBIR	10	Do.
	WLEX	18	Lexington.
	WKYT	17	Do.
	WBLG	62	Do.
Robertson	WLWT	5	Cincinnati.
	WCPO	9	Do.
	WKRC	12	Do.
	WLEX	18	Lexington.
Rockcastle	WLEX	18	Lexington.
	WKYT	27	Do.
	WBLG	62	Do.
Rowan	WSAZ	3	Charleston-Huntington.
Russell	WATE	6	Knoxville.
	WBIR	10	Do.
	WHAS	11	Louisville.
	WSM	4	Nashville.
	WLAC	5	Do.
	WSIX	8	Do.
Scott	WLEX	18	Lexington.
	WKYT	27	Do.
	WBLG	62	Do.
	WLWT	5	Cincinnati.
	WCPO	9	Do.
	WKRC	12	Do.
Shelby	WAVE	3	Louisville.
	WHAS	11	Do.
	WLKY	32	Do.
Simpson	WSM	4	Nashville.
	WLAC	5	Do.
	WSIX	8	Do.
Spencer	WAVE	3	Louisville.
	WHAS	11	Do.
	WLKY	32	Do.
Taylor	WAVE	3	Louisville.
	WHAS	11	Do.
Todd	WSM	4	Nashville.
	WLAC	5	Do.
	WSIX	8	Do.
Trigg	WSM	4	Do.
	WLAC	5	Do.
	WSIX	8	Do.
	WPSD	6	Paducah-Cape Girardeau-Harrisburg.
Trimble	WAVE	3	Louisville.
	WHAS	11	Do.
	WLKY	32	Do.
	WLWT	5	Cincinnati.
	WCPO	9	Do.
	WKRC	12	Do.
	WXIX	19	Do.
	WTTV	4	Indianapolis.
Union	WTVW	7	Evansville.
	WFIE	14	Do.
	WEHT	25	Do.
	WSIL+	3	Paducah-Cape Girardeau-Harrisburg.
	WPSD	6	Do.
Warren	WSM	4	Nashville.
	WLAC	5	Do.
	WSIX	8	Do.
	WBKO	13	Bowling Green.
Washington	WAVE	3	Louisville.
	WHAS	11	Do.
	WLKY	32	Do.
Wayne	WATE	6	Knoxville.
	WBIR	10	Do.

SIGNIFICANTLY VIEWED SIGNALS—Continued

County	Call letters, channel number and market name		
	KENTUCKY—continued		
Webster	WTVW	7	Evansville.
	WFIE	14	Do.
	WEHT	25	Do.
Whitley	WATE	6	Knoxville.
	WBIR	10	Do.
Wolfe	WSAZ	3	Charleston-Huntington.
	WCHS	8	Do.
	WLEX	18	Lexington.
	WKYT	27	Do.
Woodford	WLEX	18	Do.
	WKYT	27	Do.
	WBLG	62	Do.
	WAVE	3	Louisville.
	WHAS	11	Do.

Parish	Call letters, channel number, and market name		
	LOUISIANA		
Acadia	KATC	3	Lafayette, La.
	KLFY	10	Do.
	KLNI	15	Do.
	KALB	5	Alexandria, La.
Allen	KATC	3	Lafayette, La.
	KLFY	10	Do.
	KALB	5	Alexandria, La.
	KPLC	7	Lake Charles.
Ascension	WBRZ	2	Baton Rouge.
	WAFB	9	Do.
	WWL	4	New Orleans.
	WDSU	6	Do.
	WVUE	8	Do.
Assumption	WBRZ	2	Baton Rouge.
	WAFB	9	Do.
	WWL	4	New Orleans.
	WDSU	6	Do.
	WVUE	8	Do.
Avoyelles	KALB	5	Alexandria, La.
	WAFB	9	Baton Rouge.
	KATC	3	Lafayette, La.
	KLFY	10	Do.
Beauregard	KJAC	4	Beaumont-Port Arthur.
	KFDM	6	Do.
	KALB	5	Alexandria, La.
	KATC	3	Lafayette, La.
	KLFY	10	Do.
	KPLC	7	Lake Charles.
Bienville	KTBS	3	Shreveport-Texarkana.
	KTAL	6	Do.
	KSLA	12	Do.
	KNOE	8	Monroe-El Dorado.
	KTVE	10	Do.
Bossier	KTBS	3	Shreveport-Texarkana.
	KTAL	6	Do.
	KSLA	12	Do.
Caddo	KTBS	3	Do.
	KTAL	6	Do.
	KSLA	12	Do.
Calcasieu	KPLC	7	Lake Charles.
	KJAC	4	Beaumont-Port Arthur.
	KFDM	6	Do.
	KBMT	12	Do.
	KATC	3	Lafayette, La.
	KLFY	10	Do.
Caldwell	KNOE	8	Monroe-El Dorado.
	KTVE	10	Do.
Cameron	KPLC	7	Lake Charles.
	KJAC	4	Beaumont-Port Arthur.
	KFDM	6	Do.
	KBMT	12	Do.
	KATC	3	Lafayette, La.
	KLFY	10	Do.
Catahoula	KNOE	8	Monroe-El Dorado.
	KALB	5	Alexandria, La.
Claiborne	KTBS	3	Shreveport-Texarkana.
	KTAL	6	Do.
	KSLA	12	Do.
	KNOE	8	Monroe-El Dorado.
	KTVE	10	Do.
Concordia	KNOE	8	Do.
	KALB	5	Alexandria, La.
De Soto	KTBS	3	Shreveport-Texarkana.
	KTAL	6	Do.
	KSLA	12	Do.
East Baton Rouge	WBRZ	2	Baton Rouge.
	WAFB	9	Do.
East Carroll	KNOE	8	Monroe-El Dorado.
	KTVE	10	Do.
	WABG	6	Greenwood-Greenville.
	WLBT	3	Jackson, Miss.
	WJTV	12	Do.
East Feliciana	WBRZ	2	Baton Rouge.
	WAFB	9	Do.

Parish	Call letters, channel number, and market name		
	LOUISIANA—continued		
Evangeline	KATC	3	Lafayette, La.
	KLFY	10	Do.
	KALB	5	Alexandria, La.
Franklin	KNOE	8	Monroe-El Dorado.
	KTVE	10	Do.
Grant	KALB	5	Alexandria, La.
	KNOE	8	Monroe-El Dorado.
Iberia	KATC	3	Lafayette, La.
	KLFY	10	Do.
	KLNI	15	Do.
	WBRZ	2	Baton Rouge.
	WAFB	9	Do.
Iberville	WBRZ	2	Do.
	WAFB	9	Do.
Jackson	KNOE	8	Monroe-El Dorado.
	KTVE	10	Do.
	KTBS	3	Shreveport-Texarkana.
	KSLA	12	Do.
Jefferson	WWL	4	New Orleans.
	WDSU	6	Do.
	WVUE	8	Do.
Jefferson Davis	KATC	3	Lafayette, La.
	KLFY	10	Do.
	KPLC	7	Lake Charles.
Lafayette	KATC	3	Lafayette, La.
	KLFY	10	Do.
	KLNI	15	Do.
	WBRZ	2	Baton Rouge.
	WAFB	9	Do.
Lafourche	WWL	4	New Orleans.
	WDSU	6	Do.
	WVUE	8	Do.
	WAFB	9	Baton Rouge.
La Salle	KNOE	8	Monroe-El Dorado.
	KALB	5	Alexandria, La.
Lincoln*	KNOE	8	Monroe-El Dorado.
	KTVE	10	Do.
	KTBS	3	Shreveport-Texarkana.
Livingston	WBRZ	2	Baton Rouge.
	WAFB	9	Do.
	WWL	4	New Orleans.
Madison	WLBT	3	Jackson, Miss.
	WJTV	12	Do.
	KNOE	8	Monroe-El Dorado.
Morehouse	KNOE	8	Do.
	KTVE	10	Do.
Natchitoches	KTBS	3	Shreveport-Texarkana.
	KSLA	12	Do.
	KALB	5	Alexandria, La.
	KNOE	8	Monroe-El Dorado.
Orleans	WWL	4	New Orleans.
	WDSU	6	Do.
	WVUE	8	Do.
	WWOM	26	Do.
Ouachita	KNOE	8	Monroe-El Dorado.
	KTVE	10	Do.
Plaquemines	WWL	4	New Orleans.
	WDSU	6	Do.
	WVUE	8	Do.
Pointe Coupee	WBRZ	2	Baton Rouge.
	WAFB	9	Do.
Rapides	KALB	5	Alexandria, La.
	KLFY	10	Lafayette, La.
	KNOE	8	Monroe-El Dorado.
Red River	KTBS	3	Shreveport-Texarkana.
	KTAL	6	Do.
	KSLA	12	Do.
Richland	KNOE	8	Monroe-El Dorado.
	KTVE	10	Do.
	WLBT	3	Jackson, Miss.
Sabine	KTBS	3	Shreveport-Texarkana.
	KTAL	6	Do.
	KSLA	12	Do.
	KALB	5	Alexandria, La.
St. Bernard	WWL	4	New Orleans.
	WDSU	6	Do.
	WVUE	8	Do.
St. Charles	WWL	4	Do.
	WDSU	6	Do.
	WVUE	8	Do.
St. Helena	WBRZ	2	Baton Rouge.
	WAFB	9	Do.
St. James	WWL	4	New Orleans.
	WDSU	6	Do.
	WVUE	8	Do.
	WBRZ	2	Baton Rouge.
	WAFB	9	Do.
St. John the Baptist	WWL	4	New Orleans.
	WDSU	6	Do.
	WVUE	8	Do.
	WAFB	9	Baton Rouge.
St. Landry	KATC	3	Lafayette, La.
	KLFY	10	Do.
	KALB	5	Alexandria, La.
	WBRZ	2	Baton Rouge.
	WAFB	9	Do.

Parish	Call letters, channel number, and market name		
	LOUISIANA—continued		
St. Martin	KATC	3	Lafayette, La.
	KLFY	10	Do.
	KLNI	15	Do.
	WBRZ	2	Baton Rouge.
	WAFB	9	Do.
St. Mary	WBRZ	2	Do.
	WAFB	9	Do.
	KATC	3	Lafayette, La.
	KLFY	10	Do.
St. Tammany	WWL	4	New Orleans.
	WDSU	6	Do.
	WVUE	8	Do.
Tangipahoa	WWL	4	Do.
	WDSU	6	Do.
	WVUE	8	Do.
	WBRZ	2	Baton Rouge.
	WAFB	9	Do.
Tensas	WLBT	3	Jackson, Miss.
	WJTV	12	Do.
	KNOE	8	Monroe-El Dorado.
Terrebonne	WWL	4	New Orleans.
	WDSU	6	Do.
	WVUE	8	Do.
Union	KNOE	8	Monroe-El Dorado.
	KTVE	10	Do.
Vermilion *	KATC	3	Lafayette, La.
	KLFY	10	Do.
	KLNI	15	Do.
Vernon	KALB	5	Alexandria, La.
	KTBS	3	Shreveport-Texarkana.
Washington	WWL	4	New Orleans.
	WDSU	6	Do.
	WVUE	8	Do.
	WLOX	13	Baton Rouge.
Webster	KTBS	3	Shreveport-Texarkana.
	KTAL	6	Do.
	KSLA	12	Do.
West Baton Rouge	WBRZ	2	Baton Rouge.
	WAFB	9	Do.
West Carroll	KNOE	8	Monroe-El Dorado.
	KTVE	10	Do.
West Feliciana	WBRZ	2	Baton Rouge.
	WAFB	9	Do.
	KATC	3	Lafayette, La.
Winn	KNOE	8	Monroe-El Dorado.
	KALB	5	Alexandria, La.

County	Call letters, channel number and market name		
	MAINE		
Androscoggin	WCSH	6	Portland-Poland Spring.
	WMTW	8	Do.
	WGAN	13	Do.
Aroostook	WAGM	8	Presque Isle.
	CHSJ	4	Canada.
Cumberland	WCSH	6	Portland-Poland Spring.
	WMTW	8	Do.
	WGAN	13	Do.
Franklin	WCSH	6	Do.
	WMTW	8	Do.
	WGAN	13	Do.
	WABI	5	Bangor.
Hancock	WLBZ	2	Do.
	WABI	5	Do.
	WEMT	7	Do.
Kennebec	WCSH	6	Portland-Poland Spring.
	WMTW	8	Do.
	WGAN	13	Do.
	WLBZ	2	Bangor.
	WABI	5	Do.
Knox	WLBZ	2	Do.
	WABI	5	Do.
	WEMT	7	Do.
	WCSH	6	Portland-Poland Spring.
	WMTW	8	Do.
	WGAN	13	Do.
Lincoln	WCSH	6	Do.
	WMTW	8	Do.
	WGAN	13	Do.
Oxford	WCSH	6	Do.
	WMTW	8	Do.
	WGAN	13	Do.
	WABI	5	Bangor.
Penobscot	WLBZ	2	Do.
	WABI	5	Do.
	WEMT	7	Do.
Piscataquis	WLBZ	2	Do.
	WABI	5	Do.
	WEMT	7	Do.

SIGNIFICANTLY VIEWED SIGNALS—Continued

County	Call letters, channel number and market name		

MAINE—continued

County	Call	Ch	Market
Sagadahoc	WCSH	6	Portland-Poland Spring.
	WMTW	8	Do.
	WGAN	13	Do.
Somerset	WLBZ	2	Bangor.
	WABI	5	Do.
	WEMT	7	Do.
	WCSH	6	Portland-Poland Spring.
	WMTW	8	Do.
	WGAN	13	Do.
Waldo	WLBZ	2	Bangor.
	WABI	5	Do.
	WDMT	7	Do.
Washington*	WLBZ	2	Do.
	WABI	5	Do.
	WEMT	7	Do.
	CHSJ	4	Canada.
York	WCSH	6	Portland-Poland Spring.
	WMTW	8	Do.
	WGAN	13	Do.
	WBZ	4	Boston.
	WHDH	5	Do.
	WNAC	7	Do.

MARYLAND

County	Call	Ch	Market
Allegany	WTTG	5	Washington, D.C.
	WMAL	7	Do.
	WTOP	9	Do.
	WJAC	6	Johnstown-Altoona.
Anne Arundel	WMAR	2	Baltimore.
	WBAL	11	Do.
	WJZ	13	Do.
	WRC	4	Washington, D.C.
	WTTG	5	Do.
	WMAL	7	Do.
	WTOP	9	Do.
	WDCA	20	Do.
Baltimore including Baltimore City.	WMAR	2	Baltimore.
	WBAL	11	Do.
	WJZ	13	Do.
	WTTG	5	Washington, D.C.
Calvert	WRC	4	Do.
	WTTG	5	Do.
	WMAL	7	Do.
	WTOP	9	Do.
	WMAR	2	Baltimore.
Caroline	WMAR	2	Baltimore.
	WBAL	11	Do.
	WJZ	13	Do.
	WTTG	5	Washington, D.C.
Carroll	WMAR	2	Baltimore.
	WBAL	11	Do.
	WJZ	13	Do.
	WRC	4	Washington, D.C.
	WTTG	5	Do.
	WMAL	7	Do.
	WTOP	9	Do.
	WDCA	20	Do.
Cecil	WMAR	2	Baltimore.
	WBAL	11	Do.
	WJZ	13	Do.
	KYW	3	Philadelphia.
	WFIL	6	Do.
	WCAU	10	Do.
	WGAL	8	Harrisburg-York-Lancaster-Lebanon.
Charles	WRC	4	Do.
	WTTG	5	Do.
	WMAL	7	Do.
	WTOP	9	Do.
	WDCA	20	Do.
Dorchester	WMAR	2	Baltimore.
	WBAL	11	Do.
	WJZ	13	Do.
	WBOC	16	Salisbury.
	WRC	4	Washington, D.C.
	WTTG	5	Do.
	WMAL	7	Do.
	WTOP	9	Do.
Frederick*	WRC	4	Washington, D.C.
	WTTG	5	Do.
	WMAL	7	Do.
	WTOP	9	Do.
	WMAR	2	Baltimore.
	WBAL	11	Do.
	WJZ	13	Do.
Garrett	KDKA	2	Pittsburgh.
	WTAE	4	Do.
	WJAC	6	Johnstown-Altoona.
Harford	WMAR	2	Baltimore.
	WBAL	11	Do.
	WJZ	13	Do.
	WTTG	5	Washington, D.C.
Howard	WMAR	2	Baltimore.
	WBAL	11	Do.
	WJZ	13	Do.

SIGNIFICANTLY VIEWED SIGNALS—Continued

County	Call letters, channel number and market name		

MARYLAND—continued

County	Call	Ch	Market
	WRC	4	Washington, D.C.
	WTTG	5	Do.
	WMAL	7	Do.
	WTOP	9	Do.
	WDCA	20	Do.
Kent	WMAR	2	Baltimore.
	WBAL	11	Do.
	WJZ	13	Do.
	WRC	4	Washington, D.C.
	WTTG	5	Do.
	WTOP	9	Do.
Montgomery	WRC	4	Do.
	WTTG	5	Do.
	WMAL	7	Do.
	WTOP	9	Do.
	WDCA	20	Do.
Prince Georges	WRC	4	Do.
	WTTG	5	Do.
	WMAL	7	Do.
	WTOP	9	Do.
	WDCA	20	Do.
Queen Annes	WMAR	2	Baltimore.
	WBAL	11	Do.
	WJZ	13	Do.
	WTTG	5	Washington, D.C.
St. Marys	WRC	4	Do.
	WTTG	5	Do.
	WMAL	7	Do.
	WTOP	9	Do.
	WMAR	2	Baltimore.
Somerset	WBOC	16	Salisbury.
	WTTG	5	Washington, D.C.
Talbot	WMAR	2	Baltimore.
	WBAL	11	Do.
	WJZ	13	Do.
	WRC	4	Washington, D.C.
	WTTG	5	Do.
	WMAL	7	Do.
	WTOP	9	Do.
Washington	WRC	4	Do.
	WTTG	5	Do.
	WMAL	7	Do.
	WTOP	9	Do.
	WMAR	2	Baltimore.
	WHAG	25	Hagerstown, Md.
Wicomico	WBOC	16	Salisbury.
	WMAR	2	Baltimore.
	WBAL	11	Do.
	WJZ	13	Do.
	WTTG	5	Washington, D.C.
Worcester	WBOC	16	Salisbury.
	WTTG	5	Washington, D.C.

MASSACHUSETTS

County	Call	Ch	Market
Barnstable*	WBZ	4	Boston.
	WHDH	5	Do.
	WNAC	7	Do.
	WTEV	6	Providence.
	WJAR	10	Do.
	WPRI	12	Do.
Berkshire	WRGB	6	Albany-Schenectady-Troy.
	WTEN+	10	Do.
	WAST	13	Do.
	WTIC	3	Hartford-New Haven.
Bristol	WTEV	6	Providence.
	WJAR	10	Do.
	WPRI	12	Do.
	WBZ	4	Boston.
	WHDH	5	Do.
	WNAC	7	Do.
	WSBK	38	Do.
	WKBG	56	Do.
Dukes	WTEV	6	Providence.
	WJAR	10	Do.
	WPRI	12	Do.
	WBZ	4	Boston.
	WHDH	5	Do.
	WNAC	7	Do.
Essex	WBZ	4	Do.
	WHDH	5	Do.
	WNAC	7	Do.
	WSBK	38	Do.
	WKBG	56	Do.
Franklin	WWLP+	22	Springfield, Mass.
	WHYN	40	Do.
	WBZ	4	Boston.
	WHDH	5	Do.
	WTIC	3	Hartford-New Haven.
Hampden*	WWLP+	22	Springfield, Mass.
	WHYN	40	Do.
	WTIC	3	Hartford-New Haven.
	WNHC	8	Do.
	WHNB	30	Do.
Hampshire	WWLP+	22	Springfield, Mass.
	WHYN	40	Do.
	WTIC	3	Hartford-New Haven.

SIGNIFICANTLY VIEWED SIGNALS—Continued

County	Call letters, channel number and market name		

MASSACHUSETTS—continued

County	Call	Ch	Market
Middlesex	WBZ	4	Boston.
	WHDH	5	Do.
	WNAC	7	Do.
	WSBK	38	Do.
	WKBG	56	Do.
Nantucket	WTEV	6	Providence.
	WJAR	10	Do.
	WPRI	12	Do.
	WBZ	4	Boston.
	WHDH	5	Do.
Norfolk	WBZ	4	Do.
	WHDH	5	Do.
	WNAC	7	Do.
	WSBK	38	Do.
	WKBG	56	Do.
Plymouth	WBZ	4	Do.
	WHDH	5	Do.
	WSBK	38	Do.
	WKBG	56	Do.
	WNAC	7	Do.
	WTEV	6	Providence.
	WJAR	10	Do.
	WPRI	12	Do.
Suffolk	WBZ	4	Boston.
	WHDH	5	Do.
	WNAC	7	Do.
	WSBK	38	Do.
	WKBG	56	Do.
Worcester	WBZ	4	Boston.
	WHDH	5	Do.
	WNAC	7	Do.
	WSBK	38	Do.
	WKBG	56	Do.
	WJAR	10	Providence.
	WPRI	12	Do.
	WSMW	27	Worcester.

MICHIGAN

County	Call	Ch	Market
Alcona	WNEM	5	Flint-Saginaw-Bay City.
	WJRT	12	Do.
Alger	WLUC	6	Marquette.
	WFRV+	5	Green Bay.
Allegan	WKZO	3	Grand Rapids-Kalamazoo.
	WOOD	8	Do.
	WZZM	13	Do.
Alpena	WPBN	7	Traverse City-Cadillac.
	WWTV+	9	Do.
Antrim	WPBN+	7	Do.
	WWTV+	9	Do.
Arenac	WNEM	5	Flint-Saginaw-Bay City.
	WJRT	12	Do.
	WWTV+	9	Traverse City-Cadillac.
Baraga*	WLUC	6	Marquette.
Barry	WKZO	3	Grand Rapids-Kalamazoo.
	WOOD	8	Do.
	WZZM	13	Do.
	WJIM	6	Lansing.
Bay	WNEM	5	Flint-Saginaw-Bay City.
	WJRT	12	Do.
	WKNX	25	Do.
Benzie	WPBN+	7	Traverse City-Cadillac.
	WWTV+	9	Do.
Berrien	WBBM	2	Chicago.
	WMAQ	5	Do.
	WLS	7	Do.
	WGN	9	Do.
	WNDU	16	South Bend-Elkhart.
	WSBT	22	Do.
	WSJV	28	Do.
Branch	WKZO	3	Grand Rapids-Kalamazoo.
	WOOD	8	Do.
	WJIM	6	Lansing.
	WILX	10	Do.
Calhoun	WKZO	3	Grand Rapids-Kalamazoo.
	WOOD	8	Do.
	WJIM	6	Lansing.
	WILX	10	Do.
Cass	WNDU	16	South Bend-Elkhart.
	WSBT	22	Do.
	WSJV	28	Do.
	WGN	9	Chicago.
	WKZO	3	Grand Rapids-Kalamazoo.
	WOOD	8	Do.
Charlevoix	WPBN+	7	Traverse City-Cadillac.
	WWTV+	9	Do.
Cheboygan	WPBN+	7	Do.
	WWTV+	9	Do.

County	Call letters, channel number and market name		
	MICHIGAN—continued		
Chippewa	WPBN+	7	Do.
	WWTV+	9	Do.
	CJIC	2	Canada.
Clare	WNEM	5	Flint-Saginaw-Bay City.
	WJRT	12	Do.
	WPBN	7	Traverse City-Cadillac.
	WWTV+	9	Do.
Clinton	WJIM	6	Lansing.
	WILX	10	Do.
	WNEM	5	Flint-Saginaw-Bay City.
	WJRT	12	Do.
	WOOD	8	Grand Rapids-Kalamazoo.
Crawford	WPBN+	7	Traverse City-Cadillac.
	WWTV+	9	Do.
Delta	WFRV+	5	Green Bay.
	WLUK	11	Do.
	WLUC	6	Marquette.
Dickinson	WBAY	2	Green Bay.
	WFRV+	5	Do.
	WLUK	11	Do.
	WLUC	6	Marquette.
Eaton	WJIM	6	Lansing.
	WILX	10	Do.
	WJRT	12	Flint-Saginaw-Bay City.
	WKZO	3	Grand Rapids-Kalamazoo.
	WOOD	8	Do.
Emmet	WPBN+	7	Traverse City-Cadillac.
	WWTV+	9	Do.
Genesee	WNEM	5	Flint-Saginaw-Bay City.
	WJRT	12	Do.
	WJBK	2	Detroit.
	WWJ	4	Do.
	WXYZ	7	Do.
	WKBD	50	Do.
	WJIM	6	Lansing.
Gladwin	WNEM	5	Flint-Saginaw-Bay City.
	WJRT	12	Do.
	WWTV+	9	Traverse City-Cadillac.
Gogebic	KDAL	3	Duluth-Superior.
	WDSM	6	Do.
	WDIO+	10	Do.
	WAEO	12	Wausau-Rhinelander.
Grand Traverse	WPBN+	7	Traverse City-Cadillac.
	WWTV+	9	Do.
Gratiot	WNEM	5	Flint-Saginaw-Bay City.
	WJRT	12	Do.
	WOOD	8	Grand Rapids-Kalamazoo.
	WJIM	6	Lansing.
Hillsdale	WJIM	6	Do.
	WILX	10	Do.
	WKZO	3	Grand Rapids-Kalamazoo.
	WOOD	8	Do.
	WTOL	11	Toledo.
	WSPD	13	Do.
Houghton	WLUC	6	Marquette.
	WFRV+	5	Green Bay.
Huron *	WNEM	5	Flint-Saginaw-Bay City.
	WJRT	12	Do.
Ingham	WJIM	6	Lansing.
	WILX	10	Do.
	WJRT	12	Flint-Saginaw-Bay City.
	WKZO	3	Grand Rapids-Kalamazoo.
	WOOD	8	Do.
Ionia	WKZO	3	Do.
	WOOD	8	Do.
	WZZM	13	Do.
	WJRT	12	Flint-Saginaw-Bay City.
	WJIM	6	Lansing.
Iosco	WNEM	5	Flint-Saginaw-Bay City.
	WJRT	12	Do.
Iron	WLUC	6	Marquette.
	WFRV+	5	Green Bay.
	WAEO	12	Wausau-Rhinelander.
Isabella	WNEM	5	Flint-Saginaw-Bay City.
	WJRT	12	Do.
	WJIM	6	Lansing.
	WWTV	9	Traverse City-Cadillac.

County	Call letters, channel number and market name		
	MICHIGAN—continued		
Jackson	WJIM	6	Lansing.
	WILX	10	Do.
	WJBK	2	Detroit.
	WWJ	4	Do.
	WXYZ	7	Do.
Kalamazoo	WKZO	3	Grand Rapids-Kalamazoo.
	WOOD	8	Do.
	WZZM	13	Do.
Kalkaska	WPBN+	7	Traverse City-Cadillac.
	WWTV+	9	Do.
Kent	WKZO	3	Grand Rapids-Kalamazoo.
	WOOD	8	Do.
	WZZM	13	Do.
Keweenaw	WLUC	6	Marquette.
	CKPR	2	Canada.
Lake	WPBN+	7	Traverse City-Cadillac.
	WWTV+	9	Do.
	WZZM	13	Grand Rapids-Kalamazoo.
Lapeer	WJBK	2	Detroit.
	WWJ	4	Do.
	WXYZ	7	Do.
	CKLW	9	Do.
	WNEM	5	Flint-Saginaw-Bay City.
	WJRT	12	Do.
	WJIM	6	Lansing.
Leelanau	WPBN+	7	Traverse City-Cadillac.
	WWTV+	9	Do.
Lenawee*	WJBK	2	Detroit.
	WWJ	4	Do.
	WXYZ	7	Do.
	CKLW	9	Do.
	WKBD	50	Do.
	WTOL	11	Toledo.
	WSPD	13	Do.
	WDHO	24	Do.
Livingston	WJBK	2	Detroit.
	WWJ	4	Do.
	WXYZ	7	Do.
	CKLW	9	Do.
	WKBD	50	Do.
	WJRT	12	Flint-Saginaw-Bay City.
	WJIM	6	Lansing.
Luce	WPBN+	7	Traverse City-Cadillac.
	WWTV+	9	Do.
	WFRV+	5	Green Bay.
	WLUC	6	Marquette.
	CJIC	2	Canada.
Mackinac	WPBN+	7	Traverse City-Cadillac.
	WWTV+	9	Do.
	WFRV	5	Green Bay.
	CJIC	2	Canada.
Macomb	WJBK	2	Detroit.
	WWJ	4	Do.
	WXYZ	7	Do.
	CKLW	9	Do.
	WKBD	50	Do.
Manistee	WPBN+	7	Traverse City-Cadillac.
	WWTV+	9	Do.
	WZZM	13	Grand Rapids-Kalamazoo.
	WBAY	2	Green Bay.
	WLUK	11	Do.
Marquette	WLUC	6	Marquette.
	WFRV+	5	Green Bay.
	WLUK	11	Do.
Mason	WPBN+	7	Traverse City-Cadillac.
	WWTV+	9	Do.
	WZZM	13	Grand Rapids-Kalamazoo.
	WBAY	2	Green Bay.
	WFRV+	5	Do.
	WLUK	11	Do.
Mecosta	WPBN+	7	Traverse City-Cadillac.
	WWTV+	9	Do.
	WZZM	13	Grand Rapids-Kalamazoo.
Menominee	WBAY	2	Green Bay.
	WFRV+	5	Do.
	WLUK	11	Do.
	WLUC	6	Marquette.
Midland	WNEM	5	Flint-Saginaw-Bay City.
	WJRT	12	Do.
	WWTV+	9	Traverse City-Cadillac.
Missaukee	WPBN+	7	Do.
	WWTV+	9	Do.

County	Call letters, channel number and market name		
	MICHIGAN—continued		
Monroe	WJBK	2	Detroit.
	WWJ	4	Do.
	WXYZ	7	Do.
	CKLW	0	Do.
	WKBD	50	Do.
	WTOL	11	Toledo.
	WDHO	24	Do.
	WSPD	13	Do.
Montcalm	WKZO	3	Grand Rapids-Kalamazoo.
	WOOD	8	Do.
	WZZM	13	Do.
	WJRT	12	Flint-Saginaw-Bay City.
	WJIM	6	Lansing.
	WWTV	9	Traverse City-Cadillac.
Montmorency	WPBN+	7	Do.
	WWTV+	9	Do.
Muskegon	WKZO	3	Grand Rapids-Kalamazoo.
	WOOD	8	Do.
	WZZM	13	Do.
Newaygo	WKZO	3	Do.
	WOOD	8	Do.
	WZZM	13	Do.
	WPBN+	7	Traverse City-Cadillac.
	WWTV+	9	Do.
Oakland	WJBK	2	Detroit.
	WWJ	4	Do.
	WXYZ	7	Do.
	CKLW	9	Do.
	WKBD	50	Do.
Oceana	WKZO	3	Grand Rapids-Kalamazoo.
	WZZM	13	Do.
	WPBN+	7	Traverse City-Cadillac.
	WWTV+	9	Do.
Ogemaw	WNEM	5	Flint-Saginaw-Bay City.
	WJRT	12	Do.
	WWTV+	9	Traverse City-Cadillac.
Ontonagon	WLUC	6	Marquette.
	KDAL	3	Duluth-Superior.
	WAEO	12	Wausau-Rhinelander.
Osceola	WPBN+	7	Traverse City-Cadillac.
	WWTV+	9	Do.
	WNEM	5	Flint-Saginaw-Bay City.
	WZZM	13	Grand Rapids-Kalamazoo.
Oscoda	WPBN+	7	Traverse City-Cadillac.
	WWTV+	9	Do.
	WNEM	5	Flint-Saginaw-Bay City.
Otsego	WPBN+	7	Traverse City-Cadillac.
	WWTV+	9	Do.
Ottawa	WKZO	3	Grand Rapids-Kalamazoo.
	WOOD	8	Do.
	WZZM	13	Do.
Presque Isle	WPBN+	7	Traverse City-Cadillac.
	WWTV+	9	Do.
Roscommon	WPBN+	7	Do.
	WWTV+	9	Do.
	WNEM	5	Flint-Saginaw-Bay City.
Saginaw	WNEM	5	Do.
	WJRT	12	Do.
	WKNX	25	Do.
St. Clair	WJBK	2	Detroit.
	WWJ	4	Do.
	WXYZ	7	Do.
	CKLW	9	Do.
	WKBD	50	Do.
St. Joseph	WKZO	3	Grand Rapids-Kalamazoo.
	WOOD	8	Do.
	WNDU	16	South Bend-Elkhart.
	WSBT	22	Do.
	WSJV	28	Do.
Sanilac	WJBK	2	Detroit.
	WWJ	4	Do.
	WXYZ	7	Do.
	WNEM	5	Flint-Saginaw-Bay City.
	WJRT	12	Do.
	CFPL	10	Canada.
Schoolcraft	WLUC	6	Marquette.
	WFRV+	1	Green Bay.
Shiawassee	WNEM	5	Flint-Saginaw-Bay City.
	WJRT	12	Do.
	WJIM	6	Lansing.
	WILX	10	Do.

SIGNIFICANTLY VIEWED SIGNALS—Continued

County	Call letters, channel number and market name		

MICHIGAN—continued

County	Call letters	Ch.	Market name
Tuscola	WNEM	5	Flint-Saginaw-Bay City.
	WJRT	12	Do.
	WKNX	25	Do.
Van Buren	WKZO	3	Grand Rapids-Kalamazoo.
	WOOD	8	Do.
	WZZM	13	Do.
Washtenaw	WJBK	2	Detroit.
	WWJ	4	Do.
	WXYZ	7	Do.
	CKLW	9	Do.
	WKBD	50	Do.
Wayne	WJBK	2	Do.
	WWJ	4	Do.
	WXYZ	7	Do.
	CKLW	9	Do.
	WKBD	50	Do.
Wexford	WPBN+	7	Traverse City-Cadillac.
	WWTV+	9	Do.

MINNESOTA

County	Call letters	Ch.	Market name
Aitkin	KDAL	3	Duluth-Superior.
	WDSM	6	Do.
	WDIO+	10	Do.
	KNMT	12	Alexandria, Minn.
Anoka	WCCO	4	Minneapolis-St. Paul.
	KSTP	5	Do.
	KMSP	9	Do.
	WTCN	11	Do.
Becker	KXJB	4	Fargo.
	WDAY	6	Do.
	KTHI	11	Do.
Beltrami	KNMT	12	Alexandria, Minn.
	KDAL	3	Duluth-Superior.
	WDSM	6	Do.
	KTHI	11	Fargo.
Benton	WCCO	4	Minneapolis-St. Paul.
	KSTP	5	Do.
	KMSP	9	Do.
	WTCN	11	Do.
	KCMT	7	Alexandria, Minn.
Big Stone*	KELO+	11	Sioux Falls-Mitchell.
Blue Earth	KEYC	12	Mankato.
	WCCO	4	Minneapolis-St. Paul.
	KSTP	5	Do.
	KMSP	9	Do.
	WTCN	11	Do.
	KAUS	6	Rochester-Mason City-Austin.
Brown	WCCO	4	Minneapolis-St. Paul.
	KSTP	5	Do.
	KMSP	9	Do.
	WTCN	11	Do.
	KEYC	12	Mankato.
	KAUS	6	Rochester-Mason City-Austin.
Carlton	KDAL	3	Duluth-Superior.
	WDSM	6	Do.
	WDIO+	10	Do.
Carver	WCCO	4	Minneapolis-St. Paul.
	KSTP	5	Do.
	KMSP	9	Do.
	WTCN	11	Do.
Cass	KCMT	7	Alexandria, Minn.
	KNMT	12	Do.
	KDAL	3	Duluth-Superior.
	WDIO+	10	Do.
Chippewa	KCMT	7	Alexandria, Minn.
	WCCO	4	Minneapolis-St. Paul.
	KMSP	9	Do.
	WTCN	11	Do.
Chisago	WCCO	4	Minneapolis-St. Paul.
	KSTP	5	Do.
	KMSP	9	Do.
	WTCN	11	Do.
Clay	KXJB	4	Fargo.
	WDAY	6	Do.
	KTHI	11	Do.
Clearwater	KXJB	4	Fargo.
	WDAY	6	Do.
	KTHI	11	Do.
Cook	KDAL	3	Duluth-Superior.
	WDSM	6	Do.
	WDIO+	10	Do.
	CKPR	2	Canada.
Cottonwood	KEYC	12	Mankato.
	KSTP	5	Minneapolis-St. Paul.
	WTCN	11	Do.
	KAUS	6	Rochester-Mason City-Austin.
Crow Wing	KCMT	7	Alexandria, Minn.
	KNMT	12	Do.
	KDAL	3	Duluth-Superior.

MINNESOTA—continued

County	Call letters	Ch.	Market name
Dakota	WCCO	4	Minneapolis-St. Paul.
	KSTP	5	Do.
	KMSP	9	Do.
	WTCN	11	Do.
Dodge	KGLO	3	Rochester-Mason City-Austin.
	KAUS	6	Do.
	KROC	10	Do.
	WCCO	4	Minneapolis-St. Paul.
	KSTP	5	Do.
Douglas	KCMT	7	Alexandria, Minn.
Faribault	KGLO	3	Rochester-Mason City-Austin.
	KAUS	6	Do.
	KROC	10	Do.
	KEYC	12	Mankato.
Fillmore	KGLO	3	Rochester-Mason City-Austin.
	KAUS	6	Do.
	KROC	10	Do.
	WKBT	8	La Crosse-Eau Claire.
Freeborn	KGLO	3	Rochester-Mason City-Austin.
	KAUS	6	Do.
	KROC	10	Do.
Goodhue	WCCO	4	Minneapolis-St. Paul.
	KSTP	5	Do.
	KMSP	9	Do.
	WTCN	11	Do.
Grant*	KCMT	7	Alexandria, Minn.
Hennepin	WCCO	4	Minneapolis-St. Paul.
	KSTP	5	Do.
	KMSP	9	Do.
	WTCN	11	Do.
Houston	WKBT	8	La Crosse, Eau Claire.
	KROC	10	Rochester-Mason City-Austin.
Hubbard	KNMT	12	Alexandria, Minn.
Isanti	WCCO	4	Minneapolis-St. Paul.
	KSTP	5	Do.
	KMSP	9	Do.
	WTCN	11	Do.
Itasca	KDAL	3	Duluth-Superior.
	WDSM	6	Do.
	WDIO+	10	Do.
	KNMT	12	Alexandria, Minn.
Jackson	KEYC	12	Mankato.
	KCAU	9	Sioux City.
	KELO+	11	Sioux Falls-Mitchell.
	KSOO+	13	Do.
Kanabec	WCCO	4	Minneapolis-St. Paul.
	KSTP	5	Do.
	KMSP	9	Do.
	WTCN	11	Do.
Kandiyohi	WCCO	4	Do.
	KSTP	5	Do.
	KMSP	9	Do.
	WTCN	11	Do.
	KCMT	7	Alexandria, Minn.
Kittson	KXJB	4	Fargo.
	WDAZ	8	Do.
	KCND	12	Pembina.
	CBWT	6	Canada.
	CJAY	7	Do.
Koochiching	KDAL	3	Duluth-Superior.
	WDSM	6	Do.
	WDIO+	10	Do.
	CBWT	6	Canada.
Lac qui Parle	KCMT	7	Alexandria, Minn.
	KELO+	11	Sioux Falls-Mitchell.
Lake	KDAL	3	Duluth-Superior.
	WDSM	6	Do.
	WDIO+	10	Do.
Lake of the Woods.	CBWT	6	Canada.
Le Sueur	WCCO	4	Minneapolis-St. Paul.
	KSTP	5	Do.
	KMSP	9	Do.
	WTCN	11	Do.
	KEYC	12	Mankato.
Lincoln	KORN	5	Sioux Falls-Mitchell.
	KELO+	11	Do.
	KSOO+	13	Do.
Lyon	KELO+	11	Do.
	KSOO+	13	Do.
	KEYC	12	Mankato.
McLeod	WCCO	4	Minneapolis-St. Paul.
	KSTP	5	Do.
	KMSP	9	Do.
	WTCN	11	Do.
Mahnomen	KXJB	4	Fargo.
	WDAY	6	Do.
	KTHI	11	Do.
Marshall	KXJB	4	Do.
	WDAZ	8	Do.
	KTHI	11	Do.
	KCND	12	Pembina.
	CBWT	6	Canada.

MINNESOTA—continued

County	Call letters	Ch.	Market name
Martin	KEYC	12	Mankato.
	KAUS	6	Rochester-Mason City-Austin.
	KROC	10	Do.
Meeker	WCCO	4	Minneapolis-St. Paul.
	KSTP	5	Do.
	KMSP	9	Do.
	WTCN	11	Do.
	KCMT	7	Alexandria, Minn.
Mille Lacs	WCCO	4	Minneapolis-St. Paul.
	KSTP	5	Do.
	KMSP	9	Do.
	WTCN	11	Do.
Morrison	KCMT	7	Alexandria, Minn.
	WCCO	4	Minneapolis-St. Paul.
	KSTP	5	Do.
	KMSP	9	Do.
	WTCN	11	Do.
Mower	KGLO	3	Rochester-Mason City-Austin.
	KAUS	6	Do.
	KROC	10	Do.
Murray	KELO+	11	Sioux Falls-Mitchell.
	KSOO+	13	Do.
Nicollet	WCCO	4	Minneapolis-St. Paul.
	KSTP	5	Do.
	KMSP	9	Do.
	WTCN	11	Do.
	KEYC	12	Mankato.
Nobles	KELO+	11	Sioux Falls-Mitchell.
	KSOO+	13	Do.
	KCAU	9	Sioux City.
Norman	KXJB	4	Fargo.
	WDAY	6	Do.
	KTHI	11	Do.
Olmsted	KGLO	3	Rochester-Mason City-Austin.
	KAUS	6	Do.
	KROC	10	Do.
Otter Tail	KXJB	4	Fargo.
	WDAY	6	Do.
	KTHI	11	Do.
	KCMT	7	Alexandria, Minn.
Pennington	KXJB	4	Fargo.
	WDAY	6	Do.
	WDAZ	8	Do.
	KTHI	11	Do.
Pine	KDAL	3	Duluth-Superior.
	WDSM	6	Do.
	WDIO+	10	Do.
	WCCO	4	Minneapolis-St. Paul.
	KSTP	5	Do.
	KMSP	9	Do.
	WTCN	11	Do.
Pipestone	KORN	5	Sioux Falls-Mitchell.
	KELO+	11	Do.
	KSOO+	13	Do.
Polk	KXJB	4	Fargo.
	WDAY	6	Do.
	KTHI	11	Do.
Pope	KCMT	7	Alexandria, Minn.
Ramsey	WCCO	4	Minneapolis-St. Paul.
	KSTP	5	Do.
	KMSP	9	Do.
	WTCN	11	Do.
Red Lake	KXJB	4	Fargo.
	WDAY	6	Do.
	WDAZ	8	Do.
	KTHI	11	Do.
Redwood	KEYC	12	Mankato.
	WCCO	4	Minneapolis-St. Paul.
	KSTP	5	Do.
	KMSP	9	Do.
	WTCN	11	Do.
Renville	WCCO	4	Minneapolis-St. Paul.
	KSTP	5	Do.
	KMSP	9	Do.
	WTCN	11	Do.
	KCMT	7	Alexandria, Minn.
	KEYC	12	Mankato.
Rice	WCCO	4	Minneapolis-St. Paul.
	KSTP	5	Do.
	KMSP	9	Do.
	WTCN	11	Do.
Rock	KELO+	11	Sioux Falls-Mitchell.
	KSOO+	13	Do.
	KCAU	9	Sioux City.
Roseau	KCND	12	Pembina.
	WDAZ	8	Fargo.
	CBWT	6	Canada.
	CJAY	7	Do.
St. Louis	KDAL	3	Duluth-Superior.
	WDSM	6	Do.
	WDIO+	10	Do.
Scott	WCCO	4	Minneapolis-St. Paul.
	KSTP	5	Do.
	KMSP	9	Do.
	WTCN	11	Do.

SIGNIFICANTLY VIEWED SIGNALS—Continued

County	Call letters, channel number and market name		

MINNESOTA—continued

County	Call letters	Channel	Market name
Sherburne	WCCO	4	Do.
	KSTP	5	Do.
	KMSP	9	Do.
	WTCN	11	Do.
Sibley	WCCO	4	Minneapolis-St. Paul.
	KSTP	5	Do.
	KMSP	9	Do.
	WTCN	11	Do.
	KEYC	12	Mankato.
Stearns	WCCO	4	Minneapolis-St. Paul.
	KSTP	5	Do.
	KMSP	9	Do.
	WTCN	11	Do.
	KCMT	7	Alexandria, Minn.
Steele	KGLO	3	Rochester-Mason City-Austin.
	KAUS	6	Do.
	KROC	10	Do.
	KEYC	12	Mankato.
	WCCO	4	Minneapolis-St. Paul.
	KSTP	5	Do.
	KMSP	9	Do.
	WTCN	11	Do.
Stevens	KCMT	7	Alexandria, Minn.
	KELO+	11	Sioux Falls-Mitchell.
Swift	KCMT	7	Alexandria, Minn.
Todd	KCMT	7	Do.
Traverse	KCMT	7	Do.
	KXJB	4	Fargo.
	KELO+	11	Sioux Falls-Mitchell.
Wabasha	WCCO	4	Minneapolis-St. Paul.
	KSTP	5	Do.
	KMSP	9	Do.
	WTCN	11	Do.
	WKBT	8	La Crosse-Eau Claire.
	WEAU	13	Do.
	KAUS	6	Rochester-Mason City-Austin.
	KROC	10	Do.
Wadena	KCMT	7	Alexandria, Minn.
	KNMT	12	Do.
Waseca	WCCO	4	Minneapolis-St. Paul.
	KSTP	5	Do.
	KMSP	9	Do.
	WTCN	11	Do.
	KEYC	12	Mankato.
	KGLO	3	Rochester-Mason City-Austin.
	KAUS	6	Do.
	KROC	10	Do.
Washington	WCCO	4	Minneapolis-St. Paul.
	KSTP	5	Do.
	KMSP	9	Do.
	WTCN	11	Do.
Watonwan	KEYC	12	Mankato.
	WCCO	4	Minneapolis-St. Paul.
	KSTP	5	Do.
	WTCN	11	Do.
	KAUS	6	Rochester-Mason City-Austin.
Wilkin	KXJB	4	Fargo.
	WDAY	6	Do.
	KTHI	11	Do.
Winona	WKBT	8	La Crosse-Eau Claire.
	KAUS	6	Rochester-Mason City-Austin.
	KROC	10	Do.
Wright	WCCO	4	Minneapllis-St. Paul.
	KSTP	5	Do.
	KMSP	9	Do.
	WTCN	11	Do.
Yellow Medicine	WCCO	4	Do.
	KMSP	9	Do.
	WTCN	11	Do.
	KCMT	7	Alexandria, Minn.
	KELO+	11	Sioux Falls-Mitchell.

MISSISSIPPI

County	Call letters	Channel	Market name
Adams	KNOE	8	Monroe-El Dorado.
	KALB	5	Alexandria, La.
	WLBT	3	Jackson, Miss.
	WJTV	12	Do.
Alcorn	WREC	3	Memphis.
	WMC	5	Do.
	WHBQ	13	Do.
Amite	WVRZ	2	Baton Rouge.
	WAFB	9	Do.
	WLBT	3	Jackson, Miss.
Attala	WLBT	3	Do.
	WJTV	12	Do.
	WABG	6	Greenwood-Greenville.
Benton	WREC	3	Memphis.
	WMC	5	Do.
	WHBQ	13	Do.
Bolivar	WABG	6	Greenwood-Greenville.
	WLBT	3	Jackson, Miss.
	WREC	3	Memphis.
	WMC	5	Do.

SIGNIFICANTLY VIEWED SIGNALS—Continued

MISSISSIPPI—continued

County	Call letters	Channel	Market name
Calhoun	WREC	3	Do.
	WMC	5	Do.
	WCBI	4	Columbus, Miss.
	WABG	6	Greenwood-Greenville.
Carroll	WABG	6	Do.
	WLBT	3	Jackson, Miss.
	WJTV	12	Do.
Chickasaw	WCBI	4	Columbus, Miss.
	WMC	5	Memphis.
	WTWV	9	Tupelo.
Choctaw	WCBI	4	Columbus, Miss.
	WABG	6	Greenwood-Greenville.
	WLBT	3	Jackson, Miss.
	WJTV	12	Do.
Claiborne	WLBT	3	Do.
	WJTV	12	Do.
Clarke	WTOK	11	Meridian.
	WDAM	7	Laurel-Hattiesburg.
Clay	WCBI	4	Columbus, Miss.
Coahoma	WREC	3	Memphis.
	WMC	5	Do.
	WHBQ	13	Do.
Copiah	WLBT	3	Jackson, Miss.
	WJTV	12	Do.
	WAPT	16	Do.
Covington	WDAM	7	Laurel-Hattiesburg.
	WLOX	13	Biloxi-Gulfport-Pascagoula.
	WLBT	3	Jackson, Miss.
	WJTV	12	Do.
De Soto	WREC	3	Memphis.
	WMC	5	Do.
	WHBQ	13	Do.
Forrest	WDAM	7	Laurel-Hattiesburg.
	WLOX	13	Biloxi-Gulfport-Pascagoula.
Franklin	WLBT	3	Jackson, Miss.
	WJTV	12	Do.
	WBRZ	2	Baton Rouge.
George	WEAR	3	Mobile-Pensacola.
	WKRG	5	Do.
	WALA	10	Do.
	WLOX	13	Biloxi-Gulfport-Pascagoula.
Greene	WEAR	3	Mobile-Pensacola.
	WKRG	5	Do.
	WALA	10	Do.
	WLOX	13	Biloxi-Gulfport-Pascagoula.
	WDAM	7	Laurel-Hattiesburg.
Grenada	WABG	6	Greenwood-Greenville.
	WREC	3	Memphis.
	WMC	5	Do.
Hancock	WWL	4	New Orleans.
	WDSU	6	Do.
	WVUE	8	Do.
	WLOX	13	Biloxi-Gulfport-Pascagoula.
Harrison	WLOX	13	Do.
	WKRG	5	Mobile-Pensacola.
	WWL	4	New Orleans.
	WDSU	6	Do.
	WVUE	8	Do.
Hinds	WLBT	3	Jackson, Miss.
	WJTV	12	Do.
	WAPT	16	Do.
Holmes	WLBT	3	Do.
	WJTV	12	Do.
	WABG	6	Greenwood-Greenville.
Humphreys	WLBT	3	Jackson, Miss.
	WJTV	12	Do.
	WABG	6	Greenwood-Greenville.
Issaquena	WLBT	3	Jackson, Miss.
	WJTV	12	Do.
	WABG	6	Greenwood-Greenville.
	KNOE	8	Monroe-El Dorado.
Itawamba	WTWV	9	Tupelo.
	WCBI	4	Columbus, Miss.
Jackson*	WEAR	3	Mobile-Pensacola.
	WKRG	5	Do.
	WALA	10	Do.
	WLOX	13	Biloxi-Gulfport-Pascagoula.
Jasper	WDAM	7	Laurel-Hattiesburg.
	WLBT	3	Jackson, Miss.
	WJTV	12	Do.
	WTOK	11	Meridian.
Jefferson	WLBT	3	Jackson, Miss.
	WJTV	12	Do.
	KNOE	8	Monroe-El Dorado.
Jefferson Davis	WLBT	3	Jackson, Miss.
	WJTV	12	Do.
	WDAM	7	Laurel-Hattiesburg.
Jones	WDAM	7	Do.
	WLOX	13	Biloxi-Gulfport-Pascagoula.
	WTOK	11	Meridian.
Kemper	WTOK	11	Do.

SIGNIFICANTLY VIEWED SIGNALS—Continued

MISSISSIPPI—continued

County	Call letters	Channel	Market name
Lafayette	WREC	3	Memphis.
	WMC	5	Do.
	WHBQ	13	Do.
Lamar	WDAM	7	Laurel-Hattiesburg.
	WLOX	13	Biloxi-Gulfport-Pascagoula.
Lauderdale	WTOK	11	Meridian.
Lawrence	WLBT	3	Jackson, Miss.
	WJTV	12	Do.
	WAPT	16	Do.
Leake	WLBT	3	Do.
	WJTV	12	Do.
Lee	WTWV	9	Tupelo.
	WCBI	4	Columbus, Miss.
	WREC	3	Memphis.
	WMC	5	Do.
	WHBQ	13	Do.
Leflore	WABG	6	Greenwood-Greenville.
	WLBT	3	Jackson, Miss.
	WJTV	12	Do.
Lincoln	WLBT	3	Do.
	WJTV	12	Do.
Lowndes	WCBI	4	Columbus, Miss.
Madison	WLBT	3	Jackson, Miss.
	WJTV	12	Do.
	WAPT	16	Do.
Marion	WLBT	3	Do.
	WJTV	12	Do.
	WLOX	13	Biloxi-Gulfport-Pascagoula.
	WDAM	7	Laurel-Hattiesburg.
Marshall	WREC	3	Memphis.
	WMC	5	Do.
	WHBQ	13	Do.
Monroe	WCBI	4	Columbus, Miss.
	WTWV	9	Tupelo.
Montgomery	WABG	6	Greenwood-Greenville.
	WCBI	4	Columbus, Miss.
	WLBT	3	Jackson, Miss.
	WJTV	12	Do.
	WMC	5	Memphis.
Neshoba	WTOK	11	Meridian.
	WLBT	3	Jackson, Miss.
Newton	WTOK	11	Meridian.
	WLBT	3	Jackson, Miss.
	WJTV	12	Do.
Noxubee	WTOK	11	Meridian.
	WCBI	4	Columbus, Miss.
Oktibbeha	WCBI	4	Do.
Panola	WREC	3	Memphis.
	WMC	5	Do.
	WHBQ	13	Do.
Pearl River	WWL	4	New Orleans.
	WDSU	6	Do.
	WVUE	8	Do.
	WLOX	13	Biloxi-Gulfport-Pascagoula.
Perry	WDAM	7	Laurel-Hattiesburg.
	WLOX	13	Biloxi-Gulfport-Pascagoula.
Pike	WKRG	5	Mobile-Pensacola.
	WLBT	3	Jackson, Miss.
	WJTV	12	Do.
	WBRZ	2	Baton Rouge.
	WAFB	9	Do.
	WWL	4	New Orleans.
Pontotoc	WREC	3	Memphis.
	WMC	5	Do.
	WHBQ	13	Do.
	WTWV	9	Tupelo.
Prentiss	WREC	3	Memphis.
	WMC	5	Do.
	WHBQ	13	Do.
	WTWV	9	Tupelo.
Quitman	WREC	3	Memphis.
	WMC	5	Do.
	WHBQ	13	Do.
Rankin	WLBT	3	Jackson, Miss.
	WJTV	12	Do.
	WAPT	16	Do.
Scott	WLBT	3	Do.
	WJTV	12	Do.
	WTOK	11	Meridian.
Sharkey	WLBT	3	Jackson, Miss.
	WJTV	12	Do.
	WABG	6	Greenwood-Greenville.
Simpson	WLBT	3	Jackson, Miss.
	WJTV	12	Do.
	WAPT	16	Do.
Smith	WLBT	3	Do.
	WJTV	12	Do.
	WDAM	7	Laurel-Hattiesburg.
Stone	WLOX	13	Biloxi-Gulfport-Pascagoula.
	WDAM	7	Laurel-Hattiesburg.
	WKRG	5	Mobile-Pensacola.
	WWL	4	New Orleans.
	WDSU	6	Do.

RULES AND REGULATIONS

SIGNIFICANTLY VIEWED SIGNALS—Continued

County	Call letters, channel number and market name		
MISSISSIPPI—continued			
Sunflower	WABG	6	Greenwood-Greenville.
	WLBT	3	Jackson, Miss.
	WJTV	12	Do.
Tallahatchie	WABG	6	Greenwood-Greenville.
	WREC	3	Memphis.
	WMC	5	Do.
	WHBQ	13	Do.
Tate	WREC	3	Do.
	WMC	5	Do.
	WHBQ	13	Do.
Tippah	WREC	3	Do.
	WMC	5	Do.
	WHBQ	13	Do.
Tishomingo	WTWV	9	Tupelo.
	WREC	3	Memphis.
	WMC	5	Do.
	WHBQ	13	Do.
Tunica	WREC	3	Do.
	WMC	5	Do.
	WHBQ	13	Do.
Union	WREC	3	Do.
	WMC	5	Do.
	WHBQ	13	Do.
Walthall	WLBT	3	Jackson, Miss.
	WJTV	12	Do.
	WLOX	13	Biloxi-Gulfport-Pascagoula.
	WDAM	7	Laurel-Hattiesburg.
	WWL	4	New Orleans.
	WDSU	6	Do.
Warren	WLBT	3	Jackson, Miss.
	WJTV	12	Do.
	KNOE	8	Monroe-El Dorado.
Washington	WLBT	3	Jackson, Miss.
	WJTV	12	Do.
	KTVE	10	Monroe-El Dorado.
	WABG	6	Greenwood-Greenville.
Wayne	WDAM	7	Laurel-Hattiesburg.
	WLOX	13	Biloxi-Gulfport-Pascagoula.
	WTOK	11	Meridian.
	WEAR	3	Mobile-Pensacola.
	WKRG	5	Do.
Webster	WCBI	4	Columbus, Miss.
	WABG	6	Greenwood-Greenville.
	WLBT	3	Jackson, Miss.
Wilkinson	WBRZ	2	Baton Rouge.
	WAFB	9	Do.
Winston	WTOK	11	Meridian.
	WCBI	4	Columbus, Miss.
	WLBT	3	Jackson, Miss.
Yalobusha	WREC	3	Memphis.
	WMC	5	Do.
	WHBQ	13	Do.
	WABG	6	Greenwood-Greenville.
Yazoo	WLBT	3	Jackson, Miss.
	WJTV	12	Do.
	WABG	6	Greenwood-Greenville.
MISSOURI			
Adair	KTVO	3	Ottumwa-Kirksville.
	KHQA	7	Quincy-Hannibal.
	WGEM	10	Do.
Andrew	WDAF	4	Kansas City.
	KCMO	5	Do.
	KMBC	9	Do.
	KQTV	2	St. Joseph.
Atchison	KMTV	3	Omaha.
	WOW	6	Do.
	KETV	7	Do.
	KQTV	2	St. Joseph.
Audrain	KOMU	8	Columbia-Jefferson City.
	KRCG	13	Do.
	KHQA	7	Quincy-Hannibal.
	WGEM	10	Do.
Barry	KYTV	3	Springfield, Mo.
	KTTS	10	Do.
	KOAM	7	Joplin-Pittsburg.
	KODE	12	Do.
Barton	KOAM	7	Do.
	KODE	12	Do.
	KUHI	16	Do.
Bates	WDAF	4	Kansas City.
	KCMO	5	Do.
	KMBC	9	Do.
Benton	KMOS	6	Columbia-Jefferson City.
	KCMO	5	Kansas City.
	KMBC	9	Do.
	KYTV	3	Springfield, Mo.

SIGNIFICANTLY VIEWED SIGNALS—Continued

County	Call letters, channel number and market name		
MISSOURI—continued			
Bollinger	WPSD	6	Paducah-Cape Girardeau-Harrisburg.
	KFVS	12	Do.
Boone	KOMU	8	Columbia-Jefferson City.
	KRCG	13	Do.
Buchanan	KQTV	2	St. Joseph.
	WDAF	4	Kansas City.
	KCMO	5	Do.
	KMBC	9	Do.
Butler	WSIL+	3	Paducah-Cape Girardeau-Harrisburg.
	WPSD	6	Do.
	KFVS	12	Do.
Caldwell	WDAF	4	Kansas City.
	KCMO	5	Do.
	KMBC	9	Do.
	KQTV	2	St. Joseph.
Callaway	KOMU	8	Columbia-Jefferson City.
	KRCG	13	Do.
Camden	KYTV	3	Springfield, Mo.
	KTTS	10	Do.
	KMTC	27	Do.
	KOMU	8	Columbia-Jefferson City.
	KRCG	13	Do.
Cape Girardeau	WSIL+	3	Paducah-Cape Girardeau-Harrisburg.
	WPSD	6	Do.
	KFVS	12	Do.
Carroll	WDAF	4	Kansas City.
	KCMO	5	Do.
	KMBC	9	Do.
Carter	WPSD	6	Paducah-Cape Girardeau-Harrisburg.
	KFVS	12	Do.
	KAIT	8	Jonesboro.
Cass	WDAF	4	Kansas City.
	KCMO	5	Do.
	KMBC	9	Do.
	KCIT	50	Do.
Cedar	KYTV	3	Springfield, Mo.
	KTTS	10	Do.
	KOAM	7	Joplin-Pittsburg.
	KODE	12	Do.
Chariton	KOMU	8	Columbia-Jefferson City.
	KRCG	13	Do.
	WDAF	4	Kansas City.
	KCMO	5	Do.
	KMBC	9	Do.
Christian	KYTV	3	Springfield, Mo.
	KTTS	10	Do.
	KMTC	27	Do.
Clark	KHQA	7	Quincy-Hannibal.
	WGEM	10	Do.
	KTVO	3	Ottumwa-Kirksville.
Clay	WDAF	4	Kansas City.
	KCMO	5	Do.
	KMBC	9	Do.
	KBMA	41	Do.
	KCIT	50	Do.
Clinton	WDAF	4	Do.
	KCMO	5	Do.
	KMBC	9	Do.
	KCIT	50	Do.
	KQTV	2	St. Joseph.
Cole	KOMU	8	Columbia-Jefferson City.
	KRCG	13	Do.
Cooper	KOMU	8	Columbia-Jefferson City.
	KRCG	13	Do.
Crawford	KTVI	2	St. Louis.
	KMOX	4	Do.
	KSD	5	Do.
	KPLR	11	Do.
	KRCG	13	Columbia-Jefferson City.
Dade	KYTV	3	Springfield, Mo.
	KTTS	10	Do.
	KOAM	7	Joplin-Pittsburg.
	KODE	12	Do.
Dallas	KYTV	3	Springfield, Mo.
	KTTS	10	Do.
	KMTC	27	Do.
Daviess	WDAF	4	Kansas City.
	KCMO	5	Do.
	KMBC	9	Do.
	KQTV	2	St. Joseph.
De Kalb	WDAF	4	Kansas City.
	KCMO	5	Do.
	KMBC	9	Do.
	KQTV	2	St. Joseph.

SIGNIFICANTLY VIEWED SIGNALS—Continued

County	Call letters, channel number and market name		
MISSOURI—continued			
Dent	KTVI	2	St. Louis.
	KMOX	4	Do.
	KSD	5	Do.
	KPLR	11	Do.
	KDNL	30	Do.
	KRCG	13	Columbia-Jefferson City.
Douglas	KYTV	3	Springfield, Mo.
	KTTS	10	Do.
	KMTC	27	Do.
Dunklin	WREC	3	Memphis.
	WMC	5	Do.
	WHBQ	13	Do.
	KAIT	8	Jonesboro.
	WPSD	6	Paducah-Cape Girardeau-Harrisburg.
	KFVS	12	Do.
Franklin	KTVI	2	St. Louis.
	KMOX	4	Do.
	KSD	5	Do.
	KPLR	11	Do.
	KDNL	30	Do.
Gasconade	KTVI	2	Do.
	KMOX	4	Do.
	KSD	5	Do.
	KPLR	11	Do.
	KRCG	13	Columbia-Jefferson City.
Gentry	KQTV	2	St. Joseph.
	WDAF	4	Kansas City.
	KCMO	5	Do.
	KMBC	9	Do.
Greene	KYTV	3	Springfield, Mo.
	KTTS	10	Do.
	KMTC	27	Do.
Grundy	WDAF	4	Kansas City.
	KCMO	5	Do.
	KMBC	9	Do.
	KTVO	3	Ottumwa-Kirksville.
	KQTV	2	St. Joseph.
Harrison	KQTV	2	Do.
	WDAF	4	Kansas City.
	KCMO	5	Do.
Henry	WDAF	4	Do.
	KCMO	5	Do.
	KMBC	9	Do.
Hickory	KYTV	3	Springfield, Mo.
	KTTS	10	Do.
	KMTC	27	Do.
Holt	KQTV	2	St. Joseph.
	WDAF	4	Kansas City.
	KCMO	5	Do.
	KMBC	9	Do.
Howard	KOMU	8	Columbia-Jefferson City.
	KRCG	13	Do.
Howell	KYTV	3	Springfield, Mo.
	KMTC	27	Do.
Iron*	KTVI	2	St. Louis.
	KMOX	4	Do.
	KSD	5	Do.
	KPLR	11	Do.
	KFVS	12	Paducah-Cape Girardeau-Harrisburg.
Jackson	WDAF	4	Kansas City.
	KCMO	5	Do.
	WMBC	9	Do.
	KBMA	41	Do.
	KCIT	50	Do.
Jasper	KOAM	7	Joplin-Pittsburg.
	KODE	12	Do.
	KUHI	16	Do.
Jefferson	KTVI	2	St. Louis.
	KMOX	4	Do.
	KSD	5	Do.
	KPLR	11	Do.
	KDNL	30	Do.
Johnson	WDAF	4	Kansas City.
	KCMO	5	Do.
	KMBC	9	Do.
Knox	KHQA	7	Quincy-Hannibal.
	WGEM	10	Do.
	KTVO	3	Ottumwa-Kirksville.
Laclede	KYTV	3	Springfield, Mo.
	KTTS	10	Do.
	KMTC	27	Do.
Lafayette	WDAF	4	Kansas City.
	KCMO	5	Do.
	KMBC	9	Do.
	KCIT	50	Do.
Lawrence	KYTV	3	Springfield, Mo.
	KTTS	10	Do.
	KOAM	7	Joplin-Pittsburg.
	KODE	12	Do.
Lewis	KHQA	7	Quincy-Hannibal.
	WGEM	10	Do.
	KTVO	3	Ottumwa-Kirksville.

Significantly Viewed Signals—Continued

County	Call letters, channel number and market name		

MISSOURI—continued

County	Call letters	Channel	Market
Lincoln	KTVI	2	St. Louis.
	KMOX	4	Do.
	KSD	5	Do.
	KPLR	11	Do.
Linn	WDAF	4	Kansas City.
	KCMO	5	Do.
	KMBC	9	Do.
	KTVO	3	Ottumwa-Kirksville.
Livingston	WDAF	4	Kansas City.
	KCMO	5	Do.
	KMBC	9	Do.
	KQTV	2	St. Joseph.
McDonald	KOAM	7	Joplin-Pittsburg.
	KODE	12	Do.
	KUHI	16	Do.
Macon	KHQA	7	Quincy-Hannibal.
	WGEM	10	Do.
	KOMU	8	Columbia-Jefferson City.
	KTVO	3	Ottumwa-Kirksville.
Madison	KTVI	2	St. Louis.
	KMOX	4	Do.
	KSD	5	Do.
	KPLR	11	Do.
	KFVS	12	Paducah-Cape Girardeau-Harrisburg.
Maries	KOMU	8	Columbia-Jefferson City.
	KRCG	13	Do.
	KTVI	2	St. Louis.
	KMOX	4	Do.
	KSD	5	Do.
Marion	KHQA	7	Quincy-Hannibal.
	WGEM	10	Do.
	WJJY	14	Jacksonville, Ill.
Mercer	KTVO	3	Ottumwa-Kirksville.
	KRNT	8	Des Moines.
	WDAF	4	Kansas City.
	KCMO	5	Do.
	KQTV	2	St. Joseph.
Miller	KOMU	8	Columbia-Jefferson City.
	KRCG	13	Do.
Mississippi	WSIL+	3	Paducah-Cape Girardeau-Harrisburg.
	WPSD	6	Do.
	KFVS	12	Do.
Moniteau	KOMU	8	Columbia-Jefferson City.
	KRCG	13	Do.
Monroe	KHQA	7	Quincy-Hannibal.
	WGEM	10	Do.
	KOMU	8	Columbia-Jefferson City.
	KRCG	13	Do.
Montgomery	KOMU	8	Do.
	KRCG	13	Do.
	KTVI	2	St. Louis.
	KMOX	4	Do.
	KSD	5	Do.
	KPLR	11	Do.
	KDNL	30	Do.
Morgan	KOMU	8	Columbia-Jefferson City.
	KRCG	13	Do.
	KYTV	3	Springfield, Mo.
New Madrid	WPSD	6	Paducah-Cape Girardeau-Harrisburg.
	KFVS	12	Do.
Newton	KOAM	7	Joplin-Pittsburg.
	KODE	12	Do.
	KUHI	16	Do.
Nodaway	KQTV	2	St. Joseph.
	WDAF	4	Kansas City.
	KCMO	5	Do.
	KMTV	3	Omaha.
	WOW	6	Do.
Oregon	KYTV	3	Springfield, Mo.
	KAIT	8	Jonesboro.
Osage	KOMU	8	Columbia-Jefferson City.
	KRCG	13	Do.
	KTVI	2	St. Louis.
Ozark	KYTV	3	Springfield, Mo.
	KTTS	10	Do.
Pemiscot	WREC	3	Memphis.
	WMC	5	Do.
	WHBQ	13	Do.
	KFVS	12	Paducah-Cape Girardeau-Harrisburg.
Perry	KTVI	2	St. Louis.
	KMOX	4	Do.
	KSD	5	Do.
	KPLR	11	Do.
	KFVS	12	Paducah-Cape Girardeau-Harrisburg.

Significantly Viewed Signals—Continued

County	Call letters, channel number and market name		

MISSOURI—continued

County	Call letters	Channel	Market
Pettis	KMOS	6	Columbia-Jefferson City.
	KOMU	8	Do.
	WDAF	4	Kansas City.
	KCMO	5	Do.
	KMBC	9	Do.
Phelps*	KOMU	8	Columbia-Jefferson City.
	KRCG	13	Do.
	KTVI	2	St. Louis.
Pike	KHQA	7	Quincy-Hannibal.
	WGEM	10	Do.
	KTVI	2	St. Louis.
	KMOX	4	Do.
	KSD	5	Do.
	KPLR	11	Do.
Platte	WDAF	4	Kansas City.
	KCMO	5	Do.
	KMBC	9	Do.
	KBMA	41	Do.
	KCIT	50	Do.
	KQTV	2	St. Joseph.
Polk	KYTV	3	Springfield, Mo.
	KTTS	10	Do.
	KMTC	27	Do.
Pulaski	KYTV	3	Do.
	KOMU	8	Columbia-Jefferson City.
	KRCG	13	Do.
Putnam	KTVO	3	Ottumwa-Kirksville.
Ralls	KHQA	7	Quincy-Hannibal.
	WGEM	10	Do.
	WJJY	14	Jacksonville, Ill.
Randolph	KOMU	8	Columbia-Jefferson City.
	KRCG	13	Do.
Ray	WDAF	4	Kansas City.
	KCMO	5	Do.
	KMBC	9	Do.
	KCIT	50	Do.
Reynolds	KFVS	12	Paducah-Cape Girardeau-Harrisburg.
	KTVI	2	St. Louis.
	KMOX	4	Do.
	KSD	5	Do.
Ripley	WPSD	6	Paducah-Cape Girardeau-Harrisburg.
	KFVS	12	Do.
	KAIT	8	Jonesboro.
St. Charles	KTVI	2	St. Louis.
	KMOX	4	Do.
	KSD	5	Do.
	KPLR	11	Do.
	KDNL	30	Do.
St. Clair	WDAF	4	Kansas City.
	KCMO	5	Do.
	KMBC	9	Do.
	KYTV	3	Springfield, Mo.
	KTTS	10	Do.
St. Francois	KTVI	2	St. Louis.
	KMOX	4	Do.
	KSD	5	Do.
	KPLR	11	Do.
St. Louis including city of St. Louis.	KTVI	2	Do.
	KMOX	4	Do.
	KSD	5	Do.
	KPLR	11	Do.
	KDNL	30	Do.
Ste. Genevieve	KTVI	2	Do.
	KMOX	4	Do.
	KSD	5	Do.
	KPLR	11	Do.
	KDLN	30	Do.
Saline	WDAF	4	Kansas City.
	KCMO	5	Do.
	KMBC	9	Do.
	KMOS	6	Columbia-Jefferson City.
	KOMU	8	Do.
	KRCG	13	Do.
Schuyler	KTVO	3	Ottumwa-Kirksville.
	KHQA	7	Quincy-Hannibal.
	WGEM	10	Do.
Scotland	KHQA	7	Do.
	WGEM	10	Do.
	KTVO	3	Ottumwa-Kirksville.
Scott	WSIL+	3	Paducah-Cape Girardeau-Harrisburg.
	WSPD	6	Do.
	KFVS	12	Do.
Shannon	KYTV	3	Springfield, Mo.
Shelby	KHQA	7	Quincy-Hannibal.
	WGEM	10	Do.
	KTVO	3	Ottumwa-Kirksville.
Stoddard	WPSD	6	Paducah-Cape Girardeau-Harrisburg.
	KFVS	12	Do.
Stone	KYTV	3	Springfield, Mo.
	KTTS	10	Do.
	KMTC	27	Do.

Significantly Viewed Signals—Continued

County	Call letters, channel number and market name		

MISSOURI—continued

County	Call letters	Channel	Market
Sullivan*	KTVO	3	Ottumwa-Kirksville.
	WGEM	10	Quincy-Hannibal.
Taney	KYTV	3	Springfield, Mo.
	KTTS	10	Do.
	KMTC	27	Do.
Texas	KTYV	3	Do.
	KTTS	10	Do.
	KMTC	27	Do.
Vernon	KOAM	7	Joplin-Pittsburg.
	KODE	12	Do.
	KUHI	16	Do.
	KCMO	5	Kansas City.
Warren	KTVI	2	St. Louis.
	KMOX	4	Do.
	KSD	5	Do.
	KPLR	11	Do.
Washington	KTVI	2	Do.
	KMOX	4	Do.
	KSD	5	Do.
	KPLR	11	Do.
Wayne	WPSD	6	Paducah-Cape Girardeau-Harrisburg.
	KFVS	12	Do.
Webster	KYTV	3	Springfield, Mo.
	KTTS	10	Do.
	KMTC	27	Do.
Worth	KQTV	2	St. Joseph.
	WDAF	4	Kansas City.
	KCMO	5	Do.
Wright	KYTV	3	Springfield, Mo.
	KTTS	10	Do.
	KMTC	27	Do.

MONTANA

County	Call letters	Channel	Market
Beaverhead	KXLF+	4	Butte.
	KGVO+	13	Missoula.
Big Horn	KOOK	2	Billings.
	KULR	8	Do.
Blaine	KRTV	3	Great Falls.
	KFBB	5	Do.
	CJLH	7	Canada.
Broadwater	KXLF+	4	Butte.
	KRTV	3	Great Falls.
	KFBB	5	Do.
Carbon	KOOK	2	Billings.
	KULR	8	Do.
Carter	KOTA+	3	Rapid City.
	KXGN	5	Glendive.
Cascade	KRTV	3	Great Falls.
	KFBB	5	Do.
Chouteau	KRTV	3	Do.
	KFBB	5	Do.
Custer	KOOK	2	Billings.
	KULR	8	Do.
	KYUS	3	Miles City, Mont.
Daniels	KUMV	8	Minot-Bismarck.
	KXMD	11	Do.
	CKCK	2	Canada.
Dawson	KXGN	5	Glendive.
	KUMV	8	Minot-Bismarck.
Deer Lodge	KXLF+	4	Butte.
	KGVO+	13	Missoula.
Fallon	KDIX	2	Dickinson, N. Dak.
	KXGN	5	Glendive.
Fergus	KOOK	2	Billings.
	KULR	8	Do.
	KFBB	5	Great Falls.
Flathead	KCFW	9	Missoula.
	KREM	2	Spokane.
	KXLY	4	Do.
Gallatin	KXLF+	4	Butte.
	KGVO+	13	Missoula.
Garfield	KOOK	2	Billings.
	KULR	8	Do.
Glacier	KRTV	3	Great Falls.
	KFBB	5	Do.
	CJLH	7	Canada.
Golden Valley	KOOK	2	Billings.
	KULR	8	Do.
Granite	KXLF+	4	Butte.
	KGVO+	13	Missoula.
Hill	KRTV	3	Great Falls.
	KFBB	5	Do.
	CFCN	4	Canada.
	CJLH	7	Do.
Jefferson	KXLF+	4	Butte.
	KFBB	5	Great Falls.
	KGVO+	13	Missoula.
Judith Basin	KOOK	2	Billings.
	KULR	8	Do.
	KRTV	3	Great Falls.
	KFBB	5	Do.
Lake	KGVO+	13	Missoula.
	KXLF+	4	Butte.
	KXLY	4	Spokane.
Lewis and Clark	KBLL	12	Helena.
	KXLF+	4	Butte.
	KFBB	5	Great Falls.

RULES AND REGULATIONS

SIGNIFICANTLY VIEWED SIGNALS—Continued

MONTANA—continued

County	Call letters	Channel	Market name
Liberty	KRTV	3	Do.
	KFBB	5	Do.
	CFCN	4	Canada.
	CJLH	7	Do.
Lincoln	KREM	2	Spokane.
	KXLY	4	Do.
	KHQ	6	Do.
	KCFW	9	Missoula.
McCone	KUMV	8	Minot-Bismarck.
	KXGN	5	Glendive.
Madison	KXLF+	4	Butte.
	KGVO	13	Missoula.
Meagher	KRTV	3	Great Falls.
	KFBB	5	Do.
	KXLF+	4	Butte.
Mineral	KXLY	4	Spokane.
	KXLF+	4	Butte.
	KGVO+	13	Missoula.
Missoula	KGVO+	13	Missoula.
	KXLF+	4	Butte.
Musselshell	KOOK	2	Billings.
	KULR	8	Do.
Park	KOOK	2	Billings.
	KULR	8	Do.
	KXLF+	4	Butte.
Petroleum	KOOK	2	Billings.
	KULR	8	Do.
Phillips	KRTV	3	Great Falls.
	KFBB	5	Do.
	KOOK	2	Billings.
Pondera	KRTV	3	Great Falls.
	KFBB	5	Do.
	CJLH	7	Canada.
Powder River	KOOK	2	Billings.
	KULR	8	Do.
	KOTA+	3	Rapid City.
Powell	KXLF+	4	Butte.
	KGVO+	13	Missoula.
Prairie	KXGN	5	Glendive.
	KYUS	3	Miles City, Mont.
Ravalli	KGVO+	13	Missoula.
	KXLF+	4	Butte.
Richland	KUMV	8	Minot-Bismarck.
	KXMD	11	Do.
	KXGN	5	Glendive.
Roosevelt	KUMV	8	Minot-Bismarck.
	KXMD	11	Do.
	CKCK	2	Canada.
Rosebud	KOOK	2	Billings.
	KULR	8	Do.
	KYUS	3	Miles City, Mont.
Sanders	KREM	2	Spokane.
	KXLY	4	Do.
	KHQ	6	Do.
	KGVO+	13	Missoula.
Sheridan	KUMV	8	Minot-Bismarck.
	KXMD	11	Do.
	CKCK	2	Canada.
Silver Bow	KXLF+	4	Butte.
	KGVO+	13	Missoula.
Stillwater	KOOK	2	Billings.
	KULR	8	Do.
Sweet Grass	KOOK	2	Do.
	KULR	8	Do.
Teton	KRTV	3	Great Falls.
	KFBB	5	Do.
Toole	KRTV	3	Do.
	KFBB	5	Do.
	CFCN	4	Canada.
	CJLH	7	Do.
Treasure	KOOK	2	Billings.
	KULR	8	Do.
Valley	KUMV	8	Minot-Bismarck.
	KXMD	11	Do.
	CKCK	2	Canada.
Wheatland	KOOK	2	Billings.
	KULR	8	Do.
Wibaux	KDIX	2	Dickinson, N. Dak.
	KXGN	5	Glendive.
	KUMV	8	Minot-Bismarck.
Yellowstone	KOOK	2	Billings.
	KULR	8	Do.

NEBRASKA

County	Call letters	Channel	Market name
Adams	KHAS	5	Lincoln-Hastings-Kearney.
	KOLN+	10	Do.
	KHOL+	13	Do.
Antelope	KHQL	8	Do.
	KOLN	10	Do.
	KTIV	4	Sioux City.
	KCAU	9	Do.
Arthur	KNOP	2	North Platte.
	KHOL+	13	Lincoln-Hastings-Kearney.
Banner	KSTF	10	Cheyenne.
	KDUH	4	Rapid City.
Blaine	KNOP	2	North Platte.

NEBRASKA—continued

County	Call letters	Channel	Market name
Boone	KHAS	5	Lincoln-Hastings-Kearney.
	KHQL	8	Do.
	KOLN+	10	Do.
Box Butte	KSTF	10	Cheyenne.
	KDUH	4	Rapid City.
Boyd	KORN	5	Sioux Falls-Mitchell.
	KELO+	11	Do.
Brown	KELO+	11	Do.
Buffalo	KHAS	5	Lincoln-Hastings-Kearney.
	KOLN+	10	Do.
	KHOL+	13	Do.
Burt	KMTV	3	Omaha.
	WOW	6	Do.
	KETV	7	Do.
Butler	KMTV	3	Do.
	WOW	6	Do.
	KETV	7	Do.
	KOLN+	10	Lincoln-Hastings-Kearney.
Cass	KMTV	3	Omaha.
	WOW	6	Do.
	KETV	7	Do.
	KOLN	10	Lincoln-Hastings-Kearney.
Cedar	KTIV	4	Sioux City.
	KCAU	9	Do.
	KMEG	14	Do.
	KELO+	11	Sioux Falls-Mitchell.
Chase	KHOL+	13	Lincoln-Hastings-Kearney.
	KOMC	8	Wichita-Hutchinson.
Cherry	KELO+	11	Sioux Falls-Mitchell.
	KNOP	2	North Platte.
	KDUH	4	Rapid City.
Cheyenne	KTVS	3	Cheyenne.
	KSTF	10	Do.
	KDUH	4	Rapid City.
Clay	KHTL	4	Lincoln-Hastings-Kearney.
	KHAS	5	Do.
	KOLN+	10	Do.
	KHOL+	13	Do.
Colfax	KMTV	3	Omaha.
	WOW	6	Do.
	KETV	7	Do.
	KOLN+	10	Lincoln-Hastings-Kearney.
Cuming	KMTV	3	Omaha.
	WOW	6	Do.
	KETV	7	Do.
	KTIV	4	Sioux City.
	KCAU	9	Do.
Custer	KHAS	5	Lincoln-Hastings-Kearney.
	KOLN+	10	Do.
	KHOL+	13	Do.
	KNOP	2	North Platte.
Dakota	KTIV	4	Sioux City.
	KCAU	9	Do.
	KMEG	14	Do.
Dawes	KDUH	4	Rapid City.
	KSTF	10	Cheyenne.
Dawson	KOLN+	10	Lincoln-Hastings-Kearney.
	KHOL+	13	Do.
	KNOP	2	North Platte.
Deuel	KTVS	3	Cheyenne.
	KHOL+	13	Lincoln-Hastings-Kearney.
	KNOP	2	North Platte.
Dixon	KTIV	4	Sioux City.
	KCAU	9	Do.
	KMEG	14	Do.
Dodge	KMTV	3	Omaha.
	WOW	6	Do.
	KETV	7	Do.
Douglas	KMTV	3	Do.
	WOW	6	Do.
	KETV	7	Do.
Dundy	KAYS+	7	Wichita-Hutchinson.
	KOMC	8	Do.
	KHOL+	13	Lincoln-Hastings-Kearney.
Fillmore	KHTL	4	Do.
	KHAS	5	Do.
	KOLN+	10	Do.
	KHOL+	13	Do.
Franklin	KHAS	5	Do.
	KOLN+	10	Do.
	KHOL+	13	Do.
Frontier	KOLN+	10	Do.
	KHOL+	13	Do.
	KNOP	2	North Platte.
	KOMC	8	Wichita-Hutchinson.

NEBRASKA—continued

County	Call letters	Channel	Market name
Furnas	KOLN+	10	Lincoln-Hastings-Kearney.
	KHOL+	13	Do.
	KOMC	8	Wichita-Hutchinson.
Gage	KOLN+	10	Lincoln-Hastings-Kearney.
	KMTV	3	Omaha.
	KETV	7	Do.
Garden	KTVS	3	Cheyenne.
	KSTF	10	Do.
	KHOL+	13	Lincoln-Hastings-Kearney.
	KNOP	2	North Platte.
	KDUH	4	Rapid City.
Garfield	KHAS	5	Lincoln-Hastings-Kearney.
	KHQL	8	Do.
	KOLN+	10	Do.
Gosper	KHAS	5	Do.
	KOLN+	10	Do.
	KHOL+	13	Do.
Grant	KDUH	4	Rapid City.
	KHOL+	13	Lincoln-Hastings-Kearney.
	KNOP	2	North Platte.
Greeley	KHAS	5	Lincoln-Hastings-Kearney.
	KHQL	8	Do.
	KOLN+	10	Do.
	KHOL+	13	Do.
Hall	KHAS	5	Do.
	KOLN+	10	Do.
	KHOL+	13	Do.
Hamilton	KHAS	5	Do.
	KOLN+	10	Do.
	KHOL+	13	Do.
Harlan	KHAS	5	Do.
	KOLN+	10	Do.
	KHOL+	13	Do.
Hayes	KNOP	2	North Platte.
	KOMC	8	Wichita-Hutchinson.
Hitchcock	KAYS	7	Do.
	KOMC	8	Do.
	KHOL+	13	Lincoln-Hastings-Kearney.
Holt	KHQL	8	Do.
	KOLN+	10	Do.
	KTIV	4	Sioux City.
	KCAU	9	Do.
	KELO+	11	Sioux Falls-Mitchell.
Hooker	KNOP	2	North Platte.
	KDUH	4	Rapid City.
Howard	KHAS	5	Lincoln-Hastings-Kearney.
	KOLN+	10	Do.
	KHOL+	13	Do.
Jefferson	KHTL	4	Do.
	KHAS	5	Do.
	KOLN+	10	Do.
Johnson	KMTV	3	Omaha.
	WOW	6	Do.
	KETV	7	Do.
	KOLN+	10	Lincoln-Hastings-Kearney.
Kearney	KHAS	5	Do.
	KOLN+	10	Do.
	KHOL+	13	Do.
Keith	KNOP	2	North Platte.
	KHOL+	13	Lincoln-Hastings-Kearney.
Keya Paha	KELO+	11	Sioux Falls-Mitchell.
Kimball	KTVS	3	Cheyenne.
	KFBC	5	Do.
	KSTF	10	Do.
	KDUH	4	Rapid City.
Knox	KTIV	4	Sioux City.
	KCAU	9	Do.
	KHQL	8	Lincoln-Hastings-Kearney.
	KORN	5	Sioux Falls-Mitchell.
	KELO+	11	Do.
	KSOO+	13	Do.
Lancaster	KOLN+	10	Lincoln-Hastings-Kearney.
	KMTV	3	Omaha.
	WOW	6	Do.
	KETV	7	Do.
Lincoln	KNOP	2	North Platte.
	KOLN+	10	Lincoln-Hastings-Kearney.
	KHOL+	13	Do.
Logan	KNOP	2	North Platte.
	KHOL+	13	Lincoln-Hastings-Kearney.
Loup	KHAS	5	Do.
	KHQL	8	Do.
	KOLN+	10	Do.
McPherson	KNOP	2	North Platte.
	KHOL+	13	Lincoln-Hastings-Kearney.

County	Call letters, channel number and market name		County	Call letters, channel number and market name		County	Call letters, channel number and market name	
NEBRASKA—continued			NEBRASKA—continued			NEW HAMPSHIRE		
Madison	KTIV	4 Sioux City.	Stanton	KTIV	4 Sioux City.	Belknap	WCSH	6 Portland-Poland Spring.
	KCAU	9 Do.		KCAU	9 Do.		WMTW	8 Do.
	KHQL	8 Lincoln-Hastings-Kearney.		KHQL	8 Lincoln-Hastings-Kearney.		WGAN	13 Do.
	KOLN+	10 Do.		KOLN+	10 Do.		WBZ	4 Boston.
	WOW	6 Omaha.		KMTV	3 Omaha.		WHDH	5 Do.
Merrick	KHAS	5 Lincoln-Hastings-Kearney.		WOW	6 Do.		WMUR	9 Manchester.
	KHQL	8 Do.		KETV	7 Do.	Carroll	WCSH	6 Portland-Poland Spring.
	KOLN+	10 Do.	Thayer	KHTL	4 Lincoln-Hastings-Kearney.		WMTW	8 Do.
	KHOL+	13 Do.		KHAS	5 Do.		WGAN	13 Do.
Morrill	KSTF	10 Cheyenne.		KOLN+	10 Do.	Cheshire	WBZ	4 Boston.
	KDUH	4 Rapid City.		KHOL+	13 Do.		WHDH	5 Do.
Nance	KHAS	5 Lincoln-Hastings-Kearney.	Thomas	KNOP	2 North Platte.		WNAC	7 Do.
	KHQL	8 Do.	Thurston	KTIV	4 Sioux City.		WTIC	3 Hartford-New Haven.
	KOLN+	10 Do.		KCAU	9 Do.		WMUR	9 Manchester.
	KMTV	3 Omaha.		KMEG	14 Do.		WWLP+	22 Springfield, Mass.
Nemaha	KMTV	3 Do.		KMTV	3 Omaha.	Coos	WCSH	6 Portland-Poland Spring.
	WOW	6 Do.		WOW	6 Do.		WMTW	8 Do.
	KETV	7 Do.		KETV	7 Do.		WGAN	13 Do.
Nuckolls	KHTL	4 Lincoln-Hastings-Kearney.	Valley	KHAS	5 Lincoln-Hastings-Kearney.		WCAX	3 Burlington-Plattsburgh.
	KHAS	5 Do.		KHQL	8 Do.	Grafton	WMTW	8 Portland-Poland Spring.
	KOLN+	10 Do.		KOLN+	10 Do.		WCAX	3 Burlington-Plattsburgh.
	KHOL+	13 Do.		KHOL+	13 Do.	Hillsborough	WBZ	4 Boston.
Otoe	KMTV	3 Omaha.	Washington	KMTV	3 Omaha.		WHDH	5 Do.
	WOW	6 Do.		WOW	6 Do.		WNAC	7 Do.
	KETV	7 Do.		KETV	7 Do.		WSBK	38 Do.
	KOLN+	10 Lincoln-Hastings-Kearney.	Wayne	KTIV	4 Sioux City.		WKBG	56 Do.
Pawnee	KMTV	3 Omaha.		KCAU	9 Do.		WMUR	9 Manchester.
	WOW	6 Do.		KMEG	14 Do.	Merrimack*	WBZ	4 Boston.
	KETV	7 Do.	Webster	KHAS	5 Lincoln-Hastings-Kearney.		WHDH	5 Do.
	KOLN+	10 Lincoln-Hastings-Kearney.		KOLN+	10 Do.		WNAC	7 Do.
Perkins	KHOL +13 Do.			KHOL+	13 Do.		WMUR	9 Manchester.
	KTVS	3 Cheyenne.	Wheeler	KHAS	5 Do.		WCSH	6 Portland-Poland Spring.
	KNOP	2 North Platte.		KHQL	8 Do.		WMTW	8 Do.
Phelps	KHAS	5 Lincoln-Hastings-Kearney.		KOLN+	10 Do.		WBZ	4 Boston.
	KOLN+	10 Do.	York	KHTL	4 Do.	Rockingham	WHDH	5 Do.
	KHOL+	13 Do.		KHAS	5 Do.		WNAC	7 Do.
Pierce	KTIV	4 Sioux City.		KOLN+	10 Do.		WSBK	38 Do.
	KCAU	9 Do.					WKBG	56 Do.
	KHQL	8 Lincoln-Hastings-Kearney.	NEVADA				WMUR	9 Manchester.
	KELO+	11 Sioux Falls-Mitchell.	Churchill	KCRL	4 Reno.	Strafford	WBZ	4 Boston.
Platte	KHQL	8 Lincoln-Hastings-Kearney.		KOLO	8 Do.		WHDH	5 Do.
	KOLN+	10 Do.	Clark	KORK	3 Las Vegas.		WNAC	7 Do.
	KMTV	3 Omaha.		KHBV	5 Do.		WMUR	9 Manchester.
	WOW	6 Do.		KLAS	8 Do.		WCSH	6 Portland-Poland Spring.
	KETV	7 Do.		KSHO	13 Do.		WMTW	8 Do.
Polk	KHAS	5 Lincoln-Hastings-Kearney.	Douglas	KTVN	2 Reno.		WGAN	13 Do.
	KHQL	8 Do.		KCRL	4 Do.	Sullivan	WBZ	4 Boston.
	KOLN+	10 Do.		KOLO	8 Do.		WHDH	5 Do.
	KMTV	3 Omaha.		KTVU	2 San Francisco.		WCAX	3 Burlington-Plattsburgh.
	KETV	7 Do.	Elko	KSL	5 Salt Lake City.		WMUR	9 Manchester.
Red Willow	KOMC	8 Wichita-Hutchinson.		KBOI	2 Boise.		WWLP+	22 Springfield, Mass.
	KOLN+	10 Lincoln-Hastings-Kearney.		KTVB+	7 Do.			
	KHOL+	13 Do.		KOLO	8 Reno.	NEW JERSEY		
Richardson	KMTV	3 Omaha.	Esmeralda	KOLO	8 Do.	Atlantic	KYW	3 Philadelphia.
	WOW	6 Do.	Eureka	KUTV	2 Salt Lake City.		WFIL	6 Do.
	KETV	7 Do.		KCPX	4 Do.		WCAU	10 Do.
	KCMO	5 Kansas City.		KSL	5 Do.		WPHL	17 Do.
	KOLN+	10 Lincoln-Hastings-Kearney.	Humboldt	KOLO	8 Reno.		WTAF	29 Do.
	KQTV	2 St. Joseph.		KBOI	2 Boise.		WKBS	48 Do.
Rock	KELO+	11 Sioux Falls-Mitchell.		KTVB+	7 Do.	Bergen	WCBS	2 New York.
Saline	KHTL	4 Lincoln-Hastings-Kearney.	Lander	KTVN	2 Reno.		WNBC	4 Do.
	KOLN	10 Do.		KOLO	8 Do.		WNEW	5 Do.
	KMTV	3 Omaha.	Lincoln	KORK	3 Las Vegas.		WABC	7 Do.
	KETV	7 Do.		KLAS	8 Do.		WOR	9 Do.
Sarpy	KMTV	3 Do.		KCPX	4 Salt Lake City.		WPIX	11 Do.
	WOW	6 Do.	Lyon	KTVN	2 Reno.	Burlington	KYW	3 Philadelphia.
	KETV	7 Do.		KCRL	4 Do.		WFIL	6 Do.
Saunders	KMTV	3 Do.		KOLO	8 Do.		WCAU	10 Do.
	WOW	6 Do.	Mineral	KTVN	2 Do.		WPHL	17 Do.
	KETV	7 Do.		KCPL	4 Do.		WTAF	29 Do.
	KOLN+	10 Lincoln-Hastings-Kearney.		KOLO	8 Do.		WKBS	48 Do.
Scotts Bluff	KSTF	10 Cheyenne.	Nye	KTVN	2 Do.	Camden	KYW	3 Do.
	KDUH	4 Rapid City.		KCRL	4 Do.		WFIL	6 Do.
Seward	KOLN+	10 Lincoln-Hastings-Kearney.		KOLO	8 Do.		WCAU	10 Do.
	KMTV	3 Omaha.		KORK	3 Las Vegas.		WPHL	17 Do.
	WOW	6 Do.	Ormsby	KTVN	2 Reno.		WTAF	29 Do.
	KETV	7 Do.		KCRL	4 Do.		WKBS	48 Do.
Sheridan	KDUH	4 Rapid City.		KOLO	8 Do.	Cape May	KYW	3 Do.
Sherman	KHAS	5 Lincoln-Hastings-Kearney.	Pershing	KTVN	2 Do.		WFIL	6 Do.
	KOLN+	10 Do.		KCRL	4 Do.		WCAU	10 Do.
	KHOL+	13 Do.		KOLO	8 Do.		WPHL	17 Do.
Sioux	KSTF	10 Cheyenne.	Storey	KTVN	2 Do.		WKBS	48 Do.
	KDUH	4 Rapid City.		KCRL	4 Do.	Cumberland	KYW	3 Do.
				KOLO	8 Do.		WFIL	6 Do.
			Washoe	KTVN	2 Do.		WCAU	10 Do.
				KCRL	4 Do.		WPHL	17 Do.
				KOLO	8 Do.		WTAF	29 Do.
			White Pine	KUTV	2 Salt Lake City.		WKBS	48 Do.
				KCPX	4 Do.			
				KSL	5 Do.			

SIGNIFICANTLY VIEWED SIGNALS—Continued

NEW JERSEY—continued

County	Call letters, channel number and market name		
Essex	WCBS	2	New York.
	WNBC	4	Do.
	WNEW	5	Do.
	WABC	7	Do.
	WOR	9	Do.
	WPIX	11	Do.
Gloucester	KYW	3	Philadelphia.
	WFIL	6	Do.
	WCAU	10	Do.
	WPHL	17	Do.
	WKBS	48	Do.
Hudson	WCBS	2	New York.
	WNBC	4	Do.
	WNEW	5	Do.
	WABC	7	Do.
	WOR	9	Do.
	WPIX	11	Do.
Hunterdon	WCBS	2	Do.
	WNBC	4	Do.
	WNEW	5	Do.
	WABC	7	Do.
	WOR	9	Do.
	WPIX	11	Do.
	KYW	3	Philadelphia.
	WFIL	6	Do.
	WCAU	10	Do.
	WTAF	29	Do.
Mercer	KYW	3	Do.
	WFIL	6	Do.
	WCAU	10	Do.
	WPHL	17	Do.
	WKBS	48	Do.
	WCBS	2	New York.
	WNBC	4	Do.
	WNEW	5	Do.
	WABC	7	Do.
	WOR	9	Do.
	WPIX	11	Do.
Middlesex	WCBS	2	Do.
	WNBC	4	Do.
	WNEW	5	Do.
	WABC	7	Do.
	WOR	9	Do.
	WPIX	11	Do.
Monmouth	WCBS	2	Do.
	WNBC	2	Do.
	WNEW	5	Do.
	WABC	7	Do.
	WOR	9	Do.
	WPIX	11	Do.
Morris	WCBS	2	Do.
	WNBC	4	Do.
	WNEW	5	Do.
	WABC	7	Do.
	WOR	9	Do.
	WPIX	11	Do.
Ocean	WCBS	2	Do.
	WNBC	4	Do.
	WNEW	5	Do.
	WABC	7	Do.
	WOR	9	Do.
	WPIX	11	Do.
	WFIL	6	Philadelphia.
Passaic	WCBS	2	New York.
	WNBC	4	Do.
	WNEW	5	Do.
	WABC	7	Do.
	WOR	9	Do.
	WPIX	11	Do.
Salem	KYW	3	Philadelphia.
	WFIL	6	Do.
	WCAU	10	Do.
	WPHL	17	Do.
	WTAF	29	Do.
	WKBS	48	Do.
Somerset	WCBS	2	New York.
	WNBC	4	Do.
	WNEW	5	Do.
	WABC	7	Do.
	WOR	9	Do.
	WPIX	11	Do.
Sussex	WCBS	2	Do.
	WNBC	4	Do.
	WNEW	5	Do.
	WABC	7	Do.
	WOR	9	Do.
	WPIX	11	Do.
Union	WCBS	2	Do.
	WNBC	4	Do.
	WNEW	5	Do.
	WABC	7	Do.
	WOR	9	Do.
	WPIX	11	Do.
Warren	KYW	3	Philadelphia.
	WFIL	6	Do.
	WCAU	10	Do.
	WCBS	2	New York.
	WNBC	4	Do.
	WNEW	5	Do.
	WABC	7	Do.
	WPIX	11	Do.

NEW MEXICO

County	Call letters, channel number and market name		
Bernalillo	KOB	4	Albuquerque.
	KOAT	7	Do.
	KGGM	13	Do.
Catron	KOB	4	Do.
	KOAT	7	Do.
	KVOA	4	Tucson.
	KGUN	9	Do.
	KOLD	13	Do.
Chaves	KBIM	10	Roswell.
	KCBD	11	Lubbock.
Colfax	KOB	4	Albuquerque.
	KOAT	7	Do.
	KGGM	13	Do.
	KRDO	13	Colorado Springs-Pueblo.
Curry	KVII	7	Amarillo.
	KFDA+	10	Do.
	KCBD+	11	Lubbock.
De Baca	KCBD	11	Do.
	KOAT	7	Albuquerque.
	KBIM	10	Roswell.
Dona Ana	KROD	4	El Paso.
	KTSM	9	Do.
	KELP+	13	Do.
Eddy	KBIM	10	Roswell.
	KELP+	13	El Paso.
	KCBD+	11	Lubbock.
Grant	KROD	4	El Paso.
	KTSM	9	Do.
	KOAT	7	Albuquerque.
Guadalupe	KOB	4	Do.
	KOAT	7	Do.
	KGGM	13	Do.
Harding	KOB	4	Do.
	KOAT	7	Do.
	KGGM	13	Do.
Hidalgo	KTVK	3	Phoenix.
	KOOL	10	Do.
	KTAR	12	Do.
	KVOA	4	Tucson.
	KGUN	9	Do.
	KOLD	13	Do.
Lea North	KBIM	10	Roswell.
	KCBD+	11	Lubbock.
Lea South	KMID	2	Odessa-Midland.
	KOSA	7	Do.
	KMOM+	9	Do.
	KCBD+	11	Lubbock.
	KBIM	10	Roswell.
Lincoln	KOB	4	Albuquerque.
	KCBD+	11	Lubbock.
	KBIM	10	Roswell.
Los Alamos	KOB	4	Albuquerque.
	KOAT	7	Do.
	KGGM	13	Do.
Luna	KROD	4	El Paso.
	KTSM	9	Do.
	KELP+	13	Do.
McKinley	KOB	4	Albuquerque.
	KOAT	7	Do.
	KGGM	13	Do.
Mora	KOB	4	Do.
	KOAT	7	Do.
	KGGM	13	Do.
Otero	KROD	4	El Paso.
	KTSM	9	Do.
	KOAT	7	Albuquerque.
Quay	KGNC	4	Amarillo.
	KVII	7	Do.
	KFDA+	10	Do.
	KOAT	7	Albuquerque.
	KCBD+	11	Lubbock.
Rio Arriba	KOB	4	Albuquerque.
	KOAT	7	Do.
	KGGM	13	Do.
Roosevelt	KFDA+	10	Amarillo.
	KCBD+	11	Lubbock.
Sandoval	KOB	4	Albuquerque.
	KOAT	7	Do.
	KGGM	13	Do.
San Juan	KOB	4	Do.
	KOAT	7	Do.
	KGGM	13	Do.
San Miguel	KOB	4	Do.
	KOAT	7	Do.
	KGGM	13	Do.
Santa Fe	KOB	4	Do.
	KOAT	7	Do.
	KGGM	13	Do.
Sierra	KOB	4	Do.
	KOAT	7	Do.
	KGGM	13	Do.
Socorro	KOB	4	Do.
	KOAT	7	Do.
	KGGM	13	Do.
Taos	KOB	4	Do.
	KOAT	7	Do.
	KGGM	13	Do.
Torrance	KOB	4	Do.
	KOAT	7	Do.
	KGGM	13	Do.

NEW MEXICO—continued

County	Call letters, channel number and market name		
Union	KGNC	4	Amarillo.
	KVII	7	Do.
	KFDA+	10	Do.
	KKTV	11	Colorado Springs-Pueblo.
	KRDO	13	Do.
Valencia	KOB	4	Albuquerque.
	KOAT	7	Do.
	KGGM	13	Do.

Census county divisions in split counties

Lea North: Lea North Central, Lovington, Tatum.
Lea South: All other.

NEW YORK

County	Call letters, channel number and market name		
Albany	WRGB	6	Albany-Schenectady-Troy.
	WTEN	+10	Do.
	WAST	13	Do.
Allegany	WGP	2	Buffalo.
	WBEN	4	Do.
	WKBW	7	Do.
Bronx	WCBS	2	New York.
	WNBC	4	Do.
	WNEW	5	Do.
	WABC	7	Do.
	WOR	9	Do.
	WPIX	11	Do.
Broome	WNBF	12	Binghamton.
	WBJA	34	Do.
	WINR	40	Do.
Cattaraugus	WGP	2	Buffalo.
	WBEN	4	Do.
	WKBW	7	Do.
Cayuga	WSYR	3	Syracuse.
	WHEN	5	Do.
	WNYS	9	Do.
	WHEC	10	Rochester, N.Y.
	WOKR	13	Do.
Chautauqua	WGR	2	Buffalo.
	WBEN	4	Do.
	WKBW	7	Do.
Chemung	WSYE	18	Elmira.
	WENY	36	Do.
	WNBF	12	Binghamton.
	WNEW	5	New York.
Chenango	WSYR	3	Syracuse.
	WHEN	5	Do.
	WNYS	9	Do.
	WNBF	12	Binghamton.
	WINR	40	Do.
Clinton	WCAX	3	Burlington-Plattsburgh.
	WPTZ	5	Do.
	WVNY	22	Do.
	CBMT	6	Canada.
	CFCF	12	Do.
Columbia	WRGB	6	Albany-Schenectady-Troy.
	WTEN	+10	Do.
	WAST	13	Do.
Cortland	WSYR	3	Syracuse.
	WHEN	5	Do.
	WNYS	9	Do.
Delaware	WNBF	12	Binghamton.
	WRGB	6	Albany-Schenectady-Troy.
	WTEN	+10	Do.
	WKTV	2	Utica.
Dutchess	WCBS	2	New York.
	WNBC	4	Do.
	WNEW	5	Do.
	WABC	7	Do.
	WOR	9	Do.
	WPIX	11	Do.
	WTEN	+10	Albany-Schenectady-Troy.
Erie	WGR	2	Buffalo.
	WBEN	4	Do.
	WKBW	7	Do.
	WUTV	29	Do.
Essex	WCAX	3	Burlington-Plattsburgh.
	WPTZ	5	Do.
Franklin	WCAX	3	Do.
	WPTZ	5	Do.
	CBOT	4	Canada.
	CBMT	6	Do.
	CJSS	8	Do.
	CFCF	12	Do.
Fulton	WRGB	6	Albany-Schenectady-Troy.
	WTEN	+10	Do.
	WAST	13	Do.
Genesee	WGR	2	Buffalo.
	WBEN	4	Do.
	WKBW	7	Do.
	WROC	8	Rochester, N.Y.
	WHEC	10	Do.
	WOKR	13	Do.

SIGNIFICANTLY VIEWED SIGNALS—Continued

County	Call letters, channel number and market name		
NEW YORK—continued			
Greene	WRGB	6	Albany-Schenectady-Troy.
	WTEN+	10	Do.
	WAST	13	Do.
Hamilton	WRGB	6	Do.
	WTEN+	10	Do.
Herkimer	WKTV	2	Utica.
	WRBG	6	Albany-Schenectady-Troy.
	WHEN	5	Syracuse.
	WNYS	9	Do.
Jefferson	WWNY	7	Watertown-Carthage.
	WSYR	3	Syracuse.
	WHEN	5	Do.
	CKWS	11	Canada.
Kings	WCBS	2	New York.
	WNBC	4	Do.
	WNEW	5	Do.
	WABC	7	Do.
	WOR	9	Do.
	WPIX	11	Do.
Lewis	WWNY	7	Watertown-Carthage.
	WSYR	3	Syracuse.
	WHEN	5	Do.
	WKTV	2	Utica.
Livingston	WROC	8	Rochester, N.Y.
	WHEC	10	Do.
	WOKR	13	Do.
	WGR	2	Buffalo.
Madison	WSYR	3	Syracuse.
	WHEN	5	Do.
	WNYS	9	Do.
	WKTV	2	Utica.
Monroe	WROC	8	Rochester, N.Y.
	WHEC	10	Do.
	WOKR	13	Do.
Montgomery	WRGB	6	Albany-Schenectady-Troy.
	WTEN+	10	Do.
	WAST	13	Do.
	WKTV	2	Utica.
Nassau	WCBS	2	New York.
	WNBC	4	Do.
	WNEW	5	Do.
	WABC	7	Do.
	WOR	9	Do.
	WPIX	11	Do.
New York	WCBS	2	Do.
	WNBC	4	Do.
	WNEW	5	Do.
	WABC	7	Do.
	WOR	9	Do.
	WPIX	11	Do.
Niagara	WGR	2	Buffalo.
	WBEN	4	Do.
	WKBW	7	Do.
	WUTV	29	Do.
	CBLT	6	Canada.
	CFTO	9	Do.
	CHCH	11	Do.
Oneida East	WKTV	2	Utica.
	WUTR	20	Do.
	WSYR	3	Syracuse.
	WHEN	5	Do.
	WNYS	9	Do.
Oneida West	WSYR	3	Syracuse.
	WHEN	5	Do.
	WNYS	9	Do.
	WKTV	2	Utica.
Onondaga	WSYR	3	Syracuse.
	WHEN	5	Do.
	WNYS	9	Do.
Ontario	WROC	8	Rochester, N.Y.
	WHEC	10	Do.
	WOKR	13	Do.
	WSYR	3	Syracuse.
	WHEN	5	Do.
	WNYS	9	Do.
Orange	WCBS	2	New York.
	WNBC	4	Do.
	WNEW	5	Do.
	WABC	7	Do.
	WOR	9	Do.
	WPIX	11	Do.
Orleans	WGR	2	Buffalo.
	WBEN	4	Do.
	WKBW	7	Do.
	WUTV	29	Do.
	WROC	8	Rochester, N.Y.
	WHEC	10	Do.
	WOKR	13	Do.
Oswego	WSYR	3	Syracuse.
	WHEN	5	Do.
	WNYS	9	Do.
Otsego	WKTV	2	Utica.
	WRGB	6	Albany-Schenectady-Troy.
	WNBF	12	Binghamton.
	WHEN	5	Syracuse.
	WNYS	9	Do.

SIGNIFICANTLY VIEWED SIGNALS—Continued

County	Call letters, channel number and market name		
NEW YORK—continued			
Putnam	WCBS	2	New York.
	WNBC	4	Do.
	WNEW	5	Do.
	WABC	7	Do.
	WOR	9	Do.
	WPIX	11	Do.
Queens	WCBS	2	Do.
	WNBC	4	Do.
	WABC	7	Do.
	WOR	9	Do.
	WPIX	11	Do.
	WNEW	5	Do.
Rensselaer	WRGB	6	Albany-Schenectady-Troy.
	WTEN	10	Do.
	WAST	13	Do.
Richmond	WCBS	2	New York.
	WNBC	4	Do.
	WNEW	5	Do.
	WABC	7	Do.
	WOR	9	Do.
	WPIX	11	Do.
Rockland	WCBS	2	Do.
	WNBC	4	Do.
	WNEW	5	Do.
	WABC	7	Do.
	WOR	9	Do.
	WPIX	11	Do.
St. Lawrence	WWNY	7	Watertown-Carthage.
	WPTZ	5	Burlington-Plattsburgh.
	CBOT	4	Canada.
	CJSS	8	Do.
	CKWS	11	Do.
Saratoga	WRGB	6	Albany-Schenectady-Troy.
	WTEN+	10	Do.
	WAST	13	Do.
Schenectady	WRGB	6	Do.
	WTEN+	10	Do.
	WAST	13	Do.
Schoharie	WRGB	6	Do.
	WTEN+	10	Do.
	WAST	13	Do.
Schuyler	WSYR	3	Syracuse.
	WHEN	5	Do.
	WNYS	9	Do.
	WROC	8	Rochester, N.Y.
	WHEC	10	Do.
Seneca	WSYR	3	Syracuse.
	WHEN	5	Do.
	WNYS	9	Do.
	WROC	8	Rochester, N.Y.
	WHEC	10	Do.
	WOKR	13	Do.
Steuben	WSYR	3	Syracuse.
	WHEN	5	Do.
	WNYS	9	Do.
	WNBF	12	Binghamton.
	WBEN	4	Buffalo.
	WKBW	7	Do.
	WSYE	18	Elmira.
Suffolk East	WCBS	2	New York.
	WNBC	4	Do.
	WNEW	5	Do.
	WABC	7	Do.
	WOR	9	Do.
	WPIX	11	Do.
	WTIC	3	Hartford-New Haven.
	WNHC	8	Do.
Suffolk West	WCBS	2	New York.
	WNBC	4	Do.
	WNEW	5	Do.
	WABC	7	Do.
	WOR	9	Do.
	WPIX	11	Do.
Sullivan	WCBS	2	Do.
	WNBC	4	Do.
	WNEW	5	Do.
	WABC	7	Do.
	WOR	9	Do.
	WPIX	11	Do.
Tioga	WNBF	12	Binghamton.
	WBJA	34	Do.
	WINR	40	Do.
Tompkins	WSYR	3	Syracuse.
	WHEN	5	Do.
	WNYS	9	Do.
	WNBF	12	Binghamton.
Ulster	WCBS	2	New York.
	WNBC	4	Do.
	WNEW	5	Do.
	WABC	7	Do.
	WOR	9	Do.
	WPIX	11	Do.
	WRGB	6	Albany-Schenectady-Troy.
	WTEN+	10	Do.
Warren	WRGB	6	Do.
	WTEN+	10	Do.
	WAST	13	Do.

SIGNIFICANTLY VIEWED SIGNALS—Continued

County	Call letters, channel number and market name		
NEW YORK—continued			
Washington	WRGB	6	Do.
	WTEN+	10	Do.
	WAST	13	Do.
Wayne	WROC	8	Rochester, N.Y.
	WHEC	10	Do.
	WOKR	13	Do.
	WSYR	3	Syracuse.
	WHEN	5	Do.
	WNYS	9	Do.
Westchester	WCBS	2	New York.
	WNBC	4	Do.
	WNEW	5	Do.
	WABC	7	Do.
	WOR	9	Do.
	WPIX	11	Do.
Wyoming	WGR	2	Buffalo.
	WBEN	4	Do.
	WKBW	7	Do.
	WROC	8	Rochester, N.Y.
	WHEC	10	Do.
	WOKR	13	Do.
Yates	WSYR	3	Syracuse.
	WHEN	5	Do.
	WNYS	9	Do.
	WROC	8	Rochester, N.Y.
	WHEC	10	Do.

Census County Divisions in Split Counties

Oneida West:
Annsville, Ava, Boonville, Camden, Florence, Forestport, Lee, Rome City, Vernon, Verona, Vienna, Western, Sherrill.
Oneida East: All other.
Suffolk West: Babylon, Huntington, Islip, Smithtown.
Suffolk East: All other.

NORTH CAROLINA

County	Call letters, channel number and market name		
Alamance	WFMY	2	Greensboro-Winston-Salem-High Point.
	WGHP	8	Do.
	WSJS	12	Do.
	WRAL	5	Raleigh-Durham.
	WTVD	11	Do.
Alexander	WBTV	3	Charlotte.
	WSOC	9	Do.
	WRET	36	Do.
	WGHP	8	Greensboro-Winston-Salem-High Point.
Alleghany	WSJS	12	Do.
	WFMY	2	Do.
	WGHP	8	Do.
	WSJS	12	Do.
	WBTV	3	Charlotte.
	WSOC	9	Do.
	WDBJ	7	Roanoke-Lynchburg.
	WSLS	10	Do.
Anson	WBTV	3	Charlotte.
	WSOC	9	Do.
	WCCB	18	Do.
	WRET	36	Do.
	WBTW	13	Florence, S.C.
	WGHP	8	Greensboro-Winston-Salem-High Point.
Ashe	WBTV	3	Charlotte.
	WCYB	5	Bristol-Kingsport-Johnson City.
	WGHP	8	Greensboro-Winston-Salem-High Point.
Avery	WBTV	3	Charlotte.
	WSOC	9	Do.
	WCYB	5	Bristol-Kingsport-Johnson City.
Beaufort	WITN	7	Greenville-New Bern-Washington.
	WNCT	9	Do.
	WCTI	12	Do.
Bertie	WITN	7	Do.
	WNCT	9	Do.
	WTAR	3	Norfolk-Portsmouth-Newport News-Hampton.
	WAVY	10	Do.
	WVEC	13	Do.
Bladen	WWAY	3	Wilmington, N.C.
	WECT	6	Do.
Brunswick	WWAY	3	Do.
	WECT	6	Do.
Buncombe	WFBC	4	Greenville-Spartanburg-Asheville.
	WSPA	7	Do.
	WLOS	13	Do.
Burke	WBTV	3	Charlotte.
	WSOC	9	Do.
	WFBC	4	Greenville-Spartanburg-Asheville.
	WSPA	7	Do.
	WLOS	13	Do.

RULES AND REGULATIONS

SIGNIFICANTLY VIEWED SIGNALS—Continued

County	Call letters, channel number and market name		
NORTH CAROLINA—continued			
Cabarrus	WBTV	3	Charlotte.
	WSOC	9	Do.
	WCCB	18	Do.
	WRET	36	Do.
Caldwell	WBTV	3	Do.
	WSOC	9	Do.
	WFBC	4	Greenville-Spartanburg-Asheville.
	WODA	7	Do.
	WLOS	13	Do.
Camden	WTAR	3	Norfolk-Portsmouth-Newport News-Hampton.
	WAVY	10	Do.
	WVEC	13	Do.
Carteret	WITN	7	Greenville-New Bern-Washington.
	WNCT	9	Do.
	WCTI	12	Do.
Caswell	WFMY	2	Greensboro-Winston-Salem-High Point.
	WGHP	8	Do.
	WSJS	12	Do.
	WRAL	5	Raleigh-Durham.
	WDBJ	7	Roanoke-Lynchburg.
	WSLS	10	Do.
	WLVA	13	Do.
Catawba	WBTV	3	Charlotte.
	WSOC	9	Do.
	WCCB	18	Do.
	WSPA	7	Greenville-Spartanburg-Asheville.
	WLOS	13	Do.
Chatham	WFMY	2	Greensboro-Winston-Salem-High Point.
	WGHP	8	Do.
	WRAL	5	Raleigh-Durham.
	WTVD	11	Do.
	WRDU	28	Do.
Cherokee	WRCB	3	Chattanooga.
	WTVC	9	Do.
	WDEF	12	Do.
	WFBC	4	Greenville-Spartanburg-Asheville.
	WLOS	13	Do.
Chowan	WTAR	3	Norfolk-Portsmouth-Newport News-Hampton.
	WAVY	10	Do.
	WVEC	13	Norfolk-Portsmouth-Newport News.
	WITN	7	Greenville-New Bern-Washington.
	WNCT	9	Do.
Clay	WRCB	3	Chattanooga.
	WTVC	9	Do.
	WDEF	12	Do.
	WSB	2	Atlanta.
	WAGA	5	Do.
	WQXI	11	Do.
Cleveland	WBTV	3	Charlotte
	WSOC	9	Do.
	WFBC	4	Greenville-Spartanburg-Asheville.
	WSPA	7	Do.
	WLOS	13	Do.
Columbus	WWAY	3	Wilmington, N.C.
	WECT	6	Do.
	WBTW	13	Florence, S.C.
Craven	WITN	7	Greenville-New Bern-Washington.
	WNCT	9	Do.
	WCTI	12	Do.
Cumberland	WRAL	5	Raleigh-Durham.
	WTVD	11	Do.
	WECT	6	Wilmington, N.C.
Currituck	WTAR	3	Norfolk-Portsmouth-Newport News-Hampton.
	WAVY	10	Do.
	WVEC	13	Do.
Dare	WTAR	3	Do.
	WAVY	10	Do.
	WVEC	13	Do.
	WITN	7	Greenville-NewBern-Washington.
Davidson	WFMY	2	Greensboro-Winston-Salem-High Point.
	WGHP	8	Do.
	WSJS	12	Do.
	WBTV	3	Charlotte.
	WSOC	9	Do.
Davie	WFMY	2	Greensboro-Winston-Salem-High Point.
	WGHP	8	Do.
	WSJS	12	Do.
	WBTV	3	Charlotte.
	WSOC	9	Do.

SIGNIFICANTLY VIEWED SIGNALS—Continued

County	Call letters, channel number and market name		
NORTH CAROLINA—continued			
Duplin	WITN	7	Greenville-New Bern-Washington.
	WNCT	9	Do.
	WRAL	5	Raleigh-Durham.
	WTVD	11	Do.
	WWAY	3	Wilmington, N.C.
	WECT	6	Do.
Durham	WRAL	5	Raleigh-Durham.
	WTVD	11	Do.
	WRDU	28	Do.
	WFMY	2	Greensboro-Winston-Salem-High Point.
Edgecombe	WITN	7	Greenville-New Bern-Washington.
	WNCT	9	Do.
	WRAL	5	Raleigh-Durham.
	WTVD	11	Do.
Forsyth	WFMY	2	Greensboro-Winston-Salem-High Point.
	WGHP	8	Do.
	WSJS	12	Do.
Franklin	WRAL	5	Raleigh-Durham.
	WTVD	11	Do.
	WNCT	9	Greenville-New Bern-Washington.
Gaston	WBTV	3	Charlotte.
	WSOC	9	Do.
	WCCB	18	Do.
	WRET	36	Do.
	WLOS	13	Greenville-Spartanburg-Asheville.
Gates	WTAR	3	Norfolk-Portsmouth-Newport News-Hampton.
	WAVY	10	Do.
	WVEC	13	Do.
Graham	WATE	6	Knoxville.
	WBIR	10	Do.
	WFBC	4	Greenville-Spartanburg-Asheville.
	WLOS	13	Do.
Granville	WRAL	5	Raleigh-Durham.
	WTVD	11	Do.
	WRDU	28	Do.
Greene	WITN	7	Greenville-New Bern-Washington.
	WNCT	9	Do.
	WRAL	5	Raleigh-Durham.
	WTVD	11	Do.
Guilford	WFMY	2	Greensboro-Winston-Salem-High Point.
	WGHP	8	Do.
	WSJS	12	Do.
Halifax	WITN	7	Greenville-New Bern-Washington.
	WNCT	9	Do.
	WRAL	5	Raleigh-Durham.
	WTVD	11	Do.
Harnett	WRAL	5	Do.
	WTVD	11	Do.
	WECT	6	Wilmington, N.C.
Haywood	WFBC	4	Greenville-Spartanburg-Asheville.
	WSPA	7	Do.
	WLOS	13	Do.
Henderson	WFBC	4	Do.
	WSPA	7	Do.
	WLOS	13	Do.
Hertford	WTAR	3	Norfolk-Portsmouth-Newport News-Hampton.
	WAVY	10	Do.
	WVEC	13	Do.
Hoke	WRAL	5	Raleigh-Durham.
	WTVD	11	Do.
	WBTW	13	Florence, S.C.
	WGHP	8	Greensboro-Winston-Salem-High Point.
	WECT	6	Wilmington, N.C.
Hyde	WITN	7	Greenville-New Bern-Washington.
	WNCT	9	Do.
	WCTI	12	Do.
Iredell	WBTV	3	Charlotte.
	WSOC	9	Do.
	WCCB	18	Do.
	WRET	36	Do.
	WGHP	8	Greensboro-Winston-Salem-High Point.
	WSJS	12	Do.
Jackson	WFBC	4	Greenville-Spartanburg-Asheville.
	WSPA	7	Do.
	WLOS	13	Do.

SIGNIFICANTLY VIEWED SIGNALS—Continued

County	Call letters, channel number and market name		
NORTH CAROLINA—continued			
Johnston	WRAL	5	Raleigh-Durham.
	WTVD	11	Do.
	WITN	7	Greenville-New Bern-Washington.
Jones	WITN	7	Do.
	WNCT	9	Do.
	WCTI	12	Do.
Lee	WRAL	5	Raleigh-Durham.
	WTVD	11	Do.
	WRDU	28	Do.
	WFMY	2	Greensboro-Winston-Salem-High Point.
Lenoir	WITN	7	Greenville-New Bern-Washington.
	WNCT	9	Do.
	WCTI	12	Do.
	WRAL	5	Raleigh-Durham.
Lincoln	WBTV	3	Charlotte.
	WSOC	9	Do.
	WCCB	18	Do.
	WRET	36	Do.
	WSPA	7	Greenville-Spartanburg-Asheville.
	WLOS	13	Do.
McDowell	WFBC	4	Do.
	WSPA	7	Do.
	WLOS	13	Do.
	WBTV	3	Charlotte.
Macon	WFBC	4	Greenville-Spartanburg-Asheville.
	WSPA	7	Do.
	WLOS	13	Do.
	WSB	2	Atlanta.
Madison	WFBC	4	Greenville-Spartanburg-Asheville.
	WLOS	13	Do.
Martin	WITN	7	Greenville-New Bern-Washington.
	WNCT	9	Do.
	WCTI	12	Do.
Mecklenburg	WBTV	3	Charlotte.
	WSOC	9	Do.
	WCCB	18	Do.
	WRET	36	Do.
Mitchell	WBTV	3	Do.
	WCYB	5	Bristol-Kingsport-Johnson City.
	WFBC	4	Greenville-Spartanburg-Asheville.
	WSPA	7	Do.
	WLOS	13	Do.
Montgomery	WFMY	2	Greensboro-Winston-Salem-High Point.
	WGHP	8	Do.
	WBTV	3	Charlotte.
	WSOC	9	Do.
Moore	WFMY	2	Greensboro-Winston-Salem-High Point.
	WGHP	8	Do.
	WRAL	5	Raleigh-Durham.
	WTVD	11	Do.
	WECT	6	Wilmington, N.C.
Nash	WITN	7	Greenville-New Bern-Washington.
	WNCT	9	Do.
	WRAL	5	Raleigh-Durham.
	WTVD	11	Do.
New Hanover	WWAY	3	Wilmington, N.C.
	WECT	6	Do.
Northampton	WTAR	3	Norfolk-Portsmouth-Newport News-Hampton.
	WAVY	10	Do.
	WVEC	13	Do.
	WITN	7	Greenville-New Bern-Washington.
	WNCT	9	Do.
Onslow	WITN	7	Do.
	WNCT	9	Do.
	WCTI	12	Do.
	WWAY	3	Wilmington, N.C.
	WECT	6	Do.
Orange	WRAL	5	Raleigh-Durham.
	WTVD	11	Do.
	WRDU	28	Do.
	WFMY	2	Greensboro-Winston-Salem-High Point.
	WGHP	8	Do.
Pamlico	WITN	7	Greenville-New Bern-Washington.
	WNCT	9	Do.
	WCTI	12	Do.
Pasquotank	WTAR	3	Norfolk-Portsmouth-Newport News-Hampton.
	WAVY	10	Do.
	WVEC	13	Do.
	WYAH	27	Do.

County	Call letters, channel number and market name		

NORTH CAROLINA—continued

County	Call letters	Ch.	Market name
Pender	WWAY	3	Wilmington, N.C.
	WECT	6	Do.
Perquimans	WTAR	3	Norfolk-Portsmouth-Newport News-Hampton.
	WAVY	10	Do.
	WVEC	13	Do.
Person	WRAL	5	Raleigh-Durham.
	WTVD	11	Do.
	WRDU	28	Do.
	WFMY	2	Greensboro-Winston-Salem-High Point.
	WDBJ	7	Roanoke-Lynchburg.
	WSLS	10	Do.
	WLVA	13	Do.
Pitt	WITN	7	Greenville-New Bern-Washington.
	WNCT	9	Do.
	WCTI	12	Do.
	WRAL	5	Raleigh-Durham.
Polk	WFBC	4	Greenville-Spartanburg-Asheville.
	WSPA	7	Do.
	WLOS	13	Do.
	WBTV	3	Charlotte.
Randolph	WFMY	2	Greensboro-Winston-Salem-High Point.
	WGHP	8	Do.
	WSJS	12	Do.
Richmond	WBTV	3	Charlotte.
	WSOC	9	Do.
	WBTW	13	Florence, S.C.
	WGHP	8	Greensboro-Winston-Salem-High Point.
Robeson	WWAY	3	Wilmington, N.C.
	WECT	6	Do.
	WBTW	13	Florence, S.C.
	WRAL	5	Raleigh-Durham.
	WTVD	11	Do.
Rockingham	WFMY	2	Greensboro-Winston-Salem-High Point.
	WGHP	8	Do.
	WSJS	12	Do.
	WDBJ	7	Roanoke-Lynchburg.
	WSLS	10	Do.
Rowan	WBTV	3	Charlotte.
	WSOC	9	Do.
	WCCB	18	Do.
	WRET	36	Do.
	WFMY	2	Greensboro-Winston-Salem-High Point.
	WGHP	8	Do.
	WSJS	12	Do.
Rutherford	WFBC	4	Greenville-Spartanburg-Asheville.
	WSPA	7	Do.
	WLOS	13	Do.
	WBTV	3	Charlotte.
	WSOC	9	Do.
Sampson	WRAL	5	Raleigh-Durham.
	WTVD	11	Do.
	WITN	7	Greenville-New Bern-Washington.
	WNCT	9	Do.
	WWAY	3	Wilmington, N.C.
	WECT	6	Do.
Scotland	WBTW	13	Florence, S.C.
	WGHP	8	Greensboro-Winston-Salem-High Point.
	WRAL	5	Raleigh-Durham.
	WTVD	11	Do.
	WECT	6	Wilmington, N.C.
Stanly	WBTV	3	Charlotte.
	WSOC	9	Do.
	WGHP	8	Greensboro-Winston-Salem-High Point.
Stokes	WFMY	2	Do.
	WGHP	8	Do.
	WSJS	12	Do.
Surry	WFMY	2	Do.
	WGHP	8	Do.
	WSJS	12	Do.
Swain	WFBC	4	Greenville-Spartanburg-Asheville.
	WSPA	7	Do.
	WLOS	13	Do.
Transylvania	WFBC	4	Do.
	WSPA	7	Do.
	WLOS	13	Do.
Tyrrell	WTAR	3	Norfolk-Portsmouth-Newport News-Hampton.
	WAVY	10	Do.
	WVEC	13	Do.
	WITN	7	Greenville-New Bern-Washington.
	WNCT	9	Do.
Union	WBTV	3	Charlotte.
	WSOC	9	Do.
	WCCB	18	Do.
	WRET	36	Do.

NORTH CAROLINA—continued

County	Call letters	Ch.	Market name
Vance	WRAL	5	Raleigh-Durham.
	WTVD	11	Do.
Wake	WRAL	5	Do.
	WTVD	11	Do.
Warren	WRAL	5	Do.
	WTVD	11	Do.
	WITN	7	Greenville-New Bern-Washington.
	WNCT	9	Do.
Washington	WITN	7	Do.
	WNCT	9	Do.
	WCTI	12	Do.
Watauga	WBTV	3	Charlotte.
	WCYB	5	Bristol-Kingsport-Johnson City.
	WGHP	8	Greensboro-Winston-Salem-High Point.
Wayne	WITN	7	Greenville-New Bern-Washington.
	WNCT	9	Do.
	WRAL	5	Raleigh-Durham.
	WTVD	11	Do.
Wilkes	WGHP	8	Greensboro-Winston-Salem-High Point.
	WSJS	12	Do.
	WBTV	3	Charlotte.
	WSOC	9	Do.
Wilson	WITN	7	Greenville-Winston-Salem-High Point.
	WNCT	9	Do.
	WRAL	5	Raleigh-Durham.
	WTVD	11	Do.
Yadkin	WFMY	2	Greensboro-Winston-Salem-High Point.
	WGHP	8	Do.
	WSJS	12	Do.
	WBTV	3	Charlotte.
	WSOC	9	Do.
Yancey	WFBC	4	Greenville-Spartanburg-Asheville.
	WSPA	7	Do.
	WLOS	13	Do.
	WCYB	5	Bristol-Kingsport-Johnson City.

NORTH DAKOTA

County	Call letters	Ch.	Market name
Adams	KDIX	2	Dickinson, N. Dak.
	KFYR	5	Minot-Bismarck.
Barnes	KXJB	4	Fargo.
	WDAY	6	Do.
	KTHI	11	Do.
Benson	KXJB	4	Do.
	WDAZ	8	Do.
	KTHI	11	Do.
	KXMC	13	Minot-Bismarck.
Billings	KDIX	2	Dickinson, N. Dak.
	KFYR	5	Minot-Bismarck.
Bottineau	KMOT	10	Do.
	KXMC	13	Do.
Bowman	KDIX	2	Dickinson, N. Dak.
	KFYR	5	Minot-Bismarck.
	KOTA+	3	Rapid City.
Burke	KUMV	8	Minot-Bismarck.
	KMOT	10	Do.
	KXMC	13	Do.
	CKOS	3	Canada.
Burleigh	KFYR	5	Minot-Bismarck.
	KXMB	12	Do.
Cass	KXJB	4	Fargo.
	WDAY	6	Do.
	KTHI	11	Do.
Cavalier	WDAZ	8	Do.
	KCND	12	Pembina.
	CJAY	7	Canada.
Dickey	KXJB	4	Fargo.
	WDAY	6	Do.
	KTHI	11	Do.
	KELO+	11	Sioux Falls-Mitchell.
	KSOO+	13	Do.
Divide	KUMV	8	Minot-Bismarck.
	KXMD	11	Do.
	CKCK	2	Canada.
Dunn	KDIX	2	Dickinson, N. Dak.
	KFYR	5	Minot-Bismarck.
	KUMV	8	Do.
Eddy	KXJB	4	Fargo.
	WDAZ	8	Do.
	KTHI	11	Do.
Emmons	KFYR	5	Minot-Bismarck.
	KXMB	12	Do.
Foster	KXJB	4	Fargo.
	WDAY	6	Do.
	WDAZ	8	Do.
	KTHI	11	Do.
Golden Valley	KUMV	8	Minot-Bismarck.
	KDIX	2	Dickinson, N. Dak.
Grand Forks*	KXJB	4	Fargo.
	WDA	6	Do.
	WDAZ	8	Do.
	KTHI	11	Do.

NORTH DAKOTA—continued

County	Call letters	Ch.	Market name
Grant	KFYR	5	Minot-Bismarck.
	KXMB	12	Do.
Griggs	KXJB	4	Fargo.
	WDAY	6	Do.
	WDAZ	8	Do.
	KTHI	11	Do.
Hettinger	KFYR	5	Minot-Bismarck.
	KDIX	2	Dickinson, N. Dak.
Kidder	KFYR	5	Minot-Bismarck.
	KXMB	12	Do.
La Moure	KXJB	4	Fargo.
	WDAY	6	Do.
	KTHI	11	Do.
Logan	KFYR	5	Minot-Bismarck.
	KXMB	12	Do.
McHenry	KMOT	10	Do.
	KXMC	13	Do.
McIntosh	KFYR	5	Do.
	KXMB	12	Do.
McKenzie	KUMV	8	Do.
	KXMD	11	Do.
McLean	KFYR	5	Do.
	KXMB	12	Do.
	KXMC	13	Do.
Mercer	KFYR	5	Do.
	KXMC	13	Do.
Morton East	KFYR	5	Do.
	KXMB	12	Do.
Morton West	KFYR	5	Do.
	KXMB	12	Do.
Mountrail	KUMV	8	Do.
	KMOT	10	Do.
	KXMC	13	Do.
Nelson	KXJB	4	Fargo.
	WDAY	6	Do.
	WDAZ	8	Do.
	KTHI	11	Do.
Oliver	KFYR	5	Minot-Bismarck.
	KXMB	12	Do.
Pembina	KCND	12	Pembina.
	KXJB	4	Fargo.
	WDAZ	8	Do.
	CBWT	6	Canada.
	CJAY	7	Do.
Pierce	KMOT	10	Minot-Bismarck.
	KXMC	13	Do.
Ramsey	KXJB	4	Fargo.
	WDAZ	8	Do.
	KTHI	11	Do.
Ransom	KXJB	4	Do.
	WDAY	6	Do.
	KTHI	11	Do.
Renville	KMOT	10	Minot-Bismarck.
	KXMC	13	Do.
Richland*	KXJB	4	Fargo.
	WDAY	6	Do.
	KTHI	11	Do.
Rolette	KMOT	10	Minot-Bismarck.
	KXMC	13	Do.
	CKX	5	Canada.
Sargent	KXJB	4	Fargo.
	WDAY	6	Do.
	KTHI	11	Do.
Sheridan	KFYR	5	Minot-Bismarck.
	KXMC	13	Do.
Sioux	KFYR	5	Do.
	KXMB	12	Do.
Slope	KDIX	2	Dickinson, N. Dak.
Stark	KDIX	2	Do.
	KFYR	5	Minot-Bismarck.
Steele	KXJB	4	Fargo.
	WDAY	6	Do.
	KTHI	11	Do.
Stutsman	KXJB	4	Do.
	WDAY	6	Do.
	KTHI	11	Do.
	KFYR	5	Minot-Bismarck.
Towner	KXJB	4	Fargo.
	WDAZ	8	Do.
Traill	KXJB	4	Do.
	WDAY	6	Do.
	KTHI	11	Do.
Walsh	KXJB	4	Do.
	WDAZ	8	Do.
	KTHI	11	Do.
	KCND	12	Pembina.
Ward	KMOT	10	Minot-Bismarck.
	KXMC	13	Do.
Wells	KFYR	5	Do.
	KXMC	13	Do.
	KXJB	4	Fargo.
	WDAZ	8	Do.
	KTHI	11	Do.
Williams	KUMV	8	Minot-Bismarck.
	KXMD	11	Do.

Census County Divisions in Split Counties

Morton East: Mandan, Mandan North, Mandan South.
Morton West: All other.

SIGNIFICANTLY VIEWED SIGNALS—Continued

County	Call letters, channel number and market name		

OHIO

County	Call letters	Ch.	Market name
Adams	WLWT	5	Cincinnati.
	WCPO	9	Do.
	WKRC	12	Do.
Allen	WIMA	35	Lima.
	WHIO	7	Dayton.
	WANE	15	Fort Wayne.
	WTOL	11	Toledo.
	WSPD	13	Do.
	WDHO	24	Do.
Ashland	WKYO	3	Cleveland.
	WEWS	5	Do.
	WJW	8	Do.
	WUAB	43	Do.
	WKBF	61	Do.
Ashtabula	WKYC	3	Cleveland.
	WEWS	5	Do.
	WJW	8	Do.
	WICU	12	Erie.
	WJET	24	Do.
	WSEE	35	Do.
Athens	WSAZ	3	Charleston-Huntington.
	WCHS	8	Do.
	WHTN	13	Do.
	WLWC	4	Columbus, Ohio.
Auglaize	WLWD	2	Dayton.
	WHIO	7	Do.
	WIMA	35	Lima.
Belmont	WTRF	7	Wheeling-Steubenville.
	WSTV	9	Do.
	KDKA	2	Pittsburgh.
	WTAE	4	Do.
Brown	WLWT	5	Cincinnati.
	WCPO	9	Do.
	WKRC	12	Do.
	WXIX	19	Do.
Butler	WLWT	5	Cincinnati.
	WCPO	9	Do.
	WKRC	12	Do.
	WXIX	19	Do.
	WLWD	2	Dayton.
	WHIO	7	Do.
Carroll	WTRF	7	Wheeling-Steubenville.
	WSTV	9	Do.
	WKYC	3	Cleveland.
	WEWS	5	Do.
	WJW	8	Do.
	KDKA	2	Pittsburgh.
	WTAE	4	Do.
	WIIC	11	Do.
Champaign	WLWD	2	Dayton.
	WHIO	7	Do.
	WLWC	4	Columbus, Ohio.
	WTVN	6	Do.
	WBNS	10	Do.
Clark	WLWD	2	Dayton.
	WHIO	7	Do.
	WKEF	22	Do.
	WLWC	4	Columbus, Ohio.
	WTVN	6	Do.
	WBNS	10	Do.
Clermont	WLWT	5	Cincinnati.
	WCPO	9	Do.
	WKRC	12	Do.
	WXIX	19	Do.
Clinton	WLWT	5	Cincinnati.
	WCPO	9	Do.
	WKRC	12	Do.
	WLWD	2	Dayton.
	WHIO	7	Do.
	WKEF	22	Do.
Columbiana	KDKA	2	Pittsburgh.
	WTAE	4	Do.
	WIIC	11	Do.
	WKYC	3	Cleveland.
	WEWS	5	Do.
	WJW	8	Do.
	WTRF	7	Wheeling-Steubenville.
	WSTV	9	Do.
	WFMJ	21	Youngstown.
	WKBN	27	Do.
	WYTV	33	Do.
Coshocton	WLWC	4	Columbus, Ohio.
	WTVN	6	Do.
	WBNS	10	Do.
	WTRF	7	Wheeling-Steubenville.
	WSTV	9	Do.
	WHIZ	18	Zanesville.
Crawford	WLWC	4	Columbus, Ohio.
	WTVN	6	Do.
	WBNS	10	Do.
	WKYC	3	Cleveland.
	WEWS	5	Do.
	WJW	8	Do.
	WTOL	11	Toledo.
	WSPD	13	Do.
Cuyahoga	WKYC	3	Cleveland.
	WEWS	5	Do.
	WJW	8	Do.
	WUAB	43	Do.
	WKBF	61	Do.

OHIO—continued

County	Call letters	Ch.	Market name
Darke	WLWD	2	Dayton.
	WHIO	7	Do.
	WKEF	22	Do.
	WCPO	9	Cincinnati.
Defiance	WTOL	11	Toledo.
	WSPD	13	Do.
	WDHO	24	Do.
	WANE	15	Fort Wayne.
	WPTA	21	Do.
	WKJG	33	Do.
Delaware	WLWC	4	Columbus, Ohio.
	WTVN	6	Do.
	WBNS	10	Do.
Erie	WKYC	3	Cleveland.
	WEWS	5	Do.
	WJW	8	Do.
	WUAB	43	Do.
	WKBF	61	Do.
	WTOL	11	Toledo.
	WSPD	13	Do.
Fairfield	WLWC	4	Columbus, Ohio.
	WTVN	6	Do.
	WBNS	10	Do.
Fayette	WLWC	4	Columbus, Ohio.
	WTVN	6	Do.
	WBNS	10	Do.
	WHIO	7	Dayton.
Franklin	WLWC	4	Columbus, Ohio.
	WTVN	6	Do.
	WBNS	10	Do.
Fulton	WTOL	11	Toledo.
	WSPD	13	Do.
	WDHO	24	Do.
Gallia	WSAZ	3	Charleston-Huntington.
	WCHS	8	Do.
	WHTN	13	Do.
Geauga	WKYC	3	Cleveland.
	WEWS	5	Do.
	WJW	8	Do.
	WUAB	43	Do.
	WKBF	61	Do.
Greene	WLWD	2	Dayton.
	WHIO	7	Do.
	WKEF	22	Do.
	WCPO	9	Cincinnati.
	WKRC	12	Do.
	WTRF	7	Wheeling-Steubenville.
Guernsey	WSTV	9	Do.
Hamilton	WLWT	5	Cincinnati.
	WCPO	9	Do.
	WKRC	12	Do.
	WXIX	19	Do.
Hancock	WTOL	11	Toledo.
	WSPD	13	Do.
	WDHO	24	Do.
Hardin	WLWC	4	Columbus, Ohio.
	WTVN	6	Do.
	WBNS	10	Do.
	WTOL	11	Toledo.
	WSPD	13	Do.
Harrison	WTRF	7	Wheeling-Steubenville.
	WSTV	9	Do.
	KDKA	2	Pittsburgh.
	WTAE	4	Do.
	WIIC	11	Do.
Henry	WTOL	11	Toledo.
	WSPD	13	Do.
	WDHO	24	Do.
Highland	WLWT	5	Cincinnati.
	WCPO	9	Do.
	WKRC	12	Do.
	WXIX	19	Do.
	WLWC	4	Columbus, Ohio.
	WTVN	6	Do.
	WBNS	10	Do.
	WHIO	7	Dayton.
Hocking	WLWC	4	Columbus, Ohio.
	WTVN	6	Do.
	WBNS	10	Do.
Holmes	WKYC	3	Cleveland
	WEWS	5	Do.
	WJW	8	Do.
Huron	WKYC	3	Cleveland.
	WEWS	5	Do.
	WJW	8	Do.
	WUAB	43	Do.
	WKBF	61	Do.
	WTOL	11	Toledo.
	WSPD	13	Do.
Jackson	WSAZ	3	Charleston-Huntington.
	WCHS	8	Do.
	WHTN	13	Do.
Jefferson	WTRF	7	Wheeling-Steubenville.
	WSTV	9	Do.
	KDKA	2	Pittsburgh.
	WTAE	4	Do.
	WIIC	11	Do.
Knox	WLWC	4	Columbus, Ohio.
	WTVN	6	Do.
	WBNS	10	Do.

OHIO—continued

County	Call letters	Ch.	Market name
Lake	WKYC	3	Cleveland.
	WEWS	5	Do.
	WJW	8	Do.
	WUAB	43	Do.
	WKBF	61	Do.
Lawrence	WSAZ	3	Charleston-Huntington.
	WCHS	8	Do.
	WHTN	13	Do.
Licking	WLWC	4	Columbus, Ohio.
	WTVN	6	Do.
	WBNS	10	Do.
Logan	WLWC	4	Columbus, Ohio.
	WTVN	6	Do.
	WBNS	10	Do.
	WLWD	2	Dayton.
	WHIO	7	Do.
Lorain	WKYC	3	Cleveland.
	WEWS	5	Do.
	WJW	8	Do.
	WUAB	43	Do.
	WKBF	61	Do.
Lucas*	WTOL	11	Toledo.
	WSPD	13	Do.
	WDHO	24	Do.
	WJBK	2	Detroit.
	WXYZ	7	Do.
Madison	WLWC	4	Columbus, Ohio.
	WTVN	6	Do.
	WBNS	10	Do.
Mahoning	WFMJ	21	Youngstown.
	WKBN	27	Do.
	WYTV	33	Do.
Marion	WLWC	4	Columbus, Ohio.
	WTVN	6	Do.
	WBNS	10	Do.
Medina	WKYC	3	Cleveland.
	WEWS	5	Do.
	WJW	8	Do.
	WUAB	43	Do.
	WKBF	61	Do.
Meigs	WSAZ	3	Charleston-Huntington.
	WCHS	8	Do.
	WHTN	13	Do.
Mercer	WLWD	2	Dayton.
	WHIO	7	Do.
	WANE	15	Fort Wayne.
	WPTA	21	Do.
	WKJG	33	Do.
	WIMA	35	Lima.
Miami	WLWD	2	Dayton.
	WHIO	7	Do.
	WKEF	22	Do.
	WKTR	16	Kettering (Dayton).
Monroe	WTRF	7	Wheeling-Steubenville.
	WSTV	9	Do.
	WDTV	5	Clarksburg-Weston.
	WTAE	4	Pittsburgh.
Montgomery	WLWD	2	Dayton.
	WHIO	7	Do.
	WKTR	16	Do.
	WKEF	22	Do.
	WCPO	9	Cincinnati.
	WKRC	12	Do.
Morgan	WLWC	4	Columbus, Ohio.
	WTVN	6	Do.
	WBNS	10	Do.
	WSAZ	3	Charleston-Huntington.
	WCHS	8	Do.
	WHTN	13	Do.
	WTAP	15	Parkersburg.
	WHIZ	18	Zanesville.
Morrow	WLWC	4	Columbus, Ohio.
	WTVN	6	Do.
	WBNS	10	Do.
Muskingum	WHIZ	18	Zanesville.
	WLWC	4	Columbus, Ohio.
	WTVN	6	Do.
	WBNS	10	Do.
Noble	WTRF	7	Wheeling-Steubenville.
	WSTV	9	Do.
Ottawa	WTOL	11	Toledo.
	WSPD	13	Do.
	WDHO	24	Do.
	WEWS	5	Cleveland.
	WJBK	2	Detroit.
Paulding	WANE	15	Fort Wayne.
	WPTA	21	Do.
	WKJG	33	Do.
Perry	WLWC	4	Columbus, Ohio.
	WTVN	6	Do.
	WBNS	10	Do.
	WHIZ	18	Zanesville.
Pickaway	WLWC	4	Columbus, Ohio.
	WTVN	6	Do.
	WBNS	10	Do.

OHIO—continued

County	Call letters, channel number and market name
Pike	WLWC 4 Do.
	WTVN 6 Do.
	WBNS 10 Do.
	WSAZ 3 Charleston-Huntington.
	WHTN 13 Do.
	WEWS 5 Do.
	WJW 8 Do.
	WUAB 43 Do.
	WKBF 61 Do.
Preble	WLWD 2 Dayton.
	WHIO 7 Do.
	WKEF 22 Do.
	WLWT 5 Cincinnati.
	WCPO 9 Do.
	WKRC 12 Do.
	WXIX 19 Do.
Putnam	WTOL 11 Toledo.
	WSPD 13 Do.
	WDHO 24 Do.
	WIMA 35 Lima.
Richland	WKYC 3 Cleveland.
	WEWS 5 Do.
	WJW 8 Do.
	WBNS 10 Columbus, Ohio.
Ross	WLWC 4 Do.
	WTVN 6 Do.
	WBNS 10 Do.
Sandusky	WTOL 11 Toledo.
	WSPD 13 Do.
	WDHO 24 Do.
Scioto	WSAZ 3 Charleston-Huntington
	WCHS 8 Do.
	WHTN 13 Do.
Seneca	WTOL 11 Toledo.
	WSPD 13 Do.
	WDHO 24 Do.
	WEWS 5 Cleveland.
Shelby	WLWD 2 Dayton.
	WHIO 7 Do.
	WKEF 22 Do.
Stark	WKYC 3 Cleveland.
	WEWS 5 Do.
	WJW 8 Do.
	WUAB 43 Do.
	WKBF 61 Do.
Summit	WKYC 3 Cleveland.
	WEWS 5 Do.
	WJW 8 Do.
	WUAB 43 Do.
	WKBF 61 Do.
	WAKR 23 Akron.
Trumbull	WFMJ 21 Youngstown.
	WKBN 27 Do.
	WYTV 33 Do.
	WKYC 3 Cleveland.
	WEWS 5 Do.
	WJW 8 Do.
Tuscarawas	WKYC 3 Cleveland.
	WEWS 5 Do.
	WJW 8 Do.
	WTRF 7 Wheeling-Steubenville.
	WSTV 9 Do.
Union	WLWC 4 Columbus, Ohio.
	WTVN 6 Do.
	WBNS 10 Do.
Van Wert	WANE 15 Fort Wayne.
	WPTA 21 Do.
	WKJG 33 Do.
	WIMA 35 Lima.
Vinton	WSAZ 3 Charleston-Huntington.
	WCHS 8 Do.
	WHTN 13 Do.
	WLWC 4 Columbus, Ohio.
	WTVN 6 Do.
	WBNS 10 Do.
Warren	WLWT 5 Cincinnati.
	WCPO 9 Do.
	WKRC 12 Do.
	WXIX 19 Do.
	WLWD 2 Dayton.
	WHIO 7 Do.
	WKEF 22 Do.
Washington	WSAZ 3 Charleston-Huntington.
	WCHS 8 Do.
	WHTN 13 Do.
	WTAP 15 Parkersburg.
	WTRF 7 Wheeling-Steubenville.
Wayne	WKYC 3 Cleveland.
	WEWS 5 Do.
	WJW 8 Do.
	WUAB 43 Do.
	WKBF 61 Do.

OHIO—continued

County	Call letters, channel number and market name
Williams	WTOL 11 Toledo.
	WSPD 13 Do.
	WDHO 24 Do.
	WANE 15 Fort Wayne.
	WPTA 21 Do.
	WKJG 33 Do.
Wood	WTOL 11 Toledo.
	WSPD 13 Do.
	WDHO 24 Do.
	WKBD 50 Detroit.
Wyandot	WTOL 11 Toledo.
	WSPD 13 Do.
	WDHO 24 Do.
	WLWC 4 Columbus, Ohio.
	WBNS 10 Do.

OKLAHOMA

County	Call letters, channel number and market name
Adair	KTEW 2 Tulsa.
	KOTV 6 Do.
	KTUL 8 Do.
Alfalfa	WKY 4 Oklahoma City.
	KOCO 5 Do.
	KWTV 9 Do.
Atoka	KTEN 10 Ardmore-Ada.
	KXII 12 Do.
Beaver	KTVC 6 Wichita-Hutchinson.
	KGLD 11 Do.
	KUPK 13 Do.
	KFDA+ 10 Amarillo.
Beckham	KSWO 7 Wichita Falls-Lawton.
Blaine	WKY 4 Oklahoma City.
	KOCO 5 Do.
	KWTV 9 Do.
Bryan	KTEN 10 Ardmore-Ada.
	KXII 12 Do.
	KDFW 4 Dallas-Fort Worth.
	WFAA 8 Do.
	KTVT 11 Do.
Caddo	WKY 4 Oklahoma City.
	KOCO 5 Do.
	KWTV 9 Do.
Canadian	WKY 4 Oklahoma City.
	KOCO 5 Do.
	KWTV 9 Do.
Carter	KTEN 10 Ardmore-Ada.
	KXII 12 Do.
	KWTV 9 Oklahoma City.
	KFDX 3 Wichita Falls-Lawton.
	KAUZ 6 Do.
Cherokee	KTEW 2 Tulsa.
	KOTV 6 Do.
	KTUL 8 Do.
Choctaw	KTVT 11 Dallas-Fort Worth.
	KTEN 10 Ardmore-Ada.
	KXII 12 Do.
Cimarron	KGNC 4 Amarillo.
	KVII 7 Do.
	KDFA+ 10 Do.
Cleveland	WKY 4 Oklahoma City.
	KOCO 5 Do.
	KWTV 9 Do.
Coal	KTEN 10 Ardmore-Ada.
	KXII 12 Do.
Comanche	WKY 4 Oklahoma City.
	KFDX 3 Wichita Falls-Lawton.
	KAUZ 6 Do.
	KSWO 7 Do.
Cotton	KFDX 3 Do.
	KAUZ 6 Do.
	KSWO 7 Do.
Craig	KTEW 2 Tulsa.
	KOTV 6 Do.
	KTUL 8 Do.
	KOAM 7 Joplin-Pittsburg.
Creek	KTEM 2 Tulsa.
	KOTV 6 Do.
	KTUL 8 Do.
Custer	WKY 4 Oklahoma City.
	KOCO 5 Do.
	KWTV 9 Do.
Delaware	KTEW 2 Tulsa.
	KOTV 6 Do.
	KTUL 8 Do.
	KOAM 7 Joplin-Pittsburg.
	KODE 12 Do.
Dewey	WKY 4 Oklahoma City.
	KOCO 5 Do.
	KWTX 9 Do.
Ellis	WKY 4 Do.
	KOCO 5 Do.
	KWTV 9 Do.
	KGNC 4 Amarillo.
	KFDA+ 10 Do.

OKLAHOMA—continued

County	Call letters, channel number and market name
Garfield	WKY 4 Oklahoma City.
	KOCO 5 Do.
	KWTV 9 Do.
Garvin	WKY 4 Do.
	KOCO 5 Do.
	KWTV 9 Do.
	KTEN 10 Ardmore-Ada.
Grady	WKY 4 Oklahoma City.
	KOCO 5 Do.
	KWTV 9 Do.
Grant	WKY 4 Do.
	KOCO 5 Do.
	KWTV 9 Do.
	KARD 3 Wichita-Hutchinson.
	KAKE 10 Do.
Greer	KFDX 3 Wichita Falls-Lawton.
	KAUZ 6 Do.
	KSWO 7 Do.
	KFDA+ 10 Amarillo.
Harmon	KFDX 3 Wichita Falls-Lawton.
	KAUZ 6 Do.
	KSWO 7 Do.
Harper	WKY 4 Oklahoma City.
	KOCO 5 Do.
	KWTV 9 Do.
	KTVC 6 Wichita-Hutchinson.
	KUPK 13 Do.
Haskell*	KTUL 8 Tulsa.
	KFSA 5 Fort Smith.
Hughes	WKY 4 Oklahoma City.
	KOCO 5 Do.
	KWTV 9 Do.
	KTEN 10 Ardmore-Ada.
	KTEW 2 Tulsa.
	KOTV 6 Do.
	KTUL 8 Do.
Jackson	KFDX 3 Wichita Falls-Lawton.
	KAUZ 6 Do.
	KSWO 7 Do.
Jefferson	KFDX 3 Do.
	KAUZ 6 Do.
	KSWO 7 Do.
	KXII 12 Ardmore-Ada.
Johnston	KTEN 10 Do.
	KXII 12 Do.
Kay	WKY 4 Oklahoma City.
	KOCO 5 Do.
	KWTV 9 Do.
	KTEW 2 Tulsa.
	KOTV 6 Do.
Kingfisher	WKY 4 Oklahoma City.
	KOCO 5 Do.
	KWTV 9 Do.
Kiowa	KFDX 3 Wichita Falls-Lawton.
	KAUZ 6 Do.
	KSWO 7 Do.
	WKY 4 Oklahoma City.
	KOCO 5 Do.
Latimer	KTUL 8 Tulsa.
	KTEN 10 Ardmore-Ada.
	KFSA 5 Fort Smith.
Le Flore	KFSA 5 Do.
	KTUL 8 Tulsa.
Lincoln	WKY 4 Oklahoma City.
	KOCO 5 Do.
	KWTV 9 Do.
Logan	WKY 4 Oklahoma City.
	KOCO 5 Do.
	KWTV 9 Do.
Love	KDFW 4 Dallas-Fort Worth.
	WFAA 8 Do.
	KTVT 11 Do.
	KXII 12 Ardmore-Ada.
	KFDX 3 Wichita Falls-Lawton.
	KAUZ 6 Do.
	KSWO 7 Do.
McClain	WKY 4 Oklahoma City.
	KOCO 5 Do.
	KWTV 9 Do.
McCurtain	KTBS 3 Shreveport-Texarkana.
	KTAL 12 Do.
	KSFA 5 Fort Smith.
McIntosh	KTEW 2 Tulsa.
	KOTV 6 Do.
	KTUL 8 Do.
Major	WKY 4 Oklahoma City.
	KOCO 5 Do.
	KWTV 9 Do.
Marshall	KTEN 10 Ardmore-Ada.
	KXII 12 Do.
	KDFW 4 Dallas-Fort Worth.
Mayes	KTEW 2 Tulsa.
	KOTV 6 Do.
	KTUL 8 Do.

SIGNIFICANTLY VIEWED SIGNALS—Continued

County	Call letters, channel number and market name		

OKLAHOMA—continued

County	Call letters	Channel	Market name
Murray	WKY	4	Oklahoma City.
	KOCO	5	Do.
	KWTV	9	Do.
	KTEN	10	Ardmore-Ada.
	KXII	12	Do.
Muskogee	KTEW	2	Tulsa.
	KOTV	6	Do.
	KTUL	8	Do.
Noble	WKY	4	Oklahoma City.
	KOCO	5	Do.
	KWTV	9	Do.
Nowata	KTEW	2	Tulsa.
	KOTV	6	Do.
	KTUL	8	Do.
Okfuskee	KTEW	2	Do.
	KOTV	6	Do.
	KTUL	8	Do.
	KTEN	10	Ardmore-Ada.
	WKY	4	Oklahoma City.
	KOCO	5	Do.
	KWTV	9	Do.
Oklahoma	WKY	4	Do.
	KOCO	5	Do.
	KWTV	9	Do.
Okmulgee	KTEW	2	Tulsa.
	KOTV	6	Do.
	KTUL	8	Do.
Osage	KTEW	2	Do.
	KOTV	6	Do.
	KTUL	8	Do.
Ottawa	KOAM	7	Joplin-Pittsburg.
	KODE	12	Do.
	KUHI	16	Do.
	KOTV	6	Tulsa.
	KTUL	8	Do.
Pawnee	KTEW	2	Do.
	KOTV	6	Do.
	KTUL	8	Do.
Payne	WKY	4	Oklahoma City.
	KOCO	5	Do.
	KWTV	9	Do.
	KTEW	2	Tulsa.
	KOTV	6	Do.
Pittsburg	KTEW	2	Do.
	KOTV	6	Do.
	KTUL	8	Do.
	KTEN	10	Ardmore-Ada.
Pontotoc	KTEN	10	Do.
	WKY	4	Oklahoma City.
	KOCO	5	Do.
	KWTV	9	Do.
Pottawatomie	WKY	4	Do.
	KOCO	5	Do.
	KWTV	9	Do.
Pushmataha	KTEN	10	Ardmore-Ada.
	KXII	12	Do.
Roger Mills	KFDA+	10	Amarillo.
	WKY	4	Oklahoma City.
Rogers	KTEW	2	Tulsa.
	KOTV	6	Do.
	KTUL	8	Do.
Seminole	WKY	4	Oklahoma City.
	KOCO	5	Do.
	KWTV	9	Do.
	KTEN	10	Ardmore-Ada.
Sequoyah	KTEW	2	Tulsa.
	KOTV	6	Do.
	KTUL	8	Do.
	KSFA	5	Fort Smith.
Stephens	KFDX	3	Wichita Falls-Lawton.
	KAUZ	6	Do.
	KSWO	7	Do.
Texas	KGNC	4	Amarillo.
	KVII	7	Do.
	KFDA+	10	Do.
	KTVC	6	Wichita-Hutchinson.
	KGLD	11	Do.
	KUPK	13	Do.
Tillman	KFDX	3	Wichita Falls-Lawton.
	KAUZ	6	Do.
	KSWO	7	Do.
Tulsa	KTEW	2	Tulsa.
	KOTV	6	Do.
	KTUL	8	Do.
Wagoner	KTEW	2	Do.
	KOTV	6	Do.
	KTUL	8	Do.
Washington	KTEW	2	Do.
	KOTV	6	Do.
	KTUL	8	Do.
Washita	WKY	4	Oklahoma City.
	KOCO	5	Do.
	KWTV	9	Do.
	KFDX	3	Wichita Falls-Lawton.
	KAUZ	6	Do.
	KSWO	7	Do.
Woods	WKY	4	Oklahoma City.
	KOCO	5	Do.
	KWTV	9	Do.
Woodward	WKY	4	Do.
	KOCO	5	Do.
	KWTV	9	Do.

SIGNIFICANTLY VIEWED SIGNALS—Continued

OREGON

County	Call letters	Channel	Market name
Baker	KBOI	2	Boise.
	KTVB+	7	Do.
Benton	KATU	2	Portland, Oreg.
	KOIN	6	Do.
	KPTV	12	Do.
	KEZI	9	Eugene.
	KVAL	13	Do.
Clackamas	KATU	2	Portland, Oreg.
	KOIN	6	Do.
	KGW	8	Do.
	KPTV	12	Do.
Clatsop	KATU	2	Do.
	KOIN	6	Do.
	KGW	8	Do.
	KPTV	12	Do.
	KING	5	Seattle-Tacoma.
Columbia	KATU	2	Portland, Oreg.
	KOIN	6	Do.
	KGW	8	Do.
	KPTV	12	Do.
Coos	KCBY	11	Eugene.
	KOBI	5	Medford.
Crook	KATU	2	Portland, Oreg.
	KOIN	6	Do.
	KGW	8	Do.
	KPTV	12	Do.
	KEZI	9	Eugene.
Curry	KIEM	3	Eureka.
	KVIQ	6	Do.
Deschutes	NA.		
Douglas	KPIC	4	Eugene.
	KEZI	9	Do.
	KOBI	5	Medford.
Gilliam	KEPR	19	Yakima.
	KNDU	25	Do.
	KOIN	6	Portland, Oreg.
	KGW	8	Do.
	KPTV	12	Do.
Grant	KBOI	2	Boise.
	KTVB+	7	Do.
Harney	NA.		
Hood River	KATU	2	Portland, Oreg.
	KOIN	6	Do.
	KGW	8	Do.
	KPTV	12	Do.
Jackson	KOBI	5	Medford.
	KMED	10	Do.
Jefferson	KATU	2	Portland, Oreg.
	KOIN	6	Do.
	KGW	8	Do.
Josephine	KOBI	5	Medford.
	KMED	10	Do.
Klamath	KOTI	2	Klamath Falls.
	KMED	10	Medford.
Lake	KOTI	2	Klamath Falls.
Lane Inner	KEZI	9	Eugene.
	KVAL	13	Do.
Lane Outer*	KEZI	9	Do.
	KVAL	13	Do.
	KOIN	6	Portland, Oreg.
Lincoln	KATU	2	Do.
	KOIN	6	Do.
	KGW	8	Do.
	KPTV	12	Do.
	KEZI	9	Eugene.
	KVDO	3	Salem, Oreg.
Linn	KATU	2	Portland, Oreg.
	KOIN	6	Do.
	KPTV	12	Do.
	KEZI	9	Eugene.
	KVAL	13	Do.
	KVDO	3	Salem, Oreg.
Malheur	KBOI	2	Boise.
	KTVB+	7	Do.
Marion	KATU	2	Portland, Oreg.
	KOIN	6	Do.
	KGW	8	Do.
	KPTV	12	Do.
	KVDO	13	Salem, Oreg.
Morrow	KEPR	19	Yakima.
	KNDU	25	Do.
	KVEW	42	Do.
	KATU	2	Portland, Oreg.
	KOIN	6	Do.
	KGW	8	Do.
Multnomah	KATU	2	Do.
	KOIN	6	Do.
	KGW	8	Do.
	KPTV	12	Do.
Polk	KATU	2	Do.
	KOIN	6	Do.
	KGW	8	Do.
	KPTV	12	Do.
	KVDO	3	Salem, Oreg.
Sherman	KATU	2	Portland, Oreg.
	KOIN	6	Do.
	KGW	8	Do.
	KPTV	12	Do.

SIGNIFICANTLY VIEWED SIGNALS—Continued

OREGON—continued

County	Call letters	Channel	Market name
Tillamook	KATU	2	Do.
	KOIN	6	Do.
	KGW	8	Do.
	KPTV	12	Do.
	KVDO	3	Salem, Oreg.
Umatilla	KEPR	19	Yakima.
	KNDU	25	Do.
	KVEW	42	Do.
Union	KREM	2	Spokane.
	KXLY	4	Do.
	KHQ	6	Do.
	KTVB+	7	Boise.
Wallowa	KREM	2	Spokane.
	KXLY	4	Do.
	KHQ	6	Do.
Wasco	KATU	2	Portland, Oreg.
	KOIN	6	Do.
	KGW	8	Do.
	KPTV	12	Do.
Washington	KATU	2	Do.
	KOIN	6	Do.
	KGW	8	Do.
	KPTV	12	Do.
Yamhill	KATU	2	Do.
	KOIN	6	Do.
	KGW	8	Do.
	KPTV	12	Do.
	KVDO	3	Salem, Oreg.
Wheeler	NA.		

Census County Divisions in Split Counties

Lane Inner: Eugene, Springfield, Eugene West.
Lane Outer: All other.

PENNSYLVANIA

County	Call letters	Channel	Market name
Adams	WGAL	8	Harrisburg-York-Lancaster-Lebanon.
	WMAR	2	Baltimore.
	WBAL	11	Do.
	WJZ	13	Do.
	WTTG	5	Washington, D.C.
Allegheny	KDKA	2	Pittsburgh.
	WTAE	4	Do.
	WIIC	11	Do.
	WPGH	53	Do.
Armstrong	KDKA	2	Do.
	WTAE	4	Do.
	WIIC	11	Do.
	WPGH	53	Do.
	WJAC	6	Johnstown-Altoona.
Beaver	KDKA	2	Pittsburgh.
	WTAE	4	Do.
	WIIC	11	Do.
	WPGH	53	Do.
	WSTV	9	Wheeling-Steubenville.
Bedford	WJAC	6	Johnstown-Altoona.
	WFBG	10	Do.
Berks	KYW	3	Philadelphia.
	WFIL	6	Do.
	WCAU	10	Do.
	WPHL	17	Do.
	WKBS	48	Do.
	WGAL	8	Harrisburg-York-Lancaster-Lebanon.
Blair	WJAC	6	Johnstown-Altoona.
	WFBG	10	Do.
Bradford	WNBF	12	Binghamton.
	WSYE	18	Elmira.
	WENY	36	Do.
Bucks	KYW	3	Philadelphia.
	WFIL	6	Do.
	WCAU	10	Do.
	WPHL	17	Do.
	WTAF	29	Do.
	WKBS	48	Do.
Butler	KDKA	2	Pittsburgh.
	WTAE	4	Do.
	WIIC	11	Do.
	WPGH	53	Do.
	WJAC	6	Johnstown-Altoona.
Cambria	WJAC	6	Do.
	WFBG	10	Do.
	KDKA	2	Pittsburgh.
	WTAE	4	Do.
Cameron	NA.		
Carbon	KYW	3	Philadelphia.
	WFIL	6	Do.
	WCAU	10	Do.
	WNEP	16	Wilkes-Barre-Scranton.
Centre	WJAC	6	Johnstown-Altoona.
	WFBG	10	Do.
Chester	KYW	3	Philadelphia.
	WFIL	6	Do.
	WCAU	10	Do.
	WPHL	17	Do.
	WTAF	29	Do.
	WKBS	48	Do.

SIGNIFICANTLY VIEWED SIGNALS—Continued

County	Call letters, channel number and market name		

PENNSYLVANIA—continued

County	Call letters	Channel	Market name
Clarion	KDKA	2	Pittsburgh.
	WTAE	4	Do.
	WIIC	11	Do.
	WJAC	6	Johnstown-Altoona.
Clearfield	WJAC	6	Do.
	WFBG	10	Do.
Clinton*	WFBG	10	Do.
Columbia	WNEP	16	Wilkes-Barre-Scranton.
	WDAU	22	Do.
	WBRE	28	Do.
Crawford	WICU	12	Erie.
	WJET	24	Do.
	WSEE	35	Do.
Cumberland	WGAL	8	Harrisburg—York-Lancaster-Lebanon.
	WHP	21	Do.
	WTPA	27	Do.
Dauphin	WGAL	8	Do.
	WHP	21	Do.
	WTPA	27	Do.
Delaware	KYW	3	Philadelphia.
	WFIL	6	Do.
	WCAU	10	Do.
	WPHL	17	Do.
	WTAF	29	Do.
	WKBS	48	Do.
Elk	WJAC	6	Johnstown-Altoona.
	WFBG	10	Do.
Erie	WICU	12	Erie.
	WJET	24	Do.
	WSEE	35	Do.
Fayette	KDKA	2	Pittsburgh.
	WTAE	4	Do.
	WIIC	11	Do.
Forest	WICU	12	Erie.
	WJAC	6	Johnstown-Altoona.
	KDKA	2	Pittsburgh.
	WTAE	4	Do.
Franklin	WRC	4	Washington, D.C.
	WTTG	5	Do.
	WMAL	7	Do.
	WTOP	9	Do.
	WMAR	2	Baltimore.
	WBAL	11	Do.
	WJZ	13	Do.
	WGAL	8	Harrisburg—York-Lancaster-Lebanon.
Fulton	WRC	4	Washington, D.C.
	WTTG	5	Do.
	WMAL	7	Do.
	WJAC	6	Johnstown-Altoona.
	WFBG	10	Do.
Greene	KDKA	2	Pittsburgh.
	WTAE	4	Do.
	WIIC	11	Do.
	WPGH	53	Do.
	WTRF	7	Wheeling-Steubenville.
Huntingdon	WJAC	6	Johnstown-Altoona.
	WFBG	10	Do.
Indiana	KDKA	2	Pittsburgh.
	WTAE	4	Do.
	WIIC	11	Do.
	WJAC	6	Johnstown-Altoona.
	WFBG	10	Do.
Jefferson	WJAC	6	Do.
	WFBG	10	Do.
	KDKA	2	Pittsburgh.
	WTAE	4	Do.
Juniata	WGAL	8	Harrisburg—York-Lancaster-Lebanon.
	WHP	21	Do.
	WTPA	27	Do.
	WFBG	10	Johnstown-Altoona.
Lackawanna	WNEP	16	Wilkes-Barre-Scranton
	WDAU	22	Do.
	WBRE	28	Do.
Lancaster	WGAL	8	Harrisburg—York-Lancaster-Lebanon.
	WLYH	15	Do.
	WTPA	27	Do.
	KYW	3	Philadelphia.
	WFIL	6	Do.
	WCAU	10	Do.
	WPHL	17	Do.
Lawrence	KDKA	2	Pittsburgh.
	WTAE	4	Do.
	WIIC	11	Do.
	WFMJ	21	Youngstown.
	WKBN	27	Do.
	WYTV	33	Do.
Lebanon	WGAL	8	Harrisburg—York-Lancaster-Lebanon.
	WLYH	15	Do.
	WHP	21	Do.
	WTPA	27	Do.
Lehigh	KYW	3	Philadelphia.
	WFIL	6	Do.
	WCAU	10	Do.
	WPHL	17	Do.

PENNSYLVANIA—continued

County	Call letters	Channel	Market name
Luzerne	WNEP	16	Wilkes-Barre-Scranton.
	WDAU	22	Do.
	WBRE	28	Do.
Lycoming	WNEP	16	Do.
	WDAU	22	Do.
	WBRE	28	Do.
	WFBG	10	Johnstown-Altoona.
McKean	WGR	2	Buffalo.
	WBEN	4	Do.
	WKBW	7	Do.
Mercer	WFMJ	21	Youngstown.
	WKBN	27	Do.
	WYTV	33	Do.
	KDKA	2	Pittsburgh.
	WIIC	11	Do.
Mifflin	WGAL	8	Harrisburg-York-Lancaster-Lebanon.
	WJAC	6	Johnstown-Altoona.
	WFBG	10	Do.
Monroe	KYW	3	Philadelphia.
	WFIL	6	Do.
	WCAU	10	Do.
	WCBS	2	New York.
	WNBC	4	Do.
	WNEW	5	Do.
Montgomery	KYW	3	Philadelphia.
	WFIL	6	Do.
	WCAU	10	Do.
	WPHL	17	Do.
	WTAF	29	Do.
	WKBS	48	Do.
Montour	Over 90 percent cable penetration.		
Northampton	KYW	3	Philadelphia.
	WFIL	6	Do.
	WCAU	10	Do.
	WNEW	5	New York.
	WOR	9	Do.
	WPIX	11	Do.
Northumberland	Over 90 percent cable penetration.		
Perry	WGAL	8	Harrisburg-York-Lancaster-Lebanon.
	WHP	21	Do.
	WTPA	27	Do.
Philadelphia	KYW	3	Philadelphia.
	WFIL	6	Do.
	WCAU	10	Do.
	WPHL	17	Do.
	WTAF	29	Do.
	WKBS	48	Do.
Pike	WCBS	2	New York.
	WNBC	4	Do.
	WNEW	5	Do.
	WABC	7	Do.
	WNEP	16	Wilkes-Barre-Scranton.
	WDAU	22	Do.
	WBRE	28	Do.
Potter	WGR	2	Buffalo.
	WBEN	4	Do.
	WKBW	7	Do.
Schuylkill	KYW	3	Philadelphia.
	WFIL	6	Do.
	WCAU	10	Do.
	WGAL	8	Harrisburg-York-Lancaster-Lebanon.
Snyder	WGAL	8	Do.
	WHP	21	Do.
	WTPA	27	Do.
	WNEP	16	Wilkes-Barre-Scranton.
	WBRE	28	Do.
Somerset	WJAC	6	Johnstown-Altoona.
	KDKA	2	Pittsburgh.
	WTAE	4	Do.
Sullivan	WNBF	12	Binghamton.
	WNEP	16	Wilkes-Barre-Scranton.
	WDAU	22	Do.
	WBRE	28	Do.
Susquehanna	WNBF	12	Binghamton.
	WNEP	16	Wilkes-Barre-Scranton.
	WDAU	22	Do.
	WBRE	28	Do.
Tioga	WSYE	18	Elmira.
	WENY	36	Do.
	WNBF	12	Binghamton.
Union	WNEP	16	Wilkes-Barre-Scranton.
	WDAU	22	Do.
	WBRE	28	Do.
Venango	KDKA	2	Pittsburgh.
	WTAE	4	Do.
	WIIC	11	Do.
	WICU	12	Erie.
	WJAC	6	Johnstown-Altoona.
Warren	WGR	2	Buffalo.
	WBEN	4	Do.
	WKBW	7	Do.
	WICU	12	Erie.

SIGNIFICANTLY VIEWED SIGNALS—Continued

County	Call letters	Channel	Market name

PENNSYLVANIA—continued

County	Call letters	Channel	Market name
Washington	KDKA	2	Pittsburgh.
	WTAE	4	Do.
	WIIC	11	Do.
	WPGH	53	Do.
	WTRF	7	Wheeling-Steubenville.
	WSTV	9	Do.
Wayne	WNEP	16	Wilkes-Barre-Scranton.
	WDAU	22	Do.
	WBRE	28	Do.
	WNBF	12	Binghamton.
	WNBC	4	New York.
	WNEW	5	Do.
Westmoreland	KDKA	2	Pittsburgh.
	WTAE	4	Do.
	WIIC	11	Do.
	WPGH	53	Do.
	WJAC	6	Johnstown-Altoona.
Wyoming	WNEP	16	Wilkes-Barre-Scranton.
	WDAU	22	Do.
	WBRE	28	Do.
York	WGAL	8	Harrisburg-York-Lancaster-Lebanon.
	WTPA	27	Do.
	WSBA	43	Do.
	WMAR	2	Baltimore.
	WBAL	11	Do.
	WJZ	13	Do.

RHODE ISLAND

County	Call letters	Channel	Market name
Bristol	WTEV	6	Providence.
	WJAR	10	Do.
	WPRI	12	Do.
	WBZ	4	Boston.
	WNAC	7	Do.
	WSBK	38	Do.
Kent	WTEV	6	Providence.
	WJAR	10	Do.
	WPRI	12	Do.
	WHDH	5	Boston.
	WNAC	7	Do.
	WSBK	38	Do.
Newport	WTEV	6	Providence.
	WJAR	10	Do.
	WPRI	12	Do.
	WNAC	7	Boston.
	WSBK	38	Do.
	WKBG	56	Do.
Providence	WTEV	6	Providence.
	WJAR	10	Do.
	WPRI	12	Do.
	WBZ	4	Boston.
	WHDH	5	Do.
	WNAC	7	Do.
	WSBK	38	Do.
	WKBG	56	Do.
Washington*	WTEV	6	Providence.
	WJAR	10	Do.
	WPRI	12	Do.

SOUTH CAROLINA

County	Call letters	Channel	Market name
Abbeville	WFBC	4	Greenville-Spartanburg-Asheville.
	WSPA	7	Do.
	WLOS	13	Do.
	WJBF	6	Augusta.
Aiken*	WJBF	6	Do.
	WRDW	12	Do.
Allendale	WJBF	6	Do.
	WRDW	12	Do.
Anderson	WFBC	4	Greenville-Spartanburg-Ashville.
	WSPA	7	Do.
	WLOS	13	Do.
Bamberg	WJBF	6	Augusta.
	WRDW	12	Do.
	WCSC	5	Charleston, S.C.
	WIS	10	Columbia, S.C.
Barnwell	WJBF	6	Augusta.
	WRDW	12	Do.
	WIS	10	Columbia, S.C.
Beaufort	WUSN	2	Charleston, S.C.
	WCIV	4	Do.
	WCSC	5	Do.
	WSAV	3	Savannah.
	WTOC	11	Do.
Berkeley	WUSN	2	Charleston, S.C.
	WCIV	4	Do.
	WCSC	5	Do.

SIGNIFICANTLY VIEWED SIGNALS—Continued

County	Call letters, channel number and market name		

SOUTH CAROLINA—continued

County	Call letters	Ch.	Market name
Calhoun	WIS	10	Columbia, S.C.
	WNOK	19	Do.
	WRDW	12	Augusta.
	WUSN	2	Charleston, S.C.
	WCSC	5	Do.
Charleston	WUSN	2	Do.
	WCIV	4	Do.
	WCSC	5	Do.
Cherokee	WFBC	4	Greenville-Spartanburg-Asheville.
	WSPA	7	Do.
	WLOS	13	Do.
	WBTV	3	Charlotte.
	WSOC	9	Do.
Chester	WBTV	3	Do.
	WSOC	9	Do.
	WIS	10	Columbia, S.C.
	WFBC	4	Greenville-Spartanburg-Asheville.
	WSPA	7	Do.
	WLOS	13	Do.
Chesterfield	WBTV	3	Charlotte.
	WSOC	9	Do.
	WCCB	18	Do.
	WRET	36	Do.
	WIS	10	Columbia, S.C.
	WBTW	13	Florence, S.C.
Clarendon	WIS	10	Columbia, S.C.
	WUSN	2	Charleston, S.C.
	WCIV	4	Do.
	WCSC	5	Do.
	WBTW	13	Florence, S.C.
Colleton	WUSN	2	Charleston, S.C.
	WCIV	4	Do.
	WCSC	5	Do.
Darlington	WBTW	13	Florence, S.C.
	WIS	10	Columbia, S.C.
Dillon*	WBTW	13	Florence, S.C.
	WWAY	3	Wilmington, N.C.
	WECT	6	Do.
Dorchester	WUSN	2	Charleston, S.C.
	WCIV	4	Do.
	WCSC	5	Do.
Edgefield	WJBF	6	Augusta.
	WRDW	12	Do.
Fairfield	WIS	10	Columbia, S.C.
	WNOK	19	Do.
	WOLO	25	Do.
	WSPA	7	Greenville-Spartanburg-Asheville.
Florence	WBTW	13	Florence, S.C.
	WIS	10	Columbia, S.C.
Georgetown	WUSN	2	Charleston, S.C.
	WCIV	4	Do.
	WCSC	5	Do.
Greenville	WFBC	4	Greenville-Spartanburg-Asheville.
	WSPA	7	Do.
	WLOS	13	Do.
Greenwood	WFDO	4	Do.
	WSPA	7	Do.
	WLOS	13	Do.
	WJBF	6	Augusta.
Hampton	WJBF	6	Do.
	WRDW	12	Do.
	WUSN	2	Charleston, S.C.
	WCIV	4	Do.
	WCSC	5	Do.
	WSAV	3	Savannah.
	WTOC	11	Do.
Horry	WWAY	3	Wilmington, N.C.
	WECT	6	Do.
	WCSC	5	Charleston, S.C.
	WBTW	13	Florence, S.C.
Jasper	WSAV	3	Savannah.
	WTOC	11	Do.
	WUSN	2	Charleston, S.C.
	WCIV	4	Do.
	WCSC	5	Do.
Kershaw	WIS	10	Columbia, S.C.
	WNOK	19	Do.
	WOLO	25	Do.
	WBTW	13	Florence, S.C.
Lancaster	WBTV	3	Charlotte.
	WSOC	9	Do.
	WCCB	18	Do.
	WRET	36	Do.
	WIS	10	Columbia, S.C.
Laurens	WFBC	4	Greenville-Spartanburg-Asheville.
	WSPA	7	Do.
	WLOS	13	Do.
Lee	WIS	10	Columbia, S.C.
	WBTW	13	Florence, S.C.
Lexington	WIS	10	Columbia, S.C.
	WNOK	19	Do.
	WOLO	25	Do.

SOUTH CAROLINA—continued

County	Call letters	Ch.	Market name
McCormick	WJBF	6	Augusta.
	WRDW	12	Do.
	WFBC	4	Greenville-Spartanburg-Asheville.
	WSPA	7	Do.
Marion	WBTW	13	Florence, S.C.
	WIS	10	Columbia, S.C.
	WWAY	3	Wilmington, N.C.
	WECT	6	Do.
Marlboro	WBTW	13	Florence, S.C.
	WIS	10	Columbia, S.C.
	WECT	6	Wilmington, N.C.
Newberry	WFBC	4	Greenville-Spartanburg-Asheville.
	WSPA	7	Do.
	WLOS	13	Do.
	WJBF	6	Augusta.
	WIS	10	Columbia, S.C.
Oconee	WFBC	4	Greenville-Spartanburg-Asheville.
	WSPA	7	Do.
	WLOS	13	Do.
Orangeburg	WIS	10	Columbia., S.C.
	WJBF	6	Augusta.
	WRDW	12	Do.
	WUSN	2	Charleston, S.C.
	WCIV	4	Do.
	WCSC	5	Do.
Pickens	WFDO	4	Greenville-Spartanburg-Asheville.
	WSPA	7	Do.
	WLOS	13	Do.
Richland	WIS	10	Columbia, S.C.
	WNOK	19	Do.
	WOLO	25	Do.
Saluda	WJBF	6	Augusta.
	WRDW	12	Do.
	WIS	10	Columbia, S.C.
	WFBC	4	Greenville-Spartanburg-Asheville.
	WSPA	7	Do.
Spartanburg	WFBC	4	Do.
	WSPA	7	Do.
	WLOS	13	Do.
	WBTV	3	Charlotte.
Sumter	WIS	10	Columbia, S.C.
	WNOK	19	Do.
	WOLO	25	Do.
	WBTW	13	Florence, S.C.
Union	WFBC	4	Greenville-Spartanburg-Asheville.
	WSPA	7	Do.
	WLOS	13	Do.
	WBTV	3	Charlotte.
Williamsburg	WUSN	2	Charleston, S.C.
	WCIV	4	Do.
	WCSC	5	Do.
	WIS	10	Columbia, S.C.
	WBTW	13	Florence, S.C.
York*	WBTV	3	Charlotte.
	WSOC	9	Do.
	WCCB	18	Do.
	WRET	36	Do.
	WSPA	7	Greenville-Spartanburg-Asheville.

SOUTH DAKOTA

County	Call letters	Ch.	Market name
Aurora	KORN	5	Sioux Falls-Mitchell.
	KELO+	11	Do.
	KSOO+	13	Do.
Beadle	KORN	5	Do.
	KELO+	11	Do.
	KSOO+	13	Do.
Bennett	KOTA	3	Rapid City.
	KDUH	4	Do.
Bon Homme	KORN	5	Sioux Falls-Mitchell.
	KELO+	11	Do.
	KSOO+	13	Do.
	KTIV	4	Sioux City.
	KCAU	9	Do.
Brookings	KORN	5	Sioux Falls-Mitchell.
	KELO+	11	Do.
	KSOO+	13	Do.
Brown	KELO+	11	Do.
	KSOO+	13	Do.
Brule	KORN	5	Do.
	KELO+	11	Do.
Buffalo	KORN	5	Do.
	KELO+	11	Do.
Butte	KOTA+	3	Rapid City.
	KRSD+	7	Do.
Campbell	KFYR	5	Minot-Bismarck.
	KXMB	12	Do.

SOUTH DAKOTA—continued

County	Call letters	Ch.	Market name
Charles Mix	KORN	5	Sioux Falls-Mitchell.
	KELO+	11	Do.
	KSOO+	13	Do.
Clark	KELO+	11	Do.
	KSOO+	13	Do.
Clay	KTIV	4	Sioux City.
	KCAU	9	Do.
	KMEG	14	Do.
	KELO+	11	Sioux Falls-Mitchell.
	KSOO+	13	Do.
Codington	KELO+	11	Do.
	KSOO+	13	Do.
Corson	KFYR	5	Minot-Bismarck.
	KXMB	12	Do.
Custer	KOTA+	3	Rapid City.
	KRSD+	7	Do.
Davison	KORN	5	Sioux Falls-Mitchell.
	KELO+	11	Do.
	KSOO+	13	Do.
Day	KELO+	11	Do.
	KSOO+	13	Do.
Deuel	KELO+	11	Do.
	KSOO+	13	Do.
Dewey	KFYR	5	Minot-Bismarck.
	KXMB	12	Do.
	KELO	11	Sioux Falls-Mitchell.
Douglas	KORN	5	Do.
	KELO+	11	Do.
Edmunds	KELO+	11	Do.
	KSOO+	13	Do.
Fall River	KOTA+	3	Rapid City.
	KDUH	4	Do.
	KSTF	10	Cheyenne.
Faulk	KELO+	11	Sioux Falls-Mitchell.
	KSOO+	13	Do.
Grant*	KELO+	11	Do.
	KCMT	7	Alexandria, Minn.
Gregory	KORN	5	Sioux Falls-Mitchell.
	KELO+	11	Do.
Haakon	KOTA+	3	Rapid City.
	KELO+	11	Sioux Falls-Mitchell.
Hamlin	KELO+	11	Do.
	KSOO+	13	Do.
Hand	KELO+	11	Do.
	KSOO+	13	Do.
Hanson	KORN	5	Do.
	KELO+	11	Do.
	KSOO+	13	Do.
Harding	KOTA+	3	Rapid City.
Hughes	KELO+	11	Sioux Falls-Mitchell.
Hutchinson	KORN	5	Do.
	KELO+	11	Do.
	KSOO+	13	Do.
Hyde	KELO+	11	Do.
	KSOO+	13	Do.
Jackson	KOTA+	3	Rapid City.
	KELO+	11	Sioux Falls-Mitchell.
Jerauld	KORN	5	Do.
	KELO+	11	Do.
	KSOO+	13	Do.
Jones	KELO+	11	Do.
Kingsbury	KORN	5	Do.
	KELO+	11	Do.
	KSOO+	13	Do.
Lake	KORM	5	Do.
	KELO+	11	Do.
	KSOO+	13	Do.
Lawrence	KOTA+	3	Rapid City.
	KRSD+	7	Do.
Lincoln	KORN	5	Sioux Falls-Mitchell.
	KELO+	11	Do.
	KSOO+	13	Do.
	KTIV	4	Sioux City.
	KCAU	9	Do.
Lyman	KELO+	11	Sioux Falls-Mitchell.
McCook	KORN	5	Do.
	KELO+	11	Do.
	KSOO+	13	Do.
McPherson	KELO+	11	Do.
	KSOO+	13	Do.
	KFYR	5	Minot-Bismarck.
Marshall	KELO+	11	Sioux Falls-Mitchell.
	KSOO+	13	Do.
	KXJB	4	Fargo.
Meade*	KOTA+	3	Rapid City.
	KRSD+	7	Do.
Mellette	KELO+	11	Sioux Falls-Mitchell.
Miner	KORN	5	Do.
	KELO+	11	Do.
	KSOO+	13	Do.
Minnehaha	KORN	5	Do.
	KELO+	11	Do.
	KSOO+	13	Do.
	KCAU	9	Sioux City.
Moody	KORN	5	Sioux Falls-Mitchell.
	KELO+	11	Do.
	KSOO+	13	Do.
Pennington	KOTA+	3	Rapid City.
	KRSD+	7	Do.

SIGNIFICANTLY VIEWED SIGNALS—Continued | SIGNIFICANTLY VIEWED SIGNALS—Continued | SIGNIFICANTLY VIEWED SIGNALS—Continued

SOUTH DAKOTA—continued

County	Call letters	Channel	Market name
Perkins	KFYR	5	Minot-Bismarck.
	KDIX	2	Dickinson, N. Dak.
	KOTA+	3	Rapid City.
Potter	KELO+	11	Sioux Falls-Mitchell.
	KSOO+	13	Do.
	KFYR	5	Minot-Bismarck.
Roberts	KELO+	11	Sioux Falls-Mitchell.
	KSOO+	13	Do.
	WDAY	6	Fargo.
Sanborn	KORN	5	Sioux Falls-Mitchell.
	KELO+	11	Do.
	KSOO+	13	Do.
Shannon	KOTA+	3	Rapid City.
	KDUH	4	Do.
	KRSD+	7	Do.
Spink	KELO+	11	Sioux Falls-Mitchell.
	KSOO+	13	Do.
Stanley*	KELO+	11	Do.
Sully	KELO+	11	Do.
Todd	KELO+	11	Do.
Tripp	KORN	5	Do.
	KELO+	11	Do.
Turner	KORN	5	Do.
	KELO+	11	Do.
	KSOO+	13	Do.
	KTIV	4	Sioux City.
	KCAU	9	Do.
Union	KTIV	4	Do.
	KCAU	9	Do.
	KMEG	14	Do.
	KELO+	11	Sioux Falls-Mitchell.
	KSOO+	13	Do.
Walworth	KFYR	5	Minot-Bismarck.
	KXMB	12	Do.
	KELO+	11	Sioux Falls-Mitchell.
Washabaugh	KOTA+	3	Rapid City.
	KELO+	11	Sioux Falls-Mitchell.
Yankton	KTIV	4	Sioux City.
	KCAU	9	Do.
	KELO+	11	Sioux Falls-Mitchell.
	KSOO	13	Do.
Ziebach	KOTA+	3	Rapid City.

TENNESSEE

County	Call letters	Channel	Market name
Anderson	WATE	6	Knoxville.
	WBIR	10	Do.
	WTVK	26	Do.
Bedford	WSM	4	Nashville.
	WLAC	5	Do.
	WSIX	8	Do.
Benton	WSM	4	Do.
	WLAC	5	Do.
	WSIX	8	Do.
Bledsoe	WRCB	3	Chattanooga.
	WTVC	9	Do.
	WDEF	12	Do.
Blount	WATE	6	Knoxville.
	WBIR	10	Do.
	WTVK	26	Do.
Bradley	WRCB	3	Chattanooga.
	WTVC	9	Do.
	WDEF	12	Do.
Campbell	WATE	6	Knoxville.
	WBIR	10	Do.
	WTVK	26	Do.
Cannon	WSM	4	Nashville.
	WLAC	5	Do.
	WSIX	8	Do.
Carroll	WSM	4	Do.
	WLAC	5	Do.
	WSIX	8	Do.
	WBBJ	7	Jackson, Tenn.
	WREC	3	Memphis.
	WPSD	6	Paducah-Cape Girardeau-Harrisburg.
Carter	WCYB	5	Bristol-Kingsport-Johnson City.
	WJHL	11	Do.
	WKPT	19	Do.
Cheatham	WSM	4	Nashville.
	WLAC	5	Do.
	WSIX	8	Do.
Chester	WREC	3	Memphis.
	WMC	5	Do.
	WHBQ	13	Do.
	WBBJ	7	Jackson, Tenn.
Claiborne	WATE	6	Knoxville.
	WBIR	10	Do.
Clay	WSM	4	Nashville.
	WLAC	5	Do.
	WSIX	8	Do.
Cocke	WATE	6	Knoxville.
	WBIR	10	Do.
	WLOS	13	Greenville-Spartanburg-Asheville.
Coffee	WSM	4	Nashville.
	WLAC	5	Do.
	WSIX	8	Do.

TENNESSEE—continued

County	Call letters	Channel	Market name
Crockett	WREC	3	Memphis.
	WMC	5	Do.
	WHBQ	13	Do.
	WBBJ	7	Jackson, Tenn.
Cumberland	WATE	6	Knoxville.
	WBIR	10	Do.
	WIVC	9	Chattanooga.
Davidson	WSM	4	Nashville.
	WLAC	5	Do.
	WSIX	8	Do.
Decatur	WSM	4	Do.
	WLAC	5	Do.
	WSIX	8	Do.
	WBBJ	7	Jackson, Tenn.
De Kalb	WSM	4	Nashville.
	WLAC	5	Do.
	WSIX	8	Do.
Dickson	WSM	4	Do.
	WLAC	5	Do.
	WSIX	8	Do.
Dyer	WREC	3	Memphis.
	WMC	5	Do.
	WHBQ	13	Do.
	WBBJ	7	Jackson, Tenn.
	KFVS	12	Paducah-Cape Girardeau-Harrisburg.
Fayette	WREC	3	Memphis.
	WMC	5	Do.
	WHBQ	13	Do.
Fentress	WATE	6	Knoxville.
	WBIR	10	Do.
Franklin	WSM	4	Nashville.
	WLAC	5	Do.
	WSIX	8	Do.
	WTVC	9	Chattanooga.
	WDEF	12	Do.
Gibson	WREC	3	Memphis.
	WMC	5	Do.
	WHBQ	13	Do.
	WBBJ	7	Jackson, Tenn.
Giles	WSM	4	Nashville.
	WLAC	5	Do.
	WSIX	8	Do.
Grainger	WATE	6	Knoxville.
	WBIR	10	Do.
Greene	WCYB	5	Bristol-Kingsport-Johnson City.
	WJHL	11	Do.
	WLOS	13	Greenville-Spartanburg-Asheville.
	WATE	6	Knoxville.
	WBIR	10	Do.
Grundy	WRCB	3	Chattanooga.
	WTVC	9	Do.
	WDEF	12	Do.
	WSM	4	Nashville.
	WLAC	5	Do.
	WSIX	8	Do.
Hamblen	WATE	6	Knoxville.
	WBIR	10	Do.
	WCYB	5	Bristol-Kingsport-Johnson City.
	WLOS	13	Greenville-Spartanburg-Asheville.
Hamilton	WRCB	3	Chattanooga.
	WTVC	9	Do.
	WDEF	12	Do.
Hancock	WATE	6	Knoxville.
	WBIR	10	Do.
	WCYB	5	Bristol-Kingsport-Johnson City.
Hardeman	WREC	3	Memphis.
	WMC	5	Do.
	WHBQ	13	Do.
Hardin	WBBJ	7	Jackson, Tenn.
	WREC	3	Memphis.
	WMC	5	Do.
Hawkins	WCYB	5	Bristol-Kingsport-Johnson City.
	WJHL	11	Do.
	WLOS	13	Greenville-Spartanburg-Asheville.
	WATE	6	Knoxville.
	WBIR	10	Do.
Haywood	WREC	3	Memphis.
	WMC	5	Do.
	WHBQ	13	Do.
	WBBJ	7	Jackson, Tenn.
Henderson	WBBJ	7	Do.
	WREC	3	Memphis.
	WMC	5	Do.
	WSM	4	Nashville.
	WLAC	5	Do.
Henry	WSM	4	Do.
	WLAC	5	Do.
	WSIX	8	Do.
	WPSD	6	Paducah-Cape Girardeau-Harrisburg.

TENNESSEE—continued

County	Call letters	Channel	Market name
Hickman	WSM	4	Nashville.
	WLAC	5	Do.
	WSIX	8	Do.
Houston	WSM	4	Do.
	WLAC	5	Do.
	WSIX	8	Do.
Humphreys	WSM	4	Do.
	WLAC	5	Do.
	WSIX	8	Do.
Jackson	WSM	4	Do.
	WLAC	5	Do.
	WSIX	8	Do.
Jefferson	WATE	6	Knoxville.
	WBIR	10	Do.
	WTVK	26	Do.
	WLOS	13	Greenville-Spartanburg-Asheville.
Johnson	WCYB	5	Bristol-Kingsport-Johnson City.
	WJHL	11	Do.
Knox	WATE	6	Knoxville.
	WBIR	10	Do.
	WTVK	26	Do.
Lake	WPSD	6	Paducah-Cape Girardeau-Harrisburg.
	KFVS	12	Do.
	WREC	3	Memphis.
	WMC	5	Do.
	WHBQ	13	Do.
Lauderdale	WREC	3	Do.
	WMC	5	Do.
	WHBQ	13	Do.
Lawrence	WSM	4	Nashville.
	WLAC	5	Do.
	WSIX	8	Do.
	WHNT	19	Huntsville-Decatur-Florence.
	WAAY	31	Do.
Lewis	WSM	4	Nashville.
	WLAC	5	Do.
	WSIX	8	Do.
Lincoln	WHNT	19	Huntsville-Decatur-Florence.
	WAAY	31	Do.
	WMSL	48	Do.
	WSM	4	Nashville.
	WLAC	5	Do.
	WSIX	8	Do.
Loudon	WATE	6	Knoxville.
	WBIR	10	Do.
	WTVK	26	Do.
McMinn	WRCB	3	Chattanooga.
	WTVC	9	Do.
	WDEF	12	Do.
	WATE	6	Knoxville.
McNairy	WREC	3	Memphis.
	WMC	5	Do.
	WHVQ	13	Do.
	WBBJ	7	Jackson, Tenn.
Macon	WSM	4	Nashville.
	WLAC	5	Do.
	WSIX	8	Do.
Madison	WBBJ	7	Jackson, Tenn.
	WREC	3	Memphis.
	WMC	5	Do.
	WHBQ	13	Do.
Marion	WRCB	3	Chattanooga.
	WTVC	9	Do.
	WDEF	12	Do.
Marshall	WSM	4	Nashville.
	WLAC	5	Do.
	WSIX	8	Do.
Maury	WSM	4	Do.
	WLAC	5	Do.
	WSIX	8	Do.
Meigs	WRCB	3	Chattanooga.
	WTVC	9	Do.
	WDEF	12	Do.
Monroe	WATE	6	Knoxville.
	WBIR	10	Do.
	WRCB	3	Chattanooga.
	WTVC	9	Do.
	WDEF	12	Do.
Montgomery	WSM	4	Nashville.
	WLAC	5	Do.
	WSIX	8	Do.
Moore	WHNT	19	Huntsville-Decatur-Florence.
	WAAY	31	Do.
	WMSL	48	Do.
	WSM	4	Nashville.
	WLAC	5	Do.
	WSIX	8	Do.
Morgan	WATE	6	Knoxville.
	WBIR	10	Do.
	WTVC	9	Chattanooga.
Obion	WPSD	6	Paducah-Cape Girardeau-Harrisburg.
	KFVS	12	Do.
	WBBJ	7	Jackson, Tenn.

SIGNIFICANTLY VIEWED SIGNALS—Continued

County	Call letters, channel number and market name		

TENNESSEE—continued

County	Call letters	Ch.	Market
Overton	WSM	4	Nashville.
	WLAC	5	Do.
	WSIX	8	Do.
Perry	WSM	4	Do.
	WLAC	5	Do.
	WSIX	8	Do.
Pickett	WSM	4	Do.
	WLAC	5	Do.
	WSIX	8	Do.
Polk	WRCB	3	Chattanooga.
	WTVC	9	Do.
	WDEF	12	Do.
	WATL	36	Atlanta.
Putnam	WSM	4	Nashville.
	WLAC	5	Do.
	WSIX	8	Do.
Rhea	WRCB	3	Chattanooga.
	WTVC	9	Do.
	WDEF	12	Do.
Roane	WATE	6	Knoxville.
	WBIR	10	Do.
	WRCB	3	Chattanooga.
	WTVC	9	Do.
	WDEF	12	Do.
Robertson	WSM	4	Nashville.
	WLAC	5	Do.
	WSIX	8	Do.
Rutherford	WSM	4	Do.
	WLAC	5	Do.
	WSIX	8	Do.
Scott*	WATE	6	Knoxville.
	WBIR	10	Do.
Sequatchie	WRCB	3	Chattanooga.
	WTVC	9	Do.
	WDEF	12	Do.
Sevier	WATE	6	Knoxville.
	WBIR	10	Do.
	WTVK	26	Do.
Shelby	WREC	3	Memphis.
	WMC	5	Do.
	WHBQ	13	Do.
Smith	WSM	4	Nashville.
	WLAC	5	Do.
	WSIX	8	Do.
Stewart	WSM	4	Do.
	WLAC	5	Do.
	WSIX	8	Do.
Sullivan	WCYB	5	Bristol-Kingsport-Johnson City.
	WJHL	11	Do.
	WKPT	19	Do.
Sumner	WSM	4	Nashville.
	WLAC	5	Do.
	WSIX	8	Do.
Tipton	WREC	3	Memphis.
	WMC	5	Do.
	WHBQ	13	Do.
Trousdale	WSM	4	Nashville.
	WLAC	5	Do.
	WSIX	8	Do.
Unicoi	WCYB	5	Bristol-Kingsport-Johnson City.
	WJHL	11	Do.
Union	WATE	6	Knoxville.
	WBIR	10	Do.
	WTVK	26	Do.
Van Buren	WSM	4	Nashville.
	WLAC	5	Do.
	WSIX	8	Do.
	WRCB	3	Chattanooga.
	WDEF	12	Do.
Warren	WSM	4	Nashville.
	WLAC	5	Do.
	WSIX	8	Do.
Washington	WCYB	5	Bristol-Kingsport-Johnson City.
	WJHL	11	Do.
	WKPT	19	Do.
Wayne	WSM	4	Nashville.
	WLAC	5	Do.
	WSIX	8	Do.
Weakley	WPSD	6	Paducah-Cape Girardeau-Harrisburg.
	KFVS	12	Do.
	WBBJ	7	Jackson, Tenn.
White	WSM	4	Nashville.
	WLAC	5	Do.
	WSIX	8	Do.
Williamson	WSM	4	Do.
	WLAC	5	Do.
	WSIX	8	Do.
Wilson	WSM	4	Do.
	WLAC	5	Do.
	WSIX	8	Do.

SIGNIFICANTLY VIEWED SIGNALS—Continued

TEXAS

County	Call letters	Ch.	Market
Anderson	KDFW	4	Dallas-Fort Worth.
	WBAP	5	Do.
	WFAA	8	Do.
	KTVT	11	Do.
	KLTV	7	Tyler.
Andrews	KMID	2	Odessa-Midland.
	KOSA	7	Do.
	KMOM+	9	Do.
Angelina	KTRE	9	Tyler.
Aransas	KIII+	3	Corpus Christi.
	KRIS	6	Do.
	KZTV	10	Do.
Archer	KFDX	3	Wichita Falls-Lawton.
	KAUZ	6	Do.
	KSWO	7	Do.
Armstrong	KGNC	4	Amarillo.
	KVII	7	Do.
	KFDA+	10	Do.
Atascosa	WOAI	4	San Antonio.
	KENS	5	Do.
	KSAT	12	Do.
Austin	KPRC	2	Houston.
	KHOU	11	Do.
	KTRK	13	Do.
	KHTV	39	Do.
Bailey	KCBD+	11	Lubbock.
	KLBK	13	Do.
	KFDA+	10	Amarillo.
Bandera	WOAI	4	San Antonio.
	KENS	5	Do.
	KSAT	12	Do.
Bastrop	KTBC	7	Austin, Tex.
	KHFI	42	Do.
	WOAI	4	San Antonio.
	KENS	5	Do.
	KSAT	12	Do.
Baylor	KFDX	3	Wichita Falls-Lawton.
	KAUZ	6	Do.
	KSWO	7	Do.
Bee	KIII+	3	Corpus Christi.
	KRIS	6	Do.
	KZTV	10	Do.
	WOAI	4	San Antonio.
	KENS	5	Do.
	KSAT	12	Do.
Bell	KCEN	6	Waco-Temple.
	KWTX	10	Do.
	KTBC	7	Austin, Tex.
Bexar	WOAI	4	San Antonio.
	KENS	5	Do.
	KSAT	12	Do.
	KWEX	41	Do.
Blanco	WOAI	4	Do.
	KENS	5	Do.
	KSAT	12	Do.
	KTBC	7	Austin, Tex.
	KHFI	42	Do.
Borden	KCBD+	11	Lubbock.
	KLBK	13	Do.
	KSEL	28	Do.
Bosque	KDFW	4	Dallas-Fort Worth.
	WBAP	5	Do.
	WFAA	8	Do.
	KTVT	11	Do.
	KCEN	6	Waco-Temple.
	KWTX	10	Do.
Bowie	KTBS	3	Shreveport-Texarkana.
	KTAL	6	Do.
	KSLA	12	Do.
Brazoria	KPRC	2	Houston.
	KHOU	11	Do.
	KTRK	13	Do.
	KHTV	39	Do.
Brazos	KBTX	3	Waco-Temple.
	KCEN	6	Do.
	KTVT	11	Dallas-Fort Worth.
Brewster	NA.		
Briscoe	KGNC	4	Amarillo.
	KVII	7	Do.
	KFDA+	10	Do.
Brooks	KIII+	3	Corpus Christi.
	KRIS	6	Do.
	KZTV	10	Do.
Brown	KRBC+	9	Abilene-Sweetwater.
	KTXS	12	Do.
	KTVT	11	Dallas-Fort Worth.
Burleson	KBTX	3	Waco-Temple.
	KCEN	6	Do.
	KTBC	7	Austin, Tex.
Burnet	KTBC	7	Do.
	KHFI	42	Do.
	KWTX	10	Waco-Temple.
Caldwell	WOAI	4	San Antonio.
	KENS	5	Do.
	KSAT	12	Do.
	KTBC	7	Austin, Tex.
	KHFI	42	Do.

SIGNIFICANTLY VIEWED SIGNALS—Continued

TEXAS—continued

County	Call letters	Ch.	Market
Calhoun	NA.		
Callahan	KRBC+	9	Abilene-Sweetwater.
	KTXS	12	Do.
Cameron	KGBT	4	McAllen-Brownsville.
	KRGV	5	Do.
Camp	KTBS	3	Shreveport-Texarkana.
	KTAL	6	Do.
	KSLA	12	Do.
	KLTV	7	Tyler.
Carson	KGNC	4	Amarillo.
	KVII	7	Do.
	KFDA+	10	Do.
Cass	KTBS	3	Shreveport-Texarkana.
	KTAL	6	Do.
	KSLA	12	Do.
Castro	KGNC	4	Amarillo.
	KVII	7	Do.
	KFDA+	10	Do.
	KLBK	13	Lubbock.
Chambers	KPRC	2	Houston.
	KHOU	11	Do.
	KTRK	13	Do.
	KHTV	39	Do.
	KJAC	4	Beaumont-Port Arthur.
	KFDM	6	Do.
	KBMT	12	Do.
Cherokee	KLTV	7	Tyler.
	KTRE	9	Do.
	KTBS	3	Shreveport-Texarkana.
Childress	NA.		
Clay	KFDX	3	Wichita Falls-Lawton.
	KAUZ	6	Do.
	KSWO	7	Do.
Cochran	KCBD+	11	Lubbock.
	KLBK	13	Do.
	KSEL	28	Do.
Coke	KRBC+	9	Abilene-Sweetwater.
	KTXS	12	Do.
	KCTV	8	San Angelo.
Coleman	KRBC+	9	Abilene-Sweetwater.
	KTXS	12	Do.
Collin	KDFW	4	Dallas-Fort Worth.
	WBAP	5	Do.
	WFAA	8	Do.
	KTVT	11	Do.
	KDTV	39	Do.
Collingsworth	KVII	7	Amarillo.
	KFDA+	10	Do.
	KFDX	3	Wichita Falls-Lawton.
	KAUZ	6	Do.
	KSWO	7	Do.
Colorado	KPRC	2	Houston.
	KHOU	11	Do.
	KTRK	13	Do.
	WOAI	4	San Antonio.
	KENS	5	Do.
	KSAT	12	Do.
Comal	WOAI	4	San Antonio.
	KENS	5	Do.
	KSAT	12	Do.
Comanche	KDFW	4	Dallas-Fort Worth.
	WBAP	5	Do.
	WFAA	8	Do.
	KTVT	11	Do.
	KRBC+	9	Abilene-Sweetwater.
Concho			Over 90 percent cable penetration.
Cooke	KDFW	4	Dallas-Fort Worth.
	WBAP	5	Do.
	WFAA	8	Do.
	KTVT	11	Do.
Coryell	KCEN	6	Waco-Temple.
	KWTX	10	Do.
	KTVC	7	Austin, Tex.
	KTVT	11	Dallas-Fort Worth.
Cottle			Over 90 percent cable penetration.
Crane	KMID	2	Odessa-Midland.
	KOSA	7	Do.
	KMOM+	9	Do.
Crockett	NA.		
Crosby	KCBD+	11	Lubbock.
	KLBK	13	Do.
	KSEL	28	Do.
Culberson	NA.		
Dallam	KGNC	4	Amarillo.
	KVII	7	Do.
	KFDA+	10	Do.
Dallas	KDFW	4	Dallas-Fort Worth.
	WBAP	5	Do.
	WFAA	8	Do.
	KTVT	11	Do.
	KDTV	39	Do.
Dawson	KCBD+	11	Lubbock.
	KLBK	13	Do.
	KSEL	28	Do.
	KMXN	34	Do.
	KMID	2	Odessa-Midland.

County	Call letters, channel number and market name		
TEXAS—continued			
Deaf Smith	KGNC	4	Amarillo.
	KVII	7	Do.
	KFDA+	10	Do.
Delta	KDFW	4	Dallas-Fort Worth.
	WBAP	5	Do.
	WFAA	8	Do.
	KTVT	11	Do.
Denton	KDFW	4	Dallas-Fort Worth.
	WBAP	5	Do.
	WFAA	8	Do.
	KTVT	11	Do.
	KDTV	39	Do.
DeWitt	WOAI	4	San Antonio.
	KENS	5	Do.
	KSAT	12	Do.
Dickens	KCBD+	11	Lubbock.
	KLBK	13	Do.
	KSEL	28	Do.
	KTXS	12	Abilene-Sweetwater.
Dimmit	NA.		
Donley	KGNC	4	Amarillo.
	KVII	7	Do.
	KFDA+	10	Do.
Duval	KIII+	3	Corpus Christi.
	KRIS	6	Do.
	KZTV	10	Do.
Eastland	KRBC+	9	Abilene-Sweetwater.
	KTXS	12	Do.
Ector*	KMID	2	Odessa-Midland.
	KOSA	7	Do.
	KMOM+	9	Do.
Edwards	WOAI	4	San Antonio.
	KENS	5	Do.
	KSAT	12	Do.
Ellis	KDFW	4	Dallas-Fort Worth.
	WBAP	5	Do.
	WFAA	8	Do.
	KTVT	11	Do.
	KDTV	39	Do.
El Paso	KROD	4	El Paso.
	KTSM	9	Do.
	KELP+	13	Do.
Erath	KDFW	4	Dallas-Fort Worth.
	WBAP	5	Do.
	WFAA	8	Do.
	KTVT	11	Do.
Falls	KCEN	6	Waco-Temple.
	KWTX	10	Do.
Fannin	KDFW	4	Dallas-Fort Worth.
	WBAP	5	Do.
	WFAA	8	Do.
	KTVT	11	Do.
	KXII	12	Ardmore-Ada.
Fayette*	KTBC	7	Austin, Texas.
	KPRC	2	Houston.
	KHOU	11	Do.
	KTRK	13	Do.
	KENS	5	San Antonio.
	KSAT	12	Do.
Fisher*	KRBC+	9	Abilene-Sweetwater.
	KTXS	12	Do.
Floyd	KCBD+	11	Lubbock.
	KLBK	13	Do.
	KSEL	28	Do.
Foard	KFDX	3	Wichita Falls-Lawton.
	KAUZ	6	Do.
	KSWO	7	Do.
Fort Bend	KPRC	2	Houston.
	KHOU	11	Do.
	KTRK	13	Do.
	KHTV	39	Do.
Franklin	NA.		
Freestone	KDFW	4	Dallas-Fort Worth.
	WBAP	5	Do.
	WFAA	8	Do.
	KTVT	11	Do.
Frio	WOAI	4	San Antonio.
	KENS	5	Do.
	KSAT	12	Do.
Gaines	KCBD+	11	Lubbock.
	KLBK	13	Do.
	WFAA	8	Dallas-Fort Worth.
	KTVT	11	Do.
	KDTV	39	Do.
Galveston	KPRC	2	Houston.
	KHOU	11	Do.
	KTRK	13	Do.
	KHTV	39	Do.
Garza*	KCBD+	11	Lubbock.
	KLBK	13	Do.
	KSEL	28	Do.
Gillespie	WOAI	4	San Antonio.
	KENS	5	Do.
	KSAT	12	Do.
	KTBC	7	Austin, Tex.
Glasscock	KMID	2	Odessa-Midland.
	KOSA	7	Do.
	KMOM+	9	Do.

County	Call letters, channel number and market name		
TEXAS—continued			
Goliad	WOAI	4	San Antonio.
	KENS	5	Do.
	KSAT	12	Do.
Gonzales	WOAI	4	San Antonio.
	KENS	5	Do.
	KSAT	12	Do.
Gray	KGNC	4	Amarillo.
	KVII	7	Do.
	KFDA+	10	Do.
Grayson	KDFW	4	Dallas-Fort Worth.
	WBAP	5	Do.
	WFAA	8	Do.
	KTVT	11	Do.
	KXII	12	Ardmore-Ada.
Gregg	KTBS	3	Shreveport-Texarkana.
	KTAL	6	Do.
	KSLA	12	Do.
	KLTV	7	Tyler.
Grimes*	KPRC	2	Houston.
	KHOU	11	Do.
	KTRK	13	Do.
	KBTX	3	Waco-Temple.
Guadalupe	WOAI	4	San Antonio.
	KENS	5	Do.
	KSAT	12	Do.
Hale*	KCBD	11	Lubbock.
	KLBK	13	Do.
	KSEL	28	Do.
Hall	Over 90 percent cable penetration.		
Hamilton	KDFW	4	Dallas-Fort Worth.
	WBAP	5	Do.
	WFAA	8	Do.
	KTVT	11	Do.
	KCEN	6	Waco-Temple.
	KWTX	10	Do.
Hansford	KGNC	4	Amarillo.
	KVII	7	Do.
	KFDA+	10	Do.
Hardeman	KFDX	3	Wichita Falls-Lawton.
	KAUZ	6	Do.
	KSWO	7	Do.
Hardin	KJAC	4	Beaumont-Port Arthur.
	KFDM	6	Do.
	KBMT	12	Do.
Harris	KPRC	2	Houston.
	KHOU	11	Do.
	KTRK	13	Do.
	KHTV	39	Do.
Harrison	KTBS	3	Shreveport-Texarkana.
	KTAL	6	Do.
	KSLA	12	Do.
Hartley	KGNC	4	Amarillo.
	KVII	7	Do.
	KFDA+	10	Do.
Haskell	KRBC+	9	Abilene-Sweetwater.
	KTXS	12	Do.
	KFDX	3	Wichita Falls-Lawton.
Hays	WOAI	4	San Antonio.
	KENS	5	Do.
	KSAT	12	Do.
	KTBC	7	Austin, Tex.
Hemphill	Over 90 percent cable penetration.		
Henderson	KDFW	4	Dallas-Fort Worth.
	WBAP	5	Do.
	WFAA	8	Do.
	KTVT	11	Do.
	KLTV	7	Tyler.
Hidalgo	KGVT	4	McAllen-Brownsville (Lwr R Grnd).
	KRGV	5	Do.
Hill	KDFW	4	Dallas-Fort Worth.
	WBAP	5	Do.
	WFAA	8	Do.
	KTVT	11	Do.
	KWTX	10	Waco-Temple.
Hockley*	KCBD+	11	Lubbock.
	KLBK	13	Do.
	KSEL	28	Do.
Hood	KDFW	4	Dallas-Fort Worth.
	WBAP	5	Do.
	WFAA	8	Do.
	KTVT	11	Do.
Hopkins	KDFW	4	Dallas-Fort Worth.
	WBAP	5	Do.
	WFAA	8	Do.
	KLVT	11	Do.
	KLTV	7	Tyler.
Houston	KTRE	9	Lufkin-Nacogdoches.
	KBTX	3	Waco-Temple.
Howard	KMID	2	Odessa-Midland.
	KWAB	4	Do.
	KOSA	7	Do.
	KMOM+	9	Do.
Hudspeth	KROD	4	El Paso.
	KTSM	9	Do.
	KELP+	13	Do.

County	Call letters, channel number and market name		
TEXAS—continued			
Hunt	KDFW	4	Dallas-Fort Worth.
	WBAP	5	Do.
	WFAA	8	Do.
	KTVT	11	Do.
Hutchinson	KGNC	4	Amarillo.
	KVII	7	Do.
	KFDA+	10	Do.
Irion	KCTV	8	San Angelo.
	KRBC+	9	Abilene-Sweetwater.
Jack	KFDX	3	Wichita Falls-Lawton.
	KAUZ	6	Do.
	KSWO	7	Do.
	KDFW	4	Dallas-Fort Worth.
	WBAP	5	Do.
	WFAA	8	Do.
	KTVT	11	Do.
Jackson	KPRC	2	Houston.
	KHOU	11	Do.
	KTRK	13	Do.
	KHT	39	Do.
Jasper	KJAC	4	Beaumont-Port Arthur.
	KFDM	6	Do.
	KBMT	12	Do.
Jeff Davis	KMID	2	Odessa-Midland.
	KOSA	7	Do.
	KMOM+	9	Do.
Jefferson North	KJAC	4	Beaumont-Port Arthur.
	KFDM	6	Do.
	KBMT	12	Do.
Jefferson South	KJAC	4	Do.
	KFDM	6	Do.
	KBMT	12	Do.
Jim Hogg	KIII+	3	Corpus Christi.
	KRIS	6	Do.
	KZTV	10	Do.
Jim Wells	KIII+	3	Corpus Christi.
	KRIS	6	Do.
	KZTV	10	Do.
Johnson	KDFW	4	Dallas-Forth Worth.
	WBAP	5	Do.
	WFAA	8	Do.
	KTVT	11	Do.
	KDTV	39	Do.
Jones	KRBC+	9	Abilene-Sweetwater.
	KTXS	12	Do.
Karnes	WOAI	4	San Antonio.
	KENS	5	Do.
	KSAT	12	Do.
Kaufman	KDFW	4	Dallas-Forth Worth.
	WBAP	5	Do.
	WFAA	8	Do.
	KTVT	11	Do.
	KDTV	39	Do.
Kendall	WOAI	4	San Antonio.
	KENS	5	Do.
	KSAT	12	Do.
Kenedy	KIII+	3	Do.
	KRIS	6	Do.
	KZTV	10	Do.
Kent	KCBD+	11	Lubbock.
	KLBK	13	Do.
	KSEL	28	Do.
	KTXS	12	Abilene-Sweetwater.
Kerr	WOAI	4	San Antonio.
	KENS	5	Do.
	KSAT	12	Do.
Kimble	WOAI	4	Do.
	KENS	5	Do.
	KSAT	12	Do.
King	KFDX	3	Wichita Falls-Lawton.
	KAUZ	6	Do.
	KSWO	7	Do.
	KTXS	12	Abilene-Sweetwater.
Kinney	Over 90 percent cable penetration.		
Kleberg	KIII	3	Corpus Christi.
	KRIS	6	Do.
	KZTV	10	Do.
Knox	KFDX	3	Wichita Falls-Lawton.
	KAUZ	6	Do.
	KSWO	7	Do.
	KTXS	12	Abilene-Sweetwater.
Lamar	KDFW	4	Dallas-Fort Worth.
	WBAP	5	Do.
	WFAA	8	Do.
	KTVT	11	Do.
	KXII	12	Ardmore-Ada.
Lamb	KCBD+	11	Lubbock.
	KLBK	13	Do.
	KSEL	28	Do.
Lampasas	KCEN	6	Waco-Temple.
	KWTX	10	Do.
	KTBC	7	Austin, Tex.
La Salle	NA.		
Lavaca	WOAI	4	San Antonio.
	KENS	5	Do.
	KSAT	12	Do.
	KPRC	2	Houston.

RULES AND REGULATIONS

County	Call letters, channel number and market name		
TEXAS—continued			
Lee	KTBC	7	Austin, Tex.
	KHFI	42	Do.
	KBTX	3	Waco-Temple.
	KCEN	6	Do.
Leon	KBTX	3	Do.
	KCEN	6	Do.
	KWTX	10	Do.
Liberty	KPRC	2	Houston.
	KHOU	11	Do.
	KTRK	13	Do.
	KHTV	39	Do.
Limestone	KDFW	4	Dallas-Fort Worth.
	WBAP	5	Do.
	WFAA	8	Do.
	KTVT	11	Do.
	KCEN	6	Waco-Temple.
	KWTX	10	Do.
Lipscomb	KGNC	4	Amarillo.
	KVII	7	Do.
	KFDA+	10	Do.
Live Oak	WOAI	4	San Antonio.
	KENS	5	Do.
	KSAT	12	Do.
Llano	KTBC	7	Austin, Tex.
	KHFI	42	Do.
Loving	KOSA	7	Odessa-Midland.
	KMOM+	9	Do.
Lubbock	KCBD+	11	Lubbock.
	KLBK	13	Do.
	KSEL	28	Do.
Lynn	KCBD+	11	Do.
	KLBK	13	Do.
	KSEL	28	Do.
	KMXN	34	Do.
McCulloch	Over 90 percent cable penetration.		
McLennan	KCEN	6	Waco-Temple.
	KWTX	10	Do.
	KDFW	4	Dallas-Fort Worth.
	WFAA	8	Do.
	KTVT	11	Do.
McMullen	WOAI	4	San Antonio.
	KENS	5	Do.
	KSAT	12	Do.
Madison	KBTX	3	Waco-Temple.
	KPRC	2	Houston.
Marion	KTBS	3	Shreveport-Texarkana
	KTAL	6	Do.
	KSLA	12	Do.
Martin	KMID	2	Odessa-Midland.
	KOSA	7	Do.
	KMOM+	9	Do.
Mason	NA.		
Matagorda	KPRC	2	Houston.
	KHOU	11	Do.
	KTRK	13	Do.
	KHTV	39	Do.
Maverick	Over 90 percent cable penetration.		
Medina	WOAI	4	San Antonio.
	KENS	5	Do.
	KSAT	12	Do.
Menard	KRBC+	9	Abilene-Sweetwater.
	KCTV	8	San Angelo.
Midland	KMID	2	Odessa-Midland.
	KOSA	7	Do.
	KMOM+	9	Do.
Milam	KCEN	6	Waco-Temple.
	KWTX	10	Do.
	KTBC	7	Austin, Tex.
Mills	KCEN	6	Waco-Temple.
	KWTX	10	Do.
	KRBC+	9	Abilene-Sweetwater.
	KTXS	12	Do.
Mitchell	KMID	2	Odessa-Midland.
	KMOM	9	Do.
	KTXS	12	Abilene-Sweetwater.
Montague	KFDX	3	Wichita Falls-Lawton.
	KAUZ	6	Do.
	KSWO	7	Do.
	KDFW	4	Dallas-Fort Worth.
	WBAP	5	Do.
	KTVT	11	Do.
	KDTV	39	Do.
Montgomery	KPRC	2	Houston.
	KHOU	11	Do.
	KTRK	13	Do.
	KHTV	39	Do.
Moore	KGNC	4	Amarillo.
	KVII	7	Do.
	KFDA+	10	Do.
Morris	KTBS	3	Shreveport-Texarkana
	KTAL	6	Do.
	KSLA	12	Do.
Motley	Over 90 percent cable penetration.		
Nacogdoches	KTBS	3	Shreveport-Texarkana.
	KSLA	12	Do.
	KTVT	11	Dallas-Fort Worth.
	KTRE	9	Tyler.
Navarro	KDFW	4	Dallas-Fort Worth.
	WBAP	5	Do.
	WFAA	8	Do.
	KTVT	11	Do.

County	Call letters, channel number and market name		
TEXAS—continued			
Newton	KJAC	4	Beaumont-Port Arthur.
	KFDM	6	Do.
	KBMT	12	Do.
Nolan	KRBC+	9	Abilene-Sweetwater.
	KTXS	12	Do.
Nueces	KIII+	3	Corpus Christi.
	KRIS	6	Do.
	KZTV	10	Do.
Ochiltree	KGNC	4	Amarillo.
	KVII	7	Do.
	KFDA+	10	Do.
Oldham	KGNC	4	Do.
	KVII	7	Do.
	KFDA+	10	Do.
Orange	KJAC	4	Beaumont-Port Arthur.
	KFDM	6	Do.
	KBMT	12	Do.
Palo Pinto	KDFW	4	Dallas-Fort Worth.
	WBAP	5	Do.
	WFAA	8	Do.
	KTVT	11	Do.
Panola	KTBS	3	Shreveport-Texarkana.
	KTAL	6	Do.
	KSLA	12	Do.
Parker	KDFW	4	Dallas-Fort Worth.
	WBAP	5	Do.
	WFAA	8	Do.
	KTVT	11	Do.
Parmer	KGNC	4	Amarillo.
	KVII	7	Do.
	KFDA+	10	Do.
	KCBD+	11	Lubbock.
Pecos	KMID	2	Odessa-Midland.
	KOSA	7	Do.
	KMOM+	9	Do.
Polk	KTRE	9	Tyler.
	KJAC	4	Beaumont-Port Arthur.
	KFDM	6	Do.
	KPRC	2	Houston.
Potter	KGNC	4	Amarillo.
	KVII	7	Do.
	KFDA+	10	Do.
Presidio	KOSA	7	Odessa-Midland.
Rains	KDFW	4	Dallas-Fort Worth.
	WBAP	5	Do.
	WFAA	8	Do.
	KTVT	11	Do.
Randall	KGNC	4	Amarillo.
	KVII	7	Do.
	KFDA+	10	Do.
Reagan	KMID	2	Odessa-Midland.
	KOSA	7	Do.
	KMOM	9	Do.
Real	WOAI	4	San Antonio.
	KENS	5	Do.
	KSAT	12	Do.
Red River	KTBS	3	Shreveport-Texarkana
	KTAL	6	Do.
	KSLA	12	Do.
Reeves	KOSA	7	Odessa-Midland.
	KMOM+	9	Do.
Refugio	KIII+	3	Corpus Christi.
	KRIS	6	Do.
	KZTV	10	Do.
Roberts	KGNC	4	Amarillo.
	KVII	7	Do.
	KFDA+	10	Do.
Robertson*	KCEN	6	Waco-Temple.
	KWTX	10	Do.
Rockwall	KDFW	4	Dallas-Fort Worth.
	WBAP	5	Do.
	WFAA	8	Do.
	KTVT	11	Do.
	KDTV	39	Do.
Runnels	KRBC+	9	Abilene-Sweetwater.
	KTXS	12	Do.
	KCTV	8	San Angelo.
Rusk	KTBS	3	Shreveport-Texarkana
	KTAL	6	Do.
	KSLA	12	Do.
	KLTV	7	Tyler.
Sabine	KJAC	4	Beaumont-Port Arthur.
	KFDM	6	Do.
	KTBS	3	Shreveport-Texarkana.
	KSLA	12	Do.
	KTRE	9	Tyler.
San Augustine	KTBS	3	Shreveport-Texarkana.
	KSLA	12	Do.
	KTRE	9	Tyler.
San Jacinto	KPRC	2	Houston.
	KHOU	11	Do.
	KTRK	13	Do.
San Patricio	KIII+	3	Corpus Christi.
	KRIS	6	Do.
	KZTV	10	Do.

County	Call letters, channel number and market name		
TEXAS—continued			
San Saba	KDFW	4	Dallas-Fort Worth.
	KRBC+	9	Abilene-Sweetwater.
	KTBC	7	Austin, Tex.
	KCEN	6	Waco-Temple.
	KWTX	10	Do.
Schleicher	KCTV	8	San Angelo.
	KRBC+	9	Abilene-Sweetwater.
Scurry	KRBC+	9	Do.
	KTXS	12	Do.
Shackelford	KRBC+	9	Do.
	KTXS	12	Do.
Shelby	KTBS	3	Shreveport-Texarkana.
	KTAL	6	Do.
	KSLA	12	Do.
Sherman	KGNC	4	Amarillo.
	KVII	7	Do.
	KFDA+	10	Do.
Smith	KLTV	7	Tyler.
	KDFW	4	Dallas-Fort Worth.
	KTVT	11	Do.
	KTBS	3	Shreveport-Texarkana.
	KSLA	12	Do.
Somervell	KDFW	4	Dallas-Fort Worth.
	WBAP	5	Do.
	WFAA	8	Do.
	KTVT	11	Do.
Starr	KGBT	4	McAllen-Brownsville (Lwr R Grnd).
	KRGV	5	Do.
Stephens	KRBC+	9	Abilene-Sweetwater.
	KTXS	12	Do.
	WFAA	8	Dallas-Fort Worth.
	KTVT	11	Do.
Sterling	KRBC+	9	Abilene-Sweetwater.
	KCTV	8	San Angelo.
Stonewall	KRBC+	9	Abilene-Sweetwater.
	KTXS	12	Do.
Sutton	NA.		
Swisher	KGNC	4	Amarillo.
	KVII	7	Do.
	KFDA+	10	Do.
Tarrant	KDFW	4	Dallas-Fort Worth.
	WBAP	5	Do.
	WFAA	8	Do.
	KTVT	11	Do.
	KDTV	39	Do.
Taylor	KRBC	9	Abilene-Sweetwater.
	KTXS	12	Do.
Terrell	Over 90 percent cable penetration.		
Terry*	KCBD+	11	Lubbock.
	KLBK	13	Do.
	KSEL	28	Do.
Throckmorton	KFDX	3	Wichita Falls-Lawton.
	KAUZ	6	Do.
	KSWO	7	Do.
Titus	KTBS	3	Shreveport-Texarkana.
	KTAL	6	Do.
	KSLA	12	Do.
Tom Green	KCTV	8	San Angelo.
	KRBC+	9	Abilene-Sweetwater
Travis	KTBC	7	Austin, Tex.
	KHFI	42	Do.
Trinity	KRTE	9	Tyler.
	KPRC	2	Houston.
	KBTX	3	Waco-Temple.
Tyler	KJAC	4	Beaumont-Port Arthur.
	KFDM	6	Do.
	KBMT	12	Do.
Upshur	KTBS	3	Shreveport-Texarkana·
	KTAL	6	Do.
	KSLA	12	Do.
	KLTV	7	Tyler.
Upton	KMID	2	Odessa-Midland.
	KOSA	7	Do.
	KMOM+	9	Do.
Uvalde	WOAI	4	San Antonio.
	KENS	5	Do.
	KSAT	12	Do.
Val Verde	N.A.		
Van Zandt	KDFW	4	Dallas-Fort Worth.
	WBAP	5	Do.
	WFAA	8	Do.
	KTVT	11	Do.
	KLTV	7	Tyler.
Victoria	WOAI	4	San Antonio.
	KENS	5	Do.
	KSAT	12	Do.
	KIII	3	Corpus Christi.
Walker	KPRC	2	Houston.
	KHOU	11	Do.
	KTRK	13	Do.
	KHTV	39	Do.
	KBTX	3	Waco-Temple.
Waller	KPRC	2	Houston.
	KHOU	11	Do.
	KTRK	13	Do.
	KHTV	39	Do.

County	Call letters, channel number and market name		County	Call letters, channel number and market name		County	Call letters, channel number and market name	
TEXAS—continued			**UTAH—continued**			**VERMONT—continued**		
Ward	KMID	2 Odessa-Midland.	Iron	KUTV	2 Do.	Orange	WCAX	3 Burlington-Plattsburgh.
	KOSA	7 Do.		KCPX	4 Do.		WMTW	8 Portland-Poland Spring.
	KMOM+	9 Do.		KSL	5 Do.	Orleans	WCAX	3 Burlington-Plattsburgh.
Washington	KPRC	2 Houston.	Juab	KUTV	2 Do.		WPTZ	5 Do.
	KHOU	11 Do.		KCPX	4 Do.		WMTW	8 Portland-Poland Spring.
	KTRK	13 Do.		KSL	5 Do.		CBMT	6 Canada.
	KHTV	39 Do.	Kane	KUTV	2 Do.		CFCF	12 Do.
	KBTX	3 Waco-Temple.		KCPX	4 Do.	Rutland	WCAX	3 Burlington-Plattsburgh.
Webb	KGNS	8 Laredo.		KSL	5 Do.		WPTZ	5 Do.
	XEFE	2 Mexico.	Millard	KUTV	2 Do.		WRGB	6 Albany-Schenectady-Troy.
Wharton	KPRC	2 Houston.		KCPX	4 Do.		WTEN+	10 Do.
	KHOU	11 Do.		KSL	5 Do.		WAST	13 Do.
	KTRK	13 Do.	Morgan	KUTV	2 Do.	Washington	WCAX	3 Burlington-Plattsburgh.
	KHTV	39 Do.		KCPX	4 Do.		WPTZ	5 Do.
Wheeler	KFDA+	10 Amarillo.		KSL	5 Do.		WMTW	8 Portland-Poland Spring.
Wichita	KFDX	3 Wichita Falls-Lawton.	Piute	KUTV	2 Do.	Windham	WMTW	8 Do.
	KAUZ	6 Do.		KCPX	4 Do.		WHDH	5 Boston.
	KSWO	7 Do.		KSL	5 Do.	Windsor	WCAX	3 Burlington-Plattsburgh.
Wilbarger	KFDX	3 Do.	Rich	KUTV	2 Do.		WMTW	8 Portland-Poland Spring.
	KAUZ	6 Do.		KCPX	4 Do.			
	KSWO	7 Do.		KSL	5 Do.	**VIRGINIA**		
Willacy	KGBT	4 McAllen-Brownsville (Lwr R Grnd).	Salt Lake	KUTV	2 Do.	Accomack	WTAR	3 Norfolk-Portsmouth-Newport News-Hampton.
	KRGV	5 Do.		KCPX	4 Do.		WAVY	10 Do.
Williamson	KTBC	7 Austin, Tex.		KSL	5 Do.		WVEC	13 Do.
	KHFI	42 Do.	San Juan	KUTV	2 Do.		WBOC	16 Salisbury.
	KCEN	6 Waco-Temple.		KCPX	4 Do.		WTTG	5 Washington, D.C.
	KWTX	10 Do.		KSL	5 Do.	Albemarle and Charlottesville City.	WTVR	6 Richmond.
Wilson	WOAI	4 San Antonio.	Sanpete	KUTV	2 Do.		WXEX	8 Do.
	KENS	5 Do.		KCPX	4 Do.		WWBT	12 Do.
	KSAT	12 Do.		KSL	5 Do.		WSVA	3 Harrisonburg.
	KWEX	41 Do.	Sevier	KUTV	2 Do.	Alleghany and Covington City including Clifton Forge City.	WDBJ	7 Roanoke-Lynchburg.
Winkler	KMID	2 Odessa-Midland.		KCPX	4 Do.		WSLS	10 Do.
	KOSA	7 Do.		KSL	5 Do.			
	KNOM+	9 Do.	Summit	KUTV	2 Do.	Amelia	WTVR	6 Richmond.
Wise	KDFW	4 Dallas-Fort Worth.		KCPX	4 Do.		WXEX	8 Do.
	WBAP	5 Do.		KSL	5 Do.		WWBT	12 Do.
	WFAA	8 Do.	Tooele	KUTV	2 Do.	Amherst	WDBJ	7 Roanoke-Lynchburg.
	KTVT	11 Do.		KCPX	4 Do.		WSLS	10 Do.
	KDTV	39 Do.		KSL	5 Do.		WLVA	13 Do.
Wood	KTBS	3 Shreveport-Texarkana.	Uintah	KUTV	2 Do.	Appomattox	WDBJ	7 Do.
	KTAL	6 Do.		KCPX	4 Do.		WSLS	10 Do.
	KSLA	12 Do.		KSL	5 Do.		WLVA	13 Do.
	KDFW	4 Dallas-Fort Worth.	Utah	KUTV	2 Do.	Arlington and Alexandria City.	WRC	4 Washington, D.C.
	WBAP	5 Do.		KCPX	4 Do.		WTTG	5 Do.
	WFAA	8 Do.		KSL	5 Do.		WMAL	7 Do.
	KTVT	11 Do.	Wasatch	KUTV	2 Do.		WTOP	9 Do.
	KLTV	7 Tyler.		KCPX	4 Do.		WDCA	20 Do.
Yoakum	KCBD+	11 Lubbock.		KSL	5 Do.	Augusta and Stronton City and Waynesboro City.	WTVR	6 Richmond.
	KLBK	13 Do.	Washington	KUTV	2 Do.		WWBT	12 Do.
	KBIM	10 Roswell.		KCPX	4 Do.		WSVA	3 Harrisonburg.
Young	KFDX	3 Wichita Falls-Lawton.		KSL	5 Do.		WTTG	5 Washington, D.C.
	KAUZ	6 Do.		KORK	3 Las Vegas.	Bath	WDBJ	7 Roanoke-Lynchburg.
	KSWO	7 Do.	Wayne	KUTV	2 Salt Lake City.		WSLS	10 Do.
Zapata	KGNS	8 Laredo.		KCPX	4 Do.		WHIS	6 Bluefield-Beckley-Oak Hill.
	KGBT	4 McAllen-Brownsville (Lwr R Grnd).		KSL	5 Do.		WSVA	3 Harrisonburg.
	XEFB	3 Mexico.	Weber	KUTV	2 Do.	Bedford	WDBJ	7 Roanoke-Lynchburg.
Zavala	WOAI	4 San Antonio.		KCPX	4 Do.		WSLS	10 Do.
	KENS	5 Do.		KSL	5 Do.		WLVA	13 Do.
						Bland	WHIS	6 Bluefield-Beckley-Oak Hill.
			VERMONT				WDBJ	7 Roanoke-Lynchburg.
			Addison*	WCAX	3 Burlington-Plattsburgh.		WSLS	10 Do.
				WPTZ	5 Do.	Botetourt	WDBJ	7 Do.
			Bennington	WRGB	6 Albany-Schenectady-Troy.		WSLS	10 Do.
				WTEN+	10 Do.		WLVA	13 Do.
				WAST	13 Do.	Buchanan	WOAY	4 Bluefield-Beckley-Oak Hill.
			Caledonia	WCAX	3 Burlington-Plattsburgh.		WHIS	6 Do.
				WMTW	8 Portland-Poland Spring.		WCYB	5 Bristol-Kingsport-Johnson City.
			Chittenden	WCAX	3 Burlington-Plattsburgh.	Buckingham	WTVR	6 Richmond.
				WPTZ	5 Do.		WXEX	8 Do.
				WVNY	22 Do.		WWBT	12 Do.
				CFCF	12 Canada.	Brunswick	WTVR	6 Do.
			Essex	WCAX	3 Burlington-Plattsburgh.		WXEX	8 Do.
				WMTW	8 Portland-Poland Spring.		WWBT	12 Do.
			Franklin	WCAX	3 Burlington-Plattsburgh.	Campbell and Lynchburg City.	WDBJ	7 Roanoke-Lynchburg.
				WPTZ	5 Do.		WSLS	10 Do.
				WVNY	22 Do.		WLVA	13 Do.
				CBMT	6 Canada.	Caroline	WTVR	6 Richmond.
				CFCF	12 Do.		WXEX	8 Do.
			Grand Isle	WCAX	3 Burlington-Plattsburgh.		WWBT	12 Do.
				WPTZ	5 Do.		WTTG	5 Washington, D.C.
				WVNY	22 Do.			
				CBMT	6 Canada.			
				CFCF	12 Do.			
			Lamoille	WCAX	3 Burlington-Plattsburgh.			
				WPTZ	5 Do.			
				WVNY	22 Do.			
				WMTW	8 Portland-Poland Spring.			
				CBMT	6 Canada.			

Census county divisions in split counties

Jefferson North: Beaumont, Nome-China.
Jefferson South: All other.

UTAH

Beaver	KUTV	2 Salt Lake City.
	KCPX	4 Do.
	KSL	5 Do.
Box Elder	KUTV	2 Do.
	KCPX	4 Do.
	KSL	5 Do.
Cache	KUTV	2 Do.
	KCPX	4 Do.
	KSL	5 Do.
Carbon	KUTV	2 Do.
	KCPX	4 Do.
	KSL	5 Do.
Daggett	KUTV	2 Do.
	KCPX	4 Do.
	KSL	5 Do.
Davis	KUTV	2 Do.
	KCPX	4 Do.
	KSL	5 Do.
Duchesne	KUTV	2 Do.
	KCPX	4 Do.
	KSL	5 Do.
Emery	KUTV	2 Do.
	KCPX	4 Do.
	KSL	5 Do.
Garfield	KUTV	2 Do.
	KCPX	4 Do.
	KSL	5 Do.
Grand	KUTV	2 Do.
	KCPX	4 Do.
	KSL	5 Do.

SIGNIFICANTLY VIEWED SIGNALS—Continued

County	Call letters, channel number and market name		

VIRGINIA—continued (column 1)

County	Call letters	Channel	Market name
Carroll	WDBJ	7	Roanoke-Lynchburg.
	WSLS	10	Do.
	WHIS	6	Bluefield-Beckley-Oak Hill.
	WFMY	2	Greensboro-Winston-Salem-High Point.
	WGHP	8	Do.
	WOJO	10	Do.
Charlotte	WDBJ	7	Roanoke-Lynchburg.
	WSLS	10	Do.
	WLVA	13	Do.
	WTVR	6	Richmond.
Charles City	WTVR	6	Do.
	WXEX	8	Do.
	WWBT	12	Do.
Chesterfield and Colonial Heights City	WTVR	6	Do.
	WXEX	8	Do.
	WWBT	12	Do.
Clarke	WRC	4	Washington, D.C.
	WTTG	5	Do.
	WMAL	7	Do.
	WTOP	9	Do.
Craig	WDBJ	7	Roanoke-Lynchburg.
	WSLS	10	Do.
Culpeper	WRC	4	Washington, D.C.
	WTTG	5	Do.
	WMAL	7	Do.
	WTOP	9	Do.
Cumberland	WTVR	6	Richmond.
	WXEX	8	Do.
	WWBT	12	Do.
Dickenson	WCYB	5	Bristol-Kingsport-Johnson City.
Dinwiddie and Petersburg City.	WTVR	6	Richmond.
	WXEX	8	Do.
	WWBT	12	Do.
Essex	WTVR	6	Do.
	WXEX	8	Do.
	WWBT	12	Do.
	WTTG	5	Washington, D.C.
Fairfax and Fairfax City and Falls Church City.	WRC	4	Do.
	WTTG	5	Do.
	WMAL	7	Do.
	WTOP	9	Do.
	WDCA	20	Do.
Fauquier	WRC	4	Do.
	WTTG	5	Do.
	WMAL	7	Do.
	WTOP	9	Do.
Floyd	WDBJ	7	Roanoke-Lynchburg.
	WSLS	10	Do.
Fluvanna	WTVR	6	Richmond.
	WXEX	8	Do.
	WWBT	12	Do.
Franklin	WDBJ	7	Roanoke-Lynchburg.
	WSLS	10	Do.
	WLVA	13	Do.
Frederick and Winchester City.*	WRC	4	Washington, D.C.
	WTTG	5	Do.
	WMAL	7	Do.
	WTOP	9	Do.
Giles	WDBJ	7	Roanoke-Lynchburg.
	WSLS	10	Do.
	WHIS	6	Bluefield-Beckley-Oak Hill.
Gloucester	WTAR	3	Norfolk-Portsmouth-Newport News-Hampton.
	WAVY	10	Do.
	WVEC	13	Do.
	WTVR	6	Richmond.
	WXEX	8	Do.
Goochland	WTVR	6	Do.
	WXEX	8	Do.
	WWBT	12	Do.
Grayson	WDBJ	7	Roanoke-Lynchburg.
	WSLS	10	Do.
	WFMY	2	Greensboro-Winston-Salem-High Point.
	WGHP	8	Do.
	WSJS	12	Do.
Greene	WTVR	6	Richmond.
	WXEX	8	Do.
	WWBT	12	Do.
	WSVA	3	Harrisonburg.
Greensville	WTVR	6	Richmond.
	WXEX	8	Do.
	WWBT	12	Do.
	WTAR	3	Norfolk-Portsmouth-Newport News-Hampton.
	WAVY	10	Do.
Halifax	WDBJ	7	Roanoke-Lynchburg.
	WSLS	10	Do.
	WLVA	13	Do.
Hanover	WTVR	6	Richmond.
	WXEX	8	Do.
	WWBT	12	Do.
Henrico and Richmond City.	WTVR	6	Do.
	WXEX	8	Do.
	WWBT	12	Do.

VIRGINIA—continued (column 2)

County	Call letters	Channel	Market name
Henry and Martinsville City.	WDBJ	7	Roanoke-Lynchburg.
	WSLS	10	Do.
	WFMY	2	Greensboro-Winston-Salem-High Point.
	WGHP	8	Do.
	WSJS	12	Do.
Highland	WDBJ	7	Roanoke-Lynchburg.
	WSLS	10	Do.
	WSVA	3	Harrisonburg.
Isle of Wight	WTAR	3	Norfolk-Portsmouth-Newport News-Hampton.
	WAVY	10	Do.
	WVEC	13	Do.
James City and Williamsburg City.	WTAR	3	Do.
	WAVY	10	Do.
	WVEC	13	Do.
	WTVR	6	Richmond.
	WXEX	8	Do.
	WWBT	12	Do.
King and Queen	WTVR	6	Do.
	WXEX	8	Do.
	WWBT	12	Do.
	WAVY	10	Norfolk-Portsmouth-Newport News-Hampton.
King George	WRC	4	Washington, D.C.
	WTTG	5	Do.
	WMAL	7	Do.
	WTOP	9	Do.
	WDCA	20	Do.
King William	WTVR	6	Richmond.
	WXEX	8	Do.
	WWBT	12	Do.
Lancaster	WTVR	6	Do.
	WXEX	8	Do.
	WWBT	12	Do.
	WTAR	3	Norfolk-Portsmouth-Newport News-Hampton.
	WAVY	10	Do.
Lee	WCYB	5	Bristol-Kingsport-Johnson City.
	WJHL	11	Do.
	WATE	6	Knoxville.
	WBIR	10	Do.
Loudoun	WRC	4	Washington, D.C.
	WTTG	5	Do.
	WMAL	7	Do.
	WTOP	9	Do.
	WDCA	20	Do.
Louisa	WTVR	6	Richmond.
	WXEX	8	Do.
	WWBT	12	Do.
Lunenburg	WTVR	6	Do.
	WXEX	8	Do.
	WWBT	12	Do.
Madison	WTVR	6	Do.
	WXEX	8	Do.
	WWBT	12	Do.
	WSVA	3	Harrisonburg.
	WRC	4	Washington, D.C.
	WTTG	5	Do.
Mathews	WTAR	3	Norfolk-Portsmouth-Newport News-Hampton.
	WAVY	10	Do.
	WEVC	13	Do.
	WTVR	6	Richmond.
	WXEX	8	Do.
Mecklenburg	WDBJ	7	Roanoke-Lynchburg.
	WSLS	10	Do.
	WLVA	13	Do.
	WRAL	5	Raleigh-Durham.
	WTVD	11	Do.
	WTVR	6	Richmond.
	WXEX	8	Do.
Middlesex	WTVR	6	Do.
	WXEX	8	Do.
	WWBT	12	Do.
	WTAR	3	Norfolk-Portsmouth-Newport News-Hampton.
	WAVY	13	Do.
Montgomery and Radford City.	WDBJ	7	Roanoke-Lynchburg.
	WSLS	10	Do.
	WLVA	13	Do.
Nansemond and Suffolk City.	WTAR	3	Norfolk-Portsmouth-Newport News-Hampton.
	WAVY	10	Do.
	WVEC	13	Do.
	WYAH	27	Do.
Nelson	WDBJ	7	Roanoke-Lynchburg.
	WSLS	10	Do.
	WLVA	13	Do.
	WTVR	6	Richmond.
	WWBT	12	Do.
New Kent	WTVR	6	Do.
	WXEX	8	Do.
	WWBT	12	Do.

VIRGINIA—continued (column 3)

County	Call letters	Channel	Market name
Norfolk and Chesapeake City and Portsmouth City and Norfolk City.	WTAR	3	Norfolk-Portsmouth-Newport News-Hampton.
	WAVY	10	Do.
	WVEC	13	Do.
Northampton	WTAR	3	Norfolk-Portsmouth-Newport News-Hampton.
	WAVY	10	Do.
	WVEC	13	Do.
Northumberland	WTVR	6	Richmond.
	WXEX	8	Do.
	WWBT	12	Do.
	WTAR	3	Norfolk-Portsmouth-Newport News-Hampton.
	WAVY	10	Do.
	WTTG	5	Do.
Nottoway	WTVR	6	Richmond.
	WXEX	8	Do.
	WWBT	12	Do.
Orange	WTVR	6	Do.
	WXEX	8	Do.
	WWBT	12	Do.
	WRC	4	Washington, D.C.
	WTTG	5	Do.
	WMAL	7	Do.
Page	WTOP	9	Do.
	WSVA	3	Harrisonburg.
	WTVR	6	Richmond.
Patrick	WFMY	2	Greensboro-Winston Salem-High Point.
	WGHP	8	Do.
	WSJS	12	Do.
	WDBJ	7	Roanoke-Lynchburg.
	WSLS	10	Do.
Pittsylvania and Danville City.	WDBJ	7	Roanoke-Lynchburg.
	WSLS	10	Do.
	WLVA	13	Do.
	WFMY	2	Greensboro-Winston Salem-High Point.
	WGHP	8	Do.
	WSJS	12	Do.
Powhatan	WTVR	6	Richmond.
	WXEX	8	Do.
	WWBT	12	Do.
Prince Edward	WTVR	6	Do.
	WXEX	8	Do.
	WWBT	12	Do.
	WLVA	13	Roanoke-Lynchburg.
Prince George and Hopewell City.	WTVR	6	Richmond.
	WXEX	8	Do.
	WWBT	12	Do.
Prince William	WRC	4	Washington, D.C.
	WTTG	5	Do.
	WMAL	7	Do.
	WTOP	9	Do.
	WDCA	20	Do.
Pulaski	WDBJ	7	Roanoke-Lynchburg.
	WSLS	10	Do.
	WHIS	6	Bluefield-Beckley-Oak Hill.
Rappahannock	WRC	4	Washington, D.C.
	WTTG	5	Do.
	WMAL	7	Do.
	WTOP	9	Do.
	WSVA	3	Harrisonburg.
Richmond	WTVR	6	Richmond.
	WXEX	8	Do.
	WWBT	12	Do.
	WTTG	5	Washington, D.C.
Roanoke and Roanoke City and Salem City.	WDBJ	7	Roanoke-Lynchburg.
	WSLS	10	Do.
	WLVA	13	Do.
	WRFT	27	Do.
Rockbridge	WDBJ	7	Do.
	WSLS	10	Do.
	WLVA	13	Do.
Rockingham and Harrisonburg City.	WSVA	3	Harrisonburg.
	WTVR	6	Richmond.
	WWBT	12	Do.
	WTTG	5	Washington, D.C.
Russell	WCYB	5	Bristol-Kingsport-Johnson City.
	WJHL	11	Do.
	WHIS	6	Bluefield-Beckley-Oak Hill.
Scott	WCYB	5	Bristol-Kingsport-Johnson City.
	WJHL	11	Do.
Shenandoah	WRC	4	Washington, D.C.
	WTTG	5	Do.
	WMAL	7	Do.
	WTOP	9	Do.
	WSVA	3	Harrisonburg.
Smyth	WCYB	5	Bristol-Kingsport-Johnson City.
	WJHL	11	Do.

SIGNIFICANTLY VIEWED SIGNALS—Continued

County	Call letters, channel number and market name		
VIRGINIA—continued			
Southampton	WTAR	3	Norfolk-Portsmouth-Newport News-Hampton.
	WAVY	10	Do.
	WVEC	13	Do.
Spotsylvania and Fredericksburg City.	WRC	4	Washington, D.C.
	WTTG	5	Do.
	WMAL	7	Do.
	WTOP	9	Do.
	WTVR	6	Richmond.
Stafford	WRC	4	Washington, D.C.
	WTTG	5	Do.
	WMAL	7	Do.
	WTOP	9	Do.
	WDCA	20	Do.
	WTVR	6	Richmond.
Surry	WTVR	6	Do.
	WXEX	8	Do.
	WWBT	12	Do.
	WTAR	3	Norfolk-Portsmouth-Newport News-Hampton.
	WAVY	10	Do.
	WVEC	13	Do.
Sussex	WTVR	6	Richmond.
	WXEX	8	Do.
	WWBT	12	Do.
	WTAR	3	Norfolk-Portsmouth-Newport News-Hampton.
	WAVY	10	Do.
Tazewell	WOAY	4	Bluefield-Beckley-Oak Hill.
	WHIS	6	Do.
	WDBJ	7	Roanoke-Lynchburg.
Virginia Beach and Virginia. Beach City,	WTAR	3	Norfolk-Portsmouth-Newport News-Hampton.
	WAVY	10	Do.
	WVEC	13	Do.
Warren	WRC	4	Washington, D.C.
	WTTG	5	Do.
	WMAL	7	Do.
	WTOP	9	Do.
	WMAR	2	Baltimore.
	WBAL	11	Do.
	WJZ	13	Do.
	WSVA	3	Harrisonburg.
Washington and Bristol City.*	WCYB	5	Bristol-Kingsport-Johnson City.
	WJHL	11	Do.
	WKPT	19	Do.
Westmoreland	WRC	4	Washington, D.C.
	WTTG	5	Do.
	WMAL	7	Do.
	WTOP	9	Do.
	WTVR	6	Richmond.
	WXEX	8	Do.
Wise	WCYB	5	Bristol-Kingsport-Johnson City.
	WJHL	11	Do.
Wythe	WDBJ	7	Roanoke-Lynchburg.
	WSLS	10	Do.
	WHIS	6	Bluefield-Beckley-Oak Hill.
Hampton-Newport News and Hampton City and Newport News City.	WTAR	3	Norfolk-Portsmouth-Newport News-Hampton.
	WAVY	10	Do.
	WVEC	13	Do.
York	WTAR	3	Do.
	WAVY	10	Do.
	WVEC	13	Do.
WASHINGTON			
Adams	KREM	2	Spokane.
	KXLY	4	Do.
	KHQ	6	Do.
	KEPR	19	Yakima.
	KNDU	25	Do.
Asotin	KREM	2	Spokane.
	KXLY	4	Do.
	KHQ	6	Do.
	KLEW	3	Yakima.
Benton	KEPR	19	Do.
	KNDU	25	Do.
	KVEW	42	Do.
Chelan	KREM	2	Spokane.
	KXLY	4	Do.
	KHQ	6	Do.
Clallam	KOMO	4	Seattle-Tacoma.
	KING	5	Do.
	KIRO	7	Do.
	KVOS	12	Bellingham.
	CBUT	2	Canada.
	CHEK	6	Do.
	CHAN	8	Do.

SIGNIFICANTLY VIEWED SIGNALS—Continued

County	Call letters, channel number and market name		
WASHINGTON—continued			
Clark	KATU	2	Portland, Oreg.
	KOIN	6	Do.
	KGW	8	Do.
	KPTV	12	Do.
Columbia	KREM	2	Spokane.
	KXLY	4	Do.
	KHQ	6	Do.
Cowlitz	KATU	2	Portland, Oreg.
	KOIN	6	Do.
	KGW	8	Do.
	KPTV	12	Do.
Douglas	KREM	2	Spokane.
	KXLY	4	Do.
	KHQ	6	Do.
Ferry	KREM	2	Do.
	KXLY	4	Do.
	KHQ	6	Do.
Franklin	KEPR	19	Yakima.
	KNDU	25	Do.
	KVEW	42	Do.
Garfield	KREM	2	Spokane.
	KXLY	4	Do.
	KHQ	6	Do.
Grant	KREM	2	Do.
	KXLY	4	Do.
	KHQ	6	Do.
Grays Harbor	KOMO	4	Seattle-Tacoma.
	KING	5	Do.
	KIRO	7	Do.
Island	KOMO	4	Do.
	KING	5	Do.
	KIRO	7	Do.
	KTNT	11	Do.
	KVOS	12	Bellingham.
	CHEK	6	Canada.
Jefferson	KOMO	4	Seattle-Tacoma.
	KING	5	Do.
	KIRO	7	Do.
	KTNT	11	Do.
King	KOMO	4	Do.
	KING	5	Do.
	KIRO	7	Do.
	KTNT	11	Do.
Kitsap	KOMO	4	Do.
	KING	5	Do.
	KIRO	7	Do.
	KTNT	11	Do.
Kittitas	KNDO	23	Yakima.
	KIMA	29	Do.
Klickitat	KATU	2	Portland, Oreg.
	KOIN	6	Do.
	KGW	8	Do.
	KPTV	12	Do.
Lewis	KOMO	4	Seattle-Tacoma.
	KING	5	Do.
	KIRO	7	Do.
	KTNT	11	Do.
	KATU	2	Portland, Oreg.
	KOIN	6	Do.
	KGW	8	Do.
	KPTV	12	Do.
Lincoln	KREM	2	Spokane.
	KXLY	4	Do.
	KHQ	6	Do.
Mason	KOMO	4	Seattle-Tacoma.
	KING	5	Do.
	KIRO	7	Do.
	KTNT	11	Do.
Okanogan	KREM	2	Spokane.
	KXLY	4	Do.
	KHQ	6	Do.
Pacific	KOMO	4	Seattle-Tacoma.
	KING	5	Do.
Pend Oreille	KREM	2	Spokane.
	KXLY	4	Do.
	KHQ	6	Do.
Pierce	KOMO	4	Seattle-Tacoma.
	KING	5	Do.
	KIRO	7	Do.
	KTNT	11	Do.
San Juan	KOMO	4	Do.
	KING	5	Do.
	KIRO	7	Do.
	KVOS	12	Bellingham.
	CBUT	2	Canada.
	CHEK	6	Do.
	CHAN	8	Do.
Skagit	KOMO	4	Seattle-Tacoma.
	KING	5	Do.
	KIRO	7	Do.
	KTNT	11	Do.
	KVOS	12	Bellingham.
	CHEK	6	Canada.
	CHAN	8	Do.
Skamania	KATU	2	Portland, Oreg.
	KOIN	6	Do.
	KGW	8	Do.
	KPTV	12	Do.
Snohomish	KOMO	4	Seattle-Tacoma.
	KING	5	Do.
	KIRO	7	Do.
	KTNT	11	Do.

SIGNIFICANTLY VIEWED SIGNALS—Continued

County	Call letters, channel number and market name		
WASHINGTON—continued			
Spokane	KREM	2	Spokane.
	KXLY	4	Do.
	KHQ	6	Do.
Stevens	KREM	2	Do.
	KXLY	4	Do.
	KHQ	6	Do.
Thurston	KOMO	4	Seattle-Tacoma.
	KING	5	Do.
	KIRO	7	Do.
	KTNT	11	Do.
	KTVW	13	Do.
Wahkiakum	KATU	2	Portland, Oreg.
	KOIN	6	Do.
	KPTV	12	Do.
Walla Walla	KEPR	19	Yakima.
	KNDU	25	Do.
	KVEW	42	Do.
	KREM	2	Spokane.
	KXLY	4	Do.
	KHQ	6	Do.
Whatcom	KVOS	12	Bellingham.
	KOMO	4	Seattle-Tacoma.
	KING	5	Do.
	KIRO	7	Do.
	CBUT	2	Canada.
	CHEK	6	Do.
	CHAN	8	Do.
Whitman	KREM	2	Spokane.
	KXLY	4	Do.
	KHQ	6	Do.
Yakima	KNDO	23	Yakima.
	KIMA	29	Do.
	KAPP	35	Do.
WEST VIRGINIA			
Barbour	WDTV	5	Clarksburg-Weston.
	WBOY	12	Do.
	KDKA	2	Pittsburgh.
	WTAE	4	Do.
Berkeley	WRC	4	Washington, D.C.
	WTTG	5	Do.
	WMAL	7	Do.
	WTOP	9	Do.
	WMAR	2	Baltimore.
Boone	WSAZ	3	Charleston-Huntington.
	WCHS	8	Do.
	WHTN	13	Do.
Braxton	WSAZ	3	Do.
	WCHS	8	Do.
	WOAY	4	Bluefield-Beckley-Oak Hill.
	WDTV	5	Clarksburg-Weston.
Brooke	WTRF	7	Wheeling-Steubenville.
	WSTV	9	Do.
	KDKA	2	Pittsburgh.
	WTAE	4	Do.
	WIIC	11	Do.
Cabell	WSAZ	3	Charleston-Huntington.
	WCHS	8	Do.
	WHTN	13	Do.
Calhoun	WSAZ	3	Charleston-Huntington.
	WCHS	8	Do.
	WHTN	13	Do.
	WDTV	5	Clarksburg-Weston.
Clay	WSAZ	3	Charleston-Huntington.
	WCHS	8	Do.
	WOAY	4	Bluefield-Beckley-Oak Hill.
Doddridge *	WDTV	5	Clarksburg-Weston.
	WBOY	12	Do.
Fayette	WOAY	4	Bluefield-Beckley-Oak Hill.
	WHIS	6	Do.
	WSAZ	3	Charleston-Huntington.
	WCHS	8	Do.
	WHTN	13	Do.
Gilmer	WDTV	5	Clarksburg-Weston.
	WBOY	12	Do.
	WOAY	4	Bluefield-Beckley-Oak Hill.
	WSAZ	3	Charleston-Huntington.
Grant	WSVA	3	Harrisonburg.
	WJAC	6	Johnstown-Altoona.
Greenbrier	WOAY	4	Bluefield-Beckley-Oak Hill.
	WHIS	6	Do.
	WDBJ	7	Roanoke-Lynchburg.
	WSLS	10	Do.
Hampshire	WRC	4	Washington, D.C.
	WTTG	5	Do.
	WTOP	9	Do.
	WMAR	2	Baltimore.
	WSVA	3	Harrisonburg.
	WJAC	6	Johnstown-Altoona.

SIGNIFICANTLY VIEWED SIGNALS—Continued

County	Call letters	channel number	and market name

WEST VIRGINIA—continued

County	Call letters	Channel	Market name
Hancock	WTRF	7	Wheeling-Steubenville.
	WSTV	9	Do.
	KDKA	2	Pittsburgh.
	WTAE	4	Do.
	WIIC	11	Do.
Hardy	WSVA	3	Harrisonburg.
	WRC	4	Washington, D.C.
	WTTG	5	Do.
	WTOP	9	Do.
Harrison	WDTV	5	Clarksburg-Weston.
	WBOY	12	Do.
	WTAE	4	Pittsburgh.
Jackson	WSAZ	3	Charleston-Huntington.
	WCHS	8	Do.
	WHTN	13	Do.
Jefferson	WRC	4	Washington, D.C.
	WTTG	5	Do.
	WMAL	7	Do.
	WTOP	9	Do.
	WMAR	2	Baltimore.
Kanawha	WSAZ	3	Charleston-Huntington.
	WCHS	8	Do.
	WHTN	13	Do.
Lewis	WDTV	5	Clarksburg-Weston.
	WBOY	12	Do.
Lincoln	WSAZ	3	Charleston-Huntington.
	WHTN	13	Do.
	WCHS	8	Do.
Logan	WSAZ	3	Do.
	WCHS	8	Do.
	WHTN	13	Do.
Marion	KDKA	2	Pittsburgh.
	WTAE	4	Do.
	WDTV	5	Clarksburg-Weston.
	WBOY	12	Do.
	WTRF	7	Wheeling-Steubenville.
	WSTV	9	Do.
Marshall	WTRF	7	Do.
	WSTV	9	Do.
	KDKA	2	Pittsburgh.
	WTAE	4	Do.
	WIIC	11	Do.
Mason	WSAZ	3	Charleston-Huntington
	WCHS	8	Do.
	WHTN	13	Do.
McDowell	WSAZ	3	Do.
	WCHS	8	Do.
	WHTN	13	Do.
	WOAY	4	Bluefield-Beckley-Oak Hill.
	WHIS	6	Do.
Mercer	WOAY	4	Do.
	WHIS	6	Do.
	WDBJ	7	Roanoke-Lynchburg.
	WSLS	10	Do.
Mineral	Over 90 percent cable penetration.		
Mingo	WSAZ	3	Charleston-Huntington.
	WCHS	8	Do.
	WHTN	13	Do.
	WHIS	6	Bluefield-Beckley-Oak Hill.
Monongalia	KDKA	2	Pittsburgh.
	WTAE	4	Do.
	WIIC	11	Do.
	WBOY	12	Clarksburg-Weston.
	WTRF	7	Wheeling-Steubenville.
Monroe	WHIS	6	Bluefield-Beckley-Oak Hill.
	WDBJ	7	Roanoke-Lynchburg.
	WSLS	10	Do.
Morgan	WRC	4	Washington, D.C.
	WTTG	5	Do.
	WMAL	7	Do.
	WTOP	9	Do.
	WMAR	2	Baltimore.
	WFBG	10	Johnstown-Altoona.
Nicholas	WSAZ	3	Charleston-Huntington.
	WCHS	8	Do.
	WHTN	13	Do.
	WOAY	4	Bluefield-Beckley-Oak Hill.
Ohio	WTRF	7	Wheeling-Steubenville.
	WSTV	9	Do.
	KDKA	2	Pittsburgh.
	WTAE	4	Do.
	WIIC	11	Do.
Pendelton	WSVA	3	Harrisonburg.
Pleasants	WTRF	7	Wheeling-Steubenville.
	WCHS	8	Charleston-Huntington.
	WDTV	5	Clarksburg-Weston.
Pocahontas	WDBJ	7	Roanoke-Lynchburg.
	WSLS	10	Do.
	WHIS	6	Bluefield-Beckley-Oak Hill.
Preston	KDKA	2	Pittsburgh.
	WTAE	4	Do.
	WIIC	11	Do.
	WDTV	5	Clarksburg-Weston.
	WTRF	7	Wheeling-Steubenville.

WEST VIRGINIA—continued

County	Call letters	Channel	Market name
Putnam	WSAZ	3	Charleston-Huntington.
	WCHS	8	Do.
	WHTN	13	Do.
Raleigh	WOAY	4	Bluefield-Beckley-Oak Hill.
	WHIS	6	Do.
	WSAZ	3	Charleston-Huntington.
	WCHS	8	Do.
	WHTN	13	Do.
Randolph	WDTV	5	Clarksburg-Weston.
	WBOY	12	Do.
	WCHS	8	Charleston-Huntington.
Ritchie	WSAZ	3	Do.
	WCHS	8	Do.
	WHTN	13	Do.
	WDTV	5	Clarksburg-Weston.
	WBOY	12	Do.
	WTRF	7	Wheeling-Steubenville.
Roane	WSAZ	3	Charleston-Huntington.
	WCHS	8	Do.
	WHTN	13	Do.
Summers	WOAY	4	Bluefield-Beckley-Oak Hill.
	WHIS	6	Do.
Taylor	WDTV	5	Clarksburg-Weston.
	WBOY	12	Do.
Tucker	KDKA	2	Pittsburgh.
	WTAE	4	Do.
	WDTV	5	Clarksburg-Weston.
	WBOY	12	Do.
	WTRF	7	Wheeling-Steubenville.
	WSTV	9	Do.
Tyler	WTRF	7	Do.
	WDTV	5	Clarksburg-Weston.
Upshur	WDTV	5	Do.
	WBOY	12	Do.
Wayne	WSAZ	3	Charleston-Huntington.
	WCHS	8	Do.
	WHTN	13	Do.
Webster	WSAZ	3	Do.
	WOAY	4	Bluefield-Beckley-Oak Hill.
	WDTV	5	Clarksburg-Weston.
Wetzel	WTRF	7	Wheeling-Steubenville.
	WSTV	9	Do.
	KDKA	2	Pittsburgh.
	WTAE	4	Do.
Wirt	WSAZ	3	Charleston-Huntington.
	WCHS	8	Do.
	WHTN	13	Do.
Wood	WSAZ	3	Do.
	WCHS	8	Do.
	WHTN	13	Do.
	WTAP	15	Parkersburg.
Wyoming	WOAY	4	Bluefield-Beckley-Oak Hill.
	WHIS	6	Do.
	WCHS	8	Charleston-Huntington.

WISCONSIN

County	Call letters	Channel	Market name
Adams	WSAU	7	Wausau-Rhinelander.
	WAOW	9	Do.
	WKBT	8	La Crosse-Eau Claire.
	WEAU	13	Do.
	WISC	3	Madison.
Ashland	KDAL	3	Duluth-Superior.
	WDSM	6	Do.
	WDIO+	10	Do.
Barron	WCCO	4	Minneapolis-St. Paul.
	KSTP	5	Do.
	KMSP	9	Do.
	WTCN	11	Do.
	WEAU	13	La Crosse-Eau Claire.
Bayfield	KDAL	3	Duluth-Superior.
	WDSM	6	Do.
	WDIO+	10	Do.
Brown	WBAY	2	Green Bay.
	WFRV+	5	Do.
	WLUK	11	Do.
Buffalo	WKBT	8	La Crosse-Eau Claire.
	WEAU	13	Do.
	KROC	10	Rochester-Mason City-Austin.
Burnett	WCCO	4	Minneapolis-St. Paul.
	KSTP	5	Do.
	KMSP	9	Do.
	WTCN	11	Do.
	KDAL	3	Duluth-Superior.
	WDSM	6	Do.
Calumet	WBAY	2	Green Bay.
	WFRV+	5	Do.
	WLUK	11	Do.
Chippewa	WKBT	8	La Crosse-Eau Claire.
	WEAU	13	Do.

WISCONSIN—continued

County	Call letters	Channel	Market name
Clark	WSAU	7	Wausau-Rhinelander.
	WAOW	9	Do.
	WKBT	8	La Crosse-Eau Claire.
	WEAU	13	Do.
Columbia	WISC	3	Madison.
	WMTV	15	Do.
	WKOW	27	Do.
Crawford	WKBT	8	La Crosse-Eau Claire.
	WMT	2	Cedar Rapids-Waterloo.
	KWWL	7	Do.
	KCRG	9	Do.
Dane	WISC	3	Madison.
	WMTV	15	Do.
	WKOW	27	Do.
Dodge	WTMJ	4	Milwaukee.
	WITI	6	Do.
	WISN	12	Do.
	WISC	3	Madison.
	WMTV	15	Do.
	WKOW	27	Do.
Door	WBAY	2	Green Bay.
	WFRV+	5	Do.
	WLUK	11	Do.
Douglas	KDAL	3	Duluth-Superior.
	WDSM	6	Do.
	WDIO+	10	Do.
Dunn	WKBT	8	La Crosse-Eau Claire.
	WEAU	13	Do.
	WCCO	4	Minneapolis-St. Paul.
	KSTP	5	Do.
	KMSP	9	Do.
	WTCN	11	Do.
Eau Claire	WKBT	8	La Crosse-Eau Claire.
	WEAU	13	Do.
Florence	WLUC	6	Marquette.
	WFRV+	5	Green Bay.
	WAEO	12	Wausau-Rhinelander.
Fond du Lac	WBAY	2	Green Bay.
	WFRV+	5	Do.
	WLUK	11	Do.
	KFIZ	34	Fond du Lac.
	WTMJ	4	Milwaukee.
	WITI	6	Do.
	WISN	12	Do.
Forest	WBAY	2	Green Bay.
	WFRV+	5	Do.
	WLUK	11	Do.
	WSAU	7	Wausau-Rhinelander.
	WAOW	9	Do.
	WAEO	12	Do.
Grant	WMT	2	Cedar Rapids-Waterloo.
	KWWL	7	Do.
	KCRG	9	Do.
	WISC	3	Madison.
Green	WISC	3	Do.
	WMTV	15	Do.
	WKOW	27	Do.
	WREX	13	Rockford.
	WTVO	17	Do.
	WCEE	23	Do.
Green Lake	WBAY	2	Green Bay.
	WFRV+	5	Do.
	WLUK	11	Do.
	WISC	3	Madison.
	WMTV	15	Do.
Iowa	WISC	3	Do.
	WMTV	15	Do.
	WKOW	27	Do.
Iron	KDAL	3	Duluth-Superior.
	WDSM	6	Do.
	WDIO+	10	Do.
Jackson	WKBT	8	La Crosse-Eau Claire.
	WEAU	13	Do.
Jefferson	WTMJ	4	Milwaukee.
	WITI	6	Do.
	WISN	12	Do.
	WISC	3	Madison.
	WMTV	15	Do.
	WKOW	27	Do.
Juneau	WKBT	8	La Crosse-Eau Claire.
	WEAU	13	Do.
	WISC	3	Madison.
	WSAU	7	Wausau-Rhinelander.
	WAOW	9	Do.
Kenosha	WBBM	2	Chicago.
	WMAQ	5	Do.
	WLS	7	Do.
	WGN	9	Do.
	WTMJ	4	Milwaukee.
	WITI	6	Do.
	WISN	12	Do.
Kewaunee	WBAY	2	Green Bay.
	WFRV+	5	Do.
	WLUK	11	Do.
La Crosse	WKPT	8	La Crosse-Eau Claire.
	WEAU	13	Do.
	WXOW	19	Do.
Lafayette	WISC	3	Madison.
	WMTV	15	Do.
	WKOW	27	Do.

SIGNIFICANTLY VIEWED SIGNALS—Continued

County	Call letters, channel number and market name		

WISCONSIN—continued

County	Call letters	Ch.	Market name
Langlade	WSAU	7	Wausau-Rhinelander.
	WAOW	9	Do.
	WAEO	12	Do.
	WBAY	2	Green Bay.
	WFRV+	5	Do.
	WLUK	11	Do.
Lincoln	WSAU	7	Wausau-Rhinelander.
	WAOW	9	Do.
	WAEO	12	Do.
Manitowoc	WBAY	2	Green Bay.
	WFRV+	5	Do.
	WLUK	11	Do.
Marathon	WSAU	7	Wausau-Rhinelander.
	WAOW	9	Do.
	WAEO	12	Do.
	WEAU	13	La Crosse-Eau Claire.
Marinette	WBAY	2	Green Bay.
	WFRV+	5	Do.
	WLUK	11	Do.
Marquette	WISC	3	Madison.
	WMTV	15	Do.
	WKOW	27	Do.
	WBAY	2	Green Bay.
	WFRV+	5	Do.
	WLUK	11	Do.
Menominee	WBAY	2	Green Bay.
	WFRV+	5	Do.
	WLUK	11	Do.
Milwaukee	WTMJ	4	Milwaukee.
	WITI	6	Do.
	WISN	12	Do.
	WVTV	18	Do.
Monroe	WKBT	8	La Crosse-Eau Claire.
	WEAU	13	Do.
Oconto	WBAY	2	Green Bay.
	WFRV+	5	Do.
	WLUK	11	Do.
Oneida	WSAU	7	Wausau-Rhinelander.
	WAOW	9	Do.
	WAEO	12	Do.
Outagamie	WBAY	2	Green Bay.
	WFRV+	5	Do.
	WLUK	11	Do.
Ozaukee	WTMJ	4	Milwaukee.
	WITI	6	Do.
	WISN	12	Do.
	WVTV	18	Do.
Pepin*	WKBT	8	La Crosse-Eau Claire.
	WEAU	13	Do.
	WCCO	4	Minneapolis-St. Paul.
	KSTP	5	Do.
Pierce	WCCO	4	Do.
	KSTP	5	Do.
	KMSP	9	Do.
	WTCN	11	Do.
Polk	WCCO	4	Do.
	KSTP	5	Do.
	KMSP	9	Do.
	WTCN	11	Do.
Portage	WSAU	7	Wausau-Rhinelander.
	WAOW	9	Do.
	WBAY	2	Green Bay.
	WFRV+	5	Do.
	WLUK	11	Do.
Price	WSAU	7	Wausau-Rhinelander.
	WAOW	9	Do.
	WAEO	12	Do.
	WEAU	13	La Crosse-Eau Claire.
Racine	WTMJ	4	Milwaukee.
	WITI	6	Do.
	WISN	12	Do.
	WVTV	18	Do.
	WLS	7	Chicago.
	WGN	9	Do.
Richland	WISC	3	Madison.
	WKBT	8	La Crosse-Eau Claire.
Rock	WREX	13	Rockford.
	WTVO	17	Do.
	WCEE	23	Do.
	WISC	3	Madison.
	WMTV	15	Do.
	WKOW	27	Do.
Rusk	WKBT	8	La Crosse-Eau Claire.
	WEAU	13	Do.
	WSAU	7	Wausau-Rhinelander.
St. Croix	WCCO	4	Minneapolis-St. Paul.
	KSTP	5	Do.
	KMSP	9	Do.
	WTCN	11	Do.
Sauk	WISC	3	Madison.
	WMTV	15	Do.
	WKOW	27	Do.
Sawyer	KDAL	3	Duluth-Superior.
	WDSM	6	Do.
	WDIO+	10	Do.
Shawano	WBAY	2	Green Bay.
	WFRV+	5	Do.
	WLUK	11	Do.
	WSAU	7	Wausau-Rhinelander.

SIGNIFICANTLY VIEWED SIGNALS—Continued

County	Call letters, channel number and market name		

WISCONSIN—continued

County	Call letters	Ch.	Market name
Sheboygan	WTMJ	4	Milwaukee.
	WITI	6	Do.
	WISN	12	Do.
	WBAY	2	Green Bay.
	WFRV+	5	Do.
	WLUK	11	Do.
Taylor	WSAU	7	Wausau-Rhinelander.
	WAOW	9	Do.
	WEAU	13	La Crosse-Eau Claire.
Trempealeau	WKBT	8	Do.
	WEAU	13	Do.
Vernon	WKBT	8	Do.
	WEAU	13	Do.
	KROC	10	Rochester-Mason City-Austin.
Vilas	WSAU	7	Wausau-Rhinelander.
	WAOW	9	Do.
	WAEO	12	Do.
Walworth	WTMJ	4	Milwaukee.
	WITI	6	Do.
	WISN	12	Do.
	WBBM	2	Chicago.
	WGN	9	Do.
	WISC	3	Madison.
	WREX	13	Rockford.
Washburn	KDAL	3	Duluth-Superior.
	WDSM	6	Do.
	WDIO+	10	Do.
Washington	WTMJ	4	Milwaukee.
	WITI	6	Do.
	WISN	12	Do.
	WVTV	18	Do.
Waukesha	WTMJ	4	Milwaukee.
	WITI	6	Do.
	WISN	12	Do.
	WVTV	18	Do.
Waupaca	WBAY	2	Green Bay.
	WFRV+	5	Do.
	WLUK	11	Do.
	WSAU	7	Wausau-Rhinelander.
Waushara	WBAY	2	Green Bay.
	WFRV+	5	Do.
	WLUK	11	Do.
	WSAU	7	Wausau-Rhinelander.
Winnebago	WBAY	2	Green Bay.
	WFRV+	5	Do.
	WLUK	11	Do.
	KFIZ	34	Fond Du Lac.
Wood	WSAU	7	Wausau-Rhinelander.
	WAOW	9	Do.
	WEAU	13	La Crosse-Eau Claire.

WYOMING

County	Call letters	Ch.	Market name
Albany	KOA	4	Denver.
	KLZ	7	Do.
	KBTV	9	Do.
	KFBC	5	Cheyenne.
Big Horn	KOOK	2	Billings.
	KULR	8	Do.
	KWRB	10	Casper-Riverton.
Campbell	Over 90 percent cable penetration.		
Carbon	KTWO	2	Casper-Riverton.
	KFBC	5	Cheyenne.
Converse	KTWO	2	Casper-Riverton.
	KSTF	10	Cheyenne.
Crook	KOTA+	3	Rapid City.
	KTWO	2	Casper-Riverton.
Fremont	KTWO	2	Do.
	KWRB	10	Do.
Goshen	KSTF	10	Cheyenne.
	KDUH	4	Rapid City.
Hot Springs	KTWO	2	Casper-Riverton.
	KWRB	10	Do.
Johnson	KTWO	2	Do.
Laramie	KFBC	5	Cheyenne.
	KWGN	2	Denver.
	KOA	4	Do.
	KLZ	7	Do.
	KBTV	9	Do.
Lincoln	KID	3	Idaho Falls-Pocatello.
	KIFI	8	Do.
	KCPX	4	Salt Lake City.
	KSL	5	Do.
Natrona	KTWO	2	Casper-Riverton.
Niobrara	KTWO	2	Do.
	KFBC	5	Cheyenne.
Park	KOOK	2	Billings.
	KULR	8	Do.
Platte	KFBC	5	Cheyenne.
	KSTF	10	Do.
	KTWO	2	Casper-Riverton.
Sheridan	KOOK	2	Billings.
	KULR	8	Do.
	KTWO	2	Casper-Riverton.
	KOTA+	3	Rapid City.
Sublette	KTWO	2	Casper-Riverton.
	KID	3	Idaho Falls-Pocatello.

SIGNIFICANTLY VIEWED SIGNALS—Continued

County	Call letters, channel number and market name		

WYOMING—continued

County	Call letters	Ch.	Market name
Sweetwater	Over 90 percent cable penetration.		
Teton	KID	3	Idaho Falls-Pocatello.
	KIFI	8	Do.
Uinta	KUTV	2	Salt Lake City.
	KCPX	4	Do.
	KSL	5	Do.
Washakie	KTWO	2	Casper-Riverton.
	KWRB	10	Do.
	KCPX	4	Salt Lake City.
	KSL	5	Do.
Weston	KTWO	2	Casper-Riverton.
	KOTA+	3	Rapid City.
Yellowstone National Park	KID	3	Idaho Falls-Pocatello.
	KULR	8	Billings.

Appendix C

Paragraphs 5–8 of memorandum opinion and order denying "Motion For Stay Pending Appeal" of Nevada Independent Broadcasting Corp., Henderson, Nev. 34 FCC 2d 165 (1972):

A detailed history of the proceedings leading to the promulgation of the rules is set forth in the report and order and will not be repeated here. See 37 F.R. at 3252–62. Suffice it to say that there were several sets of proposed rules offered in the years preceding adoption. These evolved into a letter of intent that was sent to Congress in August 1971 (31 FCC 2d 115) that outlined the proposed final shape of the rules. The letter, in turn, was tempered in some respects by a consensus agreement entered into by principal industry groups before the rules were finaly adopted and released in February 1972. What ultimately emerged in terms of specific rules was admittedly different from what was initially proposed. But such is the nature of the rule making proceedings. The APA does not demand that the rules, as finally articulated, be identical to those first proposed. What the APA demands is that the notice of proposed rule making include "the terms or substance of the proposed rule or a description of the subjects and issues involved," (5 U.S.C. 553(b)(3)) and that "the agency shall give interested persons an opportunity to participate in the rule making through submission of written data, views, or arguments with or without opportunity for oral presentation." (5 U.S.C. 555(c)). It is unquestionable that few administrative proceedings have been so open or so subject to written comment and oral presentation. More than 700 separate substantive comments were received in the six dockets comprising this rulemaking, 175 persons appeared before the Commission in oral argument, and scores of persons participated in panel discussions on various aspects and issues pertaining to the rules. Every conceivable point of view, both public and private, was represented.

The public was given ample notice of the "subjects and issues involved" in the rulemaking. Petitioner argues that the Commission's letter of intent "was specific enough for the purposes of rule making but it was not used for that purpose (because) (n)o comments on the proposals were invited," and that the final rules contain a copyright exclusivity provision that is "novel and previously undiscussed." But an examination of the subjects and issues (rather than the precise rules) and their genesis demonstrates that the dictates of the APA have been honored.

Part 76 of the rules, challenged by petitioner, regulates (a) the signals that may be carried by cable systems, depending on the size of the television market in which the

system is located and the nature and source of the television signal to be carried, (b) program exclusivity and nonduplication protection to be granted to certain broadcasters, depending on location of the broadcaster and cable system and the exclusivity contracts involved, and (c) cablecasting, public access, and minimum channel requirements, depending on the size of the television market in which the cable system is located.

The rules respond basically to three broad issues: What is the permissible degree of cable penetration in each of the various size television markets, what safeguards are necessary to ensure the healthy maintenance of broadcast television, and to what degree should a cable system operate as an outlet for local community expression? Each of the above issues, as stated, was discussed at length in the rule making proceedings. The enunciation of these issues gave reasonable notice that the Commission was contemplating the adoption of rules addressed to these issues that might involve carriage requirements and restrictions, exclusivity and nonduplication protection as a means of implementing the desired safeguards, and minimum technical requirements. Being aware of these issues, interested parties had the opportunity to comment on them freely and to offer their own solutions, proposals and counterproposals. Indeed, it was some of these suggestions that eventually found acceptance in the rules. The Commission was under no obligation, however, to put out a new notice of proposed rule making each time it received a proposal from one of the participants that the Commission found convincing. Having received comments from all quarters on all subject areas, and thereby being as fully informed as practical, the law leaves to Commission discretion the structuring of the final rules. See, e.g., Owensboro on the Air, Inc. v. United States, 262 F. 2d 702 (D.C. Cir. 1958), cert. denied, 360 U.S. 911; Buckeye Cablevision, Inc. v. FCC, 387 F. 2d 220 (D.C. Cir. 1967); Mount Mansfield Television Inc. v. FCC, 442 F. 2d 470 (2d Cir. 1971).

[FR Doc.72–10100 Filed 7–13–72;8:45 am]

Before the
FEDERAL COMMUNICATIONS COMMISSION
Washington, D.C. 20554

In the Matter of   )
)
Amendment of Part 74, Subpart K, of the Commission's   ) Docket No. 18397
Rules and Regulations Relative to Community Antenna   ) Docket No. 18397-A
Television Systems; and Inquiry into the Development   )
of Communications Technology and Services to Formu-   )
late Regulatory Policy and Rulemaking and/or Legisla-   )
tive Proposals.   )
)
Amendment of Section 74.1107 of the Commission's Rules   )
and Regulations to Avoid Filing of Repetitious Requests.   ) Docket No. 18373
)
Amendment of Section 74.1031(c) and 74.1105(a) and (b)   ) Docket No. 18416
of the Commission's Rules and Regulations as they Re-   )
late to Addition of New Television Signals.   )
)
Amendment of Part 74, Subpart K, of the Commission's   ) Docket No. 18892
Rules and Regulations Relative to Federal-State or   )
Local Relationships in the Community Antenna Television   )
System Field; and/or Formulation of Legislative Pro-   )
posals in this Respect.   )
)
Amendment of Subpart K of Part 74 of the Commission's   ) Docket No. 18894
Rules and Regulations with Respect to Technical Stand-   )
ards for Community Antenna Television Systems.   )

E R R A T U M

Released: July 13, 1972

Appendix A of the _Memorandum Opinion and Order on Reconsideration_
of the Cable Television Report and Order in the above entitled matters,
FCC 72-530, released June 26, 1972, is corrected with respect to §76.13
(a) and (b), to read as follows:

§76.13 Filing of applications.

*   *   *   *   *

(a) *   *   *

(4) A statement that explains how the proposed system's
franchise and its plans for availability and administration of access
channels and other nonbroadcast cable services are consistent with the
provisions of §§76.31, 76.201, and 76.251;

> NOTE: If the proposed system's franchise was issued
> prior to March 31, 1972, only substantial consistency
> with the provisions of §76.31 need be demonstrated in
> the statement required in subparagraph (4), until the
> end of the current franchise period, or March 31, 1977,
> whichever occurs first.

A   A   A   N   N

(b) * * *

(3) If the system has not commenced operations but has been authorized to carry one or more television signals, a copy of the franchise, license, permit, or certificate granted to construct and operate a cable television system, and a statement that explains how the system's franchise is substantially consistent with the provisions of §76.31;

> NOTE: If only substantial consistency with the provisions of §76.31 is demonstrated in the statement required in subparagraph (3), a certificate of compliance that is granted pursuant to §76.11 shall be valid only until the end of the system's current franchise period, or March 31, 1977, whichever occurs first.

*   *   *   *   *

FEDERAL COMMUNICATIONS COMMISSION

Ben F. Waple
Secretary

NOTE: Rules changes herein will be covered by the 1972 Edition of Volume III.

37 Fed. Reg. 23104-23150 (October 28, 1972)

Title 47--TELECOMMUNICATION

Chapter 1--Federal Communications Commission

[FCC 72-943]

PART 76--CABLE TELEVISION SERVICES

Procedures for Filing of Applications

Order. In the matter of amendment of Part 76,
Subpart B, of the Commission's rules and regulations concerning procedures in the Cable Television Service.
   1. In paragraph 117, "Reconsideration of Cable Television Report and Order." FCC-72-530, '36
FCC 2d 326, 366, we stated that when there is a
dispute as to whether the appropriate franchising
authority is on the State of local level notice
of filing of an application for a certificate of
compliance should be served on all authorities
that are claiming jurisdiction. In addition, in
several States a State regulatory body has jurisdiction to confirm or deny franchises granted by
local authorities, or otherwise regulate cable
television. In either case, both the local and
State authorities should be served with copies of
the application for certificate of compliance.
We have examined the Cable Television Rules and
have concluded that certain editorial revisions
are necessary to emphasize this procedural point.
   2. First, we are amending the first part of
§ 76.13 (a)(7), (b)(7), and (c)(5), of the Commission's rules to insure that in cases where there
are State and local authorities asserting jurisdiction over cable television, even where such
jurisdiction is pendente lite, both are served
with copies of the application for certificate of
compliance.
   3. Second, we are amending the second part of
§ 76.13 (a)(7), (b)(7), and (c)(5) of the Commission's rules to make it clear that, unless either
the State or local body makes a copy of the application for certificate of compliance available for
public inspection in the community of the system,
the applicant will provide for public inspection
of the application in said community.
   4. Since these amendments are interpretive or
relate to Commission procedure, the prior notice
provisions of section 4 of the Administrative Procedure Act, 5 U.S.C. section 653, do not apply.
   Authority for the rule amendments adopted herein is contained in sections 2, 3, 4 (i), and (j)
of the Communications Act of 1934, as amended.
   Accordingly, it is ordered, That effective October 31, 1972, Part 76 of the Commission's rules
and regulations is amended as set forth below.

(Secs. 2, 3, 4, 43 Stat., as amended, 1064, 1065,
1066; 47 U.S.C. 152, 153, 154)

Adopted: October 18, 1972.

Released: October 24, 1972.

FEDERAL COMMUNICATIONS COMMISSION,

[SEAL]   Ben F. Waple, Secretary.

   Chapter I of Title 47 of the Code of Federal
Regulations is amended as follows:
   In § 76.13, paragraphs (a)(7), (b)(7), and (c)
(5) are revised as follows:

§ 76.13 Filing of applications.

   *   *   *   *   *

   (a) * * *
   (7) A statement that a copy of the completed
application has been served on any local or State
agency or body asserting authority to franchise,
license, certify, or otherwise regulate cable
television, and that if such application is not
made available by any such authority for public
inspection in the community of the system, the applicant will provide for public inspection of the
application at any accessible place (such as a
public library, public registry for documents, or
an attorney's office) in the community of the system at any time during regular business hours;

   *   *   *   *   *

   (b) * * *
   (7) A statement that a copy of the completed
application has been served on any local or State
agency or body asserting authority to franchise,
license, certify, or otherwise regulate cable
television, and that if such application is not
made available by any such authority for public
inspection in the community of the system, the applicant will provide for public inspection of the
application at any accessible place (such as a
public library, public registry for documents, or
an attorney's office) in the community of the system at any time during regular business hours;

   *   *   *   *   *

   (c) * * *
   (5) A statement that a copy of the completed
application has been served on any local or State
agency or body asserting authority to franchise,
license, certify, or otherwise regulate cable
television, and that if such application is not
made available by any such authority for public
inspection in the community of the system, the
application at any accessible place (such as a
public library, public registry for documents or
an attorney's office) in the community of the
system at any time during regular business hours;

   *   *   *   *   *

[FR Doc. 72-18435 Filed 10-27-72; 8:49 am]

Cable Policy

[In re: Docket Numbers 18397, 18397-A, 18373, 18416, 18892 and 18894]

Opinion of Commissioner Nicholas Johnson, Concurring in Part and Dissenting in Part

Foreword

On Thursday, February 3, 1972, I issued a preliminary opinion in this matter. The text of that opinion follows:

The much-heralded new dawn for cable turns out to be a cold and smog-filled day.

The White House interference in the process makes a mockery of the FCC's independence and role as an arm of Congress.

The Commission's about-face accommodation of the desires of the largest broadcasters, cable companies and copyright interests-- after long hearings and the declaration of its August 5th policy as in the "public interest"--makes a shambles of the spirit of the Administrative Procedure Act. This failing is so severe that even issuing this document as a proposed rulemaking for public comment would not cure it.

For FCC Chairman Burch to engage in secret bargaining sessions designed to bind his fellow Commissioners to policies in which they have had no participation is an affront to a multi-man Commission.

The hurried issuance of today's document means that few of the full opinions of the six Commissioners will be available, and the only people to get copies of the document for a matter of days will be a few favored Congressmen, lobbyists, trade magazines and press. The use of the Federal Register will take a week, and may also preclude publication of separate statements.

The substance is little better than the procedure. It is not true, as the majority states, that the compromise "does not disturb the basic structure of our August 5 plan." Unlike the August 5th rules, at least 40% of the American people, those who live in the largest cities, will not not get cable. This serves no one's interests-- save the most powerful broadcasters and program owners who now get their way. The multi-million-dollar big city corporate owners, whose "National Association of Broadcasters" exacted the added protectionism for them, don't need it. Small broadcasters may-- but don't get it. If cable is to grow, it must be in the big cities--where it's precluded. If the potential need and demand for leased channels, public access channels, and minority program- ming are to be served it must be in the big cities. It won't be.

The limitations on what even small-town cable can carry are ridicu- lous. With all its capacity to bring the American people dozens of signals from thousands of miles, the FCC rules won't even let cable systems carry some signals that its subscribers can pick up off the air with rabbit ears! There are severe limitations on the cities from which signals can be imported. An elaborate, almost unintelligible section (inserted by the richest program owners after the August 5th policy excluded it) prohibits the showing of the programs most desired by the public. The FCC agrees, more- over, to tie its hands and never make future changes in part of this arrangement.

A fuller opinion will follow.

Now, two working days later, the fuller opinion promised has been pre- pared, and follows.

Introduction

In future years, when students of law or government wish to study the decision making process at its worst, when they look for examples of industry domination of government, when they look for Presidential interference in the operation of an agency responsible to Congress, they will look to the FCC handling of the never-ending saga of cable tele- vision as a classic case study. It is unfortunate, if not fatal, that the decision must be described in these terms, for of the national com- munications policy questions before us, none is more important to the country's future than cable television.

The Commission has promulgated rules for cable television which are designed to introduce, in a conservative fashion, the benefits of cable to some of the people of this country. To the extent they will, to some extent, achieve that purpose, I concur with the majority. Because they are substantially different from the rules I would have preferred to adopt, and because the Commission arrived at those rules through a process I find wholly inconsistent with the spirit of the Administrative Procedure Act, the concept of independent regulatory agencies, and possibly the due process clause of the Fifth Amendment, I am compelled to dissent in part, as well.

I. Cable Development: A Model

Unencumbered by political and vested economic pressures, cable television would develop like any new technology--in the market place. Systems would be built in markets in which consumer demand made building profitable. These systems would import distant signals to the extent of market demand. One could expect that after providing all markets with affiliates of the three national commercial networks, local independents and public broadcasting stations, perhaps two to five independent stations from various parts of the country, regional networks or some additional out of market non-commercial stations, cable systems would have little incentive to import more signals. Indeed, a system would probably be commercially more attractive if it provided additional channels with nonbroadcast ("cablecasting") services rather than additional channels of commercial television. In any event, the market-- the cost of importing additional signals compared with the additional income they would provide the cable entrepreneur--would seek its own level. And, I would guess, that level would be somewhere between eight to fifteen signals, depending upon the region of the country involved.

I would impose limited regulations on this basic marketplace system. I would require all systems in the larger cities to have a minimum

capacity of 40 channels, half of which would be dedicated to other than over-the-air broadcast services. Of the one half of the channels reserved for purposes other than over-the-air broadcast signals, at least one would be dedicated to state and local government use, one would be dedicated to educational use, one would be dedicated to the public use (all on a first come-first serve basis, free of charge), and the others would be leased to all comers at fixed rates. Systems would be required to expand channel capacity in accordance with demand, in the manner set out in the August 5 letter and these rules.

Because a national cable network could develop under this system, some protection would have to be afforded the public as well as the systems from anti-competitive agreements between microwave systems and "overzealous" independents. To this end microwave systems could be required to offer all independent signals along their routes to cable systems, to prevent the aggressive independent from using anticompetitive methods to achieve the network result.

The over-the-air broadcast system as we know it is an important element of our society and is entitled to some protection. No one wants massive numbers of over-the-air stations suddenly to go bankrupt and leave the air because of cable. Cable is currently almost wholly dependent upon over-the-air stations for its programming; there are many homeowners who can't or won't have cable; and the continued competition and choice for the viewer between cable and over-the-air signals is his only ultimate protection against cable abuses. The question is only how much protectionism is warranted and necessary at a time when no station has yet gone off the air becuase of cable. I would provide, for starters, only that no cable system could simultaneously duplicate a local station's program with that of an imported station. Then, if a local station could demonstrate that (1) it is deteriorating substantially (i.e., a steady decline of gross revenue), and (2) that such deterioration is a result of the existence of cable television in its market, special relief could then be made available. If the problem became widespread, new general protection could be fashioned at that time.

I would regulate to prevent further concentration of control of the mass media. Our rules prevent cross ownership of broadcast stations and cable systems in the same market, and common ownership of national networks and cable systems. I would also consider rules prohibiting any single company from owning more than one cable system in the top 50 markets, and any single system from reaching more than one percent of the country's television households. I would consider prohibiting cross-ownership of newspapers and cable television in the same market.

As a matter of principle, I believe copyright holders should be compensated for the use of their products by cable systems. But regulations implementing that right need not take the form of exclusivity (prohibiting a cable system from carrying the program at all), as they do in these rules. Regulations could simply require the automatic payment of fees to the copyright holders, through a mechanism similar to that used by ASCAP for song writers. However, I am not convinced that the FCC is the appropriate forum in which such decisions should be made-- any more than the FCC should attempt to legislate minimum wage legislation for cable systems, or zoning restrictions. As I discuss more fully later in this opinion, the CBS v Tele-Prompter suit may clarify the copyright situation beyond the Fortnightly set of facts. Fortnightly Corporation v. United Artists, 392 U.S. 390 (1968). In any event, I would expect either the courts or Congress to adjust the interests of the competing parties--not the FCC--as the Commission indicated it believed on August 5.

Finally, I would support regulations limiting subscriber charges, lease prices for leased channels, and rates charged by utilities for the use of their poles.

The model I have outlined ought to have the support of most people of independent mind--"free entrepreneurs" and "regulators" alike. It serves the "public interest" and is wholly consistent with the profit motive. The problem, of course, is that it does not have the support of the most powerful broadcasters--a group whose political influence is unrivaled in our time.

The rules we adopt today vary from this model; in some cases they are quite similar, while in others they are based on a wholly different

philosophical premise. But a persistent current, running throughout the rules, is an absence of adequate rationale, satisfactory justifications for departures from this model.

II. August 5, 1971 and its Aftermath

On August 5, 1971, the Commission, in a 6 to 1 decision, transmitted to Congress a "letter of intent," outlining its proposed rules for cable television. These rules were the result of exhaustive public hearings at which all positions were aired. The result reached was a far cry from the free enterprise model described above; it was itself a compromise, intended to adjust and protect various economic interests, and to accommodate "political realities." But it was a compromise we agree was feasible, and one under which cable could at least get started.

Subsequent to our adoption of the August 5 letter, apparently not satisfied with the concessions made to each of them, broadcasters and copyright owners, with the support and encouragement of the White House and Chairman Burch (and the participation of cable interests), carved up the cable pie in a manner more to their liking. In its rules the Commission puts its stamp of approval on the results of these closed door sessions by implementing the precise terms of the industry's agreement.

The new rules graphically demonstrate what economic protectionism can do to a sound regulatory scheme. In our August 5 letter of intent, we recognized that the big city markets, more than the others, needed both the additional entertainment programming and the nonbroadcast benefits of cable television. Thus, while we held back cable development in the big cities in some respects (for example, only three independents' signals were permitted), we provided sufficient benefits to stimulate its beginning. The regulatory scheme permitted cable systems in the top 50 markets to distribute, as a minimum, three network stations and three independent stations. Systems in the second 50 markets were

permitted three network and two independent stations. Those systems in markets below the top 100 were permitted three network affiliates and one independent station. And systems in cities without any television stations were permitted an unlimited number of independents. (All systems could also carry non-commercial and foreign language television stations, and radio signals.)

We never felt that cable's future was tied to distant broadcast signals; if that were all that was involved, it is doubtful that we would have spent one-tenth the effort we have expended. With this scheme, we hoped that systems in the larger markets, where diversity of interests most required the nonbroadcast advantages of cable—such as access, leased channels for community groups, and educational channels—would have sufficient attractiveness to subscribers so as to provide the economic base necessary for the development of these services.

Markets 51 to 100 were given fewer distant signals on the theory that the over-the air stations there had less in the way of both revenue and audience to support much imported competition. Systems in the smallest cities, those located more than 35 miles from any television station, were to be permitted unlimited distant signals on the theory that, barring any rationale for station protectionism, there was no reason not to revert to the model of unlimited signal importation.

This was the state of affairs on August 5, 1971. Thereafter, the vested economic interests—broadcasters (who felt threatened by this new technological competition), copyright holders (who were afraid cable systems would diminish the value of their products), and the cable industry (who felt threatened by the political power of the broadcasters—once joined by Chairman Burch and the President—to stop our August 5 policy entirely in Congress)— met with the representatives of the White House and with FCC Chairman Burch and finally agreed to the compromise that the majority refers to as the "consensus agreement."

The compromise carved up the action among the three industries, at the expense of the viewing public, by making three changes in the policy we announced on August 5. Despite the majority's assurances that its "incorporation into our new rules for cable does not disturb the basic structure of our August 5 plan," the compromise was, of course, designed to disturb the basic structure and succeeded in doing so.

215

III. Policy and Protectionism

The compromise and the rules promulgated by the Commission are a far cry from the free enterprise model of cable television. They are a patchwork of protectionism, designed to foster the interests of vested economic institutions at the expense of the public. Admittedly, under these rules cable will be able to make a very modest start in some of the smallest markets. It will not, however, grow with the speed and the impact it would have under less restrictive rules. The major failings of the compromise and the rules, as I see them, involve the exclusivity protection, the viewing standard, and leapfrogging.

Exclusivity protection. The rules provide for "run of the contract exclusivity" to stations in the top 50 markets, and two year exclusivity to stations in markets 51-100. That is, a program supplier can sell, and a station can buy, an "exclusive" right to a given program, and gain thereby the legally enforceable right to keep any other station in the market from showing it. Now, says the FCC, the station can use that "exclusivity" to keep a cable system from importing that program from an out-of-market station as well. In other words, if a station in one of these markets has a contractual right to show David Frost or The Pawnbroker, no cable system in that market can import it from another city. Thus, although top 100 market systems are "permitted" to import distant signals, these signals will have to be blacked out whenever they carry programs covered under exclusive contracts. One of the principal services offered by cable--not just different programming, but alternative schedules for the same programming--is hereby simply wiped out. Further, programs or films subject to local "exclusivity" may not be imported by cable even though the local station may not show them for years.

Translated into concrete examples, based on current programming and currently existing contractual arrangements, a cable system in Charlotte, North Carolina, the forty-second market, would have to black out over 16 hours a day of programming from WTCG-TV, Atlanta, Georgia, if it chose to import that station. A system in Fort Wayne, Indiana,

the eighty-second ranking market, would have to black out WGN-TV, Chicago, should it chose to import it, for over eight hours daily. Obviously, we can expect to find a rush to exclusive contracts in the future to permit local stations to take advantage of the FCC-sanctioned anti-competitive device.

Viewing standard. Television signals can often be picked up off the air from 60 to 100 miles distance in proper terrain with a good antenna. The advantage of cable is that it can bring subscribers more signals than they can get off the air. That's because the cable system has a taller, more powerful receiving antenna than most homeowners, and because it can relay signals by microwave over long distances (the same way the networks relay their signals from New York around the country to affiliates). Even with a little "rabbit ears" antenna, however, I can, for example, pick up Baltimore signals on my home receiver in Washington. One would assume, therefore, that cable systems would be permitted by the FCC to provide their subscribers at least what the subscribers can already pick up off the air. Right? Wrong. The rules contain a unique concept known as the "viewing standard." Cable systems in all cities with television stations are required to carry all stations licensed to cities within a 35 mile circle around them. That's no problem; most cable systems would want to do that anyway. The problem comes in defining what additional signals the cable may carry as, in effect, "local signals"--that is, signals that will not count as "distant" imported signals. I would define that as "viewable" signals, whether technically defined as "predicted Grade B," actual Grade B, or most pragmatically, what the cable operator can, in fact, pick up with his antenna. In my case, for example, those Baltimore signals would be considered "viewable," even though, in fact, one would generally watch the Washington signals whenever the same network program is being shown by both. (By contrast, the same network's news may be shown at different times in Washington and Baltimore, and being able to watch both cities' signals thereby increases the number of networks' news shows that may be watched.) This is decidedly not the FCC/industry "viewing standard." Its standard is not whether the station can be watched, but whether it is, in fact, watched. Such an inquiry is, of

217

course, directed solely at protection of the local station's market revenues, not to the technological capabilities of cable. The details of "share" and "net weekly circulation" are spelled out in the majority's document and are not necessary to our discussion. It's sufficient to note that the August 5 policy was that any station actually viewed by 1% of the local homes could be carried and that the "compromise" raises that to 2%--and thereby cuts in about half the number of stations that may be carried. (For example, none of those Baltimore signals I can now watch could be carried by a Washington cable system.)

Leapfrogging. The rules provide for the importation of a limited number of distant signals. However, although technologically capable of bringing in distant signals from anywhere in the country, if a cable system wants to bring in a signal from a city in one of the top 25 markets--obviously, the most desirable stations--it must reach out only to the closest two top 25 cities. Only when forced to black out one or both of those signals can a system go nationwide for programming. That is, it may not "leapfrog" closer stations in order to reach out for more distant (and desirable) stations.

The net result of this compromise--exclusivity, viewing standard, and leapfrogging--is to reverse the priorities we established in August. The exclusivity provisions in the top 50 markets were designed to protect the copyright holders, who derive over 80% of their profits from sales to stations in the top 50 markets. Under these provisions, virtually all attractive programming will be unavailable to cable systems during terms of contracts that theoretically can exist forever. (The Commission promises to study the question of the length of exclusive contracts, but bare promises are a far cry from operating rules. And even if the Commission were to someday limit contracts, say, to five years, a term of this length will in many cases make the program highly unattractive.) This resulting lack of available programming will doom cable in the top 50 markets. It will literally have nothing to sell.

The exclusivity provisions in markets 51 to 100 are designed to protect broadcasters. The copyright holders don't really care about these markets, as they earn less than 20% of their revenues there. The broadcasters, vicariously protected in the top 50 markets by the interests

218

of the copyright holders, managed to negotiate two year exclusivity in the remaining markets. Thus, cable systems will not be able to show popular programs until two years after they are available to broadcasters. Granted, cable may still begin, but its attractiveness will be limited.

There is no exclusivity in the small markets and nonmarket areas. These were the cities "given" to the cable industry by broadcasters and copyright holders.

The compromise agreement not only makes little sense from a sound regulatory point of view, it's not even very sensible selfish protectionism. While, on the one hand, our August 5 plan expressly provided benefits to the big city systems by permitting them to import some signals, the compromise burdens these systems by imposing prohibitive exclusivity, viewing standard, and leapfrogging requirements.

There may be some truth to the argument that television stations in small markets can be injured economically through audience fragmentation when even one additional competitive station comes to town via cable. But it should be clear that stations in the major markets, already competing with large numbers of other television stations and other entertainment and news outlets, are less likely to be injured by an additional station or two. Yet it is in these major markets where the regulations inhibit cable, and the smaller ones where cable is free to develop. This result can only be explained in terms of the sheer political power that the history of the compromise represents.

IV. History and Failings of the "Consensus Agreement"

It is impossible to have a full understanding of the significance of the Commission's adoption of the consensus without first fully exploring the background of both the consensus and the rules.

In 1968 we imposed what amounted to a freeze on cable television development in the major cities--even though never denominated as such.

We adopted procedures that we said would enhance the growth of cable, and which I believed would actually work. Under these procedures, no cable system in a top 100 market would be permitted to import distant signals unless it received retransmission consent from that station. This never worked.

The battle lines reformed around the issue of distant signals. Most broadcasters were perfectly happy to permit passive cable systems—systems which only transmitted local signals. Some broadcasters and copyright holders argued that even these passive systems should be required to pay copyright fees for local programs that showed on their systems. The Supreme Court rejected this argument in Fortnightly Corporation v. United Artists, 392 U.S. 390 (1968).

This did not, however, necessarily settle the question of a system's authority to carry distant signals without paying copyright. Fortnightly was read narrowly by the FCC and limited to its facts: that is, no copyright fee would be required for the showing on cable systems of local stations, but the question of distant signals remained unsettled.

The parties refused to budge. Broadcasters and copyright holders threatened to block any cable rules that permitted the importation of distant signals until copyright legislation was adopted—by exerting their impressive political influence in Congress, forcing Congressional hearings. Cable owners refused to support copyright legislation until the cable rules were adopted. The Senate Copyright Subcommittee refused to pass a copyright revision until the question of cable was settled, and it refused to enact a separate copyright law for cable. The process ground to a halt.

Finally, the Commission, after months of thorough study, acting precisely as one would hope a quasi-legislative body should act, promulgated its August 5 letter. For one of the few times in my tenure as an FCC Commissioner, I was able to join with a near-unanimous majority on a major issue of communications policy.

Unfortunately, our historic example was not to be. Three months later, the industries had used their White House leverage to fashion their own cable policy, and the consensus agreement was born.

The implications of the Commission's decision to adopt the compromise are as serious a threat to the democratic system of government as any we have witnessed in almost 200 years of our history. While the majority goes to great lengths to describe how our accepting the compromise was really in the public interest because it facilitated the promulgation of these rules and the passage of copyright legislation, it utterly fails to take into consideration the threat to the public interest posed by setting the precedent of deferring to big business whenever it possesses the power to impede the development of a regulatory scheme (or legislation or an executive decision). We, as a society, profess to abhor political blackmail, and struggle to insulate our decision making process from the influence of those who would sacrifice the common good for greedy self interest. Yet here we find a Commission, made up of public citizens appointed by a president, agreeing that this method of decision making is in the public interest. I am not naive enough to think that this process has not been repeated hundreds of time prior to this occasion by this and other agencies; but I am shocked when, rather than try to hide the reality, we applaud it as an appropriate method of doing the people's business.

This procedure is rendered even more abhorrent when one sees it in the perspective of the industry power over regulatory agencies that already exists. Industries have often written the legislation under which the agencies act. They may have veto power over the Commissioners appointed. Their knowledge of the working of the agency is enhanced by their hiring away the ablest of its employees. (Most former FCC Commissioners are now working, in one way of another, for one of the industries they were formerly responsible for regulating.) The potential of such future employment (at much higher pay) has been characterized by Ralph Nader as "the deferred bribe." The "regulated" industry influences the agency's appropriations, even its forms and inquiries (through OMB "industry advisory committees"). The industry has the money to contract for any study, hire any consultant, and file whatever legal briefs and other documentation may be necessary to influence the decision "on the merits." It can send representatives to walk the halls of the agency, and provide luncheons for Commissioners and employees.

It fights at every turn (generally with agency backing) any participation by public interest law firms in matters before the agency. Now, on top of all this, what the FCC seems to be saying is that if, notwithstanding this stacked deck, the industry still loses, we will then let it win because it's so politically powerful it can get its way anyway. The whole sordid story doesn't augur well for those who are urging the disaffected to "work for change within the system."

The value we have trampled on comes to us from at least three different sources: the Administrative Procedure Act (APA), the philosophical concept of independent Congressional agencies, and the due process clause of the fifth amendment. The APA was designed to establish an orderly procedure by which adminstrative agencies can collect information necessary for them to make intelligent decisions. It provides an opportunity for all interested parties to comment on a proposal (in this case, cable television regulation), reply comments from those who wish to dispute what others have said, and public hearings in the event the agency feels they are desirable. After this process, the agency is free to consult or use any source it wishes. Thus, although adoption of the consensus agreement may not be prohibited by the APA, such an action is clearly inconsistent with the spirit of an Act which attempts to set out an orderly public procedure by which decisions of this nature are made. The FCC often issues proposed rule makings which are little more than superficial rewrites of the requests of one special interest or another. That is not the point. In this instance we went out of our way to canvas the full range of public and industry opinion before issuing our August 5 policy. For Chairman Burch subsequently to go into secret sessions with industry spokesmen, and accept their rewrite of the rules, and then force the industry version down the throats of his fellow Commissioners, Congress and public alike makes an unnecessarily cruel hoax of what started out as a fairly commendable undertaking.

Perhaps more serious is the fact that one major party to the compromise (described by some as the "glue" that holds the compromise together) was the Director of the President's Office of Telecommunications Policy. His participation, indeed the very existence of his Office, looms large as a threat to the independence of the FCC as an agency

responsible only to Congress. This alternative voice tends to turn the Commission into a partisan body, by causing it to react on political rather than sound policy grounds; further, it tends to increase the rivalry between the President and Congress, a rivalry which is healthy only when it results in constructive dialogue as opposed to destructive bickering. And, no less serious, it legitimizes the Administration's carrot/stick approach to broadcasters, serving as it does as an ambiguous, fear-inducing institutional outlet for the President's attacks and rewards to the media.

Finally, the history of this proceeding, beginning as it did with an honest and good-faith effort to develop the best possible cable television rules, and ending with complete and utter deference to the demands of the most powerful elements of the industry, may have left us with a legacy that cannot withstand Constitutional scrutiny. In 1934, and again in 1935, the Supreme Court had occasion to address a markedly similar question in the context of New Deal legislation. Under laws subsequently struck down by the Court, industry committees were given the authority to promulgate binding regulations on their entire industry. In striking this legislation on several grounds (some of which are not applicable here, and in any event have been reversed by later Court decisions), the Supreme Court said:

> But would it be seriously contended that Congress could delegate its legislative authority to trade or industrial associations or groups so as to empower them to enact the laws they deem to be wise and beneficent for the rehabilitation and expansion of their trade or industries? Could trade or industrial associations or groups be constituted legislative bodies for that purpose because such associations or groups are familiar with the problems of their enterprises? And, could an effort of that sort be made valid by such a preface of generalities as to permissible aims as we find in section 1 of title I? The answer is obvious. Such a delegation of legislative power is unknown to our law and is utterly inconsistent with the constitutional prerogatives and duties of Congress. Schechter Corp. v. United States, 295 U.S. 495, 537 (1934).

In a later case the Court made a similar declaration:

The power conferred upon the majority [of the industry to establish binding wage and hour laws] is, in effect, the power to regulate the affairs of an unwilling minority. This is legislative delegation in its most obnoxious form; for it is not even delegation to an official or an official body, presumptively disinterested, but to private persons whose interests may be and often are adverse to the interests of others in the same business. The record shows that the conditions of competition differ among the various localities. . . . The difference between producing coal and regulating its production is, of course, fundamental. The former is a private activity; the latter is necessarily a governmental function, since, in the very nature of things, one person may not be entrusted with the power to regulate the business of another, especially a competitor. . . . The delegation is so clearly arbitrary, and so clearly a denial of rights safeguarded by the due process clause of the Fifth Amendment. . . . Carter v. Carter Coal Co., 298 U.S. 238, 311 (1935).

No one would contend that these cases are "on all fours" with the case before us. In the NRA cases the Court was concerned with a direct, statutory delegation of decision making and regulatory power by Congress to an industry; here the "delegation" resulted from the FCC's capitulation to the sheer power of the industry, and does not involve continuing regulatory responsibility. Further, these cases have been overruled on many other grounds, and it is difficult to say with certainty that this aspect of the cases is as vital today as it was in 1935, even though they have never been overruled on these grounds. But the fact remains that the Supreme Court has addressed the underlying issues present here and has found the procedures wanting.

The very existence of this compromise, and the fact that as a practical matter the Commission was obliged to either accept it in its entirety or not at all (with the necessary result of eliminating the prospects of any cable for months or years), made the act of putting out the rules based on this compromise as a Further Notice of Proposed Rule Making for public comment an exercise in futility. I tried to offer modest revisions of some of the compromise provisions to make them a wee bit more palatable; Chairman Burch would not budge. It was fait accompli or nothing. It would have been hypocrisy in the extreme to solicit comments suggesting changes we were not free to make. The only question that we, as Commissioners, had to decide, was whether we

were willing to sacrifice a fundamental value of a democratic society--
the independence of government officials from the influence of big
business--in exchange for some cable television. The majority concluded
that it was in the public interest to do so. I could not. No amount
of comment could expand our ability to resolve this fundamental juris-
prudential question, and asking for public comment would have been nothing more than a cheap attempt to camouflage what, in my view, is a fatal
flaw in our procedure.

V.   Conclusion: The Politics of Cable

In view of the fact that the FCC has, in effect, abandoned its role
as the formulator of policy and the interpreter of law for that of the
political pundit, perhaps I am obliged to engage in a little political
comment myself.

The wisdom and validity of the FCC's acceptance of industry rules
in place of its own turns on one issue--accepting the majority's interpretation. Put most bluntly, had we held firm to the August 5 policy,
could we have brought it off? The majority thinks not; I think we could
have done it.

I say "accepting the majority's interpretation" because it is, itself, a questionable assumption. The majority is saying, in effect,
that a regulatory commission must consider not just the legitimate interests of all parties but also their political power. Its responsibility, says the FCC, goes beyond simply finding and promulgating the
policy most "in the public interest." It must also consider the power
of any of the powers before it to use political influence with the
White House or Congress to render its policy ineffective.

The contrary position, of course is that a regulatory commission
should simply declare the policy as it sees it and let the chips fall
where they may in terms of subsequent actions by Congress, White House,
or courts. (One might observe, for example, that the FCC has seemingly

225

given little consideration in recent months to the likelihood that its decisions might be overturned by the courts.)

Since the latter position seems to have few adherents, I will simply offer it without stating a personal preference, and proceed to taking on the majority on its own ground. What were the politics of the August 5 policy?

Chairman Burch at one point declared to a House Committee that we could have a cable policy by the end of May 1971. Hearings on Federal Communications Commission Activities (1971) before the Subcommittee on Communications and Power of the House Committee on Interstate and Foreign Commerce, 92nd Cong., 1st Sess., ser. 92-8 at 20 (1971). (This was later changed to August 5, December 31, March 1, 1972, and finally the date selected, March 31--which ultimately may have to be extended for petitions for reconsideration.) That declaration prompted an immediate reaction from broadcasters, pressuring their Senators to hold up the policy one way or another. The Senators, in turn, communicated their constituent problems to Senator Pastore, Chairman of the Subcommittee on Communications of the Senate Commerce Committee. Hearings on Community Antenna Television Problems before the Subcommittee on Communications of the Senate Committee on Commerce, 92nd Cong., 1st Sess., ser. 92-12 at 1-2 (1971). Senator Pastore, for whatever reasons, called the FCC before his Subcommittee in June 1971. At that time Chairman Burch outlined the substance of what became the August 5 policy. Senator Pastore indicated his desire to know the details of the policy before it was released. Senate Hearings at 107. Commissioner Bartley and I complained on the record that this was contributing to the delay sought by the broadcasters. Senate Hearings at 72 and 107. Chairman Burch's testimony seemed to Commissioner Bartley and me to be an adequate preview of the policy for Congress. Indeed, I argued within the Commission at the time that even that testimony may have been going too far. (My own view is that Congress established the FCC to formulate communications policy, and that, in general, it ought to leave it alone to do its job, subject to two exceptions: general "oversight" hearings to review what the agency has done after the fact, and subject matter legislative hearings that necessarily preempt the FCC's authority to act on the issue

under review. This was neither. This is a view which Senator Hart supported during the Hearing. <u>Senate Hearings</u> at 57-58.)

Even accepting for sake of argument that the FCC is obliged to comply with every Congressman's every wish, it seemed to me that our participation in the hearing had achieved that purpose. Chairman Burch further promised that the Committee could get an advance look at the final policy (which I also felt to be unnecessary), and that the policy would be out before Congress adjourned (August 5, which I felt to be later than necessary). In no event do I think Senator Pastore's requests (for the hearing, and for the advance look at the policy) required that the August 5 policy be issued in anything other than final form. And so it was that I, once again, protested the additional delay when Chairman Burch indicated to his fellow Commissioners that the August 5 policy was not going to be issued as final rule making, but as some kind of an unprecedented "letter" to the Chairmen of the Senate and House Communications Subcommittees. In any event, at that time we were promising the policy would be finally issued by December 31, 1971.

The question is, what would have happened had we issued that August 5 policy as final rule making sometime between August 5 and December 31? Bear in mind that those who voted for it on August 5 felt morally obliged to stick with it, notwithstanding the fact that each of us had some misgivings about various parts of the document. Bear in mind also that Commissioner Robert Wells, the only Commissioner not to vote for the policy, had left; Commissioner Wiley, who took his place, and Commissioner Reid, who replaced Commissioner Houser, might well have voted for the August 5 policy (based upon their votes and opinions today).

We had discussed the policy in open hearings with both sides of Congress. We had given them the document in advance, in effect, with the August 5 letter. No Senator or Congressman could have made any resonable argument that he was caught unaware, or that more time was necessary to evaluate the matter. (Indeed, Senator Pastore was on record as hoping the policy would <u>not</u> change: "I hope we don't end up with one resolution and then have to chase another idea, because that has happened time and time again." <u>Senate Hearings</u> at 37).

Most significantly, Chairman Burch would have been going forward with a unanimous (or, at worst, nearly unanimous) Commission—something he clearly doesn't have for his current industry policy. He and I, and the others, would be declaring to Congress, the industry, and the public, with a single harmonious voice, that we were in agreement on a policy that was, indeed, in the public interest.

No dissatisfied industry spokesmen could have argued to us, or to Congress, that they had not had an adequate opportunity to be heard—fully and fairly. Our 1971 hearings were widely known to have been among the best in the agency's history.

As for national Presidential politics, our rules make absolutely no sense at all given the current state of our economy. The installation of cable systems in our largest cities would require capital expenditures in the millions of dollars. Thousands of people would be put to work building the facility, laying the cable and making the connections to the subscribing homes. In short, cable could provide a shot in the arm for our ailing economy where it is needed most—our cities. If our sole purpose for taking this action is to protect broadcasters and copyright holders, it would be far more beneficial to all concerned simply to subsidize them directly, perhaps from the taxes paid by cable systems, than to deprive the people of our major cities of both the economic growth and the technological development that cable could bring. Politics involves more than campaign contributions from the wealthy, and media exposure by broadcasters. It also involves the ability to marshall evidence of having done something for the people. How can the FCC's decision possibly be squared with the President's recent State of the Union message?

> We also will help meet our goal of full employment in peacetime with a set of major initiatives to stimulate more imaginative use of America's great capacity for technological advance, and to direct it toward improving the quality of life for every American.
>
> In reaching the moon, we demonstrated what miracles American technology is capable of achieving. Now the time has come to move more deliberately toward making full use of that technology here on earth, in harnessing the wonders of science to the service of man. 118 Cong. Rec. H 146-47 (daily ed. January 20, 1972).

The only miracle with cable technology is that it still exists at all.

No one, of course, can <u>know</u> what is going to happen to any policy in Washington. One often suspects that "D.C." stands for the Delay Capital of the world. Broadcasters and copyright owners (and possibly even some cable operators) would have attempted to stop the policy. So what's new? They are trying to stop today's so-called "consensus" policy, too--giving further proof to the fact that there just ain't no such thing as a consensus between <u>all</u> the economic interests that are involved in this policy (as distinguished from those segments of industry represented at the closed White House meetings with Chairman Burch). What we're engaged in is predictions, game theory. So that's why I put all the chess men on the board. And when I look at them, and consider all the plays I've watched (and participated in) during the past 10 years in this town, what I <u>think</u> would have happened is that--after a few abortive phone calls and letters from the Hill, a threatened White House "task force," and some faulty court suits--the August 5 policy would have become the law of the land.

And that, at least, is a good deal more than the likelihood of a lived-happily-ever-after ending for the policy we're throwing up on the table today.

NEWS   Federal Communications Commission

1919 M Street, NW.

Washington, D.C. 20554

Public Notice

                              FOR RELEASE
                              12:00 NOON
                              CENTRAL DAYLIGHT TIME
                              MAY 15, 1972

NATIONAL CABLE TELEVISION ASSOCIATION

Chicago, Illinois

May 15, 1972

                              REMARKS BY:

                              DEAN BURCH, CHAIRMAN
                              Federal Communications Commission

Every time we get together, it seems to be the first day of the Year One. There is always some major milestone just behind us or just ahead. Today is no exception.

(Particularly today, as a matter of fact. The NCTA itself is marking a change in dynasty. It's been a pleasure doing business with John Gwin; I anticipate the same with Bill Bresnan and David Foster. All three, it seems to me, represent cable's own best image: equal measures of energy, imagination and responsibility--with a little prayer thrown in. Clearly, the NCTA is alive and well.)

A year ago, the Commission was hammering out its Letter of Intent to the Congress. The essence of that document was an "adequate service formula", tailored to market size, that would permit just enough broadcast signal carriage to give cable a leg up--that would at the same time hold the impact on the existing broadcast system to a tolerable minimum--and that would quite deliberately keep cable lean and mean. We were seeking, in effect, a "competitive entry" formula. And the only guarantee we offered anyone, cable or broadcasting, was an honest chance to serve the public.

The core of our policy was to make cable's evolution dependent on those nonbroadcast, broadband services that are unique to cable technology. It still is the core of our policy.

A year ago, I said to you at your Washington meeting that now the ball is in cable's court--that it's up to you whether cable is going to be just another way of moving broadcast signals around (hardly worth the ulcers involved) or whether it is going to become a genuinely new and competitively different medium of communications, offering everything from entertainment and sports and movies to classroom instruction and commercial services and public "rap" sessions. That was up to you. It still is.

Now ten months later, the Letter of Intent represents the Dead Sea Scrolls of cable—simply the first draft of the Bible. The Consensus Agreement was entered into, and in my view is binding on all the principals. The Cable Television Report and Order, the true gospel, is in force. Round One of reconsideration involving petitions for stay is behind us. Round Two is close to completion.

(And without presuming on the other Commissioners' prerogatives, I do not now anticipate radical revisions in the March 31 rules.) The "freeze" and "regulatory lag" are, hopefully, nothing but reminders of the old days, gone forever.

It is not, of course, quite that simple. That there will be a cable industry in the nation's future seems to me no longer open to question. Competitive coexistence is a fact of life for you and broadcasting. But there are bugs in the new rules—some of them the size of elephants—and the precise shape of that future is dim. It is filled with imponderables.

Cable copyright legislation fits under both the "bug" and the "imponderable" category. Ending open warfare and moving to the negotiating table, and ultimately to legislation, was the whole thrust behind the Consensus Agreement. It was a reasonable compromise for which no apologies are necessary. That was why the Commission decided explicitly to implement its key provisions in the new rules. That is why I continue to call for hard, good faith bargaining on a fee schedule that all the parties can live with.

I haven't been in on any of the sessions, I don't intend to get directly involved, and I'm not going to engage in any speculation about "who shot John". I'm not even going to speculate about the effects of the District Court decision in CBS v. TelePrompTer. You can read it many ways. But I do say—not for the first time and not for the last—that unless and until cable is brought within the competitive program distribution market and pays a fair price for the product it uses, you

232

will be operating under a cloud and on shaky foundations. And that is no way to build an industry.

Turning to another area of perplexity, the Commission opted for dual Federal/State-local jurisdiction because cable clearly calls some of both (and maybe all three) into play. We are attempting to integrate cable into a national communications structure. At the same time, it is clearly a local "convenience and necessity"--in both the technical and substantive senses of that phrase. We left the line of demarcation hazy because, quite candidly, we were not yet ready to draw it with precision. That had to be left to experience and experimentation. All of which leaves you in something of a no-man's-land. (Here in Illinois, as in a few other states where even intrastate jurisdiction is up for grabs, it's virtually a "free fire zone".)

This afternoon we will convene the steering group of a Federal/State Advisory Committee--along with one on Technical Standards--and, hopefully, these committees will help in time to reduce the ambiguities. But there is one way that all of you can help right now--and that is by _not_ playing games with the spirit of the rules. The same goes for state and local franchising authorities.

The rules set out basic access requirements, as you know--one public channel in perpetuity, and an educational and a governmental channel free for an initial five-year experimental period. The rules also suggest reasonable franchise fees--in the range of three to five percent of gross revenues. Deliberately, the rules do not tie down all the loose ends. That would freeze cable into a single mold, and it would represent a case of "decision first, evidence later". (We'll permit special showings and grant special exceptions from the start, but only on the basis of compelling presentations that the public will benefit.)

The obvious danger--and we can see it coming even if we're not sure what to do about it--is that both free-channel bonanzas and excessive franchise fees are likely to become bargaining chips in the bidding _and_

233

franchising process. The tempation will be "promise now", whatever it may take, and worry later about the consequences. But, as usual, it will be the public who pays--either in the form of cross-subsidy from subscriber fees, or loading the burdens on cable to the point that the bird never will fly.

Don't count on the Commission to bail you out. Indeed, I cannot urge too strongly--and again the message is directed as much to state and local franchising authorities as it is to cable operators--don't <u>force</u> the Commission to bail you out. Game-playing, whether in the form of over-promise or under-performance, is not the mark of a responsible industry. And the simple solution is "don't play games".

I'm frank to confess that the rules are terribly complicated. Cable begins its new era in a regulatory maze. Because cable and broadcasting are so closely linked--in carriage and exclusivity rules, grandfathering, leapfrogging, viewing standards, you name it--I doubt that it could have been otherwise. The access requirements add another element of complexity. All we can promise is to be reasonable in applying the rules--and prepared to revisit them as we accumulate experience. At least we all begin life equal; we're as perplexed as you are.

The biggest of the imponderables, quite candidly, is that we don't yet know if the rules will "work" at all. On the other hand, it may turn out that even the very limited amount of distant signal importation we're permitting will have unforeseen impact on over-the-air broadcasting. On the other hand, in a abundance of caution, we may have kept cable not so much "lean and hungry" as starved to death.

Nor do we really know how the exclusivity rules will work out in practice. They may leave little programming available to cable beyond tired re-runs and Charlie Chan movies. The leapfrogging rules may give cable in some parts of the country not much in the way of attractive distant independents to choose from. The access requirements may turn out to be one big public yawn--and the experimental channels may languish for want of imagination on the part of educational and governmental entities.

234

I've been deliberately laying out the worst possible prognosis, vastly overstated on the side of pessimism. I hope and expect that it won't come to pass. But the blunt fact is: we're not yet certain.

All I do know is that we are as anxious to find out as you are. We want cable to get moving--and that means we want the rules to work as intended. Otherwise it's two long, arduous years down the tube--and the great promise of cable, all that it may mean for new horizons of public service, still just a promise.

I want to stress that the rulebook is not carved in stone. Except for the complement of signals to which compulsory licenses will be applicable, it is open to change and refinement. We retain all our options.

Another thing about which I feel very confident is that the core of our policy is sound; that cable's future is and ought to be tied to those nonbroadcast services that are, in the last analysis, what cable is all about.

Last year I ran down the list of what some of these broadband services might be--indeed, what they already are in certain communities. I don't intend today to repeat that drill. Certainly I'm not going to stand up here and lecture you about the potentialities of your own industry.

There is no Burch Plan for the guaranteed future of cable television. For that matter, there can be no NCTA plan either. But I feel confident--for each community, for the New Yorks and Chicagos, as well as the East Overshoes and Splitlips--there is some combination of services that will get cable under way, sell it and establish it as a permanent feature of the landscape. Your job, and mine, is to find and effectuate that winning combination.

If we're able to do that, then the rules will work, bugs and all. And the Commission's effort will have paid off. The matter was expressed rather well, if I do say so myself, in the Cable Television Report and Order:

For more than three years we have been gathering data, soliciting
views, hearing arguments, evaluating studies, examining alternatives, authorizing experiments--turning finally to public panel
discussions unique in communications rule making--and, in this
effort, have necessarily postponed the substantial public benefits
that cable promises. In these circumstances, we do not see that
there can be any case for further delay. It is time to act.

We did act, and our commitment is for the long haul. The rest, ladies
and gentlemen, is up to you.

UNITED STATES et al., Petitioners,

v

MIDWEST VIDEO CORPORATION

— US —, 32 L Ed 2d 390, 92 S Ct —

[No. 71-506]

Argued April 19, 1972.   Decided June 7, 1972.

SUMMARY

The Federal Communications Commission adopted a rule providing that no cable television system having 3,500 or more subscribers shall carry the signal of any television broadcast station unless the system also operates to a significant extent as a local outlet by cablecasting and has available facilities for local production and presentation of programs other than automated services (20 FCC2d 201).   An operator of cable television systems with more than 3,500 subscribers petitioned for review in the United States Court of Appeals for the Eighth Circuit, which set aside the order (441 F2d 1322).

On certiorari, five members of the United States Supreme Court, although not agreeing on an opinion, agreed that the judgment be reversed.

BRENNAN, J., announced the judgment of the court, and in an opinion joined by WHITE, MARSHALL, and BLACKMUN, JJ., expressed the view that the FCC's rule was valid as reasonably ancillary to the effective performance of its various responsibilities for the regulation of television broadcasting.

BURGER, Ch. J., concurring in the result, expressed the view that while the FCC's position strained the outer limits of its jurisdiction, it should be allowed wide latitude.

DOUGLAS, J., joined by STEWART, POWELL, and REHNQUIST, JJ., dissented on the ground that the FCC lacks authority to compel cable television systems to originate programs.

HEADNOTES

Classified to U. S. Supreme Court Digest, Annotated

Communications § 15 — cable television — program origination

1. A Federal Communications Commission rule providing that no cable television system having 3,500 or more subscribers shall carry the signal of any television broadcast station unless the system also operates to a significant extent as a local outlet by cablecasting and has available facilities for local production and presentation of programs other than automated services is valid as reasonably ancillary to the effective performance of the FCC's various responsibilities for the regulation of television broadcasting.

Communications § 15 — FCC jurisdiction — cablecasting

2. The devotion of cable television systems to broadcast transmission—together with the interdependencies between that service and cablecasts, and the necessity for unified regulation—suffices to bring cablecasts within the jurisdiction of the Federal Communications Commission under § 2(a) of the Communications Act of 1934 (47 USCS § 152(a)), which makes the Act applicable to all interstate and foreign communication by wire or radio and all interstate and foreign transmission of energy by radio.

Points from Separate Opinions

Communications § 15 — cable television — rules

3. Federal Communications Commission regulations are valid in requiring cable television systems (1) to carry, upon request and in a specified order of priority within the limits of their channel capacity, the signals of broadcast stations into whose service area they bring competing signals; (2) to avoid, upon request, the duplication on the same day of local station programming; and (3) to refrain from bringing new distant signals into the 100 largest television markets except upon a prior showing that that service will be consistent with the public interest. [From separate opinion by Brennan, White, Marshall, and Blackmun, JJ.]

TOTAL CLIENT-SERVICE LIBRARY® REFERENCES

Am Jur 2d, Telecommunications (1st ed, Radio and Television § 4)

US L Ed Digest, Communications § 15

ALR Digests, Radio and Television § 4

L Ed Index to Anno, Federal Communications Commission; Radio and Television

ALR Quick Index, Radio and Television

Federal Quick Index, Federal Communications Commission; Radio and Television

ANNOTATION REFERENCES

Licensing and control of telecast facilities. 95 L Ed 1075.

Validity and construction of municipal ordinances regulating community antenna television service (CATV). 41 ALR3d 384.

Legal aspects of television. 15 ALR2d 785.

Points from Separate Opinions—Continued

Communications § 15 — cable television — regulatory authority

4. The Federal Communications Commission's power to regulate cable television is not limited to controlling the competitive impact that cable television may have on broadcast services, but extends also to requiring cable television affirmatively to further statutory policies. [From separate opinion by Brennan, White, Marshall, and Blackmun, JJ.]

Communications § 15 — television regulation — goals

5. The goals of increasing the number of outlets for community self-expression and augmenting the public's choice of programs and types of services are within the Federal Communications Commission's mandate for the regulation of television broadcasting. [From separate opinion by Brennan, White, Marshall, and Blackmun, JJ.]

Evidence § 961 — sufficiency — FCC rule

6. A Federal Communications Commission rule requiring cable television systems with 3,500 or more subscribers to originate programs is supported by substantial evidence that it will promote the public interest where the 3,500-subscribers standard encompasses less than 10 percent of the existing cable television systems; there is no data tending to demonstrate that systems with 3,500 subscribers cannot cablecast without impairing their financial stability, raising rates, or reducing the quality of service; all systems may apply for ad hoc waiver of the rule; and systems with less than 10,000 subscribers are excused from originating programs pending action on their waiver requests. [From separate opinion by Brennan, White, Marshall, and Blackmun, JJ.]

SYLLABUS BY REPORTER OF DECISIONS

The Federal Communications Commission (FCC) promulgated a rule that "no CATV system having 3,500 or more subscribers shall carry the signal of any television broadcast station unless the system also operates to a significant extent as a local outlet by cablecasting [i.e., originating programs] and has available facilities for local production and presentation of programs other than automated services." Upon challenge of respondent, an operator of CATV systems subject to the new requirement, the Court of Appeals set aside the regulation on the ground that the FCC had no authority to issue it. *Held:* The judgment is reversed.

441 F2d 1322, reversed.

Mr. Justice Brennan, joined by Mr. Justice White, Mr. Justice Marshall, and Mr. Justice Blackmun, concluded that:

1. The rule is within the FCC's statutory authority to regulate CATV at least to the extent "reasonably ancillary to the effective performance of the Commission's various responsibilities for the regulation of television broadcasting." United States v Southwestern Cable Co., 392 US 157, 178, 20 L Ed 2d 1001, 1016, 88 S Ct 1994.

2. In the light of the record in this case, there is substantial evidence that the rule, with its 3,500 standard and as it is applied under FCC guidelines for waiver on a showing of financial hardship, will promote the public interest within the meaning of the Communications Act of 1934.

The Chief Justice concluded that until Congress acts to deal with the problems brought about by the emergence of CATV, the FCC should be allowed wide latitude.

Brennan, J., announced the Court's judgment and delivered an opinion in which White, Marshall, and Blackmun, JJ., joined. Burger, C. J., filed an opinion concurring in the result. Douglas, J., filed a dissenting opinion, in which Stewart, Powell, and Rehnquist, JJ., joined.

APPEARANCES OF COUNSEL

Lawrence G. Wallace argued the cause for petitioners.

Harry M. Plotkin argued the cause for respondent.

OPINION OF THE COURT

Mr. Justice **Brennan** announced the judgment of the Court and an opinion in which Mr. Justice **White,** Mr. Justice **Marshall,** and Mr. Justice **Blackmun** joined.

Community antenna television (CATV) was developed long after the enactment of the Communications Act of 1934, 48 Stat 1064, as amended, 47 USC § 151, as an auxiliary to broadcasting through the transmission of radio signals by wire to viewers otherwise unable to receive them because of distance or local terrain.[1] In United States v Southwestern Cable Co. 392 US 157, 20 L Ed 2d 1001, 88 S Ct 1994 (1968), where we sustained the jurisdiction of the Federal Communications Commission to regulate the new industry at least to the extent "reasonably ancillary to the effective performance of the Commission's various responsibilities for the regulation of television broadcasting," id., at 178, 20 L Ed 2d at 1016, we observed that the growth of CATV since the establishment of

the first commercial system in 1950 has been nothing less than "'explosive.'" Id., at 163, 20 L Ed 2d at 1008.[2] The potential of the new industry to augment communication services now available is equally phenomenal.[3] As we said in Southwestern, id., at 164, 20 L Ed 2d at 1008, CATV "[promises] for the future to provide a national communications system, in which signals from selected broadcasting centers would be transmitted to metropolitan areas throughout the country." Moreover, as the Commission has noted, "the expanding multichannel capacity of cable systems could be utilized to provide a variety of new communications services to homes and businesses within a community," such as facsimile reproduction of documents, electronic mail delivery, and information retrieval. Notice of Proposed Rulemaking and Notice of Inquiry, 15 FCC2d 417, 419–420 (1968). Perhaps more important, CATV systems can themselves originate pro-

---

1. "CATV systems receive the signals of television broadcasting stations, amplify them, transmit them by cable or microwave, and ultimately distribute them by wire to the receivers of their subscribers." United States v Southwestern Cable Co., 392 US 157, 161, 20 L Ed 2d 1001, 1007, 88 S Ct 1994 (1968). They "perform either or both of two functions. First, they may supplement broadcasting by facilitating satisfactory reception of local stations in adjacent areas in which such reception would not otherwise be possible; and second, they may transmit to subscribers the signals of distant stations entirely beyond the range of local antennae." Id., at 163, 20 L Ed 2d at 1008.

2. There are now 2,678 CATV systems in operation, 1,916 CATV franchises outstanding for systems not yet in current operation, and 2,804 franchise applications pending. Weekly CATV Activity Addenda, 12 Television Digest, at 9 (Feb. 28, 1972).

3. For this reason the Commission has recently adopted the term "cable television" in place of CATV. See Report and Order on Cable Television Service; Cable Television Relay Service, 37 Fed Reg 3252 n 9 (1972) (hereinafter cited as Report and Order on Cable Television Service).

grams, or "cablecast"—which means, the Commission has found, that CATV can "[increase] the number of local outlets for community self-expression and [augment] the public's choice of programs and types of services, without use of broadcast spectrum . . . ." Id., at 421.

Recognizing this potential, the Commission, shortly after our decision in Southwestern, initiated a general inquiry "to explore the broad question of how best to obtain, consistent with the public interest standard of the Communications Act, the full benefits of developing communications technology for the public, with particular immediate reference to CATV technology . . . ." Id., at 417. In particular, the Commission tentatively concluded, as part of a more expansive program for the regulation of CATV,[4] "that, for now and in general, CATV program origination is in the public interest," id., at 421, and sought comments on a proposal "to condition the carriage of television broadcast signals (local or distant) upon a requirement that the CATV system also operate to a significant extent as a local outlet by originating." Id., at 422. As for its authority to impose such a requirement, the Commission stated that its "concern with CATV carriage of broadcast signals is not just a matter of avoidance of adverse effects, but extends also to requiring CATV affirmatively to further statutory policies." Ibid.

On the basis of comments received, the Commission on October 24, 1969, adopted a rule providing that "no CATV system having 3,500 or more subscribers shall carry the signal of any television broadcast station unless the system also operates to a significant extent[5] as a local outlet by cablecasting[6] and has

---

4. The early regulatory history of CATV, canvassed in Southwestern, need not be repeated here, other than to note that in 1966 the Commission adopted rules, applicable to both microwave and non-microwave CATV systems, to regulate the carriage of local signals, the duplication of local programing, and the importation of distant signals into the 100 largest television markets. See p —, 32 L Ed 2d p 397, infra. The Commission's 1968 notice of proposed rulemaking addressed, in addition to the program origination requirement at issue here, whether advertising should be permitted on cablecasts and whether the broadcast doctrines of "equal time," "fairness," and sponsorship identification should apply to them. Other areas of inquiry included the use of CATV facilities to provide common carrier service; federal licensing and local regulation of CATV; cross-ownership of television stations and CATV systems; reporting and technical standards; and importation of distant signals into major markets. The notice offered concrete proposals in some of these areas, which were acted on in the Commission's First Report and Order, 20 FCC 2d 201 (1969) (hereinafter cited as First Report and Order), and Report and Order on Cable Television Service. See also Memorandum Opinion and Order, 23 FCC 2d 825 (1970) (hereinafter cited as Memorandum Opinion and Order). None of these regulations, aside from the cablecasting requirement, is now before us, see n 14, infra, and we, of course, intimate no view on their validity.

5. "By significant extent [the Commission indicated] we mean something more than the origination of automated services (such as time and weather, news ticker, stock ticker, etc.) and aural services (such as music and announcements). Since one of the purposes of the origination requirement is to insure that cablecasting equipment will be available for use by others originating on common carrier channels, 'operation to a signficant extent as a local outlet' in essence necessitates that the CATV operator have some kind of video cablecasting system for the production of local live and delayed programing (e.g., a camera and a video tape recorder, etc.)." First Report and Order 214.

6. "Cablecasting" was defined as "pro-

available facilities for local production and presentation of programs other than automated services." 47 CFR § 74.1111(a).[7] In a report accompanying this regulation, the Commission stated that the tentative conclusions of its earlier notice of proposed rulemaking: "recognize the great potential of the cable technology to further the achievement of long-established regulatory goals in the field of television broadcasting by increasing the number of outlets for community self-expression and augmenting the public's choice of programs and types of services . . . . They also reflect our view that a multi-purpose CATV

operation combining carriage of broadcast signals with program origination and common carrier services,[8] might best exploit cable channel capacity to the advantage of the public and promote the basic purpose for which this Commission was created: 'regulating interstate and foreign commerce in communication by wire and radio so as to make available, so far as possible, to all people of the United States a rapid, efficient, nationwide, and worldwide wire and radio communication service with adequate facilities at reasonable charges . . .' (sec. 1 of the Communications Act).[9] After

---

gramming distributed on a CATV system which has been originated by the CATV operator or by another entity, exclusive of broadcast signals carried on the system." 47 CFR § 74.1101(j). As this definition makes clear, cablecasting may include not only programs produced by the CATV operator, but "films and tapes produced by others, and CATV network programing." First Report and Order 214. See also id., at 203. The definition has been altered to conform to changes in the regulation, see n 7, infra, and now appears at 47 CFR § 76.5(w). See Report and Order on Cable Television Service 3279. Although the definition now refers to programing "subject to the exclusive control of the cable operator," this is apparently not meant to effect a change in substance or to preclude the operator from cablecasting programs produced by others. See id., at 3271.

7. This requirement, applicable to both microwave and nonmicrowave CATV systems without any "grandfathering" provision, was originally scheduled to go into effect on January 1, 1971. See First Report and Order 223. On petitions for reconsideration, however, the effective date was delayed until April 1, 1971, see Memorandum Opinion and Order 827, 830, and then, after the Court of Appeals decision below, suspended pending final judgment here. See 36 Fed. Reg. 10876 (1971). Meanwhile, the regulation has been revised and now appears at 47 CFR § 76.201(a). The revision has no significance for this case. See Memorandum Opinion and

Order 827, 830 (revision effective Aug. 14, 1970); Report and Order on Cable Television Service 3271, 3277, 3287 (revision effective March 31, 1972).

8. Although the Commission did not impose common carrier obligations on CATV systems in its 1969 report, it did note that "the origination requirement will help ensure that origination facilities are available for use by others originating on leased channels." First Report and Order 209. Public access requirements were introduced in the Commission's Report and Order on Cable Television Service, although not directly under the heading of common carrier service. See Report and Order on Cable Television Service 3277.

9. Section 1 of the Act, 48 Stat 1064, as amended, 47 USC § 151, states:

"For the purpose of regulating interstate and foreign commerce in communication by wire and radio so as to make available, so far as possible, to all the people of the United States a rapid, efficient, Nation-wide, and world-wide wire and radio communication service with adequate facilities at reasonable charges, for the purpose of the national defense, for the purpose of promoting safety of life and property through the use of wire and radio communication, and for the purpose of securing a more effective execution of this policy by centralizing authority heretofore granted by law to several agencies and by granting additional authority with respect to interstate and foreign commerce in wire and radio communication, there is created a commission to be known as the

full consideration of the comments filed by the parties, we adhere to the view that program origination on CATV is in the public interest."[10] First Report and Order, 20 FCC2d 201, 202 (1969).

The Commission further stated, id., at 208–209:

"The use of broadcast signals has enabled CATV to finance the construction of high capacity cable facilities. In requiring in return for these uses of radio that CATV de-

vote a portion of the facilities to providing needed origination service, we are furthering our statutory responsibility to 'encourage the larger and more effective use of radio in the public interest' (§ 303(g)).[11] The requirement will also facilitate the more effective performance of the Commission's duty to provide a fair, efficient, and equitable distribution of television service to each of the several States and communities (§ 307(b)),[12] in areas where we have been unable to accomplish this through broadcast media."[13]

'Federal Communications Commission' which shall be constituted as hereinafter provided, and which shall execute and enforce the provisions of this chapter."

10. In so concluding, the Commission rejected the contention that a prohibition on CATV originations was "necessary to prevent potential fractionalization of the audience for broadcast services and a siphoning off of program material and advertising revenue now available to the broadcast service." First Report and Order 202. "[B]roadcasters and CATV originators . . . ," the Commission reasoned, "stand on the same footing in acquiring the program material with which they compete." Id., at 203. Moreover, "a loss of audience or advertising revenue to a television station is not in itself a matter of moment to the public interest unless the result is a net loss of television service," ibid.—an impact that the Commission found had no support in the record and that, in any event, it would undertake to prevent should the need arise. See id., at 203–204. See also Memorandum Opinion and Order 826 n 3, 828–829.

11. Section 303(g), 48 Stat 1082, 47 USC § 303(g), states that "[e]xcept as otherwise provided in this chapter, the Commission from time to time, as public convenience, interest, or necessity requires, shall" "[s]tudy new uses for radio, provide for experimental uses of frequencies, and generally encourage the larger and more effective use of radio in the public interest . . . ."

12. Section 307(b), 48 Stat 1084, as amended, 47 USC § 307(b), states:

"In considering applications for licenses [for the transmission of energy, communications, or signals by radio], and

modifications and renewals thereof, when and insofar as there is demand for the same, the Commission shall make such distribution of licenses, frequencies, hours of operation, and of power among the several States and communities as to provide a fair, efficient, and equitable distribution of radio service to each of the same."

13. The Commission added: "[I]n authorizing the receipt, forwarding, and delivery of broadcast signals, the Commission is in effect authorizing CATV to engage in radio communication, and may condition this authorization upon reasonable requirements governing activities which are closely related to such radio communication and facilities." First Report and Order 209 (citing, inter alia, § 301 of the Communications Act, 48 Stat 1081, 47 USC § 301 (generally requiring licenses for the use or operation of any apparatus for the interstate or foreign transmission of energy, communications, or signals by radio)). Since, as we hold, infra, the authority of the Commission recognized in Southwestern is sufficient to sustain the cablecasting requirement at issue here, we need not, and do not, pass upon the extent of the Commission's jurisdiction over CATV under § 301. See e.g., FCC v Pottsville Broadcasting Co., 309 US 134, 138, 84 L Ed 656, 659, 60 S Ct 437 (1940); General Telephone Co. of Cal. v FCC, 413 F2d 390, 404–405 (CADC 1969); Philadelphia Television Broadcasting Co. v FCC, 359 F2d 282, 284 (CADC 1966): "In a statutory scheme in which Congress has given an agency various bases of jurisdiction and various tools with which to protect the public interest, the agency is entitled to some leeway in

[1] Upon the challenge of respondent Midwest Video Corporation, an operator of CATV systems subject to the new cablecasting requirement, the United States Court of Appeals for the Eighth Circuit set aside the regulation on the ground that the Commission "is without authority to impose" it. 441 F2d 1322, 1328 (1971).[14] "The Commission's power [over CATV] . . . ," the court explained, "must be based on the Commission's right to adopt rules that are reasonably ancillary to its responsibilities in the broadcasting field," id., at 1326—a standard that the court thought the Commission's regulation "goes far beyond." Id., at 1327.[15] The court's opinion may also be understood to hold the regulation invalid as not supported by substantial evidence that it would serve the public interest. "The Commission report itself shows," the court said, "that upon

the basis of the record made, it is highly speculative whether there is sufficient expertise or information available to support a finding that the origination rule will further the public interest." Id., at 1328. "Entering into the program origination field involves very substantial expenditures," id., at 1327, and "[a] high probability exists that cablecasting will not be self-supporting," that there will be a "substantial increase" in CATV subscription fees, and that "in some instances" CATV operators will be driven out of business. Ibid.[16] We granted certiorari. 404 US 1014, 30 L Ed 2d 661, 92 S Ct 676 (1972). We reverse.

I

[3] In 1966 the Commission promulgated regulations that, in general, required CATV systems (1) to carry, upon request and in a specified

---

choosing which jurisdictional base and which regulatory tools will be most effective in advancing the Congressional objective."

14. Although this holding was specifically limited to "existing cable television operators," the court's reasoning extended more broadly to all CATV systems, and, indeed, its judgment set aside the regulation in all its applications. See 441 F2d, at 1328.

Respondent also challenged other regulations, promulgated in the Commission's First Report and Order and Memorandum Opinion and Order, dealing with advertising, "equal time," "fairness," sponsorship identification, and per-program or per-channel charges on cablecasts. The Court of Appeals, however, did not "[pass] on the power of the FCC . . . to prescribe reasonable rules for such CATV operators who voluntarily choose to originate programs," id., at 1326, since respondent acknowledged that it did not want to cablecast and hence lacked standing to attack those rules. See id., at 1328.

15. The court held, in addition, that the Commission may not require CATV operators "as a condition of [their] right to use . . . captured [broadcast]

signals in their existing franchise operation to engage in the entirely new and different business of originating programs." Id., at 1327. This holding presents no separate question from the "reasonably ancillary" issue that need be considered here. See n 22, infra.

16. Concurring in the result in a similar vein, Judge Gibson concluded that although "the FCC has authority over CATV systems," "the order under review is confiscatory and hence arbitrary," 441 F2d, at 1328, for the regulation "would be extremely burdensome and perhaps remove from the CATV field many entrepreneurs who do not have the resources, talent and ability to enter the broadcasting field." Id., at 1329. If this is to suggest that the regulation is invalid merely because it burdens CATV operators or may even force some of them out of business, the argument is plainly incorrect. See n 31, infra. The question would still remain whether the Commission reasonably found on substantial evidence that the regulation on balance would promote policy objectives committed to its jurisdiction under the Communications Act, which, for the reasons given infra, we hold that it did.

order of priority within the limits of their channel capacity, the signals of broadcast stations into whose service area they brought competing signals; (2) to avoid, upon request, the duplication on the same day of local station programing; and (3) to refrain from bringing new distant signals into the 100 largest television markets except upon a prior showing that that service would be consistent with the public interest. See Second Report and Order, 2 FCC2d 725 (1966). In assessing the Commission's jurisdiction over CATV against the backdrop of these regulations,[17] we focused in Southwestern chiefly on § 2(a) of the Communications Act, 48 Stat 1064, as amended, 47 USC § 152(a), which provides in pertinent part: "The provisions of this [Act] shall apply to all interstate and foreign communication by wire or radio . . . , which originates and/or is received within the United States, and to all persons engaged within the· United States in such communication . . . ." In view of the Act's definitions of "communication by wire" and "communi-

cation by radio,"[18] the interstate character of CATV services,[19] and the evidence of congressional intent that "[t]he Commission was expected to serve as the 'single Government agency' with 'unified jurisdiction' and 'regulatory power over all forms of electrical communication, whether by telephone, telegraph, cable, or radio,' " 392 US, at 167–168, 20 L Ed 2d 1010, 1011 (footnotes omitted), we held that § 2(a) amply covers CATV systems and operations. We also held that § 2(a) is itself a grant of regulatory power and not merely a prescription of the forms of communication to which the Act's other provisions governing common carriers and broadcasters apply:

"We cannot [we said] construe the Act so restrictively. Nothing in the language of § [2(a)], in the surrounding language, or in the Act's history or purposes limits the Commission's authority to those activities and forms of communication that are specifically described by the Act's other provisions. . . . Certainly Congress could not in 1934 have foreseen the development of

---

[3] 17. Southwestern reviewed, but did not specifically pass upon the validity of, the regulations. See 392 US, at 167, 20 L Ed 2d at 1010. Their validity was, however, subsequently and correctly upheld by courts of appeals as within the guidelines of that decision. See, e.g., Black Hills Video Corp. v FCC, 399 F2d 65 (CA8 1968).

18. Sections 3(a), (b), 48 Stat 1065, 47 USC § 153(a), (b), define these terms to mean "the transmission" "of writing, signs, signals, pictures, and sounds of all kinds," whether by cable or radio, "including all instrumentalities, facilities, apparatus, and services (among other things, the receipt, forwarding, and delivery of communications) incidental to such transmission."

19. "Nor can we doubt that CATV systems are engaged in interstate communication, even where . . . the inter-

cepted signals emanate from stations located within the same State in which the CATV system operates. We may take notice that television broadcasting consists in very large part of programming devised for, and distributed to, national audiences; [CATV operators] thus are ordinarily employed in the simultaneous retransmission of communications that have very often originated in other States. The stream of communication is essentially uninterrupted and properly indivisible. To categorize [CATV] activities as intrastate would disregard the character of the television industry, and serve merely to prevent the national regulation that 'is not only appropriate but essential to the efficient use of radio facilities.' Federal Radio Comm'n v Nelson Bros. Co. 289 US 266, 279 [77 L Ed 1166, 1175, 53 S Ct 627, 89 ALR 406]" 392 US, at 168–169, 20 L Ed 2d at 1011.

community antenna television systems, but it seems to us that it was precisely because Congress wished 'to maintain, through appropriate administrative control, a grip on the dynamic aspects of radio transmission,' FCC v Pottsville Broadcasting Co., [309 US], at 138, [84 L Ed at 659] that it conferred upon the Commission a 'unified jurisdiction' and 'broad authority.' Thus, '[u]nderlying the whole [Communications Act] is recognition of the rapidly fluctuating factors characteristic of the evolution of broadcasting and of the corresponding requirement that the administrative process possess sufficient flexibility to adjust itself to these factors.' [Ibid.] Congress in 1934 acted in a field that was demonstrably 'both new and dynamic,' and it therefore gave the Commission 'a comprehensive mandate,' with 'not niggardly but expansive powers.' National Broadcasting Co. v United States, 319 US 190, 219, [87 L Ed 1344, 1364, 63 S Ct 997]. We have found no reason to believe that § [2] does not, as its terms suggest, confer regulatory authority over 'all interstate . . . communication by wire or radio.' " Id., at 172–173, 20 L Ed 2d at 1013 (footnotes omitted).

This conclusion, however, did not end the analysis, for § 2(a) does not in and of itself prescribe any objectives for which the Commission's regulatory power over CATV might properly be exercised. We accordingly went on to evaluate the reasons for which the Commission had asserted jurisdiction and found that "the Commission has reasonably concluded that regulatory authority over CATV is imperative if it is to perform with appropriate effectiveness certain of its other responsibilities." Id., at 173, 20 L Ed 2d at 1014. In particular, we found that the Commission had reasonably determined that " 'the unregulated explosive growth of CATV,' " especially through "its importation of distant signals into the service areas of local stations" and the resulting division of audiences and revenues, threatened to "deprive the public of the various benefits of [the] system of local broadcasting stations" that the Commission was charged with developing and overseeing under § 307(b) of the Act.[20] Id., at 175, 20 L Ed 2d at 1015. We therefore concluded, without expressing any view "as to the Commission's authority, if any, to regulate CATV under any other circumstances or for any other purposes," that the Commission does have jurisdiction over CATV "reasonably ancillary to the effective performance of [its] various responsibilities for the regulation of television broadcasting . . . [and] may, for these purposes, issue 'such rules and regulations and prescribe such restrictions and conditions, not inconsistent with law,' as 'public convenience, interest, or necessity requires.' " Id., at 178, 20 L Ed 2d at 1016 (quoting § 303(r) of the Act, 50 Stat 191, 47 USC § 303(r)).

[1, 2] The parties now before us do not dispute that in light of Southwestern CATV transmissions are subject to the Commission's jurisdiction as "interstate . . . com-

---

20. See n 12, supra. See also § 303 (f), (h), 48 Stat 1082, 47 USC § 303(f), (h) (authorizing the Commission to prevent interference among stations and to establish areas to be served by them respectively). "In particular, the Commission feared that CATV might . . . significantly magnify the characteristically serious financial difficulties of UHF and educational television broadcasters." 392 US, at 175–176, 20 L Ed 2d at 1015.

munication by radio or wire" within the meaning of § 2(a) even insofar as they are local cablecasts.[21] The controversy instead centers on whether the Commission's program

origination rule is "reasonably ancillary to the effective performance of [its] various responsibilities for the regulation of television broadcasting."[22] We hold that it is.

---

21. This, however, is contested by the State of Illinois as amicus curiae. It is, nevertheless, clear that cablecasts constitute communication by wire (or radio if microwave transmission is involved), as well as interstate communication if the transmission itself has moved interstate, as the Commission has authorized and encouraged. See First Report and Order 207–208 (regional and national interconnections) and n 6, supra. The capacity for interstate nonbroadcast programing may in itself be sufficient to bring cablecasts within the compass of § 2(a). In Southwestern we declined to carve CATV broadcast transmissions, for the purpose of determining the extent of the Commission's regulatory authority, into interstate and intrastate components. See n 19, supra. This result was justified by the extent of interstate broadcast programing, the interdependencies between the two components, and the need to preserve the " 'unified and comprehensive regulatory system for the [broadcasting] industry.' " 392 US, at 168, 20 L Ed 2d at 1011 (quoting FCC v Pottsville Broadcasting Co., n 13, supra, at 137, 84 L Ed at 659). A similar rationale may apply here, despite the lesser "interstate content" of cablecasts at present.

But we need not now decide that question because, in any event, CATV operators have, by virtue of their carriage of broadcast signals, necessarily subjected themselves to the Commission's comprehensive jurisdiction. As Mr. Chief Justice (then Judge) Burger has stated in a related context:

"The Petitioners [telephone companies providing CATV channel distribution facilities] have, by choice, inserted themselves as links in this indivisible stream and have become an integral part of interstate broadcast transmission. They cannot have the economic benefits of such carriage as they perform and be free of the necessarily pervasive jurisdiction of the Commission." General Telephone Co. of Cal. v FCC, n 13, supra, at 401.

[?] The devotion of CATV systems to broadcast transmission—together with the interdependencies between that service and cablecasts, and the necessity for uni-

fied regulation—plainly suffices to bring cablecasts within the Commission's § 2(a) jurisdiction. See generally Barnett, State, Federal, and Local Regulation of Cable Television, 47 Notre Dame L 685, 721–723, 726–734 (1972).

22. Since "[t]he function of CATV systems has little in common with the function of broadcasters," Fortnightly Corp. v United Artists Television, Inc., 392 US 390, 400, 20 L Ed 2d 1176, 1183, 88 S Ct 2084 (1968), and since "[t]he fact that . . . property is devoted to a public use on certain terms does not justify . . . the imposition of restrictions that are not reasonably concerned with the proper conduct of the business according to the undertaking which the [owner] has expressly or impliedly assumed," Nor. Pac. Ry. v North Dakota, 236 US 585, 595, 59 L Ed 735, 741, 35 S Ct 429 (1915), respondent also argues that CATV operators may not be required to cablecast as a condition for their customary service of carrying broadcast signals. This conclusion might follow only if the program origination requirement is not reasonably ancillary to the Commission's jurisdiction over broadcasting. For, as we held in Southwestern, CATV operators are, at least to that extent, engaged in a business subject to the Commission's regulation. Our holding on the "reasonably ancillary" issue is therefore dispositive of respondent's additional claim. See pp 20–22, infra.

It should be added that Fortnightly Corp. v United Artists Television, Inc., supra, has no bearing on the "reasonably ancillary" question. That case merely held that CATV operators who retransmit, but do not themselves originate copyrighted works do not "perform" them within the meaning of the Copyright Act, 61 Stat 652, as amended, 17 USC § 1, since "[e]ssentially, [that kind of] a CATV system no more than enhances the viewer's capacity to receive the broadcaster's signals . . . ." 392 US, at 399, 20 L Ed 2d at 1183. The analogy thus drawn between CATV operations and broadcast viewing for copyright purposes obviously does not dictate the extent of the Commission's authority to regulate CATV under the Communications Act. Indeed,

[4] At the outset we must note that the Commission's legitimate concern in the regulation of CATV is not limited to controlling the competitive impact CATV may have on broadcast services. Southwestern refers to the Commission's "various responsibilities for the regulation of television broadcasting." These are considerably more numerous than simply assuring that broadcast stations operating in the public interest do not go out of business. Moreover, we must agree with the Commission that its "concern with CATV carriage of broadcast signals is not just a matter of avoidance of adverse effects, but extends also to requiring CATV affirmatively to further statutory policies." P —, supra, 32 L Ed 2d p 394. Since the avoidance of adverse effects is itself the furtherance of statutory policies, no sensible distinction even in theory can be drawn along those lines. More important, CATV systems, no less than broadcast stations, see, e.g., Federal Radio Comm'n v Nelson Bros. Co., 289 US 266, 77 L Ed 1166, 53 S Ct 627, 89 ALR 406 (1933) (deletion of a station), may enhance as well as impair the appropriate provision of broadcast services. Consequent-

ly, to define the Commission's power in terms of the protection, as opposed to the advancement, of broadcasting objectives would artificially constrict the Commission in the achievement of its statutory purposes and be inconsistent with our recognition in Southwestern "that it was precisely because Congress wished 'to maintain, through appropriate administrative control, a grip on the dynamic aspects of radio transmission,' . . . that it conferred upon the Commission a 'unified jurisdiction' and 'broad author-ity.'" P —, supra, 32 L Ed 2d p 399.[23]

The very regulations that formed the backdrop for our decision in Southwestern demonstrate this point. Those regulations were, of course, avowedly designed to guard broadcast services from being undermined by unregulated CATV growth. At the same time, the Commission recognized that "CATV systems . . . have arisen in response to public need and demand for improved television service and perform valuable public services in this respect." Second Report and Order, 2 FCC2d 725, 745 (1966).[24]

Southwestern, handed down only a week before Fortnightly, expressly held that CATV systems are not merely receivers, but transmitters of interstate communication subject to the Commission's jurisdiction under that Act. See 392 US, at 168, 20 L Ed 2d at 1010.

23. See also General Telephone Co. of Cal. v FCC, n 13, supra, at 398:

"Over the years, the Commission has been required to meet new problems concerning CATV and as cases have reached the courts the scope of the Act has been defined, as Congress contemplated would be done, so as to avoid a continuing process of statutory revision. To do otherwise in regulating a dynamic public service function such as broadcasting would place an intolerable regulatory bur-

den on the Congress—one which it sought to escape by delegating administrative functions to the Commission."

24. The Commission elaborated: "CATV . . . has made a significant contribution to meeting the public demand for television service in areas too small in population to support a local station or too remote in distance or isolated by terrain to receive regular or good off-the-air reception. It has also contributed to meeting the public's demand for good reception of multiple program choices, particularly the three full network services. In thus contributing to the realization of some of the most important goals which have governed our allocations planning, CATV has clearly served the public interest 'in the larger and more effective

[32 L Ed 2d]—26

Accordingly, the Commission's express purpose was not: "to deprive the public of these important benefits or to restrict the enriched programing selection which CATV makes available. Rather, our goal here is to integrate the CATV service into the national television structure in such a way as to promote maximum television service to all people of the United States (secs. 1 and 303(g) of the act [nn. 9 and 11, supra]), both those who are cable viewers and those dependent on off-the-air service. The new rules . . . are the minimum measures we believe to be essential to insure that CATV continues to perform its valuable supplementary role without unduly damaging or impeding the growth of television broadcast service." Id., at 745–746.[25]

In implementation of this approach CATV systems were required to

carry local broadcast station signals to encourage diversified programing suitable to the community's needs as well as to prevent a diversion of audiences and advertising revenues.[26] The duplication of local station programing was also forbidden for the latter purpose, but only on the same day as the local broadcast so as "to preserve, to the extent practicable, the valuable public contribution of CATV in providing wider access to nationwide programing and a wider selection of programs on any particular day." Id., at 747. Finally, the distant importation rule was adopted to enable the Commission to reach a public interest determination weighing the advantages and disadvantages of the proposed service on the facts of each individual case. See id., at 776, 781–782. In short, the regulatory authority asserted by the Commission in 1966 and generally sustained

---

use of radio.' And, even in the major market, 'where there may be no dearth of service . . . , CATV may . . . increase viewing opportunities, either by bringing in programing not otherwise available or, what is more likely, bringing in programing locally available but at times different from those presented by the local stations." Second Report and Order, 2 FCC2d 725, 781 (1966). See also id., at 745.

25. This statement, made with reference only to the local carriage and non-duplication requirements, was no less true of the distant importation rule. See id., at 781–782.

26. The regulation, for example, retained the provision of the Commission's earlier rule governing CATV microwave systems under which a local signal was not required to be carried "if (1) it substantially duplicates the network programing of a signal of a higher grade, and (2) carrying it would—because of limited channel capacity—prevent the system from carrying a nonnetwork signal, which would contribute to the diversity of its service." First Report and Order, 38 FCC 683, 717 (1965). See Second Report and

Order, n 24, supra, at 752–753. Moreover, CATV operators were warned that, in reviewing their discretionary choice of stations to carry among those of equal priority in certain circumstances, the Commission would "give particular consideration to any allegation that the station not carried is one with closer community ties." Second Report and Order, supra, at 755. In addition, operators were required to carry the signals of local satellite stations even if they also carried the signals of the satellites' parents; otherwise, "the satellite [might] lose audience for which it may be originating some local programing and [find] its incentive to originate programs [reduced]." Id., at 755–756. Finally, the Commission indicated that, in considering waivers of the regulation, it would "[accord] substantial weight" to such considerations as whether "the programing of stations located within the State would be of greater interest than those of nearer, but out-of-State stations [otherwise required to be given priority in carriage]—e.g., covering of political elections and other public affairs of statewide concern." Id., at 753.

by this Court in Southwestern was authority to regulate CATV with a view not merely to protect but to promote the objectives for which the Commission had been assigned jurisdiction over broadcasting.

[1] In this light the critical question in this case is whether the Commission has reasonably determined that its origination rule will "further the achievement of long-established regulatory goals in the field of television broadcasting by increasing the number of outlets for community self-expression and augmenting the public's choice of programs and types of services . . . ." P —, supra, 32 L Ed 2d p 395. We find that it has.

[5] The goals specified are plainly within the Commission's mandate for the regulation of television broadcasting.[27] In National Broadcasting Co. v United States, 319 US 190, 87 L Ed 1344, 63 S Ct 997 (1943), for example, we sustained Commission regulations governing relations between broadcast stations and network organizations for the purpose of preserving the stations' ability to serve the public interest through their programing. Noting that "[t]he facilities of radio are not large enough to accommodate all who wish to use them," id., at 216, we held that the Communications "Act does not restrict the Commission merely to supervision of [radio] traffic. It puts upon the Commission the burden of determining the composition of that traffic." Id., at 215–216, 87 L Ed at 1362. We then upheld the Commission's judgment that: " '[w]ith

the number of radio channels limited by natural factors, the public interest demands that those who are entrusted with the available channels shall make the fullest and most effective use of them.' " Id., at 218, 87 L Ed at 1363.

" 'A station licensee must retain sufficient freedom of action to supply the program . . . needs of the local community. Local program service is a vital part of community life. A station should be ready, able, and willing to serve the needs of the local community by broadcasting such outstanding local events as community concerts, civic meetings, local sports events, and other programs of local consumer and social interest.' " Id., at 203, 87 L Ed at 1356.

[1] Equally plainly the broadcasting policies the Commission has specified are served by the program origination rule under review. To be sure, the cablecasts required may be transmitted without use of the broadcast spectrum. But the regulation is not the less, for that reason, reasonably ancillary to the Commission's jurisdiction over broadcast services. The effect of the regulation, after all, is to assure that in the retransmission of broadcast signals viewers are provided suitably diversified programing—the same objective underlying regulations sustained in National Broadcasting Co. v United States, supra, as well as the local carriage rule reviewed in Southwestern and subsequently upheld. See p — and nn 17 and 26, supra, 32 L Ed 2d p 402. In es-

---

27. As the Commission stated, "it has long been a basic tenet of national communications policy that 'the widest possible dissemination of information from diverse and antagonistic sources is essential to the welfare of the public.' As-

sociated Press v United States, 326 US 1, 20, [89 L Ed 2013, 2030, 65 S Ct 1416;] Red Lion Broadcasting Co., Inc. v Federal Communications Commission, 395 US 367 [23 L Ed 2d 371, 89 S Ct 1794] . . . ." First Report and Order 205.

sence the regulation is no different from Commission rules governing the technological quality of CATV broadcast carriage. In the one case, of course, the concern is with the strength of the picture and voice received by the subscriber, while in the other it is with the content of the programing offered. But in both cases the rules serve the policies of §§ 1 and 303(g) of the Communications Act on which the cablecasting regulation is specifically premised, see pp — – —, supra, 32 L Ed 2d pp 395-396,[28] and also, in the Commission's words, "facilitate the more effective performance of [its] duty to provide a fair, efficient, and equitable distribution of television service to each of the several States and communities" under § 307(b). P —, supra, 32 L Ed 2d p 396.[29] In sum, the regulation preserves and enhances the integrity of broadcast signals and therefore is "reasonably ancillary to the effective performance of the Commission's various responsibilities for the regulation of television broadcasting."

Respondent, nevertheless, maintains that just as the Commission is powerless to require the provision of television broadcast services where there are no applicants for station licenses no matter how important or desirable those services may be, so, too, it cannot require CATV operators unwillingly to engage in cablecasting. In our view, the analogy respondent thus draws between entry into broadcasting and entry into cablecasting is misconceived. The Commission is not attempting to compel wire service where there has been no commitment to undertake it. CATV operators to whom the cablecasting rule applies have voluntarily engaged themselves in providing that service, and the Commission seeks only to ensure that it satisfactorily meets community needs within the context of their undertaking.

For these reasons we conclude that the program origination rule is within the Commission's authority recognized in Southwestern.

II

The question remains whether the regulation is supported by substantial evidence that it will promote the public interest. We read the opinion of the Court of Appeals as holding that substantial evidence to that effect is lacking because the regulation creates the risk that the added burden of cablecasting will result in increased subscription rates and even the termination of CATV services. That holding is patently incorrect in light of the record.

In first proposing the cablecasting requirement, the Commission noted

28. Respondent apparently does not dispute this, but contends instead that §§ 1 and 303(g) merely state objectives without granting power for their implementation. See Brief for Midwest Video Corporation, at 24. The cablecasting requirement, however, is founded on those provisions for the policies they state and not for any regulatory power they might confer. The regulatory power itself may be found, as in Southwestern, see pp 11, 13, supra, in 47 USC §§ 152(a), 303(r).

29. Respondent asserts that "it is difficult to see how a mandatory [origination] requirement . . . can be said to aid the Commission in preserving the availability of broadcast stations to the several states and communities." Brief for Midwest Video Corporation, at 24. Respondent ignores that the provision of additional programing outlets by CATV necessarily affects the fairness, efficiency, and equity of the distribution of television services. We have no basis, it may be added, for overturning the Commission's judgment that the effect in this regard will be favorable. See pp — – — and n 10, supra, 32 L Ed 2d pp 395-396.

that "[t]here may . . . be practical limitations [for compliance] stemming from the size of some CATV systems" and accordingly sought comments "as to a reasonable cutoff point [for application of the regulation] in light of the cost of the equipment and personnel minimally necessary for local originations." Notice of Proposed Rulemaking and Notice of Inquiry, 15 FCC2d 417, 422 (1968). The comments filed in response to this request included detailed data indicating, for example, that a basic monochrome system for cablecasting could be obtained and operated for less than an annual cost of $21,000 and a color system, for less than $56,000. See First Report and Order, 20 FCC2d 201, 210 (1969). This data, however, provided only a sampling of the experience of the CATV systems already engaged in program origination. Consequently, the Commission: "decided not to prescribe a permanent minimum cutoff point for required origination on the basis of the record now before us. The Commission intends to obtain more information from originating systems about their experience, equipment, and the nature of the origination effort. . . . In the meantime, we will prescribe a very liberal standard for required origination, with a view toward lowering this floor in . . . further proceedings, should the data obtained in such proceedings establish the appropriateness and desirability of such action." Id., at 213.

On this basis the Commission chose to apply the regulation to systems with 3,500 or more subscribers, effective January 1, 1971.

"This standard [the Commission explained] appears more than reasonable in light of the [data filed], our decision to permit advertising at natural breaks . . . , and the 1-year grace period. Moreover, it appears that approximately 70 percent of the systems now originating have fewer than 3,500 subscribers; indeed, about half of the systems now originating have fewer than 2,000 subscribers. . . . [T]he 3,500 standard will encompass only a very small percentage of existing systems at present subscriber levels, less than 10 percent." Ibid.

On petitions for reconsideration the Commission observed that it had "been given no data tending to demonstrate that systems with 3,500 subscribers cannot cablecast without impairing their financial stability, raising rates or reducing the quality of service." Memorandum Opinion and Order, 23 FCC2d 825, 826 (1970). The Commission repeated that "[t]he rule adopted is minimal in the light of the potentials of cablecasting,"[30] but, nonetheless, on its own motion postponed the effective date of the regulation to April 1, 1971, "to afford additional preparation time." Id., at 827.

This was still not the Commission's final effort to tailor the regulation to the financial capacity of CATV operators. In denying respondent's motion for a stay of the effective date of the rule, the Commission reiterated that "there has been no showing made to support the view that compliance . . . would be an unsustainable burden." Memorandum Opinion and Order, 27 FCC2d 778, 779 (1971). On the other hand, the Commission recog-

30. Commissioner Bartley, however, dissented on the ground that the regulation should apply only to systems with over 7,500 subscribers. Memorandum Opinion and Order 831.

nized that new information suggested that CATV systems of 10,000 ultimate subscribers would operate at a loss for at least four years if required to cablecast. That data, however, was based on capital expenditure and annual operating cost figures "appreciably higher" than those first projected by the Commission. Ibid. The Commission concluded:

"While we do not consider that an adequate showing has been made to justify general change, we see no public benefit in risking injury to CATV systems in providing local origination. Accordingly, if CATV operators with fewer than 10,000 subscribers request ad hoc waiver of [the regulation], they will not be required to originate pending action on their waiver requests. . . . Systems of more than 10,000 subscribers may also request waivers, but they will not be excused from compliance unless the Commission grants a requested waiver . . . .

[The] benefit [of cablecasting] to the public would be delayed if the . . . stay [requested by respondent] is granted, and the stay would, therefore, do injury to the public's interest." Ibid.

[6] This history speaks for itself. The cablecasting requirement thus applied is plainly supported by substantial evidence that it will promote the public interest.[31] Indeed, respondent does not appear to argue to the contrary. See Tr. of Oral Arg., at 43–44. It was, of course, beyond the competence of the Court of Appeals itself to assess the relative risks and benefits of cablecasting. As we said in National Broadcasting Co. v United States, 319 US 190, 224, 87 L Ed 1344, 1367, 63 S Ct 997 (1943):

"Our duty is at an end when we find that the action of the Commission was based upon findings supported by evidence, and was made pursuant to authority granted by Congress.

---

31. Nor is the regulation infirm for its failure to grant "grandfather" rights, see n 7, supra, as the Commission warned would be the case in its Notice of Proposed Rulemaking and Notice of Inquiry, 15 FCC2d 417, 424 (1968). See, e. g., Federal Radio Comm'n v Nelson Bros. Co., 289 US 266, 282, 77 L Ed 1166, 1176, 53 S Ct 627, 89 ALR 406 (1933) ("the power of Congress in the regulation of interstate commerce is not fettered by the necessity of maintaining existing arrangements which would conflict with the execution of its policy"). Judge Tuttle has elaborated, General Telephone Co. of Southwest v United States, 449 F2d 846, 863–864 (CA5 1971):

"In a complex and dynamic industry such as the communications field, it cannot be expected that the agency charged with its regulation will have perfect clairvoyance. Indeed as Justice Cardozo once said, 'Hardship must at times result from postponement of the rule of action till a time when action is complete. It is one of the consequences of the limitations of the human intellect and of the

denial to legislators and judges of infinite prevision.' Cardozo, The Nature of the Judicial Process 145 (1921). The Commission, thus, must be afforded some leeway in developing policies and rules to fit the exigencies of the burgeoning CATV industry. Where the on-rushing course of events have outpaced the regulatory process, the Commission should be enabled to remedy the [problem] . . . by retroactive adjustments, provided they are reasonable. . . .

"Admittedly the rule here at issue has an effect on activities embarked upon prior to the issuance of the Commission's Final Order and Report. Nonetheless the announcement of a new policy will inevitably have retroactive consequences. . . . The property of regulated industries is held subject to such limitations as may reasonably be imposed upon it in the public interest and the courts have frequently recognized that new rules may abolish or modify pre existing interests."

With regard to federal infringement of franchise rights, see generally Barnett, n 21, supra, at 703–705 and n 116.

It is not for us to say that the 'public interest' will [in fact] be furthered or retarded by the . . . [regulation]."

See also, e.g., United States v Storer Broadcasting Co., 351 US 192, 203, 100 L Ed 1081, 1091, 76 S Ct 763 (1956); General Telephone Co. of Southwest v United States, 449 F2d 846, 858–859, 862–863 (CA5 1971).

Reversed.

Mr. Chief Justice **Burger,** concurring in the result.

This case presents questions of extraordinary difficulty and sensitivity in the communications field as the opinions of the divided Court of Appeals and our own divisions reflect. As Mr. Justice Brennan has noted, Congress could not anticipate the advent of CATV when it enacted the regulatory scheme nearly 40 years ago. Yet that statutory scheme plainly anticipated the need for comprehensive regulation as pervasive as the reach of the instrumentalities of broadcasting.

In the four decades spanning the life of the Communications Act, the courts have consistently construed the Act as granting pervasive jurisdiction to the Commission to meet the expansion and development of broadcasting. That approach was broad enough to embrace the advent of CATV, as indicated in the plurality opinion. CATV is dependent totally on broadcast signals and is a significant link in the system as a whole and therefore must be seen as within the jurisdiction of the Act.

Concededly the Communications Act did not explicitly contemplate either CATV or the jurisdiction the Commission has now asserted. However Congress was well aware in the 1930's that broadcasting was a dynamic instrumentality, that its future could not be predicted, that scientific developments would inevitably enlarge the role and scope of broadcasting and that in consequence regulatory schemes must be flexible and virtually open-ended.

Candor requires acknowledgment, for me at least, that the Commission's position strains the outer limits of even the open-ended and pervasive jurisdiction that has evolved by decisions of the Commission and the courts. The almost explosive development of CATV suggests the need of a comprehensive re-examination of the statutory scheme as it relates to this new development, so that the basic policies are considered by Congress and not left entirely to the Commission and the courts.

I agree with the plurality's rejection of any meaningful analogy between requiring CATV operators to develop programing and the concept of commandeering someone to engage in broadcasting. Those who exploit the existing broadcast signals for private commercial surface transmission by CATV—to which they make no contribution—are not exactly strangers to the stream of broadcasting. The essence of the matter is that when they interrupt the signal and put it to their own use for profit, they take on burdens, one of which is regulation by the Commission.

[1] I am not fully persuaded that the Commission has made the correct decision in this case and the thoughtful opinions in the Court of Appeals and the dissenting opinion here reflect some of my reservations. But the scope of our review is limited and does not permit me to resolve this issue as perhaps I would were I a member of the Fed-

eral Communications Commission. That I might take a different position as a member of the Commission gives me no license to do so here. Congress has created its instrumentality to regulate broadcasting, has given it pervasive powers, and the Commission has generations of experience and "feel" for the problem. I therefore conclude that until Congress acts, the Commission should be allowed wide latitude and I therefore concur in the result reached by this Court.

Mr. Justice **Douglas**, with whom Mr. Justice **Stewart**, Mr. Justice **Powell**, and Mr. Justice **Rehnquist** concur, dissenting.

The policies reflected in the plurality opinion may be wise ones. But whether CATV systems should be required to originate programs is a decision that we certainly are not competent to make and in my judgment the Commission is not authorized to make. Congress is the agency to make the decision and Congress has not acted.

CATV captures TV and radio signals, converts the signals, and carries them by microwave relay transmission or by coaxial cables into communities unable to receive the signals directly. In United States v Southwestern Cable Co., 392 US 157, 20 L Ed 2d 1001, 88 S Ct 1994, we upheld the power of the Commission to regulate the transmission of signals. As we said in that case:

"CATV systems perform either or both of two functions. First, they may supplement broadcasting by facilitating satisfactory reception of local stations in adjacent

areas in which such reception would not otherwise be possible; and second, they may transmit to subscribers the signals of distant stations entirely beyond the range of local antennae. As the number and size of CATV systems have increased, their principal function has more frequently become the importation of distant signals." Id., at 163, 20 L Ed 2d at 1008.

CATV evolved after the Communications Act of 1934, 48 Stat 1064 was passed. But we held that the reach of the Act which extends "to all interstate and foreign communication by wire or radio," 47 USC § 152(a), was not limited to the precise methods of communication then known. 392 US, at 173, 20 L Ed 2d at 1013.

Compulsory origination of programs is, however, a far cry from the regulation of communications approved in Southwestern Cable. Origination requires new investment and new and different equipment, and an entirely different cast of personnel.[1] See 20 FCC2d 201, 210–211. We marked the difference between communication and origination in Fortnightly Corp. v United Artists, 392 US 390, 20 L Ed 2d 1176, 88 S Ct 2084, and made clear how foreign the origination of programs is to CATV's traditional transmission of signals. In that case, CATV was sought to be held liable for infringement of copyrights of movies licensed to broadcasters and carried by CATV. We held CATV not liable, saying:

"Essentially, a CATV system no more than enhances the viewer's capacity to receive the broadcaster's

---

1. In light of the striking difference between origination and communication, the suggestion that "the regulation is no different from Commission rules governing the technical quality of CATV broadcast carriage," ante, at —, 32 L Ed 2d at 403, appears misconceived.

signals; it provides a well-located antenna with an efficient connection to the viewer's television set. It is true that a CATV system plays an 'active' role in making reception possible in a given area, but so do ordinary television sets and antennas. CATV equipment is powerful and sophisticated, but the basic function the equipment serves is little different from that served by the equipment generally furnished by a television viewer. If an individual erected an antenna on a hill, strung a cable to his house, and installed the necessary amplifying equipment, he would not be 'performing' the programs he received on his television set. The result would be no different if several people combined to erect a cooperative antenna for the same purpose. The only difference in the case of CATV is that the antenna system is erected and owned not by its users but by an entrepreneur.

"The function of CATV systems has little in common with the function of broadcasters. CATV systems do not in fact broadcast or rebroadcast. Broadcasters select the programs to be viewed; CATV systems simply carry, without editing, whatever programs they receive. Broadcasters procure programs and propagate them to the public; CATV systems receive programs that have been released to the public and carry them by private channels to additional viewers. We hold that CATV operators, like viewers and unlike broadcasters, do not perform the programs that they receive and carry." Id., at 400–401, 20 L Ed 2d at 1183, 1184.

The Act forbids any person from operating a broadcast station without first obtaining a license from the Commission. 47 USC § 301. Only

qualified persons may obtain licenses and they must operate in the public interest. 47 USC §§ 308, 309. But nowhere in the Act is there the slightest suggestion that a person may be compelled to enter the broadcasting or cablecasting field. Rather, the Act extends "to all interstate and foreign communication by wire or radio . . . which *originates and/or is received* within the United States." 47 USC § 152(a) (emphasis added). When the Commission jurisdiction is so limited, it strains logic to hold that this jurisdiction may be expanded by requiring someone to "originate" or "receive."

The Act, when dealing with broadcasters, speaks of "applicants," "applications for licenses," see 47 USC §§ 307, 308, and "whether the public interest, convenience and necessity will be served by the granting of such application." 47 USC § 309(a). The emphasis on the Committee Reports was on "original applications" and "application for the renewal of a license." HR Rep No 1918, 73d Cong, 2d Sess, p 48; S Rep No 781, 73d Cong, 2d Sess, pp 7, 9. The idea that a carrier or any other person can be drafted against his will to become a broadcaster is completely foreign to the history of the Act, as I read it.

CATV is simply a carrier having no more control over the message content than does a telephone company. A carrier may of course seek a broadcaster's license; but there is not the slightest suggestion in the Act or in its history that a carrier can be bludgeoned into becoming a broadcaster while all other broadcasters live under more lenient rules. There is not the slightest cue in the Act that CATV carriers can

be compulsorily converted into broadcasters.

The plurality opinion performs the legerdemain by saying that the requirement of CATV origination is "reasonably ancillary" to the Commission's power to regulate television broadcasting.[2] That requires a brand new amendment to the broadcasting provisions of the Act which only the Congress can effect. The Commission is not given carte blanche to initiate broadcasting stations; it cannot force people into the business. It cannot say to one who applies for a broadcast outlet in city A that the need is greater in city B and he will be licensed there. The fact that the Commission has authority to regulate origination of programs if CATV decides to enter the field does not mean that it can compel CATV to originate programs. The fact that the Act directs the Commission to encourage the larger and more effective use of radio in the public interest, 47 USC § 303 (g), relates to the objectives of the Act and does not grant power to compel people to become broadcasters any more than it grants the power to compel broadcasters to become CATV operators.

The upshot of today's decision is to make the Commission's authority over activities "ancillary" to its responsibilities greater than its authority over any broadcast licensee. Of course, the Commission can regulate a CATV that transmits broadcast signals. But to entrust the Commission with the power to force some, a few, or all CATV operators into the broadcast business is to give it a forbidding authority. Congress may decide to do so. But the step is a legislative measure so extreme that we should not find it interstitially authorized in the vague language of the Act.

I would affirm the Court of Appeals.

---

2. The separate opinion of The Chief Justice reaches the same result by saying "CATV is dependent totally on broadcast signals and is a significant link in the system as a whole and therefore must be seen as within the jurisdiction of the Act." Ante, at —, 32 L Ed 2d at 407. The difficulty is that this analysis knows no limits short of complete domination of the field of communications by the Commission. This reasoning—divorced as it is from any specific statutory basis—could as well apply to the manufacturers of radio and television broadcasting and receiving equipment.

NEWS   Federal Communications Commission
        1919 M Street, NW.
        Washington, D.C. 20554
        Public Notice

For information on releases and texts call 632-0002

87959

August 22, 1972 - G

FRANCHISE PROVISIONS AT VARIANCE WITH FCC CABLE TELEVISION RULES

The following letter has been sent to Western Communications, Inc.
by Sol Schildhause, Chief of the Cable Television Bureau, in response
to an inquiry about the extent to which Federal-State/local franchising authorities can establish regulations which are different from
those established by the FCC for cable television systems:

> This is in reply to your letter of August 3, 1972, in which you
> question the extent to which franchising authorities can establish
> regulations in excess of or different than the regulations estab-
> lished by the Commission in the Cable Television Report and Order.
> I have taken the liberty of combining some of your questions and
> re-phrasing others for purposes of clarity.

> 1. Q. May a franchising authority in a major television
>   market specify a minimum channel capacity in excess of
>   20 channels?

>   A. In footnote 25 or the Memorandum Opinion and Order on
>   Reconsideration of the Cable Television Report and Order,
>   the Commission stated that while it has preempted the
>   area of channel capacity, it would not foreclose a sys-
>   tem from meeting more stringent local requirements. . .
>   "upon a demonstration of need for such channel capacity
>   and the system's ability to provide it." (emphasis sup-
>   plied) Also see paragraph 132 and footnote 70 of the
>   Cable Television Report and Order.

2. Q. May a franchising authority outside a major market specify a minimum channel capacity and, if so, can this minimum channel capacity be in excess of what this Commission requires for a major market?

   A. Cities outside major markets may specify a minimum channel capacity, but such capacity may not be in excess of what the Commission requires for systems in major markets. See Section 76.251(b) of the Rules and paragraph 148 of the Cable Television Report and Order.

3. Q. May a franchising authority located outside a major television market require a cable system to maintain a plant having the technical capacity for nonvoice return communications?

   A. Yes - see Section 76.251(b) of the Commission's Rules and paragraphs 132 and 148 of the Cable Television Report and Order.

4. Q. May any franchising authority require a more sophisticated form of return communications?

   A. In footnote 25 of the Memorandum Opinion and Order on Reconsideration of the Cable Television Report and Order the Commission states, "Where a franchising authority has a plan for actual use of a more sophisticated two-way capability and the cable operator can demonstrate its feasibility both practically and economically, we will consider, in the certificating process, allowing such equipment". (emphasis supplied)

5. Q. Can a franchising authority require all access services to be made available at no charge?

   A. No - The Commission will consider in the certificating process, however, requirements that additional public access channels or some educational channels be offered at no charge or a reduced cost on an experimental basis. See paragraph 132 of the Cable Television Report and Order.

6. Q. Can a franchising authority require the franchisee to make available more access channels than those specified by the Commission?

   A. No - unless during the certificating process the Commission is shown that such additional channels are necessary and capable of being used according to an existing, viable plan. See Section 76.251(a)(11)(iv) of the Rules, and paragraph 132 of the Cable Television Report and Order.

7. Q. Can a franchising authority require a franchisee to provide access services outside major markets?

   A. Yes - but to no greater extent than the Commission requires for systems in major markets. See Section 76.251 b) and paragraphs 132 and 148 of the Cable Television Report and Order.

8. Q. May a franchising authority impose a franchise fee based upon revenues derived from "auxiliary" services such as advertising revenues, leased channel revenues, pay cable revenues, etc.?

   A. No - Subscriber revenues are considered to be those revenues derived from regular subscriber services - i.e., the carriage of broadcast signals and required non-broadcast services.

9. Q. May a franchising authority insist on a franchise fee higher than 3% if the excess fee is to be used for funding public access services?

   A. There is no hard and fast answer to this question at present. Clearly, however, the factors that would bear heavily in the Commission's consideration of any such scheme would include the amount of excess fee, the danger that, through funding, local governments would control public access programming, and the possibility of other alternatives.

10. Q. May a franchising authority require a faster construction schedule than that suggested by the Commission?

    A. Yes - See Section 76.31(a)(2) of the Commission's Rules.

11. Q. May a franchising authority require systems with fewer than 3500 subscribers to engage in local origination?

    A. The Commission has preempted this field. See paragraph 48, First Report and Order, 20 FCC 2d 201, at 223. See also "Clarification of CATV First Report as to Scope of Federal Pre-Emption," 20 FCC 2d 741. Under these circumstances, I believe the Commission would reject such a requirement. Further, the Commission preemption extends to policy concerning any waiver of the origination rule.

12. Q. May a franchising authority establish technical standards in excess of those required by the Commission?

    A. Yes - (See paragraph 91 of the Memorandum Opinion and Order on Reconsideration of the Cable Television Report and Order.) The Commission will not, however, assume responsibility for enforcement of more stringent technical

standards. Local authorities should therefore be prepared
to assume the burden of such enforcement.

13. Q. May a franchising authority limit a franchisee to provid-
       ing services that can be performed only by the franchisee
       itself?

    A. No - Clearly the concept of access services is to offer
       the benefits of a multiplicity of channels to the public.
       Thus in Sections 76.251(a)(11)(i) and (iii), system opera-
       tors are specifically forbidden to exercise control over
       the program content of public and leased access channels.

The foregoing responses to your questions should indicate the degrees
to which the Commission will sanction franchise provisions at variance with its cable regulatory program. Where variances are sought,
as for instance where a franchise calls for extra access channels,
greater channel capacity, or a higher franchise fee, detailed showings will be required during the certificating process. If such a
showing is inadequate, the Commission will not issue a certificate
of compliance.

It is recommended, therefore, that franchises be drawn to include
severability clauses that will enable the Commission to authorize
system operations without the delay that might be created by the
necessity for franchise amendments.

I hope the foregoing is responsive to your inquiry. If I can be of
further assistance to you, do not hesitate to call on me.

(Slip Opinion)

SUPREME COURT OF THE UNITED STATES

Syllabus

TELEPROMPTER CORP. ET AL. *v.* COLUMBIA BROADCASTING SYSTEM, INC., ET AL.

CERTIORARI TO THE UNITED STATES COURT OF APPEALS FOR THE SECOND CIRCUIT

No. 72-1628. Argued January 7, 1974—Decided March 4, 1974*

Several creators and producers of copyrighted television programs brought this suit claiming that defendants had infringed their copyrights by intercepting broadcast transmissions of copyrighted material and rechanneling these programs through various community antenna television (CATV) systems to paying subscribers. The District Court dismissed the complaint on the ground that the cause of action was barred by this Court's decision in *Fortnightly Corp.* v. *United Artists Television Corp.*, 392 U. S. 390. On appeal, the Court of Appeals divided CATV systems into two categories for copyright purposes: (1) those where the broadcast signal was already "in the community" served by the system, and could be received there either by a community antenna or by standard rooftop or other antennae belonging to the owners of television sets; and (2) those where the systems imported "distant signals" from broadcasters so far away from the CATV community that the foregoing local facilities could not normally receive adequate signals. Holding that CATV reception and retransmission of non-"distant" signals do not constitute copyright infringement, but that reception and retransmission of "distant" signals amount to a "performance" and thus constitute copyright infringement, the court affirmed as to those systems in the first category, but reversed and remanded as to the remaining systems. *Held:*

1. The development and implementation, since the *Fortnightly* decision, of new functions of CATV systems—program origination,

---

*Together with No. 72-1633, *Columbia Broadcasting System, Inc., et al.* v. *Teleprompter Corp. et al.*, also on certiorari to the same court.

sale of commercials, and interconnection with other CATV systems—even though they may allow the systems to compete more effectively with the broadcasters for the television market, do not convert the entire CATV operation, regardless of distance from the broadcasting station, into a "broadcast function," thus subjecting the CATV operators to copyright infringement liability, but are extraneous to a determination of such liability, since in none of these functions is there any nexus with the CATV operators' reception and rechanneling of the broadcasters' copyrighted materials. Pp. 7–10.

2. The importation of "distant" signals from one community into another does not constitute a "performance" under the Copyright Act. Pp. 10–19.

(a) By importing signals which could not normally be received with current technology in the community it serves, a CATV system does not, for copyright purposes, alter the function it performs for its subscribers, but the reception and rechanneling of these signals for simultaneous viewing is essentially a viewer function, irrespective of the distance between the broadcasting station and the ultimate viewer. P. 13.

(b) Even in exercising its limited freedom to choose among various "distant" broadcasting stations, a CATV operator cannot be viewed as "selecting" broadcast signals, since when it chooses which broadcast signals to rechannel, its creative function is then extinguished and it thereafter "simply carr[ies], without editing, whatever programs [it] receive[s]," *Fortnightly Corp.* v. *United Artists Television Corp., supra,* at 400. Nor does a CATV system importing "distant" signals procure and propagate them to the public, since it is not engaged in converting the sights and sounds of an event or a program into electronic signals available to the public, the signals it receives and rechannels having already been "released to the public" even though not normally available to the specific segment of the public served by the CATV system. Pp. 14–15.

(c) The fact that there have been shifts in current business and commercial relationships in the communications industry as a result of the CATV systems' importation of "distant" signals, does not entail copyright infringement liability, since by extending the range of viewability of a broadcast program, the CATV systems do not interfere in any traditional sense with the copyright holders' means of extracting recompense for their creativity

or labor from advertisers on the basis of all viewers who watch the particular program. Pp. 15–19.

476 F. 2d 338, affirmed in part, reversed in part, and remanded to District Court.

STEWART, J., delivered the opinion of the Court, in which BRENNAN, WHITE, MARSHALL, POWELL, and REHNQUIST, JJ., joined. BLACKMUN, J., filed an opinion dissenting in part. DOUGLAS, J., filed a dissenting opinion, in which BURGER, C. J., joined.

SUPREME COURT OF THE UNITED STATES

Nos. 72–1628 and 72–1633

Teleprompter Corporation et al.,
Petitioners,

72–1628   v.

Columbia Broadcasting System,
Inc., et al.

Columbia Broadcasting System,
Inc., et al., Petitioners,

72–1633   v.

Teleprompter Corporation et al.

On Writs of Certiorari
to the United States
Court of Appeals for
the Second Circuit.

[March 4, 1974]

Mr. Justice Stewart delivered the opinion of the
Court.

The plaintiffs in this litigation, creators and producers
of televised programs copyrighted under the provisions of
the Copyright Act of 1909, as amended, 17 U. S. C. § 1
et seq., commenced suit in 1964 in the United States
District Court for the Southern District of New York,
claiming that the defendants had infringed their copyrights by intercepting broadcast transmissions of copyrighted material and rechanneling these programs
through various community antenna television (CATV)
systems to paying subscribers.[1] The suit was initially

---

[1] The exclusive rights of copyright owners are specified in § 1
of the Copyright Act:

"Any person entitled thereto, upon complying with the provisions
of this title, shall have the exclusive right:

stayed by agreement of the parties, pending this Court's decision in *Fortnightly Corp.* v. *United Artists Television Corp.,* 392 U. S. 290. In that case, decided in 1968, we

---

"(a) To print, reprint, publish, copy, and vend the copyrighted work;

"(b) To translate the copyrighted work into other languages or dialects, or make any other version thereof, if it be a literary work; to dramatize it if it be a nondramatic work; to convert it into a novel or other nondramatic work if it be a drama; to arrange or adapt it if it be a musical work; to complete, execute, and finish it if it be a model or design for a work of art;

"(c) To deliver, authorize the delivery of, read, or present the copyrighted work in public for profit if it be a lecture, sermon, address or similar production, or other nondramatic literary work; to make or procure the making of any transcription or record thereof by or from which, in whole or in part, it may in any manner or by any method be exhibited, delivered, presented, produced, or reproduced; and to play or perform it in public for profit, and to exhibit, represent, produce, or reproduce it in any manner or by any method whatsoever. The damages for the infringement by broadcast of any work referred to in this subsection shall not exceed the sum of $100 where the infringing broadcaster shows that he was not aware that he was infringing and that such infringement could not have been reasonably foreseen; and

"(d) To perform or represent the copyrighted work publicly if it be a drama or, if it be a dramatic work and not reproduced in copies for sale, to vend any manuscript or any record whatsoever thereof; to make or to procure the making of any transcription or record thereof by or from which, in whole or in part, it may in any manner or by any method be exhibited, performed, represented, produced, or reproduced; and to exhibit, perform, represent, produce, or reproduce it in any manner or by any method whatsoever; and

"(e) To perform the copyrighted work publicly for profit if it be a musical composition; and for the purpose of public performance for profit, and for the purposes set forth in subsection (a) hereof, to make any arrangement or setting of it or of the melody of it in any system of notation or any form of record in which the thought of an author may be recorded and from which it may be read or reproduced . . . ." 17 U. S. C. § 1.

held that the reception and distribution of television broadcasts by the CATV systems there involved did not constitute a "performance" within the meaning of the Copyright Act, and thus did not amount to copyright infringement.[2] After that decision the plaintiffs in the present litigation filed supplemental pleadings in which they sought to distinguish the five CATV systems challenged here from those whose operations had been found not to constitute copyright infringement in *Fortnightly*.[3] The District Court subsequently dismissed the complaint on the ground that the plaintiffs' cause of action was barred by the *Fortnightly* decision. 355 F. Supp. 618. On appeal to the United States Court of Appeals for the Second Circuit, the judgment was affirmed in part and reversed in part, and the case was remanded to the District Court for further proceedings. 476 F. 2d 338. Both the plaintiffs and the defendants petitioned for certiorari, and, because of the seemingly important ques-

---

[2] Although the Copyright Act does not contain an explicit definition of infringement, it is settled that unauthorized use of copyrighted material inconsistent with the "exclusive rights" enumerated in § 1, constitutes copyright infringement under federal law. See M. Nimmer, Copyright § 100, at 376 (1972). Use of copyrighted material not in conflict with a right secured by § 1, however, no matter how widespread, is not copyright infringement. "The fundamental [is] that 'use' is not the same thing as 'infringement,' that use short of infringement is to be encouraged . . . ." B. Kaplan, An Unhurried View of Copyright 57 (1967).

It appears to be conceded that liability in this case depends entirely on whether the defendants did "perform" the copyrighted works. Teleprompter has not contended in this Court that, if it did "perform" the material, its performance was not "in public" within the meaning of § 1 (c) of the Act (nondramatic literary works) or "publicly" under § 1 (d) (dramatic works). Cf. *Fortnightly Corp.* v. *United Artists Television, Inc.*, 392 U. S. 390, 395 n. 13.

[3] The plaintiffs' amended complaints also contained allegations of additional copyright infringements on various dates in 1969 and 1971.

tions of federal law involved, we granted both petitions. 414 U. S. 817.

I

The complaint alleged that copyright infringements occurred on certain dates at each of five illustrative CATV systems located in Elmira, New York; Farmington, New Mexico; Rawlins, Wyoming; Great Falls, Montana; and New York City. The operations of these systems typically involved the reception of broadcast beams by means of special television antennae owned and operated by Teleprompter, transmission of these electronic signals by means of cable or a combination of cable and point-to-point microwave [4] to the homes of

---

[4] The Court of Appeals in this case described the differences between point-to-point microwave transmission and broadcasting in the following terms:

"A microwave link involves the transmission of signals through the air. However, microwave transmission in itself is not broadcasting. A broadcast signal, according to 47 U. S. C. § 153 (o), is transmitted by a broadcaster for '[reception] by the public.' In the case of microwave, the signal is focused and transmitted in a narrow beam aimed with precision at the receiving points. Thus, microwave transmission is point-to-point communication. The receiving antenna must be in the path of the signal beam. If the transmission must cover a considerable distance, the microwave signal is transmitted to the first receiving point from which it is retransmitted to another receiving point, and this process is repeated until the signal reaches the point from which it is distributed by cable to subscribers." 476 F. 2d 338, 343 n. 6.

The plaintiffs argued in the District Court and in the Court of Appeals that "the use of microwave, in and of itself, is sufficient to make a CATV system functionally equivalent to a broadcaster and thus subject to copyright liability . . . ." *Id.*, at 348–349. This contention was rejected by the Court of Appeals on the ground that microwave transmission "is merely an alternative, more economical in some circumstances, to cable in transmitting a broadcast signal from one point in a CATV system to another," *id.*, at 349, and the argument has not been renewed in this Court.

subscribers, and the conversion of the electromagnetic signals into images and sounds by means of the subscribers' own television sets.[5] In some cases the distance between the point of original transmission and the ultimate viewer was relatively great—in one instance more than 450 miles—and reception of the signals of those stations by means of an ordinary rooftop antenna, even an extremely high one, would have been impossible because of the curvature of the earth and other topographical factors. In others, the original broadcast was relatively close to the customers' receiving sets and could normally have been received by means of standard television equipment. Between these extremes were systems involving intermediate distances where the broadcast signals could have been received by the customers' own television antennae only intermittently, imperfectly, and sporadically.[6]

Among the various actual and potential CATV operations described at trial the Court of Appeals discerned, for copyright purposes, two distinct categories. One category included situations where the broadcast signal was already "in the community" served by a CATV system, and could be received there either by standard rooftop or other antennae belonging to the owners of

---

[5] For general descriptions of CATV systems and their operation, see *United States* v. *Southwestern Cable Co.*, 392 U. S. 157; M. Seiden, An Economic Analysis of Community Antenna Television Systems and the Television Broadcasting Industry (1965); Note, Regulation of Community Antenna Television, 70 Col. L. Rev. 837 (1970); Note, The Wire Mire: The FCC and CATV, 79 Harv. L. Rev. 366 (1965).

[6] In two of the cities involved in this suit signals not normally receivable by household sets because of distance or terrain could be received by rooftop antennae because of the use by the broadcasting stations of "translators," under license from the Federal Communications Commission, which rebroadcast a specific station's signals. See 476 F. 2d, at 344 & n. 7.

television sets or by a community antenna erected in or adjacent to the community. Such CATV systems, the court found, performed essentially the same function as the CATV systems in *Fortnightly* in that they "no more than enhance the viewer's capacity to receive the broadcaster's signals," 392 U. S., at 399. The second category included situations where the CATV systems imported "distant signals" from broadcasters so far away from the CATV community that neither rooftop nor community antennae located in or near the locality could normally receive signals capable of providing acceptable images.

The Court of Appeals determined that "[w]hen a CATV system is performing this second function of distributing signals that are beyond the range of local antennas, . . . to this extent, it is functionally equivalent to a broadcaster and thus should be deemed to 'perform' the programming distributed to subscribers on these imported signals." 476 F. 2d, at 349. The Court of Appeals found that in two of the operations challenged in the complaint—those in Elmira and New York City—the signals received and rechanneled by the CATV systems were not "distant signals," and as to these claims the court affirmed the District Court's dismissal of the complaint. As to the three remaining systems, the case was remanded for further findings in order to apply the appellate court's test for determining whether or not the signals were "distant." [7] In No. 72–1633 the plaintiffs

---

[7] The Court of Appeals acknowledged that a determination of what is a "distant signal" was "difficult," and "that a precise judicial definition of a distant signal is not possible." *Id.*, at 350. FCC regulations at one time provided that for regulatory purposes a distant signal was one "which is extended or received beyond the Grade B contour of the station." 47 CFR § 74.1101 (i), (1971) (removed 37 Fed. Reg. 3278, Feb. 12, 1972). A Grade B contour was defined as a line along which good reception may be expected 90% of the time at 50% of the locations. *United States* v. *South-*

TELEPROMPTER CORP. *v.* CBS

ask this Court to reverse the determination of the Court of Appeals that CATV reception and retransmission of signals that are not "distant" do not constitute copyright infringement. In No. 72–1628, the defendants ask us to reverse the appellate court's determination that reception and retransmission of "distant" signals amount to a "performance," and thus constitute copyright infringement on the part of the CATV systems.

II

We turn first to the assertions of the petitioners in No. 72–1633 that irrespective of the distance from the broadcasting station, the reception and retransmission of its signal by a CATV system constitute a "performance" of a copyrighted work. These petitioners contend that a number of significant developments in the technology and actual operations of CATV systems mandate a reassessment of the conclusion reached in *Fortnightly* that CATV systems act only as an extension of a television set's function of converting into images and sounds the signals made available by the broadcasters to the public. In *Fortnightly* this Court reviewed earlier cases in the federal courts and determined that while analogies to the functions of performer and viewer envisioned by the Congress in 1909—that of live or filmed performances

---

western Cable Television Co., supra, 392 U. S., at 163, n. 16. The Court of Appeals recognized that "this definition [is] unsuitable for copyright purposes because . . . any definition phrased in terms of what can be received in area homes using rooftop antennas would fly in the face of the mandate of *Fortnightly.*" 476 F. 2d, at 350. The court found instead that "it is easier to state what is not a distant signal than to state what is a distant signal. Accordingly, we have concluded that any signal capable of projecting, without relay or retransmittal an acceptable image that a CATV system receives off-the-air during a substantial portion of the time by means of an antenna erected in or adjacent to the CATV community is not a distant signal." *Id.,* at 351 (footnote omitted).

watched by audiences—were necessarily imperfect, a simple line could be drawn: "Broadcasters perform. Viewers do not perform." 392 U. S., at 398 (footnotes omitted). Analysis of the function played by CATV systems and comparison with those of broadcasters and viewers convinced the Court that CATV systems fall "on the viewer's side of the line." *Id.,* at 399 (footnote omitted).

> "The function of CATV systems has little in common with the function of broadcasters. CATV systems do not in fact broadcast or rebroadcast. Broadcasters select the programs to be viewed; CATV systems simply carry, without editing, whatever programs they receive. Broadcasters procure programs and propagate them to the public; CATV systems receive programs that have been released to the public and carry them by private channels to additional viewers. We hold that CATV operators, like viewers and unlike broadcasters, do not perform the programs that they receive and carry." *Id.,* at 400–401 (footnotes omitted).

The petitioners claim that certain basic changes in the operation of CATV systems that have occurred since *Fortnightly* bring the systems in question here over to the broadcasters' "side of the line." In particular, they emphasize three developments that have taken place in the few years since the *Fortnightly* decision. First, they point out that many CATV systems, including some of those challenged here, originate programs wholly independent of the programs that they receive off-the-air from broadcasters and rechannel to their subscribers.[8]

---

[8] Program origination initially consisted of simple arrangements on spare channels using automated cameras providing time, weather, news ticker, or stock ticker information, and aural systems with music or news announcements. The function has been expanded

It is undisputed that such CATV systems "perform" those programs which they produce and program on their own; but it is contended that, in addition, the engagement in such original programming converts the entire CATV operation into a "broadcast function," and thus a "performance" under the Copyright Act. Second, these petitioners assert that Teleprompter, unlike the CATV operators sued in *Fortnightly*, sells advertising time to commercial interests wishing to sell goods or services in the localities served by its CATV systems. The sale of such commercials, they point out, was considered in the *Fortnightly* opinion as a function characteristically performed by broadcasters. 392 U. S., at 400 n. 28, citing *Intermountain Broadcasting & Television Corp.* v. *Idaho Microwave, Inc.*, 196 F. Supp. 315, 325. Finally, they contend that by engaging in interconnection with other CATV systems—whereby one CATV system that originates a program sells the right to redistribute it to other CATV systems that carry it simultaneously to their own subscribers—the CATV operators have similarly transferred their functions into that of broadcasters, thus subjecting themselves to copyright infringement liability.[9]

The copyright significance of each of these functions—program origination, sale of commercials, and interconnection—suffers from the same logical flaw: in none of

---

to include coverage of sports and other live events, news services, moving picture films, and specially created dramatic and nondramatic programs. See FCC, First Report and Order, 20 FCC 2d 201 (1969); *United States* v. *Midwest Video Corp.*, 406 U. S 649

[9] The Court of Appeals limited its discussion of interconnection among CATV systems to two instances of live coverage of championship heavyweight boxing contests. While the respondents contend that additional examples of interconnection were presented in the trial testimony, they do not suggest that material copyrighted by anyone other than the CATV operators was carried by any such interconnection, and thus the exact number of such instances is of no significance.

these operations is there any nexus with the defendants' reception and rechannelling of the broadcasters' copyrighted materials. As the Court of Appeals observed with respect to program origination, "[e]ven though the origination service and the reception service are sold as a package to the subscribers, they remain separate and different operations, and we cannot sensibly say that the system becomes a 'performer' of the broadcast programming when it offers both origination and reception services, but remains a non-performer when it offers only the latter." 476 F. 2d, at 347. Similarly, none of the programs accompanying advertisements sold by CATV or carried via an interconnection arrangement among CATV systems involved material copyrighted by the petitioners.[10]

For these reasons we hold that the Court of Appeals was correct in determining that the development and implementation of these new functions, even though they may allow CATV systems to compete more effectively with the broadcasters for the television market, are simply extraneous to a determination of copyright infringement liability with respect to the reception and retransmission of broadcasters' programs.

III

In No. 72–1628 Teleprompter and its subsidiary, Conley Electronics Corp., seek a reversal of that portion of the Court of Appeals' judgment that determined that the importation of "distant" signals from one community into another constitutes a "performance" under the Copy-

---

[10] While the technology apparently exists whereby a CATV system could retransmit to its subscribers broadcast programs taken off-the-air but substitute its own commercials for those appearing in the broadcast, none of the instances of claimed infringement involved such a process.

right Act. In concluding that rechanneling of "distant" signals constitutes copyright infringement while a similar operation with respect to more nearby signals does not, the court relied in part on a description of CATV operations contained in this Court's opinion in *United States* v. *Southwestern Cable Co.,* 392 U. S. 157, announced a week before the decision in *Fortnightly:*

> "CATV systems perform either or both of two functions. First, they may supplement broadcasting by facilitating satisfactory reception of local stations in adjacent areas in which such reception would not otherwise be possible; and second, they may transmit to subscribers the signal of distant stations entirely beyond the range of local antennae." 392 U. S., at 163.

The Court in *Southwestern Cable,* however, was faced with conflicting assertions concerning the jurisdiction of the Federal Communications Commission to regulate in the public interest the operations of CATV systems. Insofar as the language quoted had other than a purely descriptive purpose, it was related only to the issue of regulatory authority of the Commission. In that context it did not and could not purport to create any separation of functions with significance for copyright purposes.[11]

---

[11] The FCC has consistently contended that it is without power to alter rights emanating from other sources, including the Copyright Act. In 1966 it indicated that its proposed rules regulating CATV operations would not "affect in any way the pending copyrights suit, involving matters entirely beyond [the FCC's] jurisdiction." Second Report and Order, Community Antenna Television Systems, 2 FCC 2d 725, 768. This position is consistent with the terms of the Communications Act of 1934, the source of the Commission's regulatory power, which provides, in part:

"Nothing in this chapter contained shall in any way abridge or alter the remedies now existing at common law or by statute, but

In the briefs and at oral argument various rationales for the distinction adopted by the Court of Appeals have been advanced. The first, on which the court itself relied, is the assertion that by importing signals from distant communities the CATV systems do considerably more than "enhance the viewer's capacity to receive the broadcaster's signals," *Fortnightly, supra,* 392 U. S., at 399, and instead "bring signals into the community that would not otherwise be receivable on an antenna, even a large community antenna, erected in that area." 476 F. 2d, at 349. In concluding that such importation transformed the CATV systems into performers, the Court of Appeals misconceived the thrust of this Court's opinion in *Fortnightly.*

In the *Fortnightly* case the Court of Appeals had concluded that a determination of whether an electronic function constituted a copyright "performance" should depend on "how much did the [CATV system] do to bring about the viewing and hearing of a copyrighted work." 377 F. 2d 872, 877. This quantitative approach was squarely rejected by this Court:

> "[M]ere quantitative contribution cannot be the proper test to determine copyright liability in the context of television broadcasting . . . . Rather, resolution of the issue before us depends upon a determination of the function that CATV plays in

---

provisions of this chapter are in addition to such remedies." 47 U. S. C. § 414.

Thus, it is highly unlikely that the "distant signal" definition adopted by the Commission or a differentiation of function based on such a definition was intended to or could have copyright significance. Indeed, as noted, the Court of Appeals in the present case found that the Commission's definition off a "distant signal" was unsatisfactory for determining if a "performance" under the Copyright Act had occurred. See n. 7, *supra.*

the total process of television broadcasting and reception." 392 U. S., at 397.

By importing signals that could not normally be received with current technology in the community it serves, a CATV system does not, for copyright purposes, alter the function it performs for its subscribers. When a television broadcaster transmits a program, it has made public for simultaneous viewing and hearing the contents of that program. The privilege of receiving the broadcast electronic signals and of converting them into the sights and sounds of the program inheres in all members of the public who have the means of doing so. The reception and rechanneling of these signals for simultaneous viewing is essentially a viewer function, irrespective of the distance between the broadcasting station and the ultimate viewer.

In *Fortnightly* the Court reasoned that "[i]f an individual erected an antenna on a hill, strung a cable to his house, and installed the necessary amplifying equipment, he would not be 'performing' the programs he received on his television set," 392 U. S., at 400, and concluded that "[t]he only difference in the case of CATV is that the antenna system is erected and owned not by its users but by an entrepreneur." *Ibid.* In the case of importation of "distant signals," the function is essentially the same. While the ability or inclination of an individual to erect his own antenna might decrease with respect to distant signals because of the increased cost of bringing the signal to his home, his status as a "non-performer" would remain unchanged. Similarly, a CATV system does not lose its status as a nonbroadcaster, and thus a non-"performer" for copyright purposes, when the signals it carries are those from distant rather than local sources.

It is further argued that when a CATV operator increases the number of broadcast signals that it may

receive and redistribute, it exercises certain elements of choice and selection among alternative sources and that this exercise brings it within scope of the broadcaster function. It is pointed out that some of the CATV systems importing signals from relatively distant sources could with equal ease and cost have decided to import signals from other stations at no greater distance from the communities they serve. In some instances, the CATV system here involved "leapfrogged" nearer broadcasting stations in order to receive and rechannel more distant programs.[12] By choosing among the alternative broadcasting stations, it is said, a CATV system functions much like a network affiliate which chooses the mix of national and local program material it will broadcast.

The distinct functions played by broadcasters and CATV systems were described in *Fortnightly* in the following terms:

> "Broadcasters select the programs to be viewed; CATV systems simply carry, without editing, whatever programs they receive. Broadcasters procure programs and propagate them to the public; CATV systems receive programs that have been released to the public and carry them by private channels to additional viewers." 392 U. S., at 400.

Even in exercising its limited freedom to choose among various broadcasting stations, a CATV operator simply cannot be viewed as "selecting," "procuring," or "propagating" broadcast signals as those terms were used in *Fortnightly*. When a local broadcasting station selects

---

[12] For example, it was represented in a brief before this Court that the Farmington, New Mexico CATV system imported signals from a Los Angeles station even though 113 other stations were closer or equidistant, including a number which, unlike the Los Angeles station, were in the same time zone as the Farmington community.

a program to be broadcast at a certain time, it is exercising a creative choice among the many possible programs available from the national network with which it is affiliated, from copyright holders of new or rerun motion pictures, or from its own facilities to generate and produce entirely original program material. The alternatives are myriad, and the creative possibilities limited only by scope of imagination and financial considerations. An operator of a CATV system, however, makes a choice as to which broadcast signals to rechannel to its subscribers, and its creative function is then extinguished. Thereafter it "merely carr[ies], without editing, whatever programs [it] receive[s]." *Ibid.* Moreover, a CATV system importing "distant" signals does not procure programs and propagate them to the public, since it is not engaged in converting the sights and sounds of an event or a program into electronic signals available to the public. The electronic signals it receives and rechannels have already been "released to the public" even though they may not be normally available to the specific segment of the public served by the CATV system.

Finally, it is contended that importation of "distant" signals should entail copyright infringement liability because of the deleterious impact of such retransmission upon the economics and market structure of copyright licensing. When a copyright holder first licenses a copyrighted program to be shown on broadcast television, he typically cannot expect to recoup his entire investment from a single broadcast. Rather, after a program has had a "first run" on the major broadcasting networks, it is often later syndicated to affiliates and independent stations for "second run" propagation to secondary markets. The copyright holders argue that if CATV systems are allowed to import programs and rechannel them into secondary markets they will dilute the profitability of

279

later syndications, since viewer appeal, as measured by various rating systems, diminishes with each successive showing in a given market. We are told that in order to ensure "the general benefits derived by the public from the labors of authors," *Fox Film Corp.* v. *Doyal,* 286 U. S. 123, 127, and "the incentive to further efforts for the same important objects," *id.,* at 128, citing *Kendall* v. *Winsor,* 62 U. S. (21 How.) 322, 328, current licensing relationships must be maintained.

In the television industry, however, the commercial relations between the copyright holders and the licensees on the one hand and the viewing public on the other are such that dilution or dislocation of markets does not have the direct economic or copyright significance that this argument ascribes to it. Unlike propagators of other copyrighted material, such as those who sell books, perform live dramatic productions, or project motion pictures to live audiences, holders of copyrights for television programs or their licensees are not paid directly by those who ultimately enjoy the publication of the material—that is, the television viewers—but by advertisers who use the drawing power of the copyrighted material to promote their goods and services. Such advertisers typically pay the broadcasters a fee for each transmission of an advertisement based on an estimate of the expected number and characteristics of the viewers who will watch the program. While, as members of the general public, the viewers indirectly pay for the privilege of viewing copyrighted material through increased prices for the goods and services of the advertisers, they are not involved in a direct economic relationship with the copyright holders or their licensees.[13]

---

[13] Some commentators have suggested that if CATV systems must pay license fees for the privilege of retransmitting copyrighted broadcast programs, the CATV subscribers will in effect be paying twice for the privilege of seeing such programs: first through in-

By extending the range of viewability of a broadcast program, CATV systems thus do not interfere in any traditional sense with the copyright holders' means of extracting recompense for their creativity or labor. When a broadcaster transmits a program under license from the copyright holder it has no control over the segment of the population which may view the program— the broadcaster cannot beam the program exclusively to the young or to the old, only to women or only to men— but rather he gets paid by advertisers on the basis of all viewers who watch the program. The use of CATV does not significantly alter this situation. Instead of basing advertising fees on the number of viewers within the range of direct transmission plus those who may receive "local signals" via a CATV system, broadcasters whose reception ranges have been extended by means of "distant signal" CATV rechanneling will merely have a different and larger viewer market.[14] From the point of view of the broadcasters, such market extension may mark a reallocation of the potential number of viewers each station may reach, a fact of no direct concern under the Copyright Act. From the point of view of the copyright holders, such market changes will mean that the

---

creased prices for the goods and services of the advertisers who pay for the television broadcasts and a second time in the increased cost of the CATV service. Note, CATV and Copyright Liability: On a Clear Day You Can See Forever, 52 Va. L. Rev. 1505, 1515 (1966); Note, CATV and Copyright Liability, 80 Harv. L. Rev. 1514, 1522–1523 (1967). See n. 15, *infra.*

[14] Testimony and exhibits introduced in the District Court indicate that the major rating services include in their compilations statistics concerning the entire number of viewers of a particular program, including those who receive the broadcast via "distant" transmission over CATV systems. The weight given such statistics by advertisers who bid for broadcast time and pay the fees which support the broadcasting industry was not, however, established. See n. 15, *infra.*

compensation a broadcaster will be willing to pay for the use of copyrighted material will be calculated on the basis of the size of the direct broadcast market augmented by the size of the CATV market.[15]

---

[15] It is contended that copyright holders will necessarily suffer a net loss from the dissemination of their copyrighted material if license-free use of "distant signal" importation is permitted. It is said that importation of copyrighted material into a secondary market will result in a loss in the secondary market without increasing revenues from the extended primary market on a scale sufficient to compensate for that loss. The assumption is that local advertisers supporting "first run" programs will be unlikely to pay significantly higher fees on the basis of additional viewers in a "distant" market because such viewers will typically have no commercial interest in the goods and services sold by purely local advertisers. For discussion of the possible impact of CATV "distant signal" importation on advertiser markets for broadcast television, see Note, *supra,* 52 Va. L. Rev., at 1513–1516; Note, *supra,* 80 Harv. L. Rev., at 1522–1525. The Court of Appeals noted that "[n]o evidence was presented in the court below to show that regional or local advertisers would be willing to pay greater fees because the sponsored program will be exhibited in some distant market, or that national advertisers would pay more for the relatively·minor increase in audience size that CATV carriage would yield for a network program," and concluded that "[i]ndeed, economics and common sense would impel one to an opposite conclusion." 476 F. 2d, at 342 n. 2. Thus, no specific findings of fact were made concerning the precise impact of "distant signal" retransmission on the value of program copyrights. But such a showing would be of very little relevance to the copyright question we decide here. At issue in this case is the limited question of whether CATV transmission of "distant" signals constitutes a "performance" under the Copyright Act. While securing compensation to the holders of copyrights was an essential purpose of that Act, freezing existing economic arrangements for doing so was not. It ·has been suggested that the best theoretical approach to the problem might be "[a] rule which called for compensation to copyright holders only for the actual advertising time 'wasted' on local advertisers unwilling to pay for the increase in audience size brought about by the cable transmission," Note, 87 Harv. L. Rev. 665, 675 n. 32 (1974). But such a rule would entail extended factfinding and a legislative, rather

These shifts in current business and commercial relationships, while of significance with respect to the organization and growth of the communications industry, simply cannot be controlled by means of litigation based on copyright legislation enacted more than half a century ago, when neither broadcast television nor CATV was yet conceived. Detailed regulation of these relationships, and any ultimate resolution of the many sensitive and important problems in this field, must be left to Congress.[16]

The judgment of the Court of Appeals is affirmed in part and reversed in part, and these cases are remanded to the District Court with directions to reinstate its judgment.

It is so ordered.

---

than a judicial, judgment. In any event, a determination of the best alternative structure for providing compensation to copyright holders, or a prediction of the possible evolution in the relationship between advertising markets and the television medium, is beyond the competence of this Court.

[16] The pre-*Fortnightly* history of efforts to update the Copyright Act to deal with technological developments such as CATV was reviewed in the *Fortnightly* opinion at 392 U. S., at 396 n. 17. At that time legislative action to revise the copyright laws so as to resolve copyright problems posed by CATV was of such apparent imminence that the Solicitor General initially suggested to this Court that it defer judicial resolution of the *Fortnightly* case in order to allow a speedy completion of pending legislative proceedings. Those legislative activities, however, did not bear fruit, apparently because of the diversity and delicacy of the interests affected by the CATV problem. See 117 Cong. Rec. 2001 (Feb. 8, 1971) (remarks of Sen. McClellan). Further attempts at revision in the 91st Congress, S. 542, and the 92d Congress, S. 644, met with a similar lack of success. At present, Senate hearings in the Subcommittee on Patents, Trademarks and Copyrights have been held on a bill that would amend the Copyright Act, S. 1361, but the bill has not yet been reported out of that subcommittee. A companion bill has been introduced in the House of Representatives, H. R. 8186 and referred to Judiciary Committee No. 3, but no hearings have yet been scheduled.

SUPREME COURT OF THE UNITED STATES

Nos. 72–1628 and 72–1633

Teleprompter Corporation et al.,
Petitioners,

72–1628   *v.*

Columbia Broadcasting System,
Inc., et al.

Columbia Broadcasting System,
Inc., et al., Petitioners,

72–1633   *v.*

Teleprompter Corporation et al.

On Writs of Certiorari to the United States Court of Appeals for the Second Circuit.

[March 4, 1974]

Mr. Justice Blackmun, dissenting in part.

I was not on the Court when *Fortnightly Corp.* v. *United Artists Television, Inc.,* 392 U. S. 390 (1968), was decided. Were that case presented for the first time today, I would be in full agreement with what Mr. Justice Fortas said in dissent. I would join his unanswered—and, for me, unanswerable—reliance on Mr. Justice Brandeis' unanimous opinion in *Buck* v. *Jewell-LaSalle Realty Corp.,* 283 U. S. 191 (1931). But *Fortnightly* has been decided, and today the Court adheres to the principles it enunciated and to the simplistic basis* on which it rests.

With *Fortnightly* on the books, I, as Mr. Justice Douglas, would confine it "to its precise facts and leave any extension or modification to the Congress." *Ante,* p. ——. The United States Court of Appeals for the Second Circuit decided the present case as best it could

---

*"Broadcasters perform. Viewers do not perform." 392 U. S., at 398 (footnotes omitted).

with the difficulties inherent in, and flowing from, *Fort-nightly* and the Copyright Act, and within such elbow-room as was left for it to consider the expanding technology of modern-day CATV. Judge Lumbard's opinion, at 476 F. 2d 338, presents an imaginative and well-reasoned solution without transgressing upon the restrictive parameters of *Fortnightly*. I am in agreement with that opinion and would therefore affirm the judgment.

SUPREME COURT OF THE UNITED STATES

Nos. 72–1628 AND 72–1633

Teleprompter Corporation et al.,
　　　　Petitioners,
72–1628　　　　*v.*
Columbia Broadcasting System,
　　Inc., et al.

Columbia Broadcasting System,
　　Inc., et al., Petitioners,
72–1633　　　　*v.*
Teleprompter Corporation et al.

On Writs of Certiorari
to the United States
Court of Appeals for
the Second Circuit.

[March 4, 1974]

Mr. Justice Douglas, with whom The Chief Justice concurs, dissenting.

The Court today makes an extraordinary excursion into the legislative field. In *United Artists Television, Inc.* v. *Fortnightly Corp.*, 392 U. S. 390, the lower courts had found infringement of the copyright, but this Court reversed holding that the CATV systems in *Fortnightly* were merely a "reception service," were "on the viewer's side of the line" *id.*, at 399, and therefore did not infringe the copyright act. They performed by cable, reaching into towns which could not receive a TV signal due, say, to surrounding mountains and expanded the reach of the TV signal within the confines of the area which a broadcaster's telecast reached.

Whatever one thinks of *Fortnightly*, we should not take the next step necessary to give immunity to the present CATV organizations. Unlike those involved in *Fortnightly*, the present CATV's are functionally the equivalent to a regular broadcaster. TV waves travel in straight

lines, thus reaching a limited area on the earth's curved surface. This scientific fact has created for regulatory purposes separate television markets.[1] Those whose telecast has covered one market or geographic area are under *Fortnightly* estopped to say that one who through CATV reaches by cable remote hidden valleys in that area, infringe the broadcaster's copyright. But the CATV's in the present case go hundreds of miles, erect receiving stations or towers that pick up the programs of distant broadcasters, and carry them by cable into a wholly different area.

In any realistic practical sense the importation of these remote programs into the new and different market is performing a broadcast function by the cable device. Respondents in the present case exercised their copyright privileges and licensed performance of their works to particular broadcasters for telecast in the distant market. Petitioners were not among those licensees. Yet they are granted use of the copyright material without payment of any fees.

The Copyright Act, 17 U. S. C. § 1 (c) and (d), gives the owner of a copyright "the exclusive right" to present the creation "in public for profit" and to control the manner or method by which it is "reproduced." A CATV that builds an antenna to pick up telecasts in Area B and then transmits it by cable to Area A is

---

[1] The Communications Act of 1934, §§ 303 (c), (d), (h), empowered the FCC to: "[A]ssign frequencies for each individual station," "determine the power which each station shall use," "[d]etermine the location of . . . individual stations," and "[h]ave authority to establish areas or zones to be served by any station." 47 U. S. C. §§ 303 (c), (d), and (h). Pursuant to these powers and others granted it by the Communications Act, the FCC has supervised the establishment and maintenance of a nationwide system of local radio and television broadcasting stations, each with primary responsibility to a particular community.

reproducing the copyright work not pursuant to a license from the owner of the copyright but by theft. That is not "encouragement to the production of literary (or artistic) works of lasting benefit to the world" that we extolled in *Mazer* v. *Stein,* 347 U. S. 201, 219. Today's decision is at war with what Chief Justice Hughes, speaking for the Court in *Fox Film Corporation* v. *Doyal,* 286 U. S. 123, 130, described as the aim of Congress:

> "Copyright is a right exercised by the owner during the term at his pleasure and exclusively for his own profit and forms the basis for extensive and profitable business enterprises. The advantage to the public is gained merely from the carrying out of the general policy in making such grants and not from any direct interest which the Government has in the use of the property which is the subject of the grants."

The CATV system involved in the present case performs somewhat like a network-affiliated broadcast station which imports network programs originated in distant telecast centers by microwave, off-the-air cable, precisely as petitioners do here.[2] Petitioner in picking up these distant signals is not managing a simple antenna reception service. It goes hundreds of miles from the community it desires to serve, erects a receiving station and then selects the programs from TV and radio stations in that distant area which it desires to distribute in its own distant market. If "function" is the key test as *Fortnightly* says, then functionally speaking petitioners are broadcasters; and their acts of piracy are flagrant

---

[2] Farmington, New Mexico, into which petitioners pipe programs stolen from Los Angeles is 600 miles away; and petitioner developed an intricate hookup "over twenty-three steps over a roundabout. 1300 mile route to establish the link." See 355 F. Supp., at 622.

violations of the Copyright Act. The original broadcaster is the licensor of his copyright and it is by virtue of that license that, say, a Los Angeles station is enabled lawfully to make its broadcasts. Petitioner receives today a license-free importation of programs from the Los Angeles market into Farmington, New Mexico, a distant second market. Petitioners not only rebroadcast the pirated copyright programs; they themselves—unlike those in *Fortnightly*—originate programs and finance their original programs [3] and their pirated programs by sales of time to advertisers. That is the way the owner of these copyrighted programs receives value for his copyrights. CATV does the same thing; but it makes its fortunes through advertising rates based in part upon pirated copyright programs. The Court says this is "a fact of no direct concern under the Copyright Act"; but the statement is itself the refutation of its truth. Rechanneling by CATV of the pirated programs robs the copyright owner of his chance for monetary rewards through advertising rates [4] on rebroadcasts in the distant area and gives those monetary rewards to the group that has pirated the copyright.

We are advised by an *amicus* brief of the Motion Picture Association that films from TV telecasts are being imported by CATV into their own markets in competition with the same pictures licensed to TV stations in the area into which the CATV—a nonpaying pirate of the films—imports them. It would be difficult to imagine a more flagrant violation of the Copyright Act. Since

---

[3] *Cable Television Report and Order*, 36 F. C. C. 2d 143, 290. And see *Rules re Micro-wave Served CATV*, 38 F. C. C. 683 (1965); *Cable Television Report and Order*, 36 F. C. C. 2d 148 (1972); *Radio Signals, Importation by Cable Television*, 36 F. C. C. 2d 630.

[4] We sustained the Commission's authority to require CATV to originate programs in a 5–4 decision in 1972. *United States* v. *Midwest Video Corp.*, 406 U. S. 649.

the Copyright Act is our only guide to law and justice in this case, it is difficult to see why CATV systems are free of copyright license fees, when they import programs from distant stations and transmit them to their paying customers in a distant market. That result reads the Copyright Act out of existence for CATV. That may or may not be desirable public policy. But it is a legislative decision that not even a rampant judicial activism should entertain.

There is nothing in the Communications Act that qualifies, limits, modifies, or makes exception to the Copyright Act.

"Nothing in this chapter contained shall in any way abridge or alter the remedies now existing at common law or by statute, but provisions of this chapter are in addition to such remedies." Moreover the Federal Communications Commission has realized that it can "neither resolve, nor avoid" the problem under the Copyright Act, when it comes to CATV.[5]

On January 14, 1974, the Cabinet Committee on Cable Communications headed by Clay T. Whitehead made its report to the President. That report emphasizes the need for the free flow of information in a society that honors "freedom of expression"; and it emphasizes that CATV is a means to that end and that CATV is so

---

[5] The Solicitor General in his brief in the *Fortnightly* case urged that the cable transmission of other station's programs into distant markets be subject to copyright protection:

". . . much of the advertising which accompanies the performance of copyrighted works, such as motion pictures, is directed solely at potential viewers who are within the station's normal service area—'local' advertising and 'national spot' advertising both fall within that category. Such advertisers do not necessarily derive any significant commercial benefit from CATV carriage of the sponsored programs outside of the market ordinarily served by the particular station, and accordingly may be unwilling to pay additional amounts for such expanded coverage."

closely "linked to electronic data processing, telephone, television and radio broadcasting, the motion picture and music industries, and communication satellites" *id.,* pp. 5–6, as to require "a consistent and coherent national policy" *id.,* 6. The Report rejects the regulatory framework of the Federal Communications Commission because it creates "the constant danger of unwarranted governmental influence or control over what people see and hear on television broadcast programming" *id.,* 7. The Report opts for a limitation of "the number of channels over which the cable operator has control of program content and to require that the bulk of channels be leased to others." *Ibid.*

The Report recognizes that "copyright liability" is an important phase of the new regulatory program the Committee envisages, *id.,* 14. The pirating of copyrights sanctioned by today's decision is anathema to the philosophy of this Report:

". . . Both equity and the incentives necessary for the free and competitive supply of programs require a system in which program retailers using cable channels negotiate and pay for the right to use programs and other copyrighted information. Individual or industry-wide negotiations for a license, or right, to use copyrighted material are the rule in all the other media and should be the rule in the cable industry.

"As a matter of communications policy, rather than copyright policy, the program retailer who distributes television broadcast signals in addition to those provided by the cable operator should be subject to full copyright liability for such retransmissions. However, given the reasonable expectations created by current regulatory policy, the cable operator should be entitled to a non-negotiated, blanket

license, conferred by statute, to cover his own retransmission of broadcast signals."

The Whitehead Commission Report has of course no technical, legal bearing on the issue before us. But it strongly indicates how important to legislation is the sanctity of the copyright and how opposed to ethical business systems is the pirating of copyright materials. The Court can reach the result it achieves today only by "legislating" important features of the Copyright Act out of existence. As stated by THE CHIEF JUSTICE in *United States* v. *Midwest Video Corp., supra,* at 676, "The almost explosive development of CATV suggests the need of a comprehensive re-examination of the statutory scheme as it relates to this new development, so that the basic policies are considered by Congress and not left entirely to the Commission and the Courts."

That counsel means that if we do not override *Fortnightly,* we should limit it to its precise facts and leave any extension or modification to the Congress.

S. 1361

IN THE SENATE OF THE UNITED STATES

MARCH 26, 1973

Mr. McCLELLAN introduced the following bill; which was read twice and
referred to the Committee on the Judiciary

A BILL

For the general revision of the Copyright Law, title 17 of the United States
Code, and for other purposes.

1   *Be it enacted by the Senate and House of Representatives of the*

2   *United States of America in Congress assembled,*

3   TITLE I—GENERAL REVISION OF COPYRIGHT LAW

4   SEC. 101. Title 17 of the United States Code, entitled "Copyrights,"

5   is hereby amended in its entirety to read as follows:

6   TITLE 17—COPYRIGHTS

§ 111. **Limitations on exclusive rights: Secondary transmissions**

(a) CERTAIN SECONDARY TRANSMISSIONS EXEMPTED.—The secondary transmission of a primary transmission embodying a performance or display of a work is not an infringement of copyright if:

(1) the secondary transmission is not made by a cable system, and consists entirely of the relaying, by the management of a hotel, apartment house, or similar establishment, of signals transmitted by a broadcast station licensed by the Federal Communications Commission, within the local service area of such station, to the private lodgings of guests or residents of such establishment, and no direct charge is made to see or hear the secondary transmission; or

(2) the secondary transmission is made solely for the purpose and under the conditions specified by clause (2) of section 110; or

(3) the secondary transmission is made by a common, contract, or special carrier who has no direct or indirect control over the content or selection of the primary transmission or over the particular recipients of the secondary transmission, and whose activities with respect to the secondary transmission consist solely of providing wires, cables, or other communications channels for the use of others: *Provided*, That the provisions of this clause extend only to the activities of said carrier with respect to secondary transmissions and do not exempt from liability the activities of others with respect to their own primary or secondary transmission; or

(4) the secondary transmission is made by a governmental body, or other nonprofit organization, without any purpose of direct or indirect commercial advantage, and without charge to the recipients of the secondary transmission other than assessments necessary to defray the actual and reasonable costs of maintaining and operating the secondary transmission service.

(b) SECONDARY TRANSMISSION OF PRIMARY TRANSMISSION TO CONTROLLED GROUP.—Notwithstanding the provisions of subsections (a) and (c), the secondary transmission to the public of a primary transmission embodying a performance or display of a work is actionable as an act of infringement under section 501, and is fully subject to the remedies provided by sections 502 through 506, if the primary transmission is not made for reception by the public at large but is controlled and limited to reception by particular members of the public.

1   (c) SECONDARY TRANSMISSIONS BY CABLE SYSTEMS.—

2   (1) Subject to the provisions of subsections (a) and (b), but not-

3 withstanding the provisions of clauses (2) and (4) of this subsection,

4 the secondary transmission to the public by a cable system of a pri-

5 mary transmission made by a broadcast station licensed by the Federal

6 Communications Commission and embodying a performance or display

7 of a work is subject to compulsory licensing under the conditions speci-

8 fied by subsection (d), in the following cases:

9   (A) Where the signals comprising the primary transmission

10   are exclusively aural; or

11   (B) Where the reference point of the cable system is within the

12   local service area of the primary transmitter; or

13   (C) Where the reference point of the cable system is outside

14   any United States television market, as defined in accordance

15   with subsection (f).

16   (2) Subject to the provisions of subsections (a), (b), and (e) and

17 of clauses (1) and (4) of this subsection, the secondary transmission

18 to the public by a cable system of a primary transmission made by a

19 broadcast station licensed by the Federal Communications Commis-

20 sion and embodying a performance or display of a work is subject to

21 compulsory licensing under the conditions specified by subsection (d),

22 in the following cases:

23   (A) Where the reference point of the cable system is within a

24   United States television market, as defined in accordance with

25   subsection (f), but the signal of the primary transmitter—

26   (i) when added to the signals of those television broadcast

27   stations whose local service areas are within that market,

28   and of any other television broadcast stations whose signals

29   are being regularly and lawfully used under this section by

30   the cable system for secondary transmissions, does not exceed

31   the number of signals of stations specified by clause (3)

32   as comprising adequate television service for that market;

33   and

34   (ii) is the signal of a television broadcast station of the

35   type whose lack deprives the market of adequate service in

36   accordance with the standards specified by clause (3), and

37   is closer to the market than the signal of any other station

38   of the same type, whose local service area is not within the

39   market; or

40   (B) Where, notwithstanding the provisions of subclause (A),

1   the cable system or its predecessor in title had, before January 1,

2   1971, in accordance with the applicable rules of the Federal

3   Communications Commission, made regular secondary trans-

4   missions of the transmissions of the primary transmitter or its

5   predecessor in title. And provided that such regular secondary

6   transmissions shall be exempt from the requirements of clauses

7   (4)(A) and (4)(B) of subsection (c).

8   (3) For the purposes of this subsection, "adequate television serv-

9   ice" within a United States television market is defined according to

10   the numerical rank of the market and the number and type of those

11   operating broadcast stations licensed by the Federal Communications

12   Commission whose local service areas are within that market. Con-

13   struction permits shall not be included in any computation for this

14   purpose.

15   (A) In markets 1 through 50, adequate television service com-

16   prises the network stations transmitting the programs of all the

17   television networks providing national transmissions, three inde-

18   pendent commercial stations, and one noncommercial educational

19   station.

20   (B) In markets 51 and below, adequate television service com-

21   prises the network stations transmitting the programs of all the

22   television networks providing national transmissions, two inde-

23   pendent commercial stations, and one noncommercial educa-

24   station.

25   (4) Subject to the provisions of subsections (a) and (b) and of

26   clause (1) of this subsection, but notwithstanding the provisions of

27   clause (2) of this subsection, the secondary transmission to the public

28   by a cable system of a primary transmission made by a broadcast

29   station licensed by the Federal Communications Commission and

30   embodying a performance or display of a work is actionable as an

31   act of infringement under section 501, and is fully subject to the

32   remedies provided by sections 502 through 506, in the following

33   cases:

34   (A) Where the cable system, at least one month before the

35   date of the secondary transmission, has not recorded the notice

36   specified by subsection (d); or

37   (B) Where the reference point of the cable system falls within

38   a circle defined by a radius of thirty-five air miles, or within a

39   radius as subsequently determined by the Federal Communica-

40   tions Commission, after notice and public hearings, from the cen-

ter of a United States television market, as defined in accordance with subsection (f), and—

(i) the primary transmission is made by a television broadcast station whose local service area is outside the market; and

(ii) a television broadcast station licensed by the Federal Communications Commission, whose local service area is within the market, has the exclusive right, under an exclusive license or other transfer of copyright, to transmit any performance or display of the same version of the work covered by the exclusive license or other transfer of copyright; and

(iii) except where the market is one of the first fifty of the United States television markets, the particular version of the work covered by the exclusive license or other transfer of copyright has never been transmitted to the public in a syndicated showing in the market by the station specified by paragraph (ii), or by any other television broadcast stations licensed by the Federal Communications Commission whose local service area is within the market; and

(iv) the station specified by paragraph (ii) has given written notice of said exclusive right to the cable system within the specified time limits and in accordance with the other requirements that the Register of Copyrights shall prescribe by regulation.

(C) Where the reference point of the cable system is within a United States television market, as defined in accordance with subsection (f), and—

(i) the content of the particular transmission program consists primarily of an organized professional team sporting event occurring simultaneously with the initial fixation and primary transmission of the program; and

(ii) the secondary transmission is made for reception wholly or partly outside the local service area of the primary transmitter; and

(iii) the secondary transmission is made for reception wholly or partly within the local service area of one or more television broadcasting stations licensed by the Federal Communications Commission, none of which has received authorization to transmit said program within such area.

(d) COMPULSORY LICENSE FOR SECONDARY TRANSMISSIONS BY CABLE SYSTEMS.—

(1) For any secondary transmission to be subject to compulsory

1 licensing under subsection (c), the cable system shall, at least one
2 month before the date of the secondary transmission, record in the
3 Copyright Office, in accordance with requirements that the Register of
4 Copyrights shall prescribe by regulation, a notice including a state-
5 ment of the identity and address of the person who owns the secondary
6 transmission service or has power to exercise primary control over it,
7 together with the name and location of the primary transmitter.

8 (2) A cable system whose secondary transmissions have been subject
9 to compulsory licensing under subsection (c) shall, during the months
10 of January, April, July, and October, deposit with the Register of
11 Copyrights, in accordance with requirements that the Register shall
12 prescribe by regulation—

13 (A) A statement of account, covering the three months next
14 preceding, specifying the number of channels on which the cable
15 system made secondary transmissions to its subscribers, the names
16 and locations of all primary transmitters whose transmissions
17 were further transmitted by the cable system, the total number
18 of subscribers to the cable system, and the gross amounts paid to
19 the cable system by subscribers for the basic service of providing
20 secondary transmissions of primary broadcast transmitters; and

21 (B) A total royalty fee for the period covered by the statement,
22 computed on the basis of specified percentages of the gross receipts
23 from subscribers to the cable service during said period, as
24 follows:

25 (i) 1 percent of any gross receipts up to $40,000;
26 (ii) 2 percent of any gross receipts totalling more than
27 $40,000 but not more than $80,000;
28 (iii) 3 percent of any gross receipts totalling more than
29 $80,000, but not more than $120,000;
30 (iv) 4 percent of any gross receipts totalling more than
31 $120,000, but not more than $160,000; and
32 (v) 5 percent of any gross receipts totalling more than
33 $160,000.

34 The total royalty fee shall include an additional 1 percent of
35 the gross receipts paid by subscribers for the basic service of
36 providing secondary transmissions of primary broadcast trans-
37 mitter for each channel on which the cable system, under a com-
38 pulsory license, is permitted by the Federal Communications
39 Commission to increase the number of signals comprising ade-
40 quate service pursuant to clause (2)(B) of subsection (e).

1   (3) The royalty fees thus deposited shall be distributed in accord-

2   ance with the following procedures:

3   (A) During the month of July in each year, every person claiming

4   to be entitled to compulsory license fees for secondary transmissions

5   made during the preceding twelve-month period shall file a claim with

6   the Register of Copyrights, in accordance with requirements that Reg

7   ister shall prescribe by regulation. Notwithstanding any provisions of

8   the antitrust laws (the Act of October 15, 1914, 38 Stat. 730, and any

9   amendments of any such laws), for purposes of this clause any claim-

10   ants may agree among themselves as to the proportionate division of

11   compulsory licensing fees among them, may lump their claims together

12   and file them jointly or as a single claim, or may designate a common

13   agent to receive payment on their behalf.

14   (B) After the first day of August of each year, the Register of

15   Copyrights shall determine whether there exists a controversy concern-

16   ing the distribution of royalty fees deposited under clause (2). If he

17   determines that no such controversy exists, he shall, after deducting

18   his reasonable administrative costs under this section, distribute such

19   fees to the copyright owners entitled, or to their designated agents.

20   If he finds the existence of a controversy he shall certify to that fact

21   and proceed to constitute a panel of the Copyright Royalty Tribunal

22   in accordance with section 803. In such cases the reasonable adminis-

23   trative costs of the Register under this section shall be deducted prior

24   to distribution of the royalty fee by the tribunal.

25   (C) After deducting the costs of administration, 15 percent of the

26   royalty fees collected shall be maintained in a special fund, and shall

27   be distributed, according to regulations prescribed by the Register of

28   Copyrights, to the copyright owners, or their designated agents, of

29   musical works.

30   (D) During the pendency of any proceeding under this subsection,

31   the Register of Copyrights or the Copyright Royalty Tribunal shall

32   withhold from distribution an amount sufficient to satisfy all claims

33   with respect to which a controversy exists, but shall have discretion to

34   proceed to distribute any amounts that are not in controversy.

35   (e) PREEMPTION OF OTHER LAWS AND REGULATIONS.—

36   (1) Except as provided by clause (2), on and after January 1, 1973,

37   all Federal, State, and local laws and regulations restricting the right

38   of a cable system to make secondary transmissions in any case made

39   subject to compulsory licensing by this section are preempted by this

40   title. Thereafter, unless specifically authorized by this subsection, the

1 Federal Communications Commission or any other governmental
2 agency or instrumentality shall not issue or enforce any order, notice,
3 rule, or regulation requiring a cable system to obtain authority of the
4 copyright owner as a condition for making any secondary transmis-
5 sion, or prohibiting a cable system from making secondary trans-
6 missions within an area where such secondary transmissions are per-
7 missible under the compulsory licensing provisions of subsection (c).
8 However, nothing in this section shall be construed to preempt the
9 authority of the Federal Communications Commission, with respect
10 to a cable system whose reference point is within a United States
11 television market—

12 (A) to prevent the cable system from further transmitting a
13 primary transmission made by a television broadcast station,
14 whose local service area is outside the market, on the same day
15 that another station licensed by the Commission, whose local serv-
16 ice area is within the market, transmits the same transmission
17 program;

18 (B) to compel the cable system to make secondary transmis-
19 sions of primary transmissions by television broadcast stations
20 licensed by the Commission, whose local service area is within the
21 market; and

22 (C) to regulate the operations of a cable system otherwise than
23 as provided by this section.

24 (2) Notwithstanding the provisions of clause (1), the Federal Com-
25 munications Commission shall have the responsibility to establish vari-
26 ous criteria and definitions as provided by subsection (f), and shall
27 have the authority in the public interest, and in accordance with re-
28 quirements that the Commission shall prescribe by regulation, to do
29 the following:

30 (A) to permit a cable system to substitute, for the signal of the
31 station specified in the compulsory licensing provisions of para-
32 graph (ii) of subsection (c) (2) (A), a more distant signal;

33 (B) to increase the number of signals of stations specified
34 in the compulsory licensing provisions of clause (3) of subsec-
35 tion (c) as comprising adequate television service for a United
36 States television market; and

37 (C) to permit a cable system that is required to delete a signal
38 under the provisions of clause (4) of subsection (c), to substitute
39 the signal of another station of the same kind and within the

300

quantitative limits specified by the compulsory licensing provisions of clause (3) of subsection (c).

(f) DEFINITIONS.—

(1) As used in this section, the following terms and their variant forms mean the following:

(A) A "primary transmission" is a transmission made to the public by the transmitting facility whose signals are being received and further transmitted by the secondary transmission service, regardless of where or when the performance or display was first transmitted.

(B) A "secondary transmission" is the further transmitting of a primary transmission simultaneously with the primary transmission.

(C) A "cable system" is a facility operated for purposes of commercial advantage that receives signals transmitted by one or more television broadcast stations licensed by the Federal Communications Commission and simultaneously makes secondary transmissions of such signals by wires, cables, or other communications channels to subscribing members of the public who pay for such service.

(2) As used in this section, the following terms and their variant forms have the meanings given to them in definitions that the Federal Communications Commission shall publish in the Federal Register during July, 1971, and annually in July thereafter. Said definitions shall have binding effect upon the 1st day of January of the year following their publication; they shall be based upon the general criteria provided by this clause, and upon specific criteria adopted by the Commission in the public interest and in the light of changing industry practices and communications technology. Annual publication of the definitions shall be accompanied by publication of lists specifying the reference points for all cable systems in the United States, the numerical rank of all United States television markets, and all network stations, independent commercial stations, and noncommercial educational stations, together with maps showing the specific geographical location of all said reference points, the area encompassed by all said United States television markets, and the local service areas of all said stations.

(A) The "reference point" of a cable system is the longitude and latitude, expressed in degrees, minutes, and seconds, of a point representing the effective center of operations of a cable system, taking into

1 account factors of geography, demography, and concentration of
2 subscribers.

3   (B) A "United States television market" is a community or group
4 of communities incorporating the local service areas of one or more
5 television broadcast stations licensed by the Federal Communications
6 Commission. The numerical ranking of such a market shall depend
7 primarily upon the number of viewers in the market receiving tele-
8 vision signals, but may be affected by other factors including the num-
9 ber of signals available in the market, concentration of population,
10 industrial development, and level of income.

11   (C) The "local service area" of a broadcast station comprises the
12 entire geographic area within the radius that the station's signal is
13 expected to reach effectively under normal conditions, including any
14 parts of the area within that radius that its signal fails to reach effec-
15 tively because of terrain, structures, or other physical or technical
16 barriers. Where the local service area of one station overlaps with that
17 of another, the overlapping area is considered within the local service
18 areas of both stations.

19   (D) A "network station" is a television broadcast station that is
20 owned or operated by, or affiliated with, one of the television networks
21 providing nationwide transmissions, and that transmits substantially
22 all of the programming supplied by such network.

23   (E) An "independent commercial station" is a television broadcast
24 station operated for commercial advantage, other than a network
25 station.

26   (F) A "noncommercial educational station" is a station operated
27 without any direct or indirect purpose of commercial advantage, whose
28 programming consists preponderantly of instructional, educational,
29 or cultural subject matter.

CBS
Columbia Broadcasting System, Inc.
51 West 52 Street
New York, New York 10019
(212) 765-4321

Frank Stanton, Vice Chairman of the Board

Dear Mr. Chairman:

The Federal Communications Commission now is considering proposed rules
for the regulation of cable television which will have a profound impact not only upon that medium but upon the structure and character of
over-the-air free broadcasting in this country.

The rules being considered by the Commission are based on those outlined
by Chairman Burch in his letter of August 5, 1971, to the Senate Commerce Committee and to the House Interstate and Foreign Commerce Committee, as amended by certain suggestions made by Clay T. Whitehead, Director of the Office of Telecommunications Policy (OTP), in an effort to
mediate differences between CATV, broadcast and copyright interests.
The amended proposal—wisely referred to as the "CATV Compromise"—
was offered by Dr. Whitehead on November 5 to certain interested private parties, namely, groups broadly representing the cable television
industry, the broadcasting industry and several major motion picture
companies. The proposal was accompanied by instructions to accept or
reject the Compromise on or before November 11.

On November 10 the Board of Directors of the National Association of
Broadcasters (NAB) considered the CATV Compromise. During these deliberations the CBS representatives argued strongly that the proposal was
not in the public interest nor in that of broadcasters. We urged, in
particular, that those portions dealing with copyright treatment of
CATV and copyright restrictions upon broadcasters, first made available
by the OTP only five days earlier, should not be acted on by broadcasters
without careful study and mature consideration. We also pointed out
that the Trial in the United States District Court for the Southern
District of New York of CBS v. Teleprompter had been concluded in September. The outcome of that lawsuit will determine the proper application of existing copyright law to cable systems in this country and,
thus, drastically affect the underlying, and we believe erroneous, premise of the CATV Compromise.

Nevertheless, a majority of the members of the NAB Board, moved in substantial part by considerations extrinsic to any issue involved in the
CATV Compromise, voted to endorse the proposal. The other groups likewise endorsed it.

Because of our disagreement with that endorsement, because of our intense interest in the proper relationship between broadcasting and CATV as evidenced by our past testimony in the Congress on these issues, and because our present position and intentions with respect to CBS v. Teleprompter may have been misinterpreted, it seems to me that the CBS position with respect to the CATV Compromise should be made as clear as possible to the Committees of Congress which will ultimately recommend legislation in this area of important public policy. Hence this letter.

First, we have no intention of abandoning CBS v. Teleprompter, even if it were appropriate, or possible, to withdraw from a case which is now before the court.

Unlike the purely passive CATV systems--relaying nearby television signals to subscribers--which were the subject of the Fortnightly case (where the Supreme Court held that there was no copyright liability), the CATV systems in the Teleprompter case import distant signals from stations hundreds of miles away, originate programming, sell advertising, engage in cable networking and otherwise conduct their business in a manner functionally equivalent to that of a broadcast station and most certainly not comparable to a home viewer's antenna. We have never regarded CBS v. Teleprompter, a suit initiated when we confidently expected to be an important factor in cable communications, as a lawsuit which is "hostile" to CATV or which is calculated to secure an unfair advantage over that medium. Rather, the purpose of the lawsuit is to provide a just copyright environment--one proper for the consideration of comprehensive copyright legislation under which free broadcasting and its potential major competitor would compete on terms fair to both media. Our position is not based upon any desire to unfairly exploit copyright. We own the copyrights only in a small proportion of the material broadcast by us, the bulk of such copyrights being owned by others. We have repeatedly expressed our intention, should we prevail in that lawsuit, to use every effort to acquire the necessary rights in copyrighted material broadcast by us to the end that cable systems could carry the signals of nearby television stations without charge or at a nominal charge. We adopt the same position with respect to any copyright legislation that may eventuate and we would note that our position on this matter is more favorable to such cable systems than are the copyright aspects of the CATV Compromise.

With respect to distant signals, however, we see no reason to grant to cable systems extraordinary copyright privileges which are not available to competitive broadcasters. It would be extremely anomalous for the Commission, which has previously eschewed interference with Congressional copyright policy, to sanction such privileges as a part of its regulatory scheme. The Commission requires that broadcast licensees meticulously canvass the needs and interests of their communities and provide programming to meet those interests. But the proposed CATV regulations would, in effect, create a new and favored class of broadcaster in this country, the independent stations in the country's major cities, whose programs will be imported by hundreds and perhaps thousands

of CATV systems, fractionalizing the local broadcasters' audiences for broadcasts of community interest. Nor do we see any justification for the provisions which curb the ability of local television stations to contract for reasonable copyright exclusivity in program material which they obtain from networks or which they purchase from program syndicators.

In any event, even if the subsidization of cable is to be accepted as a legitimate objective of public policy, there is a notable omission in the proposal. While undertaking to give cable special copyright privileges at the expense of broadcast licensees, it does not contain any provision which would protect broadcasters and the public at such time as the growth of cable makes it economically possible for CATV operators to bid sports and other attractions away from free television. In Chairman Burch's August 5 letter he notes that the Commission is "not unmindful of the possibility that a nationwide interconnected cable network, whether achieved by terrestrial or satellite technology, could remove sports programming from conventional broadcast television by offering sports teams more favorable terms than broadcast interests might be willing to pay." He characterizes this problem as "far from imminent," but expresses an intention to "keep a close watch on this question" and states that he would "welcome Congressional guidance in this area of national concern."

The risk that cable television at a relatively early date could siphon off sports and other events from over-the-air television is now much more imminent. We had not previously thought the protective anti-siphoning legislation would be necessary or desirable if cable systems were subjected to the same copyright obligations as their broadcast competitors. However, the inequality in copyright treatment provided under the CATV Compromise gives a distinct economic advantage to CATV in competition with free broadcasting. It is not difficult to foresee where this advantage would lead. It is the stated and evident objective of CATV to use the imported programs to increase the number of CATV subscribers and take away the audience of local broadcasters. With the loss of audience, of course, would come a loss in revenue for the broadcasters.

The Compromise agreement thus would set in motion a continuing buildup in the economic base for CATV and a continuing erosion of the economic base for free broadcasting stations. All concerned should understand clearly the serious consequences of this course. CATV eventually would be in a position to outbid free broadcasting not only for copyrighted programs but for major attractions such as outstanding live sports events, entertainment features and the services of the most popular and expensive entertainment stars. This could be true for CATV networks outbidding broadcasting networks for programs of national interest; it could be equally true for local CATV systems outbidding local broadcasters for the rights to carry local sports and other local events.

As this process develops, the erosion of free broadcasting would gather accelerated momentum. If, as we are repeatedly told by protagonists for cable television, it is to become the dominant mode of communications

in this country, supplanting over-the-air broadcasters, we see no reason why broadcasters should be compelled to provide a massive subsidy to help to bring this about.

What is at stake, however, is much more than the supplanting of one communications medium by a competing medium. There is a crucial public interest. For CATV would bring these attractions only to those who are able to pay for them, and for the most part only to those who live in populated areas where CATV is feasible.

It should be pointed out that the typical monthly charge for CATV equals or exceeds the current monthly average household expenditure for all spectator and participatory amusements. Millions of families who are poor, especially those who live in the inner cities and in rural areas, would lose the sports and entertainment and public affairs features which they now receive free over-the-air.

Accordingly, if the regulatory structure is to provide a forced subsidy for CATV at the expense of broadcasting, certainly this structure should provide protection against the siphoning of major attractions from free broadcasting to CATV. The FCC, in its rules for that form of cable television which imposes additional program charges, provides reasonable protection against such siphoning. The absence of protection in the Compromise agreement on regular CATV, which is only another form of pay television, is a glaring failure to safeguard the broadest public interest.

Finally, there is a remaining major question of public policy, presented by the proposed CATV rules, which has not received Congressional consideration and imperatively requires it. Although the FCC holds broadcasters to standards of accountability to the public interest, the CATV Compromise proposes that the cable operator "must not censor or exercise program content control of any kind." Instead of requiring CATV operators to exercise responsibility over programming, it suggests that the Commission should "explore whether it would be feasible or desirable to provide subscribers with a locked switch to cut off the public access or leased channels, should parents wish to control children's viewing."

The implications seem clear. Cable operators would not only be at liberty but compelled to present any kind of program material now seen in any medium of communications, including those media which limit exposure of certain material to adult audiences. Presumably, the only limitation would be material prohibited by criminal law.

This seems to us a startling departure from the historic public interest tradition of American broadcasting. It would be one thing for the FCC to reaffirm that the Commission, under Section 326 of the Communications Act, will not censor or otherwise interfere with the content of program material carried by cable television. It is quite another thing, however, for the Commission to require by regulation that a cable operator may not exercise editorial responsibility over what he offers. To do so is to posit that the counterpart function of a publisher or editor is superfluous in cable television, to be exercised by an individual transiently

willing to pay for a channel in order to transmit whatever material he chooses. This deficiency is especailly disturbing in view of the commit-ment of many CATV systems to more and more program origination. Here again, the Commission would be creating a double standard: one for over-the-air television and one for cable systems.

No other medium of public communication in this country--no newspaper, no magazine, no television or radio station or network--whether licensed or not by the government, is prohibited from exercising responsibility over the content of its material. We believe that such a prohibition involves a corruption of the First Amendment and a disservice to the public interest that should be corrected by Congress.

$$* \quad * \quad * \quad * \quad * \quad * \quad *$$

In summary, the Compromise agreement was accepted by the private organi-zations involved under a deadline that did not permit adequate consider-ation of the complexities and the potentially serious long-term portents. These matters should now be reviewed carefully by the Commission, by the Congress, and by all those concerned with the future of television ser-vice to the public.

What is at stake in the pending regulatory proposals is no more and no less than the question whether government policy should be directed at preserving and enhancing, or diminishing and destroying, free over-the-air broadcasting in this country. Those who see cable television as the wave of the future are not, of course, concerned about the consequences to free television from a policy favoring cable television. Thus, for example, the Sloan Commission on Cable Communications (On The Cable: The Television of Abundance, McGraw-Hill, 1971), blandly dismisses such consequences for free television in these words:

> "But in any case, if over-the-air television is to fall victim, in some degree or another, to technological change, it is in no dif-ferent position from any other enterprise in which investments have been made, and possesses no greater right than other industries to protection from technological change. It does not appear to the Commission that the industry needs or warrants further protection by regulatory agencies."

If the current threat posed to over-the-air television were only that of "technological change," we would have little proper concern. What does concern us, and what moves us to write this letter, is the risk that free television will fall victim not to technological change, but to a deliberate and, we think, mistaken public policy which would prevent free and fair competition between these media and favor cable television at the expense of free television. In this respect we urge, contrary to the Sloan Commission, that over the air television is in a "different position" from other enterprises. It is in a different position by vir-tue of the fact that it is the means by which the American public receives most of its news and information as well as its entertainment, and does so without distinctions based on ability to pay or on geographical sep-aration.

Over-the-air television, whatever its faults, is a means of communication which deserves much of the American people and should not be dismissed as merely another "enterprise in which investments have been made." It warrants, if not "protection" in the narrow sense referred to by the Sloan Commission, every fair opportunity to continue and to strengthen its service to the American people.

With all good wishes.

Sincerely,

The Honorable John O. Pastore
Chairman
Subcommittee on Communications
United States Senate
Washington, D.C.   20510

The Honorable John L. McClellan
Chairman
Subcommittee on Patents, Trademarks
   and Copyrights
Unites States Senate
Washington, D.C.   20510

The Honorable Torbert H. Macdonald
Chairman
Subcommittee on Communications and
   Power
House of Representatives
Washington, D.C.   20515

The Honorable Robert W. Kastenmeier
Chairman
Subcommittee No. 3
Judiciary Committee
House of Representatives
Washington, D.C.   20515

January 4, 1972

FEDERAL COMMUNICATIONS COMMISSION FCC 71-946
Washington, D.C. 20554   66781
September 8, 1971

W. Theodore Pierson, Jr.
Pierson, Ball & Dowd
1000 Ring Building
Washington, D.C. 20036

Dear Mr. Pierson:

This is in reply to your letter of July 7, 1971, written on behalf of
Time-Life Broadcast, Inc., and Sterling Manhattan Cable Television,
Inc., in which you ask the Commission to issue an interpretive ruling
that pay television cablecasting operations by Sterling Manhattan Cable
Television are "affirmatively authorized." You state that affirmative
authorization by the Commission is necessary for CATV systems presently
franchised in New York City because Section 4(1) of the New York City
Franchise, issued to Sterling Manhattan Cable Television, states in
part that, "The Company shall not engage in Pay Television nor shall
it deliver signals of any person engaged in Pay Television, unless and
until affirmatively authorized by the FCC."

In paragraph 17 of the Notice of Proposed Rule Making in Docket 18397,
15 FCC 2d 417 (1968), the Commission indicated that CATV's could make
per-program charges. More recently, in Clarification of CATV First
Report, 20 FCC 2d 741 (1969), this Commission ruled that local authorities are pre-empted from interfering with federally authorized CATV
origination and advertising. Accordingly, the Commission has pre-empted
the field of pay television cablecasting so that local franchise terms
are inoperative and no further affirmative authorization is required.
In any event, the cited actions make clear the Commission's recognition
that pay television on cable may serve the public interest and it has,
to the extent indicated, already authorized such operations. And see
"Letter of Intent of August 5, 1971", FCC 71-787.

Any such operation must be consistent with the Commission's present
regulations (See section 74.1121 of the Commission's Rules) and any
revision thereof (e.g., see Notice of Proposed Rule Making in Docket
No. 18893, FCC 70-678). This present letter is thus not to be construed
to sanction, authorize, or encourage the carriage of any specific program on pay cablevision. As you know, the Commission has expressed its
continued concern that programming now presented on broadcast television
might be siphoned off to cable, and it intends to keep a close watch
on this question and to take whatever action is necessary to protect
the public interest.

We trust the foregoing adequately supplies the information sought in your letter of July 7, 1971.

Commissioner Robert E. Lee not participating; Commissioner Johnson concurring in part and dissenting in part and issuing a statement; Commissioner Wells concurring in the result.

BY DIRECTION OF THE COMMISSION*

Ben F. Waple
Secretary

*See attached statement of Commissioner Johnson.

AGREEMENT FORM FOR POLITICAL BROADCASTS

STATION and LOCATION _____ _____ 19____

I, _____, (being)
(supporting) _____,

a legally qualified candidate for the office of _____ in the

_____ election, do hereby request station time as follows:

—LENGTH OF BROADCAST—	—HOUR—	DAYS	TIMES PER WEEK	—TOTAL NO. WEEKS—	—RATE—

DATE OF FIRST BROADCAST	DATE OF LAST BROADCAST

The broadcast time will be used by _____

I represent that the advance payment for the above-described broadcast time has been furnished by

_____ and you are authorized to so describe the sponsor in your

log, or otherwise, and to announce the program as paid for by such person(s).

The entity furnishing the payment, if other than an individual person, is: ( ) (1) a corporation; ( ) (2) a committee; ( ) (3) an association; or ( ) (4) other unincorporated group.

   (a) The corporation or other entity is organized under the laws of _____ .

   (b) The officers, board of directors and chief executive officers of the entity are: _____

   _____

   _____

It is my understanding that: The above is the same uniform rate for comparable station time charged all such other candidates for the same public office described above; the charges above do not exceed the charges made for comparable use of said station for other purposes; and the same is agreeable to me.

In the event that the facilities of the station are utilized for the above-stated purpose, I agree to abide by all provisions of the Communications Act of 1934, as amended, and rules and regulations of the Federal Communications Commission governing such broadcasts, in particular those provisions reprinted on the back hereof, which I have read and understand. I further agree to indemnify and hold harmless the station for any damages or liability that may ensue from the performance of the said broadcasts.

* For the above broadcast, I agree to prepare a script or transcription, which will be delivered to the station at least _____ before the time of the scheduled broadcast.

_____
(Candidate, Supporter or Agent)

Accepted)
Rejected) by _____ Title _____

If rejected, the reasons therefor are as follows:

This application, whether accepted or rejected, will be available for public inspection for a period of two years, in accordance with FCC Regulations (AM, Section 73.120; FM, Section 73.290; TV, Section 73.657).

* This statement is not applicable if the candidate is personally using the time.

LAWS AND REGULATIONS GOVERNING POLITICAL BROADCASTERS*

From the Communications Act of 1934, as amended:

Section 315. (a) If any licensee shall permit any person who is a legally qualified candidate for any public office to use a broadcasting station, he shall afford equal opportunities to all other such candidates for that office in the use of such broadcasting station: Provided, That such licensee shall have no power of censorship over the material broadcast under the provisions of this section. No obligation is hereby imposed upon any licensee to allow the use of its station by any such candidate. Appearance by a legally qualified candidate on any—

(1) bona fide newscast.

(2) bona fide news interview.

(3) bona fide news documentary (if the appearance of the candidate is incidental to the presentation of the subject or subjects covered by the news documentary), or

(4) on-the-spot coverage of bona fide news events (including but not limited to political conventions and activities incidental thereto),

shall not be deemed to be use of a broadcasting station within the meaning of this subsection. Nothing in the foregoing sentence shall be construed as relieving broadcasters, in connection with the presentation of newscasts, news interviews, news documentaries, and on-the-spot coverage of news events, from the obligation imposed upon them under this Act to operate in the public interest and to afford reasonable opportunity for the discussion of conflicting views on issues of public importance.

(b) The charges made for the use of any broadcasting station for any of the purposes set forth in this section shall not exceed the charges made for comparable use of such station for other purposes.

(c) The Commission shall prescribe appropriate rules and regulations to carry out the provisions of this section.

From the Rules of the Commission Governing Radio Broadcast Services:

Section 73.119. Sponsored programs, announcements of.—(a) When a standard broadcast station transmits any matter for which money, services, or other valuable consideration is either directly or indirectly paid or promised to, or charged or received by, such station, the station shall broadcast an announcement that such matter is sponsored, paid for, or furnished, either in whole or in part, and by whom or on whose behalf such consideration was supplied: Provided, however, that "services or other valuable consideration" shall not include any service or property furnished without charge or at a nominal charge for use on, or in connection with, a broadcast unless it is so furnished in consideration for an identification in a broadcast of any person, product, service, trademark, or brand name beyond an identification which is reasonably related to the use of such service or property on the broadcast.

(b) The licensee of each standard broadcast station shall exercise reasonable diligence to obtain from its employees, and from other persons with whom it deals directly in connection with any program matter for broadcast, information to enable such licensee to make the announcement required by this section.

(c) In any case where a report (concerning the providing or accepting of valuable consideration by any person for inclusion of any matter in a program intended for broadcasting) has been made to a standard broadcast station, as required by Section 508 of the Communications Act of 1934, as amended, of circumstances which would have required an announcement under this section had the consideration been received by such standard broadcast station, an appropriate announcement shall be made by such station.

(d) In the case of any political program or any program involving the discussion of public controversial issues for which any records, transcriptions, talent, scripts, or other material or services of any kind are furnished, either directly or indirectly, to a station as an inducement to the broadcasting of such program, an announcement shall be made both at the beginning and conclusion of such program on which such material or services are used that such records, transcriptions, talent, scripts, or other material or services have been furnished to such station in connection with the broadcasting of such program: Provided, however, that only one such announcement need be made in the case of any such program of 5 minutes' duration or less, which announcements may be made either at the beginning or conclusion of the program.

(e) The announcement required by this section shall fully and fairly disclose the true identity of the person or persons by whom or in whose behalf such payment is made or promised, or from whom or in whose behalf such services or other valuable consideration is received, or by whom the material or services referred to in paragraph (d) of this section are furnished. Where an agent or other person contracts or otherwise makes arrangements with a station on behalf of another, and such fact is known to the station, the announcement shall disclose the identity of the person or persons in whose behalf such agent is acting instead of the name of such agent.

(f) In the case of any program, other than a program advertising

commercial products or services, which is sponsored, paid for, or furnished, either in whole or in part, or for which material or services referred to in paragraph (d) of this section are furnished, by a corporation, committee, association, or other unincorporated group the announcement required by this section shall disclose the name of such corporation, committee, association, or other unincorporated group. In each such case, the station shall require that a list of the chief executive officers or members of the executive committee or of the board of directors of the corporation, committee, association or other unincorporated group shall be made available for public inspection at the studios or general offices of one of the standard broadcast stations carrying the program in each community in which the program is broadcast.

(Corresponding Rules—FM, 73.289; TV, 73.654.)

Section 73.120. Broadcasts by candidates for public office.

(a) Definitions. A "legally qualified candidate" means any person who has publicly announced that he is a candidate for nomination by a convention of a political party or for nomination or election in a primary, special, or general election, municipal, county, state or national, and who meets the qualifications prescribed by the applicable laws to hold the office for which he is a candidate, so that he may be voted for by the electorate directly or by means of delegates or electors, and who:

(1) has qualified for a place on the ballot or

(2) is eligible under the applicable law to be voted for by sticker, by writing in his name on the ballot, or other method, and

(i) has been duly nominated by a political party which is commonly known and regarded as such, or

(ii) makes a substantial showing that he is a bona fide candidate for nomination or office, as the case may be.

(b) General requirements. No station licensee is required to permit the use of its facilities by any legally qualified candidate for public office, but if any licensee shall permit any such candidate to use its facilities, it shall afford equal opportunities to all other such candidates for that office to use such facilities: Provided, That such licensee shall have no power of censorship over the material broadcast by any such candidate.

(c) Rates and practices. (1) The rates, if any, charged all such candidates for the same office shall be uniform and shall not be rebated by any means direct or indirect. A candidate, shall, in each case, be charged no more than the rate the station would charge if the candidate were a commercial advertiser whose advertising was directed to promoting its business within the same area as that encompassed by the particular office for which such person is a candidate. All discount privileges otherwise offered by a station to commercial advertisers shall be available upon equal terms to all candidates for public office. (2) In making time available to candidates for public office no licensee shall make any discrimination between candidates in charges, practices, regulations, facilities, or services for or in connection with the service rendered pursuant to this part, or make or give any preference to any candidate for public office or subject any such candidate to any prejudice or disadvantage; nor shall any licensee make any contract or other agreement which shall have the effect of permitting any legally qualified candidate for any public office to broadcast to the exclusion of other legally qualified candidates for the same public office.

(d) Records; inspection. Every licensee shall keep and permit public inspection of a complete record of all requests for broadcast time made by or on behalf of candidates for public office, together with an appropriate notation showing the disposition made by the licensee of such requests, and the charges made, if any, if request is granted. Such records shall be retained for a period of two years.

(e) A request for equal opportunities must be submitted to the licensee within one week of the day on which the prior use occurred

(f) A candidate requesting such equal opportunities of the licensee or complaining of non-compliance to the Commission shall have the burden of proving that he and his opponent are legally qualified candidates for the same public office. (Corresponding Rules—FM 73.290; TV, 73.657.)

Section 73.112 Program Log:

(a) the following entries shall be made in the program log: * * *

(1) (v) An entry for each program presenting a political candidate, showing the name and political affiliation of such candidate. * * *

(2) (iv) An entry showing that the appropriate announcement(s) (sponsorship, furnishing material or services, etc.) have been made as required by Section 317 of the Communications Act and §73.119. A check mark will suffice but shall be made in such a way as to indicate the matter to which it relates. * * *

(4) (ii) An entry for each announcement presenting a political candidate, showing the name and political affiliation of such candidate.

(Corresponding Rules—FM, 73.282; TV, 73.663.)

* For further details see NAB's "A Political Broadcast Catechism" (5th Ed.). Available on request.

§ 73.603 Numerical designation of television channels.

(a)

(b) In Alaska and Hawaii, the frequency bands 76-82 Mc/s and 82-88 Mc/s are allocated for non-broadcast use. These frequency bands (Channels 5 and 6) will not be assigned in Alaska or Hawaii for use by television broadcast stations.

Channel No.	Frequency band (mega-cycles)	Channel No.	Frequency band (mega-cycles)
2.........	54-60	43........	644-650
3.........	60-66	44........	650-656
4.........	66-72	45........	656-662
5.........	76-82	46........	662-668
6.........	82-88	47........	668-674
7.........	174-180	48........	674-680
8.........	180-186	49........	680-686
9.........	186-192	50........	686-692
10........	192-198	51........	692-698
11........	198-204	52........	698-704
12........	204-210	53........	704-710
13........	210-216	54........	710-716
14........	470-476	55........	716-722
15........	476-482	56........	722-728
16........	482-488	57........	728-734
17........	488-494	58........	734-740
18........	494-500	59........	740-746
19........	500-506	60........	746-752
20........	506-512	61........	752-758
21........	512-518	62........	758-764
22........	518-524	63........	764-770
23........	524-530	64........	770-776
24........	530-536	65........	776-782
25........	536-542	66........	782-788
26........	542-548	67........	788-794
27........	548-554	68........	794-800
28........	554-560	69........	800-806
29........	560-566	70........	806-812
30........	566-572	71........	812-818
31........	572-578	72........	818-824
32........	578-584	73........	824-830
33........	584-590	74........	830-836
34........	590-596	75........	836-842
35........	596-602	76........	842-848
36........	602-608	77........	848-854
37........	608-614	78........	854-860
38........	614-620	79........	860-866
39........	620-626	80........	866-872
40........	626-632	81........	872-878
41........	632-638	82........	878-884
42........	638-644	83........	884-890

(c) Channel 37, 608-614 Mc/s, is not available for assignment prior to January 1, 1974.

[28 F.R. 13660, Dec. 14, 1963, as amended at 35 F.R. 11179, July 11, 1970]

47 C.F.R. § 1.51

§ 1.51. Number of copies of pleadings, briefs and
other papers.

Except as otherwise specifically provided in
the Commission's rules and regulations, the number of copies of pleadings, briefs and other papers to be filed is as follows:

(a) In hearing proceedings, the following number of copies shall be filed:

(1) If the paper filed relates to a matter to
be acted on by the presiding officer or the Chief
Hearing Examiner, an original and 6 copies shall
be filed.

(2) If the paper filed relates to matters to
be acted on by the Review Board, an original and
12 copies shall be filed.

(3) If the paper filed relates to matters to
be acted on by the Commission, an original and
19 copies shall be filed.

(4) If more than one person presided (is presiding) at the hearing, an additional copy shall
be filed for each such additional person.

(5) In CATV hearing cases, two additional copies of all pleadings shall be filed.

(b) In rule making proceedings which have not
been designated for hearing, an original and 14
copies of all papers shall be filed.

(c) In matters other than rule making and hearing cases, an original and 11 copies of all papers
shall be filed.

(d) Where statute or regulation provides for
service by the Commission of papers filed with the
Commission, an additional copy of such papers
shall be filed for each person to be served.

(e) The parties to any proceeding may, on notice, be required to file additional copies of any
or all papers filed in that proceeding.

[36 F.R. 7423, Apr. 20, 1971]

FEDERAL COMMUNICATIONS COMMISSION
Washington, D. C. 20554

Form Approved
OMB No. 52-R0197

Code No.

ANNUAL REPORT OF CABLE TELEVISION SYSTEMS

If this report does not cover the full calendar year, indicate the period covered:

Period Beginning _____ , 19___ , and Ending _____ , 19____ .

A separate FCC Form 325 must be filed for each cable television system. Each separate and distinct community or municipal entity (including single, discrete, unincorporated areas) served by cable television facilities constitutes a separate cable television system, even if there is a single head end and identical ownership of facilities extending into several communities. Unless otherwise indicated, information should be stated as of the last day of the calendar year.

NOTE. Where cable television systems, as defined above, are under common ownership and *all* of the information requested in Sections 2 and 3 of this form is identical for each of them, one fully completed copy of this form may be filed for all of the systems, *with a completed copy of Section 1 of this form attached for each of the communities or municipal entities covered by the composite report.* If such a composite report is filed, Question 9 below should be answered.

SECTION 1. GENERAL INFORMATION

(If no label appears in this block,
fill in Questions - 4 below)

IF THE INFORMATION ON THE ABOVE LABEL IS COMPLETELY CORRECT, DO NOT ANSWER QUESTIONS 1-4 BELOW. IF THERE IS NO LABEL, OR IF ANY OF THE INFORMATION ON THE LABEL IS INCORRECT, ANSWER THE PARTS OF QUESTIONS 1-4 NECESSARY TO CORRECT THE LABEL OR TO SUPPLY MISSING INFORMATION.

1. _____
Full and exact name of cable television system (See Instruction 2)

2. _____
Mailing address

3. _____
City   State   ZIP Code

4. _____
Community Served   County   State

5. Number of subscribers on December 31 (See Instruction 3): .... _____

6. Population of community served (See Instruction 4): .... _____

7. Name and address of person to receive communications concerning this system, in addition to above (e.g., attorney or engineering consultant):

     a. _____
                                     Name

     b. _____
                               Mailing address

     c. _____
             City   *State*   *ZIP Code*

8. Date cable television service began ...._____
                                                     Month   *Year*

9. If a composite FCC Form 325 is being filed which is applicable to the reporting cable television system, provide the following information:

     ┌──────────────┐   a. _____
     │   │   *Name of cable television system given on composite report*
     │   Code No.   │   b. _____
     └──────────────┘   *Community served*   *County*   *State*

10. Estimated number of homes passed by cable on December 31 ...._____

11. Estimated number of strand or street miles of cable on December 31 .... _____

This report must be certified by the individual owning the reporting cable television system, if individually owned; by a partner, if a partnership; by an officer of the corporation, if a corporation; or by a representative holding power of attorney in case of physical disability of an individual owner or his absence from the United States.

CERTIFICATION

I certify that I have examined this report, and that all statements of fact contained therein are true, complete, and correct to the best of my knowledge, information, and belief, and are made in good faith.

_____   _____
        (Signature)   *(Title)*

_____   _____
   (Printed name of person signing)   *(Date signed)*

WILLFUL FALSE STATEMENTS MADE ON THIS FORM ARE PUNISHABLE BY FINE OR IMPRISONMENT. U.S. CODE, TITLE 18, SECTION 1001.

1. Provide the following information concerning television stations carried by the reporting cable television system:

	Call letters	City	State	Channel Designation		Method of Reception 1/
				Broadcast	Cable	
1.						
2.						
3.						
4.						
5.						
6.						
7.						
8.						
9.						
10.						
11.						
12.						
13.						
14.						
16.						
17.						
18.						
19.						
20.						

1/ Use the following code to indicate the method of reception of the particular television station by the cable television system:

O - off-the-air; CC - common carrier microwave; P - private microwave company (Business Radio or Community Antenna Relay Service).

2. Provide the following information concerning Standard or FM radio stations carried by the reporting cable television system (If allband FM is carried, write "allband" instead of listing each station).

	Call letters	City	State
1.			
2			
3.			
4.			
5.			
6.			
7.			
8.			
9.			
10.			

	City letters	City	State
11.			
12.			
13.			
14.			
15.			
16.			
17.			
18.			
19.			
20.			

3. Cable channel utilization. (See Instruction 5)

   a. Number of channels used for television signal carriage............_____

   b. Number of channels used for automated (time and weather, news ticker, etc.) program originations without "background music"....._____

   c. Number of channels used for automated program originations with "background music"..._____

   d. Number of channels used by the cable television operator or by others for non-automated program originations...................._____

   e. Number of channels used to deliver AM or FM broadcast signals to radio receivers.._____

   f. Number of channels available for television signal carriage but not in use..._____
      Reason(s) for non-use:_____

      _____

NOTE: Questions 4-9 of this Section should be answered with reference to the specified week for reporting, which is defined in Instruction 6. If it is not possible for the reporting cable television system to use the specified week, explain reasons in Exhibit No.____ . In such instances, the system may use any seven consecutive days out of the 60-day period immediately preceding the date of filing of this report; the dates used should be stated below.

IF THE REPORTING CABLE TELEVISION SYSTEM CARRIED NO AUTOMATED OR NON-AUTOMATED PROGRAM ORIGINATIONS DURING THE SPECIFIED WEEK, CHECK THIS BOX / / AND CONTINUE WITH QUESTION 10.

    / / Specified week used.   / / Week of _____, 19___
                        through _____, 19___ used.

4. Automated program originations and other automated services. (See Instruction 7)

Type	No. of cable television channels used	Hours during specified week;	Type
a. Time and weather			f. Burglar alarm system Yes / / No / /
b. News ticker			g. Fire alarm system Yes / / No / /
c. Sports ticker			h. Police surveillance Yes / / No / /
d. Stock market ticker			i. Facsimile reproduction Yes / / No / /
e. Other automated programming (Specify): _____ _____ _____ _____			j. Preference polling Yes / / No / /
			k. Utility meter reading Yes / / No / /
			l. Other services (Specify): _____ _____

318

5. Sources of non-automated program originations (See Instructions 7
   and 8. Include programming presented on free or leased channels.)

Type	No. of cable tele-vision channels used	Total hours during specified week
a. Local		
b. CATV Network or Series of System Interconnections		
c. Film		
d. Video Tape 1/		
e. Other (Specify): _____ _____		

     1/ If video tape was used, check appropriate box(es): ____
       ½-inch tape /__/   ¼-inch tape /__/   other /__/

6. Types of non-automated programming (Indicate the approximate number of
   hours of non-automated program originations during the specified week
   involving each of the following types of programming. Include programming
   presented on free or leased channels. Submit as Exhibit No. ___ a copy of a
   typical week's program originations schedule, if said schedules are
   published. See Instruction 9 for definitions of the programming categories.)

    a. News Programs (N)..............................._____hours
    b. Sports Programs (S)............................_____hours
    c. Public Affairs (Including Political and Editorial)
       Programs (P).................................._____hours
    d. Religious Programs (R).........................._____hours
    e. Instructional (Including Educational Institutions)
       Programs (I).................................._____hours
    f. Entertainment Programs (E)....................._____hours
    g. Other programs (O)............................._____hours

7. Use of cable television channels for non-automated program originations.

    The following questions concern use of channel time for non-automated
    program originations by the reporting cable television system's operator
    or by others. If a single cable channel was used for more than one
    purpose on a share-time basis during the specified week (for example,
    local government and leased uses), write "shared" and indicate how
    many channels were shared.

    a. Total number of channels used for non-automated program originations
       by the cable television operator or others during the specified
       week ...._____

7. (Continued)

b. Number of channels, if any, used by state or local government during the specified week.,.._____

c. Number of channels, if any, used by local or regional educational agencies, including schools, during the specified week..............._____

d. Number of channels, if any, used without charge by private groups or individual members of the public during the specified week......._____

e. Number of channels leased or subject to other sale of time during the specified week..._____

f. Number of channels used by the cable television operator during the specified week ...._____

g. What charges are assessed for use of the channel(s) described in (e) above? (As applicable, distinguish between lengths of leases or amounts of time sold, amount of equipment required, etc. A schedule of charges may be submitted as Exhibit No. ___ in lieu of a written response.) _____

_____

_____

8. Public service announcements. If the cable television system originates public service announcements (announcements for which no charge is made and which promote programs, activities, or services of governmental, community service, charitable, or similar organizations), indicate the total number of such announcements originated during the specified week...................._____

9. Commercial advertising in conjunction with program originations.

a. During the specified week, approximately how many minutes of commercial matter were carried on the cable television system per hour of program origination (including programming presented on leased channels)?

NOTE: If the commercial matter is continuously visible throughout the program (e.g. use of split-screen technique in conjunction with news ticker), the entire length of the program should be counted as commercial time. If the commercial matter is visible at regular intervals but not continuously (e.g. advertising messages scanned at intervals in a time-weather service), only the amount of time that the commercial matter is visible should be counted as commercial time.

(1) Automated originations......................................_____minut

(2) Non-automated originations..................................._____minutes

320

9. (Continued)

b. Purchasers of commercial advertising during the specified week. (Indicate the approximate number of purchasers in each category, and the proportion of total advertising revenues, to the nearest whole per cent, derived from each category of purchasers during the specified week.):

	Number of Purchasers	Per Cent of Total Advertising Revenue
(1) Local advertisers............................	_____	_____ %
(2) Advertisers in other nearby communities..	_____	_____ %
(3) Regional advertisers......................	_____	_____ %
(4) National advertisers......................	_____	_____ %
(5) Other (Specify): _____.	_____	_____ %

c. What advertising rates are charged? (A rate card may be submitted as Exhibit No. ___ in lieu of a written response.)

d. Are lessees of channels or purchasers of channel time permitted to solicit advertising for themselves.   Yes___ No___

10. If the reporting cable television system is operating under a franchise, submit as Exhibit No. ___ a copy of such franchise, unless one is already on file, in which case only a copy of changes (if any) since the last filing need be filed.

NOTE: As used in Question 10, "franchise" means any instrument --local ordinance, franchise, certificate, permit, or license -- by which the granting municipal or state authority authorizes construction and operation of a cable television system or use of public streets and ways in connection therewith.

321

NOTE: Pages 3-7 of this Section   must be submitted for each corporation or other business entity named in Question 1 below, as well as for the reporting cable television system, except as explained below.

If the reporting cable television system is unable to supply all of the information requested in this Section for itself and all of the corporations or other business entities named in Question 1 below, it should request the appropriate persons or business entities to supply the missing information directly to the Commission.

If the corporation or other business entity concerning which information is sought is a broadcast licensee or permittee and has filed ownership reports required by the Commission's Rules, then the information requested in Questions 3, 4, 5, 6 and 7 (a) of this Section need not be filed. Instead, indicate that such information is on file with the Commission, and specify the call letters under which the information was filed, the date of the last complete report and the dates of any subsequent supplements thereto, and the file number(s), if any.

If several cable television systems are under common ownership, so that one or more of the corporations or other business entities named in Question 1(c) below are the same for them, such cable television systems need not file separate copies of pages 3-7 of this Section for those corporations or other business entities. Instead, the reporting cable television system may indicate in Question 2 below the names of the corporations or other business entities which are the same, and the name and location of the cable television system whose annual report includes a fully completed copy of pages 3-7 of this Section for each of these corporations or other business entities.

In Question 1 (c) and (d), the heading preceding Question 3, and Questions 4 and 6, where a social security or employer identification (E.I.) number is requested, individuals should use their social security number and business entities should use the number used to identify themselves when filing reports with the Social Security Administration.

1. Business structure

    a. Check appropriate box to indicate business format of reporting cable television system:

      /¯/ Individually owned   /¯/ Partnership

      /¯/ Corporation   /¯/ Other

    b. If the reporting cable television system is individually owned, give the name, city and state of residence, and social security number of the owner in the space provided, and then continue with Question 7:

_____

Name (last, first)   Social Security No.

_____

City   State

1. (Continued)

    c.

        (i) If any corporation or other business entity owns 25% or more of the voting stock of the reporting cable television system, furnish the name, city and state of principal office, and employer identification number of each such corporation or other business entity on a separate sheet of paper in accordance with the "MODEL" attached to the Instructions.

       (ii) If 25% or more of the voting stock of an entity named above is owned by another corporation or other business entity, furnish the name, city and state of principal office, and employer identification number of each such corporation or other business entity on a separate sheet of paper in accordance with the "MODEL" attached to the Instructions. If 25% or more of the voting stock of an entity just named is owned by another corporation or other business entity, furnish the name, city and state of principal office, and employer identification number of each such corporation on a separate sheet of paper in accordance with the "MODEL".

    d. (i) If the reporting cable television system owns 25% or more of the voting stock of any corporation or other business entity, furnish the name, city and state of principal office, and employer identification number of each such corporation or other business entity on a separate sheet of paper in accordance with the "MODEL" attached to the Instructions.

       (ii) If the entity or entities named above own 25% or more of the voting stock of a corporation or other business entity, furnish the name, city and state of principal office, and employer identification number of each such corporation or other business entity on a separate sheet of paper in accordance with the "MODEL" attached to the Instructions.

2. If another FCC Form 325 is being filed which contains Section 3 ownership information for one or more of the corporations or other business entities named in Question 1(c) above, and if the reporting cable television system wishes to incorporate that information into this form, the following information should be provided:

_____

Name of corporation or other business entity (from Question 1(c))

_____

Name of cable television system whose Form 325 includes ownership dat

_____

Community served   County   State

323

Name of business entity reporting below	City	State	E.I. No.

3. Capitalization (Only applies to corporations)

Class of stock (preferred, common, or other)	Voting or non-voting	Number of shares		
		Authorized	Issued and Outstanding	Treasury	Unissued

(Number of shares sub-columns: Authorized, Issued and Outstanding, Treasury, Unissued)

4. Officers, directors, and ownership interests. (For partnership, list the name, social security number, city and state of residence, and per cent of interest in the partnership of each general or limited partner. For corporations, list all officers and directors (whether or not they own stock), and stockholders who own 3% or more of the voting stock of the corporation. If an ownership interest exists, record this to the nearest whole per cent based on the total number of outstanding shares of voting stock of the corporation, exclusive of treasury stock. Where stock is held **by a stockholder in a street name, this fact should be noted, but no further information concerning such stockholder need be furnished.**)

Name		City	State	Social security or E.I. number	Corporate position 1/	Percent interest
Last	First					

1/ Use code for corporate position: P- president; T-treasurer; S-secretary; V-vice-president; O-other officers: D-director

324

5. If there is any close family relationship (i.e., husband-wife, parent-offspring, brothers, sisters, or brother-sister) between any of the officers, directors, or stockholders listed in Question 4 of this Section, list below the names of the persons and the relationship.

Name of related person				Relationship
Last	First	Last	First	

6. If any of the corporate stock listed in Question 4 of this Section is held for any other person who is the beneficial owner of the stock, list below the name of the beneficial owner and the name of the person who votes the stock (e.g., trustee, executor, or custodian).

Name of person voting stock		Name of beneficial owner		City	State	Social security or E.I. number of beneficial owner
Last	First	Last	First			

325

7. Other communications interests

If any of the persons, corporations, or other business entities named in Questions 1, 4, or 6 of this Section is a stockholder owning 5% or more of the voting stock of any communications entity of the type described below, or is an officer, director, partner, or individual owner of such an entity, fill in the appropriate information. If the interest is a fiduciary one, e.g., trustee, check column F. Record ownership interests to the nearest whole per cent (based on the total number of outstanding shares of voting stock, exclusive of treasury stock, in the case of corporations).

a. Interests in any AM, FM, or television broadcast licensee or permittee.

Name of individual or entity having ownership interest		Name of broadcast entity	Call letters	AM, FM, or TV	Nature of interest 1/	% Voting interest	F
Last	First						

b. Interests in other cable television systems.

Name of individual or entity having ownership interest		Name of cable television system	City	State	Nature of interest 1/	% Voting interest	F
Last	First						

1/ Use code for nature of interest: O-officer; D-director; S-stockholder (other than officer or director); P-partner; I-owned as an individual.

7. (Continued)

c. Interests in manufacturers of cable television equipment.

Name of individual or entity having ownership interest		Name of cable television equipment manufacturer	City	State	Nature of interest 1/	% Voting interest	F
Last	First						

d. Interests in communications common carriers.

Name of individual or entity having ownership interest		Name of communications common carrier	City	State	Nature of interest 1/	% Voting interest	F
Last	First						

e. Interests in daily newspapers.

Name of individual or entity having ownership interest		Publisher's Name	Published for		Nature of interest 1/	% Voting interest	F
Last	First		City	State			

1/ Use code for nature of interest: O - officer; D - director; S - stockholder (other than officer or
director); P - partner; I - owned as an individual.

8. Has any person named in Questions 4, 5, or 6 of this Section been found guilty of any felony in any federal or state court within the past 10 years?

    /__/ Yes   /__/ No

If the answer is "Yes", submit as Exhibit No. ___ a statement disclosing the person and matters involved and identifying the court and proceeding by date and file numbers.

9. If any of the persons listed in Questions 4, 5, or 6 of this Section are aliens, submit as Exhibit No. ____ a list of their names, addresses, and nationalities.

10. If the reporting cable television system is unable to supply all of the information requested in this Section for itself and all of the corporations or other business entities named in Question 1, submit as Exhibit No. ___ a list of those persons or business entities for which any of the requested information is not being furnished, and include a detailed explanation of why the omitted material is unavailable.

If information requested in this Section is being submitted directly by any person or business entity other than the reporting cable television system, this page should be signed and dated in the space indicated below:

_____ (Name of individual submitting information, if submitted on behalf of that individual)	_____ (Signature of individual submitting information, if submitted on behalf of that individual)
	_____ Date
_____ (Name of business entity supplying information on behalf of itself or others)	_____ (Signature of person signing for business entity supplying information)
	_____ Title Date

WILLFUL FALSE STATEMENTS MADE ON THIS FORM ARE PUNISHABLE BY FINE OR IMPRISONMENT. U. S. CODE, TITLE 18, SECTION 1001.

DO NOT RETURN THESE INSTRUCTIONS TO THE COMMISSION

G E N E R A L   I N S T R U C T I O N S

(Annual Report of Cable Television Systems (FCC Form 325))

1. This form is to be filed as follows for all cable television systems (See Section 74.1101(a) for definition of a cable television system):

   a. Annually on or before March 1, for the preceding calendar year.

   b. File one copy of this form with the Federal Communications Commission, Washington, D. C. 20554.

   c. This form must be filled out completely. Typewriter or ink must be used. The expression "none" or "not applicable" should be given as the answer to any particular inquiry where it truly and completely states the fact.

   d. If the space provided for any information is insufficient or if it is necessary to insert additional statements, the insert pages shall be securely fastened in the report and shall be of durable paper conforming to this form in size and width of margin. Each insert shall bear the Section and Question number to which it pertains and the name of the cable television system.

INSTRUCTIONS FOR SECTION 1

2. The name of the cable television system in Question 1 shall be its full and exact name, and should be that of the business entity directly responsible for operation of the system.

3. In determining the number of subscribers to a cable television system (Question 5), where the system offers bulk rates to multiple-outlet subscribers, such as apartment house or motel operators, each bulk-rate contract is viewed as a number of subscriptions to be calculated by dividing the total annual charge for the bulk-rate contract by the system's basic annual subscription rate for an individual household. (Thus, for example, if a cable television system charges an apartment house operator $1000 a year for a bulk-rate contract and charges individual households a basic rate of $50 per year, the bulk-rate contract is counted as 20 subscriptions (i.e., 1000 ÷ 50 = 20)). Where a variety of "annual subscription rates" for individual households exists (e.g., $50 per year, if paid in one sum, or $60 per year, if paid on a per-month basis), the rate used in the subscriber formula should be the lowest annual rate which is offered to individual subscribers ($50 here). Likewise, if the bulk-rate contract is on a monthly basis, it should be divided by the lowest monthly rate which is offered. In the preceding example, a $50 per year charge should be viewed as a charge of $4.17 per month.

4. Indicate in Question 6 the population of the community served, according to the most recent U.S. Census. Estimate the community's population if it did not exist at the time of the last Census or if no Census figure is available for the community.

INSTRUCTIONS FOR SECTION 2

5. In Question 3, if a single cable channel is used for more than one purpose on a share-time basis (for example, automated and non-automated program originations are carried on the same cable channel), write "shared" and indicate how many channels are being shared.

   In Question 3(e), if allband FM is delivered by the reporting cable television system, write "allband".

6. The "specified week" for purposes of completing Questions 4-9 is the seven-day period commencing at 12:01 a.m. on December 1, 1971, and ending at midnight on December 7, 1971.

7. In Questions 4 and 5, if a single cable channel is used to provide more than one type of automated program origination (Question 4) or more than one mode of non-automated origination (Question 5) on a share-time basis, write "shared" in the "channels" column and indicate how many channels were shared.

8. Definitions for sources of programming in Question 5:

   (a) A local program (L) is any program originated or produced by a **cable television system, a channel lessee or other user, or an agent of either,** the content or setting of which is geographically local, primarily involves appearances by local persons, or primarily focuses on issues or subjects of local interest. A local program fed to a cable television network or series of interconnected cable television systems shall be classified by the originating system as local.

   (b) A cable television network program (NET) is any program furnished to the system by a cable television network (national, regional, or special) or by a series of interconnected cable television systems. Delayed cablecasts of programs originated by cable television networks or series of interconnected cable television systems are classified as network.

   (c) A film program (F) is any program consisting of the presentation of feature or syndicated films.

   (d) A video-taped program (VID) is any program not defined in (a), (b), or (c) above, presented via video tape.

330

9. Definitions for types of programming in Question 7:

(a) News programs (N) include reports dealing with current local, national, and international events, including weather and stock market reports, and, when an integral part of a news program, commentary, analysis, and sports news.

(b) Sports programs (S) include play-by-play and pre- or post-game related activities and separate programs of sport instructions, news or information (e.g., fishing opportunities, golfing instructions, etc.).

(c) Public Affairs (Including Political and Editorial) programs (P) include talks, commentaries, discussions, speeches, editorials, political programs (including those which present candidates for public office), documentaries, forums, panels, round tables, and similar programs primarily concerning local, national, and international public affairs.

(d) Religious programs (R) include sermons or devotionals, religious news, and music, drama, and other types of programs designed primarily for religious purposes.

(e) Instructional (Including Educational Institution) programs (I) include any program involving the discussion of, or primarily designed to further an appreciation or understanding of, literature, music, fine arts, history, geography, and the natural and social sciences; and programs devoted to occupational and vocational instruction, instruction with respect to hobbies, and similar programs intended primarily to instruct. It also includes any programs prepared by, in behalf of, or in cooperation with, educational institutions, educational organizations, libraries, museums, PTA's, or similar organizations. Sports programs shall not be included.

(f) Entertainment programs (E) include all programs intended primarily as entertainment, such as music, drama, variety, comedy, quiz, etc.

(g) Other programs (O) include all programs not falling within definitions (a) through (f).

SECTION 74.1101(a) OF THE COMMISSION'S RULES AND REGULATIONS

§74.1101 Definitions.

(a) Community antenna television system. The term "community antenna television system" ("CATV system") means any facility which, in whole or in part, receives directly or indirectly over the air and amplifies or otherwise modifies the signals transmitting programs broadcast by one or more television stations and distributes such signals by wire or cable to subscribing members of the public who pay for such service, but such term shall not include (1) any such facility which serves fewer than 50 subscribers, or (2) any such facility which serves only the residents of one or more apartment dwellings under common ownership, control, or management, and commercial establishments located on the premises of such an apartment house.

MODEL

OWNERSHIP INFORMATION TO BE FURNISHED IN ACCORDANCE WITH QUESTIONS 1(c) and 1(d)
OF SECTION 3 OF FCC FORM 325

c. (i) If any corporation or other business entity owns 25% or more of the voting stock of the reporting cable television system, enter the name, city and state of principal office, and employer identification number of each such corporation or other business entity in (A)-(D) below:

(A) _____
| Name | City | State | E.I. Number |

(B) _____
| Name | City | State | E.I. Number |

(C) _____
| Name | City | State | E.I. Number |

(D) _____
| Name | City | State | E.I. Number |

(ii) If 25% or more of the voting stock of an entity named in (A) above is owned by another corporation or other business entity, enter the name, city and state of principal office, and employer identification number of each such corporation or other business entity in (A)(1)-(A)(4) below. If 25% or more of the voting stock of an entity named in (A)(1)-(A)(4) below is owned by another corporation or other business entity, enter the name, city and state of principal office, and employer identification number of each such corporation or other business entity in the sublines of (A)(1)-(A)(4) below. If the names of corporations or other business entities are listed in (B), (C), and/or (D) above, use the same format that is provided below for (A). DO NOT provide information beyond the (A)(1)(a)-(A)(1)(d), etc. level.

(A)(1) _____
| Name | City | State | E.I. Number |

(A)(1)(a) _____
| Name | City | State | E.I. Number |

(A)(1)(b) _____
| Name | City | State | E.I. Number |

(A)(1)(c) _____
| Name | City | State | E.I. Number |

(A)(1)(d) _____
| Name | City | State | E.I. Number |

(A)(2) _____
| Name | City | State | E.I. Number |

(A)(2)(a) _____
| Name | City | State | E.I. Number |

(A)(2)(d) _____
| Name | City | State | E.I. Number |

(A)(3) _____
| Name | City | State | E.I. Number |

(A)(3)(a) _____
| Name | City | State | E.I. Number |

(A)(3)(d) _____
| Name | City | State | E.I. Number |

(A)(4) _____
| Name | City | State | E.I. Number |

(A)(4)(a) _____
| Name | City | State | E.I. Number |

(A)(4)(d) _____
| Name | City | State | E.I. Number |

d. (i) If the reporting cable television system owns 25% or more of the voting stock of any corporation or other business entity, enter the name, city and state of principal office, and employer identification number of each such corporation or other business entity in (A)-(D) below:

(A) _____
| Name | City | State | E.I. Number |

(B) _____
| Name | City | State | E.I. Number |

(C) _____
| Name | City | State | E.I. Number |

(D) _____
| Name | City | State | E.I. Number |

(ii) If the entity named in (A) above owns 25% or more of the voting stock of a corporation or other business entity, enter the name, city and state of principal office, and employer identification number of each such corporation or other business entity below, in such a way that (A)(1), (A)(2), ...etc. are subsidiaries of (A). If the names of corporations or other business entities are listed in (B), (C), and/or (D) above, use the same format that is provided below for (A). DO NOT provide information beyond the (A)(1)-(A)(4), etc. level.

(A)(1) _____
| Name | City | State | E.I. Number |

(A)(2) _____
| Name | City | State | E.I. Number |

(A)(3) _____
| Name | City | State | E.I. Number |

(A)(4) _____
| Name | City | State | E.I. Number |

FEDERAL COMMUNICATIONS COMMISSION
Washington, D. C. 20554

Form Approved
OMB No. 52-R0220

Code No.

CABLE TELEVISION ANNUAL FINANCIAL REPORT

For Period Beginning _____ , 19___ , and Ending _____ , 19___ .

A separate FCC Form 326 must be filed for each cable television system. Each separate and distinct community or municipal entity (including single, discrete unincorporated areas) served by cable television facilities constitutes a separate cable television system, even if there is a single head end and identical ownership of facilities extending into several communities. Cable television systems which commenced operations prior to December 1, 1971 may report on a fiscal year basis. All other systems must use a calendar year basis. Unless otherwise indicated, all information should be stated as of the last day of the fiscal or calendar year. See Instructions for definitions of terms used in the Schedules.

NOTE: Where cable television systems, as defined above, are under common ownership and normally keep a consolidated set of bookkeeping records, one fully completed copy of this form may be filed for all of the systems, *with a completed copy of Pages 1 and 2 of this form attached for each of the communities or municipal entities covered by the composite report.* If such a composite report is filed, Question 8 below should be answered.

(If no label appears in this block,
fill in Questions 1-4 below)

IF THE INFORMATION ON THE ABOVE LABEL IS COMPLETELY CORRECT, DO NOT ANSWER QUESTIONS 1-4 BELOW. IF THERE IS NO LABEL, OR IF ANY OF THE INFORMATION ON THE LABEL IS INCORRECT, ANSWER THE PARTS OF QUESTIONS 1-4 NECESSARY TO CORRECT THE LABEL OR TO SUPPLY MISSING INFORMATION.

1. _____
   Full and exact name of cable television system (See Instruction 2)

2. _____
   Mailing address

3. _____
   City   *State*   *ZIP Code*

4. _____
   Community served   *County*   *State*

ITEM 16   333

5. Number of subscribers on December 31 (See Instruction 3): .... _____

6. Population of community served (See Instruction 4): .... _____

7. Subscriber fees:

         a. Installation fee: $_____ ;   b. Monthly subscriber fee: $_____ ;

         c. Per program or per channel charges (A rate card may be submitted as Exhibit No.____ in lieu of a written response.): _____

8. If a composite FCC Form 326 is being filed which encompasses the finances of the reporting cable television system, provide the following information:

         a. _____
              Name of cable television system given on composite report

         b. _____
              Community served   *County*   *State*

This report must be certified by the individual owning the reporting cable television system, if individually owned; by a partner, if a partnership; by an officer of the corporation, if a corporation; or by a representative holding power of attorney in case of physical disability of an individual owner or his absence from the United States.

CERTIFICATION

I certify that I have examined this report, and that all statements of fact contained therein are true, complete, and correct to the best of my knowledge, information, and belief, and are made in good faith.

_____   _____
      (Signature)   *(Title)*

_____   _____
(Printed name of person signing)   *(Date signed)*

WILLFUL FALSE STATEMENTS MADE ON THIS FORM ARE PUNISHABLE BY FINE OR IMPRISONMENT. U.S. CODE, TITLE 18, SECTION 1001.

Line No.	SCHEDULE I. CABLE TELEVISION REVENUES AND EXPENSES	AMOUNT (Omit Cents)
	CABLE TELEVISION REVENUES	$
	SUBSCRIBER REVENUES:	
1	Installation charges	
2	Regular subscription charges	
3	Per program or per-channel charges	
4	Other subscriber revenues	
5	TOTAL SUBSCRIBER REVENUES	
	NON-SUBSCRIBER REVENUES:	
6	Advertising on cable channels programmed by reporting cable television system	
7	Leasing or other sale of channel time and/or facilities	
8	Other non-subscriber revenues	
9	TOTAL NON-SUBSCRIBER REVENUES	
10	TOTAL CABLE TELEVISION REVENUES (lines 5 + 9)	
	CABLE TELEVISION EXPENSES	
	SERVICE EXPENSES:	
11	Technical and maintenance payroll	
12	Pole and duct rentals	
	Microwave service costs:	
13	Community Antenna Relay Service (CARS) or Business Radio Service	
14	Domestic Public Point-to-Point Microwave Radio Service	
	Tariff (Leaseback) Charges /Lines 15-18 apply only to cable television systems receiving telephone company channel service/	
15	Cable	
16	Subscriber drops	
17	Input equipment	
18	Other	
19	All other service expenses	
20	TOTAL SERVICE EXPENSES	
	PROGRAM ORIGINATION EXPENSES:	
21	Program origination payroll	
22	Cost of outside program services (other than line 23)	
23	Performance or program rights	
24	All other program origination expenses	
25	TOTAL PROGRAM ORIGINATION EXPENSES	
	SELLING, GENERAL, AND ADMINISTRATIVE EXPENSES:	
26	Selling, general, and administrative payroll	
27	Franchise fees	
28	All other selling, general, and administrative expenses (other than Federal income taxes)	
29	TOTAL SELLING, GENERAL, AND ADMINISTRATIVE EXPENSES	
30	INTEREST EXPENSES	
31	DEPRECIATION OF TANGIBLE ASSETS, AND AMORTIZATION OF INTANGIBLE ASSETS AND LEASEHOLD IMPROVEMENTS	
32	TOTAL CABLE TELEVISION EXPENSES (lines 20 + 25 + 29 + 30 + 31)	

Line No.	SCHEDULE 2. CABLE TELEVISION INCOME	AMOUNT (Omit Cents)
1	Cable Television Revenues (from Schedule 1, line 10)	$
2	Cable Television Expenses (from Schedule 1, line 32)	
3	Cable Television Operating Income or Loss (line 1 minus line 2)	

SCHEDULE 3. TANGIBLE AND INTANGIBLE PROPERTY

Line No.	Class of Property	(1) Original Cost	(2) Depreciation to Date	(3) Depreciated Value (Col. (1) minus Col.(2))	(4) Property Life (in Years) Used for Depreciation	(5) Depreciation Method Used (Code)1/
1	Land and Buildings	$	$	$		
2	Head End					
3	Trunk and Distribution System					
4	Program Origination Equipment					
5	All Other Tangible Property					
6	TOTAL TANGIBLE PROPERTY (lines 1-5)				/////////	/////////
7	TOTAL INTANGIBLE PROPERTY					

1/ Use the following code to indicate depreciation method used (Column 5):

Code No.	Method
1	Straight Line
2	Declining Balance
3	Sum-of-the-years' Digits
4	Other

SCHEDULE 4. NON-RECURRING TELEPHONE COMPANY CHARGES
(TO BE COMPLETED ONLY BY CABLE TELEVISION SYSTEMS OPERATING UNDER A TELEPHONE COMPANY TARIFF (LEASEBACK))

Line	Class of Charge	(1) Original Cost	(2) Amortization to Date	(3) Unamortized Value (Col. (1) minus Col.(2))	(4) Amortization Term (in Years)	(5) Amortization Method Used (See Schedule 3 for Code)
1	Subscriber Drop Installations	$	$	$		
2	Cable					
3	Other Non-Recurring Charges					
4	TOTAL (lines 1-3)				/////////	/////////

SCHEDULE 5. EMPLOYMENT
Indicate the number of employees for the workweek in which December 31 falls:
Full time _____   Part-time _____

336

G E N E R A L   I N S T R U C T I O N S

(Cable Television Annual Financial Report (FCC Form 326))

1. This form is to be filed as follows for all cable television systems
    (See Section 74.1101(a) for definition of a cable television system):

    a. Annually on or before April 1, for the preceding calendar year.
        NOTE: A cable television system which commenced operations prior
        to December 1, 1971 may report on a fiscal year basis, in which
        case the form shall be filed annually no more than ninety (90)
        days after the close of the system's fiscal year, for the pre-
        ceding fiscal year.

    b. File one copy of this form with the Federal Communications Com-
        mission, Washington, D. C. 20554.

    c. The report must cover the full calendar or fiscal year to which
        it refers. If the system was operated for part of the year under
        other ownership, the present and former owners must file reports
        covering their respective periods of operation.

    d. Use dollar figures only. Cents should not be shown.

2. The name of the cable television system in Question 1, Page 1, shall
    be its full and exact name, and should be that of the business entity
    directly responsible for operation of the system.

3. In determining the number of subscribers to a cable television system
    (Question 5, Page 2), where the system offers bulk rates to multiple-
    outlet subscribers, such as apartment house or motel operators, each
    bulk-rate contract is viewed as a number of subscriptions to be cal-
    culated by dividing the total annual charge for the bulk-rate con-
    tract by the system's basic annual subscription rate for an indivi-
    dual household. (Thus, for example, if a cable television system
    charges an apartment house operator $1000 a year for a bulk-rate
    contract and charges individual households a basic rate of $50 per
    year, the bulk-rate contract is counted as 20 subscriptions (i.e.,
    1000 ÷ 50 = 20). Where a variety of "annual subscription rates" for
    individual households exists (e.g., $50 per year, if paid in one sum,
    or $60 per year, if paid on a per-month basis), the rate used in the
    subscriber formula should be the lowest annual rate which is offered
    to individual subscribers ($50 here). Likewise, if the bulk-rate
    contract is on a monthly basis, it should be divided by the lowest
    monthly rate which is offered. In the preceding example, a $50 per
    year charge should be viewed as a charge of $4.17 per month.

4. Indicate in Question 6, Page 2, the population of the community served,
    according to the most recent U.S. Census. Estimate the community's
    population if it did not exist at the time of the last Census or if
    no Census figure is available for the community

DEFINITIONS AND INSTRUCTIONS FOR SCHEDULES

Schedule 1: CABLE TELEVISION REVENUES AND EXPENSES

Lines 11, 21, and 26. Salaries: Apportion among these line entries the salary or wages of persons working in more than one job area.

Lines 21-25. Program Origination Expenses: Includes automated program originations (e.g., news or stock market ticker) as well as non-automated program originations.

Line 22. Cost of outside program services: Includes film and tape rentals, program packages, and program formats. If appropriate, apportion part of these costs to Line 23 (Performance or program rights).

Line 23. Performance or program rights: Includes copyright fees, music license fees, cost of rights to show live performances, and payments, if any, for television station programming carried by the cable television system.

Line 27. Franchise fees: Includes any franchise, license, or permit fees payable during the reporting year to local or state authorities.

Schedule 3: TANGIBLE AND INTANGIBLE PROPERTY

Original Cost (Column 1) is the cost of all tangible property used for cable television operations. If the property was acquired by the purchase of a going business, the cost reported should be that part of the total purchase price assignable to tangible property.

Property Life (in Years) Used for Depreciation (Column 4) is the number of years over which depreciation is to be taken on a class of property (e.g. head end). If the number of years of life varies among different items within a class of property, indicate the number of years used for the largest proportion (by value) of the property in that class.

Depreciation Method Used (Column 5). If more than one depreciation method is used within a class of property, indicate the method used for the largest proportion (by value) of the property in that class.

Line 1. Land and Buildings: Includes real property and buildings except the head end shack.

Line 2. Head End: Includes tower, antennae, electronic equipment (but not program origination equipment), head end shack, and all other equipment associated with the head end facility.

Line 3. Trunk and Distribution System: Includes equipment and
property that are associated with the distribution from the head
end to the subscriber, such as poles, cable, amplifiers, hardware,
taps, blocks, transformers, and other subscriber connection devices.

Line 4. Program Origination Equipment: Includes cameras, lights,
audio equipment, film projectors, video tape recorders, monitoring
and test equipment, film, tapes, etc.

Schedule 4.   NON-RECURRING TELEPHONE COMPANY CHARGES

This Schedule applies only to cable television systems receiving
telephone company channel service under a tariff (leaseback arrangements). Generally the non-recurring charges will be one-time
installation fees.

SECTION 74.1101(a) OF THE COMMISSION'S RULES AND REGULATIONS:

74.1101 Definitions

(a) Community antenna television system. The term "community antenna
television system" ("CATV system") means any facility which, in whole or
in part, receives directly or indirectly over the air and amplifies or
otherwise modifies the signals transmitting programs broadcast by one
or more television stations and distributes such signals by wire or cable
to subscribing members of the public who pay for such service, but such
term shall not include (1) any such facility which serves fewer than 50
subscribers, or (2) any such facility which serves only the residents of
one or more apartment dwellings under common ownership, control, or management, and commercial establishments located on the premises of such an
apartment house.

1. On May 3, 1971, the Commission issued a notice of proposed rule making in this docket,[1] in which we proposed the adoption of new rules, analogous to those already adopted for the broadcast and common carrier services, which would require—

(1) The adoption of equal employment opportunity (EEO) policies and practices by all cable television (cable) system operators and community antenna relay (CAR) station licensees and permittees; and

(2) Filing of—

(i) EEO program statements by—

(a) All CAR station permittees and licensees and cable system operators with five or more full-time employees; and

(b) All applicants for CAR and cable authorizations (except where five or more full-time employees at any time during the first 12 months of operation are not anticipated);

(ii) Annual EEO employment reports by all CAR permittees and licensees and cable operators with five or more full-time employees; and

(iii) Annual EEO complaint reports, by all CAR licensees and cable operators, of EEO complaints against them which have been submitted to appropriate governmental bodies.

2. Comments were received from the National Cable Television Association (NCTA); Gulf Communicators, Inc. (Gulf); Allen's TV Cable Service, Inc., and 66 other parties, filing jointly (Allen); the Jerrold Corp., NewChannels Corp., and Cox Cable Communications, Inc., filing jointly (Jerrold); and the Office of Communication, the Board for Homeland Ministries, and the Commission for Racial Justice, of the United Church of Christ, filing jointly (United Church). NCTA is a national association of cable system operators; the 67 parties to the Allen comments are all cable system operators and/or CAR station licensees; Jerrold, NewChannels, and Cox Cable are also cable operators; and the three parties to the United Church comments are instrumentalities of a 2,000,000-member religious denomination.

3. In their comments, the parties argue variously—

(1) Regarding the appropriateness of the proceeding—

(i) That the cable EEO rule making proceeding should be abandoned altogether (NCTA, Allen) on the grounds that any cable EEO regulation (a) would exceed the Commission's jurisdiction, (b) is unnecessary, and (c) would unreasonably and needlessly burden the Commission and the industry; and

(ii) That EEO regulations applicable to the cable industry should be adopted and enforced because, "[i]n light of the potential impact of CATV systems on private and business life, it is most essential that those charged with delivering such services be broadly representative of their communities" (United Church);

(2) Regarding required filing of EEO statements and reports—

(i) That the employment unit size (five full-time employees) proposed by the Commission as the cutoff point for exemptions from some of the proposed requirements is too low (NCTA, Allen, Jerrold);

(ii) That the proposed requirement of detailed annual employment reports should be abandoned, and replaced by a single line with a checkbox in the general cable annual report form to the effect that "FCC requirements with respect to nondiscrimination in employment have been fulfilled" (NCTA);

(iii) That, with respect to EEO-violation complaints, (i) the employer should not be required to submit any report at all unless an EEO-violation complaint was actually filed during the 12-month period in question (Jerrold, Gulf); and (ii) if an EEO-violation complaint was filed against the employer, space could be provided in the EEO annual employment report in which the employer could indicate the relevant information (Jerrold);[2] and

(iv) That the Commission's proposed definition of a "single employment unit" as potentially comprehending a cable operation serving more than one political subdivision (see proposed § 74.1125(b)(3))[3] is "totally inconsistent with the heretofore uniform definition of a CATV system as each separate political subdivision in which a cable operation is located" and that the Commission should "state the reason for and justify by explanation this seemingly arbitrary action" (Allen);

(3) Regarding the definition of "employee"—

(i) Personnel with ownership interests in the cable or CAR operation should not be counted as employees (Allen); and

(ii) That clarification is needed as to whether a person should be counted as an employee if—

(a) He is a director who (1) is nominally on the payroll, but has primary employment or obligations elsewhere, or (2) has a place of business or residence in a location removed from that of the reporting unit; or if

[Docket No. 19246; FCC 72–275]

PART 76—CABLE TELEVISION SERVICE

PART 78—CABLE TELEVISION RELAY SERVICE

Nondiscrimination in Employment Practices

Report and order. In the matter of amendment of the Commission's rules to require operators or community antenna television systems and community antenna relay station licensees to show nondiscrimination in their employment practices, Docket No. 19246.

[4] Petitioner, Emmett Valley Broadcasteres, did not file a supporting comment in this proceeding to make a public interest showing as to the need of Emmett for a first radio service. This indicates a possible loss of interest by petitioner in establishing a station at Emmett. However, since allocation proceedings are based on the public interest value of meeting the needs of communities, rather than meeting the private interests of specific proponents, we are assigning Channel 209A to Emmett in hope that it will be applied for by petitioner or some other interested party.

[5] Commissioners Robert L. Lee and Johnson absent.

[1] FCC 71–455, 29 FCC 2d 18, 36 F.R. 8457.

[2] Jerrold suggests, in this connection, that the rules could require the filing of a separate EEO-violation complaint report where an employment unit (i) is not required to file an annual employment report, but (ii) has been the subject of an EEO-violation complaint.

[3] In the Cable Television Report and Order, FCC 72–108, released Feb. 3, 1972, the Commission revised and renumbered the cable television rules in the new Part 76. Under this scheme, § 74.1125 became § 76.311. In referring to the rules adopted herein, we shall use the new numbering for easy reference to the attached appendix which contains the full text of the rules. However, the section numbers utilized in the proposed rules appended to the notice of proposed rule making in Docket No. 19246 will be used in those cases where the reference or quotation relates to the originally proposed rules rather than to the rules as finally adopted herein.

(b) He works for the system or station on a voluntary or experimental basis and receives no pay (Jerrold);

(4) Regarding other contents of material filed by the employer—

(i) That cable systems and CAR stations should be required to include, in their annual reports, showings of affirmative action with respect to the employment of women (United Church); and

(ii) That the Commission provide clarification as to how formal a "training program" must be to be so characterized in the employer's filings to the Commission (Jerrold); and

(5) Regarding filing deadlines, that a common date, e.g., May 31 of each year, should be designated as the EEO filing deadline for (a) reporting changes in, or submitting new, employment programs, (b) filing complaint reports, and (c) filing annual employment reports; and that employment programs should be required to be submitted either (i) on the first May 31 after service commences, or (ii) within 30 days after the initial commencement of service (Jerrold); [4]

(6) Regarding employer maintenance of EEO records for public inspection—

(i) That the employer should be permitted to require of persons wishing to examine his EEO records on his premises that they make a reasonable showing of need for such examination (Gulf);

(ii) That he should not be required to keep EEO records, for such inspection, at each place of employment reported on, but should be permitted instead to keep the records (a) at a central office (Gulf), or (b) "in the same place as other material which must be kept available for public inspection" (Jerrold);

(iii) That he should be required to retain such records for public inspection not for 2 years but for 5 (United Church); and

(iv) That he should be required to provide any member of the public who asks for it with a quick-copy of any EEO material kept by the employer for public inspection, and charge no more than 5 cents per page for the service" (United Church);

(7) Regarding the posting of EEO notices at the place of employment (see proposed § 74.1125(c)(2)(i)(a)), that the notice of EEO rights should be in one or more languages in addition to English "where 5 percent of the population or service area consists of Spanish-surnamed Americans or other groups whose original or native language is not English" (United Church); and

---

[4] In connection with this recommendation, Jerrold suggests that the proposed §74.1025(c)(1)(i)(d) be amended to read, "If pursuant to § 74.1025(c)(1)(i)(b) or § 74.1031(f), a station has been exempted from the requirement that it file an equal employment opportunity statement, but has failed to satisfy the conditions of that exemption at any time for 6 months of the past calendar year, it shall file the statement on or before May 31 of the following calendar year"; and that § 74.1125(c)(1)(i)(d) be amended similarly.

(8) Regarding liaison with other governmental agencies, that the Commission should itself review alleged EEO violations instead of referring them to the U.S. Equal Employment Opportunities Commission or to other State or local authorities for determination (United Church).

4. In support of the contention that the Commission lacks jurisdiction to regulate the EEO aspects of cable system operation, it is argued (by Allen) that: (1) "in United States v. Southwestern Cable Co., 392 U.S. 157 (1968), * * * the [Supreme] Court held that the Commission's authority to regulate CATV 'is restricted to that reasonably ancillary to the effective performance of the Commission's responsibilities for the regulation of television broadcasting'" (id. at 178); (2) in Midwest Video Corporation v. United States, 441 F. 2d 1322 (1971), the U.S. Court of Appeals, 8th Circuit, held that the Commission's mandatory program origination rule is invalid in that it "'goes far beyond the regulatory power approved in Southwestern Cable Co.'" (id. at 1325); and (3) EEO regulation of cable systems is "far less related to effective broadcast regulation than [is] enactment of program origination rules": In view of the economic interplay and service competition between broadcasting and cablecasting, "it is at least arguable—if not tenable—that prescription of cablecast rules is reasonably related to effective broadcast regulation," but "imposition of employment discrimination rules has not the slightest bearing on effective broadcast regulation."

5. The foregoing jurisdictional contention and argument are rejected for reasons spelled out in the text of the "Petition for a Writ of Certiorari * * *" recently submitted by the Solicitor General to the Supreme Court in response to the 8th Circuit's Midwest Video decision. As noted in the certiorari petition, the Supreme Court in Southwestern did not restrict the Commission's authority to regulate CATV "to that reasonably ancillary to the effective performance of the Commission's responsibilities for regulation of television broadcasting"; it merely restricted its holding to that question. At the same time, the text of the Supreme Court opinion in Southwestern indicated that the Communications Act confers on the Commission essentially the same jurisdiction over CATV that it has over the broadcasting industry. That jurisdiction includes the authority to condition FCC authorization of cable systems upon their service to the public interest via their compliance with reasonable regulations by the Commission (and other governmental units) which both (i) further the National policy against discrimination in employment and, (ii) in so doing, increase the likelihood that such systems and stations will operate generally in a non-discriminatory fashion, beneficial to all segments of the communities served.

6. Measured in the light of such considerations as the importance of the objectives sought, the practical relationship between those ends and the regulatory means in question, and the extent of the "burden" which would be imposed upon cable systems, we believe that both the proposed rules and the rules finally adopted herein are reasonable indeed: The objectives are important; the regulatory means set forth are not excessive and do further those objectives; and the "burden" placed upon cable system operators is in fact quite slight.

Moreover, as stated in the certiorari petition—

* * * even if Commission regulation of CATV must be "reasonably ancillary" to regulation of broadcasting, there is no reason to believe that this standard would restrict the Commission to regulation designed to prevent deleterious competition to the broadcasting industry. Having become "an integral part of interstate broadcast transmission," CATV operators "cannot have the economic benefits of such carriage as they perform and be free of the necessarily pervasive jurisdiction of the Commission." * * * Thus, even if there must be some connection between the Commission's regulation of CATV and its responsibilities in the broadcasting area, such connection is present when, as here, the Commission attempts to require CATV operators, whose principal product is the retransmission of broadcast signals, and who serve the same functions in many areas as broadcasters, to meet some of the same basic standards of reponsibility to the public that are imposed on broadcasters. (Citations deleted.)

7. Although none of the respondents question the importance of the Commission's objectives in this proceeding, two of the comments on file (NCTA, Allen) generally question the need for adoption of EEO regulations in the cable field to achieve those ends. In the NCTA pleading, it is flatly asserted that "cable systems do not practice discrimination * * * [regarding] race, color, religion, national origin, or sex" in hiring practices, and that "An industry which probably represents the last great chance for minorities to get into realistic communications media will not tolerate discriminatory employment practices." Although we hope that this is so, we cannot, in the light of our responsibility to the public, rely upon such representations alone. Facts are needed, regularly brought up to date, to provide continuing assurance that such representations are generally correct, and to draw our attention to those cable employers (hopefully few) whose employment practices are unsatisfactory in this regard. Moreover, in view of the importance of cable's present and prospective roles in the American communications system, and in the lives of the American people, we are persuaded that a mere abstinence from overt discriminatory practices is not sufficient. The effects of past discrimination in our country are such that a positive continuing program of specific practices designed to assure equal opportunity in every aspect of CATV employment is required if the term "equal opportunity employer" is to be truly meaningful. Elements of such a program are noted at

§ 76.311 (b)(2) and (c)(2). The program items set forth therein are not intended to be fully applicable to every situation. As stated in the introductory words of § 76.311(c)(2): "The program should reasonably address itself to such specific areas as set forth below, to the extent that they are appropriate in terms of employment unit size, location, etc." [5]

8. NCTA suggests that the only reason for the application of EEO rules to cable systems is that the Commission has previously applied such rules to broadcast and common carrier facilities. Although that assertion is incorrect, it is of value nonetheless in that it draws attention to the fact that cable is quite similar in its major aspects to both broadcasting and common carrier service, and that both of these are already subject to EEO rules. Cable—which already serves close to 10 percent of the American people, and may be only at the threshold of its major growth—is ancillary to broadcast service in that it retransmits the signals of broadcast stations. It may exercise some discretion as to which broadcast signals, and programs, it will carry. To the extent that it engages directly in program origination, it performs for its subscribers the basic functions of a broadcast station itself. Like a common carrier, it is almost certain, in any given service area, to be a monopoly (or, at best, a semimonopoly) whose customers have no recourse to the normal operation of the market place if they desire cable service but are not satisfied with the quality of that service, or with the willingness of their local cable system to meet the specialized programing and other needs of particular segments of the population. Like a common carrier, the local cable system either is, or will soon become, a contractor for delivery of communications (cablecast, etc.) by others on its leased, or free-of-charge, "access" channels.

9. Cable, by virtue of its multichannel capacity, is uniquely capable of serving the special programing, and other communications needs of discriminated-against minority groups. But a company which is not an equal opportunity employer is less likely than it otherwise would be, to recognize and respond to those needs. In the light of this fact, among others, it would certainly be improper for the Commission to countenance discriminatory employment practices by cable systems at the same time as it forbids such practices by broadcast and common carrier facilities.

10. The Allen pleading contends that the rules proposed by the Commission would result in an unnecessary duplication of regulatory functions entrusted to other agencies, notably the U.S. Equal Employment Opportunity Commission. EEOC is an agency with limited resources which functions in part as a direct instrument of the National policy against discrimination in employment, and in part as a facilitator of the efforts of other Governmental units. EEOC has welcomed, encouraged, and assisted this Commission in the development of our own EEO-regulatory program, and we look forward to further cooperation with EEOC in the implementation of our program. Moreover, because of the special nature of cable, its special impact upon the community, and our special responsibilities as guardians of the public interest in the wire and radio communications field, the reporting requirements adopted herein will affect many cable employment units which EEOC could not reach.

11. Having considered the contention, in three of the pleadings, that the employment size proposed by the Commission as the cutoff point for exemption from certain of the filing requirements (five full-time employees), is too low, we have decided, for the present, to stay with that figure. We are not persuaded that the filing requirements adopted herein will impose a significant burden upon employment units with five or more full-time employees. However, we will reexamine the matter in the future, on the basis of further information and experience.

12. As previously noted, one of the comments (Allen) questions the Commission's definition of an "employment unit" as potentially including more than one cable system, on the ground that deviation from the Commission's more typical unit of cable regulation (i.e., an individual cable system whose facilities serve only one political subdivision) is a "seemingly arbitrary action." That comment presumably refers to the following provisions:

(1) In § 74.1125(c)(3):

Where two or more community antenna relay stations and/or community antenna television stations, under common ownership and/or control, are so interrelated in their management, operations, and utilization of employees as to constitute a single employment unit, the [equal employment opportunity] program shall be jointly established, maintained, and controlled by them.

And (2) in § 74.1125(f)(2):

Where, pursuant to § 74.1125(c)(3), an equal employment opportunity program is jointly established by two or more community antenna relay stations and/or community antenna television systems with an aggregate total of * * * [10] or more full-time employees, a combined employment report shall be filed.[6]

These provisions are not "arbitrary," either "seemingly" or in fact. Many functionally unitary cable operations serve two or more political subdivisions from a single headend. In such cases, the Commission's characterization of the operation as consisting of two or more "systems" each serving a different political subdivision is a "legal fiction," useful for many purposes, which takes into account the separate franchising powers of separate political subdivisions. In certain aspects of day-to-day cable activity, however—including employment practices—such usage does not sufficiently reflect the business realities referred to in § 74.1125(c)(3). In determining the appropriate unit of regulation, we have borrowed from the architectural profession the precept that form should follow function. In doing so, no "arbitrary action" is involved; just plain good sense.

13. Upon consideration of the comments cited at paragraph 3(3), supra, concerning the definition of "employee," we have concluded that, within the meaning of the rules adopted by this order, (1) membership on the employer's corporate board of directors is not a basis for designation as an employee even if the person in question receives honoraria for attendance at board meetings; (2) a cablecast-program director who works for the employer only a few hours per week and receives compensation therefor is a part-time employee; (3) employment unit personnel who have ownership interests in the company (e.g., proprietor, member of his immediate family, stockholder) and who are paid for their services or who may, directly or indirectly, obtain significant benefit from such ownership interest, are employees; and (4) other unpaid personnel are not employees. Where a reporting employer believes that preparation of his annual employment report in accordance with this may create a distorted impression of his employment practices, he is certainly welcome to avoid that possibility by appending an explanatory statement to his report.

14. We have not adopted the recommendation that cable employers be required to include, in their annual reports, showings of affirmative action with respect to the employment of women (United Church). Such a provision would require employers to provide more annual information regarding their employment of women than is required with respect to their employment of Negroes, Orientals, American Indians, and Spanish-surnamed Americans. It should be noted, however, (1) that new § 76.311(c) (1) and (2) require the employer to include in its equal employment opportunity program statement "specific practices to be followed in order to assure equal employment on the basis of sex * * *"; and (2) the employer is required to set forth, in its annual employment report, both (i) the total numbers of males and females, respectively, in each job category, and (ii) the total numbers of males and females in each job category per minority group.

15. Also not adopted is the recommendation that an employer be exempted from the requirement that he submit an EEO-complaints report for any year in which no EEO complaints have been filed (Gulf, Jerrold). In support of that recommendation, it is argued that "The fact

---

[5] In response to Jerrold's inquiry as to how formal a "training program" must be to be so characterized in the employer's filings, we will not attempt here to prescribe the employer's language usage in his equal employment opportunity program statement—except to call upon him to avoid ambiguity and vagueness if possible; e.g., by use of examples and explanations when appropriate.

[6] For the reasons stated in paragraph 11 supra, the provision as adopted herein differs from the originally proposed language in that the number of full-time employees referred to has been changed from 5 to 10.

that no filing was made by a * * * cable system would indicate that no such complaints were filed during the preceding reporting period" (Gulf). Unfortunately, however, the absence of an annual EEO-complaints report from an employer may just as easily be the result of a clerical error on his part. If in fact no EEO-violation complaints were filed against an employer during the reporting period, the requirement that he mail a simple statement to that effect to the Commission surely places no burden of any consequence upon him.

16. We have adopted the recommendation by Jerrold that the time limits for filing of annual employment reports, EEO-violation complaint reports, and changes in EEO program statements be changed to "on or before May 31 of each year." Also, proposed § 74.1105(e) has been amended (and is renumbered § 76.13(a)(8)) to permit omission of an EEO program statement from a new-system application for certificate of compliance, if the proposed system operator (1) believes that there will not be 10 full-time employees at any time during the first 3 months of the year following commencement of operation of the CATV system, and (2) submits a statement justifying that conclusion.[7]

17. With respect to the employer's duty to maintain copies of his EEO records for a period of time, and to make them available for inspection by members of the public—

(a) We do not agree with Gulf that the employer should be permitted to inquire of persons wishing to inspect such records that they make "a reasonable showing of need" for such inspection. Our position in this matter is consistent with our adoption, on December 15, 1971, of a memorandum opinion and order concerning the duties of a broadcast station under § 1.526(d) of the Commission's rules.[8] In that ruling, we permitted broadcast licensees to require persons wishing to inspect their files to give their names and addresses; but prohibited broadcasters from requiring of such a person that he identify the organization on whose behalf he is requesting such inspection. We pointed out that any member of the public is entitled to inspect the files regardless of membership in an organization, and said that a requirement of organization disclosure would be likely to do more harm than good in that it might discourage some persons from seeking access to the licensee's public files. See also our "By Direction" letter of November 3, 1971, to

WBRN, Inc., 32 FCC 2d 474, FCC 71–1141, in which we stated, inter alia, that "those letters which are required to be made available [by broadcast licensees] for public inspection must be accessible to the public during regular business hours and such records should be provided to members of the public on request without requiring that they identify * * * the particular documents they wish to inspect."

(b) The Gulf and Jerrold comments regarding the location(s) of EEO records maintained by the employer for public inspection reflect a need for amendment of proposed § 74.1125(f) (2)(i) so that they will more clearly reflect the Commission's intent, which is that—

(i) Each employment unit shall maintain for public inspection such portions of the employer's EEO file as pertain to the employment practices and policies of that employment unit.[9]

(ii) An employer who is required to file a consolidated annual employment report, shall maintain an adequately indexed consolidated EEO file, containing copies of all the material included in the EEO files of the employment units reported upon in his consolidated annual employment report.

(iii) The EEO file for a system (or a single employment unit including that system) shall be maintained at the principal work-place of the employment unit, or at any accessible location (such as a public registry for documents or an attorney's office) in the principal community served by the employment unit. The headquarters office EEO file and the consolidated EEO file shall be maintained, (a) respectively, at the headquarters office, and the principal office of the employer, or (b) at any accessible place (such as a public registry for documents or an attorney's office) in the community in which the office is located. The employer shall provide reasonable accommodations at these locations for undisturbed inspection of his EEO records by members of the public during regular business hours.[10]

We do not believe (nor have the respondents in this proceeding submitted any evidence tending to support the belief) that the maintenance by individual employment units of EEO records for public inspection will impose a significant burden upon cable systems.

(c) United Church states, in support of its recommendation that employers be required to retain EEO records for public inspection for at least 5 years (instead

of the 2-year retention period set forth in the notice of proposed rule making):

It is extremely difficult to establish patterns of discrimination under any circumstances, especially if the staff involved is small and there are few turnovers. It would be impossible to establish such patterns if records are available only for 2 years. The retention of the records for an additional 3 years would not impose a substantial burden on an operator, licensee or permittee. In the case of a small enterprise with limited storage space, the records will be quite brief unless a large number of discrimination complaints have been filed. If in a rare case they have, it is absolutely essential that such records be available for public inspection for a period longer than 2 years.

We are persuaded by that argument, and have amended the proposed rules accordingly. (See § 78.311(f)(2)(ii).)

(d) We are not persuaded, however, by United Church's further recommendation that the employer be required to provide any member of the public who asks for it, with a quick-copy (Xerox or other) of any EEO material kept by the employer for public inspection, and charge no more than 5 cents per page for the service. Although the provision of such service by an employer would certainly be an expression of good faith on his part, the requirement that employers provide such service would, in our judgment, place an unreasonable burden upon many of them, particularly in light of (1) the availability, to persons wishing to copy EEO records, of quite inexpensive portable quick-copy equipment; and (2) the employer's obligation to provide adequate facilities for examination of EEO records.[11]

18. Upon consideration of United Church's comment regarding bulletin board EEO notices (see paragraph 3(7) supra), we have amended and renumbered proposed § 74.1125(c)(2)(i)(a) to read:

Posting notices in the cable operator's offices and places of employment informing employees, and applicants for employment, of their equal employment rights and their right to notify the Equal Employment Opportunity Commission, the Federal Communications Commission, or other appropriate agency. Where a significant percentage of employees, employment applicants, or residents of the community of a cable television system are Spanish-surnamed Americans, such notice should be posted in Spanish and English. Similar use should be made of other languages, in such posted equal employment opportunity notices, where appropriate.

A fixed mathematical definition of "significant percentage" (e.g., the 5 percent figure recommended by United Church) has been avoided pending further Commission experience. However, we are confident that employers will recognize the desirability of bi- (or multi-) lingual EEO notices, where appropriate, both as aids in communication of rights and as earnests of the employer's good faith.

19. In paragraph five of the notice of proposed rule making in this docket we indicated our intention to refer EEO complaints within the jurisdiction of the U.S. Equal Employment Opportunity

---

[7] The originally proposed language of this section would permit omission of the EEO program statement only if the cable certificate applicant believed that the system would not have 10 or more full-time employees at any time during the first 12 months of operation of the system. The substituted language is more harmonious with the references to January, February, and March in newly added § 76.611(e)(8). (These new provisions eliminate needless payroll-period differences between cable annual employment reports and broadcast and common carrier annual employment reports.)

[8] 32 FCC 2d 729.

[9] Including copies of the applicable EEO program statement and changes therein, annual reports to the Commission, etc., as specified in the above referenced section of the rules, now renumbered as § 76.311(f) (2)(i).

[10] Including, e.g., the use of a table and chair, adequate lighting, and a wall electric outlet for copying equipment which the EEO-record examiner may bring along. See also the Commission's Public Notice of February 23, 1971 (28 FCC 2d 71), regarding the availability of locally maintained records for inspection by members of the public.

[11] See n. 10, supra.

Commission to that agency, and thereafter to maintain appropriate liaison in the matter with EEOC and the Department of Justice; and to act similarly with respect to EEO complaints within the jurisdiction of State and local government authorities. United Church objects to such referrals on the grounds that: (1) They would yield inconsistent and uneven enforcement of the rules; (2) agencies receiving such referrals would typically lack the Commission's enforcement powers; and (3) Commission action (investigation etc.) in a matter after referral of it to another governmental agency [12] would both (i) undermine that other agency, if done frequently, and (ii) subject complainants and employers to a double procedural burden.

20. In this, as in all other aspects of its EEO programs, the Commission intends to keep its options open, and to learn from experience. It should be noticed, in this connection, that the procedure in question is described not in the new rules themselves, but only in the text of our notice. We recognize the possibility of problems arising from referral of complaints to other agencies for their investigation and action. We are prepared to modify our procedures, both generally and on a case-by-case basis, when necessary to avoid such problems. In any case, it should be perfectly clear that the Commission, in referring complaints to other agencies, would not be abdicating its responsibility to take appropriate action on such cases itself.

21. Finally, we note that although the adoption of two almost identical sets of equal employment opportunity rules—one for CATV systems and one for CAR stations—was originally proposed, we are now persuaded that since CAR stations are always auxiliary facilities to cable systems, and are not separate employment units as a practical matter, no useful purpose would be served by the application of separate rules to these related entities. [13] Hence, we are placing the body of the equal employment opportunity rules in Part 76 (Cable Television Service) with a cross-reference to them in Part 78 (Cable Television Relay Service) and an introductory paragraph in the rules making it clear that they apply to operators of cable systems, both in that capacity and as licensees or permittees of CAR stations. Where a cable system or a headquarters office has employees whose duties are related to the operation of a CAR station, these employees shall be considered employees of the cable system or headquarters office employment unit for purposes of the rules. This approach to CAR stations parallels the rules in the broadcast area, where no separate provisions apply to auxiliary stations.

22. For the reasons set forth in the foregoing paragraphs, *It is ordered,* Pur-

[12] Because of dissatisfaction with the results of such referral.

[13] Under § 78.13 of the rules, a CAR license will be issued only to the owner of a CATV system or to a cooperative enterprise wholly owned by CATV owners or operators.

suant to the authority contained in sections 2(a), 3 (h) and (d), 4(i), 301, 303, and 403 of the Communications Act of 1934, as amended, that effective May 9, 1972, Parts 76 and 78 of the Commission's Rules are amended as set forth below, and that the proceedings in Docket No. 19246 are terminated. [14]

(Secs. 2, 3, 4, 301, 303, 403, 48 Stat., as amended, 152, 1064, 1066, 1081, 1082, 1094; 47 U.S.C. 152, 153, 154, 301, 303, 403)

Adopted: March 23, 1972.

Released: March 29, 1972.

FEDERAL COMMUNICATIONS
COMMISSION, [15]
[SEAL]   BEN F. WAPLE,
                                Secretary.

Chapter I of Title 47 of the Code of Federal Regulations is amended as follows:

A. In Part 76, Cable Television Service:

1. In 76.13, paragraph (a) (8) is added, to read as follows:

§ 76.13   Filing of applications.

*   *   *   *   *

(a) * * *

(8) A statement of the proposed system's equal employment opportunity program, as described in § 76.311. However, if the operator of the proposed system believes that the system will (continuously during January, February, and March of the year following commencement of operations) satisfy the conditions in § 76.311(c) (1) (i) (b), he may submit a statement justifying that conclusion in lieu of a statement of the proposed system's equal employment opportunity program.

*   *   *   *   *

2. In § 76.305, paragraph (f) is added, to read as follows:

§ 76.305   Logging and recordkeeping requirements.

*   *   *   *   *

(f) *Equal employment opportunities.* See § 76.311(f).

3. Section 76.311 is added, to read as follows:

§ 76.311   Equal employment opportunities.

The following provisions apply to all operators of cable television systems, both in that capacity and as licensees or permittees of cable television relay stations. Where a cable system or a headquarters office has employees whose duties are related to the operation of a cable television relay station, these employees shall be considered employees of

[14] An "Annual Employment Report" (FCC Form 395) having appropriate cable television references will be published after clearance of the necessary changes in the existing form with the Office of Management and Budget.

[15] Commissioner Bartley concurring in part and dissenting in part and issuing a statement, filed as part of the original document; Commissioners Robert E. Lee and Johnson absent.

the cable system or headquarters office employment unit for purposes of this section.

(a) *General policy.* Equal opportunity in employment shall be afforded by all operators of cable television systems to all qualified persons, and no person shall be discriminated against in employment because of race, color, religion, national origin, or sex.

(b) *Equal employment opportunity program.* (1) Each cable television system shall establish, maintain, and carry out a positive continuing program of specific practices designed to assure equal opportunity in every aspect of system employment policy and practice.

(2) Under the terms of its program, a system shall:

(i) Define the responsibility of each level of management to insure a positive application and vigorous enforcement of the policy of equal opportunity, and establish a procedure to review and control managerial and supervisory performance;

(ii) Inform its employees and recognized employee organizations of the positive equal employment opportunity policy and program and enlist their cooperation;

(iii) Communicate the system's equal employment opportunity policy and program and its employment needs to sources of qualified applicants without regard to race, color, religion, national origin, or sex, and solicit their recruitment assistance on a continuing basis;

(iv) Conduct a continuing program to exclude every form of prejudice or discrimination based upon race, color, religion, national origin, or sex from the system's personnel policies and practices and working conditions;

(v) Conduct continuing review of job structure and employment practices and adopt positive recruitment, training, job design, and other measures needed to assure genuine equality of opportunity to participate fully in all organizational units, occupations, and levels of responsibility in the system.

(3) Where two or more cable television systems under common ownership or control are so interrelated in their management, operations, and utilization of employees as to constitute a single employment unit, the program shall be jointly established, maintained, and carried out by them. (Under other circumstances, the term "single employment unit" refers to an individual cable television system or to a headquarters office.)

(c) *Additional information to be furnished to the Commission*—(1) *Equal employment programs to be filed by operators of systems.* (i) The operator of each cable television system shall file a statement of its equal employment opportunity program not later than June 30, 1972, indicating specific practices to be followed in order to assure equal employment opportunity for females, Negroes, Orientals, American Indians, and Spanish-surnamed Americans in such aspects of employment practices as recruitment, selection, training, placement,

promotion, pay, working conditions, demotion, layoff, and termination.

(a) Any changes or amendments to existing programs shall be filed with the Commission on or before May 31 of each year thereafter.

(b) If the system (1) has fewer than five full-time employees, and (2) does not (within the meaning of paragraph (b) (3) of this section together with other cable television systems constitute a single employment unit with an aggregate total of five or more full-time employees, an equal employment opportunity program statement need not be filed for the employment unit which consists of or includes the system.

(c) (1) Where, pursuant to paragraph (b) (3) of this section, a program is jointly established by two or more cable systems with an aggregate total of 10 or more full-time employees, a multiple cable operator shall file a combined statement. (2) A multiple cable operator shall file a separate equal employment opportunity program statement for each headquarters office if that office has five or more full-time employees, and its work is primarily related to the operation of more than one cable television system under common ownership or control.

(d) If, pursuant to (b) of this subdivision or § 76.13(a) (8), a cable operator has been exempted from the requirement that it file an equal employment opportunity program statement, but has failed to satisfy the conditions of that exemption at any time during the first 3 months of a calendar year, it shall file the statement on or before May 31 of that year.

(2) Contents of the equal employment program statement. The program should reasonably address itself to such specific areas as set forth below, to the extent that they are appropriate in terms of employment unit size, location, etc.

(i) To assure nondiscrimination in employment. (a) Posting notices in the cable operator's offices and places of employment informing employees, and applicants for employment, of their equal employment opportunity rights, and their right to notify the Equal Employment Opportunity Commission, the Federal Communications Commission, or other appropriate agency if they believe they have been discriminated against. Where a significant percentage of employees, employment applicants, or residents of the community of a cable television system are Spanish-surnamed Americans, such notice should be posted in Spanish and English. Similar use should be made of other languages in such posted equal employment opportunity notices, where appropriate;

(b) Placing a notice in bold type on the employment application informing prospective employees that discrimination because of sex, race, color, religion, or national origin is prohibited and that they may notify the Equal Employment Opportunity Commission, the Federal Communications Commission, or other appropriate agency if they believe they have been discriminated against;

(c) Placing employment advertisements in media that have significant circulation among minority-group people in the recruiting area;

(d) Recruiting through schools and colleges with significant minority-group enrollments;

(e) Maintaining systematic contacts with minority and human relations organizations, leaders, and spokesmen to encourage referral of qualified minority or female applicants;

(f) Encouraging present employees to refer minority or female applicants;

(g) Making known to the appropriate recruitment sources in the employer's immediate area that qualified minority members and females are being sought for consideration whenever the cable operator hires.

(ii) To assure nondiscrimination in selection and hiring. (a) Instructing personally those on the staff of the system who make hiring decisions that all applicants for all jobs are to be considered without discrimination;

(b) Where union agreements exist, cooperating with the union or unions in the development of programs to assure qualified minority persons or females of equal opportunity for employment, and including an effective nondiscrimination clause in new or renegotiated union agreements;

(c) Avoiding use of selection techniques or tests that have the effect of discriminating against minority groups or females;

(iii) To assure nondiscriminatory placement and promotion. (a) Instructing personally those of the system's staff who make decisions on placement and promotion that minority employees and females are to be considered without discrimination, and that job areas in which there is little or no minority or female representation should be reviewed to determine whether this results from discrimination;

(b) Giving minority groups and female employees equal opportunity for positions which lead to higher positions. Inquiring as to the interest and skills of all lower paid employees with respect to any of the higher paid positions, followed by assistance, counselling, and effective measures to enable employees with interest and potential to qualify themselves for such positions;

(c) Reviewing seniority practices to insure that such practices are nondiscriminatory and do not have a discriminatory effect;

(d) Avoiding use of selection techniques or tests that have the effect of discriminating against minority groups or females.

(iv) To assure nondiscrimination in other areas of employment practices. (a) Examining rates of pay and fringe benefits for present employees with equivalent duties and adjusting any inequities found;

(b) Providing opportunity to perform overtime work on a basis that does not discriminate against qualified minority group or female employees.

(d) Report of complaints filed against operators of systems. (1) All operators of cable television systems shall submit an annual report to the Commission no later than April 1 of each year indicating whether any complaints regarding violations by the operator of equal employment provisions of Federal, State, territorial, or local law have been filed before any body having competent jurisdiction.

(i) The report shall state with respect to each such complaint: The parties involved, the date filed, the courts or agencies before which the matter has been heard, the appropriate file number (if any), and the respective disposition or current status of the complaint.

(ii) Any cable operator who has filed such information with the Equal Employment Opportunity Commission need not do so with the Federal Communications Commission, if such previous filing is indicated.

(e) Report of annual employment. (1) Each operator of a cable television system with five or more full-time employees (including those whose duties are related to the operation of a cable television relay station) shall file with the Commission, on or before May 31 of each year, on FCC Form 395, an annual employment report.

(2) (i) Where pursuant to paragraph (b) (3) of this section, an equal employment opportunity program is jointly established by two or more cable television systems with an aggregate total of five or more full-time employees, a combined (single employment unit) annual employment report shall be filed.

(ii) A multiple cable operator shall file a separate annual employment report for each headquarters office if that office has five or more full-time employees, and its work is primarily related to the operation of more than one cable television system under common ownership or control.

(iii) Where, pursuant to subdivisions (i) and (ii) of this subparagraph, if more than one annual employment report is filed with respect to (a) cable television systems under common ownership or control, or (b) headquarters offices performing work related to such systems, a multiple cable operator shall also file a consolidated report, covering all system and headquarters office employees included in those reports.

(3) The data contained in each annual employment report required by subparagraphs (1) and (2) (i) and (ii) of this paragraph shall reflect the figures from any one payroll period in January, February, or March of the year during which the report is filed. The same payroll period should be used in each year's annual employment report.

(4) Annual employment reports required by this paragraph shall be filed on or before May 31 of each year.

(f) Records available to the public—(1) Commission records. A copy of every annual employment report, equal employment opportunity program, and reports on complaints regarding violation

of equal employment provisions of Federal, State, territorial, or local law, and copies of all exhibits, letters, and other documents filed as part thereof, all amendments thereto, all correspondence between the cable operator and the Commission pertaining to the reports after they have been filed and all documents incorporated therein by reference, are open for public inspection at the offices of the Commission.

(2) *Records to be maintained locally for public inspection by operators—*(i) *Records to be maintained.* Each operator of a cable television system required to file annual employment reports, equal employment opportunity programs, and annual reports on complaints regarding violations of equal employment provisions of Federal, State, territorial, or local law shall maintain, for public inspection, a file containing a copy of each such report and copies of all exhibits, letters, and other documents filed as part thereto, all correspondence between the cable operator and the Commission pertaining to the reports after they have been filed and all documents incorporated therein by reference. An employer who is required to file a consolidated annual employment report shall maintain an adequately indexed consolidated equal employment opportunity file, containing copies of all the material included in the equal employment opportunity files of the headquarters offices and other employment units reported upon in his consolidated annual employment report.

(ii) *Period of retention.* The documents specified in subdivision (i) of this subparagraph shall be maintained for a period of 5 years.

(iii) *Where maintained.* The equal employment opportunity file for a system (or a single employment unit including that system) shall be maintained at the principal workplace of the employment unit, or at any accessible location (such as a public registry for documents or an attorney's office) in the principal community served by the employment unit. The headquarters office equal employment opportunity file and the consolidated equal employment opportunity file shall be maintained (*a*) respectively, at the headquarters office and the principal office of the employer, or (*b*) at any accessible place (such as a public registry for documents or an attorney's office) in the community in which the office is located. The employer shall provide reasonable accommodations at these locations for undisturbed inspection of his equal employment opportunity records by members of the public during regular business hours.

4. Section 76.409 is added, to read as follows:

§ 76.409 **Annual employment report.**

An "Annual Employment Report" (FCC Form 395) shall be filed with the Commission for each cable television system, as defined in § 76.5, on or before May 31 of each year, in accordance with the provisions of § 76.311.

B In Part 78, Cable Television Relay Service.

1. Section 78.75 is added, to read as follows:

§ 78.75 **Equal employment opportunities.**

See § 76.311 of this chapter.

[FR Doc.72–4971 Filed 3–30–72;8:50 am]

Report to the President
The Cabinet Committee on Cable Communications

The following sections A–E constitute a summary outline of the Committee's long-range recommendations (Chapter III) as they affect cable operators, channel users, telephone common carriers, the FCC, and the franchising authorities. The exceptions to those recommendations, which would apply during the transition period (Chapter IV), are summarized in section F.

A. Policies Affecting Cable System Operators

1. Operators should be **required** to:

a. Offer their channels, or time on their channels, for lease to others for any lawful purpose, and without discrimination among comparable uses and users (pp. 29–30, 44–45),[1] with the exception of the channels used for retransmission of the broadcast signals authorized for carriage by the FCC's cable rules, plus one or two additional channels. The FCC's rules regarding broadcast signal carriage will apply to channels used for retransmission of the broadcast signals (note 2, pp. 29–30).

b. Comply with Federal and franchising authority requirements to construct cable systems with adequate channel capacity (p. 44).

c. Comply with the minimum technical standards established for cable distribution by the FCC (p. 41).

d. Offer customers a selective means to control or prevent reception of programming or information services which the customer does not wish to receive, and to prevent interception of personal or confidential information distributed over cable (pp. 38, 41).

2. Operators should be **allowed** to:

a. Own and operate other media outlets such as newspapers, magazines, or broadcast stations or networks including those within the same market area as the cable system (p. 32).

3. Operators should be **prohibited** from:

a. Having any financial or ownership interest in, or any control of, the production, selection, financing or marketing of the program or information services supplied by channel users leasing the operators' distribution facilities (pp. 29–30); with the exception noted in section A.1.a.

---

[1] All page references are to Chapter III, except where otherwise indicated. [Report to the President of The Committee, Executive Office of the President, Washington, D. C., January 14, 1974.]

b. Participating in the joint ownership or control of cable systems, interconnection facilities, and program supply services (p. 31).

B. Policies Affecting Program Retailers and Other Channel Users

1. Channel users should be **required** to:

a. Adhere to all applicable provisions of copyright laws and accept full liability for any program materials or information services they may supply (p. 39).

2. Channel users should be **allowed** to:

a. Lease channels or obtain other distribution services from any cable system with which they have no financial relationship or other form of common interest or control—with the exception noted in section A.1.a.—and offer to the public any lawful program materials or information services via such system (pp. 29–30, 37–39).

b. Establish such charges as they consider appropriate for the programming or information services they supply, without regulation by Federal, state, or local authorities (pp. 38–39).

c. Have legal recourse against any cable system operator: (1) who denies access or discriminates against the channel user by reason of the content of the user's message or the user's race, religion, nationality, or beliefs; or (2) who otherwise engages in practices that violate the requirement of non-discriminatory channel lease rates (p. 44).

3. Channel users should be **prohibited** from:

a. Providing any information or taking any action in violation of relevant laws and statutes protecting privacy and governing dissemination of obscene, libelous, or otherwise illegal material, as well as material the cable customer has indicated he does not wish to receive (p. 38).

b. Requiring viewers to pay a fee for professional sports programming unless consistent with the FCC's anti-siphoning restrictions (p. 37).

C. Policies Affecting Telephone Common Carriers

1. Common carriers should be **required** to:

a. Provide pole, conduit, or other right-of-way access to any franchised cable system operator at reasonable rates and without discrimination among users or uses (p. 34).

2. Common carriers should be **allowed** to:

a. Offer local cable distribution service on a "lease-back" basis to any franchised cable system operator (p. 34).

b. Obtain franchises to operate as cable system operators outside

of any area in which they have exclusive authority to provide telephone service (p. 34).

3. Common carriers should be **prohibited** from:

a. Owning, controlling or operating any cable system within their telephone service areas, i.e., performing any function not associated with actual signal distribution, such as the operation of cable system "head-ends" used for information origination, reception, conversion, switching, or other processing functions (p. 34).

D. Policies Affecting the Federal Communications Commission (FCC)

1. FCC shuld be **permitted** only to:

a. Establish minimum technical standards for cable distribution systems, only as needed to ensure compatibility, interoperability, privacy and security of cable systems (p. 41).

b. Require that cable systems be constructed with adequate channel capacity (p. 44).

c. Apply restrictions to the presentation for a fee of professional sports programs (pp. 37, 41).

2. FCC should **not be permitted** to:

a. Regulate in any way the information content of any services carried by cable systems including any regulations as to the balance or "fairness" of such information (p. 38).

b. Require minimum channel capacity to be leased to others; designate special purpose channels; require expansion of channel capacity or construction of two-way capacity (Chapter IV, pp. 9–10).

c. Regulate the rates or earnings of cable operators or channel users, or require any free service (pp. 38–39).

d. Limit, by regulation or policy, the ownership of cable systems by broadcast stations or networks, or by newspapers, magazines, or other media outlets, or limit the number of cable systems to be owned by one firm or the number of customers to be served by one firm (p. 32).

E. Policies Affecting Franchising Authorities

1. Franchising authorities should be **required** to:

a. Award non-exclusive franchises for the use of public rights-of-way by cable systems, and collect franchise fees for such use to the extent the fees merely compensate for the costs of regulation or costs incurred in the use of the public rights-of-way (p. 43).

b. Require that the rates, terms, and conditions, for channel leasing, not unreasonably discriminate among comparable channel uses and users (pp. 44–45).

c. Require that the cable operator make available one channel to be used for public access purposes (note 9, p. 11).

d. Require, through negotiations with prospective cable operators, that cable systems be constructed with adequate channel capacity (p. 44).

2. Franchising authorities should be **permitted** to:

a. Set maximum limits on the rates or charges imposed on customers for cable installation (p. 45).

b. Establish franchising conditions dealing with the cable system operator's qualifications; construction timetables; extension of service to all portions of the franchise area; handling of service complaints; and other conditions not expressly forbidden to franchising authorities (p. 45).

3. Franchising authorities should **not be permitted** to:

a. Regulate the information content of any service carried by a cable operator including any regulation as to the balance or "fairness" of such information (p. 38).

b. Award exclusive franchises for cable systems or require dedicated free channels for special purposes (pp. 43–44).

c. Impose franchise fees on cable systems, when the primary purpose is to raise revenues (p 43).

d. Regulate the rate of return or earnings of cable operators or the rates charged by program or information suppliers to their subscribers (pp. 42–43).

F. Transition Policies

The following exceptions to the long-range policy recommendations would apply during the transition period, which would end when 50 percent of the nation's households were connected to cable systems (p. 52).

1. Cable operators would be exempt from the prohibition on offering programming directly or having financial or other interests in the programming and other services offered over their systems (p. 53).

2. Franchising authorities would have to require cable operators to:

a. Make available for lease to others at least one equivalent channel for every channel used by the cable operator for retransmission of broadcast signals or for program originations (p. 53).

b. Establish a pattern of gradual lessening of the cable operator's control of channels by increasing the proportion of channels to be leased to others (p. 53).

3. The Federal Communications Commission would continue to:

a. Prohibit future ownership of cable systems by television broadcast networks and by television broadcast stations in their station service areas (p. 53).

b. Apply restrictions on the type of entertainment programming that can be offered to cable system customers for a fee and adapt such restrictions to changing conditions in the broadcast, cable, and programming industries (p. 54).

MONDAY, APRIL 22, 1974

WASHINGTON, D.C.

Volume 39 ■ Number 78

PART II

FEDERAL COM- MUNICATIONS COMMISSION

CABLE TELEVISION

■

Clarification of Rules and Notice of Proposed Rulemaking

FEDERAL COMMUNICATIONS COMMISSION

[47 CFR Part 76]

[FCC 74–384; 11288; Docket Nos. 20018–20024]

CABLE TELEVISION

Proposed Clarification of Rules

In the matter of:

Amendment of Part 76 of the Commission's rules and regulations relative to the advisability of Federal preemption of cable television technical standards or the imposition of a Moratorium on non-Federal standards (Docket No. 20018).

Amendment of Part 76 of the Commission's rules and regulations relative to an inquiry on the need for additional rules in the area of public proceedings and qualifications for franchisees, § 76.31(a)(1) (Docket No. 20019).

Amendment of Part 76 of the Commission's rules and regulations relative to requiring additional assurances on the establishment of line extension provisions in franchises—§ 76.31(a) (1), (2) (Docket No. 20020).

Amendment of Part 76 of the Commission's rules and regulations relative to amending existing franchise duration rules—§ 76.31(a)(3) to lengthen maximum term and impose a minimum term (Docket No. 20021).

Amendment of Part 76 of the Commission's rules and regulations relative to an inquiry on the advisability of adding specific rules to § 76.31(a)(3) regarding franchise expiration, cancellation and continuation of service (Docket No. 20022).

Amendment of Part 76 of the Commission's rules and regulations relative to an inquiry on the need for new regulations in the area of transfers of control of cable television franchises (Docket No. 20023).

Amendment of Part 76 of the Commission's rules and regulations relative to a specific requirement in § 76.31 (a)(5) that the local official responsible for subscriber complaints be identified in the franchise (Docket No. 20024).

I. INTRODUCTION

1. On February 2, 1972, the Commission adopted the Cable Television Report and Order (37 FR 3252, 36 FCC 2d 143). Reconsideration of Report and Order (37 FR 13848, 36 FCC 2d 326). In that report we adopted a comprehensive set of new rules for most aspects of cable television operation. The report was separated into four main categories:

a. Television broadcast signal carriage;
b. Access to and use of non-broadcast cable channels, including minimum channel capacity;
c. Technical standards;
d. The appropriate division of regulatory jurisdiction between the Federal and state-local levels of government.

Particularly as to the last three categories, we stated repeatedly that new regulatory concepts and procedures were being employed and that many of these rules were experimental in nature and would be clarified, modified, or changed as the situation warranted. The rules were an attempt to create a flexible regulatory framework that took into account the constant and necessary flux inherent in any emerging industry such as cable television. The time has come, after two years of operational experience, to make some modifications and clarifications of our rules to keep pace with the changing picture presented by cable's development and to resolve whatever ambiguities may exist.

2. Our interest in the development of cable television is not passive. While the bedrock of our regulatory authority over cable clearly derives from its use of broadcast signals (see "U.S. v. Southwestern Cable Co.", 392 U.S. 157, "Midwest Video v. U.S.", 406 U.S. 649), this is not where our concern ends. This Commission is primarily responsible for the development and maintenance of a nationwide communication system (Communications Act of 1934 as amended, sec. 1). Cable television is undeniably part of that system and presumably will become a major and integrally vital element of what many see as the broadband communications system of the future. We are concerned that we do not, in our efforts to mold the communications structure of the future, unduly hamper the developing structure of today. Over-expectation and anticipatory regulation can be just as damaging, if not more damaging, than no regulation at all.

3. The need for flexibility in our rules and a willingness to modify them as needed is best illustrated by the technological changes that have occurred within the past two years. In this relatively short time span, we have seen the development of cable television converters that have nearly doubled the maximum channel capacity. Satellite transmission to cable systems has become a technical reality. Two-way subscriber response systems have moved from the drawing boards to test installations. Any regulations of cable television must be designed with enough flexibility to allow for these changes.

4. Two years of experience in administering our rules has also given us the opportunity to pinpoint the weaknesses, identify the areas creating undue confusion or misinterpretation, and catalogue our own mistakes. This process of refining our rules was significantly aided by the reports submitted to us by the special Federal/State-Local Advisory Committee [FSLAC] that was established for this purpose when we adopted the Cable Television Report and Order 37 FR 3252 at 3277, Paragraph 188. That Committee spent more than 250 hours in public meetings debating many of the issues we will deal with here. In many cases, the clarification we are providing today is in response to the confusion or need for more specificity highlighted by those meetings. The final report of the FSLAC Steering Committee [1] has been thoroughly reviewed by this Commission prior to the preparation of this document. The review included a special meeting held between the Steering Committee and the full Commission in public session on December 11, 1973. The actions we are taking today are not intended to be dispositive of the FSLAC report. That report did provide valuable guidance, however, in the prepartion of this document. We expect to continue work that has already been initiated relating to the FSLAC recommendations, and future actions based on the FSLAC report will be so noted.

5. We are issuing this clarification and suggesting modifications only after a great deal of careful study and two years of experience with the present rules. Many interrelated rule making proceedings and requests for waivers, special relief, or declaratory rulings have been received during that time. Some of those pending requests will be either resolved or modified by our action today.

6. This document is intended to both clarify our existing rules and policies and at the same time open new inquiries where appropriate. In areas where a new rule is proposed or the change suggested goes beyond clarification or non-substantive modification, we have so noted it by specifically inviting comments and assigning a docket number to the issue. As in all other notices of proposed rule making and inquiry, comments are invited from all interested parties. We emphasize in this regard, however, that we intend to act expeditiously on these matters. While many of the issues considered today cross the subject matter categories employed in the Cable Television Report and Order, we will attempt to deal with them within that framework to maintain continuity.

II. TELEVISION BROADCAST SIGNAL CARRIAGE

7. We do not intend to suggest any modifications in our signal carriage rules at this time. Several rule makings are outstanding (i.e., non-duplication RM–2275, Docket No. 19995) and will be dealt with in due course. However, some general comments on signal carriage, particularly as it relates to other issues in this report, are appropriate.

SIGNAL CARRIAGE JURISDICTION

8. The fact that this Commission has pre-empted jurisdiction of any and all signal carriage regulation is unquestioned. Nonetheless, occasionally we receive applications for certificates of compliance which enclose franchises that attempt to delineate the signals to be carried by the franchisee cable operator. Franchising authorities do not have any jurisdiction or authority relating to signal carriage. While the franchiser might want to include a provision requiring the operator to carry all signals allowable under our rules, that is

---

[1] The final report of the Steering Committee of the FCC Cable Television Advisory Committee on Federal/State-Local Regulatory Relationships is available for $6.50 from the National Technical Information Service, 5285 Port Royal Road, Springfield, Virginia 22151, Order No. PB 223–147.

as far as the franchiser can or should go. In fact, because of the complexities of our signal carriage rules, even that statement in a franchise could be troublesome. We have been faced in some instances with the unfortunate situation where, because the franchise included signal carriage requirements inconsistent with our rules, we were forced to delay the grant of a certificate awaiting amendment of the franchise. In other cases, where the franchise included a severability clause, we were able to grant the certificate. Even in those instances, it would have been preferable had the franchising authority omitted the signal carriage clauses altogether.

LEAPFROGGING

9. We note that a further suggestion on signal carriage was made by the Federal/State-Local Advisory Committee final report submitted by its Steering Committee (hereinafter referred to as the FS LAC Report). The report designates over-the-air signal carriage as Issue #19 and states:

Signal carriage requirements are and should remain in exclusively federal jurisdiction. Additionally, the Committee recommends that when there is a joint petition by the cable operator and the franchising authority for a waiver of the leapfrogging rules based on a showing of community interest, the Commission should give additional weight to such petitions in considering the waiver request.

We agree with this position and have adopted it in some cases presented to us. (See Commission on Cable Television of the State of New York, 43 FCC 2d 826, FCC 73-1148, CSR-342). We intend to continue investigating such waiver requests on an ad hoc basis, and as noted in the above-cited case, as we gain more experience in this area, we may consider appropriate amendments of our leapfrogging rules (§ 76.59, 61 et seq.) to accommodate the carriage of in-state signals in some or all situations.

SIGNAL DELETION

10. Several procedural changes have also been suggested in this area, particularly as they relate to applications for certificates of compliance. In § 76.13 (a) (1) and (b)(1), we require indication of the signals an operator is authorized to carry as well as specification of the signals requested to be added to that authorization. In many instances, this has led to situations where there are clearly many more signals authorized than could technically be carried or are desired. We intend to amend this rule to require that the applicant indicate, when applicable, what signals should be deleted from the authorization as well as added.

11. We recognize that, in many cases, the reason there are more signals authorized than can technically be carried is that some of those signals are only carried in part. This is consistent with § 76.55(b) which simply requires that a particular program may not be altered or deleted in part. The carriage of signals not required by our rules is left to the discretion of the cable operator. In those cases, however, where signals are going to be dropped completely, we want to be apprised. A procedural change in § 76.55 (b), should be sufficient to accomplish that result.[2]

III. ACCESS TO AND USE OF NON-BROADCAST CHANNELS

12. A comprehensive and innovative set of new rules regarding cable television access channels was adopted in our 1972 regulations. In the Report and Order in Docket No. 18396 et al., we clearly stated the basis and rationale for these new rules:

Broadcast signals are being used as a basic component in the establishment of cable systems, and, it is therefore appropriate that the fundamental goals of a national communications structure be furthered by cable—the opening of new outlets for local expression, the promotion of diversity in television programming, the advancement of educational and instructional television and increased informational services of local government. (Para. 121.)

13. We reiterated this over-all concern for the development of cable television in the reconsideration of the Cable Television Report and Order, 37 FR 13848, 36 FCC 2d 326:

* * * Cable Television as it grows, must be integrated into a nationwide communications structure. Were we to permit an uncontrolled development of cable we would be breaking our obligations under the Communications Act of 1934, as amended. This Commission was created, amid the chaotic development in the field of radio, * * * to make available, so far as possible, to all the people of the United States a rapid, efficient, nationwide, and worldwide wire and radio communications service. * * * (Section 1, 47 USC 151). As an integral part of interstate broadcast transmission, cable operators "cannot have the economic benefits of such carriage as they perform and be free of the necessary and pervasive jurisdiction of the Commission" (General Telephone of California v. FCC, 413 F 2d 390, 401 (C.A.D.C.) (1969), Cert. denied, 396 U.S. 888. Thus, we conceive it to be our obligation to consider the actual and potential services of cable television and create a Federal policy which insures that these services can be distributed equitably, on a nationwide basis as merely one link in our communications systems * * *. (Para. 74.)

From watching the development of our access program, we are now, more than ever, convinced of the propriety and need for such a program. Access is still in its infancy and it has a long, hard struggle ahead before it becomes an accepted part of the communication process in this country. We knew this would be the case when we instituted the rules noting:

* * * We recognize that in any matter involving future projections, there are necessarily certain imponderables. These access rules constitute not a complete body of detailed regulations but a basic framework within which we may measure cable's technological promise, assess its role in our nationwide scheme of communications, and learn how to adapt its potential for energetic growth to serve the public. (Para. 117.)

---

[2] Formal action to effect this procedural amendment will be announced in a separate Commission document.

14. We believe that the access channels we have required will eventually serve the public in many ways. However, we are also aware that the requirement for providing these channels imposes a burden on the cable operator, particularly on the small, older systems now required to provide access channels and the new large systems that provide services to many small communities. We also note that many franchisors outside the major markets are now including access requirements in their renewal proceedings.[3]

ACCESS ON CONGLOMERATE SYSTEMS

15. For the most part, our access channel requirements do not appear to be overly burdensome. To date we find no reason to alter the rule requiring at least four access channels (public, educational, government, and leased). The application of that regulation, however, must stand on a flexible and reasonable basis. One issue that is being raised in this regard, and which we wish to clarify here, is the effect of the rule in multi-jurisdictional systems. In the Cable Television Report and Order, we stated that "* * * To the extent that the access requirements pose problems for systems operating in small communities in major markets, such systems are free to meet their obligations through joint building and related programs * * *." Our intent here is to make clear that we have and will continue to entertain petitions and special showings to allow the joint use of access channels and facilities. (e.g., Century Cable Communications, Inc., CAC-1914, FCC 74-63.) There is no need, as we see it, to require a system providing service to a large number of small suburban communities to have a separate public access channel for each one of those communities when in reality none of those access channels is or would likely be fully utilized. In fact, in such a situation, it might be better, in terms of fostering public access channel use, to have one or two channels significantly used and "lit" rather than a multiplicity of channels "dark" for a major portion of the time because of scarcity of programming. On the other hand, we want again to put all cable operators on notice that although we may grant waivers of immediate provisions for access channels we still expect and will require operators to have sufficient channel capacity to meet any reasonable demand.

CHANNEL CAPACITY

16. Questions arising out of our channel capacity rules (§ 76.251(a)(1)) also indicate that clarification is necessary. Our efforts to establish minimum/maximum channel capacity requirements were based on a study of the existing technology at the time of the adoption of

---

[3] We allow the addition of such requirements in smaller market franchises so long as they are consistent with and no greater than our rules for the major markets. See Cable Television Report and Order, 37 FR 3252, at 3272, Para. 148, § 76.251(b).

those rules. We were attempting to indicate to the industry that they must have sufficient channel capacity to meet foreseeable future demands, and, at the same time, we were cautioning franchising authorities that requiring excessive technological capacity was detrimental to our overall program. A "20-channel" system, in essence, requires construction that is sufficient for any currently foreseeable demand; that is, single cable with converter, dual cable, or eventually dual cable with converter. We continue to be of the opinion that this is sufficient. We note that some communities have contemplated requiring massive extra bandwidth provisions, such as operational capacity for 120 video channels. The present need or value of such excess has yet to be proved. Apparently the theory is that many discrete groups could thereby each have their own separate access channel. However, it appears from current experience that, for now, the more successful access experiments are those where a cooperative effort is made by many groups to fill an access channel. The advantage of such cooperation is that it results in the channel's use for a substantial portion of the day so that viewers become accustomed to seeing programming originate on the channel as a normal course of events rather than as an occasional special event. The provision for special access channels for various discrete groups may, we fear work to their detriment in that rather than pooling their efforts to program one channel, each will go its separate way and ultimately none may succeed. We envisioned and continue to promote the concept of pooled facilities. For instance, the school systems in a community should be able to cooperate to program an educational channel. Their time and resources would be better spent and more effectively utilized by joint effort than by each demanding his own channel and then not being able to fully utilize it.

FACILITY REQUIREMENTS

17. Our access program, and the burden it imposes on the cable operator, has been carefully weighed and we consider it to be both reasonable and in the public interest. We are requiring the provision of free access channels and some facilities to utilize them. We envision this access program as an opportunity for a multiplicity of persons and groups to become active in the use of the communications media for the first time. For access channels to work the individuals and groups being offered access must design their own programs, develop their own resources, and foster the use and value of the channels. This is not accomplished by demanding that the cable operator, having provided the free channels, should now also pay to program the channels. An unfortunate misconception seems to have developed because of some over-expectations at the prospect of free access channels. Demands are being made not only for excessive amounts of free equipment but also free programming and engineering personnel to man the equipment. Cable sub-

scribers are being asked to subsidize the local school system, government, and access groups. This was not our intent and may, in fact, hamper our efforts at fostering cable technology on a nationwide scale. Too often these extra equipment and personnel demands become franchise bargaining chips rather than serious community access efforts. We are very hopeful that our access experiment will work. We recognize the difficulties inherent in developing access programming and will have more to say on the subject later. We do not think, however, that simply putting more demands on the cable operator will make public access a success. Access will only work, we suspect, when the rest of the community assumes its responsibility to use the opportunity it has been provided.

18. In order to clarify the meaning and intent of our access requirements, we will review them here as they appear in our rules.

19. Sections 76.251(a) (1) and (2), as noted earlier, are meant to assure that any new cable system being built is designed with sufficient capacity for any foreseeable future demand. We think these rules adequately meet that goal and see no need to modify them. It should be noted, however, that we recognize that in some cases strict application of these rules would not be reasonable. This is particularly true where, because an older system is already carrying a great number of grandfathered signals, or a new system must carry a large number of "local" stations, a system would have to have an inordinately large channel capacity in order to double its bandwidth pursuant to § 76.251(a) (2). We will continue to entertain waiver requests in such circumstances. This does not mean, however, that a waiver will be granted to allow a system to continue operating without any extra capacity. All systems covered by our rules will have to have sufficient capacity to meet their access obligations and have some capacity left over for future use. Waivers will be granted in instances where the extra capacity required by the rule would appear to have no foreseeable relationship to future demand.

BANDWIDTH ACTIVATION REQUIREMENTS

20. Some questions have been raised as to when the extra bandwidth must be activated. Some systems claim 20- or 24- or 26-channel capacity by having the capability of installing converters on a single trunk system. We have occasionally been asked when that converter must be installed. Our application of this rule is purely pragmatic. The rule requires bandwidth "* * * available for immediate or potential use * * *." No system will receive a certificate of compliance if its activated capacity is insufficient to meet our access requirements (including at least one channel available for leased use). So long as the system always has that much immediately available and usable capacity, it will be considered in compliance with our rules, assuming, of course, that the remaining capacity can be activated without significant rebuilding or delay.

CHANNEL ACTIVATION

21. In this regard, we believe it is necessary to clarify the language of the channel expansion formula in § 76.251 (a) (8) of the rules. This Section requires that a new designated access channel be made available when the first channel is in use for a specified period of time. The "time trigger" (channel use for 80 percent of the time during any consecutive three-hour period for six consecutive weeks) applies to each channel individually. For instance, if the public access channel is filled to that degree, a new public access channel must be designated upon request regardless of the amount of use being made of the other access channels. Additional special designated channels need not be provided free of charge. Reasonable charges consistent with our access policy can be assessed so long as the free channel in each category remains available an a non-discriminatory basis.

TWO-WAY

22. In § 76.251(a) (3), we require that the technical capacity for non-voice return communication be designed into any new cable facility affected by the rule. We fully explain the rationale for this requirement in Paras. 128, and 129 of the Report and Order. This rule does not require that the cable system be operational in the return mode. Once again, as in the case of channel capacity, we want to make sure that new systems being built will be able to meet all present and foreseeable future service obligations without the need for significant rebuilding or delay. We are aware that at present there are few, if any, proven, economically viable uses for two-way cable communications. To require operational two-way systems at this time, therefore, might impose unreasonable costs on the cable operator. In some cases, we have noted that franchising authorities are requiring the immediate operational installation of two-way facilities. Before a certificate of compliance is granted in any such case, we require a showing of the intended use of such facilities and a showing that such a requirement will not adversely affect the system's viability or otherwise inhibit it from complying with the federal goal of a nationwide cable communications grid.

PRIVACY

23. Many questions and fears have developed about the use of two-way equipment. In this regard, the statement made in the FSLAC Report is most appropriate:

The issue of privacy and its relationship to the legitimate uses or potential uses of cable television is a highly emotional one. The fears of many, that cable television will bring with it "1984-type" surveillance and monitoring is in the public mind regardless of the technological factors that argue against such uses. These fears must be met. At the moment, the potential for over-reaction to such fears and the inclusion of impractical and prohibitive allegedly protective requirements in franchises prompts the Committee to suggest that: Protection of subscriber privacy may take the form of regulation and judicially enforceable sanctions, and may be ad-

dressed at federal, state or local levels. The Committee believes that the principal problem area relates to the individualized monitoring of subscriber viewing habits, without explicit advance consent, and the disclosure of such information. Restraints on such activity should not impede systemwide, non-individually addressed "sweeps," or the operator's acquisition of information for purposes of verifying system integrity, controlling return path transmissions, or billing for pay services.

24. We agree fully with the Committee on this point. Without denigrating the well-intentioned pleas for caution voiced by many groups, we feel that there has been much misinformed over-reaction to this problem. Some franchises have included provisions to guard against monitoring that are not only impractical but often impossible to comply with. Other provisions have been included which purport to prohibit activities by the cable operator, such as generalized performance "sweeps," which are necessary to assure system integrity. Equipment to "monitor monitoring" has been required that does not even exist. It should be sufficient at this time to caution franchising authorities against excessive regulation in this regard. We are watching this situation carefully and will take any action necessary to protect the privacy of cable subscribers. Such action may take the form of added regulations at the agency level to assure privacy or possibly even Congressional action.[4] All governmental jurisdictions should be on guard to guarantee that the right of privacy is maintained. As we noted when we instituted the two-way requirement, any use of two-way communications, any activation of return service must always be at the subscriber's option (Para. 129).

FREE CHANNELS

25. In § 76.251(a)(4), (5), and (6), we require the provision of public, educational, and governmental access channels. We continue to view these channels as experimental. After only two years of experience with these rules, it would be premature to characterize the experiment as a success or failure. We would prefer more experience before significantly changing these requirements. Once again, however, it appears necessary to reiterate that until we can gain more experience with the experiment already under way, we are reluctant to allow major alterations by individual franchising authorities without good cause. Unquestionably, in some areas, because of particular local needs and facilities, different access programs might be useful. In those cases, we will entertain petitions for waiver of our general rule. To date, however, we have received several applications for extra access channels and equipment on the "more is better" concept rather than on any actual need or plan for use. As we have

already noted, "more" may not be better, and, indeed, may be worse. Any proposals in franchises requiring access channels or facilities in excess of what is required in our rules must be shown to be reasonable and necessary for a planned local program of use. A showing in the application for a certificate of compliance must be made that indicates what the nature of the added requirement is, how it will be implemented, who will pay for the extra services and equipment, how much they will cost, and how the costs, if borne by the cable operator, will add to rather than detract from his overall service offering.

ACCESS CHANNEL REGULATION

26. As to the actual plans for use of the access channels we have required, we want to emphasize that there is a great deal of flexibility. Different communities, operators, and access groups will find various ways of utilizing their channels most effectively. We expect that many variations will be tried. It would be a mistake for any regulatory authority or board to attempt, at this formative stage, to delimit too particularly how the access channels should work.

27. Our effort at creating a public access channel was meant to give the maximum access possible to local groups. It is for this reason that we initially described the channel as one that should be available on a "first-come, first-served non-discriminatory" basis. The best example of why we say that it is premature to establish firm rules for the access channel is the myriad number of questions we have been asked arising from that statement. By attempting to answer some of them here, we hope to clarify the policy considerations behind our access rules.

28. Some have questioned whether our rules would allow a particular person or group to reserve access channel time on a long-term basis, e.g., every Thursday night from 8 to 9 p.m. We did not intend that our rule would prohibit an access programmer from developing a viewership at a particular time by consistent programming. Therefore, this type of reserved time would be consistent with our rules. However, we also want to assure that all desirable time slots are not "frozen" and thereby monopolized or not available to the occasional programmer. Some balance is necessary. We are allowing cable operators to design their access channel rules to accommodate both interests and shall remain sensitive to the possibility that abuses might develop.

EDUCATIONAL ACCESS

29. Our educational access channel rules were designed to promote the use of that channel by educational authorities in the community. Much was claimed in the original dockets which led to the adoption of this rule about the potential for educational channels on cable. Little has developed. In retrospect, it appears that our limitation of one free educational access channel was wise. Designating vast channel capacity for

education only to see it lie fallow serves no purpose. Two questions have repeatedly been raised about our educational access rules: (1) Who qualifies as an "educational authority" to use the channel, and (2) what extra equipment, assistance, etc., can be demanded or offered for educators in a franchise agreement?

30. Our concept of "educational authority" was not meant to restrict the use of this channel to the local public school board. Any school, college, or university, public or private, formal or informal, should have the opportunity to air programming on this channel. The one exception to this interpretation would be commercial educational enterprises (computer schools, beauty schools, etc.) that would in essence be using the channel for advertising which we have specifically disallowed on the educational access channel. Any bona fide educational interest should have access to the educational channel. We envision a working educational channel as one where the programmers work out a reasonable ability to program this channel nor should cable operator to utilize this opportunity offered to them. It might be possible, for instance, for a high school and a college to produce complementary instructional programming of benefit to both. It is not the cable operator's responsibility to program this channel nor should he be expected to.

31. The problem of increasing demands in franchises for extra channels, money, equipment, personnel, etc., will be dealt with in section V of this document.

LEASED CHANNELS

32. It is too early to discern any trends regarding our leased access channel rules (§ 76.251(a)(7)). It remains our intent to keep these channels as free as possible from any regulation that might restrict or artificially alter their growth. This is particularly true in the area of rate regulation. We have pre-empted this area with the explicit purpose of allowing the market place to function freely. We note that many authorities are already talking about regulating leased channel rates and/or rates for pay cable services. It is premature to regulate along these lines. Such regulation might destroy any chance for this emerging communications service by stifling competition, setting incorrect rates, and establishing an atmosphere which deters experimentation, innovation, or speculation. We have pre-empted this area to avoid those pitfalls. It is unclear how a regulatory body could now establish reasonable rates for services that are untested, unproven, and which have not even established a consistent record as to costs, expenses, subscription, etc.

33. As we noted, in the Cable Television Report and Order, Para. 130, 131, dual jurisdictional regulation of access channels would cause great confusion and might inhibit their growth on a nationwide basis. Different regulation, rate structures, etc., for instance, on channels where a par program or per channel charge is made might unduly

hamper the. obviously interstate effort involved by cable operators and programmers to secure a large enough audience to make this new communications. medium a viable economic success. We cannot allow such a multiplicity of regulation to detract from our national program.

34. While we have decided to prohibit non-federal rate regulation. of leased channel uses or users at this time and have further announced our intention of refraining from imposing any federal regulations now, some guidelines regarding leased channel operation might be helpful. We recognize that many of the early efforts at rate regulation were motivated by concerns over potential abuse of the cable operator's position. We noted such a potential in the 1972 rules (Para. 126). To date there has been little evidence that the cable operators are hoarding capacity for their own uses or are setting preferential or prohibitive rates to maintain a monopoly position. Should such a situation develop, we will, of course, stop it. It is in the cable operator's best interest for this not to happen. All parties must be given access to the leased channels at rates not designed to prohibit entry. This is especially true in the area of pay cable. Evidence that cable operators are restricting entry would obviously lead to demands that cable be re-defined as a common carrier. We do not think this would be a good idea at this time. In fact, it would probably be detrimental.[5] But abuse, particularly of leased channel access, will surely result in far more restrictive regulation.

35. Some cable operators and franchising authorities have suggested a program whereby preferential rates for leased public, educational, and governmental channels are offered to non-commercial users. Thus, when the free channels are filled, or when, for instance, an educational user wants to put specialized programming on a separate educational channel, he could lease a channel at a lower rate than would be available to a commercial user. This concept appears sound, and we do not discourage cable operators from experimenting with such preferential rate structures. Specific franchise requirements or controls of this nature, however, remain pre-empted. We favor a market place experimentation in this area for now.

IV. TECHNICAL STANDARDS

36. We repeatedly stated in the original cable rules and in the reconsideration of them that our technical standards were only a first step in what we expected would be a long process of refin-

ing the technical parameters of cable television. In the Reconsideration of Cable Television Report and Order, we indicated that franchising authorities could also promulgate technical standards. That decision has now been brought into question.

37. The FSLAC report, while acknowledging an apparent problem regarding unrealistic standards being developed at state and local levels, recommends that this dual jurisdictional approach be maintained at least until the completion of the FCC Cable Television Technical Advisory Committee's (CTAC) work. However, the FSLAC report also recommends (Issue #4) that we issue cautionary advice to franchising authorities noting that our rules should suffice in a majority of cases and that any more stringent standards must be enforced locally. This recommendation also urges that we retain oversight authority to deal with any unrealistic standards that may be promulgated.

38. We recognize that this is an area of significant conflict. The experience we have already gained from CTAC's preliminary work and the confusion engendered by some of our original rules indicates that much more work needs to be done. Our technical advisory committee is making progress in this direction.[6] Most State Governors have already named liaisons with the Committee at our request so that we may coordinate as much of this activity as possible. The question now arises as to whether we should institute, at the least, a moratorium on the promulgation of non-federal technical standards until the completion of CTAC's work.

39. A petition for rule making has already been received from the National Cable Television Association regarding technical standards, pre-emption,[7] and the opinion of our own Office of Chief Engineer suggests that the multiplicity of conflicting technical standards has become a problem and might not be in the public interest. There has been considerable comment on the desirability of uniform standards and argument that the lack of such uniform standards could conceivably hamper the development of cable television because technical equipment could not be manufactured for nationwide use.

40. Understandably, the imposition of a moratorium or the complete subject matter pre-emption of technical standards are issues of considerable debate. For this reason, we invite interested parties to submit comments on the question of whether cable television technical standards should be totally pre-empted or a moratorium on additional non-federal technical standards should be a imposed until the completion of

the technical advisory Committee's work.[8]

V. FEDERAL/STATE-LOCAL RELATIONSHIPS

41. In our 1972 rules we adopted an ambitious program of creative federalism in the area of cable television franchising. In essence, we developed an approach of dualism toward the granting of cable franchises. We recognized that the complexities and national character of cable television called for nationwide rules and guidelines. At the same time, we acknowledged that the essentially local service offered by cable television, at least in its formative stages, could best be developed through local partcipation and enforcement. Our rules attempted to blend these needs into a cohesive, cooperative program between federal and local authorities. This effort appears to have been basically successful.

42. One significant new development, however, has become a complicating factor. State governments have begun asserting a regulatory role in cable television, thus adding a third-tier to the regulatory scheme. When we adopted our rule we envisioned a system whereby federal rules and guidelines would be complemented by one other regulatory authority—the so-called "local" level of government. We did not specify cities or municipalities because we recognized that in some states the state government would serve as the "local" authority rather than some smaller political subdivision. Indeed, this was the case in 1972, since several states had already asserted state jurisdiction over cable franchising (e.g., Connecticut, Nevada, Rhode Island, and Vermont). However, at that time there were no states asserting an additional regulatory function while leaving other regulatory and franchising matters to localities. It is this latter development that concerns us. A major portion of the FSLAC report deals with this "three-tier" problem (see Part II, FSLAC Final Report). In our December meeting with the FSLAC Steering Committee this was also a prime topic of discussion. We intend, in the near future, to deal with this question specifically. For the purposes of this document, however, it should be sufficient to caution all regulatory bodies involved or considering involvement in cable television that we are concerned about the developing duplicative and burdensome overregulation of cable television.

43. The purpose of this notice of proposed rulemaking and memorandum opinion and order is to clarify and in some cases modify the existing rules. Our experience to date indicates that one of the areas most in need of clarification is our franchise standards and their relationship to the rest of our rules.

44. Once again, we think it would be easiest to review all of our franchising

---

[5] This, of course, is consistent with the position we took in the Cable Television Report and Order, Para. 146. It does not mean that at some future time, once cable technology has sufficiently matured, that common carrier status would necessarily still be inappropriate. We note that the same position has now been taken by the President's Cabinet Committee on Cable Communications.

[6] The Committee expects to complete the first phase of its work by late this year.
[7] RM–2196 *Petition for Rule Making To Standardize Technical Standards* filed May 23, 1973, by the National Cable Television Association.

[8] Comments filed in response to this *Notice of Proposed Rule Making* should be referred to Docket No. 20018.

standards here in the order that they appear in the rules. For present purposes, the use of the term "local" authority should be read as referring to the local or state authority, whichever is appropriate in the particular jurisdiction. Generally, we assume that whichever non-federal authority grants the franchise will also be responsible for complying with all the other franchise-related aspects of those rules.

FRANCHISING AUTHORITY

45. In § 76.31(a), we require that cable operators in order to receive a certificate of compliance, must have a "* * * franchise or other appropriate authorization." In most cases, this has not caused any difficulties. It is not necessary that the document in question be called a "franchise." Depending on the laws of the particular jurisdiction, the authorization may take the form of a franchise, franchise and ordinance, license, permit, certificate of convenience and necessity, etc. The point is that documents must be provided showing that authorization from the appropriate local authority or authorities has been granted to the applicant to build a cable television system. This authorization must be complete before we will process an application for a certificate of compliance. The applicant must be in the position of being able to begin operation or construction immediately upon receipt of a certificate of compliance. All local and state processes (if any) must be completed before we will certify that an applicant has complied with our rules. It would be administratively burdensome and unnecessarily time consuming for us to process applications only to find that the applicant failed to secure full local approval to build and operate a proposed system. We will not process an application at this time which contains only the municipal franchise if the applicant is required by state law to also have state approval of the franchise.

46. We have had some difficulty when there is apparently no appropriate authority in the state empowered to grant a franchise. To date, we have, consistent with paragraph 116 of Reconsideration, granted certificates in such cases when an appropriate alternative proposal is supplied. We do not like this procedure but see no way around it so long as some states delay designating the appropriate local jurisdictional authority. It should be reiterated, however, that before we will proceed with an application claiming that there is no local authority capable of issuing a franchise or other appropriate authorization, we expect formal statements to that effect from the local authorities. We have no desire to become involved in the interpretation of state laws. We assume the regularity and accuracy of local official interpretation of state law unless specifically shown otherwise. We would urge, however, that in the few remaining areas where this problem still exists it be clarified at the state level in the near future.

47. Another, although less frequent problem, has come to our attention where

a franchising authority, while apparently having the authority to grant a franchise or other appropriate authorization consistent with our rules, declines to do so. Alternative proposals by the cable operator in such cases will not be accepted. Where a franchising authority has the power to comply with our rules but does not, a certificate of compliance will not be issued.

FRANCHISE STANDARDS

48. Section 76.31(a) also requires that the franchise contain "* * * recitations and provisions consistent with the following requirements." This has caused problems in cases where, although our rules were indeed followed, the fact that they were was not stated in the franchise. We have allowed applicants to remedy minor deficiencies by official communications from franchising authorities, thus avoiding the time consuming process of franchise amendments. Such a less formal process has allowed us to administer our rules with flexibility and will be continued. This is not to say that substantive omissions can be corrected in this manner.

49. In cases where, for instance, a statement that a full public proceeding was held was not included in the franchise and we find that such a proceeding was in fact held, we will not reject the application. Of course, franchising authorities and applicants would be wise to comply totally with the letter of our rules. We are simply stating here that we intend to apply our rules reasonably and see that their intent is followed even if, in some instances, their particulars are not. As always, such decisions will be made on a case by case basis, and, of necessity, such consideration will unavoidably slow the certificating process.

FRANCHISEE SELECTION—PUBLIC PROCEEDINGS

50. We think that the intent of § 76.31(a)(1) is clear. Prior to the selection of a franchisee, we expect the franchising authority to investigate the applicants' legal, character, financial, technical, and other pertinent qualifications. We also require that the public be given the opportunity to become involved in this process. There are many ways that this can be done. Many of the larger cities have had comprehensive hearings on the design of a cable ordinance. Others have established citizens' committees which held open publicized meetings and reported back their findings to the local authorities. Smaller localities, as a rule, have confined the process to their regular city council meetings. All of these methods are presently acceptable.

51. The purpose of our present rule is to assure that the public has been given notice and a right to be heard regarding the development of cable television in any particular area. We, of course, cannot guarantee nor would it be possible to require that all public input be heeded or adopted. We do not intend to act as a "court of last resort" for those who disagree with the decisions of their

elected officials. Our present requirement for public proceedings is administered on the basis of a "reasonable man" standard. So long as the public has been given a reasonable opportunity to participate in the franchising process, we currently consider our "public proceeding" requirement as having been met. We presume the regularity of action by local officials. Except in the extraordinary case, if local officials assure us that they have made appropriate investigations of the franchisee's qualifications and that the public has had an opportunity to participate in the process we will not delve further into the particular methodology or decision factors in any specific franchise grant.

52. Some have argued that we should strengthen these rules. The FSLAC report (Issue No. 2) recommends that we articulate minimum due process standards. The National Black Media Coalition (RM–2278 filed November 12, 1973) makes a similar request and suggests further that we adopt very specific notice and time requirements for public meetings. There have also been suggestions that we require specific information that should be requested or given to franchising authorities prior to the selection of a franchise.

53. We are not sure that such an approach is practical in the dual jurisdictional program we have set out. We recognize that the procedures for granting franchises differ in the various jurisdictions. There are questions whether such procedures are susceptible to nationwide regulation or whether the procedures as well as the general inquiry into the qualification of franchise applicants would better be left to local officials. We invite any interested party to comment on these questions.[9] Should more specific rules be adopted to articulate the appropriate public proceedings required prior to the selection of a franchisee (type and length of notice, etc.)? Should franchising authorities be given specific guidelines and requirements on the information to be considered prior to the selection of a franchisee? And in either case, if the answer is yes, what should the guidelines or recommendations be and how should they be enforced?

54. Of course, while we proceed in this inquiry, our "reasonable man" doctrine will remain in effect. By way of advice to authorities preparing to embark on the franchising process, particularly in the larger urban areas where there is a great deal of citizen interest in cable, experience suggests the following:

a. Publicly announced meetings specifically on the topic of cable television are most helpful. These meetings can be used to educate both the citizens and the city officials on precisely what cable is and is not and how cable relates to the needs of the particular locality.

b. Specific procedures for granting the franchise should be established, published,

---

[9] Comments filed pursuant to this Notice of Proposed Rule Making and Inquiry should be referred to Docket No. 20019.

and followed. It should be noted that, as has already happened in several cases, not following to the letter a municipality's own rules can cause considerable delay and acrimony.

c. Cities that have initially established an ordinance on cable television and have approved it without looking first to who will receive the franchise have found this to be a beneficial procedure avoiding many of the pitfalls involved in an unrealistic bidding contest on a combined franchise and ordinance. Such bidding contests, with cable operators and city officials offering or demanding provisions unrelated to the actual needs of the city or viable operation of the system, are harmful to all parties.

d. An open, written bid proposal by all applicants is helpful but care should be taken lest this become another for of bidding contest. It should be noted by cities and franchise alike that whenever a franchise application is incorporated by reference in a franchise it must be made part of the application for certificate of compliance from us. It will be reviewed for compliance with our rules in such a situation.

55. The process of soliciting bids for a cable television franchise often leads to excesses in both demands and offers. As we just noted, any bid application incorporated by reference in the franchise will be reviewed for consistency with the cable television regulations we have established. The fact that an "offer" was tendered and accepted by the franchising authority rather than demanded by that authority makes no practical difference in the administering of our rules. We look at all provisions, particularly for extra services or equipment, that are enforceable against the franchisee regardless of how they originated.

56. We do not mean to imply that any of the particular suggestions mentioned above are necessary to comply with our present requirements. They are simply illustrations of some successful approaches to the problem. In essence we anticipate for now that the franchising process includes open access to the decisionmaking process both for citizens and applicants, fairness to all parties, and consistency in the administration of any rules adopted to grant the franchise.

57. Many parties have asked whether we intended our current rules on public proceedings to include franchise renewal proceedings. The simple answer is yes. We have made no specific requirements either in the initial grant procedures or in renewal procedures. We do not require that there be written bids or even that there must be competitive procedures. In some cases, negotiated bids with selected applicants may be appropriate. These are matters for the franchising authority to decide. Particularly in renewals, which we will discuss more fully later, there may be no reason for competitive bidding. In both initial and renewal proceedings, however, we do require open access, consistency, and overall fairness. We may add new requirements as a result of the inquiry we have just initiated in this area but these minimums, we are confident, will not change.

CONSTRUCTION—LINE EXTENSION

58. In both § 76.31(a) (1) and (2), we refer to the "* * * adequacy and feasibility of * * * construction arrangements" and that the cable operator must "* * * equitably and reasonably extend energized trunk cable * * *." Confusion arising from these requirements prompts further clarification.

59. It was our intent that all parts of a franchise area that could reasonably be wired would be wired. The initial problem we were trying to cope with was the "hole in the donut" situation that could have developed in larger markets, that is, the wiring of the more affluent outlying areas of a city while ignoring the center city or the wiring of the "desirable" section of town and not providing the communications benefits of cable to the poorer areas. It now develops that in most instances this is not as much of a problem as was feared. In fact, the problem is reversed. The high density areas are being wired but the outlying, less populated suburbs are not.

60. Clearly, this problem can best be dealt with at the local level since every community presents unique demographic vagaries. Some over-all guidelines, however, should be set out. Obviously, the ideal case is where a franchisee is required to wire the entire franchise area. This is our present rule. The purpose of the rule was to assure that no "cream-skimming", wiring just the economically lucrative portions of a franchise area, would take place. We are aware, however, that many franchises are being granted that do not encompass the entire political subdivision of the grantor. Such grants are appropriate so long as they are not used as a device to deprive certain portions of the population of service. In some cases, cities decide to grant multiple franchises to different franchisees for various discrete sections of the franchise area. This is acceptable so long as the ultimate result is complete coverage of the area. Clearly, if the area was subdivided in such a way that one area would be highly lucrative while another was marginal and not sought after, the result would be "cream-skimming." This would be unacceptable. Other jurisdictions define the franchise area by way of a so-called "line extension" clause, that is where the cable operator is only required to wire those parts of the political subdivision that contain a specified number of homes per mile measured on some stated formula or base. The numbers we have seen range generally from 30 to 60 homes per mile. In some cases, we acknowledge such a formula is justified. The potential subscribership in a particular community may be marginal in terms of system viability, and the extension of lines to citizens in outlying areas or pockets might spell the difference between success and failure of the system. In other cases, however, systems have apparently

sought to maximize profits by only serving densely populated areas even though an averaging of the density figures to include those miles of cable plant in the sparsely populated areas indicated that the system would still be viable.

61. A middle course has been adopted in some instances whereby a formula is established in the franchise so that if outlying pockets of viewers wish the cable extended to them they must pay the specified costs involved in extending the trunk line.

62. We can see reasonable justifications in all of these approaches. They point up the necessity of local involvement in the cable process to deal with the unique problems presented by various communities. We think it would be a mistake to attempt to specify a nationwide rule on this point. Indeed, it might be very difficult to create any such rule even on a state by state level. This is a job for the localities.

63. Because we recognize this problem, we have and will continue to grant certificates of compliance to applicants whose franchises do not require our ideal, the wiring of the entire community. However, before we do, we want assurances in the application and from the franchisor that the public, and particularly those citizens directly affected by the exclusions or conditional wiring provisions, are informed of the effect of such provisions before they are adopted. In at least some cases such notification has been accomplished by local newspaper articles including maps indicating the specifically affected areas. In others, local officials directly contacted affected homeowners. Unfortunately, however, in many cases line extension policies were set without any consultation with the citizens involved, and at least a few instances have been found where even the franchising authority did not fully comprehend the effect of its actions. We are not prohibiting line extension provisions in franchises, but we do intend to require that there be a showing that such provisions were developed knowledgeably and publicly. Any line extension formulas arrived at under these conditions are likely to be reasonable, having taken into consideration costs, population density and averages, terrain problems, long range land development plants, etc. under public scrutiny.

64. Since the assurances we are requesting do not presently appear in our rules, we plan to make the appropriate amendments consistent with the proposal outlined above. We invite any interested party to comment on this suggested addition to our rules. Any other proposals submitted aimed at remedying this problem will also be considered prior to the adoption of any specific new rule or filing procedure.[10]

[10] Comments filed pursuant to this notice of proposed rule making should be referred to Docket No. 20020.

COUNTY-WIDE FRANCHISES

65. This entire discussion of franchise area delineation takes on even more immediate importance in the many incorporated, county-regulated areas of the country. Clearly a large county with many non-contiguous-population pockets does not expect one franchise to wire the entire county, particularly at the rate of construction we have recommended. For this reason, we have consistently contacted applicants for certificates of compliance with blanket county franchises and requested more specific information on what areas the system plans to serve. Certificates of compliance will only be granted for those specified areas, not for an entire county, unless the applicant truly intends to serve the entire area within a specified construction time schedule.

66. In most instances, county franchisees are in fact developing systems for particular unincorporated communities within the county. It would be a significant help to us if county governments designated what they considered to be the discrete communities within their jurisdiction. Such delineations are, after all, uniquely a part of the responsibility of local officials. Their conclusions will have significant impact on the applicability of our rules (for example, in the area of our filing and access channel requirements which apply to each discrete community).

EXTENSION OF SERVICE

67. One of the most common complaints about cable television received by this Commission is a potential subscriber's inability to obtain service. This generally is caused by one of three situations: there is no cable television system in the locality; there is a system, but it will not extend its lines; or there is a system in an adjacent jurisdiction, and it is unable to extend beyond its franchise area.

68. In the first instance, of course, there is little that can be said other than that the community should, if there is substantial interest, seek a franchise. The second case, refusal to extend service, relates directly to our previous discussion of line extension policies. The third problem, however, is more difficult.

69. In an increasing number of cases, we are finding that newly developed areas, housing developments and the like, find themselves unable to obtain service because they are located either at an extreme fringe or outside the franchise area. This is an unavoidable and vexing problem that can only be remedied by cooperation and planning. The FSLAC report (Issue #15) comments on this situation in detail and we think their conclusions are a helpful guideline to all regulatory authorities:

Extensions of service, both within a given franchise area and into adjacent areas, have and will continue to be made voluntarily by operators in response to economic and public relations forces. Jurisdiction for mandatory (involuntary) extensions may be applied only to new franchises and renewals.

Discussion: The subject of mandating extensions either within the franchise area or to contiguous areas prompted considerable debate within the Committee.

Relating to extension of existing systems, the Committee opposes forced extensions into areas not specifically contemplated or required by the existing franchise. The Committee believes that in most cases an operator's failure to make an apparently feasible extension (i.e., to respond voluntarily to normal market forces) is evidence of economic and engineering problems faced in effecting extensions which were not contemplated when the system was originally designed. In this area there also appear to be valid legal concerns about modifications of existing contracts.

The Committee holds a different view in the case of new franchises where the franchisee has of his own free will accepted line extension requirements as an initial condition of doing business. In such cases we urge both the franchisee and the franchising authority to seriously think and plan the area's development pattern over the term of the franchise so that future engineering problems can be avoided. It would seem appropriate, and consistent with our views relating to the definitions of the franchise area for the franchising authority and franchisee to agree upon an expandable definition of the required service area including a clear statement of the condition under which extensions could be mandated.

For these purposes renewals can be viewed as new franchises, assuming * * * that a non-renewed operator who has faithfully performed his expired franchise is assured of realizing fair value for his property. Subject to this condition, as in the case of a new franchise, whatever requirements are imposed would be the result of an arms-length agreement.

Finally this position on renewals would tolerate the imposition of extension requirements on existing franchisees in the extraordinary cases where, by state action establishing an overriding public interest in receipt of cable service, all existing franchises were terminated and reissued.[11]

70. In this regard, particularly in areas where there are pockets of population or growing suburban subdivisions, we would urge that franchising authorities in contiguous communities join together in planning for future cable development. We have seen several cases in which a new housing development was unable to get cable service because it was on the extreme edge of its community but the adjoining jurisdiction's cable system was readily available. A 'joint powers' agreement or other type of cooperative arrangement between the communities could easily solve these problems.

71. We are treating this subject in considerable detail because we consider it one of the most important factors in local and regional franchising. Service extension and the delineation of the franchise should be one of the primary concerns of local regulatory authorities. It has received too little attention in the past.

---

[11] See also Minority Comment, FSLAC Report Appendix A, which argues that state action may be the most appropriate or effective method of establishing nonconfiscatory required service extensions.

FRANCHISE LENGTH

72. Our rules limit the length of a new franchise to a maximum of 15 years (Section 76.31(a)(3)). This rule was prompted by the initial trend in franchising that led to extremely long (i.e., 99-year) franchises which afforded local authorities no opportunity to review and modify the franchise agreement if necessary. Lengthy franchise grants, we noted in our 1972 report, "* * * are an invitation to obsolescence in light of the momentum of cable technology" (par. 182). We also stated, in the reconsideration of cable television report and order, Para. 111, that there might be some instances where longer franchises are warranted and that we would entertain waiver requests in those cases. The FSLAC report recommends that this rule be changed in favor of a more flexible approach. They argue that, particularly in the larger cities, 15 years may not be sufficient time to develop and make profitable the advanced and complex broadband communications systems being contemplated. The Committee Report states:

"* * * It is our feeling that a fifteen year maximum period does not sufficiently deal with the difficulties of financing modern systems in cities of widely varying size. Accordingly we recommend that the maximum franchise period be redefined as a range of fifteen to twenty-five years, with specific periods within that range to be determined by individual franchising authorities. As an integral part of this recommendation provision should be made by the franchising authority for review at least every five years, commencing at most ten years after the franchise grant.

The central purpose of such reviews would be to consider such issues as system performance, design modifications, and the possible need for changes in franchise terms. Such reviews might result in alterations in the basic franchise, franchise extensions, and other possible changes in the agreements between the parties. In no case would such review periods preclude proceedings by the franchising authority at any time for termination of the franchise for cause."

73. The problem of minimum franchise terms has also been raised. In some cases, certificates of compliance are being sought for franchises with a one-year term. We question the advisability of this short a franchise duration. The capital costs and commitments involved in building a cable television system would seem to dictate against entrepreneurs accepting such short terms. We understand that in some states a year-to-year franchise is easier to secure than a term franchise necessitating a public referendum. However, such year-to-year franchises impose significant risks and increased administrative burdens. We intend to consider a rule imposing some minimum franchise term, possibly between 5 and 7 years, to remedy this problem.

74. We invite any interested party to submit comments on both the proposal made in the FSLAC report and our suggestion for a minimum term requirement as well as any other suggestions

for modifications of our rules on franchise duration. Of particular interest would be any cash flow figures supporting contentions for the need for longer franchise terms.[12]

FRANCHISE MODIFICATION AND RENEWAL

75. The entire subject of franchise duration, modification, renewal, expiration, and cancellation is one that is fraught with difficulties. First, a few points should be made to clarify our own filing requirements in this area. While we are considering propsals to change the duration rules (§ 76.31(a)(3)), the existing 15-year maximum will remain in effect. This maximum applies to both the initial grant and any renewals. A franchise calling for a 15-year term with a renewal option at the sole discretion of the franchisee does not comply with the rule. The franchisor must at least review, in a public proceeding, the performance of the system operator and the adequacy of the franchise as well as its consistency with our rules prior to renewal. This is not to say that any bid procedures are required or that any new franchise offering must be made, but simply that a public review of the franchise must be held with the opportunity for citizen input prior to renewal. In this regard, it should be noted that our rules, and the certificates of compliance we grant, are based in part on the franchises included in the application. The certificate does not apply to renewal franchises or to the terms of franchises significantly amended in any way, such as a change in termination date, service obligations,[13] or franchise fees. Any such substantial change or renewal we consider to effectively terminate the existing franchise and that termination (or in effect the granting of a new franchise) requires recertification (or certification in the first instance on grandfathered franchises). An exception to this doctrine is a change in subscriber rates. Such a change is consistent with our rules so long as it is done in a public meeting and will not be considered to have terminated the existing, certified franchise.[14]

_____
[12] Comments filed pursuant to this Notice of Rulemaking should be referred to Docket No. 20021.

[13] The term "service obligation" or "service package" as used herein includes generally all requirements imposed on the franchisee relating to local origination or access programming, equipment, personnel, or any other purported obligations relating to programming or any other special benefits required for specified programmers or subscribers.

[14] We, of course, are primarily concerned with changes that have some relationship to our rules. We are not including in this interpretation of when recertification is required changes in such areas are indemnity or bonding requirements, specific construction or safety alterations, reporting and enforcement procedures and the like. We recognize that the franchise is a "living" document and changes must be made from time to time to reflect current situations and practices.

76. The reason we are taking this approach to substantial changes in franchises should be obvious. Our entire program of certification would be meaningless if significant alterations, potentially contrary to our rules, could be made in a franchise after we had certified that it complied with federal regulations. Any substantial change in a franchise, of course, would automatically end any "grandfathering" rights regarding other provisions in the same franchise. Our "grandfather" of pre-March 31, 1972 franchises was meant to give franchising authorities a reasonable amount of time in which to bring their franchises into compliance. If they are now changing provisions in the franchises, they also have had ample opportunity to acquaint themselves with the new rules and will be expected to comply with all of our franchise requirements. In dealing with previously certified applicants, we will assume that they are operating pursuant to the already certified franchising during the certificating process for the new franchise.

FRANCHISE EXPIRATION AND CANCELLATION

77. In Reconsideration we expressed concern over situations where franchise renewal applicants threaten to terminate service to the public rather than reach an accord with the franchising authority. Once again, the comments in the FSLAC Report (Issue #10) are helpful by way of clarification:

* * * [T]wo * * * problems in this area * * * bear mentioning. First, as the franchise term draws to a close with no assured renewal or fair compensation in sight, the cable operator acquires a strong disincentive to invest in needed new equipment that he cannot be certain of amortizing over the remaining term; the result, obviously, is a deterioration of service. Second, unfortunately, this situation has in the past created extreme and sometimes unwarranted pressures on franchise authorities and system operators to reach renewal agreements. Both these excessive pressures and the disincentive should be removed.

First, the Committee feels there should be no cancellation or expiration of the franchise without fair procedures and fair compensation. The existing franchisee should be given adequate notice and opportunity to be heard. Furthermore, we suggest that if the decision is adverse to the existing franchisee, the franchisor should have some provisions for an assignable obligation to acquire the system at a predetermined compensation formula. In the case of non-renewal this formula should call for payment of fair market value of the system as a going concern; whereas in the case of cancellation of the franchise for material breach of its terms, the compensation criterion might call for depreciated original cost with no value assigned to the franchise. In either case, the Committee would suggest that there be provision for impartial arbitration if the negotiators fail to agree on a price. The franchisor's obligation should be fully assignable to a successor franchisee selected by the franchisor.

It is also advisable, we believe, that there be a requirement that, during the reasonable interim period while transfer of the system is being arranged, the original franchisee be required to continue service to the public as a trustee for his successor in interest, subject to an accounting for net earnings or losses during the interim period.

All of these provisions should be included in the franchise itself so that the parties to the franchise know their respective rights and obligations and can plan their operations accordingly.

78. We think the Committee's advice is well taken. All the provisions mentioned are of utmost importance to the orderly process of renewal or transfer of system control. The public is directly and potentially severely affected if these provisions, or ones like them, are not contained in the franchise. We strongly suggest that all franchising authorities include such provisions.

79. Our concern in this area is so great, particularly as to guaranteed continuation of service to the public, that we are considering adopting rules requiring franchises to contain specific provisions and procedures relating to expiration, cancellation, and continuation of service. We invite all interested parties to comment on this proposal.[15] Particular attention in the comments should be given to whether the rule should be a general one, simply requiring that franchises contain such provisions or whether the rules should be more specific as to the type of safeguards we should require to protect the public interest in this area. We would reiterate, however, that regardless of the outcome of this proceeding, franchising authorities would be wise to adopt the type of provisions discussed above. Too many instances have already come to our attention of threats to cut off service to the public.

TRANSFERS

80. In a related notice of rulemaking and inquiry, we would like to explore the difficult problem of transfers or assignments of control of franchises.[16] At the moment we have no firm rules in this area and many questions have arisen. We note, for instance, that most new franchises require prior local approval before a transfer can take place. We assume that such approvals are given only after full public proceedings. We do not, however, require the inclusion of such provisions in the franchise at this time. Comments are invited on whether such requirements should be added to § 76.31 of our Rules.

81. Unfortunately, this is not as simple as it may appear. What, for instance, constitutes a transfer of control? If corporate ownership changes by acquisition, merger, etc., yet the local franchisee remains the same, should this trigger a public proceeding with all that entails? What effect would such an interpretation have on the ability of multiple system operators to consolidate, merge, etc., in the open market place? Should franchising authorities even be concerned with

_____
[15] Comments filed pursuant to this notice of proposed rulemaking should be referred to Docket No. 20022.

[16] Comments filed pursuant to this notice of proposed rulemaking and inquiry should be referred to Docket No. 20023.

this type of "transfer" so long as the negotiated terms of their franchise are enforceable? Clearly it is time for us to inquire into these areas and adopt appropriate regulations to deal with them where necessary.

82. A question has been raised as to whether franchise transfers constitute a "significant change" so as to require recertification. At the moment, they do not. It would seem that so long as the franchise terms comply with our rules and the franchise is so certified, it is unnecessary for us to require recertification of the same document. The selection of the franchise holder is, after all, a local matter under our rules.

83. While we do not consider recertification necessary because the terms of operation in a simple transfer or assignment remain the same, we are considering adding a provision to our filing requirements for the submission of a new Form 325 for any transferred system. Such a rule would assure us that our files are always updated on transfers of ownership as soon as they occur. When we receive this information, it would be checked for compliance with our cross-ownership rules. A statement of such compliance accompanying the submission might also be required. Comments on this proposal or any other recommendations on dealing with the complex problems involved with transfers of control as well as franchise expiration, cancellation, and termination are invited. The discussion of these problems above should put all parties on notice that we consider this a particularly difficult area that requires careful study and perhaps additional regulation.

SUBSCRIBER RATE REGULATION

84. In § 76.31(a)(4) we require that cable systems, in order to receive a certificate of compliance, must have a franchise providing for franchisor approval of initial charges for installation and regular subscriber service. We have intentionally and specifically limited rate regulation responsibilities to the area of regular subscriber service, and we will continue to do so. We have defined "regular subscriber service" as that service regularly provided to all subscribers. This would include all broadcast signal carriage and all our required access channels including origination programming. It does not include specialized programming for which a per-program or per-channel charge is made. The purpose of this rule was to clearly focus the regulatory responsibility for regular subscriber rates. It was not meant to promote rate regulation of any other kind.

85. After considerable study of the emerging cable industry and its prospects for introducing new and innovative communications services, we have concluded that, at this time, there should be no regulation of rates for such services at all by any governmental level. Attempting to impose rate regulation on specialized services that have not yet developed would not only be premature but would in all likelihood have a chilling effect on the anticipated develop-

ment. This is precisely what we are trying to avoid. The same logic applies to all other areas of rate regulation in cable, i.e., advertising, pay services, digital services, alarm systems, two way experiments, etc. No one has any firm idea of how any of these services will develop or how much they will cost. Hence, for now we are preempting the field and have decided not to impose restrictive regulations. Of course, at such time as clear trends develop and if we find that the free market place does not adequately protect the public interest, we will act, but not until then.

SUBSCRIBER COMPLAINTS

86. Assuring that subscribers receive quality service and quick resolution of any complaints is one of the most important regulatory functions to be performed at all levels of government. The primary locus of responsibility, however, must be at the local level, where the service is. For this reason we stated in § 76.31(a)(5) that specific procedures for the resolution of subscriber complaints shall be included in the franchise and that there shall be a local business office or agent available to subscribers to remedy complaints. Many franchises are now being reviewed which have full statements of the franchisee's obligations to resolve subscriber complaints but no indication whether the franchisor has any responsibilities. We wish to make it clear, therefore, that this obligation was meant to cover both parties.

87. If no specific franchise statement indicates with whom or where to register complaints at the local level and what will be done with them once received, the public is not well served. The result of this information gap to subscribers is that local complaints often are sent to this Commission.[17] We, in turn, inform the correspondent that his complaint is within the purview of local not federal officials, and he should contact them. Much time and effort is thus wasted.

88. In order to fully comply with § 76.31(a)(5), therefore, we expect that franchising authorities from this point on will include specific provisions in the franchise on what government official will be directly responsible for receiving and acting upon subscriber complaints.[18] We would also urge that this information along with the specified procedure for reporting trouble to the cable operator be given to all subscribers as they are hooked into the system. Some communities have required that a card with this information on it be given to each new subscriber. It seems to have worked

---

[17] This Commission has established a subscriber complaint service to aid the public. Its efforts are primarily aimed at clearing up misunderstandings between subscribers and systems with regard to our rules and the informal resolution of complaints.

[18] We propose changing our rules to make this requirement clear. Any interested party may file comments on this notice of proposed rulemaking. Comments on this subject should be referred to Docket No. 20024.

well, and we would encourage adoption of this approach.

89. Some questions have been raised regarding the meaning of our requirement for a "local" business office. In most cases, this is a clear requirement. A system serving one city should have a business office or other means in that community to receive and act on subscriber complaints. However, we will be flexible in the interpretation of this rule as it relates to a single head-end multi-community system or a "county" system. The operator of a single-plant multi-community system need not have a business office in each of the communities served so long as subscribers can call a local telephone number to register complaints and personnel are available to act on those complaints. On the other hand, we will not accept a situation where there is only one business office in a large county necessitating long-distance telephone calls for some subscribers to register their complaints.

FRANCHISE CONSISTENCY

90. As we have said throughout the period of developing these rules for cable television, the process is evolutionary. We expect to continue to modify, clarify, add, or eliminate provisions as the need arises. We intend to remain flexible in this regard and franchising authorities should be on notice that this is the case. For this reason, we included the requirement § 76.31(a)(6) that a franchise should specifically contain provisions allowing for amendments to comply with our rules. Unfortunately, although this rule appears to be clear on its face, many franchises have not included such a provision. It should be understood that any required modifications would have to be made even where a franchise does not specifically state that it is amendable to comply with our changes within one year. However, we would prefer a clear statement in the franchise to that effect to make sure all parties are aware of the possible need for modifications.

FRANCHISE FEES

91. In § 76.31(b) a limitation is imposed on the franchise fee deemed acceptable in an application for a certificate of compliance. Many questions have been raised about the perimeters of this limitation (see, e.g., FSLAC final report (Appendix B) "memorandum regarding clarification of § 76.31(b) and related matters" and the associated minority opinion.)

92. The purpose of the limitation we imposed was clearly stated in the Cable Television Report and Order:

* * * We are seeking to strike a balance that permits the achievement of federal goals and at the same time allows adequate revenues to defray the costs of local regulation.

We have found no reason to change our position on this matter. The use of the franchise fee mechanism as a revenue raising device frustrates our efforts at developing a nationwide broadband

communications grid. Excessive fees or other demands in effect create an obstruction to interstate commerce which must be avoided.

93. The figure of three percent of gross subscriber revenues seems to more than adequately compensate the average franchising authority for actual regulatory costs. We have provided a waiver mechanism for fees up to five percent of gross subscriber revenues in those cases where an unusual or experimental regulatory program is proposed that can be shown to need the extra revenue.

94. Because of the many questions raised regarding this rule we will review the reasoning, intent, and scope of § 76.31(b) as it relates to the rest of our rules. First, some definitions appear necessary.

"GROSS SUBSCRIBER REVENUES"

95. The term "gross subscriber revenues" is meant to include only those revenues derived from the supplying of regular subscriber service, that is, the installation fees, disconnect and reconnect fees, and fees for regular cable benefits including the transmission of broadcast signals and access and origination channels if any. It does not include revenues derived from per-program or per-channel charges, leased channel revenues, advertising revenues, or any other income derived from the system.

FEES FROM AUXILIARY SERVICES

96. We recognize that the income derived from auxiliary cable services may at some future time constitute the bulk of a cable system's receipts. We have no intention of depriving the franchising authority of a reasonable percentage of those receipts at that time. But for now, the monies derived from ancillary services are best used to support the development of those experimental and largely unprofitable services. We encourage experimentation in ancillary services. Any funds that can be freed to support those services will ultimately benefit the community of the system and aid our efforts at seeing these services develop nationwide.

97. Because we are presently imposing a "gross subscriber revenues" limit on franchise fees which may, at some future date, be lifted, we suggest that franchising authorities write their franchise fee provision flexibly, that is, using a "gross subscriber revenues" base for now but including a provision for the base to change to "gross revenues" automatically in the event that this Commission changes its rules.

98. There have been several cases where a franchise fee was based on something other than gross subscriber revenues. Generally, such instances arise when the fee is based on a specific monetary figure per year per subscriber. In those cases, the percentage is figured based on the subscriber rate and an average penetration estimate. Regardless of how the fee is stated, however, we will attempt to translate the fee into a percentage of gross subscriber revenues to

see if it reasonably complies with our rules.

STATE AND LOCAL FEES

99. It should be noted that we include all non-Federal regulatory fees in our limitation. The purposes stated in the *Report and Order* would clearly be circumvented if we interpreted the rule otherwise. Our concern that "* * * high local regulatory fees may burden cable television to the extent that it will be unable to carry out its part in our national communications policy * * *" (Para. 185) is just as valid if the burdensome fees are imposed by a combination of local authorities. Accordingly, both local franchise fees and state fees, if any, will be added together to determine compliance with our fee limitations.

100. Another related problem has recently been brought to our attention in this area of fees. Several jurisdictions are now attempting to impose a "use tax" as well as a fee for cable television service. It would appear that such a tax, particularly when its purpose is described as general revenue raising, results in the same potential harm we are attempting to avoid by imposing a franchise fee limitation. While the particular cases before us (CSR–479, Stockton, Cal., and CSR–499, State of Florida) will be dealt with in separate actions, we think it is necessary to express our concern about this development. The burdens and obstructions to the growth of a viable nationwide communications grid remain the same whether imposed via a fee or a tax mechanism.

FRANCHISE FEE WAIVERS

101. As we noted earlier and made clear in the report and order, waiver of the three percent ceiling is available. Indeed, even our rules indicate that up to five percent could be considered a reasonable fee depending on specific showings. Many have asked what exactly need be shown to allow a fee between three and five percent.

102. While each case, of necessity, is different and must be handled on an individual basis, some general guidelines can be given. The bulk of the regulatory burden at the local level comes in the first few years of cable development. The creation of a cable ordinance and the granting of a franchise as well as supervision of construction all occur in this period. Aside from normal franchise enforcement and review, very little actual regulation on a day-to-day basis goes on after this initial surge of activity. The number of franchises now being adopted with our fee limitation intact indicates that three percent of gross subscriber revenues does cover these costs.

103. It is the rare case where a more comprehensive regulatory program is contemplated that extra fees might be justified. Such programs are usually in the larger markets or where experimental applications of cable are being attempted. In these cases, we recognize that our three percent fee limit might not cover the costs incurred. Where it can be shown that the three percent

figure will not be adequate and that the specific contemplated costs of the specific regulatory program require extra input in the form of fees up to five percent of gross subscriber revenues, we will entertain waiver requests.

104. Petitions to justify fees in excess of three percent should include both a full description of the special regulatory program contemplated and a full accounting of estimated costs. Such petitions should also contain information on the estimated subscriber penetration and the derived figures on revenue anticipated from the franchise fee. It is only with a complete showing of this nature that we can realistically determine if the extra fee request is justified and that it will not adversely affect the operator's ability to accomplish federal objectives.

105. The recitation of the normal obligations to oversee a franchisee assumed by the local authority is not sufficient to warrant extra fees. Justifications that simply allocate a portion of the time and salary of various city officials to cable regulation without a full explanation of the special regulatory program to be carried out will also not be considered sufficient. Such an allocation, without amplification, would only confirm that the fee is being used to augment the general treasury as a revenue raising device.[19]

106. The reason we have allowed for extra fees despite our concern over the possible strain such fees impose on our nationwide program is to maintain flexibility. In those cases where a special office of telecommunications (such as in New York City) is warranted by unique circumstances or special personnel is hired to handle cable television regulation and complaints, the new costs could in part be covered by the higher franchise fee. Very few situations of this type have come to our attention.

LUMP SUM PAYMENTS

107. Included in our fee limitation is a notation on lump sum payments or payments-in-kind. It is important that everyone understand the ramifications of this notation. Were we to allow a large initial lump sum payment for securing the franchise it would negate the effort we have made to limit the franchise fee. Bidding contests would continue unabated. The public would be the ultimate loser since the franchising authorities and bidders would focus on bidding rather than how and by whom the best service can be provided to the community. We therefore include any lump sum payments in our injunction on the ultimate size and effect of the fee. Such payments are amortized over the term of the franchise to determine their effect on the percentage figure. One exception to this method is stated consulting fees and expenses incurred in the granting or renewal of the franchise. If these fees

---

[19] We note that the Report to the President by the Cabinet Committee on Cable Communications (1974) has adopted our view against the use of cable franchise fees for such purposes. Recommendation 9(c).

are not excessive and can be shown as direct costs to the franchising authority, we think they should be recoverable from the ultimate franchisee or from all franchise applicants as has been done in some cases. It is not unusual for the franchising authority to spend several thousand dollars for an independent survey or consultant to aid in developing the cable ordinance. So long as these expenses do not become a new form of bidding we will not include them in our calculation of franchise fees. A specific showing of the expenses, however, should be made. Ideally, the expenses should be calculated and set prior to franchise bidding and the established costs either allocated among the bidders or applicable to any franchisee. Of course, we will continue to watch such charges for any evidence of abuse.

EXTRA SERVICE PACKAGE REQUIREMENTS

108. Another area that we closely monitor in relation to the franchise fee is the rather all-encompassing problem of "extra services". This has included everything from the free wiring of entire school systems to the building of television studios attached to the local high school, extra free channels, fees for access groups, and even free television sets for city officials. This is a very difficult problem to deal with, as can be seen from the Federal/State-Local Advisory Committee's rather lengthy discussion of the topic, *supra*. It is precisely because these "extra services" take such diverse forms that specific guidelines are almost impossible to enunciate. We will attempt to discuss some of the more commonly requested extra services and their relationship to our overall policy. In that way we would hope that franchisors and applicants can be more sensitive and responsive to the problems we see developing.

109. In many if not most franchises, the franchisee is required to install one free "tap" or "drop" in each local school and often in every other government building (city hall, firehouse, etc.). We have no objection to such a provision. In a few instances, however, the free extra service has been much greater. Some franchises have required the cable operator, for instance, to wire each room in all the local public schools. This in essence requires the operator to internally wire the school system free of charge. Such an expense can be considerable, especially when several hundred rooms might be involved. The cost of equipment and materials alone could amount to more than the revenue derived from the franchise fee. It is this sort of indirect "payment-in-kind" that we are watching very closely and will not allow without justification. This type of expense is just as real and has just as much of an effect on the franchisee as a simple fee. All parties must begin to recognize that when such costs are incurred they of necessity often become trade-offs on service provided elsewhere to the community at large. In this example we merely have the cable operator subsidizing the school system. That is not his function.

110. A trend seems to be developing where franchising authorities specify in the franchise the production equipment to be made available. Some franchises have become so technical that they even include the model numbers of particular microphones and cables. While such "service package" requirements are not prohibited by our rules, we do not think it is a particularly good idea. Technology in the area of low-cost video production equipment is advancing so rapidly that such specifications are likely to be an invitation to planned obsolescence. We only repeat, in this regard, that origination and access will not work because of anything written in a franchise. It is far more important for the franchising authority to assure itself of the character, responsiveness, and interest of the potential operator than it is to write strict franchise provisions in this area. The mere requiring of specific cameras and equipment will not guarantee successful community access. Real commitment and interest cannot be required in any legal document.

111. As was noted earlier, if the franchising authority wishes to specify the service package it expects from the operator in the franchise, we will not stop it from doing so. Reasonable service offerings can and are being made in the franchising process. Both franchising authorities and franchise applicants must recognize, however, that any specification of services will reflect on the costs of the over-all service to the community. Excessive service demands or offers will affect the viability of the system. Cable operators must learn that accepting such demands simply to secure a franchise may not be in their or the cities best interest. Similarly, franchise authorities must be cautious in accepting high priced extra service offerings on the basis of bid procedures. The net effect of some superficially attractive offerings might be a basic system that does not find it possible economically to serve the community properly.

112. It has been our policy to date to view any service package requirements in relation to our franchise fee limitation. We plan to relax this approach experimentally. The service package—so long as it is directly related to services and equipment which can potentially benefit all cable users—will now be treated as a contractual question and, so long as the package is not clearly excessive, solely up to the discretion of the franchisor and franchisee.[20] We wish to emphasize, however, that we are relaxing the effect of our rules experimentally. Any evidence that cable operators or franchisors are using this relaxation to return to the damaging process of simple "bidding contests" will result in the immediate reinstitution of our former procedures.

---

[20] In this context we are discussing "service packages" only as they relate to equipment, personnel, etc. This does not include pre-empted services such as extra channels, origination programming, etc.

113. It should be noted that we are making a distinction on what will or will not be viewed as part of the franchise fee "payment-in-kind" limitation. Required extra services that benefit only one group of special users is still considered a type of cross-subsidy that will be viewed in relation to the franchise fee. As an example, the operator being required to wire the entire local school system for closed circuit cable use would still be considered payment-in-kind. Specific equipment or personnel requirements where the benefits are available to all cable users would not.

114. Our purpose, in part, in imposing a franchise fee or payment-in-kind limitation was to prevent the siphoning of the limited available capital for cable development for other uses, thereby threatening the success of our overall national goals. We intend to maintain that limitation. Reasonable service requirements that directly benefit cable development and use by all parties is compatible with that purpose.

115. Another reason for this adjustment in our review policy is that the complexities involved in any service package offering and the innumerable variations result in an ad hoc administrative process that cannot be effectively carried out with any consistency. We are, however, sensitive to our obligation to insure that abuses do not arise that will threaten our nationwide program. For this reason, we expect to issue a Notice of Proposed Rulemaking in the near future that will suggest revisions in our filing and reporting procedures so that we can get more specific data on the costs of special service packages.

116. The information we will be seeking is also information that any responsible franchising authority should demand prior to accepting any applicant's proposal, i.e., what are the expected expenses involved in the service offering; how will those expenses contribute to the quality of cable services in the community; what will be the effect of those expenses on the financial viability of the system, etc.

117. We will no longer attempt to "second guess" the franchising authority on the answers to those types of questions. It is hoped that all parties will realize that decisions made in the area of required services may well have a major impact on the development of cable in any particular locale. We will, however, continue to monitor such agreements. If we find that serious abuses are arising that could effect our national goals we stand ready to re-establish procedures to remedy the problem.

118. Once again, it should be emphasized that the flexibility we are encouraging in service packages is restricted to services, equipment or personnel available to all cable users. Proposals that would benefit only one class of cable users would not be acceptable. Studios, equipment, or mobile vans designated for use or given specifically to one group such as the educational authority or a public access group would not be reasonable.

Such equipment, etc., must inure to the benefit of all users, including the cable operator, for his own origination programming, if any. As was explained in detail earlier in this document, guidelines, and procedures for waivers will remain in force regarding channel capacity, extra access channel demands, etc.

USE OF FEES FOR OTHER PURPOSES

119. In yet another area where the franchise fee limitation has come into question, we have received many inquiries regarding the use of the fee for purposes other than to defray regulatory costs. Proposals have been made, for instance, to use a portion of the franchise fee to pay for access programming or to aid local educational broadcast facilities. As a general rule we have stated that the franchise fee should be based on regulatory costs. It should not be used for revenue raising purposes. We continue to hold this position at this time.

120. As with most of our other regulations in this field, we intend to maintain flexibility. We will entertain waiver requests for the use of franchise fees for non-regulatory purposes. Such requests, however, must be very specific. Information on how the funds will be used, distributed, and accounted for must be included. A showing that the proposed use of the fee is consistent with our regulatory program and will benefit the development of a broadband communications system will also be necessary. In carefully reviewed cases where a specific experimental program designed for a particular community is presented we will consider granting waivers of our rules. Generally speaking the use of these "extra" fees will be limited to the same maximum now imposed for regulatory purposes, five percent of gross subscribed revenues. In most cases that have come to our atten-

tion the special uses fees are limited to the two percent "pad" between three and five percent. It is unlikely that we will allow waivers for any proposal that exceeds a total of five percent for regulatory and non-regulatory purposes.

121. Proposals to use the two percent "pad" in the franchise fee rules for public access purposes pose several significant problems for us. While we recognize the need for additional funding for access, there are serious difficulties, we feel, with governmental funding of programming. These difficulties exist regardless of the mechanism for distribution. We intend to issue a separate document shortly that will address this specific issue.

CONCLUSION

122. In summation, on the question of franchise fees and extra services or other obligations, we intend to be vigilant and monitor any such requirements thoroughly to assure that no undue burdens are being imposed that would result in a diminution of the overall goals we have set for cable television. Reasonability is the keynote to any such program, and we will remain flexible and open to any thoroughly considered proposals. Our rules and the service requirements we impose on cable operators are intended to provide a solid base for the development of a nationwide means of broadband communications. In most instances, no more is required or, indeed, desirable at this time. It is unreasonable to expect an infant industry to be able to start where we all hope it will eventually end—as a truly new and innovative highly complex broadband network. It must be allowed to grow in stages or it will be killed by overexpectation and excessive demands.

123. The clarifications and guidance we have provided in this document will hopefully aid all parties in our effort to

develop responsive and flexible regulations for an emerging industry. The announcement of several new rule making inquiries herein is yet another testimonial to the fact that we intend to continue to investigate, clarify, modify or change our regulations as the situation warrants. The regulatory concepts we have adopted are new and many of our rules are experimental. We welcome any supported recommendations aimed at improving them.

124. Authority for the rule makings proposed herein is contained in sections 4(i), 303, and 403 of the Communications Act of 1934, as amended. All interested parties are invited to file written comments on these rule making proposals on or before June 7, 1974, and reply comments on or before June 21, 1974. Please note that separate docket numbers have been assigned to individual rulemaking inquiries initiated herein. Comments should also be filed separately. In reaching a decision on these matters, the Commission may take into account any other relevant information before it, in addition to the comments invited by this Notice.

125. In accordance with the provisions of § 1.419 of the Commission's rules and regulations, an original and 14 copies of all comments, replies, pleadings, briefs or other documents filed in this proceeding, shall be furnished to the Commission. Responses will be available for public inspection during regular business hours in the Commission Public Reference Room at its Headquarters in Washington, D.C.

Adopted: April 15, 1974.

Released: April 17, 1974.

FEDERAL COMMUNICATIONS COMMISSION

[SEAL] VINCENT J. MULLINS,
Secretary.

[FR Doc.74-9141 Filed 4-19-74;8:45 am]

GLOSSARY

NOTE: Except for occasional direct quotations, the following definitions are paraphrases of terminology used by the Federal Communications Commission. Other definitions are presented in the text where specially relevant.

Access: Use by programers other than the operator (and excluding broadcast television stations carried on cable) of cable television channels to reach subscribers; FCC rules require "access channels" for public, educational, local government, and leased use in major television markets.

Affiliate: A television broadcast station under a regular contractual arrangement with, or owned by, a television network.

Cablecasting: Programing on cable television not transmitted over the air; FCC regulations require origination cablecasting by cable operators on systems with 3500 or more subscribers and capability for access cablecasting in major television markets.

Cable television: Apparatus for distributing video programing by wire connections to individual television receivers; "cable television service" includes carriage of video television signals locally available, carried by microwave from distant points, and originated solely for wire distribution, as well as nonvideo programing and return service.

CBS v. TelePrompTer: The District Court for the Southern District of New York on May 2, 1972, Civ. No. 64-3814 (S.D.N.Y.), held that cable systems may import, via microwave relays, distant television programs that are broadcast beyond the range of local reception and distribute them to subscribers by cable without liability for copyright payments under the Copyright Act of 1909. The decision of Judge Constance Baker Motley is being appealed. See also "*Fortnightly.*"

Certificate of compliance: FCC regulations require issuance of a federal certificate of compliance before a cable television system may "commence operations

[1] All page references are to Chapter III, except where otherwise indicated. [Report to the President of The Committee, Executive Office of the President, Washington, D. C., January 14, 1974.]

[carrying broadcast television] or add a television broadcast signal to existing operations." 47 C.F.R. § 76.11.

Common carrier A privately owned economic entity of high public importance granted monopoly rights under various state and federal legislation, subject to public regulation to insure adequate and nondiscriminatory service at "reasonable rates." See "Public Utility."

Communications Act of 1934: Enacted by the Congress long before the development of cable television, and not specifically amended with respect to cable television, the Communications Act grants regulatory authority to the Federal Communications Commission and provides specific standards for licensing of broadcast stations and regulation of interstate common carriers.

Consensus Agreement: An agreement negotiated among representatives of broadcast stations, cable television operators, and copyright owners to provide for the limited growth of cable television systems; the Consensus Agreement, which serves as the basis for the FCC's cable rules, stipulates principles applicable to protect "program exclusivity" (which see) and presumes enactment of complementary copyright legislation.

Distant signals: Television broadcast over the air not receivable by local television sets but which may be carried (via microwave) and distributed by cable television systems under strict FCC quotas.

Equal time: The requirement enforced by the FCC that broadcast stations, and cable systems on origination channels, furnish opportunities to publicize opposing candidates for public office.

Fairness doctrine: The requirement enforced by the FCC that broadcast stations, and cable systems on origination channels, present opportunities for airing conflicting views on issues of public importance.

Federal Communications Commission (FCC): The federal regulatory body, consisting of seven members appointed by the President and confirmed by the Senate, which regulates interstate communications under the Communications Act of 1934.

"Footnote 69": Where two major television markets are adjacent to each other and some viewers in each can receive broadcast television signals from stations licensed to the other market, the FCC has established special rules for distributing broadcast signals via cable; the classification refers to a footnote reference in a 1966 FCC order, "Second Report and Order in Docket 15971."

Fortnightly: In *Fortnightly Corp. v. United Artists Television, Inc.,* 392 U.S. 390 (1968), the Supreme Court held that carriage by cable television of locally available broadcast signals does not give rise to copyright liability under the Copyright Act of 1909, leaving liability for carriage of distant signals unresolved; see also *CBS v. TelePrompTer.*

Franchise: The legal authorization required under applicable state law from state and/or local governments for establishment of cable television service; made a precondition for FCC issuance of a certificate of compliance under rules respecting scope and procedure. 47 C.F.R. § 76.31.

"Freeze": Beginning in 1966 and continuing under proposed new regulations issued in 1968, the FCC imposed severe limitations on use by cable television systems of additional distant signals in major markets, thus effectively curbing cable television growth in those areas.

Grade-A contour: The line demarcating an area within which broadcast signals have sufficient strength to enable viewers to receive them at 90 percent of the locations embraced for at least 70 percent of the time. 47 C.F.R. § 73.683.

Grade-B contour: The line demarcating an area within which broadcast signals have sufficient strength to enable viewers to receive them at 90 percent of the locations embraced for at least 50 percent of the time. 47 C.F.R. § 73.683.

"Grandfathering": The vesting of preexisting rights by regulatory action as of a particular time, exempting such rights from prospective policy changes, to achieve fairness and avoid constitutional questions of deprivation of property; applied in regulatory changes affecting, e.g., carriage of broadcast television signals, program exclusivity, and required channel capacity.

Independent: Description applicable to ownership and control of television broadcast stations not affiliated with a television network.

"Leapfrogging": Carriage of distant television signals by cable television systems, whereby a geographically more distant signal is carried in preference to a nearer signal with the same programing, subject to restriction by FCC rules.

Local signals: Signals of broadcast stations locally receivable by viewers on television receivers; classification of such local signals for carriage by cable television systems is subject to special rules established by FCC.

Major television markets: The 100 largest concentrations of television viewers in the United States, ranging from New York City to Columbia, South Carolina, as listed in 47 C.F.R. § 76.51; differentiated between the "first 50 major markets" and the "second 50 major markets," and distinguished overall from "smaller" television markets. See also "Specified Zone" for definition applicable to market contours.

Midwest Video: In *United States v. Midwest Video Corporation*, 406 U.S. 649, 32 L Ed 2d 390 (1972), the Supreme Court upheld the FCC's authority to order cable television systems with 3500 or more subscribers to originate programing, on the grounds that such authority is "reasonably ancillary" to the agency's responsibilities for broadcast television under the Communications Act of 1934.

Minimum channel capacity: FCC rules require cable systems in major television markets to have available a number of channels for access and origination equivalent to the number of broadcast signals carried, subject to an overall minimum requirement of twenty (20) television channels of usable bandwidth.

Network: Originator and distributor of commercial television programing on a national basis, specifically, American Broadcasting Corporation (ABC), Columbia Broadcasting System (CBS), and National Broadcasting Corporation (NBC).

Opposition: Process by which interested persons may object to the grant of federal certification to cable television system; they must do so within 30 days after a petition for certification has been filed.

Origination: Cablecasting, defined by the FCC as "programing (exclusive of broadcast signals) carried on a cable television system over one or more channels and subject to the exclusive control of the cable operator." 47 C.F.R. § 76.5.

Pay-cablevision: The use of cable systems of one or more origination channels to distribute video programing for which a per-program or per-channel reception charge is made in addition to regular service; subject to FCC rules with respect to type and frequency of programing under 47 C.F.R. § 76.225.

Personal attack rule: Requirement that, when a verbal attack has been made an identified person or group, the cable system provide a reasonable opportunity to respond; applicable to broadcast television and origination channels, but not to access channels.

Petition for Special Relief: Mechanism by which waiver of FCC rules is sought by a cable system, franchising authority, station licensee, permitee, applicant, or other interested person.

Preemption: Occupation of a field of activity by federal authority through regulatory, legislative, or Constitutional provision; inconsistent state and local regulation is Constitutionally prohibited.

Program exclusivity: Protection imposed against cable carriage of broadcast programing, requiring that the cable system "black out" specific programs on distant signals carried where sole distribution rights exist by agreement between program suppliers and local broadcast stations.

Program origination: Programing required to be offered by cable systems with 3500 or more subscribers, in fulfillment of an FCC rule that such systems must serve "to a significant extent as a local outlet." 47 C.F.R. § 76.201(a).

Public utility: Economic entity providing vital public service subject to public regulation, such as telephone, telegraph, distribution of energy, etc., under private or public ownership; usually regulated to permit only "reasonable" returns on invested capital.

Reconsideration: The FCC's *Memorandum, Opinion and Order of Reconsideration of the Cable Television and Order,* 37 Fed. Reg. 13848 (July 14, 1972).

Report and Order: The FCC's *Cable Television Report and Order* (Docket Nos. 18397, 18397-A, 18416, 18892, and 18894), 37 Fed. Reg. 3251 (February 12, 1972), effective March 31, 1972.

"Significant viewing": Measurement standard of extent and intensity of local viewing of television broadcast stations used by FCC as alternative basis for requiring signal carriage by cable systems, according to standards set forth in 47 C.F.R. § 76.5; see also, "specified zone."

Southwestern Cable: In *U.S. v. Southwestern Cable Company,* 392 U.S. 157 (1968), the Supreme Court ruled that the FCC has the authority to prohibit carriage of distant signals, such authority being "reasonably ancillary to the effective performance of the Commission's various responsibilities for the regulation of television broadcasting" under the Communications Act of 1934.

Specified Zone: The area "extending 35 air miles from the reference point in the community to which" a television broadcast station is licensed or authorized by the FCC; used by the FCC as the radius for determining television markets. 47 C.F.R. § 76.5(f).

Syndicated programing: Programing distributed to independent broadcast stations.

Translator: A television broadcast facility utilized to increase the effective range of television broadcast signals.

Two-way capability: Technical capability of cable television system to carry "return" signals from dispersed terminals to any other point in a cable television system.

Waiver: See "Petition for Special Relief."

"Wild cards": Broadcast signals of independent television stations which cable systems are permitted to add to the quotas of distant signals they are authorized to carry in major television markets.

"Zapple ruling": FCC provision extending equal-time rules to supporters of or spokesmen for a candidate during an election campaign: named after the congressional aide seeking the ruling.

SELECTED RAND BOOKS

Bagdikian, Ben H. *The Information Machines: Their Impact on Men and the Media.* New York: Harper and Row, 1971.

Bretz, Rudy. *A Taxonomy of Communication Media.* Englewood Cliffs, N. J.: Educational Technology Publications, 1971.

Bruno, James E. (ed.). *Emerging Issues in Education: Policy Implications for the Schools.* Lexington, Mass.: D. C. Heath and Company, 1972.

Cohen, Bernard, and Jan M. Chaiken. *Police Background Characteristics and Performance.* Lexington, Mass.: D. C. Heath and Company, 1973.

Coleman, James S., and Nancy L. Karweit. *Information Systems and Performance Measures in Schools.* Englewood Cliffs, N. J.: Educational Technology Publications, 1972.

Dalkey, Norman C. (ed.). *Studies in the Quality of Life: Delphi and Decision-making.* Lexington, Mass.: D. C. Heath and Company, 1972.

DeSalvo, Joseph S. (ed.). *Perspectives on Regional Transportation Planning.* Lexington, Mass.: D. C. Heath and Company, 1973.

Downs, Anthony. *Inside Bureaucracy.* Boston, Mass.: Little, Brown and Company, 1967.

Fisher, Gene H. *Cost Considerations in Systems Analysis.* New York: American Elsevier Publishing Co., 1971.

Haggart, Sue A. (ed.). *Program Budgeting for School District Planning.* Englewood Cliffs, N. J.: Educational Technology Publications, 1972.

Harman, Alvin. *The International Computer Industry: Innovation and Comparative Advantage.* Cambridge, Mass.: Harvard University Press, 1971.

Levien, Roger E. (ed.). *The Emerging Technology: Instructional Uses of the Computer in Higher Education.* New York: McGraw-Hill Book Company, 1972.

Meyer, John R., Martin Wohl, and John F. Kain. *The Urban Transportation Problem.* Cambridge, Mass.: Harvard University Press, 1965.

Nelson, Richard R., Merton J. Peck, and Edward D. Kalachek. *Technology, Economic Growth and Public Policy.* Washington, D.C.: The Brookings Institution, 1967.

Novick, David (ed.). *Current Practice in Program Budgeting (PPBS): Analysis and Case Studies Covering Government and Business.* New York: Crane, Russak and Company, Inc., 1973.

Park, Rolla Edward. *The Role of Analysis in Regulatory Decisionmaking.* Lexington, Mass.: D. C. Heath and Company, 1973.

Pascal, Anthony H. (ed.). *Racial Discrimination in Economic Life.* Lexington, Mass.: D. C. Heath and Company, 1972.

Pascal, Anthony H. *Thinking about Cities: New Perspectives on Urban Problems.* Belmont, Calif.: Dickenson Publishing Company, 1970.

Quade, Edward S., and Wayne I. Boucher. *Systems Analysis and Policy Planning: Applications in Defense.* New York: American Elsevier Publishing Company, 1968.

Sharpe, William F. *The Economics of Computers.* New York: Columbia University Press, 1969.

Williams, John D. *The Compleat Strategyst: Being a Primer on the Theory of Games of Strategy.* New York: McGraw-Hill Book Company, 1954.

A NEW VARIORUM EDITION

OF

SHAKESPEARE

EDITED BY

HORACE HOWARD FURNESS

HON. PH. D. (HALLE), HON. L. H. D. (COLUMB.), HON. LL.D. (PENN. ET HARV.)
HON. LITT. D. (CANTAB.)

A MIDSOMMER NIGHTS DREAME

[*SEVENTH EDITION*]

PHILADELPHIA
J. B. LIPPINCOTT COMPANY
LONDON: 5, HENRIETTA STREET, COVENT GARDEN

Copyright, 1895, by H. H. FURNESS.

WESTCOTT & THOMSON,   PRESS OF J. B. LIPPINCOTT COMPANY,
Electrotypers and Stereotypers, Phila.   Phila.

IN MEMORIAM

PREFACE

'I KNOW not,' says Dr JOHNSON, 'why SHAKESPEARE calls this play "A *Midsummer* Night's Dream," when he so carefully informs us that it happened on the night preceding *May-day*.'

'The title of this play,' responds Dr FARMER, 'seems no more intended to denote the precise *time of the action* than that of *The Winter's Tale*, which we find was at the season of sheep-shearing.'

'In *Twelfth Night*,' remarks STEEVENS, 'Olivia observes of Malvolio's seeming frenzy, that "it is a very *Midsummer* madness." That time of the year, we may therefore suppose, was anciently thought productive of mental vagaries resembling the scheme of SHAKESPEARE's play. To this circumstance it might have owed its title.'

'I imagine,' replies the cautious MALONE, 'that the title was suggested by the time it was first introduced on the stage, which was probably at Midsummer: "A Dream for the *entertainment* of "a Midsummer night." *Twelfth Night* and *The Winter's Tale* had probably their titles from a similar circumstance.'

Here the discussion of the *Title of the Play* among our forbears closed, and ever since there has been a general acquiescence in the reason suggested by MALONE: however emphatic may be the allusions to May-day, the play was designed as one of those which were common at Midsummer festivities. To the inheritors of the English tongue the potent sway of fairies on Midsummer Eve is familiar. The very title is in itself a charm, and frames our minds to accept without question any delusion of the night; and this it is which shields it from criticism.

Not thus, however, is it with our German brothers. Their native air is not spungy to the dazzling spells of SHAKESPEARE's genius. Against his wand they are magic-proof; they are not to be hugged into his snares; titles of plays must be titles of plays, and indicate what they mean.

Accordingly, from the earliest days of German translation, this discrepancy in the present play between festivities. with the magic

rites permissible only on *Walpurgisnacht*, the first of May, and a
dream seven weeks later on *Johannisnacht*, the twenty-fourth of June,
was a knot too intrinse to unloose, and to this hour, I think, no German
editor has ventured to translate the title more closely than by *A
Summernight's Dream*.   In the earliest translation, that by WIELAND
in 1762, the play was named, without comment as far as I can discover,
Ein St. Johannis Nachts-Traum.   But then we must remember that
WIELAND was anxious to propitiate a public wedded to French dramatic laws and unprepared to accept the barbarisms of *Gilles* SHAKE-
SPEARE.   Indeed, so alert was poor WIELAND not to offend the purest
taste that he scented, in some incomprehensible way, a flagrant impro-
priety in ' Hence, you long-legged *spinners*, hence ;' a dash in his
text replaces a translation of the immodest word ' spinner,' which is
paraphrased for us, however, in a footnote by the more decent word
' spider,' which we can all read without a blush.

  ESCHENBURG, VOSS, SCHLEGEL, TIECK, BODENSTEDT, SCHMIDT (to
whom we owe much for his *Lexicon*), all have *Ein Sommer Nachts
Traum*.   RAPP follows WIELAND, but then RAPP is a free lance ; he
changes Titles, Names, Acts, and Scenes at will ; *The Two Gentlemen
of Verona* becomes *The Two Friends of Oporto*, with the scene laid in
Lisbon, and with every name Portuguese.   But SIMROCK, whose *Plots of
Shakespeare's Plays*, translated and issued by *The Shakespeare Society*
in 1840, is helpful,—SIMROCK boldly changed the title to *Walpurgisnachtstraum*, and stood bravely by it in spite of the criticisms of
KURZ in the *Shakespeare Jahrbuch* (iv, 304).   SIMROCK's main difficulty seems to me to be one which he shares in common with many
German critics, who apparently assume that SHAKESPEARE's ways
were their ways, and that he wrote with the help of the best *Conver-
sations-Lexicon* within his reach ; that at every step SHAKESPEARE
looked up historical evidence, ransacked the classics, and burrowed
deeply in the lore of Teutonic popular superstitions ; accordingly,
if we are to believe SIMROCK, it was from the popular superstitions of
Germany that SHAKESPEARE, in writing the present play, most largely
drew.

  TIECK, in a note to SCHLEGEL's translation in 1830, had said that the
Johannisnacht, the twenty-fourth of June, was celebrated in England,
and indeed almost throughout Europe, by many innocent and super-
stitious observances, such as seeking for the future husband or sweet-
heart, &c.   This assertion SIMROCK (p. 436, ed. Hildburghausen,
1868) uncompromisingly pronounces false ; because the only custom mentioned by GRIMM in his *Mythologie*, p. 555, as taking place
on Midsummer Eve is that of wending to neighboring springs,

there to find healing and strength in the waters. On Midsummer Night there were only the Midsummer fires. When, however, TIECK goes on to say that 'many herbs and flowers are thought to 'attain only on this night their full strength or magical power,' he takes SIMROCK wholly with him; here at last, says the latter, in this fact, 'that the magic power of herbs is restricted to certain tides 'and times, lies the source of all the error in the title of this play, 'a title which cannot have come from SHAKESPEARE's hands.' All the blame is to be laid on the magic herbs with which the eyes of the characters in the play were latched. SHAKESPEARE, continues SIMROCK, must have been perfectly aware that he had represented this drama as played, not at the summer solstice, but on the Walpurgis night,—Theseus makes several allusions to the May-day observances; and inasmuch as this old symbolism was vividly present to the poet, we may assume that he placed the marriage of Theseus and Hippolyta on the first of May, because the May King and May Queen were wont to be married within the first twelve days of that month. Even Oberon's and Titania's domestic quarrel over the little changeling 'is founded on the German legends of the gods'; Frea and Gwodan quarrel in the same way over their devotees, and Frigga and Odin, in the *Edda*, over Geirröd and Agnar. 'The commentators,' complains SIMROCK, 'are profuse enough with their explanations where 'no explanations are needed, but not a hint do they give us of the 'reason why Puck is called a "wanderer," whereas it is an epithet 'which originated in the wanderings of Odin.' This *Germanising* of SHAKESPEARE is, I think, pushed to its extreme when SIMROCK finds an indication of Puck's high rank among the fairies in the mad sprite's 'other name, Ruprecht, which is *Ruodperacht*, the Glory-glittering.' It is vain to ask where SHAKESPEARE calls Puck 'Ruprecht;' it is enough for SIMROCK that Robin Goodfellow's counterpart in German Folk lore is Ruprecht, and that he chooses so to translate the name Robin. As a final argument for his adopted title, *Walpurgisnachtstraum*, SIMROCK (p. 437) urges that Oberon, Titania, and Puck could not have had their sports on Midsummer's Eve, because this is the shortest night in the year and it was made as bright as day by bonfires. In reply to KURZ's assertion that WIELAND's *Oberon* suggested GOETHE's *Intermezzo* (that incomprehensible and ineradicable defect in GOETHE's immortal poem), SIMROCK replies (*Quellen des Shakespeare*, 2d ed. ii, 343, 1870) that GOETHE took no hint whatever from WIELAND's *Oberon*, but named his *Intermezzo—A Walpurgisnachts Traum* 'in 'deference to SHAKESPEARE, just as SHAKESPEARE himself would have 'named his own play, knowing that the mad revelry of spirits, for

'which the night of the first of May is notorious, then goes rushing
'by like a dream.'

This brief account of a discussion in Germany is not out of place
here. From it we learn somewhat of the methods of dealing with
SHAKESPEARE in that land which claims an earlier and more intimate appreciation of him than is to be found in his own country—a
claim which, I am sorry to say, has been acknowledged by some of
SHAKESPEARE'S countrymen who should have known better.

The discrepancy noted by Dr JOHNSON can be, I think, explained
by recalling the distinction, always in the main preserved in England,
between festivities and rites attending the May-day celebrations and
those of the twenty-fourth of June: the former were allotted to the
day-time and the latter to the night-time.* As the wedding sports
of Theseus, with hounds and horns and Interludes, were to take place
by daylight, May day was the fit time for them; as the cross purposes
of the lovers were to be made straight with fairy charms during slumber, night was chosen for them, and both day and night were woven
together, and one potent glamour floated over all in the shadowy realm
of a midsummer night's dream.

The text of the First Folio, the *Editio Princeps*, has been again
adopted in the present play, as in the last four volumes of this
edition. It has been reproduced, from my own copy, with all the
exactitude in my power. The reasons for adopting this text are
duly set forth in the Preface to *Othello*, and need not be repeated.
Time has but confirmed the conviction that it is the text which a
student needs constantly before him. In a majority of the plays it is
the freshest from SHAKESPEARE'S own hands.

As in the case of fifteen or sixteen other plays of SHAKESPEARE, *A
Midsummer Night's Dream* was issued in Quarto, during SHAKESPEARE'S
lifetime. In this Quarto form there were two issues, both of them
dated 1600. To only one of them was a license to print granted by
the Master Wardens of the Stationers' Company—the nearest approach
in those days to the modern copyright. The license is thus reprinted
by ARBER in his *Transcript of the Stationers' Registers*, vol. iii, p. 174:†

* How many, how various, how wild, and occasionally how identical these festivities were, the curious reader may learn in Brand's *Popular Antiquities*, i, 212–
247, 298–337, Bohn's ed., or in Chambers's *Book of Days*.

† In Malone's reprint of this entry, the title reads a 'Mydsomer Nyghte Dreame.'
It may be worth while to mention what, I believe, has been nowhere noticed, the
variation in the title as it stands in the Third and Fourth Folios: 'A Midsummers
nights Dreame.

ɔ. octobris [1000]

Thomas ffyssher Entred for his copie vnder the handes of master
RODES | and the Wardens A booke called *A*
mydsommer nightes Dreame . . . . . vj^d

The book thus licensed and entered appeared eventually with
the following title page :—'A | Midſommer nights | dreame. | As it
'hath beene ſundry times pub- | *lickely aɛted, by the Right honoura-* |
'ble, the Lord Chamberlaine his | *ſeruants.* | *Written by William*
'*Shakeſpeare.* | [Publishers punning device of a king-fisher, with a
'reference, in the motto, to the old belief in halcyon weather:
'*motos ſalea companere fluctus*] ¶ Imprinted at London, for *Thomas*
'*Fisher,* and are to | be ſoulde at his ſhoppe, at the Signe of the
'White Hart, | in *Fleeteſtreete.* 1600.'

The Quarto thus authorised is called the First Quarto (Q₁), and
sometimes Fisher's Quarto.

No entry of a license to print the other Quarto has been found
in the *Stationers' Registers.* Its title is as follows:—'A | Midſom-
'mer nights | dreame. | As it hath beene ſundry times pub- | *likely*
'*aɛted, by the Right Honoura-* | ble, the Lord Chamberlaine his |
'*ſeruants.* | *VVritten by VVilliam Shakeſpeare.* | [Heraldic device, with
'the motto *Post Tenebras Lvx.*] *Printed by Iames Roberts,* 1600.'

This is termed the Second Quarto (Q₂) or Roberts's Quarto. The
second place is properly allotted to it, because, apart from the plea that
an unregistered edition ought not, in the absence of proof, to take precedence of one that is registered, it is little likely, so it seems to me, that
Fisher would have applied for a license to print when another edition
was already on the market ; and he might have saved his registration fee.
There are, however, two eminent critics who are inclined to give the
priority to this unregistered Quarto of Roberts. 'Perhaps,' says HAL-
LIWELL,* 'Fisher's edition, which, on the whole, seems to be more cor-
'rect than the other, was printed from a corrected copy of that published
'by Roberts It has, indeed, been usually supposed that Fisher's edition
'was the earliest, but no evidence has been adduced in support of this
'assertion, and the probabilities are against this view being the correct
'one. Fisher's edition could not have been published till nearly the
'end of the year ; and, in the absence of direct information to the
'contrary, it may be presumed that the one printed by Roberts is
'really the first edition.' If the 'probabilities,' thus referred to, are
the superiority of Fisher's text and the lateness in the year at which it
was registered, both may be, I think, lessened by urging, first, that

---

* *Memoranda on the Midsummer Night's Dream*, p. 34, 1879.

the excellence of the text is counterbalanced by the inferiority of the typography, a defect little likely to occur in a second edition; and, secondly, in regard to the 'end of the year,' HALLIWELL, I cannot but think, overlooked the fact that the year began on the 25th of March; the 8th of October was therefore only a fortnight past the middle of the year.

The other critic who does not accept Fisher's registered copy (Q₁) as earlier than Roberts's unregistered copy (Q₂) is FLEAY, to whom 'it seems far more likely' (*The English Drama*, ii, 179) that 'Roberts 'printed the play for Fisher, who did not, for some reason unknown to 'us, care to put his name on the first issue; but finding the edition 'quickly exhausted, and the play popular, he then appended his 'name as publisher.' Furthermore, FLEAY makes the remarkable assertion that 'printer's errors are far more likely to have been intro- 'duced than corrected in a second edition.' From FLEAY's hands we have received such bountiful favours in his *Chronicle History of the London Stage* and in his *Biographical Chronicle of the English Drama* that it seems ungracious to criticise. Shall we not, like Lokman the Wise, 'accept one bitter fruit'? and yet this bitter fruit is elsewhere of a growth which overruns luxuriantly all dealings with the historical SHAKESPEARE, where surmise is assumed as fact, and structures are reared on imaginary foundations. Does it anywhere stand recorded, let me respectfully ask, that Thomas Fisher 'found that the edition 'was quickly exhausted'?

Thus, then, with these two texts and the Folio we have our critical apparatus for the discovery, amid misprints and sophistications, of SHAKESPEARE's own words, which is the butt and sea-mark of our utmost sail. To enter into any minute examination of the three texts is needless in an edition like the present. It is merely forestalling the work, the remunerative work, of the student; wherefor all that is needed is fully given in the TEXTUAL NOTES, which therein fulfill the purpose of their existence. Results obtained by the student's own study of these Textual Notes will be more profitable to him than results gathered by another, be they tabulated with ultra-German minuteness. It is where only one single text is before him that a student needs another's help. This help is obtrusive when, as in this edition, there are prac- tically forty texts on the same page. All that is befitting here, at the threshold of the volume, is to set forth certain general conclusions.

In the Folio, the Acts are indicated. In none of the three texts is there any division into Scenes.

In Fisher's Quarto (Q₁), although the entrances of the characters are noted, the exits are often omitted, and the spelling throughout is

archaic, for instance, *shee, bedde, dogge,* &c., betraying merely a compositor's peculiarity; to this same personal equation (to borrow an astronomical phrase) may be attributed such spellings as *bould,* I, i, 68; *chaunting,* I, i, 82; *graunt,* I, i, 234; *daunce,* II, i, 90; *Perchaunce,* II, i, 144; *ould,* v, i, 273, and others elsewhere. Its typography when compared with that of the Second Quarto is inferior, the fonts are mixed, and the type old and battered. On the other hand, the Second Quarto, Roberts's, has the fairer page, with type fresh and clear, and the spelling is almost that of to-day. The exits, too, are more carefully marked than in what is assumed to be its predecessor. Albeit the width of Roberts's page is larger than Fisher's, the two Quartos keep line for line together; where, now and then, there happens to be an overlapping, the gap is speedily spaced out. In both Quartos the stage directions are, as in copies used on the stage, in the imperative, such as '*wind horns,*' '*sleep,*' &c. Both Quartos have examples of spelling by the ear. In ' Ile watch Titania when she is ' asleepe' (II, i, 184) Roberts's compositor, following the sound, set up ' Ile watch Titania whence she is asleepe.' In the same way the compositors of both Quartos set up : ' Dians bud, or Cupids flower,' instead of: ' Dian's bud *o'er* Cupid's flower.' Again, it is the similarity of sound which led the compositors to set up : ' When the Wolf beholds the Moon,' instead of *behowls.* And, indeed, I am inclined to regard all the spelling in Fisher's Quarto, archaic and otherwise, as the result of composing by the ear from dictation, instead of by the eye from manuscript; hence the spelling becomes the compositor's personal equation. Moreover, many of the examples of what is called the ' absorption' of consonants are due, I think, to this cause. Take, for instance, a line from the scene where Bottom awakes. Roberts's Quarto and the Folio read : ' if he go about to expound this dream.' Fisher's compositor heard the sound of ' to' merged in the final *t* of ' about,' and so he set up, ' if he go about expound ' this dream.' The same absorption occurs, I think, in a line in *The Merchant of Venice,* which, as it has never, I believe, been suggested, and has occurred to me since that play was issued in this edition, I may be pardoned for inserting here as an additional instance of the same kind. Shylock's meaning has greatly puzzled editors and critics where he says to the Duke at the beginning of the trial : ' I'll not answer that : But say it is my humour, Is it an- ' swered?' Thus read, the reply is little short of self-contradiction. Shylock says that he will not answer, and yet asks the Duke if he is answered. Grant that the conjunction *to* was heard by the compositor in the final *t* of ' But,' and we have the full phrase ' I'll

'not answer that but *to* say it is my humour,' that is, 'I'll answer that 'no further than to say it is my humour. Is it answered?'

In the discussion of misprints in general, and especially of these instances of absorption—and these instances are numberless—not enough allowance has been made, I think, for this liability to compose by sound to which compositors even at the present day are exposed when with a retentive memory they carry long sentences in their minds, and to which compositors in the sixteenth and seventeenth centuries were most especially exposed, when, as we have reason to believe, they did not, as a rule, compose by the eye from a copy before them, but wholly by the ear from dictation.* Furthermore, it is not impossible that many of the examples adduced to prove that the text of sundry Quartos was obtained from hearing the play on the stage may be traced to hearing the play in the printer's office. Be this as it may, it is assuredly more likely that such blunders as 'Eagles' for *Ægle*, or 'Peregenia' for *Perigouna* (of North's *Plutarch*), in II, i, 82, are due to the deficient hearing of a compositor, than that they were so written by a man of as accurate a memory as SHAKESPEARE, whose 'less Greek' was ample to avoid such misnomers.

In the address 'To the great Variety of Readers' prefixed by HEMINGE and CONDELL to the First Folio, we are led by them to infer that the text of that edition was taken directly from SHAKESPEARE'S own manuscript, which they had received from him with 'scarse a blot.' Unfortunately, in the present case, this cannot be strictly true. The proofs are only too manifest that the text of the Folio is that of Roberts's Quarto (Q_2). Let us not, however, be too hasty in imputing to HEMINGE and CONDELL a wilful untruth. It may be that in using a printed text they were virtually using SHAKESPEARE'S manuscript if they knew that this text was printed directly from his manuscript, and had been for years used in their theatre as a stage copy, with possibly additional stage-business marked on the margin for the use of the prompter, and here and there sundry emendations, noted possibly by the author's own hand, who, by these changes, theoretically authenticated all the rest of the text.

* Conrad Zeltner, a learned printer of the 17th century, said . . . . 'that it was customary to employ a reader to read aloud to the compositors, who set the types from dictation, not seeing the copy. He also says that the reader could dictate from as many different pages or copies to three or four compositors working together. When the compositors were educated, the method of dictation may have been practised with some success; when they were ignorant, it was sure to produce many errors. Zeltner said he preferred the old method, but he admits that it had to be abandoned on account of the increasing ignorance of the compositors.'—*The Invention of Printing*, &c. by T. L. DE VINNE, New York, 1876, p. 524.

The Folio was printed in 1623. We know that *A Midsummer Night's Dream* was in existence in 1598. Is it likely that during the quarter of a century between these two dates, many leaves of legible manuscript would survive of a popular play, which had been handled over and over again by indifferent actors or by careless boys? That many and many a play did really survive in manuscript for long years, we know, but then they had not, through lack of popularity, probably been exposed to as much wear and tear of stage use as *A Midsummer Night's Dream*, wherein, too, about a third of the actors were boys.

Be this, however, as it may, in those days when an editor's duty, hardly to this hour fully recognised, of following the *ipsissima verba* of his author, was almost unknown, it is an allowable supposition that HEMINGE and CONDELL, unskilled editors in all regards, believed they were telling the substantial truth when they said they were giving us as the copy of SHAKESPEARE'S own handwriting, that which they knew was printed directly from it, and which might well have been used many a time and oft on the stage by SHAKESPEARE himself.

Let us not be too hasty in condemning SHAKESPEARE'S two friends who gathered together his plays for us. To be sure, it was on their part a business venture, but this does not lessen our gratitude. Had HEMINGE and CONDELL foreseen, what even no poet of that day, however compact of all imagination, could foresee, 'the fierce light' which centuries after was destined 'to beat' on every syllable of every line, it is possible that not even the allurements of a successful stroke of business could have induced them to assume their heavy responsibility; they might have 'shrunk blinded by the glare,' the world have lacked the Folio, and the current of literature have been, for all time, turned awry.

The reasons which induced SHAKESPEARE'S close friends and fellow-actors to adopt Roberts's Quarto (Q$_2$) as the Folio text, we shall never know, but adopt it they did, as the Textual Notes in the present edition make clear, with manifold proofs. It is not, however, solely by similarity of punctuation, or even of errors, that the identity of the two texts is to be detected; these might be due to a common origin; but there are ways more subtle whereby we can discover the 'copy' used by the compositors of the Folio. Should a noteworthy example be desired, it may be found in III, i, 168–170, where Titania calls for Pease-blossom, Cobweb, Moth, and Mustard-seed, and the four little fairies enter with their 'Ready,' 'And I,' 'And I,' 'And I.' In the Folio, Titania's call is converted into a stage-direc-

tion, with *Enter* before it, and the little fairies as they come in respond ' Ready ' without having been summoned. Had the Folio been our only text, there would have been over this line much shedding of Christian and, I fear it must be added, unchristian ink. But by referring to the Quartos we find that it is in obedience to Titania's call that the atomies enter, and that *Enter foure Fairyes* is the only stage-direction there. Like all proper names in both Quartos and Folios, the names Peaseblossom and the others are in Italics, as are also all stage-directions. In Fisher's Quarto (Q_1) Titania's summons is correctly printed as the concluding line of her speech, thus : —' *Peafe-bloffome, Cobweb, Moth,* and *Muftard-feede ?*' In Roberts's Quarto (Q_2) the line is also printed as of Titania's speech, but the compositor carelessly overlooked both the ' and ' in Roman, which he changed to Italic, and the interrogation at the end, which he changed to a full stop, thus converting it apparently into a genuine stage-direction, and as such it was incontinently accepted by his copyist the compositor of the Folio, who prefixed *Enter* and changed *Enter foure Fairies* into *and foure Fairies*, thereby making the number of Fairies eight in all ; and he may have thought himself quite ' smart,' as the Yankees say, in thus clearing up a difficulty which was made for him by Roberts's compositor, through the printing in Italic of ' and ' and through the change of punctuation. Thus it is clear, I think, that in this instance there can be little doubt that Roberts's Quarto was the direct source of the text of the First Folio.

There are, however, certain variations here and there between the Quartos and Folio which indicate in the latter a mild editorial supervision. For instance, in II, i, 95 both Quartos read ' euerie pelting riuer ;' the Folio changes ' pelting ' to ' petty,' an improvement which bears the trace of a hand rather more masterful than that of a compositor who elsewhere evinces small repugnance at repeating errors. In III, i, 90, after the *exit* of Bottom, Quince says, according to the Quartos, 'A stranger Pyramus than e'er played ' here '—a remark impossible in Quince's mouth. The Folio corrects by giving it to Puck. In III, ii, 227, in the Quartos, Hermia utters an incurably prosaic line, 'I am amazed at your words ;' the Folio, with a knowledge beyond that of a mere compositor, prints, ' I ' am amazed at your *passionate* words.'

Again, there is another class of variations which reveal to us that the copy of the Quarto, from which the Folio was printed, had been a stage-copy. In the first scene of all, Theseus bids Philostrate, as the Master of the Revels, ' go stir up the Athenian youth to merriments.' Philostrate retires and immediately after

Egeus enters. In no scene throughout the play, except in the very last, are Philostrate and Egeus on the stage at the same time, so that down to this last scene one actor could perform the two parts, and this practice of 'doubling' must have been frequent enough in a company as small as at The Globe. In the last scene, however, it is the duty of Philostrate to provide the entertainment, and Egeus too has to be present. There can be no 'doubling' now, and one of the two characters must be omitted. Of course it is the unimportant Philostrate who is stricken out; Egeus remains, and becomes the Master of the Revels and provides the entertainment. In texts to be used only by readers any change whatever is needless, but in a text to be used by actors the prefixes to the speeches must be changed, and '*Phil.*' must be erased and '*Egeus*' substituted. And this, I believe, is exactly what was done in the copy of the Quarto from which the Folio was printed,—but in the erasing, one speech (V, i, 84) was accidentally overlooked, and the tell-tale '*Phil.*' remained. This, of itself, is almost sufficient proof that the Folio was printed from a copy which was used on the stage.

Furthermore, cumulative proofs of this stage-usage are afforded both by the number and by the character of the stage-directions. In Fisher's Quarto (Q₁) there are about fifty-six stage-directions; in Roberts's (Q₂), about seventy-four; and in the Folio, about ninety-seven, not counting the division into Acts. Such minute attention to stage-business in the Folio as compared with the Quartos should not be overlooked.

There remain in the Folio two other traces of a stage copy which, trifling though they may be, add largely, I cannot but think, to the general conclusion. In V, i, 134, before Pyramus and the others appear, we have the stage-direction '*Tawyer with a Trumpet before them.*' In 'Tawyer' we have the name of one of the company, be it Trumpeter or Presenter, just as in *Romeo and Juliet* we find '*Enter Will Kempe.*' The second trace of the prompter's hand is to be found, I think, in III, i, 116, where Pyramus, according to the stage-direction of the Folio, enters '*with the Asse head.*' In all modern editions this is of course changed to '*an* Ass's head,' but the prompter of SHAKESPEARE's stage, knowing well enough that there was among the scanty properties but one Ass-head, inserted in the text 'with *the* Asse head'—the only one they had.

In any review of the text of the Folio one downright oversight should be noted. It is the omission of a whole line, which is given in both Quartos. The omission occurs after III, ii, 364, where the omitted line as given by the Quartos is:—

' *Her.* I am amaz'd, and know not what to fay.   *Exeunt.*'

Had the Folio omitted Hermia's speech while retaining the *Exeunt,* we might infer that the omission was intentional; but, as there is no *Exeunt* in the Folio where it is needed, the conclusion is inevitable that the omission of the whole line is merely a compositor's oversight, and not due to an erasure by the prompter or the author, who had the line before him in his Quarto.

To sum up the three texts:—Fisher's registered Quarto, or The First Quarto, has the better text, and inferior typography. Roberts's unregistered Quarto, or The Second Quarto, corrects some of the errors in Fisher's, is superior to it in stage-directions; in spelling; and, occasionally, in the division of lines; but is inferior in punctuation. The First Folio was printed from a copy of Roberts's Quarto, which had been used as a prompter's stage copy. Thus theoretically there are three texts; virtually there is but one. The variations between the three will warrant scarcely more than the inference that possibly in the Folio we can now and then detect the revising hand of the author. In any microscopic examination of the Quartos and Folios, with their commas and their colons, we must be constantly on our guard lest we fall into the error of imagining that we are dealing with the hand of SHAKESPEARE; in reality it is simply that of a mere compositor.

The stories of the texts of *A Midsummer Night's Dream* and of *The Merchant of Venice* are much alike. In both there are two Quartos, and in both a Quarto was the ' copy ' for the Folio, and in both the inferior Quarto was selected; both plays were entered on the *Stationers' Registers* in the month of October, of the same year; both were the early ventures of young stationers (*The Merchant of Venice* was Thomas Heyes's second venture, and *A Midsummer Night's Dream* Thomas Fisher's first), and in both of them James Roberts figures as the almost simultaneous printer of the same play. And it is this James Roberts who is, I believe, the centre of all the entanglement over these Quartos of *The Merchant of Venice* and of *A Midsummer Night's Dream,* just as I have supposed him to be in the case of *As You Like It* (see *As You Like It,* p. 296, and *Merchant of Venice,* p. 271 of this edition). I will here add no darker shadows to the portrait of James Roberts, which, in the *Appendix* to *As You Like It,* was painted ' from the depths of my consciousness.' I will merely emphasize the outlines by supposing that young Thomas Heyes and young Thomas Fisher were the victims of the older, shrewder James Roberts, who in some unknowable way was close enough to

the Lord Chamberlain's Servants to obtain, honestly or, I fear me, dishonestly, manuscript copies of SHAKESPEARE's plays, and, unable, through ill-repute with the Wardens, to obtain a license to print, he sold these copies to two inexperienced young stationers; and then, after his victims' books were published, in one case actually printing the Quarto for one of them, he turned round and issued a finer and more attractive edition for his own benefit. Then, after the two rival editions were issued, the same friendship or bribery, which obtained for him a copy taken from the manuscript of SHAKESPEARE, led the actors to use James Roberts's clearly printed page in place of the worn and less legible stage manuscript. Hence it may be that HEMINGE and CONDELL, knowing the craft whereby the text of Roberts's Quarto was obtained, could with truth refer to it as 'stolne and surreptitious,' and yet at the same time adopt a copy of it which had been long in use on the stage, worn and corrected perchance by the very hand of the Master, as the authentic text for the Folio; and in announcing that they had used SHAKESPEARE's own manuscript, their assertion was a grace not greatly 'snatched beyond the bounds of truth.'

Thus, by the aid of that pure imagination which is a constant factor in the solution of problems connected with SHAKESPEARE as a breather of this world, we may solve the enigma of the Quartos and Folio of this play and of the others where James Roberts figures.

It is perhaps worth while to note the ingenuity, thoroughly German, with which Dr ALEXANDER SCHMIDT converts the heraldic device on the title-page of James Roberts's Quarto into an example of punning arms. 'The crowned eagle,' says the learned lexicographer (*Program*, &c. p. 14), 'on the left of the two compartments into 'which the shield is divided, probably indicates King *James*, Eliza 'beth's successor, and gives us the printer's surname. The key, with 'intricate wards, on the right, is the tool and arms of a "*Roberts*man," 'as a burglar was then termed.' If my having in Heraldry is a younger brother's revenue, Dr SCHMIDT's having in that intricate department of *gentilesse* is apparently that of a brother not appreciably older, most probably a twin. According to my ignorance, the shield is an achievement, where the husband's and the wife's arms are impaled. If this be so, leaving out of view the extreme improbability of any reference in the 'crowned eagle to Elizabeth's successor' three years before Elizabeth's death, the key in the sinister half of the shield is Mrs Roberts's arms; and though my estimate of her husband's honesty is small, I am not prepared to brand the wife as a burglar. James Roberts printed several other Quartos,

B

and whether or not he was unwilling to give further publicity to his wife's burglarious propensity, and thereby disclose the family skeleton, it is impossible to say; but certain it is that he did not afterward adopt these *armes parlantes*, as they were termed, but used innocent and misleading flourishes calculated to baffle detectives.

No commentary on a play of Shakespeare's is now-a-days complete without a discussion of the DATE OF ITS COMPOSITION. Could we be content with dry, prosaic facts, this discussion in the present play would be brief. MERES mentions *A Midsummer Night's Dream* among others, in 1598. This is all we know. But in a discussion over any subject connected with SHAKESPEARE, who ever heard of resting content with what we know? It is what we do not know that fills our volumes. MERES's *Wits Commonwealth* was entered in the *Stationers' Registers* in September, 1598, when the year, which began in March, was about half through. MERES must have composed his book before it was registered. This uncertainty as to how long before registration MERES wrote, added to the uncertainty as to how long before the writing by MERES the play of *A Midsummer Night's Dream* had been acted, leaves the door ajar for speculation; critics have not been slow to see therein their opportunity, and, flinging the door wide open, have given to surmises and discursive learning a flight as unrestricted as when ' wild geese madly ' sweep the sky.' Of course it can be only through internal evidence in the play itself that proof is to be found for the *Date of Composition* before 1598. This evidence has been detected at various times by various critics in the following lines and items :—

' *Thorough bush, thorough briar.*'—II, i, 5 ;

Titania's description of the disastrous effects on the weather and harvests caused by the quarrel between her and Oberon.—II, i, 94–120 ;

' *And hang a pearl in every cowslip's ear.*'—II, i, 14 ;

' *One sees more devils than vast Hell can hold.*'—V, i, 11 ;

A poem of Pyramus and Thisbe ;

The date of Spenser's *Faerie Queene* ,

The ancient privilege of Athens, whereby Egeus claims the disposal of his daughter, either to give her in marriage or to put her to death.—I, i, 49 ;

' *The thrice three Muses, mourning for the death of learning, late deceast in beggerie.*' —V, i, 59 ;

And finally, the whole play being intended for the celebration of some noble marriage, it is only necessary to find out for whose marriage it was written, and we have found out the *Date of Composition*.

If this array of evidence pointed to one and the same date, it would be fairly conclusive of that date. But the dates are as manifold as their advocates; and there is not one of them which has not been, by some critic or other, stoutly denied, and all of them collectively by DYCE. Of some of them it may be said that they are apparently founded on two premises: First, that although SHAKE-SPEARE'S vocation was the writing of plays, yet his resources were so restricted that his chief avocation lay in conveying lines and ideas from his more original and vigorous contemporaries. And secondly, that although SHAKESPEARE could show us a bank whereon the wild thyme grows and fill our ears with Philomel's sweet melody, yet he could not so depict a season of wet weather that his audience would recognise the picture unless they were still chattering with un-timely frosts. (It has always been a source of wonder to me that the thunderstorm in *Lear* is not used to fix the date.)

The last item in this list, namely that which assumes the play to have been written for performance at some noble wedding, is one of the chiefest in determining the year of composition. From our know-ledge of the stage in those days this assumption may well be granted. But we must be guarded lest we assume too much. To suppose that Shakespeare could not have written his play for an imaginary noble marriage is to put a limitation to his power, on which I for one will never venture. And, furthermore, knowing that SHAKESPEARE wrote to fill the theatre and earn money for himself and his fellows, to suppose that he could not, without a basis of fact, write a play with wooing and wedding for its theme, which should charm and fas-cinate till wooing and wedding cease to be, is to impute to him a dis-trust of his own power in which I again, for one, will bear no share. How little he wrote for the passing hour, how fixedly he was grounded on the 'eternal verities,' how small a share in his plays trifling, local, and temporary allusions bear, is shown by the popularity of these plays, now at this day when every echo of those allusions has died away. If the plays were as saturated with such allusions as the critics would fain have us believe, if all his chief characters had pro-totypes in real life, then, with the oblivion of these allusions and of these prototypes, there would also vanish, for us, the point and meaning of his words, and SHAKESPEARE'S plays would long ago have ceased to be the source of 'tears and laughter for all time.' No noble marriage

was needed **as an occa**sion to bring out within SHAKESPEARE's century
that witless opera *The Fairy Queen*, and yet almost all the allusions
to a marriage to be found in *A Midsummer Night's Dream* are there
repeated.   I have given a short account of this opera in the *Appendix*, page 340, partly to illustrate this very point.   Moreover, this same
denial of SHAKESPEARE's dramatic power is everywhere thrust forward.   It is pushed even into his *Sonnets*, and for every sigh there
and for every smile we must needs, forsooth, fit an occasion.   SHAKE-
SPEARE cannot be permitted to bewail his outcast state, but we must
straight sniff a peccadillo.   We deny to SHAKESPEARE what we grant
to every other poet.   Had he written *The Miller's Daughter* of TENNY-
SON, the very site of the mill-dam would have been long ago fixed, the
stumps of the ' three chestnuts ' discovered, and probably fragments of
the ' long green box ' wherein grew the mignonette.   Probably no
department of literature is more beset than the Shakespearian with
what WHATELY happily terms the ' Thaumatrope fallacy.'   It is in constant use in demonstrating allusions in the plays, and pre-eminently in
narrating the facts of his most meagre biography.   On one side of a
card is set forth theories and pure imaginings interspersed with ' of
' course,' ' it could not be otherwise,' ' natural sequence,' &c., &c.,
and on the other side SHAKESPEARE; and, while the card is rapidly
twirled, before we know it we see SHAKESPEARE firmly imbedded in the
assumption and are triumphantly called on to accept a proven fact.

In the *Appendix* will be found a discussion of the items of internal
evidence which bear upon THE DATE OF COMPOSITION.   In this whole
subject of fixing the dates of these plays I confess I take no atom
of interest, beyond that which lies in any curious speculation.   But
many of my superiors assert that this subject, to me so jejune, is of keen
interest, and the source of what they think is, in their own case, refined
pleasure.   To this decision, while reserving the right of private judgement, I yield, at the same time wishing that these, my betters,
would occasionally go for a while ' into retreat,' and calmly and
soberly, in seclusion, ask themselves what is the chief end of man
in reading SHAKESPEARE.   I think they would discern that not by
the discovery of the dates of these plays is it that fear and compassion, or the sense of humor, are awakened: the clearer vision
would enable them, I trust, to separate the chaff from the wheat ;
and that when, before them, there pass scenes of breathing life,
with the hot blood stirring, they would not seek after the date of
the play nor ask SHAKESPEARE how old he was when he wrote it.
' The poet,' says LESSING, ' introduces us to the feasts of the gods, and

'great must be our *ennui* there, if we turn round and inquire after the
' usher who admitted us.'   When, however, between every glance we
try to comprehend each syllable that is uttered, or strain our ears to
catch every measure of the heavenly harmony, or trace the subtle
workings of consummate art,—that is a far different matter; therein
lies many a lesson for our feeble powers; then we share with SHAKE-
SPEARE the joy of his meaning.   But the dates of the plays are purely
biographical, and have for me as much relevancy to the plays themselves as has a chemical analysis of the paper of the Folio or of the
ink of the Quartos.

Due explanations of THE TEXTUAL NOTES will be found in the
Appendix, page 344.   It has been mentioned in a previous volume of
this edition—and it is befitting that the statement should be occasionally recalled—that in these Textual Notes no record is made of the
conjectural emendations or rhythmical changes proposed by ZACHARY
JACKSON, or by his copesmates BECKETT, SEYMOUR, and Lord CHED-
WORTH.   The equable atmosphere of an edition like the present must
not be rendered baleful by exsufflicate and blown surmises.   It is
well to remember that this play is a ' Dream,' but, of all loves, do not
let us have it a nightmare.   It is painful to announce that in succeeding volumes of this edition to these four criticasters must be added
certain others, more recent, whose emendations, so called, must be left
unrecorded here.

There is abroad a strange oblivion, to call it by no harsher
name, among the readers of SHAKESPEARE, of the exquisite nicety
demanded, at the present day, in emending SHAKESPEARE's text—a
nicety of judgement, a nicety of knowledge of Elizabethan literature,
a nicety of ear, which alone bars all foreigners from the task, and,
beyond all, a thorough mastery of SHAKESPEARE's style and ways of
thinking, which alone should bar all the rest of us.   Moreover, never
for a minute should we lose sight of that star to every wandering
textual bark which has been from time immemorial the scholar's surest
guide in criticism: *Durior lectio preferenda est.*   The successive win-
nowings are all forgot, to which the text has been subjected for nigh
two hundred years.   Never again can there be such harvests as were
richly garnered by ROWE, THEOBALD, and CAPELL, and when to these
we add STEEVENS and MALONE of more recent times, we may rest
assured that the gleaning for us is of the very scantiest, and reserved
only for the keenest and most skilful eyesight.   At the present day
those who know the most venture the least.   We may see an example
of this in *The Globe* edition, where many a line, marked with an

obelus as incorrigible, is airily emended by those who can scarcely
detect the difficulty which to the experienced editors of that edition
was insurmountable. Moreover, by this time the text of SHAKE-
SPEARE has become so fixed and settled that I think it safe to
predict that, unless a veritable MS of SHAKESPEARE'S own be discovered, not a single future emendation will be generally accepted
in critical editions. Indeed, I think, even a wider range may
be assumed, so as to include in this list all emendations, that
is, substitutions of words, which have been proposed since the
days of COLLIER. Much ink, printer's and other, will be spared
if we deal with the text now given to us in *The Globe* and in the
recent (second) *Cambridge Edition*, much in the style of NOLAN'S
words to Lord LUCAN: 'There is the enemy, and there are your
'orders.' There is the text, and we must comprehend it, if we can.
But if, after all, in some unfortunate patient the *insanabile cacoethes
emendi* still lurk in the system, let him sedulously conceal its products from all but his nearest friends, who are bound to bear a friend's
infirmities. Should, however, concealment prove impossible, and
naught but publication avail, no feelings must be hurt if we sigh
under our breath, 'Why will you be talking, Master Benedict?
'Nobody minds ye.'

The present play is one of the very few whereof no trace of the
whole Plot has been found in any preceding play or story; but that
there was such a play—and it is more likely to have been a play than a
story which SHAKESPEARE touched with his heavenly alchemy—is, I
think, more probable than improbable. I have long thought that
hints (hints, be it observed) might be found in that lost play of *Huon
of Burdeaux* which HENSLOWE records (*Shakespeare Society*, p. 31)
as having been performed in 'desembr' and 'Jenewary, 1593,' and
called by that thrifty but illiterate manager '*hewen of burdokes.*'
Be this as it may, all that is now reserved for us in dealing with the
SOURCE OF THE PLOT is to detect the origin of every line or thought
which SHAKESPEARE is supposed to have obtained from other writers.
The various hints which SHAKESPEARE took here, there, and everywhere in writing this play will be found set forth at full length in the
Appendix, p. 268. Among them I have reprinted several which could
not possibly have been used by SHAKESPEARE, because of the discrepancy
in dates; but as they are found in modern editions, and have arguments based on them, I have preferred to err on the side of fulness.
I have not reprinted DRAYTON'S *Nymphidia*, which is in this list of
publications subsequent in date to *A Midsummer Night's Dream;*

first, because of its extreme length; and secondly, because it is access
ible in the popular, and deservedly popular, edition of the present
play set forth by the late Professor MORLEY, at an insignificant cost.
The temptation to reprint it, nevertheless, was strong after reading an
assertion like the following: 'Shakespeare unquestionably borrowed
' from DRAYTON's *Nymphidia* to set forth his "Queen Mab," and enrich
' his fairy world of the *Midsummer Night's Dream.*' * The oversight
here in regard to the date of the *Nymphidia* is venial enough. It is
not the oversight that astonishes: it is that any one can be found to assert
that SHAKESPEARE 'borrowed' from the *Nymphidia*, and that the loan
'enriched' his fairy world. HALLIWELL (*Fairy Mythology*, p. 195)
speaks of the *Nymphidia* as 'this beautiful poem.' To me it is dull,
commonplace, and coarse. There is in it a constant straining after a
light and airy touch, and the poet, as though conscious of his failure,
tries to conceal it under a show of feeble jocosity, reminding one of
the sickly smile which men put on after an undignified tumble. Do
we not see this forced fun in the very name of the hero, ' Pigwiggen'?
When Oberon is hastening in search of Titania, who has fled to 'her
' dear Pigwiggen,' one of the side-splitting misadventures of the Elfin
King is thus described :—

> ' A new adventure him betides:
> He met an ant, which he bestrides,
> And post thereon away he rides,
>   Which with his haste doth stumble,
> And came full over on her snout;
> Her heels so threw the dirt about,
> For she by no means could get out,
>   But over him doth tumble.'

Moreover, is it not strange that the borrower, SHAKESPEARE, gave
to his fairies such names as *Moth, Cobweb, Peaseblossom*, when he
might have 'enriched' his nomenclature from such a list as this?—

> ' Hop, and Mop, and Dryp so clear,
> Pip, and Trip, and Skip that were
> To Mab, their sovereign ever dear,
>   Her special maids of honour;
> Fib, and Tib, and Pinck, and Pin,
> Tick, and Quick, and Jil, and Jin,
> Tit, and Nit, and Wap, and Win,
>   The train that wait upon her.'

HALLIWELL-PHILLIPPS † mentions a manuscript which he had seen

* GERALD MASSEY: *Shakespeare's Sonnets*, p. 573, ed. 1866; *ib.*, ed. 1872.
† *Memoranda on the Midsummer Night's Dream*, p. 13, 1879.

of CHARLES LAMB, wherein LAMB 'speaks of SHAKESPEARE 'as having
'"invented the fairies."' No one was ever more competent than
LAMB to pronounce such an opinion, and nothing that LAMB ever said
is more true. There were no real fairies before SHAKESPEARE'S. What
were called 'fairies' have existed ever since stories were told to wide-
eyed listeners round a winter's fire. But these are not the fairies of
SHAKESPEARE, nor the fairies of today. They are the fairies of Grimm's
Mythology. Our fairies are spirits of another sort, but unless they
wear SHAKESPEARE'S livery they are counterfeit. The fairies of Folk
Lore were rough and repulsive, taking their style from the hempen
homespuns who invented them; they were gnomes, cobbolds, lubber-
louts, and, descendants though they may have been of the Greek
Nereids, they had lost every vestige of charm along their Northern
route.

Dr JOHNSON's final note on the present play is that 'fairies in
'[Shakespeare's] time were much in fashion, common tradition had
'made them familiar, and Spenser's poem had made them great.' If
the innuendo here be that SPENSER's fairies and SHAKESPEARE's fairies
were allied, the uncomfortable inference is inevitable that Dr JOHN-
SON's reading of his *Faerie Queene* did not extend to the Tenth
Canto of the Second Book, where 'faeryes' are described and the
descent given of the Faerie Queene, Gloriana. Along the line of
ancestors we meet, it is true, with Oberon; but, like all his progenitors
and descendants, he was a mortal, and with no attributes in common
with SHAKESPEARE's Oberon except in being a king. To save the
student the trouble of going to SPENSER, the passages referred to
are reprinted in the Appendix, p. 287. Merely a cursory glance at
these extracts will show, I think, that as far as proving any real connection between the two Oberons is concerned, they might as well
have been 'the unedifying Tenth of Nehemiah.'

Reference has just been made to HENSLOWE'S *hewen of burdokes*,
with the suggestion that it may have supplied SHAKESPEARE with some
hints when writing the present comedy. One of the hints which I
had in mind is the name Oberon, and his dwelling in the East. No
play founded on the old romance of *Huon of Burdeaux* could have
overlooked the great *Deus ex machinâ* of that story, who is almost as
important a character as Huon himself, so that HENSLOWE's 'hewen'
must have had an Oberon, and as 'hewen' was acted in 1593, we get
very close to the time when MERES wrote his *Wits Commonwealth* and
extolled SHAKESPEARE's *Midsummer Night's Dream*, in 1598. It may
be interesting to note that although the character, Oberon, appears for

the first time in this old French romance of *Huon*, KEIGHTLEY has shown that the model is the dwarf Elberich in Wolfram von Eschenbach's ballad of 'Otnit' in the *Heldenbuch*. Furthermore, the names Elberich and Oberon are the same. ' From the usual change of *l* into ' *u* (as *al* = *au*, *col* = *cou*, &c.) in the French language, Elberich or Albe- 'rich (derived from Alp, Alf) becomes Auberich; and *ich* not being 'a French termination, the usual one of *on* was substituted, and so it 'became Auberon, or Oberon.' *

There is one point, however, which certainly yields a strong presumption that Huon's Oberon was, directly or indirectly, the progenitor of Shakespeare's Oberon. Attention was called to it by Mr. S. L. LEE (to whom we are indebted for the valuable excursus in *The Merchant of Venice* on the 'Jews in England') in his *Introduction* to *Duke Huon of Burdeaux*.† 'The Oberon of the great poet's fairy- 'comedy,' says Mr LEE, 'although he is set in a butterfly environment, 'still possesses some features very similar to those of the romantic 'fairy king. . . . The mediæval fairy dwells in the East, his kingdom 'is situated somewhere to the east of Jerusalem, in the far-reaching 'district that was known to mediæval writers under the generic name 'of India. Shakespeare's fairy is similarly a foreigner to the western 'world. He is totally unlike Puck, his lieutenant, "the merry wan- '"derer of the night," who springs from purely English superstition, 'and it is stated in the comedy that he has come to Greece "from the '"farthest steep of India." Titania, further, tells her husband how 'the mother of her page-boy gossiped at her side in their home, "in '"the spiced Indian air by night-fall." And it will be remembered 'that an Indian boy causes the jealousy of Oberon.'

It is, however, quite possible to account for these coincidences on the supposition that there was an Oberon on the English stage, intermediate between Huon's and SHAKESPEARE's. It is difficult to believe that if SHAKESPEARE went direct to *Duke Huon* no trace of the progenitor should survive in the descendant other than in the Eastern references, striking though they are, just pointed out by Mr LEE. The two Oberons do not resemble each other in person, for, although Huon's Oberon 'hathe an aungelyke vysage,' yet is he 'of heyght but 'of .iii. fote, and crokyd shulderyd' (p. 63). Again, 'the dwarfe of 'the fayre, kynge Oberon, came rydynge by, and had on a gowne so 'ryche that it were meruayll to recount the ryches and fayssyon thereof 'and it was so garnyshyd with precyous stones that the clerenes of them 'shone lyke the sone. Also he had a goodly bow in hys hande so

* *Fairy Mythology*, ii, 6, foot-note, 1833.
† *Early English Text Society*, Part i, p. l.

' ryche that it coude not be esteemyde, and hys arrous after the same
' sort and they had suche proparte that any beest in the worlde that he
' wolde wyshe for, the arow sholde areste hym.   Also he hade about
' hys necke a ryche horne hangyng by two lases of golde, the horne
' was so ryche and fayre, that there was neuer sene none suche ' (p. 65).

It may be also worth while to remark that the parentage of Huon's
Oberon was, to say the least, noteworthy.   His father was Julius
Cæsar, and his mother by a previous marriage became the grandmother
of Alexander the Great (p. 72).   It was this strain of mortality derived
from his father that made Oberon, although king of ye fayrey, mortal.
' I am a mortall man as ye be,' he said once to Charlemagne (p. 265),
and shortly after he added to his dear friend, the hero of the romance,
' Huon,' quod Oberon, ' know for a truth I shal not abyde longe in
' this worlde, for so is the pleasure of god. it behoueth me to go in to
' paradyce, wher as my place is apparelled ; in ye fayrye I shal byde
' no longer ' (p. 267).

Unquestionably, this Oberon of *Huon of Burdeaux* is a noble
character, brave, wise, of an infinite scorn of anything untrue or
unchaste, and of an aungelyke visage withal, but except in name and
dwelling he is not SHAKESPEARE's Oberon.

When we turn to Puck the case is altered.   We know very well
all his forbears.   About him and his specific name Robin Goodfellow has been gathered by antiquarian and archæological zeal a
greater mass of comment than about any other character in the play.
The larger share of it is Folk Lore, but beyond the proofs of the
antiquity of the name and of his traditional mischievous character
little needs either revival or perpetuation in the present edition.   The
sources of the knowledge of popular superstitions were as free to
SHAKESPEARE as to the authors whose gossip is cited by the anti-
quarians,—all had to go to the stories at a winter's fire authorised by
a grandam.

Sundry ballads are reprinted in the *Appendix*, for which the claim
is urged that they have influenced, or at least preceded, SHAKE-
SPEARE.   There also will be found the extracts from Chaucer's
Knight's Tale which have been cited by many editors as the story
to which the present play owes much.   It is difficult to understand the grounds for this belief.   There is no resemblance between
the tale and the drama beyond an allusion to the celebration of May
day, and the names Theseus and Philostrate.   For the name Hippolyta,
SHAKESPEARE must have deserted *Chaucer*, who gives it ' Ipolita,' and

resorted to his *Plutarch*. STAUNTON truly remarks that 'the persist-
'ence [of the commentators] in assigning the groundwork of the
'fable to Chaucer's *Knight's Tale* is a remarkable instance of the
'docility with which succeeding writers will adopt, one after another,
'an assertion that has really little or no foundation in fact.'

No little space in the *Appendix* is allotted to the extracts from
Greene's *Scottish History of James IV*. This was deemed necessary,
because of the great weight of any assertion made by Mr W. A. WARD,
who thinks that to this drama Shakespeare was 'in all probability'
indebted for the entire machinery of Oberon and his fairy-court.
With every desire to accept Mr WARD's view, I am obliged to acknowledge that I can detect no trace of the influence of Greene's drama on
A Midsummer Night's Dream.

In the *Appendix* will be found the views of various critics concerning the DURATION OF THE ACTION. This Duration is apparently set
forth by SHAKESPEARE himself with emphatic clearness in the opening
lines of the play. Theseus there says that 'four happy days bring in
'another moon,' and Hippolyta replies that 'four nights will quickly
'dream away the time.' When, however, it is sought to compute
this number of days and nights in the course of the action, difficulties
have sprung up of a character so insurmountable that a majority of the
critics have not hesitated to say that SHAKESPEARE failed to fulfill this
opening promise, and that he actually miscalculated, in such humble
figures, moreover, as three and four, and mistook the one for the other.
Nay, to such straits is one critic, FLEAY, driven in his loyalty to SHAKE-
SPEARE that, rather than acknowledge an error, he very properly prefers
to suppose that some of the characters sleep for twenty-four consecutive
hours—an enviable slumber, it must be confessed, when induced by
SHAKESPEARE's hand and furnished by that hand with dreams.

That SHAKESPEARE knew 'small Latin and less Greek' is sad
enough. It is indeed depressing if to these deficiencies we must add
Arithmetic. Is there no evasion of this shocking charge? Is there
not a more excellent way of solving the problem?

The great event of the play, the end and aim of all its action, is
the wedding of Theseus and Hippolyta. Why did SHAKESPEARE
begin the play four days before that event? If the incidents were to
occur in a dream, one night is surely enough for the longest of dreams;
the play might have opened on the last day of April, and as far as the
demands of a dream were concerned the *dramatis personæ* have all waked
up, after one night's slumber, bright and fresh on May-day morning

Why then, was the wedding deferred four days? It is not for us to 'ha'e the presoomption' to say what was in SHAKESPEARE's mind, or what he thought, or what he intended. We can, in a case like this, but humbly suggest that as a most momentous issue was presented to Hermia, either of being put to death, or else to wed Demetrius, or to abjure for ever the society of men, SHAKESPEARE may have thought that in such most grave questions the tender Athenian maid was entitled to at least as much grace as is accorded to common criminals; to give her less would have savoured of needless harshness and tyranny on the part of Theseus, and would have been unbecoming to his joyous marriage mood. Therefore to Hermia is given three full days to pause, and on the fourth, the sealing day 'twixt Theseus and Hippolyta, her choice must be announced. Three days are surely enough wherein a young girl can make up her mind; our sense of justice is satisfied; a dramatic reason intimated for opening the play so long before the main action; and the 'four happy days' of Theseus are justified.

The problem before us, then, is to discover any semblance of probability in the structure of a drama where to four days there is only one night. Of one thing we are sure: it is a midsummer night, and therefore full of enchantment. Ah, if enchantment once ensnares us, and SHAKESPEARE's enchantment at that, day and night will be alike a dream after we are broad awake. To the victims of fairies, time is nought, divisions of day and night pass unperceived. It is not those inside the magic circle, but those outside— the spectators or the audience—for whom the hours must be counted. It is we, after all, not the characters on the stage, about whom SHAKESPEARE weaves his spells. It is our eyes that are latched with magic juice. The lovers on the stage pass but a single night in the enchanted wood, and one dawn awakens them on May day. We, the onlookers, are bound in deeper charms, and must see dawn after dawn arise until the tale is told, and, looking back, be conscious of the lapse of days as well as of a night.

If 'four happy days,' as Theseus says, 'bring in another moon' on the evening of the first of May, the play must open on the twenty-seventh of April, and as, I think, it is never the custom when counting the days before an event to include the day that is passing, the four days are: the twenty-eighth, the twenty-ninth, the thirtieth of April, and the first of May. Hippolyta's four nights are: the night which is approaching—namely, the twenty-seventh, the twenty-eighth, the twenty-ninth, and the thirtieth of April. The evening of the first of May she could not count; on that evening she was married. (We must count thus

on our fingers, because one critic, Mr DANIEL, has said that Hippolyta should have counted *five* nights.)

The play has begun, and SHAKESPEARE'S two clocks are wound up; on the face of one we count the hurrying time, and when the other strikes we hear how slowly time passes. But before we really begin to listen, SHAKESPEARE presents to us ' one fair enchanted cup,' which we must all quaff. It is but four days before the moon like to a silver bow will be new bent in heaven, and yet when Lysander and Hermia elope on the morrow night, we find, instead of the moonless darkness which should enshroud the earth, that ' Phœbe' is actually beholding ' her silver visage in the watery glass,' and ' decking with liquid pearl ' the bladed grass.' It is folly to suppose that this can be our satellite— our sedate Phœbe hides her every ray before a new moon is born. On Oberon, too, is shed the light of this strange moon. He meets Titania ' by moonlight,' and Titania invites him to join her ' moon- ' light revels.' Even almanacs play us false. Bottom's calendar assures us that the moon will shine on the ' night of the play.' Our new moon sets almost with the sun. In a world where the moon shines bright in the last nights of her last quarter, of what avail are all our Ephemerides, computed by purblind, star-gazing astronomers? And yet in the agonising struggle to discover the year in which SHAKESPEARE wrote this play this monstrous moon has been over- looked, and dusty Ephemerides have been exhumed and bade to divulge the Date of Composition, which will be unquestionably divulged can we but find a year among the nineties of the sixteenth century when a new moon falls on the first of May. But even here, I am happy to say, Puck rules the hour and again misleads night-wanderers. There is a whole week's difference between the new moons in Germany and in England in May, 1590, and our ears are so dinned with Robin Goodfellow's ' Ho! ho! ho!' over the discrepancy that we cannot determine whether Bottom's almanac was in German or in English. (I privately think that, as befits Athens and the investigators, it was in Greek, with the Kalends red-lettered.) Into such dilemmas are we led in our vain attempts to turn a stage moon into a real one, and to discover the Date of Composition from internal evidence.

In *Othello* many days are compressed into thirty-six hours; in *The Merchant of Venice* three hours are made equivalent to three months. In the present play four days are to have but one night, and I venture to think that, thanks to the limitations of SHAKESPEARE'S stage, this was a task scarcely more difficult than those in the two plays just mentioned.

Grant that the play opens on Monday, Hippolyta's four nights are

then, Monday night, Tuesday night, Wednesday night, and Thursday
night. Why does Lysander propose to elope with Hermia ' *to-morrow*
' night,' and Hermia agree to meet him ' *morrow* deep midnight ' ?
One would think that not only a lover's haste but a wise prudence
would counsel flight that very night. Why need we be told with so
much emphasis that the Clowns' rehearsal was to be held ' *to-morrow*
' night ' ? Is it not that both by the specified time of the elopement
and by the specified time of the rehearsal we are to be made conscious
that Monday night is to be eliminated? If so, there will then remain
but three nights to be accounted for before the wedding day, and these
three nights are to be made to seem as only one. If while this long
night is brooding over the lovers we can be made to see two separate
dawns, the third dawn will be May day and the task will be done.
We must see Wednesday's dawn, Thursday's dawn, and on Friday
morning early Theseus's horns must wake the sleepers.

It is not to be expected that these dawns and the days following
them will be proclaimed in set terms. That would mar the impression
of one continuous night. They will not be obtruded on us. They
will be intimated by swift, fleeting allusions which induce the belief
almost insensibly that a new dawn has arisen. To be thoroughly re-
ceptive of these impressions we must look at the scene through the
eyes of SHAKESPEARE'S audience, which beholds, in the full light of
an afternoon, a stage with no footlights or side-lights to be darkened
to represent night, but where daylight is the rule ; night, be it remembered, is to be assumed only when we are told to assume it.

The Second Act opens in the wood where Lysander and Hermia
were to meet at ' deep midnight ' ; they have started on their journey to
Lysander's aunt, and have already wandered so long and so far that
Demetrius and Helena cannot find them, and they decide to ' tarry
' for the comfort of the day.' This prepares us for a dawn near at
hand. They must have wandered many a weary mile and hour since
midnight. Oberon sends for the magic flower, and is strict in his
commands to Puck after anointing Demetrius's eyes to meet him ' ere
' the first cock crow.' Again an allusion to dawn, which must be
close at hand or the command would be superfluous. Puck wanders
' through the forest ' in a vain search for the lovers. This must
have taken some time, and the dawn is coming closer. Puck finds
the lovers at last, chants his charm as he anoints, by mistake,
Lysander's eyes, and then hurries off with ' I must now to Oberon.'
We feel the necessity for his haste, the dawn is upon him and the cock
about to crow. To say that these allusions are purposeless is to believe
that SHAKESPEARE wrote haphazard, which he may believe who lists.

This dawn, then, whose streaks we see lacing the severing clouds, is that of Wednesday morning. We need but one more dawn, that of Thursday, before we hear the horns of Theseus. Lest, however, this impression of a new day be too emphatic, SHAKESPEARE artfully closes the Act with the undertone of night by showing us Hermia waking up after her desertion by Lysander. Be it never forgotten that while we are looking at the fast clock we must hear the slow clock strike.

The Third Act begins with the crew of rude mechanicals at their rehearsal. If we were to stop to think while the play is going on before us, we should remember that rightfully this rehearsal is on Tuesday night; but we have watched the events of that night which occurred long after midnight; we have seen a new day dawn; and this is a new Act. Our consciousness tells us that it is Wednesday. Moreover, who of us ever imagines that this rehearsal is at night? As though for the very purpose of dispelling such a thought, Snout asks if the moon shines the night of the play, which is only two or three nights off. Would such a question have occurred to him if they had then been acting by moonlight? Remember, on SHAKESPEARE's open-air stage we must assume daylight unless we are told that it is night. Though we assume daylight here at the rehearsal, we are again gently reminded toward the close of the scene, as though at the end of the day, that the moon looks with a watery eye upon Titania and her horrid love.

The next scene is night, Wednesday night, and all four lovers are still in the fierce vexation of the dream through which we have followed them continuously, and yet we are conscious, we scarcely know how, that outside in the world a day has slipped by. Did we not see Bottom and all of them in broad daylight? Lysander and Demetrius *exeunt* to fight their duel; Hermia and Helena depart, and again a dawn is so near that darkness can be prolonged, and the starry welkin covered, only by Oberon's magic 'fog as black as Acheron,' and over the brows of the rivals death-counterfeiting sleep can creep only by Puck's art. So near is day at hand that this art must be plied with haste, 'for night's swift dragons cut the clouds full fast, And 'yonder shines Aurora's harbinger.' Here we have a second dawn, the dawn of Thursday morning. All four lovers are in the deepest slumber—a slumber 'more dead than common sleep,' induced by magic. And the First Folio tells us explicitly before the Fourth Act opens that '*They ſleepe all the Act.*'

Wednesday night has passed, and this Act, the Fourth, through which they sleep, befalls on Thursday, after the dawn announced by Aurora's harbinger has broadened into day. Surely it is only on a

midsummer noon that we can picture Titania on a bed of flowers, coying Bottom's amiable cheeks and kissing his fair large ears.   Never could Bottom even, with or without the ass's nowl, have thought of sending Cavalery Cobweb to kill a red-hipt humble-bee on the top of a thistle at night, when not a bee is abroad.   It must be high noon.   But Bottom takes his nap with Titania's arms wound round him; the after noon wanes; Titania is awakened and disenchanted; she and Oberon take hands and rock the ground whereon the lovers still are lying, and then, as though to settle every doubt, and to stamp, at the close, every impression ineffaceably that we have reached Thursday night, Oberon tells his Queen that they will dance in Duke Theseus's house ' *to-* ' *morrow* midnight.'   But before the Fairy King and Queen trip away, Puck hears the morning lark, the herald of Friday's dawn, and almost mingling with the song we catch the notes of hunting horns. So the scene closes, with the mindful stage-direction that the *Sleepers Lye ſtill.*   It was not a mere pretty conceit that led SHAKESPEARE to lull these sleepers with fairy music and to rock the ground; this sleep was thus charmed and made 'more dead than common sleep' to reconcile us to the long night of Thursday, until early on Friday morning the horns of Theseus's foresters could be heard.   The horns are heard; the sleepers 'all start up'; it is Friday, the first of May, and the day when Hermia is to give answer of her choice.

The wheel has come full circle.   We have watched three days dawn since the lovers stole forth into the wood *last night*, and four days since we first saw Theseus and Hippolyta *yesterday*.   The lovers have quarrelled, and slept not through one night, but three nights, and these three nights have been one night.   Theseus's four days are all right, we have seen them all; Hippolyta's four nights are all right, we have seen them all.

There are allusions in the Second Act, undeniably, to the near approach of a dawn, and again there are allusions in the Third Act undeniably to the near approach of a dawn; wherefore, since divisions into Acts indicate progress in the action or they are meaningless, I think we are justified in considering these allusions, in different Acts, as referring to two separate dawns; that of Wednesday and that of Thursday, the only ones we need before the May-day horns are heard on Friday.

For those who refuse to be spellbound it is, of course, possible to assert that these different allusions refer to one and the same dawn, and that the duration of the action is a hopeless muddle.   If such an attitude toward the play imparts any pleasure, so be it; one of the objects of all works of art is thereby attained, and the general sum of

happiness of mankind is increased. For my part, I prefer to submit myself an unresisting victim to any charms which SHAKESPEARE may mutter; should I catch him at his tricks, I shall lift no finger to break the spell; and that the spell is there, no one can deny who ever saw this play performed or read it with his imagination on the wing.

Thus far we have been made by SHAKESPEARE to condense time; we are equally powerless when he bids us expand it. Have these days after all really passed so swiftly? Oberon has just come from the farthest steep of India on purpose to be present at this wedding of Hippolyta. We infer that he takes Titania by surprise by the suddenness of his appearance, and yet before the first conference of these Fairies is half through we seem to have been watching them ever since the middle summer's spring, and we are shivering at the remembrance of the effect of their quarrel on the seasons. Oberon knows, too, Titania's haunts, the very bank of wild thyme where she sometimes sleeps at night. He cannot have just arrived from India. He must have watched Titania for days to have found out her haunts. Then, too, how long ago it seems since he sat upon a promontory and marked where the bolt of Cupid fell on a little Western flower!—the flower has had time to change its hue, and for maidens to give it a familiar name. It is not urged that these allusions have any connection with Theseus's four days; it is merely suggested that they help to carry our imaginations into the past, and make us forget the present, to which, when our thoughts are again recalled, we are ready to credit any intimation of a swift advance, be it by a chance allusion or by the sharp division of an Act.

These faint scattered hints are all near the beginning of the Play: it is toward the close, after we have seen the time glide swiftly past, that the deepest impressions of prolonged time must be made on us. Accordingly, although every minute of the dramatic lives of Oberon and Titania has been apparently passed in our sight since we first saw them, yet Oberon speaks of Titania's infatuation for Bottom as a passion of so long standing that at last he began to pity her, and that, meeting her *of late* behind the wood where she was seeking sweet favours for the hateful fool, he obtained the little changeling child. Again, when Bottom's fellows meet to condole over his having been transported, and have in vain sent to his house, Bottom appears with the news that their play has been placed on the list of entertainments for the Duke's wedding. We do not stop to wonder when and where this could have been done, but at once accept a conference and a discussion with the Master of the Revels. Finally, it is in the last Act that the weightiest impression is made of time's slow passage and that many a

c

day has elapsed. When Theseus decides that he will hear the tragical mirth of 'Pyramus and Thisbe,' Egeus attempts to dissuade him, and says that the play made his eyes water *when he saw it rehearsed*. When and where could he have seen it rehearsed? We witnessed the first and only rehearsal, and no one else was present but ourselves and Puck; immediately after the rehearsal Bottom became the god of Titania's idolatry, and fell asleep in her arms; when he awoke and returned to Athens his comrades were still bewailing his fate; he enters and tells them to prepare for an immediate performance before the Duke. Yet Egeus saw a rehearsal of the whole play with all the characters, and laughed till he cried over it.

Enthralled by SHAKESPEARE's art, and submissive to it, we accept without question every stroke of time's thievish progress, be it fast or slow; and, at the close, acknowledge that the promise of the opening lines has been redeemed. But if, in spite of all our best endeavours, our feeble wits refuse to follow him, SHAKESPEARE smiles gently and benignantly as the curtain falls, and begging us to take no offence at shadows, bids us think it all as no more yielding than a dream.

H. H. F.

March, 1895.

A Midsommer Nights Dreame

Dramatis Personæ

Theſeus, *Duke of* Athens.
Egeus, *an* Athenian *Lord.*
Lyſander, *in Love with* Hermia.
Demetrius, *in Love with* Hermia.   **5**
Quince, *the Carpenter.*
Snug, *the Joiner.*
Bottom, *the Weaver.*
Flute, *the Bellows-mender.*
Snout, *the Tinker.*   **10**
Starveling, *the Tailor.*

Hippolita, *Princeſs of the* Amazons, *betrothed to* Theſeus.
Hermia, *Daughter to* Egeus, *in love with* Lyſander.
Helena, *in love with* Demetrius.

Attendants.   **15**
Oberon, *King of the Fairies.*
Titania, *Queen of the Fairies.*

---

1. First given by Rowe.   5. in Love with *Hermia.*] belov'd of *Helena.* Cap.

---

2. **Theseus**] Throughout the play, a trisyllable: Thesēus.

6. **Quince**] BELL (iii, 182, note), letting the cart, as Lear's Fool says, draw the horse, asserts that Shakespeare adopted this name from the old German comedy *Peter Squenz.*

8. **Bottom**] HALLIWELL. Nicholas was either a favourite Christian name for a weaver, or a generic appellation for a person of that trade. Bottom takes his name from a bottom of thread. '*Anguinum,* a knotte of snakes rolled together lyke a bottome of threede.'—Elyot's *Dictionarie,* 1559. ['*Botme* of threde.'—*Promp. Parv.* In a footnote WAY gives ' "A bothome of threde, *filarium.*"—*Cath. Angl.* "Bottome of threde, *gliceaux, plotton de fil.*"—*Palsg.* Skinner derives it from the French *boteau, fasciculus.*' In *Two Gent.* III, ii, 53, Shakespeare uses it as a verb meaning to *wind,* to *twist.* For an example of its modern use by Colman, *The Gentleman,* No. 5: 'Give me leave to wind up the bottom of my loose thoughts on conversation,' &c., and references to Bentley, *Works,* iii, 537, and to Charles Dibdin, *The Deserter,* I, i, see FITZ-EDWARD HALL'S *Modern English,* 1873, p. 217.—ED.]

16, 17. MALONE (ii, 337, 1821): Oberon and Titania had been introduced in a

1

[Oberon . . . Titania]

dramatic entertainment before Queen Elizabeth in 1591, when she was at Elvetham in Hampshire; as appears from *A Description of the Queene's Entertainment in Progress at Lord Hartford's*, &c. in 1591. Her majesty, after having been pestered a whole afternoon with speeches in verse from the three Graces, Sylvanus, Wood Nymphs, &c., is at length addressed by the Fairy Queen, who presents her majesty with a chaplet, ' Given me by Auberon the fairie king.' [Malone does not mention, but W. ALDIS WRIGHT does (*Preface*, p. xvi), that the name of the Fairy who thus addressed her majesty was not Titania, but 'Aureola, the Queene of Fairyland.' For the derivation of the name Oberon, see KEIGHTLEY's note in Preface to this volume, p. xxv.—ED.]

17. **Titania**] KEIGHTLEY (*Fairy Myth.* ii, 127): It was the belief of those days that the Fairies were the same as the classic Nymphs, the attendants of Diana: ' That fourth kind of spiritis,' says King James, ' quhilk be the gentilis was called Diana, and her wandering court, and amongs us called the *Phairie*.' The Fairy-queen was therefore the same as Diana, whom Ovid frequently styles Titania.

HUNTER (*New Illust.* i, 285): We shall be less surprised to find Diana in such company when we recollect that there is much in the Fairy Mythology which seems but a perpetuation of the beautiful conceptions of primeval ages, of the fields, woods, mountains, rivers, and the margin of the sea being haunted by nymphs, the dryades and hamadryades, oreades and naiades.

SIMROCK (*Die Quellen des Sh.* 2te Aflge, ii, 344): The *Handbook of German Myth.* (p. 414, § 125) gives us an explanation of the name of Titania, in that it shows how elvish spirits, and Titania is an elfin queen, steal children, and children are called *Titti*, whence the name of *Tittilake*, wherefrom, according to popular belief, children are fetched. . . . The name does not come from classic mythology, which knows no Titania; nor is it of Shakespeare's coinage, who had enough classic culture to know that the Titans were giants, not elves. [It is rare, indeed, to catch a German napping in the classics, but, *aliquando dormitat*, &c. Almost any Latin Dictionary would have given Simrock the reference to Ovid, *Meta.* iii, 173: ' Dumque ibi perluitur solita Titania lympha,' where ' Titania ' is Diana, who is about to be seen by Actæon. Golding, with whose translation of Ovid we suppose that Shakespeare was familiar, gives us no help here; in the three other places where Ovid uses the name Titania as an epithet of Latona, of Pyrrha, and of Circe, Golding does not use that name, but a paraphrase.—ED.]

BAYNES (*Fraser's Maga.* Jan. 1880, p. 101, or *Shakespeare Studies*, 1894, p. 210) [Keightley's] statement is that Titania occurs once in the *Metamorphoses* as a designation of Diana. [A remarkable and, I think, unusual oversight on the part of Prof. Baynes. Vide Kéightley, *supra.*—ED.] But in reality the name occurs not once only, but several times, not as the designation of a single goddess, but of several female deities, supreme or subordinate, descended from the Titans. . . . Diana, Latona, and Circe are each styled by Ovid ' Titania.' . . . Thus used [the name] embodies rich and complex associations connected with the silver bow, the magic cup, and the triple crown. . . . Diana, Latona, Hecate are all goddesses of night, queens of the shadowy world, ruling over its mystic elements and spectral powers. The common name thus awakens recollections of gleaming huntresses in dim and dewy woods, of dark rites and potent incantations under moonlit skies, of strange aërial voyages, and ghostly apparitions of the under-world. It was, therefore, of all possible names, the one best fitted to designate the queen of the same shadowy empire, with its

phantom troops and activities, in the Northern mythology. And since Shakespeare, with prescient inspiration, selected it for this purpose, it has naturally come to represent the whole world of fairy beauty, elfin adventure, and goblin sport connected with lunar influences, with enchanted herbs, and muttered spells. The Titania of Shakespeare's fairy mythology may thus be regarded as the successor of Diana and other regents of the night belonging to the Greek Pantheon. [It is not easy to over-estimate the value of what Prof. BAYNES now proceeds to note. Not since MAGINN's day has so direct an answer been given to FARMER with his proofs that SHAKESPEARE knew the Latin authors only through translations.—ED.] Reverting to the name Titania, however, the important point to be noted is that Shakespeare clearly derived it from his study of Ovid in the original. It must have struck him in reading the text of the *Metamorphoses*, as it is not to be found in the only translation which existed in his day. Golding, instead of transferring the term Titania, always translates it in the case of Diana by the phrase 'Titan's daughter,' and in the case of Circe by the line: 'Of Circe, who by long descent of Titans' stocke, am borne.' Shakespeare could not therefore have been indebted to Golding for the happy selection. On the other hand, in the next translation of the *Metamorphoses* by Sandys, first published ten years after Shakespeare's death, Titania is freely used. . . . But this use of the name is undoubtedly due to Shakespeare's original choice, and to the fact that through its employment in the *Midsummer Night's Dream* it had become a familiar English word. Dekker, indeed, had used it in Shakespeare's lifetime as an established designation for the queen of the fairies. It is clear, therefore, I think, that Shakespeare not only studied the *Metamorphoses* in the original, but that he read the different stories with a quick and open eye for any name, incident, or allusion that might be available for use in his own dramatic labours.

18. **Puck**] R. GRANT WHITE (ed. i): Until after Shakespeare wrote this play 'puck' was the generic name for a minor order of evil spirits. The name exists in all the Teutonic and Scandinavian dialects; and in New York [and Pennsylvania.—ED.] the Dutch have left it *spook*. The name was not pronounced in Shakespeare's time with the *u* short. Indeed, he seems to have been the first to spell it 'puck,' all other previous or contemporary English writers in whose works it has been discovered spelling it either *powke*, *pooke*, or *pouke*. There seems to be no reason to doubt that Shakespeare and his contemporaneous readers pronounced it *pook*. The fact that it is made a rhyme to 'luck' is not at all at variance with this opinion, because it appears equally certain that the *u* in that word, and in all of similar orthography, had the sound of *oo*. My own observation had convinced me of this long before I met with the following passages in Butler's *English Grammar*, 1633: '. . . for as *i* short hath the sound of *ee* short, so hath *u* short of *oo* short.' p. 8. 'The Saxon *u* wee have in sundry words turned into *oo*, and not onely *u* short into *oo* short (*which sound is all one*),' &c. p. 9.

W. A. WRIGHT (*Preface*, xvi): Puck is an appellative and not strictly a proper name, and we find him speaking of himself, 'As I am an honest Puck,' 'Else the Puck a liar call.' In fact, Puck, or pouke, is an old word for devil, and it is used in this sense in the *Vision of Piers Ploughman*, 11345 (ed. T. Wright): 'Out of the poukes pondfold No maynprise may us fecche.' And in the *Romance of Richard Coer de Lion*, 4326 (printed in Weber's *Metrical Romances*, vol. ii): 'He is no man be is a pouke.' The Icelandic *púki* is the same word, and in Friesland the kobold

Peaſebloſſom, ⎱
Cobweb,   ⎰ *Fairies.*
Moth,   ⎰
Muſtardſeed, ⎰

Other Fairies attending on the King and Queen.

SCENE Athens, *and a Wood not far from it.*

[Theobald added :]
Philostrate, *Master of the Sports to the Duke.*
Pyramus, ⎱
Thisbe,   ⎰
Wall,   ⎰ *Characters in the* Interlude *perform'd by*
Moonshine, ⎰   *the* Clowns.
Lyon,   ⎰

---

or domestic spirit is called Puk. In Devonshire, pixy is the name for a fairy, and in Worcestershire we are told that the peasants are sometimes *poake ledden*, that is, misled by a mischievous spirit called *Poake*. 'Pouk-laden' is also given in Hartshorne's *Shropshire Glossary*. [The inquisitive student, the *very* inquisitive student, is referred to BELL's *Shakespeare's Puck*, 3 vols. 1852–64, where will be found a mass of Folk-lore of varying value, whereof the drift may be learned from an assertion by the author (vol. iii, p. 176) to the effect that 'unless this entire work hitherto is totally valueless, it must follow that our poet's original view of this beautiful creation [*A Midsummer Night's Dream*] is entirely owing to foreign support.'—ED.]

26. **Philostrate**] FLEAY (*Life and Work*, p. 185) says that Shakespeare got this name from Chaucer's *Knighte's Tale*.

MALONE in his *Life of Shakespeare* (Var. '21, ii, 491) suggests that not a journey between London and Stratford was made by Shakespeare which did not probably supply materials for subsequent use in his plays; 'and of this,' he goes on to say 'an instance has been recorded by Mr. Aubrey: "The humour of . . . the cunstable in a Midsomer's Night's Dreame, he happened to take at Grenden in Bucks (I thinkè it was Midsomer Night that he happened to lye there) which is the roade from London to Stratford, and there was living that constable about 1642, when I first came to Oxon: Mr. Jos. Howe is of the parish, and knew him" [Halliwell, *Memoranda*, &c. 1879, p. 31]. It must be acknowledged that there is here a slight mistake, there being no such character as a constable in *A Midsummer Night's Dream*. The perſon in contemplation probably was Dogberry in *Much Ado*.'

A
MIDSOMMER
Nights Dreame.

Actus primus. [*Scene I.*]

Enter Theſeus, Hippolita, with others.

Theſeus.

Ow faire Hippolita, our nuptiall houre
Drawes on apace : foure happy daies bring in   5
Another Moon : but oh, me thinkes, how flow
This old Moon wanes ; She lingers my deſires   7

Midſommer Nights] Midſummers nights F_3F_4 (thus also throughout in running title). Midsummer - Night's Rowe.

1. Actus primus.] Om. Qq.

[Scene, *the Duke's Palace in* Athens. Theob. A State-Room in Theſeus's Palace. Cap.

2. with others.] with Attendants.

Rowe. Philostrate, with **Attendants**. Theob.

4. *houre*] *hower* Q_1.

5. *apace*] *apaſe* Q_1.
foure] *fower* Q_1.

6. *Another*] *An other* Q_1.
me thinkes] *me-thinks* Q_2.

7. *wanes ;*] *wanes !* Q_1. *wanes :* Q_2. *wanes ?* Ff. *wanes !* Rowe et seq.

7. *deſires*] *deſires,* Q_1.

---

1. **Actus primus**] The division into Acts is marked only in the Folios; neither in the Quartos nor in the Folios is there any division into Scenes. The division into Scenes which has most generally obtained is that of CAPELL, which I have followed here, with the exception of the last Act, wherein I have followed the CAMBRIDGE EDITION. Albeit Capell's division is open to criticism, particularly in the Second Act, the whole subject is, I think, a matter of small moment to the student, and more concerns the stage-manager, who, after all, will make his own division to suit his public, regardless of the weight of any name or text, wherein he is quite right. For the student it is important that there should be some standard of Act, Scene, and Line for the purpose of reference. This standard is supplied in *The Globe* edition.—ED.

7. **lingers**] For other instances of this active use, see SCHMIDT *s. v.*, or ABBOTT, § 290.

Like to a Step-dame, or a Dowager,   8
Long withering out a yong mans reuennew.

   Hip. Foure daies wil quickly fteep thẽfelues in nights   10
Foure nights wil quickly dreame away the time:
And then the Moone, like to a filuer bow,
Now bent in heauen, fhal behold the night   13

---

8. *Step-dame*] *Stepdame* Q₁. *Step-dam* Q₂.

   withering out] *wintering on* Warb. *withering-out* Cap. *widowing on* Gould.

9. *yong*] *young* Q₂F₃F₄.

10, 11. *Foure*] *Fower* Q₁.

10. *nights*] *night :* Q₁, Theob. Warb.

Johns. Cam. Wr. Wh. ii. *nights*, Ff et cet. (subs.).

11. *nights*] *daies* Q₂.

13. *Now bent*] QqFf, Coll. i. *Never bent* Johns. *New bent* Rowe et cet. (hyphened by Dyce.)

   night] *height* Daniel.

---

   8. **Dowager**] CAPELL: Dowagers that are long-lived wither out estates with a witness, when their jointures are too large, and what remains too little for the heir's proper supportance; whose impatience to bury them must (in that case) be of the strongest degree.

   9. **withering out**] STEEVENS: Thus, 'And there the goodly plant lies withering out his grace.'—Chapman, *Iliad*, iv, 528. [This is quoted in reply to Warburton's assertion that 'withering out' is not good English.]—WHALLEY (p. 55): Compare, 'Ut piger annus Pupillis, quos dura premit custodia matrum; Sic mihi tarda fluunt ingrataque tempora.'—*Horace, Epist.* I, i. 21.

   10. **nights**] Independently of the avoidance of the repetition of the word in the next line, and of sibilants, I prefer the abstract *night* of Q₁.—ED.

   13. **Now bent**] ROWE changed this to 'new bent,' and has been followed, I think, by every subsequent editor, except by Dr Johnson, and by Collier in his First Edition. Johnson's 'never bent' must be, of course, a misprint, although no correction of it is made in his *Appendix*, where similar misprints are corrected. The CAM. ED. does not note it.—KNIGHT, while accepting *new*, believes that it was used in the sense of 'now,' a belief which probably arose from the very common misprint of the one word for the other.—DYCE (*Rem.* p. 44) says that this misprint of 'now' for *new* is 'one of the commonest.'—'However graceful as the opening of the play,' says HUNTER (*Illust.* i, 287), 'and however pleasing these lines may be, they exhibit proof that Shakespeare, like Homer, may some-times slumber; for, as the old moon had still four nights to run, it is quite clear that at the time Hippolyta speaks of there would be no moon, either full-orbed or "like to a silver bow," to beam on their solemnities, or to make up for the deficient properties of those who were to represent Pyramus and Thisbe, by moonlight, at the tomb of Ninus.'—COLLIER, in his first ed. believes that the difficulty may be solved by restoring the original reading, whereof the meaning is that 'then the moon, which is *now* bent in heaven like a silver bow, shall behold the night of our solemnities.' This is specious, but on reflection I think we shall find that DYCE (*Rem.* p. 44) puts it none too strongly when he says: 'If Shakespeare had written "Now," intending the passage to have the meaning which Mr Collier gives it, I feel convinced that he would have adopted a different collocation of words.'—COLLIER in his next edition adopted *New* on the authority of his 'old annotator.'—FLEAY (*Life and Work*, p.

Of our folemnities.

The. Go *Philoſtrate*,   15
Stirre vp the Athenian youth to merriments,
Awake the pert and nimble ſpirit of mirth,
Turne melancholy forth to Funerals:
The pale companion is not for our pompe,   19

16. *the*] *th'* Pope, Theob. Han. Warb.   19. *pompe,*] *pompe.* Qq.
17. *pert*] *peart* Qq.   [Exit *Phil.* Theob.
18. *melancholy*] *melancholly* F_3F_4.

---

185) : The time-analysis of this play has probably been disturbed by omissions in pro-
ducing the Court ...... I, i, 196-265 ought to form, and probably did, in the
original play, a separate scene; it certainly does not take place in the palace. To
the same cause must be attributed the confusion as to the moon's age; cf. I, i, 222
with the opening lines; the new moon was an after-thought, and evidently derived
from a form of the story in which the first day of the month and the new moon were
coincident, after the Greek time-reckoning.

14. **solemnities**] Just as *solemn* frequently means *formal, ceremonious*, so here
'solemnities' refers, I think, to the ceremonious celebration of the nuptials, and is
used more in reference to the idea of ceremony than of festivity. Theseus afterwards
uses it (IV, i, 203) again in the same sense, 'We'll hold a feast in great solemnity.'
—ED.

15. **Philostrate**] A trisyllable, see V, i, 43, where the Qq give *Philostrate* instead
of 'Egeus,' and where the scanning proves that it is trisyllabic.—ED.

16. **merriments**] I think the final *s* is as superfluous here as just above in
'nights.'—ED.

17. **pert**] SKEAT (*Dict. s. v.*) : In Shakespeare [this] means *lively, alert*. Middle
English, *pert*, which, however, has two meanings and two sources, and the meanings
somewhat run into one another. 1. In some instances *pert* is certainly a corruption
of *apert*, and *pertly* is used for 'openly' or 'evidently,' see *Will. of Palerne*, 4930, &c.
In this case the source is the French *apert*, open, evident, from Lat. *apertus*. 2. But
we also find 'proud and pert,' Chaucer, *Cant. T.* 3948; 'Stout he was and *pert*,' *Li
Beaus Disconus*, l. 123 (Ritson). There is an equivalent form, *perk*, which is really
older; the change from *k* to *t* taking place occasionally, as in Eng. *mate* from Mid. Eng.
make. ['Pert' is still a common word in New England, used exactly in the Shake-
sperian sense and pronounced as it is spelled in the Qq, *peart*, i. e. *peert*.—ED.]

19. **The**] GREY (1, 41) : I am apt to believe that the author gave it, '*That* pale
companion,' which has more force. And, besides the moon, another pale companion
was to be witness to the marriage pomp and solemnity, as Hippolyta had said just
before. 'The moon,' &c.—*Anon.*

19. **companion**] W. A. WRIGHT: That is, fellow. These two words have completely exchanged their meanings in later usage. 'Companion' is not now used con-
temptuously as it once was, and as *fellow* frequently is. [SCHMIDT's examples are
not appropriately distributed under the several shades of meaning of this word; the
contemptuous tone in many of them is not caught.—ED.]

19. **pompe**] 'Funerals,' with its imagery of long processions, suggested here, I
think, this word 'pompe' in its classic sense. See note on line 23 below.—ED.

Hippolita, I woo'd thee with my fword,   20
And wonne thy loue, doing thee iniuries :
But I will wed thee in another key,
With pompe, with triumph, and with reuelling.

Enter Egeus and his daughter Hermia, Lyſander,
and Demetrius.   25
Ege. Happy be *Theſeus,* our renowned Duke.

23. *reuelling*] *revelry* T. White, Coll.   *Helena*, Q₁. and Lyſander, Helena, Q₂.
MS, Ktly.   26. *Duke.*] *duke* Q₁ (Ashbee). *duke.*
24. Lyſander] and *Lyſander.* and   Q₁ (Griggs).

---

19. WHITE (ed. i) : At the end of Theseus's address to Philostrate it has been the
practice in modern editions to mark his exit. But such literalism is almost puerile.
Theseus surely did not mean that Philostrate should then rush out incontinent, and
begin on the moment to awake 'the pert and nimble spirit of mirth' in the Athenian
youth. [Philostrate must leave at once, if he is the 'double' of Egeus.—ED.]

20. Hippolita, &c.] GREY (i, 41), followed by KNIGHT, here quotes a long passage from Chaucer's *Knighte's Tale,* beginning at line 860: 'Whilom as olde stories
tellen us, There was a duk that highte Theseus,' &c. See Appendix, 'Source of the
Plot.'—ED.

23. pompe,] WARTON (quoted by W. A. WRIGHT) in a note on Milton's *Sampson Agonistes,* 1312: 'This day to Dagon is a solemn feast, With sacrifices, triumph,
pomp, and games,' suggests that Milton applied 'pomp' to the appropriated sense
which it bore to the Grecian festivals, where the πομπή, a principal part of the ceremony, was the spectacular procession. Shakespeare, adds WRIGHT, in *King John,*
III, i, 304, also has the word with a trace of its original meaning : 'Shall braying
trumpets and loud churlish drums, Clamours of hell, be measures of our pomp ?'

23. triumph] MALONE : By triumph, as Mr Warton has observed, we are to
understand *shows,* such as masks, revels, &c.—STEEVENS : In the *Duke of Anjou's
Entertainment at Antwerp,* 1581 : 'Yet notwithstanding their triumphes [those of
the Romans] have so borne the bell above all the rest, that the word *triumphing,*
which cometh thereof, *hath beene applied to all high, great, and statelie dooings.'*—
W. A. WRIGHT : The title of Bacon's 37th *Essay* is 'Of Masques and Triumphs,'
and the two words appear to have been synonymous, for the Essay treats of masques
alone. [Falstaff says of Pistol : 'O, thou art a perpetual triumph, an everlasting
bonfire-light!'—*1 Hen. IV :* III, iii, 46.]

23. reuelling] T. WHITE (ap. Fennell) : There is scarcely a scene in this play
which does not conclude with a rhyming couplet. I have no doubt, therefore,
Shakespeare wrote 'revel*ry.*' [Before this emendation can be considered we must
know the pronunciation both of 'key' and of 'revelry' in Shakespeare's time. It is
by no means impossible that 'revelry,' where the *y* final is unaccented, was pronounced
revelrei. If the word be spelled *revelrie,* then it may rhyme with 'key,' if we were
sure that Shakespeare did not pronounce that word *kay.* Dryden (Ellis, i, 87) rhymes
key with *lay, sway, prey.*—KEIGHTLEY'S positive assertion that *revelry* is the 'right
word' alone justifies any extended notice of White's emendation, which happens to
be also one of Collier's 'Old Corrector's.'—ED.]

26. Duke.] The notes in the Variorum, 1821, afford abundant examples, if any be

The. Thanks good *Egeus*: what's the news with thee ?   27

Ege. Full of vexation, come I, with complaint

Againſt my childe, my daughter Hermia.

Stand forth Dometrius.   30

My Noble Lord,

This man hath my conſent to marrie her.

Stand forth Lyſander.

And my gracious Duke,

This man hath bewitch'd the boſome of my childe: 35

Thou, thou *Lyſander*, thou haſt giuen her rimes,

And interchang'd loue-tokens with my childe :

Thou haſt by Moone-light at her window ſung,

With faining voice, verſes of faining loue,   39

---

27. Egeus :] Egeus. Qq.
    what's] *Whats* Q₁.
30. As beginning line 31, Rowe et seq.
    Dometrius] F₁.
33. As beginning line 34, Rowe et seq.
    Lyſander] Liſander Q₁.
35. *This man*] *This* Ff, Rowe, Pope,
Cap. Mal. Steev. Var.

35. *bewitch'd*] *witch'd* Theob. Warb.
Johns. Dyce ii, iii, Ktly, Huds.
36. *Thou, thou*] *Thou*, Gould.
37. *loue-tokens*] *loue tokens* Qq. *love-
token* F₄.
38. *haſt...light*] *haſt,...light*, Q₁.
39. *faining loue*] *feigned love* Han.
Walker (*Crit*. iii, 46).

---

needed, of the uſe of this title, in our early literature, applied to any great leader,
such as 'Duke Hamilcar,' 'Duke Hasdrubal,' 'Duke Æneas,' and, in Chaucer's
Knight's Tale, cited above, 'Duk Theseus,' where, it has been suggested, Shakespeare found it.—ED.

27. Egeus] As has been already noted this is a trisyllable, with the accent on
the middle syllable.   The Second Folio spells it 'Egæus.'

30, 33. These lines are clearly part of the text, but being in the imperative mood,
so familiar in stage-copies, the compositor mistook them for stage-directions, and set
them up accordingly.—ED.

35. The Textual Notes show the editorial struggles to evade what has been deemed
the defective metre of this line.   It is needful to retain 'man' as an antithesis to
'man' in line 32 ; and the change of ' bewitch'd ' into *witch'd* has only THEOBALD for
authority   To my ear the line is rendered smooth by reducing ' hath ' to '*th* ; ' This
man 'th bewitch'd,' &c.—just as in the next line ' thou 'st given her rhymes ' better
accords with due emphasis than ' thou hast giv'n her rhymes.'—ED.

39. faining voice . . . faining loue] It is not easy to see why every editor,
without exception, I believe, should have followed ROWE'S change to *feigning*,
a change which HUNTER (*Illust*. i, 287) characterises, properly I think, as
' injudicious.'   Surely there was nothing feigned nor false in Lysander's love, nor
any discernible reason why he should sing in a falsetto voice.   His love was
sincere, and because it was outspoken Demetrius's wrath was stirred.   HALLIWELL
says that probably ' Egeus intended to imply that the love of Lysander was assumed
and deceptive,' but there is no intimation of this anywhere except in this change by
Rowe.   I cannot but think that the original word of the QqFf is here correct, and

And ſtolne the impreſſion of her fantaſie,   40
With bracelets of thy haire, rings, gawdes, conceits,
Knackes, trifles, Noſe-gaies, ſweet meats (meſſengers
Of ſtrong preuailment in vnhardned youth)
With cunning haſt thou filch'd my daughters **heart**,
Turn'd her obedience (which is due to me)   45
To ſtubborne harſhneſſe.   And my gracious **Duke**,
Be it ſo ſhe will not heere before your Grace,
Conſent to marrie with *Demetrius*,
I beg the ancient priuiledge of Athens ;
As ſhe is mine, I may diſpoſe of her ;   50
Which ſhall be either to this Gentleman,
Or to her death, according to our Law,
Immediately prouided in that caſe.   53

42, 43. (*meſſengers ... youth*)] No
parenthesis, Rowe.
  42. *Noſe-gaies*] *noſegaies* Qq.
  43. *vnhardned*] *vnhardened* Qq. *unharden'd* Rowe.
  44. *filch'd*] *filcht* Qq.
  46. *harſhneſſe*] *hardness* Coll. (MS).
  47. *Be it*] *Be't* Pope+, Dyce iii.

47. *ſo...heere*] *ſo,...heere*, Q$_r$.
48. Demetrius,]   Demetrius.   Q$_r$
(Griggs).
  49. *ancient*] *auncient* Q$_r$.
     Athens ;] *Athens :* Q$_r$.   *Athens,*
Ff.
  52. *death,*] *death ;* Q$_r$.

---

that it is used in its not unusual sense of *loving, longing, yearning.* So far from
feigning being the true word, I think a better paraphrase of 'faining' would be
love-sick.—ED.

  40. stolne the impression of her fantasie] W. A. WRIGHT: That is, secretly
stamped his image on her imagination. [This 'impression,' taken, as it were, on yielding wax, *may* have suggested the use of the word 'unhardened' in line 43, and
Theseus's words in 57, 58.—ED.]

  41. gawdes] W. A. WRIGHT: Trifling ornaments, toys. Both 'gawd' and *jewel*
are derived from the Latin *gaudium ;* the latter coming to us immediately from the
Old French *joel,* which is itself *gaudiale.*

  41. conceits] *Gentileſſes ·* Prettie conceits, deuiſes, knacks, feats, trickes.—Cotgrave.

  47. Be it so] ABBOTT, § 133: 'So' seems to mean *in this way, on these
terms,* and the full construction is, 'be it (if it be) *so* that.' See 'so,' III, ii, 329,
post.

  52. to her death] WARBURTON: By a law of Solon's, parents had an absolute
power of life and death over their children. So it suited the poet's purpose well
enough to suppose the Athenians had it before. Or perhaps he neither thought nor
knew anything of the matter.

  53. Immediately, &c.] STEEVENS: Shakespeare is grievously suspected of having
been placed, while a boy, in an attorney's office. The line before us has an undoubted
smack of legal common-place. Poetry disclaims it.

The. What fay you Hermia? be aduis'd faire Maide,
To you your Father fhould be as a God ;   55
One that compos'd your beauties ; yea and one
To whom you are but as a forme in waxe
By him imprinted : and within his power,
To leaue the figure, or disfigure it :
Demetrius is a worthy Gentleman.   60
    Her. So is *Lyfander.*
    The. In himfelfe he is.
But in this kinde, wanting your fathers voyce.
The other muft be held the worthier.
    Her. I would my Father look'd but with my eyes.   65
    The. Rather your eies muft with his iudgment looke.
    Her. I do entreat your Grace to pardon me.
I know not by what power I am made bold,
Nor how it may concerne my modeftie
In fuch a prefence heere to pleade my thoughts :   70
But I befeech your Grace, that I may know
The worft that may befall me in this cafe,
If I refufe to wed *Demetrius.*
    The. Either to dye the death, or to abiure   74

---

54. *Maide,*] *maid.* Q,Ff.   66. *looke.*] *looke,* Q₁.
55. *To you*] *To you,* Q₁.   67. *me.*] *me,* Ff.
59. *leaue*] *'leve* Warb.   68. *bold*] *bould* Q₁.
61. *Lyfander*] *Lifander* Q₁.   70. *prefence*] *prefence,* Qq.
63. *voyce.*] *voice,* Qq. *voice* Ff.

---

58. **power**] For other examples of an ellipsis of *it is*, see ABBOTT, § 403.

59. **leaue**] Warburton's emendation, *'leve*, is incomprehensible without a word of explanation. It stands for '*releve*, to heighten or add to the beauty of the figure, which is said to be *imprinted by him*. 'Tis from the French *relever*.'—JOHNSON : The sense is,—you owe to your father a being which he may at pleasure continue or deftroy.

63. **in this kinde**] This phrase, like Hermia's 'in this case,' line 72, refers to the present question of marriage.—ED.

69. **concerne my modestie**] W. A. WRIGHT : That is, nor how much it may affect my modesty. [Is it not rather, how much it may affect my reputation for modesty ?—ED.]

74. **dye the death**] JOHNSON : This seems to be a solemn phrase for death inflicted by law.—Note on *Meas. for Meas.* II, iv, 165.—W. A. WRIGHT : Generally, but not uniformly, applied to death inflicted by law ; for instance, it is apparently an intensive phrase in Sackville's *Induction*, line 55 : ' It taught mee well all earthly things be borne To dye the death.' Shakespeare, however, uses the expression always of a judicial punishment. Cf. *Ant. and Cleop.* IV, xiv, 26 : ' She hath be-

For euer the fociety of men.   75
Therefore faire Hermia queftion your defires,
Know of your youth, examine well your blood,
Whether (if you yeeld not to your fathers choice)
You can endure the liuerie of a Nunne,
For aye to be in fhady Cloifter mew'd,   80
To liue a barren fifter all your life,
Chanting faint hymnes to the cold fruitleffe Moone,
Thrice bleffed they that mafter fo their blood,
To vndergo fuch maiden pilgrimage,
But earthlier happie is the Rose diftil'd,   85

---

77. *blood.*] *blood.* F₃F₄.
78. *if you yeeld not*] *not yielding* Pope, Han.
81. *barren*] *barraine* Q₁.
82. *Chanting*] *Chaunting* Q₁.
83. *their*] *there* Q₁.
84. *pilgrimage,*] *pilgrimage.* F₃F₄.

85. *earthlier happie*] *earthlyer happy* Q₁. *earthlier happy* Q₂. *earlier happy* Rowe ii. *earthly happier* Cap. Knt, Coll. i, ii, Sing. Sta. *earthlier-happy* Walker, Dyce, Huds.
diftil'd] *diftol'd* Gould (p. 56).

---

tray'd me and shall die the death.' Even when Cloten says (*Cym.* IV, ii, 96) to Guiderius, ' Die the death,' he looks upon himself as the executioner of a judicial sentence in killing an outlaw. See *Matthew* xv, 4.

77. **Know**] STAUNTON : That is, ascertain from your youth.

77. **blood**] DYCE : That is, disposition, inclination, temperament, impulse.—W. A. WRIGHT : Passion as opposed to reason. See below, line 83, and *Ham.* III, ii, 74 : ' Whose blood and judgement are so well commingled.'

78. **Whether**] For multitudinous instances of this monosyllabic pronunciation, see WALKER, *Vers.* 103, or ABBOTT, § 466, or Shakespeare *passim.*

79. **Nunne**] W. A. WRIGHT : For the word ' nun,' applied to a woman in the time of Theseus, see North's *Plutarch* (1631), p. 2 : ' But Egeus desiring (as they say) to know how he might haue children, went into the city of Delphes, to the Oracle of Apollo : where, by a Nunne of the temple, this notable prophecie was giuen him for an answer.' ' Livery,' which now denotes the dress of servants, formerly signi-fied any distinctive dress, as in the present passage. Cf. *Pericles*, II, v, 10; and III, iv, 10.

82. **faint**] ROLFE : That is, without feeling or fervour. [But is such an impu-tation of insincerity, almost of hypocrisy, in keeping with the dignified seriousness of the Duke's adjuration ? May it not be that midnight hymns chanted by nuns within a convent's walls must always sound ' faint' to the ears of men outside ?—ED.]

83, 84. **so . . . To**] For instances of the omission of *as* after *so*, see ABBOTT, § 281.

84. **pilgrimage**] W. A. WRIGHT : This sense of ' pilgrimage' is in accordance with the usage of Scripture. Compare *Genesis* xlvii, 9 : ' The days of the years of my pilgrimage are an hundred and thirty years.' And *As You Like It*, III, ii, 138 : how brief the life of man Runs his erring pilgrimage.'

85. **earthlier happie**] JOHNSON : ' Earthlier' is so harsh a word, and ' earthlier

Then that which withering on the virgin thorne,   86

86. *Then*] *Than* F₄.

---

happy,' for *happier earthly,* a mode of speech so unusual, that I wonder none of the editors have proposed *earlier happy* [see Textual Notes].—STEEVENS: We might read, *earthly happy.*—KNIGHT (who follows Capell): If, in the orthography of the Folio, the comparative had not been used, it would have been *earthlie happie;* and it is easy to see, therefore, that the *r* has been transposed.—HUNTER (i, 288): This is perhaps one of Shakespeare's 'unfiled expressions,' one which he would have a little polished had he ever 'blotted a line,' and yet the words after all convey their meaning with sufficient clearness. The virgin is thrice blessed, as respects the heaven for which she prepares herself; but, looking only to the present world, the other is the happier lot. [The objections to Capell's reading] are, 1st, that it is against authority; 2d, that nothing is gained by it; 3d, that if there is any difference in the meaning it is a deterioration, not an improvement; and 4th, that it spoils the melody.—R. G. WHITE (ed. i): Capell's change substitutes a comparison of degree for one of kind, impairs the rhythm of the line, gives a weak thought for a strong one, is based on a limitation of the flexibility of the language even in the hands of Shakespeare, and, in short, is little less than barbarous. There is no better adjective than *earthly,* and none which can be better made comparative or superlative.—WALKER (*Crit.* i, 27): If, indeed, it be not too obvious, this means *more earthly-happy.* [Both WALKER (*Crit.* iii, 46) and HALLIWELL (*ad loc.*) cite Erasmus's *Colloquies, Colloq. Proci et Puellæ,*—'Ego rosam existimo feliciorem, quæ marescit in hominis manu, delectans interim et oculos et nares, quam quæ senescit in frutice.'—DYCE: *Earthy happier* is a more correct expression, doubtless; but Shakespeare (like his contemporaries) did not always write *correctly.*—J. F. MARSH (*Notes & Qu.* 5th, x, 243, 1878) asserts that it is impossible to make sense of this passage. 'Happiness is predicated of both roses. The earthliness only of their happiness is the subject of comparison. The distilled rose enjoys a more earthly, and the withered rose a less earthly, happiness, and the more earthly happiness is assumed to be the preferable state. This, the only possible construction, is a *reductio ad absurdum.*' [Marsh hereupon suggests that *eathlier* is a word which differs from the text by the omission of only a single letter. '"Uneath" is found in *2 Hen. VI:* II, iv, 8; Spenser in many places has *eath* as an adjective; Fairfax's *Tasso* has *eathest;* and Peele, *Honour of the Garter,* has *eathly* as an adverb, of which the word now proposed would be the regular comparative form. . . . True, I find no authority for the exact word; but the very fact of its being unusual would increase its liability to be misprinted by the substitution of a word so very like it in appearance.' It is proper to add that Marsh would not disturb the present text, because sanctioned by the authority of the QqFf, but where sense is impossible he holds conjectures to be legitimate. At one time he was 'half inclined to suggest the possibility that *rathelier* was the original word.' Marsh is the only critic, I believe, who finds the meaning obscure; it is the 'unusual mode of speech' which has given rise to discussion. Theseus's meaning is clear, however much we may disagree with the sentiment, that in an earthly sense the married woman is happier than the spinster.—ED.]

85. *distil'd*] MALONE: This is a thought in which Shakespeare seems to have much delighted. We meet with it more than once in the *Sonnets.* See *Sonnet* 5: 'But flowers distill'd, though they with winter meet, Leese but their show; their substance still lives sweet.' So also *Sonn.* 54.

Growes, liues, and dies, in fingle bleffedneffe.   87

　Her.   So will I grow, fo liue, fo die my Lord,
Ere I will yeeld my virgin Patent vp
Vnto his Lordfhip, whofe vnwifhed yoake,   90
My foule confents not to giue foueraignty.

　The.   Take time to paufe, and by **the next new Moon**
The fealing day betwixt my loue and me,
For euerlafting bond of fellowfhip :
Vpon that day either prepare to dye,   95
For difobedience to your fathers will,
Or elfe to wed *Demetrius* as hee would,
Or on *Dianaes* Altar to proteft
For aie, aufterity, and fingle life.

　Dem.   Relent fweet *Hermia*, and *Lyfander*, **yeelde**   100
Thy crazed title to my certaine right.

　Lyf.   You haue her fathers loue, *Demetrius* :   102

90. *whofe vnwifhed*] *to whofe vn-wifhed* F₂F₃.   *to whofe unwifhd* F₄, Rowe+, Cap. Steev. Mal. Coll.

96. *your*] *you* F₂.
97 transpose to follow 99, Wagner conj.

---

89. **virgin Patent**] That is, my patent to be a virgin.

90. **Lordship**] KNIGHT : That is, authority.   The word *dominion* in our present translation of the Bible (*Romans* vi) is *lordship* in Wicklif's translation.

90. **whose**] The instances given by ABBOTT, § 201, of the omission of the preposition before the indirect object of some verbs, such as *say, question,* and, in the present instance, *consent,* show that the insertion of 'to' in F₂ was needless.

91. After this line, Hermia, in Garrick's Version, 1763, sings the following song, the music by ' Mr Smith ' :—

> ' With mean disguise let others nature hide,
> 　And mimick virtue with the paint of art ;
> I scorn the cheat of reason's foolish pride,
> 　And boast the graceful weakness of my heart ;
> The more I think, the more I feel my pain,
> 　And learn the more each heav'nly charm to prize ;
> While fools, too light for passion, safe remain,
> 　And dull sensation keeps the stupid wise.'

93, 94. **sealing . . . bond**] Again legal phraseology.

101. **crazed title**] W. A. WRIGHT : That is, a title with a flaw in it.   Compare Lyly's *Euphues* (ed. Arber), p. 58 : ' Yes, yes, *Lucilla*, well doth he knowe that the glasse once crased, will with the least clappe be cracked.'—D. WILSON (*Caliban*, &c., p. 242) : Query, *razed* title.   The decision of Theseus has just been given, by which all claim or title of Lysander to Hermia's hand is erased.   The word *razed* repeatedly occurs in this sense in the dramas.

Let me haue *Hermiaes* : do you marry him.   103
   Egeus.   Scornfull *Lyſander*, true, he hath my **Loue**;
Aud what is mine, my loue ſhall render him.   105
And ſhe is mine, and all my right of her,
I do eſtate vnto *Demetrius*.
   Lyſ.   I am my Lord, as well deriu'd as **he**,
As well poſſeſt : my loue is more then his :
My fortunes euery way as fairely ranck'd   110
(If not with vantage) as *Demetrius* :
And (which is more then all theſe boaſts can **be**)
I am belou'd of beauteous *Hermia*.
Why ſhould not I then proſecute my right?
Demetrius, Ile auouch it to his head,   115
Made loue to *Nedars* daughter, *Helena*,
And won her ſoule : and ſhe (ſweet Ladie) dotes,
Deuoutly dotes, dotes in Idolatry,
Vpon this ſpotted and inconſtant man.   119

---

103. Hermiaes] *Hermia* Tyrwhitt.
104. Lyſander,] Lyſander: F_3F_4. Lyſander! Rowe.
106. *her*,] F_2.   *her* QqF_3F_4.
109. *then*] *than* Q_1F_4.

110. *fortunes*] *Fortune*'s Rowe, Pope Theob. Warb. Johns.
111. Demetrius] Demetrius' Har. Demetrius's Johns.
113. *beauteous*] *beautious* Qq.
115. *Ile*] *I'le* F_3F_4.

---

107. **estate vnto**] If Shakespeare elsewhere discloses the lawyer, he betrays the layman here.   A lawyer would, instinctively almost, say 'estate *upon*' or '*on*,' as, indeed, Shakespeare has done elsewhere, in the only two places, I believe, in which he has used the verb: *Temp.* IV, i, 85, and *As You Like It*, V, ii, 13.   HANMER incontinently changed it to *upon*.—ED.

113. **beauteous**] The spelling 'beautious' in the two Quartos may possibly indicate a pronunciation of *ti* like *sh*.   If so, it is possibly the pronunciation of merely the compositors, and it is somewhat strange that both of them should here agreu.   This is another reminder of the gap which lies between Shakespeare and us, and of the futility of examining microscopically the spelling or even the punctuation of his plays as they have been transmitted to us.—ED.

115. **to his head**] W. A. WRIGHT: That is, before his face, openly and unreservedly.   Compare *Meas. for Meas.* IV, iii, 147; *Much Ado*, V, i, 62.

116. **Nedars**] WALKER (*Crit.* ii, 30): Perhaps a mistake of the printer's for *Nestor*,—of course not the Pylian.   'Very unlikely, I think,' adds Dyce (ed. ii). [*If* this play is founded on an older play, we have here, perchance, a reminiscence of the original, or, which I think more likely, this familiar reference is designed merely to give vividness.—ED.]

119. **spotted**] JOHNSON: As *spotless* is innocent, so 'spotted' is wicked.—D. WILSON (p. 243): No one would venture to disturb the text.   But I may note here

The. I muſt confeſſe, that I haue heard ſo much,   120
And with *Demetrius* thought to haue ſpoke thereof :
But being ouer-full of ſelfe-affaires,
My minde did loſe it.   But *Demetrius* come,
And come *Egeus*, you ſhall go with me,
I haue ſome priuate ſchooling for you both.   125
For you faire *Hermia*, looke you arme your ſelfe,
To fit your fancies to your Fathers will ;
Or elſe the Law of Athens yeelds you vp
(Which by no meanes we may extenuate)
To death, or to a vow of ſingle life.   13o
Come my *Hippolita*, what cheare my loue ?
Demetrius and *Egeus* go along :
I muſt imploy you in ſome buſineſſe
Againſt our nuptiall, and conferre with you
Of ſomething, neerely that concernes your ſelues.   135
  Ege.   With dutie and deſire we follow you.   *Exeunt*
            Manet Lyſander and Hermia.   137

123. *loſe*] *looſe* Q₁.   137. Manet...] Om. Qq.
127. *fancies*] *fancy* Ktly conj.   [Scene II. Pope, Han. Warb.
133. *imploy*] *employ* Q₁F₃F₄.   Fleay.
134. *nuptiall*] *nuptialls* Ff, Rowe+.

a conjectural change as harmonising, by antithesis with Helena's ' devout idolatry '
to her forsworn lover : ''*Pon this apostate and*,' &c.
    122. **selfe-affaires**] For similar compounds with *self*, see ABBOTT, § 20.
    126. **For**] For other instances of this use in the sense of *as regards*, see ABBOTT,
§ 149.
    131. **Hippolita**] WARBURTON : Hippolita had not said one single word all this
while.   Had a modern poet had the teaching of her, we should have found her the
busiest amongst them ; and, without doubt, the Lovers might have expected a more
equitable decision.   But Shakespeare knew better what he was about, and observed
decorum.
    134. **nuptiall**] W. A. WRIGHT : Shakespeare, except in two instances [*Othello*,
II, ii. 8, and *Pericles*, V, iii, 80], employs the singular form of this word.   In the
same way we have ' funeral ' and ' funerals.'   Compare *Jul. Cæs.* V, iii, 105 : ' His
funerals shall not be in our camp '; although in this case it is the singular form that
has survived.   [As long as the source of our knowledge of Shakespeare's language is
a text transmitted to us by several compositors, it is hazardous to assert that Shakespeare employs any special form of a word.   In the instance from *Othello*, the Qq,
it is true, have the plural, ' nuptialls,' but the word in the Ff is in the singular, as
Wright himself notes, *Tempest*, V, i, 362, of this edition.—ED.]
    135. **neerely**] For other transpositions of adverbs, see ABBOTT, § 421.
    137. **Manet, &c.**] W. A. WRIGHT : It was a strange oversight on the part of

Lyſ.   How now my loue? Why is your cheek ſo pale?   138
How chance the Roſes there do fade ſo faſt?

Her.   Belike for want of raine, which I could well   140
Beteeme them, from the tempeſt of mine eyes.

141. *Beteeme*] *Bestream* or *Bestow*   141. *mine*] *my* Qq, Cam. Wh. ii.
D. Wilson (withdrawn).

Egeus to leave his daughter with Lysander. VERITY: The plot requires this private
conference between Hermia and Lysander, at which the scheme to leave Athens may
be arranged. Shakespeare's device to bring about the conference is . . . artificial. . . .
In his later plays, when he is more experienced in stage-craft, Shakespeare so contrives
his plot that one event springs naturally from another, in accordance with probability. [As the *Text. Notes* show, POPE, followed by HANMER and WARBURTON, began
here a new scene, but as these editors are wont to begin new scenes whenever there
is any shifting of characters, small attention need be paid to their divisions. Yet, at
the same time, a new scene, in spite of the *Manent*, &c. of F_I, would certainly help
to remove the objections urged by WRIGHT and VERITY; and, indeed, such a division
was proposed by FLEAY (*Robinson' Epit. of Lit.* Apr. 1879), on the ground that it
is unlikely that Lysander and Hermia would indulge in confidential conversation in
Theseus's palace, and that when Helena enters Hermia should say, 'God speed, fair
Helena! *whither away?*'—this new scene, says FLEAY, ' is clearly in a street.' This
last assertion reveals a difficulty in the way of adopting Fleay's proposed division.
It is perhaps a little less likely that Lysander and Hermia would indulge in a confidential conversation in the open street than in an empty room of Theseus's palace.
Finally, it is hard utterly to ignore the grey authority of the Folio with its *Manet*,
when we are almost sure that the copy from which the Folio was printed was a
stage-copy.—ED.]

139. **chance**] The full phrase would be, ' How chances it,' as in *Hamlet*, II, ii,
343: ' How chances it they travel?' See also *post*, V, i, 315; or ABBOTT, § 37.

140. **Belike**] W. A. WRIGHT: This word is unusual if not singular in form. It
is recorded in Nodal and Milner's *Lancashire Glossary* as still in use.

141. **Beteeme**] POPE: Beteem, or pour down upon 'em. JOHNSON: Give them,
bestow upon them. The word is used by Spenser. CAPELL: The word which
Skinner explains—*effundere seu ab uno vase in aliud transfundere* is—teem; and is
(it seems) a local word only, proper to Lincolnshire: so that the *particula otiosa*
before it should be Shakespeare's; and he a user of other liberties with it, making
' beteem them ' stand for ' beteem *to* them,' *i. e.* the roses: If the passage be uncor-
rupted, and this the sense of ' beteem ' (of both which there is some suspicion), he
must have us'd it that his verb might suit the strength of his substantive, ' tempest,'
requiring a pouring out. STEEVENS: ' So would I' (said th' enchaunter), ' glad
and faine Beteeme to you this sword, you to defend.'—*Fairie Queene* [Bk II, canto
viii, 19]. But I rather think that to ' beteem ' in this place signifies (as in the northern counties) to *pour out*. [In a note on ' beteem ' in *Hamlet*, I, ii, 141, Steevens
says]: This word occurs in Golding's *Ovid*, 1587, and from the corresponding Latin
word (*dignatur*, bk x, line 157) must necessarily mean to *vouchsafe, deign, permit*, or
suffer. KNIGHT: That is, pour forth. COLLIER: To ' teem ' is certainly to pour
out, but that sense is hardly wanted here. [STAUNTON, R. G. WHITE, and W. A.
WRIGHT all give the meaning *afford, yield, allow*. The last says there is ' probably

2

Lyf.   For ought that euer I could reade,   142
Could euer heare by tale or hiſtorie,
The courſe of true loue neuer did run ſmooth,
But either it was different in blood.   145
    Her.   O croſſe! too high to be enthral'd to loue.

142. *For*] *Eigh me : for* Qq. *Hermia, for* Ff, Rowe+, Cap. Wh. i. *Ah me, for* Johns. Steev. Mal. Knt, Coll. Sing. Hal. *Ay me! for* Dyce, Sta. Cam. Wh. ii.
        ought] *aught* Q$_1$, Warb. Johns. Steev. et seq.

142. *euer I could*] *I could euer* Qq, Cap. Coll. Hal. Sta. Cam. Wh. ii.
143. *heare*] *here* Q$_1$.
145-147. *blood. ...yeares.*] *bloud;... yeares;* or *blood—...years—* Qq, Rowe et cet.
146. *enthral'd*] *inthrald* Qq.
        loue] *low !* Theob. Warb. et seq.

a reference to the other meaning of the word, *to pour.*' DYCE (*Gloss.*) gives a happy and concise paraphrase : 'to give in streaming abundance,' but even here it is not absolutely necessary to add the idea of abundance. 'Beteem' is here used, I think, exactly as it is asserted to be by Pope and suggested by Capell. The tempest of Hermia's eyes could readily pour down the rain to revive the roses in her cheeks.—ED.]

142. **For**] HUNTER (*Illust.* i, 288) finds in the 'Hermia' of the Second Folio (see Textual Notes) 'a point and pathos even beyond what the passage, as usually printed, possesses. A skilful actor might give great effect to the name ; and we ought always to remember, what Shakespeare never forgot, that he was writing for spokesmen, not in the first instance for students in their closets.' R. G. WHITE (ed. i) : The exclamation ['Ay me !'] is unsuited to Lysander and to his speech ; and I believe that it was an error of the press, or of the transcribers, for the proper name, and that its absence in the Folio is the result of its erasure in the Quarto stage-copy, the interlineation of the correct word having been omitted by accident. [White's objections were removed before he printed his second edition. The line as it stands in the Folio is certainly deficient, and although I agree both with Hunter, that the direct personal address is more impressive, and with White, that 'Ay me' seems out of character and is somewhat lackadaisical, yet the authority of the Quartos greatly outweighs that of the Second Folio, and we cannot quite disregard it.—ED.]

144. **The course, &c.**] W. A. WRIGHT : Bishop Newton, in his edition of Milton [1749], called attention to the resemblance between Lysander's complaint and that of Adam in *Paradise Lost*, x, 898-906.

146, 148. COLERIDGE (p. 101) : There is no authority for any alteration,—but I never can help feeling how great an improvement it would be, if the two former of Hermia's exclamations were omitted [lines 146 and 148];—the third and only appropriate one would then become a beauty, and most natural. HALLIWELL (*Introd.* p. 70) goes further, and thinks 'it cannot be denied' that Lysander's speech would be improved by the omission of all of Hermia's interpolations, and adds that Dodd and Planché have so printed it. This HALLIWELL afterwards modified by the reflection (p. 36, folio ed.) that 'the author evidently intended both the speakers should join in passionately lamenting the difficulties encountered in the path of love.'

146. **loue**] THEOBALD'S reasons for his change to *low*, which has been uniformly adopted from the days of Warburton, are that Hermia, if she undertakes to answer Lysander's complaint of the difference in blood, 'must necessarily say *low*. So the

Lyf.   Or elfe mifgraffed, in refpeƈt of yeares.   147
Her.   O fpight! too old to be ingag'd to yong.
Lyf.   Or elfe it ftood vpon the choife of merit.   149

148. *to yong*] *too young* F₄, Rowe i.   149. *merit.*] Ff. *merit—* Rowe, Wh.
                                                 i. *friends;* Qq et cet. *men* Coll. (MS)

---

antithesis is kept up in the terms; and so she is made to condole the disproportion of blood and quality in lovers.   And this is one of the curses, that Venus, on seeing Adonis dead, prophecies shall always attend love, in our author's *Venus and Adonis*, lines 1136–1140.'

147. **misgraffed**] That is, ill-grafted.   SKEAT (s. v. *graff*): The form *graft* is corrupt, and due to a confusion with *graffed*, originally the past participle of *graff*. Shakespeare has 'grafted,' *Macb.* IV, iii, 51; but he has rightly also 'graft' as a past participle, *Rich. III:* III, vii, 127.   The verb is formed from the substantive *graff*, a scion.   Old French, *graffe, grafe*, a style for writing with a sort of pencil, whence French *greffe*, 'a graff, a slip, or young shoot.'—Cotgrave; so named from the resemblance of the cut slip to the shape of a pointed pencil.   [See *As You Like It*, III, ii, 116, of this edition.]

147. **in respect**] The COWDEN-CLARKES (*Sh. Key*, p. 627): We have discovered recurrent traces of special features of style marking certain plays by Shakespeare, which lead us to fancy that he thought in that particular mode while he was writing that particular drama.   Sometimes it is a peculiar word, sometimes a peculiar manner of construction, sometimes a peculiar fashion of employing epithets or terms in an unusual sense.   Throughout [this present] play the word 'respect' is used somewhat peculiarly; so as to convey the idea of *regard* or *consideration*, rather than the more usually assigned one of *reverence* or *deference*, as in the present line; see also line 170, just below, II, ii, 217, and 232, V, i, 98.

149. **merit**] As the Folio was printed from the Second Quarto, and presumably a stage-copy at that, the substitution of the word 'merit' for 'friends' of the Quarto can hardly be deemed either a compositor's sophistication or an accident.   A change so decided must have been made with authority; it is a change, moreover, not from an obscure word to a plainer word, but from a plain word to one more recondite in meaning.   A 'choice of merit' is a choice enforced through desert or as a reward, qualities with which true love or 'sympathy in choice' can have nothing in common.   It is a choice good enough in itself, but worldly-wise, calculating, one of the roughest of obstructions to the course of true love, in that it may be urged by parents so plausibly; and this very urging is implied in Hermia's phrase of choosing 'by another's eye,' and possibly the vehemence of her expletive indicates that this obstruction is the worst of the three.   But with the exception of ROWE and R. G. WHITE (in his first edition) all editors have adopted 'friends' of the Quartos, and only two have any remarks on it.   'The alteration in the Folio,' says KNIGHT, 'was certainly not an accidental one, but we hesitate to adopt the reading, the meaning of which is more recondite than that of *friends*.   The "choice of merit" is opposed to the "sympathy in choice,"—the merit of the suitor recommends itself to "another's eye," but not to the person beloved.'—R. G. WHITE says, 'the "choice of merit" is, plainly enough, not the spontaneous, and at first unconscious, preference of the lover.'   This is in his first edition; the second edition is silent.—The *Cambridge Editors* (vol i, *Preface*, xii) pronounce 'the reading of the Folios certainly wrong.'   And yet, in spite of all,

Her.   O hell ! to choofe loue by anothers eie.   150
Lyf.   Or if there were a fimpathie in choife,
Warre, death, or fickneffe, did lay fiege to it ;
Making it momentarie, as a found :
Swift as a fhadow, fhort as any dreame,
Briefe as the lightning in the collied night,   155
That (in a fpleene) vnfolds both heauen and earth ;

150. *eie.*] *eyes !* Q₁, Coll. Wh. i, Dyce iii. *eyes.* Q₂, Cam. Wh. ii. *eye.* Ff, Rowe et cet.

153. *momentarie*] *momentany* Qq, Mal. Steev. Coll. Hal. Dyce, Sta. Cam.
156. *fpleene*] *sheen* Han. MS conj. ap. Cam.

---

after a careful review, as the Duke says in *As You Like It*, 'I would not change it.' —ED.]

153, &c. CAPELL: This passage rises to a pitch of sublimity that is not exceeded by any other in Shakespeare.

153. **momentarie**] JOHNSON: [*Momentany* of the Qq] is the old and proper word.—HENLEY: 'That short momentany rage' is an expression of Dryden.— KNIGHT: *Momentany* and 'momentary' were each indifferently used in Shakespeare's time. We prefer the reading of the Folio, because *momentary* occurs in four other passages of our poet's dramas; and this is a solitary example of the use of *momentany*, and that only in the Quartos. The reading of the Folio is invariably 'momentary.'—COLLIER: Stubbes, in 1593, preferred *momentany* to 'momentary,' where in the list of errors of the press, before his *Motive to Good Works*, he enumerated the misprinting of 'momentary,' instead of *momentany*, in the following passage, p. 188 : 'this life is but momentary, short and transitory; no life, indeed, but a shadow of life.'—STAUNTON: We have improvidently permitted too many of our old expressions to become obsolete.—HALLIWELL: 'Momentary' is hardly to be considered a modernisation; in *Meas. for Meas.* III, i, 114, 'momentary' in F₁ and F₂ is altered to *momentany* in F₃ [and F₄.—ED.].—WALKER (*Crit.* iii, 46): With *momentany* compare the old adjective *miscellany*, e. g. *miscellany poems*. Donne has *momentane, Sermon* cxlviii, ed. Alford,—'a single, and momentane, and transitory man.'—W. A. WRIGHT: *Momentany* seems to have been the earlier form, from Fr. *momentaine*, Lat. *momentaneus*.

154. **swift as a shadow**] Compare 'love's heralds should be thoughts, Which ten times faster glide than the sun's beams, Driving back shadows over louring hills.' —*Rom. and Jul.* II, v, 4.—ED.

155. **collied**] STEEVENS: That is, *black, smutted with coal.* A word still used in the Midland counties.—HALLIWELL: 'I colowe, I make blake with a cole, *je charbonne.*'—Palsgrave, 1530. 'Colwyd, *carbonatus.*'—*Prompt. Parv.* ['Charbonné. Painted, marked, written, with a coale, collowed, smeered, blacked with coales; (hence) also, darkened.'—Cotgrave.]

156. **spleene**] WARBURTON: Shakespeare, always hurried on by the grandeur and multitude of his ideas, assumes, every now and then, an uncommon license in the use of his words. Particularly in complex moral modes it is usual with him to employ one only to express a very few *ideas* of that number of which it is composed. Thus, wanting here to express the ideas—*of a sudden*, or—*in a trice*, he uses the

And ere a man hath power to fay, behold,   157
The iawes of darkneffe do deuoure it vp :
So quicke bright things come to confufion.

  Her. If then true Louers haue beene euer croft,   160
It ftands as an edict in deftinie :
Then let vs teach our triall patience,
Becaufe it is a cuftomarie croffe,
As due to loue, as thoughts, and dreames, and fighes,
Wifhes and teares ; poore Fancies followers.   165
  Lyf. A good perfwafion ; therefore heare me *Hermia*,

---

157. *behold*] *hehold* F$_2$.   161. *It ftands*] *If't stand* Rann conj.
158. *do*] *to* F$_3$F$_4$.   164. *due*] *dewe* Q$_1$.

---

word 'spleen,' which, partially considered, signifying *a hasty sudden fit*, is enough
for him, and he never troubles himself about the further or fuller use of the word.
Here he uses 'spleen' for *a sudden hasty fit ;* so, just the contrary, in *The Two Gent.*
he uses 'sudden' for *splenetic :* 'sudden quips.' And it must be owned this sort of
conversion adds a force to the diction.—NARES : In this sense of *violent haste* we do
not find the word so used by other writers.—HUNTER (i, 289) : This is a mistake ;
and it will be seen that a happier choice could not have been made than the poet has
made of this word. ' Like winter fires that with disdainful heat The opposition of the
cold defeat ; And in an angry spleen do burn more fair The more encountered by the
frosty air.'—*Verses* by Poole, before his *England's Parnassus*, 1637. So in Lithgow's
Nineteen Years Travels, 1632, p. 61 : 'All things below and above being cunningly perfected, . . . we recommend ourselves in the hands of the Almighty, and in the meanwhile attended their fiery salutations. In a *furious spleen*, the first holla of their courtesies, was the progress of a martial conflict,' &c. [This note of Hunter has been
quoted by Staunton and by Halliwell, yet, as both Poole and Lithgow are post-Shakespearian, and possibly may have drawn the phrase from this very passage, its value
as an illustration is doubtful.—ED.]

 157. **say, behold**] Compare ' like the lightning which doth cease to be, Ere one
can say " It lightens." '—*Rom. and Jul.* II, ii, 119.

 161. **edict**] For a list of words in which the accent was formerly nearer the end
than at present, see ABBOTT, § 490. W. A. WRIGHT notes that 'edict' has the
accent on the penultimate in *1 Hen. IV :* IV, iii, 79.

 165. **Fancies**] It is scarcely necessary to remark that in Shakespeare 'fancy'
means *love ;* see 'fancy free,' II, i, 170; 'fancy-sick,' III, ii, 99; and ' Helena, in
fancy followed me,' IV, ii, 181. ARBER (*Introd. to Dryden's Essay on Dramatic
Poesie.—Eng. Garner*, iii, 502) notes four changes of the meaning of ' fancy.' First,
in the Elizabethan Age it was but another word for personal *Love* or *Affection.* Second, the Restoration Age understood by it, *Imagination, the mental power of pic-
turing forth.* Third, Coleridge endeavoured yet further to distinguish between
Imagination and *Fancy.* Fourth, it is now used in another sense, ' I do not *fancy*
that,' equivalent to ' I do not *like* or *prefer* that.'

 166. **perswasion**] SCHMIDT defines this as *opinion, belief.* W. A. WRIGHT suggests that as persuasion ' signifies a *persuasive argument*, it may perhaps have that

I haue a Widdow Aunt, a dowager,   167
Of great reuennew, and fhe hath no childe,
From Athens is her houfe remou'd feuen leagues,
And fhe refpeéts me, as her onely fonne :   170
There gentle *Hermia*, may I marrie thee,
And to that place, the fharpe Athenian **Law**
Cannot purfue vs.   If thou lou'ft me, then
Steale forth thy fathers houfe to morrow night :
And in the wood, a league without the towne,   175
(Where I did meete thee once with *Helena*,
To do obferuance for a morne of May)   177

---

167. *Aunt*] *Ant* Q₂.
169. *remou'd*] *remote* Qq, Cap. Steev.
Mal. Coll. Hal. Dyce, Sta. Cam.
170. Transposed to follow line 168,

Johns. conj. Ktly, Huds.
173. *lou'ft*] *loueft* Qq.
177. *for a*] Ff, Rowe, Wh. i.   *to the*
Pope +.   *to a* Qq, Cap. et cet.

---

sense here.   Hermia's words have carried conviction to Lysander and persuaded him.—ED.

169, 170. JOHNSON proposed to transpose these lines, reading in line 169, 'Her house from Athens is,' &c.—KEIGHTLEY (p. 130) : Common sense dictates this transposition.   Line 170, it is evident, has been an addition made by the poet in the margin.

169. **remou'd**] A change to the 'remote' of the Qq is unnecessary.   Familiarity has reconciled us to this word in *Hamlet*, ' It waves you to a more removed ground.' Again, *As You Like It*, III, ii, 331 : ' Your accent is something finer, than you could purchase in so remoued a dwelling.'—ED.

174. **forth**] For other examples of 'forth,' used as a preposition equivalent to *from*, see ABBOTT, § 156.

175. **the wood, a league**] HALLIWELL : This wood in the next scene is called the ' Palace wood,' and is there described as being ' a mile without the town.'   It appears that Shakespeare, in this and other instances, made a league and a mile synonymous.   The league was certainly variously estimated.   In Holland's translation of *Ammianus Marcellinus* it is reckoned as a mile and a half.

177. **obferuance**] KNIGHT : See Chaucer, *Knight's Tale*, 1500, where the very expression occurs : 'And for to doon his observance to May.'   [I doubt if there be a breather of the world, whose native speech is English, who does not know that Mayday is welcomed with more or less festivity.   As W. A. WRIGHT says, ' scarcely an English poet from Chaucer to Tennyson is without a reference to the simple customs by which our ancestors celebrated the advent of the flowers.'   Details of these customs, which are endless, can scarcely be said to be strictly illustrative of Shakespeare.   To mention Brand's *Popular Antiquities*, Strutt's *Sports and Pastimes*, Stubbes's *Anatomie of Abuses*, or Chambers's *Book of Days* will be quite sufficient, and no student of Folk-lore will be at a loss for other quarters into which to pursue his enquiry.—ED.]

177. **for a**] That Chaucer, in the line quoted above, has the expression ' observance *to* May,' has been, I suppose, a sufficing reason for following the Quartos here, but the improvement is scarcely appreciable.—ED.

There will I ftay for thee.   178
  Her. My good *Lyfander,*
I fweare to thee, by Cupids ftrongeft bow,   180
By his beft arrow with the golden head,

---

178. Hereupon, in Garrick's *Version*, Lysander sings as follows. (May we not assume that, foreseeing the inspiration which Milton would draw from this play, Lysander deems it no felony to convey freely from *L'Allegro*?)

> ' When that gay season did us lead
>   To the tann'd hay-cock in the mead,
>   When the merry bells rung round,
>   And the rebecks brisk did sound,
>   When young and old came forth to play
>   On a sunshine holyday;
>
> ' Let us wander far away,
>   Where the nibbling flocks do stray
>   O'er the mountains barren breast,
>   Where labouring clouds do often rest,
>   O'er the meads with daisies py'd,
>   Shallow brooks and rivers wide.'

179, &c. WARBURTON : Lysander does but just propose her running away from her father at midnight, and straight she is at her oaths that she will meet him at the place of rendezvous. Not one doubt or hesitation, not one condition of assurance for Lysander's constancy. Either she was nauseously coming, or she had before jilted him, and he could not believe her without a thousand oaths. But Shakespeare observed nature at another rate. The speeches are divided wrong. [Hereupon Warburton gives to Lysander lines 180–187 and to Hermia lines 188 and 189. This reading attracted but little attention in Warburton's own day, and still less since. If any answer be needed, it is sufficiently given by HEATH, who says (p. 42)]: No doubt [Hermia's] conduct is not to be justified according to the strict rules of prudence. But when it is considered that she is deeply in love, and a just allowance is made for the necessity of her situation, being but just sentenced either to death, a vow of perpetual virginity, or a marriage she detested, every equitable reader, and I am sure the fair sex in general, will be more inclined to pity than to blame her. . . . Lysander asks no oaths of her. They are the superfluous, but tender effusion of her own heartfelt passion. . . . Would any man in his senses, when he is giving the strongest assurances of his fidelity to his mistress, endeavour at the same time to defeat the purpose, and destroy the effect of them, by expressly reminding her how often her sex had been deceived and ruined by trusting to such security? Whereas in her mouth these expressions have the greatest beauty. She finely insinuates to her lover that she is not insensible of the hazard she runs from the entire confidence she reposes in him; but at the same time she lets him see that she loves him with a passion above being restrained by this or any other consideration. This excess of tenderness, expressed with so much delicacy, must very strongly affect every mind that is susceptible of a sympathy with these generous sentiments.

181. best arrow] HALLIWELL : An allusion to the two arrows mentioned in Ovid's *Metamorphoses*, i, 466 : ['tone causeth Loue, the tother doth it slake. That

By the fimplicitie of Venus Doues,   182
By that which knitteth foules, and profpers loue,
And by that fire which burn'd the Carthage Qucene,
When the falfe Troyan vnder faile was feene,   185
By all the vowes that euer men haue broke,
(In number more then euer women fpoke)
In that fame place thou haft appointed me,
To morrow truly will I meete with thee.

   Lyf. Keepe promife loue : looke here comes *Helena.* 190

      Enter Helena.

  Her. God fpeede faire *Helena*, whither away ?
  Hel. Cal you me faire *?* that faire againe vnfay,
Demetrius loues you faire : O happie faire !   194

---

183. *loue*] *loues* Q₁, Pope et seq.   192. *fpeede faire*] *speed, fair* Theob.
185. *Troyan*] *Trojan* F₄.   Warb. Johns. Cap. Steev. Mal.
191. [Scene III. Pope+.   194. *you*] Ff, Wh. ii. *you*, Rowe ii+,
             Cap. Wh. i. *your* Qq et cet.

---

causeth loue, is all of golde with point full sharpe and bright, That chaseth loue is blunt, whose steele with leaden head is dight.'—Golding's trans.]

 181. **golden head**] GREEN (*Emblem Writers*, p. 401) suggests that Shake-speare might have derived this epithet, 'golden,' quite as well from Alciat's 154th and 155th Emblem, ed. 1581, or from Whitney, p. 132, 1586, as from Golding's *Ovid.*

 182, 183, 186–189. 'These six lines,' says ROFFE (p. 53), 'have been excellently set by Sir Henry Bishop as a solo, which was sung by Miss Stephens, as Hermia, in the operatised *Midsummer Night's Dream.*'

 183. This line is transposed to follow line 181 in SINGER'S second edition. This edition derives its chief value from the contributions to it of W. W. LLOYD. This transposition is probably an emendation by the latter; he proposed it in *Notes and Queries*, 6th ser. vol xi, p. 182, 1878, which he would not have done had it not been his own. HUDSON adopted this transposition, which KEIGHTLEY (*Exp.* 130) says is unnecessary, because the allusion in line 183 is not to the arrows, but 'most prob-ably to the *Cestus* of Venus.'—ED.

 184. **Carthage Queene**] For many another noun-compound, see ABBOTT, § 430. STEEVENS: Shakespeare had forgot that Theseus performed his exploits before the Trojan war, and consequently long before the death of Dido.—W. A. WRIGHT: But Shakespeare's Hermia lived in the latter part of the sixteenth century, and was con-temporary with Nick Bottom the weaver.

 194–197, 204, 205. In Garrick's Version these six lines are sung by Helena. The air by Mr. Christopher Smith. Line 194 reads : 'O Hermia fair, O happy, happy fair,' and the last line : 'You sway the motions of your lover's heart.' In the *List of All the Songs and Passages in Shakspere which have been set to Music*, issued by the *New Shakspere Society*, p. 35, three other compositions adapted to these lines are noted ;

Your eyes are loadftarres, and your tongues fweet ayre   195
More tuneable then Larke to fhepheards eare,
When wheate is greene, when hauthorne buds appeare,
Sickneffe is catching : O were fauor fo,
Your words I catch, faire *Hermia* ere I go,   199

198. *fo,*] *so!* Theob.+, Cap. Steev.   *words Ide* Ff, Rowe, Pope, Theob. Mal.
Var.   '90, Sta. *Your worth Ild* Wagner conj.
  199. *Your words I*] Qq, Coll. i. *Your*   *Your's would I* Han. et cet.

see also ROFFE's *Handbook*, p. 54. Hermia in turn sings lines 217–220; again the
air is by Smith, who has also, set to music, lines 248–253.

194. *you faire*] In the Folio 'you' and 'your' are so frequently confounded (for
many examples, see WALKER, *Crit.* ii, 190) that the choice here may well depend on
personal preference. Those who prefer 'your fair' of the Qq take 'fair' as a noun
(for which there is abundant authority, see ABBOTT, § 5); and take it again as a
noun also in 'O happie faire !' For my part, I prefer to take it as a noun only in the
latter phrase. 'Demetrius loves you, it is you who are fair. Ah, happy fairness, that
can bring such blessings !'—ED.

195. **loadstarres**] JOHNSON: This was a compliment not unfrequent among the
old poets. The *lode-star* is the *leading* or *guiding* star, that is, the *pole-star*. The
magnet is, for the same reason, called the *lode-stone*, either because it leads iron or
because it guides the sailor. Milton has the same thought in *L'Allegro*, 80: 'Where
perhaps some beauty lies, The cynosure of neighbouring eyes' [κυνόσουρα being the
Greek name for the constellation Ursa Minor, in which is the pole-star.—W. A.
WRIGHT.] Davies calls Queen Elizabeth : 'Lode-stone to hearts, and lode-stone to
all eyes.'—GREY (i, 44) : Sir John Maundevile, in his *voiages* and *travailes*, ch. 17,
speaking of *Lemery*, saith : 'In that Lond, ne in many othere beȝonde that, no man
may see the Sterre transmontane, that is clept the Sterre of the See, that is unmevable,
and that is toward the Northe that we clepen the Lode Sterre.'—HALLIWELL, as an
aid to our imaginations, gives us a wood-cut of a six-pointed star.

198. **fauor**] STEEVENS: That is, *feature, countenance.*—HALLIWELL (*Introd.* p.
72, 1841) : 'Favour' is not here used, as all editors and commentators have supposed,
in the sense of *countenance*, but evidently in the common acceptation of the term—
'O, were favour so,' *i. e.* favour in the eyes of Demetrius ; a particular application of
a wish expressed in general terms.—STAUNTON : Sometimes in Shakespeare it means
countenance, features, and occasionally, as here, *good graces* generally. [Whether
'favor' refers to the qualities of mind or of person is decided, I think, by the enumeration which follows.—ED.]

199. **Your words I**] KNIGHT, albeit adopting Hanmer's emendation, says that
the text of the Folio will give an intelligible meaning if we include in a parenthesis
'Your words I catch, fair Hermia,' adding 'it is in the repetition of the word *fair*
that Helena catches the words of Hermia ; but she would also catch her voice, her
intonation, and her expression as well as her words.'—COLLIER, in his first edition, is
the only editor who adopts the text of the Folio, and justifies it ; 'the meaning is,' he
says, 'that Helena only catches the words and not the voice of Hermia.' In his second edition he followed Hanmer.—The text of the Second Folio, 'Your words I'd
catch,' MALONE pronounces 'intelligible,' and STAUNTON, who also adopts it, remarks
that 'Helena would catch not only the beauty of her rival's aspect and the melody

My eare fhould catch your voice, my eye, your eye,   200
My tongue fhould catch your tongues fweet melodie,
Were the world mine, *Demetrius* being bated,
The reft Ile giue to be to you tranflated.   203

---

203. *Ile*] *ile* Q₁.   *I'le* F₃F₄.   *I'd* Han. Cam. Wh. ii, Ktly, Huds.

---

of her tones, but her language also,' which applies quite as well to Hanmer's emendation.—'But,' says W. A. WRIGHT, 'Hanmer's correction gives a better sense.'
However reluctant we may be to desert the QqFf, I am afraid we must submit.—ED.

200. **eare . . . voice**] DYCE (ed. ii): Mr W. N. Lettsom would read, 'My *hair*
should catch your *hair*, my eye your eye,' and defends the alteration thus: 'As the
passage stands at present, Helena wishes her *ear* may resemble the *voice* of Hermia!
I conceive that, in the first place, "*heare*"—"*heare*" [a common old spelling of
'*hair*'] was transformed into "*eare*"—"eare" by the blunder of a transcriber. The
verse was then operated upon by a sophisticator, who regarded nothing but the line
before him, and was not aware of the true meaning of "*my eye your eye*," but took
"*catch*" in the ordinary sense, not in the peculiar sense of contracting a disease,
which it bears throughout this passage.'—DEIGHTON: If any change were allowable,
I should be inclined to read: 'My *fair* should catch your *fair*,' *i. e.* the personal
beauty you have ascribed to me should catch your personal beauty, . . . *fair* being the
general term including the particulars 'eye' and 'tongue.' 'Voice' seems clearly
wrong, . . . and with my conjecture we have in these two lines a complete corre-
spondency with lines 194, 195.—[HUDSON adopted Lettsom's emendation, wherein, I
think, the fact is overlooked that, while it is quite possible for Helena's eyes to catch
the love-light that lies in Hermia's, and for Helena's tongue to catch the melody of
her rival's, by no possibility can Helena's hair be made to resemble Hermia's, short
of artificial means. Deighton's emendation is certainly more plausible than Lettsom's. Both of them, however, are, I think, needless. To a compositor, 'eare'
might be mistaken for *fair* or *hair*, but it is unlikely that for either of these words
he should mis-read or mis-hear 'voice.'—ED.]

200. **my**] ABBOTT, § 237: *Mine* is almost always found before *eye*, *ear*, &c. where
no emphasis is intended.   But where there is antithesis we have *my*, *thy*.   See, also,
III, ii, 230: 'To follow *me* and praise *my* eies and face?'

200, 201. **eye . . . melodie**] I cannot believe that to Elizabethan ears the rhyme
here was imperfect.   It was as perfect as are all the others in this scene.   'Melody,'
therefore, must have been pronounced then as it is in German at this day: *melodei*.
If additional proof be needed, compare the Fairy's song in II, ii, 15, 16: 'Philomele with melodie, Sing in your sweet Lullaby,' where the music is marred if the
rhyme be not perfect.—ED.

202. **bated**] That is, excepted.

203. **Ile**] LETTSOM: Read *I'd*. I cannot but think that the frequent confusion
of 'Ile' and 'Ide' is a misprint, not an idiom.—DYCE (ed. ii, where the foregoing
note is found): But it certainly appears that our ancestors frequently used '*will*'
where we now use 'would,' *e. g.* 'If I *should* pay your worship those again, Perchance
you *will* not bear them patiently.'—*Com. of Err.* I, ii, 85; 'I *would* bend under
any heavy weight That *he'll* enjoin me to.'—*Much Ado*, V, i, 286.

203. **translated**] That is, transformed, as in Quince's 'Bottom, bless thee; thou
art translated,' III, i, 124.

O teach me how you looke, and with what art
you fway the motion of *Demetrius* hart.   205
 Her. I frowne vpon him, yet he loues me ftill.
 Hel. O that your frownes would teach my fmiles
fuch skil.
 Her. I giue him curfes, yet he giues me loue.
 Hel. O that my prayers could fuch affeＣtion mooue. 210
 Her. The more I hate, the more he followes me.
 Hel. The more I loue, the more he hateth me.
 Her. His folly Helena is none of mine.
 Hel. None but your beauty, wold that fault wer mine
 Her. Take comfort : he no more fhall fee my face, 215
Lyfander and my felfe will flie this place.
Before the time I did *Lyfander* fee,   217

---

213. *folly Helena*] *fault, oh Helena* Coll. (MS). *no fault* Q₁ et cet.
Han. *folly, Helen* Dyce ii, iii, Huds. 214. *None...wold*] *None.—But your*
fault, fair Helena Coll. (MS). *beauty ;—'would* Henderson ap. Var.
  none] Q₂Ff, Rowe, Pope, Han. *beauty*] *beauty's* Daniel, Huds.

213, 214. It is by no means easy to decide between the text as we have it above
in the Folio, and the text of Q₁ (which has been adopted by a majority of editors) :
'His folly, Helena, is no fault of mine.' If we assume that Hermia is trying to comfort her dear friend with assurances of her enduring love, then there is a charm in
this asseveration, in the Folio, that she does not share in Demetrius's folly, which
gives hate for love, but that she returns love for love ; and her words become sympathetic and caressing. But if we adopt the text of Q₁, Hermia's words have a faint
tinge of acerbity (which, it must be confessed, is not altogether out of character), as
though she were defending herself from some unkind imputation, and wished to close
the discussion (which would also be not unnatural). It is again in favour of the
Quarto that Helena replies ' would *that* fault were mine.' The demonstrative ' that '
seems clearly to refer to a ' fault ' previously expressed. This weighs so heavily with
Capell that he says the word ' fault ' must ' of necessity have a place ' in Hermia's
line. Lastly, it is in favour of the Folio that Helena's first words are Hermia's last.
' It is *none* of mine,' says Hermia, ' It is *none* of yours,' assents Helena. On the
whole, therefore, I adhere to the text of the Folio.—ED.

215, &c. JOHNSON : Perhaps every reader may not discover the propriety of these
lines. Hermia is willing to comfort Helena, and to avoid all appearance of triumph
over her. She therefore bids her not to consider the power of pleasing as an advantage to be much envied or much desired, since Hermia, whom she considers as pos-
sessing it in the supreme degree, has found no other effect of it than the loss of happiness.—DEIGHTON : How powerful must be the graces of my beloved one, seeing
that they have made Athens a place of torture to me ; *i. e.* since so long as she
remained in it she could not marry Lysander. [According to Johnson's interpretation,
' he,' in the phrase ' he hath turn'd,' refers, not to Lysander, but to ' love,' Hermia's
own love, which is doubtful.—ED.]

Seem'd Athens like a Paradife to mee.   218
O then, what graces in my Loue do dwell,
That he hath turn'd a heauen into hell.   220
 Lyf. *Helen,* to you our mindes we will vnfold,
To morrow night, when *Phœbe* doth behold
Her filuer vifage, in the watry glaffe,
Decking with liquid pearle, the bladed graffe
(A time that Louers flights doth ftill conceale)   225
Through *Athens* gates, haue we deuis'd to fteale.
 Her. And in the wood, where often you and I,
Vpon faint Primrofe beds, were wont to lye,
Emptying our bofomes, of their counfell fweld :

---

218. *like a*] *as a* Q₁, Cap. Steev. Mal. '90, Coll. Dyce, Cam. Wh. ii, Ktly.
219. *do*] *must* Coll. (MS).
220. *into*] Q₂Ff, Rowe, Pope, Han. Johns. Hal. *vnto a* Q₁, Theob. Warb. Cap. Steev. Mal. Knt, Dyce, Sta. Cam.

unto Var.'03, '13, '21.   *into a* White.
226. *gates*] *gate* F₃F₄, Rowe+.
229. *counfell fweld*] QqFf, Rowe i, Hal. *counsells swell'd* Rowe ii, Pope, Warb. *counsells sweet* Theob. Han. Johns. Ktly. *counsel sweet* Cap. et cet.

---

220. **into**] DYCE (*Rem.* 44) : The context, 'a heaven,' is quite enough to deter-mine that the reading of Fisher's 4to [Q₁], 'unto *a* hell,' is the right one, excepting that 'unto' should be '*into*.' Compare a well-known passage of Milton : 'The mind is its own place, and in itself Can make *a* heaven of hell, *a* hell of heaven.'—*Par. Lost,* i, 254.

225. **still**] Constantly, always. See Shakespeare *passim.*

228. **faint Primrose beds**] STEEVENS : Whether the epithet 'faint' has reference to the colour or smell of primroses, let the reader determine. [I think it refers to the colour. Twice (in *Winter's Tale,* IV, iv, 122, and in *Cym.* IV, ii, 221) Shake-speare speaks of '*pale* primroses.'—DELIUS supposes that 'faint' is here used pro-leptically, and refers to 'beds for those who are weary. Compare "lazy bed," *Tro. & Cres.* I, iii.'—ED.]

229. **sweld**] THEOBALD : This whole scene is strictly in rhyme, and that it devi-ates [here and in line 232], I am persuaded is owing to the ignorance of the first, and the inaccuracy of the later, editors ; I have, therefore, ventured to restore the rhymes, as, I make no doubt, but the poet first gave them. *Sweet* was easily cor-rupted into 'sweld,' because that made an antithesis to 'emptying'; and 'strange companions' [line 232] our editors thought was plain English; but *stranger companies* a little quaint and unintelligible. Our author elsewhere uses the substantive *stranger* adjectively, and *companies* to signify 'companions.' See *Rich. II:* I, iii, 143 : 'But tread the stranger paths of banishment'; and in *Hen. V:* I, i, 53 : 'His companies unletter'd, rude and shallow.' And so in a parallel word : 'My riots past, my wild societies,' *Merry Wives,* III, iv, 8.—HEATH (p. 44) : It is evident, as well from the dissonance of the rhyme as from the absurdity and false grammar of the expression, 'bosoms swell'd of their counsels,' that 'swell'd' is corrupt. Mr Theobald hath by a very happy conjecture corrected this wrong reading ; [the meaning then is] emptying our bosoms of those secrets upon which we were wont to consult each other with so

There my *Lyſander,* and my ſelfe ſhall meete,   230
And thence from *Athens* turne away our eyes
To ſeeke new friends and ſtrange companions,
Farwell ſweet play-fellow, pray thou for vs,
And good lucke grant thee thy *Demetrius.*
Keepe word *Lyſander* we muſt ſtarue our ſight,   235
From louers foode, till morrow deepe midnight.
                                      Exit Hermia.
    Lyſ.   I will my *Hermia.*   *Helena* adieu,
As you on him, *Demetrius* dotes on you.   *Exit Lyſander.*   239

---

232. *ſtrange companions*] *ſtranger*
companies Theob. Han. Johns. Mal.
Steev. Knt, Coll. White, Dyce, Sta.
Cam. Ktly.

234. *grand*] *grand* Q,
    thy] *thine* Rowe ii.
239. *dotes*] *dote* Qq, Pope et seq.

---

sweet a satisfaction.   The poet seems to have had in his eye *Psalm* lv, 14: 'We took
sweet counsel together.'—STEEVENS adheres to the Folio, because 'a *bosom swell'd
with secrets* does not appear as an expression unlikely to have been used by our author
who speaks of a *stuff'd bosom* in *Macbeth.*   In *Rich. II:* IV, i, 298, we have "the
unseen grief That swells with silence in the tortured soul." "*Of* counsels swell'd"
may mean, swell'd *with* counsels.'—HALLIWELL also defends the Folio, and pronounces Theobald's emendation 'unnecessary' (*Introd.* 73): 'If Shakespeare had
written *sweet* and *stranger companies,* it is very improbable that these words could
have been so changed either by the actors or printers.'   In his Folio edition, fifteen
years later than his *Introduction,* Halliwell is still of the same mind: 'Theobald
in each instance sacrifices the sense to the ear, the participle "emptying" corrobo-
rating the old reading "swell'd," and the comparative, as applied to companions or
companies, being pointless.'   He then adds: 'In a previous speech of Hermia's all the
lines rhyme with the exception of the three commencing ones.   If Theobald's theory
be correct, the two lines in that speech ending with the words "bow" and "head"
should be altered so as to rhyme.'—COLLIER (ed. ii): The (MS) amends 'swell'd'
and 'companions' [as Theobald amends them], though, somewhat to our surprise,
no change is made in the epithet 'strange.'—DYCE (ed. i): I give here Theobald's
emendations, . . . and I give them in the belief that more certain emendations were
never made.—W. A. WRIGHT: The rhyme is decisive in favour of Theobald's conjecture.   [In a modernised text Theobald's emendations should be adopted unquestionably.   See the following note by Walker.—ED.]

232. **strange**] It is noteworthy as a corroboration of Theobald's emendation that
WALKER (*Crit.* ii, 53) cites this present word among his many examples of the confusion of final *e* and *er.*   See II, ii, 81.

239. **dotes**] A clear instance of the interpolation of the final *s,* early recognised
by POPE as an error, and acknowledged by every subsequent editor.—WALKER'S
article, dealing with this final *s* (*Crit.* i, 233), is one of the most valuable of his many
valuable articles.   'The interpolation of an *s* at the end of a word—generally, but
not always, a noun substantive—is remarkably frequent in the Folio.   Those who are
conversant with the MSS of the Elizabethan Age may perhaps be able to explain its

Hele.   How happy fome, ore otherfome can be *?*   240
Through *Athens* I am thought as faire as fhe.
But what of that *?* *Demetrius* thinkes not fo :
He will not know, what all, but he doth know,
And as hee erres, doting on *Hermias* eyes ;
So I, admiring of his qualities :   245
Things bafe and vilde, holding no quantity,

---

243. *he doth*] Ff, Rowe, White. *hee*   246. *vilde*] F_2F_3, Knt, Hal. *vile*
doe Q_1. *he do* Q_2, Pope et cet. QqF_4 et cet.

---

origin.   Were it not for the different degree of frequency with which it occurs in different parts of the Folio,—being comparatively rare in the Comedies (except perhaps
in *The Winter's Tale*), appearing more frequently in the Histories, and becoming
quite common in the Tragedies,—I should be inclined to think it originated in some
peculiarity of Shakespeare's hand-writing.'   There is another example of it in this
play, cited as such by Walker (IV, i, 208, 'every things seemes double '), but which
might possibly receive a different explanation.   There are several examples in *As You
Like It*, cited, in this edition, at I, iii, 60, together with instances from other plays not
noticed by Walker ; I can recall no single example in *The Tempest*.   We know that
the Folio was printed at the charges of four Stationers.   May not this interpolated *s*,
which is local in its frequency, be due, not to Shakespeare's handwriting, but to the
compositors in the different printing-offices ?—ED.

   240. **othersome**] HALLIWELL : A quaint but pretty phrase of frequent occurrence
in early works.   It is found in the Scripture, *Acts* xvii, 18.—ABBOTT (p. 5) gives
an example from Heywood, who, 'after dividing human diners into three classes,
thus : " Some with small fare they be not pleased, Some with much fare they be dis-
eased, Some with mean fare be scant appeased," adds, with truly Elizabethan freedom, " But of all *somes* none is displeased To be welcome." '—W. A. WRIGHT refers
to *Two Noble Kinsmen*, IV, iii ; *Meas. for Meas.* III, ii, 94 ; also *2 Esdras* xiii, 13.
[See also Lily's *Love's Meta.* III, i, p. 232, ed. Fairholt.]

   245. **admiring of**] See ABBOTT, § 178, for other examples of verbal nouns.—W.
A. WRIGHT : In this construction ' admiring ' is a verbal noun, originally governed
by a preposition, *in* or *on*, which has disappeared, but which exists sometimes in the
degraded form *a*, in such words as ' a hunting,' ' a building.'—VERITY : I take ' admiring ' as a present participle, and ' of ' as the redundant preposition found in Elizabethan English with many verbs ; cf. Bacon, *Advancement of Learning*, II, xxiii,
13 : ' Neither doth learning *admire* or *esteem of* this architecture.'   So, in the same
work (II, xxv, 7), ' define of ' and ' discern of ' (II, xxi, 1).

   246, 247.   GREEN (*Emblem Writers*, p. 349) finds a parallel to the sentiment in
these lines in an emblem, engraved by De Passe in 1596, illustrating the apothegm :
' Perpolit incultum paulatim tempus amorem.'   The illustration represents Cupid
watching a bear which is licking her cub into shape, and is accompanied by Latin
and French stanzas.   As the present is, I think, one of the happiest examples of Green's
theory, the space is well bestowed in giving these stanzas in full : ' Ursa novum fertur
lambendo fingere foetum Paulatim et formam, quae decet, ore dare ; Sic dominam, ut
valde sic cruda sit aspera Amator Blanditiis sensim mollet et obsequio.'   '*Peu à peu.
Ceste masse de chair, que toute ourse faonne [*sic*] En la leschant se forme à son com-

Loue can tranfpofe to forme and dignity,   247
Loue lookes not with the eyes, but with the minde,
And therefore is wing'd *Cupid* painted blinde.
Nor hath loues minde of any iudgement tafte :   250
Wings and no eyes, figure, vnheedy hafte.
And therefore is Loue faid to be a childe,
Becaufe in choife he is often beguil'd,
As waggifh boyes in game themfelues forfweare ;
So the boy Loue is periur'd euery where.   255
For ere *Demetrius* lookt on *Hermias* eyne,
He hail'd downe oathes that he was onely mine.
And when this Haile fome heat from *Hermia* felt,
So he diffolu'd, and fhowres of oathes did melt,
I will goe tell him of faire *Hermias* flight :   260

251. *figure,*] *figure* Rowe et seq.   256. *eyne*] Q₂ (Ashbee) F₂F₃. *eyen*
     hafte] *haft* F₄.   Q₁Q₂ (Griggs). *eyn* F₄.
 253. *is often*] *is oft* Q₂. *often is* Ff, 257. *onely*] *only* F₂F₄.
Rowe, Pope, Han. White. *is fo oft* Q₁,   258. *this*] *his* Q₂.
Theob. et cet.   259. *So he*] *Lo, he* Cap. *Soon it*
 254. *in game themfelues*] *themfelues*   Rann. *Soon he* Daniel.
in game F₃F₄, Rowe+.

---

mencement.   Par servir : par flatter, par complaire en aymant, L'amour rude à l'abord,
à la fin se façonne.'—ED.

   246. **no quantity**] JOHNSON : *Quality* seems a word more suitable to the sense
than 'quantity,' but either may serve.—STEEVENS : 'Quantity' is our author's word.
So in *Hamlet*, III, ii, 177 : 'For women's fear and love hold quantity.'—SCHMIDT :
That is, bearing no proportion to what they are estimated by love.

   254. **game**] JOHNSON : This signifies here, not contentious play, but *sport, jest.*

   256. **eyne**] W. A. WRIGHT : This Old English plural is used by Shakespeare
always on account of the rhyme, except in *Lucrece*, 1229, and *Pericles*, III, *Gower*, 5.

   259. **So**] ABBOTT, § 66 : 'So' (like the Greek οὖτω δή) is often used where we
should use *then.*

   260. **goe tell**] See ABBOTT, § 349.   Also 'go seeke,' II, i, 13.

   260, &c. COLERIDGE (p. 101) : I am convinced that Shakespeare availed himself
of the title of this play in his own mind, and worked upon it as a dream throughout,
but especially, and perhaps unpleasingly, in this broad determination of ungrateful
treachery in Helena, so undisguisedly avowed to herself, and this, too, after the witty,
cool philosophising that precedes.   The act itself is natural, and the resolve so to act
is, I fear, likewise too true a picture of the lax hold which principles have on a
woman's heart, when opposed to, or even separated from, passion and inclination.
For women are less hypocrites to their own minds than men are, because in general
they feel less proportionate abhorrence of moral evil in and for itself, and more of its
outward consequences, as detection and loss of character, than men,—their natures
being almost wholly extroitive.   Still, however just in itself, the representation of this
is not poetical ; we shrink from it, and cannot harmonise it with the ideal.

Then to the wood will he, to morrow night   261
Purſue her ; and for his intelligence,
If I haue thankes, it is a deere expence :
But heerein meane I to enrich my paine,
To haue his ſight thither, and backe againe.   *Exit.*   265

262. *his*] *this* Qq, Rowe et seq.

262. his] This is one of WALKER's instances (IV, i, 88 is another) where, in this play, *his* and *this* have supplanted one another (*Crit.* ii, 221).

263. deere expence] STEEVENS: That is, it will *cost him much* (be a severe constraint on his feelings,) to make even so slight a return for my communication.—COLLIER (ed. ii): This reading may be reconciled to meaning, but the alteration of the MS at once claims our acceptance; *it is dear recompense* can mean nothing but the expression of great satisfaction on the part of Helena at the reward she hopes to receive for her intelligence.—LETTSOM (*Blackwood*, Aug. 1853): The Old Corrector [*i. e.* Collier's MS] is an old woman who, in this case, has not merely mistaken, but has directly reversed, Shakespeare's meaning. So far from saying that Demetrius's thanks will be any 'recompense' for what she proposes doing, Helena says the very reverse, that they will be a severe aggravation of her pain. 'A dear expense' here means a painful purchase, a bitter bargain. 'If I have thanks, the sacrifice which I make in giving Demetrius this information will be doubly distressing to me.' Of course she would much rather that Demetrius, her old lover, did not thank her for setting him on the traces of his new mistress. Thanks would be a mockery in the circumstances, and this is what Helena means to say. Such is manifestly the meaning of the passage, as may be gathered both from the words themselves and from the connection with the context. The *sight* of Demetrius, and not his *thanks*, was to be Helena's *recompense*.—DYCE (ed. i): The MS Corrector was evidently in total darkness as to the meaning of the passage; nor could Mr Collier himself have paid much attention to the context, when he recommended so foolish an alteration as a singular improvement.—STAUNTON: Does it not mean that, as to gratify her lover with this intelligence, she makes the most painful sacrifice of her feelings, his thanks, even if obtained, are dearly bought?—DELIUS: Helena assuredly means that she purchases even the thanks of Demetrius at a high price, namely, at the price of fostering and furthering Demetrius's love for Hermia, and therefore of her own harm.—W. A. WRIGHT: That is, it will cost me dear, because it will be in return for my procuring him a sight of my rival.

265. In Garrick's Version, Helena, before she departs, sings as follows:—

> ' Against myself why all this art,
>   To glad my eyes, I grieve my heart;
>   To give him joy, I court my bane !
>   And with his sight enrich my pain.'

The Air is by ' Mr. Burney.'

[*Scene II.*]

Enter Quince the Carpenter, Snug the Ioyner, Bottome the Weauer, Flute the bellowes-mender, Snout the Tinker, and Starueling the Taylor.

Quin.   Is all our company heere?

Bot.   You were beſt to call them generally, man by   5
man, accoᵣding to the ſcrip.

Qui.   Here is the ſcrowle of euery mans name, which
is thought fit through all *Athens*, to play in our Enterlude before the Duke and the Dutches, on his wedding
day at night.   10

[Scene IV. Pope+.   Scene III. Fleay.
Scene II. Cap. et seq.   Scene changes to
a Cottage. Theob.   A Room in Quince's
House. Cap.

1, 2. Snug...Snout] and *Snugge*, the
Ioyner, and *Bottom*, the Weauer, and
Flute, the Bellowes mender, & *Snout*, Q₁.

2. Snout] Snowt F₃F₄, Rowe, Pope.

6. *to*] Om. Q₂.

8. *Enterlude*] *interlude* Theob. et
seq.

9. *the Dutches*] *Dutchess* Pope ii,
Theob. Warb. Johns. Steev. Mal. Var.
Coll. Sing. Ktly.

---

1. JOHNSON: In this scene Shakespeare takes advantage of his knowledge of the
theatre to ridicule the prejudices and the competitions of the players.   Bottom, who
is generally acknowledged the principal actor, declares his inclination to be for a
tyrant, for a part of fury, tumult, and noise, such as every young man pants to perform
when he first steps upon the stage.   The same Bottom, who seems bred in a tiring-
room, has another histrionical passion.   He is for engrossing every part, and would
exclude his inferiors from all possibility of distinction.   He is therefore desirous to
play Pyramus, Thisbe, and the Lion, at the same time.—STAUNTON suggests the possibility that 'in the rude dramatic performance of these handicraftsmen of Athens,
Shakespeare was referring to the plays and pageants exhibited by the trading companies of Coventry, which were celebrated down to his own time, and which he
might very probably have witnessed.'   This is not impossible, especially in view of
the fact, which I do not remember to have seen noticed in connection with the present
play, that midsummer eve was especially chosen as the occasion for a 'showe' or
'watche,' performed by various companies of handicraftsmen.   'Heare we maye note
that yᵉ showe or watche, on midsomer eaue, called "midsomer showe," yearely now
vsed within yᵉ Citti of Chester, was vsed in yᵉ tyme of those whitson playes &
before,' so says David Rogers, in 1609, *Harl.* MS, 1944, quoted by F. J. FURNIVALL
in Appendix to 'Forewords' of *The Digby Mysteries*, p. xxiii, *New Sh. Soc.*—ED.

For remarks on Bottom's character, see Appendix.

5. **you were best**] For this substitution for the full phrase *to you it were best*, see
ABBOTT, § 230.

5. **generally**] W. A. WRIGHT: This, in Bottom's language, means *particularly,
severally.*

6. **scrip**] GREY (i, 45): Formerly used in the same sense with *script*, and signified a scrip of paper or any manner of writing.

3

Bot.   Fiıſt, good *Peter Quince,* ſay what the play treats   11
on : then read the names of the Actors : and ſo grow on
to a point.

Quin.   Marry our play is the moſt lamentable Comedy, and moſt cruell death of *Pyramus* and *Thisbie.*   15

Bot.   A very good peece of worke I aſſure you, and a

11, 17, 40. Peter] Peeter Q_1.
12, 13. *grow...point*] *grow to a point*
Qq, Cap. Steev. Mal. Var. Coll. Sing.
Sta. Dyce, Ktly, Cam. *grow on to ap-*

point F_4. *go on to a point* Warb. *go
on to appoint* Coll. MS.
14. *Marry*] *Mary* Q_1.

---

9. **his wedding**] R. G. WHITE (ed. i) : This use of ' his ' is in conformity to the
usage of educated persons in Shakespeare's day.

12, 13. **grow . . . point**] JOHNSON : ' Grow ' is used in allusion to his name,
Quince.—STEEVENS : It has, I believe, no reference to the name.   I meet with the
same kind of expression in *Wily Beguiled,* 'As yet we are grown to no conclusion.'
[I do not think this is to be found in *Wily Beguiled.*—ED.]   Again, in *The Arraignment of Paris,* 1584 : ' Our reasons will be infinite, I trow, Unless unto some other
point we grow ' [II, i].—WARNER upholds, as an original emendation, the reading
' appoint ' of F_4, and explains : ' Quince first tells them the name of the play, then
calls the actors by their names, and after that tells each of them what part is set down
for him to act.   Perhaps Shakespeare wrote "to *point,*" i. e. to appoint.'—HALLI-
WELL : Warner's suggestion was probably derived from the Opera of *The Fairy
Queen,* 1692, where the sentence is thus given :—' and so go on to appoint the parts.'
THOMAS WHITE (p. 29) : Does not this mean *draw to a conclusion,* alluding to Bottom's trade of a weaver ?   In a tract in the public library at Cambridge, with the following title—*The Reformado precisely characterised by a modern Churchman*—occurs
this passage : ' Here are mechanicks of my profession who can seperate the pieces of
salvation from those of damnation, *measure out* the thread, substantially *pressing the
points,* till they have fashionably filled up their work with a *well-bottomed* conclusion.'
—STAUNTON : That is, and so to business.   A common colloquial phrase formerly.—
R. G. WHITE : The speech as it stands is good colloquial Bottom-ese.—W. A.
WRIGHT : It is not always quite safe to interpret Bottom, but he seems to mean
' come to the point.'

14. **lamentable Comedy**] STEEVENS : This is very probably a burlesque on the
title-page of Cambyses, '*A lamentable Tragedie, mixed full of pleasant Mirth, con-
teyning the life of Cambises, King of Percia,* &c. by Thomas Preston ' [1561 ?   It is,
I think, very doubtful if any burlesque of a particular play was meant.   At any rate,
Shakespeare's audiences probably were not so learned that they could at once appreciate the fling at a tragedy in all likelihood thirty years old.   Moreover, even in Dryden's time the limits of Tragedy and Comedy were vague.   *Cymbeline* is still classed
among Tragedies.—ED.]

15. **Pyramus**] See Appendix, *Source of the Plot.*

16. **worke**] KNIGHT : Bottom and Sly both speak of a theatrical representation as
they would of a piece of cloth or a pair of shoes.   [Perhaps the antithesis may be in
calling a ' play ' a ' work.'   Ben Jonson was the first, I believe, to call his Plays
Works.—ED.]

merry.   Now good *Peter Quince,* call forth your Actors   17
by the fcrowle. Mafters fpread your felues.

Quince.   Anfwere as I call you.   *Nick Bottome* the
Weauer.   20

Bottome.   Ready ; name what part I am for , and
proceed.

Quince.   You *Nicke Bottome* are fet downe for *Py-*
ramus.

Bot.   What is *Pyramus,* a louer, or a tyrant ?   25

Quin.   A Louer that kills himfelfe moft gallantly for
loue.

Bot.   That will aske fome teares in the true perfor-
ming of it *:* if I do it, let the audience looke to their eies :
I will mooue ftormes ; I will condole in fome meafure.   30
To the reft yet, my chiefe humour is for a tyrant.   I could

---

20. *Weauer.*] *Weauer ?* Q$_1$.
25. Pyramus,] Pyramus ? Q$_1$.
26. *gallantly*] *gallant* Qq, Cap. Coll.
Sing. Sta. Ktly, Cam.
29. *it : if*] *it.   If* Q$_1$.
30. *teares*] *stones* Coll. MS.

31. *reft yet,*] QqFf, Rowe, Pope, Sta.
Dyce ii, iii.   *rest ;—yet,* Theob. et cet.
(subs.).
      To the reft] As a stage direction,
Opera, 1692, Deighton conj.

---

20. **Weauer**] In the *Transactions* of *The New Shakspere Soc.* 1877–79, p. 425,
G. H. OVEREND describes and transcribes a bill, addressed to Cardinal Wolsey as
Chancellor, wherein is contained the 'complaint of one George Maller, a glazier,
against Thomas Arthur, a tailor, whom he had undertaken to train as a player.'

26. **gallantly**] COLLIER : This improves the grammar [of the Quartos], but renders the expression less characteristic.—R. G. WHITE (ed. I) : On the contrary, it
makes the speech quite unsuited to good Peter Quince, who always speaks correctly.
Indeed, it should be observed that purely grammatical blunders are rarely or never
put into the mouths of Shakespeare's characters; probably because grammatical
forms, in minute points at least, were not so fixed and so universally observed in his
day as to make violations of them very ridiculous to a general audience.   He depends
for burlesque effect upon errors more radically nonsensical and ludicrous.

30. **condole**] W. A. WRIGHT : Bottom, of course, blunders, but it is impossible to
say what word he intended to employ.   Shakespeare uses 'condole' only once besides, and he then puts it into the mouth of Ancient Pistol, who in such matters is as
little of an authority as Bottom.   See *Hen. V:* II, i, 133 : 'Let us condole the
knight,' that is, mourn for him.   In *Hamlet,* I, ii, 93, 'condolement' signifies the
expression of grief.

31. **rest yet,**] STAUNTON : The colon after 'rest' in modern editions is a deviation
which originated perhaps in unconsciousness of one of the senses Shakespeare attributes to the word 'yet.'   'To the rest yet,' is simply, 'To the rest now,' or, as he shortly
after repeats it, '*Now,* name the rest of the players.'—W. A. WRIGHT gives two instances of the use of 'yet' in this unemphatic position : Lord Herbert of Cherbury's

play *Ercles* rarely, or a part to teare a Cat in, to make all 32
fplit the raging Rocks ; and fhiuering fhocks fhall break
the locks of prifon gates, and *Phibbus* carre fhall fhine
from farre, and make and marre the foolifh Fates. This 35

32. *Cat*] *Cap* Warb.

    in, to] *in. To* Pope, Han. *in
and to* Ktly. *in two*, Bottom the Weaver,
1661. *in : To* Theob. et seq. (subs.).

32, 33. *to make all fplit*] Separate
line, Cap.

33–35. *the raging ... Fates*] QqFf,
Rowe+, Sta. Eight lines, Johns. et.

cet.

33. *fplit the*] QqFf, Rowe ii, Pope
Han. Sta. *fplit to* F₄, Rowe i. *split—
"the* Theob. et cet. (subs.).

    and fhiuering] *With fhivering*
Farmer, Steev.'85, '93.

34. Phibbus] *Phibbus's* Rowe i. *Phib
bus'* Theob. ii et seq.

Life, p. 57 : ' Before I departed yet I left her with child of a son' ; and *Meas. for
Meas.* III, ii, 187 : ' The duke yet would have dark deeds darkly answered.'

32. **Ercles**] MALONE : In Greene's *Groat's-worth of Wit*, 1592, a player who is
introduced says : ' The twelue labors of *Hercules* haue I terribly thundered on the
stage.'—HALLIWELL : Henslowe, in his Diary, mentions ' the firste parte of Hercu-
lous,' a play acted in 1595, and afterwards, in the same manuscript, the ' two partes
of Hercolus ' are named as the work of Martin Slather or Slaughter. In Sidney's
Arcadia : ' leaning his hands vpon his bill, and his chin vpon his hands, with the
voyce of one that playeth *Hercules* in a play ' [Lib. i, p. 50, ed. 1598].—W. A.
WRIGHT : The part of Hercules was like that of Herod in the Mysteries, one in
which the actor could indulge to the utmost his passion for ranting.

32. **teare a Cat**] EDWARDS (p. 52) : A burlesque upon Hercules's killing a lion.
—HEATH (p. 45) takes Warburton's emendation, *cap*, seriously, and supposes ' it
might not be unusual for a player, in the violence of his rant, sometimes to tear his
cap.'—And CAPELL takes Bottom seriously and supposes ' he might have seen
" Ercles " acted, and some strange thing *torn* which he mistook for a cat.'—STEE-
VENS : In Middleton's *The Roaring Girl*, 1611, there is a character called ' Tearcat,'
who says : ' I am called by those who have seen my valour, Tearcat ' [V, i]. In an
anonymous piece, called *Histriomastix*, 1610, a captain says to a company of players :
' Sirrah, this is you would rend and tear the cat upon a stage.' [Act V, p. 73, ed.
Simpson, who attributes large portions of the play to Marston, and places the date
circa 1599, but a few years later, therefore, than the *Mid. N. Dream.*—ED.]

33. **all split**] FARMER : In *The Scornful Lady*, II, iii, by Beau. and Fl. we meet
with ' Two roaring boys of Rome, that made all split.' DYCE : The phrase was a
favourite expression with our old dramatists.—In his *Few Notes*, p. 61, Dyce observes
that he believes ' it has not been remarked ' that the expression is properly a ' nautical
phrase : " He set downe this period with such a sigh, that, *as the Marriners say*, a man
would haue thought *al would haue split againe.*"—Greene's *Neuer too late*, sig. G3,
ed. 1611.']—W. A. WRIGHT : Compare with all this, which it illustrates, Hamlet's
advice to the players, III, ii, 9, &c : ' to hear a robustious periwig-pated fellow tear a
paffion to tatters, to very rags,' &c.

33–35. **the raging . . . Fates**] THEOBALD : I presume this to be either a quotation from some fustian old play, or a ridicule on some bombastic rants, very near
resembling a direct quotation.—R. G. WHITE (ed. i) : Does not Bottom's expression
in line 35, ' This *was* lofty,' make it certain that it is a quotation ?—STAUNTON : The

was lofty. Now name the reſt of the Players. This 36
is *Ercles* vaine, a tyrants vaine : a louer is more condoling.

Quin. *Francis Flute* the Bellowes-mender.

Flu. Heere *Peter Quince.* 40

Quin. You muſt take *Thisbie* on you.

Flut. What is *Thisbie,* a wandring Knight ?

Quin. It is the Lady that *Pyramus* muſt loue.

Flut. Nay faith, let not mee play a woman, I haue a
beard comming. 45

---

37. Ercles] *Ercles's* Opera, 1692;
Ercles' Theob. et seq.
    vaine...vaine] *veine...veine* Ff.
reign...reign Bottom the Weaver, 1661.
    louer] *lover's* Opera, 1692, Dan-
iel, Huds.

39. *mender.*] *mender ?* Q₁.
41. *You*] *Flute, you* Q₁, Cap. Sta.
Cam.
42. Flut.] Fla. Q₁.
    Thisbie,] Thiſby ? Q₁.

---

chief humour of Bottom's 'lofty' rant consists in the speaker's barbarous disregard
of sense and rhythm; yet, notwithstanding this, and that the whole is printed as
prose, carefully punctuated to be unintelligible in all the old copies, modern editors
will persist in presenting it in good set doggerel rhyme. [I think Staunton somewhat
exaggerates the 'careful' mispunctuation of the old copies; there is but one instance
of mispunctuation, namely in 'to make all split the raging rocks,' which, after all,
might be due to the compositor, a second Bottom perchance. As W. A. Wright says,
it is not always quite safe to interpret Bottom, but I am inclined to think that 'raging'
should be pronounced *ragging*, which will better indicate the word *ragged*, which
was, *perhaps*, the true word, than 'raging.'—ED.]

39. **Bellowes-mender**] STEEVENS: In Ben Jonson's *Masque of Pan's Anniver-
sary* a man of the same profession is introduced. I have been told that a 'bellows-
mender' was one who had the care of organs, regals, &c. [But from the context in
Ben Jonson's masque the 'bellows' were of the ordinary, domestic kind.—ED.]

44. **woman**] JOHNSON: This passage shows how the want of women on the old
stage was supplied. If they had not a young man who could perform the part, with
a face that might pass for feminine, the character was acted in a mask, which was at
that time a part of a lady's dress, so much in use that it did not give any unusual
appearance to the scene; and he that could modulate his voice in a female tone,
might play the woman very successfully. It is observed in Downes's *Roscius Anglicanus* [(p. 26, ed. Davies) of Kynaston that he 'made a compleat Female Stage
Beauty; performing his parts so well . . . that it has since been disputable among the
judicious, whether any woman that succeeded him so sensibly touched the audience
as he ']. Some of the catastrophes of the old comedies, which make lovers marry
the wrong women, are, by recollection of the common use of masks, brought nearer
to possibility.—HALLIWELL: Previously to the Restoration, the parts of women were
usually performed by boys or young men. 'In stage plays, for a boy to put one the
attyre, the gesture, the passions of a woman; for a meane person to take upon him
the title of a Prince with counterfeit porte and traine, is by outwarde signes to shewe

Qui. That's all one, you fhall play it in a Maske, and   46
you may fpeake as fmall as you will.

Bot. And I may hide my face, let me play *Thisbie* too :
Ile fpeake in a monftrous little voyce ; *Thifne, Thifne,* ah   49

---

48. *And*] *An* Pope et seq. *(An'*   49. Thifne, Thifne] *Thisby, Thisby*
Johns.).   Han. *Listen, listen!* White ii.

    too] *to* Qq.

---

themselves otherwise then they are.'—Gosson's *Playes Confuted in five Actions,* n. d.
Occasional instances, however, of women appearing on the London stage occurred
early in the seventeenth century. Thus says Coryat, in his *Crudities,* 1611, p. 247,
speaking of Venice,—' here I observed certaine things that I never saw before, for I
saw women acte, a thing that I never saw before, though I have heard that it hath
beene sometimes used in London ; and they performed it with as good a grace, action,
gesture, and whatsoever convenient for a player, as ever I saw any masculine actor.'
According to Prynne, some women acted at The Blackfriars in the year 1629, and
one in the previous year. It appears from the passage in the text, and from what follows, that the actor's beard was concealed by a mask, when it was sufficiently prominent to render the personification incongruous ; but a story is told of Davenant stating
as a reason why the play did not commence, that they were engaged in 'shaving the
Queen.' The appearance of female actors was certainly of very rare occurrence
previously to the accession of Charles II. The following is a clause in the patent
granted to Sir W. Davenant :—' That, the women's parts in plays have hitherto been
acted by men in the habits of women, at which some have taken offence, we do permit, and give leave, for the time to come, that all women's parts be acted by women.'
Langbaine in his *Account of the English Dramatic Poets,* 1691, p. 117, speaking of
Davenport's King John and Matilda, observes that the publisher, Andrew Penny-
cuicke, acted the part of Matilda, ' women in those times not having appear'd on the
stage.' Hart and Clun, according to the *Historia Histrionica,* 1699, ' were bred up
boys at The Blackfriars, and acted women's parts ;' and the same authority informs
us that Stephen Hammerton ' was at first a most noted and beautifull woman-actor.'
An actor named Pate played a woman's part in the Opera of *The Fairy Queen,*
1692. [According to Malone (Var. '21, iii, 126), it is the received tradition that
Mrs Saunderson, who afterwards married Betterton, was the first English actress.
Unmarried women were not styled ' Miss ' until towards the close of the seventeenth
century. For a discussion of the earliest appearance of actresses on the English
stage, see notes on pp. 288, 289 of *As You Like It,* and p. 397 of *Othello,* in this
edition.—ED.]

    47. **small**] HALLIWELL : That is, low, soft, feminine. Slender, describing Anne
Page (*Mer. Wives,* I, i, 49), observes that ' she has brown hair and speaks small like
a woman.' The expression is an ancient one, an example of it occurring in Chaucer,
The Flower and the Leaf, line 180, ' With voices sweet entuned and so smalle.'
[Many other examples are given by Halliwell, dating from 1552 to 1638, but the
phrase in the present passage is amply explained by Bottom's ' monstrous little voice,'
if any explanation be at all required.—ED.]

    49. **Thisne, Thisne**] W. A. WRIGHT : These words are printed in italic in the
old copies, as if they represented a proper name, and so ' Thisne ' has been regarded
as a blunder of Bottom's for Thisbe. But as he has the name right in the very next

Pyramus my louer deare, thy *Thisbie* deare, and Lady   50
deare.

   Quin.   No no, you muſt play *Pyramus*, and *Flute*, you
Thisby.

   Bot.   Well, proceed.

   Qu.   *Robin Starueling* the Taylor.   55

   Star.   Heere *Peter Quince.*

   Quince.   *Robin Starueling*, you muſt play *Thisbies*
mother?

   Tom Snowt, the Tinker.

   Snowt.   Heere *Peter Quince.*   60

   Quin.   You, *Pyramus* father ; my ſelf, *This bies* father ;
Snugge the Ioyner, you the Lyons part : and I hope there
is a play fitted.   63

52, 53. *you* Thisby] *your Thisby*
Rowe i.
   55. *Taylor.*] *Tailer ?* Q₁.
   58. *mother ?*] *mother :* Qq.
   59 closes line 58, Qq, Cap. et seq.
      Tinker.] *Tinker ?* Q₁.

60. Peter] *Peeor* F₂.
62. *and I hope*] *I hope* Rowe ii+.
      there] *here* Qq, Cap. Mal. Var.
Knt, Coll. Sing. Hal. Sta. Dyce, Cam.
White ii.

---

line, it seems more probable that ' Thiſne ' ſignifies *in this way ;* and he then gives a
specimen of how he would aggravate his voice.   *Thissen* is given in Wright's *Provincial Dictionary* as equivalent to *in this manner ;* and *thissens* is so used in Norfolk.—R. G. WHITE (ed. ii) says that Bottom did not use ' in this way such words as
thissen.'—VERITY : Probably a mistake for ' Thisbe,'—but whose ?   Most likely not
the printer's (contrast the next line).   And if Bottom's, why does he make it only
here ?   Perhaps the reason is that the name is the first word that he has to utter in
this his first attempt to speak in a ' monstrous little voice.'   For an instant, may be, it
plays him false, then by the next line he has recovered himself.   [W. A. Wright's
note carries conviction.   It is not impossible that Capell also thus interpreted the
words, which he prints in Roman, with a dash before and after, whereas proper
names he invariably prints in Italics.   In Mrs Centlivre's *Platonick Lady*, IV, i,
1707, Mrs Dowdy ' enters drest extravagantly in French Night cloaths and Furbe-
lows,' and says : ' If old Roger Dowdy were alive and zeen me thisen, he wou'd
zwear I was going to fly away.'—ED.]
   58. **mother**] THEOBALD : There seems a double forgetfulness of our poet in relation to the Characters of this Interlude.   The father and mother of Thisbe, and the
father of Pyramus, are here mentioned, who do not appear at all in the Interlude ;
but ' Wall ' and ' Moonshine ' are both employed in it, of whom there is not the least
notice taken here.—CAPELL : What the moderns call a forgetfulness in the poet was,
in truth, his judgement : [these parts] promised little, and had been too long in ex-
pectance ; whereas Quince's ' Prologue ' and the other actors, ' Moon-shine ' and
' Wall,' elevate and surprise.—STEEVENS : The introduction of Wall and Moonshine
was an afterthought ; see III, i, 59 and 67.

Snug.   Haue you the Lions part written? pray you if
be, giue it me, for I am flow of ftudie.   65

Quin.   You may doe it *extemporie*, for it is nothing
but roaring.

Bot.   Let mee play the Lyon too, I will roare that I
will doe any mans heart good to heare me.   I will roare,
that I will make the Duke fay, Let him roare againe, let   70
him roare againe.

Quin.   If you fhould doe it too terribly, you would
fright the Dutcheffe and the Ladies, that they would
fhrike, and that were enough to hang vs all.

All.   That would hang vs euery mothers fonne.   75

Bottome.   I graunt you friends, if that you fhould
fright the Ladies out of their Wittes, they would
haue no more difcretion but to hang vs : but I will aggrauate my voyce fo, that I will roare you as gently as
any fucking Doue ; I will roare and 'twere any Nightin-   80
gale.

64. *if*] *if it* QqFf.
72. *If*] *And* Q₁. *An* Cap. et seq.
76. *friends*] *friend* F₄, Rowe i.
    if that] *if* Qq, Pope+, Cap. Cam.

80. *roare*] *roare you* Qq, Pope+,
Steev. Mal. Var. Knt, Coll. Hal. Sta.
Dyce, Cam.
    and] *an* Rowe ii et seq.

---

65. **studie**] STEEVENS: 'Study' is still the cant term used in a theatre for getting
any nonsense by heart.   Hamlet asks the player if he can 'study a speech.'—MA-
LONE: Steevens wrote this note to vex Garrick, with whom he had quarreled.
'Study' is no more a 'cant term' than any other word of art, nor is it applied necessarily to 'nonsense.'

71. **againe**] COWDEN-CLARKE: Not only does Bottom propose to play every part
himself, but he anticipates the applause, and encores his own roar.

78. **aggrauate**] W. A. WRIGHT: Bottom, of course, means the very opposite, like
Mrs Quickly, in *2 Hen. IV:* II, iv, 175: 'I beseek you now, aggravate your choler.'

80. **sucking Doue**] W. A. WRIGHT: Oddly enough, Bottom's blunder of 'suck
ing dove' for 'sucking lamb' has crept into Mrs Clarke's *Concordance*, where *2 Hen.
VI:* III, i, 71 is quoted, 'As is the sucking dove or,' &c.—BAILEY (*Received Text*,
&c. ii, 198): 'Sucking dove' is so utterly nonsenical that it is marvellous how it has
escaped criticism and condemnation.   So far from suffering such a fate, it continues
to be quoted as if it were some felicitous phrase.   The plea can scarcely be set up
that it is humorous, for the humour of the passage lies in Bottom's undertaking to
roar gently and musically, although acting the part of a lion, and is not at all dependent on the incongruity of representing a dove as sucking.   The blunder, which is
whimsical enough, may be rectified by the smallest of alterations—by striking out a
single letter from 'dove,' leaving the clause 'as gently as any sucking *doe*.'   [Had
Bailey no judicious friend?—ED.]

80. **and 'twere**] STEEVENS: As if it were.   Compare *Tro. & Cres.* I, ii, 188:

Quin. You can play no part but *Piramus*, for *Pira-* 82
mus is a ſweet-fac'd man, a proper man as one ſhall ſee in
a ſummers day; a moſt louely Gentleman-like man, therfore you muſt needs play *Piramus*. 85

Bot. Well, I will vndertake it. What beard were I
beſt to play it in?

Quin. Why, what you will.

Bot. I will diſcharge it, in either your ſtraw-colour
beard, your orange tawnie beard, your purple in graine 90

---

84. *Gentleman-like man*] *Gentleman-like-man* F$_3$F$_4$, Rowe.

'He will weep you, an 'twere a man born in April.' [For many examples
where *an* and *and* have been confounded, see WALKER, *Crit.* ii, 153, or ABBOTT,
§ 104.]

89. **straw-colour beard**] HALLIWELL: The custom of dyeing beards is frequently referred to. 'I have fitted my divine and canonist, dyed their beards and
all.'—*Silent Woman*. Sometimes the beards were named after Scriptural personages,
the colours being probably attributed as they were seen in old tapestries. 'I ever
thought by his red beard he would prove a Judas.'—*Insatiate Countess*, 1613. 'That
Abraham-coloured Trojon' is mentioned in *Soliman and Perseda*, 1599; and 'a
goodly, long, thick Abraham-colour'd beard' in *Blurt, Master Constable*, 1602.
Steevens has conjectured that *Abraham* may be a corruption of *auburn*. A 'whay-
coloured beard' and 'a kane-coloured beard' are mentioned in the *Merry Wives*,
1602, the latter being conjectured by some to signify a beard of the colour of cane,
which would be nearly synonymous with the straw-coloured beard alluded to by
Bottom.

90. **purple in graine**] MARSH (*Lectures*, &c. p. 67): The Latin *granum* signifies
a seed, and was early applied to all small objects resembling seeds, and finally to all
minute particles. A species of oak or ilex (*Quercus coccifera*) is frequented by an
insect of the genus *coccus*, which, when dried, furnishes a variety of red dyes, and
which, from its seed-like form, was called in Later Latin *granum*, in Spanish, *grana*,
and *graine* in French; from one of these is derived the English word *grain*, which,
as a coloring material, strictly taken, means the dye produced by the coccus insect,
often called in the arts *kermes*; this dye (like the murex of Tyre) is capable of
assuming a variety of reddish hues, whence Milton and other poets often use *grain*
as equivalent to *Tyrean purple*, as in *Il Penseroso*: 'All in a robe of darkest grain.'
[Marsh here gives many instances from Milton, Chaucer, and others showing that, in
the use of the word *grain*, color is denoted.] The phrase 'purple-in-grain' in Bottom's
speech signifies a color obtained from kermes, and doubtless refers to a hair-dye of
that material. The color obtained from kermes or grain was peculiarly durable, that
is, *fast*, which word in this sense is etymologically the same as *fixed*. When, then, a
merchant recommended his purple stuffs as being dyed in *grain*, he originally meant
that they were dyed with *kermes*, and would wear well, and this phrase was afterwards applied to other colors as expressing their durability. Thus, in *The Com. of
Err.* III, ii, 107, when Antipholus says, 'That's a fault that water will mend,' 'No,
sir,' Dromio replies, ''tis in grain; Noah's flood could not do it.' And again in

beard, or your French-crowne colour'd beard, your per-   91
fect yellow.

Quin. Some of your French Crownes haue no haire
at all, and then you will play bare-fac'd. But mafters here
are your parts, and I am to intreat you, requeft you, and   95
defire you, to con them by too morrow night : and meet
me in the palace wood, a mile without the Towne, by
Moone-light, there we will rehearfe : for if we meete in
the Citie, we fhalbe dog'd with company, and our deui-
fes knowne.   In the meane time, I wil draw a bil of pro-   100

---

91. *colour'd*] *colour* Qq, Cap. Steev.
Mal. Var. Coll. Sing. Hal. Sta. Dyce,
Cam. White ii.

91, 103. *perfect*] *perfit* Qq.

96. *too morrow*] Q₂.
98. *we will*] *will wee* Q₁, Cap. Steev.
Mal. Var. Knt, Coll. Hal. Sta. Dyce,
Cam. White ii.

Twelfth Night, I, v, 253, when Viola insinuates that Olivia's complexion had been
improved by art, the latter replies, ''Tis in grain, sir; 'twill endure wind and weather.'
In both these examples it is the sense of permanence, a well-known quality of the
color produced by *grain* or *kermes,* that is expressed.   It is familiarly known that if
wool be dyed before spinning, the color is usually more permanent than when the
spun yarn or manufactured cloth is first dipped in the tincture.   When the original
sense of *grain* grew less familiar, and it was used chiefly as expressive of *fastness* of
color, the name of the effect was transferred to an ordinary known cause, and *dyed in
grain,* originally meaning dyed with kermes, then dyed with fast color, came at last
to signify dyed in *the wool,* or raw material.   The verb *ingrain,* meaning to incorporate a color or quality with the natural substance, comes from *grain* used in this
last sense.   *Kermes* is the Arabic and Persian name of the coccus insect, and occurs
in a still older form, *krmi,* in Sanscrit.   Hence come the words *carmine* and *crimson.*
The Romans sometimes applied to the coccus the generic name *vermiculus,* a little
worm or insect, the diminutive of *vermis,* which is doubtless cognate with the Sanscrit *krmi,* and from which comes *vermilion,* erroneously supposed to be produced by
the kermes, and it may be added that *cochineal,* as the name both of the dye, which
has now largely superseded *grain,* and of the insect which produces it, is derived,
through the Spanish, from *coccum,* the Latin name of the Spanish insect.

91. **French-crowne colour'd**] It is manifest that this means the yellowish color
of a gold coin.   In Quince's reply there is a reference to the baldness which resulted
from an illness supposed to be more prevalent in France than elsewhere.

97. **a mile**] See note on 'league,' in I, i, 175.

97. **without**] See IV, i, 171, 'where we might be Without the perill of the Athenian Law,' where 'without' is used locatively, as here.—ED.

100. **properties**] From 1511, when the Church-wardens of Bassingborne, for a
performance of the play of *Saint George,* disbursed 'xx, s' 'To the garnement-man
for garnements and *propyrts*' (Warton's *Hist. of Eng. Poetry,* iii, 326, cited by
Steevens), to the present day, the 'properties' are the stage requisites of costume
or furniture.   In Henslowe's *Diary* (p. 273, *Sh. Soc.*) there is an 'Enventary
tacker of all the properties for my Lord Admiralles men, the 10 of Marche 1598'

perties, ſuch as our play wants.   I pray you faile me not.   101

Bottom.   We will meete, and there we may rehearſe more obſcenely and couragiouſly.   Take paines, be per-fe&ct, adieu.

Quin.   At the Dukes oake we meete.   105

---

103. *more*] *moſt* Q₁, Cap. Sta. Cam. White ii.

103-105. *Take paines...meete*] Given

to Quince, Coll. ii, iii (MS), Sing. Dyce ii, iii, Ktly, Huds.

103. *paines*] *paine* Ff, Rowe.

---

wherein we find such items as 'j rocke, j cage, j tombe, j Hell mought (*i. e.* mouth).' Again, 'Item, ij marchpanes, & the sittie of Rome.' 'Item, j wooden canepie; owld Mahemetes head,' &c. Halliwell, *ad loc.* and Collier's *Eng. Dram. Poetry*, iii, 159, give abundant references to the use of the word.—ED.

103. **obscenely**] GREY (i, 47): I should have imagined that Shakespeare wrote 'more obscurely,' had I not met with the following distinction in Randolph's *Muses Looking-Glass*, IV, ii (p. 244, ed. Hazlitt): '*Kataplectus.* Obscenum est, quod intra scænam agi non opportuit.' [The point is scarcely worth noting, but I think that 'scænam' is here used not as 'on the stage,' but merely as 'in public,' and the whole phrase is only an ordinary definition of 'obscenum.'—SCHMIDT (*Lex.*) gives a mis-use of 'obscenely' by Costard similar to Bottom's: 'When it comes so smoothly off, so obscenely, as it were, so fit.'—*Love's Lab. L.* IV, i, 145; from which example DEIGHTON infers that Bottom meant 'more *seemly*.'—ED.]

103, 105. **Take pains . . . meete**] COLLIER (*Notes*, p. 100): These words are given to Quince by the Old Corrector, and they seem to belong to him, as the manager of the play, rather than to Bottom. [This plausible suggestion was adopted by Dyce and Hudson with due acknowledgement, by Singer and Keightley without acknow-ledgement: the latter is excusable because he printed from Singer, and more than once expressed his regret that he had followed Singer's text without more careful thought, but Singer has less excuse. I know of no editor who more freely made use, without acknowledgement, of his fellow editors' notes, than Singer, and no one was more bitter than he in denunciation of what he assumed to be Collier's literary dishonesty. Plausible though this present emendation be, it is doubtful if an assumption of the manager's duty be not characteristic of Bottom.—ED.]

105. **Dukes oake**] HALLIWELL: The conjecture is, perhaps, a whimsical one, but the localities here mentioned, 'the Palace Wood' and the 'Duke's Oak,' bear some appearance of being derived from English sources, and, in a certain degree, support an opinion that they were either taken from an older drama, or were names familiar to Shakespeare as belonging to real places in some part of his own country.

105. Garrick thus ended the scene:—

Bot. But hold ye, hold ye, neighbours; are your voices in order, and your tunes ready? For if we miss our musical pitch, we shall be all sham'd and abandon'd.

Quin. Ay, ay! Nothing goes down so well as a little of your sol, fa, and long quaver; therefore let us be in our airs—and for better assurance I have got the pitch pipe.

Bot. Stand round, stand round! We'll rehearse our eplog—Clear up your pipes, and every man in his turn take up his stanza-verse,—Are you all ready?

All. Ay, ay!—Sound the pitch-pipe, Peter Quince.   [Quince *blows.*

Bot. Now make your reverency and begin.

Bot.   Enough, hold or cut bow-ſtrings.   *Exeunt*   106

106. *cut*] *break* or *not* Han. conj. MS ap. Cam.

Song—*for Epilogue.*
By Quince, Bottom, Snug, Flute, Starveling, Snout.

Quin.   Most noble Duke, to us be kind;
Be you and all your courtiers blind,
That you may not our errors find,
        But smile upon our sport.
For we are simple actors all,
Some fat, some lean, some short, some tall;
Our pride is great, our merit small;
        Will that, pray, do at court?

Starv.   The writer too of this same piece,
Like other poets here of Greece,
May think all swans, that are but geese,
        And spoil your princely sport.
Six honest folk we are, no doubt,
But scarce know what we've been about,
And tho' we're honest, if we're out,
        That will not do at court.

[Bottom and Flute in turn continue the song, but the foregoing is as much as need be here repeated.]

Bot. Well said, my boys, my hearts! Sing but like nightingales thus when you come to your misrepresentation, and we are made forever, you rogues! So! steal away now to your homes without inspection, meet me at the Duke's oak—by moonlight—mum's the word.

All. Mum!   [*Exeunt all stealing out.*

106. **hold or cut bow-strings**] CAPELL (*Notes*, p. 102) : This phrase is of the proverbial kind, and was born in the days of archery: when a party was made at butts, assurance of meeting was given in the words of that phrase; the sense of the person using them being that he would 'hold' or keep promise, or they might '*cut his bowstrings,*' demolish him for an archer.—STEEVENS: In *The Ball,* by Chapman and Shirley, 1639 : '*Scutilla.* Have you devices To jeer the rest? *Lucina.* All the regiment of them, or I'll break my bowstrings.'—[II, iii]. The 'bowstring' in this instance may mean only the *strings* which make part of the bow of a musical instrument. [It is quite possible, but there is nothing in the context of the play to lead us to the inference. A 'kit' is mentioned in the preceding act.]—MALONE : To meet, *whether bowstrings hold or are cut,* is to meet in all events. 'He hath twice or thrice cut Cupid's bowstring,' says Don Pedro, in *Much Ado,* III, ii, 10, 'and the little hangman dare not shoot at him.'—STAUNTON and W. A. WRIGHT approve of Capell's explanation; DYCE is unable to determine whether it be true or not.

Actus Secundus. [*Scene I.*]

Enter a Fairie at one doore, and Robin good-
fellow at another.

Rob.   How now ſpirit, whether wander you ?

Fai.   Ouer hil, ouer dale, through buſh, through briar,   5

Ouer parke, ouer pale, through flood, through fire,

I do wander euerie where, ſwifter then y̆ Moons ſphere ;   7

1. Om. Qq.

[Scene I. Rowe et seq. Scene, a
Wood. Theob. A Wood near *Athens*.
Cap.

2. Enter...doore] Enter, from opposite
sides, a Fairy, Cap.

    Fairie] fairy Q₂.

    and] and Puck, or Rowe.

3. at another.] Om. Cap.

4. Rob.] Puck. Rowe et seq.

4. whether] Q₂, F₃.

5–9. *Ouer*...(*green*) Eight lines, Pope
et seq.

5, 6. *through*] *thorough* Q₁, Cap. et
seq.

7. *then*] *than* Q₁.

    Moons] *moones* Steev. Mal. Var.
White ii. *moony* Steev. conj. White i
Huds. *moonĕ's* Ktly.

2, 4, 17, &c. **Robin**] See FLEAY, V, i, 417.

2. **doore**] DYCE (*Rem.* p. 45): The ' doors ' refer to the actual stage-locality, not
to the scene supposed to be represented. . . . More than one editor of early dramas
has mistaken the meaning of *door* in the stage-directions. According to the old copies
of Beau. and Fl.'s *Wit without Money*, III, iv, Luce enters, and ' lays a suit and letter
at the door' (i. e. at the stage-door, at the side of the stage) ; according to Weber's ed.
she ' lays a suit and letter *at a house door*' ! !

4. To read this line rhythmically we must, according to Walker (see note, line 32
of this scene, and *Vers.* 103) and Abbott (§ 466), contract ' spirit ' into *sprite*, and
' whither ' into *whi'er*, thus ; ' Hów now | sprite, whi'er | wánder | yóu.' I am not
sure, however, that the ear is not quite as well satisfied with the line as it stands.—ED.

5, 6. According to GUEST (i, 172), the sameness of rhythm in these lines calls up in
the mind the idea of ' a *multitudinous* succession.'—COLERIDGE, as quoted by Collier,
said that ' the measure had been invented and employed by Shakespeare for the sake
of its appropriateneʃs to the rapid and airy motion of the Fairy by whom the passage
is delivered.' In line 110 of this scene we again have ' through,' where, as here, the
First Quarto has ' thorough,' and is followed by every editor. ' Thorough ' is merely
a mode of spelling of the Early English *thurh*, to indicate the pronunciation of *r*
final, which ABBOTT, § 478, calls ' a kind of " burr." ' Drayton imitated these lines
in his *Nymphidia*, 1627.

7. **Moons**] STEEVENS : Unless we suppose this to be the Saxon genitive case,
moones, the metre will be defective. So in Spenser, *Fairie Queene*, III, i, 15 : 'And
eke through fear as white as whales bone.' Again, in a letter from Gabriel Harvey
to Spenser, 1580 : ' Have we not *God hys wrath* for *Goddes* wrath, and a thousand
of the same stampe, wherein the corrupte orthography in the most, hath been the sole
or principal cause of corrupte prosodye in over-many ?' The following passage

[7. Moons sphere ;]

however, in Sidney's *Arcadia* [Lib. III, p. 262, 1598] may suggest a different read-' ing: ' Diana did begin. What mov'd me to invite Your presence (sister deare) first to my Moony spheare.'—COLLIER : It has been usual to print ' moons ' as two syllables, as if it were to be pronounced like ' whales ' in *Love's Lab. Lost*, V, ii, 332, ' To show his teeth as white as whale's bone,' but all that seems required for the measure is to dwell a little longer than usual upon the monosyllable ' moons.'—With Collier, ABBOTT agrees, and in § 484 gives a long list of examples where ' monosyllables con- taining diphthongs and long vowels are so emphasized as to dispense with an unaccented syllable;' among them is the present line, as well as line 58, ' But *room* Fairy, heere comes Oberon.'—R. G. WHITE (ed. i) and HUDSON adopt ' moony sphere ' on the ground not only that it is a common poetical phrase, but that it is certain Shakespeare would not have allowed, among lines of exquisite music, a line so unrhythmical as this as it stands in the Folio.—W. A. WRIGHT : ' Moon's ' is a disyllable, as ' Earth's ' in *The Tempest*, IV, i, 110 : ' Earth's increase, foison plenty.' Compare, also, IV, i, 107, of the present play, where the true reading is that of the First Quarto : ' Trip we after night's shade.' The Second Quarto and the Folios read ' the night's,' but this disturbs the accent of the verse.—Finally, we have GUEST, whose rhythmical solution differs from all others, and is to me the true one. ' Steevens,' says GUEST (i, 294), ' with that mischievous ingenuity which called down the happy ridicule of Gif- ford, thought fit to *improve* the metre of Shakespeare [by reading *moones*. But the Qq and Ff are] against him. The flow of Shakespeare's line is quite in keeping with the peculiar rhythm which he has devoted to his fairies. It wants nothing from the critic but his forbearance. Burns, in his *Lucy*, has used this section [viz. 5. *p.* of two accents] often enough to give a peculiar charm to his metre :

> " O wat ye wha's : in yon || town | ,
> Ye see the e'enin sun upon ?
> The fairest dame's : in yon || town | ,
> The e'enin sun is shining on."

Moore also, in one of his beautiful melodies, has used a compound stanza, which opens with a stave, like Burns's :

> " While gazing on : the moon's || light | ,
> A moment from her smile I turn'd
> To look at orbs : that, more || bright, |
> In lone and distant glory burn'd." ' '

To those who are familiar with Guest's volumes the concise formula ' 5. *p.*' needs no explanation, but to others it may be as well to explain, in fewest possible words, that it designates a section of a verse composed of two iambs, where a pause takes the place of the second unaccented syllable. As an illustration of ' 5.' alone, without the '*p.*', take the first section of the line, ' I'll loòk | tŏ lìke : if looking liking move '; or take the second section in one of the lines before us : ' I do wan : dĕr èv' | rў whère.' If now '*p.*' be added to ' 5.', we have the scansion of the line under discus- sion, as well as the lines from Burns and Moore : ' Swifter than : thĕ moòns || ⌣ sphère '; ' While gazing on : thĕ moòn's || ⌣ lìght, | &c. In the line in *The Tem- pest*, IV, i, 110 (IV, i, 122 of this ed.; which see, with the notes), this same rule could be applied, were it not that there is authority in the Folios for the insertion of a syl- lable : ' Eàrth's ìncreàse : ⌣fòi | zŏn plèn | ty.' The $F_2F_3F_4$ inserted ' and,' ' Earth's

And I ſerue the Fairy Queene, to dew her orbs vpon the　　8
The Cowſlips tall, her penſioners bee,　　　　　(green.

---

8. *orbs*] *herbs* Grey.　*cups* Wilson.　　　9. *tall*] *all* Coll. MS.

---

increase and foizon plenty,' an addition which is as harmless as it is needless. It is
important, I think, to emphasise this use of these *moræ vacuæ*, or, as Guest calls
them, 'the pauses filling the place of an unaccented syllable,' so familiar to us in
Greek and Latin, especially in Plautus; a neglect of them is a serious defect, I think,
in much of the scansion of Shakespeare's verse.—ED.

7. **sphere**] FURNIVALL (*New Sh. Soc. Trans.* 1877–79, p. 431): At the date of
this play the Ptolemaic system was believed in, and the moon and all the planets and
stars were supposed to be fixed in hollow crystalline spheres or globes. These spheres
were supposed to be swung bodily round the earth in twenty-four hours by the top
sphere, the *primum mobile*, thus making an entire revolution in one day and night.
[Furnivall reprints from Batman on Bartholomeus *de Proprietatibus Rerum*, the
following sections: 'What is the World'; 'Of the distinction of heauen'; 'Of
heauen Emperio'; 'Of the sphere of heauen'; 'Of double mouing of the Planets';
'Of the Sunne'; 'Of the Moone'; Of the starre Comets'; and 'Of fixed Starres.'
For the 'music of the spheres,' see notes, *Mer. of Ven.* V, i, 74, of this edition.—ED.]

8. **dew her orbs**] JOHNSON: The 'orbs' are circles supposed to be made by the
fairies on the ground, whose verdure proceeds from the fairies' care to water them.
Thus, Drayton [*Nymphidia*, p. 162, ed. 1748]: 'And in their courses make that
round, In meadows and in marshes found, Of them so call'd the Fairy ground.'—
STEEVENS: Thus, in Olaus Magnus *de Gentibus Septentrionalibus :* '—similes illis
spectris, quæ in multis locis, præsertim nocturno tempore, suum saltatorium orbem cum
omnium musarum concentu versare solent.' It appears from the same author that these
dancers always parched up the grass, and therefore it is properly made the office of
the fairy to refresh it.—DOUCE (i, 180): When the damsels of old gathered the May
dew on the grass, and which they made use of to improve their complexions, they
left undisturbed such of it as they perceived on the fairy rings; apprehensive that the
fairies should in revenge destroy their beauty. Nor was it reckoned safe to put the
foot within the rings, lest they should be liable to the fairies' power.—HALLIWELL:
These 'orbs' are the well-known circles of dark-green grass, frequently seen in old
pasture-fields, generally called 'fairy-rings,' and supposed to be created by the growth
of a species of fungus, *Agaricus orcades*, Linn. These circles are usually from four to
eight feet broad, and from six to twelve feet in diameter, and are more prominently
marked in summer than in winter.—BELL (*Puck*, &c. iii, 193): The intention seems
rather to point to gathering the dew for the queen to wash her face in; a powerful
means of continual youth. [See Brand's *Popular Antiq.* ii, 480, ed. Bohn; or Dyer,
Folk-lore of Sh. p. 15; see also *The Tempest*, V, i, 44, of this ed.—CAPELL gives what
he terms 'a reverie of long standing' as to the origin of these fairy-rings: in substance it is that if air from the earth rises into the vapours hanging over a meadow
a bubble must be the consequence, and when the bubble breaks the matter of which
it was composed is deposited in a circular form; and as this matter is prolific, the grass
of these circles is more verdant than elsewhere. Evidently Banquo had convinced
Capell that the earth hath bubbles as the water hath. The latest explanation of
these 'fairy-rings' is contained in an Address delivered by J. SIDNEY TURNER at the
Fiftieth Annual Meeting of the South-Eastern Branch of the Brit. Med. Assoc., and

In their gold coats, fpots you fee,   10
Thofe be Rubies, Fairie fauors,

10. *coats*] *cups* Coll. MS.

---

reported in the *Brit. Med. Journ.* 28 July, '94, wherein it is noted that the 'so-called
" fairy-rings " on hills and downs were produced by the better and more vigorous
growth of the grass, owing to the excess of nitrogen afforded by the fungi, which
composed the ring of the previous year.'—ED.]

9. **Cowslips . . . pensioners**] JOHNSON: The cowslip was a favorite among the
fairies. Thus, Drayton, *Nymphidia :* 'And for the Queen a fitting bower, Quoth he
is that fair cowslip-flower, On Hipcut-hill that groweth; In all your train there's not
a fay That ever went to gather May, But she hath made it in her way The tallest
there that groweth.'—T. WARTON: This was said in consequence of Queen Elizabeth's fashionable establishment of a band of military courtiers, by the name of *pensioners.* They were some of the handsomest and *tallest* young men, of the best families and fortune that could be found. Hence, says Mrs Quickly, *Merry Wives,* II,
ii, 79, ' and yet there has been earls, nay, which is more, pensioners.' They gave the
mode in dress and diversions.—KNIGHT: They were the handsomest men of the first
families,—tall, as the cowslip was to the fairy, and shining in their spotted gold coats
like that flower under an April sun.—HALLIWELL: Holles, in his life of the first
Earl of Clare, says : ' I have heard the Earl of Clare say, that when he was pensioner
to the Queen, he did not know a worse man of the whole band than himself; and
that all the world knew he had then an inheritance of 4000*l.* a year.' ' In the month
of December,' 1539, says Stowe, *Annals,* p. 973, ed. 1615, ' were appointed to waite
on the king's person fifty gentlemen, called *Pensioners* or *Speares,* like as they were
in the first yeare of the king; unto whom was assigned the summe of fiftie pounds,
yerely, for the maintenance of themselves, and everie man two horses, or one horse
and a gelding of service.'—W. A. WRIGHT: See Osborne's *Traditional Memoirs of
Queene Elizabeth* (in *Secret History of the Court of James the First,* i, 55). When
Queen Elizabeth visited Cambridge in 1564, she was present at a performance of the
Aulularia of Plautus in the ante-chapel of King's College, on which occasion her
gentlemen pensioners kept the stage, holding staff torches in their hands (Cooper's
Annals of Cambridge, ii, 193).—WALKER (*Crit.* iii, 47): The passage in Milton's
Penseroso, l. 6, alludes to the pensioners' dress: ' —*gaudy shapes*—As thick and numberless As the *gay* motes that people the sunbeams, Or likest hovering dreams, The
fickle pensioners of Morpheus' train.' In those times pensioners, like pursuivants,
progresses, &c., were still things familiar, and naturally suggested themselves as subjects for simile or metaphor. [In 1598 Paul Hentzner saw these pensioners guarding
the queen on each side; they were still ' fifty in number, with gilt halberds.' See
Rye's *England as seen by Foreigners,* p. 105.]

10. **spots**] PERCY: There is an allusion in *Cymbeline* to the same red spots, 'A
mole cinque-spotted, like the crimson drops I' th' bottom of a cowslip.'—HALLI-
WELL: Parkinson, speaking of this species of cowslip (the *Primula veris,* the common
cowslip of the fields), mentions its ' faire yellow flowers, with spots of a deeper yellow at the bottome of each leafe.'—*Paradisus Terrestris,* 1629, p. 244. Collier's
MS Corrector, in altering ' coats ' to *cups* was probably thinking of one of the names
of the crowfoot, which was *golde cup*; but the flowers of the cowslip are not, strictly
speaking, cups.

In thofe freckles, liue their fauors,   12
I muft go feeke fome dew drops heere,
And hang a pearle in euery cowflips eare.
Farewell thou Lob of fpirits, Ile be gon, .   `   15

---

13. *heere*] *here and there* Han. Cap. **clear** Daniel.

---

13. **go seeke**] Cf. 'goe tell,' I, i, 260.

14. **hang a pearle**] For the similarity of this line to 'Hanging on every leaf an orient pearl,' in *Doctor Dodypoll*, and for the inferences thence drawn, see Appendix, *Date of Composition*.—W. A. WRIGHT: There are numberless allusions to the wearing of jewels in the ear, both by men and women, in Shakespeare and in contemporary writers. Cf. *Rom. and Jul.* I, v, 48: 'like a rich jewel in an Ethiope's ear.' Also Marlowe, *Tamburlane*, First Part, I, i; Ben Jonson, *Every Man in his Humour*, IV, vii; *Every Man out of his Humour*, Induction.—HALLIWELL: There are two allusions in this line—first, to the custom of wearing a pearl in the ear; second, to the notion that the dewdrop was the commencing form of the pearl. 'If we believe the naturalists, Pearl is ingendred of the dew of Heaven in those parts of the earth where it is most pure and serene, and the cockle opening at the first rayes of the sun to receive those precious drops, plungeth into the sea with its booty, and conceives in its shell the pearl which resembles the heavens, and imitateth its clearness.'—*The History of Jewels*, &c. 1675. [One of the 'naturalists' just referred to, who assert that pearls originate from dew, is probably Pliny; see Holland's trans. Ninth Booke, cap. xxxv.]

14. After this line, in Garrick's Version, the Fairy sings as follows. The Air is by 'Mr Mich. Arne:'—

  'Kingcup, daffodil and rose,
   Shall the fairy wreath compose;
   Beauty, sweetness, and delight,
   Crown our revels of the night:
    Lightly trip it o'er the green
    Where the Fairy ring is seen;
    So no step of earthly tread,
    Shall offend our Lady's head.

   Virtue sometimes droops her wing,
   Beauty's bee, may lose her sting;
   Fairy land can both combine,
   Roses with the eglantine:
    Lightly be your measures seen,
    Deftly footed o'er the green;
    Nor a spectre's baleful head
    Peep at our nocturnal tread.'

15. **Lob**] JOHNSON: *Lob, lubber, looby, lobcock*, all denote inactivity of body and dulness of mind.—WARTON (*Obs. on Spenser*, i, 120, 1762), in a note on the 'lubbar-fiend' in *L'Allegro*, remarks that this 'seems to be the same traditionary being that is mentioned by Beaumont and Fletcher: "—There's a pretty tale of a witch, that had the devil's mark about her (God bless us!), that had a giant to her son, that was

4

Our Queene and all her Elues come heere anon.   16
    Rob. The King doth keepe his Reuels here to night,
Take heed the Queene come not within his fight,
For *Oberon* is paſsing fell and wrath,
Becauſe that ſhe, as her attendant, hath   20
A louely boy ſtolne from an Indian King,
She neuer had ſo ſweet a changeling,   22

---

16. *her*] *our* Globe (misprint).   21. *boy ſtolne*] *boy ſtollen*, Q₁. *boy, stol'n* Theob. et seq. (except Knt).

---

called Lob-lie-by-the-fire."—*The Knight of the Burning Pestle*' [III, iv, p. 191, ed. Dyce, who says that this remark of Warton that 'Milton confounded the "lubbar fiend" with the sleepy giant in *The Knight of the Burning Pestle* is erroneous.']—COLLIER: The fairy, by this word 'lob,' reproaches Puck with heaviness, compared with his own lightness.—STAUNTON: 'Lob' here, I believe, is no more than another name for *clown* or *fool;* and does not necessarily denote inactivity either of body or mind.—THOMS (*Three Notelets*, p. 89): Dr Johnson's observation in the present place is altogether misplaced. For here the name 'Lob' is doubtless a well-established fairy epithet; and the passage from *The Knight of the Burning Pestle* confirms this. Grimm mentions a remarkable document, dated 1492, in which Bishop Gebhard of Halberstadt, complains of the reverence paid to a spirit called *den guden lubben*, and to whom bones of animals were offered on a mountain.—R. G. WHITE: 'Lob' is here used by the fairy as descriptive of the contrast between Puck's squat figure and the airy shapes of the other fays.—DYCE: R. G. White is probably right. As Puck could fly 'swifter than arrow from the Tartar's bow,' and 'could put a girdle round about the earth in forty minutes,' the Fairy can hardly mean, as Collier supposes, 'to reproach Puck with heaviness.' [Why should a merry wanderer of the night be 'squat'? Omitting this epithet, I think White's and Staunton's explanation the true one. Any elf taller than a cowslip would be a lubber to a fairy that could creep into an acorn-cup. Many references to the use of the word 'lob' will be found in Nares and Halliwell.—ED.]

16. According to the *List of Songs*, &c of the *New Shakspere Soc.*, the foregoing sixteen lines have been set to music by no less than seven different composers.

19. **fell and wrath**] W. A. WRIGHT: 'Fell' is from the Old French *fel*, Italian *fello*, with which *felon* is connected. 'Wrath' is so written for the sake of the rhyme. In Anglo-Saxon wráð is both the substantive 'wrath' and the adjective 'wroth.'

22. **changeling**] JOHNSON: This is commonly used for the child supposed to be left by the fairies, but here for the child taken away. [The *e* mute in this word is pronounced; for other examples, see Abbott, § 487, or Walker, *Crit.* iii, 47.]—DRAKE (*Sh. and His Times*, ii, 325): The Beings substituted [by the Fairies] for the healthy offspring of man were apparently idiots, monstrous and decrepid in their form, and defective in speech. . . . The cause assigned for this evil propensity on the part of the Fairies was the dreadful obligation they were under of sacrificing the tenth individual to the Devil every, or every seventh, year. . . . For the recovery of the unfortunate substitutes thus selected for the payment of their infernal tribute, various charms and contrivances were adopted, of which the most effectual, though the most horrible, was

And iealous *Oberon* would haue the childe   23
Knight of his traine, to trace the Forrefts wilde.
But fhe (perforce) with-holds the loued boy,   25
Crownes him with flowers, and makes him all her ioy.
And now they neuer meete in groue, or greene,
By fountaine cleere, or fpangled ftar-light fheene,   28

24. *of his*] *of this* F₃F₄.

---

the assignment to the flames of the supposed changeling, which it was firmly believed would, in consequence of this treatment, disappear, and the real child return to the lap of its mother. 'A beautiful child of Ouerhaueroe, in Nithsdale,' relates Mr Cromek from tradition, 'on the second day of its birth, and before its baptism, was changed, none knew how, for an antiquated elf of hideous aspect. It kept the family awake with its nightly yells, biting the mother's breasts, and would be neither cradled nor nursed. The mother, obliged to be from home, left it in charge to the servant girl. The poor lass was sitting bemoaning herself,—"Wer't nae for thy girning face I would knock the big, winnow the corn, and grun the meal!"—"Lowse the cradle band," quoth the Elf, "and tent the neighbours, and I'll work yere wark." Up started the elf, the wind arose, the corn was chaffed, the outlyers were foddered, and the hand-mill moved around, as by instinct, and the *knocking mell* did its work with amazing rapidity. The lass and her elfin servant rested and diverted themselves, till, on the mistress's approach, it was restored to the cradle, and began to yell anew. The girl took the first opportunity of slyly telling her mistress the adventure. "What'll we do wi' the wee diel?" said she. "I'll wirk it a pirn," replied the lass. At the middle hour of the night the chimney-top was covered up, and every inlet barred and closed. The embers were blown up until glowing hot, and the maid, undressing the elf, tossed it on the fire. It uttered the wildest and most piercing yells, and, in a moment, the Fairies were heard moaning at every wonted avenue, and rattling at the window-boards, at the chimney-head, and at the door. "In the name o' God bring back the bairn," cried the lass. The window flew up; the earthly child was laid unharmed in the mother's lap, while its grisly substitute flew up the chimney with a loud laugh." '—*Remains of Nithsdale and Galloway Song*, p. 308.

24. **to trace**] This has here, I think, a more restricted meaning than 'to walk over, to pace,' as SCHMIDT defines it, or than 'to traverse, wander through,' as defined by W. A. WRIGHT. There is an intimation here of hunting, of tracing the tracks of game (a tautological expression, but which illustrates the meaning). Spenser thus uses it transitively: 'The Monster swift as word, that from her went, Went forth in hast, and did her footing trace,' *Faerie Queene*, III, vii, line 209; in the present passage it is used intransitively, as in Milton's *Comus*, also with the idea of hunting, although this meaning was not attached to it by HOLT WHITE, who first cited the passage: 'And like a quiver'd Nymph with arrows keen May trace huge forests.'—line 422.—ED.

28. **sheene**] JOHNSON: Shining, bright, gay.—W. A. WRIGHT: Milton, with the passage in his mind, uses 'sheen' as a substantive. See Comus, 1003: 'But far above in spangled sheen, Celestial Cupid, her fam'd son, advanc'd.' [If Milton, at the time of his writing *Comus* had been blind, which he was not, and had listened to

But they do fquare, that all their Elues for feare
Creepe into Acorne cups and hide them there.   30
   Fai.   Either I miftake your fhape and making quite,
Or elfe you are that fhrew'd and knauifh fpirit   32

---

29. *fquare*] *quarrel* Wilson.   32. *fpirit*] *fprite* Q₁, Rowe et seq.
31. *Either*] *Or* Pope +.

---

the reading of *A Mid. N. Dream*, he might have readily accepted 'sheen' as a
noun, with 'starlight' in the genitive, 'starlight's sheen.'—ED.]

   29. **square**] PECK (p. 223): I fancied our author wrote *jar* (a word which sounds
very like *squar*), but then a neighbour of mine, on my showing him the passage,
guessed *squall* to be the true reading. And I should like *squall* as well as *jar*. . . .
Yet, upon the whole, perhaps Shakespeare never wrote 'square' to express a *quarrel*.
For I am sometimes inclined to think he wrote, in most of these places, *sparre.*—
HALLIWELL: 'I square, I chyde or vary, *je prens noyse;* of all the men lyvyng, I
love not to square with hym.'—Palsgrave, 1530. 'To square' was, therefore, properly, to quarrel noisily, to come to high words; but in Shakespeare's time the term
was applied generally in the sense of *to quarrel*, and it was also in common use as a
substantive.—W. A. WRIGHT: In his description of the singing in the church at
Augsburg, Ascham uses the word 'square' in the sense of *jar* or *discord:* 'The præ-
centor begins the psalm, all the church follows without any square, none behind, none
before, but there doth appear one sound of voice and heart amongst them all.'—
Works, ed. Giles, i, 270. [Cotgrave gives: '*Se quarrer.* To strout, or square it,
looke big on 't, carrie his armes a kemboll braggadochio-like.' The examples in
Nares and Dyce (*Gloss.*), which it is needless to repeat here, adequately prove the
meaning *to quarrel.*—ED.]

   29. **that**] For instances of 'that' equivalent to *so that*, see, if need be, ABBOTT,
§ 283.

   31. **Either**] See WALKER (*Vers.* 103) or ABBOTT, § 466, for instances of the contraction, in pronunciation, into monosyllables of such words as *either, neither, whether,
mother, brother, even, heaven,* &c. Another instance is in II, ii, 162.

   32. **spirit**] See Q₁ in Textual Notes. WALKER (*Crit.* i, 193): It may safely be
laid down as a canon that the word 'spirit,' in our old poets, wherever the metre
does not compel us to pronounce it disyllabically, is a monosyllable. And this is
almost always the case. The truth of this rule is evident from several considerations.
In the first place, we never meet with other disyllables—such, I mean, as are incapable
of contraction—placed in a similar situation; the apparent exceptions not being really
exceptions (see *Vers. passim*). Another argument is founded on the unpleasant rip-
ple which the common pronunciation occasions in the flow of numberless lines, interfering with the general run of the verse; a harshness which, in some passages, must
be evident to the dullest ear. Add to this the frequent substitution of *spright* or
sprite for 'spirit' (in all the different senses of the word, I mean, and not merely in
that of *ghost*, in which *sprite* is still used); also *spreet*, though rarely (only in the ante-
Elizabethan age, I think, as far as I have observed); and sometimes *sp'rit* and *sprit*.
For the double spelling, *spright* and *sprite*, one may compare *despight* and *despite;*
which in like manner subsequently assumed different meanings, *despight* being used
for *contempt, despectus*. . . . Perhaps it would be desirable, wherever the word occurs
as a monosyllable, to write it *spright*, in order to ensure the proper pronunciation of

Cal'd Robin Good-fellow.   Are you not hee,   33
That frights the maidens of the Villagree,
Skim milke, and fometimes labour in the querne,   35
And bootleffe make the breathleffe hufwife cherne,

<hr/>

33. *you not*] *not you* Q₁, Cap. Sta.
Cam. White ii.

34. *frights*] *fright* F₃F₄, Rowe+,
Mal. Steev. Var. White i.

　　Villagree] Q₂F₂F₃. *Villageree* Q₁,
Rowe, Theob. Warb. Johns. *Vilagree*

F₄. *villag'ry* Cap. Steev. *villagery*
Han. et cet.

35–38. *Skim...labour...make...make
...Mifleade*] *Skims...labours...makes...
makes...misleads* Mal. conj. Coll. Dyce,
Huds.

35. *fometimes*] *sometime* Dyce ii, iii.

<hr/>

the line. I prefer *spright* to *sprite*, inasmuch as the latter invariably carries with it a spectral association. [See also *Macbeth*, IV, i, 127, or *Mer. of Ven.* V, i, 96, of this edition.]

33–40. In Garrick's *Version* these lines are sung by the Fairy to an Air by Mr Mich. Arne. Many liberties are taken with the text which are not worth reprinting here.

33. **Robin Good-fellow**] See Appendix, *Source of the Plot.*

34, 35, &c. **frights . . . Skim . . . labour**] The Textual Notes will show the grammatical changes adopted by editors in order to give a uniformity which is, after all, needless. Abbott, § 224, after several examples of 'he' and 'she' used for *man* and *woman*, adds that 'this makes more natural the use [in the present line] of "he that," with the third person of the verb.' See also 'are you he that hangs?' —*As You Like It*, III, ii, 375, of this ed. Again, in § 415, after sundry examples of a change of construction caused by a change of thought, Abbott says of the present passage that 'the transition is natural from "Are not you the person who frights?" to "Do not you skim?"'—W. A. Wright: We have in English both constructions. For instance, in *Exodus* vi, 7: 'And ye shall know that I am the Lord your God, which bringeth you out from under the burdens of the Egyptians.' And in *Samuel* v, 2: 'Thou wast he that leddest out and broughtest in Israel.'

34. **Villagree**] W. A. Wright: That is, village population, and so peasantry. Johnson defines it as a district of villages, but it denotes rather a collection of villagers than a collection of villages. No other instance of the word is recorded.

35, 37. **sometimes . . . sometime**] R. G. White (ed. i): Both forms of the word were used indifferently; and in the present case the instinctive perception of euphony, which was so constant a guide of Shakespeare's pen, and in this play, perhaps, more so than in any other, seems to have determined the choice.

35, 36. Johnson: The sense of these lines is confused. Are not you he (says the fairy) that fright the country girls, that skim milk, work in the hand-mill, and make the tired dairy-woman churn without effect? The mention of the mill seems out of place, for she is not now telling the good, but the evil, that he does. I would regulate the lines thus: 'And sometimes make the breathless housewife churn Skim milk, and bootless labour in the quern.' [Rann adopted this 'regulation.'] Or by a simple transposition of the lines. Yet there is no necessity of alteration.—Ritson: Dr Johnson's observation will apply with equal force to his 'skimming the milk,' which, if it were done at a proper time and the cream preserved, would be a piece of service. But we must understand both to be mischievous pranks. He skims the milk

And fometime make the drinke to beare no barme,   37
Mifleade night-wanderers, laughing at their harme,
Thofe that Hobgoblin call you, and fweet Pucke,
You do their worke, and they fhall haue good lucke.   40
Are not you he?
    Rob.   Thou fpeak'ft aright;   42

42, 43. One line, Qq.
42. *Thou*] *The same, thou* Han. *I am
—thou* Johns.   *Fairy, thou* Coll. ii, iii
(MS), Dyce ii, iii, Huds. *Indeed, thou*
Schmidt.

42. *fpeak'ft*] *fpeakeft* Q₁. *speakest me*
Cap.

   fpeak'ft aright] *speakest all aright*
Wagner conj.

---

when it ought not to be skimmed, and grinds the corn when it is not wanted.—HAL-
LIWELL: 'Labour in' is equivalent to 'labour with.' In the old ballad of *Robin
Goodfellow* he is described as working at a malt-quern for the benefit of the maids.
[See Appendix.]

   35. querne] HALLIWELL: A hand-mill for grinding corn; *cwéorn*, Anglo-Saxon.
In its most primitive form it consisted merely of one revolving stone, worked by a
handle, moving in the circular cup of a larger one. Boswell, in his *Tour to the
Hebrides*, speaks of its being in use there: 'We saw an old woman grinding corn
with the *quern*, an ancient Highland instrument, which, it is said, was used by the
Romans'; and Dr Johnson, in his *Tour* to the same place, says, 'when the water-mills
in Skye and Raasa are too far distant, the housewives grind their oats with a *quern*,
or hand-mill.' See Chaucer, *Monke's Tale*, where Sampson is described, 'But now
he is in prifoun in a cave, Ther as thay made him at the querne grynde' [l. 83, ed.
Morris]. In Wiclif's translation of the New Testament a passage is thus rendered:
'tweine wymmen schulen ben gryndynge in o querne, oon schal be taken and the
tother lefte.'—DELIUS unaccountably prefers to interpret 'quern' not as a hand-mill,
but as the ordinary *churn*, 'in which,' he adds, 'milk is turned into butter.'

   37. barme] STEEVENS: A name for *yeast*, yet used in our Midland counties, and
universally in Ireland.—HALLIWELL: This provincial term is still in use in Warwickshire, and in 1847 I observed a card advertising 'fresh barm' in Henley Street, at
Stratford-on-Avon, within a few yards of the poet's birth-place.

   38. Misleade] HALLIWELL: This line was remembered by Milton, 'a wand'ring
fire. . . . Hovering and blazing with delusive light, Misleads th' amaz'd night-wan-
derer from his way.'—*Par. Lost*, ix, 634.

   39. sweet Pucke] TYRWHITT: The epithet is by no means superfluous, as
'Puck' alone was far from being an endearing appellation. It signified nothing
better than *fiend* or *devil*. [See p. 3, *antè*, or Appendix, *Source of the Plot*.]

   42. Thou] JOHNSON: I would fill up the verse which, I suppose, the author left
complete—'*I am*, thou speak'st aright.'—COLLIER (ed. ii): *Fairy* [see Text. Notes]
is from the MS. Some word of two syllables is wanting to complete the line. (Ed.
iii): Here, we may be pretty sure, we have the poet's own word.—DYCE: *Fairy* is
far better than the other attempts that have been made to complete the metre.—R. G.
WHITE (ed. i): Collier's MS is probably correct. But as the pause naturally made
before the reply to the fairy's question may have been intended to take the place of
the missing foot, I have made no addition to the text of the Qq and Ff. ABBOTT
§ 506, agrees with R. C. White, as also the present ED.

I am that merrie wanderer of the night :   43
I ieft to *Oberon,* and make him fmile,
When I a fat and beane-fed horfe beguile,   45
Neighing in likeneffe of a filly foale,
And fometime lurke I in a Goffips bole,
In very likeneffe of a roafted crab :
And when fhe drinkes, againft her lips I bob,
And on her withered dewlop poure the Ale.   50
The wifeft Aunt telling the faddeft tale,

---

46. *of a*] *like a* F₃F₄, Rowe.   47. *bole*] *bowl* F₄.
filly] Q₂Ff, Rowe+, Hal. *filly*   49. *bob*] *bub* Gould.
Q₁ et cet.   50. *withered*] QqFf, Rowe, Cam. ii.
47. *fometime*] *fometimes* F₃F₄,   *wither'd* Pope et cet.
Rowe+.   *dewlop*] *dewlap* Rowe ii.

---

43. See Delius's note on line 154, below.

46. **filly foale**] HALLIWELL: 'Silly' is probably the right reading, in the sense of *fimple*. [For the folk-lore in reference to the various animals whereof the shapes were assumed by fairies, see THOMS's *Three Notelets*, p. 55. I can see no reason for deferting the Folio.—ED.]

47. **Goffips bole**] W. A. WRIGHT: Originally a christening-cup; for a gossip or godsib was properly a sponsor. Hence, from signifying those who were associated at the feftivities of a chriftening, it came to denote generally thofe who were accuf tomed to make merry together. Archbishop Trench mentions that the word retains its original signification among the peasantry of Hampshire. He adds, 'Gossips are first, the sponsors, brought by the act of a common sponsorship into affinity and near familiarity with one another; secondly, these sponsors, who, being thus brought together, allow themselves one with the other in familiar, and then in trivial and idle, talk; thirdly, any who allow themselves in this trivial and idle talk, called in French *commérage*, from the fact that *commère* has run through exactly the same stages as its English equivalent.'—*Eng. Past and Present*, pp. 204-5, 4th ed. Warton, in his note on Milton's *L'Allegro*, 100, identifies 'the spicy nut-brown ale' with the gossip's bowl of Shakespeare. 'The composition was ale, nutmeg, sugar, toast, and roasted crabs or apples. It was called Lambs-wool.' See Breton's *Fantastickes*, *January* : 'An Apple and a Nutmeg make a Gossip's cup.'

48. **very**] That is, true, exact.

48. **crab**] STEEVENS: That is, a wild apple of that name.—HALLIWELL: 'The crabbe groweth somewhat like the apple-tree, but full of thornes, and thicker of branches; the flowers are alike, but the fruite is generally small and very sower, yet some more than others, which the country people, to amend, doe usually roft them at the fire, and make them their winter's junckets.'—Parkinson's *Theat. Botanicum*, 1640.

51. **Aunt**] Unquestionably 'aunt' was at times applied to a woman of low charac ter (see the examples cited by NARES, *s. v.*), but here the adjective 'wisest' shows that it means merely 'the most sedate old woman.' R. G. WHITE calls attention to the common use of 'aunt' as well as 'uncle,' as applied to 'good-natured old people' at

Sometime for three-foot ftoole, miftaketh me,		52
Then flip I from her bum, downe topples fhe,
And tailour cries, and fals into a coffe.
And then the whole quire hold their hips, and loffe,		55

54. *tailour*] *rails, or* Han. Warb. Cap.		55. *loffe*] *laugh* Coll. Cam.
tail-sore Anon ap. Cap.

---

the North and to the old negroes at the South; HALLIWELL cites Pegge as authority for a similar usage in Cornwall.

54. **tailour**] JOHNSON: The custom of crying *tailor* at a sudden fall backwards I think I remember to have observed. He that slips beside his chair falls as a tailor squats upon his board.—HALLIWELL: This explanation by Dr Johnson has not been satisfactorily supported. The expression is probably one of contempt, equivalent to *thief*, and possibly a corruption of the older word *taylard*, which occurs in the Romance of *Richard Cœur de Lion*, where two French justices term that sovereign, when reviling him, a 'taylard,' upon which the choleric monarch instantly clove the skull of the first and nearly killed the second. The Elizabethan use of the term, as one of contempt, appears to be confirmed by the following passage in *Pasquil's Night Cap*, 1612: 'Theeving is now an occupation made, Though men the name of tailor doe it give.'—BELL (iii, 194): It may be thought fanciful, but not altogether improbable, to explain this custom by one equally low at the present day, as when black-guards press rudely the hats of passengers over their eyes; and of a female's cry: *bonnet her.* So that I should read: *tail her.*—PERRING (p. 113) would read *traitor*, on the score that it would be much more consistent with the aunt's 'disposition, her age, her dignity, and, I may add, with the serious nature of her story, to raise against her invisible foe that fierce cry of "traitor," which was wont to be raised against suspected political malcontents, . . . in using which the "wisest aunt" associated herself with kings and queens and empresses of the earth.' [It is difficult to believe that this is put forth seriously. A discussion was started in *Notes & Queries* (7th S. ii, 385, 1886) by J. BOUCHIER asking 'Why tailor any more than cobbler, hosier, or barber?' To which A. H. (7th S. iii, 42) replied that a tailor's assistance would be needed when 'a sudden tumble eventuates in the rent of a necessary garment.' This interpretation was pronounced untenable by C. F. S. WARREN, M. A. (Ib. p. 264), 'because a sudden fall backwards will not split petticoats as it will trousers.'—HYDE CLARKE adds, with more truth than appositeness, that 'there were tailors for women in most countries of the West and East, as there still are in many. In London tailors make riding breeches for women.' In this diverting discussion, from Halliwell downwards, it needs scarcely an ounce of civet to sweeten the imagination, if it be suggested that the slight substitution of an *e* for an *o* in the word 'tailor' will show that, as boys in swimming take a '*header*,' the wisest Aunt was subjected to the opposite.—ED.]

55. **quire**] DYCE: A company, an assembly. [With a suggestion here of its meaning of acting in concert.—ED.]

55. **loffe**] CAPELL (104): A rustic sounding of *laugh*, to whose spelling all the elder editions assimilate 'cough,' and its sound should incline to it.—HALLIWELL: This is the ancient pronunciation of the word. Ben Jonson, in *The Fox*, makes *daughter* rhyme with *laughter;* and in the old nursery ballad of *Mother Hubbard*, after she had bought her dog a 'coffin' she came home and found he was *loffing!* In

And waxen in their mirth, and neeze, and fweare,   56
A merrier houre vvas neuer wafted there.
But roome Fairy, heere comes *Oberon.*   58

56. *waxen*] *yexen* Farmer, Sing.   *room, room* Marshall.
58. *roome*] *make room* Pope+, Cap.   58. *Fairy*] *Faery* Johns. conj. Steev.
Ktly. *room now* Dyce ii, iii, Huds.   Mal. Knt. *Faëry* Sing. i, ii, Sta.

---

some line in Harrington's *Most Elegant and Wittie Epigrams*, 1633, *lafter* (laughter) rhymes with *after*. There appears to have been some variation as to the pronunciation of the word. Marston, in *The Parasitaster*, 1606, mentions a critic who vowed 'to leve to posteritie the true orthography and pronunciation of laughing.' [I doubt if Halliwell's quotation from Marston be exactly germane. The 'critique' to whom it refers was in 'the Ship of Fools,' and his puzzle was, I think, not the mere spelling or pronunciation of the word *laugh* or *laughter*, but what combination of letters would express the sound of *laughing*, a puzzle which need not be restricted to the days of Elizabeth. It is almost impossible to fix the exact pronunciation, in the XVIth or XVIIth century, of *laugh* or *laughter*, especially as there are indications of a change which was at this time creeping over these words as well as such words as *daughter*, *slaughter*, and the like. See ELLIS (*Early Eng. Pronunciation*, p. 963). As a boy of 16, in Warwickshire, Shakespeare may have heard a pronunciation of these words quite different from that which he heard in his mature years, in London. See *Ibid.* p. 144. In the present spelling I think we have, as Capell suggests, a phonetic attempt to reproduce the 'robustious' laughter of boors, just as, nowadays, Chawbacon's laughter is spelled 'Haw! haw!' and 'loffe' should be retained in the text. WHALLEY refers to Milton's *L'Allegro:* 'And Laughter holding both his sides,' line 32.—ED.]

56. **waxen**] JOHNSON: That is, *increase*, as the moon *waxes*.—STEEVENS: Dr Farmer observes to me that 'waxen' is probably corrupted from *yoxen* or *yexen*, to hiccup. It should be remembered that Puck is at present speaking with an affectation of ancient phraseology. SINGER pronounces Farmer's needless emendation to be 'undoubtedly the true reading,' and adopts, without acknowledgement, *more suo*, Steevens's remark about the affectation of ancient phraseology, of which affectation I see no proof.—ED.

56. **neeze**] W. A. WRIGHT: That is, *sneeze;* A.-S. *niesan;* Germ. *niesen.* Similarly, we find the two forms of the same word : 'knap' and 'snap'; 'top' and 'stop'; 'cratch' and 'scratch'; 'lightly' and 'slightly'; 'quinsy' and 'squinancy.' In *2 Kings* iv, 35, the text originally stood, 'And the child neesed seven times,' but the word has been altered in modern editions to 'sneezed.' In *Job* xli, 18, however, 'neesings' still holds its place. Compare *Homilies* (ed. Griffiths, 1859), p. 227 : 'Using these sayings: such as learn, God and St. Nicholas be my speed; such as neese, God help and St. John; to the horse, God and St. Loy save thee.' Cotgrave gives both forms, 'Esternuer. To neeze or sneeze.'

58. **roome Fairy**] JOHNSON: Fairy, or Faery, was sometimes of three syllables, as often in Spenser.—DYCE (ed. ii) : I have inserted *now* for the metre's sake, which is surely preferable to the usual modern emendation, '*make* room.' To print 'But room *Faëry*' is too ridiculous.—NICHOLSON (*N. & Qu.* 3d Ser. V, 49, 1864) suggests *·oomer*, a sea-phrase, 'which, in speaking of the sailing of ships, meant to alter the course, and go free of one another.' Thus, in Hakluyt, Best, narrating how in

Fair. And heere my Miſtris :
Would that he vvere gone.   60

> *Enter the King of Fairies at one doore with his traine,*
> *and the Queene at another with hers.*

Ob. Ill met by Moone-light,
Proud *Tytania.*
    Qu. What, iealous *Oberon?* Fairy skip hence.   65

59, 60. One line, Qq, Pope et seq.

60. *he*] *we* Ff, Rowe, Johns.

[Scene II. Pope+, Var. Knt, Sing. Ktly.

61. the King] King F_3F_4.

63, 64. One line, Qq, Pope et seq.

64. Tytania] Titania F_3F_4, Rowe et seq.

65. Qu.] Tit. Cap. et seq. (subs.).

    Fairy skip] *fairies, skip* Theob Han. Warb. Johns. Coll. Sing. White. Sta. Dyce, Cam. Ktly.

Frobisher's second voyage the ships were caught in a storm amidst drifting icebergs, says : 'We went roomer [off our course, and more before the wind] for one (iceberg), and loofed [luffed up in the wind] for another.' Hence *roomer* aptly expresses one of two courses which must be adopted by an inferior vessel when it meets another, whose sovereignty entitles her to hold on her way unchecked. The fairy had luffed, and so stayed her course to speak with Puck. Having interchanged civilities, Here, says Puck, comes Oberon, bearing down upon you full sail; do you, vassal as you are of a power that he is unfriends with, alter your course; go off before the wind, and free of him. In a word, *roomer.* If objection be made to the use, by Puck, of a sea-phrase, I would quote the inlander Romeo, who speaks of the high top-gallant of his joy. ABBOTT, § 484, who gives more than twenty pages to examples of the lengthening of words in scanning, has 'room' in the present passage among them. [No change is absolutely necessary. The break in the line affords, I think, sufficient pause to fill up the metre.—ED.]

63. See Delius's note on line 154, below.

65. **Fairy skip**] THEOBALD silently changed this to *Fairies skip,* and the Text. Notes show how generally he has been followed by the best editors, who have urged as their plea : first, the ease with which the final *s* of *Fairies* might have been lost to the ear in the first *s* of 'skip.'—WALKER (*Crit.* i, 265) cites this passage in his *Article* on the omission of the *s,* and says the words are 'surely' '*Fairies* skip.'—COLLIER finds no reason why a particular fairy should be addressed unless we suppose that Oberon is referred to; but this DYCE (ed. i) disproves by citing the following line : ' I have forsworn *his* bed and company.' Secondly, Titania evidently wishes her whole train to withdraw, because at line 149 she distinctly says, ' Fairies away.'—B. NICHOLSON (*N. & Qu.* 4th Ser. V, 56) questions the conclusiveness of this last command, because the circumstances may have changed, and while the king and queen have been wrangling the attendant courtiers and maids of honour may have been frisking, flirting, intermingling, and have become scattered, and her majesty wishes to recall them.—CAPELL (p. 104) is the only editor who justifies the Folio, and, I think, with adequate reason for so trifling a question, which, after all, is mainly for the eye; Capell says that the fairy thus addressed is Titania's ' leading fairy, her gentleman-usher, whose moving-off would be a signal for all the rest of the train.'—COLLIER

I haue forſworne his bed and companie.   66

   Ob.   Tarrie raſh Wanton ; am not I thy Lord ?

   Qu.   Then I muſt be thy Lady : but I know

When thou vvaſt ſtolne away from Fairy Land,

And in the ſhape of *Corin,* ſate all day,   70

Playing on pipes of Corne, and verſing loue

To amorous *Phillida.*   Why art thou heere

Come from the fartheſt ſteepe of *India* ?   73

---

69. *vvaſt*] Ff, Rowe, Pope, Ktly.   70. *ſate*] *ſat* QqF₄.
haſt Qq, Theob. et seq.   73. *ſteepe*] *ſteppe* Q₁.   *step* Cap.

reports an emendation by HARNESS: 'Fairies *keep*'; and DYCE adds one of his
own: 'Fairies *trip.*'

  69. **vvast**] KEIGHTLEY (*N. & Qu.* 2d Ser. IV, 262; *Exp.* 131) is the only editor
who upholds the reading of the Ff.   He maintains that by 'wast' Titania means that
Oberon 'stole away' only once, whereas 'hast' of the Qq implies a habit.   'Moreover, Shakespeare invariably employs the verb substantive with "ſtolen away," except
in the case of a doubly-compound tense.'

  71. Corne] RITSON: The shepherd boys of Chaucer's time had '—many ſlowte
and liltyng horne, And pipes made of grene corne.'—[*House of Fame,* iii, 133, ed
Morris.   Albeit that 'corn' is, in England, applied to any cereal, yet the 'pipes of
corn' on which Corin played were probably the same as the '*oaten* straws' on which
'the ſhepherds pipe' in *Love's Lab. Lost,* V, ii, 913; *avena* is used in Latin in
the same way.   The 'corne' mentioned in line 98, below, is, of course, not oats,
but wheat.—ED.]

  72. Phillida] F. A. MARSHALL (p. 369): Do not these lines rather militate
against the idea of Oberon and Titania being such very diminutive people ?   Could
a manikin hope to impress the 'amorous Phillida' ?   Again, Oberon's retort on
Titania seems to imply that she was capable of inspiring a passion in that prototype
of all Don Juans, Theseus.   Perhaps these fairies were supposed to possess the power
of assuming the human shape and size, or, what is more likely, to Shakespeare they
were so entirely creatures of the imagination that they never assumed, to his mind's
eye, any concrete form.   [In the first place, if we must resort to a prosaic interpretation, Marshall's query is answered by the fact that Oberon assumed 'the shape of
Corin'; in the second place, one of the strokes of humour in this whole scene, between atomies who can creep into acorn-cups, and for whom the waxen thigh of a bee
affords an ample torch, lies in the assumption by them of human powers and of super-
human importance.   Not only is Titania jealous of the bouncing Amazon, but this
their quarrel influences the moon in the sky, changes the seasons, and affects disas-
trously the whole human race.   There is a touch of the same humour, but deeply
coarsened, in the scandal which Gulliver's conduct started when he was at the court
of Laputa.—ED.]

  73. steepe] WHITE (ed. i): *Steppe,* of the first Quarto, is 'but a strange accident,
for the word was not known in Shakespeare's day.'—W. A. WRIGHT: It is danger-
ous to assert a proposition which may be disproved by a single instance of the contrary.   There is certainly no *a priori* reason why the present passage should not furnish that instance, inasmuch as a word of similar origin, 'horde,' was perfectly well

But that forſooth the bouncing *Amazon*
Your buskin'd Miſtreſſe, and your Warrior loue,   75
To *Theſeus* muſt be Wedded ; and you come,
To giue their bed ioy and proſperitie.
      Ob.   How canſt thou thus for ſhame *Tytania*,
Glance at my credite, vvith *Hippolita* ?
Knowing I knovv thy loue to *Theſeus* ?   80
Didſt thou not leade him through the glimmering night
From *Peregenia,* whom he rauiſhed ?
And make him vvith faire Eagles breake his faith   83

---

75. *buskin'd*] *bukskined* so quoted many times by Hermann.
81. *through the glimmering night*] *glimmering through the night* Warb.

82. Peregenia] *Perigune* Theob. Pope ii. *Periguné* Theob. ii. *Perigynè* Han. *Perigouna* White.
83. *Eagles*] *Ægle* Rowe et seq.

---

known in England at the beginning of the 17th century. On the other hand, too much weight must not be attached to the spelling of Q_I, for in III, ii, 88, 'sleep' is misprinted *slippe*. [It is almost needless to restrict to Q_I this variation in spelling; it applies to the Folios as well; in the very passage referred to by W. A. Wright, *sleep* is printed 'slip' in all the Folios, and was first corrected by Rowe. Accordinging to the *Century Dictionary, steppe* was introduced into the scientific literature of Western Europe by Humboldt, and in popular use it is nowhere applied but to regions dominated by Russia; there is no need of its use, I think, in the present passage.—ED.]

76. **must**] Simply definite futurity, as in Portia's, 'Then must the Jew be merciful.' For other instances, see ABBOTT, § 314.

79. **Glance**] W. A. WRIGHT: That is, *hint at, indirectly attack.* Thus, in Bacon's *Advancement of Learning,* i, 7, § 8 (p. 57, ed. Wright) : ' But when Marcus Philosophus came in, Silenus was gravelled and out of countenance, not knowing where to carp at him; save at the last he gave a glance at his patience towards his wife.'

81. **glimmering**] WARBURTON upholds his wanton emendation by asserting that Titania conducted Theseus 'in the appearance of fire through the dark night.' Had he forgotten 'The west yet glimmers with some streaks of day,' *Macb.* III, iii, 5 ? —ED.

82. **Peregenia**] STAUNTON : ' This Sinnis had a goodly faire daughter called Perigouna, which fled away when she saw her father slaine. . . . But Theseus finding her, called her, and sware by his faith he would use her gently, and do her no hurt, nor displeasure at all.'—North's *Plutarch* [p. 279, ed. Skeat. MALONE thinks that Shakespeare changed the name for the sake of rhythm, but the rhythm remains the same with either spelling, and we are by no means certain that Shakespeare took the name from Plutarch, or that he ever saw the name as it is thus spelled by the printer.—ED.]

83. **Eagles**] STAUNTON : ' For some say that Ariadne hung herself for sorrow, when she saw that Theseus had cast her off. Other write, that she was transported by mariners into the ile of Naxos, where she was married unto Œnarus, the priest of Bacchus; and they think that Theseus left her, because he was in love with another

With *Ariadne*, and *Atiopa*?

   Que. Thefe are the forgeries of iealoufie,   85
And neuer fince the middle Summers fpring
Met vve on hil, in dale, forreft, or mead,
By paued fountaine, or by rufhie brooke,
Or in the beached margent of the fea,   89

---

84. Atiopa] Antiopa QqFf.
86. *the*] *that* Han. Warb. Cap.
   fpring] *prime* D. Wilson.

89. *in the*] QqFf, Rowe, Hal. Sta.
Dyce, Cam. Wh. ii. *on the* Pope et
cet.

---

as by these verses should appear: Ægles, the nymph, was loved of Theseus, Who was the daughter of Panopeus.'—North's *Plutarch* [p. 284, ed. Skeat] DYCE (*Remarks*, p. 46): In Shakespeare's time it was not uncommon to use the genitive of proper names for the nominative. At an earlier period this practice prevailed almost universally. Even in a modern book, and the work of a scholar, we find, 'a natural grotto, more beautiful than Ælian's description of Atalanta's, or that in Homer, where *Calypsos* lived.'—Amory's *Life of John Buncle*, i, 214, ed. 1756. [Is it not a little misleading to call this added final *s* the sign of the 'genitive case'? Walker's long list (*Crit.* i, 233) shows the frequency with which the final *s* was added, not only to proper names, but to all words. If it be the genitive case in 'Eagles,' why should this solitary genitive be surrounded by the nominative forms 'Peregenia,' 'Ariadne,' and 'Antiopa'? We need some other cause than inflection, I think, to explain this sibilant tendency, be it in some peculiar flourish in writing, or be it in some delicate phonetic demand, which our modern ears have lost.—ED.]

   84. **Atiopa**] STAUNTON: 'Philochurus, and some other hold opinion, that [Theseus] went thither with Hercules against the Amazons: and that to honour his valiantness, Hercules gave him Antiopa the Amazone. . . . Bion . . . saith that he brought her away by deceit and stealth, . . . and that Theseus enticed her to come into his ship, who brought him a present; and so soon as she was aboord, he hoysed his sail, and so carried her away.'—North's *Plutarch* [p. 286, ed. Skeat].

   86. **the**] WARBURTON: We should read *that*. It appears to have been some years since the quarrel first began.—CAPELL adopts this emendation, and also believes that the midsummer was 'a distant one'; it is not easy to see on what ground. Perhaps on the supposition that the quarrel began at the birth of the little Indian boy, or when Oberon piped to amorous Phillida. But there is no intimation of it in the text. — ED.

   86. **middle Summers fpring**] CAPELL (*Notes*, ii, 104) understands this as the spring preceding the 'midsummer in which the quarrel took place.'—But STEEVENS shows that it means 'the beginning of *middle* or *mid* summer.' 'Spring,' for *beginning* is used in *2 Hen. IV:* IV, iv, 35: 'As flaws congealed in the spring of day.' Also in *Luke* i, 78: 'Whereby the dayspring from on high hath visited us.'

   88. **paued fountaine**] HENLEY: That is, fountains whose beds were covered with pebbles, in opposition to those of the rushy brooks, which are oozy.—KNIGHT: 'Paved' is here used in the same sense as in the 'pearl-paved ford' of Drayton, the 'pebble-paved channel' of Marlowe, and the 'coral-paven bed' of Milton.

   89. **in**] HALLIWELL: That is, *within;* unnecessarily changed by Pope.—DYCE (ed. i): 'In' was often used for *on*. So in *Cymb.* III, vi, 50: 'Gold strew'd i' the

To dance our ringlets to the whiſtling Winde,   90
But vvith thy braules thou haſt diſturb'd our ſport.
Therefore the Windes, piping to vs in vaine,   92

---

floor' (where Boswell cites, from the Lord's Prayer, 'Thy will be done in earth').—
1863. Mr W. N. Lettsom observes to me: 'Is it not hazardous to retain "*in* the
beachèd margent," when Shakespeare has written, in *A Lover's Complaint*, "*Upon*
whose margent weeping she was set"? It is true that *in* is frequently used before
earth, mountain, hill, and the like; but this scarcely warrants "*in* the floor," for the
word *floor* seems to give exclusively the notion of *surface*, while the other words
express also *abode* or *locality*. It is, besides, not merely more or less probable, but
positively certain, that printers confound these prepositions, as, for instance, in *Rich.
III:* V, i, "To turn their own points *on* their masters' bosoms," where the Ff have
in, the Qq *on.*' [See 'falling in the Land,' line 94, below. Mrs Furness's *Concordance* gives many instances where 'in' is used where we should use *on.* The question
of changing the present text to *on* should be weighed only by an editor of a modern text, for the use of young beginners.—ED.]

89. **beached**] W. A. WRIGHT: That is, formed by a beach, or which serves as a
beach. Cf. *Timon,* V. i, 219: 'Upon the beached verge of the salt flood.' For similar instances of adjectives formed from substantives, see 'guiled,' *Mer. of Ven.* III,
ii, 97; 'disdain'd,' *1 Hen. IV:* I, iii, 183; 'simple-answer'd,' that is, simple in your
answer, furnished with a simple answer, which is the reading of the Ff in *Lear,* III,
vii, 43; 'the caged cloister,' the cloister which serves as a cage, *Lover's Com.* 249;
'ravin'd,' for ravenous, *Macb.* IV, i, 24; 'poysened,' for poisonous, Lily, *Euphues,*
p. 196 (ed. Arber): 'Nylus breedeth the precious stone and the poysened serpent.'
[Also 'the delighted spirit,' *Meas. for Meas.* III, i, 121.]

89. **margent**] HALLIWELL: One of the old forms of *margin,* of so exceedingly
common occurrence as merely to require a passing notice. It seems to have first
come in use in the sixteenth century, and has only become obsolete within the past
generation, many instances of it occurring in writers of the time of the first Georges.
—W. A. WRIGHT: Shakespeare never uses *margin*.

90. **ringlets**] W. A. WRIGHT refers these 'ringlets' to the 'orbs' in line 8, above.
Can they be the same? The fairy rings 'whereof the ewe not bites' are found where
grass grows green in pastures, but not by the paved fountain nor by rushy brook, and
never in the beached margent of the sea, on those yellow sands where, of all places,
from Shakespeare's day to this, fairies foot it featly, and toss their gossamer ringlets
to the whistling and the music of the wind.—ED.

91. **braules**] W. A. WRIGHT: That is, quarrels. Originally, a *brawl* was a
French dance, as in *Love's Lab. L.* III, i, 9: 'Will you win your love with a French
brawl?' And it was a dance of a violent and boisterous character, as appears by the
following extract from Cotgrave: 'Bransle: m. A totter, swing, or swidge; a shake,
shog, or shocke; a stirring, an vncertain and inconstant motion; ... also, a brawle,
or daunce, wherein many (men and women) holding by the hands sometimes in a
ring, and other whiles at length, moue altogether.' It may be, however, that there is
no etymological connexion between these two words, which are the same in form.—
MURRAY (*New Eng. Dict.*) separates this word from *brawl,* a French dance; the
origin and primary sense of the former are uncertain.

92. **piping to us in vain**] 'We have piped unto you, and ye have not danced.'
—*Matt.* xi, 17.

As in reuenge, haue fuck'd vp from the fea   93
Contagious fogges : Which falling in the Land,
Hath euerie petty Riuer made fo proud,   95
That they haue ouer-borne their Continents.
The Oxe hath therefore ftretch'd his yoake in vaine,
The Ploughman loft his fweat, and the greene Corne
Hath rotted, ere his youth attain'd a beard :
The fold ftands empty in the drowned field,   100
And Crowes are fatted vvith the murrion flocke,
The nine mens Morris is fild vp with mud,   102

---

95. *Mud*] QqFf, Rowe i, Kal, *Marr* Theob. i, Cam. *murrain* Theob. ii et
Rowe ii et cet.   cet.
    petty] Ff, White. *paltry* Bell.   102. *nine mens Morris*] *nine mens
pelting Qq et cet.   morris* F_3. *Nine-mens-morris* F_4, Rowe,
99. *his youth*] *its youth* Pope, Han.   Dyce ii, iii. *nine-mens morris* Pope.
Warb.   *nine-mens morrice* Cap.
101. *murrion*] QqFf, Rowe, Pope,

---

95. **Hath**] For other examples of singular verbs following relatives, when the ante-
cedents are plural, see ABBOTT, § 247.—W. A. WRIGHT : ' Hath,' following ' Land,'
is here singular by attraction.

95. **petty**] I can see no reason why we should here desert the Folio, especially as
there is, according to all authorities, from Dr Johnson down, a tinge of contempt in
the ' pelting ' of the Qq, which is here needless; insignificance is all-sufficient.—ED.

96. **they**] W. A. WRIGHT : The plural follows loosely, as representing the collection of individual rivers.

96. **Continents**] JOHNSON : Borne down the banks that contain them. So in
Lear, III, ii, 58 : ' —close pent-up guilts Rive your concealing continents.'

97, &c. WARBURTON maintains that the assertion that Shakespeare borrowed the
description of the miseries of the country from Ovid (*Met.* V, 474–484) will admit
of no dispute. No editor, as far as I know, has taken any notice of this indisputable
instance of Shakespeare's thieving propensity, except HALLIWELL, who gives at
length Golding's translation, which he who has time to waste may read on p. 64 of
that Translation, ed. 1567.—ED.

101. **murrion**] No one familiar with the Old Testament needs to be told the
meaning of this word; see *Exodus* ix, 3 ' For the variety of the spelling', says W
A. WRIGHT, ' compare *Lear*, I, i, 65, where the Ff are divided between " champains "
and " champions." '

102. **nine mens Morris**] JAMES : In that part of Warwickshire where Shakespeare was educated, and in the neighbouring parts of Northamptonshire, the shepherds and other boys dig up the turf with their knives to represent a sort of imperfect
chess-board. It consists of a square, sometimes only a foot in diameter, sometimes
three or four yards. Within this is another square, every side of which is parallel to
the external square, and these squares are joined by lines drawn from each corner of
both squares, and the middle of each line. One party, or player, has wooden pegs,
the other stones, which they move in such a manner as to take up each other's men as
they are called, and the area of the inner square is called the pound, in which the

And the queint Mazes in the wanton greene ,   103

103. *queint*] *quaint* Johns.   103. *in*] *on* Coll. MS.

men taken up are impounded.   These figures are always cut upon the green turf or
leys, as they are called, or upon the grass at the end of ploughed lands, and in rainy
seasons never fail to be choked up with mud.—ALCHORNE: A figure is made on the
ground by cutting out the turf, and two persons take each nine stones, which they
place by turns in the angles, and afterwards move alternately, as at chess or draughts.
He who can place three in a straight line may then take off any one of his adversary's,
where he pleases, till one, having lost all his men, loses the game.   [This variety of
the game corresponds with what W. A. WRIGHT says he has seen in Suffolk : 'Three
squares, instead of two, are drawn one within the other, and the middle points of the
parallel sides are joined by straight lines, leaving the inmost square for the pound.
But the corners of the squares are not joined.   The corners of the squares and the
middle points of the sides are the places where the men may be put, and they move
from place to place along the line which joins them.'—COTGRAVE gives s. v. *Merelles,*
' The boyish game called Merills, or fiue-pennie Morris; played here most commonly
with stones, but in France with pawnes, or men made of purpose, and tearmed
Merelles.'—DOUCE (i, 184): This game was sometimes called *the nine mens merrils,*
from *merelles* or *mereaux,* an ancient French word for the jettons or counters, with
which it was played.   The other term, *morris,* is probably a corruption suggested by
the sort of *dance* which in the progress of the game the counters performed.   In the
French *merelles* each party had three counters only, which were to be placed in a line
in order to win the game.   It appears to have been the *Tremerel* mentioned in an old
fabliau. . . . Dr Hyde thinks the morris or merrils was known during the time that
the Normans continued in possession of England, and that the name was afterwards
corrupted into *three mens morals* or *nine mens morals.*   If this be true, the conversion
of *morals* into *morris,* a term so very familiar to the country people, was extremely
natural.   The doctor adds that it was likewise called *nine-penny,* or *nine-pin miracle,*
three-penny morris, five-penny morris, nine-penny morris. or *three-pin, five-pin,* and
nine-pin morris, all corruptions of *three-pin,* &c. merels.—Hyde, *Hist. Nerdiludii,* p.
202 —STAUNTON: Whether the game is now obsolete in France, I am unable to say ;
but it is still practised, though rarely, in this country, both on the turf and on the
table, its old title having undergone another mutation and become ' Mill.'   [See also
Nares, *Glossary ;* Strutt's *Sports and Pastimes,* p. 279, sec. ed. ; Halliwell ad loc.
&c., &c.]

     103. queint Mazes] STEEVENS: This alludes to a sport still followed by boys,
i. e. what is now called *running the figure of eight.*—W. A. WRIGHT: But I have
seen very much more complicated figures upon village greens, and such as might
strictly be called mazes or labyrinths.   On St. Catherine's Hill, Winchester, ' near the
top of it, on the north-east side, is the form of a labyrinth, impressed upon the turf,
which is always kept entire by the coursing of the sportive youth through its mean-
derings.   The fabled origin of this Dædalæan work is connected with that of the
Dulce Domum song.'—Milner, *Hist. of Winchester,* ii, 155.—HALLIWELL gives a
wood-cut from an old print of *The Shepherd's Race* or *Robin Hood's Race,* ' a maze
which was formerly on the summit of a hill near St. Ann's Well, about one mile from
Nottingham.   The length of the path was 535 yards, but it was all obliterated by the
plough in the year 1797, on the occasion of the enclosure of the lordship of Sneinton.'

For lacke of tread are vndiſtinguiſhable.
The humane mortals want their winter heere, 105

<table>
<tr><td>105, 106. Transposed to follow line 112. Elze (Notes, 1880, p. 41).</td><td>105. winter heere,] winter here. Q₁. winter chear [i. e. cheer] Theob. conj.</td></tr>
</table>

105, 106. Transposed to follow line
112. Elze (*Notes*, 1880, p. 41).
  105. *want ... heere,*] *want, ... here ;*
White ii. *wail...here ;* Kinnear.

105. *winter heere,*] *winter here.* Q₁.
winter chear [i. e. *cheer*] Theob. conj.
Han. Sing. ii, Coll. ii, Hal. Dyce ii, iii.
winter hoar ; Herr. *winter hire* D. Wilson. *winter gear* Brae ap. Cam.

---

105. humane mortals] That is, mankind as distinguished from fairies ; Titania, herself immortal, afterwards (line 140) refers to the mother of her changeling as ' being mortal ' ; and a fairy addresses Bottom with, ' Hail, mortal, hail !' thus indicating that fairies were not mortal. But STEEVENS, unmindful of the fact that Shakespeare's fairies are unlike all other fairies, especially unlike the fairies of *Huon of Bordeaux*, or of Spenser, started a controversy by asserting that ' fairies were not *human*, but they were yet *subject to mortality*,' and ' that " human " might have been here employed to mark the difference between *men* and *fairies*.' The controversy which followed, which may be found in the *Variorum* of 1821, and in Ritson's *Quip Modest*, p. 12, it would be a waste of time to transfer to these pages, and which, since Ritson was one of the disputants, it would be superfluous to characterise as acrimonious.—ED.

105. want their winter heere] THEOBALD : I once suspected it should be ' want their winter chear,' *i. e.* their jollity, usual merry-makings at that season.—WARBURTON : It seems to me as plain as day that we ought to read ' want their winters heried,' *i. e.* praised, celebrated ; an old word, and the line that follows shows the propriety of it here.—CAPELL (*Notes*, ii, 104) : That is, their accustomed winter, in a country thus afflicted ; to wit, a winter enlivened with mirth and distinguished with grateful hymns to their deities.—JOHNSON proposed that we should read ' want their *wonted year*,' and transposed the lines as follows : 105, 111–118, 106, 107, 108, 110, 109, 119. His conjecture re-appeared only in the Variorums of 1773, 1778, and 1785 ; it was omitted, after his death, from the Variorum of 1793.—MALONE'S note in the Variorum of 1790, which is sometimes quoted as ' Malone's own,' is merely a combination of the note of Theobald and Capell.—KNIGHT : The ingenious author of a pamphlet, *Explanations and Emendations*, &c., Edinburgh, 1814, would read : ' The human mortals want ; their winter here, No night,' &c. The writer does not support his emendation by any argument, but we believe that he is right. [Knight adopted this punctuation in his text.] The swollen rivers have rotted the corn, the fold stands empty, the flocks are murrain, the sports of summer are at an end, the human mortals *want*. This is the climax. Their winter is *here*—is come—although the season is the latter summer [how does this accord with the title of the play ?—ED.] or autumn ; and in consequence the hymns and carols which gladdened the nights of a seasonable winter are wanting to this premature one.—R. G. WHITE (ed. i) : It is barely possible that ' want ' is a misprint for *chant*, and that Titania, wishing to contrast the gloom of the spurious, with the merriment of the real, Winter, says, ' when their Winter is here, the human mortals chant ; but *now* no night is blessed with hymn or carol ' ; and that we should read : ' The human mortals *chant*,—their Winter here ;'—STAUNTON : ' Want,' in this passage does not appear to mean *need, lack, wish for*, &c., but to be used in the sense of *be without*. The human mortals are *without* their winter here. It occurs, with the same meaning, in a well-known passage in *Macb.* III, vi : ' Men must not

5

No night is now with hymne or caroll bleſt ;   106
Therefore the Moone (the gouerneſſe of floods)
Pale in her anger, waſhes all the aire ;
That Rheumaticke diſeaſes doe abound.   109

---

walk too late Who cannot *want* the thought,' &c.—KEIGHTLEY (*Exp.* 131) : I should
prefer *summer* for 'winter,' for in Dr Forman's *Diary* of the year 1594—which year
Shakespeare had certainly in view—we read : ' This monethes of June and July were
very wet and wonderfull cold, like winter, that the 10 dae of Julii many did syt by
the fyer, yt was so cold ; and soe was it in Maye and June ; and scarse too fair dais
together all that tyme, but it rayned every day more or lesse. Yf it did not raine then
was it cold and cloudye. . . . There were many gret fludes this sommer.' It is possible, however, that the error may lie in ' want,' for which we might read *have*, or
some such word.—HUDSON (ed. ii) : ' Want their winter here ' cannot possibly be
right ; it gives a sense all out of harmony with the context. I think the next line naturally points out *minstrelsy* as the right correction. [And so Hudson's text reads.]—
DYCE (ed. ii) : ' Heere ' is proved to be nonsense by the attempts to explain it. [This
puzzling line R. G. White, in his first edition, pronounces ' unless greatly corrupted,
one of the most obscure and unsatisfactory in all Shakespeare's works.' Whether
' want ' mean *to lack*, or *to desire*, or *to be without*, it cannot be satisfactorily interpreted
in connection with ' here ' in the sense of time. ' Here ' and now, while Titania is
talking, is either April or midsummer, and although at this season in the course of
nature winter is assuredly *lacking*, it is erroneous to suppose that human mortals are
now *desiring* its presence ; in fact, it is because there are signs of winter at midsummer that the world is mazed. The only solution which I can find is to take ' here,'
not in the sense of time, but of place. Here in Warwickshire, says Titania, in effect
(for of course she and Oberon are in the Forest of Arden, with never a thought of
Athens ; whoever heard of the nine mens morris on the slopes of Pentelicus ?), ' here
the poor human mortals have no summer with its sports, and now they have had no
winter with its hymns and carols.' With this interpretation of ' here,' which Capell
was the first to suggest, and whose words, ' in this country,' seem to have been overlooked by recent editors, the line scarcely needs emendation.—ED.]

   107. Therefore] To JOHNSON this passage ' remained unintelligible,' most probably because he misinterpreted, I think, this ' therefore.' He says, ' Men find no
winter, therefore they sing no hymns, the moon provoked by this omission alters the
seasons : That is, the alteration of the seasons produces the alteration of the seasons.'
—MALONE points out that there is a succession of ' therefores,' all pointing to the fairy
quarrel as the cause of the war of the elements : ' *Therefore* the winds,' &c. ; ' the ox
hath *therefore*,' &c., and the present line, which is not logically connected with the
omission of hymns and carols.

   108. Pale] Because it can shine but dimly through the contagious fogs.—ED.

   109. Rheumaticke] Again used with the accent on the first syllable in *Ven. and
Ad.* 135.—MALONE : Rheumatic diseases signified, in Shakespeare's time, not what
we now call rheumatism, but distillations from the head, catarrhs, &c. So, in the
Sydney Memorials, i, 94 (1567), we find : ' he hath verie much distemporid divers
parts of his bodie ; as namelie, his hedde, his stomack, &c. And therby is always
subject to distillacions, coughes, and other rumatick diseases.—W. A. WRIGHT adds
that it would be ' more correct to say that the term included all this in addition to

And through this diftemperature, we fee   110
The feafons alter ; hoared headed frofts
Fall in the frefh lap of the crimfon Rofe,
And on old *Hyems* chinne and Icie crowne,   113

---

110. *through*] *thorough* Q₁F₂F₃, Rowe
ii et seq.

111. *hoared headed*] Q₂. *hoared-*
-headed F₂. *hoary-headed* F₃. *hoary*

headed Q₁F₄ et cet.

113. Hyems] *Adam's* Herr.

chinne] *thin* Tyrwhitt, Hal.
White, Dyce, Sta. Cam.

---

what is now understood by it. Cotgrave has "Rumatique : com. Rhewmaticke ; troubled with a Rhewme," and he defines "Rume : f. A Rhewme, Catarrhe ; Pose, Murror"' Drou gives a somewhat different meaning, defining it : 'splenetic, humoursome, peevish,' and cites *2 Hen. IV:* II, iv, 62, 'as rheumatic as two dry toasts,' which JOHNSON explains by 'which cannot meet but they grate one another.'

109, 110. Johnson's suggestion (see note on line 105, *supra*) to transpose these two lines, HUDSON adopts; an emendation as harmless as it is needless, if 'distemperature' refers to the washing of the air by the moon, to which it is quite possible it may refer.—But W. A. WRIGHT, following Malone, says that 'distemperature' refers to the 'disturbance between Oberon and Titania, not to the perturbation of the elements,' and cites *Per.* V, f, 27 : 'Upon what ground is his distemperature?' 'where it is used of the disturbance of mind caused by grief. Again, *Rom. and Jul.* II, iii, 40 : "Thou art uproused by some distemperature."' On the other hand, SCHMIDT (*Lex.*) gives an example from *1 Hen. IV:* V, i, 3, quite parallel to the present line, where 'distemperature' refers not to mental, but to physical disturbance : 'how bloodily the sun begins to peer above yon bosky hill ! the day looks pale at his distemperature.' It must be confessed that the reiterated reference to a personal quarrel between atomies as the cause of elemental and planetary disturbances is in accord with the whole passage and to be preferred ; but at the same time it cannot be denied that the 'Therefore' in line 107 may contain a sufficient reference to the fairy brawl, and that 'distemperature' may mean the anger of the moon.—ED.

110. **through**] See II, i, 5.

113. **chinne**] The earliest critic who, in print, suggested *chill* is GREY (i, 49, 1754,), but in 1729 THEOBALD wrote to WARBURTON (Nichols, *Lit. Hist.* ii, 232) : 'it staggered me to hear of a chaplet or garland on the "chin." I therefore conjectured it should be "*chill* and icy crown." But upon looking into Paschalius *de Coronis*, I find many instances of the ancients having chaplets on their necks, as well as temples ; so that, if we may suppose Hyem is represented here as an old man bending his chin towards his breast, then a chaplet round his neck may properly enough be said to be on his chin. So I am much in doubt about my first conjecture.'—To CAPELL also (*Notes*, p. 104) the same emendation occurred independently, and he, too, was restrained from adopting it in his text by his classical knowledge ; he had a 'distant remembrance of the *incana barba* of a Silenus, or some such person, having a "chaplet" put on it by nymphs that are playing with him.'—In support of the text, however, or rather in what they considered support of the text, WESTON and MALONE adduced passages from Virgil (*Æneid*, iv, 253) and Golding's *Ovid* (Seconde Booke, p. 15) which have no parallelism with the present phrase, but contain merely a description of Winter with his 'hoarie beard' and 'snowie frozen crown.'—It wa reserved for TYRWHITT to suggest an emendation which has been since adopte

An odorous Chaplet of ſweet Sommer buds
Is as in mockry ſet.   The Spring, the Sommer,   115
The childing Autumne, angry Winter change

116. *childing*] *chiding* F₄, Pope, Han. Cap.   *chilling* or *churlish* Herr.

by many of the ablest editors; he remarked ' I should rather be for *thin*, i. e. thin-haired.'—In support, STEEVENS cites *Lear*, IV, vii, 36: ' To watch—poor perdu!—With this thin helm;' and *Rich. II:* III, ii, 112: ' White-beards have arm'd their thin and hairless scalps Against thy majesty.'—And W. A. WRIGHT adds *Timon*, IV, iii, 144: ' Thatch your poor thin roofs With burthens of the dead.'—DYCE (*Remarks*, p. 46), after giving in full the citations of Weston and Malone just mentioned, ob-serves: ' Now, in good truth, there is not the slightest resemblance between these two quotations and the absurdity which they are adduced to illustrate and defend.   When Virgil describes Atlas with rivers streaming from his chin, and when Ovid paints Winter with icicles dangling on his beard and crown, we have such pictures pre sented to us as the imagination not unwillingly receives; but *Hyems with a chaplet of summer buds on his* CHIN is a grotesque which must surely startle even the dullest reader.'—In deference to Dyce's opinion, HALLIWELL adopted *thin* in his text, but confesses that he is ' not quite convinced that " chin " is incorrect,' ' the author evi-dently intended a grotesque contrast,—" is, *as in mockery*, set;" the proper appendage being ice.'—' What was a chaplet doing on old Hyems's " *chin* "?' asks R. G. WHITE, ' How did it get there? and when it got there, how did it stay?'—Lastly, WALKER (*Crit*. ii, 275) in an Article on the confusion of *c* and *t*, pronounces *thin* clearly right. [I cannot but think that there is some slight corroboration of Tyrwhitt's emendation in the use of the word ' chaplet,' which is almost restricted to the head.   Would not the word have been *garland* had it been meant to have the summer buds about old Hyems's neck and resting in mockery on his chin or beard?—ED.]

116. *childing*] STEEVENS: This is the *frugifer autumnus*.—HOLT WHITE: Thus in Fairfax's *Tasso*, xviii, 26: ' An hundreth plants beside (euen in his sight) Childed an hundreth nymphes, so great, so dight.'   *Childing* is an old term in botany, when a small flower grows out of a large one; ' the childing autumn ' therefore means the autumn which unseasonably produces flowers on those of summer.—W. A. WRIGHT: It means the autumn which seasonably produces its own fruits.   It is the change of seasons which makes it abnormal.—KNIGHT: ' The childing autumn ' is the ' *teem ing* autumn ' of our poet's 97th *Sonnet*.—ABBOTT, § 290: That is, autumn pro-ducing fruits as it were children.—J. B. NOYES (*Poet-Lore*, p. 531, Oct. 1892): No passage has yet been produced from any writer to justify the definition of ' childing ' as *fruitful*, and it is presumed that none fairly can be.   I believe the word ' childing ' to be a corrupt spelling of the ignorant compositor, a vulgar and strong form of the true reading *chilling*. [See HERR's conj., Text. Notes.]   Edward Coote, in *The English Schoole-Master*, p. 19, 1624, 15th ed., writes: ' But it is both unusual and needlesse to write *bibbl* and *childd*, to make them differ from *bible* and *child*.'   It therefore seems extremely probable that ' childing ' or *chillding* is simply a corrupt spelling of *chilling*, formed in the same manner as ' oilde ' from ' oile ' [where?—ED.], and ' beholds ' from *behowls*, which corrupt spellings are found in the Folio text of this play.   A passage from Greene's *Orpharion*, 1599, p. 20 [p. 37, ed. Grosart], would seem to dispel any lingering doubt as to the proposed emendation: ' for the childing colde of Winter, makes the Sommers Sun more pleasant.'—[In his *Glossarial*

Their wonted Liueries, and the mazed world ,   117
By their increaſe, now knowes not which is which ;
And this ſame progeny of euills,
Comes from our debate, from our diſſention,   120
We are their parents and originall.
     Ober. Do you amend it then, it lies in you,   122

---

117. *mazed*] *amazed* F$_3$F$_4$, Rowe +.   Warb. *encrease* Cap.
mazed Johns. Steev. Mal. Sing. ii, Ktly.   119, 120. *And...Comes*] One line, F1
118. *increaſe*] *inverse* Han. *inchase*   et seq. *And...evil comes* F$_4$, Rowe +.

...x, OKOSAKI anticipates Noyes in the correction of 'childing' to *chiding* in this
passage from Greene.—In MURRAY'S *N. E. Dict.* there are the following citations,
in addition to the present passage, in support of the meaning *fertile, fruitful,* and also
of the botanical meaning of 'childing,' noted by Holt White : '1609, HEYWOOD,
Brit. Troy, V, xix, 111, By him (Saturn) . . . Childing Tellus beares. 1636, GERARD'S
Herbal, II, cciii, 635, Another pretty double daisie, which . . . puts forth many foot-
stalkes carrying also little double floures . . . whence they haue fitly termed it the
childing Daisie. 1688, R. HOLME, *Armoury,* II, 64/2 : The Childing Pink groweth
. . . on upright stalks. 1776, WITHERING, *Bot. Arrangem.* (1830), II, 539 : *Dian
thus prolifer,* Childing or Proliferous Pink. 1879, PRIOR, *Plant-n.,* Childing Cud
weed, *Gnaphalium germanicum.*' Surely the text of the Folio may stand. From
time immemorial Autumn has been symbolised by harvests and by fruits. If there
be any virtue in illustrating Shakespeare by himself, we cannot overlook the parallel
passage cited by Knight : 'The teeming autumn, big with rich increase, Bearing the
wanton burthen of the prime.' In each of my three copies of F$_4$ 'childing' is spelled
chiding, yet it would be unsafe to assert that this is the reading in all copies. Neither
CAPELL nor the CAMBRIDGE EDITOR makes any mention of it, but both credit it to
Pope. Capell adopted it in his text, and justifies it in his notes by saying that he
could not see 'how the epithet " angry " could well have presented itself to the poet,
if " chiding " had not preceded.'—R. G. WHITE supposed that the change was original with him. 'I am so sure,' he says (ed. i), 'that " childing " is a misprint for
chiding (in allusion to the lowering skies and harsh winds of Autumn, as the next
epithet figures the increased inclemency of Winter,) . . . that I wonder that the suggestion has not been made before.'—ED.]

   117. **mazed**] That is, confuſed, bewildered; it is not an abbreviation for *amazed,*
as it is sometimes printed in modern editions. See Text. Notes.

   118. **increase**] Warburton's substitution *inchase* is unintelligible without his explanation that it refers to the temperature in which the seasons are *set* or *inchased* like
jewels.—Whereupon HEATH (p. 47) observes, none too strongly, that 'a season set
in a warm or cold temperature borders very nearly upon downright nonsense.' 'If
[Warburton] had recollected the Psalm he every day repeats in the evening service
of the Common Prayer, he would have found that " increase " signifies product,
growth.' 'The seasons had so changed their wonted liveries that it was no longer
possible to distinguish them one from another by their products.'

   119. **progeny of euills**] For contemporary references to these meteorological disturbances, see Appendix, *Date of Composition.*

Why fhould *Titania* croffe her *Oberon* ?   123
I do but beg a little changeling boy,
To be my Henchman.   125
    Qu.   Set your heart at reft,
The Fairy land buyes not the childe of me,
His mother was a Votreffe of my Order,
And in the fpiced *Indian* aire, by night
Full often hath fhe goffipt by my fide,   130
And fat with me on *Neptunes* yellow fands,
Marking th'embarked traders on the flood,
When we haue laught to fee the failes conceiue,
And grow big bellied with the wanton winde :
Which fhe with pretty and with fwimming gate,   135
Following (her wombe then rich with my yong fquire)

123. Oberon] Orberon F₄.

128. *Votreffe*] *votaress* Dyce, Coll. ii, Cam.

130. *hath fhe*] *fhe hath* F₃F₄, Rowe +.

131. *And fat*] *And fat*, Q₁.

132. *on the*] *of the* F₃F₄, Rowe, Pope, Han.

133. *we haue*] *we* F₃F₄.

135. *gate*] *gait* Cap. et seq.

136. *Following (her…fquire)*] *Follying (her…squire)* Warb. Theob. Han. *(Following her. …squire)* Kenrick, Farmer, Steev. Rann. Mal. *Following her womb, …squire,* Hal. White i (subs.). *Following her womb…squire.*— White ii.

---

124. In this contest over a boy, BELL (ii, 207) detects the contest of Jupiter over Hercules.

125. **Henchman**] The meaning of this word is given as concisely as may be in Sherwood's *French-English Dictionary*, appended to Cotgrave : 'A hench-man, or hench boy. Page d'honneur ; qui marche devant quelque Seigneur de grand authoritie.' Its derivation is still somewhat in doubt. SKEAT derives it from *hengst-man*, horse-man, groom ; Anglosaxon *hengest =* horse. For a prolonged discussion wherein many examples are cited, one as early as 1415, see *Notes and Queries*, 8th Ser. III, 478, 1893, where references are given to all the preceding communications in that periodical. HALLIWELL devotes more than two folio pages, with a wood-cut, to the elucidation of the word ; but for all purposes of present illustration, Sherwood's definition appears to be ample.—ED.

127. **The Fairy land**] COLLIER (ed. ii) : The MS has *Thy ;* and as Titania afterwards speaks to Oberon of '*thy* fairy kingdom,' it is probably right. [If improvement be justifiable, this trivial emendation is harmless.—ED.]

135. **swimming**] Of course this refers to a gliding motion on or in the water ; at the same time, it is well to remember that to Elizabethan ears there may have been here the suggestion of a graceful dance. That there was a step in dancing called *the swim* we know, but of its style we are ignorant. DANIEL (see note, *As You Like It,* V, iv, 73, of this ed.) collected references to this dance from Beau. & Fl., Massinger, and Steele ; ELZE added another from Chapman ; to them may be added, from Jonson's *Cynthia's Revels :* '*Moria.* You wanted the swim in the turn. *Philautia.* Nay, … the swim and the trip are properly mine ; everybody will affirm it that has any

Would imitate, and faile vpon the Land,   137
To fetch me trifles, and returne againe,
As from a voyage, rich with merchandize.
But fhe being mortall, of that boy did die,   140
And for her fake I do reare vp her boy,
And for her fake I will not part with him.
    Ob. How long within this wood intend you ftay *?*
    Qu. Perchance till after *Thefeus* wedding day.
If you will patiently dance in our Round,   145
And fee our Moone-light reuels, goe with vs;
If not, fhun me and I will fpare your haunts.
    Ob. Giue me that boy, and I will goe with thee.
    Qu. Not for thy Fairy Kingdome. Fairies away: 149

---

139. *rich with*] *ripe with* Coll. MS.   **144.** Thefeus] *Theseus's* Rowe i. *The-*
    merchandize] *marchandife* Q₁.   *seus'* Rowe ii et seq.
141. *I doe*] *doe I* Qq, Cap. Mal.'90,   **149.** *Fairies*] *Elves* Pope+.
Sta. Cam. White ii.

---

judgement in dancing.'—II, i, p. 270, ed. Gifford, 1816. Unfortunately, Gifford has
no note on it.—ED.

136. Following] WARBURTON'S emendations, not unfrequently, as in the present
instance, composed of words coined by himself, need explanation; a bare record in the
Text. Notes is almost unintelligible. 'Following' he changes to *follying*, and says it
means 'wantoning in sport and gaiety,'—HEATH rightly explained that the little
mother 'followed on the land the ship which sailed on the water, . . . and that she
continued following it for some time, . . . and would then pick up a few trifles, and
"return again, As from a voyage, rich with merchandise."' Bad as is WARBURTON'S
change, which, by the way, Dr JOHNSON pronounced 'very ingenious,' it is to me preferable to KENRICK'S repulsive punctuation (*Rev.* p. 19). He removes the excellent
parentheses of the Folio, and puts a comma after 'wombe'; having thus coarsened
Titania's sweet picture and degraded her words to the slang level of 'following one's
nose,' he complacently adds: 'this is the method a critic should take with the poets.
Trace out their images, and you will soon find how they expressed themselves.' It is
to be regretted that Kenrick has, substantially, so good a following; it is incomprehensible that LETTSOM (ap. Dyce, ed. ii) should say he was right.—ED.

137. imitate] C. C. HENSE (*Sh.'s Sommernachtstraum Erläutert*, 1851, p. 7):
Shakespeare's fairies delight in whatsoever is comic, hence it is thoroughly characteristic that Titania in recalling the loveliness of her friend should dwell with fondest
recollection on the laughter called forth by the imitation of the embark'd traders.

143. stay] For other examples of the omission *to* before the infinitive, see ABBOTT,
§ 349.

145. Round] HALLIWELL: '*Orbis saltatorius,* the round danse, or the dansing of
the rounds.'—*Nomenclator,* 1585. So in Elyot's *Boke of the Governour,* 1537: 'In
stede of these we haue nowe base daunsis, bargenettes, pauions, turgions, and roundes'
[i, 230, ed. Croft]. The round was, in fact, what is now called the country-dance.

149. Fairy] 'By the advice of Dr Farmer,' STEEVENS 'omitted this useless adjec-

We ſhall chide downe right, if I longer ſtay.   *Exeunt.*   150
    Ob.   Wel, go thy way : thou ſhalt not from this groue,
Till I torment thee for this iniury.
My gentle *Pucke* come hither ; thou remembreſt
Since once I ſat vpon a promontory,   154

---

153. *remembreſt*] *rememberest* Cam.   *that I* Rowe.   *Since I once* Coll. MS
154. *Since once I*] *Since I* Ff.   *Since*   ap. Cam.

---

tive as it spoils the metre.'   And then, can it be believed ? pronounced the following
' Fairies ' as a trisyllable !—ED.

152. **iniury**] W. A. WRIGHT : This word has here something of the meaning of
insult, and not of wrong only.   Compare III, ii, 153, and the adjective ' injurious ' in
the sense of ' insulting, insolent ' in III, ii, 202.   In the Authorised Version of *1 Tim-
othy* i, 13, ' injurious ' is the rendering of *ὑβριστής.*

153–175.   For notes on this passage, see p. 75.

154. **Since**] For other examples of the use of ' since ' for *when*, see ABBOTT,
§ 132, where it is said that this meaning arises from the ' omission of " it is " in
such phrases as " it is long *since* I saw you," when condensed into " long *since*, I
saw you."   Thus *since* acquires the meaning of " ago," " in past time," adverbially,
and hence is used conjunctively for " when, long ago." '—VERITY gives a refined
analysis of this usage : ' " Since " is used by Shakespeare as equivalent to *when* only
after verbs denoting recollection.   Perhaps this use comes from the meaning *ever
since ;* if you recollect a thing *ever since* it occurred, you must recollect *when* it
occurred.'   In *2 Hen. VI :* III, i, 9, the Queen says, ' We know the time since he
was mild and affable ' ; at first sight, the use of ' since ' appears here to disprove
Verity's rule, but in reality it conforms to it.   In ' we know the time ' there is
involved the idea of recollection.—ED.

154. **Since once I sat, &c.**] DELIUS (*Sh. Jahrbuch,* vol. xii, p. 1, 1877) has collected examples of what he ' ventures to term ' ' the epic element ' in Shakespeare's
dramas.   By this ' epic element ' is meant those passages where the poet, through
the mouth of one of his characters, lets those circumstances be narrated or described
which might have been presented scenically.   It is needless to call attention to the
important bearing of this subject on Shakespeare's dramatic art.   Of the present play
Delius says (p. 4) : The previous quarrel between Oberon and Titania, which has
such disastrous consequences for all nature and for mankind, Shakespeare describes
at length through the mouths of the Fairy King and Queen themselves ; just as he
had shortly before made the roguish Puck boast of his own knavish tricks in order to
prepare the audience for those tricks which he was afterwards to play in the drama.
A third descriptive or epic element is in the present passage, where Oberon describes
the magic properties of the little western flower.   Be the meaning of this much-vexed
passage what it may, this much is certain, that a visible scenic representation of it was
precluded by the meagre theatrical resources of the day ; and yet so essential to the
developement of the action is this magic flower that a picture of it must be drawn as
vividly and as visibly as possible before the mind's eye.   And here it is where Shakespeare has completely succeeded.   While listening in the theatre to Oberon's words
the spectators *saw* Oberon himself on the promontory.   With Oberon's eyes they *saw*
Cupid's love-shaft miss the fair vestal throned by the west, and fall upon the little

And heard a Meare-maide on a Dolphins backe,   155
Vttering such dulcet and harmonious breath,
That the rude sea grew ciuill at her song,
And certaine starres shot madly from their Spheares,
To heare the Sea-maids musicke.

   Puc.   I remember.   160

   Ob.   That very time I say (but thou couldst not)
Flying betweene the cold Moone and the earth,
Cupid all arm'd ; a certaine aime he tooke   163

---

155. *Meare-maide*] *mermaid* Rowe.   161. *I say*] *I saw* Q₁, Rowe et seq.
156. *harmonious*] *hermonious* Q₁.   163. *all arm'd*] *alarm'd* Warb.
158. *Spheares*] *Shpeares* F₃.   Theob.   *all-arm'd* Johns.

---

flower before milk-white, now purple with love's wound.   They *saw* the siren, as a
contrast to the invulnerable chastity of that vestal, control the sea with her seductive
songs, and entice the stars, maddened with love, from their spheres.   [If the spectators saw this, did they see what Shakespeare intended?   Delius speaks of a 'siren';
a mermaid was not necessarily a 'siren,' nor is 'dulcet and harmonious breath' necessarily 'seductive.'   Moreover, does not Delius overshoot the mark when he represents
Shakespeare as resorting to the epic element here, not from artistic reasons, but
because of the poverty of his stage?   Delius's Essay has been translated in the *New
Shakspere Society's Transactions*, Part ii, pp. 207, 232.—ED.]

   158. **certaine**] W. A. WRIGHT: Here used of an indefinite number, as in *Temp.*
V, i, 53: 'I'll break my staff, Bury it certain fathoms in the earth.'   [This interpretation is, of course, allowable, but I am by no means sure that there is not an added
beauty in taking 'certain' in the meaning of *sure, fixed ;* does it not heighten the
power of the mermaid's song, that it could bring down the very stars, fixed in the
sky.   Schmidt (*Lex.*) furnishes a parallel example from the *R. of L.* where the skies
were sorry at the burning of Ilion, 'And little stars shot from their fixed places.'—
l. 1525.   That this interpretation is hostile to the theory that the 'certain stars' were
the Duke of Norfolk and the Earls of Northumberland and Westmoreland, is possibly an additional reason why it should be preferred.—ED.]

   157. Prof. A. S. COOK (*Academy*, 30 Nov. 1889) calls attention to the parallelism
of this line to the description, in the Sixth Canto of the *Orlando*, of 'una Sirena Che
col suo dolce canto accheta il mare.'

   158. **Spheares**] See note on 'moon's sphere' in line 7 of this scene.

   163. **all arm'd**] WARBURTON, on the supposition that the beauty of the passage
would be heightened if Cupid were represented as frightened at the Queen's declaration for a single life, changed this to *alarm'd*, and Dr JOHNSON gravely defended
the original text, and explained that 'it does not signify dressed in panoply.'   Earlier
than Johnson, however, GREY (i, 52) had rightly remarked that 'all arm'd' means
nothing more 'than being arm'd with bow and quiver, the proper and classical arms
of Cupid, which yet he sometimes feigned to lay aside.'—And CAPELL, too, came to
the rescue of a phrase that would have needed no comment had not the perverse and
ingenious Warburton given it a twist, whereof the effects have more or less endured
until now.—W. A. WRIGHT observes that 'all' is merely emphatic,—'not in full
armour, but with all his usual weapons.'

At a faire Veſtall, throned by the Weſt,
And loos'd his loue-ſhaft ſmartly from his bow,   165
As it ſhould pierce a hundred thouſand hearts,
But I might ſee young *Cupids* fiery ſhaft
Quencht in the chaſte beames of the watry **Moone**;
And the imperiall Votreſſe paſſed on,
In maiden meditation, fancy free.   170
Yet markt I where the bolt of *Cupid* fell.
It fell vpon a little weſterne flower;
Before, milke-white; now purple with loues wound,
And maidens call it, Loue in idleneſſe.   174

---

164. *by the*] *by* Qq.   169. *Votreſſe*] *votaress* Knt, Coll.
166. *ſhould*] *would* F₄, Rowe i.   Dyce, Sta. Cam. White ii.
168. *Quencht*] *Quench* F₃F₄.   170. *fancy free*] *fancy-free* Ff et seq.

---

164. **by**] For other examples of a similar use of 'by,' see ABBOTT, § 145.

165. **loos'd**] DYCE: The technical term in archery. See Puttenham's *Arte of Poesie*, 1589, p. 145: 'th' Archer's terme, who is not said to finish the feate of his shot before he give the loose, and deliuer his arrow from his bow.' Compare, in the excellent old ballad of *Adam Bell, Clim of the Clough, and William of Cloudesly,* 'They loused theyr arowes bothe at ones.'—[Child's *Eng. and Scot. Popular Ballads,* V, 26.]

166. **As**] For other instances where 'as' is equivalent to *as if,* see ABBOTT, § 107; and see § 312 for examples of 'might,' in the next line, used in the sense of *was able, could.*

170. **fancy free**] STEEVENS: That is, exempt from the power of love.

173. **Before, milke-white**] HUNTER (i, 293): The change of the flower from white to purple was evidently suggested by the change of the mulberry in Ovid's story of Pyramus. HALLIWELL: Shakespeare was so minute an observer of nature, it is possible there is here an allusion to the changes which take place in the colours of plants arising from solar light and the character of the soil. [Lyte, in his *Nieuve Herball,* 1578, p. 147, speaking of the different kinds of violets (and Love-in-idleness is the *viola tricolor,* see next note), says: 'There is also a thirde kinde, bearing floures as white as snow. And also a fourth kinde (but not very common), whose floures be of a darke Crymsen, or old reddish purple colour, in all other poyntes like to the first, as in leaues, seede, and growing.' If any appeal to Botany be needed, which I doubt, we appear to have here a sufficing response.—ED.]

174. **Loue in idlenesse**] In his Part II, chap. ii, *Of Pances or Hartes ease,* Lyte says: 'This floure is called . . . in Latine . . . Viola tricolor, Herba Trinitatis, Iacea, and Herba Clauellata: in English Pances, Loue in idlenes, and Hartes ease' (p. 149, ed. 1578). W. A. WRIGHT quotes Gerard (*Herball,* p. 705, ed. 1597) as calling the flower 'Harts ease, Pansies, Liue in Idlenes, Cull me to you, and three faces in one hood.'—ELLACOMBE (p. 151) has added from Dr Prior more common names, such as: 'Herb Trinity, Fancy, Flamy, Kiss me, Cull me or Cuddle me to you, Tickle my fancy, Kiss me ere I rise, Jump up and kiss me, Kiss me at the garden gate, Pink of my John, &c.' I think the commonest name in this country is Johnny-jump-up.—ED.

Fetch me that flower ; the hearb I ſhew'd thee once,   175

175. *ſhew'd*] *ſhewed* Q₁.

153–175. My gentle Pucke . . . that flower] This speech of Oberon has been the subject of more voluminous speculation than any other twenty-five lines in Shakespeare.   Perhaps not unnaturally.   Let an allegory be once scented and the divagations are endless.   That there is an allegory here has been noted from the days of Rowe, but how far it extended and what its limitations and its meanings have since then proved prolific themes.   According to Rowe, it amounted to no more than a compliment to Queen Elizabeth, and this is the single point on which all critics since his day are agreed.   In his *Life of Shakespear* (p. viii, 1709) ROWE says that ' Queen Elizabeth had several of [Shakespear's] plays acted before her, and without doubt gave him many gracious marks of her favour.   It is that maiden Princess, plainly, whom he intends by a " fair veſtal throned by the Weſt " ; and that whole passage is a Compliment very properly brought in, and very handsomely apply'd to her.'   The next advance was made by Warburton, and however unwilling we may be to accept instruction from his dogmatic lips, and however much he may have been derided and mangled by Ritson, it still remains that his interpretation has been accepted by one, at least, of the able critics of our day.—' The first thing,' says WARBURTON, ' observable in these words [the first seven lines of Oberon's speech] is that this action of the *Mermaid* is laid in the same time and place with Cupid's attack upon the *vestal*.   By the vestal every one knows is meant Queen Elizabeth.   It is very natural and reasonable then to think that the Mermaid stands for some eminent personage of her time.   And if so, the allegorical covering, in which there is a mixture of satire and panegyric, will lead us to conclude that this person was one of whom it had been inconvenient for the author to speak openly, either in praise or disparaise.   All this agrees with Mary Queen of Scots, and with no other.   Queen Elizabeth could not bear to hear her commended ; and her successor would not forgive her satirist.   But the poet has so well marked out every distinguished circumstance of her life and character in this beautiful allegory, as will leave no room to doubt about his secret meaning.   She is called a *Mermaid*—1, to denote her reign over a kingdom situate in the sea, and 2, her beauty and intemperate lust, " Ut turpiter atrum Desinat in piscem mulier formosa supernè," for as Elizabeth, for her chastity, is called a *Vestal*, this unfortunate lady, on a contrary account, is called a *Mermaid*.   3. An ancient story may be supposed to be here alluded to.   The emperor Julian tells us, *Epistle* 41, that the Sirens (which, with all the modern poets, are mermaids) contended for precedency with the Muses, who, overcoming them, took away their wings.   The quarrels between Mary and Elizabeth had the same cause and the same issue.

' " On a dolphin's back ".   This evidently marks out that distinguishing circumstance of Mary's fortune, her marriage with the Dauphin of France, son of Henry II.

' " Uttering such dulcet and harmonious breath " : This alludes to her great abilities of genius and learning, which rendered her the most accomplished Princess of her age. . . .

' " That the rude sea grew civill at her song " : By " rude sea " is meant Scotland encircled with the ocean ; which rose up in arms against the Regent, while she was in France.   But her return home presently quieted those disorders. . . . There is the greater justness and beauty in this image, as the vulgar opinion is, that the mermaid always sings in storms.

' " And certaine starres shot madly from their spheares, To heare the Sea-maids

---

[153–175. **My gentle Pucke . . . that flower**]

musicke ": Thus concludes the description, with that remarkable circumstance of this unhappy lady's fate, the destruction she brought upon several of the English nobility, whom she drew in to support her cause. This, in the boldest expression of the sublime, the poet images by *certain stars shooting madly from their spheres.* By which he meant the earls of Northumberland and Westmoreland, who fell in her quarrel; and principally the great duke of Norfolk, whose projected marriage with her was attended with such fatal consequences. Here, again, the reader may observe a peculiar justness in the imagery. The vulgar opinion being that the mermaid allured men to destruction by her songs. . . . On the whole, it is the noblest and justest allegory that was ever written. The laying it in *fairy land*, and out of nature, is in the character of the speaker. And on these occasions Shakespeare always excels himself.'

This interpretation of the 'noblest and justest allegory' (Warburton's innocent way of praising his own ingenuity) was accepted for forty years, and duly appeared in each succeeding edition of the *Variorum* down to 'Steevens's Own,' in 1793, when that editor found he could not 'dissemble his doubts concerning it.' 'Why,' he asks, 'is the *thrice-married* Queen of Scotland styled a Sea-*maid?* and is it probable that Shakespeare (who understood his own political as well as poetical interest) should have ventured such a panegyric on this ill-fated Princess during the reign of her rival, Elizabeth? If it was unintelligible to his audience, it was thrown away; if obvious, there was danger of offence to her majesty. . . . To these remarks may be added those of a like tendency which I met with in *The Edinburgh Magazine*, Nov. 1786: "That a complement to Queen Elizabeth was intended in the expression of the 'fair Vestal throned in the West' seems to be generally allowed; but how far Shakespeare designed, under the image of the mermaid, to figure Mary, Queen of Scots, is more doubtful. If by the 'rude sea grew civil at her song' is meant, as Dr Warburton supposes, that the tumults of Scotland were appeased by her address, the observation is not true; for that *sea* was in a storm during the whole of Mary's reign. Neither is the figure just, if by the 'stars shooting madly from their spheres' the poet alluded to the fate of the Earls of Northumberland and Westmoreland, and particularly of the Duke of Norfolk, whose projected marriage with Mary was the occasion of his ruin. It would have been absurd and irreconcileable to the good sense of the poet to have represented a nobleman *aspiring* to marry a queen, by the image of a star *shooting* or *descending* from its sphere." '

The doubts merely hinted at by Steevens become withering sneers from RITSON. 'I shall not dispute,' says he, 'that by "the fair vestal" Shakspeare intended a compliment to Queen Elizabeth, who, I am willing to believe, at the age of sixty-eight, was no less chaste than beautiful; but whether any other part of Oberon's speech have an allegorical meaning or not, I presume, in direct opposition to Dr Warburton, to contend that it agrees with any other rather than with Mary, Queen of Scots. The "mixture of satire and panegyric" I shall examine anon. I only wish to know, for the present, why it should have been "inconvenient for the author to speak openly" in "dispraise" of the Scottish queen. If he meant to please "the imperial votress," no incense could have been half so grateful as the blackest calumny. But, it seems, "her successor would not forgive her satirist." Who then was her "successor" when this play was written? Mary's son, James? I am persuaded that, had Dr Warburton been better read in the history of those times, he would not have found this monarch's succession quite so certain, at that period, as to have prevented Shakspeare. who was by no means the refined speculatist he would induce one to suppose, from

[153 175. My gentle Pucke . . . that flower]

gratifying the "fair vestal" with sentiments so agreeable to her. However, if "the poet has so well marked out every distinguishing circumstance of her life and character, in this beautiful allegory, as will leave no room to doubt about his secret meaning," there is an end of all controversy. For, though the satire would be cowardly, false, and infamous, yet, since it was couched under an allegory, which, while perspicuous as glass to Elizabeth, would have become opake as a mill-stone to her successor, Shakspeare, lying as snug as his own Ariel in a cowslip's bell, would have had no reason to apprehend any ill consequences from it. Now, though our speculative bard might not be able to foresee the sagacity of the Scotish king in smelling out a plot, as I believe it was some years after that he gave any proof of his excellence that way, he could not but have heard of his being an admirable witch-finder, and, surely, the skill requisite to detect a witch must be sufficient to develope an allegory; so that I must needs question the propriety of the compliment here paid to the poet's prudence. Queen Mary "is called a *Mermaid*—1, to denote her reign over a kingdom situate in the sea." In that respect, at least, Elizabeth was as much a mermaid as herself. "And 2, her beauty and intemperate lust; for as Elizabeth, for her chastity, is called a Vestal, this unfortunate lady, on a contrary account, is called a *mermaid*." All this is as false as it is foolish: The mermaid was never the emblem of lust; nor was the "gentle Shakspeare" of a character or disposition to have insulted the memory of a murdered princess by so infamous a charge. The most abandoned libeller, even Buchanan himself, never accused her of "intemperate lust"; and it is pretty well understood at present that, if either of these ladies were remarkable for her purity, it was *not* Queen Elizabeth. "3. An ancient story may be supposed to be here alluded to: the Emperor Julian tells us that the *Sirens* (which, with all the modern poets, are *mermaids*) contended for precedency with the Muses, who, overcoming them, took away their wings." Can anything be more ridiculous? *Mermaids* are half women and half *fishes*: where then are their wings? or what possible use could they make of them if they had any? The *Sirens* which Julian speaks of were partly women and partly *birds*; so that "the pollusion," as good-man Dull hath it, by no means "holds in the exchange." [Florio gives: 'Sirena, *a Syren, a Mermaide*,' and Cotgrave: 'Serene: f. *A Syren, or Mermaid*.' Hence it seems that the words were to a certain extent interchangeable in Shakespeare's day, and Ritson's sneers in this regard must be tempered.] "The quarrels between Mary and Elizabeth had the same cause and the same issue." That is, they contended for precedency, and Elizabeth, overcoming, took away the other's *wings*. The secret of their contest for precedency should seem to have been confined to Dr Warburton. It would be in vain to enquire after it in the history of the time. The Queen of Scots, indeed, flew for refuge to her treacherous rival (who is here again the mermaid of the allegory, alluring to destruction, by her songs or fair speeches, and wearing, it should seem, like a cherubim, her wings on her neck), Elizabeth, who was determined she should fly no more, and in her eagerness to tear them away, happened, inadvertently, to take off her head. The situation of the poet's mermaid, *on a dolphin's back*, "evidently marks out that distinguishing circumstance in Mary's fortune, her marriage with the dauphin of France." A mermaid would seem to have but a strangely aukward seat on the back of a dolphin, but that, to be sure, is the poet's affair, and not the commentator's; the latter, however, is certainly answerable for placing a Queen on the back of her husband—a very extraordinary situation, one would think, for a married lady; and of which I only recollect a single instance, in the common print, of "a poor man

[153-175. **My gentle Pucke** . . . that flower]

loaded with mischief." Mermaids are supposed to sing, but their *dulcet and harmonious breath* must, in this instance, to suit the allegory, allude to " those great abilities of genius and learning," which rendered Queen Mary " the most accomplished princess of her age." This compliment could not fail of being highly agreeable to the " fair Vestal." " By the rude sea is meant Scotland *incircled with the ocean*, which rose up in arms against the regent, while she [Mary] was in France. But her return home quieted these disorders; and had not her strange ill conduct afterwards more violently inflamed them, she might have passed her whole life in peace." Dr Warburton, whose skill in geography seems to match his knowledge of history and acuteness in allegory, must be allowed the sole merit of discovering Scotland to be an *island*. But, as to the disorders of that country being quieted by the Queen's return, it appears from history to be full as peaceable before as it is at any time after that event. Whether, in the revival or continuance of these disorders, she, or her idiot husband, or fanatical subjects, were most to blame, is a point upon which doctors still differ; but, it is evident, that if the enchanting song of the commentator's mermaid civilized the rude sea for a time, it was only to render it, in an instant, more boisterous than ever; those great abilities of genius and learning, which rendered her the most accomplished princess of her age, not availing her among a parcel of ferocious and enthusiastic barbarians, whom even the lyre of Orpheus had in vain warbled to humanize. Brantome, who accompanied her, says she was welcomed home by a mob of five or six hundred ragamuffins, who, in discord, with the most execrable instruments, sung *psalms* (which she was supposed to dislike) under her chamber window: "*He!*" adds he, "*quelle musique et quelle repos pour sa nuit!*" However, it seems "there is great justness and beauty in this image, as the vulgar opinion is that the mermaid always sings in storms." " The vulgar opinion," I am persuaded, is peculiar to the ingenious commentator; as, if the mermaid is ever supposed to sing, it is in *calms* which presage storms. I can perceive no propriety in calling the insurrection of the Northern earls the quarrel of Queen Mary, unless in so far as it was that of the religion she professed. But this, perhaps, is the least objectionable part of a chimerical allegory of which the poet himself had no idea, and which the commentator, to whose creative fancy it owes its existence, seems to have very justly characterised in telling us it is " out of nature "; that is, as I conceive, perfectly groundless and unnatural.'

Warburton may have urged inappropriate reasons for representing Mary as a mermaid, but history, it must be confessed, bears him out so far as to show that she was caricatured under this shape in her own day. In *Notes & Qu.* (3d Ser. V, 338, 1864) W. PINKERTON quotes the following from Strickland's *Queens of Scotland*, V, 231 : Among other cruel devices practised against Mary at this season by her cowardly assailants was the dissemination of gross personal caricatures; which, like the placards charging her as an accomplice in her husband's murder, were fixed on the doors of churches and other public places in Edinburgh. . . . Mary was peculiarly annoyed at one of these productions, called " The Mermaid," which represented her in the character of a crowned siren, with a sceptre [" formed of a hawk's lure "—Pinkerton], and flanked with the regal initials " M. R." This curious specimen of party malignity is still preserved in the State Paper Office.'

In 1794, WHITER (*A Specimen of a Commentary*, &c. p. 186) gave a wholly new turn to the discussion when he observed that the whole passage 'is very naturally derived from the *Masque* or the *Pageant*, which abounded in the age of Shakespeare; and which would often quicken and enrich the fancy of the poet with wild and orig-

[153–175. My gentle Pucke . . . that flower]

inal combinations.' To prove that a representation of a dolphin bearing a singer on his back was not uncommon at these spectacles, Whiter cites the anecdote about Harry Goldingham, given by Malone (see III, i, 44), and then concludes: 'In the present example we may perhaps be inclined to suspect that Shakespeare, in this whole description of the mermaid, the dolphin, the vestal, and Cupid, directly alludes to some actual exhibition which contained all these particulars, and which had been purposely contrived and presented before Elizabeth to compliment that princess at the expense of her unfortunate rival. So favorite a representation does the *riding on a dolphin* appear to have been in the time of our poet, that it was sometimes introduced among the quaint devices in the art of cookery,' whereof Whiter cites an example from Jonson's Masque of *Neptune's Triumph*, and from his *Staple of News;* as an illustration that the *sea-maid's music* is to be referred to the same source he cites a passage from Jonson's *Masque*, performed on *Twelfth Night*, 1605.

These examples are eminently useful, I think, as evidence of the small likelihood there is that any one in Shakespeare's audience attached any allegorical significance to Oberon's description, beyond his allusion to the 'fair Vestal throned by the West.'

In 1797, PLUMPTRE (*Appendix to Obs. on Hamlet*, p. 61) feebly answered Ritson's criticisms; for instance, it does not strike him 'as necessary that the Queen should be placed on the *back* of her husband. The word "back" might suggest to the Poet merely the idea of her being united to him, or *backing him*, i. e. their interests strengthening (or seconding, or supporting) each other by their union.' His only contribution to the discussion is his supposition that by 'Cupid's attack upon the Vestal' was meant 'the accomplishments of the Earl of Leicester.'

The pageant which Whiter supposed to have been the groundwork of Oberon's description, BOADEN found, as he believed, in '*The Princelie Pleasures*,' which Leicester devised for the entertainment of the Queen at Kenilworth in 1575, when Shakespeare was a boy. 'Where is the improbability,' he asks (*On the Sonnets*, p. 8, 1837), 'that Shakespeare in his youth should have ventured, under the wing of Greene, his townsman, even to Kenilworth itself? It was but fourteen miles distant from Stratford. Nay, that he should at eleven years of age have personally witnessed the reception of the great Queen by the mighty favourite, and perhaps have even discharged some youthful part in the pageant written by Mr Ferrers, sometime lord of misrule in the Court? Was there nothing about the spectacle likely to linger in one of "imagination all compact," a youth of singular precocity, with a strong devotion to the Muses, and little inclined, as we know, to "drive on the affair of wool at home with his father"? Nay, is there no part of his immortal works which bears *evidence* upon the question of his youthful visit? We should expect to find such graphic record in a composition peculiarly devoted to *Fancy*, and there, if I do not greatly err, we undoubtedly find it.' Boaden hereupon proceeds to show that this 'composition' is the *Midsummer Night's Dream*, and the 'graphic record' is Shakespeare's description from memory, in this speech of Oberon, of what Gascoigne calls *The Princely Pleasures at Kenilworth Castle*, and, as a corroboration of his interpretation, briefly cites certain passages from Gascoigne and from Laneham's *Letter;* as these passages are given with greater fullness by Halpin, the next commentator, it is not worth while to give their abridgement here. Let it be noted, however, that to Boaden belongs the credit of first calling attention to them. He continues:—

'Shakespeare's impression of the scene was strong and general; he does not write

[153–175. **My gentle Pucke . . . that flower**]

as if the tracts of Gascoigne and Laneham lay upon his table. His description is exactly such as, after seventeen years had elapsed, a reminiscence would suggest to a mind highly poetical.' After referring to Leicester as 'Cupid,' 'who then, or never, expected to carry his romantic prize,' and to the Queen as the 'fair vestal,' Boaden concludes:—'But the splendid captivations of Leicester were not disdained by all female minds, and the bolt of Cupid is seldom discharged in vain. Shakespeare has told us where it fell, "upon a little western flower." Why, alas! can we not ask the kindred spirit, Sir Walter Scott, whether he can conceive his own Amy Robsart more beautifully and touchingly figured than she appears to be in this exquisite metaphor?'

Doubtless Sir Walter's 'kindred spirit,' when in the flesh, would have smilingly answered his questioner that no fairer description could be anywhere found of 'his own Amy Robsart,' but that the Earl of Leicester's Amy Robsart had been dead fifteen years when The Princely Pleasures took place at Kenilworth.

The Rev. N. J. HALPIN next takes up the wondrous tale, and in a remarkaole Essay, printed by *The Shakespeare Society* (*Oberon's Vision*, &c, 1843), followed Boaden (unwittingly, as he claims) in identifying the scene of Oberon's vision with Leicester's entertainment of Elizabeth at Kenilworth; but he carries the allegory much farther than it had ever been carried before, and finds an explanation for Oberon's every phrase. His one hundred and eight octavo pages must be greatly condensed here.

However refined may be the interpretation, and however sure the elucidation of certain portions of Oberon's speech, one thing, it seems to me, is beyond all allegorical explanation, and that is 'the little western flower'; it is a genuine flower that Oberon wishes, and it is a genuine flower that Puck brings him. Let imagination run riot in a south sea of discovery with regard to every other detail—this little flower is a fact, and its magic properties must be put to use. But Halpin scouts the idea that this little flower is to be taken literally, oblivious of the difficulty into which his theory leads him, when it comes to squeezing this flower on the lover's eyelids.

'It is obvious,' says Halpin, p. 11, 'that throughout the passage under consideration the little flower is the leading object, the principal figure, to whose development all the rest—the mermaid and her dolphin, the music and the stars, Cupid and his quiver, the vestal and her moonbeams—are but accessories; intimating the time, the place, and the occasion, of its investment of its singular properties. The language throughout, with the exception of *the little flower*, is admitted to be allegorical. If this be really the case—if we are to take the little flower in its literal meaning, as a little western flower and "nothing more"—we have then, instead of a poetical beauty, a poetical anomaly, of which it would be difficult to find another example in the whole range of literature—an allegory, to wit, in which all the accessories are allegorical, but the principal figure real and literal! [Does not Halpin here forget that this elaborate allegory in all its accessories is of his own creation?] . . . I therefore infer that our "little western flower" is also an allegorical personage. . . . I conclude also that this personage is a female; not only because the delicate flower is an appropriate image of feminine beauty, but because the shaft levelled at a female bosom penetrates its heart and influences its destinies.' Halpin digresses for short space to explain that 'Dian's bud,' which has power to dispel the charm of the little flower, is Queen Elizabeth; and by way of proof cites a passage from Greene's *Friar Bacon*, where she is styled '*Diana's Rose*.' [Is it not clear, therefore, that when Greene, in acknowledged adulation of the Queen, styles her Diana's *Rose*, that Shakespeare, who had

[153–175. My gentle Pucke . . . that flower]

no connection with Greene's play, can have no other reference when he too speaks of Diana's *bud?* If we refuse to accept a conclusion like this, there will soon be an end to all Shakespearian explanations.] Halpin disposes of the assumption that the 'little western flower' was Mary, Queen of Scots, by maintaining that, with reference to Elizabeth, 'Mary was neither a *little* flower nor a *western* flower. She was Elizabeth's equal, and her kingdom lay *north* of her rivals' (p. 15). Due acknowledgment is given to Boaden for his discovery that in Oberon's first speech the *time* and *place* of the action is intimated—namely, the 'princely pleasures' at Kenilworth; and in Oberon's second speech the *persons* engaged in it, although, of course, Halpin was too well read to accept Amy Robsart as the 'little western flower.' It is clear that Leicester-Cupid was carrying on a double intrigue—with the fair Vestal on the one hand, and the little western flower on the other; and that when his bolt missed one it fell upon the other; the task now is to discover the identity of the latter, but before entering on it Halpin discusses more fully than had been hitherto discussed: first, the several features of 'the princely pleasures' to which Oberon referred; and, secondly, Boaden's conjecture that Shakespeare had himself witnessed those pleasures under the escort of his townsman, Greene.

First, in regard to the princely pleasures there are three authorities: Laneham's *Letter: whearin Part of the Entertainment untoo the Queenz Majesty, at Killingwoorth Castl in Warwick Sheer, in this Soommerz Progrest 1575, iz signified;* Gascoigne's *Princely Pleasures, with the Masque, intended to have been presented before Queen Elizabeth at Kenilworth Castle;* and Dugdale's *Antiquities of Warwickshire.* It will be well to give Halpin's collation of the three authorities unabridged, that the reader may judge how closely the scene is reproduced in Oberon's description.

'*Shakespeare* "A mermaid on a dolphin's back."

'*Laneham.* " Her Highnesse returning, cam thear, upon a swimming *mermayd*, *Triton*, Neptune's blaster," &c. [The italics throughout are, of course, Halpin's.]

'*Gascoigne.* " *Triton*, in the *likenesse of a mermaide*, came towards the Queen's Majestie as she passed over the bridge."

'*Laneham* (again). " *Arion*, that excellent and famouz muzicien, in tyre and appointment straunge, ryding alofte upon hiz old freend the *dolphin*," &c.

'*Gascoigne* (again). " From thence her Majestie passing yet further on the bridge, *Protheus* appeared sitting *on a dolphin's back.*" (The very words, as Mr. Boaden observes, of Shakespeare.)

'*Dugdale:* " Besides all this, he had upon the pool a *Triton riding on a mermaid* 18 foot long; as also *Arion on a dolphin.*"

'From this collation it appears that the impressions made on the eye-witnesses of the spectacle did not exactly correspond. The mythological figure that to Laneham appeared to be " Triton *upon a swimming mermaid*," to Gascoigne seemed to be " Triton *in the likeness of a mermaid.*" Again: the group that Gascoigne thought to be " *Protheus* on a dolphin's back" was taken by Laneham and Dugdale's informant for "*Arion* on the back of his old friend, the dolphin." Who can wonder, then, that to a more imaginative fancy the group should present the idea of " a mermaid on a dolphin's back " ? But to proceed :

'*Shakespeare.* " Uttering such dulcet and harmonious breath."

'*Laneham :* " Heerwith *Arion*, after a feaw well-coouched words unto her Majesty, beegan a delectabl ditty of a song well apted to a melodious noiz; compounded of

6

[153–175. My gentle Pucke . . . that flower]

six severall instruments, al coovert, casting soound from the dolphin's belly within; *Arion*, the seaventh, sitting thus singing (az I say) without.''

'*Gascoigne:* '' And the dolphyn was conveyed upon a boate, so that the owers seemed to be his fynnes. Within the which dolphyn, a consort of musicke was secretly placed; the which sounded; and Protheus, clearing his voyce, sang this song of congratulation,'' &c.

'*Dugdale:* ''*Arion* on a dolphin with rare musick.'' Here, too, we observe a similar discrepancy between the two eye-witnesses, touching the musician which sang upon the dolphin's back. Gascoigne supposed it to be *Protheus;* Laneham (and Dugdale's informant) thought it *Arion.* Laneham and Gascoigne were of the household of Leicester; if they could not agree what to make of this figure '' in its tyre and appointment straunge,'' surely the mere spectator may be pardoned for the mistake (if it were one) which transformed it into a mermaid. . . .

'*Shakespeare:* '' That the rude sea grew civill at her song.''

'*Laneham:* '' Mooving heerwith from the bridge, and fleeting more into the pool, chargeth he [*Triton* on his mermaid] in Neptune's name both Eolus and al his windez, the waters with hiz springs, hiz fysh, and fooul, and all his clients in the same, that they ne be so hardye in any fors to stur, but keep them calm and quiet while this Queen be prezent.''

'*Gascoigne:* '' Triton, in the likenesse of a mermaide, came towards the Queene's Majestie as she passed over the bridge, and to her declared that Neptune had sent him to her Highnes'' (and here he makes a long speech, partly in prose, partly in verse, declaring the purport of his message:) '' furthermore commanding both the waues to be calme, and the fishes to giue their attendance.'' '' And herewith,'' adds Gascoigne, '' Triton soundeth his trompe, and spake to the winds, waters, and fishes, as followeth:

" You windes, returne into your caues,   and silent there remaine,
   You waters wilde, suppress your waues,   and keep you calm and plaine;
   You fishes all, and each thing else   that here haue any sway,
   I charge you all, in Neptune's name   you keep you at a stay.''

' Here, again, we have the same slight variations which characterise the preceding parallels. In Laneham, it is '' Triton, on a swimming Mermaid,'' that calms the waves; in Gascoigne, '' Triton, in the likenesse of a Mermaid ''; and in Shakespeare, the '' Mermaid '' herself.

' We come now to the last particular of the pageant:

'*Shakespeare:* ''And certain stars shot madly from their spheres,
   To hear the sea-maid's music.''

'*Laneham:* ''At last the Altitonant displaz me hiz mayn poour; with blaz of burning darts, flying too and fro, leams of starz corruscant, streamz and hail of firie sparkes, lightninges of wildfier a-water and lond; flight and shoot of thunderboltz, all with such continuans, terror and vehemencie, that the heavins thundred, the waters scourged, the earth shooke.''

'*Gascoigne:* '' There were fireworks shewed upon the water, the which were both strange and well-executed; as sometimes passing under water a long space; when all men thought they had been quenched, they would rise and mount out of the water againe, and burn very furiously untill they were entirely consumed.''

' We have now, perhaps, sufficient evidence before us to identify the time and place of Oberon's Vision with the Princely Pleasures of Kenilworth.'

[153-175. My gentle Pucke . . . that flower]

Secondly, Boaden's surmise, that it was under the wing of a poor player that the boy, William Shakespeare, witnessed the festivities at Kenilworth, arouses Halpin's gentle indignation; it was under no such humble escort that the little boy of eleven went thither, but 'as a capable and gratified spectator in the suite of his high-minded kinsman, the head of the Arden family, and in the company of his father and mother,' among the nobility and gentry.   For, according to Halpin, 'Shakespeare was of gentle birth on both sides of the house,' and, following Malone, he connects the Ardens of Wilnecote with Robert Arden, Groom of the Chamber to Henry VII, and hereby makes Shakespeare of near kinship to the Edward Arden who incurred Leicester's implacable hate (by what he said and did at these very festivities, according to Halpin), and was put to death in 1583.   As this Edward Arden knew the secret history of Leicester's amours, it was from his lips, so Halpin conjectures (p. 46), that Shakespeare, who was nineteen years of age when Arden was executed, may have learned the mystery of the Kenilworth festivities.   This explains, so thinks Halpin, what Oberon means when he says, '*I* could see, but thou could'st not.'

But ('which doth allay the good precedence') HALLIWELL (*Life*, p. 17) says there is 'no good proof' that Robert Arden, Groom of the Chamber to Henry VII, and ancestor of Edward Arden, was 'related to the Ardens of Wilnecote'; and that 'we find the poet of nature rising where we would wish to find him rise, from the inhabitants of the valley and woodland.'   If the relationship between Oberon and Edward Arden vanishes into air, into thin air, then much of Halpin's insubstantial pageant fades with it and leaves but a wreck behind.

Halpin now addresses himself (p. 25) to the discovery of the 'little western flower': It is clear that the entertainment at Kenilworth was Leicester's 'bold stroke for a wife'; it was certainly an expensive one, it cost him £60,000, it is said; and the stroke failed.   Halpin thinks that from Laneham and Gascoigne we can learn the very day when the Earl's plans were frustrated.   There certainly appears to have been one day during which the Queen remained indoors, and the pageants prepared for that day were postponed.   Both Laneham and Gascoigne attribute the Queen's seclusion to the weather, but Halpin prefers to believe that it was due to a cause, which Sir Walter Scott imagined and made use of, in *Kenilworth;* 'or to an event of a similar kind, *in offence*, to wit, *arising out of female jealousy*.   And such precisely is the transaction which—visible to Oberon and the superior intelligences—was indiscernible to Puck and the meaner spirits in attendance.'   Of course the object of Elizabeth's jealousy was the little western flower, and Leicester's history must be scanned to find her out.   'Leicester,' says Halpin, p. 30, 'was, in fact, married (whether lawfully or otherwise) to three wives: first, Amy Robsart, in the year 1550; secondly, to Douglas, widow of the Earl of Sheffield, in or about 1572; and lastly, to Lettice, widow of Walter, Earl of Essex, 1576.   This last date brings us so close upon the royal visit to Kenilworth and to the disturbance of its festivities, that whatsoever were the embarrassments ascribed to Leicester by Sir Walter Scott, or whatever the incident alluded to by Shakespeare in the line—" before milk-white, now purple with Love's wound "—I cannot withhold my belief that they bear true reference to the Lady Lettice, Countess of Essex and none other.'

It is not worth while to follow Halpin in his history of Leicester, especially as his statements by no means tally in all particulars with the facts set forth in DEVEREUX's *Lives and Letters of the Earls of Essex*, 1853.   I am here giving Halpin's conclusions drawn from other sources.   At the time of the Princely Pleasures, Leicester's wife

[153–175. **My gentle Pucke . . . that flower**]

was Lady Douglas, Countess of Sheffield, but he was having an intrigue with Lady Essex, whose husband was in Ireland. ' Doubtless the ladies of the court attended their mistress on her Summer Progress; doubtless the wives of her principal officers of state and of her chief nobility either attended in her suite or were invited to grace her reception.   Amongst one or other of these classes it is but natural to suppose that the wife of a nobleman so high as Essex in the confidence and employment of the Queen, and a mistress so dear to the heart of her Majesty's princely entertainer, would not have been omitted.   We may then safely conclude that the Countess of Essex was a partaker of these splendid festivities; and as lovers are known to think themselves most unobserved when most in a crowd of company, no occasion can be imagined more likely to encourage those petty indiscretions which would betray their secret to the keen-sighted few than the crowded and bustling scenes of pleasure in which they were engaged.   "*I* saw, but *thou* couldst not," is the sly remark of Oberon' (pp. 42, 43).

Among these 'keen-sighted few' was Edward Arden, Shakespeare's 'distinguished kinsman,' and his informant.   When, eight years afterwards, Arden fell a victim to Leicester's vengeance, although the ostensible cause of his condemnation to death was high treason, the chief cause was, according to Dugdale, for 'certain harsh expressions touching his [Leicester's] private accesses to the Countess of Essex before she was his wife.'   As Leicester was married to Lady Essex 'soon after' the death of the Earl of Essex in 1576, and as the princely pleasures took place in 1575, Halpin thinks it is clear that Arden's 'harsh expressions' must have been uttered at Kenilworth during the festivities.   In regard to the time that elapsed between Essex's death and the marriage of his widow to Leicester, Halpin's 'soon after' is in reality two years.   Essex died in September, 1576, and the marriage took place in September, 1578, three years after the Princely Pleasures.   'Shakespeare was nineteen years of age at the death of his kinsman; he may, therefore, have heard the story from his own lips. . . . Have we not, then, in the connection between the death of Edward Arden and the guilty secret of the Lady Essex the grounds of a probable conclusion that her Ladyship is the person intended to be designated under the allegory of the "little western flower?"' (p. 46).   So varied is taste in such matters that I cannot presume to decide whether or not it detracts from the sentiment of the occasion, to reflect that the 'little western flower,' at the time of the festivities of Kenilworth, was between thirty-five and forty years old.

Halpin now turns to one of Lylie's court-plays, called *Endymion*, wherein he finds such collateral evidence of his theory as will bring satisfaction to 'the most incredulous minds.'   The earliest known edition of *Endymion* is dated 1591, 'though probably written and performed (if not published) some years before.'   It will not prove worth the labour to enter here into all the details of Halpin's analysis of this play, which fills nigh thirty of his hundred pages; it is sufficient to accept his conclusions, viz. that *Endymion* is an allegory from beginning to end, veiling Leicester's clandestine marriage with Lady Douglas Sheffield, pending his suit for the hand of his royal mistress, and the consequences of that hazardous engagement; it is parallel to Shakespeare's allegory, except that instead of the little western flower, we have the Countess of Sheffield.   If here and there known facts belie the allegory, such as where the Lady Douglas, under the name of Tellus, represents herself as a 'poor credulous virgin,' we can always apply the reflection that 'in works of fiction we must not expect a rigid conformity with the facts they shadow forth.'   Halpin concludes that *Endymion*

[153-175. My gentle Pucke . . . that flower]

is the Earl of Leicester; *Cynthia*, Queen Elizabeth; *Tellus*, the Countess of Sheffield,
and so on.   There is also another character in Lylie's allegory which finds its parallel in Oberon's vision, and this is the 'unobtrusive *Floscula*, who contributes nothing
to the action, and but little to the dialogue.'   In her, Halpin recognises the *little* western *flower*, the Countess of Essex; and finding that, in this instance, Shakespeare's
English is a translation of Lylie's Latin, he observes that the same holds good in the
case of Lylie's *Cynthia*, who is Shakespeare's *Moon*, i. e. Queen Elizabeth; and
Lylie's *Tellus*, who is Shakespeare's *Earth*, i. e. the Countess of Sheffield.   Oberon
says that he saw ' Cupid ' ' Flying between the *cold moon* and the *earth* ' ; ' it is necessary to observe,' says Halpin (p. 89), 'how accurately, discriminately, and delicately
the nice, descriptive touches of the poet are adapted to the rank, family, and misfortunes of the unhappy lady who is shadowed out under the allegory of "the Little
Flower."   1. She is a "little" flower, as compared with the royal vestal—she a
countess, Elizabeth a queen.   [As a fact, the Countess of Essex's grandmother and
Anne Bulleyn were sisters; her mother and Queen Elizabeth were therefore cousins.]
2. She is a "western" flower, that is, an English flower—an Englishwoman, a member of the English court.   If, beyond this, the epithet have a special signification, it
may refer to the office and residence of her noble husband, the Earl of Essex, who
was warden of Wales, the most western part of Britain, and *she*, therefore, *par excellence*, a *western* flower, i. e. a *western* lady.   [Halpin forgets that relatively to Oberon
and the scene of *A Midsummer Night's Dream* the whole British isle was in the
west—the fair vestal herself was throned by the west.]   3. She was once "milk-
white," indicating her purity and reputation while true to the nuptial bond with
Essex; but, 4, has become "purple with Love's wound," signifying either the shame
of her fall from virtue, or the deeper crimson of a husband's blood.   Finally, her
name is " Love in idleness," one of the many fanciful names of the *Viola tricolor*—
all indicative of the tender passion accompanied with concealment—such as " Pansies " (*pensées*, thoughts), " Cuddle-me-close," " Kiss-at-the-garden-gate," " Two-
faces-under-a-hood," &c.   But there is a peculiar elegance and significancy in the
synonym which Shakespeare has selected—" Love-in-Idleness."   It indicates the
occasion of her fall,—the absence of her lord, the waste of her affections, the " idleness," as it were, of her heart, unoccupied with domestic duties, and left a prey to
the sedulous villany of a powerful and crafty betrayer. . . . The story is an eventful
one.   It involves the fate of princes, statesmen, and nobles, and is therefore fitly
ushered in with portents, which, in the universal belief of the time, omened the fortunes of the great.   The mermaid singing her enchantments—a superstition descended
from the ancient fable of the sirens—was the old and apposite type of those female
seductions generally so fatal to their objects.   The " stars shooting madly from their
spheres " were, in that stage of the march of intellect, the prodigies which foreboded
disasters to the great.   The whole literature of that period abounds with allusions to
those " skiey influences."   On this occasion, the phenomenon seems to have signified
a *Star*—a high and mighty potentate—wildly rushing from the sphere of the bright
and lofty *Moon*—a princess of the highest rank—darting beneath the attractions of
the *Earth*—another lady, but of inferior grade—and falling in a jelly, as falling stars
are apt to do, on the lap of *Love in idleness*, an emblematic *flower*, signifying, in the
typical language of the day, a mistress in concealment. . . . Let us now compare the
poetical allegory (in juxtaposition) with a simple paraphrase of the literal meaning
which has been assigned to it. . . .

[153–175. My gentle Pucke . . . that flower]

Text	Paraphrase
OBERON.	OBERON.
My gentle Puck, come hither.	Come hither, Puck. You doubtless
Thou rememberest,	remember when, once upon a time, sit-
When once I sat upon a pro-	ting together on a rising ground, or *bray* *
montory *	by the side of a piece of water, we saw
And saw	what to us appeared (though to others it
	might have worn a different semblance)
a mermaid on a dolphin's back,	a mermaid sitting on a dolphin's back,
Uttering such dulcet and harmoni-	and singing so sweetly to the accompani-
ous breath	ment of a band of music placed inside of
	the artificial dolphin that one could very
That the rude sea grew civil at her	well imagine the waves of the mimic sea
song;	before us would, had they been ruffled,
	have calmed down to listen to her mel-
And certain stars shot madly	ody; and at the same time, there was
	a flight of artificial fireworks resembling
from their spheres	stars, which plunged very strangely out of
	their natural element into the water, and,
	after remaining there a while, rose again
	into the air, as if wishing to hear once
To hear the sea-maid's music.	more the sea-maid's music.
PUCK.	PUCK.
I remember.	I remember such things to have been
	exhibited amongst the pageantry at Kenil-
	worth Castle, during the Princely Plea-
	sures given on the occasion of Queen Eliz-
	abeth's visit in 1575.
OBERON.	OBERON.
	You are right. Well, at that very time
That very time I saw—	and place, I (and perhaps a few other of
	the choicer spirits) could discern a circum-
(*but* thou *couldst not,*)	stance that was imperceptible to you (and
	the meaner multitude of guests and visit-
Flying	ants): in fact, I saw—wavering in his
	passion
between the cold moon	between (Cynthia, or) Queen Elizabeth,
and the Earth,	and (Tellus, or) the Lady Douglas, Count-
Cupid	ess of Sheffield, (Endymion, or) the Earl
	of Leicester,
all-armed.	all-armed, in the magnificence of his prep-
	arations for storming the heart of his Royal
	Mistress.

---

* Probably "the Brayz" mentioned by Laneham as "linking a fair park with the castle on the South," and adjacent to the "goodly pool of rare beauty, breadth, length, and depth."—See Nichols's *Progresses.*

[153–175. My gentle Pucke . . . that flower]

Text	Paraphrase
A certain aim he took	He made a pre-determined and a well-directed effort for the hand of Elizabeth,
At a fair Vestal	
throned by the West;	the Virgin Queen of England;
And loosed a love-shaft madly [sic] *from his bow,*	and presumptuously made such love to her—rash under all the circumstances—
As it should pierce	as if he fancied that neither she nor any
a hundred thousand hearts;	woman in the world could resist his suit;
But I might see	but it was evident to me (and to the rest
young Cupid's fiery shaft	of the *initiated*), that the ardent Leicester's desperate venture
Quenched in the chaste beams	was lost in the pride, prudery, and jealousy of ~~passion, which invariably swayed~~
of the wat'ry Moon;	the tide of Elizabeth's passions; and the
And the imperial Votaress	Virgin Queen
passed on,	finally departed from Kenilworth Castle un-shackled with a matrimonial engagement,
in maiden meditation	
fancy-free.	and as heart-whole as ever.
Yet	And yet (continues Oberon) curious to observe the collateral issues of this amorous preparation, I watched (whatever others
marked I	may have done) and discovered the person
where the bolt of Cupid fell:	on whom Leicester's irregular passion was secretly fixed:
It fell	it was fixed
upon a little western flower,	upon Lettice, at that time the wife of Walter, Earl of Essex, an Englishwoman of rank inferior to the object of his great ambition; who, previous to this unhappy
Before milk-white;	attachment, was not only pure and innocent in conduct, but unblemished also in reputation; after which she became not only deeply inflamed with a criminal pas-
now purple with Love's wound;	sion, and still more deeply (perhaps) stained with a husband's blood, but the subject, also, of shame and obloquy.
And maidens	Those, however, who pity her weakness, and compassionate her misery, still offer a
call it	feeble apology for her conduct, by calling it the result of her husband's voluntary absence, of the waste of affections naturally
Love in Idleness.	tender and fond, and of the idleness of a heart that might have been faithful if busied with honest duties, and filled with domestic loves.
	You cannot mistake, after all I have said—
Fetch me that flower.	Go—fetch me that flower.

[153-175. My gentle Pucke . . . that flower]

Such is Halpin's explanation of 'Oberon's vision.' It does not appear, despite its ingenuity, to have made any impression on some of the best Shakespearian editors; it may well be that they were appalled by its intricacy and length. It is not even alluded to by DYCE, COLLIER, or STAUNTON. Possibly they were repelled by the cruel conclusion that it was not a flower, but Lettice Knollys, that was to be squeezed in Titania's eyes. However, Halpin has one staunch follower, one who with a greedy ear will devour up any discourse which aims at identifying Shakespeare's characters with that group around Southampton, to whose loves, to whose jealousies, to whose hates he would fain have us believe Shakespeare crammed his plays to bursting with allusions.

Mr GERALD MASSEY (*The Secret Drama of Shakespeare's Sonnets*, 1888) asserts that Halpin has 'conclusively shown the "little western flower"' to be Lettice Knollys, but on one or two minor points Halpin does not take Massey with him. 'My interpretation,' says MASSEY (p. 446), 'of Oberon's remark, "That very time I saw, *but thou couldst not*," is to this effect: Shakespeare is treating Puck, for the moment, as a personification of his own boyhood. "*Thou rememberest* the rare vision we saw at the 'Princely Pleasures' of Kenilworth?" "I remember," replies Puck. So that he was then present, and saw the sights and all the outer realities of the pageant. But the Boy of eleven could not see what Oberon saw—the matrimonial mysteries of Leicester; the lofty aim of the Earl at a Royal prize, and the secret intrigue then pursued by him and the Countess of Essex. Whereupon, the Fairy King unfolds in Allegory what he before saw in vision, and clothes the naked skeleton of fact in the very bloom of beauty. My reading will dovetail with the other to the strengthening of both. But Mr Halpin does not explain *why* this "little flower" should play so important a part; why it should be the chief object and final cause of the whole allegory, so that the royal range of the imagery is but the mere setting; why it should be the only link of connection betwixt the allegory and the play. My rendering alone will show why and how. The allegory was introduced on account of these two cousins; [it should be here observed that, according to Mr Massey, the *causa causans* of the present play was the jealousy of Elizabeth Vernon, and her bickerings with her cousin Lady Rich, who are, respectively, Helena and Hermia]; the "little western flower" being mother to Lady Rich and aunt to Elizabeth Vernon. The Poet pays the Queen a compliment by the way, but his allusion to the love-shaft loosed so impetuously by Cupid is only for the sake of marking where it fell, and bringing in the Flower. It is the little flower alone that is necessary to his present purpose, for he is entertaining his "Private Friends" more than catering for the amusement of the Court. This personal consideration will explain the tenderness of the treatment. Such delicate dealing with the subject was not likely to win the Royal favour; the "imperial votaress" never forgave the "little western flower," and only permitted her to come to Court once, and then for a private interview, after her Majesty learned that Lettice Knollys had really become Countess of Leicester. Shakespeare himself must have had sterner thoughts about the lady, but this was not the time to show them; he had introduced the subject for poetic beauty, not for poetic justice. He brings in his allegory, then, on account of those who are related to the "little western flower," and in his use of the flower he is playfully tracing up an effect to its natural cause. The mother of Lady Rich is typified as the flower called "Love-in-idleness." . . . And the daughter was like the mother. "It comes from his mother," said the Queen, with a sigh, speaking of the dash of wilful devilry and the Will-o'-

[153-175. My gentle Pucke . . . that flower]

the wisp fire in the Earl of Essex's blood! Shakespeare, in a smiling mood, says the very same of Lady Rich and her love-in-idleness. " It comes from her mother!" She, too, was a genuine " light-o'-love," and possessed the qualities attributed to the " little western flower "—the vicious virtue of its juice, the power of glamourie by communicating the poison with which Cupid's arrow was touched when dipped for doing its deadliest work. These she derives by inheritance; and these she has tried to exercise in real life on the lover of her cousin. The juice of " love-in-idleness " has been dropped into Southampton's eyes, and in the Play its enchantment has to be counteracted. And here I part company with Mr Halpin. "*Dian's bud*," the " *other herb*," does not represent *his* Elizabeth, the Queen, but *my* Elizabeth, the " faire Vernon." It cannot be made to fit the Queen in any shape. If the herb of more poten tial spell, " whose liquor hath this *virtuous* property " that it can correct all errors of sight, and " undo this hateful imperfection " of the enamoured eyes—" Dian's bud, o'er Cupid's flower, Hath such force and blessed power,—" were meant for the Queen, it would have no application whatever in life, and the allegory would not *impinge* on the Play. Whose eyes did this virtue of the Queen purge from the grossness of wanton love? Assuredly not Leicester's, and as certainly not those of the Lady Lettice. The facts of real life would have made the allusion a sarcasm on the Queen's virgin force and " blessed power," such as would have warranted Iago's expression, " *blessed fig's end!*" If it be applied to Titania and Lysander, what had the Queen to do with them, or they with her? The allegory will not go thus far; the link is missing that should connect it with the drama. No. " Dian's bud " is not the Queen. It is the emblem of Elizabeth Vernon's true love and its virtue in restoring the "precious seeing " to her lover's eyes, which had in the human world been doating wrongly. It symbols the triumph of love-in-earnest over love-in-idleness; the influence of that purity which is here represented as the offspring of Dian. Only thus can we find that the meeting-point of Queen and Countess, of Cupid's flower and Dian's bud, in the Play, which is absolutely essential to the existence and the oneness of the work; only thus can we connect the cause of the mischief with its cure. The allusion to the Queen was but a passing compliment; the influence of the " *little western flower* " and its necessary connection with persons in the drama are as much the *sine quâ non* of the Play's continuity and developement as was the jealousy of Elizabeth Vernon a motive-incident in the poetic creation.'

Warburton's explanation that by the mermaid the Queen of Scots was meant, was silently adopted by JOHNSON, and was praised by CAPELL. I have said that one of our best modern critics had also accepted it.—HUNTER (*New Illust.* i, 291) observes, as follows: I profess at once my adherence to the interpretation which Bishop Warburton has given of the allegorical portion of this celebrated passage, so far as to the mermaid representing the Queen of Scots; and I think I can perceive some reasons for this, which were not adverted to by himself and which have been left unnoticed by Ritson, [by Boaden, and by Halpin]. . . . It may be admitted that to place a mermaid on the back of a dolphin is perhaps not the happiest conception that might have been formed, and there have been found critics who have scoffed at it; but this has nothing to do with the question whether the mermaid had any counterpart in the allegory, and whether that counterpart was the Queen of Scots. . . . Seeing the large space which the mermaid occupies, it can hardly be that, if there is an allegory at all, she does not bear a part in it; and, seeing how everything said of the mermaid has its counterpart in the Queen of Scots, and not in any other person, it can hardly be that

[153–175. **My gentle Pucke . . . that flower**]

the mermaid was not intended to represent her. She has the dolphin with her, which may certainly seem very well to arise out of the fact that she had been married to the Dauphin of France; she utters 'dulcet and harmonious breath'; and, beside the general charm which surrounded this royal lady, . . . if we must interpret the allegory in a literal spirit, we know on the best authority that she had an 'alluring Scottish accent,' which, with the agreeableness of her conversation, fascinated all that approached her, and subdued even harsh and uncivil minds. But some were touched by it more than others. She had not been long in England when two Northern earls broke out in open rebellion, and would have made her queen. . . . Here, at least, it must be admitted that we have what answers very well to stars that 'shot madly from their spheres To hear the sea-maid's music.' There is not indeed a circumstance about the mermaid to which we do not find something correspondent in the Scottish Queen. Now proceed to the other half of the allegory. 'That very time I saw (but thou could'st not).' *That very time* :—These words are most important. At the very time when the Duke of Norfolk was aspiring to the hand of the Queen of Scots, and so, shooting from his sphere, the Queen of England was herself strongly solicited to marry. [See lines 161–165.] Halpin would give Cupid a counterpart. The Earl of Leicester, according to his theory, is Cupid. This never could have been the intention of the poet, who uses one of the most ordinary of all figures, supplied from the store-house of the ancient mythology, to represent the advances which were made to Elizabeth. The expression *at that very time* appears to have escaped the notice of the learned commentator who shewed the true interpretation of this passage, and yet it appears to me to connect the two parts and to leave no shadow of doubt that his hypothesis is the right one. The identity in respect of time happens to be very distinctly marked in a few lines in Camden's *Annals :* ' Non majorem curam et operam ad has nuptias conficiendas adhibuerunt Galli, quam Angli nonnulli ad alias accelerandas inter Scotorum Reginam et Norfolchium.' The suitor to Queen Elizabeth was, of course, the Duke of Anjou. At the very time when at the sea-maid's music certain stars shot from their spheres, the strong dart aimed by Cupid against Elizabeth fell innocuous; and she passed on ' In maiden meditation fancy-free.' The allegory ends here, according to all just rule, when the flower is introduced. This flower was a real flower about to perform a conspicuous part in the drama, and the allegory is written expressly to give a dignity to the flower; it is the splendour of preparation intended to fix attention on the flower, whose peculiar virtues were to be the means of effecting some of the most important purposes of the drama. The passage resembles, in this respect, one a little before, in which there is an interest given to the little henchman by the recital of the gambols of Titania with his mother on the sea-shore of India, and the interest thrown around Othello's handkerchief. The allegory has been complete, and has fulfilled its purpose when we come to the flower, which in the hands of the poet undergoes a beautiful metamorphose, and has now acquired all the interest which it was desirable to give it, and poetically and dramatically necessary, considering the very important part which was afterwards to be performed by it.

In the copy of Hanmer's *Shakespeare*, which Mrs F. A. KEMBLE used in her Public Readings, and which she gave to the present Editor, there is in the margin opposite this passage the following MS note by that loved and venerated hand :—' It always seems to me the crowning hardship of Mary Stuart's hard life to have had this precious stone thrown at her by the hand of Shakespeare—it seems to me most miserable, even

The iuyce of it, on fleeping eye-lids laid,   176
Will make or man or woman madly dote
Vpon the next liue creature that it fees.
Fetch me this hearbe, and be thou heere againe,
Ere the *Leuiathan* can fwim a league.   180
   Pucke.   Ile put a girdle about the earth, in forty mi-
nutes.   182

177. *or man*] *a man* F₃F₄, Rowe.
178. *it fees*] *is seen* Coll. MS.
181. *Ile...earth*] One line, Pope et
seq.

181. *about*] *roūd about* Q₁.   *round about* Pope et seq.
182. [Exit. Ff.

when I think of all her misery, that she should have had this beautiful, bad record from the humanest man that ever lived, and, for her sins, the greatest poet—and she that was wise (not good) and prosperous, to have this crown of stars set on her narrow forehead by the same hand.'

Apart from the impossibility, which Hunter sees, but Halpin and Massey do not see, of including in the allegory 'the little western flower,' there is to me in the acceptance of Halpin's whole theory one obstacle which is insurmountable, and this is, the length of time which had elapsed between the festivities at Kenilworth and the date of this play. To suppose that Shakespeare's audience, whether at court or at the theatre, would at once, on hearing Oberon's vision, recall Leicester's intrigue of twenty years before, is to assume a capacity for court-scandal which verges on the supernatural, and a memory for it which could be regarded only with awe. Moreover, taking the very earliest date ascribed by any critic to this play, 1590, at that time 'Cupid' had been dead two years, and 'the little western flower' was living with her third husband. Finally, KURZ has pointed out (*Sh. Jahrbuch*, 1869, p. 295) that as far as the Princelie Pleasures were concerned the age was so accustomed to such performances that any reference to these particular festivities would be understood by no one but the poet himself; 'they were a drop, glittering 'tis true, but yet a mere drop in a sea of similar festivals, with pageants and plays wherein there was a deadly sameness of subjects drawn from the mythology of the Renaissance-Antique. Nay, a glance at the various Courts of the Continent enlarges this sea to an ocean; such revelries were everywhere, and all of them described and printed and engraved and passed on from Court to Court—from highest Jove to the latest sea-monsters, all hackneyed alike.' ED.

180. **Leuiathan**] W. A. WRIGHT: The margins of the Bibles in Shakespeare's day explained leviathan as a whale, and so no doubt he thought it.

181. **Ile**] COLLIER'S MS changed this to *I'd*, which LETTSOM (ap. Dyce, ed. ii) says the sense requires. Collier, however, did not adopt it; HUDSON did.

181. **girdle**] STEEVENS: Perhaps this phrase is proverbial. Compare Chapman's *Bussy d'Ambois*, 1607: 'To put a girdle round about the world.'—*Works*, ii, 6.—HALLIWELL: This metaphor is not peculiar to Shakespeare. The idea and expression were probably derived from the old plans of the world, in which the Zodiac is represented as 'a girdle round about the earth.' Thus, says the author of *The Compost of Ptolomeus*, 'the other is large, in maner of a girdle, or as a garland of flowers, which they doe call the Zodiack.' [Halliwell cites several other examples to the

Ober.   Hauing once this iuyce,   183
Ile watch *Titania*, when fhe is afleepe,
And drop the liquor of it in her eyes :   185
The next thing when fhe waking lookes vpon,
(Be it on Lyon, Beare, or Wolfe, or Bull,
On medling Monkey, or on bufie Ape)
Shee fhall purfue it, with the foule of loue.
And ere I take this charme off from her fight,   190
(As I can take it with another hearbe)
Ile make her render vp her Page to me.
But who comes heere? I am inuifible,
And I will ouer-heare their conference.

Enter Demetrius, Helena following him.   195

Deme.   I loue thee not, therefore purfue me not,

184. *when*] *whence* Q₂.   188. *On medling*] *Or medling* Rowe,
     afleepe] *a fleepe* Q₁F₃.   Pope.
185. *in her*] *on her* Han.   190. *off from*] *from of* Q₁. *from off*
186. *when*] *which* Rowe+. *then* Q₁,   Theob. Cap. Sta. Cam. White ii.
Cap. et seq.   194. [Scene III. Pope+.

---

same effect, and Staunton, who says that the phrase seems to have been a proverbial mode of expressing a voyage round the world, adds another from Shirley's *Humourous Courtier*, I, i : ' Thou hast been a traveller, and convers'd With the Antipodes, almost put a girdle About the world.'   See also, to the same purpose, Walker, *Crit.* iii, 48.—Green (*Emblem Writers*, p. 413) gives an Emblem by Whitney, 1586, representing a globe whereon rides Drake's ship, which first circumnavigated the earth ; to the prow of this ship is attached a girdle which goes round the world, while the other end is held by the hand of God, issuing from the clouds.—Ed.]

181. **forty**] Elze (*Notes*, &c. 1889, p. 230) has collected a large number of instances of the use of ' forty' as an indefinite number, in German as well as in English, from the ' forty days and forty nights' of the Deluge to Whittier's *Barbara Frietchie*, 1879 : ' Forty flags with their silver stars, Forty flags with their crimson bars.'

184. **when she**] Note how the ear of the compositor of Q₂ misled him when he set up *whence she* for ' when she.'—Ed.

185. **drop the liquor**] See the extract from the *Diana* of George of Montemayor, in Appendix, *Source of the Plot.*

193. **inuisible**] Theobald : As Oberon and Puck may be frequently observed to speak when there is no mention of their entering, they are designed by the poet to be supposed on the stage during the greatest part of the remainder of the play, and to mix, as they please, as spirits, with the other actors, without being seen or heard, but when to their own purpose.—Collier (ed. ii) : Among the ' properties' enumerated in Henslowe's *Diary* is ' a robe for to go invisible.'   Possibly Oberon wore, or put on, such a robe, by which it was understood that he was not to be seen.

196. **pursue me not**] Mrs F. A. Kemble [*MS note*] : Was it not well devised

Where is *Lysander,* and faire *Hermia*?   197
The one Ile ftay, the other ftayeth me.
Thou toldft me they were ftolne into this wood;
And heere am I, and wood within this wood,   200
Becaufe I cannot meet my *Hermia.*

198. *ftay...ftayeth*] QqFf, Knt, Hal.
flay...flayeth Thirlby, Theob. et cet.
199. *into*] *vnto* Qq, Cap. Steev.'85,
Sta. Cam. White ii.

200. *wood...wood*] *wodde,...wood* Q₁.
wode...wood Han. Cap. Cam.
201. *my*] *with* Mal. Steev.'93, Var.
Sing. i.

to make the timid, feminine Helena the pursuer of her indifferent, inconstant lover?
We know how she looked—tall and slender, fair, delicate, and fragile. If the short,
round, dark-eyed Hermia had thus wooed a man, it would have been unlovely.
Shakespeare has wonderfully given this bold position to a ' maiden never bold '; and
the pale, pathetic figure imploring vainly a man's love, and enduring patiently his
contemptuous refusal, still represents a more tender and feminine idea than the bloom-
ing, well-beloved maiden pointing to the remote turf where she will have her lover
lie that he may not offend her by his nearness while they sleep together in the wood.

198. **stay . . . stayeth**] At an early date, 1729, the Rev. STYAN THIRLBY, in a
letter to Theobald, proposed, without comment, the change of 'stay . . . stayeth' to
slay . . . slayeth, and this excellent emendation has commended itself to almost every
editor since then. As far as I know, the only defenders of the original text are
HEATH, KNIGHT, and HALLIWELL. The first urges (p. 50) that 'there is not the
least foundation for imputing this bloody disposition [expressed by Thirlby's change]
to Demetrius. His real intention is sufficiently expressed by [the Folio, viz:] " I will
arrest Lysander, and disappoint his scheme of carrying off Hermia; for 'tis upon the
account of this latter that I am wasting away the night in this wood." I believe, too,
another instance cannot be given, wherein a lady is said to *slay* her lover by the slight
she expresses for him.' [*Aliquando dormitat,* &c. The truly admirable Heath quite
forgot the song in *Twelfth Night :* ' I am slain by a fair, cruel maid,' II, iv, 55. He
properly referred, however, 'stay' to Lysander, and 'stayeth' to Hermia. But
KNIGHT, who adds no new argument, confuses them. HALLIWELL merely reprints
Heath's note, and adds two needless instances, where 'stay' means *to arrest.* ZACH-
ARY JACKSON, who, with his tribesmen, BECKET and LORD CHEDWORTH, is never
quoted in these pages, upholds the Folio, so says Knight; this is quite sufficient to
condemn it.—R. G. WHITE (ed. i), in reference to the plea urged by Heath, that it
is unnecessary to attribute murderous designs to Demetrius, properly calls attention to
Demetrius's wish (III, ii, 67) to give Lysander's carcase to his hounds, and he might
have added Hermia's fear, expressed more than once, that her lover had been slain
by Demetrius.—ED.]

200. **wood . . . wood**] Of course, a play upon words, where the former 'wood'
means *enraged,* and, as it is the Anglosaxon *wód,* examples of it may be found in our
earliest literature. It is worth considering whether, in a modernised text, it would
not be well to indicate the difference in meaning by spelling the former *wode,* as has
been done by HANMER, CAPELL, and by W. A. WRIGHT, in *The Cambridge Edition.*
A slight objection to it lies in the fact that we are by no means sure that there was a
distinction between the words in general pronunciation. The *wodde* of Q₁ may be a
mere misprint, or the peculiar spelling of a single compositor.—ED.

Hence, get thee gone, and follow me no more.   202

    Hel.   You draw me, you hard-hearted Adamant,
But yet you draw not Iron, for my heart
Is true as fteele. Leaue you your power to draw,   205
And I fhall haue no power to follow you.

    Deme.   Do I entice you? do I fpeake you faire?
Or rather doe I not in plaineft truth,
Tell you I doe not, nor I cannot loue you?

    Hel.   And euen for that doe I loue thee the more;   210
I am your fpaniell, and *Demetrius*,
The more you beat me, I will fawne on you.   212

---

202. *thee*] *the* Q₁F₃.   209. *nor*] *not* Qq. *and* Pope, Han.
204. *Iron, for*] *Iron for* Gould.   210. *thee*] Q₂Ff, Rowe+, White i.
205. *you*] Om. F₃F₄.   *you* Q₁ et cet.

---

203. **You**] If Shakespeare indicated shades of meaning by the use of *thou* and *you* (and sometimes I am inclined, so difficult or so fanciful is the analysis, to think he did not always so indicate them), it would be interesting to note in this dialogue the varying emotions of love, contempt, respect, and anger that flit over the speakers and find expression in these personal pronouns.—ED.

203. **Adamant**] Cotgrave gives. 'Aimant: m. *A louer, a seruant, a sweet-heart, also, the Adamant, or Load-stone.*' Again, 'Calamite: m. *The Adamant, Loadstone, or Magnes-stone.*' The qualities of the lodestone are well known at the present day, and as they were no less well known in Shakespeare's day, examples of their use in poetry or prose are superfluous. It is sufficient to know that lodestone and 'adamant' were formerly synonymous.—ED.

204. **for**] LETTSOM (ap. Dyce, ed. ii) queries if this should not be *though*, and HUDSON suspects that 'he is right, as he is apt to be.'—MARSHALL (*Henry Irving Sh.* p. 372) adopts *though*, and says 'for' in the sense of *because* is nonsense. 'If we retain "for,"' he urges, 'we must take it as equivalent to *for all*, i. e. *in spite of all.*'—D. WILSON (p. 248): In the Ff 'Iron' is printed with a capital, which, in F₂ is somewhat displaced and separated from the *ron*. This has apparently suggested to the former possessor of my copy an ingenious emendation, which he has written on the margin, thus: 'You draw, not I run, for, &c. Among my own annotations are [*sic*] included this conjectural reading, 'you draw *no truer;* for,' &c. [There is no need of change if we take 'draw not' in the sense of the opposite of drawing, namely, of repulsion, which is not logical, it must be granted, but then Helena was not logical; 'you are,' she says, in effect, 'adamant only as far as I am concerned; you repel iron, as is shown by your repelling my heart, which is true steel'; or there may have been the image in Helena's mind of a piece of lodestone, such as all of us have often seen, encrusted with bits of iron, which have been drawn to it, and she says to Demetrius, in effect, 'You do not draw iron, because if you did, my heart, which is the truest steel, would be close to your heart, and I should be folded in your arms.'—ED.]

209. **nor I cannot**] For examples of this common double negative, see ABBOTT § 406, and for 'euen,' in the next line, see line 31 of this scene.

Vſe me but as your ſpaniell ; ſpurne me, ſtrike me,   213
Negleƈt me, loſe me ; onely giue me leaue
(Vnworthy as I am)to follow you.   215
What worſer place can I beg in your loue,
(And yet a place of high reſpeƈt with me)
Then to be vſed as you doe your dogge.
   Dem.   Tempt not too much the hatred of my ſpirit,
For I am ſicke when I do looke on thee.   220
   Hel.   And I am ſicke when I looke not on you.
   Dem.   You doe impeach your modeſty too much,
To leaue the Citty, and commit your ſelfe
Into the hands of one that loues you not,
To truſt the opportunity of night,   225
And the ill counſell of a deſert place,
With the rich worth of your virginity.
   Hel.   Your vertue is my priuiledge : for that
It is not night when I doe ſee your face.
Therefore I thinke I am not in the night,   230
Nor doth this wood lacke world's of company,
For you in my reſpeƈt are nll the world.
Then how can it be ſaid I am alone,
When all the world is heere to looke on me ?   234

---

214. *loſe*] *looſe* Q₁. *loathe* Anon. ap.
Hal.
  216. *can*] *can can* F₂.
  218. *doe*] Ff, Rowe, White i. *do use*
Var.'21, Sing. i. *vſe* Qq et cet.
   dogge.] *dog ?* Rowe.

228. *priuiledge : for that*] *privilege
for that.* Tyrwhitt, Steev.'78, Rann.
Mal. Sing. Knt, Coll. Dyce, Hal. White
i, Ktly, C. Clarke, Huds. Rolfe.
  232. *nll*] F₁.

---

214. **lose**] HALLIWELL: Perhaps this means blot me out of your memory, lose all remembrance of me.

222. **impeach**] STEEVENS: That is, bring it into question, as in *Mer. of Ven.* III, ii, 280: 'doth impeach the freedom of the state.'

228. **for that**] TYRWHITT'S punctuation (see Text. Notes), which makes 'that' refer to Helena's leaving the city, has been adopted by all the best editors down to STAUNTON, who returned to the Ff and Qq. Every editor, without exception I think, has substituted a comma at the end of the next line, after 'face,' instead of the full stop. Staunton has a respectable following in the CAMBRIDGE EDITORS.—ABBOTT § 287, expresses no preference, and, indeed, the present question is one of the many instances where the scales are so nicely balanced that a transient mood may decide it.—ED.

229. **It is not night, &c.**] JOHNSON: Compare '—Tu nocte vel atra Lumen, et in solis tu mihi turba locis.'—Tibullus, *Carm.* IV, xiii, 11.

232. **respect**] That is, as far as I am concerned.

Dem.   Ile run from thee, and hide me in the brakes,   235
And leaue thee to the mercy of wilde beaſts.

Hel.   The wildeſt hath not ſuch a heart as you ;
Runne when you will, the ſtory ſhall be chang'd :
Apollo flies, and *Daphne* holds the chaſe ;
The Doue purſues the Griffin, the milde Hinde   240
Makes ſpeed to catch the Tyger.   Bootleſſe ſpeede,
When cowardiſe purſues, and valour flies.

Demet.   I will not ſtay thy queſtions, let me go ;
Or if thou follow me, doe not beleeue,
But I ſhall doe thee miſchiefe in the wood.   245

Hel.   I, in the Temple, in the Towne, and Field
You doe me miſchiefe.   Fye *Demetrius,*
Your wrongs doe ſet a ſcandall on my ſexe :
We cannot fight for loue, as men may doe ;
We ſhould be woo'd, and were not made to wooe.   250
I follow thee, and make a heauen of hell,

---

243. *queſtions*] *question* Steev. conj.
Dyce ii, iii, Walker, Huds.
244. *thou*] *you* Rowe, Pope, Han.
246, 257. *I*] *Ay* Rowe et seq.
246. *and*] Q₂Ff, Rowe, Pope, Han.

Var. Knt, Hal. White i, Sta.   *the* Q₁ et
cet.
250. [Demetrius breaks from her, and
Exit. Cap. et seq. (subs.).
251. *I*] *Ile* Qq, Cap. et seq.

---

240. **Griffin**] WAY (*Prompt. Parv.* s. v. *Grype*, footnote): This fabulous animal
is particularly described by Sir John Maundevile, in his account of Bacharie. ' In
that contree ben many griffounes, more plentee than in ony other contree.   Sum men
seyn that thei han the body upward as an eagle, and benethe as a lyoune, and treuly
thei seyn sothe that thei ben of that schapp.   But o griffoun hathe the body more
gret, and is more strong thanne viij. lyouns, of suche lyouns as ben o this half, and
more gret and strongere than an c. egles, suche as we han amonges us.'   He further
states that a griffin would bear to its nest a horse, or a couple of oxen yoked to the
plough ; its talons being like horns of great oxen, and serving as drinking cups ; and
of the ribs and wing feathers strong bows were made.

240. **the milde**] For other examples of unemphatic monosyllables, like the present ' the,' standing in an emphatic place, see ABBOTT, § 457.

243. **questions**] STEEVENS : Though Helena certainly puts a few insignificant
' questions ' to Demetrius, I cannot but think our author wrote *question*, i. e. discourse,
conversation.   So in *As You Like It*, III, iv, 39 : ' I met the duke yesterday, and
had much question with him.'   [The same emendation occurred to WALKER, *Crit.* i,
248.]—W. A. WRIGHT : The plural may denote Helena's repeated efforts at inducing
Demetrius to talk with her.

245. **But**] For many other passages illustrating the 'preventive meaning' of *but*,
see ABBOTT, § 122.

251. **I follow**] There is really no reason for deserting the Ff here.—ED.

To die vpon the hand I loue ſo well.   *Exit.*   252

    Ob. Fare thee well Nymph, ere he do leaue this groue,

Thou ſhalt flie him, and he ſhall ſeeke thy loue.

Haſt thou the flower there? Welcome wanderer.   255

Enter Pucke.

    Puck. I, there it is.

    Ob. I pray thee giue it me.

I know a banke where the wilde time blowes,   259

---

252. Exit.] Om. Q₁. Exeunt. Rowe +.   257. *there*] *here* Lettsom, Huds.
254. [Re-enter Puck. Cap. et seq.   259. *where*] *whereon* Pope +, Cap.
256. [Scene IV. Pope +.   Steev. Rann, Sing. i, Dyce ii, iii, Huds.

---

252. To die] That is, *in dying*, not *in order to die.* For similar instances of this
gerundial usage, see Abbott, § 356.

252. die vpon the hand] W. A. Wright : ' Upon' occurs in a temporal sense
in some phrases, where it is used with the cause of anything. In such cases the consequence follows ' upon ' the cause. For instance, in *Much Ado*, IV, i, 225 : ' When
he shall hear she died upon his words.' Again, in the same play, IV, ii, 65 : 'And
upon the grief of this suddenly died.' Also ' on ' is used in a local sense with the
instrument of an action. See below, II, ii, 112 : ' O how fit a word, Is that vile name
to perish on my sword !' And *Jul. Cæs.* V, i, 58 : ' I was not born to die on Brutus'
sword.' Hence, metaphorically, it occurs in *Lear*, II, iv, 34 : ' On whose contents
They summoned up their meiny.' None of these instances are strictly parallel to the
one before us, but they show how ' upon the hand ' comes to be nearly equivalent to
' by the hand,' while with this is combined the idea of local nearness to the beloved
object which is contained in the ordinary meaning of ' upon.' A better example is
found in Fletcher's *Chances*, I, ix : ' Give me dying, As dying ought to be, upon mine
enemy, Parting with mankind by a man that's manly.'

255–258. Hast . . . me.] Dyce (ed. ii) : ' The first part of each of these two
verses,' says Mr W. N. Lettsom, ' is inconsistent with the second part. Should we
not read and point ? " Hast thou the flower there, welcome wanderer ? *Puck.* Ay,
here it is. *Obe.* I pray thee give it me." ' Mr Swynfen Jervis proposes : ' Welcome, wanderer. Hast thou the flower there ?' [Lettsom's punctuation of line 255
is certainly good, but the change of ' there ' to *here* seems needless ; in either case
the word would be uttered with a gesture. According to the footnotes in the Cam.
Ed., Zachary Jackson anticipated Swynfen Jervis. The reason is given in the *Preface*
to this volume for the exclusion from these Textual Notes of Jackson's conjectures.
—Ed.]

259. where] Malone, Keightley, Abbott (§ 480), and W. A. Wright pronounce this as a disyllable.—R. G. White (ed. i) says that ' Malone reasonably supposed ' it to be ' used as a disyllable,' and added, ' it may, at least, very properly have
a disyllabic quantity,'—a distinction which it is somewhat difficult to comprehend ;
it is even more difficult to comprehend what rhythmical advantage these eminent editors imagine has been gained by this conversion of a monosyllable into a disyllable,
when by its position in the verse the ictus must fall on its manufactured second syl-
lable. Can it be that their ears are pleased by ' I knòw | a bànk | whe-ère | the

7

Where Oxflips and the nodding Violet growes,	260
Quite ouer-cannoped with lufcious woodbine,
With fweet muske rofes, and with Eglantine;	262

260. *Oxflips*] *Oxlips* Q₁. *the Oxflips*
F₄, Rowe. *oxslip* Pope, Han. *ox-lip*
Theob. Warb. Johns.
261. *Quite ouer-cannoped*] *Quite ouer-cannopi'd* Q₁. *O'er-cannopy'd* Pope,

Theob. Warb. Johns. Cap. *White clover canopied* Bulloch.
261. *lufcious*] *lufhious* Qq. *lush* Theob. conj. Steev.'93, Coll. ii (MS), Dyce ii, Huds.

---

wìld | thyme blòws. | ' ? Unless the ictus be preserved the disyllable has been made in vain. To me, it would be better ignominiously to adopt Pope's *whereòn*. But there is no need of appealing either to Pope or to Malone. Let a pause before ' where' take the place of a syllable, as in 'swifter than the moon's sphere' in line 7 of this scene; which see. With my latest editorial breath I will denounce these disyllables devised to supply the place of a pause.—ED.

260. **Oxslips**] ' The Oxelip, or the small kinde of white Mulleyn, is very like to the Cowslippe aforesaide, sauing that his leaues be greater and larger, and his floures be of a pale or faynt yellow colour, almost white and without sauour.'—Lyte, p. 123, ed. 1578.—KEIGHTLEY (*Exp.* 132, and *N. & Qu.* 2d Ser. xii, 264) transposes 'oxlip' and ' violet,' because, as he alleges, the former 'nods' and the latter does not. This wanton change in the character of the *ox*lip he justifies by a line from *Lycidas* about the *cows*lip, a different plant : ' With cowslips wan, that hang the pensive head.'— v. 14. Unquestionably the violets in this country nod, whatever their British brothers may do.—ED.

260. **grows**] Either the singular by attraction, or from the image in the mind of one bed of oxlips and violets growing together.—ED.

261. **luscious**] JOHNSON : On the margin of one of my Folios an unknown hand has written ' *lush* woodbine,' which, I think, is right. This hand I have since discovered to be Theobald's.—RITSON : *Lush* is clearly preferable in point of sense, and absolutely necessary in point of metre.—STEEVENS : Compare *Temp.* II, i, 52 : How lush and lusty the grass looks !'—W. A. WRIGHT : That is, sweet-scented ; generally sweet to the taste. [It can be no disgrace to accept this line as an Alexandrine : ' Quite ò | ver-càn | opèd | with lùs | ciòus | woodbìne,' where the resolved syllables of ' lus-ci-ous ' need not be harshly nor strongly emphasised.—ED.]

261. **woodbine**] 'Woodbine or Honysuckle hath many small branches, whereby it windeth and wrappeth it selfe about trees and hedges. . . . Woodbine groweth in all this Countrie in hedges, about inclosed feeldes, and amongst broome or firres. It is founde also in woodes. . . . This herbe, or kinde of Bindeweede, is called . . . in Englishe Honysuckle, or Woodbine, and of some Caprifoyle.'—Lyte, p. 390, ed. 1578. [See IV, i, 48.]

262. **muske roses . . . Eglantine**] ' The sixth kinde of Roses called Muske Roses, hath slender springes and shutes, the leaues and flowers be smaller then the other Roses, yet they grow vp almost as high as the Damaske or Prouince Rose. The flowers be small and single, and sometimes double, of a white colour and pleasant sauour, in proportion not muche vnlyke the wilde Roses, or Canel Roses. . . . The Eglantine or sweete brier, may be also counted of the kindes of Roses, for it is lyke to the wilde Rose plante, in sharpe and cruel shutes, springes, and rough branches.'— Lyte, p. 654.

There fleepes *Tytania*, fometime of the night,   263
Lul'd in thefe flowers, with dances and delight:
And there the fnake throwes her enammel'd skinne,   265
Weed wide enough to rap a Fairy in.
And with the iuyce of this Ile ftreake her eyes,
And make her full of hatefull fantafies.
Take thou fome of it, and feek through this groue;   269

263. *fometime* QqFf, Dyce, Sta. Cam.   266. *rap*] *wrappe* Q$_1$.   *wrap* Ff.
White ii.   *some time* Rowe et cet.   267. *And*] *There* Han.   *Then* Ktly.
264. *flowers*] *bowers* Coll. MS, White i.   *Now* Lettsom.
   with] *from* Han.

---

263. **sometime of the night**] ABBOTT, § 176: That is, sometimes during the
night.—W. A. WRIGHT: The accent shows that 'sometime' should not be separated
into two words.

264. **these flowers**] COLLIER (ed. ii): Where the MS substitutes *bowers* for
'flowers,' we refuse the emendation, because it is not required.—R. G. WHITE (ed.
i): The context plainly shows that 'flowers' is a misprint. 'A *bank*' ' *oercanopied*'
with woodbine, musk-roses, and eglantine is certainly a bower; and, says Oberon,
'*there* sleeps Titania,' and '*there* the snake throws her enamell'd skin.' Finally,
Puck says, III, ii, 9, 'near to her close and consecrated *bower.*'—DYCE (ed. ii):
' Oddly enough, Knight has attacked the MS Corrector's reading *bowers* with a string
of absurdities; while R. G. White, who adopts it, makes a remark that is conclusive
against it, viz. that "a bank overcanopied with woodbine, musk-roses, and eglantine
is certainly a bower." I strongly suspect that the genuine reading is "this bower." '
—W. N. LETTSOM. [Hudson adopted this conjecture of Lettsom. I do not know
where to find Knight's attack on Collier's MS to which Lettsom refers, and I cannot see why R. G. White's remark, which Lettsom quotes, is conclusive against the
adoption of *bowers*. Hudson adds another reference, III, i, 205, 'lead him to my
bower.'—ED.]

265, 266. **And . . . in**] KEIGHTLEY (*Exp.* 132, and *N. & Qu.* 2d Ser, xii, 264)
transposes these two lines so as to follow line 262, a transposition which is, so he says,
'imperatively demanded by the sequence of ideas'; he also suggests that these two
lines 'may have been an addition made by the poet or transcriber in the margin, and
taken in in the wrong place.'—HUDSON adopted this transposition, which certainly
has much in its favour, and reads, 'And *where* the snake' instead of 'And there the
snake.' 'With the old order,' says Hudson, 'it would naturally seem that Oberon
was to streak the snake's eyes instead of Titania's,' especially, he might have added,
since 'snake' is, as W. A. WRIGHT points out, feminine, see *Macb.* III, ii, 13: 'We
have scotch'd the snake. . . . She'll close,' &c.—J. CROSBY (*Lit. World*, Boston, 1
June, '78) anticipated Hudson in substituting *where* for 'there.'

266. **Weed**] A garment; the word now survives in 'widows' weeds.'

267. **And**] KEIGHTLEY: If this be the right word, something must have been
lost, *e. g.* 'Upon her will I steal there as she lies'; but the poet's word may have
been what I have given, *Then*, strongly emphaticized, and written *Than*, the two
first letters of which having been effaced, the printer made it 'And.'

267. **streake**] W. A. WRIGHT: That is, stroke, touch gently.

A fweet *Athenian* Lady is in loue   270
With a difdainefull youth : annoint his eyes,
But doe it when the next thing he efpies,
May be the Lady.   Thou fhalt know the man,
By the *Athenian* garments he hath on.
Effect it with fome care, that he may proue   275
More fond on her, then fhe vpon her loue ;
And looke thou meet me ere the firft Cocke crow.
    Pu. Feare not my Lord, your feruant fhall do fo. *Exit.*   278

276. *on her*] *of her* Rowe, Pope,   277. *thou*] *you* Rowe+.
Theob. Han. Warb.   278: Exit.] Exeunt. Qq.
    her loue] *his love* Han.

273, 274. **man . . . on**] STEEVENS : I desire no surer evidence to prove that the
broad Scotch pronunciation once prevailed in England, than such a rhyme as the first
of these words affords to the second.—W. A. WRIGHT : In an earlier part of the
scene ' crab' rhymes to ' bob,' and ' cough' to ' laugh' ; but from such imperfect
rhymes, of which other examples occur in III, ii, 369, 370 [where the present rhyme
of *man, on*, is repeated] ; III, ii, 435, 436 [*there, here*] ; Ib. 484, 486 [*ill, well,*—is
any rhyme here intended ? Wright's last reference is to ' V, i, 267, 268 ' of his own
text (corresponding to V, i, 289, 290 of the present text), which must be, of course,
a misprint ; the two words are *here* and *see*. Wright then continues] it is unsafe to
draw any inference as to Shakespeare's pronunciation. [But is it not begging the
question to call these rhymes ' imperfect' ? The presumption is that they are perfect,
and to say that they are not, assumes a complete knowledge of Shakespeare's pronunciation. If Shakespeare again and again rhymes short *a* with short *o*, and Ellis
(*E. E. Pronun.* p. 954) gives ten or a dozen instances, is it unfair to infer that to his
ear the rhyme was perfect ? may we not thus approximate to his pronunciation ? Of
course, the standard which Ellis derived from certain lists in Salesbury is not here
involved. I am merely urging a gentle plea against a general condemnation of
Steevens's remark, which, when it was made, indicated, I think, that Steevens's face
was turned in the right direction.—ED.]

    276. **on**] For numerous examples of this construction with ' on,' see ABBOTT, §§
180. 181 ; and for the subjunctive ' meet,' in the next line, see Ib. § 369.

[*Scene II.*]

Enter Queene of Fairies, with her traine.

Queen. Come, now a Roundell, and a Fairy fong;
Then for the third part of a minute hence,
Some to kill Cankers in the muske rofe buds,
Some warre with Reremife, for their leathern wings,   5
To make my fmall Elues coates, and fome keepe backe
The clamorous Owle that nightly hoots and wonders   7

[Scene V. Pope+. Scene III. Steev.
Mal. Sing. Knt. Dell. ii, Kly. **Act III,**
sc. i. Fleay. Scene II. Cap. et cet.
[Another Part of the Wood. Cap.
1. Enter] Enter *Titania* Q₁.

3. *for*] '*fore* Theob. Han. Johns.
Huds. ... Huds. ... *Kinnear.*
a minute] *the midnight* Warb. *the*
Minuit Id. conj.
6. *fome keepe*] *keep fome* F₄.

---

2. **Roundell**] See note on line 10.

3. **for**] THEOBALD thus explains his text '*fore*: The Poet undoubtedly intended
Titania to say, Dance your Round, and sing your song, and then instantly (*before* the
third part of a minute) begone to your respective duties.—HEATH (p. 51): I should
rather incline to read: *in.* That is, after your song and dance have ended vanish in
the third part of a minute, and leave me to my rest.—CAPELL: It rather seems that
the queen's command is expressive of the short time in which she should be asleep
after their song and dance; that absence is enjoined, but 'till she were asleep; after
which, they might return if they pleased and follow the tasks she set them even
about her 'cradle' as Puck calls it, her sleep's soundness would not be disturb'd by
them; and this hint of its soundness is not unnecessary: for we see presently that it
is not broke by the persons that enter next, nor by the clowns 'till Bottom brays-out
his song.

3. **a minute**] WARBURTON pronounces this 'nonsense,' and actually substituted
in his text *the midnight.*—STEEVENS: But the persons employed are *fairies*, to whom
the third part of a *minute* might not be a very short time to do such work in. The
critic might as well have objected to the epithet 'tall,' which the fairy bestows on the
cowslip. But Shakespeare, throughout the play, has preserved the proportion of other
things in respect of these tiny beings, compared with whose size a cowslip might be
tall, and to whose powers of execution a minute might be equivalent to an age.—
HALLIWELL: This quaint subdivision of time exactly suits the character of the fairy
speaker and her diminutive world.

4. **Cankers**] PATTERSON (p. 34): This larva, *Lozotænia Rosana*, passes by the
'smirch'd tapestry,' and chooses for its domicile 'the fresh lap of the crimson rose.'
It there lives among the blossoms, and prevents the possibility of their further development.—HALLIWELL says that this name is applied to almost any kind of destructive
caterpillar. [Here in this country a popular distinction is drawn, I think, between
cankers and caterpillars. The former are small and hairless, the latter may be large
or small, but always hairy.—ED.]

5. **Reremise**] W. A. WRIGHT: That is, bats; A.-S. *hrère-mús*, from *hreran*, to
stir, to agitate, and so equivalent to the old name *flittermouse.* Cotgrave has, 'Chau-
vesouris: m. A Batt, Flittermouse, Reremouse.'

At our queint ſpirits : Sing me now aſleepe,   8
Then to your offices, and let me reſt.

Fairies Sing.   10

You ſpotted Snakes with double tongue,
Thorny Hedgehogges be not ſeene,
Newts and blinde wormes do no wrong,   13

8 *ſpirits*] *sports* Han. Warb.   10. Fairies Sing] Song.   First Fairy.
Sing] *Come, sing* Han.   Cap. et seq. (subs.).
                                          11–27. In Roman, Q₁.

7. **clamorous**] WALKER (*Crit.* i, 157) concludes that this word, in many places in Shakespeare, evidently signifies *wailing*.

8. **queint**] Cotgrave has, 'Coint: m. Quaint, compt, neat, fine, spruce, briske, smirke, smug, daintie, trim, tricked vp.'

10. **Fairies Sing**] CAPELL was the first to divide this song into two stanzas of four lines each, with a chorus of six lines, from line 15 to line 20 inclusive In the stanzas we have the 'Fairy Song' which the Queen calls for, and in the Chorus we have the 'Roundell,' which was 'danc'd-to as well as sung.' [This solves the difficulty of combining a dance and that which the text tells us was a song. *Rondel*, says Skeat, is an older form of *rondeau*, which Cotgrave explains as 'a rime or sonnet which ends as it begins.' Tyrwhitt cites a passage from Jonson's *Tale of a Tub*, II, i, which shows that *rondel* was a dance: 'You'd have your daughter and maids Dance o'er the fields like faies to church, this frost. I'll have no rondels, I, in the queen's paths.'—p. 154, ed. Gifford. Staunton says that a 'roundel' is 'a *dance*, where the parties joined hands and formed a ring.' He gives no authority, but adds, 'this kind of dance was sometimes called a *round*, and a *roundelay* also, according to Minshew, who explains: "Roundelay, *Shepheards daunce*."']

13. **Newts**] '*Of the Nevte or Water Lizard.* This is a little blacke Lizard, called *Wassermoll* or *Wasseraddex*, that is, a Lizard of the Water. : . . They liue in standing water or pooles, as in ditches of Townes and Hedges. . . . There is nothing in nature that so much offendeth it as salt, for so soone as it is layde vpon salt, it endeauoureth with all might & maine to runne away. . . . Beeing moued to anger, it standeth vpon the hinder legges, and looketh directlie in the face of him that hath stirred it, and so continueth till all the body be white, through a kind of white humour or poyson, that it swelleth outward, to harme (if it were possible) the person that did prouoke it.'—Topsell, p. 212.—W. A. WRIGHT: 'A newt' is an evet or eft (A.-S. *efete*), the *n* of the article having become attached to the following word, as in 'nonce,' 'noumpere' = umpire, and others. In 'adder' the opposite practice has taken place, and 'a nadder' (A.-S. *næddre*) has become 'an adder'; so 'an auger' is really 'a nauger' (A.-S. *nafegár*). ['Orange' may be also added.]

13. **blinde wormes**] '*Of the Slovv-Worme.* This Serpent was called in auncient time among the Græcians *Tythlops* and *Typhlines*, and *Cophia*, because of the dimnes of the sight thereof, and the deafenes of the eares and hearing. . . . It beeing most euident that it receiueth name from the blindnes and deafenes thereof, for I haue often prooued, that it neither heareth nor seeth here in England, or at the most it

Come not neere our Fairy Queene.
Philomele with melodie, 15
Sing in your ſweet Lullaby.
Lulla, lulla, lullaby, lulla, lulla, lullaby,
Neuer harme, nor ſpell, nor charme,
Come our louely Lady nye,
So good night with Lullaby. 20
    2. Fairy. Weauing Spiders come not heere,
Hence you long leg'd Spinners, hence:
Beetles blacke approach not neere;
Worme nor Snayle doe no offence.
Philomele with melody, &c. 25
    1. Fairy. Hence away, now all is well;
One aloofe, ſtand Centinell.   *Shee ſleepes.* 27

15, 25, Philomele] Chorus. *Philomel*
Cap. et seq. (subs.).
    16. Sing in your] *Singing her* Rann.
    in your] in our Q₁, Cap. et seq.
now your Coll. MS.
    21. 2. Fairy] 1. Fai. Qq (subs.), Cam.

22. Spinners] Spinders Qₙ.
26. 1. Fairy] 2. Fai. Qq (subs.), Coll.
Sta. Cam.
27. Shee ſleepes] Om. QqF₃F₄. Exeunt Fairies. Rowe.

---

seeth no better then a Mole. . . . They love to hide themselues in Corne-fieldes vnder the rype corne when it is cut downe. It is harmlesse except being prouoked, yet many times when an Oxe or a Cow lieth downe in the pasture, if it chaunce to lye vppon one of these Slow-wormes, it byteth the beast, & if remedy be not had, there followeth mortalitie or death, for the poyson thereof is very strong.—Topsell, p. 239. Marshall (*Irving Sh.* p. 374) says that it is impossible to imagine two animals more harmless than newts and blind-worms. Topsell, who was translating Gesner probably at the very hour Shakespeare was writing this play, gives us the belief, not only of the common folk, but of the naturalists of the time.—ED.

    15, 16. melodie . . . Lullaby] See I, i, 200.

    21. **Spiders**] It is not necessary to suppose that any deadly or even venomous qualities are here attributed to spiders, any more than to beetles, worms, or snails. It is enough that they are repulsive. Albeit, Topsell (p. 246), at the beginning of his long chapter on 'Spyders,' says: 'All spyders are venomous, but yet some more, and some lesse. Of Spyders that neyther doe nor can doe much harm, some of them are tame, familiar, and domesticall, and these be cōmonly the greatest among the whole packe of them. Others againe be meere wilde, liuing without the house abroade in the open ayre, which by reason of their rauenous gut, and greedy deuouring maw, haue purchased to thēselues the name of wolfes and hunting Spyders.' At the close, however, of his chapter (p. 272) he acknowledges that 'Our Spyders in England are not so venomous as in other parts of the world. . . . We cannot chuse but confesse that their byting is poysonlesse, as being without venome, procuring not the least touch of hurt at all to any one whatsoeuer.'—ED.

    11–25. No less than eight musical settings of this song are recorded in the *List*, &c, issued by *The New Shakspere Soc.*

Enter Oberon.   28

Ober. What thou feeft when thou doft wake,
Doe it for thy true Loue take: 30
Loue and languifh for his fake.
Be it Ounce, or Catte, or Beare,
Pard, or Boare with briftled haire,
In thy eye that fhall appeare,
When thou wak'ft, it is thy deare,   35
Wake when fome vile thing is neere.

Enter Lifander and Hermia.

Lif. Faire loue, you faint with wandring in ẙ woods,
And to fpeake troth I haue forgot our way:
Wee'll reft vs *Hermia,* if you thinke it good,   40
And tarry for the comfort of the day.
Her. Be it fo *Lyfander* ; finde you out a bed,
For I vpon this banke will reft my head.
Lyf. One turfe fhall ferue as pillow for vs both,
One heart, one bed, two bofomes, and one troth.   45

---

29. [to Tit. squeezing the flower upon her eyelids. Cap.
30. *thy*] *thy thy* Q₂.
    true Loue] *true-love* Harness, Knt, Dyce, Sta. Cam.
33. *haire*] *hear* Ktly.
34. *that*] *what* Pope, Han.

36. Exit Oberon. Rowe.
37. [Scene VI. Pope+.
38. *woods*] Q₂Ff, Rowe i, Sta. *wood* Q₁ et cet.
41. *comfort*] *comfor* Q₁.
42. *Be it*] *Bet it* Q₁. *Be't* Pope+, Dyce ii, iii.

26, 27. Capell was the first to indicate that these two lines are not a part of the song; he has been followed, of course, by all the editors since his day.—ED.

30. **true Loue**] W. A. WRIGHT: Possibly a corruption. In Icelandic, *trú-lofa* is to betroth. [Is not the hyphen (see Text. Notes) a corruption?—ED.]

32. **Catte**] W. A. WRIGHT: This must be the wild cat.

33. **haire**] KEIGHTLEY (*Exp.* 133): The rhyme demands the old form, *hear.* [Keightley is right, as far as he goes, but if we are to adopt the Shakespearian pronunciation in this word we must go further, and not only pronounce 'hair' *hear*, but 'bear' *beer*, which was also right. It seems scarcely worth while to adopt Shakespeare's pronunciation in isolated instances, unless there is a decided need, as in 'melody' and 'lullaby.' Although these five lines were probably perfect rhymes originally, yet as 'bear' and 'hair' are perfect rhymes at present, no change seems necessary.—ED.]

38. **with**] For other examples of 'with' thus used, see ABBOTT, § 193.

45. **one troth**] W. A. WRIGHT: One faith or trust, pledged to each other in betrothal.

45. After this line, in Garrick's *Version*, the lovers sing a duet. It is scarcely

Her. Nay good *Lyſander*, for my ſake my deere 46
Lie further off yet, doe not lie ſo neere.
Lyſ. O take the ſence ſweet, of my innocence,
Loue takes the meaning, in loues conference,
I meane that my heart vnto yours is knit, 50
So that but one heart can you make of it.
Two boſomes interchanged with an oath , 52

---

46. *good*] *god* Q₁.
48, 49. *innocence...conference*] *conference...innocence* Warb. Theob. *inno-
....nf..... Coll. ii (MS)*
49. *takes*] *take* Tyrwhitt, Rann.
50. *is*] *it* Q₁.

51. *can you*] Ff, White i. *wee can*
Q₁. *can we* Cap. Sta. *you can* White
ii. *we can* Q₂ et cet.
50 *interchanged*] Ff, White i *inter
chained* Qq et cet.

---

worth while to cumber these pages with the words either of this song or of the fifteen
others scattered through the rest of the play. They are all weak variations of the
same weak theme—reflections from the 'tea-cup times of hood and hoop While yet
the patch was worn.' The specimens already given will prove, I am sure, quite
sufficient.—ED.

48. **innocence**] WARBURTON's needless emendation called forth JOHNSON's
almost needless paraphrase : ' Understand the meaning of my innocence, or my innocent meaning. Let no suspicion of ill enter thy mind.'

49. **conference**] JOHNSON : In the conversation of those who are assured of each
other's kindness, not *suspicion* but *love* takes the meaning. No malevolent interpretation is to be made, but all is to be received in the sense which *love* can find and
which *love* can dictate.—TYRWHITT : I would read : ' Love *take* the meaning,' &c,
that is, '*Let* love take the meaning,' &c.—COLLIER (ed. ii) : *Confidence* is a happy
emendation of the MS. What Lysander means is that Hermia should take the
innocence of his intentions in the *confidence* of his love, and thence he proceeds to
explain the fulness, fidelity, and purity of his attachment.—LETTSOM (*Blackwood's
Maga.* Aug. 1853) : The alteration of 'conference' into *confidence* is an *improvement*,
most decidedly *for the worse*. What Lysander says is, that love puts a good construction on all that is said or done in the 'conference' or intercourse of love. *Confidence* makes nonsense. [To this Dyce (ed. i) gives a hearty assent.]

51. **can you**] R. G. WHITE (ed. i) : The reading of F₁ is not only authoritative
in this essential change, but far more significant than that of the Quartos. Lysander
in his attempt to meet the objections which Hermia makes to his proposition, may,
with much more propriety and effect, attribute to his mistress alone the desire of separating him from her, than to make himself a party to such an endeavour.

52. **interchanged**] R. G. WHITE (ed. i) : *Interchained* of the Qq conveys the
comparatively commonplace thought that the lovers' hearts were bound together ;
' interchanged ' represents them as having been given each to the other, as the most
solemn instruments are made, interchangeably.—MARSHALL : The considerations
which have induced us to adopt *interchained* are these : (1) it is more consonant in
sense with line 50, '—my heart unto yours is *knit*' ; and (2) 'bosom,' though used
as *desire* (*Meas. for Meas.* IV, iii, 139), or as *inmost thoughts* (*Oth.* III, i, 58), seems
never to be used for 'the affections' themselves. Shakespeare would scarcely have

So then two bofomes, and a fingle troth.   53
Then by your fide, no bed-roome me deny,
For lying fo, *Hermia*, I doe not lye.   55
   Her. *Lyfander* riddles very prettily;
Now much befhrew my manners and my pride,
If *Hermia* meant to fay, *Lyfander* lied.
But gentle friend, for loue and courtefie
Lie further off, in humane modefty,   60
Such feparation, as may well be faid,
Becomes a vertuous batchelour, and a maide,
So farre be diftant, and good night fweet friend;
Thy loue nere alter, till thy fweet life end.
   Lyf. Amen, amen, to that faire prayer, fay I,   65
And then end life, when I end loyalty:
Heere is my bed, fleepe giue thee all his reft.
   Her. With halfe that wifh, the wifhers eyes be preft.
Enter Pucke.   *They fleepe.*
   Puck. Through the Forreft haue I gone,   70
But *Athenian* finde I none,

---

55. *lying fo,* Hermia] *Hermia, lying so* Schmidt.

  lye] *lie* Cap.

60. *off, in...modefty,*] Q₂F₂. *off, in... modesty:* Q₁, Han. *off in...modesty,* F₃F₄. *off; in...modesty,* Theob. et cet. (subs.).

60. *humane*] *human* F₄.

67. *my*] *thy* Rowe i.

69. *They fleepe*] Om. Qq.

71. *finde*] Q₂Ff, Knt, Hal. White i. *found* Q₁ et cet.

---

faid, ' We have *interchanged* bosoms.' The objection to *interchained* is, not that it occurs only in this passage, but that it is not to be found in any other writer, ancient or modern, as far as I can discover.

57. **beshrew**] STEEVENS expresses it a little too strongly when he says that this word ' implies a sinister wish.'—DYCE defines it more correctly, I think, as ' a mild form of imprecation, equivalent to " a mischief on." ' Pronounced *beshrow*, as Walker (*Crit.* i, 158) has shown; it is thus spelled in several instances in the Folio, as well as *shrowd* and *shrode* for ' shrewd.' ' Shrewsbury ' is still pronounced by some *Shrowsbury*.—ED.

60–63. **in humane modesty . . . distant**] W. A. WRIGHT: The sense is clear, though the syntax is imperfect. Delius connects ' as may well be said ' with ' in human modesty,' but the construction is rather ' in human modesty (let there be) such separation,' &c., and ' So far be distant ' is merely a repetition of the same thing. —L. WILSON (p. 248): Titania's use of the phrase ' human mortals ' is very expressive but ' human modesty ' seems a needless pleonasm. . . . If any change be made, ' *common* modesty ' would better suit the context.

68. **be**] For other examples of the subjunctive used optatively, see ABBOTT, § 365.

71. **finde**] By the sequence of tenses this should be as it is in Q₂, *found.* It is

One whofe eyes I might approue   72
This flowers force in ftirring loue.
Night and filence : who is heere ?
Weedes of *Athens* he doth weare :   75
This is he (my mafter faid)
Defpifed the *Athenian* maide :
And heere the maiden fleeping found,
On the danke and durty ground.
Pretty foule, fhe durft not lye   80
Neere this lacke-loue, this kill-curtefie.

---

72. *One*] *On* QqFf et cet.

81. *Neere...curtefie*] *Near to this lack-love, this kill-curtesie* Pope, Steev.'73, '78. *Near to this kill-curtesie* Theob. Han. Cap. *Near to this lack-love kill-curtesie* Warb. *Near this lack-love kill-courtesie* Johns. *Near this lack-love, kill courtesy* Steev.'85, '93, Coll. ii (MS). *Nearer this lack-love, this kill-courtesy* Walker, Dyce ii, iii, Huds. *Near... court'sy* Sta.

---

therefore an instance of an error the opposite to that of which WALKER (*Crit.* ii, 271) gives an example, where *finde* was printed 'found'; Lettsom, in a footnote, calls attention to the present passage.

75. **Weedes**] That is, garments; see II, i, 266.

81. **Neere . . . curtesie**] THEOBALD: This verse, as Ben Jonson says, is broke loose from his fellows, and wants to be tied up. I believe the poet wrote: 'Near to this kill-courtesie.' And so the line is reduced to the measure of the other. But this term being somewhat quaint and uncommon, the Players, in my opinion, officiously clapped in the other as a Comment; and so it has ever since held possession.—MALONE: If we read 'near' as a disyllable, like many other similar words, we shall produce a line of ten syllables, a measure which sometimes occurs in Puck's speeches: 'I must go seek some dew drops here; And hang a pearl in every cowslip's ear.' Again, 'I go, I go; look how I go; Swifter than arrow from a Tartar's bow.'—KNIGHT agrees with Malone that it is 'evidently intended for a long line amidst those of seven or eight syllables.'—WALKER (*Crit.* ii, 52): Read *Nearer* for 'Neere.' The force of *nearer* and Lysander's discourtesy (as it appeared to our friend Puck) are explained by the scene immediately preceding between Lysander and Hermia. . . . I suspect that *e* for *er* in the terminations of words is not an infrequent error in the old editions of our poets. . . . See I, i, 232, '*strange* companions'; though this perhaps might be accounted for otherwise. . . . The converse error also appears sometimes in the Folio, though, I think, less frequently. See III, i, 209: 'Tye vp my *louers* tongue, bring him silently.'—ABBOTT, § 504: There is difficulty in scanning this line. It is of course possible that 'kill-curt'sy' may have the accent on the first, but thus we shall have to accent the first 'this' and 'love' with undue emphasis. It is also more in Shakespeare's manner to give 'courtesy' its three syllables at the end of a line. I therefore scan: '(Near this) láck-love, | thís kill | cóurte | sý.' Perhaps, however, as in *Macb.* III, v, 34, 35, and ? 21, a verse of five accents is purposely introduced.—VERITY: Best scan the line as four iambic feet, thus: 'Near thís | lack-lòve | , this kíll- | court'sẏ.' The first *this* may be accented because said with emphatic contempt—Puck pointing at Lysander. The syllable that immediately

Churle, vpon thy eyes I throw   82
All the power this charme doth owe :
When thou wak'ſt, let loue forbid
Sleepe his ſeate on thy eye-lid.   85

82. *thy*] *the* F₃F₄.   82. [Squeezes the flower on Lysander's
                                           eyelids. Dyce.

---

follows a strongly-accented syllable is liable to lose its own stress : hence the stress
on *lòve*, not *lack*. Where a word occurs twice in the same line it is generally accented
differently : hence the second *this* is unaccented, the stress falling on *kill* (which
accentuation has also the merit that it varies the accent of the two compounds, *lack
lòve—kìll-court'sy*). The last foot is simple. Shakespeare often introduces an iambic
rhythm into a trochaic passage for the sake of variety; and this line treated thus as
iambic will correspond with line 78, also four iambics. [I cannot believe that any
scansion is worthy of consideration which subordinates to the rhythm the meaning
and the force of words. The rhythm must emphasize the idea, not neglect it, still
less mar it. In this line there are two compound words of emphatic vituperation,
and in both the force lies in the first syllable, which must be accented, unless we are
to make the rhythm superior to the sense. There is no necessity to convert, with
Walker, ' Near ' into *Nearer;* the sense does not demand it; but even if the sense does
demand the comparative degree, we have that degree already in the very word itself,
or with the *er* lying perdue if necessary in the final *r*, just as *This is* is delicately
heard in ' This' a dull sight' (*Lear*, V, iii, 283), which is one of Walker's own
excellent suggestions. Taking, therefore, the text as it stands, the rhythm and the
sense are, in the first half of the line, with the strong accent where it belongs : ' Nèar
this | làck-love.' The difficulty, then, is to scan the second half, which, if the tro-
chaic measure is to be kept up, will bring the emphasis, or arsis, on ' this,' which is
all right, but the thesis on ' kill,' which is all wrong. The solution which I find here
is that neither from Puck's tongue nor from any one else's would these vehement
compounds, ' lack-love ' and ' kill-courtesy,' glide off glibly. No intelligent reader
of the line but would instinctively pause before each of them, and in that pause before
the second we may find the thesis of the foot of which ' this ' is the arsis; and, after
the pause, be ready for a new and emphatic arsis in ' kill.' If there be, after all, a
certain harshness in thus reading the line, is it not in keeping ? May we not imagine
the indignant little sprite as uttering the words through almost clenched teeth, and
with a spite to which the reduplicated *k*-sound in ' kill-curtesy,' corresponding to the
pitying liquids in ' lack-love,' lends an emphasis ? Wherefore the text of the Folio
is right, I think, and waits for its harmony on the reader's voice.—ED.]

83. owe] Where this word occurs in *Othello*, STEEVENS observed that it means to
own, to *possess*, whereupon PYE (p. 330) remarked, ' Very true; but do not explain it
so often '; and I think Pye takes us all with him.—ED.

84, 85. When . . . eye-lid] DANIEL (p. 31) : The only meaning that can attach
to these lines, as they at present stand, is that when Lysander awakes, Love is to forbid Sleep to occupy his (Love's or Sleep's ?) seat on Lysander's eye-lid. In other
words, when Lysander awakes, he is no longer to be asleep ! . . . Puck's intention in
anointing the sleeper's eyes is clearly to make him fall in love with her whom he had
hitherto contemned. Read, therefore, ' let love forbid *Keep* his seat,' &c. ' Forbid '
here has the meaning of accursed, placed under an interdict, as in *Macbeth*, ' He

So awake when I am gone: 86
For I muſt now to *Oberon*.   *Exit.*

Enter Demetrius and Helena running.

Hel.   Stay, though thou kill me, ſweete *Demetrius.*
De.   I charge thee hence, and do not haunt me thus.   90
Hel.   O wilt thou darkling leaue me? do not ſo.
De.   Stay on thy perill, I alone will goe.
Exit Demetrius.

Hel.   O I am out of breath, in this fond chace,
The more my prayer, the leſſer is my grace,   95
Happy is *Hermia*, whereſoere ſhe lies;
For ſhe hath bleſſed and attractiue eyes.
How came her eyes ſo bright? Not with ſalt teares.
If ſo, my eyes are oftner waſht then hers.
No, no, I am as vgly as a Beare;   100
For beaſts that meete me, runne away for feare,
Therefore no maruaile, though *Demetrius*
Doe as a monſter, flie my preſence thus.
What wicked and diſſembling glaſſe of mine,   104

---

88. [Scene VII. Pope+.   93. Om. Qq.
89. *Stay*] *Say* Ff.   96. *whereſoere*] *wherefore* F₄.
91. *darkling*] *Darling* F₄, Rowe.   102. *maruaile*] *mavaile* F₂.

---

shall live a man forbid'; and the sense of the passage is that love, which was *forbid*, should, when the sleeper awoke, *keep his seat* or enthrone himself on his eye-lid. Compare *King John*, III, iii, 45: 'Making that idiot laughter *keep* men's eyes.' [I cannot think that emendation is necessary. Puck's charm is to awaken in Lysander such a feverish love that sleep will be banned from his eyes, a symptom of the passion common enough. If we adopt Daniel's change, Love must be exiled from its consecrated home, the heart, and seated, of all places in the world, on an eye-lid. —ED.]

91. darkling] STEEVENS: That is, in the dark. The word is likewise used by Milton [*Par. Lost*, iii, 39: 'As the wakeful bird Sings darkling.'—W. A. WRIGHT.] The COWDEN-CLARKES (*Sh. Key*, p. 545): Besides its direct meaning of *in the dark*, 'darkling,' as Shakespeare employs it, includes the meaning of *baffled, deserted, bereft of light and help*. [Note the not unnatural—nay, almost plausible—sophistication,—*darling* of F₄ followed by Rowe, which is here recorded, I believe, for the first time.—ED.]

94. fond] W. A. WRIGHT: That is, foolish, with perhaps something of the other meaning which the word now has.

100, 101. Beare . . . feare] Note again this rhyme.—ED.

103. as a monster] This refers not to Demetrius, but to Helena herself.

Made me compare with *Hermias* fphery eyne?   105
But who is here?   *Lyfander* on the ground;
Deade or afleepe?   I fee no bloud, no wound,
Lyfander, if you liue, good fir awake.

   Lyf.   And run through fire I will for thy fweet fake.
Tranfparent *Helena*, nature her fhewes art,   110
That through thy bofome makes me fee thy heart.
Where is *Demetrius*? oh how fit a word
Is that vile name, to perifh on my fword!

   Hel.   Do not fay fo *Lyfander*, fay not fo:
What though he loue your *Hermia*? Lord, what though?   115
Yet *Hermia* ftill loues you; then be content.

   Lyf.   Content with *Hermia*? No, I do repent
The tedious minutes I with her haue fpent.   118

---

106. Lyfander] *Lysander!* Cap. et
seq. (subs. except Coll. White i).
   ground;] *ground?* Q₁, Coll.
grouud? Q₂. ground! Cap. et seq.
   107. *Deade*] *Dead!* Cap. et seq.
(subs.).
   109. *fake.*] *sake,* Cap. (in Errata).
   [Waking. Rowe et seq. (subs.).
   110. Helena,] *Helen*, Rowe ii+, Dyce
ii, iii. *Helena!* Cap. et cet.

110. *nature her fhewes*] *nature fhewes
Qq*, Cap. Mal.'90, Cam. White ii, Rolfe.
nature here fhews Ff, Rowe+, Steev.
Coll. Dyce i. *Nature shows her* Var.'21,
Knt, Hal. Sing. White i, Sta. Dyce ii, iii,
Huds. Ktly.
   111. *thy heart*] *my heart* Walker,
Dyce ii, iii, Huds.
   112. *is*] Om. Ff.

---

  105. **sphery**] W. A. WRIGHT: 'Sphere' is used by Shakespeare to denote first
the orbit in which a star moves, and then the star itself.

  110. **Helena**] WALKER (*Crit.* i, 230): Read *Helen* [See Text. Notes], as in half
a dozen other passages in the play. [So also, nine lines below Walker would read
Helen; and again, 'to avoid the trisyllabic termination,' in III, ii, 337.]

  110. **her shewes**] MALONE: Probably an error of the press for *shews her.*—R. G.
WHITE (ed. i): Plainly but an accidental transposition. [Both of these remarks
seem to me wrong; they quite remove the astonishment which Lysander expresses at
the fact that *Nature* can show *art.* To me it is clear that we must read either with
the Qq and retain 'Helena,' or hold 'her' to be a misprint (corrected in the following Ff) for *here*, and, with Walker, read 'Helen.'—ED.]

  111. **thy heart**] WALKER (*Crit.* i, 300): Read, '*my* heart.' The old poetical
commonplace; e. g. *As You Like It,* V, iv, 120: 'That thou mightst join her hand
with his, Whose heart within her bosom is.' Compare *Sonnet* 133: 'Prison my heart
in thy steel bosom's ward.'

  112. **Demetrius**] TIESSEN (*Archiv f. n. Sp.*, &c., vol. lviii, p. 4, 1877): We
would be grateful to editors if they would only tell us why the '*name*' of Demetrius
should be thus referred to. Is there a covert reference to *demit*, i. e. to humble, to
subject, or to *meat* which is stuck on a spit? [*i. e.* 'De-meat-rius,' I suppose. This
insight of the way in which a learned German reads his Shakespeare would be interesting if it were not so depressing.—ED.]

Not *Hermia,* but *Helena* now I loue;
Who will not change a Rauen for a Doue?   120
The will of man is by his reafon fway'd:
And reafon faies you are the worthier Maide.
Things growing are not ripe vntill their feafon;
So I being yong, till now ripe not to reafon,
And touching now the point of humane skill,   125
Reafon becomes the Marfhall to my will,
And leades me to your eyes, where I orelooke
Loues ftories, written in Loues richeft booke.
    Hel. Wherefore was I to this keene mockery borne?
When at your hands did I deferue this fcorne?   130
Ift not enough, ift not enough, yong man,
That I did neuer, no nor neuer can,
Deferue a fweete looke from *Demetrius* eye,
But you muft flout my infufficiency?
Good troth you do me wrong(good-footh you do)   135
In fuch difdainfull manner, me to wooe.
But fare you well; perforce I muft confeffe,
I thought you Lord of more true gentleneffe.   138

---

119. Helena *now*] Q₂Ff, Var.'21, Sing. Knt, Hal. White i. *Helena* Q₁, Pope, Theob. Han. Warb. Cap. Steev. Rann, Mal.'90, Sing. Coll. Dyce i, Sta. Cam. Ktly, White ii, Rolfe. *Helen now* Johns. Walker, Dyce ii, iii, Huds.
  124. *ripe not*] *not ripe* Rowe ii, Pope,
Han. *riped not* Schmidt.
  125. *humane*] *human* Rowe et seq.
  128. *Loues ftories*] *Love-stories* Walker, Dyce ii, iii, Huds.
  133. Demetrius] *Demetrius's* Rowe i. *Demetrius'* Rowe ii et seq.
  134. *infufficiency*] *infufficency* Q₂.

---

124. **ripe not**] STEEVENS: 'Ripe' is here a verb, as in *As You Like It*, II, vii, 26, 'And so from hour to hour, we ripe and ripe.'

125. **touching now the point**] STEEVENS: That is, my senses being now at the utmost height of perfection.—W. A. WRIGHT: Having reached the height of discernment poffible to man.

126. **the Marfhall**] JOHNSON: That is, my will now follows reason.

128. **Loues richest booke**] STEEVENS: So in *Rom. & Jul.* I, iii, 86: 'And what obscured in this fair volume lies, Find written in the margent of his eyes.'

131. It is not easy to decide whether these repetitions here, in the next line, and in line 135 are characteristic of Helena (in Shakespearian phrase, 'tricks' of hers) or are the effects of sobbing. I think that when Helena finds that to the scorn of Demetrius is added the scorn of Lysander (she has just said, 'Wherefore was I to *this* keen mockery born? When at *your* hands did I deserve this scorn?'), she bursts into uncontrollable tears. And yet there are somewhat similar repetitions in lines 114, 115, above, where is no question of tears, which sound weak, unless they be a trait of character.—ED.

Oh, that a Lady of one man refus'd,

Should of another therefore be abus'd.   *Exit.* 140

  Lyf.   She fees not *Hermia* : *Hermia* fleepe thou there,

And neuer maift thou come *Lyfander* neere ;

For as a furfeit of the fweeteft things

The deepeft loathing to the ftomacke brings :

Or as the herefies that men do leaue,   145

Are hated moft of thofe that did deceiue :

So thou, my furfeit, and my herefie,

Of all be hated ; but the moft of me ;

And all my powers addreffe your loue and might,

To honour *Helen,* and to be her Knight.   *Exit.* 150

  Her.   Helpe me *Lyfander,* helpe me ; do thy beft

To plucke this crawling ferpent from my breft.

Aye me, for pitty ; what a dreame was here ?

Lyfander looke, how I do quake with feare :

Me-thought a ferpent eate my heart away,   155

And yet fat fmiling at his cruell prey.

Lyfander, what remoou'd ?   *Lyfander,* Lord,

What, out of hearing, gone ? No found, no word ?

Alacke where are you ? fpeake and if you heare :   159

---

141. *Hermia : Hermia*] *Hermia* F₃F₄.
Hermia.—Hermia Coll.

 144. *the*] *a* Ff, Rowe, Pope, Han.

 146. *that*] *they* Qq, Rowe et seq.

 149. *And all my powers*] *And, all my
powers,* Han. Cap. et seq. (subs.).

   your] *their* Coll. MS.

 151. [Starting. Cap.

 153. *Aye*] QqFf, Rowe+, White,
Dyce, Cam. (subs.).   *Ah* Cap. et cet.

 154. *I do*] *do I* Pope, Han.

 155. *eate*] *ate* Knt.

 156. *yet fat*] Ff, Rowe.   *you fate* Qq
(*fat* Q₂) et cet. (subs.).

 157. *Lyfander, what*] *Lyfander what,*
Q₁.   *Lyfander ! what* Rowe ii.   *Lyfander ! what,* Han. et seq.

 158. *hearing, gone ? No found,*] *hearing gone ? No sound,* Theob. Warb. Johns.
hearing ? gone ?   No sound ? Cap. (Errata) et seq. (subs.).

 159. *and if*] *an if* Cap. et seq.

---

 155. **eate**] WHITE (ed. i): The same form as here of the verb, and the same
orthography is given elsewhere, which not only forbids us to read *ate,* but accords
with the supposition that the present and preterite tenses were not distinguished even
in pronunciation, but both had the pure sound of *e.*   And yet the strong preterite—
ate, is, of course, the older form.

 156. **prey**] W. A. WRIGHT: Here used for the act of preying, as in *Macb.* III,
ii, 53: 'Whiles nights black agents to their preys do rouse.'

 159. **and if**] This is, I think, equivalent to something more than simply *if*; it is,
at least, a strongly emphasized *if.*   See ABBOTT, § 105, which assuredly applies to
the present passage.—ED.

Speake of all loues ; I found almoſt with feare.　　　　160
No , then I well perceiue you are not nye,
Either death or you Ile finde immediately.　　　　*Exit.* 162

Actus Tertius. [*Scene I.*]

Enter the Clownes.

Bot.　Are we all met?

Quin.　Pat, pat, and here's a maruailous conuenient
place for our rehearſall.　This greene plot ſhall be our　　5
ſtage, this hauthorne brake our tyring houſe, and we will
do it in action, as we will do it before the Duke.　　7

---

160. *Speake of*] *Speake, of* Q₁, Cap. et
seq.

　ſound] *ſwoune* Q₁. *ſwound* Q₂
Ff, Rowe i, Hal.　*swoon* Rowe ii et cet.

161. *No,*] *No ?* Theob. Warb. et seq.

162. *Either*] *Or* Pope+, Cap. Steev.
'85.

I. Om. Qq.　Act III, Scene i. Rowe
et seq. Act III, Sc. ii. Fleay.　The Wood.
Pope.　The Same. Cap.

2. Enter...] Enter Quince, Snug, Bottom, Flute, Snowt, and Starveling.　The
Queen of Fairies lying asleep. Rowe et
seq. (subs. asleep, but invisible. Hal.).

4. *Pat*] *Par* F₂F₄.

　maruailous] *maruailes* Q₁. *marvels*
Cap.

5. *plot*] *plat* F₄, Rowe i.

6. *tyring houſe*] *'tiring-house* Coll.

---

160. **of all loues**] ABBOTT, § 169, 'of' is used in adjurations and appeals to signify *out of.*　'Of charity, what kin are you to me?'—*Twelfth Night,* V, i, 237.
Hence, the sense of *out of* being lost, it is equivalent to *for the sake of, by.*　[As in
the present instance.　HALLIWELL says that the phrase is of very common occurrence; he gives eight or nine examples, and the references to as many more.]

160. **sound**] As the Folio was set up by at least four different sets of compositors,
it is irrational to expect any uniformity of spelling.　Accordingly we find this word,
besides its present form, spelled 'swoon,' 'swoone,' 'swowne.'—ED.

160. **almost**] For examples of similar transposition, see ABBOTT, § 29.　The
idiom of the language has somewhat changed since Shakespeare's day in regard to
the position of this adverb.　Again and again it is placed after the word it qualifies,
when we should now place it before it; as here, where the position is quite independent of rhythm.—ED.

162. **Either**] See II, i, 31.

4. **maruailous**] CAMBRIDGE EDD.: Capell appears to have considered the reading of Q₁ as representing the vulgar pronunciation of 'marvellous,' and he therefore
printed it 'marvels,' as in IV, i, 27.

6. **hauthorne-brake**] See line 75 *post.*

6. **tyring house**] COLLIER: That is, '*Attiring*-house,' the place where the actors
attired themselves.　Every ancient theatre had its 'tiring-room or 'tiring-house,

8

Bot.   *Peter quince?*   8

Peter.   What faift thou, bully *Bottome?*

Bot.   There are things in this Comedy of *Piramus* and   10
Thisby, that will neuer pleafe.   Firft, *Piramus* muft draw a
fword to kill himfelfe; which the Ladies cannot abide.
How anfwere you that?

Snout.   Berlaken, a parlous feare.

Star.   I beleeue we muft leaue the killing out, when   15
all is done.

Bot.   Not a whit, I haue a deuice to make all well.
Write me a Prologue, and let the Prologue feeme to fay,
we will do no harme with our fwords, and that *Pyramus*
is not kill'd indeede : and for the more better affurance,   20
tell them, that I *Piramus* am not *Piramus*, but *Bottome* the
Weauer ; this will put them out of feare.

Quin.   Well, we will haue fuch a Prologue, and it fhall
be written in eight and fixe.   24

---

8. Peter quince ?] Q₂. Peeter Quince ?
Q₁. Peter Quince ? Ff. Peter Quince—
Theob. et seq. (subs.).

14. *Berlaken*]*Berlakin* Q₁. *By'rlaken*
Pope. *By'r-lakin* Cap. *By'r lakin* Dyce.

14. *parlous*] *par'lous* Cap.

17. *deuice*] *deuife* Q₁.

18. *feeme*] *serve* Gould.

20. *the more better*] *the better* Rowe ii.
more better Pope +.

---

9. bully] MURRAY (*N. E. D.* s. v.): Etymology obscure; possibly an adaptation of the Dutch *boel*, 'lover (of either sex),' also 'brother'; compare Middle High
German *buole*, modern German *buhle*, 'lover,' earlier also 'friend, kinsman.' . . . A
term of endearment and familiarity, originally applied to either sex; sweetheart,
darling.   Later, to men only, implying friendly admiration; good friend, fine fellow,
'gallant.'   Often prefixed as a sort of title to the name or designation of the person
addressed, as in 'bully Bottom,' 'bully doctor.'   1538, BALE, *Thre Lawes*, 475 :
'Though she be sumwhat olde It is myne owne swete bullye, My muskyne and my
mullye.'

10. There are things] WALKER (*Crit.* ii, 256): *Qu.* 'There are *three* things,'
&c.   See what follows.   I think, indeed, it is required.   [If anything may be said to
be required in dealing with Bottom's logic or language.—ED.]

14. Berlaken] STEEVENS: That is, by our Ladykin, or little Lady.   [The spelling is, probably, true to the pronunciation.]

14. parlous] STEEVENS: Corrupted from *perilous*.—HALLIWELL: It is used in
the generic sense of *excessive*, and sometimes with the signification of *wonderful*.
[See ABBOTT, § 461, for examples of many other words similarly contracted.]

17. Not a whit] W. A. WRIGHT: As 'not' is itself a contraction of *nâwiht* or
nawhit, 'not a whit' is redundant.

18. seeme to say] W. A. WRIGHT: Compare Launcelot's language in *Mer. of
Ven.* II, iv, 11 : 'An it shall please you to break up this, it shall seem to signify.'

20. more better] For double comparatives, see ABBOTT, § 11.

Bot. No, make it two more, let it be written in eight   25
and eight.

  Snout. Will not the Ladies be afear'd of the Lyon?

  Star. I feare it, I promiſe you.

  Bot. Maſters, you ought to conſider with your ſelues, to
bring in (God ſhield vs) a Lyon among Ladies, is a moſt   30
dreadfull thing. For there is not a more fearefull wilde
foule then your Lyon liuing: and wee ought to looke
to it.

  Snout. Therefore another Prologue muſt tell he is not
a Lyon.   35

  Bot. Nay, you muſt name his name, and half his face
muſt be ſeene through the Lyons necke, and he himſelfe
muſt ſpeake through, ſaying thus, or to the ſame defeēt;   38

27. *afear'd*] *afraid* Rowe ii+.
29. *Maſters*] *Maiſters* Ff.
    your ſelues,] *your ſelfe,* Qq. *your-
ſelves;* Rowe.

33. *to it*] *toote* Q₁. *to't* Cap. Sta. Cam.
White ii.
37. *necke*] *mask* Gould.
38. *defect*] *deffect* Q₂.

  24. **eight and ſixe**] CAPELL refers this to the number of lines, fourteen, 'which,'
as he says, 'is the measure of that time's ſonnets; all Shakeſpeare's are writ in it.'
'Bottom wants it writ in "two more"; instead of which, when we come to 't, we
find it just the same number less.'—MALONE interprets it as referring to the common
ballad metre of 'alternate verses of eight and six syllables,' and this interpretation
has been adopted. Capell assumes that we have this Prologue in Act V. Whereas,
this special Prologue which Bottom calls for nowhere appears. It seems almost
needless to call attention to the fact that this rehearsal does not correspond to the
play as it is acted before the Duke. See note on line 84 below. If this were a
genuine rehearsal of the play, its repetition at the public performance would be
wearisome.—ED.

  25, 26. **eight and eight**] HALLIWELL: An anonymous MS annotator alters this
to *eighty-eight*, an evident blunder.

  28. **I fear it**] It is almost foolish to attempt any emendation in the language
of these clowns, but it seems not unlikely that this should be 'I, I fear it,' that is,
'Ay, I fear it.' ED.

  29. **ſelues, to bring**] W. A. WRIGHT: The construction here, with only a comma
instead of a colon, is 'You ought to consider with yourselves (that) to bring in,' &c.

  31. **dreadful thing**] MALONE finds 'an odd coincidence' here between this
remark and an incident which happened, not in London, nor even in England, but
in Scotland in 1594, at the christening of the eldest son of James the First. 'While
the king and queen were at dinner a chariot was drawn in by "a black-moore. This
chariot should have been drawne in by a lyon, but because his presence might have
brought some feare to the nearest, or that the sights of the lights and the torches
might have commoved his tameness, it was thought meete that the Moor should ſup-
ply that room."' [—Reprinted in Somers's *Tracts*, ii, 179, W. A. Wright.]

Ladies, or faire Ladies, I would wifh you, or I would
requeft you, or I would entreat you, not to feare, not to   40
tremble : my life for yours.   If you thinke I come hither
as a Lyon, it were pitty of my life.   No, I am no fuch
thing, I am a man as other men are ; and there indeed let
him name his name, and tell him plainly hee is *Snug* the
ioyner.   45

Quin.   Well, it fhall be fo ; but there is two hard
things, that is, to bring the Moone-light into a chamber : for you know, *Piramus* and *Thisby* meete by Moone-
light.

Sn.   Doth the Moone fhine that night wee play our   50
play ?

41. *hither*] *hether* Q₂.
42. *pitty*] *pittty* F₂.
44. *tell him*] *tell them* Qq, Rowe et

seq.   *'em* Anon. ap. Cam.
50. Sn.] Qq. Snout. Cam. Rife, White
ii.   Snug. Ff et cet.

42. **of my life**] ABBOTT, § 174: 'Of' passes easily from meaning *as regards* to
concerning, about [as here, and also in line 188 of this scene : 'I desire you of more
acquaintance,' and again in IV, i, 145 : 'I wonder of there being here.']—W. A.
WRIGHT : That is, it were a sad thing for my life, that is, for me.   See V, i, 239.   It
would seem that in this expression 'of my life' is either all but superfluous or else a
separate exclamation, as in *Merry Wives*, I, i, 40 : 'Ha ! o' my life, if I were young
again, this sword should end it.'   The phrase occurs again in *Meas. for Meas.* II, i,
77 : 'It is pity of her life, for it is a naughty house.'   And in the same play, II, iii,
42, compare ''Tis pity of him,' equivalent to, it is a sad thing for him.

44. **name his name**] MALONE : I think it not improbable that Shakespeare meant
to allude to a fact which happened in his time at an entertainment exhibited before
Queen Elizabeth.   It is recorded in a MS collection of anecdotes, &c., entitled
Merry Passages and Jeasts, MS Harl. 6395: 'There was a spectacle presented to
Q : Elizabeth vpon the water and amongst others, Harr. Golding : was to represent
Arion vpon the Dolphin's backe, but finding his voice to be very hoarse and vnpleas-
ant when he came to performe it, he teares of his Disguise, and swears he was none
of Arion not he, but eene honest Har. Goldingham ; which blunt discoverie pleasd
the Queene better, then if it had gone thorough in the right way ; yet he could order
his voice to an instrument exceeding well.'   [I have followed, in spelling and punctuation, W. A. Wright, who is here presumably more accurate than either Malone or
Halliwell.—ED.]   The collector appears to have been nephew to Sir Roger L'Es-
trange.—KNIGHT : This passage will suggest to our readers Sir Walter Scott's description of the pageant at Kenilworth, when Lambourne, not knowing his part, tore off
his vizard and swore 'he was none of Arion or Orion either, but honest Mike Lambourne, that had been drinking her Majesty's health from morning till midnight.'

50. **Sn.**] Throughout this scene there appears to be but little uniformity in the
spelling of the names of the characters.   Quince is sometimes '*Quin.*' and sometimes
'*Pet.*'   Thisby is sometimes '*This.*' and sometimes '*Thys.*'   At line 54 we have
Enter Pucke, and at line 77 '*Enter Robin,*' as though it were another character

Bot.   A Calender, a Calender, looke in the Almanack,   52
finde out Moone-fhine, finde out Moone-fhine.

Enter Pucke.

Quin.   Yes, it doth fhine that night.   55

Bot. Why then may you leaue a cafement of the great
chamber window (where we play) open, and the Moone
may fhine in at the cafement.

Quin. I, or elfe one muft come in with a bufh of thorns
and a lanthorne, and fay he comes to disfigure, or to pre-   60
fent the perfon of Moone-fhine.   Then there is another
thing, we muft haue a wall in the great Chamber ; for *Pi-*
ramus and *Thisby* (faies the ftory) did talke through the
chinke of a wall.

Sn. You can neuer bring in a wall.   What fay you   65
Bottome ?

Bot. Some man or other muft prefent wall, and let
him haue fome Plafter, or fome Lome, or fome rough
caft about him, to fignifie wall ; or let him hold his fin-   69

---

54. Enter Pucke] Ff, Om. Qq et cet.
56. Bot.] Cet. Q₁.
57. *great chamber window*] *great*
chamber-window Knt.   *great-chamber*
Anon. ap. Cam.
59. *I,*] *Ay,* Rowe et seq.

65. Sn.] Q₂. Sno. Q₁. Snu. F₂. Snout.
Cam. Rlfe, White ii.   Snug. F₃F₄ et cet.
68. *Lome*] *lime* Coll. MS.
69. *or let*] *and let* Coll. MS, Dyce,
Huds. Rlfe, White ii.

---

and as though Puck were not already there.   Even the running title is '*A Midsomer*
nights Dreame.'   And there are trifling variations in the spelling of other names.
Wherefore, when we have, as in the present instance, merely '*Sn.*' we are free to
choose between *Snug* and *Snowt.*   The F₂F₃F₄ adopted *Snug*, and nearly every
editor has followed them.   The CAMBRIDGE EDD. elected *Snowt.*   It is a matter of
small importance; indeed, the very word 'importance' is almost too strong to apply
to the subject.—ED.

52. **Calender**] HALLIWELL asserts, but without giving his authority, that the calendars of Shakespeare's time were in 'even greater use than the almanacs of the
present day, and were more frequently referred to.'—KNIGHT: The popular almanac
of Shakespeare's time was that of Leonard Digges, the worthy precursor of the
Moores and the Murphys.   He had a higher ambition than these his degenerate
descendants; for, while they prophecy only by the day and the week, he prognosti-
cated *for ever*, as his title-page shows : 'A Prognostication *euerlastinge* of right good
effect, fruictfully augmented by the auctour, contayning plain, briefe, pleasaunte,
chosen rules to iudge the Weather by the Sunne, Moone, Starres, Comets, Rainebow,
Thunder, Cloudes, with other extraordinarye tokens, not omitting the Aspects of the
Planets, with a briefe iudgement *for euer*, of Plenty, Lucke, Sickenes, Dearth, Warres,
&c., opening also many natural causes worthy to be knowen' (1575).

69. **or let him**] DYCE (ed. i): This mistake of 'or' for *and* was occasioned by

gers thus ; and through that cranny, ſhall *Piramus* and   70
Thisby whiſper.

   Quin. If that may be, then all is well. Come, ſit
downe euery mothers ſonne, and rehearſe your parts.
Piramus, you begin ; when you haue ſpoken your ſpeech,
enter into that Brake, and ſo euery one according to his   75
cue.

Enter Robin.

   Rob. What hempen home-ſpuns haue we ſwagge-
    ring here,
So neere the Cradle of the Faierie Queene?   80
What, a Play toward?   Ile be an auditor,
An Actor too perhaps, if I ſee cauſe.

   Quin.   Speake *Piramus* : *Thisby* ſtand forth.

   Pir.   *Thisby*, the flowers of odious ſauors ſweete.   84

---

70. *that cranny*] *the cranny* Rowe+.	81. *toward*] *tow'rd* Pope+.
74. *Your*] *Yonr* Q$_1$.	82. *too perhaps*] *to perhappes* Q$_1$.
77. Scene II. Pope+.	84, 86, 106. Pir.] Bot. Cam. Rlfe,
Enter Robin.] Enter Puck. Rowe	White ii.
et seq. (subs.). Enter Puck behind.	84. *flowers*] *flower* Pope+, Cap. Steev.
Theob.	'73, '78, '85.
78, 79. *ſwaggering*] *ſwaggring* Qq.	*ſauors*] *savour's* Rowe, Pope.
80. *Faierie*] *Fairy* Qq.	*savour* Hal.

---

or ' having occurred twice before.   (It is but fair to Mr Collier's MS Corrector to
mention that this mistake did not escape him.)

   75. **Brake**] In defining this to be a 'thicket or furze-bush,' Steevens evidently
supposed that it was different from the hawthorn brake before mentioned.—HUNTER
(i, 295): Brake has many different senses.   Here it is used for what was otherwise
called a *frame*, a little space with rails on each side, which, in this instance, were
formed or at least intertwined with hawthorn. . . . See notice of the 'frame or brake'
in Barnaby Googe's *Book of Husbandry*, 1614, p. 119.—HALLIWELL: Kennett, MS
Lansd. 1033, defines *brake*, 'a small plat or parcel of bushes growing by themselves.'
This seems to be the right meaning here, although a single bush is also called a *brake*.
. . . The *brake* mentioned by Barnaby Googe would only be found in cultivated land,
not in the centre of the 'palace wood.'

   76. **cue**] MURRAY (*N. E. D.* s. v.): Origin uncertain.   It has been taken as
equivalent to French *queue*, on the ground that it is the tail or ending of the preceding
speech; but no such use of *queue* has ever obtained in French (where 'cue' is called
réplique), and no literal sense of *queue* or *cue* leading up to this appears in 16th century English.   On the other hand, in the 16th and early 17th centuries it is found written Q, *q*, *q*., or *qu*, and it was explained by 17th century writers as a contraction for
some Latin word (sc. *qualis, quando*), said to have been used to mark in actors'
copies of plays the points at which they were to begin.   But no evidence confirming
this has been found.

   84, &c. The speeches delivered at this rehearsal do not afterwards appear when

Quin. Odours, odours. 85

Pir. Odours fauors fweete,

So hath thy breath, my deareft *Thisby* deare.

But harke, a voyce : ftay thou but here a while,

And by and by I will to thee appeare. *Exit. Pir.*

Puck. A ftranger *Piramus,* then ere plaid here. 90

85. *Odours, odours*] Odours, odorous Qq.

87. *hath*] *that* Rowe i. *doth* Rowe ii+, Cap. Steev.

After this, a line lost. Wagner conj.

88. *a while*] *a whit* Theob. Han.

Warb. Johns. Cap.

89. Exit Pir.] Exit. Qq.

90. Puck.] Quin. Qq.

[Aside. Pope+, Cap. Steev. Mai. Var. Hal. Coll. (MS).

[Exit. Cap. et seq.

---

the play is performed before the Duke.—SIMPSON (*School of Shakspere,* ii, p. 88) finds in this lack of correspondence a precedent for the same lack in the Play within the Play of *Histrio-Mastix* (pp. 32–39, ed. Simpson), and asks, 'Was the *Midsummer Night's* the provocative of the *Histrio-Mastix?* Who was the author of the *Pyramus and Thisbe* there parodied?'

84. **of odious sauors**] COLLIER (ed. i) : Possibly we ought to read 'the flowers *have* odours, savours sweet, or '*odorous* savours sweet.'—*Ib.* (ed. ii) : The MS has 'flowers *have* odious savours sweet,' and rightly, as the next line of the supposed tragedy demonstrates, 'So *hath* thy breath,' &c. The corruption has been 'of' for *have ;* unless we are to suppose it to be one of the blunders of the 'hempen-home-spuns.'

84. **sauors**] This singular here used after a plural nominative, may have been perhaps intended, says ABBOTT, § 333, to be a sign of low breeding and harsh writing in this play of *Pyramus and Thisbe.* See III, ii, 466 : 'Two of both kindes makes up foure.' [But compare R. G. White's note on 'gallantly,' I, ii, 26 ; and also the next note below by the learned German to whom we owe the *Lexicon.*]

84. **sweete**] SCHMIDT (*Programm,* &c., p, 4) : However absurd may be the poesy of these Clowns, in rhythm and grammar it is irreproachable, therefore 'hath' in line 87 cannot be right. In Shakespearian dialogue (*dialogue,* be it observed) it is an inviolable rule that in alternate rhymes, when the second and fourth verses rhyme, the first and the third rhyme likewise. A sequence of endings like *sweet . . . dear . . . while . . . appear* violates Shakespear's use and wont. Wherefore, either *sweet* or *awhile* must be corrupt, probably the former. It is conceivable that Peter Quince, presumably the author of this tragedy of 'Pyramus and Thisbe,' wanted to say more, in his hyperbolic style, than that Thisbe's breath equalled in sweetness the odours of flowers,—odour did not amount to much, it is too commonplace ; we shall enter into his spirit if we read : 'Thisbe, the flowers of odours' savour's *vile* (or : the odorous flowers' savour's vile), So *not* thy breath,' &c.

88. **while**] THEOBALD changed this to *whit,* in order to rhyme with 'sweete,' and the change is harmless enough if there be a single uncouthness here which is not intentional.— MALONE goes even further, and supposes that two lines have been lost, one to rhyme with 'sweete' and another with 'while.'—ED.

89, 90. **And . . . here**] JULIUS HEUSER (*Sh. Jahrbuch,* xxviii, p. 207) : These two lines form a so-called *capping verse,* that is, a verse which contains a response to what precedes, although the speaker has not been directly questioned. They are

Thif. Muſt I ſpeake now? 91

Pet. I marry muſt you. For you muſt vnderſtand he goes but to ſee a noyſe that he heard, and is to come a-gaine.

Thyf. Moſt radiant *Piramus,* moſt Lilly white of hue, 95
Of colour like the red roſe on triumphant bryer,
Moſt brisky Iuuenall, and eke moſt louely Iew,
As true as trueſt horſe, that yet would neuer tyre,
Ile meete thee *Piramus,* at *Ninnies* toombe.

Pet. Ninus toombe man: why, you muſt not ſpeake 100
that yet; that you anſwere to *Piramus*: you ſpeake all
your part at once, cues and all. *Piramus* enter, your cue is
paſt; it is neuer tyre. 103

91, 95, 104. Thif.] Flu. Cam. Rlfe, White ii.
92, 100, 107. Pet.] Quin. Q$_1$, Rowe et seq.
97. *brisky Iuuenall*] *brisky Juvenile* Rowe ii+. *Briskly Juvenile* Han.

100. *why,*] *Why?* Q$_1$.
103. [Enter Pyramus. Rowe, Pope. Re-enter Bottom with an Ass-head. or, Puck and Bottom... Theob. Warb. Johns. Steev. Mal. Knt, Coll. White i, Sta.

---

generally in rhyme and are supposed to have a comic effect. [For this 'so-called *capping verse*' which, I think, appears here in literature for the first time, Simpson is indirectly responsible; its definition is Heuser's own. In Simpson's edition of *Faire Em* (*School of Sh.* ii, 422) he gives a collation with the Bodleian text of certain rhymes made by Fair Em and Trotter, and remarks that they are defective 'according to all rules of capping verses.' This remark ELZE quoted (*Sh. Jahrbuch,* xv, 344) in his notes on *Faire Em,* and added humourously that in Rowley's *When You See me You Know me* we had to deal with *rime couée.* This 'capped rhyme,' I am afraid, misled Heuser, to whom apparently the phrase 'to cap verses' was unfamiliar, and hence he supposed that there is a certain style of verse called 'capping.'—ED.]

90. Puck] Note that the Qq have *Quin.,* a serious blunder, whereof the correction adds much to the value which we should attach to the text of F$_1$. In a modernised text, I think, a period and a dash should close the preceding line, and a dash commence the present, so as to join the two speeches, and make Puck's the continuation, in sense, of Pyramus's: 'And by and by I will to thee appear,——a stranger Pyramus than e'er play'd here!' adds Puck in anticipation of the Ass-head which he was about to apply. (I find, by a MS marginal note, that I am herein anticipated by ALLEN.)—ED.

97. Iuuenall] W. A. WRIGHT: See *Love's Lab. L.* I, ii, 8, where this word again occurs; it was affectedly used, and appears to have been designedly ridiculed by Shakespeare.

97. eke] HALLIWELL: This word was becoming obsolete, and is used by Shakespeare only in burlesque passages.

102. cues and all] STAUNTON: To appreciate the importance of *cues* it must be borne in mind that when the 'parts' or written language of a new play are distributed, each performer receives only what he has himself to recite; consequently, if this

Thyſ. O, as true as trueſt horſe, that yet would neuer
tyre : 105

Pir. If I were faire, *Thisby* I were onely thine.

Pet. O monſtrous. O ſtrange. We are hanted; pray
maſters, flye maſters, helpe.

The Clownes all Exit.

Puk. Ile follow you, Ile leade you about a Round, 110

---

104. *O, as*] *O,—As* Theob. et seq.
(subs.).

105. *tyre :*] *tyre.* Qq.

[Re-enter Bottom with an Asse's
head. Han. Re-enter Puck and Bottom…
Cap. Dyce, Cam. White ii.

106. *I were faire, Thisby*] Q₂Ff. *I
were, fair Thisby,* Mal. conj. Coll. Hal.
I were fair Thisby, White i. *I were so,*

fair *Thisby*, Ktly. *I were fair, fair
Thisby* Anon. ap. Cam. *I were fairer
Thisby* Schmidt. *I were faire, Thisby,* Q₁ et
cet

107. *hanted*] *haunted* Qq.

109. *The…Exit.*] Om. Qq. The…
Exeunt. F₃F₄.

110. *Puk.*] Rob. Qq.

about] *'bout* Walker, Dyce ii, iii.

---

were unaccompanied by *cues* or catchwords from the other parts, he would be utterly
at a loss to know either when to make his entrance on the scene or to join in the
dialogue.

106. **I were faire, Thisby**] MALONE: Perhaps we ought to point thus : ' If I
were, [*i. e.* as true, &c.] fair Thisbe, I were only thine.'—STAUNTON, after quoting
this remark of Malone, replies : There cannot be a doubt of it, if we absolutely insist
upon making bully Bottom speak sensibly, which Shakespeare has taken some pains
to show he was never designed to do.—HUDSON (p. 121) even mends the metre, and
reads : '*An* if I were,' &c. He thinks the punctuation of the Folio is ' rather too
fine-drawn to be appreciated on the stage. Perhaps we ought to read, " If I were
true, fair Thisbe," &c., which is the meaning, either way, as the words are spoken in
reply to Thisbe's "As true as truest horse," &c.'

110. **a Round**] That is, a dance, but probably of a more fantastic and less orderly
style than that to which Titania invites Oberon when she asks him to ' dance patiently in our round,' II, i, 145. The phrase ' to lead about a Round ' has, however,
an uncouth sound ; ' about ' certainly seems superfluous, or almost tautological. Is it
permissible to suppose that ' a round ' is one word, *around*, and that in view of the
enumeration in the next five lines of the separate distresses, may not Puck have begun
this enumeration here : ' I'll follow you—I'll lead you—about—around—.' ? The
objection, almost a fatal one, to this reading is that nowhere is this word *around* to
be found, either in Shakespeare or in the Bible, 1611. But, as W. A. Wright says in
regard to *steppe*, II, i, 73, ' there is certainly no *a priori* reason why' the present passage ' should not furnish' an instance of it ; the word itself, although not in the sense
which I here ascribe to it, is, according to Murray (*N. E. D.* s. v.), as old as *c.* 1300,
and is used by Spenser, ' The fountaine where they sat arounde.'—*Shep. Cal.* June
30, and elsewhere. Wherefore the word itself, as an adverb, is not an anomaly. As
a preposition it is used by Milton in the sense here claimed for it as an adverb, and
the following example is given by Murray under the definition ' On all sides of, in all
directions from ' ; ' They around the flag Of each his faction . . . Swarm populous.'—
Par. Lost. II, 900. That there is need of such an adverb is proved by the examples

Through bogge, through bufh, through brake, through   111
Sometime a horfe Ile be, fometime a hound :   (bryer,
A hogge, a headleffe beare, fometime a fire,
And neigh, and barke, and grunt, and rore, and burne,
Like horfe, hound, hog, beare, fire, at euery turne.   *Exit.*   115
        Enter Piramus with the Affe head.
    Bot. Why do they run away ?   This is a knauery of
them to make me afeard.   *Enter Snowt.*   118

---

111. *Through bogge, through bush,*]
Thro' brook, thro bog, Peck. *Through
bog, through mire, through bush,* Johns.
conj.   *Through bog, through burn,
through bush,* Ritson.   *Through bog,
through brook, through bush* Lettsom
ap. Dyce, Marshall.
    bufh] *bnfh* F₃.

112. *Sometime*]*Sometimes* F₄, Rowe + .
    fometime]   *fometimes*   F₃F₄,
Rowe + .
113. *headleffe*] *heedless* Del. conj.
curbless Gould.
    fometime] *fometimes* F₄, Johns.
116. Enter...] Om. Qq. Enter Bottom with an Ass Head. Rowe, Pope.

---

of its use by eminent modern writers, as collected in the *N. E. D.* All that is
humbly urged for it here is that it may receive the stamp of respectability by admission
to Shakespeare's vocabulary.—ED.

    112, 113. COLLIER and HALLIWELL appeal to sundry popular ballads as authority
for these transformations.

    114, 115. Note the pelting, rattling staccato, which sounds like the explosion of a
pack of Chinese firecrackers, at the heels of the flying clowns ?—ED.

    116. Enter, &c.] It is needless to call attention to the patent dislocation of this
stage-direction.—B. NICHOLSON (*N. & Qu.* 4th Ser. V, 56) justifies its present position on the ground that according to line 109 all the clowns, Pyramus included, had
rushed off, and for '*Enter*' we should here read *Re-enter.* But no trust is to be
placed in the stage-directions on this imperfectly printed page of the Folio, where, at
line 54, we have '*Enter Pucke,*' who says no word for more than twenty lines nor
goes out, and yet, at line 77, we have '*Enter Robin.*' It is, however, a simple matter
to arrange the present action; we have Puck's account of it all in III, ii, 21, and by
it we know that Pyramus enters with the ass's head after line 105.—ED.

    116. the Asse head] I cannot but think that this trifling expression stamps this
stage-direction as taken from a play-house copy. See *Preface.*—ED.

    116. Asse head] 'If I affirme, that with certeine charmes and popish praiers I
can set an horsse or an asses head upon a mans shoulders, I shall not be beleeved;
or if I doo it, I shall be thought a witch. And yet if *J. Bap. Neap.* experiments be
true, it is no difficult matter to make it seeme so; and the charme of a witch or papist
joined with the experiment, will also make the woonder seeme to proceed thereof.
The words used in such case are vncerteine, and to be recited at the pleasure of the
witch or cousener. But the conclusion is this : Cut off the head of a horsse or an
asse (before they be dead), otherwise the vertue or strength thereof will be the lesse
effectuall, and make an earthen vessell of fit capacitie to conteine the same, and let
it be filled with the oile and fat thereof : cover it close, and dawbe it over with lome :
let it boile over a soft fier three daies continuallie, that the flesh boiled may run into
oile, so as the bare bones may be seene : beate the haire into powder, and mingle the

Sn. O *Bottom,* thou art chang'd; What doe I ſee on thee? 120

Bot. What do you ſee? You ſee an Aſſe-head of your owne, do you? 122

---

120. *thee?*] *thee? An Ass's head?* Johns. conj.
 [Exit. Cap. Exit frightened. Coll.

121. *Aſſe-head*] *ass's head* Var.'03, '13, '21, Sing. i.

122. [Exit Snout. Dyce, Cam.

---

same with the oile; and annoint the heads of the standers by, and they shall seeme to have horsses or asses heads.'—Scot's *Discovery of Witchcraft,* 1584, p. 315, ed. Nicholson.—That this was the passage whence Shakespeare took the idea of fixing an ass's head on Bottom was suggested first by DOUCE, i, 192, and the suggestion has been since then generally adopted.—B. NICHOLSON, however, is inclined to think (*N. & Qu.* 6th Ser. IV, 2) that a previous passage (p. 99, ed. Nicholson) gave the first and greater foundation to work upon. "The bodie of man is subject to . . . sicknesses and infirmities whereunto an asses body is not inclined; and man's body must be fed with bread, &c, and not with hay. *Bodins* asseheaded man must either eat hale or nothing; as appeareth by the storie." Nicholson thinks that this eating hay is very likely to have suggested Bottom's 'great desire to a bottle of hay'; and furthermore, both passages from Scot, especially the former, 'show that Shakspeare here introduced no unknown creature of his imagination, but brought before his audiences one which they had known by report. It was not the creature so much as his walking and talking as set forth, that made it supremely ridiculous.'—THOMS, also (*Three Notelets,* p. 68), infers from Scot that 'the possibility of such transformations was in Shakespeare's day an article of popular belief.' Bodin's story is to be found on p. 94 of Scot, ed. 1584, wherein a young man, as in Apuleius, was changed completely into an ass.—STEEVENS: The metamorphosis of Bottom's head might have been suggested by a trick mentioned in the *History of the Damnable Life and Deserved Death of Dr. John Faustus,* chap. xliii:—'The guests having sat, and well eat and drank, Dr. Faustus made that every one had an ass's head on, with great and long ears, so they fell to dancing, and to drive away the time until it was midnight, and then every one departed home, and as soon as they were out of the house, each one was in his natural shape, and so they ended and went to sleep.'—DOUCE refers to a receipt for this metamorphosis in Albertus Magnus *de Secretis Naturæ,* of which there was an English translation printed at London by William Copland. This receipt is thus given by W. A. WRIGHT (it is much less elaborate than Scot's, and really places the experiment within reach of the humblest): 'It thou wilt that a mans head seeme an Asse head. Take vp the couering of an Asse & anoint the man on his head.'

120. **thee?**] JOHNSON: It is plain by Bottom's answer, that Snout mentioned an *ass's head.* Therefore we should read: 'what do I see on thee? *An ass's head?*'— HALLIWELL: This suggestion by Dr Johnson is not necessary, the phrase being a vernacular one of the day, and originally in the present place created probably great amusement when thus spoken by Bottom in his translated shape. Mrs Quickly, in the *Merry Wives,* says, 'You shall have a fool's head of your own.' According to Pinkerton, 'The phrase—You see an ass's head of your own; do you?—is a trite vulgarism, when a person expresses a foolish amazement at some trifling oddity in another's dress or the like.'

Enter Peter Quince.   123

Pet. Bleſſe thee *Bottome*, bleſſe thee; thou art tranſlated.   *Exit.*   125

Bot. I ſee their knauery; this is to make an aſſe of me,
to fright me if they could; but I will not ſtirre from
this place, do what they can.   I will walke vp and downe
here, and I will ſing that they ſhall heare I am not a-
fraid.   130
The Wooſell cocke, ſo blacke of hew,
With Orenge-tawny bill.
The Throſtle, with his note ſo true,
The Wren and little quill.

Tyta.   What Angell wakes me from my flowry bed?   135
Bot.   The Finch, the Sparrow, and the Larke,
The plainſong Cuckow gray ;   137

---

125. Exit.] Exit frightened. Coll. Exeunt Snout and Quince. Sta.
129. *I will*] *will* F₃F₄, Rowe i.
130. [Sings. Pope et seq.
131. *Wooſell cocke*] *Wooſel cock* F₄,
Rowe. *Ousel cock* Pope +. *ouzel cock*
Cap. *oosel-cock* Steev.

132. *Orenge*] *Orange* Qq, Rowe ii et
seq.
133. *with*] *will* F₄, Rowe i.
134. *and*] *with* Qq, Pope et seq.
quill.] *quill ;* Cap. et seq.
135. [waking. Rowe.   Sings waking.
Pope.
136. Sings. Theob. et seq.

---

129. **they shall**] For other examples of the future where we should use the infinitive or subjunctive, see ABBOTT, § 348.

131. **Woosel Cocke**] W. A. WRIGHT: The male blackbird.   The word in the
Ff and Qq is probably the same as French *oiseau*, of which the old form was *oisel*.
Cotgrave gives, 'Merle : m. A Mearle, Owsell, Blackbird.   Merle noir.   The Blackbird, or ordinarie Owsell.'   [For further ornithological discussion, of great interest,
doubtless, to British naturalists, the student is referred to the voluminous notes of
HALLIWELL, STEEVENS, DOUCE, and COLLIER.   HARTING'S decision (p. 139) that
the owzel-cock is the *Turdus merula*, and Cotgrave's definition, are ample for us in
this country, and perhaps for all others elsewhere.—ED.]

133. **Throstle**] HARTING (p. 137): It is somewhat singular that the thrush (*Turdus musicus*), a bird as much famed for song as either the nightingale or the lark, has
been so little noticed by Shakespeare.   We have failed to discover more than three
passages in which this well-known bird is mentioned.   [The spelling 'Trassell,' in
the Qq and F₁ of *The Mer. of Ven.* I, ii, 58 (of this ed.), probably, with a broad *a*,
gives the pronunciation.—ED.]

134. **and little quill**] Remembering that it is Bottom who is singing, I cannot
but think it needless to change 'and' to *with*, as the Qq read.   Of course, 'quill'
here means *pipe* or *note*.—ED.

137. **plainsong**] CHAPPELL (p. 51, footnote): Prick-song meant harmony written
or pricked down, in opposition to plain-song, where the descant rested with the will

Whofe note full many a man doth marke,   138
And dares not anfwere, nay.
For indeede, who would fet his wit to fo foolifh a bird?   140
Who would giue a bird the lye, though he cry Cuckow,
neuer fo?

   Tyta. I pray thee gentle mortall, fing againe,
Mine eare is much enamored of thy note;
On the firft view to fay, to fweare I loue thee.   145
So is mine eye enthralled to thy fhape.
And thy faire vertues force (perforce) doth moue me.

   Bot. Me-thinkes miftreffe, you fhould haue little
reafon for that: and yet to fay the truth, reafon and
loue keepe little company together, now-adayes.   150
The more the pittie, that fome honeft neighbours will

---

139. *nay.*] *nay ;*— Cap. et seq.
  144. *enamored*] *enamoured* Q₁F₄.
enamour'd Rowe et seq.
  145. Transposed to follow line 147,
Q₁, Theob. et seq.

145. *to fweare*] *to swear*, Theob. et
seq.
  147. *vertues*] *vertue's* Rowe ii et seq.
virtue, Coll. conj.
      doth] *do* Thirlby.
  148. *miftreffe*] *maiftreffe* Ff.

---

of the singer. Thus the florid counterpoint in use in churches is slyly reproved in *The Four Elements*, circa 1517: '*Humanity.* Peace, man, prick-song may not be despised For therewith God is well pleased, In the church oft times among. *Ignorance.* Is God well pleased, trow'st thou, thereby? Nay, nay, for there is no reason why, For is it not as good to say plainly, Give me a spade, As give me a spa, ve, va, ve, va, ve, vade?'[—p. 49, ed. Hazlitt. T. WARTON, apparently misled by the word 'plain,' supposed that 'plain-song' meant 'having no variety of strains,' or having 'the uniform modulation of the chant,' and herein he is followed by DYCE and R. G. WHITE. HARTING, however, gives a different character to the Cuckoo's song; of this present line he says, p. 150:] The cuckoo, as long ago remarked by John Heywood, begins to sing early in the season with the interval of a minor third; the bird then proceeds to a major third, next to a fourth, then to a fifth, after which its voice breaks, without attaining a minor sixth. It may, therefore, be said to have done much for musical science, because from this bird has been derived the minor scale, the origin of which has puzzled so many; the cuckoo's couplet being the minor third sung downwards.

  139. nay] HALLIWELL: Bottom here refers to an opinion very prevalent in Shakespeare's time that the unfaithfulness of a wife was always guided by a destiny which no human power could avert.

  140. set his wit to] W. A. WRIGHT: That is, would match his wit against. So *Tro. and Cres.* II, i, 94: 'Will you set your wit to a fool's?'

  145–147. See Text. Notes for the proper order of these lines.

  149. reason and loue] VERITY: Compare the old proverb that 'a man cannot love and be wise,' from the maxim, *amare et sapere vix deo conceditur.*

not make them friends.   Nay, I can gleeke vpon occa-   152
fion.

 Tyta. Thou art as wife, as thou art beautifull.

 Bot. Not fo neither : but if I had wit enough to get   155
out of this wood, I haue enough to ferue mine owne
turne.

 Tyta. Out of this wood, do not defire to goe,
Thou fhalt remaine here, whether thou wilt or no.
I am a fpirit of no common rate :   160
The Summer ftill doth tend vpon my ftate,

---

152. gleeke] POPE: Joke or scoff.—BOSWELL: See Jamieson's *Scottish Diction-ary,* s. v. *Glaik, s.* [where the first meaning is : 'The reflection of the rays of light on the roof or wall of a house, or on any other object, from a lucid body in motion. Hence, to *cast the glaiks on* one, to make the reflection fall on one's eyes so as to con-found and dazzle.' The third meaning is : 'A deception or trick. *To play the glaiks with* one, to gull, to cheat. . . . This sense would suggest that it is radically the same with North of England *gleek,* to deceive, to beguile, as it is used by Shakespeare, " I can *gleek* upon occasion "; Lambe thinks it has been improperly rendered *joke* or *scoff.*' Jamieson's definition of the verb, however, viz. 'to trifle with, to spend time idly or playfully,' does not greatly vary from that of POPE, NARES, DYCE, STAUNTON, COLLIER, W. A. WRIGHT, and others, who define 'gleek' as *scoffing, jesting,* &c., a meaning which is certainly borne out in the only other passage where it is used as a verb in Shakespeare.   Gower, in referring to Pistol's treatment of Fluellen, says to the former, 'I have seen you gleeking and galling at this gentleman twice or thrice.' —*Hen. V:* V, i, 78.   The COWDEN-CLARKES (*Sh. Key,* p. 39) thus define the word : 'That is, *gibe, jeer;* in modern slang, *chaff.*   The expression originated in the name for a game of cards, called " gleek," in which game " a gleek " was the term for a set of three particular cards ; " to gleek," for gaining an advantage over ; and " to be gleeked," for being tricked, cheated, duped, or befooled.   Hence the words " gleek " and " gleeking " became used for being tauntingly or hectoringly jocose.'   But, after all, is it worth while to strain after any exact meaning in Bot-tom's words ?   Did he, more than nebulously, know his own meaning ? STAUNTON says : 'The all-accomplished Bottom is boasting of his versatility.   He has shown. by his last profound observation on the disunion of love and reason, that he pos-sesses a pretty turn for the didactic and sententious ; but he wishes Titania to under-stand that upon fitting occasion he can be as waggish as he has just been grave.'   To which W. A. WRIGHT replies : 'But a " gleek " is rather a satirical than a waggish joke, and in this vein Bottom flatters himself he has just been rather successfully indulging.'   Whatever the meaning of 'gleek,' I think it is clear that Bottom refers to what he has just said, not to what he may say in the future.   It is perhaps worth while merely to note that in the Opera of *The Fairy-Queen,* 1692, Bottom says here, instead of 'gleek,' ' Nay I can break a Jest on occasion.'   Garrick in his version, 1763, retained 'gleek.'—ED.]

160, 161. I am . . . state] FLEAY (*Life & Work,* p. 181): These lines are so closely like those in Nash's *Summer's Last Will,* where Summer says : 'Died had I indeed unto the earth, But that Eliza, England's beauteous Queen, *On whom all*

And I doe loue thee ; therefore goe with me,   162
Ile giue thee Fairies to attend on thee ;
And they fhall fetch thee Iewels from the deepe,
And fing, while thou on preffed flowers doft fleepe :   165
And I will purge thy mortall groffeneffe fo,
That thou fhalt like an airie fpirit go.

Enter Peafe-bloffome, Cobweb, Moth, Muftard-
feede, and foure Fairies.

Fai.   Ready ; and I, and I, and I, Where fhall we go ?   170

---

165. d[...]] ab[...] Γ₃Γ₄, Rowe i.

167. [Scene III. Pope +.

168, 169. Enter Peafe-bloffome...Muf-
tard-feede,] Peafe-bloffome...*and* Muftard feede ? (Continued to Titania.) Qq,
Theob. et seq.

168. Moth] *Mote* White.

169. and four Fairies.] Enter four
Fairies. Qq, Theob. et seq.

170. Fai.] Fairies. Qq.
Fai. *Ready ; and I, and I,*] 1.

Fai. Ready. 2. Fai. And I. 3. Fai
And I. Rowe et seq.   Peas-blossom.
Ready. Cobweb. *And I.* Mote. *And I.*
White, Dyce (subs.).

170. *and I, Where fhall we go ?*] 4.
Fair. *And I. Where shall we go?* Rowe +.
4. Fai. *Where shall we go?* Farmer, Steev.
'93, Coll. Mustard-seed. *And I.* All.
Where shall we go ? White, Dyce (subs.).
4. *And I.* All. *Where shall we go ?* Cap.
et cet. (subs.).

---

seasons prosperously attend, Forbad the execution of my fate,' &c., that I think they
are alluded to by Shakespeare.

161. still] Always.

168. **Moth**] R. G. WHITE : This is the invariable spelling of *mote* in the old
copies, as, for instance, in this play, V, i, 322. The editors, not having noticed this
orthography or that 'moth' was pronounced *mote* in Shakespeare's day, Fairy *Mote*
has been hitherto presented as Fairy *Moth*. [In his *Introduction* to *Much Ado*, and
in his note on 'Enter Armado and Moth,' in *Love's Lab. L.* I, ii, R. G. White has
gathered the following instances in proof of the old pronunciation of *th :* 'I am here
with thee and thy goats as the most capricious poet, honest Ovid, was among the
Goths.'—*As You Like It*, III, iii, 7 ; 'You found his Moth, the King your Moth did
see ; but I a beame doe finde in each of three.'—*Love's Lab. L.* IV, iii, 161 ; 'O
heaven, that there were but a moth in yours [*sc.* eye].'—*King John*, IV, i, 92. Wic-
liff wrote, in *Matthew* vi, 'were ruft and *mought* diftryeth.' To these examples he
adds in the prefent note:] From Withal's *Shorte* [Latin] *Dictionarie for Young*
Beginners, London, 1568 : 'A moth or motte that eateth clothes, *tinea.*'—fol. 7 a ; 'A
barell or greate bolle, *Tina, næ. Sed tinea, cum e, vermiculus est, anglicè*, A mought.'
—fol. 43 a ; and this from Lodge's *Wits Miserie*, 'They are in the aire like *atomi in*
sole, mothes in the sun.' [In his *Memorandums of Eng. Pronunciation*, &c., Shakespeare's Works, xii, p. 431, White has collected many more examples, such as : *nos-*
trils, nosethrills ; *apotecary*, apothecary ; *autority*, authority ; *t'one*, the one ; *t'other*,
the other ; *swarty*, swarthy ; *fift*, fifth ; *sixt*, sixth ; *Sathan*, Satan ; *Antony*, Anthony ;
wit, withe [an interesting example, by which alone can be explained the pun
in *Love's Lab. L.* I, ii, 94, 'green wit'] ; *pother*, pudder, potter ; *noting*, nothing
[White contends that the title of the play should be pronounced *Much Ado about*

Tita.   Be kinde and curteous to this Gentleman,   171
Hop in his walkes, and gambole in his eies,
Feede him with Apricocks, and Dewberries,
With purple Grapes, greene Figs, and Mulberries,
The honie-bags fteale from the humble Bees,   175
And for night-tapers crop their waxen thighes,
And light them at the fierie-Glow-wormes eyes,   177

172. *gambole*] *gambol* Cap.   177. *wormes*] *worms'* Kinnear.
175. *The*] *Their* Coll. MS.

Noting]; *With Sundayes,* Whit Sundays, &c., &c.—A. J. ELLIS, after a thorough
discussion of this memorandum of White, comes to this temperate general conclusion (*Early Eng. Pronun.* p. 972): 'There does not appear to be any reason for
concluding that the genuine English *th* ever had the sound of *t*, although some final
t's have fallen into *th*. As regards the alternate use of *d* and *th* in such words as
murther, further, father, &c., there seems reason to suppose that both sounds existed,
as they still exist, dialectically, vulgarly, and obsolescently.' As regards the name
of the little Fairy now present, however, I have no doubt that R. G. White is
entirely right.—ED.

170. R. G. WHITE was the first to substitute the fairies' names, instead of
numerals, before each repetition of 'and I.'—CAPELL was the first editor to mark
that 'All' united in the question 'Where shall we go?' Chronologically, he was
anticipated in *The Fairy-Queen, An Opera,* 1692.

173. **Apricocks**] W. A. WRIGHT: This is the earlier and more correct spelling
of *apricots*. The word has a curious history. In Latin the fruit was called *praecoqua*
(Martial, *Epig.* xiii, 46) or *praecocia* (Pliny, *H. N.* xv, 11), from being early ripe ;
Dioscorides (i, 165) called it in Greek πραικόκια. Hence, in Arabic, it became *bar-
quq* or *birquq*, and with the article *al-barquq* or *al-birquq ;* Spanish, *albarcoque ;* Italian, *albricocco* (Torriano) ; French, *abricot ;* and English, *abricot, abricoct* (Holland's
Pliny, xv, 11), *apricock*, or *apricot*.

173. **Dewberries**] HALLIWELL cites Parkinson's *Theatrum Botanicum,* 1640,
wherein the 'Deaw-berry or Winberry' is the *Rubus tricoccos*, and quotes a long
description. 'Other writers,' he adds, 'make it synonymous with the dwarf mulberry or knotberry, *Rubus chamæmorus*, and it is worth remarking that this fruit is
still called the dewberry by the Warwickshire peasantry. It is exceedingly plentiful
in the lanes between Stratford-on-Avon and Aston Cantlowe.'—W. A. WRIGHT says
its 'botanical name is *Rubus caesius*.' But of what avail are botanical names for
fruits of autumn and for flowers of spring which are not only in bloom but are ripe
in a dream on a midsummer night ?—ED.

177. **eyes**] JOHNSON: I know not how Shakespeare, who commonly derived his
knowledge of nature from his own observation, happened to place the glow-worm's
light in his eyes, which is only in his tail.—HALLIWELL, with greater entomological
accuracy, describes the light as 'emanating from the further segments of the abdo-
men,' and he might also have caught tripping even Dr Johnson himself for referring
to the glow-worm as masculine.—M. MASON: Dr Johnson might have arraigned
Shakespeare with equal propriety for sending his fairies to *light* their tapers at the fire
of the glow-worm, which in *Hamlet* he terms *uneffectual :* 'The glow-worm . . . gins

To haue my loue to bed, and to arife :   178
And plucke the wings from painted Butterflies,
To fan the Moone-beames from his fleeping eies.   180
Nod to him Elues, and doe him curtefies.

    1. *Fai.*   Haile mortall, haile.

    2. *Fai.*   Haile.

    3. *Fai.*   Haile.

    Bot.   I cry your worfhips mercy hartily ; I befeech   185
your worfhips name.

    Cob.   *Cobweb.*

    Bot.   I fhall defire you of more acquaintance, good
Mafter *Cobweb* : if I cut my finger, I fhall make bold
with you.   190
Your name honeft Gentleman ?

    Peaf.   *Peafe bloffome.*

    Bot.   I pray you commend mee to miftreffe *Squafh* ,   193

---

178. *haue*] *show* Gould.

179. *plucke*] *pluke* F$_2$.

182, 184. 1. Fai. ...*Haile*.] 1. F. *Hail mortal !* 2. *hail !* 3. *hail !* 4. *hail !* Cap. et seq. (subs.). Peas. *Hail mortal !* Cob. *Hail !* Moth. *Hail !* Mus. *Hail !* Dyce,

White.

185. *worfhips*] *worship's* Rowe+, Steev.'85, Var.'21, Knt, Coll. Hal. Ktly.

186. *worfhips*] *worship's* Rowe et seq.

187, 189. Cobweb.] Cobwed. F$_2$.

188. *you of*] *of you* Rowe+.

---

to pale his uneffectual fire.'   As we all know, and as Monk Mason himself probably knew, 'uneffectual' in *Hamlet* does not mean incapable of imparting fire, but of showing in the matin light.   But Dr Johnson, of all men, could not complain at being 'knocked down with the butt of a pistol.'   Indeed, he is sufficiently answered by the line in Herrick's *To Julia*, familiar as a household word : ' Her eyes the glow-worm lend thee.'—ED.—HAZLITT (*Characters*, &c., p. 130) : This exhortation is remarkable for a certain cloying sweetness in the repetition of the rhymes.

188. **you of**] STEEVENS, MALONE, STAUNTON, and HALLIWELL give examples from old authors of this construction, which may be termed common.   It is quite sufficient to refer to the note on line 42 of this scene, where ABBOTT, § 174, is cited, who gives additional examples, if even a single one be needed.   The modern phrase in line 195. 'I shal desire of you more acquaintance,' is possibly a misprint.—ED.

189. **if I, &c.**] MALONE notes that there is a dialogue ' very similar to the present ' in *The Mayde's Metamorphosis*, by Lilly.   This play was published anonymously in 1600, possibly after Lilly's death, and so little resembles in style all of the other plays by that author that Fairholt does not even include it in Lilly's *Works*.—ED.

193. **Squash**] SKEAT (*Dict.* s. v. *to squash*) : To crush, to squeeze flat.   No doubt commonly regarded as an intensive form of *quash ;* the prefix *s-* answering to Old French *es-* = Latin *ex-*.   But it was originally quite an independent word, and even now there is a difference in sense ; to *quash* never means to squeeze flat. . . . Derivative : *squash*, substantive, a soft unripe peascod [whereof Shakespeare himself gives the best definition in *Twel. N.* I, v, 165 : ' Not yet old enough for a man, nor young

9

your mother, and to mafter *Peafcod* your father.   Good
mafter *Peafe-bloffome*, I fhal defire of you more acquain-   195
tance to.   Your name I befeech you fir ?

Muf.   *Muftard-feede.*

Peaf.   *Peafe-bloffome.*

Bot.   Good mafter *Muftard feede*, I know your patience well : that fame cowardly gyant-like Oxe beefe   200
hath deuoured many a gentleman of your houfe.   I promife you, your kindred hath made my eyes water ere
now.   I defire you more acquaintance, good Mafter
Muftard-feede.

Tita.   Come waite vpon him, lead him to my bower.   205
The Moone me-thinks, lookes with a watrie eie,
And when fhe weepes, weepe euerie little flower,   207

195. *of you more*] *you of more* Qq, Cap. et seq.

acquaintance to.] *acquaintance, to.* Q₁. *acquaintance too.* Ff et seq.

198. *Peaf. Peafe-bloffome.*] Om. Qq Ff et seq.

199, 200. *your patience*] *your parentage* Han. Warb. *your puissance* Rann

conj. *your passions* Farmer. *you passing* Mason.

202. *hath*] *have* Cap. conj.

203. *you more*] *your more* F₃F₄, Cam. White ii. *more of your* Rowe+. *you, more* Cap. Steev.'85, Mal.'90. *you of more* Dyce, White i, Coll. ii.

207. *weepe*] *weepes* Q₁, Han. Cap. et seq.

enough for a boy; as a squash is before 'tis a peascod.' Our American vegetable, *squash*, is, according to the *Century Dict.*, an abbreviation of *squanter-squash*, a corruption of the American Indian *asqútasquash*. The authorities are Roger Williams, *Key to Lang. of America*, ed. 1643, and Josselyn, *N. E. Rarities*, 1672, Amer. Antiq. Soc. iv, 193.—ED.]

198. This is merely a compositor's negligent repetition of line 192, and was, of course, corrected in the next Folio.

199. **patience**] JOHNSON approved of HANMER's change to *parentage ;* FARMER fancied the true word was *passions*, i. e. sufferings.—CAPELL: 'Patience' is put for *impatience, hotness ;* applicable, to a proverb, to the gentleman the speech addresses; and that this is its ironical sense, the ideas that follow after seem to confirm; insinuating that this hotness, being hereditary in the family, had been the cause that many of them had been 'devour'd' in their quarrels with 'ox-beef,' and of his crying for them.—REED: These words are spoken ironically. According to the opinion prevailing in our author's time, mustard was supposed to excite choler —KNIGHT: The *patience* of the family of Mustard in being devoured by the ox-beef is one of those brief touches of wit, so common in Shakespeare, which take him far out of the range of ordinary writers.—HALLIWELL: Bottom is certainly speaking ironically, thinking perhaps of the old proverb—as hot as mustard. [Can there be a better proof of Mustard-seed's long suffering patience than that, being strong enough to force tears from Bottom's eyes, he permits himself to be devoured by a big cowardly Ox-beef ?—ED.]

207. **she weepes**] WALKER (*Crit.* iii, 48): Alluding to the supposed origin of

Lamenting fome enforced chaftitie.   208
Tye vp my louers tongue, bring him filently.   *Exit.*

[*Scene II.*]

Enter King of Pharies, folus.

Ob. I wonder if *Titania* be awak't;
Then what it was that next came in her eye,
Which fhe muft dote on, in extremitie.

Enter Pucke.   5

Here comes my meffenger : how now mad fpirit,
What night-rule now about this gaunted groue?   7

209. *louers*] *love's* Pope+, Cap. Steev. Mal. Knt, White, Dyce, Sta. Cam.

louers tongue] *louer's tongue and* Coll. ii (MS).

Exit.] Exeunt. Rowe.

Scene IV. Pope+. Scene II. Cap. et seq. Act IV, Sc. i. Fleay. Another Part of the Wood. Cap.

1. Enter...] Enter Oberon. Cap. et seq. folus.] and Robin Goodfellow. Qq, Om Theob. Warb. et seq.

4. *extremitie*] *extreamitie* Q₁.

5. Om. Qq. After *meffenger*, line 6, Dyce.

6. *fpirit*] *sprite* Pope+.

7. *gaunted*] *haunted* QqFf.

dew in the moon. *Macb.* III, v: 'Upon the corner of the moon There hangs a vaporous drop profound.' Fletcher, *Faithful Shepherdess*, iv, 4, Moxon, vol. i, p. 279: 'Showers of more price, more orient, and more round Than those that hang upon the moon's pale brow.'

209. **louers**] MALONE: Our poet has again used 'lover' as a monosyllable in *Twelfth N.* II, iv, 66: 'Sad true lover never find my grave.'—STEEVENS: In the passage quoted from *Twelfth N.* 'true lover' is evidently a mistake for *true love*, a phrase which occurs in the next scene, line 92. How is 'louer' to be pronounced as a monosyllable? [See WALKER (*Crit.* ii, 55), cited at II, ii, 81. There can be, I think, no doubt that *love* is the true word here. Is it insinuated that however deeply Titania may be enamoured with Bottom's fair large ears, and her eye enthralled to his shape, she can find no corresponding charm in his talk? There is a limit even to the powers of the magic love-juice; Bottom's tongue must be tied.—ED.]

4. **must**] Compelled by the love-juice.

6. **spirit**] See II, i, 32.

7. **night-rule**] STEEVENS: This should seem to mean here, what frolic of the night, what revelry is going forward?—NARES: Such conduct as generally *rules* in the night.—HALLIWELL quotes from the *Statutes of the Streets of London*, ap. Stowe, p. 666: 'No man shall, after the houre of nine at the night, keep any rule whereby any such sudden outcry be made in the still of the night,' &c. [Dyce's definition of 'rule' applies to this quotation from Stowe, and to other examples given by Halliwell, as well as to the present 'night-rule.' After quoting Nares's definition of 'rule,' viz. that it is apparently put for behaviour or conduct; with some allusion perhaps to the frolics called *mis-rule*,' DYCE adds: 'I believe it is equivalent to "revel, noisy sport";' Coles has "Rule (stir), *Tumultus*."—*Lat. and Eng. Dict.*' Whereby we come round

Puck.   My Miſtris with a monſter is in loue,   8
Neere to her cloſe and conſecrated bower,
While ſhe was in her dull and ſleeping hower,   10
A crew of patches, rude Mcehanicals,
That worke for bread vpon *Athenian* ſtals,
Were met together to rehearſe a Play,
Intended for great *Theſeus* nuptiall day :
The ſhalloweſt thick-skin of that barren ſort,   15
Who *Piramus* preſented, in their ſport,
Forſooke his Scene, and entred in a brake,
When I did him at this aduantage take,
An Aſſes nole I fixed on his head.
Anon his *Thisbie* muſt be anſwered,   20
And forth my Mimmick comes : when they him ſpie,

8, 9. *loue,...bower,*] *loue,...bower.* Q₁.
love. ...bower, Rowe et seq.
    11. *Mcehanicals*] F₁.
    14. Theſeus] *Theseus'* Rowe ii.
    15. *thick-skin*] *thick-skull* Han.
    16. *preſented, in their ſport,*] QqFf.
presented in their sport, Coll. Hal. Wh. i,

Sta. Dyce ii, iii.   *presented, in their sport*
Rowe et cet.
    19. *nole*] *nowl* Johns.   *nose* ' Bottom
the Weaver.'
    21. *Mimmick*] F₂F₃.   *Minnick* Q₁.
Minnock Q₂, Pope, Theob. Warb. Johns.
Steev.'85.   *Mimick* F₄ et cet. (subs.).

---

pretty nearly to Steevens's definition of 'night-rule' just given.—W. A. WRIGHT's
note here reads : 'Night-order, revelry, or diversion. "Rule" is used in the sense
of conduct in *Twelfth N.* II, iii, 132: "Mistress Mary, if you prized my lady's
favour at anything more than contempt, you would not give means for this uncivil
rule."' It is quite possible, I think, that here too Dyce's definition will apply, and
that 'rule' means something more than simply *conduct.* Malvolio certainly intends
to use vigorous language, and Sir Toby's conduct was extremely boisterous.—ED.]

    11. **patches**] Elsewhere in Shakespeare, e. g. *Tempest,* III, ii, 66, and *Mer. of
Ven.* II, v, 49 (of this ed.) this word has some reference, from the parti-coloured
dress, to the domestic fool, but here it means, I think, merely *ill-dressed* fellows, or as
Johnson has it, *tatterdemalions.*—ED.

    15. **thick-skin**] STEEVENS [note, *Mer. Wives,* IV, v, 2]: Thus, Holland's *Pliny,*
p. 346: 'Some measure not the finenesse of spirit and wit by the puritie of bloud,
but suppose creatures are brutish, more or lesse, according as their skin is thicker or
thinner.'—HALLIWELL: A common term of contempt for a stupid country bumpkin.

    15. **barren sort**] STEEVENS: Dull company.

    17. **in**] For other instances where 'in' is equivalent to *into,* see ABBOTT, § 159.

    19. **nole**] W. A. WRIGHT: A grotesque word for head, like pate, noddle. In
the Wicliffite versions of *Genesis,* xlix, 8, where the earlier has 'thin hondis in the
skulles of thin enemyes'; the later has 'thin hondis schulen be in the nollis of thin
enemyes'; the Latin being *cervicibus.* Probably 'nole,' like 'noddle,' was the back
part of the head, and so included the neck. Cotgrave has 'Occipital, . . . belonging
to the noddle, or hinder part of the head.'

    21. **Mimmick**] JOHNSON, on the ground that *minnock* was 'apparently a word of

As Wilde-geefe, that the creeping Fowler eye,   22
Or ruffed-pated choughes, many in fort

23. *ruffed-pated*] *ruffei pated* Q₁. *ruffed pated* Q₂. *ruffet-pated* F₄ et seq.

contempt,' believed that this misprint of Q₂ was right.—RITSON (p. 44) conjectured
mammock, which 'signifies a *huge misshapen thing;* and is very properly applied
by a Fairy to a clumsy over-grown clown.'—MALONE: ' Mimmick ' is used as synonymous to *actor* in Decker's *Guls Hornebooke*, 1609: 'and draw what troope
you can from the stage after you: the *Mimicks* are beholden to you, for allowing
them elbow roome' [—p. 253, ed. Grosart].—W. A. WRIGHT cites a passage
from Herrick's *The Wake*, ii, 62, where, again, the word has the same meaning,
actor.

23. **russed-pated choughes**] Whether or not by the name ' chough,' one species
of bird, and that the ' Cornish ' or ' Red-legged Crow,' was always meant is doubtful.
—HARTING (p. 118) says that we may infer the existence of ' various choughs ' from
a passage in O'Flaherty's *West or H' Iar Connaught*, 1684, p. 13 :—' I omit other
ordinary fowl and birds, as bernacles, wild geese, swans, cocks-of-the-wood, wood-
cocks, *choughs*, rooks, *Cornish choughs, with red legs and bills,*' &c. ' Here,' adds
Harting, ' the first-mentioned choughs were in all probability jackdaws.' Furthermore, ' the jackdaw, though having a grey head, would more appropriately bear the
designation " russet-pated " than any of its congeners. We may presume, therefore,
that this is the species to which Shakespeare intended to refer. The head of the
chough, like the rest of its body, is perfectly black.'—The difficulty of reconciling
the colour ' russet ' with what is perfectly black is so grave that W. A. WRIGHT
changed the text to ' russet-patted,' and remarked : ' I have not hesitated to adopt
Mr. Bennett's suggestion (*Zoological Journal*, v, 496), communicated to me by Professor Newton, to substitute *russet-patted* or red-legged (Fr. *à pattes rousses*) for the
old reading, which is untrue of the chough, for it has a russet-coloured bill and feet,
but a perfectly black head.' Hereupon followed a discussion in *Notes & Queries*
(5th Ser. xii, 444; 6th Ser. ix, 345, 396, 470; x, 499), whereof the substance is as
follows : B. NICHOLSON maintains that change is needless ; whatever be the colour of
' russet ' it is properly applied to the chough; and in confirmation cites N. Breton,
Strange Newes, &c. [p. 12, ed. Grosart], where the ' Russet-coate ' of the chough is
twice referred to.—F. A. MARSHALL adopts Harting's interpretation that the choughs
here mentioned are jackdaws, but finds it difficult even then to account for the epithet
russet in the sense of ruddy-brown as applied to them. As to the emendation proposed by Bennett and adopted by W. A. Wright, Marshall maintains that there is no
such word as *patted*, and even if there were Shakespeare would not have applied to
the claws what was distinctive of the whole leg; moreover, he would not have called
that ' russet ' which is scarlet or vermilion. Hereupon it became necessary to determine what the colour really is which ' russet ' represents. From the seven or eight
references supplied by Richardson's *Dict. s. v.* ' Russet,' Marshall thinks that his
own suggestion is perfectly justified, that ' *russet* might apply to the *grey* colour of
the jackdaw's head,' but never to the bright red of the Cornish chough's feet and
legs. Moreover he is confirmed, by a reconsideration of all the passages in Shakespeare where ' chough ' occurs, in the belief that it ' never meant anything else but
jackdaw.'—The discussion was closed by W. A. Wright, who, with a magnanimity
unfortunately rare, acknowledged that Marshall was ' perfectly right in his suggestion

(Riſing and cawing at the guns report)
Seuer themſelues, and madly ſweepe the skye :   25
So at his ſight, away his fellowes flye,
And at our ſtampe, here ore and ore one fals ;   27

---

that *russet* in Shakespeare's time described the *grey*-coloured head of the jackdaw; I have, therefore, restored the old reading.   I was induced to adopt Mr Bennett's conjecture, perhaps too hastily, from the feeling that the epithet " russet " as usually understood was inappropriate, and from the absence of any satisfactory evidence for another meaning.   Lately, however, on looking into the question afresh, I have found proof that " russet," although rather loosely used, did bear the meaning of grey or ash-coloured, and I now give the evidence for the benefit of others.   In the *Prompt. Parv.* (*cir.* 1440) we find, "*Russet*, Gresius," which is the French *gris.*—Junius's *Nomenclator*, trans. Higins (ed. Fleming, 1587), p. 178, gives :—"*Rauus . . . Faune, tané, rosset*, russet or tawnie colour."—*Rava* in Horace (*Od.* iii, 27, 3) is an epithet of the she-wolf.—"*Grigietto*, a fine graie or sheepes *russet.*"—Florio, *A Worlde of Wordes*, 1598.   "*Gris.* m. *ise.* f. Gray, *light-russet*, grizle, ash-coloured, hoarie, whitish."—Cotgrave, *Fr. Dict.* 1611.—"Also, whosoever have about him hanging to anie part of his bodie the heart of a toad, enfolded within a peece of cloth of a *white russet* colour (*in panno leucophœo*), hee shall be delivered from the quartane ague."—Holland's *Pliny*, 1601, xxxii, 10.   " Contrariwise, that which is either purple or ash-coloured and *russet* to see too, &c. (*Purpurea aut leucophœa*)."—*Ibid.*, xxiv, 12.   In the last passage *ash-coloured* and *russet* are evidently synonymous, and equivalent to *leucophœa*.   But to show that *russet* was rather loosely applied it is sufficient to quote another instance from the same volume.   In Holland's *Pliny*, xi, 37 (vol. i, p. 335), the following is the translation of " aliis nigri, aliis ravi, aliis glauci coloris orbibus cirumdatis " :—" This ball and point of the sight is compassed also round about with other circles of sundry colours, black, blewish, tawnie, *russet*, and red ;" the last three epithets being to all appearance alternative equivalents of *ravi*.   *Russet*, so far as one can judge, described a sad colour, and was applied to various shades both of grey and brown.   That *chough* and *jackdaw* were practically synonymous may be inferred from Holland also.   In his translation of *Pliny*, x, 29 (vol. i, p. 285) we find :— " And yet in the neighbor quarters of the Insubrians neere adjoining, we shall have infinite and innumerable flockes and flights of *choughes and jack dawes* (*gracculorum monedularumque*)."   Here *gracculus* is the chough, and *monedula* the jackdaw ; but in xvii, 14 (vol. i, p. 516), where the Latin has only *monedula*, the translator renders, " It is said moreover, that the *Chough* or *Daw* hath given occasion hereof by laying up for store seeds and other fruits in crevises and holes of trees, which afterwards sprouted and grew."   If *monedula*, therefore, can be rendered in one passage by " jackdaw " and in another by " chough or daw," it is not too much to assume that in the mind of the translator, who was a physician at Coventry in Shakespeare's own county, the chough and the jackdaw were the same bird.' [See ' gray light,' line 443, *post.*]

    23. sort] Company ; see line 15.

    27. stampe] THEOBALD (*Nichols*, 233) : Perhaps ' at our *stump* here,'—pointing to the stump of some tree, over which the frighted rustics fell.—JOHNSON : Fairies are never represented stamping, or of a size that should give force to a stamp, nor

He murther cries, and helpe from *Athens* cals. 28
Their fenſe thus weake, loſt with their tears thus ſtrong,
Made fenfeleſſe things begin to do them wrong. 30
For briars and thornes at their apparell ſnatch,
Some ſleeues, ſome hats, from yeelders all things catch,
I led them on in this diſtracted feare,
And left ſweete *Piramus* tranſlated there :
When in that moment(ſo it came to paſſe) 35
Tytania waked, and ſtraightway lou'd an Aſſe.
 Ob. This fals out better then I could deuiſe : 37

32. ſtampe] ſtampe Γ₃Γ₄

could they have distinguished the stamps of Puck from those of their own compan-ions. I read, 'at *a stump.*' So Drayton : 'A pain he in his head-piece feels, Against a *stubbed tree* he reels, And up went poor Hobgoblin's heels, &c. . . . A *stump* doth *trip him* in his pace, Down fell poor Hob upon his face,' &c.—[*Nymphidia*, p. 166, ed. 1748. The CAMBRIDGE EDITORS record this conj. as adopted in Johnson's text, and also as anticipated by Theobald. They were possibly misled by the 'I read' in Johnson's note, which means merely that he conjectures ; the original 'stamp' is retained in Johnson's text ; and they overlooked that Theobald's conj. is '*our* stump.' —ED.]—RITSON : Honest Reginald Scott says : 'Robin Goodfellow . . . would chafe exceedingly if the maid or good wife of the house . . . laid anie clothes for him bee-sides his messe of white bread and milke, which was his standing fee. For in that case he saith, What have we here ? Hemton, hamten, here will I never more tread nor stampen.'—*Discoverie of Witchcraft*, 1584, p. 85.—STEEVENS : The *stamp* of a fairy might be efficacious though not loud ; neither is it necessary to suppose, when supernatural beings are spoken of, that the size of the agent determines the force of the action. See IV, i, 97 : 'Come, my queen, take hand with me, And *rock* the ground,' &c.—ALLEN (MS) : It cannot be 'our' ; there was no *we* in the case ; no fairy but Puck alone ; and it was nobody's stamp that made the boors scatter ; it was merely the sight of Bottom's new head. *Perhaps :* 'at *one* stamp,'— as we might say : at one bound, at one rush ; for they started so instantly, all together, that all their feet struck the ground, on starting to run, with one stamp, one noise (Anticipa-tive of *stampede !*). [If change be needed, Allen's conj. is worthy of adoption. That Shakespeare has nowhere else thus used 'stamp' amounts to but little. Puck's sud-den change to 'our,' when he was the sole agent, is somewhat unaccountable. W. A. WRIGHT interprets the phrase 'at hearing the footsteps of the fairies,' but we have no authority for the presence of any other fairy than Puck, who says, '*I* did him at this advantage take,' '*I* fixed an asses nole,' and '*I* led them on,' &c. The misprint of 'our' for *one* is of the simplest. Since the foregoing note was written, the Second Edition of the *Cambridge Edition* has appeared ; in it 'our stamp' is duly credited as Theobald's conj., but 'a stamp,' as Johnson's reading, is still retained.—ED.]

28. He] ABBOTT, § 217 : Used like *hic* (in the antithesis between *hic* . . . *ille*).

30. senselesse] DYCE (*Rem.* 47) asks why Collier has a comma after this word. It was probably an oversight ; it is corrected in Collier's third edition. —ED

But haſt thou yet lacht the *Athenians* eyes,   38

38. *lacht*] *latcht* Q₁F₃F₄. *lech'd* Han.   *washed* Orger.
Cap.   *streak'd* or *bath'd* D. Wilson.   38. Athenians] *Athenian* F₄, Rowe i.

38. **lacht**] HANMER: Or letch'd, lick'd over, *lecher,* Fr. to lick.—STEEVENS: In
the North it signifies to *infect.*—STAUNTON, referring to Hanmer's note, says that he
has found no instance of the word thus used.—DYCE, however, gives no other meaning than this of Hanmer, and cites Richardson's *Dict.* as adopting it.—HALLIWELL
gives the meaning *to catch.* 'Hence, metaphorically,' he continues, 'to infect.
"*Latching,* catching, infecting," Ray's *English Words,* ed. 1674, p. 29. The word
occurs in the first sense in *Macbeth* [IV, iii, 196]. I believe the usual interpretation,
licked over, is quite inadmissible; but it is to be observed that the direction was to
anoint the eyes. The love-juice literally caught the Athenian's eyes.'—W. A.
WRIGHT: In the other passages where 'latch' is used by Shakespeare it has the
sense of *catch,* from A.-S. *læccan,* or *gelæccan.* See *Macbeth,* and *Sonn.* 113, 6, of
the eye: 'For it no form delivers to the heart Of bird, of flower, or shape, which it
doth latch.' Compare also Holland's *Pliny,* viii, 24, of the Ichneumon: 'In fight
he sets up his taile, & whips about, turning his taile to the enemie, & therein latcheth
and receiveth all the strokes of the Aspis.' In the present passage 'latch'd' must
signify caught and held fast as by a charm or spell, like the disciples going to Emmaus
(*Luke* xxiv, 16): 'their eyes were holden, that they should not know him.' There
appears to be no evidence for Hanmer's interpretation. On the other hand, a 'latch-
pan' in Suffolk and Norfolk is a dripping-pan, which catches the dripping from the
meat; and Bailey gives 'latching' in the sense of catching, infectious; as it is still
used in the North of England.—DANIEL (p. 32): Perhaps the right word should be
hatch'd. In Beaumont and Fletcher it is a word of frequent occurrence, meaning
generally to cover thinly, as in gilding, lackering, varnishing, or staining. [Here
follow seven or eight examples of the use of *hatch,* all of which corroborate Gifford's
definition: 'Literally, to hatch is to inlay; metaphorically, it is to adorn, to beautify,
with silver, gold, &c.'—Note on 'thy chin is hatched with silver,' Shirley, *Love in a
Maze,* II, ii, cited by Dyce. Daniel's suggestion is upheld by Deighton.]—W. W.
SKEAT (*Academy,* 11 May, 1889): The word here used has nothing to do with
'latch,' to catch. Mr W. A. Wright cites *latch-pan,* so called because it '*catches* the
dripping'; and the Prov. English *latching,* catching. Halliwell remarks on *latch-
pan* that 'every cook in Suffolk could settle the dispute,' and adds, 'the Athenian's
eyes were Puck's latch-pans.' The fact is that the whole trouble has arisen from this
etymology of 'latch-pan.' The explanation depends upon the fact that there are two
distinct verbs, both spelt 'latch,' which are wholly unrelated to each other. Shakespeare's 'latch' is related to 'latch-pan' precisely because a *latch-pan* is totally
unconnected with 'latch,' to catch. It correctly means *dripping-pan,* because 'latch'
means to drip, or to cause to drop or to dribble. To '*latch* with love-juice' is to
drop love-juice upon, to distil upon, to dribble on, or simply to moisten. If we will
give up the Anglo-Saxon *gelæccan,* and consider the common Eng. verb 'to leak,' we
shall soon come to a satisfactory result. To 'leak' means to admit drops of water;
and 'latch' is practically the causal form. The use of the latter occurs in Prov. Eng.
latch on, 'to put water on the mash when the first wort is run off,' says Halliwell. It
means merely to dribble on, to pour on slowly. The Swedish has the very phrase.
Widegren's *Swedish Dict.* (1788) gives us '*Laka,* to distil, to fall by drops.' This

With the loue iuyce, as I did bid thee doe?

Rob. I tooke him fleeping (that is finifht to)   40
And the *Athenian* woman by his fide,
That when he wak't, of force fhe muft be eyde.

Enter Demetrius and Hermia.

Ob. Stand clofe, this is the fame *Athenian.*
Rob. This is the woman, but not this the man.   45
Dem. O why rebuke you him that loues you fo?
Lay breath fo bitter on your bitter foe.
Her. Now I but chide, but I fhould vfe thee worfe,
For thou (I feare) haft giuen me caufe to curfe,
If thou haft flaine *Lyfander* in his fleepe,   50
Being ore fhooes in bloud, plunge in the deepe, and kill
          me too: 52

40, 45. Rob.] Puck. Rowe et seq.
40. *fleeping (that...to)*] *fleeping; that
.too;* Rowe+.
    to] *too* Ff.
42. *wak't*] *wakes* Pope+.
43. Scene V. Pope+.
44, 45. Aside. Cap. They stand apart.

Coll. ii (MS).
    51. *the deepe*] *knee-deep* Coleridge (ap.
Walker), Maginn, Phelps, Dyce ii, iii,
Ktly, Huds.
    51, 52. *and kill me too*] Sep. line,
Rowe ii et seq.
    52. *too*] *to* Qq.

---

laka gives us the original *a;* the mutated *a* occurs in Swed. lāka, 'to leak.' Icelandic has the strong verb *leka*, 'to drip, to dribble, also to leak.' Koolman's *E.
Friesic Dict.* also helps us. He gives: *lek*, 'a drop, a dripping from a roof'; *lek-bēr*,
'drop-beer,' *i. e.* beer caught by standing a vessel under a leaky cock of a cask; *lek-
fat*, 'a drop-vessel,' *i. e.* a vessel in which drops are collected. The connexion of
the latter with 'a *latch-pan*' is obvious. The nearest-related Anglo-Saxon word is
leccan, 'to moisten, wet, irrigate.' This would have given a verb to *letch*, with the
sense 'to moisten.' The Prov. Eng. *latch* seems to be due to some confusion between
this form and the base *lak*, which appears in the Swedish *laka*, Danish *lage*, and in
the past tense of the Icel. strong verb; or else, as is common in English, 'latch,' to
catch, and the less-known 'letch,' to moisten, were fused under one (viz. the com-
moner) form. Whatever the true history of the form of the word may be, I think we
need have no doubt now as to its true sense.

  46, 48. **you . . . thee**] Note that Demetrius uses the respectful 'you,' while Hermia replies with the contemptuous 'thou.'—ED.

  51. **bloud**] STEEVENS: So in *Macb.* III, iv, 136: 'I am in blood Stepp'd in so
far,' &c.

  51. **the deepe**] WALKER (*Crit.* iii, 49): Read, with Coleridge, 'knee-deep.'
Compare Heywood, *Woman Killed with Kindness*, Dodsley, vii, 268: 'Come, come,
let's in; Once over shoes, we are straight o'er head in sin.' *Qu.* Is it a proverbial
phrase?—HALLIWELL quotes a note by Phelps in which this emendation 'knee-deep'
is given, but no reference to Coleridge as the author. If Coleridge be the author, he

The Sunne was not fo true vnto the day,   53
As he to me.   Would he haue ftollen away,
From fleeping *Hermia*? Ile beleeue as foone   55
This whole earth may be bord, and that the Moone
May through the Center creepe, and fo difpleafe
Her brothers noonetide, with th'*Antipodes*.
It cannot be but thou haft murdred him,
So fhould a mutrherer looke, fo dead, fo grim.   60
    Dem.   So fhould the murderer looke, and fo fhould I,
Pierft through the heart with your ftearne cruelty :
Yet you the murderer looks as bright as cleare,
As yonder *Venus* in her glimmering fpheare.   64

54. *away*,] *away* Rowe et seq.
55. *From*] *Frow* Q₁.
57. *difpleafe*] *disease* Han. *displace* D. Wilson. *disseise* Annandale ap. Marshall.
58. *with th'*] *i' th'* Warb. *with the* Cap. Steev. Mal. Knt, Dyce, Cam. Wh. ii.

60. *mutrherer*] F₁. *murderer* Q₄ *dead*] *dread* Pope+.
61. *murderer*] F₂F₃. *murtherer* F₄ Rowe. *murthered* Q₁. *murdered* Q₂ *murther'd* or *murder'd* Pope et cet.
63. *looks*] *looke* Qq, Rowe et seq.

must antedate Phelps; I am unable, however, to say where in Coleridge's notes the emendation is to be found.   DYCE, who adopts it, states no more than the fact that it is Coleridge's, and that Walker approved of it.   The instances are extremely rare where Dyce does not cite volume and page, and his omission to cite them in regard to Coleridge leads me to think that Walker alone was his authority.   I strongly suspect that it was not Coleridge, after all, who proposed the amendment, but Maginn.   In a foot-note (*Shakespeare Papers*, p. 138, ed. 1860) MAGINN says: 'Should we not read "*knee* deep"? As you are already over your shoes, wade on until the bloody tide reaches your knees.   In Shakespeare's time *knee* was generally spelt *kne;* and between *the* and *kne* there is not much difference in writing.'   In Phelps's note, quoted by Halliwell, this last sentence of Maginn is repeated word for word.   The objection to this emendation, not absolutely fatal, but still serious, is one that Maginn evidently felt when he substituted *wade* for 'plunge'; in water knee-deep we can certainly wade, but it can hardly be said that we can *plunge* into it.—ED.

    51, 52. **and kill me too**] Of course ROWE was right in making a separate line of these words.   Probably some dramatic action, such as offering her breast to him to strike, completed the line.—SCHMIDT, however, conjectures (*Programm*, &c., p. 5) that some words have dropped out, because 'even in a tragedy, where there is talk of real killing, Shakespeare would not have laid so strong an emphasis on such a phrase as "And kill me too" as to let it interpose between two rhyming couplets.'   The cheap plea of an omission should be our very last resort.—ED.

    56. **whole**] W. A. WRIGHT: Solid. Compare *Macb.* III, iv, 22 : 'Whole as the marble.'

    60. **dead**] STEEVENS: Compare *2 Henry IV:* I, i, 71 : 'Even such a man, so faint, so spiritless, So dull, so dead in look, so woe-begone.'—CAPELL : Pope's change to *dread* is implied in 'grim'; by 'dead' is meant *pale*.

    61, 63. **murderer . . . looks**] Corrected in the Qq.

Her. What's this to my *Lyfander* ? where is he *?* 65
Ah good *Demetrius,* wilt thou giue him me ?
 Dem. I'de rather giue his carkaffe to my hounds.
 Her. Out dog, out cur, thou driu'ft me paft the bounds
Of maidens patience. Haft thou flaine him then?
Henceforth be neuer numbred among men. 70
Oh, once tell true, euen for my fake,
Durft thou a lookt vpon him, being awake ?
And haft thou kill'd him fleeping ? O braue tutch :
Could not a worme, an Adder do fo much ?
An Adder did it ; for with doubler tongue 75
Then thine(thou ferpent) neuer Adder ftung.
 Dem. You fpend your paffion on a mifpri'sd mood,
I am not guiltie of *Lyfanders* blood :
Nor is he dead for ought that I can tell.
 Her. I pray thee tell me then that he is well. 80

---

67. *I'de*] Ff. *Ide* Q₂. *I'd* Rowe, Hal. Wh. i, Sta. *I'ad* Pope+. *I had* Q₁ et cet.

68. *bounds*] *bonds* Q₂.

71. *tell true*] *tell true, and* Ff, Rowe, Pope, Theob. Han. Warb. *tell true :* *tell true* Q₁, Johns. et seq. (subs.).

72. *a*] *haue* Qq, Rowe ii et seq.

73. *tutch*] *touch* Rowe et seq.

75. *An*] *And* F₂.

77. *on a...mood*] *in a...flood* Coll. MS.

79. *ought*] *aught* Theob. ii, Warb. Johns. Mal. Steev. Knt, Coll. Dyce et seq.

---

64. **glimmering**] W. A. WRIGHT: Faintly shining; this epithet seems in contradiction to 'bright' and 'clear' of the previous line.

69. **him then ?**] Does not the wildness of Hermia's grief suggest that we should thus punctuate: 'Hast thou slain him ? Then Henceforth be never,' &. ?—ED.

71. **tell true**] We must again look to the Quartos for the rhythmical completion of this line.

72. **thou a lookt**] I am not sure that this 'a,' the mere suggestion of *have*, does not permit an increased emphasis of scorn to be thrown on 'looked.' I am quite sure, however, that Capell did not improve the vigour of the line when he took away the interrogation mark at the end and substituted a comma, wherein he has been generally followed.—ED.

73. **tutch**] JOHNSON: The same with our *exploit*, or rather *stroke*. A brave touch, a noble stroke, *un grand coup*. 'Mason was verie merie, . . . pleasantlie playing, both, with the shrewde touches of many courste boyes, and with the small discretion of many leude Scholemasters.'—Ascham [*The Scholemaster*, p. 18, ed. Arber].

77. **mispris'd mood**] JOHNSON: That is, mistaken; so below [line 93], 'misprision' is *mistake*.—MALONE: 'Mood' is anger, or perhaps rather, in this place, *capricious fancy*.—STEEVENS: I rather conceive that '*on* a mispris'd *mood*' is put for *in* a mistaken *manner*.' See ABBOTT, § 180, for instances of the use of 'on' for *in*. —ALLEN (MS): It might be 'on a mispris'd *word*,'—you have mistaken the meaning of my word 'murder'd' or 'carcase.'

Dem. And if I could, what fhould I get therefore ?   81
Her. A priuiledge, neuer to fee me more ;
And from thy hated prefence part I : fee me no more.
Whether he be dead or no.   *Exit.*

Dem. There is no following her in this fierce vaine,   85
Here therefore for a while I will remaine.
So forrowes heauineffe doth heauier grow :
For debt that bankrout flip doth forrow owe,
Which now in fome flight meafure 'it will pay,
If for his tender here I make fome ftay.   *Lie downe.*   90

---

81. *And*] QqFf, Rowe+. *And,* Coll. Wh. i.   *An* Cap. et cet.
82. *fee me*] *see him* Steev.'85 (misprint ?).
83. *part I:*] *part I so :* Pope et seq.
83, 84. *fee...no.*] Sep. line, Pope et seq.

84. *he be*] *he's* Pope+.
88. *bankrout flip*] *bankrout flippe* Q₁. *bankrupt sleep* Rowe et seq.
90. Lie downe.] Ly doune. Q₁. Lies down. Rowe.
[Scene VI. Pope, Han.

---

81. **And if**] The rule is so uniform in the Ff and Qq that 'and if' is 'an if,' that any exception must find unusual support in the meaning or force of the phrase. 'An if' is not a mere reduplication of 'if'; it adds much to the uncertainty of the doubt. Wherefore, I think, before we can decide that 'and if' is equivalent to *an if* in any given example, we must be sure that this added doubt is intended. Is this the case here? The emphatic thought in this line is '*what* should I get therefor?' and the emphatic word is 'what.' There is no such emphasis on the doubt that the 'if' need be duplicated. The sense would be quite as good, perhaps even better, if a comma were placed after 'And,' a shade of contempt might be then detected: 'And, if I could, *what* should,' &c. Wherefore, if an exception to the rule is to be made, I should make it here. It is in such cases as this that we feel the need of the Greek Moods and Particles.—ED.

83. **part I :**] Every editor, I believe, since Pope has adopted the latter's change for rhyme's sake, 'part I *so*.' That *so* is the word which the compositor has omitted I have no doubt, but whether or not we should adopt Pope's punctuation I have strong doubts. Hermia is at the height of her passion, and I cannot imagine her as using a phrase like 'part I *so*!' where *so* has really not only little meaning, but actually detracts from the force of her vigorous determination to part. I prefer a full stop, and read, 'from thy hated presence part I. So, See me no more,' &c.—ED.

84. **Whether**] For instances of the very common contraction in scanning into *Whe'er*, see Walker, *Vers.* 103; Abbott, § 466; it is certainly better to make this contraction than to change 'he be' into *he's*, with Pope.—ED.

87. **So**] DEIGHTON: 'So' seems out of place here, it not being correlative to anything; possibly it is a mistake for *since*, the *so* of 'sorrow' being caught by the transcriber's eye.

88–90. **debt . . . bankrout . . . tender**] MARSHALL thinks that the 'prosaic and legal character' of these words 'smells' of an attorney's office. The fondness of Shakespeare for similes drawn from bankruptcy, even in the most impassioned pas-

Ob.   What haſt thou done? Thou haſt miſtaken quite   91
And laid the louc iuyce on ſome truc loues ſight :
Of thy miſpriſion, muſt perforce enſue
Some true loue turn'd, and not a falſe turn'd true.

Rob.   Then fate ore-rules, that one man holding troth,   95
A million faile, confounding oath on oath.

Ob.   About the wood, goe ſwifter then the winde,
And *Helena* of *Athens* looke thou finde.
All fancy ſicke ſhe is, and pale of cheere,
With ſighes of loue, that coſts the freſh bloud deare.   100

---

91. [Coming forward with Puck. Coll.
ii.
   92. *the*] *thy* F₄, Rowe+, Steev.'73.
     louc] F₁ (ap. Editor's copy).
     true loues] *true-love's* Cap. et seq.
   94. *turn'd, and*] *turn'd falſe* Han.
     true loue] *true-love* Var.'21 et seq.
   95. Rob.] Puck. Rowe et seq.

95. *that*] *for* Han.
96. *A million*] *And million* Del. (mis-print?).
   97. Ob.] Rob. F₂.
   98. *looke*] *see* Rowe+.
   100. *coſts*] *cost* Theob. ii +, Steev. Mal. Knt, Coll. Sing. Hal. Dyce, Sta.

---

sages, may be learned from Mrs Cowden-Clarke's, and Mrs Furness's *Concordances.*
—ED.

88. slip] COLLIER calls attention to a similar spelling, which sometimes occurs, of 'ship' for *sheep.*

90. **Lie downe**] Another stage-direction in the imperative, betraying the stage-house copy.—ED.

93. Of] For instances where 'of,' meaning *from,* passes naturally into the meaning *resulting from, as a consequence of,* see ABBOTT, § 168.

93. misprision] Mistake. See 'mispris'd,' line 77.

95, 96. **Then . . . oath**] DEIGHTON : Puck's excuse for his carelessness does not seem to be very logical. Possibly the meaning is ; Then, if that happens, the fault is fate's, who so often is too strong for men's intentions that, for one man who keeps faith, a million, whatever their intentions, give way and break oath after oath, *i. e.* any number of oaths.—GERVINUS (p. 196, trans.) : The poet further depicts his fairies as beings of no high intellectual development. Whoever attentively reads their parts will find that nowhere is reflection imparted to them. Only in one exception does Puck make a sententious remark upon the infidelity of man, and whoever has pene-trated into the nature of these beings will immediately feel that it is out of harmony. [Or, in other words, it does not happen to fadge with the scheme of fairydom which the learned German has evolved ; and christened Shakespeare's.—ED.]

95. that] For instances where 'that' means *in that,* see ABBOTT, § 284.

96. confounding] SCHMIDT (*Lex.*) will supply many examples where 'confound' means *to ruin, to destroy.* Here the meaning is 'breaking oath upon oath.'

99. fancy] That is, love. See I, i, 165.

99. cheere] SKEAT (*Dict.*) : Middle English *chere,* commonly meaning the face ; hence mien, look, demeanour. Old French *chere, chiere,* the face, look.

100 , costs] Many excellent modern editors follow Theobald in needlessly

By fome illufion fee thou bring her heere,   101
Ile charme his eyes againſt ſhe doth appeare.
   Robin.   I go, I go, looke how I goe,
Swifter then arrow from the *Tartars* bowe.   *Exit.*
   Ob.   Flower of this purple die,   105
Hit with *Cupids* archery,
Sinke in apple of his eye,
When his loue he doth eſpie,
Let her ſhine as glorioufly
As the *Venus* of the sky.   110
When thou wak'ſt if ſhe be by,
Beg of her for remedy.

Enter Pucke.

   Puck.   Captaine of our Fairy band,
Helena is heere at hand,   115
And the youth, miſtooke by me,

---

102. *doth*] *doe* Qq, Cap. Steev. Mal. Coll. Sing. Dyce, Cam. Wh. ii.

103. Robin.] Rob. Ff. Puck. Rowe et seq.

   looke] *look, master,* Han.

104. Exit.] Om. Q$_1$.

106. [Squeezes the flower on Demetrius's eyelids. Dyce.

112. *of her*] *of her,* Q$_1$.

---

changing 'costs' into 'cost.' W. A. WRIGHT explains the singular here as by attraction, but ABBOTT, § 247, gives so many examples of *that* with a plural antecedent followed by a verb in the singular, where attraction cannot apply, that it is perhaps better to explain examples like the present as the result of an idiom, and that the principle of attraction applies when the clause is not dependent.—ED.

100. **dear**] STEEVENS: So in *2 Hen. VI:* III, ii, 61 : 'Might liquid tears or heart-offending groans, Or blood-consuming sighs recall his life, I would be blind with weeping, sick with groans, Look pale as primrose with blood-drinking sighs.' Again, *3 Hen. VI:* IV, iv, 22 : 'Ay, ay, for this I draw in many a tear And stop the rising of blood-sucking sighs.' All alluding to the ancient supposition that every sigh was indulged at the expense of a drop of blood. [See also to the same effect: ' Dry sorrow drinks our blood.'—*Rom. & Jul.* III, v, 59; ' Like a spendthrift sigh That hurts by easing.'—*Ham.* IV, vii, 123; 'let Benedick, like cover'd fire, Consume away in sighs.'—*Much Ado,* III, i, 78.]—STAUNTON: The notion that sighing tends to impair the animal powers is still prevalent.

104. **Tartars**] DOUCE: So in Golding's *Ovid,* Bk 10: 'And though that she Did fly as swift as Arrow from a Turkye bowe.'—W. A. WRIGHT: Compare *Rom. & Jul.* I, iv, 5 : ' Bearing a Tartar's painted bow of lath.' Also Bacon's *Advancement of Learning,* Bk II, xiv, 11 : ' Yet certain it is that words, as a Tartar's bow, do shoot back upon the understanding of the wisest.'

106. See II, i, 171.

107. **in apple**] For similar omissions of the article, see ABBOTT, § 89.

Pleading for a Louers fee.   117
Shall we their fond Pageant fee?
Lord, what fooles thefe mortals be!

   Ob. Stand afide: the noyfe they make,   120
Will caufe *Demetrius* to awake.

   Puck. Then will two at once wooe one,
That muft needs be fport alone:
And thofe things doe beft pleafe me,
That befall prepofteroufly.   125

Enter Lyfander and Helena.

   Lyf. Why fhould you think ẏ I fhould wooe in fcorn?   127

---

125. *prepofteroufly*] *prepoft'roufly* Q₁,   125. [Scene VII. Pope, Han. Scene
Theob.+, Cap.   VI. Warb. Johns.
   [They stand apart. Coll. ii.

---

117. **Louers fee**] HALLIWELL: Three kisses were properly a lover's fee. 'How many, saies Batt; why, three, saies Matt, for that's a mayden's fee,' MS Ballad, circa 1650. [No great weight can be attached, I think, to post-Shakespearian quotations, especially when there is but a single one. Moreover, I doubt if 'lover's fee' here means an *honorarium*, but its meaning is rather, *estate, right by virtue of his title as lover.*—ED.]

123. **sport alone**] COLLIER: A coarse character, under the name of Robin Good-fellow, is introduced into the play of *Wily Beguiled*, the first edition of which is dated 1606, but which must have been acted perhaps ten years earlier; there one of Robin Goodfellow's frequent exclamations is, 'Why this will be sport alone,' meaning such excellent sport that nothing can match it.—HALLIWELL: A vernacular phrase signifying excellent sport. 'This islande were a place alone for one that were vexed with a shrewd wyfe.'—Holinshed, 1577. 'Now, by my sheepe-hooke, here's a tale alone.' —Drayton's *Shepherd's Garland*, 1593. [Collier's interpretation is the better. 'Sport alone' means sport all by itself, that is, unparalleled. ABBOTT, § 18, gives as its equivalent *above all things*, and cites in addition to the present passage, 'I am alone the villain of the earth.'—*Ant. & Cleop.* IV, vi, 30; 'So full of shapes is fancy That it alone is high fantastical.'—*Twelfth Night*, I, i, 15.—ED.]

125. **preposterously**] STAUNTON [Note on *Tam. of the Shr.* III, i, 9]: Shake-speare uses 'preposterous' closer to its primitive and literal sense of *inverted order*, ὕστερον πρότερον, than is customary now. With us, it implies *monstrous, absurd, ridiculous*, and the like; with him it meant *misplaced, out of the natural* or *reasonable course.*

127. **should wooe**] ABBOTT, § 328, thinks that there is no other reason for the use of 'should' here than that it denotes, like *sollen* in German, a statement not made by the speaker. It may be so, and yet the idea of *ought to*, equally with *sollen*, may be imputed to it here. 'Why should you think that I *ought to* woo in scorn?' As was said in *The Tempest* on the phrase 'where should he learn our language?' the use of 'should' in Shakespeare is of the subtlest.—ED.

Scorne and derifion neuer comes in teares :   128
Looke when I vow I weepe ; and vowes fo borne,
In their natiuity all truth appeares.   130
How can thefe things in me, feeme fcorne to you?
Bearing the badge of faith to proue them true.
   Hel.   You doe aduance your cunning more & more,
When truth kils truth, O diuelifh holy fray!
Thefe vowes are *Hermias.* Will you giue her ore?   135
Weigh oath with oath, and you will nothing weigh.
Your vowes to her, and me, (put in two fcales)
Will euen weigh, and both as light as tales.
   Lyf.   I had no iudgement, when to her I fwore.
   Hel.   Nor none in my minde, now you giue her ore.   140
   Lyf.   *Demetrius* loues her, and he loues not you. *Awa.*

---

128. *comes*] *come* Qq, Rowe et seq.
129. *borne*] *born* F$_3$F$_4$.
134. *truth kils truth*] *trueth killes truth* Q$_1$.

134. *diuelifh holy*] *devilish-holy* Cap. et seq.
141. Awa.] Om. Qq. Awaking. Rowe. Starting up. Coll.

---

128. **comes**] Is there any necessity to change this to the plural, with the Qq? Cannot 'scorn-and-derision' be conceived of as one mingled emotion of the mind? —ED.

129, 130. **vowes so borne . . . appears**] WALKER (*Crit.* i, 56) thinks that there is here 'an instinctive striving after a natural arrangement of words inconsistent with modern English grammar'; and ABBOTT, §§ 417, 376, classes 'vows so born' either as a 'noun absolute' or as a 'participle used with a Nominative Absolute.' I cannot but think that both critics, misled by the singular 'appears,' have mistaken the construction. 'Appears' should be, according to modern grammar, in the plural; its subject is 'vows,' it is singular merely by attraction; 'all truth' is the predicate, not the subject. My paraphrase, therefore, is: 'vows, thus born, appear, from their very nativity, to be all pure truth.' The next lines seem to confirm it. It can hardly be supposed that Lysander means to assert that 'all truth,' universal truth, is to be found in such vows.—ED.

132. **badge**] STEEVENS: This is an allusion to the 'badges' (*i. e.* family crests) anciently worn on the sleeves of servants and retainers. So in *Temp.* V. i, 267: 'Mark but the badges of these men, my lords, Then say if they be true.'

134. **When . . . fray**] W. A. WRIGHT: If Lysander's present protestations are true, they destroy the truth of his former vows to Hermia, and the contest between these two truths, which in themselves are holy, must in the issue be devilish and end in the destruction of both.

138. **tales**] W. A. WRIGHT: Or idle words. There is the same contrast between truths and tales in *Ant. & Cleop.* II, ii, 136: 'Truths would be tales, Where now half tales be truths.' [May not 'tales' here mean *stories of the imagination, pure fiction?*—ED.]

141. WALKER (*Crit.* iii, 49): There is *perhaps* a line lost after this line.—SCHMIDT

   Dem. O *Helen,* goddeſſe, nimph, perfect, diuine,   142
To what my, loue, ſhall I compare thine eyne!
Chriſtall is muddy, O how ripe in ſhow,
Thy lips, thoſe kiſſing cherries, tempting grow!   145
That pure congealed white, high *Taurus* ſnow,
Fan'd with the Eaſterne winde, turnes to a crow;
When thou holdſt vp thy hand.   O let me kiſſe
This Princeſſe of pure white, this ſeale of bliſſe.

   Hell. O ſpight! O hell! I ſee you are all bent   150
To ſet againſt me, for your merriment:
If you were ciuill, and knew curteſie,
You would not doe me thus much iniury.
Can you not hate me, as I know you doe,   154

142. *perfect, diuine*] *perfect diuine* Q₁.
143. *To what my,*] *To what? my* F₃F₄.
146. *congealed*] *coniealed* Q₁.
149. *Princeſſe*] *pureneſs* Han. Warb. *impreſs* Coll. ii (MS), Sta. *purest* Lett-som (ap. Dyce) *Empreſs* Marshall conj.
149. *Princeſſe of pure*] *quintessence of* Bailey (withdrawn).
    white] *whites* Bailey.
150. *are all*] *all are* Qq, Pope et seq.

(*Programm,* &c., p. 5) makes the same conjecture, which is, I think, needless. The emphasis with which Lysander pronounces the name Demetrius may have awakened the bearer of it, and in the new turn given to the dramatic action the loss of a rhyming line was not felt.—ED.

141. **Awa.**] Evidently the abbreviation of *Awake;* another mandatory stage-direction of a play-house copy.—ED.

145. **kissing cherries**] KNIGHT: These 'kissing cherries' gave Herrick a stock in trade for half a dozen poems. We would quote the 'Cherry Ripe,' had it not passed into that extreme popularity which almost renders a beautiful thing vulgar. [Knight here quotes 'The Weeping Cherry,' which the inquisitive reader may find in Herrick's *Hesperides,* &c., vol. i, p. 10, ed. Singer.]

146. **Taurus**] JOHNSON: The name of a range of mountains in Asia.

149. **Princesse**] HEATH (p. 53): I can see no objection to this reading. 'Tis not an unusual expreſſion to call the moſt excellent and perfect in any kind the prince of the kind. [This note Capell properly quotes with approval.]—COLLIER (ed. i): It may be doubted from the context whether *impress* were not Shakespeare's word.—IB. (ed. ii): This emendation [*impress*] of the MS can hardly be wrong; the old reading, 'princess,' cannot be right. *Impress* and 'seal' are nearly the same thing; and, in consistency with this alteration, it may be observed that in Beaumont and Fletcher's *Double Marriage,* IV, iii, Virolet calls Julianna's hand 'white seal of virtue.'—DYCE (*Rem.* p. 48): When Mr Collier offered [his] very unnecessary conjecture, *impress,* he did not see that these two rapturous encomiums on the hand of Helena have no connexion with each other. Demetrius terms it 'princess of pure white,' because its whiteness exceeded all other whiteness; and 'seal of bliss,' because it was to confirm the happiness of her accepted lover.

10

But you muſt ioyne in ſoules to mocke me to?   155
If you are men, as men you are in ſhow,
You would not vſe a gentle Lady ſo ;
To vow, and ſweare, and ſuperpraiſe my parts,
When I am ſure you hate me with your hearts.
You both are Riuals, and loue *Hermia* ;   160
And now both Riuals to mocke *Helena*.
A trim exploit, a manly enterprize ,
To coniure teares vp in a poore maids eyes,
With your deriſion ; none of noble ſort,
Would ſo offend a Virgin, and extort   165

---

155. *ioyne in ſoules*] ioyne, in ſoules, Q₁. *join in flouts* Han. *join in scorns* or *scoffs* Johns. conj. (withdrawn). *join, ill souls*, Tyrwhitt. *join in scouls* Blackstone (ap. *Var.*'85). *join in shoals* T. H. W. (*Gent. Mag.* lv, p. 278, 1785). *join in soul* Mason, Rann. *join, in sooth* Bailey (ii, 202). *join in taunts* Elze (*Athen.* 26 Oct. '67). *join in sport* Wetherell (*Athen.* 2 Nov. '67). *join in sports* D. Wilson. *join insults* Spedding (ap. *Cam.*), Leo (*Athen.* 27 Nov.'80).

155. *to ?*] *too ?* Q₂Ff.
156. *are men*] *were men* Qq, Han. Cap. et seq.
157. *ſo ;*] *ſo ?* Ff.
160, 161. *Riuals*] *Riuals…Riualles* Q₁.
164. *deriſion ; none*] *deriſion None,* Q₁. *deriſion, none* Q₂. *derision ! None* Theob. +, Steev. et seq. (subs.).
        noble] *nobler* Rowe i, Theob. ii, Warb. Johns. Steev.'85.

---

155. **in soules**] WARBURTON: This line is nonsense. It should read thus : ' But muſt join insolents to mock me too ?'—STEEVENS : ' Join in souls ' is to join heartily, unite in the same mind. [See Text. Notes for sundry emendations of a phrase which needs no help whatsoever. The notes attending these emendations are not here recorded ; having no obscurity in the text to explain, they amount to but little else than an announcement by their authors of a preference of their own words to Shakespeare's.—ED.]

160, 161. As a warning against rearing any theory based on the spelling in the old eds., note the different spelling of ' rivals ' in two consecutive lines in Q₁.

162. **trim**] SCHMIDT (*Lex.*) says that as an adjective this is ' mostly used with irony.' ' Mostly ' is, I think, a little too comprehensive ; that ' trim ' is sometimes used ironically is true, but the same may be said of *fine, pretty*, and of many another adjective.—ED.

164. **sort**] MALONE : Here used for *degree* or *quality*. [Not necessarily referring to *rank*, although W. A. WRIGHT quotes Cotgrave : ' Gens de mise. Persons of worth, sort, qualitie.'—ED.]

165. **extort**] SCHMIDT (*Lex.*) defines this by *To wring, wrest*, and calls attention to the parallel meaning of *to move* or *wake* a person's patience, and therefore *to make impatient*, in *Much Ado*, V, i, 102 : ' We will not wake your patience ' ; and in *Rich. III:* I, iii, 248 : ' end thy frantic curse, Lest to thy harm thou move our patience.'— ALLEN (MS) : May this not possibly mean : to produce by *torture* the *suffering* of a poor soul. To take away from a poor soul her patience, seems to me commonplace. For ' patience ' compare ' I know your patience well,' III, i, 199.

A poore foules patience, all to make you fport.   166

   Lyſa. You are vnkind *Demetrius*;be not fo,
For you loue *Hermia*; this you know I know;
And here with all good will, with all my heart,
In *Hermias* loue I yeeld you vp my part;   170
And yours of *Helena*, to me bequeath,
Whom I do loue, and will do to my death.

   Hel. Neuer did mockers waſt more idle breth.

   Dem. *Lyſander*, keep thy *Hermia*, I will none:
If ere I lou'd her, all that loue is gone.   175
My heart to her, but as gueſt-wiſe ſoiourn'd,
And now to *Helen* it is home return'd,
There to remaine.

   Lyſ. It is not fo.   179

169. *here*] *heare* Q₁. *heere* Q₂.

171. *yours of*] *your's of* Rowe i. *your's in* Coll. ii (MS).

172. *will do*] *will love* Cam. Edd. conj.
to my] *till my* Q₂, Coll. White, Cam.

173. *waſt*] *waſte* QqFf.

176. *to her*] *with her* Johns. Steev. Mal. Var. Knt, Sing. Hal. Coll. ii, Dyce ii, iii, Ktly.

177. *it is*] *is it* Q₁, Cap. Mal. Var. Coll. Dyce, White, Sta. Cam.

178. *There*] *There ever* Pope+.

179. *It is*] *Helen, it is* Q₁, Cap. et seq.

172. **will do**] The CAM. EDD. conjecture 'will *love*,' which is certainly an improvement, but then—

174. **none**] ABBOTT, § 53: 'None' is still used by us for *nothing*, followed by a partitive genitive, 'I had none of it'; and this explains the Elizabethan phrase, 'She will none of me.'—*Twelfth Night*, I, iii, 113.

176. **to her**] COLLIER 'reluctantly abandoned' this 'to' for JOHNSON's emendation *with*, because 'the phrase is *sojourned with*, not *sojourned to*, although there was formerly great license in the use of prepositions.'—DYCE adopted *with* because the 'to' in this line was 'an error occasioned by the "to" immediately below.'—R. G. WHITE refused to change because it does not appear sufficiently clear that 'to' was not the old idiom.—DELIUS interprets 'to her' as generally equivalent to *as to her*, and in the present instance, by attraction from 'guestwise,' the phrase is equivalent to *as a guest to her.*—W. A. WRIGHT: There are other instances of 'to' in Shakespeare in a sense not far different from that in the present passage. Compare *Meas. for Meas.* I, ii, 186: 'Implore her in my voice that she make friends To the strict deputy.' *Two Gent.* I, i, 57: 'To Milan let me hear from thee by letters.' *Com. of Err.* IV, i, 49: 'You use this dalliance to excuse Your breach of promise to the Porpentine.' In all these cases the sense is quite clear, but there is a confusion in the construction. In the Devonshire dialect 'to' is frequently used for 'at,' and it is a common Americanism.—ALLEN (MS): May not this be like a familiar Greek construction? My heart [*went away* from its proper home] *to* her, and sojourned [*with her*] merely as a *guest*. Confirmed by: Now it has *returned* to me. Cf. Robert Browning's *Strafford* (p. 309), V, ii: 'You've been *to* Venice, father?'

179. **It is not so**] If one likes the pronunciation of 'Helen' with the accent on

De. Diſparage not the faith thou doſt not know,   180
Leſt to thy perill thou abide it deare.
Looke where thy Loue comes, yonder is thy deare.

Enter Hermia.

Her. Dark night, that from the eye his function takes,
The eare more quicke of apprehenſion makes ,   185
Wherein it doth impaire the ſeeing ſenſe ,
Ir paies the hearing double recompence.
Thou art not by mine eye, *Lyſander* found ,
Mine eare (I thanke it) brought me to that ſound.
But why vnkindly didſt thou leaue me ſo ?   (to go?   190
Lyſan. Why ſhould hee ſtay whom Loue doth preſſe
Her. What loue could preſſe *Lyſander* from my ſide ?
Lyſ. *Lyſanders* loue (that would not let him bide)
Faire *Helena* ; who more engilds the night,
Then all yon fierie oes, and eies of light.   195

---

181. *Leſt*] *Leaſt* Qq.

   abide] *aby it* Q₁, Cap. Steev. Mal. Knt, Coll. Dyce, White, Sta. Cam. Ktly.

182. Scene VIII. Pope, Han. Scene VII. Warb. Johns.

187. *Ir*] F₁.

188. Lyſander] Lyſander, Q₁.

189. *brought*] *broogbt* F₃.

   that] *thy* Qq, Pope et seq.

193. (*that...bide*)] No parenthesis, Rowe et seq.

   bide] *'bide* Theob. ii, Warb. Johns.

195. *oes*] *o's* F₄, Rowe+. *orbs* Grey.

   eies] *eyes* F₃F₄.

---

the last syllable, there can be no objection to following the Q₁ here. But where a line is divided between two speakers, the inevitable pause is, I think, to be preferred in scansion to the stop-gap of an ill-accented word.—ED.

181. abide] The First Quarto's *aby* is here correct, the form 'abide' in the present phrase, according to SKEAT, is 'a mere corruption.'—W. A. WRIGHT [reading '*aby* it,' thus interprets :] That is, pay for it, atone for it. See below, line 353, and Spenser, *Faerie Queene*, IV, i, 53 : 'Yet thou, false squire, his fault shalt deare aby.' The Ff read 'abide' in both passages, as does Q₂ here. There is another word *aby*, in an entirely different sense, which is etymologically the same as 'abide'; but our word is from the A.-S. *abicgan*, to redeem. And 'abide,' which is synonymous with the former, is often confounded with the latter. [See also line 452, below.]

181. it deare] WALKER (*Crit.* i, 307): Possibly *here ; (heere—deare*).

195. oes] STEEVENS: Shakespeare uses O for a circle. So in *Hen. V*, Prol. 13 : may we cram Within this wooden O the very casques That did affright the air at Agincourt.' Again, in John Davies of Hereford's *Microcosmos*, 1605, p. 233 : ' Which silver oes and spangles over-ran.'—STAUNTON : ' Oes' were small circular bosses of shining metal.—HALLIWELL cites : ' and oes, or spangs, as they are of no great cost, so are they of most glory.'—Bacon's *Essay*, xxxvii, p. 157, ed. Wright.

Why feek'ft thou me? Could not this make thee know,   196
The hate I bare thee, made me leaue thee fo?

   Her.   You fpeake not as you thinke; it cannot be.

   Hel.   Loe, fhe is one of this confederacy,
Now I perceiue they haue conioyn'd all three,   200
To fafhion this falfe fport in fpight of me.
Iniurious Hermia, moft vngratefull maid,
Haue you confpir'd, haue you with thefe contriu'd
To baite me, with this foule derifion?
Is all the counfell that we two haue fhar'd,   205
The fifters vowes, the houres that we haue fpent,
When wee haue chid the hafty footed time,
For parting vs; O, is all forgot?
All fchooledaies friendfhip, child-hood innocence?
We Hermia, like two Artificiall gods,   210

197. *bare*] bear F₄, Rowe+, Dyce, Coll. Sta. Cam. i, Ktly, White ii.

201. *of me*] to me Johns.

206. *fifters vowes*] QqFf, Rowe+. *sister vows* Cap. *sister-vows* Dyce ii, iii. *sisters' vows* Steev. et cet.

208. *O, is all*] O and is all Ff, Rowe+, Cap. Steev. Knt, Hal. Sta. Dyce ii, iii, Huds. *O, is all now* Mal. *O, now, is all* Var. *Oh, is this all* Ktly. *Oh, is*

this then Ktly conj. *O, is it all* Spedding (ap. Cam.), Glo. White ii. *O, is all this* Huds. conj.

209. *fchooledaies*] school-day Cap. Steev.'85, Dyce ii, iii, Huds.

child-hood] child-hoods F₃F₄, Rowe i.

210. *two Artificiall gods*] to artificer gods or two artificial buds D. Wilson.

[Here, at least, we have a word which our German brothers must paraphrase. They cannot translate it literally, albeit Schlegel ventured it. The German capital O is apparently a circle drawn from the depths of the German consciousness; of course there had to be an æsthetic flourish in it. Is the supposition too fanciful that the punning on *o*'s and *i*'s begins with '*en*gilds'?—ED.

206. **sisters vowes**] DYCE (ed. ii): Here the old eds. have '*sisters* vowes,' and a little below, '*schoole daies* friendship' (though in the same line with 'childhood innocence').

208. **O, is all forgot**] The Text Notes show the harmless attempts to bring this line into the right butter-woman's rank to market. The break in the line gives ample pause for supplying a lost syllable. Moreover, the emotion expressed by 'O' can easily prolong the sound enough to fill the gap, and that, too, without lengthening it into an 'Irish howl,' as Steevens, with a malicious glance at Malone's nationality, once termed a similar suggestion by the latter.—ED.

208. **forgot**] REED: Mr Gibbon observes that in a poem of Gregory Nazianzen, on his own life, are some beautiful lines which burst from the heart, and speak the pangs of injured and lost friendship, resembling these. He adds, 'Shakespeare had never read the poems of Gregory Nazianzen; he was ignorant of the Greek language; but his mother-tongue, the language of nature, is the same in Cappadocia and in Britain.'—Gibbon's *Hist.* iii, 15.

IIaue with our needles, created both one flower,   211
Both on one fampler, fitting on one cufhion,
Both warbling of one fong, both in one key ;
As if our hands, our fides, voices, and mindes
Had beene incorporate.   So we grew together,   215
Like to a double cherry feeming parted ,
But yet a vnion in partition ,   217

211. *Haue...both*] *Created with our needles both* Pope+.
needles] *neelds* Rann, Mal.'90, Steev.'93, Var. Knt, Sta. Dyce ii, iii.
214. *our fides*] *and sides* Cap.

215. *beene*] *bin* Qq.
217. *yet*] Om. F_3F_4.
a vnion] *an vnion* QqF₄, Rowe ⌐ , Coll. Hal. White, Cam.

---

210. **Artificiall**] WALKER (*Crit.* i, 96): This is here used with reference to the agent ; *deabus artificibus similes*.—WALKER (*Ib.* i, 154) in his valuable chapter on ' Ovid's influence on Shakespeare ' suggests that there is in these lines an unconscious allusion to the story of Arachne and Minerva (' with a variety ') which had impressed Shakespeare in reading.—For a list of adjectives which have both an active and a passive meaning, see ABBOTT, § 3.—GEO. GOULD (p. 15): Read ' artificial *girls*,' viz. Helena and Hermia, who are like a pair of girls in waxwork. [Gifford's vocation of censor is as necessary as it is unenviable. Gifford should have died hereafter.—ED.]

211. **needles**] STEEVENS: This was probably written by Shakespeare *neelds* (a common contraction in the Inland counties at this day), otherwise the verse would be inharmonious.—ABBOTT, § 465 : ' Needle,' which in *Gammer Gurton* rhymes with ' feele,' is often pronounced as a monosyllable. ' Deep clerks she dumbs, and with her need*le* composes.'—*Per.* V, Gower, 5 ; ' I would they were in Afric both together ; myself by with a need*le* that I might prick.'—*Cym.* I, i, 168 ; ' Or when she would with sharp need*le* wound.'—*Per.* IV, Gower, 23. In the latter passage ' needle wound ' is certainly harsh, though Gower does bespeak allowance for his verse. A. J. ELLIS suggests *'ld* for ' would,' which removes the harshness. ' And grì | ping ìt | the nèed*le* | his fìn | ger pricks.'—*R. of L.* 319 ; ' Their nèed*les* | to làn | ces, ànd | their gènt | le hèarts.'—*King John*, V, ii, 157 ; ' To threàd | the pòst | ern òf | a smàll | need*le*'s èye.'—*Rich. II:* V, v, 17. ' Needle's ' seems harsh, and it would be more pleasing to modern readers to scan ' the pòst | ern òf a | small neè | dle's èye.' But this verse, in conjunction with *Per.* IV, Gower 23, may indicate that ' needle ' was pronounced as it was sometimes written, very much like *neeld*, and the *d* in *neeld*, as in *vild* (vile), may have been scarcely perceptible.—CAMBRIDGE EDITORS: Pope's reading is rendered extremely improbable by the occurrence of the word ' Have ' at the beginning of the line in all the old copies, and could only have been suggested by what Pope considered the exigencies of the metre. ' Needles ' may have been pronounced as Steevens writes it, *neelds ;* but, if not, the line is harmonious enough. [One instance of ' needle ' no one, I believe, has noticed, where it must be pronounced as a disyllable. It occurs in *R. of L.*, within two lines, strangely enough, of the line cited by Abbott: ' Lucretia's glove, wherein the needle sticks,' line 217. This proves, I think, that the word was pronounced by Shakespeare either as a monosyllable or as a disyllable, according to the needs of his rhythm.—ED.]

Two louely berries molded on one ſtem,   218
So with two ſeeming bodies, but one heart,
Two of the firſt life coats in Heraldry,   220

218. *louely*] *loving* Coll. ii, iii (MS).   220. *firſt life*] *first life*, Ff, Rowe,
219. *So*] *Or* Han.   Pope. *first*, *like* Folks, Theob. et seq.
      220, 221. Om. Coll. MS.

---

218. **louely**] COLLIER (ed. ii): It is unlikely that Helen would call herself a *lovely* berry. The change to *loving* is in the MS, and it is precisely the thought which the speaker is carrying on; we have no doubt Shakespeare wrote *loving*. Elsewhere the same misprint occurs.—DYCE (ed. ii): But was not 'lovely' sometimes used as equivalent to *loving*? Compare our author's *Tam. of the Shr.* III, ii: 'And seal the title with a *lovely* kiss'; also, 'And I will give thee many a *lovely* kiss.'—Peele's *Arraignment of Paris—Works*, p. 358, ed. Dyce, 1861. 'A father, brother, and a vowèd friend. *K. of Eng.* Link all these *lovely* styles, good king, in one.'—Greene's *James IV—Works*, p. 189, ed. Dyce, 1861. [Collier might not unreasonably answer Dyce, that all these three examples are exactly the misprints which he said might be found elsewhere, and that they corroborate the emendation of the MS, which seems, it must be confessed, unusually happy to the present ED.]

220. **of the first life**] THEOBALD: The true correction of this passage [the change of 'life' to *like*] I owe to the friendship and communication of the ingenious Martin Folks, Esq. Two of the first, second, &c. are terms peculiar to Heraldry to distinguish the different Quarterings of Coats.—M. MASON: Every branch of a family is called 'a house,' and none but the 'first' of the 'first house' can bear the arms of a family without some distinction. 'Two of the first,' therefore, means *two coats of the first house*, which are properly 'due but to one.' [This explanation seems to have satisfied no subsequent editor except KNIGHT.]—RITSON (*Cursory Crit.* 44): The two 'seeming bodies' united by 'one heart' are resembled to *coats in heraldry, crowned with one crest.* And this happens either where the *heir* keeps his *paternal* and *maternal* coats, or *the husband his own* and *his wife's* in *separate shields*, as is done on the Continent; or, as at present with us, in the quarterings of the same shield, in both cases there are 'two coats, due but to one, and crowned with one crest,' which is clearly the author's allusion. But I am sorry to add that he must have entirely misunderstood, since he has so strangely misapplied, the expression 'Two of the first,' which, in heraldical jargon, always means *two* objects *of the first* colour mentioned, that is, the *field*. For instance, in blazoning a coat they will say, *Argent*, upon a fesse *gules*, *two* mullets *of the first*, that is, *argent*, the colour of the *field*. These words are, therefore, a melancholy proof that our great author sometimes retained the phrase after he had lost the idea or [applied] the former without sufficient precaution as to the latter. [If the 'heraldical jargon' of the whole passage is confined to these two lines, and if 'first' is a technical term, which can refer only to colour, then Ritson is technically right, and the greatness of a name cannot excuse a blunder. But DOUCE (i, 194) thinks that a deeper heraldic meaning is here imputed to Shakespeare than he intended, and that 'first' does not refer to colour. 'Helen,' says Douce, 'exemplifies her position by a simile,—" we had *two of the first*, i. e. *bodies*, like the double coats in heraldry that belong to man and wife as *one person*, but which, like our *single heart*, have but *one crest*." ' This is certainly a common-sense explanation. W. A. WRIGHT says it is 'the correct one.' Staunton,

Due but to one and crowned with one creſt.   221
And will you rent our ancient loue aſunder,
To ioyne with men in ſcorning your poore friend?
It is not friendly, 'tis not maidenly.
Our ſexe as well as I, may chide you for it,   225

---

221. *creſt*] *creaſt* Q₁.   222. *rent*] *rend* Rowe +, Coll. White i.

however, shows that there is more 'heraldical jargon' in the passage than had been hitherto supposed, and that 'first' may perhaps apply neither to 'colour' nor to 'bodies,' but to heraldical 'partitions.']—STAUNTON: The plain heraldical allusion is to the simple impalements of two armorial ensigns, as they are marshalled side by side to represent a marriage; and the expression 'Two of the First' is to *that particular form of dividing the shield, being the first in order of the nine ordinary partitions of the Escutcheon.* These principles were familiarly understood in the time of Shakespeare by all the readers of the many very popular heraldical works of the period, and an extract from one of these will probably render the meaning of the passage clear. In *The Accidence of Armorie*, by Gerard Leigh, 1597, he says, 'Now will I declare to you of IX sundrie Partitions: the *First whereof is a partition from the highest part of the Escocheon to the lowest. And though it must be blazed so, yet it is a joining together.* It is also as a mariage, that is to say, *two cotes;* the man's on the right side, and the woman's on the left; as it might be said that Argent had married with Gules.' In different words, this is nothing else than an amplification of Helena's own expression,—'seeming parted; But yet a union in partition.' The shield bearing the arms of two married persons would of course be surmounted by one crest only, as the text properly remarks, that of the husband. In Shakespeare's day the only pleas for bearing two crests were ancient usage or a special grant. The modern practice of introducing a second crest by an heiress has been most improperly adopted from the German heraldical system; for it should be remembered that as a female cannot wear a helmet, so neither can she bear a crest. [The solitary objection which I can see to Staunton's explanation, and it is one of small moment, is that 'partition' is in the singular. Had Helen's phrase been 'a union in partitions,' Staunton's argument would be, I think, indisputable. As the text stands, however, I doubt if Shakespeare's thoughts were turned thus early to heraldry; 'partition' was the logical word to use after 'parted' in the preceding line; but the very sound of the word in Shakespeare's mental ear may have started a train of heraldical imagery which found expression later on. Although 'partition' is a technical term, I do not think the real heraldry begins until we come to 'Two of the first,' when, having mentioned 'partition' and referred to bodies before he referred to hearts, he used 'first' as satisfying the former, 'partition,' and as pointing to the latter, 'bodies.' So that Douce and Staunton may be measurably harmonised, and Ritson is wrong in thinking that Shakespeare blundered. So far from being remiss in his heraldry, he was so at home in it that he could play with its terms. DYCE merely quotes Douce and Staunton at length, but expresses no opinion.—ED.]

222. **rent**] W. A. WRIGHT: The old form of *rend.* Compare *A Lover's Complaint*, 55: 'This said, in top of rage the lines she rents.' It occurs also in several passages of *The Authorised Version*, but has been modernised in later editions, and is left only in *Jer.* iv, 30.

Though I alone doe feele the iniurie.   226

   Her. I am amazed at your paſſionate words,
I ſcorne you not; It ſeemes that you ſcorne me.

   Hel. Haue you not ſet *Lyſander,* as in ſcorne
To follow me, and praiſe my eies and face?   230
And made your other loue, *Demetrius*
(Who euen but now did ſpurne me with his foote)
To call me goddeſſe, nimph, diuine, and rare,
Precious, celeſtiall?   Wherefore ſpeakes he this
To her he hates?   And wherefore doth *Lyſander*   235
Denie your loue (ſo rich within his ſoule)
And tender me (forſooth) affeċtion,
But by your ſetting on, by your conſent?
What though I be not ſo in grace as you,
So hung vpon with loue, ſo fortunate?   240
(But miſerable moſt, to loue vnlou'd)
This you ſhould pittie, rather then deſpiſe.

   Her. I vnderſtand not what you meane by this.   243

---

227. *I am*] *Helen, I am* Pope, Han.
paſſionate] Om. Qq, Pope, Han.
240. *loue*] *loves* Cap.

240, 241. *fortunate?...vnlou'd*] *fortunate;...unlov'd?* Theob. *fortunate, ... unlov'd!* Knt. *fortunate, ...unlov'd,* Coll. *fortunate, ...unlov'd...* Ktly.

---

225. **for it**] WALKER (*Vers.* 79): It may be remarked that *on't, for't,* and the like, at the end of verses, have in many instances been corrupted into *of it, for it,* &c. So with *it,* in general, at the end of a line.   An ear properly imbued with the Shake-spearian rhythm in general, and with certain plays in particular,—I mean the earlier dramas (the *Mid. N. D.* for instance) in which double endings to the lines occur comparatively seldom,—invariably detects the fault.   [In the present line ' for it '] sensibly infringes on the ' *monosyllabo-teleutic* ' flow of the poem.   Read *for't.*

227. **passionate**] The omission of this emphatic word in Q_2, from which the Folio was printed, is another cumulative proof that this Qto had been a play-house copy, and had in it omissions supplied and corrections made, before it came to be used as the original from which the Folio was set up.—ED.

230. **me . . . my**] See ' my,' I, i, 200.

232. **euen but now**] ABBOTT, § 38: '*Even* now' with us is applied to an action that has been going on for some long time and *still* continues, the emphasis being laid on ' now.'   In Shakespeare the emphasis is often to be laid on *even*, and ' *even* now ' means ' *exactly* or *only* now,' *i. e.* scarcely longer ago than the present; hence ' *but* now.'   We use ' *just* now ' for the Shakespearian ' *even* now,' laying the emphasis on ' just.'   [See *Mer. of Ven.* III, ii, 176, and *As You Like It,* II, vii, 4 of this ed., where this same note of Abbott is quoted.]

240. **hung vpon**] ALLEN (MS): May not this be here used as it is in *Sonn.* xxxi, 10: ',Hung with the trophies of my lovers gone,' and in *Jul. Cæs.* I, i, 74: ' let no images Be hung with Cæsar's trophies ' ?

Hel.   I, doe, perſeuer, counterfeit ſad lookes,
Make mouthes vpon me when I turne my backe,   245
Winke each at other, hold the ſweete ieſt vp :
This ſport well carried, ſhall be chronicled.
If you haue any pittie, grace, or manners,
You would not make me ſuch an argument :
But fare ye well, 'tis partly mine owne fault,   250
Which death or abſence ſoone ſhall remedie.
   Lyſ.   Stay gentle *Helena,* heare my excuſe,
My loue, my life, my ſoule, faire *Helena.*
   Hel.   O excellent !
   Her.   Sweete, do not ſcorne her ſo.   255
   Dem.   If ſhe cannot entreate, I can compell.
   Lyſ.   Thou canſt compell, no more then ſhe entreate.
Thy threats haue no more ſtrength then her weak praiſe.
Helen, I loue thee, by my life I doe ;   259

---

244. *I, doe, perſeuer*] *I doe. Perſeuer*
Q₁. *I do, perſeuer* F₄. *Ay, do, perſevere*
Rowe, Johns. *Ay do, perſever* Pope. *I
do ;—perceive* D. Wilson. *Ay, do, perſever* Theob. et seq.
245. *mouthes*] *mows* Steev. Var. Knt.
246. *ieſt*] *ieaſt* Qq.
248. *haue*] *had* Coll. ii (MS), Huds.
250. *fare ye well*] *faryewell* Q₂.
   mine] *my* Q₁, Cam. White ii.

253. *my life*] Om Ff, Rowe.
255. [To Lys. Cap.
256. *cannot*] *can not* Cap. (Errata).
257. *compell, no more*] *compell no
more,* Q₁. *compell no more* F₃F₄, Rowe
et seq.
258. *praiſe*] *prays* Cap. Mal. '90.
prayers Theob. et cet.
259. Helen,] Helen. F₄ (as though
Helena were the speaker).

---

244. **I, doe,**] HUNTER (*Illust.* i, 296) pronounces the usual reading, 'Ay, do,'
'bad,' and upholds Q₁, wherein he hears the 'grave and serious tone' in which Helen
replies to Hermia's assertion : 'I understand not what you mean by this.'

244. **perſeuer**] For other examples of this same accent, see ABBOTT, § 492.

246. **hold . . . vp**] W. A. WRIGHT: That is, keep it going, carry it on. Compare *Merry Wives,* V, v, 109 : 'I pray you, come, hold up the jest no higher.' And
Much Ado, II, iii, 126 : 'He hath ta'en the infection ; hold it up' ; that is, keep up
the sport.

249. **argument**] JOHNSON : Such a *subject* of light merriment.

258. **praise**] THEOBALD : In the preceding line there is an antithesis betwixt
'compel' and 'entreat' ; this contrast is wanting in 'threats' and 'praise' ; wherefore
we need make no difficulty of substituting *prayers.* Indeed, my suspicion is that the
poet might have coined a substantive plural (from the verb *to pray*), *prays,* i. e. *pray-
ings, entreaties, beseechings ;* and the identity of sound might give birth to the corruption of it into 'praise.'—CAPELL (who adopted Theobald's conjecture) : 'Prays' (a
nomen verbale) is a bold coinage, but proper ; has the sense of *prayers,* but with more
contempt in it ; the sound perfectly of the word it gave birth to, and its form nearly
when that word was writ—*prayse.* [Theobald's conjecture is plausible. It is quite

I fweare by that which I will lofe for thee,   260
To proue him falfe, that faies I loue thee not.

    Dem. I fay, I loue thee more then he can do.

    Lyf. If thou fay fo, with-draw and proue it too.

    Dem. Quick, come.

    Her. *Lyfander*, whereto tends all this?   265

    Lyf. Away, you *Ethiope.*

    Dem. No, no, Sir, feeme to breake loofe;

Take on as you would follow,   268

---

260. *lofe*] *loofe* Q₄.

263. *too*] *to* Qq. *true* Anon. conj.

264. *come.*] *come,—* Cap. *come!* Dyce.

266. Ethiope] *Ethiop, you* Heath. [Holding him. Coll.

267. *No, no, Sir, feeme*] *No, no ; heele Seeme* Q₁. *No, no, hee'l feeme* Q₂. *No no, he'll feem* Pope +, Steev.'85, Hal. *No, no ; he'll not come.—Seem* Cap. Rann. *No no ; he'll—sir, Seem* Mal. Var. *No, no, sir :—he will Seem* Steev.'93. *No, no, sir :—seem* Knt, Sing. ii, Dyce i, White i, Rolfe. *No, no, he'll—Seem* Coll. Sta. White ii. *No, no ; he'll... Seem* Cam. Cla. *No, no, sir ; you Seem*

Lettsom, Dyce ii, iii. *No, no, sir :—do ; Seem* Huds. *No, no ; he'll but Seem* Nicholson (ap. Cam.). *No ! no, sir ; thou'lt Seem* Kinnear. *No, no : he'll not stir* (or *not budge*) *Seem* or *No, no, sir, no : Seem* Schmidt. *Her. No, no ; he'll—* Dem. *Seem* Joicey (*N. & Qu.* 11 Feb.'93).

267, 268. *feeme...follow*] One line, Q₁, Cap. et seq.

    to...follow] One line, Pope +, Ktly (the latter reading *you'd follow me*).

267. *to break loofe*] *To break away* Pope +.

268. *you*] *he* Pope +, Coll. iii.

---

in Shakespeare's manner to form such nouns from verbs, and in the present case, as Theobald says, *prays* is *idem sonans* with the text.—ED.]

266. **Ethiope**] From this we learn that Hermia is a brunette, just as we are shortly told that she is low of stature.—ED.

267. **No . . . seeme**] MALONE: This passage, like almost all in which there is a sudden transition or the sense is hastily broken off, is much corrupted in the old copies. . . . Demetrius, I suppose, would say, No, no; he'll not have the resolution to disengage himself from Hermia. But, turning abruptly to Lysander, he addresses him ironically: 'Sir, seem to break loose,' &c. [See Text. Notes for Malone's composite text.]—HALLIWELL [who follows the Qq]: The opening of this speech seems to be in relation, very ironically, to Lysander's previous one, implying that he is making no real effort to detach himself from the lady. Demetrius then personally addresses Lysander in the most provoking language that presents itself.—HUDSON modifies Lettsom's conjecture, adopted by Dyce, by substituting *do* for *you*, and thus justifies it: Demetrius is taunting Lysander, as if the latter were making believe that he wants to break loose from Hermia, who is clinging to him, and go apart with Demetrius and fight it out. This sense, it seems to me, is much better preserved by *do* than by *you*. We have had a like use of *do* a little before: 'Ay, do, persever,' &c. Also in *King Lear*, I, i: 'Do; kill thy physician,' &c.—W. A. WRIGHT: Unless a line has fallen out, this reading [see Text. Notes] gives as good a sense as any. Demetrius first addresses Hermia, and then breaks off abruptly to taunt Lysander with not showing much eagerness to follow him.—D. WILSON (p. 255): A pair of distracted

But yet come not : you are a tame man, go.

Lyf.   Hang off thou cat, thou bur ; vile thing let loofe,   270
Or I will fhake thee from me like a ferpent.

Her.   Why are you growne fo rude ?
What change is this fweete Loue ?

Lyf.   Thy loue ? out tawny *Tartar*, out ;
Out loathed medicine ; O hated poifon hence.   275

Her.   Do you not ieft ?

Hel.   Yes footh, and fo do you.

Lyf.   *Demetrius* : I will keepe my word with thee.

Dem.   I would I had your bond : for I perceiue
A weake bond holds you ; Ile not truft your word.   280

Lyf.   What, fhould I hurt her, ftrike her, kill her dead ?
Although I hate her, Ile not harme her fo.

Her.   What, can you do me greater harme then hate ?   283

269. *tame man*] *tameman* Walker (*Crit.* ii, 136).

270. *off*] *of* Q₁.
    bur] *but* Ff.

272, 273. *Why...this*] One line, Q₁, Pope et seq.

273. *this fweete Loue ?*] *this ? Sweet love !* Pope+. *this ? Sweet love,—* Cam. White ii.

275. *O*] *δ* Qq. Om. Pope+, Cap. Steev. Mal. Knt, Cam. Dyce ii, iii, Ktly, White ii.
    poifon] *potion* Q₁, Cap. Steev. Mal. Coll. Dyce, Sta. Cam. Ktly, White ii.

281, 283. *What,*] *What ?* Q₁. *What !* Coll. ii, iii.

283. *What...harme*] *What greater harm can you do me* Han.
    hate] *harm* F₄.

lovers, set at cross purposes by Puck's knavish blundering, are giving vent to the most extravagant violence of language. Helena says, a very little before, ' O spite ! O hell ! I see you all are bent,' &c. In like fashion, as it appears to me, Demetrius now exclaims, in language perfectly consistent with the rude epithets Lysander is heaping on Hermia, ' No, no ; hell Seems to break loose ; take on as you would, fellow !'—BULLOCH (p. 62) : The utterances of Demetrius at what is passing are astonishment, interpretation of it, sarcastic advice, a summons to a challenge, and an ironical compliment, ending with a contemptuous dismissal. [Therefore read] ' *Now, now,* Sir ! *Hell's abyss Seems* to break loose ; take on as you would *flow,* But yet come *on.*' Lysander would appear to be as Sebastian, in *The Tempest, standing water ;* and Demetrius as Antonio would excite him to action and teach him how to *flow.* [With the majority of editors I think the whole line is addressed to Lysander, but I do not think that ' No, no, Sir ' has any reference to Hermia's having been called an ' Ethiop.' Demetrius shows no such zeal when Lysander afterward showers opprobrious epithets on the damsel. To my ears ' No, no, Sir ' is a taunting sneer, in modern street-language, ' No you don't ! You can't come that game over me !' and Lettsom's emendation follows well : ' You merely seem to break loose,' &c.—ED.]

274. **tawny**] Another reference to Hermia's brunette complexion.—ED.

280. **weake bond**] Alluding to Hermia's arms, which were clinging around Lysander. Demetrius scornfully intimates that Lysander, from cowardice, does not really wish to be free. This explains Lysander's vehement reply—ED.

Hate me, wherefore? O me, what newes my Loue?
Am not I *Hermia*? Are not you *Lyſander*? 285
I am as faire now, as I was ere while.
Since night you lou'd me; yet ſince night you left me.
Why then you left me (O the gods forbid
In earneſt, ſhall I ſay? 289

---

284. *newes*] *means* Coll. ii, iii (MS),   288. *forbid*] *forbid*) QqFf. *forbid!*
Sing. ii, Ktly, Marshall.   Rowe. *forbid it!* Theob. Warb. Johns.

281, 283. In the way of punctuation, I prefer the interrogative 'What?' of Q₁ to the 'What!' of Collier and the 'What' of all the rest.—ED.

284. wheretore] For other instances where the stronger accent is on the ſecond syllable, see WALKER (*Vers.* 111), or ABBOTT, § 490.

284. newes] COLLIER (ed. ii): For more than two hundred years the text here was the ridiculous question 'what *news*, my love?' It has been repeated in edition after edition, ancient and modern; and so it might have continued but for the discovery of the MS, which ſhows that *means* has always been misprinted 'news'.—LETTSOM (*Blackwood's Maga.* Aug. 1853) thinks that this change of the MS 'seems to be right.' —HALLIWELL thinks it 'very plausible, but unnecessary. "What news?" here means What *novelty* is this?'—DYCE (ed. ii): We have a passage in *Tam. of the Shr.* I, i, which makes the alteration of Collier's MS a doubtful one: there Lucentio exchanges dress with his servant Tranio; presently Lucentio's other servant, Biondello, enters, and exclaims in great surprise, 'Master, has my fellow Tranio stol'n your clothes? Or you stol'n his? or both? pray, what's the news?'—R. G. WHITE (ed. i): Collier's MS substitution is one of the most plausible readings [in the list]. But when we also consider that as this is Hermia's first interview with her lover since Puck's application of the flower to his eyes, she may well express surprise at the novelty of his declaration that he hates her; and when, besides, we find the same word, 'newes,' in the QqFf, there does not seem to be sufficient warrant for a change in the authentic text. —MARSHALL (*Irving Sh.*): I cannot find a single instance in which 'What news?' or 'What news with you?' is not addressed to some person who has only just appeared on the scene. . . . But Hermia is here under the influence of strong emotion. Is it likely, under such circumstances, that she would employ such a colloquial phrase? Were she less in earnest, less deeply wounded, and playing the part of an indignant coquette, whose philanderings had been discovered, she might say, 'What new-fangled notion is this of your hating me?' But she is too much in earnest to play with words. The exclamation 'O me!' is not one of skittish and affected suspence; it is a cry of real mental anguish; and I cannot think any one with a due sense of dramatic fitness would admit the reading 'what news?' in the sense accepted by all the commentators. [We must doggedly shut our eyes to the substitution of any phrase, which is merely an alleged improvement where the sense of the original texts is clear. It seems to have been generally supposed that 'What news?' can be uttered only in an idle, indifferent way, but it is conceivable that very tragic pathos can be imparted to the word 'news.' Moreover, the continuity of thought upholds the original text in contrasting the new present with the old past: 'I am as fair *now* as I *was*,' &c. Above all, the sound rule that *durior lectio preferenda est* should be ever present. —ED.]

Lyf. I, by my life ;   290
And neuer did defire to fee thee more.
Therefore be out of hope, of queftion, of doubt ;
Be certaine, nothing truer : 'tis no ieft,
That I doe hate thee, and loue *Helena.*

  Her. O me, you iugler, you canker bloffome,   295
You theefe of loue ; What, haue you come by night,
And ftolne my loues heart from him ?

  Hel. Fine yfaith :
Haue you no modefty, no maiden fhame,   299

---

292. *of doubt*] *doubt* Pope+, Cap.
Steev. Mal. Sta. Dyce ii, iii, Coll. iii.
Om. Anon. (ap. Cam.).

293. *certaine,*] *certaine :* Qq.

295. *iugler, you*] *jugler, oh you* Pope+,
Steev.'85. *jugler, you ! you* Cap. *juggleēr,
you* Ktly.

298. *yfaith*] *Ifaith* Q₁. *ifaith* Q₂.

---

292. Therefore ... doubt] To cure this Alexandrine, Pope omitted 'of' before
'doubt'; which is effective if 'question' be pronounced as a disyllable, as is allowable.—WALKER (*Crit.* iii, 49) proposed to print 'Therefore' as a separate line, which
is merely a deference paid to the eye.—In support of Pope, LETTSOM (ap. Dyce)
cites : 'Ay, in the temple, in the town, the field.'—SCHMIDT (*Programm*, p. 6) transposed the words, so as to read, 'Therefore be out of hope, of doubt, of question,'
which is good. But, after all, it seems to me to be better to accept it as an incorrigible Alexandrine, necessitated by the need that each clause should have its fullest
effect and be cumulative up to the climax.—ED.

   295. iugler] MALONE, WALKER (*Vers.* 8), ABBOTT, § 477, all pronounce this
word *juggeler*—a needless deformity, when an exclamation-mark can take the place
of a syllable.—ED.

   295. canker-blossom] STEEVENS : This is not here the blossom of the *canker* or
wild rose, alluded to in *Much Ado*, I, iii, 28 : 'I had rather be a canker in a hedge
than a rose in his grace,' but a worm that preys on the buds of flowers. So in II, ii,
4 of this play : 'Some to kill cankers in the musk-rose buds.' [Albeit there is abundant evidence to show that Steevens was acquainted with Capell's *Notes*, no blame
can attach to him for overlooking explanations imbedded in that gnarled and almost
unwedgeable mass. Witness the following, on the present line : 'Judges of nature's
language in situations like that of the speaker will be at no loss to decide instantan-
eously which line should have preference, theirs [i. e. other editors], or that of this
copy : The first component of the word it [*i. e.* the line] concludes with is a verb; the
compound was overlook'd, or had had a place in the *Glossary* [*i. e.* Capell's own Glos-
sary] ; what is said of it now will make it clear to all Englishmen.' In reference to
these notes well did Lettsom parody Johnson's panegyric on Addison : 'Whoever
wishes to attain an English style uncouth without simplicity, obscure without concise-
ness, and slovenly without ease, must give his nights and days to the Notes of Capell.'
The provoking part of it is that Capell's meaning is too good to be disregarded. We
cannot afford to overlook it. In the present instance he is exactly right. 'You canker-
blossom' is not 'you blossom eaten by a canker,' but 'you who cankers blossoms.'—
ED.]

No touch of bafhfulneffe *?* What, will you teare   300
Impatient anfwers from my gentle tongue?
Fie, fie, you counterfeit, you puppet, you.
   Her. Puppet? why fo? I, that way goes the game.
Now I perceiue that fhe hath made compare
Betweene our ftatures, fhe hath vrg'd her height,   305
And with her perfonage, her tall perfonage,
Her height (forfooth) fhe hath preuail'd with him.
And are you growne fo high in his efteeme,
Becaufe I am fo dwarfifh, and fo low *?*
How low am I, thou painted May-pole? Speake,   310
How low am I? I am not yet fo low,
But that my nailes can reach vnto thine eyes.
   Hel. I pray you though you mocke me, gentlemen,
Let her not hurt me; I was neuer curft: 314

---

302. *counterfeit*] *counterfait* Q₁. *coun-
terfet* Q₂.

303. *why fo ?*] *why, so :* Theob. Warb.
Johns. Dyce.

303. *way goes*] *ways go* Rowe, Pope.

306. *tall perfonage*] *tall parfonage* Q₂.

313. *gentlemen*] *gentleman* Q₁.

---

301. **tongue ?**] Note the genesis of a sophistication. This interrogation mark became in F₃, by accident, a parenthesis: 'tongue)' This caught the eye of the compositor of F₄ in setting up from F₃, and supposing that the preceding half of the parenthesis had been omitted, supplied it, and enclosed the whole line in parentheses, to the confusion of the sense.—ED.

304. **compare**] For other instances of the conversion of one part of speech into another, see ABBOTT, § 451.

306. **And . . . personage**] ABBOTT, § 476, thus scans: 'And wĭth | her pèrson | age, hèr | tall pèr | sonàge,' as an illustration of his rule that when a word is repeated twice in a verse, and increases in emphasis, it receives one accent the first time and two accents the second. The result here is, I think, neither smoothness nor due emphasis. I prefer, 'And wĭth | her pèr | sŏnăge | her tăll | pĕrsŏnăge,' that is, the two strongly emphasized words are, the first 'personage' and 'tall.'—ED.

310. **painted May-pole**] STEEVENS: So in Stubbes's *Anatomie of Abuses,* 1583 [p. 149, ed. *New Sh. Soc.*]: 'They haue twentie or fortie yoke of Oxen, euery Oxe hauing a sweet nose-gay of flouers placed on the tip of his hornes; and these Oxen drawe home this May-pole (this stinking Ydol, rather) which is couered all ouer with floures and hearbs, bound round about with strings from the top to the bottome, and sometime painted with variable colours.'—HALLIWELL gives many extracts to show the antiquity and fashion of painted May-poles, and quotes an observation by Fairholt that 'the term applied by Hermia to Helena is a sort of inseparable conjunction, when the old custom of painting the May-pole is duly considered, and conveys a deeper satire than that applied to her height alone.' [This is doubtless true, but, at the same time, it is possible that in the epithet 'painted' there may be an allusion to the clear red and white of Helena's blonde complexion.—ED.]

I haue no gift at all in ſhrewiſhneſſe ;   315
I am a right maide for my cowardize ;
Let her not ſtrike me : you perhaps may thinke,
Becauſe ſhe is ſomething lower then my ſelfe.
That I can match her.
    Her.   Lower ? harke againe.   320
    Hel.   Good *Hermia*, do not be ſo bitter with me,
I euermore did loue you *Hermia*,
Did euer keepe your counſels, neuer wronged you,
Saue that in loue vnto *Demetrius*,
I told him of your ſtealth vnto this wood.   325
He followed you, for loue I followed him,
But he hath chid me hence, and threatned me
To ſtrike me, ſpurne me, nay to kill me too ;
And now, ſo you will let me quiet go,
To *Athens* will I beare my folly backe,   330
And follow you no further.   Let me go.
You ſee how ſimple, and how fond I am.
    Her.   Why get you gone : who iſt that hinders you ?
    Hel.   A fooliſh heart, that I leaue here behinde.
    Her.   What, with *Lyſander* ?   335
    Her.   With *Demetrius*.
    Lyſ.   Be not afraid, ſhe ſhall not harme thee *Helena*.
    Dem.   No ſir, ſhe ſhall not, though you take her part.   338

---

318. *ſhe is*] QqF₂F₃, Coll. White, Cam. *ſhe's* F₄ et cet.

323. *wronged*] *wrongd* Q₁, Rowe et seq. (subs.).

325. *vnto this*] *into this* F₄, Rowe i. *into the* Rowe ii, Pope, Han.

326. *followed you*] *follow'd you* Rowe et seq.

326. *followed him*] *followd him* F₂, Rowe et seq.

328. *too*] *to* Qq.

333. *Why*] *Why ?* Q₁.

336. Her.] F₂.   Hel. QqF₃F₄ et cet.

337. *ſhall*] *will* F₄, Rowe, Pope, Han. *Helena*] *Helen* Walker, Dyce ii iii.

---

314. curst] STEEVENS: That is, shrewish or mischievous.—W. A. WRIGHT: So in *Tam. of the Shr.* I, i, 186 : 'Her eldest sister is so curst and shrewd '; in *Much Ado*, II, i, 22 : 'For it is said, God sends a curst cow short horns.' Cotgrave defines, 'Meschant.   Wicked, impious, vngracious, . . . also curst, mischievous, harsh, froward.'

316. right] ABBOTT, § 19: This was used by Shakespeare with the indefinite article, to mean *real*, down-*right*.

329. so] See I, i, 47.

332. fond] That is, foolish.   See Shakespeare *passim*.

337. Helena] See II, ii, 110.

Hel. O when ſhe's angry, ſhe is keene and ſhrewd,
She was a vixen when ſhe went to ſchoole,   340
And though ſhe be but little, ſhe is fierce.

Her. Little againe ? Nothing but low and little ?
Why will you ſuffer her to flout me thus ?
Let me come to her.

Lyſ. Get you gone you dwarfe,   345
You *minimus,* of hindring knot-graſſe made,
You bead, you acorne.

Dem. You are too officious,
In her behalfe that ſcornes your ſeruices.
Let her alone, ſpeake not of *Helena,*   350
Take not her part. For if thou doſt intend

---

339. *ſhe's] ſhe is* Q₁.   346. minimus] *Minim, you,* Theob. ii,
                                                  Han.

---

340. **vixen**] W. A. WRIGHT: Properly, a she-fox. The form of the word is
especially interesting as being the only instance in which the feminine termination *-en*
has been preserved. See Morris, *English Accidence,* c. x, § 73. It occurs in Anglo-
Saxon as *fixen,* and in German as *füchsin.*

346. **minimus**] THEOBALD: This is no term of art, that I can find; and I can
scarce be willing to think that Shakespeare would use the masculine of an adjective
to a woman. I doubt not but he might have wrote, 'You *Minim, you,*' i. e. you
diminutive of the creation, you reptile.—NARES: The word came into use probably
from the musical term *minim,* which, in the very old notation, was the shortest note,
though now one of the longest.

346. **knot-graſſe**] STEEVENS: It appears that 'knot-grass' was anciently supposed to prevent the growth of any animal or child. See Beaumont & Fletcher's
The Knight of the Burning Pestle [II, ii, p. 157, ed. Dyce]: 'Should they put him
into a straight pair of gaskins, 'twere worse than knot-grass; he would never grow
after it.' Again, in *The Coxcomb* [II, ii, p. 150, ed. Dyce]: 'We want a boy
extremely for this function, Kept under for a year with milk and knot-grass.'—ELLA-
COMBE (p. 101): The *Polygonum aviculare,* a British weed, low, straggling, and
many-jointed, hence its name of Knot-grass. There may be another explanation of
'hindering' than that given by Steevens. Johnstone tells us that in the North, 'being
difficult to cut in the harvest time, or to pull in the process of weeding, it has obtained
the sobriquet of the Deil's-lingels.' From this it may well be called 'hindering,'
just as the *Ononis,* from the same habit of catching the plough and harrow, has
obtained the prettier name of 'Rest-harrow.' [To the same effect GREY (i, 61).
'Hindering' applies not only to 'knot-grass,' but also to Hermia; hence it becomes,
in reality, a botanical pun.—ED.]

347. **bead**] W. A. WRIGHT: As beads were generally black, there is a reference
here to Hermia's complexion as well as to her size.

351. **intend**] STEEVENS: That is, pretend. So in *Much Ado,* II, ii, 35: 'Intend
a kind of zeal both to the Prince and Claudio.'

11

Neuer ſo little ſhew of loue to her,   352
Thou ſhalt abide it.
    Lyſ. Now ſhe holds me not,
Now follow if thou dar'ſt, to try whoſe right,   355
Of thine or mine is moſt in *Helena.*
    Dem. Follow ? Nay, Ile goe with thee cheeke by
iowle.   *Exit Lyſander and Demetrius.*
    Her. You Miſtris, all this coyle is long of you.
Nay, goe not backe.   360
    Hel. I will not truſt you I,
Nor longer ſtay in your curſt companie.
Your hands then mine, are quicker for a fray,
My legs are longer though to runne away.   364
[**Her.* I am amaz'd, and know not what to ſay. *Exeunt.*] 364

---

353. *abide*] *abie* Q₂. *aby* Q₁, Pope et
seq.
    356. *Of*] *Or* Theob. Warb. Johns.
Steev. Mal. Var. Knt, Sing. Coll. ii, iii
(MS), Sta.
    358. Exit...] Exit. Q₂. Om. Q₁.
    359. *long*] *'long* Cap. et seq. (except

Hal. Dyce).
    361. *you I,*] *you, I,* Q₁. *you* Rowe i.
    364. *away.*] *away.* Exeunt. Ff. *away.*
Exeunt: Herm. pursuing Helena. Theob.
        * Her. *I am amaz'd, and know
not what to ſay.* Qq, Pope, Han. Cap. et
seq. (except White i).

---

353. **abide**] See line 181, *supra.*
    356. **Of thine or mine**] MALONE: If the line had run *Of mine or thine*, I should
have suspected that the phrase was borrowed from the Latin: Now follow, to try
whose right of *property*—of *meum* or *tuum*—is greatest in Helena. [See *The Tempest*, II, i, 32 of this edition, where is given the following note:] WALKER (*Crit.* ii,
353), in a paragraph on the use of *former*, the comparative, to which *foremost* is the
superlative, quotes this passage from Sidney's *Arcadia*, B. i, p. 63: 'the question
arising, who should be the former against Phalantus, of the blacke, or the ill-appar-
elled knight,' &c., ' *i. e.*' explains Walker, ' whether the blacke or the, &c. should be
the first to wage combat with Phalantus.' Whereupon LETTSOM, Walker's editor,
remarks that this example 'shows that the First Folio is right in " Which *of* he, *or*
Adrian." '
    358. **iowle**] W. A. WRIGHT: Side by side, close together, as the cheek to the
jole or jaw.
    359. **coyle**] That is, confusion, turmoil. See Shakespeare *passim.*
    364.* THEOBALD'S stage-direction ' Exit Hermia pursuing Helena ' cannot be right.
That this line was accidentally omitted by the printers of F₁ is clear, I think, from the
fact that there is no *Exit* or *Exeunt* for the two girls.—R. G. WHITE, in his first edition, justified the omission, but in his second edition inserted the line, without a note.
In the first edition it stands: ' The line is so unsuited to Hermia's quickness of temper and tongue, to the state of her mind, and to the situation, and so uncalled for by
Helena's speech, which elicits it, that we should gladly accept the testimony of the
authentic copy, that it is either the interpolation of some player who did not want to

Enter Oberon and Pucke. 365

Ob. This is thy negligence, ſtill thou miſtak'ſt,
Or elſe committ'ſt thy knaueries willingly.

Puck. Beleeue me, King of ſhadowes, I miſtooke,
Did not you tell me, I ſhould know the man,
By the *Athenian* garments he hath on ? 370
And ſo farre blameleſſe proues my enterpize,
That I haue nointed an Athenians eies,
And ſo farre am I glad, it ſo did ſort,
As this their iangling I eſteeme a ſport.

Ob. Thou ſeeſt theſe Louers ſeeke a place to fight, 375
Hie therefore *Robin*, ouercaſt the night,
The ſtarrie Welkin couer thou anon,
With drooping fogge as blacke as *Acheron*, 378

---

365. [Scene IX. Pope, Han. Scene VIII. Warb. Johns.
Enter...] Om. Qq.
367. *willingly*] *wilfully* Qq, Cap. Mal. Steev.'93, Var. Coll. Sing. Hal. Dyce, Sta. Cam. Ktly, White ii.
368. *ſhadowes*] *fairies* Gould.
370. *garments*] *garment* Glo. (misprint).

370. *hath*] *had* Q₁, Theob. et seq.
371. *enterpize*] F₁ (Editor's copy), Vernor & Hood's *Repr.*, Staunton's *Photolith.* *enterprize* Booth's *Repr.*
372. *nointed*] *'nointed* Rowe et seq.
373. *ſo did*] *did so* Rowe+. *did not* Steev.'85 (misprint).
378. *fogge*] *fogs* Theob. ii, Warb. Johns.

---

leave the stage without a speech, or a piece of the author's work which he cancelled as unsatisfactory or superfluous.' [See *Preface* to this volume, p. xv.—ED.]

371. **enterpize**] See Text. Notes for a variation in *Reprints* of F₁.—ED.

372. **nointed**] For a list of words whose prefixes are dropped, see ABBOTT, § 460.

373. **sort**] An allusion to fate. 'All the forms of *sort*,' says SKEAT (*Dict.* s. v.), ' are ultimately due to Lat. *sortem*, acc. of *sors*, lot, destiny, chance, condition, state.'—ED.

374. **As**] I am not sure that in a modern text there should not be a semicolon after ' sort ' in the previous line, to indicate that this 'As' does net follow the ' so ' in that line (unlike the ' so ' and 'As' in lines 379, 380), but means *oecause, since.*—ED.

378. **Acheron**] W. A. WRIGHT: The river of hell in classical mythology, supposed by Shakespeare to be a pit or lake. Compare *Macb.* III, v, 15 : 'And at the pit of Acheron Meet me,' &c.; *Tit. And.* IV, iii, 44 : 'I'll dive into the burning lake below And pull her out of Acheron by the heels,'—R. G. WHITE (ed. ii) : A river in Hades, which Shakespeare mistook to be a pit. [That Shakespeare in *Macbeth* may have supposed Acheron to be a pit is quite likely, but he made no mistake in the present passage. The rivers of hell were black, and it is with this blackness alone that comparison is here made. In Shakespeare's contemporary, Sylvester, there is the same simile : ' In *Groon-land* field is found a dungeon, A thousand-fold more dark than *Acheron.*'—*The Vocation*, line 532, ed. Grosart. And if it be urged that Sylvester has here fallen into the same error, and overlooked the fact that Acheron is a river, so be it. Shakespeare has a good companion, then, to bear half the disgrace of his oversight in *Macbeth.*—ED.]

And lead thefe teftie Riuals fo aftray,
As one come not within anothers way.   38c
Like to *Lyfander*, fometime frame thy tongue,
Then ftirre *Demetrius* vp with bitter wrong;
And fometime raile thou like *Demetrius*;
And from each other looke thou leade them thus,
Till ore their browes, death-counterfeiting, fleepe   385
With leaden legs, and Battie-wings doth c reepe;
Then crufh this hearbe into *Lyfanders* eie,
Whofe liquor hath this vertuous propertie,
To take from thence all error, with his might,
And make his eie-bals role with wonted fight.   390
When they next wake, all this derifion
Shall feeme a dreame, and fruitleffe vifion,
And backe to *Athens* fhall the Louers wend
With league, whofe date till death fhall neuer end.
Whiles I in this affaire do thee imply,   395
Ile to my Queene, and beg her *Indian* Boy;
And then I will her charmed eie releafe
From monfters view, and all things fhall be peace.
     Puck. My Fairie Lord, this muft be done with hafte,
For night-fwift Dragons cut the Clouds full faft,   400

---

385. *counterfeiting, fleepe*] *counterfeit-ing fleep* Ff.
386. *legs*] *ledgs* Q₂.
     Battie] *Batty* Qq.
389. *his might*] *its might* Rowe +.

395. *imply*] *imploy* Q₁F₄. *apply* Q₂.
400. *night-fwift*] *nights fwift* Q₁
     night fwift Q₂. *nights-fwft* F₂. *nights-fwift* F₃F₄. *night's swift* Rowe et seq.

---

388. **vertuous**] JOHNSON: Salutiferous. So he calls, in *The Tempest*, *poisonous* dew, 'wicked dew.'—R. G. WHITE (ed. i): 'Virtue' was used of old, and is some-times now used, for *power*, especially in the sense of healing or corrective power; as in the Gospels: 'I perceive some virtue has gone out of me.'—*Luke* viii, 16.

392. **shall seeme a dreame**] GUEST (i, 130) gives other examples from Shake-speare of this effective 'middle-sectional rhyme,' *e. g.* 'He hath won With *fame* a *name* to Caius Martius; these.'—*Cor.* II, i; 'With *cuffs* and *ruffs*, and farthingales and things.'—*Tam. of the Shr.* V, iii; 'Or *groan* for *Joan?* or spend a minute's time.'—*Love's L. L.* IV, iii.

391, 392. **derision ... vision**] To be pronounced *dissolutè*.

395. **imply**] The Q₁ corrects this compositor's error.

400. **night-swift**] This word, instead of *night's-swift*, may be accounted for, if the printers of F₁ composed from dictation.—ED.

400. **Dragons**] STEEVENS: So in *Cymb.* II, ii, 48: 'Swift, swift you dragons of the night.' The task of drawing the chariot of the night was assigned to dragons on account of their supposed watchfulness.—MALONE: This circumstance Shakespeare

And yonder fhines *Auroras* harbinger ; 401
At whofe approach Ghofts wandring here and there,
Troope home to Church-yards ; damned fpirits all,
That in croffe-waies and flouds haue buriall,
Alreadie to their wormie beds are gone ; 405
For feare leaft day fhould looke their fhames vpon,
They wilfully themfelues dxile from light,
And muft for aye confort with blacke browd night.
 Ob. But we are fpirits of another fort :
I, with the mornings loue haue oft made fport, 410

---

403. *Church - yards*]   church - yard
Theob. ii, Johns.
407. *themfelues dxile*] F₁. exile them-
felues F₃F₄, Rowe +.

408. *black browd*] black browed Q₁.
black-browd F₃F₄.
410. *mornings loue*] morning loue Ff.
Morning-Love Rowe i. *Morning-Light*
Rowe ii +. *morning's love* Cap. et seq.

---

might have learned from a passage in Golding's *Ovid*, which he has imitated in *The Tempest :* 'And brought asleep the dragon fell, whose eyes were never shet.'—W. A. WRIGHT : Milton perhaps had this passage in his mind when he wrote *Il Penseroso*, 59 : 'While Cynthia checks her dragon-yoke Gently o'er the accustom'd oak.' On which Keightley remarks it is wrong mythology, 'for Demeter, or Ceres, alone had a dragon yoke.' Drayton also (*The Man in the Moon*, 431) says that Phœbe 'Calls downe the Dragons that her chariot drawe.'

401. **harbinger**] I suppose this must have had two accents, on the first and on the last syllable, and the latter pronounced to rhyme with ' there.'—ED.

404. **crosse-waies and flouds**] STEEVENS : The ghosts of self-murderers, who are buried in cross-roads ; and of those, who being drowned, were condemned (accord-ing to the opinion of the ancients) to wander for a hundred years, as the rites of sepul-ture had never been regularly bestowed on their bodies. That the waters were some-times the place of residence for ' damned spirits ' we learn from the ancient bl. l. romance of *Syr Eglamoure of Artoys*, no date : ' Let some preest a gospel saye, For doute of fendes in the flode.'

405. **wormie**] STEEVENS : This has been borrowed by Milton in his *On the death of a Fair Infant :* ' Or that thy beauties lie in wormy bed.'

406. **vpon**] For other examples of the transposition of prepositions, see ABBOTT, § 203 ; and for examples of an accent nearer to the end than with us, like ' exile,' in the next line, see *Ib.* § 490.

407. **dxile**] THIRLBY (Nichols, *Illust.* ii, 224) : I read *exiled*, and incline to think Oberon's speech should begin here.

408. **black-browd**] STEEVENS : So in *King John*, V, vi, 17 : ' here walk I in the black brow of night To find you out.'

410. **mornings loue**] There has been some difficulty in determining the refer-ence here.—CAPELL suggests that it may mean ' the star Phosphorus ; possibly the sun ; and the sense be that the speaker had sported with one or other of these, *i. e.* wanton'd in them ; but the simpler sense is that he had courted the morning, made her his love-addresses ; the lady's name is Aurora.'—STEEVENS takes it for granted

And like a Forrefter, the groues may tread,   411
Euen till the Eafterne gate all fierie red,
Opening on *Neptune*, with faire bleffed beames,
Turnes into yellow gold, his falt greene ftreames.
But notwithftanding hafte, make no delay :   415
We may effect this bufineffe, yet ere day.
    Puck. Vp and downe, vp and downe, I will leade
them vp and downe : I am fear'd in field and towne.
Goblin, lead them vp and downe : here comes one.   419

---

413. *faire bleſſed*] *far-blessing* Han.
Warb. *fair-blessed* Walker, Dyce ii.
415. *notwithſtanding*] *notwiſtanding*,
Q₁. *notwithstanding*, Theob. et seq.

416. [Exit Oberon. Rowe.
417–419. *Vp...downe*] Two lines, Q₁.
Four lines, Pope et seq.
417. *downe, vp*] *down then, up* Han.

---

that it is Tithonus, the husband of Aurora.—HOLT WHITE thinks, and DYCE and
W. A. WRIGHT agree with him, that 'Cephalus, the mighty hunter and paramour of
Aurora, is intended. The context, "And like a forester," &c. seems to show that the
chase was the "sport" which Oberon boasts he partook with the "morning's love."'
—HALLIWELL says that 'Oberon merely means to say metaphorically that he has
sported with Aurora, the morning's love, the first blush of morning; and that he is
not, like a ghost, compelled to vanish at the dawn of day.' [This interpretation is
to me the most natural, and more in harmony than the others with the drift of Oberon's speech, which is to contrast with the fate of the damned spirits, who must consort with black-browed night, his liberty in the fair blessed beams of day, and not to
boast that he is privileged to sport with Phosphorus, or Tithonus, or Cephalus.—ED.]
    413. beames,] I believe that DYCE (ed. ii) and HUDSON, who printed from him,
are the only editors who have here followed WALKER's convincing suggestion (*Crit.*
iii, 49) that the comma after 'beams' be erased. It is with these beams that the
streams are turned to gold. Compare 'gilding pale streams with heavenly alchemy'
—*Sonn.* 33.—ED.
    414. salt greene] TATHWELL (ap. Grey, i, 62): Qu. *sea*-green. But perhaps
the contrast is intended between 'yellow gold' and 'salt green.' [Undoubtedly
'salt green' is *sea green.*—ED.]
    415. notwithstanding] In this word occurs one of those insignificant variants in
different copies of the same edition. The CAM. ED. records as in Q₁ (Fisher's) *not-
wistandiug*, and the same is recorded in HENRY JOHNSON's microscopically minute
collation, whereas Ashbee's Facsimile and Griggs's Photo-lithographic Facsimile both
have *notwistanding*. But this minute collation of what is not Shakespeare's work,
but that of a printer, in whom we take no atom of interest, leads, I am afraid,
nowhither.—ED.
    417. Vp and downe, &c.] COLLIER: These four lines [according to Pope's divis-
ion] are possibly a quotation from some lost ballad respecting Puck and his pranks;
he would otherwise hardly address himself as 'Goblin.' The exit of Oberon is not
marked in the old copies, and the last line [419] might belong to him, if we suppose
him to have remained on the stage.
    419. Goblin, lead] THIRLBY (Nichols, *Illust.* ii, 224) conjectured *Goblin'll lead*
—an *emendatio certissima*, I think; a clear case of absorption. STAUNTON, however,

Enter Lyſander. 420

Lyſ. Whcrc art thou, proud *Demetrius* ?
Speake thou now.

Rob. Here villaine, drawne & readie. Where art thou ?

Lyſ. I will be with thee ſtraight.

Rob. Follow me then to plainer ground. 425

Enter Demetrius.

Dem. Lyſander, ſpeake againe ;
Thou runaway, thou coward, art thou fled ?
Speake in ſome buſh : Where doſt thou hide thy head ? 429

420. Enter] Re-enter Cap.
421, 422. One line, Qq, Pope et seq.
423, 425, &c. Rob.] Puck. Rowe et seq.
425. *to plainer ground*] Separate line, Theob. et seq. (except Hal).
[Lys. goes out, as following Dem.

Theob. Exit Lys. as following the Voice, which seems to go off. Cap.
429. *Speake in ſome buſh :*] Speak. In some bush ? Cap. et seq. (subs.).
Speake...head ?] Speak in some bush, where thou dost hide thy head. Han.

in a note on ' Sicilia is a so-forth ' (*Wint. T.* I, ii, 218, contributed to *The Athenæum,* 27 June, '74), gives a strikingly novel interpretation of the whole line. It is not a happy interpretation, it must be confessed, but it has a sad interest as being one of the very last notes which sprang from that fertile and learned mind, and one which, alas, its writer never saw in print. It is as follows : ' There can be no doubt with those well read in our old drama that *et cetera* in like manner, from being used to express vaguely what a writer or speaker hesitated to call by its plain name, came at length to signify the object itself. " Yea, forsooth " is possibly another case in point. The Puritanical citizens, who were afraid of a good air-splitting oath, and indulged only in mealy-mouthed protestations, got the name of " yea-forsooths " [see *2 Hen. IV :* I, ii, 41]. I am not sure but that in the same way we get the meaning of [the present line, which is], perhaps, no other than a nickname given to the mischievous sprite to indi-cate his will-o'-the-wisp propensities, and to be read : " *Goblin-lead-them-up-and-down.*" Still more curious, there is some reason for believing that what has always been regarded as a harmless exclamation of Master Flute : "A paramour is, *God bless us,* a thing of nought," was really meant as a term of reproach. Compare V, i, 323 : " He for a man, *God warrant us ;* she for a woman, *God bless us,*" expressions which have hitherto defied explanation, but which are quite intelligible as terms of oppro-brium. The one being a male *God warrant us ;* the other a female *God bless us.* The rationale of these latter expressions being so employed must be gathered, I appre-hend, from the all-prevalent fear of witchcraft formerly. When a suspected person came in presence, or was even spoken of, it was customary to invoke the protection of Heaven, and the usual form of invocation was " God bless us !" In the course of time this formula was used to denominate the individual whose malice was depre-cated, and finally became a by-name for any one of ill-omened repute.' It is only Staunton's interpretation of the present line that is to be deprecated in the foregoing note.—ED.

423. drawne] That is, with sword drawn.

429. Speake . . . bush] CAPELL : Very nature and knowledge of what is acting

Rob. Thou coward, art thou bragging to the ſtars,   430
Telling the buſhes that thou look'ſt for wars,
And wilt not come ? Come recreant, come thou childe,
Ile whip thee with a rod. He is defil'd
That drawes a ſword on thee.

    Dem. Yea, art thou there ?   435
    Ro. Follow my voice, we'l try no manhood here. *Exit.*
    Lyſ. He goes before me, and ſtill dares me on,
When I come where he cals, then he's gone:
The villaine is much lighter heel'd then I :
I followed faſt, but faſter he did flye ; *ſhifting places.*   440
That fallen am I in darke vneuen way,
And here wil reſt me. Come thou gentle day : *lye down.*
For if but once thou ſhew me thy gray light,   443

---

430. *bragging*] *begging* F₃F₄, Rowe.
436. Exit.] Exeunt. Qq.
437. [Lys. comes back. Theob. Re-
-enter Lys. Cap.
438. *cals,*] *cals me,* Ff, Rowe+.

438. *he's*] *he is* Q₁.
440. *followed*] *follow'd* Rowe et seq.
  ſhifting places.] Om. Qq.
442. lye down.] Om. Qq. Lyes down.
Rowe.

---

will tell us, the line is spoke with great pauses; its sense this, indicated by the tone,
Speak. Are you crept into *some bush ?*

  440. **shifting places**] R. G. WHITE (ed. i) : This stage-direction is misplaced, as
it plainly refers to Puck, Lysander, and Demetrius, and belongs several lines above.
[R. G. White is the only editor, I believe, who has done more than merely mention
that this puzzling stage-direction is to be found in the Folio; his suggestion is not
altogether satisfactory. Just below Demetrius accuses Lysander of 'shifting every
place,' which certainly seems to refer to this stage-direction, and may indicate some
unusual alacrity on the part of Lysander in his attempts in the dense darkness to find
Demetrius. It is clear that Demetrius follows Puck's voice off the stage at line 436.
To make Demetrius enter and fall asleep and then Lysander enter and fall asleep,
would have smacked of tameness in the repetition, and we should have had but little
proof that the two men were really in bitter earnest. Whereas if Demetrius plunges
into the darkness and we lose sight of him mad in the pursuit of Puck's voice, and
then see Lysander enter, rush hither and thither, half frenzied, shifting his place
every minute, then the conviction is forced on us that this is a fight to the death, and
the somnolent power of Puck's charm in allaying the fury is heightened. There is
anotner point which adds somewhat to the belief that this stage-direction is correctly
placed : it is not mandatory, as are many other stage-directions in this play, or as that
two lines lower, 'lye down'; it does not tell the actor what to do, but describes
what he does. Hence I adhere to the Folio, both as to the propriety of this 'shifting
places' and as to its location.—ED.]

  443. **gray**] MARSHALL : Compare *Ham.* I, i, 166, 'But look, the morn, in *russet*
mantle c'ad,' where 'russet,' as has been pointed out in line [23 of this scene], means
grey.

Iie finde *Demetrius,* and reuenge this fpight.

Enter Robin and Demetrius. 445

Rob. Ho, ho, ho ; coward, why com'ft thou not?

Dem. Abide me, if thou dar'ft. For well I wot,

Thou runft before me, fhifting euery place, 448

444. [sleeps. Cap.
445. Enter Robin] Robin Qq.
446. *Ho, ho, ho ;*] *Ho, ho ; ho, ho!* Cap.
Steev.'93, Var. Knt, Dyce ii, iii, Ktly.

446. *why com'ft*] *why then com'st*
Han. *why comest* Johns. Steev.'85, Rann,
Mal. *wherefore comest* Schmidt.

446. **Ho, ho, ho**] RITSON: This exclamation would have been uttered by Puck with greater propriety if he were not now playing an assumed character, which he, in the present instance, seems to forget. In the old song, printed by Peck and Percy [see 'Robin Goodfellow,' Percy's *Reliques,* &c. in Appendix], in which all his gambols are related, he concludes every stanza with *Ho, ho, ho!* So in *Grim the Collier of Croydon* [Robin Goodfellow says], 'Ho, ho, ho, my masters! No good fellowship!' [V, i, p. 459, Hazlitt's *Dodsley*]. Again, in Drayton's *Nymphidia* [p. 164, ed. 1748], 'Hoh, hoh, quoth Hob, God save thy grace.' It was not, however, as has been asserted, the appropriate exclamation, in our author's time, of this eccentric character; the devil himself having, if not a better, at least an older, title to it. So in *Histriomastix* (as quoted by Mr Steevens in a note on *Rich. III*), 'a *roaring Devil* enters, with the *Vice* on his back, *Iniquity* in one hand, and *Juventus* in the other, crying, "Ho! ho! ho! these babes mine are all." '—[p. 40, ed. Simpson]. Again, in *Gammer Gurton's Needle,* 'But Diccon, Diccon, did not the devil cry ho, ho, ho?' [II, iii]. And, in the same play, 'By the mass, ich saw him of late cal up a great blacke devill, O, the knave cryed, ho, ho, he roared and he thundered' [III, ii]. So in the Epitaph attributed to Shakespeare: 'Hoh! quoth the devil, 'tis my John o' Coombe.' Again, in Goulart's *Histories,* 1607: 'the *Diuills* in horrible formes . . . assoone as they beheld him ran unto him, crying *Hoh, Hoh,* what makest thou here?' Again, in the same book, 'The *blacke guests* . . . roared and cryed out, *Hoh,* sirra, let alone the child.' Indeed, from a passage in *Wily Beguiled,* 1606, I suspect that this same 'knavish sprite' was sometimes introduced on the stage as a demi-devil: 'I'll rather,' it is Robin Goodfellow who speaks, 'put on my flashing red nose and my flaming face, and come wrap'd in a calf's skin, and cry ho, ho.'—[p. 319, ed. Hawkins, and p. 256, ed. Hazlitt's *Dodsley,* in both places it is printed *bo, bo.*—ED.].— STAUNTON. There is an ancient Norfolk proverb, 'To laugh like Robin Goodfellow,' which means, we presume, to laugh in mockery or scorn. This derision was always expressed by the exclamation in the text, . . . which seems with our ancestors always to have conveyed the idea of something fiendish and unnatural, and is the established burden of the songs which describe the frolics of Robin Goodfellow.—W. A. WRIGHT: There is nothing so exceptional in the cry as to make it inappropriate [as Ritson suggested] to Puck in an assumed character.—BELL (ii, 121), whose 'humour' was Teutonic folk-lore, connects by this exclamation, Puck with *The Wild Huntsman.*

447. **Abide**] W. A. WRIGHT: Wait for me, that we may encounter. [It is possible that 'me' may be merely the ethical dative, and thus 'abide' may be relieved from any unusual meaning, and the phrase be equivalent merely to 'Stand still.'—ED.]

And dar'ft not ftand, nor looke me in the face.
Where art thou?   450
   Rob. Come hither, I am here.
   Dem. Nay then thou mock'ft me; thou fhalt buy this
     deere,
If euer I thy face by day-light fee.
Now goe thy way : faintneffe conftraineth me,   455
To meafure out my length on this cold bed,
By daies approach looke to be vifited.
Enter Helena.

   Hel. O weary night, O long and tedious night,
Abate thy houres, fhine comforts from the Eaft,   460
That I may backe to *Athens* by day-light,
From thefe that my poore companie deteft ;
And fleepe that fometime fhuts vp forrowes eie,   463

---

450. *thou ?*] *thou now ?* Q$_2$, Cap. Coll.
Sing. Hal. Dyce, White, Sta. Cam. Ktly.
  451. *Come*] *Come thou* Pope+.
  452. *fhalt*] *fhat* Q$_1$.
    buy] *'by* Johns. conj. Coll. Sing.
Dyce, Sta.
  455. *faintneffe*] *faitnneffe* F$_2$.
  457. [Lyes down. Rowe.
  [Scene X. Pope, Han. Warb.

Scene IX. Johns.
  458. Enter...] Enter... and throws
herself down. Cap.
  460. *fhine comforts*] *fhine comforts,*
Q$_1$. *shine, comforts,* Theob. Warb. Johns.
Cap. Mal. Steev.'93, Knt, White i, Sta.
  463. *fometime*] *fometimes* QqF$_3$F$_4$,
Rowe+, Knt, Coll. i, ii, Sing. Cam. Ktly.
White ii.

---

452. **buy**] JOHNSON : That is, thou shalt dearly pay for this. Though this is
sense and may well enough stand, yet the poet perhaps wrote ' thou shalt 'by it dear.'
—STAUNTON : There can be little doubt the true word was *'by.*—W. A. WRIGHT :
The phrase [' buy it dear '], if a corruption, was so well established in Shakespeare's
time as to make a change unnecessary. Compare *I Hen. IV :* V, iii, 7 : 'The Lord
of Stafford dear to-day hath bought Thy likeness.' And *2 Hen. VI :* II, i, 100 : ' Too
true ; and bought his climbing very dear.' Besides, the two words were etymologically connected. [See line 181, above.]

  460. **comforts**] This may be an accusative, the object of ' shine '; it may be a
vocative, like ' night '; or it may be a nominative, with ' shine ' as its verb ; whichever the reader may think the most pathetic.—ED.

  462. **detest**] WALKER (*Crit.* ii, 311) : In writers of [Shakespeare's] age *detest* is
used in the sense which as then it still retained from its original *detestari,* being indicative of something spoken, not of an affection of the mind ; compare *attest, protest,* which
still retain their etymological meaning. Bacon, *Advancement of Learning,* B. ii, speaking of secrecy in matters of government, 'Again, the wisdom of antiquity . . . in the
description of torments and pains . . . doth detest the offence of facility.' Thus, *Ant.
and Cleop.* IV, xiv, 55, 'Since Cleopatra died I've liv'd in such dishonour, that the
gods Detest my baseness.' [Walker gives several other examples, besides the present
passage, which justify his observation.—ED.]

Steale me a while from mine owne companie. *Sleepe.*

   Rob. Yet but three *?* Come one more,   465

Two of both kindes makes vp foure.

Here fhe comes, curft and fad,

Cupid is a knauifh lad,

Enter Hermia.

Thus to make poore females mad.   470

   Her. Neuer fo wearie, neuer fo in woe,

Bedabbled with the dew, and torne with briars,

I can no further crawle, no further goe;

My legs can keepe no pace with my defires.

Here will I reft me till the breake of day,   475

Heauens fhield *Lyfander*, if they meane a fray.

   Rob. On the ground fleepe found,

Ile apply your eie gentle louer, remedy.   478

---

465. *three ?*] *three here ?* Han.
466. *makes*] *make* F₃, Pope+, Coll. Hal. White i.
467. *comes*] *cometh* Han.
469. Enter Hermia.] Om. Qᵣ. After line 470, Rowe et seq.
473. *further*] *farther* Coll. White i.
475. [lies down. Cap.
476. *Heauens*] *Heaven* Anon. (ap. Cam.).
476. [Lyes down. Rowe.
477–480. Six lines, Coll. Sing. Ktly. Ten lines, Warb. et cet.
477. *fleepe*] *sleep thou* Han. Cap.
478. *your*] QqFf, Hal. *to your* Rowe et cet.
[squeezing the Juice on Lysander's eye. Rowe.

465–470. VERITY: A trochaic measure of three feet with extra syllable at the end. Scan 'three' as a disyllable; likewise 'comes,' thus: 'Yèt but | thrȇe ? | Cŏme one | mòre,' and ' Hère she | cŏmes | cùrst and | sad.' [Why not say that these two lines are made up of amphimacers, and so avoid any barbarous prolongation of syllables? Thus: ' Yĕt bŭt thrēe | Cōme ŏne mōre,' and ' Hēre shĕ cōmes. | Cūrst ănd sād.' Or even why give technical terms, which are merely to guide us when in doubt, to lines which no English tongue can possibly pronounce other than rhythmically?—ED.]

466. **makes**] See III, i, 84.

477, 478. On . . . eie] TATHWELL (ap Grey, i, 63) would read as two lines; 'because verses with the middle rhyme, which were called *leonine* or *monkish* verces, seem to have been the ancient language of *charms* and incantations.'

477–480. **On . . . eye**] GUEST (i, 185): A section of two accents is rarely met with as an independent verse. The cause was evidently its shortness. Shakespeare, however, has adopted it into that peculiar rhythm in which are expressed the wants and wishes of his *fairy-land*. Under Shakespeare's sanction it has become classical, and must now be considered as the *fairy-dialect* of English literature.

478. **your eie**] HALLIWELL, who alone of all editors follows the QqFf here in the omission of the preposition *to*, asserts that '"apply" did not necessarily require the addition of the preposition. The verb occurs without it in *The Nice Wanton*, 1560. The versification is irregular.' The versification is irregular only when we

When thou wak'ſt, thou tak'ſt

True delight in the ſight of thy former Ladies eye,   480

And the Country Prouerb knowne,

That euery man ſhould take his owne.

In your waking ſhall be ſhowne.

Iacke ſhall haue *Iill*, nought ſhall goe ill,

The man ſhall haue his Mare againe, and all ſhall bee   485

well.

They ſleepe all the Act.   487

---

479. *wak'ſt, thou tak'ſt*] *wakest next, thou takest* Han. *wak'st Next, thou tak'st* Cap. *wak'st See thou tak'st* Tyrwhitt, Coll. ii (MS).

 tak'ſt] *rak'ſt* F_2F_3.

484. Two lines, Johns. et seq.

485. *Mare*] *mate* Gould.

485, 486. *and...well.*] Separate line, Coll. Sing. White i, Ktly.

485. *all ſhall bee*] *all be* Rowe+.

486. *well*] *still* Steev. conj.

487. They...] Om. Qq. They sleep. Rowe.

---

count the syllables on our fingers; a solitary example, and that too, not quoted in full, is hardly sufficient to make a rule, especially in days of careless printing.—ED.

479. **thou tak'st**] TYRWHITT: The line would be improved, I think, both in its measure and construction, if it were written ' *see* thou tak'st.'—DYCE: But *see* would require *take*. Compare above, ' sleep sound.'—GUEST (i, 292): The propriety of the rhythm will be better understood if we suppose (what was certainly intended) that the fairy is pouring the love-juice on the sleeper's eye while he pronounces the words ' thou tak'st.' The words form, indeed, the fairy's ' charm,' and the rhythm is grave and emphatic as their import. I cannot see how the construction is bettered [by Tyrwhitt's emendation], and the correspondence, no less than the fitness of the numbers, is entirely lost.

484. **Iacke . . . Iill**] STEEVENS: This is to be found in Heywood's *Epigrammes upon Proverbes*, 1567: 'All shalbe well, Iacke shall haue Gill: Nay, nay, Gill is wedded to Wyll.'—GREY: Jill seems to be a nickname for Julia or Julianna.—HALLIWELL: The nicknames of Jack and Jill, as generic titles for a man and woman, are of great antiquity.—STAUNTON cites instances of this phrase from Skelton's *Magnyfycence*, Dyce's ed. i, 234; from Heywood's *Dialogue*, 1598, sig. F 3; *Love's Lab. L.* V, ii, 305.

485, 486. **The . . . well**] W. A. WRIGHT: This seems to have been a proverbial expression, implying that all would be right in the end. Compare Fletcher, *The Chances*, III, iv: '*Fred.* How now? How goes it? *John.* Why, the man has his mare again, and all's well, Frederic.'

487. Another descriptive stage-direction, if such an expression be allowable, like 'shifting places,' above.—ED.

Actus Quartus. [*Scene I.*]

Enter Queene of Fairies, and Clowne, and Fairies, and the
    King behinde them.

Tita. Come, fit thee downe vpon this flowry bed,
While I thy amiable cheekes doe coy,   5
And fticke muske rofes in thy fleeke fmoothe head,
And kiffe thy faire large eares, my gentle ioy.

Clow. Where's *Peafe bloffome?*

Peaf. Ready.

Clow. Scratch my head, *Peafe-bloffome.* Wher's Moun-   10
fieuer *Cobweb.*

Cob. Ready.

Clowne. Mounfieur *Cobweb,* good Mounfier get your   13

---

1. A&us Quartus.] Om. Qq. Act IV,
Scene i. Rowe et seq. Act IV, Sc. ii.
Fleay.

[The Wood. Pope. The same. The
Lovers at a distance asleep. Cap.

2. and Clowne,] Bottom, Rowe et seq.
Fairies,] Faieries: Q₁.

2, 3. the King...] Oberon, behind,
unseen. Cap.

4. [seating him on a bank. Cap.

6. *fleeke fmoothe*] *sleek-smooth'd* Pope,
Han. *sleek, smooth'd* Theob. Warb.
Johns.

8, 10, &c. Clow.] Bot. Rowe.

10. *Mounfieuer*] *Mounfieur* QqFf,
Cap. White, Cam. Rolfe (throughout).
Monsieur Rowe et cet. (throughout).

13. *get your*] *get you your* Q₁, Sta.
Cam. White ii.

---

1. **Actus Quartus**] JOHNSON: I see no reason why the Fourth Act should begin
here, when there seems no interruption of the action. The division of acts seems to
have been arbitrarily made in F₁, and may therefore be altered at pleasure. [It is
precisely because there is so little 'interruption of the action' that it is necessary to
have an interruption of time, which this division supplies. At the close of the last
scene the stage is pitch-dark, doubly black through Puck's charms, and a change to
daylight is rendered less violent by a new Act. See *Preface*, p. xxxi.—ED.]

2, 8, 10, &c. **Clowne**] See FLEAY, V, i, 417.

5. **amiable**] W. A. WRIGHT: That is, lovely. Compare *Psalm* lxxxiv, 1: 'How
amiable are thy tabernacles.' And Milton, *Par. Lost*, iv, 250: 'Others whose fruit,
burnished with golden rind, Hung amiable.'

5. **coy**] STEEVENS: That is, to soothe, caress. So in Warner's *Albion's England*,
1602, vi, 30 [p. 148]: 'And whilst she coyes his sooty cheekes, or curles his sweaty
top.' Again, in Golding's *Ovid*, vii [p. 82, ed. 1567]: 'Their dangling Dewlaps
with his hand he coyd vnfearfully.'—W. A. WRIGHT: The verb is formed from the
adjective, which is itself derived from the French *coy* or *quoy*, the representative of
the Lat. *quietus*.

13. **Mounsieur**] CAMBRIDGE EDITORS: We have retained throughout this scene

weapons in your hand, & kill me a red hipt humble-Bee,
on the top of a thiſtle ; and good Mounſieur bring mee   15
the hony bag.   Doe not fret your ſelfe too much in the
action, Mounſieur; and good Mounſieur haue a care the
hony bag breake not, I would be loth to haue yon ouer-
flowne with a hony-bag ſigniour.   Where's Mounſieur
Muſtardſeed?   20
      Muſ.   Ready.
      Clo.   Giue me your neaſe, Mounſieur *Muſtardſeed.*
Pray you leaue your courteſie good Mounſieur.
      Muſ.   What's your will?
      Clo.   Nothing good Mounſieur, but to help Caualery   25
Cobweb to ſcratch.   I muſt to the Barbers Mounſieur, for
me-thinkes I am maruellous hairy about the face.   And I   27

---

18. *would*] *should* Pope ii, Theob.
Warb. Johns.
    loth] *loath* Q₁.
    yon] F₁.
    ouerflowne] *overflowed* Mal. '90
conj.
  22, 23. Prose, Q₁, Pope et seq.
  22. *your neaſe*] *thy neaſe* Pope, Theob.
Han. Warb.   *thy neife* Johns.
    neaſe] *newſe* F₂, Rowe ii.   *newſe*
F₃.   *news* F₄, Rowe i.

22. *Muſtardſeed*] *Muſtard*   F₃F₄,
Rowe i.
  23. *courteſie*] *curtſie* Q₁.   *curteſie* F₃F₄.
  25. *Caualery*] Qq, Coll. Hal. Dyce,
White, Cam. Ktly.   *Cavalero* Ff, Rowe
et cet.
  26. Cobweb] *Pease-blossom* Rann,
Hal. Dyce ii, iii.
  27. *maruellous*] *maruailes* Q₁.   *maruailous* Q₂.   *marvels* Cap.

---

the spelling of the old copies, as representing a pronunciation more appropriate to
Bottom, like 'Cavalery,' a few lines lower down.   We are aware, however, that the
word was generally so spelt.—ROLFE: It should be noted, however, that 'Monsieur,'
'Mounsieur,' 'Mounsier,' &c. are forms quite promiscuously used by the printers of
that time.   [Any indication whatever which tends to differentiate Bottom's pronunciation from Theseus's should be by all means retained.—ED.]

  22. **neaſe**] GREY: That is, fist.   So in *2 Hen. IV:* II, iv, 200: 'Sweet knight,
I kiss thy neif.'   [See Text. Notes for its evolution into *news*.—ED.]

  23. **courtesie**] SCHMIDT: That is, put on your hat.   Compare *Love's Lab. L.* V
i, 103: 'remember thy courtesy; I beseech thee, apparel thy head.'

  26. **Cobweb**] ANON. (ap. Grey, i, 64): Without doubt it should be Cavalero
Pease-blossom; as for Cavalero Cobweb, he had been just dispatched upon a perilous
adventure.—CAPELL: Unless you will solve it this way, that Cobweb laughs and goes
out, but joins the other in scratching; and this, indeed, is the likeliest, for Pease-blossom would stand but sorrily there.—HUDSON: Bottom is here in a strange predica-
ment, and has not had time to perfect himself in the nomenclature of his fairy attendants, and so he gets the names somewhat mixed.   Probably he is here addressing
Cavalery Pease-blossom, but gives him the wrong name.

  27. **marvellous**] See III, i, 4.

am fuch a tender Affe, if my haire do but tickle me, I muft   28
fcratch.

Tita.   What, wilt thou heare fome muficke, my fweet   30
loue.

Clow.   I haue a reafonable good eare in muficke.   Let
vs haue the tongs and the bones.

Muficke Tongs, Rurall Muficke.

Tita.   Or fay fweete Loue, what thou defireft to eat.   35

Clowne.   Truly a pecke of Prouender ; I could munch
your good dry Oates.   Me-thinkes I haue a great defire
to a bottle of hay : good hay, fweete hay hath no fellow.   39

28. *do but*] *doth but* Rowe ii+.
30. *fome*] *fome fome* Q₂.
32, 33. *Let vs*] *Let's* Q₁.
33. *tongs*] *tongues* F₂.   *tonges* F₃.
34. Muficke...] Ff.   Music. Tongs, ...
Pope+.   Rustic music. White i.   Rough
music. Dyce ii, Om. Qq et cet.

35. *defireft*] *desir'st* Rowe et seq. (except Cam.).
36. *Prouender*] *prouander* Q₁.
   could] *would* F₃F₄, Rowe i.
   munch] *mounch* Q₁.
38. *fweete hay*] *sweet hay*, Cap. et seq.
(except Dyce ii).

33. **tongs . . . bones**] COLLIER : Such music seems to have been played out of
sight, at this desire from Bottom.—PLANCHÉ (ap. Halliwell) : In the original sketches
of Inigo Jones, preserved in the library of the Duke of Devonshire, are two figures
illustrative of the rural music here alluded to.   'Knackers' is written by Inigo Jones
under the first figure, and 'Tonges and Key' under the second ; the 'knackers' were
usually made of bone or hard wood, and were played between the fingers, in the same
way as we still hear them every day among boys in the streets, and it is a very ancient
and popular kind of music ; the 'tongs' were struck by the 'key,' and in this way
the discordant sounds were produced that were so grateful to the ear of the entranced
Weaver.—STAUNTON : These instruments [mentioned by Planché] must be regarded
as the immediate precursors of the more musical marrow-bones and cleavers, the
introduction of which may, with great probability, be referred to the establishment of
Clare Market, in the middle of the seventeenth century ; since the butchers of that
place were particularly celebrated for their performances.   In Addison's description
of John Dentry's remarkable 'kitchen music' (*Spectator*, No. 570, 1714), the marrow-
bones and cleavers form no part of the Captain's harmonious apparatus, but the tongs
and key are represented to have become a little unfashionable some years before.   By
the year 1749, however, the former had obtained a considerable degree of vulgar
popularity, and were introduced in Bonnell Thornton's burlesque 'Ode on St. Cecil-
ia's Day, adapted to the Ancient British Musick.'   Ten years afterwards this poem
was recomposed by Dr Burney, and performed at Ranelagh, on which occasion
cleavers were cast in bell-metal to accompany the verses wherein they are mentioned.

34. **Musicke, &c.**] CAPELL : This scenical direction is certainly an interpolation
of the players, as no such direction appears in either Qto, and Titania's reply is a
clear exclusion of it.   [See Collier's suggestion noted above.]

38. **bottle**] HALLIWELL : A 'bottle of hay' was not a mere *bundle*, but some

Tita.   I haue a venturous Fairy,   40
That ſhall ſeeke the Squirrels hoard ,
And fetch thee new Nuts.   42

40, 41. One line, Q$_r$.   41. *Squirrels*] *Squirils* Q$_r$.
40, 42. Prose, Pope, Theob. Two lines,   42. *thee*] *thee thence* Han. Warb. Cap.
ending *ſeeke...nuts,* Han. et seq.   Rann, Dyce ii. iii.
40. *venturous*] *vent'rous* Cap.   *new*] *newest* Kinnear.

---

measure of that provender; by it, is now understood such a moderate bundle **as** may serve for one feed, twisted somewhat into the shape of a bottle, but in earlier times the bottles were of stated weights.   In a court-book, dated 1551, the half-penny bottle of hay is stated to weigh two pounds and a half, and the penny bottle five pounds. Cotgrave has ' Boteler, to botle or bundle up, to make into botles or bundles.'   To look for a needle in a bottle of hay is a common proverb, which occurs in Taylor's *Workes,* 1630, &c.

38. **bottle of hay**] HUNTER (i, 296): We have here an instance how imperfectly any printing can convey with fulness and precision all that a dramatist has written to be spoken on the stage.   Bottom, half man, half ass, is for a bottle of *a; hay,* or *ale,* for the actor was no doubt to speak in such a manner that both these words should be suggested.   The snatch of an old song that follows is in praise of *ale,* not ' hay.' Bottom sings, stirred to it by the rural music, the *rough music,* as it is called, which we learn from the Folio was introduced when Bottom had said ' Let us have the tongs and the bones !'   [It is to be feared that this a little too fine-spun.   First, it is extremely difficult to know when the dropping of the aspirate began to be the shibboleth of society; and secondly, I can find no trace of any song such as Hunter thinks that Bottom quotes; ' sweet' seems scarcely a fit adjective for ale.   That Bottom talks with the rudest intonation of the clowns of the day is likely.—ED.]

38. **good hay, &c.**] COLLIER: This is consistent with the notion that Bottom really partakes of the nature of the ass; not so his declaration,—I must to the barber's, &c.   He confuses his two conditions.—HALLIWELL: Bottom's desire for hay is, of course, involuntary, and has no connexion with any knowledge of his condition. It may be here remarked that it requires a close examination to enable us to reconcile the discourse of Bottom, in the present scene, with the conclusions that have been generally drawn from his language in the earlier part of the drama.   Here he is a clever humourist, and although, as throughout the play, exhibiting a consciousness of superiority, yet he is without his former absurdities.   Is it quite certain that his wrongly applied phrases in I, ii are not intended to proceed from his whimsical humour ? [See Puck's and Philostrate's description of Bottom and his fellows.—ED.]

40–42. As Titania always speaks rhythmically, these lines have proved obstinate in all endeavors to reduce them to rhythm.   The division into two lines, the first ending ' seeke,' was made by Hanmer, and he has been universally followed.   I think it not unlikely that some word has here been lost; experience has taught me that towards the foot of a column, where these present lines happen to be in the Folio, the compositors, for typographical reasons, were apt to lengthen or shorten lines, regardless of rhythm, and in this process phrases became sophisticated.   Hanmer divided the lines rightly, and I think that he was equally fortunate in supplying the word that had been probably omitted : ' The squirrel's hoard, and fetch thee *thence* new nuts.'   COLLIER supposed that *for* is the omitted word: ' and fetch *for* thee

Clown. I had rather haue a handfull or two of dried 43
peafe. But I pray you let none of your people ftirre me, I
haue an expofition of fleepe come vpon me. 45
    Tyta. Sleepe thou, and I will winde thee in my arms,
Fairies be gone, and be alwaies away.
So doth the woodbine, the fweet Honifuckle,
Gently entwift; the female Iuy fo
Enrings the barky fingers of the Elme. 50

43. *or two*] Om. Rowe i.
46. transposed to follow 47, Lettsom (ap. Dyce), Huds.
47. *alwaies*] Qq. *alwayes* F₂F₃. *always* F₄, Rowe, Pope. *a while* Han.

White, Coll. ii, iii (MS), Huds. *all ways* Theob. et cet.
49. *entwift; the female*] *entwift the Maple* Warb. Theob.
49, 50. *entwift; ...Enrings*] *entwist, ...Enring*, Han. Cap.

new nuts.' But to me the similarity between 'thee' and *thence* is the more likely source of the omission. WALKER (*Crit.* ii, 257) suggests that there has been an absorption of the definite article, the full text being 'fetch thee *the* new nuts.' But this is harsh to my ears. BULLOCH (p. 63) supposes that we have here only three-fourths of a stanza; he therefore supplies a rhyme to 'fairy' and a rhyme to 'hoard,' thus: 'And fetch the new nuts *wary To furnish forth thy board.*'—ABBOTT, § 484, says that either 'and' must be accented and 'hoard' prolonged, as Steevens asserted, or we must scan as follows: 'The squìr | rel's hòard, | and fètch | thee nèw | ˋ nùts.' I doubt if Titania's meaning demands such an emphasis on 'new,' and the prolongation of the word so as to supply the missing rhythm, which is what Abbott intends, gives a sound perilously similar to the characteristic cry of a cat. —ED.

46, 47. **Sleepe thou,** &c.] Dyce records a suggestion of Lettsom that these two lines should be transposed, which seems to me a needless change. Titania's 'Sleep thou' follows naturally after Bottom's wish, and line 47 might very well be printed in a parenthesis.—ED.

47. **alwaies**] THEOBALD, to whom we owe so much, here rightly divided this word into *all ways*, i. e. as he says, 'disperse yourselves, that danger approach us from no quarter.'—UPTON (241): 'Read "and be *away.—Away.*" [Seeing them loiter.']—HEATH (55): As the fairies here spoken to are evidently those whom the Queen had appointed to attend peculiarly on her paramour, I am inclined to think the true reading may be 'and be *always i' th'* way,' i. e. be still ready at a call.

48, 49. **woodbine, ... Gently entwist**] WARBURTON: What does the 'woodbine' entwist? The honeysuckle. But the woodbine and honeysuckle were, till now, but two names for one and the same plant. Florio interprets *Madre selva* by 'woodbinde or honniesuckle.' We must therefore find a support for the *woodbine* as well as for the *ivy*. Which is done by reading [line 49], 'Gently entwist the *Maple*, Ivy so,' &c. The corruption might happen by the first blunderer dropping the *p* in writing *maple*, which word thence became *male*. A following transcriber thought fit to change this *male* into *female*, and then tacked it as an epithet to Ivy.—UPTON (242): Read *wood rine*, i. e. the honey-suckle entwists the rind or bark of the trees:
12

[48, 49. woodbine, . . . entwist]

So doth the *wood rine* the sweet honey-suckle gently entwist.—JOHNSON: Shakespeare perhaps only meant, so the leaves involve the flower, using 'woodbine' for the plant, and 'honeysuckle' for the flower; or perhaps Shakespeare made a blunder.— STEEVENS: Baret, in his *Alvearie*, 1580, enforces the same distinction that Shake-speare thought it necessary [according to Johnson] to make: 'Woodbin that beareth the Honie-suckle.'—CAPELL, following Hanmer's text, which he says 'merits great commendation,' observes: 'honisuckle and woodbine are one, and "entwist" and "enring" are both predicated of the elm's "barky fingers."'—HEATH (55): A comma after 'entwist,' and another after 'enrings' will render any further change unnecessary. Thus:—'So the woodbine, the sweet honeysuckle doth gently entwist the barky fingers of the elm, so the female ivy enrings the same fingers.'—FARMER: It is certain that the 'woodbine' and the 'honey-suckle' were sometimes considered as different plants. In one of Taylor's poems, we have—'The woodbine, primrose, and the cowslip fine, The honisuckle, and the daffadill.'—STEEVENS: Were any change necessary I should not scruple to read the *weedbind*, i. e. smilax; a plant that twists round every other that grows in its way. In a very ancient translation of Macer's *Herball practised by Doctor Lynacre* is the following: 'Caprifolium is an herbe called woodbynde or withwynde, this groweth in hedges or in woodes, and it wyll beclyp a tre in her growynge, as doth yvye, and hath white flowers.'—GIF-FORD, in a note (referred to by Boswell) on '—— behold! How the blue bindweed doth itself infold With honey-suckle, and both these intwine Themselves with bryony and jessamine,' &c.—Jonson's *Vision of Delight—Works*, vii, 308, thus observes:— This settles the meaning of [Titania's speech]. The woodbine of Shakespeare is the blue bindweed of Jonson: in many of our counties the woodbine is still the name for the great convolvulus.—NARES: The 'blue bindweed' [of Jonson, *ut supra*] is the blue convolvulus (Gerard, 864), but the calling it 'woodbine' [in the present passage] has naturally puzzled both readers and commentators; as it seems to say that the honeysuckle entwines the honeysuckle. Supposing convolvulus to be meant all is easy, and a beautiful passage preserved. . . . The name *woodbine* has been applied to several climbing plants, and even to the ivy. In a word, if we would cor-rect the author himself, we should read: So doth the *bind-weed* the sweet honeysuckle gently entwist, &c. Otherwise we must so understand 'woodbine,' and be contented with it as a more poetical word than *bind-weed*, which probably was the feeling that occasioned it to be used.—HUNTER (i, 297): In fact woodbine and honeysuckle are but two names for one and the same plant, or, at most, the honeysuckle is but the flower of the woodbine. . . . The identity of the two is put beyond doubt by the fol-lowing passage in Googe's *Book of Husbandry:* 'The other, *the honeysuckle or the woodbine*, beginneth to flower in June.'—p. 180. All notion, therefore, of the wood-bine entwisting the honeysuckle is excluded. . . . It seems to me that the woodbine and the sweet honeysuckle are here in apposition.—R. G. WHITE (ed. i): There are few readers of Shakespeare, in America at least, who have not seen the woodbine and the honeysuckle growing together, and twining round each other from their very roots to the top of the veranda on which they are trained; and to such persons this passage is simple and plain. . . . [The flowers] of the honeysuckle are long unbroken tubes of deep scarlet, somewhat formally grouped; those of the woodbine shorter, deeply indented from the edge, of a pale buff colour, and irregularly disposed. [It is to be feared that few American readers will recognise these flowers from this description. I suppose that White refers to what is commonly called 'the coral hon-

O how I loue thee! how I dote on thee! 51

Enter Robin goodfellow and Oberon.

Ob. Welcome good *Robin*:

Seeſt thou this ſweet ſight?

Her dotage now I doe begin to pitty. 55

For meeting her of late behinde the wood,

Seeking ſweet ſauors for this hatefull foole, 57

---

51. [they sleep. Cap.

52. Enter...] Enter Puck. Rowe.
Oberon advances. Cap. ...
and Oberon.] Om. Qq.

53, 54. One line, Qq, Pope et seq.

57. *ſauors*] *fauours* Q₁, Rowe, Rann,
Hal. Dyce, White, Sta. Cam. *favors* F₄.

---

eysuckle,' to distinguish it from the 'trumpet honeysuckle,' or *tecoma;* and by woodbine he means the 'evergreen' variety. It is really, however, of small consequence,
as long as White makes it clear that he here discriminates between 'woodbine' and
'honeysuckle.'—ED.]—DYCE: My friend, the late Rev. John Mitford, an excellent
botanist, who at one time had maintained in print that Gifford's explanation of 'woodbine' was wrong, acknowledged at last that it was the only true one. (What an odd
notion of poetic composition must those interpreters have who maintain that here
woodbine and honeysuckle are put in apposition as meaning the same plant, and who,
of course, consider 'entwine' to be an intransitive verb!)—W. A. WRIGHT: The
word 'entwist' seems to describe the mutual action of two climbing plants, twining
about each other, and I therefore prefer to consider the woodbine and the honeysuckle
as distinct, the former being the convolvulus, rather than to adopt a construction and
interpretation which do violence to the reader's intelligence. [The question, reduced
to its simplest terms, is: Are there here two plants referred to, or only one? If there
are two plants, then either one or both of them bears a name which belonged to the
common speech of Shakespeare's day, and which we can now discover only by a
resort to literature, an unsure authority when it deals with the popular names of wild
flowers. To me it makes little difference what specific flower Titania calls the 'woodbine'; she means herself by it just as she designates the repulsive Bottom with two
fairies busy scratching his head, under the name of that sweet, lovely flower, the
honeysuckle; and as these two distinct vines entwist each other, so will she wind
him in her arms. As will be seen by the foregoing notes, the consensus of opinion
inclines to Gifford's interpretation of woodbine.—ED.]

49. **female**] STEEVENS: That is, because it always requires some support, which
is poetically called its husband. So Milton, *Par. Lost,* V, 215–217: 'they led the
vine To wed her elm; she spoused, about him twines Her marriageable arms.' So
Catullus, lxii, 54: 'Ulmo conjuncta marito.'

57. **savors**] STEEVENS: *Favours* of Q₁, taken in the sense of ornaments, such as
are worn at weddings, may be right.—DYCE (*Notes,* 62): I think *favours* decidedly
right. Titania was seeking flowers for Bottom to wear as *favours;* compare Greene:
'These [fair women] with syren-like allurement so entised these quaint squires, that
they bestowed all their *flowers* vpon them for *fauours.'—Quip for an Vpstart Courtier,* Sig. B 2, ed. 1620.—R. G. WHITE was at first (*Sh. Scholar,* 217) inclined to
think that 'savours' is the true word because Bottom expresses a wish for the 'sweet

I did vpbraid her, and fall out with her.	58
For fhe his hairy temples then had rounded,
With coronet of frefh and fragrant flowers.	60
And that fame dew which fomtime on the buds,
Was wont to fwell like round and orient pearles ;
Stood now within the pretty flouriets eyes,
Like teares that did their owne difgrace bewaile.
When I had at my pleafure taunted her,	65
And fhe in milde termes beg'd my patience,
I then did aske of her, her changeling childe,
Which ftraight fhe gaue me, and her Fairy fent
To beare him to my Bower in Fairy Land.
And now I haue the Boy, I will vndoe	70
This hatefull imperfection of her eyes.
And gentle *Pucke* , take this transformed fcalpe,
From off the head of this *Athenian* fwaine ;
That he awaking when the other doe,
May all to *Athens* backe againe repaire,	75
And thinke no more of this nights accidents,

61. *fomtime*] *sometimes* Johns.

63. *flouriets*] *flouret's* Johns. Mal. *flourets'* Steev.'93, Var. *flow'rets'* Knt et seq. (subs.).

68. *Fairy*] *fairies* Dyce, Ktly.

72. *transformed*] *transforming* D. Wilson.

73. *off*] *of* Q₁.

73. *this*] *the* Johns. Steev.'85, Rann.

74. *That he*] *That hee*, Q₁, Theob. Warb. Johns. Coll. Hal. Dyce. *That, he* Cam. White ii.

other] *others* Rowe+, Steev.'85, Mal.'90.

75. *May all*] *All may* Grey ap Cam.

savour' of a honey-bag, but he recanted in his subsequent edition, and decided that 'favours' is surely right, wherewith agrees the present ED.

60. **With**] ABBOTT, § 89, refers the omission of the definite article here to that class of cases where it is omitted before a noun already defined by another noun. It seems to me, however, that it is, possibly, a case of absorption in the *th* of 'With.' —ED.

62. **orient**] HALLIWELL: Sparkling, pellucid. Compare, 'His orient liquor in a crystal glass.'—*Comus* [65].—W. A. WRIGHT: Compare *Par. Lost*, i, 546: 'Ten thousand banners rise into the air, With orient colours waving.'

63. **flouriets**] CAPELL: *Flourets'* is recommended by [Heath, 56], and is indeed a word of more proper and more analogous formation; but the other ['flouriet'] was the word of the time, as this editor thinks, but has no examples at hand.

68. **Fairy**] DYCE here reads *fairies*. See II, i, 65.

74. **other**] For examples of 'other' used as a plural, see ABBOTT, § 12.

75. **May all**] ABBOTT, § 399: This might be explained by transposition, 'may all' for *all may*, but more probably *they* is implied.

But as the fierce vexation of a dreame.   77
But firſt I will releaſe the Fairy Queene.

> *Be thou as thou waſt wont to be;*
> *See as thou waſt wont to ſee.*   80
> *Dians bud, or Cupids flower,*
> *Hath ſuch force and bleſſed power.*

Now my *Titania* wake you my ſweet Queene.
　Tita. 　My *Oberon,* what viſions haue I ſeene!
Me-thought I was enamoured of an Aſſe.   85
　Ob. 　There lies your loue.
　Tita. How came theſe things to paſſe?
Oh, how mine eyes dɔth loath this viſage now!
　Ob. Silence a while. *Robin* take off his head:
Titania, muſick call, and ſtrike more dead   90

---

78. *releaſe*] *relaſe* F₄.
79–82. Roman, Q₁.
79. Be thou] *Be,* Qq, Pope et seq.
79, 80. waſt] *was* Knt.
79. [touching her Eyes with an herb. Cap.
81. bud, or] *bud o'er* Thirlby, Theob. et seq.

84, &c. Tita.] Queen. Rowe.
88. doth] *doe* Q₁Ff, Rowe et seq.
　loath this] *loathe this* Q₂. *loath his* Q₁, Cap. Mal. Var. Knt, Coll. Hal. Dyce Sta. Cam.
89. off his] *off this* Q₁, Cap. Steev. Mal. Var. Knt, Coll. Hal. Dyce, Sta. Cam. *of this* Q₂.

---

79. **Be thou**] R. G. WHITE (ed. i): In this 'thou' there is one of the instances in which it seems proper to allow strong probability and the authority of other editions to outweigh the dictum of the Folio. There is a change of rhythm for this little incantation, and that Shakespeare should have vitiated it in the very first line is improbable to the verge of impossibility; whereas the insertion of 'thou' in such a place by a transcriber or printer is an accident of a sort that frequently happens.

81. **Dians bud**] STEEVENS: This is the bud of the *Agnus Castus* or *Chaste Tree.* Thus, in Macer's *Herball,* 'The vertue of this herbe is, that he wyll kepe man and woman chaste,'—W. A. WRIGHT: It is more probably a produst of Shakespeare's imagination, which had already endued 'Cupid's flower,' the Heart's Ease, with qualities not recognised in botany.   [Was it the Heart's Ease in general which possessed these qualities, or only one particular 'little Western flower'?—ED.] Steevens's suggestion is, indeed, supported by Chaucer; see *The Flower and the Leaf,* 472–5: 'That is Diane, the goddesse of chastitie, And for because that she a maiden is, In her hond the braunch she beareth this, That *agnus castus* men call properly.'

81, 88. **or . . . loath this**] Here, within a few lines, we have two sophistications, which may be explained by the supposition that the compositors set up at dictation. —ED.

88. **this**] For other instances where *this* and *his* have supplanted one another, see WALKER (*Crit.* ii, 219, et seq.).   The same interchange seems to have taken place with 'his' in the next line. See 'his intelligence,' I, i, 262.

Then common ſleepe ; of all theſe, ſine the ſenſe.   91

 Tita. Muſicke, ho muſicke, ſuch as charmeth ſleepe.

           Muſick ſtill.   93

---

91. *common*] *cammon* F$_2$.

 ſleepe ; of all theſe, fine] *ſleep ; of all theſe find* F$_3$F$_4$. *sleep. Of all these find* Rowe i. *sleep of all these five* Thirlby, Theob. et seq.

92. *ho*] *howe* Q$_1$.

93. Muſick ſtill.] Om. Qq, Coll. Music, still. Cam. Still music. Theob. et cet.

---

91. **fine**] See Text. Note for the correction of the punctuation by THEOBALD, whose note is : This most certainly is both corrupt in the text and pointing. Would music, that was to strike them into a deeper sleep than ordinary, contribute to *fine* (or *refine*) their senses? My emendation [*five*] needs no justification. The 'five' that lay asleep were Demetrius, Lysander, Hermia, Helen, and Bottom. I ought to acknowledge that Dr Thirlby likewise started and communicated this very correction.— ANON. [ap. Halliwell]: The word 'fine' here signifies *mulctare*, and consequently Titania does the very thing Oberon desires. She *fines* or *deprives* them of their sense. —HALLIWELL : The last-quoted observations show how very difficult it is to establish the propriety of any emendation to the satisfaction of every mind. Bottom must be presumed to be at some little distance from the other sleepers, and concealed from the observation of Theseus and his train, but, on the whole, the correction [of Theobald] is to be preferred to the above subtle explanation of the original text.

93. **Musick still**] COLLIER (ed. i) : This means, probably, that the music was to cease before Puck spoke, as Oberon afterwards exclaims 'Sound music !' when it was to be renewed.—DYCE (*Remarks*, 48) : 'Music still' is nothing more than *Still music ;* compare a stage-direction in Beaumont and Fletcher's *Triumph of Time* (*Four Plays in One*), where, according to the old eds., the epithet applied to '*Trumpets*' is put last : '*Jupiter and Mercury descend severally.* Trumpets small *above.*' The music, instead of 'ceasing before Puck spoke,' was not intended to commence at all till Oberon had said 'Sound music !' The stage-direction here (as we frequently find in early eds. of plays) was placed prematurely, to warn the musicians to be in readiness.—COLLIER (ed. ii) : If, as Mr Dyce (*Remarks*, 48) suggests, 'still music' had been meant, the direction would not have been 'music still.' He evidently does not understand the force of the adverb; he mistakes it for the adjective, which occurs afterwards.—DYCE (ed. ii) : Yes, Mr Collier ventures so to write, *trusting that none of his readers will take the trouble to refer to my Remarks*, where I have quoted [a stage-direction] in which the epithet applied to '*Trumpets*' IS PUT LAST.—STAUN- TON : We apprehend that by 'Music still' or *Still music* was meant *soft, subdued music*, such music as Titania could command, 'as charmeth sleep'; the object of it being to 'strike more dead Than common sleep.' This being effected, Oberon himself calls for more stirring strains while he and the Queen take hands, 'And rock the ground whereon these sleepers be.'—DYCE (ed. ii) : I am glad to find that Mr Staunton agrees with me as to the meaning of the words '*Music still.*' I cannot, however, agree with him in the rest of his explanation. I believe that the music is not heard till Oberon echoes Titania's call for it; and that to the said *still* or *soft* music (the sole object of which is to lull the five sleepers) some sort of a *pas de deux* is danced by the fairy king and queen.

Rob.   When thou wak'ſt, with thine owne fooles eies

    peepe.   (me   95

   Ob.   Sound muſick; come my Queene, take hands with

And rocke the ground whereon theſe ſleepers be.

Now thou and I are new in amity,

And will to morrow midnight, ſolemnly

Dance in Duke *Theſeus* houſe triumphantly,   100

And bleſſe it to all faire poſterity.

There ſhall the paires of faithfull Louers be

Wedded, with *Theſeus*, all in iollity.

   Rob.   Faire King attend, and marke,

I doe heare the morning Larke.   105

   Ob.   Then my Queene in ſilence ſad,

Trip we after the nights ſhade ;   107

---

94. *When thou wak'ſt*] Q₂, Knt. *When thou awak'ſt* Ff, Rowe+, Steev.'85, Hal. *Now, when thou wak'ſt* Q₁ et cet.

96. *hands*] *hand* F₃F₄, Rowe+.

101. *faire*] *far* Han. Warb.

   poſterity] *proſperitie* Q₁, Cap. Mal. Var. Coll. Sing. Dyce i, Cam. Ktly.

102. *the*] *theſe* Ff, Rowe, Theob.+.

104. *Faire*] *Fair* F₃F₄, Rowe. *Fairy* Qq, Pope et seq.

106. *ſad,*] *fade ;* Theob. *ſtaid* Daniel.

107. *the nights*] Q₂Ff. *nights* Q₁. *nightès* Ktly. *night's* Cam. ii. *the night's* Rowe et cet.

---

98. **new**] W. A. WRIGHT: It is difficult to say whether 'new' is here an adjective or an adverb. Probably the latter, as in *Ham.* II, ii, 510, 'Aroused vengeance sets him new a-work.'

101. **faire posterity**] WARBURTON: We should read '*far* posterity,' *i. e.* to the *remotest* posterity.—HEATH (p. 56): That is, 'And bestow on it the blessing of a fair fortune to all posterity,' or, to come nearer the literal construction: 'And bless it so that the fortunes of all posterity who shall enjoy it may be fair.' Thus by this beautiful figure the two parts or branches of the blessing are united and consolidated into one expression: its extent, 'to all posterity'; and its object, 'that all that posterity may be fair,' that is, both deserving and fortunate.—MONK MASON: In the concluding song, where Oberon blesses the nuptial bed, part of his benediction is that the posterity of Theseus shall be fair. See V, i, 403.—MALONE preferred *prosperity*, induced thereto by II, i, 77.—R. G. WHITE (ed. i): *Prosperity* is a tame word here, especially as coming after 'fair.' [I prefer the present text. It involves a larger blessing. To Theseus's marriage the fairies bring present triumph, but on his house they confer the blessing of a fair posterity.—ED.]

106. **sad**] WARBURTON: This signifies only *grave, sober,* and is opposed to their dances and revels, which were now ended at the singing of the morning lark.— BLACKSTONE: A statute, 3 Henry VII, c. xiv, directs certain offences . . . 'to be tried by twelve sad men of the king's household.' [Theobald's emendation (see *Text. Notes*) was well meant, but it is not a success. The defective rhyme certainly exposes 'sad' to suspicion.—ED.]

107. **the nights**] KEIGHTLEY (p. 135): Of 'nights' I have made a disyllable [*nightès*], as being more Shakespearian than 'the night's,' which most feebly and

We the Globe can compaſſe ſoone,   108
Swifter then the wandring Moone.
   Tita. Come my Lord, and in our flight,   110
Tell me how it came this night,
That I ſleeping heere was found,
                             Sleepers Lye ſtill.
With theſe mortals on the ground.   *Exeunt.*
                        Winde Hornes.   115
    Enter Theſeus, Egeus, Hippolita and all his traine.
   Theſ. Goe one of you, finde out the Forreſter,
For now our obſeruation is perform'd;
And ſince we haue the vaward of the day,
My Loue ſhall heare the muſicke of my hounds.   120
Vncouple in the Weſterne valley, let them goe;

113. Sleepers Lye ſtill.] Om. Qq, Cap. et seq.
115. Winde Hornes.] Horns wind within. Cap. Horns winded within. Dyce. Scene II. Pope+. Act V, Sc. i. Fleay.
116. Egeus, Hippolita] Om. Qq. Egæus, Hippolita Ff.
121. *Vncouple*] *Uncoupl'd* Rann. conj. *Weſterne*] Om. Marshall. *let them*] Om. Pope+, Cap. Steev. Mal. Var. Knt, Dyce ii, iii.

---

inharmoniously throws the emphasis on 'the.' This genitive occurs more than once in our poet's earlier plays.—W. A. WRIGHT: 'Night's' is a disyllable, as 'moon's,' in II, i, 7, and 'earth's,' in *Temp.* IV, i, 110: 'Earth's increase, foison plenty.' [If the pause in these lines be observed, there will be, I think, no need of any barrel-organ regularity. 'Then my queen ‖ in silence sad, Trip we after ‖ the night's shade; We the Globe ‖ can compass soon, Swifter than ‖ the wandring moon.' As far as 'the night's shade' is concerned, the necessity of making 'night's' a disyllable is removed by the slight pause which we are forced to make between 'night's' and 'shade,' to avoid the conversion of the two words into one: *nightshade.*—ED.]

115. **Winde Hornes**] Again the mandatory direction of a stage-copy.—ED.

117. **Forrester**] KNIGHT calls attention to the fact that the Theseus of Chaucer was also a mighty hunter. The extract from Chaucer may be found in the Appendix, on the *Source of the Plot.*

118. **obseruation**] Of the rites of May, see 'obseruance for a morne of May,' I, i, 177.

119. **vaward**] DYCE: The forepart (properly of an army, 'The Vaward, *Prima acies.*'—Coles's *Lat. and. Eng. Dict.*).

120–140. HAZLITT (*Characters*, &c., p. 132): Even Titian never made a hunting-piece of a *gusto* so fresh and lusty, and so near the first ages of the world, as this.

121. **Vncouple, &c.**] CAPELL: Might not the author's copy run thus: 'Let them uncouple in the western valley; | Go; Dispatch, I say, and find the forrester.' | ? where 'Go' is no part of the verse, but a redundance, like 'Do' in this line in *Lear:* Do; kill thy physician and the fee bestow,' &c.

Difpatch I fay, and finde the Forrefter. 122

We will faire Queene, vp to the Mountaines top,

And marke the muficall confufion

Of hounds and eccho in coniunction. 125

   Hip. I was with *Hercules* and *Cadmus* once,

When in a wood of *Creete* they bayed the Beare 127

---

122. [Exit an Attend. Dyce.   127. *Beare*] *boar* Han. Cap. Dyce ii,

127. *bayed*] *bay'd* Rowe et seq. *chas'd*   iii, Coll. iii.

Rann. conj. (?)

---

126. **Hercules**] THEOBALD (Nichols, *Illust.* ii, 235): Does not the poet forget the truth of fable a little here? Hippolyta was just brought into the country of the Amazons by Theseus, and how could she have been in Crete with Hercules and Cadmus?

127. **Beare**] THEOBALD (Nichols, *Illust.* ii, 235): Should it not be *Boar?* The Erymanthian Boar, you know, is famous among the Herculean Labours.—CAPELL: The 'bear' is no animal of such a warm country as Crete; and, besides, in penning this passage the poet appears evidently to have had in his eye the *boar* of Theooaly, and to have picked up some ideas from the famous description of that hunting.—STEEVENS refers to the painting, in the temple of Mars, of 'The hunte strangled with the wilde beres,' Chaucer, *Knightes Tale*, line 1160, ed. Morris, and observes: Bear-baiting was likewise once a diversion esteemed proper for royal personages, even of the softer sex. While the princess Elizabeth remained at Hatfield House, under the custody of Sir Thomas Pope, she was visited by Queen Mary. The next morning they were entertained with a grand exhibition of bear-baiting, 'with which their high-nesses were right well content.'—*Life of Sir Thomas Pope*, cited by Warton, *Hist. Eng. Poetry*, ii, 391.—MALONE: In *The Winter's Tale* Antigonus is destroyed by a bear, who is chased by hunters. See also *Venus and Adonis*, 883: 'For now she knows it is no gentle chase, But the blunt boar, rough bear, or lion proud.'—TOLLET: Holinshed, with whose histories our poet was well acquainted, says: 'the beare is a beast commonlie hunted in the East countrie.' Pliny, Plutarch, &c. mention *bear-hunting.* Turberville, in his *Book of Hunting*, has two chapters on hunting the *bear.* —DYCE (*Remarks*, 49): In spite of what the commentators say [as just quoted], I am strongly inclined to think that 'bear' is a misprint for *boar.*—WALKER (*Crit.* iii, 50): Dyce's conjecture, *boar* (or is he referring to another critic who proposed it?), deserves attention. The story of Meleager would be sufficient to suggest it to Shakespeare.—R. G. WHITE (ed. I): Passages in Chaucer's *Knightes Tale*, Holinshed's *Chronicles*, Pliny, and Plutarch so justify 'bear' that it must remain undisturbed, but I believe that the easiest of all misprints in Shakespeare's time was made, and that we should read *boar.* This is also Mr Dyce's opinion.—DYCE (ed. ii), after quoting the notes of Walker and R. G. White, just given, adds: The 'passages' above mentioned formerly weighed little with me; now they weigh nothing.—W. A. WRIGHT: The references to 'bear' and 'bear-hunting' in Shakespeare are sufficiently numerous to justify the old reading, without going into the naturalist's question whether there are bears in Crete. Besides, according to Pliny (viii, 83), there were neither bears nor boars in the island. We may therefore leave the natural history to adjust itself, as well as the chronology which brings Cadmus with Hercules and Hippolyta into the hunting-field together.

With hounds of *Sparta* ; neuer did I heare   128
Such gallant chiding.   For befides the groues,
The skies, the fountaines, euery region neere,   130
Seeme all one mutuall cry.   I neuer heard
So muficall a difcord, fuch fweet thunder.

 Thef. My hounds are bred out of the *Spartan* kinde,
So flew'd, fo fanded, and their heads are hung   134

---

130. *fountaines*] *mountains* Anon. ap.   131. *Seeme*] *Seem'd* Ff, Rowe et seq.
Theob.

---

128. **Sparta**] W. A. WRIGHT: The Spartan hounds were celebrated for their
swiftness and quickness of scent. Compare Virgil, *Georgics*, iii, 405: 'Veloces
Spartæ catulos acremque Molossum Pasce sero pingui.'—HALLIWELL: See 'This
latter was a hounde of Crete, the other was of Spart,' in the description of Actæon's
dogs in Golding's *Ovid* [fol. 33, ed. 1567].

129. **chiding**] STEEVENS: 'Chiding' in this instance, means only *sound*. So in
Hen. VIII: III, ii, 197: 'As doth a rock against the chiding flood.'

130. **fountaines**] THEOBALD: It has been proposed to me that the author probably wrote *mountains*, from whence an echo rather proceeds than from 'fountains,'
but we have the authority of the ancients for Lakes, Rivers, and Fountains returning
a sound. See Virgil, *Æneid*, xii, 756: 'Tum vero exoritur clamor; ripaeque lacus-
que Responsant circa, et coelum tonat omne tumultu.' Propertius, *Eleg.* I, xx, 49:
'Cui procul Alcides iterat responsa; sed illi Nomen ab extremis fontibus aura refert.'
—DYCE (ed. ii) quotes the foregoing lines from Virgil, and adds, in effect, that after
all he is 'by no means sure that our author did not write *mountains*.'

131. **Seeme**] One of the many examples collected by WALKER (*Crit.* ii, 61)
where final *d* and final *e* are confounded in the Folio, 'arising in some instances, perhaps, from the juxtaposition of *d* and *e* in the compositor's case, but far oftener—as is
evident from the frequency of the erratum—from something in the old method of
writing the final *e* or *d*, and which those who are versed in Elizabethan MSS may
perhaps be able to explain.' In a footnote Walker's editor, LETTSOM, says: 'Walker's
sagacity, in default of positive knowledge, has led him to the truth. The *e*, with the
last upstroke prolonged and terminated in a loop, might easily be taken for *d*. It is
frequently found so written.'

133. **My hounds**] BAYNES (*Edin. Rev.* Oct. 1872): Shakespeare might probably
enough, as the commentators suggest, have derived his knowledge of Cretan and
Spartan hounds from Golding's *Ovid*. . . . But in enumerating the points of the slow,
sure, deep-mouthed hound it can hardly be doubted he had in view the celebrated
Talbot breed nearer home.

134. **flew'd**] WARTON: Hanmer justly remarks that 'flews' are the large chaps
of a deep-mouthed hound. See Golding's *Ovid*, iii [fol. 33, b. 1567]: 'And shaggie
Rugge with other twaine that had a Syre of Crete, And Dam of Sparta: Tone of
them callde Iollyboy, a great And large flewd hound.'

134. **sanded**] JOHNSON: So marked with small spots.—STEEVENS: It means of a
sandy colour, which is one of the true developments of a blood-hound.—COLLIER
(ed. i): This may refer to the sandy marks on the dogs, or possibly it is a misprint
for *sounded*, in allusion to their mouths. [This conjecture is omitted in Collier's ed.

With eares that fweepe away the morning dew,   135
Crooke kneed, and dew-lapt, like *Theffalian* Buls,
Slow in purfuit, but match'd in mouth like bels,
Each vnder each.   A cry more tuneable
Was neuer hallowed to, nor cheer'd with horne,
In *Creete*, in *Sparta*, nor in *Theffaly*;   140
Iudge when you heare.   But foft, what nimphs are thefe?
  Egeus. My Lord, this is my daughter heere afleepe,
And this *Lyfander*, this *Demetrius* is,
This *Helena*, olde *Nedars Helena*,   144

---

136. Theffalian] Theffalonian F₄.   142, 150, &c. Egeus.] Egæ. Ff (through-
  139. *hallowed*] hollawed F₂F₃. *hol-*   out).
lowd Qq. *hallow'd* Rowe. *hallo'd* Theob.   142. *this is*] this Q₁.
halloo'd Cap.   *holla'd* Mal.

---

ii, but it reappears in ed. iii.   In the mean time Dyce (*Remarks*, 49) had asked: 'Did
Mr Collier really believe that *sounded* could be used in the sense of "having, or giving forth, a sound"?   Besides, the earlier portion of this speech is entirely occupied
by a description of the *appearance and make* of the hounds ("sanded" denoting
their general colour); in a later part of it, Theseus describes their *cry*—"match'd
in mouth like bells."'

  137. like bels] BAYNES (*Edin. Rev.* Oct. 1872): It is clear that in Shakespeare's
day the greatest attention was paid to the musical quality of the cry.   It was a ruling
consideration in the formation of a pack that it should possess the musical fulness and
strength of a perfect canine quire.   And hounds of good voice were selected and
arranged in the hunting chorus on the same general principles that govern the formation of a cathedral or any other more articulate choir.   Thus: 'If you would have
your kennell for sweetnesse of cry, then you must compound it of some large dogges,
that have deepe solemne mouthes, and are swift in spending, which must, as it were,
beare the base in the consort, then a double number of roaring, and loud ringing
mouthes, which must beare the counter tenour, then some hollow, plaine, sweete
mouthes, which must beare the meane or middle part; and soe with these three parts
of musicke you shall make your cry perfect.'—[Markham's *Country Contentments*,
p. 6, W. A. Wright.   Down even to the days of Addison, and it may be down even
to this day, for aught I know, this tuneableness was sought after in a pack of hounds.
We all remember good old Sir Roger de Coverley's pack of *Stop-hounds*: 'what
these want in Speed, he endeavours to make amends for by the Deepness of their
mouths and the Variety of their notes, which are suited in such manner to each other,
that the whole cry makes up a complete consort.   He is so nice in this particular that
a gentleman having made him a present of a very fine hound the other day, the
Knight returned it by the Servant, with a great many expressions of civility, but
desired him to tell his Master that the dog he had sent was indeed a most excellent
Bass, but that at present he only wanted a *Counter-Tenor*.   Could I believe my friend
had ever read Shakespeare, I should certainly conclude he had taken the hint from
Theseus in the "Midsummer Night's Dream."'—ED.]

I wonder of this being heere together. 145

The. No doubt they rofe vp early, to obferue
The right of May; and hearing our intent,
Came heere in grace of our folemnity.
But fpeake *Egeus,* is not this the day
That *Hermia* fhould giue anfwer of her choice? 150
Egeus. It is, my Lord.
Thef. Goe bid the huntf-men wake them with their
hornes.

Hornes and they wake.
Shout within, they all ftart vp. 155
Thef. Good morrow friends : Saint *Valentine* is paft,
Begin thefe wood birds but to couple now?
Lyf. Pardon my Lord.
Thef. I pray you all ftand vp.
I know you two are Riuall enemies. 160
How comes this gentle concord in the world,
That hatred is is fo farre from iealoufie,
To fleepe by hate, and feare no enmity. 163

---

145. *of this*] Q₂Ff, Rowe i. *at their* Pope+, Cap. Steev.'85. *of their* Q₁ et cet.

147. *right*] *Rite* Pope et seq. (subs.).
148. *grace*] *gracc* F₄.

154, 155. Shoute within : they all starte vp. Winde hornes. Qq.
158. [He, and the rest, kneel to Theseus. Cap.
162. *is is*] F₁.

---

145. *of*] See ''Twere pity of my life,' III, i, 42, and ABBOTT, § 174, for many other examples of this usage, where we should now use a different preposition. See too, five lines lower down, 'answer of her choice.'

147. *right*] From the apparent confusion in the spelling of the words 'right' and 'rite,' we are hardly justified, I think, in imputing ignorance to the compositors. They spelled for the ear (and probably by the ear), and not, as we spell, for the eye.—ED.

150. *That*] For other examples where 'that' is equivalent to *at which time, when,* see ABBOTT, § 284; also V, i, 373 : '*That* the graves,' &c.

156. *Valentine*] STEEVENS : Alluding to the old saying that birds begin to couple on St Valentine's day. [Shakespeare knew quite as well as we know that Theseus lived long before St Valentine. But what mattered it to him, any more than it matters to us?—ED.]

158. CAPELL here added a very superfluous stage-direction, which few editors after him have had the courage to reject. Whoever is so dull as not to see the meaning in Theseus's '1 pray you all stand up,' had better close his Shakespeare and read no more that day—nor any other day. Why did not Capell further instruct us by adding *Theseus looks at them ?*—ED.

162, 163. *so farre . . . To*] For other examples of the omission of *as* after *so,* see ABBOTT, § 281.

Lyf. My Lord, I ſhall reply amaſedly,
Halfe ſleepe, halfe waking. But as yet, I ſweare, 165
I cannot truly ſay how I came heere.
But as I thinke (for truly would I ſpeake)
And now I doe bethinke me, ſo it is ;
I came with *Hermia* hither.   Our intent
Was to be gone from *Athens,* where we might be 170
Without the perill of the *Athenian* Law.

165. *ſleepe*] *'sleep* Cap. Steev. Mal.'90,
Knt, Sing. Hal. Sta. Ktly.
167, 168. (*for...is*] In parenthesis,
Cap. et seq. (subs.).
168. *I doe*] *do I* Glo. (misprint ?).
bethinke] *methink* Pope, Han.
170, 171. *Athens, where we might be
Without...Law*] Q₁Ff, Rowe+, Cap.

Steev. Mal. Knt, Hal. Sta. *Athens.
where we might Without...lawe*, Q₂.
*Athens, where we might Be without
peril...law.* Han. *Athens, where we
might Without...law*— Coll. Sing. White
i, Ktly. *Athens, where we might, Without...law,*— Dyce, White ii. *Athens,
where we might. Without...law.* Cam.

---

165. **Halfe sleep, halfe waking**] W. A. WRIGHT: Some editors regard 'sleep'
and 'waking' as adjectives, and print the former *'sleep.* Schmidt (*Lex.* p. 1419 *a*)
gives this as an instance of the same termination applying to two words, so that 'sleep
and waking' are equivalent to *sleeping and waking.* He quotes, as a possibly parallel case, *Tro. & Cres.* V, viii, 7 : 'Even with the vail and darking of the sun.' In
this case, however, 'vail' may be a substantive formed from a verb, of which there
are many instances in Shakespeare. I am inclined to think that both 'sleep' and
'waking' are here substantives, and are loosely connected with the verb 'reply'; just
as we find in *Merry Wives,* III, ii, 69 : 'He speaks holiday'; *Twelfth Night,* I, v,
115 : 'He speaks nothing but madman'; *King John,* II, i, 462 : 'He speaks plain
cannon-fire'; and as the Ff read in *As You Like It,* III, ii, 226 : 'Speak sad brow
and true maid.' [When Schmidt, in the note just cited by Wright, says of the example from *Tro. & Cres.,* 'It would not, therefore, be safe to infer the existence [here]
of a substantive *vail,*' it seems to me that he considers the passage as more than 'a
possibly parallel case.' I quite agree with Wright in his explanation, not only of the
present line, but also of the line from *Tro. & Cres.,* and I would further extend the
criticism to almost all the examples collected by Schmidt in his section on 'Suffixes
and Prefixes Omitted.'—ED.]

170, 171. **Athens, where . . . Law.**] COLLIER: The reading of Q₁ is beyond
dispute correct [viz. a comma after 'Law,' which Collier holds to be equivalent to
his dash], Lysander being interrupted by the impatience of Egeus, with 'Enough,
enough !'—DYCE (ed. ii) : Q₂ and the Ff complete the sentence very awkwardly by
adding 'be' to the reading of Q₁. Perhaps Hanmer was right in his text.—R. G.
WHITE (ed. i) : The 'be' is fatal to the rhythm of the line, and not only so, but to
the sense of the passage. For, as others have remarked, it is plain that Egeus inter-
rupts Lysander with great impetuosity; and, beside, he adds the explanation, 'They
would have stolen away,' &c., which would have been entirely superfluous had Lysander completed the expression of his intent.—STAUNTON: 'Without the peril' is
'*beyond* the peril,' &c. 'Without,' in this sense, occurs repeatedly in Shakespeare
and the books of his age. There is a memorable instance of it in *The Temp.* V, i.

Ege.   Enough, enough, my Lord : you haue enough;   172
I beg the Law, the Law, vpon his head :
They would haue ftolne away, they would *Demetrius,*
Thereby to haue defeated you and me :   175
You of your wife, and me of my confent ;
Of my confent, that fhe fhould be your wife.
 Dem.   My Lord, faire *Helen* told me of their ftealth,
Of this their purpofe hither, to this wood,
And I in furie hither followed them ;   180
Faire *Helena,* in fancy followed me.
But my good Lord, I wot not by what power,
(But by fome power it is) my loue
To *Hermia* (melted as the fnow)
Seems to me now as the remembrance of an idle gaude,   185
Which in my childehood I did doat vpon :

---

179. *this wood*] *the wood* Rowe.
180. *followed*] *follow'd* Rowe et seq.
181. *followed*] Q₂Ff, Rowe, Pope, Han. White i. *following* Q₁, Theob. et cet.

183-185. (*But...gaude*] Lines end, *Hermia...now...gaude* Pope et seq.
184. *melted as*] *Is melted as* Pope+.

Melted as doth Cap. Mal. Steev.'93, Knt, White, Hal. Coll. iii. *Melted as is* Steev. '85, Rann. *Melted as melts* Dyce ii, iii, Huds. *Melted e'en as* Ktly. *All melted as* Sta. conj. *Immaculate as* Bulloch. *Melted as thaws* Kinnear. *So melted as* or *Being melted as* Schmidt.
186. *doat*] *dote* Qq.

---

271 : 'a witch . . . That could control the moon . . . And deal in her command without her power.'   Here 'without her power' means *beyond* her power or *sphere,* as I am strongly inclined to think the poet wrote.   Thus, too, in Jonson's *Cynthia's Revels,* I, iv : 'now I apprehend you; your phrase was *Without* me before.'—W. A. WRIGHT : We cannot lay much stress on the comma at 'law' in Q₁.   'Where we might' is simply *wheresoever we might.*   [Unquestionably Staunton's interpretation of 'without' is correct; it is used locatively, in the same way, in I, ii, 97.   I prefer to retain the 'be,' notwithstanding its rhythmical superfluity.—ED.]

181. **fancy**] That is, love.

182. **wot**] W. A. WRIGHT : This is properly a preterite (A.-S. *wát,* from *witan,* to know), and is used as a present, just as *olða* and *novi.*   And not only is it used as a present in sense, but it is inflected like a present tense, for we find the third person singular 'wots' or 'wotteth.'

184. **melted**] The irregularity of the lines possibly indicates an obscurity in the MS.   Some monosyllable has been lost, and the Text. Notes show the editorial gropings for it.   Of Capell's *loth,* R. G. White says that the line is prose without it, and Staunton says it is ungrammatical with it.   Abbott, § 486, suggests that perhaps 'melted' was prolonged in pronunciation, which is doubtful, I think, because meaningless.   I prefer Dyce's 'Melted as *melts,*' it is smooth, and the iteration may possibly have led to the sophistication.—ED.

185. **gaude**] See I, i, 41.

And all the faith, the vertue of my heart,   187
The object and the pleasure of mine eye,
Is onely *Helena*.   To her, my Lord,
Was I betroth'd, ere I see *Hermia*,   190
But like a sickenesse did I loath this food,

190. betroth'd] betrothed Q₁, Rowe+, Cap. see Hermia] QqFf. *did see Hermia* Rowe i, Cap. Mal.'90. *Hermia saw* Rowe ii+. *saw Hermia* Steev. et cet.

191. *But like a*] *But like in* Steev.'93 et seq. (except Sta.). *Belike as* Bulloch. *When, like in* Kinnear.

---

190. **see**] HENRY JOHNSON (p. xv): 'See' for *saw* occurs very commonly in dialect usage in Maine, and presumably in Northern New England generally, 'Soons he see me cummin, he run.'

191. **like a sickness**] 'A sickness,' says Capell, means 'a sick thing or one sick; a common metonymy of the abstract for the concrete.'—STEEVENS changed the phrase from a preposition to a conjunction, and read 'like *in* sickness,' and owed the correction, as he said, to Dr Farmer; but HALLIWELL quotes a passage from *The Student* Oxford, 1750, where this same correction is made on the ground that 'it is little better than nonsense to make Demetrius say that he loathed the food like as he loathed a sickness.'—W. A. WRIGHT adopts Farmer's correction, but says he is 'not satisfied' with it, and the repetition of 'But,' he continues, 'inclines me to suspect that there is a further corruption.'   [I agree with Wright in thinking that there is corruption here, and that it lies in the repetition of 'But.'   That there was *a* repetition seems to me not unlikely, but it originally lay in a repetition of 'Now.'   Lettsom (*Walker's Crit.* ii, 115) supposes that the former 'But' has intruded into the place of *Then*.   I suppose that the latter 'But' has intruded into the place of 'Now.'   The strong contrast between his former and his present state, which Demetrius emphasises, warrants the repetition: '*Now*, as in health, come to my natural taste, *Now* do I,' &c.   As for Farmer's change, it is as harmless as it is needless.   I see no nonsense in saying that a man loathes a sickness.   We all do.   Had the word been *poison*, we should have been spared all notes.   Farmer's change, however, serves to show us how little repugnance there was, to cultivated ears of that day, to the use of 'like' as a conjunction. In this connection see a valuable article by WALKER (*Crit.* ii, 115), where many instances are given of the use of 'like' in 'the sense of *as*—perhaps for *like as*, as *where* for *whereas*; *when* for *whenas*.'   The present passage heads the list, with Steevens's text, 'like *in* sickness,' which apparently both Walker and his editor, Lettsom, assumed to be the original reading.   See, too, as supplementary to this article, *The Nation*, New York, 4 Aug. 1892, where Dr F. HALL, of great authority in English, has given many additional examples, and whose conclusion is as follows: 'The antiquity [of the conjunction *like*] proves to be very considerable; few good writers have ever lent it their sanction; at one stage of its history it was confined mostly to poetry, and its repute, as literary or formal English, is now but indifferent.   Yet, as a colloquialism, it is in our day, here in England, widely current in all ranks of society, from the highest to the lowest. . . . Against no one, therefore, can the charge be brought, otherwise than arbitrarily, of committing an absolute and indefensible solecism, if he chooses, in his talk, to say, for instance, "I think *like* you do."'—ED.]

But as in health, come to my naturall taſte,   192
Now doe I wiſh it, loue it, long for it ,
And will for euermore be true to it.

   Theſ.   Faire Louers, you are fortunately met ;   195
Of this diſcourſe we ſhall heare more anon.
Egeus, I will ouer-beare your will ;
For in the Temple, by and by with vs,
Theſe couples ſhall eternally be knit.
And for the morning now is ſomething worne,   200
Our purpos'd hunting ſhall be ſet aſide.
Away, with vs to *Athens* ; three and three,
Wee'll hold a feaſt in great ſolemnitie.
Come *Hippolitæ*.   *Exit Duke and Lords.*

   Dem.   Theſe things ſeeme ſmall & vndiſtinguiſhable,   205
Like farre off mountaines turned into Clouds.

   Her.   Me-thinks I ſee theſe things with parted eye,
When euery things ſeemes double.

   Hel.   So me-thinkes :
And I haue found *Demetrius*, like a iewell,   210

---

192. *But*] *Yet* Han.

193. *doe I*] *I doe* Q₁, Cam. White ii.

196. *we ſhall heare more*] *we more will here* Q₁, Steev.'93, Var. Coll. Sing. Dyce, Hal, Sta. Cam. Ktly, White ii (all reading *hear*). *we will heare more* Q₂, Cap. Mal. Knt.

203, 204. *Wee'll...Hippolitæ*] One line, Qq.

204. *Come*] *Come, my* Han. Cap. Rann, Dyce ii, iii, Huds.

204, 216. *Hippolitæ*] Q₂.

204. *Exit...Lords.*] Om. Q₁.   Exit. Q₂.

210. *found*] *fonnd* Q₁.

   like] Om. Han.

   iewell] *Gemell* Theob. Warb. Cap. *gimmal* Anon. (ap. Sing. i).

---

196. **we shall heare more**] WALKER (*Crit.* iii, 50) : I *somewhat* suspect the inversion [of Q₁].—LETTSOM (in a foot-note to this) : Here we have three authorities [F₁, Q₁ and Q₂] at variance, and who knows but Shakespeare wrote *more will we hear*?

205. **Dem.**] CAPELL (114, *b*) queries if this speech should not be given to Lysander, but gives no reason. Probably, however, for the sake of a more even distri bution of speeches.—ED.

207, 209. **Me-thinkes**] WALKER (*Vers.* 279) is undoubtedly right in surmising that in both these instances the accent is on ' Me.'

207. **parted eye**] DEIGHTON : As one would if one's eyes were not in focus with each other.

208. **things seemes**] The *s* in ' things' probably comes under Walker's rule (given at length at I, i, 239) of an interpolated *s*, but it is possible that the ear of the compositor was deceived by the *s* immediately following in ' seemes.'—ED.

210. **iewell**] WARBURTON : Hermia had observed that things appeared *double* to her. Helena replies, *so, methinks ;* and then subjoins that Demetrius was like a *jewel,*

[210. iewell]

ner own and not her own. He is here, then, compared to something which had the property of appearing to be one thing when it was another. Not the property sure of a jewel; or, if you will, of none but a false one. We should read 'like a *gemell*.' From *gemellus*, a twin. For Demetrius had that night acted two such different parts that she could hardly think them both played by one and the same Demetrius, but that there were twin Demetriuses, like the two Sosias in the farce.—THEOBALD inconti-nently adopted this emendation in his text, and observed in a note: 'If some over-nice spirits should object to *Gemell* wanting its authorities as an English word, I think fit to observe, in aid of my friend's fine conjecture, that it is no new thing with Shake-speare to coin and enfranchise words fairly derived.' Furthermore, Theobald finds both in Blount's *Glossographia* and in Philips's *World of Words*, 'Geminels,' *i. e.* Twins; and lastly, that there are 'other passages where Shakespeare uses the same manner of thought,' namely, in the case of twins in the *Comedy of Err.* and in *Twelfth Night.*—CAPELL, the editor to whom of all others we are most indebted for the text of to-day, was beguiled by the glitter of Warburton's tinsel, and also adopted it, and not only finds Warburton's reasons satisfactory in them-selves, but 'that there is in *gemell* a pleasantry, and in 'jewel' a vulgarity, that is a further recommendation of *gemell*.' The pleasantry arises, he says, 'from Helena's being now in good spirits, and able to treat her lover in the vein of her sister Hermia, her friendship's sister.'—JOHNSON: This emendation is ingenious enough to be true.—HEATH (p. 57), after denouncing the emendation as neither English nor French, gives his own paraphrase of the passage, but is not as successful therein as were Ritson and Malone subsequently. 'I have found Demetrius,' thus paraphrases Heath, 'but I feel myself in the same situation as one who, after having long lost a most valuable jewel, recovers it at last, when he least hoped to do so. The joy of this recovery, succeeding the despair of ever finding it, together with the strange circumstances which restored it to his hands, make him even doubt whether it be his own or not. He can scarcely be persuaded to believe his good fortune.' In support of Warburton's *gemell*, FARMER and STEEVENS both cite examples of its use in Drayton's *Barons Wars.*—RITSON (*Remarks*, p. 46): The learned critic [War-burton] wilfully misstates Helena's words to found his *ingenious emendation* (as every foolish and impertinent proposal is, by the courtesy of editors, intitled); she says that she has found Demetrius as a person finds a jewel or thing of great value, in which his property is so precarious as to make it uncertain whether it belongs to him or not. —MALONE: Helena, I think, means to say that having *found* Demetrius *unexpectedly*, she considered her property in him as insecure as that which a person has in a jewel that he has *found* by *accident*: which he knows not whether he shall retain, and which, therefore, may properly enough be called *his own and not his own*. She does not say, as Warburton has represented, that Demetrius *was like a jewel*, but that she had *found him* like a jewel, &c. [This explanation is to me entirely satisfactory. Of recent editors, STAUNTON has a good word for *gemell*, which, he says, 'is prefer-able to any explanation yet given of the text as it stands.']—C. BATTEN (*The Acad-emy*, 1 June, '76) suggests *double*, which 'in the jewellery trade means "a counterfeit stone composed of two pieces of crystal, with a piece of foil between them, so that they have the same appearance as if the whole substance of the crystal were col-oured." Of course the use of the word in this sense would require the knowledge of an expert, and this Shakespeare had, as is evident from his frequent use of the word "foil."'

13

Mine owne, and not mine owne.   21 ı

   Dem. It ſeemes to mee ,

That yet we ſleepe, we dreame. Do not you thinke,

The Duke was heere, and bid vs follow him ?

   Her. Yea, and my Father.   215

   Hel. And *Hippolitæ.*

   Lyſ. And he bid vs follow to the Temple.

   Dem. Why then we are awake ; lets follow him, and

by the way let vs recount our dreames.

Bottome wakes.   *Exit Louers.*   220

   Clo. When my cue comes, call me, and I will anſwer.

My next is, moſt faire *Piramus.* Hey ho. *Peter Quince* ?

Flute the bellowes-mender ? *Snout* the tinker ? *Starue-*

ling ? Gods my life ! Stolne hence, and left me aſleepe : I

haue had a moſt rare viſion. I had a dreame, paſt the wit   225

of man, to ſay, what dreame it was.   Man is but an Aſſe,

---

   212. Dem. *It*] Ff, Rowe+, Steev.'93, Knt, White i. Dem. *Are you ſure That we are awake ? It* Qq, Steev.'85, Mal. Var. Coll. Dyce i, Hal. Sta. Cam. Dem. *But are you ſure That we are well awake ? it* Cap. Rann, Dyce ii, iii. Dem. *But are you ſure That we are yet awake ? It* Ktly. Dem. *Are you ſure that we're awake ? It* White ii. Dem. *But are you ſure That now we are awake ? It* Schmidt.

     ſeemes] seems so Rowe i.

   213. *That yet*] That F_3F_4, Rowe i.

   217. *he bid*] *he did bid* Q_1, Theob. Warb. et seq.

     follow] *to follow* Pope, Han.

   218, 219. Two lines, ending *him... dreames* Rowe ii et seq.

   219. *let vs*] *lets* Q_1.

   220. Scene III. Pope+.

     Bottome...Louers] Om. Q_1. Exit. Q_2. As they go out Bottom wakes. Theob.

   222. Peter] Peeter Q_1.

   225. *I had*] *I haue had* Qq, Cap. et seq.

---

   212. Dem. It] See Text. Notes for a sentence to be found only in the Qq. ' I had once injudiciously restored these words,' says STEEVENS, ' but they add no weight to the sense of the passage, and create such a defect in the measure as is best remedied by their omission.'—DYCE (ed. ii) quotes LETTSOM as saying that ' Capell's insertions seem to me to improve the sense as well as restore the metre. I had hit upon the same conjectures long before I became acquainted with Capell.'—R. G. WHITE : Every reader with an ear and common sense must be glad that words sc superfluous and so fatal to the rhythm of two lines do not appear in F_1. But although there omitted, they have been industriously recovered from the Qq by those who consider that antiquity, not authenticity, gives authority.   [R. G. White joined the band of the industrious when putting forth his second edition.—ED.]—KEIGHTLEY : The poet's words may have been, 'Are you sure we are awake ? it seems to me.' But that would make the preceding speech terminate in a manner that does not occur in this play.

   215. Yea] W. A. WRIGHT : ' Yea ' is here the answer to a question framed in the negative, contrary to the rule laid down by Sir Thomas More, according to which it should be ' yes.'

if he goe about to expound this dreame.   Me-thought I   227
was, there is no man can tell what.   Me thought I was,
and me-thought I had.   But man is but a patch'd foole,
if he will offer to fay, what me-thought I had. The eye of   230
man hath not heard, the eare of man hath not feen, mans
hand is not able to tafte, his tongue to conceiue, nor his
heart to report, what my dreame was.   I will get *Peter*
Quince to write a ballet of this dreame, it fhall be called
Bottomes Dreame, becaufe it hath no bottome; and I will   235
fing it in the latter end of a play, before the Duke.   Per-
aduenture, to make it the more gracious, I fhall fing it
at her death.   *Exit.*   238

227. *to expound*] *expound* Q₁.
227, 229, 230. *Me-thought*] *Me thought*
Q₄.
229. *a patch'd*] *patcht a* Qq.
234. *ballet*] *Ballad* F₄.
236. *a play*] *the play* Han. Rann, Hal.

Coll. MS. *our play* Walker Dyce ii, iii,
Hudson.
238 *at her*] *after* Theob.+, Cap.
Rann, Sta. Dyce ii, Coll. iii. *at Thisby's*
Coll. MS.

---

229. **patch'd foole**] JOHNSON: That is, a fool in a parti-coloured coat.—STAUN-
TON: I have met with a remarkable proof of the supposed connexion between the
term *patch*, applied to a fool, and the garb such a character sometimes wore, in a
Flemish picture of the sixteenth century.   In this picture, which represents a grand
al fresco entertainment of the description given to Queen Elizabeth during her ' Progresses,' there is a procession of masquers and mummers, led by a fool or jester, whose
dress is covered with many-coloured coarse patches from head to heel.

230. **The eye of man, &c.**] HALLIWELL: Mistaking words was a source of merriment before Shakespeare's time. . . . This kind of humour was so very common, it
is by no means necessary to consider, with some, that Shakespeare intended Bottom
to parody Scripture.

236. **a play**] WALKER (*Crit.* ii, 320) has collected several instances of the confusion of *a* and *our;* he therefore conjectures '*our* play' here; DYCE (ed. ii) and
HUDSON adopted the conjecture.

238. **at her death**] THEOBALD: At *her* death?   At *whose?*   In all Bottom's
speech there is not the least mention of any she-creature to whom this relative can be
coupled.   I make not the least scruple, but Bottom, for the sake of a jest and to render his *Voluntary*, as we may call it, the more gracious and extraordinary, said, ' I
shall sing it *after* death.'   He, as Pyramus, is killed upon the scene, and so might
promise to rise again at the conclusion of the Interlude and give the duke his dream
by way of a song.   The source of the corruption of the text is very obvious.   The *f*
in *after* being sunk by the vulgar pronunciation, the copyist might write it from the
sound, *a'ter*, which, the wise editors not understanding, concluded two words were
erroneously got together; so splitting them, and clapping in an *h*, produced the present reading, ' at her.'—CAPELL: The singing after death does not allude to Pyramus'
death, but a death in some other play, ' a play' generally; opportunities of which the
speaker was very certain of, from the satisfaction he made no question of giving in

[*Scene II.*]

Enter Quince, Flute, Thisbie, Snout, and Starueling.

Quin. Haue you ſent to *Bottomes* houſe? Is he come home yet?

Staru. He cannot be heard of. Out of doubt hee is tranſported.   5

Thiſ. If he come not, then the play is mar'd. It goes not forward, doth it?   7

Scene IV. Pope+. Act V, Sc. ii.
Fleay. Scene II. Cap. et seq.
[Changes to the Towne. Theob.
Athens. Han. A Room in Quince's
House. Cap.
  1. Thisbie] Om. Rowe ii et seq.

1. Snout, and Starueling] and the rab-
ble. Qq.
  4. Staru.] Flut. Qq.
  6, 10, 14, &c. Thiſ.] Flute. Rowe ii et
seq.
  7. *not*] Om. F₃F₄, Rowe i.

---

discharging his present part; perhaps, too, there is a wipe in these words upon some play of the poet's time, in which a singing of this sort had been practised.—STAUN-TON: Theobald's explanation is extremely plausible. From the old text no ingenuity has ever succeeded in extracting a shred of humour or even meaning.—W. A. WRIGHT: Theobald's conjecture is certainly ingenious, and may be right. [It is an *emendatio certissima* to the present ED.]

  1. THEOBALD (Nichols, *Illust.* ii, 237) conjectured that the Fifth Act should begin here, and was the first to point out that the scene must be shifted from the Palace Wood to Athens.

  4. Staru.] COLLIER: In the Ff, as in the Qq, there is some confusion of persons, owing, perhaps, to the actor of the part of Thisbe being called *This.* in the prefixes.

  5. transported] STAUNTON: Or, as Snout expressed it when he first saw Bottom, adorned with an ass's head, *translated*, that is, *transformed*.—SCHMIDT (*Lex.*) in his third section of the meanings of this word, defines the present passage by ' to remove from this world to the next, to kill (euphemistically) '; and cites, in confirmation, *Meas. for Meas.* IV, iii, 72, where the Duke says of Barnardine ' to transport him in the mind he is were damnable.' Of course it would be temerarious to say outright that Schmidt is downright wrong, but I submit that it does not follow that a meaning which is appropriate in the Duke's mouth is appropriate in Starveling's. The presumption is strong that if 'transported' means *killed*, Starveling would not have used it. It is the mistakes of these rude mechanicals which, as Theseus says, we must take. Therefore, Starveling's 'transported' means Snout's 'translated,' which means our 'transformed.'—ED.

  6. This.] EBSWORTH (*Introd. to Griggs's Roberts's Qto*, p. xi): The first error of the Qq was the omission to mark (not *Thisbie*, but) *Thisbie's mother;* a character that had been allotted to the timid Robin Starveling, although she does not speak when the Interlude is afterwards acted. Her part is dumb-show, and therefore especially suited to the nervous tailor, who fears his own voice and shadow.

Quin.   It is not poſſible : you haue not a man in all   8
Athens, able to diſcharge *Piramus* but he.

Thiſ.   No, hee hath ſimply the beſt wit of any handy-   10
craft man in *Athens*.

Quin.   Yea, and the beſt perſon too, and hee is a very
Paramour, for a ſweet voyce.

Thiſ.   You muſt ſay, Paragon.   A Paramour is (God
bleſſe vs) a thing of nought.   15

Enter Snug the Ioyner.

Snug.   Maſters, the Duke is comming from the Temple, and there is two or three Lords & Ladies more married : If our ſport had gone forward, we had all bin made
men.   20

Thiſ.   O ſweet bully *Bottome :* thus hath he loſt ſixe-
pence a day, during his life ; he could not haue ſcaped ſix-   22

---

12. Quin.] Snout. Phelps, Hal. White ii.

  too] *to* Q₁.
14. Thiſ.] Quince. Phelps, Hal.
15. *nought*] *naught* Ff, Rowe, Theob.

Han. Warb. Cap. Knt, Hal. Dyce, Sta. Cam. White ii.

16. the Ioyner] Om. Rowe et seq.
19. *bin*] *beene* Qq.   *been* Ff.
22. *ſcaped*] *scraped* Grey.
22, 24, 25. *a day*] *a-day* Pope.

---

12. **Quin.**] PHELPS (ap. Halliwell): We give this speech to Snout, who has otherwise nothing to say, and to whom it is much more appropriate than to Quince.   Quince,
the playwright, manager, and ballad-monger, himself corrects the pronunciation of
Bottom in III, i.   The next speech by Flute [line 14] should also, we think, be given
to Quince, as the best informed of the party.   [As far as *Snout* is concerned, R. G.
WHITE, in his first edition, agreed with Phelps, and in his second edition followed
him.]—EBSWORTH (*Introd. to Griggs's Roberts's Qto,* p. xii): It is Flute who habit-
ually mistakes his words (witness his repetition of ' Ninny's tomb,' despite the correction earlier administered to him by Quince).   Therefore we may be sure that the
awkward misreading of ' Paramour ' for ' Paragon ' comes from Flute, and not from
the sensible manager, Quince.   Can we restore the right [rubric in line 14] ?   It may
have been either *Quince* or *Snout*, or even *Thisbie's Mother,* otherwise *Starveling.*
Certainly not ' *Thisbie,*' i. e. Flute.

  14, 15. **God bleſſe vs**] See Staunton's note on III, ii, 419.

  15. **nought**] W. A. WRIGHT: The two words, ' naught,' signifying worthlessness,
good-for-nothingness, and ' nought,' nothing, are etymologically the same, but the different senses they have acquired are distinguished in the spelling.—M. MASON: The
ejaculation ' God bless us !' proves that Flute imagined he was saying a naughty word
[and that the true spelling here is *naught*].

  18. **there is two or three**] For examples of ' there is ' preceding a plural subject,
see Shakespeare *passim*, or ABBOTT, § 335.

  19. **made men**] JOHNSON: In the same sense as in *The Tempest*, II, ii, 31 : ' any
strange beast there makes a man.'

pence a day.   And the Duke had not giuen him fixpence   23
a day for playing *Piramus,* Ile be hang'd.   He would haue
deferued it.   Sixpence a day in *Piramus,* or nothing.   25
                    Enter Bottome.

Bot.   Where are thefe Lads *?* Where are thefe hearts *?*
Quin.   *Bottome,* ô moft couragious day ! O moft happie houre !

Bot.   Mafters, I am to difcourfe wonders ; but ask me   30
not what.   For if I tell you , I am no true *Athenian.*   I
will tell you euery thing as it fell out.
Qu.   Let vs heare, fweet *Bottome.*
Bot.   Not a word of me : all that I will tell you, is, that
the Duke hath dined.   Get your apparell together, good   35

---

23. *And*] *An* Pope et seq.
25. *in* Piramus] *for Pyramus* Hal.
conj.
27. *hearts*] *harts* Q$_1$.
28. *Bottome,*] *Bottom !*— Theob.

29. [All croud about him. Cap.
31. *no true*] *not true* Qq.
32. *thing as*] *thing right as* Qq, Cap.
et seq. (subs.).
34. *all that*] *all* Rowe +.

---

25. **Sixpence a day**] STEEVENS : Shakespeare has already ridiculed the title-page
of *Cambyses,* by Thomas Preston, and here he seems to allude to him or some other
person who, like him, had been pensioned for his dramatic abilities.   Preston acted a
part in John Ritwise's play of *Dido* before Queen Elizabeth, at Cambridge, in 1564 ;
and the Queen was so well pleased that she bestowed on him a pension of *twenty*
pounds a year, which is little more than *a shilling a day.*—R. G. WHITE (ed. i) :
This [sixpence] seems like a jest, but is not one.   Sixpence sterling, in Shakespeare's
time, was equal to about eighty-seven and a half cents now—no mean gratuitous addition to the daily wages of a weaver during life.   See the following extract from a very
able little tract on political economy : 'And ye know xii. d. a day now will not go so
far as viii. pence would aforetime. . . . Also where xl. shillings a yere was honest
wages for a yeoman afore this time, and xx. *pence a week borde wages* was sufficient, now double as much will skante beare their charge.'—*A Conceipt of English
Pollicy,* 1581, fol. 33 b.   [That any ridicule on Preston or on any one else was here
cast by Shakespeare is, I think, extremely improbable.   It is attributing too much
intelligence to Shakespeare's audience on the one hand, and too little to Shakespeare
on the other.—ED.]

28. **cour;agious**] W. A. WRIGHT : It is not worth while to guess what Quince
intended to say.   He used the first long word that occurred to him, without reference
to its meaning ; a practice which is not yet altogether extinct.

30. **I am to discourse**] For many examples of the various ellipses after *is,* see
ABBOTT, § 405, where it is noted that ' we still retain an ellipsis of *under necessity* in
the phrase, "I *am* (yet) to learn."—*Mer. of Ven.* I, i, 5.   But we should not say :
" That ancient Painter who *being* (under necessity) to represent the griefe of the by-
standers," &c.—*Montaigne,* 3.   We should rather translate literally from Montaigne :
"Ayant à représenter."   So Bottom says to his fellows : " I am (ready) to difcourfe,"
&c.'

ſtrings to your beards, new ribbands to your pumps,   36
meete preſently at the Palace, euery man looke ore his
part : for the ſhort and the long is, our play is preferred :
In any caſe let *Thisby* haue cleane linnen : and let not him
that playes the Lion, paire his nailes, for they ſhall hang   40
out for the Lions clawes.   And moſt deare Aĉtors, eate
no Onions, nor Garlicke ; for wee are to vtter ſweete
breath, and I doe not doubt but to heare them ſay, it is a
ſweet Comedy.   No more words : away, go away.

                                             Exeunt.   45

---

38. *preferred*] *preferd* Qq. *proffer'd* Theob. conj. (Nichols, ii, 237).

43. *doubt but to*] *doubt to* F₃F₄, Rowe +.

44. *ſweet*] *most sweet* Theob. ii, Warb.

Johns.

44. *go away*] *go, away* Theob. i et seq. (subs.). *go; away* Coll. Dyce White.

45. Exeunt.] Om. Qq.

---

36. strings] MALONE: That is, to prevent the false beards, which they were to wear, from falling off.—STEEVENS: I suspect that the 'good strings' were ornamental or employed to give an air of novelty to the countenances of the performers. [As the only authority given by Steevens to support his suspicion is where the Duke, in *Meas. for Meas.* IV, ii, 187, tells the Provost to shave the head of Barnardine, and 'tie the beard,' we may not unreasonably question his interpretation.—ED.]

38. preferred] THEOBALD: This word is not to be understood in its most common acceptation here, as if their play was chosen in *preference* to the others (for that appears afterwards not to be the fact), but means that it was given in among others for the duke's option.   So in *Jul. Cæs.* III, i, 28: 'Let him go And presently prefer his suit to Cæsar.'—W. A. WRIGHT: That is, offered for acceptance; if Bottom's words have a meaning, which is not always certain.—F. A. MARSHALL queries if it has not more probably the sense of '*preferred* to the dignity (of being acted before the Duke).'   [Assuredly no one can be accused of inordinate self-conceit who asks for an explanation of Bottom's phrases which were intelligible to Snug, Flute, and Snout.—ED.]

Actus Quintus. [*Scene I.*]

Enter Thefeus, Hippolita, Egeus and his Lords.

Hip.   'Tis ftrange my *Thefeus*, y̆ thefe louers fpeake of.
The.   More ftrange then true.   I neuer may beleeue   4

1. Om. Qq.   Act V, Sc. iii. Fleay.   2. Egeus and his Lords.] and Philo-
[The Palace. Theob.   The Same. A   strate. Qq.
State-Room in Theseus's Palace. Cap.   Egeus] Egæus Ff (throughout).
                                        3. y̆] *what* Pope +.

3. y̆] For examples of the omission of the relative, see Shakespeare *passim*, or
ABBOTT, § 244; and see § 307 for examples of 'may' in the sense of *can*, as Theseus uses it the next line.

4-23. ROFFE (*Ghost Belief*, &c., p. 40): [In this speech every line] is sceptical,
yet *the conduct* of the play falsifies the Duke's reasonings, or, as they should rather
be called, his assertions.   Hippolyta having observed to him, ' 'Tis strange, my Theseus, that these lovers speak of,' he replies, *paying no attention*, be it observed, to the
fact that Hippolyta is speaking from the testimony of four persons; a very artful
stroke on the part of Shakespeare at the sceptics.   To this speech [ll. 4-23] Hippolyta very justly answers that [ll. 24-28].   Here again Shakespeare shows his nice
observation of the sceptical mind.   Every one who has conversed on any subject
with persons *predetermined, on that subject, not to believe*, must have observed how
common it is for the latter, when fairly brought to a stand-still, to lapse into a dead
silence, instead of saying, as the lover of truth would do, 'What you have alleged is
very reasonable, and I will now examine.'   *They* can say no more, nor may *you.*
Accordingly, to the incontrovertible speech of Hippolyta, Theseus makes no reply.
It is a truly noteworthy and significant fact that to the sceptical Theseus should have
been allotted by Shakespeare the sceptical idea concerning the poet, namely, as being
the embodier of the unreal, and not as being the copyist of what is true.   It is exactly
in character that the doubting Theseus should thus speak of the poetic art, and *thence
we may be sure that the poet who wrote the lines for him, thought precisely the very
reverse.*   Owing, however, to the general doubt concerning the supernatural, and the
consequent assumption of Shakespeare's disbelief [in it], this point seems never to
have been considered, and it may be safely affirmed that nine hundred and ninety-
nine readers out of every thousand would gravely quote the lines upon the poet *as
containing Shakespeare's own idea*, although, only five lines previously, *Theseus has
placed the poet in the same category with the lunatic.*   From the purely dramatic character of his works, Shakespeare can never *speak* in his own person, but he can
always *act;* that is, so frame his story as that scepticism shall be shown to be entirely
at fault.   [Be it observed that the essay, privately printed in 1851, from which the
foregoing is extracted, was written on the assumption that 'ghost-belief, rightly understood, is most rational and salutary,' and that 'the ghost-believing student' will deem
that ' it must have had the sanction of such a thinker as Shakespeare.'—ED.]—JULIA
WEDGWOOD (*Contemporary Rev.* Apr. 1890, p. 583): In the attitude of Theseus

Thefe anticke fables, nor thefe Fairy toyes,   5
Louers and mad men haue fuch feething braines,
Such fhaping phantafies, that apprehend more
Then coole reafon euer comprehends.
The Lunaticke, the Louer, and the Poet,
Are of imagination all compact.   10
One fees more diuels then vafte hell can hold ;
That is the mad man.   The Louer, all as franticke,
Sees *Helens* beauty in a brow of *Egipt*.   13

---

5. *anticke*] *antique* Q₁, Cap. Dyce, Sta. Cam. White ii. *antick* F₃F₄, Rowe +. *antic* Coll. Hal. White i, Ktly.

7. *more*] Transposed, to begin the next line, Theob. et seq.

8–10. Two lines, ending *lunatick*...

compact. Q₁.

8. *coole*] *cooler* Pope.

12. *That is the mad man*] *The mad-man. While* Pope +.   *That is, the mad-man* Cap. et seq.

13. Egipt] *Ægypt* Q₁.

---

towards the supernatural there is something essentially modern.   It is very much in the manner of Scott, or rather, there is something in it that reminds one of Scott him-self. . . . Scott thought that any contemporary who believed himself to have seen a ghost must be insane ; yet when he paints the appearance of the grey spectre to Fear-gus MacIvor, or, what seems to us his most effective introduction of the supernatural, that of Alice to the Master of Ravenswood, we feel that something within him believes in the possibility of that which he paints, and that this something is deeper than his denial, though that be expressed with all the force of his logical intellect. . . . Theseus explaining away the magic of the night is Scott himself when he drew Dousterswivel, or when he describes the Antiquary scoffing at a significant dream. . . . To paint [the supernatural] most effectually it should not be quite consistently either disbelieved or believed.   Perhaps Shakespeare was much nearer an actual belief in the fairy mythology he has half created than seems possible to a spectator of the nineteenth century.   And yet Theseus expresses exactly the denial of the mod-ern world.   And we feel at once how the introduction of such an element enhances the power of the earlier views; the courteous, kindly, man-of-the-world scepticism somehow brings out the sphere of magic against which it sets the shadow of its demand.   The belief of the peasant is emphasised and defined, while it is also inten-sified by what we feel the inadequate confutation of the prince.

6, &c. SIGISMUND (' Uebereinstimmendes zwischen Sh. und Plutarch,' *Sh. Jahr-buch*, xviii, p. 170) refers to the 'noteworthy' correspondence between this passage and the comparison of love to madness in Plutarch's *Morals*, where the resemblance, as he thinks, is too marked to be overlooked.

6. **seething**] STEEVENS : So in *The Temp.* V, i, 59 : 'thy brains, Now useless, boil'd within thy skull.'—MALONE : So also in *Wint. Tale*, III, iii, 64 : 'Would any but these boiled brains of nineteen and two-and-twenty hunt this weather ?'—DELIUS : See also *Macbeth*, II, i, 39 : 'A false creation, Proceeding from the heat-oppressed brain.'

11. **One sees, &c.**] For Chalmers's theory that in this line there is a sarcasm on Lodge's *Wits Miserie*, see Appendix, *Date of Composition*.

The Poets eye in a fine frenzy rolling, doth glance
From heauen to earth, from earth to heauen.   15
And as imagination bodies forth the forms of things
Vnknowne; the Poets pen turnes them to fhapes,
And giues to aire nothing, a locall habitation,
And a name. Such tricks hath ftrong imagination,
That if it would but apprehend fome ioy,   20
It comprehends fome bringer of that ioy.
Or in the night, imagining fome feare,
How eafie is a bufh fuppos'd a Beare?
  Hip.   But all the ftorie of the night told ouer,
And all their minds transfigur'd fo together,   25
More witneffeth than fancies images,
And growes to fomething of great conftancie;   27

---

14. *frenzy rolling,*] *frenzy, rolling,* Q₁.
14, 15. *doth glance…to heauen*] One line, Rowe et seq.
15, 16. *From…And as*] One line, Q₁.
16–19. Lines end *forth…pen…nothing …name…imagination.* Rowe ii et seq.
17. *Vnknowne;*] *unknown,* Pope et seq.

17. *fhapes*] *shape* Pope+, Dyce ii, iii.
18. *aire*] F₃. *ayery* Q₁. *ayre* F₂. *air* F₄. *aiery* Pope+. *airy* Q₂, Rowe et cet.
20. *it would*] *he would* Rowe ii, Pope, Theob.
22. *Or*] *So* Han. *For* Anon. ap. Cam.

---

13. **Egipt**] STEEVENS: By 'a brow of Egypt' Shakespeare means no more than the *brow of a gypsy.*

18. **aire**] An instance, cited by WALKER (*Crit.* ii, 48), of the confusion of *e* and *ie* final.

22, 23. R. G. WHITE (ed. i): Who can believe that these two lines are genuine? … The two preceding lines are doubtless genuine. They close the speech appropriately with a clear and conclusive distinction between the apprehensive and the comprehensive power of the imaginative mind. Where, indeed, in the whole range of metaphysical writing is the difference between the two so accurately stated and so forcibly illustrated? And would Shakespeare, after thus reaching the climax of his thought, fall a twaddling about bushes and bears? Note, too, the loss of dignity in the rhythm. I cannot even bring myself to doubt that these lines are interpolated. [This last sentence White repeats in his second edition.]—The COWDEN-CLARKES: This concluding couplet, superficially considered, has an odd, bald, flat effect, as of an anti-climax, after the magnificent diction in the previous lines of the speech; but viewed dramatically they serve to give character and naturalness to the dialogue. The speaker is carried away by the impulse of his thought and nature of his subject into lofty expression, ranging somewhat apart from the matter in hand; then, feeling this, he brings back the conversation to the point of last night's visions and the lovers' related adventures by the two lines in question.

22. **imagining**] That is, if one imagines; for examples of participles without nouns, see ABBOTT, § 378.

27. **constancie**] JOHNSON: Consistency, stability, certainty.

But howſoeuer, ſtrange, and admirable.   28

> *Enter louers , Lyſander, Demetrius, Hermia,*
> *and Helena.*   30

The.   Heere come the louers, full of ioy and mirth :
Ioy, gentle friends, ioy and freſh dayes
Of loue accompany your hearts.
   Lyſ.   More then to vs, waite in your royall walkes,
your boord, your bed.   35
   The.   Come now, what maskes, what dances ſhall
we haue,
To weare away this long age of three houres,
Between our after ſupper, and bed-time ?
Where is our vſuall manager of mirth ?   40
What Reuels are in hand ? Is there no play,
To eaſe the anguiſh of a torturing houre ?
Call *Egeus*   43

---

28. *But*] *Be't* Han.

32, 33. *Ioy...Of loue*] One line, Ff, Rowe et seq.

34, 35. *waite...bed*] One line, Ff, Rowe et seq.

34. *waite in*] *wait on* Rowe+, Cap.

38–43. Four lines, ending *betweene*...

manager...play...Philoſtrate. Q₁.

39. *our after*] *or after* Qq.

after ſupper] *after-supper* F₄, Rowe et seq.

43. Egeus.] Philoſtrate. Qq, Pope et seq.

[Enter Philostrate. Pope+.

---

28. *howsoeuer*] ABBOTT, § 47 : For 'howsoe'er it be,' ' in any case.'

28. *strange*] The COWDEN-CLARKES : Shakespeare uses this word with forcible and extensive meaning. Here, and in the opening lines of the scene, he uses it for *marvellous, out of nature, anomalous*. See also line 66, below.

28. *admirable*] That is, to be wondered at.

39. *after supper*] STAUNTON : The accepted explanation of an 'after-supper' conveys but an imperfect idea of what this refection really was. '*A rere-supper*,' says Nares, 'seems to have been a late or second supper.' Not exactly. The *rere-supper* was to the supper itself what the *rere-banquet* was to the dinner—*a dessert*. On ordinary occasions the gentlemen of Shakespeare's age appear to have dined about eleven o'clock, and then to have retired either to a garden-house or other suitable apartment and enjoyed their *rere-banquet* or dessert. Supper was usually served between five and six ; and this, like the dinner, was frequently followed by a collation consisting of fruits and sweetmeats, called, in this country, the *rere-supper ;* in Italy, Pocenio, from the Latin *Pocoenium*.

43. *Egeus*] CAPELL (p. 115 *b*) : The player editors' error in making Egeus enterer in an act he has no concern in, arose (probably) from their laying Philostrate's character in this act upon the player who had finished that of Egeus. [Which is another proof that the Folio was printed from a prompter's copy. The Qq here have, correctly, Philostrate, who was the master of the revels; and so, too, has the Folio, at

Ege.   Heere mighty *Theseus.*

The.   Say, what abridgement haue you for this eue-   45
ning?

What maske? What musicke? How shall we beguile
The lazie time, if not with some delight?

Ege.   There is a breefe how many sports are rife: 49

44, 49, 68, 79. Ege.] Philostrate Q₁.
Philo. Q₂.
  44. Theseus.] *Theseus, here* Han.
  49. *There*] *Here* Anon. ap. Hal.
    rife] *ripe* Q₁, Theob.+, Cap.

Steev. Mal. Var. Coll. Dyce, White, Sta.
Cam. Ktly.
  49. *breefe*] *briefe* QqF₂. *brief* F₃F₄.
    [presenting a Paper. Cap. Giving
a paper which Theseus hands to Lysander to read. Hal.

---

line 84,—an oversight on the part of the prompter who adapted for the stage the copy
of Q₂ from which the Folio was subsequently printed.—ED.]

  45. abridgement] STEEVENS: By 'abridgement' our author may mean a dramatic performance, which crowds the events of years into a few hours. It may be
worth while to observe that in the North the word *abatement* had the same meaning
as *diversion* or *amusement.* So in the Prologue to the Fifth book of Gawin Douglas's
version of the Æneid: 'Ful mony myrry abaytmentis follows heir.'—HENLEY: Does
not 'abridgement,' in the present instance, signify amusement to beguile the tediousness of the evening? or, in one word, *pastime?*—W. A. WRIGHT: An entertainment
to make the time pass quickly. Used in *Hamlet*, II, ii, 439, in a double sense, the
entry of the players cutting short Hamlet's talk: 'look, where my abridgement
comes.' In Steevens's quotation from Gawin Douglas, 'abaytment' is clearly the
same as the French 'esbatement,' which Cotgrave defines, 'A sporting, playing, dal-
lying, ieasting, recreation.'—[In an article on the etymology of the word 'merry,'
ZUPITZA (*Englische Studien*, 1885, vol. 8, p. 471) shows that this word originally
bore the meaning of *short* (like Old High German *murg*), and thence followed the
meaning of *that which makes the time seem short;* that is, *pleasant, agreeable, entertaining, delightful.* Hence by a parallel process 'abridgement' is used thus poetically by Shakespeare in [the present passage] as that which abridges time—namely,
pastime, diversion, amusement. 'With this poetic use of "abridgement," Vigfusson
(*Sturlunga saga*, Oxford, 1878, i, Note xxiii) compares the Old Norse *skemtan* and
skemta. The noun *skemtan* means entertainment, pastime, especially the entertainment derived from telling stories; the verb *skemta* means to entertain, to pass the
time. The Danish thus use *skjemt*, a joke, fun; *skjemte*, to joke, to amuse, &c. The
etymon of the words is Old Norse—*skammr*, short. . . . There is a development of
the same idea in Scotch, as was observed long ago by Jamieson, which corresponds to
Shakespeare's "abridgement"; we find in the Scotch the word *schorte* or *short*, equivalent to entertain, to pass the time; and *schortsum* or *shortsum*, meaning cheerful,
merry. . . . In fine, the signification of *merry* does not debar us from referring it to
the Gothic *gamaurgian*, to shorten, and Old High German *murg*, short, inasmuch as
the Old Norse *skemtan* and *skemta* from *skammr*, and the Scotch *schorte* and *schortsum*, reveal a corresponding development of meaning, and Shakespeare uses "abridgement" in the sense of amusement, pastime, diversion.' For the reference to this
article by Dr. Zupitza, I am indebted to the learning and courtesy of Prof. Dr. J. W.
BRIGHT of the Johns Hopkins University.—ED.]

Make choife of which your Highneffe will fee firft.   50

   Lif.   The battell with the Centaurs to be fung

By an Athenian Eunuch, to the Harpe.

   The.   Wee'l none of that.   That haue I told my Loue

In glory of my kinfman Hercules.

   Lif.   The riot of the tipfie Bachanals,   55

Tearing the Thracian finger, in their rage *?*

---

51. Lif.] Ff, Rowe, Pope.   The. or
Thef. Qq, Cap.   Lys. [reads] Knt, Hal.
White i, Sta.   Thes. [reads] Theob. et
cet

51–67.   Given   to   Theseus.   Qq,
Theob. +, Cap. Steev. Mal. Var. Coll.
Dyce, Cam. White ii.

51. *Centaurs*] *Centaur* F₄, Rowe i.
52. *Harpe.*] *Harpe ?* Q₁.
53. *haue I*] *I have* Theob. Warb.
Johns.

55, 59, 63. [Reads. Han. Dyce, Cam.
56. *Thracian*] *Thraſian* F₃F₄.
    rage ?] *rage.* F₄ et seq.

---

49. breefe] STEEVENS : That is, a short account or enumeration.

49. rife] THEOBALD corrected this manifest misprint, but STEEVENS dallied with it by citing examples from Sidney and from Gosson of its use (which is beside the mark.   Does any question that ' rife ' is a good word in its proper place ?), and HAL-LIWELL retained it and sustained it.   *Ripe*, of course, means ready.—ED.

51. Lis.] THEOBALD : What has Lysander to do in the affair ?   He is no courtier of Theseus's, but only an occasional guest, and just come out of the woods, so not likely to know what sports were in preparation.   I have taken the old Qq. for my guides.   Theseus reads the titles of the sports out of the list, and then alternately makes his remarks upon them.—KNIGHT : The lines are generally printed as in the Qq, but the division of so long a passage is clearly better, and is perfectly natural and proper.   'And the dignity of the monarch,' adds HALLIWELL, ' is better sustained by this arrangement.'—WHITE (ed. i) : It seems natural that, under the circumstances, a sovereign should hand such a paper to some one else to read aloud.   [In his second edition White follows the Qq.]—F. A. MARSHALL : The arrangement in the Ff is much more effective as far as the stage requirements are concerned.   COLLIER : The more natural course seems to be for Theseus both to read and comment.   [We have had so many proofs that F₁ was printed from a stage-copy that, I think, it is safest to follow it here.—ED.]

51. Centaurs] This, and the reference to Orpheus in line 56, are among the many proofs collected by WALKER (*Crit.* i, 152) of Ovid's influence on Shakespeare.   The story of the Centaurs is in Book xii of the *Metamorphoses*, and of the ' Thracian singer ' in Book xi.

52. Harpe] HALLIWELL : It is a singular circumstance that the harp is not found in any of the known relics of the ancient Greeks, so that the poet has probably unwit-tingly fallen into an anachronism.

54. Hercules] KNIGHT : Shakespeare has given to Theseus the attributes of a real hero, amongst which modesty is included.   He has attributed the glory to his ' kinsman Hercules.'   The poets and sculptors of antiquity have made Theseus him-self the great object of their glorification.—W. A. WRIGHT : The version by Theseus was different from that told by Nestor ; the latter, in Ovid, purposely omitted all men-tion of Hercules.

The.   That is an old deuice, and it was plaid   57
When I from *Thebes* came laft a Conqueror.

Lif.   The thrice three Mufes, mourning for the death
of learning, late deceaft in beggerie.   60

The.   That is fome Satire keene and criticall,
Not forting with a nuptiall ceremonie.

Lif.   A tedious breefe Scene of yong *Piramus*,
And his loue *Thisby* ; very tragicall mirth.

The.   Merry and tragicall? Tedious, and briefe? That   65
is, hot ice, and wondrous ftrange fnow.   How fhall wee
finde the concord of this difcord?   67

---

60. *of*] *Of* Qq, Pope et seq.
  beggerie.] *beggery ?* Q₁.
64. *mirth*] *mirth ?* Qq.
65–67. Prose, Q₂Ff. Three lines, ending *ice...concord...difcord.* Q₁. Three lines, ending *briefe... fnow...difcord.* Theob. et seq.
65, 66. *That...fnow*] Om. Pope.
66. *ice*] *Ife* Q₁.
  and wondrous ftrange fnow] Qq Ff, Rowe, Theob. i, Coll. i, Hal. White i, Sta. Dyce iii.   *and wonderous strange snow* Theob. ii.   *and wondrous scorching snow* Han.   *a wondrous strange shew* Warb.   *and wondrous strange black snow* Upton, Cap.   *and wondrous seething snow* Coll. ii, iii (MS).   *and wondrous swarthy snow* Sta. conj.   Dyce ii.   *and wondrous*

swarte snow Sta. conj. Kinnear.   *and wondrous sable snow* Bailey, Ktly, Elze.   *and wondrous orange* (or *raven*, or *azure*) *snow* Bailey.   *and wondrous strange in hue* Bulloch.   *and wondrous sooty snow* Herr.   *and wind-restraining snow* Wetherell (*Athen.* 2 Nov.'67).   *and ponderous flakes of snow* Leo (*Athen.* 27 Nov. '80).   *and wondrous flakes of snow* Ibid.   *and wondrous staining snow* Nicholson (ap. Cam.).   *and wondrous flaming snow* Joicey (*N. & Qu.* 11 Feb.'93).   *and wondrous fiery snow* Orger.   *and wondrous scaldinge snow* Ebsworth.
  66. *wondrous*] *wodrous* Q₁. *wonderous* Theob. ii, Johns. Steev. Rann, Mal. Var. Knt, Dyce i, White ii.

59, 60. For the various references supposed to be lying concealed in these lines, see Appendix, *Date of Composition.*

62. **ceremonie**] This example may be added to the many collected by Walker (*Crit.* ii, 73) of the trisyllabic pronunciation of *ceremony.*—Ed.

63. **Piramus**] For Golding's translation of this story from Ovid, see Appendix, *Source of the Plot.*

66. **hot ice, ... snow**] Steevens: The meaning of the line is 'hot ice, and *snow* of as strange a quality.'—M. Mason: As there is no antithesis between 'strange' and 'snow' as there is between 'hot' and 'ice,' I believe we should read, 'and wonderous *strong* snow.'—Knight: Surely, snow is a *common* thing, and, therefore, 'wonderous strange' is sufficiently antithetical—'hot ice, and snow as strange.'—Halliwell: In other words, ice and snow, wonderous hot and wonderous strange; or hot ice, and strange snow as wonderful.—Collier (ed. ii): The MS has fortunately supplied us with what must have been the language of the poet—'and wondrous *seething* snow.' *Seething* is boiling, as we have already seen at the beginning of this act; and *seething* and 'snow' are directly opposed to each other, like 'hot' and 'ice.' Thus metre and meaning are both restored, and it is not difficult to see

Ege. A play there is, my Lord, fome ten words long, 68
Which is as breefe, as I haue knowne a play;
But by ten words, my Lord, it is too long; 70
Which makes it tedious.   For in all the play,
There is not one word apt, one Player fitted.
And tragicall my noble Lord it is : for *Piramus*
Therein doth kill himfelfe.   Which when I faw
Rehearſt, I muſt confeſſe, made mine eyes water : 75
But more merrie teares, the paſſion of loud laughter
Neuer ſhed.
   Theſ. What are they that do play it ? 78

---

68. *there is*] *it is* Han. Cap. Dyce ii,
iii, Coll. iii.   *this is* Coll. ii (MS).

73, 77. Lines end, *it is...himſelfe...
confeſſe...teares...ſhed.* Ff, Rowe et seq.
74. *I ſaw*] *I saw't* Han.

---

how the misprint occurred.   Here again the corr. fo., 1630, has been of most essential service.—R. G. WHITE (ed. i) : Collier's MS emendation seems preferable to all
the others, but there is hardly sufficient ground for making so great a change in a
word which is found in the Qq and Ff.—STAUNTON : Upton's '*black* snow' comes
nearest to the sense demanded, but 'strange' could hardly have been a misprint for
black.   Perhaps we should read '*swarthy* snow.'   *Swarte*, as formerly spelt, is not so
far removed from the text as *black, scorching*, or *seething.*—WALKER (*Crit.* iii, 51) :
Perhaps *scorching* [Hanmer's] might serve as a bad makeshift.—BAILEY'S prismatic
conjectures (*The Text*, &c. i, 196) were suggested by the colours of the polar snow
as described by Arctic voyagers.—PERRING (p. 116) : The word, which has no doubt
been lost in transcription, was probably a very small one, perhaps with letters or a
sound corresponding to the termination of the word preceding it.   The final letters
of 'strange' are *ge ;* what word more fully and fairly satisfies the conditions required
than the little word *jet*, used by Shakespeare in *2 Hen. VI :* II, i, in three consecutive lines ?   Perhaps, however, it would be too much to expect editors boldly to print
'and, wondrous strange ! jet snow.'—R. G. WHITE (ed. ii) : The original text is
unsatisfactory, but not surely corrupt.—The COWDEN-CLARKES : 'Strange,' as Shakespeare occasionally uses it (in the sense of *anomalous, unnatural, prodigious*), presents sufficient image of contrast in itself.   See note on line 28, above.   [Surely there
is no need of change.   The mere fact that any child can suggest an appropriate
adjective is a reason all-sufficient for retaining Shakespeare's word, especially when
that word bears the meaning given to it by the Cowden-Clarkes.—ED.]
   68. **there is**] COLLIER (ed. ii) : We need not hesitate here to receive *this* for
'there' of the old copies.   Philostrate evidently speaks of the particular play of
Pyramus and Thisbe, which is 'some ten words long.'—DYCE (ed. ii) : Collier's MS
correction, *this*, is objectionable on account of the 'this' immediately above.
   78. **play it**] SCHMIDT (*Programm*, p. 7) finds in these lines two difficulties which
could not have been in the original MS.   The first is the incomplete verse of line 78,
and the second is the blunt answer which, so he says, no Englishman would ever think
of giving to a prince.   He, therefore, thus emends : 'What are they that do *play't ?*
Hard-handed men, | *My noble Lord* (or *My gracious Duke*) that work in Athens here.'

Ege.   Hard handed men, that worke in Athens heere,
Which neuer labour'd in their mindes till now ;   80
And now haue toyled their vnbreathed memories
With this fame play, againft your nuptiall.
   The.   And we will heare it.
   Phi.   No, my noble Lord, it is not for you. I haue heard
It ouer, and it is nothing, nothing in the world ;   85
Vnleffe you can finde fport in their intents,
Extreamely ftretcht, and cond with cruell paine ,   87

---

82. *nuptiall*] *nuptialls* Ff, Rowe+.   et seq.
84, 85. *it is…ouer*] One line, Rowe ii   86, 87. Transpose, Gould.

---

81. **vnbreathed**] STEEVENS: That is, unexercised, unpractised.

82. **nuptiall**] W. A. WRIGHT: With only two exceptions Shakespeare always uses the singular form of this word [viz. in *Othello*, II, ii. 9, where the Ff have 'nuptiall' and the Qq 'nuptialls'; and *Per.* V, iii, 80].

86. **intents**] JOHNSON: As I know not what it is to 'stretch' and 'con' an 'intent,' I suspect a line to be lost.—KENRICK (*Rev.* 19): By 'intents' is plainly meant the design or scheme of the piece intended for representation; the conceit of which being far-fetched or improbable, it might be with propriety enough called 'extremely stretched.'   As to this scheme or design being 'conn'd' (if any objection be made to the supposition of its having been written, *penn'd*), it is no wonder such players as these are represented to be 'should *con* their several parts with cruel pain.' —DOUCE (i, 196): It is surely not the 'intents' that are 'stretched and conn'd,' but the *play*, of which Philostrate is speaking. If the line 86 (' Unlesse you can,' &c.) were printed in a parenthesis all would be right.—KNIGHT and DELIUS follow Douce's suggestion, the former exactly, the latter, Delius, substituting commas for the marks of parenthesis.—R. G. WHITE (ed. i): 'Intents' here, as the subject of the two verbs, 'stretched' and 'conn'd,' is used both for *endeavour* and for *the object of endeavour*, by a license which other writers than Shakespeare have assumed.—DANIEL (p. 35): Qy. arrange and read thus: ' No, my noble lord, it is not for you, | Unless you can find sport in their intents | To do you service. I have heard it *o'er*, | And it is nothing, nothing in the world, | Extremely stretch'd and conn'd with cruel pain.' [To me, Grant White's is the right interpretation, and renders any change unnecessary. Is it any more violent to say that my intents, my endeavors, to do you service shall be stretched to my utmost ability, than it is to say, as Antonio says in *The Mer. of Ven.*, that 'my credit [for your sake] shall be rack'd to the uttermost'? —ED.]

87. **stretcht**] ULRICI (Ed. *Deut. Sh. Gesellschaft*, trans. by Dr A. Schmidt, p. 428): I cannot avoid the conclusion that there is here a misprint, albeit no objection to the phrase has hitherto been made. 'Extremely stretch'd' can by no means apply to the 'tedious brief scene' which the rude mechanicals are to perform; their 'merry tragedy,' on the contrary, is 'extremely' short. Wherefore I believe that the phrase originally stood, in Shakespeare's handwriting, not 'extremely stretch'd,' but 'extremely *wretch'd.*' [Shall we not all fervently thank the Goodness and the Grace that on our birth has smiled, and permitted us to read Shakespeare as an inheritance,

To doe you feruice. 88

Thef. I will heare that play. For neuer any thing
Can be amiffe, when fimpleneffe and duty tender it. 90
Goe bring them in, and take your places, Ladies.

89, 90. *For...it*] Two lines, ending   90, 93. *duty*] *duety* Q₁.
amiffe...it Rowe ii et seq.   91. [Exit Phil. Pope.

---

instead of having to look at him through a medium which presents fantastic distortions? Let the grateful English-speaking reader consider for a moment what would be his enjoyment of Shakespeare were he to read his verses stript of all charm of melody, of humour, and sometimes even of sense. What a tribute it is to the intelligence of our German brothers that under such disadvantages they have done what they have done!—ED.]

87. **cruell**] HALLIWELL quotes from an anonymous writer the remark that '*cruel*, among the Devonshire peasantry, is synonymous with *monstrous* in fashionable circles. The person whom the latter would denominate monstrous handsome, monstrous kind, or monstrous good-tempered, the other will style, with equal propriety, cruel handsome, cruel kind, or cruel good-tempered. The word, however, was formerly in more general use to signify anything in a superlative degree.' [It is not at all likely that this Devonshire use rules here; 'cruel' has here its ordinary meaning.—ED.]

89, 90. **For never, &c.**] STEEVENS: Ben Jonson, in *Cynthia's Revels* [V, iii], has employed this sentiment of humanity on the same occasion, when Cynthia is preparing to see a masque: 'Nothing which duty and desire to please, Bears written in the forehead, comes amiss.'

91, &c. JULIA WEDGWOOD (*Contemporary Rev.* Apr. '90, p. 584): The play of the tradesmen, which at first one is apt to regard as a somewhat irrelevant appendix to the rest of the drama, is seen, by a maturer judgement, to be, as it were, a piece of sombre tapestry, exactly adapted to form a background to the light forms and iridescent colouring of the fairies as they flit before it. But this is not its greatest interest to our mind. It is most instructive when we watch the proof it gives of Shakespeare's strong interest in his own art. It is one of three occasions in which he introduces a play within a play, and in all three the introduction, without being unnatural, has just that touch of unnecessariness by means of which the productions of art take a biographic tinge, and seem as much a confidence as a creation. How often must Shakespeare have watched some player of an heroic part proclaim his own prosaic personality, like Snug, the Joiner, letting his face be seen through the lion's head! . . . In the speech of Theseus, ordering the play, we may surely allow ourselves to believe that we hear not only the music, but the voice of Shakespeare, pleading the cause of patient effort against the scorn of a hard and narrow dilettantism. . . . 'This is the silliest stuff I ever heard,' says Hippolyta, and Theseus's answer, while it calls up deeper echoes, is full of the pathos that belongs to latent memories. 'The best in this kind are but shadows, and the worst are no worse, if imagination amend them.' Here the poet is speaking to the audience; in Hamlet, when he addresses the players, his sympathy naturally takes the form of criticism; what the Athenian prince would excuse the Danish prince would amend. But in both alike we discern the same personal interest in the actor's part, and we learn that the greatest genius who ever lived was one who could show most sympathy with incompleteness and failure.

14

Hip.   I loue not to fee wretchedneffe orecharged ;   92
And duty in his feruice perifhing.

Thef. Why gentle fweet, you fhall fee no fuch thing.

Hip.   He faies, they can doe nothing in this kinde.   95

Thef. The kinder we, to giue them thanks for nothing
Our fport fhall be, to take what they miftake ;
And what poore duty cannot doe, noble refpeƈt
Takes it in might, not merit.   99

---

92. *orecharged*] *o'ercharg'd* Rowe et seq.

97. *fport*] *sports* Steev.'85.

98. *poore duty*] *poor (willing) duty* Theob. Han. Warb. Cap. Dyce ii, iii, Coll. iii (subs.). *poor faltering duty* Ktly. *poor duty meaning* Spedding (ap. Cam.). *cannot doe*] *cannot aptly do* Bailey,

Schmidt. *cannot nobly do* Wagner. *can but poorly do* Tiessen.

98, 99. *noble...merit*] One line, Theob. et seq. (except Sta. Cam. White ii). *refpect Takes it in noble might, not noble merit.* Bulloch.

99. *might*] *mind* Bailey, Spedding (ap. Cam.).

---

97. **Our sport,** &c.] EDINBURGH MAGA. (Nov. 1786): That is, We will accept with pleasure even their blundering attempts. [Quoted by Steevens.]

98, 99. **And what,** &c.] JOHNSON: The sense of this passage as it now stands, if it has any sense, is this : What the inability of duty cannot perform, regardful generosity receives as an act of ability, though not of merit. The contrary is rather true : What dutifulness tries to perform without ability, regardful generosity receives as having the merit, though not the power, of complete performance. We should therefore read 'takes not in might, but merit.'—STEEVENS: 'In *might*' is, perhaps, an elliptical expression for *what might have been.*—HEATH (p. 58): Whatever failure there may be in the performance attempted by poor willing duty, the regard of a noble mind accepts it in proportion to the ability, not to the real merit.—KENRICK (p. 21): That is, in consequence of 'poor duty's' inability, taking the will for the deed, viz. accepting the best in its *might* to do for the best that might be done ; rating the merit of the deed itself as nothing, agreeable to the first line of Theseus's speech, 'The kinder we to give them thanks for *nothing.*'—COLERIDGE (p. 103), referring to Theobald's insertion, for the sake of rhythm, of *willing* before 'duty,' says, 'to my ears it would read far more Shakespearian thus : 'what poor duty cannot do, *yet would*, Noble,' &c.—ABBOTT, § 510, evidently unwitting that he had been anticipated by both Johnson and Coleridge, says : 'I feel confident that *but would* must be supplied, and we must read : " what poor duty cannot do, *but would*, Noble respect takes *not* in might *but* merit." '—WALKER (*Crit.* iii, 51): Something evidently has dropped out. [HALLIWELL quotes 'another editor' as proposing to read : 'what poor duty *would, but* cannot do.' This is practically the same as Coleridge's emendation, but who this 'other editor' is I do not know, and he is apparently unknown to the Cam. Ed. In the textual notes of that edition this emendation is given as 'quoted by Halliwell.'—F. A. MARSHALL adopted it.—ED.] R. G. WHITE (ed. i): The only objection to Theobald's *willing* before 'duty' is that *simple, eager, struggling*, or one of many other disyllabic words might be inserted with equal propriety.—W. A. WRIGHT: There is no need for change ; the sense being, noble respect or consideration accepts the effort to please without regard to the merit of the performance. Compare *Love's*

Where I haue come, great Clearkes haue purpofed   100
To greete me with premeditated welcomes;
Where I haue feene them fhiuer and looke pale,
Make periods in the midft of fentences,
Throttle their practiz'd accent in their feares,
And in conclufion, dumbly haue broke off,   105
Not paying me a welcome.   Truft me fweete,
Out of this filence yet, I pickt a welcome:
And in the modefty of fearefull duty,
I read as much, as from the ratling tongue
Of faucy and audacious eloquence.   110
Loue therefore, and tongue-tide fimplicity,
In leaft, fpeake moft, to my capacity.
     Egeus. So pleafe your Grace, the Prologue is addreft.   113

---

100. *Clearkes*] *Clerkes* Q₁.
102. *Where*] *When* Han. Dyce ii, iii.
105. *haue*] *th' aue* White i conj.
107. *filence yet,*] Q₂Ff. *filence, yet,* Q₁,
Cap.

112. [Enter Philomon. Pope.   Re-
enter Philostrate. Cap. et seq. (subs.).
113. Egeus.] Philoft. Qq. Phil. Pope.
     your] *you* Pope i.

Lab. L. V, ii, 517: 'That sport best pleases that doth least know how,' &c. [The
difficulty here has arisen, I think, in taking 'might' in the sense of *power, ability,*
rather than in the sense of *will;* Kenrick states the meaning concisely when he says
it is about the same as taking 'the will for the deed.'—ED.]

100. **Clearkes**] BLAKEWAY: An allusion, I think, to what happened at Warwick,
where the recorder, being to address the Queen, was so confounded by the dignity of
her presence as to be unable to proceed with his speech. I think it was in Nichols's
Progresses of Queen Elizabeth that I read this circumstance, and I have also read
that her Majesty was very well pleased when such a thing happened. It was, therefore, a very delicate way of flattering her to introduce it as Shakespeare has done
here.—WALKER (*Crit.* iii, 51) calls attention to a parallel passage in Browne's *Brittania's Pastorals*, B. ii, Song i, but as *Brittania's Pastorals* were not published until
1613, they are not of the highest moment in illustrating this present play. It is more
to the point to cite, as MALONE cites, 'Deep clerks she dumbs.'—*Pericles*, V, Prologue 5.

105. **haue**] R. G. WHITE (ed. i): As 'have' has no nominative except 'I,' three
lines above, it may be a misprint for *th' aue;* but it is far more probable that *they* is
understood; for such license was common in Shakespeare's day, or rather, it was
hardly license then.

112. It is noteworthy, as tending to show the futility of almost all collation beyond
that of specified copies, even in the case of modern editions, that the CAM. ED. here
records 'Enter Philostrate. Pope (ed. 2).   Enter Philomon. Pope (ed. 1).' In my
copies of the first and second editions of Pope, it is 'Enter Philomon' in both
instances.—ED.

113. **addreft**] STEEVENS: That is, ready.

Duke.   Let him approach.   *Flor. Trum.*

Enter the Prologue.   *Quince.*   115
Pro.   If we offend, it is with our good will.

114. Flor. Trum.] Om. Qq.   115. Enter...] Enter Quince for the
Pyramus and Thisbe. An Inter-   prologue. Rowe.
lude. Cap.   Quince] Om. Qq.
Scene II. Pope+.

114, 220, 224, &c. **Duke**] See FLEAY, line 417, below.

114. **Flor. Trum.**] STEEVENS: It appears from Dekker's *Guls Hornbook*, 1609
[chap. vi, p. 250, ed. Grosart], that the prologue was anciently ushered in by trumpets. 'Present not your selfe on the Stage (especially at a new play) vntill the quak-
ing prologue hath (by rubbing) got culor into his cheekes, and is ready to giue the
trumpets their Cue, that hees vpon point to enter.'

115. **Enter the Prologue**] MALONE (*Hist. of Eng. Stage*, Var. 1821, vol. iii,
115): The person who spoke the prologue, who entered immediately after the third
sounding, usually wore a long black velvet cloak, which, I suppose, was best suited
to a supplicatory address. Of this custom, whatever may have been its origin, some
traces remained until very lately; a black coat having been, if I mistake not, within
these few years, the constant stage-habiliment of our modern prologue-speakers. The
complete dress of the ancient prologue-speaker is still retained in the play exhibited
in *Hamlet*, before the king and court of Denmark.—COLLIER (*Dram. Hist.* iii, 245,
ed. ii): In the earlier period of our drama the prologue-speaker was either the author
in person or his representative. . . . From the Prologue to Beaumont & Fletcher's
Woman Hater, 1607, we learn that it was, even at that date, customary for the person
who delivered that portion of the performance to be furnished with a garland of bay,
as well as with a black velvet cloak. . . . The bay was the emblem of authorship, and
the use of this arose out of the custom for the author, or a person representing him,
to speak the prologue. The almost constant practice for the prologue-speaker to be
dressed in a black cloak or in black, perhaps, had the same origin. [In the light of
this statement by Collier, the appearance here in the Folio of ' Quince' is noteworthy
as an indication that the Duke was to accept Quince as the author of the play.—ED.]
KNIGHT (*Introd.* p. 331): One thing is perfectly clear to us—that the original of
these editions [the two Quartos], whichever it might be, was printed from a genuine
copy and carefully superintended through the press. The text appears to us as perfect as it is possible to be, considering the state of typography in that day. There is
one remarkable evidence of this. The prologue to the interlude of the Clowns is
purposely made inaccurate in its punctuation throughout. . . . It was impossible to
have effected the object better than by the punctuation of Roberts's edition [Q₂]; and
this is precisely one of those matters of nicety in which a printer would have failed,
unless he had followed an extremely clear copy or his proofs had been corrected by
an author or an editor.

116–125. CAPELL: In this prologue a gentle rub upon players (country ones, we'll
suppose) seems to have been intended; whose deep knowledge of what is rehears'd
by them is most curiously mark'd in the pointing of this prologue; upon which must
have been taken some pains by the poet himself when it pass'd the press; for its
punctuation, which is that of his First Quarto, can be mended by nobody. In read-

That you fhould thinke, we come not to offend,   117
But with good will.   To fhew our fimple skill ,
That is the true beginning of our end.
Confider then, we come but in defpight.   120
We do not come, as minding to content you,
Our true intent is.   All for your delight,
We are not heere.   That you fhould here repent you,
The Actors are at hand ; and by their fhow ,
You fhall know all, that you are like to know.   125

Thef. This fellow doth not ftand vpon points.

Lyf. He hath rid his Prologue, like a rough Colt : he
knowes not the ftop. A good morall my Lord.   It is not
enough to fpeake, but to fpeake true.

Hip. Indeed hee hath plaid on his Prologue , like a   130
childe on a Recorder, a found, but not in gouernment.

---

122. *is. All*] *is all* Pope.
123. *heere. That*] *here that* Pope.
125. [Exit. Dyce ii.
126. *points*] *his points* Rowe i, Coll. ii
(MS). *this points* Rowe ii.

128. *A good*] Dem. *A good* Cam. conj.
130. *his*] *this* Qq, Cap. Steev. Mal.'90,
Coll. Ktly.
131. *a Recorder*] *the Recorder* Ff,
Rowe +.

---

ing it, we apprehend we see something, and so there is; for it is just possible to point
it into meaning (not sense), and that's all; an experiment we shall leave to the
reader.—KNIGHT has kindly performed for the reader this task which Capell says
'nobody' can do : 'Had the fellow stood "upon points," it would have run thus:
"If we offend, it is with our good will That you should think we come not to offend ;
But with good will to show our simple skill.   That is the true beginning of our end.
Consider then.   We come : but in despite We do not come.   As, minding to content
you, Our true intent is all for your delight.   We are not here that you should here
repent you.   The actors are at hand ; and, by their show, You shall know all that
you are like to know."   We fear that we have taken longer to puzzle out this enigma
than the poet did to produce it.'—STAUNTON calls attention to a similar distortion by
mis-punctuation in Roister Doister's letter to Dame Custance, beginning 'Sweete mis
tresse, where as I love you nothing at all, Regarding your substance and richesse
chiefe of all,' &c.—*Ralph Roister Doister*, III, ii.

128. **the stop**] W. A. WRIGHT : A term in horsemanship, used here in a punning
sense.   Compare *A Lover's Complaint*, 109: 'What rounds, what bounds, what
course, what stop he makes!'

131. **Recorder**] CHAPPELL (*Pop. Music*, &c., 246): Old English musical instruments were made of three or four different sizes, so that a player might take any of
the four parts that were required to fill up the harmony. . . . Shakespeare speaks in
Hamlet [III, ii, 329 of this ed., which see, if needful.—ED.] of the recorder as a
little pipe, and in [the present passage says] 'like a *child* on a recorder,' but in an
engraving of the instrument it reaches from the lip to the knee of the performer. . . .
Salter describes the *recorder*, from which the instrument derives its name, as situate

Thef.   His ſpeech was like a tangled chaine: nothing   132
impaired, but all diſordered.   Who is next?
        Tawyer with a Trumpet before them.

Enter Pyramus and Thisby, Wall, Moone-ſhine, and Lyon.   135
Prol.   Gentles, perchance you wonder at this ſhow,

132, 133. *His...diſordered*] As verse.
First line, ending *chaine* (reading *impair'd*) Coll. White i, Ktly (Ktly reading *like unto*).

132. *chaine*] *skein* Anon. ap. Cam.

133. *impaired...diſordered*] *impair'd
...disorder'd* Rowe+.

        next] the next Ff, Rowe+.

134. Tawyer...] Om. Qq, Pope et seq.

135. Enter] Enter the Presenter Coll.

ii, iii (MS), Dyce ii.   Enter, with a
Trumpet and the Presenter before them.
White.

    135. Wall, Moone-ſhine] and Wall,
and Moonſhine, Q₁ — *Lyon.] Lion, as in dumb shew.*
Theob.

    136. Prol.] Presenter. White, Coll. ii,
iii (MS), Dyce ii, iii.

in the upper part of it, *i. e.* between the hole below the mouth and the highest hole
for the finger.   He says: 'Of the kinds of music, vocal has always had the preference in esteem, and in consequence the recorder, as *approaching nearest to the sweet
delightfulness of the voice,* ought to have the first place in opinion, as we see by the
universal use of it confirmed.'—SINGER (ed. ii): To *record* anciently signified to
modulate. . . . In modern cant *recorders* of corporations are called *flutes,* an ancient
jest, the meaning of which is perhaps unknown to those who use it.

    131. gouernment] M. MASON: Hamlet says, '*Govern* these ventages with your
fingers and thumb'—[III, ii, 372].

    134. Tawyer, &c.] COLLIER (ed. ii): In the MS 'Tawyer' and his trumpet are
erased, and 'Enter Presenter' is made to precede the other characters.   Such, no
doubt, was the stage-arrangement when this play was played in the time of the old
annotator, and we may presume that it was so in the time of Shakespeare.   In the
early state of our drama a *Presenter,* as he was called, sometimes introduced the characters of a play, and as Shakespeare was imitating this species of entertainment, we
need entertain little doubt that 'Tawyer with a trumpet,' of F₁, was, in fact, the Presenter, a part then filled by a person of the name of Tawyer.   In the MS also the
Presenter is made to speak the argument of the play.   This was to be made intelligible with a due observation of points, and could not properly be given to the same
performer who had delivered the prologue, purposely made so blunderingly ridiculous.
In the Qq and Ff, both the prologue and the argument, containing the history of the
piece, are absurdly assigned to one man.   Perhaps such was the case when the number of the company could not afford separate actors.—R. G. WHITE (ed. i) and DYCE
(ed. ii) adopted this plausible 'Presenter' of Collier's MS.   The former says that
'the error in the prefix ['*Prol.*' in line 136] arose from the similarity of *Pref.* and
Prol., which in the old MS could hardly be distinguished from each other.'—W. A.
WRIGHT: 'Tawyer' looks like a misprint for *Players,* unless it is the name of the
actor who played the part of Prologue.   [All doubt, however, is set at rest, and proof
afforded not only that the Folio was printed from a stage-copy, but that 'Tawyer' is
neither a misprint nor a substitution for 'Presenter,' through the discovery by HALLI-
WELL (*Outlines,* p. 500) that Tawyer 'was a subordinate in the pay of Hemmings,

But wonder on, till truth make all things plaine.   137
This man is *Piramus,* if you would know ;
This beauteous Lady, *Thisby* is certaine.
This man, with lyme and rough-caft, doth prefent   140
Wall, that vile wall, which did thefe louers funder :
And through walls chink (poor foules) they are content
To whifper.   At the which, let no man wonder.
This man, with Lanthorne, dog, and bufh of thorne,   144

139. *beauteous*] *beautious* Qq.   143. *whiſper. At*] *whisper, at* Theob.
141. *that vile*] *the vile* Ff, Rowe +.   *whisper ; at* Cap.
                                            144. *Lanthorne*] *lanterne* Q₁.

his burial at St Saviour's in June, 1625, being thus noticed in the sexton's MS note-book : " William Tawier, Mr. Heminges man, gr. and cl., xvj. *d.*" ']

139. **Thisby**] Hanmer uniformly retains this spelling where the clowns are the speakers ; elsewhere, in stage-directions, &c. his spelling is the correct, *Thisbe.* The inference is that he intends *Thisby* to be phonetic, and herein I quite agree with him. In the mouths of the clowns 'Thisbe' was pronounced, I doubt not, *Thisbei,* and ' Pyramus,' *Peiramus.*   See next note and line 170, *post.*—ED.

139. **certaine**] STEEVENS : A burlesque was here intended in the frequent recur-rence of *certain* as a bungling rhyme in poetry more ancient than the age of Shake-speare.   Thus in a short poem entitled *A lytell Treatise called the Disputacyon or the Complaynte of the Herte through perced with the Lokynge of the Eye.*   Imprynted at Lōdon in Flete-strete at the Sygne of the Sonne by Wynkyn de Worde : 'And houndes syxescore and mo certayne—To whome my thought gan to strayne certayne—Whan I had fyrst syght of her certayne—In all honoure she hath no pere certayne—To loke upon a fayre Lady certayne—As moch as is in me I am contente certayne—They made there both two theyr promysse certayne—All armed with margaretes certayne,' &c.   Again, in *The Romaunce of the Sowdone of Babylone,* ' He saide " the xij peres bene alle dede, And ye spende your goode in vayne, And therfore doth nowe by my rede, Ye ſhalle ſee hem no more certeyn." '—[ll. 2823-6, ed. E. E. Text. Soc.]. Again, ' The kinge turned him ageyn, And alle his Ooste him with, Towarde Mount-rible certeyne.'—[*Ib.* ll. 2847-9.   In the search through this Romaunce to verify Steevens's quotations I found three other examples, in lines 567, 570, and 1453, of this ' most convenient word,' as W. A. Wright says, ' for filling up a line and at the same time conveying no meaning.'—WALKER (*Crit.* i, 114) cites this ' certain' among other words as of ' a peculiar mode of rhyming—rhyming to the eye as at first sight appears.'   In this particular passage ' it is,' he says, ' of a piece with the purposely *incondite* composition of this *dramaticle.*'   Wherein, I think, he is right as far as he goes, but he does not go far enough.   Not only was this ' dramaticle ' ' incondite,' but it is meant to be thoroughly burlesque, where words are mispronounced and accents misplaced.   See lines 170, 171, below.—ED.]

140. **lyme**] HUDSON [reading *loam*]: In Wall's speech, a little after, the old copies have ' This loame, this rough-cast,' &c.   So also in III, i : 'And let him have some plaster, or some *Lome,* or some rough-cast about him.'—R. G. WHITE reverses the misprint, and thinks that ' lome ' is a misprint for ' lime.'   The Cam. Ed. notes that *loam* is als a conjecture of Capell in MS.

Prefenteth moone-fhine.   For if you will know,   145
By moone-fhine did thefe Louers thinke no fcorne
To meet at *Ninus* toombe, there, there to wooe :
This grizy beaft (which Lyon hight by name)
The trufty *Thisby*, comming firft by night,
Did fcarre away, or rather did affright :   150
And as fhe fled, her mantle fhe did fall ;
Which Lyon vile with bloody mouth did ftaine.
Anon comes *Piramus*, fweet youth and tall,
And findes his *Thisbies* Mantle flaine ;
Whereat, with blade, with bloody blamefull blade,   155
He brauely broacht his boiling bloudy breaft,

148. *grizy*] F₁. *grizly* QqFf.
      Lyon hight by name] *by name*
Lion hight Theob. Warb. Johns. Cap.
Steev. Mal. Var. Knt, Hal. Sta. Dyce ii,
iii.   *lion by name hight* Coll. iii.

149. Line marked as omitted, Ktly,
Malone conj.
150. *fcarre*] *fcare* F₃F₄.
151. *did fall*] *let fall* Pope +.
154. *his*] *his gentle* Ff, Rowe.   *his
trusty* Qq, Pope et seq.

147. wooe] R. G. WHITE (ed. i): It may be remarked here upon the rhyme of
'woo' with 'know' that the former word seems to have had the pure vowel sound of
o.   It was spelled *wooe* or *woe*, and as often in the latter way as the former.

148. hight by name] THEOBALD: As all the other parts of this speech are in
alternate rhyme, excepting that it closes with a *couplet;* and as no rhyme is left to
'name,' we must conclude either a verse is slipt out, which cannot now be retrieved;
or by a transposition of the words, as I have placed them, the poet intended a *triplet.*
[See Text. Notes.]—The COWDEN-CLARKES (*Sh. Key*, p. 674) : We believe that the
defective rhyming was intentional, to denote the slipshod style of the doggerel that
forms the dialogue in the Interlude, which we have always cherished a conviction Shakespeare intended to be taken as written by Peter Quince himself; because
in the Folio we find '*Enter the Prologue Quince,*' and because in IV, i, Bottom says,
'I will get Peter Quince *to write a ballad* of this dream,' showing that Quince is an
author as well as stage-manager and deliverer of the Prologue.   [The present Editor
wholly agrees with the foregoing.   In any attempt to improve the language of the
rude mechanicals the critic runs a perilous risk of becoming identified with them.
—ED.]

151. fall] For other examples where this verb and other intransitive verbs are used
transitively, see ABBOTT, § 291.

152, 155, 157. Lyon . . . blade . . . Mulberry] ABBOTT, § 82: Except to ridicule it, Shakespeare rarely indulges in this archaism of omitting *a* and *the.*

155, 156. JOHNSON: Upton rightly observes that Shakespeare in these lines ridi-
cules the affectation of beginning many words with the same letter.   He might have
remarked the same of 'The raging rocks And shivering shocks.'   Gascoigne, contemporary with our poet, remarks and blames the same affectation.—CAPELL descries
in these lines 'a particular burlesque of passages,' which he reprints in his *School*,
from *Sir Clyomon and Sir Chlamydes,* and refers to *Gorboduc* as 'blemished with one

And *Thisby*, tarrying in Mulberry fhade,   157
His dagger drew, and died.   For all the reft,
Let *Lyon*, *Moone-fhine*, *Wall*, and Louers twaine,
At large difcourfe, while here they doe remaine.   160
Exit all but Wall.

    Thef.   I wonder if the Lion be to fpeake.

    Deme.   No wonder, my Lord : one Lion may, when
many Affes doe.
Exit Lyon, Thisbie, and Moonefhine.   165

    Wall.   In this fame Interlude, it doth befall,
That I, one *Snowt* (by name) prefent a wall :
And fuch a wall, as I vvould haue you thinke,
That had in it a crannied hole or chinke :
Through which the Louers, *Piramus* and *Thisbie*   170

---

157. *And Thisby,...fhade*] *And* (*This-by...shade,*) Steev.'85, Mal. Steev.'93, Var. Knt, Hal. Sta. (subs.).

    in] *in the* F₃F₄, Rowe+.

161. Om. Qq.   Exeunt... Rowe+. Exeunt Prologue, Thisbe, Lion and Moonshine. Cap. Steev. Mal.   Exeunt Pies. Thisbe, Lion and Moonshine. Coll. Exeunt Prologue, Presenter, Pyramus,

Thisbe, Lion and Moonshine. White.

    163, 164. *one...doe*] Separate line, Coll. White i.

165. Om. Rowe et seq.

166. *Interlude*] *enterlude* Q₁.

167. Snowt] Flute Qq, Pope.

170. Piramus] Pyr'mus Theob. Warb. Johns.

    Thisbie] This-be Theob. i.

---

affectation, an almost continual alliteration, which Shakespeare calls " affecting the letter," and has exposed to ridicule in *Love's L. L.* IV, ii, 57 : " I will something affect the letter, for it argues facility.   The preyful princess pierced and prick'd a pretty pleasing pricket," &c.'   Steevens gives several examples of alliteration from early literature, Halliwell adds more, and Staunton still others, but as I can discern no possible light in which they illustrate Shakespeare, they are not here repeated.—W. A. WRIGHT says of this alliteration that ' it was an exaggeration of the principle upon which Anglo-Saxon verse was constructed.'

    167. Snowt] Here again is an instance of the greater accuracy for stage purposes of the Folio.   The Qq have ' Flute,' who was to act Thisby.

    169. crannied] See the extract from Golding's *Ovid*, in the Appendix.—CAPELL, who, as an actor, was, I fear, a case of arrested developement, tells us that ' the reciter who would give a comic expression to " crannied " and to " cranny " must make both vowels long.'

    170. Thisbie] GUEST (i, 91) thus scans : ' Through which | these lov | ers : Pyr | amus and | *Thisby* | ,' and adds, 'Shakespeare elsewhere accents it *This* | *by ;* he doubtless put the old and obsolete accent into the mouth of his " mechanicals " for the purposes of ridicule.'   As I understand Guest, ' the old and obsolete accent' is *Thisbee*, to rhyme with ' secretlee.'—WALKER (*Crit.* i, 114) here, as in line 139, suggests that there is a rhyme for the eye, and likewise proposes the same scansion as that just given by Guest, but adds ' this is not likely.'   I cannot wholly agree with either Guest or Walker   That ' Thisbie ' must rhyme with ' secretly ' is clear, and

Did whifper often, very fecretly.   171
This loame, this rough-caft, and this ftone doth fhew,
That I am that fame Wall ; the truth is fo.
And this the cranny is, right and finifter,
Through which the fearefull Louers are to whifper.   175

  Thef. Would you defire Lime and Haire to fpeake
better ?

  Deme. It is the vvittieft partition, that euer I heard
difcourfe, my Lord.

  Thef. Pyramus drawes neere the Wall, filence.   180

      Enter Pyramus.

  Pir. O grim lookt night, ô night with hue fo blacke,   182

---

172. *loame*] *lome* Qq. *loam* F_3F_4.
lime Cap. conj. Var.'21, Coll. Dyce i, ii,
White i.

174. [holding up one hand with a
finger expanded. Rann.

179. *difcourfe*] *difcourfed* F_3F_4.

180. *Wall, filence*] *Wall : filence* Q_1F_4,
Rowe et seq.

181. Om. Qq.

---

that in the mouth of rude mechanicals there must be an uncouth or an absurd pronunciation seems to me equally clear. ' Secretly,' like the majority of words ending
in an unaccented final *y*, was probably pronounced *secretlei* (see Ellis, *Early Eng.
Pron.* pp. 959, 977, 981) by everybody, whether mechanicals or not. The absurdity
then comes in by making ' Thisbie ' rhyme with it : *Thisbei.* See line 139, above.—ED.

172. **loame . . . shew**] The VAR. 1821 (cited by CAM. ED. as ' Reed,' which is
not, I think, strictly accurate) here reads *lime*, and notes 'so folio; quartos *lome*,' a
mis-statement which, in a note, the CAM. ED. corrects, but fails to detect what is, I
believe, the source of Boswell's or Malone's error. Either the one or the other of
these latter editors had been examining CAPELL'S *Various Readings*, where occurs
the following : ' This lime, | shew, F^s. | ,' which those who are schooled in the ' an-
fractuosities ' of the Capellian mind understand as meaning that ' This lime ' is a conjectural emendation, and that the Folios read ' shew ' instead of the *show* of Capell's
own text. Boswell or Malone overlooked the conjectural emendation and supposed
that ' F^s ' referred to *lime*, and hence, I think, the tears.—ED.

174. **sinister**] Elsewhere in *Hen. V:* II, iv, 85, this word is accented on the
middle syllable, as given by ABBOTT, § 490, but here, as Abbott says, this accent is
used comically.—W. A. WRIGHT says that ' sinister ' is used by Snout for two reasons
—first, because it is a long word, and then because it gives a sort of rhyme to
' whisper.'

178. **partition**] FARMER : I believe the passage should be read, This is the wit-
tiest *partition* that ever I heard *in discourse.* Alluding to the many stupid *partitions*
in the argumentative writings of the time. Shakespeare himself, as well as his contemporaries, uses ' discourse ' for *reasoning ;* and he here avails himself of the double
sense, as he had done before in the word ' partition.'

182. **lookt**] For examples of passive participles used not passively, see ABBOTT,
§ 374; albeit it is hardly worth while to attempt an explanation of any grammatical
anomaly in the speeches of these ' mechanicals.'—ED.

O night, which euer art, when day is not :   183
O night, ô night, alacke, alacke, alacke,
I feare my *Thisbies* promife is forgot.   185
And thou ô vvall, thou fweet and louely vvall,
That ftands betweene her fathers ground and mine,
Thou vvall, ô vvall, ô fwect and louely vvall,
Shew me thy chinke, to blinke through vvith mine eine.
Thankes courteous vvall. *Ioue* fhield thee vvell for this.   190
But vvhat fee I ? No *Thisbie* doe I fee.
O vvicked vvall, through vvhom I fee no bliffe ,
Curft bc thy ftoncs for thus deceiuing mee.

   Thef.   The vvall me-thinkes being fenfible , fhould
curfe againe.   195
   Pir.   No in truth fir, he fhould not. *Deceiuing me,*
Is *Thisbies* cue ; fhe is to enter, and I am to fpy
Her through the vvall.   You fhall fee it vvill fall.

                    Enter Thisbie.

Pat as I told you ; yonder fhe comes.   200
   Thif.   O vvall, full often haft thou heard my mones,
For parting my faire *Piramus,* and me.
My cherry lips haue often kift thy ftones ;
Thy ftones vvith Lime and Haire knit vp in thee.   204

---

186. *thou fweet and*] Ff, Rowe, White i. *O fweet and* Pope+, Ktly. *ô fweete, ô* Qq, Cap. et cet.

187. *ftands*] *ftandes* F₄. *ftandft* Q₁, Cap. Steev. Mal. Var. Coll. Dyce, White, Sta. Cam. Ktly (subs.).

189. [Wall holds up his fingers. Cap.
196–200. Prose, Pope et seq.

197. *enter,*] *enter now,* Qq, Cap. et seq.

198. *fall.*] *fall* QqF₃, Pope et seq.

199. Enter Thisbie.] After line 200 Qq, Pope et seq.

203. *haue*] *hath* F₄, Rowe.

204. *Haire*] *hayire* Q₁.
    vp in thee] *now againe* Qq.

182, 184, 186, &c. ô] I suppose that this circumflexed *o* is used merely to avoid confusion with the *o* which is an abbreviation of *of.* It is scarcely likely that it has any reference to pronunciation.—ED.

188. ô vvall, ô sweet] HALLIWELL: The repetition of the vocative case is of frequent occurrence in Elizabethan writers. Thus Gascoigne, in his translation of the *Jocasta* of Euripides, 1566, paraphrases this brief sentence of the original, 'O mother, O wife most wretched,' into : 'O wife, O mother, O both wofull names, O wofull mother, and O wofull wyfe ! O woulde to God, alas ! O woulde to God, Thou nere had bene my mother, nor my wyfe !'   Compare also the following : 'Oh ! Love, sweet Love, oh ! high and heavenly Love, The only line that leades to happy life.'—Breton's *Pilgrimage to Paradise,* 1592.

204. in thee] See Text. Notes.—WHITE (ed. i) : A variation of this kind between

Pyra. I fee a voyce ; now vvill I to the chinke ,   205
To fpy and I can heare my *Thisbies* face.   *Thisbie*?

Thif. My Loue thou art, my Loue I thinke.

Pir. Thinke vvhat thou vvilt, I am thy Louers grace,
And like *Limander* am I trufty ftill.

Thif.   And like *Helen* till the Fates me kill.   210

Pir. Not *Shafalus* to *Procrus*, was fo true.

Thif. As *Shafalus* to *Procrus*, I to you.

Pir. O kiffe me through the hole of this vile wall.

Thif. I kiffe the wals hole, not your lips at all.

Pir. Wilt thou at *Ninnies* tombe meete me ftraight   215
way ?

Thif.   Tide life, tide death, I come without delay.

Wall.   Thus haue I *Wall*, my part difcharged fo ;
And being done, thus *Wall* away doth go.   *Exit Clow.*

Du. Now is the morall downe betweene the two   220
Neighbors.

205, 206. *fee...heare*] *heare...fee* Ff,
Rowe.

206. *and I*] *an I* Pope et seq.

    Thisbie] Separate line, Rowe ii
et seq.

207. *Loue thou art, my Loue*] QqFf,
Cam. White ii.   *Love thou art, my love,*
Rowe, Pope.   *Love ! thou art, my love,*
Theob. Warb. Johns.   *Love ! thou art
my love,* Han. et cet.

209. Limander] Limandea Pope.

210. *And like*] *And I like* QqF₂, Rowe
et seq.

213. *vile*] *vilde* Q₁.

217. *Tide...tide*] ' *Tide...'tide* Cap. et
seq.

    [Exeunt Pyra. and Th. Dyce.

219. Exit Clow.] Om. Qq.   Exeunt
Wall, Pyra. and Th. Cap.

220, 225, &c. Du.] Duk. Q₁.   Thes.
Rowe et seq.

220. *morall downe*] *Moon vfed* Qq,
Pope i.   *moral down* Rowe, White i.
mure all down Theobald conj. Han. Coll.
ii.   *wall downe* Coll. MS, White ii.
mural obstacle (or *partition*) *down* Wagner conj.   *Mural down* Pope ii et cet.

F₁ and the Qq is not worthy of notice, save for the evidence it affords that the copy
of Q₂, which Heminge and Condell furnished as copy to the printers of F₁ had been
corrected either by Shakespeare or some one else in his theatre.

209, 210, 211. **Limander . . . Helen . . . Shafalus to Procrus**] CAPELL (116
a) : This 'Limander' should be Paris, by the lady he is coupl'd with ; and he is
call'd by his other name, Alexander, corrupted into 'Alisander' (as in *Love's Lab. L.*
V, ii, 567, et seq.) and 'Lisander,' which master Bottom may be allow'd to make
' Limander' of.—JOHNSON : Limander and Helen are spoken by the blundering
player for Leander and Hero. Shafalus and Procrus, for Cephalus and Procris.—
MALONE : *Procris and Cephalus,* by Henry Chute, was entered on the Stationers'
Registers by John Wolff in 1593, and probably published in the same year. It was
a poem, but not dramatic, as has been suggested.—HALLIWELL : Chute's poem is
alluded to in Nash's *Have with You to Saffron Walden,* 1596.—BLACKSTONE :

Dem. No remedie my Lord, when Wals are fo wil-   223
full, to heare without vvarning.

223. *heare*] *rear* Han. Warb. Cap. *sheer* Han. conj. MS (ap. Cam.). *leave* Gould.

Limander stands evidently for Leander, but how came 'Helen' to be coupled with
him? Might it not have originally been wrote *Heren*, which is as ridiculous a corruption of Hero as the other is of her lover?

220. **morall**] THEOBALD (*Sh. Rest.* p. 142): I am apt to think the poet wrote
'now is the *mure all* down,' and then Demetrius's reply is apposite enough.—R. G.
WHITE (ed. i) : *Mural* for *wall* is an anomaly in English, and is too infelicitous to
be regarded as one of Shakespeare's daring feats of language. . . . 'Moon used' of
the Qq could not be a misprint for 'moral down.' . . . It should be remembered that
the moon figures in the interlude, as the spectators knew; and as to the use that the
two neighbours were to make of the moon, the remark of Demetrius indicates it
plainly enough : '*No remedy*, my lord, when walls are so wilful to *hear without warning.*' But Shakespeare evidently thought that it would be plainer if the wall were
represented both as the restraint upon the passions of the lovers and as a pander to
them, and so he changed 'moon used' to 'moral down.' He did this, I believe, with
the more surety of attaining his point, because 'moral' was then pronounced *mo-ral*,
and 'mural,' as I am inclined to think, *moo-ral*. [In his ed. ii, WHITE adopts Collier's *wall* without comment.]—COLLIER (ed. ii) : It would seem that in the time of
the old MS neither 'moral' nor *mure all* were the words on the stage; he inserts
wall.—W. A. WRIGHT: Pope's emendation, so far as I am aware, has no evidence
in its favour. Perhaps the Qq reading 'Now is the Moon vsed' is a corruption of a
stage-direction, and the reading of the Ff may have arisen from an attempt to correct
in manuscript the words in a copy of the Qto by turning 'Moon' into 'Wall,' the
result being a compound having the beginning of one word and the end of the other.
If there were any evidence of the existence of such a word as *mural* used as a substantive, it would be but pedantic and affected, and so unsuited to Theseus. Having
regard, therefore, to the double occurrence of the word 'wall' in the previous speech,
and its repetition by Demetrius, I cannot but think that [Collier's *wall* is right], just
as Bottom says 'the wall is down,' line 344.—HENRY JOHNSON (p. xvi) : The agreement of the Qq gives a strong presumption in favour of the correctness of a reading.
Something besides can be said for the reasonableness of this passage. The Prologue
had announced ['moone-shine,' see lines 144–147]. The Enterlude then proceeded
as far as this agreement of Pyramus and Thisbie to meet at the tomb, and Wall, who
had served between the two neighbors, makes his explanation and leaves the stage.
Thereupon the Duke says that now, in accordance with the statement of the Prologue, the Moon will be used between the two neighbors, probably in some such ingen-
uous way as the Wall had been. [The objection to Collier's *wall* is, I think, that it
makes Theseus's remark so very tame, not far above the level of a remark by Bottom.
Perhaps it may receive a little force if we suppose that Wall suddenly drops to his
side his extended arm. I am inclined to accept White's explanation that in the old
pronunciation lay a pun, now lost, and for a pun, as Johnson said, Shakespeare would
lose the world, and be content to lose it.—ED.]

223. **to heare**] For 'to hear,' equivalent to *as to hear*, see ABBOTT, § 281.

223. **to heare**] WARBURTON: Shakespeare could never write this nonsense; we
should read : 'to *rear* without warning,' *i. e.* it is no wonder that walls should be

Dut. This is the fillieft ftuffe that ere I heard.   224

224, 227, &c. Dut.] Hip. Rowe et seq.   224. *ere*] *euer* Q₁, Cap. Steev. Mal.
Var. Dyce i, Sta. Cam. Ktly, White ii.

---

suddenly down, when they were as suddenly up; *rear'd without warning.*—HEATH: Perhaps the reader may be pleased to think the poet might possibly have written, ' to *disappear* without warning,' and in that case the words ' without warning ' must be understood to refer solely to the neighbours whose dwellings the wall in question parted.—KENRICK (*Rev.* p. 22): The interview between Pyramus and Thisbe is no sooner over than Wall, apparently without waiting for his cue, as nobody speaks to him and he speaks to no person in the drama, takes his departure. When, therefore, Demetrius replies to Theseus ' when walls are so wilful to hear without warning ' he means ' are so wilfull as to take their *cue* before it is given to them.' That the expression, however, may bear some latent meaning, I do not deny; possibly it may refer to a custom practised by the magistrates in many places abroad, of sticking up a notice or warning on the walls of ruinated or untenanted houses, for the owners either to repair or pull them quite down.—FARMER: Demetrius's reply alludes to the proverb, ' Walls have ears.' A *wall* between almost any *two neighbours* would soon be *down*, were it to exercise this faculty, without previous *warning*. [This is, perhaps, the correct interpretation.—ED.]

224. **This is,** &c.] MAGINN (p. 119): When Hippolyta speaks scornfully of the tragedy, Theseus answers that the best of this kind (scenic performances) are but shadows, and the worst no worse if imagination amend them. She answers that it must be *your* imagination then, not *theirs.* He retorts with a joke on the vanity of actors, and the conversation is immediately changed. The meaning of the Duke is that, however we may laugh at the silliness of Bottom and his companions in their ridiculous play, the author labours under no more than the common calamity of dramatists. They are all but dealers in shadowy representations of life; and if the worst among them can set the mind of the spectator at work, he is equal to the best. The answer to Theseus is that none but the best, or, at all events, those who approach to excellence, can call with success upon imagination to invest their shadows with substance. Such playwrights as Quince the carpenter,—and they abound in every literature and every theatre,—draw our attention so much to the absurdity of the performance actually going on before us that we have no inclination to trouble ourselves with considering what substance in the background their shadows should have represented. Shakespeare intended the remark as a compliment or as a consolation to less successful wooers of the comic or the tragic Muse, and touches briefly on the matter; but it was also intended as an excuse for the want of effect upon the stage of some of the finer touches of such dramatists as himself, and an appeal to all true judges of poetry to bring it before the tribunal of their own imagination; making but a matter of secondary inquiry how it appears in a theatre as delivered by those who, whatever others may think of them, would, if taken at their own estimation, ' pass for excellent men.' His own magnificent creation of fairy land in the Athenian wood must have been in his mind, and he asks an indulgent play of fancy not more for Oberon and Titania, the glittering rulers of the elements, than for the shrewd and knavish Robin Goodfellow, the lord of practical jokes, or the dull and conceited Bottom, ' the shallowest thickskin of the barren sort.'—DOWDEN (p. 70): Maginn has missed the more important significance of the passage. Its dramatic appropriateness is the essential

Du. The beſt in this kind are but ſhadowes, and the   225
worſt are no worſe, if imagination amend them.

Dut. It muſt be your imagination then, & not theirs.

Duk. If wee imagine no worſe of them then they of
themſelues, they may paſſe for excellent men. Here com
two noble beaſts, in a man and a Lion.   230

---

229. *com*] comes Ff, Rowe i. *come*   *beasts in a man* Warb. *beasts in, a moon*
Qq, Rowe ii et seq.   Han. Johns. Steev. Sing. Dyce, Ktly.
   230. *beaſts, in a man*] QqFf, Rowe i,   *beasts in, a man* Rowe ii et cet.
W. A. Wright. *beasts in a moon* Theob.

---

point to observe. To Theseus, the great man of action, the worst and the best of
these shadowy representations are all one. He graciously lends himself to be amused,
and will not give unmannerly rebuff to the painstaking craftsmen who have so labor-
iously done their best to please him. But Shakespeare's mind by no means goes
along with the utterance of Theseus in this instance any more than when he places
in a single group the lover, the lunatic, and the poet. With one principle enounced
by the Duke, however, Shakespeare evidently does agree, namely, that it is the business of the dramatist to set the spectator's imagination to work, that the dramatist
must rather appeal to the mind's eye than to the eye of sense, and that the co-opera-
tion of the spectator with the poet is necessary. For the method of Bottom and his
company is precisely the reverse, as Gervinus has observed, of Shakespeare's own
method. They are determined to leave nothing to be supplied by the imagination.
Wall must be plastered; Moonshine must carry lanthorn and bush. And when Hippolyta, again becoming impatient of absurdity, exclaims, ' I am aweary of this moon !
would he would change !' Shakespeare further insists on his piece of dramatic criticism by urging, through the Duke's mouth, the absolute necessity of the man in the
moon being *within* his lanthorn. Shakespeare as much as says, ' If you do not
approve my dramatic method of presenting fairy-land and the heroic world, here is a
specimen of the rival method. You think my fairy-world might be amended. Well,
amend it with your own imagination. I can do no more unless I adopt the artistic
ideas of these Athenian handicraftsmen.'

   230. **in a man**] THEOBALD : Immediately after Theseus's saying this, we have
' Enter Lyon and Moonshine.' It seems very probable, therefore, that our author
wrote ' in a *moon* and a lion.' The one having a crescent and a lanthorn before him,
and representing the man in the moon; and the other in a lion's hide.—MALONE :
Theseus only means to say that the ' man ' who represented the moon, and came in at
the same time, with a lanthorn in his hand and a bush of thorns at his back, was as
much a beast as he who performed the part of the lion.—FARMER : Possibly ' man '
was the marginal interpretation of *moon-calf*, and, being more intelligible, got into
the text.—W. A. WRIGHT adheres to the punctuation of the QqFf, although he
deserted it in the second edition of the *Cam. Ed.* His note is that the change of the
comma from before ' in ' to after it is unnecessary. ' " In " here signifies " in the character of," see IV, ii, 25 : " sixpence a day *in* Piramus, or nothing." Theobald, with
great plausibility, reads *moon*.' [WALKER (*Crit.* i, 315) also conjectured *moon*,
independently. Possibly the choice between ' man ' and *moon* will lie in the degree
of absurdity which strikes us in calling either the one or the other a beast.—HARNESS
has the shrewd remark, which almost settles the question in favour of ' man,' to the

Enter Lyon and Moone-ſhine.   231

Lyon. You Ladies, you (whoſe gentle harts do feare
The ſmalleſt monſtrous mouſe that creepes on floore)
May now perchance, both quake and tremble heere,
When Lion rough in wildeſt rage doth roare.   235
Then know that I, one *Snug* the Ioyner am
A Lion fell, nor elſe no Lions dam :   237

236. *one* Snug] *as* Snug Qq, Steev.'85.   237. *A Lion fell*] *No lion fell* Rowe +,
   236, 237. *one...dam*] *am Snug the*   Cap. Dyce ii, Coll. iii. *A lion-fell* Sing.
joiner in A Lion-fell, or else a Lion's   ii, White, Cam. Ktly. *A lion's fell* Field,
skin. Daniel.   Dyce i, Coll. ii.
                                               elſe] *eke* Cap. conj.

---

effect that Theseus saw merely 'a man with a lantern, and could not possibly conceive
that he was intended to " disfigure moonshine." '—ED.]

237. **Lion fell, nor else**] MALONE: That is, that I am Snug, the joiner, and
neither a lion nor a lion's dam. Dr Johnson has justly observed in a note on *All's
Well* that *nor*, in the phraseology of our author's time, often related to two members
of a sentence, though only expressed in the latter. So, in the play just mentioned,
' contempt *nor* bitterness Were in his pride or sharpness.'—I, ii, 36.—BARRON FIELD
(*Sh. Soc. Papers*, ii, 60): I would observe upon [this note of Malone] that where the
verb *follows* the negative nominatives, as in the passage quoted by Malone, this is the
phraseology not only of Shakespeare's, but of the present time, as in Gray : ' Helm
nor hauberk's twisted mail, Nor ev'n thy virtues, tyrant, shall avail,' &c., but I defy
any commentator to produce an instance of such a construction where the verb *precedes*
the nominatives. In that case, the verb has already affirmed before the word of negation comes, and the negative cannot relate back, to make the verb deny. In other
words, it is impossible that ' I am a lion, nor a lion's dam ' can mean ' I am *not* a lion,
nor a lion's dam,' or ' I am *neither* a lion nor a lion's dam.' I boldly say there is no
instance in the English language at any time of such a phraseology. And what does
Malone do with the word ' else ' ? He gives it no meaning. And why say a fell or
cruel lion ? Or introduce a lion's dam or mother ? I will now show how one little
letter shall light up the whole passage with natural meaning and give a sense to every
word : 'A lion*s* fell, nor else no lion's dam.' ' I, Snug, the joiner, am only a lion's
skin ; nor any otherwise than as a lion's skin may be said to be pregnant with a lion,
am I the mother of one.' *Fell* is a word scarcely yet obsolete for *skin*, and now the
words ' else ' and ' dam ' have a meaning ; and all this sense is obtained by only supposing that the letter *s* has dropped from the text. It might, indeed, be done without
any other alteration than that of a hyphen, *lion-fell ;* but, as we find, in other parts of
Shakespeare the words *calf's skin* and *lion's skin* with the genitive, I have thought it
better to insert the *s*.—COLLIER (ed. ii) : This judicious change of Field is doubtless
correct, as it is the reading of the MS.—LETTSOM (*Blackwood*, Aug. 1853) : Field's
excellent emendation ought to go into the text, if it has not done so already.—R. G.
WHITE (ed. i) : Field's change is the minutest ever proposed for the solution of a
real difficulty.—HALLIWELL [substantially following RITSON, p. 48] : Snug means to
say, ' I am neither a lion fell, nor in any respect a lion's dam,' that is, I am neither a
lion nor a lioness. The conjunction *nor* frequently admitted of *neither* being pre-

For if I fhould as Lion come in ftrife   238
Into this placc, 'twcre pittie of my life.
    Du. A verie gentle beaft, and of a good confcience.   240
    Dem. The verie beft at a beaft, my Lord, ỳ ere I faw.
    Lif. This Lion is a verie Fox for his valor.
    Du. True, and a Goofe for his difcretion.
    Dem. Not fo my Lord : for his valor cannot carrie
his difcretion, and the Fox carries the Goofe.   245
    Du. His difcretion I am fure cannot carrie his valor :
for the Goofe carries not the Fox. It is well ; leaue it to
his difcretion, and let vs hearken to the Moone.
    Moon. This Lanthorne doth the horned Moone pre-
fent.   250

239. *of my*] *on my* Qq, Cap. Steev. Mal. Var. Coll. Sta. Dyce ii, Ktly, Cam. White ii. *o'* Cap. conj. MS (ap. Cam.).
248. *hearken*] *liften* Q₁, Cap. Steev. Mal. Var. Coll. Dyce, Cam. Ktly.

248. *Moone*] *man* Anon. ap. Cam.
249, &c. *Lanthorne*] *lantern* Steev. '93, Mal. Reed, Knt, Sing. Dyce, Coll. Sta.

---

viously understood, and two negatives often merely strengthened the negation. Barron Field ingeniously avoided the grammatical difficulty.—STAUNTON : Field's emendation is extremely ingenious ; but in the rehearsal of this scene Snug is expressly enjoined to show his face through the lion's neck, tell his name and trade, and say : ' If you think I am come hither as a lion, it were pity of my life ; *No*, I am no such thing.' I am disposed, therefore, if *nor* is not to be taken as relating to both members of the sentence, to read [with Rowe] *i. e.* neither lion nor lioness.—WALKER (*Crit.* i, 262) : Field's emendation is perhaps right, if *A* can be tolerated. But surely Shakespeare wrote and pointed [as in Rowe]. [All appeals to grammar in the interpretation of the speeches of these clowns seem to me superfluous ; its laws are here suspended. The change of 'A' into *No* is, therefore, needless. Since 'A lion fell' (with or without a hyphen) may mean *A lion's skin*, no change whatever is required. Barron Field's high deserving lies in his discerning that ' fell ' is a noun and not an adjective ; and that by this interpretation point is given to ' lion's dam.' For Snug to say that he is ' neither a lion nor a lioness ' is, to me, pointless, but all is changed if we suppose him to say that he is a lion's skin, and only because, as such, he encloses a lion, can he be a lioness.—ED.]

239. **of my**] COLLIER (ed. ii) : ' On *your* life ' is the reading of the MS. We follow the older reading, but it is questionable. [The very fact that it is ' questionable ' makes it, in Snug's mouth, the more probable.—ED.]

241. **beft at a beaft**] WHITE (ed. i) : From the nature of this speech it is plain that ' best ' and ' beast ' were pronounced alike. [This is stated, I think, a little too strongly in a matter which is difficult of proof. Compositors, we know, were apt to spell phonetically, accordingly we find them spelling *least*, *lest*, which is a pretty good guide to the pronunciation of that word. But I can recall no instance where *beast* is spelled *best*. There may be such. Age and familiarity with the old compositors make one extremely cautious.—ED.]

15

De. He fhould haue worne the hornes on his head.   251

Du. Hee is no crefcent, and his hornes are inuifible, within the circumference.

Moon. This lanthorne doth the horned Moone pre-fent: My felfe, the man i'th Moone doth feeme to be.   255

Du. This is the greateft error of all the reft; the man fhould be put into the Lanthorne. How is it els the man i'th Moone?

Dem. He dares not come there for the candle. For you fee, it is already in fnuffe.   260

Dut. I am vvearie of this Moone; vvould he would change.   262

---

251. *on his*] *upon his* Han.
252. *no*] *not* Coll. ii, iii (MS), Dyce ii, iii.
254, 255. Two lines of verse, QqF₃F₄, Rowe et seq.
255, 268. *man i'th Moone*] *man-i'--the-moon* Dyce ii, iii.

255. *doth*] Ff, Rowe+, White i, Sta. *doe* Qq, Cap. et cet.
259, 260. Prose, Q₁, Pope et seq.
261. *vvearie*] *aweary* Q₁, Cap. Steev. Mal. Var. Coll. Dyce, White, Sta. Cam. Ktly.
vvould] *'would* Theob.

---

249. **Lanthorne**] STEEVENS needlessly modernised this word into *lantern*, and has been followed by many of the best editors, thereby obliterating the jingle, if there be one, in ‘This Lant*horne* doth the *horned* moone present.’ The CAMBRIDGE EDI-TION, both first and second, nicely discriminates between the pronunciation of Snug and of Theseus by giving *lanthorn* to the former and *lantern* to the latter. This dis-tinction W. A. WRIGHT overlooked or disregarded in his own *Clarendon Edition.* —ED.

252. **no crescent**] COLLIER [reading *not*]: The *t* most likely dropped out in the press.

255. **the man i'th Moone**] As an illustration of the text the voluminous mass of folk-lore which has gathered around this ‘man’ seems no more appropriate here than in Caliban’s allusion to him in *The Tempest.* The zealous student is referred to the two or three folio pages in Halliwell *ad loc.* or to Grimm’s *Deutsche Mythologie* there cited. From tender years every English-speaking child knows that there is a man in the moon, and is familiar with his premature descent and with his mysterious desire to visit the town of Norwich. Which is all we need to know here.—ED.

256. **greatest error of all the rest**] ABBOTT, § 409, cites this, among others, as an instance of ‘the confusion of two constructions (a thoroughly Greek idiom, though independent in English),’ and illustrates it by Milton’s famous line: ‘The fairest of her daughters, Eve,’ where the two confused constructions are ‘Eve fairer *than* ali her daughters’ and ‘Eve fairest *of* all women.’—W. A. WRIGHT cites Bacon’s Essay *Of Envy* (ed. Wright, p. 35): ‘Of all other Affections, it is the most importune and continuall.’

260. **snuffe**] JOHNSON: ‘Snuff’ signifies both the cinder of a candle and hasty anger.—STEEVENS: Thus also, in *Love’s Lab. L.* V, ii, 22, ‘You’ll mar the light by taking it in snuff.’

Du.   It appeares by his fmal light of difcretion, that   263
he is in the wane : but yet in courtefie, in all reafon, vve
muft ftay the time.   265

Lyf.   Proceed Moone.

Moon.   All that I haue to fay, is to tell you, that the
Lanthorne is the Moone; I, the man in the Moone; this
thorne bufh, my thorne bufh ; and this dog, my dog.

Dem.   Why all thefe fhould be in the Lanthorne: for   270
they are in the Moone. But filence, heere comes *Thisby.*

Enter Thisby.

Thif.   This is old *Ninnies* tombe : where is my loue ?
Lyon.   Oh.

The Lion roares, Thisby runs off.   275

Dem,   Well roar'd Lion.
Du.   Well run *Thisby.*
Dut.   Well fhone Moone.
Truly the Moone fhines with a good grace.
Du.   Wel mouz'd Lion.   280
Dem.   And then came *Piramus.*
Lyf.   And fo the Lion vanifht.   282

263. *his*] *this* Pope, Han. Mal.
268. *in the*] *ith* Q₁. *i'the* Cap. Hal. Cam.
270. *Why all*] *Why ? all* Q₁.
270, 271. *for they*] *for all thefe* Q₁, Coll. Sing. Dyce, Cam. Ktly.
273. *old...tombe*] *ould...tumbe* Q₁.
where is] *wher's* Q₂.
274. *Oh.*] *Oh. Ho. Ho.*— Han.
275. Om. Qq.
278. *fhone*] *fhoone* Q₂.

278, 279. As prose, Qq, Cap. et seq.
279. *with a*] *with* Rowe i.
[Lion shakes Thisbe's mantle, and Exit. Cap.
280. *mouz'd*] QqFf, Theob. Warb. Johns. *mouth'd* Rowe, Pope, Han. *mous'd* Cap. et cet.
281, 282. *then came...fo the Lion vanifht*] *fo comes...fo the moon vanishes* Steev.'85. *fo comes...then the moon vanishes* Farmer, Steev.'93, Var. Sing. i.

263. *fmal light of difcretion*] STAUNTON: So in *Love's Lab. L.* V, ii, 734, ' I have seen the day of wrong *through the little hole of discretion.*' The expression was evidently familiar, though we have never met with any explanation of it.

280. *mouz'd*] STEEVENS: Theseus means that the lion has well tumbled and bloodied the veil of Thisby.—MALONE: That is, to mammock, to tear in pieces, as a cat tears a mouse.

281, 282. **And . . . vanisht**] FARMER thus emended these lines: 'And *so comes* Pyramus. And *then* the *moon vanishes.*' Of this emendation STEEVENS remarks that ' it were needless to say anything in its defence. The reader, indeed. may ask why this glaring corruption was suffered to remain so long in the text.'—HARNESS: I have restored the text of F₁. Farmer's alteration on the last line, ' and so the moon vanishes,' cannot be right, for the very first lines of Pyramus on entering eulogise its

Enter Piramus.   283

Pyr. Sweet Moone, I thank thee for thy funny beames,
I thanke thee Moone, for fhining now fo bright: 285
For by thy gracious, golden, glittering beames,

286. *beames*] Qq. *gleams* Knt conj. Sta. i, White, Sing. ii, Cam. Dyce ii, iii, Marshall. *ftreames* Ff, Rowe et cet.

beams, and his last words are addressed to it as present. [To the same effect, substantially, COLLIER, ed. i.]—KNIGHT [who also returns to the QqFf]: Farmer makes
this correction, because, in the mock-play, the moon vanishes after Pyramus dies. Bu'
Demetrius and Lysander do not profess to have any knowledge of the play; it is
Philostrate who has 'heard it over.' They are thinking of the classical story, and,
like Hamlet, they are each 'a good chorus.'—DYCE (ed. i) [in answer to Knight]:
Now, if Demetrius and Lysander had *no knowledge of the play*, they must have been
sound asleep during the Dumb-show and the laboured exposition of the Prologue-
speaker. And if they were 'thinking of the classical story,' they must have read it
in a version different from that of Ovid; for, according to his account, the 'lea saeva'
had returned 'in silvas' *before the arrival of Pyramus*, who, indeed, appears to have
been somewhat slow in keeping the assignation, 'Serius egressus,' &c. (Compare,
too, the long and tedious *History of Pyramus and Thisbie* in the *Gorgious Gallery
of Gallant Inventions*, 1578, p. 171 of the reprint.) [To the foregoing Dyce adds in
his ed. ii]: Mr W. N. LETTSOM observes, 'Should not we transpose these lines, and
read, "And so the *lion's* vanished. *Now* then *comes* Pyramus"?'—Mr Swynfen Jervis would transpose the lines without altering the words. [Herein Jervis was anticipated by SPEDDING, whose emendation is recorded in the first CAM. ED., 1863, and
is adopted by HUDSON, by W. A. WRIGHT, and by WAGNER.]

286. glittering beames] KNIGHT: If the editor of F_2 had put *gleams* [instead
of *streames*] the ridicule of excessive alliteration would have been carried further.—
COLLIER: The editor of F_2 substituted *streams*, perhaps, upon some then existing
authority which we have no right to dispute.—DYCE (*Rem.* p. 49): The editor of F_2
gave here what Shakespeare undoubtedly wrote. Neither Knight nor Collier appears
to recollect that from the earliest times *stream* has been frequently used in the sense
of *ray*. [Here follow eight examples of the use of *stream* in this sense from Chaucer
to Beaumont and Fletcher, to which might be added another given by CAPELL, from
Sackville's *Induction* in the *Mirror of Magistrates*, all valuable, but superfluous here.
—STAUNTON (ed. i) adopted Knight's conj., but in his *Library Edition* returns to
'streams,' which he says he prefers.]—WALKER (*Crit.* iii, 52): I think the alliteration
requires *gleams*.—LETTSOM (footnote to Walker): I must confess I should prefer *gleams*,
but for one reason. If I may trust Mrs Cowden-Clarke, this common and convenient
word never once appears in so voluminous a writer as Shakespeare. Even its kinsman, *gloom*, is also an exile from his pages. *Glooming* or *gloomy* has slipped in at
the close of *Rom. and Jul* ; otherwise it is confined to *1 Hen. VI* and *Tit. And.* It
really looks as if Shakespeare had an objection to these words; still, for that very
reason, he may have put *gleams* into the mouth of Bottom. [Mrs Furness's *Concordance* gives an instance of *gleam'd* from the *R. of L.* 1378: 'And dying eyes gleam'd
forth their ashy lights'; and of *gloomy*, from the same, line 803: 'Keep still possession of thy gloomy place.' The unanimity of the Quartos and First Folio cannot be
lightly whistled down the wind. The fact that 'beams' is wrong and *streams* or

I truſt to taſte of trueſt *Thisbies* ſight.   287
But ſtay : O ſpight ! but marke, poore Knight,
What dreadful dole is heere ?
Eyes do you ſee ! How can it be !   290
O dainty Ducke : O Deere !
Thy mantle good ; what ſtaind with blood !
Approch you Furies fell :
O Fates *!* come, come : Cut thred and thrum,
Quaile, cruſh, conclude, and quell.   295

---

287. *taſte*] *take* Qq, Coll. Cam. White ii.   Han. Warb.   *deare* Qq, Johns. et seq.
    Thisbies] Thiſby Q₁, Coll. Cam.   292. *good ; what*] good, what, Q₁.
White ii.   Thisbie Q₂.   good, what Q₂.
288–295. Twelve lines, Pope et seq.   293. *you*] Ff, Rowe+, White i.   *ye*
291. *Deere*] Ff, Rowe, Pope, Theob.   Qq, Cap. et cet.

---

gleams manifestly right, seems to me the very reason why it should be retained in the speech of one whose eye had not heard, nor his ear seen, nor his hand tasted a dream which he had in the wood where he had gone to rehearse obscenely.—ED.]

287. **taste**] W. A. WRIGHT : This is quite in keeping with 'I see a voice,' line 205. [And yet, after this true note, Wright, in his text, follows the correct but incor rect Qq.—ED.]

293. **Approch you Furies, &c.**] MALONE : In these lines and in those spoken by Thisbe, 'O sisters three,' &c., lines 334, *et seq.* the poet probably intended, as Dr Farmer observed to me, to ridicule a passage in *Damon and Pythias*, by Richard Edwards. 1582 : ' Ye *furies*, all at once On me your torments trie : Gripe me, you greedy griefs, And present pangues of death, *You sisters three, with cruel handes With speed come stop my breath !*' [p. 44, ed. Hazlett's *Dodsley*].—W. A. WRIGHT (p. xx) : Certainly in this play [just cited] and in the tragical comedy of *Appius and Virginia*, printed in 1575, may be found doggerel no better than that which Shakespeare puts into the mouth of Bottom. See, for example, the speech of Judge Appius to Claudius, beginning, 'The furies fell of Limbo lake My princely days do short,' &c. [p. 131, ed. Hazlett's *Dodsley*]. It is also worth while to notice that the song quoted in *Rom. and Jul.* IV, v, 128, ' When griping grief the heart doth wound,' &c. is by the author of *Damon and Pythias.*

294. **thrum**] NARES : The tufted part beyond the tie, at the end of the warp, in weaving; or any collection or tuft of short thread.—WARNER : It is popularly used for very coarse yarn. The maids now call a mop of yarn a *thrum mop*.—STEEVENS : So in Howell's *Letter to Sir Paul Neale :* ' Translations are like the wrong side of a Turkey carpet, which useth to be full of thrums and knots, and nothing so even as the right side.' The thought is borrowed from Don Quixote.—HALLIWELL : So in Herrick, ' Thou who wilt not love, doe this; Learne of me what Woman is. Something made of thred and thrumme; A meere Botch of all and some.'—*Poems*, p. 84 [vol. i, p. 100, ed. Singer].

295. **quell**] JOHNSON : Murder; *manquellers* being, in the old language, the term for which *murderers* is now used.—NARES : Hence ' Jack the giant-queller' was once used [Notes on *Macbeth*, I, vii, 72].

Du. This paſſion, and the death of a deare friend,   296
Would go neere to make a man looke ſad.
    Dut. Beſhrew my heart, 'but I pittie the man.
    Pir. O wherefore Naturr, did'ſt thou Lions frame?
Since Lion vilde hath heere deflour'd my deere :   300
Which is : no, no, which was the faireſt Dame
That liu'd, that lou'd, that lik'd, that look'd with cheere.
Come teares, confound : Out ſword, and wound
The pap of *Piramus* :
I, that left pap, where heart doth hop ;   305
Thus dye I, thus, thus, thus.
Now am I dead, now am I fled, my ſoule is in the sky,
Tongue loſe thy light, Moone take thy flight,   308

296, 297. As prose, Qq, Johns. et seq.
297. *neere*] *well near* Ktly.
298. *I*] *I do* Ktly.
300. *vilde*] Qq. *vild* Ff, Hal. White i.
wild Rowe. *vile* Pope et cet.
    deere] *deare* Qq, Rowe et seq.
302. *lik'd, that look'd*] *lik't, that look't*
Qq.

303–309. Twelve lines, Johns. et seq.
305. *hop*] *rap* Gould.
    [Stabs himself. Dyce.
308. *Tongue*] *Sunne* or *sun* Capell
conj. *Moon* Elze.
    loſe] *looſe* Q₁.
    Moone] *Dog* Elze.

---

296. **This passion**] COLLIER (*Notes*, &c., p. 109) : This 'passion' has particular
reference to the 'passion' of Pyramus on the fate of Thisbe, and therefore the MS
properly changes 'and' to *on*, and reads : 'This passion *on* the death,' &c. [Collier
did not afterwards, in his ed. ii, refer to this correction.]—R. G. WHITE (*Putnam's
Maga.* Oct. 1853, p. 393) : The humour of the present speech consists in coupling
the ridiculous fustian of the clown's assumed passion with an event which would, in
itself, make a man look sad. Collier's MS extinguishes the fun at once by reading
on.—STAUNTON : This reading *on* by the MS is one proof among many of his inability to appreciate anything like subtle humour. Had he never heard the old proverbial
saying, ' He that loseth his wife and sixpence, *hath lost a tester* ' ?—W. A. WRIGHT :
For ' passion,' in the sense of violent expression of sorrow,' see line 319, and *Hamlet*,
II, ii, 587 : ' Had he the motive and the cue for passion.'
    303. **confound**] Both STEEVENS and W. A. WRIGHT cite examples to elucidate
the meaning of this word. Where is the British National Anthem ?—ED.
    305. **pap**] STEEVENS : It ought to be remembered that the broad pronunciation,
now almost peculiar to the Scotch, was anciently current in England. ' Pap,' therefore, was sounded *pop*. [See ELLIS, *Early Eng. Pron.* p. 954, where the rhyme in
these lines is noted.]
    306. **thus, thus, thus**] COLLIER (ed. ii) : Modern editors give no cause for the
death of Pyramus, but the MS places these words in the margin : *Stab himself as
often*, meaning, no doubt, every time he utters the word ' thus.'
    308. **Tongue**] CAPELL : Bottom's ' Tongue,' instead of *Sunne* or *Sun*, is a very
choice blunder.—HALLIWELL : The present error of ' tongue ' for *sun* appears too
absurd to be humorous, and it may well be questioned whether it be not a misprint.

Now dye, dye, dye, dye, dye.

 Dem. No Die, but an ace for him ; for he is but one. 310

 Liſ. Leſſe then an ace man. For he is dead, he is nothing.

 Du. With the helpe of a Surgeon, he might yet reco-
uer, and proue an Aſſe.

 Dut. How chance Moone-ſhine is gone before ? 315
Thisby comes backe, and findes her Louer.

Enter Thisby.

 Duke. She wil finde him by ſtarre-light.
Heere ſhe comes, and her paſſion ends the play.

 Dut. Me thinkes ſhee ſhould not vſe a long one for 320
ſuch a *Piramus* : I hope ſhe will be breefe.

 Dem. A Moth wil turne the ballance, which *Piramus*
which *Thisby* is the better.   (eyes.

 Lyſ. She hath ſpyed him already, with thoſe ſweete 324

---

309. [Dies. Theob. dies. Exit Moonshine. Cap.

314. *and proue*] *and yet proue* Q₁,
White i.

315, 316. Prose, Q₁, Pope et seq.

315. *chance*] *chance the* F₃F₄, Rowe+.
before ? Thisby...Louer.] *before*
Thisby...Lover ? Rowe et seq.

316. *comes*] *come* Cap. (corrected in
Errata).

317. Om. Qq. After *comes* line 319,
Cap. After line 319, Steev.

318, 319. Prose, Qq, Cap. et seq.

322. *Moth*] QqFf, Rowe+, Cap.
Steev.'85, Mal. *mote* Heath, Steev.'93
et seq.

323. *better.*] *better : he for a man ;*
God warnd vs : ſhe, for a woman ; God
bleſſe vs. Qq (subs.), Coll. Sing. Hal.
Dyce, White, Cam. Ktly (all reading
warrant), Sta. (reading *warn'd*).

---

310. **Die**] CAPELL (117 *b*) : To make even a lame conundrum of this, you are to
suppose that ' die ' implies two, as if it came from *duo.*

315. **chance**] See I, i, 139.

317. **Enter**] In this command to the actor to be ready to enter before he has to
make his actual appearance on the stage, we have another proof that the Folio was
printed from a stage-copy.—ED.

319. **Heere she comes, &c.**] THEOBALD (Nichols, *Illust.* ii, 240) : This, I think,
should be spoken by Philostrate, and not by Theseus ; for the former had seen the
interlude rehearsed and consequently knew how it ended. [This was not repeated
in Theobald's subsequent edition. He probably remembered that Theseus had seen
the Dumb-show.—ED.]

322. **Moth**] See III, i, 168.

323. **better**] See Text. Notes for a line in the Qq here omitted. We have already
had a similar omission after III, ii, 364, which was there clearly due to carelessness,
inasmuch as the necessary stage-direction ' Exeunt ' was included in the omission.
But here there is no such proof of carelessness ; and the only explanation advanced is

Dem.   And thus fhe meanes, *videlicit*.   325
This.   Afleepe my Loue?   What, dead my Doue?
O *Piramus* arife :
Speake, Speake.   Quite dumbe? Dead, dead? A tombe
Muft couer thy fweet eyes.
Thefe Lilly Lips, this cherry nofe,   330

---

325. *meanes*] QqFf, Rowe, Pope, Cam. *moans* Theob. et cet.
325–341. Twenty-three lines, Pope, Han.   Twenty-four lines, Theob. et seq.
328. *tombe*] *tumbe* Q$_1$.   *toombe* Q$_2$.
329. *thy fweet*] *my fweet* F$_3$F$_4$.

330. *Thefe...nofe*] *This lily lip, This cherry tip* Coll. ii, iii (MS).   *This lily brow, This cherry mow* Kinnear.   *These ...With cherry tips* Gould.
Lips] *brows* Theob. Warb. Johns. Steev. Sing. Ktly.   *toes* Bulloch.

---

that given first by COLLIER, that the omission was 'possibly on account of the statute against using the name of the Creator, &c., on the stage, 3 Jac. I, ch. 21, which had not passed when the original editions were printed.'   This statute, passed in 1605, imposed a penalty of ten pounds on any player who should 'jestingly or prophanely speak or use the holy name of God.'   It was, however, so easy to convert 'God bless us' into 'Lord bless us,' and was frequently so converted withal, that this explanation seems hardly adequate, and yet, until a better offers, it must suffice.—STAUNTON conjectures that for *warn'd* we should probably read *ward*, and interprets : ' From such a man, God defend us ; from such a woman, God save us.'   See Staunton's later note contributed to *The Athenæum*, cited at III, ii, 419.—ED.

324. Does not this remark of Lysander's give us an insight of the way in which Thisbe, like any amateur actor, ran at once to Pyramus's body, without looking to the right or left ?—ED.

325. **meanes**] THEOBALD : It should be *moans*, i. e. laments over her dead Pyramus.—STEEVENS : ' Lovers make moan ' (line 332) appears to countenance the alteration.—RITSON : But ' means ' had anciently the same signification as *moans*.   Pinkerton observes that it is a common term in the Scotch law, signifying to *tell*, to *relate*, to *declare ;* and the petitions to the lords of session in Scotland run : ' To the lords of council and session humbly means and shows your petitioner.'   Here, however, it evidently signifies *complains*.   Bills in Chancery begin in a similar manner : ' Humbly complaining sheweth unto your lordship,' &c.—STAUNTON : Theobald's change is, perhaps, without necessity, as ' means ' appears formerly to have sometimes borne the same signification.   Thus in *Two Gent.* V, iv, 136 : ' The more degenerate and base art thou, To make such means for her as thou hast done.'—DYCE (ed. ii) : But in this passage [cited by Staunton] ' To *make* such *means* ' surely signifies (as Steevens explains it) ' to make such interest for, take such pains about.'—W. A. WRIGHT : *Moans* does not fit in well with ' videlicet.' . . . The old word *mene* is of common occurrence.   [Jamieson, *Scotch Dict.*, gives : To Mene, Meane, To utter complaints, to make lamentations.   ' If you should die for me, sir knight, There's few for you will meane ; For mony a better has died for me, Whose graves are growing green.'—*Minstrelsy Border*, iii, 276.   Knowing the propensity which apparently, according to the critics, characterised Shakespeare, how is it that a modern poet has escaped the same condemnation ?   With this stanza from the *Border Minstrelsy* still in our ears, recall the exquisite line in ANDREW LANG'S *Helen of Troy :* ' O'er Helen's shrine the grass is growing green In desolate Therapnae.'—ED.]

Thefe yellow Cowflip cheekes   331
Are gone, are gone : Louers make mone :
His eyes were greene as Leekes.
O fifters three, come, come to mee,
With hands as pale as Milke,   335
Lay them in gore, fince you haue fhore
With fheeres, his thred of filke.
Tongue not a word : Come trufty fword :   338

336. *Lay*] *Lave* Theob. Warb. Johns.   337. *thred*] *threede* Q₁.
337. *his*] *this* F₃F₄, Rowe, Pope, Han.

---

330. **Lilly . . . nose**] THEOBALD : All Thisby's lamentation till now runs in regular rhyme and metre.   I suspect, therefore, the poet wrote 'These lilly *brows*.'   Now *black* brows being a beauty, *lilly* brows are as ridiculous as a *cherry* nose, *green* eyes, or *cowslip* cheeks.—MALONE : 'Lips' could scarcely have been mistaken, either by the eye or ear, for *brows*.—FARMER : Theobald's change cannot be right.   Thisbe has before celebrated her Pyramus as 'Lilly white of hue.'   It should be 'These lips lilly, This nose cherry.'   This mode of position adds not a little to the burlesque of the passage.—STEEVENS : We meet with somewhat like this passage in George Peele's *Old Wives Tale*, 1595 : '*Huanebango*. Her coral lips, her crimson chin. . . . *Zantippa*. By gogs-bones, thou art a flowting knave : her coral lips, her crimson chin !'—[p. 239, ed. Dyce.   I can really see no parallelism here.   Huanebango is in earnest ; he goes on to speak of her 'silver teeth,' 'her golden hair,' &c., and Zantippa is merely a coarse scold who rails at everybody ; had not this citation been repeated in modern editions, it would not have been included here.—ED.]—COLLIER (ed. ii) adopts the change of his MS, 'This lily lip, This cherry tip,' and notes that this was 'in all probability Shakespeare's language, which would have additional comic effect if Thisbe at the same time pointed to the nose of the dead Pyramus.'—R. G. WHITE : Farmer's emendation was ingenious at least.   But *nip*, a term which is yet applied to the nose in the nursery, might be mistaken for 'nofe,' written with a long *s*, and it seems to me not improbable that it was so mistaken in this instance.   [Of all tasks, that of converting the intentional nonsense of this interlude into sense seems to me the most needless.—ED.]

332. **green as leeks**] In a private letter to Lady Martin, which I am permitted to quote, Mrs ANNA WALTER THOMAS writes : 'I was interested when in Southern Wales to hear an old woman praising the beautiful blue eyes of a child in these words, "maé nhw'n lãs fel y cenin," *i. e.* they are as green as leeks, green and blue having the same word (*glas*, from the same root as our *glaucous*) in Welsh. So Thisbe must have borrowed her phrase from Welsh.'—ED.

334. **O sisters three**] See Malone's note on l. 293, above.

338. **sword**] HALLIWELL (*Memm.* 1879, p. 35) : There are reasons for believing that, notwithstanding the general opinion of the unfitness of the *Mid. N. D.* for representation, it was a successful acting play in the seventeenth century.   An obscure comedy, at least, would scarcely have furnished Sharpham with the following exceedingly curious allusion, evidently intended as one that would be familiar to the audience, which occurs in his play of *The Fleire*, published in 1607: '*Kni.* And how lives he with 'am ? *Fle.* Faith, like Thisbe in the play, 'a has almost kil'd himselfe

Come blade, my breſt imbrue :
And farwell friends, thus *Thisbie* ends ;   340
Adieu, adieu, adieu.
    Duk.   Moon-ſhine & Lion are left to burie the dead.
    Deme.   I, and Wall too.
    Bot.   No, I aſſure you, the wall is downe, that parted
their Fathers.   Will it pleaſe you to ſee the Epilogue, or   345
to heare a Bergomask dance, betweene two of our com-
pany ?
    Duk.   No Epilogue, I pray you ; for your play needs
no excuſe.   Neuer excuſe ; for when the plaiers are all
dead, there need none to be blamed.   Marry, if hee that   350
writ it had plaid *Piramus*, and hung himſelfe in *Thisbies*
garter, it would haue beene a fine Tragedy : and ſo it is
truely, and very notably diſcharg'd.   But come, your
Burgomaske ; let your Epilogue alone.
The iron tongue of midnight hath told twelue.   355
Louers to bed, 'tis almoſt Fairy time.
I feare we ſhall out-ſleepe the comming morne,   357

---

339. [Stabs herself. Dyce.
340. *farwell*] *farewell* QqFf.
341. [Dies. Theob.
344. Bot.] Lyon. Qq.
    [Starting up. Cap.
346. *Bergomask*] *Bergomaske* Q_1F_2.
350. *need*] *be* Cap. conj.

350. *Marry*] *Mary* Q_1.
   351. *hung*] Ff, Rowe +,White. *hangd*
or *hang'd* or *hanged* Qq, Cap. et cet.
   354. *Burgomaske*] *Burgomask* F_3F_4,
Rowe.   *Bergomask* Pope et seq.
    [Here a dance of Clowns. Rowe.
A dance by two of the Clowns. White.

---

with the scabberd,'—a notice which is also valuable as recording a fragment belonging to the history of the original performance of Shakespeare's comedy, the interlude
of the clowns, it may be concluded, having been conducted in the extreme of burlesque,
and the actor who represented Thisbe, when he pretends to kill himself, falling upon
the scabbard instead of upon the sword. [See C. A. BROWN in *Appendix*.]

   344. Bot.] COLLIER (ed. ii): The Qq give this speech to *Lion*.   Perhaps such
was the original distribution, but changed before F_1 was printed, to excite laughter on
the resuscitation of Pyramus.

   346. **Bergomask**] HANMER (*Gloss.*): A dance after the manner of the peasants
of Bergomasco, a country in Italy belonging to the Venetians.   All the buffoons in
Italy affect to imitate the ridiculous jargon of that people ; and from thence it became
a custom to mimic also their manner of dancing.—W. A. WRIGHT: If we substitute
Bergamo for Bergomasco, Hanmer's explanation is correct.   Alberti (*Dizion. Uni-
vers.*) says that in Italian 'Bergamasca' is a kind of dance, so called from Bergamo,
or from a song which was formerly sung in Florence.   The Italian Zanni (our 'zany')
is a contraction for Giovanni in the dialect of Bergamo, and is the nickname for a
peasant of that place.

As much as we this night haue ouer-watcht, 358
This palpable groffe play hath well beguil'd
The heauy gate of night Sweet friends to bed, 360
A fortnight hold we this folemnity.
In nightly Reuels ; and new iollitie. *Exeunt.*

Enter Pucke.

Puck Now the hungry Lyons rores,
And the Wolfe beholds the Moone : 365

---

359. *palpable groffe*] QqFf, Rowe+, Coll. Hal. White i. *palpable-gross* Cap. et cet.

360. *gate*] *gaite* Rowe ii, Pope. *gait* Johns. et seq.

362. *Reuels*] *Revel* Rowe, Pope, Theob. Han. Warb.

Scene III. Pope+. Scene II.

Cap. Steev. Mal. Var. Knt, Coll. Sing. Hal. White i, Sta. Ktly. Scene continued, Dyce, Cam. White ii, Huds. Rlfe.

364. *hungry*] *Hungarian* so quoted by Grey i, 78.

Lyons] *lion* Rowe et seq.

365. *beholds*] QqFf, Rowe, Pope, Steev.'73, '78, '85. *behowls* Warb. et cet.

---

360. gate] HEATH: I believe our poet wrote *gait*, that is, the tediousness of its progression.—STEEVENS: That is, slow progress. So in *Rich. II:* III, ii, 15: 'And heavy-gaited toads lie in their way.' [*Gait* is here applied metaphorically to hours, as in line 410 it is applied without metaphor to fairies.—ED.]

363. Enter Pucke] COLLIER (ed. ii) adds, from his MS, 'with a broom on his shoulder.' 'A broom,' says Collier, ' was unquestionably Puck's usual property on the stage, and as he is represented on the title-page of the old history of his *Mad Pranks*, 1628.'

364. Now, &c.] COLERIDGE (p. 104): Very Anacreon in perfectness, proportion, grace, and spontaneity! So far it is Greek; but then add, O! what wealth, what wild ranging, and yet what compression and condensation of, English fancy! In truth, there is nothing in Anacreon more perfect than these thirty lines, or half so rich and imaginative. They form a speckless diamond.

364. Lyons] MALONE: It has been justly observed by an anonymous writer that ' among this assemblage of familiar circumstances attending midnight, either in England or its neighbouring kingdoms, Shakespeare would never have thought of intermixing the exotic idea of the "hungry lions roaring," which can be heard no nearer than the deserts of Africa, if he had not read in the 104th Psalm: "Thou makest darkness that it may be *night*, wherein all the beasts of the forest do move; the *lions roaring* after their prey, do seek their meat from God." '—STEEVENS: I do not perceive the justness of the foregoing anonymous writer's observation. Puck, who could ' encircle the earth in forty minutes,' like his fairy mistress, might have snuffed ' the spiced Indian air;' and consequently an image, foreign to Europeans, might have been obvious to him. . . . Our poet, however, inattentive to little proprieties, has sometimes introduced his wild beasts in regions where they are never found. Thus in Arden, a forest in French Flanders, we hear of a lioness, and a bear destroys Antigonus in Bohemia.

365. beholds] WARBURTON: I make no question that it should be *behowls*, which is the wolf's characteristic property.—THEOBALD (*Letter to Warburton*, May, 1730,

Whileſt the heauy ploughman ſnores,   366

All with weary taske fore-done.

Now the waſted brands doe glow,

Whil'ſt the ſcritch-owle, ſcritching loud,

Puts the wretch that lies in woe,   370

In remembrance of a ſhrowd.

Now it is the time of night,

That the graues, all gaping wide,

Euery one lets forth his ſpright,   374

---

366. *Whileſt*] *Whilſt* Qq, Rowe et seq.
367. *fore-done*] *foredoone* Q₁.
369. *ſcritch-owle*] *ſcriech-owle* Q₁.
screech-owl Coll. Dyce, Hal. White, Cam.

369. *ſcritching*]   *ſcrieching*   Q₁.
schrieking Johns. *screeching* Coll. Dyce,
Hal. White, Cam.

Nichols, ii, 603) : I am prodigiously struck with the justness of your emendation [*be-howls*]. I remember no image whatever of the wolf *simply* gazing on the moon; but of the *night-howling* of that beast we have authority from the poets. Virgil, *Georgics*, i, 486: again, *Æneid*, vii, 16. [In Theobald's edition he added] So in Marston's *Antonio and Mellida* [Second Part, III, iii], where the whole passage seems to be copied from this of our author : ' Now barkes the wolfe against the full cheekt moon ; Now lyons half-clamd entrals roare for food ; Now croakes the toad, and night crowes screech aloud, Fluttering 'bout caſements of departed soules ; Now gapes the graves, and through their yawnes let loose Imprison'd spirits to revisit earth.'—JOHNSON : The alteration is better than the original reading, but perhaps the author meant only to say that the wolf *gazes* at the moon.—MALONE : The word ' beholds ' was, in the time of Shakespeare, frequently written *behoulds* (as, I suppose, it was then pro- nounced), which probably occasioned the mistake. These lines also in Spenser's *Fairie Queene*, Bk i, Canto v, 30, which Shakespeare might have remembered, add support to Warburton's emendation : 'And, all the while she [*Night*] stood upon the ground, The wakefull dogs did never cease to bay ; As giving warning of th' un- wonted sound, With which her yron wheeles did them affray, And her darke griesly looke them much dismay : The messenger of death, the ghastly owle, With drery shriekes did also her bewray ; And hungry wolves continually did howle At her abhorred face, so filthy and so fowle.' [If it be assumed that the compositors set up at dictation, the mishearing of ' beholds ' for *behowls* is not difficult of comprehension. —ED.]

367. **fore-done**] DYCE : That is, overcome.—ABBOTT, § 441 : *For-* is used in two words now disused, ' Forslow no longer.'—*3 Hen. VI :* II, iii, 56; ' She fordid her- self.'—*Lear*, V, iii, 256. In both words the prefix has its proper sense of *injury*.— W. A. WRIGHT : ' For ' in composition is like the German *ver-*, and has sometimes a negative and sometimes an intensive sense.

369. **scritch-owle**] DYCE (ed. ii) : I cannot but wonder that any editor should print here, with Q₂ and Ff, ' scritch ' and ' scritching,' when the best of the old eds., Q₁, has *scriech-owle* and *scrieching*.

272. **Now it is, &c.**] STEEVENS : So in *Hamlet*, III, ii, 406 : ' 'Tis now the very witching time of night When church-yards yawn.'

In the Church-way paths to glide.   375
And we Fairies, that do runne,
By the triple *Hecates* teame,
From the prefence of the Sunne,
Following darkeneffe like a dreame,
Now are frollicke ; not a Moufe   380
Shall difturbe this hallowed houfe.
I am fent with broome before,
To fweep the duft behinde the doore.

Enter King and Queene of Fairies, with their traine.
Ob.   Through the houfe giue glimmering light,   385

375. *Church-way*] *church-yard* Poole's   385, 386. *houfe giue...light, ...fier,*]
Eng. Parnassus (ap. Hal.).   *house, giv'n...light...fire,* Orger.
    381. *hallowed*] *hallow'd* Theob. Warb.   385. *the*] *this* Theob. ii, Warb. Johns.
et seq.   Steev. Var. Sing.
    384. *with*] *with all* Q$_x$.

---

373. **That**] See IV, i, 150.

377. **triple Hecates teame**] Douce: The chariot of the moon was drawn by
two horses, the one black, the other white. ' Hecate ' is uniformly a disyllable in
Shakespeare, except in *1 Hen. VI:* III, ii, 64. In Spenser and Ben Jonson it is
rightly a trisyllable. But Marlowe, though a scholar, and Middleton use it as a disyllable, and Golding has it both ways. [The daughter of Jupiter and Latona was
called Luna and Cynthia in heaven, Diana on earth, and Proserpine and Hecate in
hell.]

382. **broome**] Halliwell: Robin Goodfellow, and the fairies generally, were
remarkable for their cleanliness. Reginald Scot thus says of Puck, ' Your grand-
dames, maid, were wont to set a boll of milk for him, for (his pains in) grinding of
malt or mustard, and sweeping the house at midnight.' Compare also Ben Jonson's
masque of *Love Restored:* ' Robin Goodfellow, he that sweeps the hearth and the
house clean, riddles for the country-maids, and does all their other drudgery.' Having recounted several ineffectual attempts he had made to gain admittance, he adds,
' I e'en went back . . . with my broom and my candles and came on confidently.'
The broom and candle were no doubt the principal external characteristics of Robin.
In the *Mad Prankes*, 1628, it is stated that he ' would many times walke in the night
with a broome on his shoulder.'

383. **doore**] Farmer says that ' To sweep the dust behind the door ' is a common
expression, and a common practice in large houses, where the doors of halls and gal-
leries are thrown backward, and seldom or never shut.—Halliwell, however, gives
a more cleanly interpretation. He says that it is ' to sweep away the dust which is
behind the door.'

385. **Through . . . light**] Johnson: Milton, perhaps, had this picture in his
thought : 'And glowing embers through the gloom Teach light to counterfeit a gloom.'
—*Il Penseroso*, 79. I think it should be read, ' Through this house *in* glimmering
light.'—R. G. White (ed. i, reading *Though*): Plainly, Oberon does not intend to

By the dead and drowfie fier,   386
Euerie Elfe and Fairie fpright,
Hop as light as bird from brier,
And this Ditty after me, fing and dance it trippinglie.
    Tita.   Firft rehearfe this fong by roate,   390
To each word a warbling note.

---

386. *fier,*] QqFf, Rowe+, Sta. White.   390. Tita.] Queen. Rowe.
fire : Cap. et cet. (subs.).   *this*] *your* Q$_r$, Cap. Coll. Dyce,
  389. Two lines, Rowe ii et seq.   Sta. Cam. Ktly, White ii.
    dance it] *dance* F$_4$.

---

command his sprites to 'give glimmering light through the house *by the dead and drowsy fire,*' but to direct every elf and fairy sprite to hop as light as bird from briar, *though* the house give glimmering light by the dead and drowsy fire.—DYCE (ed. ii) : A most perplexing passage. R. G. White's reading and note, I must confess, are to me not quite intelligible. Lettsom conjectures, ' Through *this hall go* glimmering light,' &c.—HUDSON : R. G. White's reading and note seem rather to darken what is certainly none too light. Lettsom's conjecture is both ingenious and poetical in a high degree. . . . I suspect that ' By ' is simply to be taken as equivalent to *by means of.* Taking it so, I fail to perceive anything very dark or perplexing in the passage. —D. WILSON (p. 260) : My conjectural reading involves no great literal variation : ' Through the *house-wives'* glimmering light.' The couplet of Puck, which immediately precedes, sufficiently harmonises with such an idea, where with broom he sweeps the dust behind the door.—KINNEAR (p. 100) would read '—the house *gives* glimmering light *Now* the dead and drowsy fire,' &c., and remarks : ' " The dead and drowsy fire " tells the hour to the fairies,—so Puck says, l. 368, " Now the wasted brands do glow." He repeats " Now " four times, *emphasizing the hour,* ending with l. 380, "And we fairies. . . . *Now* are frolic." Oberon himself repeats the word, l. 395, "*Now,* until the break of day," &c. The whole context indicates that *Now* is the true reading. [I think it escaped the notice of Dyce and Hudson that R. G. White, in his text, restores the punctuation of the QqFf, and that it was Capell who first closed, more or less, the sentence at ' fire,' which I think is wrong; it increases the obscurity, which will still remain in spite of Hudson's interpretation of ' by,' its commonest interpretation, and it will still be perplexing to know how it is the fairies who give the glimmering light when it is given by means of the drowsy fire, unless the fairies carry the fire about with them, which is not likely. R. G. White's emendation, obtained by an insignificant change, is to me satisfactory : 'Albeit there is but a faint, glimmering light throughout the house, yet there is enough by means of the dead and drowsy fire for every Elf and Fairy to hop and sing and dance.'—ED.]

    388. **brier**] STEEVENS : This comparison is a very ancient one, being found in one of the poems of Lawrence Minot, p. 31—[ed. Ritson, ap. W. A. Wright] : ' That are was blith als brid on brere.'

    389. **it trippinglie**] This ' it ' may be, as ABBOTT, § 226 says, used indefinitely, like ' daub it,' or ' queen it,' or ' prince it '; but here it is not impossible that it refers to the ditty, which was to be both sung and danced.—KNIGHT calls attention to the use by Shakespeare of ' trip ' as the fairies' pace; it is so used in IV, i, 107. Milton's use of it for the dances of the Nymphs and the Graces in *L'Allegro* and *Comus* will occur to every one.—ED.

Hand in hand, with Fairie grace,   392
Will we fing and bleffe this place.

The Song.

Now vntill the breake of day ,   395
Through this houfe each Fairy ftray.
To the beft Bride-bed will we ,
Which by vs fhall bleffed be :   398

---

394. Om. Qq. Song and Dance. Cap.   395–416. In Roman, and given to
                                           Oberon, Qq, Johns. et seq.

---

394. The Song] JOHNSON: [This Song] I have restored to Oberon, as it apparently contains not the blessing which he intends to bestow on the bed, but his
declaration that he will bless it, and his orders to the Fairies how to perform the
necessary rites. But where then is the Song?—I am afraid it is gone with many
other things of greater value. The truth is that two songs are lost. The series of the
Scene is this: after the speech of Puck, Oberon enters and calls his Fairies to a song,
which song is apparently wanting in all the copies. Next Titania leads another song,
which is indeed lost like the former, though the Editors have endeavoured to find it.
Then Oberon dismisses his Fairies to the despatch of the ceremonies. The songs I
suppose were lost, because they were not inserted in the players' parts, from which
the drama was printed.—CAPELL [whose *Notes* were written before he had read
Johnson's edition]: That [lines 395–416] cannot be a Song is clear, even to demonstration, from the measure, the matter, and very air of every part of it; on the other
hand, it is as clear that a song, or something in nature of a song, must have come in
here; but, if this is not it, what are we to do for it? The manner in which Oberon
in his first speech, and the queen in her reply, express themselves, may incline some
to conjecture that this, which is at present before us, was designed by its Author to be
delivered in a kind of recitative, danced to by Titania and her train, and accompanied with their voices; but the arguments against its being a song are almost
equally forcible against its being recitative; and the word 'Now' seems to argue a
song preceding. Possibly such a one did exist; but Shakespeare, not being pleased
with it, nor yet inclined to mend it, scratched it out of his copy, and printed off the
play without one, as we see in the Qq; and his friends, the players—sensible of the
defect, but having nothing at hand to mend it—supplied it injudiciously in the manner
above recited. If this simple but beautiful play should ever be brought on the stage,
the insertion of some light song—in character and suited to the occasion—would do
credit to a manager's judgment, and honour to the poet who should compose it. [This
last remark is noteworthy as a revelation of the influence, even on so conservative an
editor as Capell, of an age which still believed that Shakespeare's 'wood-notes' were
'wild,' and that they could be not only improved by cultivation, but so successfully
imitated as to elude detection. See Fleay's note, line 417 below, where another
explanation of this discrepancy between the Qq and Ff is given.—ED.]

398. **blessed be]** STEEVENS: So in Chaucer's *Marchantes Tale*, line 9693, ed.
Tyrwhitt [line 575, ed. Morris]: 'And whan the bed was with the prest i-blessid.'
We learn also from 'Articles ordained by King Henry VII. for the Regulation of his
Household' that this ceremony was observed at the marriage of a Princess: 'All men
at her comming to be voided, except woemen, till she be brought to her bedd; and

And the iſſue there create,
Euer ſhall be fortunate :   400
So ſhall all the couples three,
Euer true in louing be :
And the blots of Natures hand,
Shall not in their iſſue ſtand.
Neuer mole, harelip, nor ſcarre,   405
Nor marke prodigious, ſuch as are
Deſpiſed in Natiuitie,
Shall vpon their children be.
With this field dew conſecrate,   409

408, 409. be. With...confecrate.] *be, With...conſecrate.* Coll. ii, iii (MS).

---

the man both; he sittinge in his bedd in his shirte, with a gowne cast aboute him.
Then the Bishoppe, with the Chaplaines, to come in, and blesse the bedd.'—DOUCE:
Blessing the bed was observed at *all* marriages. This was the form, copied from the
Manual for the use of Salisbury: 'Nocte vero sequente cum sponsus et sponsa ad
lectum pervenerint, accedat sacerdos et benedicat thalamum, dicens: Benedic, Domine, thalamum istum et omnes habitantes in eo; ut in tua pace consistant, et in tua
voluntate permaneant: et in amore tuo vivant et senescant et multiplicentur in longi-
tudine dierum. Per Dominum.—Item benedictio super lectum. Benedic, Domine,
hoc cubiculum, respice, quinon dormis neque dormitas. Qui custodis Israel, custodi
famulos tuos in hoc lecto quiescentes ab omnibus fantasmaticis demonum illusionibus:
custodi eos vigilantes ut in preceptis tuis meditentur dormientes, et te per soporem
sentiant: ut hic et ubique defensionis tuae muniantur auxilio. Per Dominum.—
Deinde fiat benedictio super eos in lecto tantum cum Oremus. Benedicat Deus cor-
pora vestra et animas vestras; et det super vos benedictionem sicut benedixit Abraham, Isaac, et Jacob, Amen.—His peractis aspergat eos aqua benedicta, et sic discedat
et dimittat eos in pace.'—W. A. WRIGHT: Compare *The Romans of Partenay,* or
Melusine (ed. Skeat), ll. 1009–11: 'Forsooth A Bisshop which that tyme ther was
Signed and blessid the bedde holyly; " In nomine dei," so said in that place.'

399. create] For a long list of participles like the present word, and 'consecrate,'
in line 409, where *-ed* is omitted after *t* or *d*, see ABBOTT, § 342.

408, 409. be. . . . consecrate,] COLLIER (*Notes,* &c., p. III): The MS puts a
comma after 'be' and a period after 'consecrate,' thus meaning that none of these
disfigurements shall be seen on the children consecrated with this field-dew. Then
begins a new sentence, which is judiciously altered in two words by the MS—namely,
in line 413 it reads: 'Ever shall *it safely* rest.' [The reading of Rowe ii.—ED.]
The question is whether the fairies or the issue of the different couples are to be 'consecrate' with the 'field-dew,' and there seems no reason why such delicate and
immortal beings should require it, while children might need it, to secure them from
'marks prodigious.'—DYCE (ed. ii): Collier altogether misunderstands the line, which
means 'with this *consecrated field-dew,*' i. e. fairy holy-water; and when he adds that
the field-dew was intended for 'the children,' he most unaccountably forgets that as
'the couples three' have only just retired to their respective bridal chambers, the
usual period must elapse before the birth of 'the children,' by which time ' THIS field

Euery Fairy take his gate,   410
And each feuerall chamber bleſſe,
Through this Pallace with ſweet peace,
Euer ſhall in ſafety reſt,
And the owner of it bleſt.
Trip away, make no ſtay;   415
Meet me all by breake of day.

---

410. *gate*] *gait* Johns. et seq.

413, 414. Transposed, White, Sta. Huds. Ktly.

413. Euer ſhall in ſafety] *Ever shall it safely* Rowe ii+, Cap. Steev.'85, Sing. ii, Coll. ii, iii (MS). *E'er shall it in*

safety Mal.'90, Steev. Sing.   *Ever shall 't in safety* Dyce ii, iii.

414. Two lines, Johns. of it] *of 't* Han.

415. away] *away then* Han.

416. [Exeunt. Qq. Exeunt King, Queen and Train. Cap.

---

dew' (so very prematurely provided) was not unlikely to lose its virtue, and even to evaporate, though in the keeping of fairies.

409, &c. D. WILSON (p. 260): Arranged in the following order, the consecutive relation of ideas seems to be more clearly expressed: 'Through this palace with sweet peace Every fairy take his gait, And each several chamber bless, With this field-dew consecrate; And the owners of it blest, Ever shall in safety rest,' &c.

409. field dew] DOUCE: There seems to be in this line a covert satire against holy water. Whilst the popular confidence in the power of fairies existed they had obtained the credit of doing much good service to mankind; and the great influence which they possessed gave so much offence to the holy monks that they determined to exert all their power to expel the imaginary beings from the minds of the people by taking the office of the fairies' benedictions entirely into their own hands. Of this we have a curious proof in the beginning of Chaucer's tale of *The Wife of Bath*.

410. gate] See line 360.

413, 414. Euer . . . blest.] STAUNTON: I at one time thought 'Ever shall' a misprint for *Every hall*, but it has since been suggested to me by Mr Singer, and by an anonymous correspondent, that the difficulty in the passage arose from the printer's having transposed the lines.—R. G. WHITE (ed. i): It was not until May, 1856, that the difficulty received its easy solution at the hands of a correspondent of the London *Illustrated News*, who signed his communication C. R. W. [Probably the 'anonymous correspondent' referred to by Staunton, who had then the charge of one of the columns in *The Illustrated News*.—ED.] This emendation is at once the simplest and the most consistent with the form and spirit of the context.—DYCE (ed. ii): I cannot agree with R. G. White in his estimate of this emendation; I must be allowed to prefer my own correction—the addition of a single letter. And compare the words of the supposed Fairy Queen concerning Windsor Castle: 'Strew good luck, ouphs, on every sacred room; *That it may stand till the perpetual doom*, In state [seat?] as wholesome as in state 'tis fit, Worthy the owner, and the owner it.'—*Merry Wives* V, v.—HALLIWELL: The original, in line 413, is probably correct, the nominative *palace*, being understood.—KEIGHTLEY (p. 137): This is the third or, rather, fourth

16

Robin.   If we ſhadowes haue offended,   417
Thinke but this (and all is mended)
That you haue but ſlumbred heere,
While theſe viſions did appeare.   420
And this weake and idle theame,

417. Epilogue. Hal.   418. *but this (and]* but (this, and
     Robin.] Puck. Rowe.   F₃F₄. *but this, and* Rowe et seq.
                                                        420. *theſe]* this Q₂.

---

transposition in this play. We may observe that twice before it was the second line
of the couplet that commenced with ' Ever.' For a fifth transposition in the original
eds., see III, i, 146.

417, &c. FLEAY (*Life and Work*, p. 182): The traces of the play having been
altered from a version for the stage are numerous. There is a double ending. Robin's final speech is palpably a stage-epilogue, while what precedes, from '*Enter Puck*'
to ' break of day—*Exeunt*,' is very appropriate for a marriage entertainment, but
scarcely suited for the stage. In Acts IV and V again we find the speech-prefixes
Duke, Duchess, Clown for *Theseus, Hippolita, Bottom;* such variations are nearly
always marks of alteration, the unnamed characters being anterior in date. In the
prose scenes speeches are several times assigned to wrong speakers, another common
mark of alteration. In the Fairies the character of Moth (Mote) has been excised
in the text, though he still remains among the *dramatis personæ.* [This statement is
to me inexplicable. When Titania summons four fairies (among them Moth) there
are four replies. In neither Quartos nor Folios is there a list of *dramatis personæ.*—
ED.] It is not, I think, possible to say which parts of the play were added for the
Court performance, but a careful examination has convinced me that wherever *Robin*
occurs in the stage-directions or speech-prefixes scarcely any, if any, alteration has
been made; *Puck*, on the contrary, indicates change. [Be it remembered that in
this allusion to ' the Court performance ' no special occasion is intended, for none has
been recorded, but FLEAY, throughout his *History of the London Stage*, is emphatic
in his assertion of ' the absolute subordination of public performances to Court presentations ' (*Introd.* p. 11). In proper obedience to this belief he assumes, therefore,
a Court performance in the present case. This opinion, that additions were made for
a Court performance, Fleay subsequently deserted. See *Date of Composition, post.*
—ED.]

417. shadowes] HUNTER (i, 298): Here we have a reference to a sentiment in
the play: ' The best in this kind are but shadows, and the worst are no worse if
imagination amend them,' an apology for the actor and a compliment for the critic.
What the poet had put into the mouth of one of the characters in respect of the poor
attempts of the Athenian clowns, he now, by the repetition of the word ' shadows,'
in effect says for himself and his companions. ' Shadows ' is a beautiful term by
which to express actors, those whose life is a perpetual personation, a semblance but
of something real, a shadow only of actual experiences. The idea of this resemblance was deeply inwrought in the mind of the poet and actor. When at a later
period he looked upon man again as but ' a walking shadow,' his mind immediately
passed to the long-cherished thought, and he proceeds: 'A poor player That struts
and frets his hour upon the stage And then is heard no more.'

No more yeelding but a dreame,   422
Centles, doe not reprehend.
If you pardon, we will mend.
And as I am an honeft *Pucke*,   425
If we haue vnearned lucke,
Now to fcape the Serpents tongue,
We will make amends ere long:
Elfe the *Pucke* a lyar call.
So good night vnto you all.   430
Giue me your hands, if we be friends,
And *Robin* fhall reftore amends.   432

---

422. *more yeelding*] *mere idling* D.   Coll. White i, Dyce ii, iii, Ktly.
Wilson.   425. *an*] Om. F$_3$F$_4$, Rowe+.
   423. *Centles*] *Gentles* QqFf.   429. *lyar*] *lyer* Q$_1$.
   425. *I am*] *I'm* Cap. Steev. Mal. Var. 432. [Exeunt omnes. Rowe.

---

422. **dreame**] Compare the Prologue to Lily's *The Woman in the Moone*, 1597:
'This but the shadow of our author's dreame, Argues the substance to be neere at
hand; At whose appearance I most humbly crave, That in your forehead she may
read content. If many faults escape in her discourse, Remember all is but a poet's
dreame.'—p. 151, ed. Fairholt.—ED.

425. **honest Pucke**] COLLIER: 'Puck' or *Powke* is a name of the devil, and as
Tyrwhitt remarks [II, i, 39] it is used in that sense in *Piers Ploughman's Vision*,
and elsewhere. It was therefore necessary for Shakespeare's fairy messenger to assert
his honesty, and to clear himself from any connexion with the 'helle Pouke.' ['Honest' here refers merely to his veracity, as is shown by line 429.—ED.]

426. **vnearned**] STEEVENS: That is, if we have better fortune than we have
deserved.

427. **Serpents tongue**] JOHNSON: That is, if we be dismissed without hisses.—
STEEVENS: So in Markham's *English Arcadia*, 1607: 'But the nymph, after the
custom of distrest tragedians, whose first act is entertained with a snaky salutation,'
&c.

431. **Giue . . . hands**] JOHNSON: That is, clap your hands. Give us your
applause. Wild and fantastical as this play is, all the parts in their various modes
are well written, and give the kind of pleasure which the author designed.

432. **amends**] Unwarrantably 'apprehending' (Theseus would say) that in the
second syllable of 'am*ends*' there is a punning allusion to the *end* of the play, SIM-
ROCK (Hildburghausen, 1868) takes the liberty thus to translate:

'Gute Nacht! Klatscht in die Hände,
      Dass den Dank euch Ruprecht sp——
                              Ende.   (Exit.)'

APPENDIX

APPENDIX

THE TEXT

THE TEXT is so fully discussed in the *Preface* to this volume that little remains to be added, except the opinions of two or three editors, and an account of an alleged *Third Quarto*. From the days of Dr JOHNSON all editors mention, with more or less fullness and accuracy, the Quartos and Folios, but KNIGHT is the earliest, I think, to express an opinion as to the degree of excellence with which the TEXT of this play has been transmitted to us. Although I have given the substance of his note at V, i, 115, I think it best to repeat it here.

'One thing is clear to us,' says KNIGHT (*Introductory Notice*, p. 331, 1840?),
'that the original of these editions [*i. e.* the two Quartos], whichever it might be, was
'printed from a genuine copy, and carefully superintended through the press. The
'text appears to us as perfect as it is possible to be, considering the state of typography
'of that day. There is one remarkable evidence of this. The Prologue to the inter-
'lude of the Clowns in the Fifth Act is purposely made inaccurate in its punctuation
'throughout. The speaker "does not stand upon points." It was impossible to have
'effected the object better than by the punctuation of [Q₂]; and this is precisely one
'of those matters of nicety in which a printer would have failed, unless he had fol-
'lowed an extremely clear copy, or his proofs had been corrected by an author or an
'editor.'

R. G. WHITE (ed. i, p. 18, 1857) : 'Fortunately, all of these editions [Q₁, Q₂, and
'F₁] were printed quite carefully for books of their class at that day; and the cases
'in which there is admissible doubt as to the reading are comparatively few, and,
'with one or two exceptions, unimportant.'

Rev. H. N. HUDSON (*Introduction*, p. 1, 1880) : 'In all three of these copies [the
'Quartos and Folio] the printing is remarkably clear and correct for the time, inso-
'much that modern editors have little difficulty about the text. Probably none of the
'Poet's dramas has reached us in a more satisfactory state.'

In 1841 HALLIWELL stated (*An Introd to Sh.'s Mid. N. D.* p. 9) that 'Chetwood,
'in his work entitled *The British Theatre*, 12mo. Dublin, 1750, has given a list of
'titles and dates of the early editions of Shakespeare's Plays, among which we find
'*A moste pleasaunte comedie, called A Midsummer Night's Dreame, wythe the freakes
'of the fayries*, stated to have been published in the year 1595. No copy either with
'this date or under this title has yet been discovered. It is, however, necessary to
'state that Steevens and others have pronounced many of the titles which Chetwood
'has given to be fictitious.'

Hunter, biased, possibly, by an innocent desire to fix the date of composition, is the only critic who has a good word for Chetwood, whose accuracy is commonly held in light esteem. HUNTER asks (*New Illust.* i, 283) : 'Have Chetwood's statements

'ever been examined in a fair and critical spirit, or do we dismiss them on the mere
'force of personal authority brought to bear against them? A copy cannot be pro-
'duced; but neither could a copy of the first edition of *Hamlet* be produced in the
'time of Steevens and Malone; yet it would have been a mistaken conclusion that
'no such edition existed because neither of those commentators had seen a copy.
'Chetwood gives the title somewhat circumstantially, as if he had seen a copy; and
'if some of his traditions may be shewn to be unfounded, if he may be proved to
'have been credulous, or even something worse, his writings contain some truth, and
'we cannot perhaps easily draw the line which shall separate that which is worthy
'of belief from that which is to be rejected without remorse.'

W. A. WRIGHT (*Preface*, iv) gives to Chetwood the *coup de grace* in the present
instance: 'the spelling of "wythe" is sufficient to condemn the title as spurious.'

DATE OF COMPOSITION

It is stated in the *Preface* that the following lines and allusions furnish internal
evidence of the *Date of Composition*:—

1. '*Thorough bush, thorough briar.*'—II, i, 5;

2. Titania's description of the disastrous effects on the weather and harvests caused
by the quarrel between her and Oberon.—II, i, 94–120;

3. '*And hang a pearl in every cowslip's ear.*'—II, i, 14;

4. '*One sees more devils than vast Hell can hold.*'—V, i, 11;

5. A poem of *Pyramus and Thisbe.*

6. The date of Spenser's *Faerie Queene.*

7. The ancient privilege of Athens, whereby Egeus claims the disposal of his
daughter either to give her in marriage or to put her to death.—I, i, 49;

8. '*The thrice three Muses, mourning for the death of learning, late deceast in
beggerie.*'—V, i, 59;

9. And, finally, that the play was intended for the celebration of a noble marriage.

These will now be dealt with in their foregoing order:

---

1. 'THOROUGH BUSH, THOROUGH BRIAR.'—II, i, 5.

CAPELL in 1767 (i, *Introd.* p. 64) said: 'if that pretty fantastical poem of Dray-
'ton's, call'd—"*Nymphidia or The Court of Fairy*," be early enough in time (as, I
'believe, it is; for I have seen an edition of that author's pastorals printed in 1593,
'quarto) it is not improbable, that Shakespeare took from thence the hint of his
'fairies: a line of that poem "Thorough bush, thorough briar" occurs also in his
'play.'

In the *Variorum* edition of 1773, STEEVENS asserted that Drayton's *Nymphidia*
'was printed in 1593,' but in the next *Variorum* the assertion was withdrawn, and
no decisive conclusion as to the priority of Drayton or Shakespeare was reached,
until MALONE, in the *Variorum* of 1821, settled the question in a note on 'Hob-

goblin,' II, i, 39, as follows :—'A copy of certain poems of this author [Drayton], '*The Batail of Agincourt, Nymphidia*, &c., published in 1627, which is in the col-
'lection of my friend, Mr. Bindley, puts the matter beyond a doubt; for in one of
'the blank leaves before the book, the author has written, as follows : " To the noble
' " Knight, my most honored ffrend, Sir Henry Willoughby, one of the selected
' " patrons of *thes my latest poems*, from his servant, Mi. Drayton." '

Drayton having been thus disposed of, a new claimant to priority was brought forward. 'There seems to be a certainty,' says HALLIWELL (*Memoranda*, 1879, p. 6),
'that Shakespeare, in the composition of the *Midsummer Night's Dream*, had in one
'place a recollection of the Sixth Book of *The Faerie Queene*, published in 1596, for
'he all but literally quotes the following [line 285] from the Eighth Canto of that
'book :—" Through hils and dales, through bushes and through breres,"—*Faerie*
'*Queene*, ed. 1596, p. 460. As the *Midsummer Night's Dream* was not printed
'until the year 1600, and it is impossible that Spenser could have been present at
'any representation of the comedy before he had written the Sixth Book of the
'*Faerie Queene*, it may be fairly concluded that Shakespeare's play was not composed
'at the earliest before the year 1596, in fact, not until some time after January the
'20th, 1595–6, on which day the Second Part of the *Faerie Queene* was entered on
'the books of the Stationers' Company. The sixth book of that poem was probably
'written as early as 1592 or 1593, certainly in Ireland, and at some considerable time
'before the month of November, 1594, the date of the entry of publication of the
'*Amoretti*, in the eightieth sonnet of which it is distinctly alluded to as having been
'completed previously to the composition of the latter work.'

This opinion Halliwell saw no reason to retract; he repeats it almost word for
word in his *Outlines* (1885, p. 500). But it does not meet FLEAY's approval. 'Mr
'Halliwell's fancy that Spenser's line . . . must have been imitated by Shakespeare
'. . . is very flimsy; hill and dale, bush and brier, are commonplaces of the time.'—
Life and Work, p. 186. They have been commonplaces ever since, unquestionably,
and doubtless FLEAY could have furnished many examples from contemporary authors
or he would not have made the assertion. 'Nor is there any proof,' Fleay goes on
to say, 'that this song could not have been transmitted to Ireland in 1593 or 1594.'
But what, we may ask, would have been the object in transmitting a 'commonplace'?
I quite agree with Fleay that there is small likelihood in HALLIWELL's suggestion,
but is it quite fair to scoff at a 'fancy,' and in the same breath propose another, such
as the 'transmission to Ireland'?

---

2. TITANIA'S DESCRIPTION OF THE PERVERTED SEASONS.—II, i, 86–120.

As this item of internal evidence still walks about the orb like the sun, it deserves
strict attention, and to that end, for the convenience of the reader, the whole passage
is here recalled :—

    'And neuer fince the middle Summers fpring
    'Met vve on hil, in dale, forreft, or mead,
      *   *   *   *   *
    'But vvith thy braules thou haft diftubrb'd our fport.
    'Therefore the Windes, piping to vs in vaine,
    'As in reuenge, haue fuck'd vp from the fea
    'Contagious fogges : Which falling in the Land,

' Hath euerie petty Riuer made fo proud,
' That they haue ouer-borne their Continents.
' The oxe hath therefore ftretch'd his yoake in **vaine,**
' The Ploughman loft his fweat, and the greene **Corne**
' Hath rotted, ere his youth attain'd a beard :
' The fold ftands empty in the drowned field,
' And Crowes are fatted vvith the murrion flocke,
' The nine mens Morris is fild vp with mud,
' And the queint Mazes in the wanton greene,
' For lacke of tread are vndiftinguifhable.
' The humane mortals want their winter heere,
' No night is now with hymne or caroll bleft ;
' Therefore the Moone (the gouerneffe of floods)
' Pale in her anger, washes all the aire ;
' That Rheumaticke difeafes doe abound.
' And through this diftemperature, we fee
' The feafons alter ; hoared headed frofts
' Fall in the frefh lap of the crimfon Rofe,
' And on old *Hyems* chinne and Icie crowne,
' An odorous Chaplet of fweet Sommer buds
' Is as in mockery fet.   The Spring, the Sommer,
' The childing Autumne, angry Winter change
' Their wonted Liueries, and the mazed world,
' By their increafe, now knowes not which is which ;
' And this fame progeny of euils,
' Comes from our debate, from our diffention.'

' The confusion of seasons here described,' said STEEVENS, in 1773, ' is no more
' than a poetical account of the weather which happened in England about the time
' when this play was first published.   For this information I am indebted to chance,
' which furnished me with a few leaves of an old meteorological history.'   This assertion that the ' old meteorological history' applied to the weather about the time this
play was published, that is, about 1600, STEEVENS repeated in 1778 and in 1785, but
in 1793, having adopted MALONE's chronology of the *Date of Composition,* which
placed this play in 1592, STEEVENS silently changed the application of his ' old
meteorological history' to the weather eight years earlier, and said that his few leaves
referred to the weather ' about the time the play was *written.*'   [Italics, mine.]   ' The
' date of the season,' STEEVENS goes on to say, ' may be better determined by a
' description of the same weather in Churchyard's *Charitie,* 1595, when, says he, ' "a
' " colder season, in all sorts, was never seene."   He then proceeds to say the same
' over again in rhyme :—

          ' " A colder time in world was neuer seene :
          ' " The skies do lowre, the sun and moone waxe dim ;
          ' " Sommer scarce knowne but that the leaues are greene.
          ' " The winter's waste driues water ore the brim ;
          ' " Upon the land great flotes of wood may swim.
          ' " Nature thinks scorne to do hir dutie right
          ' " Because we haue displeasde the Lord of Light."

'Let the reader compare these lines with Shakespeare's, and he will find that they
'are both descriptive of the same weather and its consequences.'

It was, however, BLAKEWAY who, in a note in the *Variorum of '21* (vol. v, p.
319), adduced yet more conclusive proofs of the extremely bad weather in 1593 and
1594, which he found in extracts, printed by Strype (*Ann.* v, iv, p. 211), from 'Dr
'King's *Lectures*, preached at York.' As W. A. WRIGHT, in his *Preface* to the
present play, has given the extracts from the Lectures themselves, I prefer, where
I can, to follow Wright, as more exact. From the second of a series of *Lectures upon Ionas*, delivered at York in 1594 and published in 1618, the following
extract, from p. 36, is given: 'The moneths of the year haue not yet gone about,
'wherein the Lord hath bowed the heauens, and come down amongst vs with more
'tokens and earnests of his wrath intended, then the agedst man of our land is able
'to recount of so small a time. For say, if euer the windes, since they blew one
'against the other, haue beene more common, & more tempestuous, as if the foure
'endes of heauen had conspired to turne the foundations of the earth vpside downe;
'thunders and lightnings neither seasonable for the time, and withall most terrible,
'with such effects brought forth, that the childe vnborne shall speake of it. The
'anger of the clouds hath beene powred downe vpon our heads, both with abundance
'and (sauing to those that felt it) with incredible violence; the aire threatned our
'miseries with a blazing starre; the pillers of the earth tottered in many whole coun-
'tries and tracts of our Ilande; the arrowes of a woeful pestilence haue beene cast
'abroad at large in all the quarters of our realme, euen to the emptying and dispeo-
'pling of some parts thereof; treasons against our Queene and countrey wee haue
'knowne many and mighty, monstrous to bee imagined, from a number of Lyons
'whelps, lurking in their dennes and watching their houre, to vndoe vs; our expecta-
'tion and comfort so fayled vs in France, as if our right armes had beene pulled from
'our shoulders.' 'The marginal note,' adds WRIGHT, 'to this passage shews the date
'to which it refers: "The yeare of the Lord 1593 and 1594."'

HALLIWELL added (*Introd. to A Mid. N. D.* 1841, p. 8) some passages from
Stowe, under date of 1594, confirming the pudder of the elements in that year: 'In
'this moneth of March **was** many great stormes of winde, which ouerturned trees,
'steeples, barnes, houses, &c., namely, in Worcestershire, in Beaudley forrest many
'Oakes were ouerturned. In Horton wood of the said shire more then 1500 Oakes
'were ouerthrowen in one day, namely, on the thursday next before Palmesunday.
'... The 11. of Aprill, a raine continued very sore more than 24. houres long and
'withall, such a winde from the north, as pearced the wals of houses, were they neuer
'so strong. ... In the moneth of May, namely, on the second day, came downe great
'water flouds, by reason of sodaine showres of haile and raine that had fallen, which
'bare downe houses, yron milles. ... This yeere in the moneth of May, fell many
'great showres of raine, but in the moneths of June and July, much more; for it
'commonly rained euerie day, or night, till S. *Iames* day, and two daies after togither
'most extreamly, all which, notwithstanding in the moneth of August there followed
'a faire haruest, but in the moneth of September fell great raines, which raised high
'waters, such as staied the carriages, and bare downe bridges, at Cambridge, Ware,
'and elsewhere, in many places. Also the price of graine grewe to be such, as a
'strike or bushell of Rie was sold for fiue shillings, a bushel of wheat for sixe, seuen,
'or eight shillings, &c., for still it rose in price, which dearth happened (after the
'common opinion) more by meanes of ouermuch transporting, by our owne merchants
'for their priuate gaine, than through the vnseasonablenesse of the weather passed.'

—*Annales*, ed. 1600, p. 1274–9. (I have added two or three sentences not given by Halliwell nor by Wright.)

Yet another testimony to these same meteorological disturbances is given by HAL-LIWELL (*Ibid.* p. 6), from Dr Simon Forman's MS (No. 384, Ashmolean Museum, Oxford), where that unabashed astrologer, who foretold the day of his own death and had the grace to fulfill the prophecy, has the following 'important observations, as Halliwell terms them, on the year 1594: 'Ther was moch sicknes but lyttle death, 'moch fruit and many plombs of all sorts this yeare and small nuts, but fewe walnuts. 'This monethes of June and July were very wet and wonderfull cold like winter, that 'the 10. dae of Julii many did syt by the fyer, yt was so cold; and soe was yt in 'Maye and June; and scarce too fair dais together all that tyme, but yt rayned every 'day more or lesse. Yf yt did not raine, then was yt cold and cloudye. Mani mur-'ders were done this quarter. There were many gret fludes this sommer, and about 'Michelmas, thorowe the abundaunce of raine that fell sodeinly; the brige of Ware 'was broken downe, and at Stratford Bowe, the water was never seen so byg as yt was; 'and in the lattere end of October, the waters burste downe the bridg at Cambridge. 'In Barkshire were many gret waters, wherewith was moch harm done sodenly.'

But the year 1594 is not to have all the bad weather; it would be poverty-stricken indeed if one and the same speech in any of Shakespeare's plays could not furnish at least two divergent opinions. Accordingly, we find CHALMERS (*Supp. Apology*, p. 368) maintaining that Titania's words refer to the fact that 'the prices of corn rose to 'a great height in 1597,' this, together with other items, to be hereafter duly mentioned, 'fixes the epoch,' according to Chalmers, 'of this fairy play to the beginning 'of the year 1598.'

As to the estimate which modern editors put on the value of these allusions by Titania in fixing the date of the play, KNIGHT, in his edition (*circa* 1840), is mildly tolerant of the weather, and thinks that the peculiarly ungenial seasons of 1593–4 'may have suggested Titania's beautiful description'; but in his *Biography* (1843, p. 360) there is the shrewd remark that 'Stowe's record that, in 1594, "notwithstanding '"in the moneth of August there followed a faire haruest," does not agree with "The '"ox hath therefore stretch'd his yoke in vain, The ploughman lost his sweat, and '"the green corn hath rotted, ere his youth attained a beard."' 'It is not necessary,' concludes KNIGHT, 'to fix Shakspere's description of the ungenial season upon 1594 'in particular.'

HALLIWELL in his *Introduction*, in 1841, set great store by his witness, Dr. Forman, and by what was to be found in the *Variorum of 1821*, but 'grizzling hair the 'brain doth clear,' and in his folio edition in 1856 he says that the 'presumed allu-'sions to contemporary events are scarcely entitled to assume the dignity of evi-'dences.' Amongst these 'presumed allusions,' however, he acknowledges that the ungenial seasons referred to in Titania's speech may be, perhaps, 'considered the 'most important.' In his *Memoranda*, 1879 (p. 5), which we may accept as his final judgment, he asserts that 'the accounts of the bad weather of 1594 are valueless in 'the question of the chronology.'

COLLIER, in both his editions, alludes to Stowe and Forman, but expresses no opinion.

DYCE in all his editions, First, Second, and Third, with outspoken British honesty (and, for that vacillating editor, extraordinary unanimity withal), pronounced the supposition that the words of Titania allude to the state of the weather in England, in 1594 'ridiculous.'

GRANT WHITE, in his First Edition (1857, p. 15), thinks that there is 'no room 'for reasonable doubt' that the date of Titania's speech is decided by the citations 'from Stowe and Forman. In his Second Edition, having in the mean time taken 'advice on the subject of Notes, as he tells us (*Preface*, p. xii), 'of his washerwoman,' he does not refer to the matter at all,—naturally, any allusion to a season when there were no 'drying days' could not but be extremely distasteful to his coadjutor.

STAUNTON (1857), while acknowledging that Titania's fine description 'is singu- 'larly applicable to a state of things prevalent in England in 1593 and 1594,' is 'not 'disposed to attach much importance to these coincidences as settling the date of 'the play.'

KURZ makes an observation which is not without weight. 'A wide-spread calam- 'ity,' he remarks (*Sh. Jahrbuch*, iv, 268, 1869), 'would have been, according to the 'ideas of those times, a topic more appropriate to the pulpit [as it really was there 'treated.—ED.] than to the stage, and, according to the ideas of *all* times, most 'inappropriate to the *comic* stage. We go to the theatre to forget our burdens; and 'he who in the midst of a gay, joyous play, without the smallest need, reminds us 'that our fields are submerged, our harvests ruined, and man and beast plague- 'stricken, may rest assured that he will not catch us again very soon seated in front 'of his stage.'

HUDSON (1880) does 'not quite see' these allusions as Dyce sees them, 'albeit I 'am apt enough to believe most of the play was written before that date [1594] 'And surely, the truth of the allusion being granted, all must admit that passing 'events have seldom been turned to better account in the service of poetry.'

W. A. WRIGHT (*Preface*, p. vi) reprints the passages from Dr. King and Stowe at length, 'if only for the purpose of showing that in all probability Shakespeare had 'not the year 1594 in his mind at all.' Notwithstanding the accounts of the direful weather in that year, there followed 'a faire harvest,' and the 'subsequent high 'prices of corn are attributed not to a deficiency of the crop, but to the avarice of 'merchants exporting it for their own gain. Now this does not agree with Titania's 'description of the fatal consequences of her quarrel with Oberon, through which '"The green corn Hath rotted, ere his youth attain'd a beard." In this point alone 'there is such an important discrepancy, that if Shakespeare referred to any particular 'season we may, without doubt, affirm it was not to the year 1594, and therefore the 'passages [from King, and Stowe, and Forman] have no bearing upon the date of 'the play. I am even sceptical enough to think that Titania's speech not only does 'not describe the events of the year 1594, or of the other bad seasons which hap- 'pened at this time, but that it is purely the product of the poet's own imagination, 'and that the picture which it presents had no original in the world of fact, any more 'than Oberon's bank or Titania's bower.'

Rev. H. P. STOKES (*Chronological Order*, &c., 1878, p. 49) thinks it 'probable' that Titania's lines refer to 'the chief dearth in Shakespeare's time in 1594-5.'

FLEAY (*Life and Work*, &c., 1886, p. 182) finds confirmation of the date 1595 in the recorded inversion of the seasons spoken of by Titania.

---

5. 'AND HANG A PEARL IN EVERY COWSLIP'S EAR.'—II, i, 14.

In the *Variorum of 1785*, STEEVENS remarked on the above line that 'the same 'thought occurs in an old comedy called *The Wisdome of Doctor Dodypoll*, 1600 'i. e. the same year in which the first printed copies of this play made their appear

ance. An enchanter says: " Twas I that lead you through the painted meadows, ' " When the light Fairies daunst upon the flowers, Hanging on every leafe an orient ' " pearle." ' [p. 135, ed. Bullen]. The author of this tiresome and mediocre comedy is unknown, and seeing that it and the present play are of the same date in publication, and that we know the latter was in existence in Meres's time, 1598, STEEVENS wisely refrained from expressing any opinion as to priority. DYCE, in 1829, discovered that a song in *Dr. Dodypoll*, ' What thing is love?' was written by Peele in *The Hunting of Cupid* (Peele's *Works*, ii, pp. 255, 260), and FLEAY (*Eng. Drama*, ii, 155) sees ' no reason for depriving him of the rest of the play,' and Fleay accordingly gives it to him. ' It was,' says FLEAY, ' most likely one of [the old plays acted by ' the children of Paul's] produced c. 1590.' Great as must be the admiration of all for Fleay's industry and almost unrivalled grasp of early dramatic history, yet not even from Fleay can we without protest accept the phrase ' most likely,' which is always, like the wrath of Achilles, the source of unnumbered woes. The present is no exception. If Fleay thought that in *Doctor Dodypoll* a line was imitated from *A Midsummer Night's Dream*, ' and spoiled in the imitation,' as he asserted in 1886 (*Life and Work*, p. 186), and that *A Midsummer Night's Dream* was ' most cer ' tainly of this date [1595] ' (*Ib.* p. 181), he would never have said in 1891 that *Doctor Dodypoll* was ' most likely ' produced ' c. 1590,' five years earlier than *A Midsummer Night's Dream*.

MALONE (ed. 1790, i, 286) observes that ' Doctor Dodipowle is mentioned by ' Nashe in his preface to Gabriel Harvey's *Hunt is Up*, printed in 1596.' Nine years later CHALMERS (*Sup. Apol.* 363) roundly asserts that *Doctor Dodypoll* ' was pub- ' lished in 1596, or before this year,' but no copy, I believe, thus dated is now known. Chalmers is, therefore, led by his premises, ' to infer that Shakespeare, according to ' the laudable practice of the bee, which steals luscious sweets from rankest weeds, ' derived his extract from Dodipol, and not Dodipol from Shakespeare.'

Malone's suggestion and Chalmers's assertion seem to have beguiled HALLIWELL into the belief that *Dr Dodypoll* was ' known to have been written as early as 1596 ' —(*Introd.* p. 10), and although he does not repeat this in his Folio Edition, but gives merely Malone's reference, in his latest *Memoranda* (1879, p. 7), we find: 'As Dr ' Dodipowle is mentioned by Nash as early as 1596, this argument would prove ' Shakespeare's comedy to have been then in existence.'

It is, however, W. A. WRIGHT (*Preface*, p. iii) who has exorcised Nash's *Dr Dodypoll* once and for ever as a factor in approximating to the date of the present play, thus: ' Nashe only mentions the name " doctor Dodypowle," without referring ' to the play, and Dodipoll was a synonym for a blockhead as early as Latimer's time.

Again, H. CHICHESTER HART (*Athenæum*, 6 Oct. 1888) points out that ' the iden- ' tical name occurs in *Hickscorner* (1552): " What, Master Doctor Dotypoll? Can- ' " not you preach well in a black boll, Or dispute any divinity?" '—Hazlitt's *Dodsley*, i, 179.

---

4. 'ONE SEES MORE DEVILS THAN VAST HELL CAN HOLD.'—V, i, 11.

In these words of Theseus, CHALMERS (*Sup. Apol.* p. 361), reading between the lines, sees something else besides ' devils ': ' plainly a sarcasm on Lodge's pamphlet, called *Wits Miserie and the Worlds Madnesse; discovering the* Incar- ' nate Devils *of this age*, which was published in 1596. Theseus had already remarked, in the same speech: " The lunatic, the Louer, and the Poet, Are of

"imagination all compact." Lodge has the same word, *compact*, as singularly 'coupled: "Heinousous thoughts *compact* them together."' This quotation from Lodge is certainly remarkable, not because Shakespeare purloined from it the common-place word 'compact,' but because he overlooked that vigorous and startling word 'Heinousous,' with its untold depths of devilish meaning. Chalmers gives no clew to the page or chapter in *Wits Miserie* where this phrase is to be found, so that many hours had to be mis-spent before I found it. It occurs in *The discouery of Asmodeus*, &c. (p. 46, ed. Hunterian Club), and let the wits' misery be imagined when the shuddering 'heinousous' stands forth as plain *heinousest ;* and 'compact,' which was the very fulcrum of Chalmers's argument, turns out to be *compacted*. Lodge's phrase is : 'Hee affembled his hainoufeft thoughts, & compacted them 'togither [*sic*].' Apart from the childishness of founding an argument on the use of one and the same word by two voluminous writers, Chalmers's quotation is apparently an example of that class, not so common now as aforetime, where a slight perversion may be ventured, in the hope that it will escape detection through lack of verification. A quotation from an author generally, without citing page or line, is suspicious.

But Chalmers is bound to prove that Theseus's line is sarcastic, and that in it Shakespeare is 'serving out' Lodge for some personal affront. This affront Chalmers detects in the omission of Shakespeare's name in the four or five 'divine wits' enumerated by Lodge : Lilly, Daniel, Spenser, Drayton, and Nash (p. 57, *ib.*). 'Owing to this preference given to other poets,' says Chalmers, p. 362, 'Shakespeare . . . now returned marked disdain for contemptuous silence.' 'There is another passage,' continues CHALMERS, still on the scent, as he believes, 'which Shakespeare may have 'felt : "They fay likewife there is a Plaier Deuil, a handfome fonne of Mammons, '"but yet I haue not feen him, becaufe he skulks in the countrie,"' &c., &c. It is not worth while to cite the rest of this long quotation (p. 40, ed. Hunterian Club), wherein the bitterest sting to Shakespeare's feelings, as is clear from Chalmers's italics, is that *he skulks in the country*.

---

5. A Poem of 'Pyramus and Thisbe.'

'There was,' according to CHALMERS (*Sup. Apol.* p. 363), 'a poem, entitled *Pyramus and Thisbe*, published by Dr. Gale in 1597 ; but Mr. Malone believed 'this to be *posterior* to The Midsummer's [*sic*] Night's Dream. On the contrary, I 'believe, that Gale's *Pyramus and Thisbe* was prior to Shakespeare's most lament-'able "Comedy of Pyramus and Thisby."' This argument was thus effectively silenced by W. A. WRIGHT (*Preface*, p. viii) : 'As no one has seen this edition of 'Gale's poem, and as the story of Pyramus and Thisbe was accessible to Shakespeare 'from other sources long before 1597, we may dismiss this piece of evidence brought 'forward by Chalmers as having no decisive weight.' See further reference to Gale in *Source of the Plot*.

---

6. The Date of Spenser's 'Faerie Queene.'

Again, CHALMERS, a commentator very fertile in resources (such as they are), says (*Ib.* p. 364) : 'It is to be remembered, that the second volume of the *Faerie 'Queene* was published in 1596 ; being entered in the Stationers' *Registers* on the '20th of January, 1595-6. This for some time furnished town talk ; which never 'fails to supply our poets with dramatical topicks. The *Faerie Queene* helped Shakespeare to many hints. In the *Midsummer's Night's Dream* the Second Act opens

' with a fairy scene : The *fairy* is forward to tell, " How I serve the *fairy queen, To*
' " *dew her orbs upon the green :* And jealous Oberon would have the child *Knight*
' " of his train, to trace the forests wild." Here, then, are obvious allusions to the
' *Faerie Queene* of 1596,' subsequent to which, be it remembered, Chalmers maintains
that *A Midsummer Night's Dream* was written.

Again, Chalmers may be safely left to W. A. WRIGHT, who replies (p. ix) to the
assertion that the second volume of *The Faerie Queene* was published in 1596 : ' To
' this I would add, what Chalmers himself should have stated, that although the
' second volume of Spenser's poem was not published till 1596, the first appeared in
' 1590, and if Shakespeare borrowed any ideas from it at all, he had an opportunity
' of doing so long before 1596. This, therefore, may be consigned to the limbo of
' worthless evidence.'

7. THE ANCIENT PRIVILEGE OF ATHENS, WHEREBY EGEUS CLAIMS TO DISPOSE
    OF HIS DAUGHTER EITHER IN MARRIAGE OR TO PUT HER TO DEATH
    I, i, 49.

CHALMERS (*Ecce, iterum Crispinus !*) urges yet other evidence to prove the late
date of the present play. ' In the first Act,' he says (p. 365), ' Egeus comes in
' *full of vexation*, with complaint against his daughter, Hermia, who had been be-
' witched by Lysander with *rhymes*, and *love tokens*, and other *messengers* of *strong*
' *prevailment in unharden'd youth ;* and claimed of the Duke the ancient privilege
' of Athens ; insisting either to dispose of her to Demetrius, or to death, " according
' " to our Law, Immediately provided in that case." . . . Our observant dramatist,
' probably, alluded to the proceedings of Parliament on this subject during the session
' of 1597. On the 7th of November of that year the bill was committed, for depriv-
' ing offenders of clergy, who, against the statute of Henry VII, should be found
' guilty of the taking away of women against their wills. On the 14th of November,
' 1597, there was a report to the House touching the abuses from *licenses for mar-*
' *riages, without bans ;* and also touching *the stealing away of men's children with-*
' *out the assent of their parents.* . . . These obvious allusions to striking transactions,
' of an interesting nature, carry the epoch of this play beyond that session of Par-
' liament, which ended on the 9th of February, 1597–8.'

Again, W. A. WRIGHT comes to the rescue (p. ix) : ' This is certainly the weak-
' est of all the proofs by which Chalmers endeavours to make out his case, for the
' law which Egeus wished to enforce was against a refractory daughter, who at the
' time at which he was speaking had not been stolen away by Lysander, and was
' only too willing to go with him.' The Parliamentary laws were directed against
the theft of heiresses, and against illegal marriages. The law Egeus invokes was
directed against disobedient daughters, whether willing victims or not.

8. ' THE THRICE THREE MUSES, MOURNING FOR THE DEATH OF LEARNING, LATE
    DECEAST IN BEGGERIE.'—V, i, 59.

In a note on ' *The thrice three Muses, mourning for the death Of learning,*
' *late deceaſt in beggerie.'*—V, i, 59, WARBURTON observed that the reference
seemed to be intended as a compliment to Spenser, who wrote a poem called *The*
' *Teares of the Muses.'* Twenty-five years later, in the Var. of 1773, WARTON makes
the same observation, and suggests that if the allusion be granted the date of the
present play might be moved somewhat nearer to 1591, the date of Spenser's poem.

In 1778 STEEVENS remarked that this 'pretended title of a dramatic performance
' might be designed as a covert stroke of satire on those who had permitted Spenser
' to die through absolute want of bread in Dublin in the year 1598—*late* deceas'd in
' beggary seems to refer to this circumstance.' In his chronology of the play, however, in this same year, MALONE says that this allusion need not necessarily be inconsistent with the early appearance of this comedy, for it might have been inserted
between the time of Spenser's death and the year 1600, when the play was published.
' Spenser, we are told by Sir James Ware, . . . did not die till 1599; ' " others " (he
' " adds), have it *wrongly*, 1598." '

Thus, this allusion to Spenser's *Tears of the Muses,* and to his death, was accepted
as evidence until KNIGHT, who found it ' difficult to understand how an elegy on the
' great poet could have been called " some satire keen and critical," ' started a new
explanation. ' Spenser's poem,' says KNIGHT (*Introductory Notice,* p. 333), ' is cer-
' tainly a satire in one sense of the word; for it makes the Muses lament that all the
' glorious productions of men that proceeded from their influence had vanished from
' the earth. . . . Clio complains that mighty peers " only boast of arms and ancestry ";
' Melpomene, that " all man's life me seems a tragedy "; Thalia is " made the servant
' " of the many "; Euterpe weeps that " now no pastoral is to be heard "; and so on.
' These laments do not seem identical with the " —mourning for the *death* Of *learn-*
' " *ing,* late deceas'd in *beggary.*" These expressions are too precise and limited to
' refer to the tears of the Muses for the decay of knowledge and art. We cannot
' divest ourselves of the belief that some real person, some real death, was alluded to.
' May we hazard a conjecture?—Greene, a man of learning, and one whom Shak-
' spere, in the generosity of his nature, might wish to point at kindly, died in 1592,
' in a condition that might truly be called beggary. But how was his death, any more
' than that of Spenser, to be the occasion of " some satire keen and critical "? Every
' student of our literary history will remember the famous controversy of Nash and
' Gabriel Harvey, which was begun by Harvey's publication in 1592 of " Four Let-
' " ters and certain Sonnets, especially touching Robert Greene, and other parties by
' " him abused." Robert Greene was dead; but Harvey came forward, in revenge of
' an incautious attack of the unhappy poet, to satirize him in his grave,—to hold up
' his vices and his misfortunes to the public scorn,—to be " keen and critical " upon
' " learning, late deceas'd in beggary." '

This conjecture of KNIGHT ' bears great appearance of probability,' says HALLI-
WELL (*Introd. Fol. Ed.* 1856, p. 5). ' The miserable death of Greene in 1592,' he continues, ' was a subject of general conversation for several years [it is to be regretted
' that no authority for this ' conversation' is given.—ED.], and a reference to the circumstance, though indistinctly expressed, would have been well understood in liter-
' ary circles at the time it is supposed the comedy was produced. " Truely I have
' " been ashamed," observed Harvey, speaking of the last days of Greene, " to heare
' " some ascertayned reportes of hys most woefull and rascall estate : how the wretched
' " fellow, or shall I say the Prince of beggars, laid all to gage for some few shillinges :
' " and was attended by lice; and would pittifully beg a penny pott of Malmesie :
' " and could not gett any of his old acquaintance to comfort, or visite him in his
' " extremity but Mistris Appleby, and the mother of Infortunatus."—*Foure Letters*
' *and certaine Sonnets,* 1592 [vol. i, p. 170, ed. Grosart]. And again, in the same
' work, " his hostisse *Isam* with teares in her eies, & sighes from a deeper fountaine
' " (for she loved him derely), tould me of his lamentable begging of a penny pott
' " of Malmesy . . . and how he was faine poore soule, to borrow her husbandes shirte,

17

' " whiles his owne was a washing: and how his dublet, and hose, and sword were
' " sold for three shillinges."—[*Ib.* p. 171]. This testimony, although emanating
' from an ill-wisher, is not controverted by the statements of Nash, who had not the
' same opportunity of obtaining correct information; and, on the whole, it cannot be
' doubted that Greene " deceas'd in beggary." His " learning " was equally notorious.
' " For judgement Jove, for learning deepe he still Apollo seemde."—*Greenes Fune-*
' *ralls*, 1594. There is nothing in the consideration that the poet had been attacked
' by Greene as the " upstart crow " to render Mr Knight's theory improbable. The
' allusion in the comedy, if applicable to Greene, was certainly not conceived in an
' unkind spirit; and the death of one who at most was probably rather jealous than
' bitterly inimical, under such afflicting circumstances, there can be no doubt would
' have obliterated all trace of animosity from a mind so generous as was that of Shake-
' speare.' The possibility that the allusion is to Spenser is precluded, so thinks Halliwell, by the date of Spenser's death, which took place early in 1599, ' unless the
' forced explanation, that the lines were inserted after the first publication, be adopted.'
This explanation is not merely ' forced.' It is impossible. ' There is greater probabil-
' ity,' continues Halliwell, ' in the supposition that there is a reference to Spenser's
' poem, *The Teares of the Muses*, which appeared in 1591, . . . but the words of
' Shakespeare certainly appear to be more positive.'

In discussing this possible allusion to *The Teares of the Muses*, COLLIER, with
more fanciful ingenuity than grave probability, detects ' a slight coincidence of expres-
' sion between Spenser and Shakespeare in the poem of the one, and in the drama of
' the other, which deserves remark: Spenser says " Our pleasant Willy, ah, is *dead*
' " *of late*." And one of Shakespeare's lines is, " Of learning, *late deceas'd* in beg-
' " gary." Yet it is quite clear, from a subsequent stanza in *The Teares of the Muses*
' that Spenser did not refer to the natural death of " Willy," whoever he were, but
' merely that he " rather chose to sit in idle cell," than write in such unfavourable
' times. In the same manner Shakespeare might not mean that Spenser (if the allu-
' sion be, indeed, to him) was actually " deceas'd," but merely, as Spenser expresses
' it in his *Colin Clout*, that he was " dead in dole." ' But by the time that COLLIER
had come to edit Spenser (1862) he had become fully persuaded [*Works*, i, xi] that
the lines in question referred ' to the death of Spenser in grief and poverty. . . . On
' the revival of plays, it was very common to make insertions of new matter especially
' adapted to the time; and this, we apprehend, was one of the additions made by
' Shakespeare shortly before his drama was published in 1600.'

R. G. WHITE, in his first edition, regards the allusion to Greene with favour,
mainly because it reveals ' the gentle and generous nature of Sweet Will ' in forgiving
and forgetting a petty wrong when the perpetrator was in the grave, and ' had been a
' fellow-labourer in the field of letters, and an unhappy one.'

STAUNTON attaches but little importance to the explanations of Titania's allusions
to the weather, and attaches still less to the present allusions to Spenser, albeit he
acknowledges that an allusion to Greene is more plausible.

DYCE regards them, one and all, as ' ridiculous.'

WARD (*Eng. Dram. Lit.* 1875, i, 380) having quoted Dyce's all-embracing
ridiculous,' and mentioned Spenser's *Teares* and his death, goes on to say that ' the
term " ridiculous " is not too strong to characterise a third supposition that [the lines
" The thrice three Muses," &c.] contain a reference to the death of Robert Greene
(1592), upon whose memory Shakespeare would certainly in that case have been
resolved to heap coals of fire.'

STOKES, however, is temerarious enough to say (*Chrono. Order*, p. 50) that he ventures to incur the ridicule [pronounced by Ward], for how can a '*satire, keen* '*and critical*, be used to "heap coals of fire"? and we know that Greene was 'regarded by Gabriel Harvey and others (including Shakespeare himself) [it is to 'be regretted that the authority for this assertion has been omitted.—ED.] with 'anything but a forgiving spirit. Surely the reference to the death "Of *learning*, '"late *deceased in beggary*," must allude to Robert Greene, "utriusque Academiæ in '"Artibus Magister" (as he styles himself on some of his title-pages), parson (miser-'abile dictu), doctor, author, who died in misery and want in a London attic.'

FLEAY (*Manual*, 1876, p. 26) says that there may be an allusion to Spenser's *Tears of the Muses*, published in 1591, or 'possibly to the death of Greene in 1592, 'or to both.'

W. A. WRIGHT (*Preface*, p. viii): 'It is difficult to see any parallel between 'Gabriel Harvey's satire and "The thrice three Muses mourning for the death of '"Of learning," which must of necessity satirize some person or persons other than 'him whose death is mourned, even supposing that any particular person is referred 'to. On the whole, I am inclined to think that Spenser's poem may have suggested 'to Shakespeare a title for the piece submitted to Theseus, and that we need not 'press for any closer parallel between them.'

To GROSART, Spenser's latest editor, it seems 'pretty clear the *Teares of the* '*Muses* ("thrice three") was intended to be designated. For only in the *Teares of* '*the Muses* is there that combination of "mourning" with satire that leads to [The-'seus's] commentary on the proposal to have such a "device" for entertainment of 'the joyous marriage-company. . . . One wishes the suggested "device . . . had 'approved itself to Theseus as it had to Philostrate For then, instead of the fooling 'of Pyramus and Thisbe . . . we might have had William Shakespeare's estimate of 'Edmund Spenser. A thousand times must [Theseus's] preference be grudged and 'lamented.'—*Spenser, Works*, i, 92.

---

9. AND, FINALLY, THAT THE PLAY WAS INTENDED FOR THE CELEBRATION OF A NOBLE MARRIAGE.

With our knowledge of the purposes for which Masques and Dramatic Entertain-ments were written, it is not improbable, from the final scene of the play, that this *Dream* was composed for the festivities of some marriage in high life, at which pos-sibly the Queen herself was present. If a noble marriage before 1598 can be found to which there are unmistakeable allusions in the play, we shall go far to confining the *Date of Composition* within narrow limits.

In the notes following Schlegel's *Translation*, in 1830, TIECK has the following (p. 353): 'Whoever understands the poet and his style must feel assured that we owe 'this work of fantasie and imagination to that same poetic intoxication which gave us '*The Merchant of Venice, Twelfth Night, As You Like It*, and *Henry V*. It was 'printed first in 1600, and we can assume that it had been already written before this 'year, for Mares [*sic*] mentions it in 1598. In this same year, 1598, the friend of 'the poet, the Earl of Southampton, espoused his beloved Mistress Varnon, to whom 'he had been long betrothed. Perhaps the germ, or the first sketch, of the drama 'was a felicitation to the newly-married pair, in the shape of a so-called Mask, in 'which Oberon, Titania, and their fairies wished and prophesied health and happi-'ness to the bridal couple. The comic antistrophe, the scene with the "rude mechan-

' " icals," formed what was termed the anti-mask. . . . Thus to this Occasional Poem
' there were added subsequently the other scenes of the comedy.   Moreover, South-
' ampton married against the wishes of the Queen, who appeared not to have known
' of it at first, because she treated it as though it had been secret.   The young Lady
' Varnon, when her lover left her to go to France, where he was presented to Henry
' IV, was an object of sympathy to all her friends.   Through this alliance Essex
' became connected with Southampton, with whom he had not been before on good
' terms.   For Southampton, as we learn from Shakespeare's *Sonnets*, many a fair one
' sighed, attracted by his charms.   Wherever we turn we meet references and allu-
' sions which, if they do not more clearly explain this wondrous poem, at least, by
' their half-glimmering explanations, tantalise the readers almost as much as Puck, in
' the play, teases the human mortals.'

ULRICI (*Shakespeare's Dram. Kunst*, 1847, p. 539; trans. by L. Dora Schmitz,
1876, ii, 81) is inclined from ' internal evidence to assume that 1596–97 was the year
' in which this piece was composed. . . . [Tieck's conjecture that it was composed for
' Southampton's marriage] I consider untenable; at all events it is not easy to see
' how the title of *A Midsummer Night's Dream* . . . could be appropriate for the
' " masque " of Oberon and Titania with its " anti-masque," the play of the mechan-
' ics, in short, for a mere epithalamium.   But, in fact, it would, in any case, be a
' strange and almost impertinent proceeding to present a noble patron with a wedding
' gift in the form of a poem where love—from its serious and ethical side—is made a
' subject for laughter and represented only from a comic aspect, in its faithlessness
' and levity, as a mere play of the imagination, and where even the marriage feast of
' Theseus appears in a comical light, owing to the manner in which it is celebrated.
' And it would have been even a greater want of tact to produce a piece, composed
' for such an occasion, on the public stage, either before or after the earl's marriage.'

GERALD MASSEY, according to whose view Shakespeare's *Sonnets*, and portions
of many of his plays, are saturated with allusions to Southampton, Essex, Lady Penelope Rich, Elizabeth Vernon, and others of that circle, discusses Oberon's command
to Puck to bring that ' little Western flower,' which, with Halpin, he believes to be
Lettice Knollys, and comes to the conclusion that ' Dian's bud' is the emblem of
Elizabeth Vernon, and, following Tieck, he has ' no doubt' (*Shakespeare's Sonnets*,
1866 and 1872, p. 481, ed. 1888, p. 443) ' that this [present] dainty drama was writ
' ten with the view of celebrating the marriage of Southampton and Elizabeth Ver-
' non; for them his Muse put on the wedding raiment of such richness; theirs was
' the bickering of jealousy so magically mirrored, the nuptial path so bestrewn with
' the choicest of our poet's flowers, the wedding bond that he so fervently blessed in
' fairy guise.   He is, as it were, the familiar friend at the marriage-feast, who gossips
' cheerily to the company of a perplexing passage in the lover's courtship, which they
' can afford to smile at now! [but that the marriage was disallowed by the Queen.—
' ed. 1888].   The play was probably composed some time before the marriage took
' place [in 1598], at a period when it may have been thought the Queen's consent
' could be obtained, but not so early as the commentators have imagined.   I have
' ventured the date of 1595.'   In a footnote there is added: ' Perhaps it was one of
' the plays presented before Mr Secretary Cecil and Lord Southampton when they
' were leaving Paris, in January, 1598, at which time, as Rowland White relates, the
Earl's marriage was secretly talked of.'

ELZE (*Jahrbuch d. deutschen Sh.-Gesellschaft*, 1869, p. 150; *Essays* trans. by L.
Dora Schmitz, 1874, p. 30) finds objections to Tieck's conjecture, in the date of

Meres's allusion in 1598, the very year of Southampton's marriage, and in the clan-destine character of that marriage, and finds allusions in the play which enforce a much earlier date. 'To state it briefly,' he says (p. 40), 'all indications point to the 'fact that [this play] was written for and performed at the marriage of the Earl of 'Essex in the year 1590.' Essex's marriage, though secret, was not clandestine, and ELZE assumes that this secrecy did not extend so far but that there could be song and music and private theatricals, and that the main thing was to keep it from the ears of the Queen until it was too late for her to refuse to sanction it; so far and no further was it secret. In Essex and his bride, the widow of Sir Philip Sidney, ELZE finds a parallel to Theseus and Hippolyta. 'Like Theseus, the bridegroom, in spite of his 'youth, was a captain and, doubtless, a huntsman as well; whether, certainly in a 'different sense from Theseus, he had won his bride by his sword could be intelligible 'only to the initiated. As a youth of seventeen he had followed his step-father, Lei-'cester, into the Netherlands, . . . and at Zutphen, in 1586, he so distinguished him-'self that Leicester knighted him.' Great clerks purposed to greet Theseus with premeditated welcomes, and when Essex returned in 1589 from his Spanish cam-paign, Peele dedicated to him his *Eclogue Gratulatory*. 'Like Theseus, he courted 'many an Aeglé and Perigenia, and then left them.' From the fact that Lady Sidney accompanied her husband to Holland and nursed him when he was mortally wounded at Zutphen, and carried him to Arnheim, ELZE thinks 'we shall scarcely be mistaken 'in conceiving her a strong heroic woman like Hippolyta—in a good sense—who in 'merry days delighted in the chase and in the barking of the hounds, like the Ama-'zon queen.' ELZE (p. 47) conceives the question, merely as a possibility, 'whether 'two of Essex's servants or officers did not enter upon their marriage at the same time 'as their master, so that the triple wedding in the play would have exactly corre-'sponded to what actually took place.' Of Puck's concluding speech, 'If we shad-'ows have offended,' &c., ELZE says that 'these lines would be flat and meaningless 'if they had not been spoken at Essex's wedding. The pardon asked for would cer-'tainly have been granted, the more readily as it could scarcely have escaped those 'interested in the play that the object of the passage in question was to put in a good 'word for them with the queen.' ELZE (p. 60) concludes: 'Thus, from whatever 'side we may view *A Midsummer Night's Dream*, and whatever points we may take 'into consideration, everything agrees with the supposition that it was written in the 'spring of the year 1590, for the wedding of the Earl of Essex with Lady Sidney.'

KURZ (*Jahrbuch d. deut. Sh.-Gesellschaft*, 1869, p. 268) upholds Elze in the sup-position that Essex's wedding was the festive occasion of the composition of this play, and suggests, as a proof, that it must have been acted before 1591; that the first three Books of Spenser's *Faery Queen*, with its idealised Queen Elizabeth, appeared in that year, and 'after that could Shakespeare let his fairy queen, albeit called Titania 'and the spouse of Oberon, fall in love with an ass? A question not to be lightly 'tossed aside. Not within half a decade at least, one would think, could he venture 'on such an incident, until the burning suspicion of an intentional allusion had cooled 'down.' KURZ has been taken seriously here. It is doubtful. There is a vein of quiet humour running through his *Essay* that makes it difficult to say whether or not he is anywhere really in earnest. From a thorough study of the *Sidney Papers* he comes to the conclusion that a certain entertainment, there mentioned, was given on the occasion of Essex's marriage, which must have taken place some time in April, 1590, either before the sixth, on which day the bride's father died, or sooner or later after it. In the latter case, her unprotected state might have accelerated the wed-

ding and justified the haste. 'There is no doubt,' says KURZ, p. 286, 'that the
'marriage itself was conducted quite privately. But the public after-celebration
'demanded a certain caution, which forsooth could not be lost sight of for months to
'come. Any unexpected festivity would arouse the curiosity and suspicion of the
'Queen, already curious and suspicious; it would be far better then to select for the
'*public* celebration some day which was a *public* festival. And such a one there was
'right off—namely, May Day, from time immemorial one of the freest festivals of the
'whole year, in city or country, by young or old, rich or poor—all was merriment.
'On this day, then, or close enough to it, a banquet [mentioned in the Sidney Let-
'ters] could take place, without exciting any comment, and afterwards a play.' This
explains the allusions to May. In short, KURZ reaches the positive conclusion
(p. 289) that the *Midsummer Night's Dream* was performed, for the first time, at a
banquet on the occasion of the unheralded festivities accompanying the marriage of
Essex, and in conjunction with the observances of May in 1590, as a masque with
significant characters, or as a masque-like comedy with a masque especially introduced, and all of it designed to conceal the object for which the festivities were
given. Hence is explained the apparent incongruity, whereby the piece seems to
have been written so emphatically for a marriage, and yet, on the other hand, does
not in some of its details seem quite appropriate thereto. Among these latter is
manifestly the allusion to Theseus's former loves; this KURZ explains (p. 291) by
supposing that, on account of the mourning for her father, the bride was not present
at the performance of the play.

The discrepancy between Hippolyta's 'new moon' and the full moon of Pyramus
and Thisbe, KURZ explains by his theory that the play was not performed at the
wedding itself, but was a part of the festivities of the following May day. 'If the
'Kalendar of 1590 gives a full moon on the first of May, then all calculations are
'upset. But be of good cheer: the old Ephemerides (*Cypr. Leovitius*, 1556–1606,
'Augsburg, 1557; *Mart. Everart*, 1590–1610, Leyden, 1597) agree in naming the
'30 April as the day whereon that May moon renewed itself.' If KURZ has rightly
understood and quoted 'the old Ephemerides,' these latter certainly corroborate,
quite remarkably, Hippolyta's words as generally adopted since Rowe's edition; but
I fail to see how they help KURZ, who says distinctly (p. 286) that Essex's marriage
(i. e. Theseus's) took place before or shortly after the sixth of April, and that it was
merely the public festivities which were held on the following May day, when the
'silver bow' must, of course, be full or gibbous if it was 'newbent' about a fortnight
or three weeks before. I am afraid no Ephemerides will reconcile Hippolyta, Quince,
and Kurz. Moreover, there is a conflict of authority. W. A. WRIGHT (*Preface*,
p. xi, footnote) took the pains to apply to Professor ADAMS, through whose kindness
he was enabled to state that 'the nearest new moon to May 1, 1590, was on April 23,
'and that there was a new moon on May 1 in 1592.' KURZ had better have left
undisturbed the dust and moonshine on the 'old Ephemerides.'

By referring *A Midsummer Night's Dream* to Essex's marriage, KURZ thinks to
solve another problem hitherto insoluble, that of accounting for Shakespeare's early
patronage by the nobility. In Theseus, the hero and statesman, lofty of manner,
appreciative of poesie, we find (p. 299) the ideal character which the popular verdict
gave to Essex; and in Hippolyta the character of Lady Frances was adequately portrayed. 'It is easy to see [p. 300] what an effect such a solution of the task must
have had on Essex, a man who could appreciate all the beauties and delicacies
of the play. . . . The performance, therefore, which so immeasurably surpassed all

demands and expectations, must have drawn, of necessity, the attention of Essex to
' the poet. . . . The Earl of three and twenty and the Poet of six and twenty
' must have become intimate as soon as they had become personally acquainted,
' Shakespeare in the inexhaustible fullness and grace of his genius; Essex with his
' captivating condescension, whereby he elevated to his own level those in a lowly
' station, and with that character so full of contradictions which offered for study at
' one and the same time a Hotspur and a Hamlet.   Whose recommendation it was,
' whereon the poet three years afterwards was introduced to Southampton, is now
' placed beyond all doubt.'

It is in reference to these speculations by Kurz that W. A. WRIGHT (*Preface*, p.
xi) caustically remarks : ' In such questions it would be well to remember the maxim
' of the ancient rabbis, " Teach thy tongue to say, I do not know." '   But is not this
a little too severe on Kurz, who is merely copying the methods of English-speaking
commentators in founding theory after theory on imaginary possibilities ?

DOWDEN (p. 67): *A Midsummer Night's Dream* was written on the occasion of
the marriage of some noble couple—possibly . . . as Mr Gerald Massey supposes;
possibly . . . as Prof. Elze supposes.

FLEAY, in his *Manual*, 1876, p. 26, gives the date as of 1592, but wider knowledge led him to the belief that this was the date of the stage-play only.   ' In its
' present form' it is of a later date.   In his *Life and Work of Shakespeare* (1886, p.
181) we find, under the year 1595, as follows : ' January 26 was the date of the mar-
' riage of William Stanley, Earl of Derby, at Greenwich.   Such events were usually
' celebrated with the accompaniment of plays or interludes, masques written specially
' for the occasion not having yet become fashionable.   The company of players
' employed at these nuptials would certainly be the Chamberlain's [the company to
' which Shakespeare belonged], who had, so lately as the year before, been in the
' employ of the Earl's brother Ferdinand.   No play known to us is so fit for the pur-
' pose as *A Midsummer Night's Dream*, which in its present form is certainly of this
' date.   About the same time Edward Russel, Earl of Bedford, married Lucy Har-
' rington.   Both marriages may have been enlivened by this performance.   This is
' rendered more probable by the identity of the Oberon story with that of Drayton's
' *Nymphidia*, whose special patroness at this time was the newly-married Countess of
' Bedford. . . . The date of the play here given is again confirmed by the description
' of the weather in II, ii. . . . Chute's *Cephalus and Procris* was entered on the Sta-
' *tioners' Registers*, 28 September, 1593; Marlowe's *Hero and Leander*, 22d October,
' 1593; Marlowe and Nash's *Dido* was printed in 1594.   All these stories are
' alluded to in the play.   The date of the Court performance must be in the winter
' of 1594–5.   But the traces of the play having been altered from a version for the
' stage are numerous [see FLEAY'S note on V, i, 417]. . . . The date of the stage-play
' may, I think, be put in the winter of 1592; and if so, it was acted, not at the Rose,
' but where Lord Strange's company were travelling.   For the allusion in V, i, 59,
' " The thrice three Muses," &c. to Spenser's *Tears of the Muses* (1591), or Greene's
' death, 3d September, 1592, could not, on either interpretation, be much later than
' the autumn of 1592, and the lines in III, i, 160, " I am a spirit of no common rate :
' " The summer still doth tend upon my state," are so closely like those in Nash's
' *Summer's Last Will* [see FLEAY'S note, *ad loc.*], that I think they are alluded to by
' Shakespeare.   The singularly fine summer of 1592 is attributed to the influence of
' Elizabeth, the Fairy Queen.   Nash's play was performed at the Archbishop's palace
' at Croydon in Michaelmas term of the same year by a " number of hammer-handed

' " clowns (for so it pleaseth them in modesty to name themselves) ;" but I believe the
' company originally satirised in Shakespeare's play was the Earl of Sussex's, Bottom,
' the chief clown, being intended for Robert Greene.' See Prof. J. M. BROWNE (*Source
of the Plot), who has in this conjecture anticipated FLEAY. In his *English Drama*,
published in 1891, FLEAY slightly modified his opinions. 'This play,' he there says
(vol. ii, p. 194), 'has certainly alternative endings : one a song by Oberon for a mar-
' riage, and then *Exeunt*, with no mark of Puck's remaining on the stage ; the other,
' an Epilogue by Puck, apparently for the Court (cf. "gentles" in l. 423). It might
' seem, as the Epilogue is placed last, that the marriage version was the earlier, and
' so I took it to be when I wrote my *Life of Shakespeare*, but the compliment to
' Elizabeth in II, i, 164, was certainly written for the Court ; and this passage is essen-
' tial to the original conduct of the play, which may have been printed from a mar-
' riage-version copy, with additions from the Court copy. This would require a date
' for the marriage subsequent to the Court performance. One version must date 1596,
' for the weather description, II, i, which can be omitted without in any way affecting
' the progress of the play, requires that date. I believe this passage was inserted for
' the Court performance in 1596, that on the public stage having taken place in 1595 ;
' but that the marriage presentation, being subsequent to this, was most likely at the
' union of Southampton and Elizabeth Vernon in 1598–9. In any case, this was
' Shakespeare's first Epilogue now extant.' FLEAY finds further confirmation of
his date (*Life and Work*, p. 185) in the lion incident noted at III, i, 31.

W. A. WRIGHT (*Preface*, ix) : If the occasion for which this play was written
' could be determined with any degree of probability, we should be able to ascertain
' within a little the time at which it was composed. But here again we embark upon
' a wide sea of conjecture, with neither star nor compass to guide us. That the *Mid-
'summer Night's Dream* may have been first acted at the marriage of some noble-
' man, and that, from the various compliments which are paid to Elizabeth, the per-
' formance may have taken place when the Queen herself was present, are no improb-
' able suppositions. But when was this conjuncture of events ? No theory which has
' yet been proposed satisfies both conditions. . . . In fact, we know nothing whatever
' about the matter, and of guesses like these [as set forth in the preceding pages]
' there is neither end nor profit.'

Here ends the discussion of the nine specified topics which are supposed to determine the *Date of Composition*. The opinions of several critics of weight, which are
general in their scope, are as follows :—

MALONE (*Variorum 1821*, ii, p. 333) : 'The poetry of this piece, glowing with all
' the warmth of a youthful and lively imagination, the many scenes which it contains
' of almost continual rhyme, the poverty of the fable, and want of discrimination among
' the higher personages, dispose me to believe that it was one of our author's earliest
' attempts in comedy.

' It seems to have been written while the ridiculous competitions prevalent among
' the histrionic tribe were strongly impressed by novelty on his mind. He would
' naturally copy those manners first with which he was first acquainted. The ambi-
' tion of a theatrical candidate for applause he has happily ridiculed in Bottom the
weaver. But among the more dignified persons of the drama we look in vain for any
traits of character. The manners of Hippolyta, the Amazon, are undistinguished
from those of other females. Theseus, the associate of Hercules, is not engaged in
any adventure worthy of his rank or reputation, nor is he in reality an agent through-

out the play. Like Henry VIII. he goes out a Maying. He meets the lovers in perplexity, and makes no effort to promote their happiness; but when supernatural ' accidents have reconciled them, he joins their company, and concludes his day's ' entertainment by uttering miserable puns at an interlude represented by a troop of ' clowns. Over the fairy part of the drama he cannot be supposed to have any influ-' ence. This part of the fable, indeed (at least as much of it as relates to the quarrels ' of Oberon and Titania), was not of our author's invention.' [This assertion rests on Tyrwhitt's remark, that ' the true progenitors of Shakespeare's Oberon and Titania ' appear to have been Pluto and Proserpine in Chaucer's *Merchant's Tale.*—ED.]. ' Through the whole piece, the more exalted characters are subservient to the interests ' of those beneath them. We laugh with Bottom and his fellows; but is a single pas-' sion agitated by the faint and childish solicitudes of Hermia and Demetrius, of Helena ' and Lysander, those shadows of each other? That a drama, of which the principal ' personages are thus insignificant, and the fable thus meagre and uninteresting, was ' one of our author's earliest compositions does not, therefore, seem a very improbable ' conjecture; nor are the beauties, with which it is embellished, inconsistent with this ' supposition; for the genius of Shakespeare, even in its minority, could embroider the ' coarsest materials with the brightest and most lasting colors.'

VERPLANCK (*Introductory Remarks*, p. 6, 1847): It seems to me very probable (though I do not know that it has appeared so to any one else) that the *Midsummer Night's Dream* was originally written in a very different form from that in which we now have it, several years before the date of the drama in its present shape—that it was subsequently remoulded, after a long interval, with the addition of the heroic personages, and all the dialogue between Oberon and Titania, perhaps with some alteration of the lower comedy; the rhyming dialogue and the whole perplexity of the Athenian lovers being retained, with slight change, from the more boyish comedy. The completeness and unity of the piece would indeed quite exclude such a conjecture, if we were forced to reason only from the evidence afforded by itself; but, as in *Romeo and Juliet* (not to speak of other dramas), we have the certain proof of the amalgamation of the products of different periods of the author's progressive intellect and power, the comparison leads to a similar conclusion here.

R. G. WHITE (ed. i, p. 16, 1857): It seems that *A Midsummer Night's Dream* was produced, in part at least, at an earlier period of Shakespeare's life than his twenty-ninth year. [That is, in 1593.] Although as a whole it is the most exquisite, the daintiest, and most fanciful creation that exists in poetry, and abounds in passages worthy even of Shakespeare in his full maturity, it also contains whole Scenes which are hardly worthy of his 'prentice hand that wrought *Love's Labour's Lost*, *The Two Gentlemen of Verona*, and *The Comedy of Errors*, and which yet seem to bear the unmistakeable marks of his unmistakeable pen. These Scenes are the various interviews between Demetrius and Lysander, Hermia and Helen, in Acts II and III. It is difficult to believe that such lines as ' Do not say so, Lysander; say not so. What though he love your Hermia? *Lord what though?*' ' When at your hands did I deserve this scorn? Is 't not enough, is 't not enough, *young man*, That I did never, no, nor never can,' &c.—Act II, Sc. i,—it is difficult to believe that these, and many others of a like character which accompany them, were written by Shakespeare after he had produced even *Venus and Adonis* and the plays mentioned above, and when he could write the poetry of the other parts of this very comedy. There seems, therefore, warrant for the opinion that this Dream was one of the very first conceptions of the young poet; that, living in a rural district where tales of house-

hold fairies were rife among his neighbors, memories of these were blended in his youthful reveries with images of the classic heroes that he found in the books which we know he read so eagerly; that perhaps on some midsummer's night he, in very deed, did dream a dream and see a vision of this comedy, and went from Stratford up to London with it partly written; that, when there, he found it necessary at first to forego the completion of it for labor that would find readier acceptance at the theatre; and that afterward, when he had more freedom of choice, he reverted to his early production, and in 1594 worked it up into the form in which it was produced. It seems to me that in spite of the silence of the Quarto title-pages on the subject, this might have been done, or at least that some additions might have been made to the play, for a performance at Court. The famous allusion to Queen Elizabeth as 'a ' fair vestal throned by the west' tends to confirm me in that opinion. Shakespeare never worked for nothing, and, besides, could he, could any man, have the heart to waste so exquisite a compliment as that is, and to such a woman as Queen Elizabeth, by uttering it behind her back? Except in the play itself I have no support for this opinion, but I am willing to be alone in it.

[In a list of Shakespeare's Works in the order in which they were probably written, R. G. WHITE (vol. i, p. xlvi, 2d ed.) gives the date of the present play as of ' 1592 (?) and 1601 (?).' The latter is an impossible date; it implies that there are additions to be found in the Folio which are not in the Quartos. There is none.—ED.]

The COWDEN-CLARKES: The internal evidence of the composition itself gives unmistakeable token of its having been written when the poet was in his flush of youthful manhood. The classicality of the principal personages, Theseus and Hippolyta; the Grecian-named characters; the prevalence of rhyme; the grace and whimsicality of the fairy-folk; the rich warmth of coloring that pervades the poetic diction; the abundance of description, rather than of plot, action, and character-developement, all mark the young dramatist. With a manifest advance in beauty beyond those which we conceive to be his earliest-written productions—*The Two Gentlemen of Verona, Comedy of Errors,* and *Love's Labour's Lost*—we believe the *Midsummer Night's Dream* to be one of his very first-written dramas after those three plays. We feel it to have been, with *Romeo and Juliet,* the work of his happy hours, when he wrote from inspiration and out of the fulness of his luxuriant imagination, between the intervals of his business-work—the adaptation of such immediately needed stage-plays as the three parts of *Henry VI.* Those, we think, he touched up for current production, for the use of the theatre at which he was employed and had a share in; but his overflowing poet-heart was put into productions like the Southern-storied *Romeo and Juliet,* and the fairy-favoured *Midsummer Night's Dream,* where every page is a forest glade flooded with golden light amid the green glooms.

According to Prof. INGRAM'S *Table of Light and Weak Endings* (*New Sh. Soc. Trans.* 1874, p. 450) the present play stands fourth in the list.

According to Dr FURNIVALL'S *Order and Groups of the Plays,* in his *Introduction* to the *Leopold Shakespeare,* this play belongs to the *First Period* or *Mistaken-Identity Group,* and its date is given '? 1590–1.'

Rev. H. P. STOKES (*Chronological Order of Sh.'s Plays,* 1878, p. 52): Mr Skeat, in his *Shakespeare's Plutarch,* speaking of the various editions of North's translation (viz. 1579, 1595, 1603, 1612, &c.), says: 'Shakespeare must certainly have known ' the work before 1603, because there is a clear allusion to it in *Midsummer Night's* ' *Dream.*' . . . Mr Skeat continues: 'Whether this play was written earlier than 1595

'I leave to the investigation of the reader.' The present investigation seems to point to that very year, and may not the re-issue of North's work in this year, after it had been so long out of print, have directed Shakespeare's attention to what so soon became his chief store-house for material to work upon?

To recapitulate, chronologically :—

MALONE	(1790)	. . . .	1592
CHALMERS	(1799)	. . . . beginning of	1598
DRAKE	(1817)	. . . .	1593
MALONE	(1821)	. . . .	1594
TIECK	(1830)	. . . .	1598
CAMPBELL	(1838)	. . . .	1594
KNIGHT	(1840)	. . . .	1594
ULRICI	(1847)	. . . .	1596–7
VERPLANCK	(1847)	. . . .	1595–6
GERVINUS	(1849)	. . . .	1594–6
W. W. LLOYD	(1856)	. . . . not before	1594
R. G. WHITE i	(1857)	. . . . Shakespeare's earliest play.	
COLLIER	(1858)	. . . . . end of 1594 or beginning of	1595
STAUNTON	(1864)	. description of seasons is singularly applicable to	1593–4
DYCE ii	(1866)	. . . . . two or three years before	1598
KEIGHTLEY	(1867)	. . . . 1594 or	1595
ELZE, KURZ	(1869)	. . . . spring of	1590
FURNIVALL	(1877)	. . . . ?	1590–1
ROLFE	(1877)	. . . . perhaps as early as	1594
W. A. WRIGHT	(1878)	. . . . before	1598
STOKES	(1878)	. . . .	1595
HALLIWELL	(1879)	. . . . after 20 January,	1595–6
HUDSON	(1880)	. . . . before	1594
R. G. WHITE ii	(1883)	. . first draft as early as 1592, if not earlier.	
FLEAY	(1886)	. . . . { Stage play, 1592	
		{ Court play, 1594–5	
MARSHALL	(1888)	. . . . approximately,	1595
MASSEY	(1888)	. . . .	1595
DEIGHTON	(1893)	. . . .	1592–1594
VERITY	(1894)	. . . . . at end of 1594 or beginning of	1595

SOURCE OF THE PLOT

CAPELL (*Introd.* vol. i, p. 64, 1767) suggested that it was 'not improbable that
' Shakespeare took a hint of his fairies' from Drayton's *Nymphidia;* 'a line of that
' poem, " Thorough bush, thorough briar," occurs also in this play.'

MALONE set at rest this suggestion by showing that the *Nymphidia* was printed
after *A Midsummer Night's Dream.* See p. 246, above.

' The rest of the play,' continues CAPELL, 'is, doubtless, invention, the names only
' of *Theseus, Hippolyta,* and *Theseus'* former loves, *Antiopa* and others, being his-
' torical; and taken from the translated *Plutarch* in the article *Theseus.*'

The passages in *Plutarch* which, as is alleged, supplied SHAKESPEARE with allusions, are as follows. They are taken from SKEAT's *Shakespeare's Plutarch,* 1875:—

' [Theseus] pricked forwards with emulation and envy of [Hercules's glory] . . .
' determined with himself one day to do the like, and the rather, because they were
' near kinsmen, being cousins removed by the mother's side.'—p. 278.

Again: 'Albeit in his time other princes of Greece had done many goodly and
' notable exploits in the wars, yet Herodotus is of opinion that Theseus was never in
' any one of them, saving that he was at the battle of the Lapithæ against the Cen-
' taurs. . . . Also he did help Adrastus, King of the Argives, to recover the bodies of
' those that were slain in the battle before the city of Thebes.'—p. 288.

Compare :—

> '*Lis.* The battell with the Centaurs to be sung
> ' By an Athenian Eunuch, to the Harpe.
> '*The.* Wee'l none of that. That haue I told my Loue
> ' In glory of my kinsman Hercules.
> '*Lis.* The riot of the tipsie Bacchanals,
> ' Tearing the Thracian singer, in their rage?
> '*The.* That is an old deuice, and it was plaid
> ' When I from *Thebes* came last a Conqueror.'

We read in Plutarch: ' This Sinnis had a goodly fair daughter called Perigouna,
' which fled away when she saw her father slain; whom [Theseus] followed and
' sought all about. But she had hidden herself in a grove full of certain kinds of
' wild pricking rushes called *stœbe,* and wild sperage which she simply like a child
' intreated to hide her, as if they had heard. . . . But Theseus finding her, called her,
' and sware by his faith he would use her gently, and do her no hurt, nor displeasure
' at all. Upon which promise she came out of the bush.'—p. 279.

Again: 'After he was arrived in Creta, he slew there the Minotaur . . . by the
' means and help of Ariadne: who being fallen in fancy with him, did give him a
' clue of thread. . . . And he returned back the same way he went, bringing with him
' those other young children of Athens, whom with Ariadne also he carried afterwards
away. . . . And being a solemn custom of Creta, that the women should be present
' to see those open sports and sights, Ariadne, being at these games among the rest,
' fell further in love with Theseus seeing him so goodly a person, so strong, and invin-
' cible in wrestling.'—p. 283. ' Some say, that Ariadne hung herself for sorrow, when
' she saw that Theseus had cast her off. Other write, that she was transported by
' mariners into the ile of Naxos, where she was married unto Œnarus the priest of
' Bacchus: and they think that Theseus left her, because he was in love with another
' as by these verses should appear :—

' Ægles, the nymph, was loved of Theseus,
' Who was the daughter of Panopeus.'—p. 284.

Again : ' Touching the voyage he made by the sea Major, Philochorus, and some
other hold opinion, that he went thither with Hercules against the Amazons : and
' that to honour his valiantness, Hercules gave him Antiopa, the Amazon.   But the
' more part of the other historiographers . . . do write, that Theseus went thither alone,
' . . . and that he took this Amazon prisoner, which is likeliest to be true. . . . Bion
' . . . saith, that he brought her away by deceit and stealth . . . and that Theseus
' enticed her to come into his ship, who brought him a present ; and so soon as she
' was aboard, he hoised his sail, and so carried her away.'—p. 286.

Again : 'Afterwards, at the end of four months, peace was taken between [the
' Athenians and the Amazons] by means of one of the women called Hippolyta.
' For this historiographer calleth the Amazon which Theseus married, Hippolyta, and
' not Antiopa.   Nevertheless some say she was slain (fighting on Theseus' side) with
' a dart, by another called Molpadia.   In memory whereof, the pillar which is
' joined to the temple of the Olympian ground was set up in her honour.   We
' are not to marvel, if the history of things so ancient be found so diversely written.'
—p. 288.

From these weeds Shakespeare gathered this honey :—

> '*Qu.* Why art thou here
> ' Come from the farthest steepe of *India* ?
> ' But that forsooth the bouncing *Amazon*
> ' Your buskin'd Mistress, and your Warrior loue,
> ' To *Theseus* must be wedded ; and you come
> ' To giue their bed ioy and prosperitie.
> '*Ob.* How canst thou thus for shame *Tytania*,
> ' Glance at my credite, with *Hippolita* ?
> ' Knowing I know thy loue to *Theseus* ?
> ' Didst thou not leade him through the glimmering night
> ' From *Peregenia*, whom he rauished ?
> ' And make him with faire Eagles breake his faith
> ' With *Ariadne*, and *Antiopa* ?'

CHAUCER'S KNIGHT'S TALE

In the *First Variorum,* 1773, STEEVENS remarked that it is ' probable that the
' hint for this play was received from Chaucer's *Knight's Tale ;* thence it is that our
' author speaks of Theseus as duke of Athens.'

This suggestion was repeated in all the Variorums down to that of 1821 ; and was
adopted by KNIGHT, in what may be fairly considered as the first critical edition
after that date.   SINGER'S edition of 1826 is little else than an abridgement, without
acknowledgement, of the Variorum of 1821 ; and HARNESS'S contribution to his edition of 1830 is mainly confined to *The Life of Shakespeare.*   KNIGHT even goes so
far as to point out the very passages ' in which, as he says, p. 343, ' it is not difficult
to trace Shakespeare.'   These passages are as follows (ed. Morris) :—

> ' Whilom, as olde stories tellen us,
> ' Ther was a duk that highte Theseus ;

'Of Athenes he was lord and governour,
'And in his tyme swich a conquerour,
'That gretter was ther non under the sonne.
'Ful many a riche contré hadde he wonne;
'That with his wisdam and his chivalrie
'He conquered al the regne of Femynye,
'That whilom was i-cleped Cithea;
'And weddede the queen Ipolita,
'And brought hire hoom with him in his contré,
'With moche glorie and gret solempnité,
'And eek hire yonge suster Emelye.
'And thus with victorie and with melodye
'Lete I this noble duk to Athenes ryde,
'And al his ost, in armes him biside.
'And certes, if it nere to long to heere,
'I wolde han told yow fully the manere,
'How wonnen was the regne of Femenye
'By Theseus, and by his chivalrye;
'And of the grete bataille for the nones
'Bytwix Athenes and the Amazones;
'And how asegid was Ypolita,
'The faire hardy quyen of Cithea;
'And of the feste that was at hire weddynge,
'And of the tempest at hire hoom comynge;
'But al that thing I most as now forbere.
'I have, God wot, a large feeld to ere.'

In a note on I, i, **177**, KNIGHT says, 'The very expression "to do observance" in
'connection with the rites of May, occurs twice in Chaucer's *Knight's Tale*:—
'This passeth yeer by yeer, and day by day,
'Til it fel oones in a morwe of May
'That Emelie, that fairer was to seene
'Than is the lilie on hire stalkes grene.
'And fresscher than the May with floures newe—
'For with the rose colour strof hire hewe,
'I not which was the fairer of hem two—
'Er it was day, as sche was wont to do,
'Sche was arisen, and al redy dight;
'For May wole have no sloggardye a nyght.
'The sesoun priketh every gentil herte,
'And maketh hin out of his sleepe sterte,
'And seith, "Arys, and *do thin observance*."'
[Page 33. The italics are Knight's.] Again :—
'And Arcite, that is in the court ryal
'With Theseus, his squyer principal,
'Is risen, and loketh on the mery day.
'And for *to doon his observance to May*.'—[p. 47].
Furthermore in a note on III, ii, 412:—'Even till the Easterne gate all fierie reu,
'Opening on *Neptune*, with faire blessed beames, Turnes into yellow gold, his salt

' greene streames,' KNIGHT says: ' This splendid passage was, perhaps, suggested
' by some line in Chaucer's *Knight's Tale*:

'The busy larke, messager of day,
' Salueth in hire song the morwe gray;
' And fyry Phebus ryseth up so bright,
' That al the orient laugheth of the light,
' And with his stremes dryeth in the greves
' The silver dropes, hongyng on the leeves.'—[p. 46].

On ' Goe one of you finde out the Forrester,' &c., IV, i, 117, KNIGHT observes:
' The Theseus of Chaucer was a mighty hunter :—

' This mene I now by mighty Theseus
' That for to honte is so desirous,
' And namely the grete hart in May
' That in his bed ther daweth him no day,
' That he nys clad, and redy for to ryde
' With hont and horn, and houndes him byside.
' For in his hontyng hath he such delyt,
' That it is al his joye and appetyt
' To been himself the grete hertes bane,
' For after Mars he serveth now Dyane.'—[p. 52].

HALLIWELL (*Introd.* p. 11, 1841) thinks that commentators have overlooked the
following passage, ' which occurs nearly at the end of *The Knight's Tale*, and may
' have furnished Shakespeare with the idea of introducing an interlude at the end of
' his play :—

' " no how the Grekes pleye
' " The wake-pleyes, kepe I nat to seye ;
' " Who wrastleth best naked, with oyle enoynt,
' " Ne who that bar him best in no disjoynt.
' " I wol not telle eek how that they ben goon
' " Hom til Athenes whan the pley is doon " [p. 91].

' The introduction of the clowns and their interlude was perhaps an afterthought.
Again, in *The Knight's Tale*, we have this passage :—

' " Duk Theseus, and al his companye,
' " Is comen hom to Athenes his cité,
' " With alle blys and gret solempnité " [p. 83],

' which bears too remarkable a resemblance to what Theseus says in the *Midsummer
' Night's Dream* to be accidental:—" Away with us to Athens: Three and three,
' " We'll hold a feast in great solempnity " [IV, i, 202].

' In the Legende of Thisbe of Babylon we read :—

' " Thus wolde they seyn :—' Allas, thou wikked walle !
' " Thurgh thyn envye thow us lettest alle !' "—[line 51],

' which is certainly similar to the following line in Pyramus's address to Wall: " O
' " wicked Wall, through whom I see no bliss !" '

The foregoing are all the extracts, I believe, which have been anywhere cited in
proof of Steevens's suggestion, the value whereof has been correctly estimated, I
think, by STAUNTON, who says (p. 476) : ' The persistence [of the commentators] in
' assigning the groundwork of the fable to Chaucer's *Knight's Tale* is a remarkable
' instance of the docility with which succeeding writers will adopt, one after the other,
' an assertion that has really little or no foundation in fact. There is scarcely any

'resemblance whatever between Chaucer's tale and Shakespeare's play, beyond that
'of the scene in both being laid at the Court of Theseus.   The Palamon, Arcite, and
'Emilie of the former are very different persons indeed from the Demetrius, Lysan-
'der, Helena, and Hermia of the latter.   Chaucer has made Duke Theseus a lead-
'ing character in his story, and has ascribed the unearthly incidents to mythological
'personages, conformable to a legend which professes to narrate events that actually
'happened in Greece.   Shakespeare, on the other hand, has merely adopted Theseus,
'whose exploits he was acquainted with through the pages of North's *Plutarch*, as a
'well-known character of romance, in subordination to whom the rest of the *dramatis*
'*personæ* might fret their hour; and has employed for supernatural machinery those
'" airy nothings " familiar to the literature and traditions of various people and nearly
'all ages.   There is little at all in common between the two stories except the name
'of Theseus, the representative of which appears in Shakespeare simply as a prince
'who lived in times when the introduction of ethereal beings, such as Oberon, Tita-
'nia, and Puck, was in accordance with tradition and romance.'

FLEAY (*Life and Work*, p. 185) says that Shakespeare got the name of Philostrate from Chaucer's *Knight's Tale*.

TYRWHITT (*Introd.* p. 97, 1798), in discussing the original of *The Marchaunde's
Tale*, says that he cannot help thinking that 'the *Pluto* and *Preserpina* in this tale
'were the true progenitors of *Oberon* and *Titania*, or rather, that they themselves
'have, once at least, deigned to revisit our poetical system under the latter names,'—
a remark which would not have been repeated here had it not been repeated, more
than once, elsewhere.

PYRAMUS AND THISBE

RITSON (*Remarks*, p. 47, 1783) in reference to *Pyramus and Thisbe* observes:
'There is an old pamphlet, containing the history of this amorous pair, in lamentable
'verse by one Dunstan Gale, which appears to have been printed in 1596; and may,
'not improbably, be found the butt of Shakespeare's ridicule in some parts of this
'interlude.'

MALONE, in a note on I, ii, **15**, gives a later date: 'A poem entitled *Pyramus and
'Thisbe*, by D. Gale, was published in 4to in 1597; but this, I believe, was posterior
'to the *Midsummer Night's Dream*.' 'On the contrary,' says CHALMERS (*Sup.
Apol.* p. 363), who also gives 1597 as the date, 'I believe that Gale's *Pyramus &*
'*Thisbe* was prior to Shakespeare's "most lamentable comedy." '

COLLIER (*Bibliog. Account*, &c., 1865, ii, 43) thus allays the breeze evoked by
Gale: 'No earlier edition [than 1617, of this poem] is known; but the dedication
'"to the worshipfull his verie friend D. B. H." is dated by the author, Dunstan Gale,
'"this 25th of November, 1596." ' From the description and specimens of this
'poem' given by Collier, we need not 'desire it of more acquaintance'; nor with
Dr. Muffet's *Silkworms and their Flies*, 1599, mentioned by COLLIER (*Ib.* i, 97) and
by HALLIWELL *ad loc.*

STEEVENS mentions a license recorded in the *Stationers' Registers* (vol. i, p. 215,
ed. Arber) as given to 'William greffeth,' in 1562, 'for pryntynge of a boke intituled
'Perymus and Thesbye.'

It appears to me to be almost childish to attempt to fix upon any single source
(except possibly *Ovid*) as the authority to which Shakespeare went for a story, with
which, in its every detail, the early literature of Europe abounds.   Would it be possible
to limit to one single writer the story of a pair of star-crost lovers, which had started in

Babylon under the shadow of the tomb of Ninus, was familiar to the Greeks and Romans, and used in the Middle Ages by pious monks as an allegory of the human soul?

The inquisitive reader is referred to a thorough and exhaustive compilation of the versions of this legend in Latin, in Greek, and in the ancient and modern literatures of France, Germany, Spain, Holland, Roumania, Italy, and England by Dr. GEORG HART (*Die Pyramus- & Thisbe-Sage*, Passau, 1889, and Part ii, 1891).

Many commentators have called attention to what they have assumed to be indications here and there of Shakespeare's having read the story of Pyramus and Thisbe in Golding's translation of *Ovid*. The story is here given from Golding (*The fourth booke*, 1567, p. 43, verso) :—

> Within the towne (of whose huge walles so monstrous high & thicke
> The fame is giuen *Semyramis* for making them of bricke)
> Dwelt hard together two yong folke in houses ioyned so nere
> That vnder all one roofe well nie both twaine conueyed were.
> The name of him was *Pyramus*, and *Thisbe* calde was she.
> So faire a man in all the East was none aliue as he,
> Nor nere a woman maide nor wife in beautie like to hir.
> This neighbrod bred acquaintance first, this neyghbrod first did stirre
> The secret sparkes, this neighbrod first an entrance in did showe,
> For loue to come to that to which it afterward did growe.
> And if that right had taken place they had bene man and wife,
> But still their Parents went about to let which (for their life)
> They could not let. For both their heartes with equall flame did burne.
> No man was priuie to their thoughts. And for to serue their turne
> In steade of talke they vsed signes, the closelier they supprest
> The fire of loue, the fiercer still it raged in their brest.
> The wall that parted house from house had riuen therein a crany
> Which shronke at making of the wall, this fault not markt of any
> Of many hundred yeares before (what doth not loue espie.)
> These louers first of all found out, and made a way whereby
> To talke toglther secretly, and through the same did goe
> Their louing whisprings verie light and safely to and fro.
> Now as a toneside *Pyramus* and *Thisbe* on the tother
> Stoode often drawing one of them the pleasant breath from other
> O thou enuious wall (they sayd) why letst thou louers thus ?
> What matter were it if that thou permitted both of vs
> In armes eche other to embrace ? Or if thou thinke that this
> Were ouermuch, yet mightest thou at least make roume to kisse.
> And yet thou shalt not finde vs churles : we think our selues in det
> For this same piece of courtesie, in vouching safe to let
> Our sayings to our friendly eares thus freely come and goe,
> Thus hauing where they stoode in vaine complayned of their woe,
> When night drew nere, they bade adew and eche gaue kisses sweete
> Vnto the parget on their side, the which did neuer meete.
> Next morning with hir cherefull light had driuen the starres aside
> And *Phebus* with his burning beames the dewie grasse had dride.
> These louers at their wonted place by foreappointment met.
> Where after much complaint and mone they couenanted to get
> Away from such as watched them, and in the Euening late

18

To steale out of their fathers house and eke the Citie gate.
And to thentent that in the fieldes they strayde not vp and downe
They did agree at *Ninus* Tumb to meete without the towne,
And tarie vnderneath a tree that by the same did grow
Which was a faire high Mulberie with fruite as white as snow,
Hard by a cool and trickling spring.   This bargaine pleasde them **both**
And so daylight (which to their thought away but slowly goth)
Did in the Ocean fall to rest, and night from thence doth rise.
Assoone as darkenesse once was come, straight *Thisbe* did deuise
A shift to wind hir out of doores, that none that were within
Perceyued hir: And muffling hir with clothes about hir chin,
That no man might discerne hir face, to *Ninus* Tumb she came
Vnto the tree, and sat hir downe there vnderneath the same.
Loue made hir bold.   But see the chance, there comes besmerde with blood
About the chappes a Lionesse all foming from the wood
From slaughter lately made of Kine to staunch hir bloudie thurst
With water of the foresaid spring.   Whome *Thisbe* spying furst
A farre by moonelight, therevpon with fearfull steppes gan flie,
And in a darke and yrkesome caue did hide hirselfe thereby.
And as she fled away for hast she let hir mantle fall
The whych for feare she left behind not looking backe at all.
Now when the cruell Lionesse hir thurst had stanched well,
In going to the Wood she found the slender weed that fell
From *Thisbe*, which with bloudie teeth in pieces she did teare
The night was somewhat further spent ere *Pyramus* came there
Who seeing in the suttle sande the print of Lions paw,
Waxt pale for feare.   But when also the bloudie cloke he saw
All rent and torne, one night (he sayd) shall louers two confounde,
Of which long life deserued she of all that liue on ground.
My soule deserues of this mischaunce the perill for to beare.
I wretch haue bene the death of thee, which to this place of feare
Did cause thee in the night to come, and came not here before.
My wicked limmes and wretched guttes with cruell teeth therfore
Deuour ye O ye Lions all that in this rocke doe dwell.
But Cowardes vse to wish for death.   The slender weede that fell
From *Thisbe* vp he takes, and streight doth beare it to the tree,
Which was appointed erst the place of meeting for to bee.
And when he had bewept and kist the garment which he knew,
Receyue thou my bloud too (quoth he) and therewithall he drew
His sworde, the which among his guttes he thrust, and by and by
Did draw it from the bleeding wound beginning for to die,   }
And cast himselfe vpon his backe, the blood did spin on hie
As when a Conduite pipe is crackt, the water bursting out
Doth shote itselfe a great way off and pierce the Ayre about.
The leaues that were vpon the tree besprincled with his blood
Were died blacke.   The roote also bestained as it stoode,
A deepe darke purple colour straight vpon the Berries cast.
Anon scarce ridded of hir feare with which she was agast,   }
For doubt of disapointing him commes *Thisbe* forth in hast,

And for hir louer lookes about, reioycing for to tell
How hardly she had scapt that night the daunger that befell.
And as she knew right well the place and facion of the tree
(As whych she saw so late before) : euen so when she did see
The colour of the Berries turnde, she was vncertain whither
It were the tree at which they both agreed to meete togither.
While in this doubtful stounde she stoode, she cast hir eye aside
And there beweltred in his bloud hir louer she espide
Lie sprawling with his dying limmes : at which she started backe,
And looked pale as any Box, a shuddring through hir stracke,
Euen like the Sea which sodenly with whissing noyse doth moue,
When with a little blast of winde it is but toucht aboue.
But when approching nearer him she knew it was hir loue.
She beate hir brest, she shricked out, she tare hir golden heares,
And taking him betweene hir armes did wash his wounds with teares,
She meynt hir weeping with his bloud, and kissing all his face
(Which now became as colde as yse) she cride in wofull case
Alas what chaunce my *Pyramus* hath parted thee and mee ?
Make aunswere O my *Pyramus*. It is thy *Thisb*, euen shee
Whome thou doste loue most heartely that speaketh vnto thee.
Giue eare and rayse thy heauie heade. He hearing *Thisbes* name,
Lift vp his dying eyes and hauing seene hir closde the same.
But when she knew hir mantle there and saw his scabberd lie
Without the swoorde : Vnhappy man thy loue hath made thee die :
Thy loue (she said) hath made thee slea thy selfe, This hand of mine
Is strong inough to doe the like. My loue no lesse than thine
Shall giue me force to worke my wound. I will pursue the dead.
And wretched woman as I am, it shall of me be sed
That like as of thy death I was the only cause and blame,
So am I thy companion eke and partner in the same,
For death which only coulde alas a sunder part vs twaine,
Shall neuer so disseuer vs but we will meete againe.
And you the Parentes of vs both, most wretched folke alyue,
Let this request that I shall make in both our names byliue
Entreate you to permit that we whome chaste and stedfast loue
And whome euen death hath ioynde in one, may as it doth behoue
In one graue be together layd. And thou vnhappie tree
Which shroudest now the corse of one, and shalt anon through mee
Shroude two, of this same slaughter holde the sicker signes for ay
Blacke be the colour of thy fruite and mourning like alway,
Such as the murder of vs twaine may euermore bewray.
This said, she tooke the sword yet warme with slaughter of hir loue
And setting it beneath hir brest, did to hir heart it shoue.
Hir prayer with the Gods and with their Parentes tooke effect.
For when the fruite is throughly ripe, the Berrie is bespect
With colour tending to a blacke. And that which after fire
Remained, rested in one Tumbe as *Thisbe* did desire.

BOSWELL (*Var.* '21, p. 193) observed that in *A Handefull of Pleasant Delites*, by Clement Robinson, 1584, there is 'A new Sonet of Pyramus and Thisbie,'—a remark which would have been scarcely worth repeating, had not FLEAY (*Life and Work*, p. 186) asserted that 'the Pyramus interlude is clearly based on C. Robinson's *Handfult* '*of Pleasant Delights*, 1584.' Boswell's allusion is clear enough : it is to the 'Sonet' signed 'I. Thomson.' But Fleay's is not so clear, inasmuch as in the 'Handfull,' besides Thomson's 'Sonet,' Pyramus is referred to by name in four other 'pleasant 'delights,' so that we might infer that it is to the number of the allusions to Pyramus that Fleay refers, and yet this would not account for employing Pyramus's story as an interlude. It is scarcely possible that Fleay could have referred, as the 'clear basis' of Shakespeare's interlude, to the following (p. 30, Arber's *Reprint*) :—

> *A new Sonet of Pyramus and Thisbie.*
> *To the, Downe right Squier.*
>
> Ou Dames (I say) that climbe the mount
>   of *Helicon*,
> Come on with me, and giue account,
>   what hath been don :
> Come tell the chaunce ye Muses all,
>   and dolefull newes,
> Which on these Louers did befall,
>   which I accuse.
> In *Babilon* not long agone,
>   a noble Prince did dwell :
> whose daughter bright dimd ech ones sight,
>   so farre she did excel.
>
> An other Lord of high renowne,
>   who had a sonne :
> And dwelling there within the towne
>   great loue begunne :
> *Pyramus* this noble Knight,
>   I tel you true :
> Who with the loue of *Thisbie* bright,
>   did cares renue :
> It came to passe, their secrets was,
>   beknowne vnto them both :
> And then in minde, their place do finde,
>   where they their loue vnclothe.
>
> This loue they vse long tract of time,
>   till it befell :
> At last they promised to meet at prime
>   by *Minus* well :
> Where they might louingly imbrace,
>   in loues delight :
> That he might see his *Thisbies* face
>   and she his sight :

In ioyfull ease, she approcht the place,
    where she her *Pyramus*
Had thought to viewd, but was renewd
    to them most dolorous.

Thus while she staies for *Pyramus*,
    there did proceed :
Out of the wood a Lion fierce,
    made *Thisbie* dreed :
And as in haste she fled awaie,
    her Mantle fine :
The Lion tare in stead of praie,
    till that the time
That *Pyramus* proceeded thus,
    and see how lion tare
The Mantle this of *Thisbie* his,
    he desperately doth fare.

For why he thought the lion had,
    faire *Thisbie* slaine.
And then the beast with his bright **blade,**
    he slew certaine :
Then made he mone and said alas,
    (O wretched wight)
Now art thou in a woful case
    For *Thisbie* bright :
Oh Gods aboue, my faithfull loue
    shal neuer faile this need :
For this my breath by fatall death,
    shal weaue *Atropos* threed.

Then from his sheathe he drew his **blade,**
    and to his hart
He thrust the point, and life did vade,
    with painfull smart :
Then *Thisbie* she from cabin came
    with pleasure great,
And to the well apase she ran,
    there for to treat :
And to discusse, with *Pyramus*
    of al her former feares.
And when slaine she, found him **truly,**
    she shed foorth bitter teares.

When sorrow great that she had **made.**
    she took in hand
The bloudie knife, to end her life,
    by fatall hand.
Yru Ladies all, peruse and **see,**
    the faithfulnesse,

How these two Louers did agree,
to die in distresse :
You Muses waile, and do not faile,
but still do you lament :
These louers twaine, who with such paine,
did die so well content.
Finis.   *I. Thomson.*

GREENE'S HISTORY OF JAMES IV.

WARD (*Eng. Dram. Hist.* 1875, i, 380) says that 'the idea of the entire machin-
' ery of Oberon and his fairy-court was, in all probability, taken by Shakespeare from
' Greene's *Scottish History of James IV* (1590 *circ.*).'

STEEVENS called attention to this drama, but he did not know at the time that
Greene was the author. WARD, to whose excellent guidance we can all trust, is so
outspoken that it behoves us to examine this play of *James IV*, and we can do no
better than to take WARD'S own account of it.

'I think,' says WARD (*Ibid.* p. 220), 'upon the whole the happiest of Greene's
' dramas is *The Scottish Historie of James IV, slaine at Flodden. Intermixed with a
' pleasant Comedie, presented by Oboram King of Fayeries* (printed in 1598). The
' title is deceptive, for the fatal field of Flodden is not included in the drama, which
' ends happily by the reconciliation of King James with his Queen Dorothea. Indeed,
' the plot of the play has no historical foundation ; James IV's consort, though of
' course she was an English princess, as she is in the play, was named Margaret, not
' Dorothea ; and King Henry VII never undertook an expedition to avenge any mis-
' deeds committed against her by her husband. But though the play is founded on
' fiction, such as we may be astonished to find applied to an historical period so little
' remote from its spectators, it is very interesting ; and, besides being symmetrically
' constructed, has passages both of vigour and pathos.' [Here follows the story,
which, as it has no alleged connection with the *Midsummer Night's Dream*, is here
omitted.] 'But though *The Scottish History of James IV* is both effective in its
' serious and amusing in its comic scenes, . . . Greene seems to have thought it neces-
' sary to give to it an adventitious attraction by what appears a quite superfluous addi-
' tion. The title describes the play as "intermixed with a pleasant Comedie pre-
' " sented by Oboram King of Fayeries," but the " pleasant comedy," in point of fact,
' consists only of a brief prelude, in which Oberon and a misanthropical Scotchman
' named Bohan introduce the play as a story written down by the latter, and of dances
' and antics by the fairies between the acts, which are perfectly supererogatory inter-
' mezzos. The " history," or body of the play itself, is represented by a set of play-
' ers, " guid fellows of Bohan's countrymen," before "Aster Oberon," who is the same
' personage as he who figures in the *Midsummer Night's Dream*, though very differ-
' ently drawn, if, indeed, he can be said to be drawn at all.'

That the reader may judge for himself how far Greene's Oberon ('Oboram' in
the title appears to be a mere misprint ; according to the texts of both Dyce and Grosart, it is uniformly 'Oberon' in the body of the play) is 'the same personage' as
Shakespeare's Oberon, and to what extent 'it is probable' that 'the entire machinery
' of Oberon and his fairy court' was taken by Shakespeare from Greene, I will here
give every line of the scenes and stage-directions wherein Oberon appears in *James
IV*. It is of small moment if they are disjointed. As we are not now concerned

with Greene, but with Shakespeare, I follow Dyce's text of the play rather than Grosart's, albeit Dyce does not apparently reproduce the original as faithfully as Grosart reproduces it; the latter says, so corrupt is the original that 'Dyce must have taken 'infinite pains in the preparation of his text.' Moreover, as Dyce's text is modernised here and there, it is all the better for present purposes :—

The Play begins : *Music playing within. Enter* ASTER OBERON, *king of fairies, and an* ANTIC, *who dance about a tomb placed conveniently on the stage, out of the which suddenly starts up, as they dance,* BOHAN, *a Scot, attired like a ridstall man, from whom the* ANTIC *flies.* OBERON *manet.*

Boh. Ay say, what's thou ?

Ober. Thy friend, Bohan.

Boh. What wot I, or reck I that ? Whay, guid man, I reck no friend, nor ay reck no loe; als ene to me. Get thee ganging, and trouble not may whayet, or ays gar thee recon me nene of thay friend, by the mary mass sall I.

Ober. Why, angry Scot, I visit thee for love; then what moves thee to wrath ?

Boh. The deil awhit reck I thy love; for I know too well that true love took her flight twenty winter sence to heaven, whither till ay can, weel I wot, ay sall ne'er find love; an thou lovest me, leave me to myself. But what were those puppets that hopped and skipped about me year whayle ?

Ober. My subjects.

Boh. Thay subjects ! whay, art thou a king ?

Ober. I am.

Boh. The deil thou art ! whay, thou lookest not so big as the king of clubs, nor so sharp as the king of spades, nor so fain as the king a' daymonds : be the mass, ay take thee to be the king of false hearts; therefore I rid thee, away or ayse so curry your kingdom, that you's be glad to run to save your life.

Ober. Why, stoical Scot, do what thou darest to me ; here is my breast, strike.

Boh. Thou wilt not threap me, this whinyard has gard many better men to lope than thou. But how now ? Gos sayds, what, wilt not out ? Whay, thou witch, thou deil ! Gads fute, may whinyard !

Ober. Why, pull, man : but what an 'twere out, how then ?

Boh. This, then, thou wear't best begone first : for ay'l so lop thy limbs, that thou's go with half a knave's carcass to the deil.

Ober. Draw it out; now strike, fool, canst thou not ?

Boh. Bread ay gad, what deil is in me ? Whay, tell me, thou skipjack, what art thou ?

Ober. Nay first tell me what thou wast from thy birth, what thou hast past hitherto, why thou dwellest in a tomb, and leavest the world ? and then I will release thee of these bonds; before, not.

Boh. And not before ! then needs must, needs sall. I was born a gentleman of the best blood in all Scotland, except the king. When time brought me to age, and death took my parents, I became a courtier, where though ay list not praise myself, ay engraved the memory of Bohan on the skin-coat of some of them, and revelled with the proudest.

Ober. But why living in such reputation, didst thou leave to be a courtier ?

Boh. Because my pride was vanity, my expense loss, my reward fair words and large promises, and my hopes spilt, for that after many years' service one outran me

and what the deil should I then do there? No, no; flattering knaves that can cog and prate fastest, speed best in the court.

Ober. To what life didst thou then betake thee?

Boh. I then changed the court for the country, and the wars for a wife: but I found the craft of swains more wise than the servants, and wives' tongues worse than the wars itself, and therefore I gave o'er that, and went to the city to dwell: and there I kept a great house with small cheer, but all was ne'er the near.

Ober. And why?

Boh. Because, in seeking friends, I found table-guests to eat me and my meat, my wife's gossips to bewray the secrets of my heart, kindred to betray the effect of my life: which when I noted, the court ill, the country worse, and the city worst of all, in good time my wife died,—ay would she had died twenty winter sooner by the mass,—leaving my two sons to the world, and shutting myself into this tomb, where if I die, I am sure I am safe from wild beasts, but whilst I live I cannot be free from ill company. Besides now I am sure gif all my friends fail me, I sall have a grave of mine own providing, this is all. Now, what art thou?

Ober. Oberon, king of fairies, that loves thee because thou hatest the world; and to gratulate thee, I brought these Antics to show thee some sport in dancing, which thou hast loved well.

Boh. Ha, ha, ha! Thinkest thou those puppets can please me? whay, I have two sons, that with one Scottish jig shall break the necks of thy Antics.

Ober. That I would fain see.

Boh. Why, thou shalt. How, boys!

Enter SLIPPER *and* NANO.

Haud your clucks, lads, trattle not for thy life, but gather opp your legs and dance me forthwith a jig worth the sight.

Slip. Why, I must talk, an I die for 't: wherefore was my tongue made?

Boh. Prattle, an thou darest, one word more, and ais dab this whinyard in thy womb.

Ober. Be quiet, Bohan. I'll strike him dumb, and his brother too; their talk shall not hinder our jig. Fall to it, dance, I say, man.

Boh. Dance Heimore, dance, ay rid thee.

[*The two dance a jig devised for the nonst.*

Now get you to the wide world with more than my father gave me, that's learning enough both kinds, knavery and honesty; and that I gave you, spend at pleasure.

Ober. Nay, for this sport I will give them this gift; to the dwarf I give a quick wit, pretty of body, and a warrant his preferment to a prince's service, where by his wisdom he shall gain more love than common; and to loggerhead your son I give a wandering life, and promise he shall never lack, and avow that, if in all distresses he call upon me, to help him. Now let them go.

[*Exeunt Slipper and Nano with courtesies.*

Boh. Now, king, if thou be a king, I will shew thee whay I hate the world by demonstration. In the year 1520, was in Scotland a king, over-ruled with parasites, misled by lust, and many circumstances too long to trattle on now, much like our court of Scotland this day. That story have I set down. Gang with me to the gallery and I'll shew thee the same in action, by guid fellows of our countrymen, and then when thou see'st that, judge if any wise man would not leave the world if he could.

Ober. That will I see: lead, and I'll follow thee.   [*Exeunt.*

[The drama of *James IV* here begins, and at the conclusion of the First Act

Bohan and Oberon again appear, and speak as follows. Of their interview Dyce says (p. 94), 'the whole of what follows, till the beginning of the next act, is a mass of 'confusion and corruption. The misprints here defy emendation.']

Enter BOHAN *and* OBERON *the Fairy-king, after the first act; to them a round of Fairies, or some pretty dance.*

 Boh. Be gad, grammercies, little king, for this;
This sport is better in my exile life
Than ever the deceitful world could yield.
 Ober. I tell thee, Bohan, Oberon is king
Of quiet, pleasure, profit, and content,
Of wealth, of honour, and of all the world;
Tied to no place, yet all are tied to me.
Live thou in this life, exil'd from world and men,
And I will shew thee wonders ere we part.
 Boh. Then mark my story, and the strange doubts
That follow flatterers, lust, and lawless will,
And then say I have reason to forsake
The world and all that are within the same.
Go, shrowd us in our harbour where we'll see
The pride of folly as it ought to be.   [*Exeunt.*

After the first Act.

 Ober. Here see I good fond actions in thy jig,
And means to paint the world's inconstant ways;
But turn thine ene, see what I can command.
  [*Enter two battles, strongly fighting, the one* SEMIRAMIS, *the other* STABROBATES : *she flies, and her crown is taken, and she hurt.*
 Boh. What gars this din of mirk and baleful harm,
Where every wean is all betaint with blood ?
 Ober. This shews thee, Bohan, what is worldly pomp :
Semiramis, the proud Assyrian queen,
When Ninus died, did tene in her wars
Three millions of footmen to the fight,
Five hundred thousand horse, of armed cars
A hundred thousand more, yet in her pride
Was hurt and conquer'd by Stabrobates.
Then what is pomp?
 Boh. I see thou art thine ene,
The bonny king, if princes fall from high :
My fall is past, until I fall to die.
Now mark my talk, and prosecute my jig.
 Ober. How should these crafts withdraw thee from the world !
But look, my Bohan, pomp allureth.
  [*Enter* CYRUS, *kings humbling themselves; himself crowned by olive Pat : at last dying, laid in a marble tomb, with this inscription*
   Whoso thou be that passest by
   For I know one shall pass, know I
   I am Cyrus of Persia,
   And, I prithee, leave me not thus like a clod of clay
   Wherewith my body is covered.   [*All exeunt.*

[Enter the king in great pomp, who reads it, and issueth, crieth vermeum.

Boh. What meaneth this?

Ober. Cyrus of Persia,
Mighty in life, within a marble grave
Was laid to rot, whom Alexander once
Beheld entomb'd, and weeping did confess,
Nothing in life could scape from wretchedness:
Why then boast men?

Boh. What reck I then of life,
Who makes the grave my tomb, the earth my wife?

Ober. But mark me more.

Boh. I can no more, my patience will not warp
To see these flatteries how they scorn and carp.

Ober. Turn but thy head.
[Enter four kings carrying crowns, ladies presenting odours to potentate enthroned, who suddenly is slain by his servants, and thrust out; and so, they eat. *[Exeunt.*

Boh. Sike is the world; but whilk is he I saw?

Ober. Sesostris, who was conqueror of the world
Slain at the last, and stamp'd on by his slaves.

Boh. How blest are peur men then that know their graves!
Now mark the sequel of my jig;
An he weele meet ends. The mirk and sable night
Doth leave the peering morn to pry abroad;
Thou nill me stay; hail then, thou pride of kings!
I ken the world, and wot well worldly things.
Mark thou my jig, in mirkest terms that tells
The loath of sins, and where corruption dwells.
Hail me ne mere with shows of guidly sights;
My grave is mine, that rids me from despights;
Accept my jig, guid king, and let me rest;
The grave with guid men is a gay-built nest.

Ober. The rising sun doth call me hence away;
Thanks for thy jig, I may no longer stay;
But if my train did wake thee from thy rest,
So shall they sing thy lullaby to nest. *[Exeunt*

[At the end of the Second Act]

Enter BOHAN *with* OBERON.

Boh. So, Oberon, now it begins to work in kind.
The ancient lords by leaving him alone,
Disliking of his humours and despite,
Let him run headlong, till his flatterers,
Sweeting his thoughts of luckless lust
With vile persuasions and alluring words,
Make him make way by murder to his will.
Judge, fairy king, hast heard a greater ill?

Ober. Nor seen more virtue in a country maid.
I tell thee, Bohan, it doth make me merry,

To think the deeds the king means to perform.

Boh. To change that humour, stand and see the rest
I trow, my son Slipper will shew's a jest.

[*Enter* SLIPPER *with a companion, boy or wench, dancing a hornpipe, and dance out again.*

Boh. Now after this beguiling of our thoughts,
And changing them from sad to better glee,
Let's to our cell, and sit and see the rest,
For, I believe, this jig will prove no jest.   [*Exeunt.*

[At the end of the Third Act Bohan appears alone, and from him we learn that the sadness of the act has put Oberon to sleep. At the conclusion of the Fourth Act]

Chorus. *Enter* BOHAN *and* OBERON.

Ober. Believe me, bonny Scot, these strange events
Are passing pleasing, may they end as well.

Boh. Else say that Bohan hath a barren skull,
If better motions yet than any past
Do not more glee to make the fairy greet.
But my small son made pretty handsome shift
To save the queen, his mistress, by his speed.

Ober. Yea, and yon laddy, for the sport he made,
Shall see, when least he hopes, I'll stand his friend,
Or else he capers in a halter's end.

Boh. What, hang my son! I trow not, Oberon;
I'll rather die than see him woe begone.

Enter a round, or some dance at pleasure.

Ober. Bohan, be pleas'd, for do they what they will,
Here is my hand, I'll save thy son from ill.   [*Exeunt.*

[In fulfillment of this promise Oberon appears towards the close of the Fifth Act, and, accompanied by Antics, silently conveys away Bohan's son, Slipper, who is in jeopardy of his life.

The foregoing extracts comprise all that Oberon does or says in the play. As far as Ward's suggestion is concerned, assent or dissent is left to the reader.]

WARD (vol. i, p. 380) says that the 'story of the magic potion [*sic*, evidently a 'mere slip of memory] and its effects Shakspeare may have found in Montemayor's '*Diana*, though the translation of this book was not published till 1598'

It is not the 'love juice,' but 'some of the fairy story,' which FLEAY (*Life and Work*, p. 186) says 'may have been suggested by Montemayor's *Diana*.' I think Fleay overlooks the fact that if, as he maintains, the date of the *Midsummer Night's Dream*, in its present shape, be 1595, it is impossible that Shakespeare could have obtained any suggestions from a book published three years later, in 1598.

I have toiled through the four hundred and ninety-six weary, dreary, falsetto, folio pages of Montemayor's *Diana*, without finding any conceivable suggestion for 'the fairy story,' other than that of the love-juice to which WARD, I think, alludes; here the hint is so broad compared with others which have been proclaimed as surely adopted elsewhere by Shakespeare, that I wonder the assertion of direct 'convey-'ance' has not been made here; to be sure we are met by the fact that Meres and

Montemayor both bear the same date; but then have we not the extremely convenient and highly accommodating refuge: that Shakespeare may have read Yong's translation in manuscript before it was published, most especially since Yong's translation is dedicated to Lady Penelope Rich, who figures, as we are assured, so freely in Shakespeare's *Sonnets*?

The passage from Yong's translation of the *Diana of George of Montemayor*, 1598, p. 123, is as follows: (it should be premised, however, that Felicia, a noble lady, 'whose course of life and onely exercise, in her stately court, is to cure and 'remedie the passions of loue,' is about to show her art to Felismena, a shepherdess temporarily blighted, and that the objects of Felicia's skill are—first, Syrenus, a shepherd immeasurably in love with a shepherdess, Diana, who in turn immeasurably loved Syrenus, but in some unaccountable way she forgot him during his temporary absence, and casually married Delius, in consequence whereof Syrenus is called 'the forgotten shepherd'; second, Silvanus, who is also in love with Diana, but by her despised, and he is called 'the despised Silvanus'; and thirdly, Silvagia, a shepherdess illimitably in love with Alanius, who, subject to his cruel father's will, cannot marry her.):—

'The Lady *Felicia* saide to *Felismena*. Entertaine this company [Syrenus, Sil-
'vanus, Silvagia and others] while I come hither againe: and going into a chamber,
'it was not long before she came out againe with two cruets of fine cristall in either
'hande, the feete of them being beaten golde, and curiously wrought and enameled:
'And coming to *Syrenus*, she saide vnto him. If there were any other remedy for
'thy greefe (forgotten Shepherd) but this, I woulde with all possible diligence haue
'sought it out, but because thou canst not now enioy her, who loued thee once so
'well, without anothers death, which is onely in the handes of God, of necessitie
'then thou must embrace another remedie, to auoide the desire of an impossible
'thing. And take thou, faire *Seluagia*, and despised *Syluanus*, this glasse, wherein
'you shall finde a soueraine remedie for all your sorrowes past & present; and a
'beginning of a ioyfull and contented life, whereof you do now so little imagine.
'And taking the cristall cruet, which she helde in her left hande, she gaue it to
'*Syrenus*, and badde him drinke; and *Syrenus* did so; and *Syluanus* and *Seluagia*
'drunke off the other betweene them, and in that instant they fell all downe to the
'ground in a deepe sleepe, which made *Filismena* not a little to woonder, ... and
'standing halfe amazed at the deepe sleepe of the shepherdes, saide to *Felicia*: If
'the ease of these Shepherds (good Ladie) consisteth in sleeping (me thinkes) they
'haue it in so ample sort, that they may liue the most quiet life in the worlde.
'Woonder not at this (saide *Felicia*) for the water they drunke hath such force, that,
'as long as I will, they shall sleepe so strongly, that none may be able to awake
'them. And because thou maist see, whether it be so or no, call one of them as
'loude as thou canst. *Felismena* then came to *Syluanus*, and pulling him by the
'arme, began to call him aloud, which did profite her as little, as if she had spoken
'to a dead body; and so it was with *Syrenus* and *Seluagia*, whereat *Felismena* mar-
'uelled very much. And then *Felicia* saide vnto her. Nay, thou shalt maruel yet
'more, after they awake, bicause thou shalt see so strange a thing, as thou didst neuer
'imagine the like. And because the water hath by this time wrought those opera-
'tions, that it shoulde do, I will awake them, and marke it well, for thou shalt heare
and see woonders. Whereupon taking a booke out of her bosome, she came to
Syrenus, and smiting him vpon the head with it, the Shepherd rose vp on his feete
in his perfect wits and judgement: To whom *Felicia* saide. Tell me *Syrenus*, if

' thou mightest now see faire *Diana,* & her vnworthy husband both togither in all the
' contentment and ioy of the worlde, laughing at thy loue, and making a sport of thy
' teares and sighes, what wouldest thou do? Not greeue me a whit (good Lady) but
' rather helpe them to laugh at my follies past. But if she were now a maide againe,
' (saide *Felicia*) or perhaps a widow, and would be married to *Syluanus* and not to
' thee, what wouldst thou then do? Myselfe woulde be the man (saide *Syrenus*) that
' woulde gladly helpe to make such a match for my friende. What thinkest thou of
' this *Felismena* (saide *Felicia*) that water is able to vnloose the knottes that peruerse
' Loue doth make? I woulde neuer haue thought (saide *Felismena*) that anie humane
' skill coulde euer attaine to such diuine knowledge as this. And looking on *Syrenus,*
' she saide vnto him. Howe nowe *Syrenus,* what meanes this? Are the teares and
' sighes whereby thou didst manifest thy loue and greefe, so soone ended? Since
' my loue is nowe ended (said *Syrenus*) no maruell then, if the effects proceeding from
 it be also determined. And is it possible now (said *Felismena*) that thou wilt loue
' *Diana* no more? I wish her as much good (answered *Syrenus*) as I doe to your
' owne selfe (faire Lady) or to any other woman that neuer offended me. But
' *Felicia,* seeing how *Felismena* was amazed at the sudden alteration of *Syrenus,* said.
' With this medicine I would also cure thy greefe (faire *Felismena*) and thine *Belisa*
' [another blighted shepherdess] if fortune did not deserue them to some greater con-
' tent, then onely to enioy your libertie. And bicause thou maist see how diuersly
' the medicines haue wrought in *Syluanus* and *Seluagia,* it shall not be amisse to
' awake them, for now they haue slept ynough: wherefore laying her booke vpon
' *Syluanus* his head, he rose vp, saying. O faire *Seluagia,* what a great offence and
' folly haue I committed, by imploying my thoughtes vpon another, after that mine
' eies did once behold thy rare beautie? What meanes this *Syluanus* (said *Felicia*).
' No woman in the world euen now in thy mouth, but thy Shepherdesse *Diana,* and
' now so suddenly changed to *Seluagia*? *Syluanus* answering her, said. As the
' ship (discreete Lady) sailes floting vp and downe, and well-ny cast away in the
' vnknowen seas, without hope of a secure hauen: so did my thoughtes (putting my
' life in no small hazard) wander in *Dianas* loue, all the while, that I pursued it.
' But now since I am safely arriued into a hauen, of all ioy and happinesse, I onely
' wish I may haue harbour and entertainment there, where my irremooueable and
' infinite loue is so firmely placed. *Felismena* was as much astonished at the seconde
' kinde of alteration of Syluanus, as at that first of *Syrenus,* and therefore saide vnto
' him laughing. What dost thou *Syluanus*? Why dost thou not awake *Seluagia*?
' for ill may a Shepherdesse heare thee, that is so fast asleepe. *Syluanus* then pull-
' ing her by the arme, began to speake out aloud vnto her, saying. Awake faire *Sel-
' uagia,* since thou hast awaked my thoughtes out of the drowsie slumber of passed
' ignorance. Thrise happy man, whom fortune hath put in the happiest estate that I
' could desire. What dost thou meane faire Shepherdesse, dost thou not heare me,
' or wilt thou not answere me? Behold the impatient passion of the loue I beare
' thee, will not suffer me to be vnheard. O my *Seluagia,* sleepe not so much, and let
' not thy slumber be an occasion to make the sleepe of death put out my vitall lightes.
' And seeing how little it auailed him, by calling her, he began to powre foorth such
' abundance of teares, that they, that were present, could not but weepe also for tender
' compassion: whereupon *Felicia* saide vnto him. Trouble not thy selfe *Syluanus,*
' for as I will make *Seluagia* answere thee, so shall not her answere be contrarie to
' thy desire, and taking him by the hand, she led him into a chamber, and said vnto
' him. Depart not from hence, vntill I call thee; and then she went againe to the

' place where *Seluagia* lay, and touching her with her booke, awaked her, as she had
' done the rest and saide vnto her.　Me thinks thou hast slept securely Shepherdesse.
' O good Lady (said she) where is my *Syluanus*, was he not with me heere ?　O God,
' who hath carried him away from hence ? or wil he come hither againe ?　Harke to
' me *Seluagia*, said *Felicia*, for me thinkes thou art not wel in thy wits.　Thy beloued
' *Alanius* is without, & saith that he hath gone wandring vp and downe in many
' places seeking after thee, and hath got his fathers good will to marrie thee : which
' shall as little auaile him (said *Seluagia*) as the sighes and teares which once in vaine
' I powred out, and spent for him, for his memorie is now exiled out of my thoughts.
' *Syluanus* mine onely life and ioy, O *Syluanus* is he, whom I loue.　O what is
' become of my *Syluanus* ?　Where is my *Syluanus* ?　Who hearing the Shepherdesse
' *Seluagia* no sooner name him, could stay no longer in the chamber, but came run-
' ning into the hall vnto her, where the one beheld the other with such apparaunt
' signes of cordiall affection, and so strongly confirmed by the mutual bonds of their
' knowen deserts, that nothing but death was able to dissolue it ; whereat *Syrenus*,
' *Felismena*, and the Shepherdesse were passing ioyfull.　And *Felicia* seeing them all
' in this contentment, said vnto them.　Now is it time for you Shepherds, and faire
' Shepherdesse to goe home to your flocks, which would be glad to heare the wonted
' voice of their knowen masters.'

　　It may be perhaps a relief to sympathetic hearts to know that Lady Felicia, as
well as Oberon, possessed an antidote, and that Syrenus did not for ever remain
insensible to Diana's charms.　The very instant that he learned that Delius was dead
and Diana a widow ' his hart began somewhat to alter and change.'　But to screen
him from any imputation of fickleness we are told (p. 466) that this change was
wrought by supernatural means, and, what is most noteworthy (I marvel it escaped
the commentators) among the means is an HERB,—beyond all question this herb is
' Dian's bud.'　Did not the Lady Felicia live at the Goddess Diana's temple ?　Any
' herb,' any ' bud ' whatsoever that she administered would be ' Dian's bud.'　It is
comfortable again to catch Shakespeare at his old tricks.　The original passage reads
thus : ' There did the secret power also of sage *Felicia* worke extraordinary effects,
' and though she was not present there, yet with her *herbes* [Italics, mine.] and
' wordes, which were of great virtue, and by many other supernaturall meanes, she
' brought to passe that *Syrenus* began now againe to renewe his old loue to *Diana*.'

―――――――

　　WARD (i, 380) says : ' I cannot quite understand whether Klein (*Gesch. des*
' *Dramas*, iv, 386) considers Shakespeare in any sense indebted to the Italian comedy
' of the *Intrighi d'Amore*, which has been erroneously attributed to Torquato Tasso.'

　　I doubt if KLEIN had that idea in his thoughts.　I think he merely holds up, in
his loyalty to Shakespeare, the *Midsummer Night's Dream* as the pattern of all comedies of *intrighi d'amore*.　KLEIN'S extraordinary command of language and vehemence of style make his purpose, at times, difficult to comprehend.　The following
is the passage referred to by Ward, and it is all the more befitting to cite it here,
because in a footnote he runs a tilt at Schöll and Ulrici :―

　　' With love-tangles, as, for example, in the scene [Klein is speaking of the Italian
' Comedy] where both Flamminio and Camillo woo Ersilia at the same time, and she,
' out of spite at the vexations she had received from her favorite Camillo, favours
' Flamminio,—with similar love-tangles and capricious waverings of heart the play of
' chance teases the lovers in *A Midsummer Night's Dream*, but with what charms,

with what poetic magic are the *intrighi d'amore* here brought into play by delicate
' fairies, like symbolically personified winks and hints of an elfin world playing among
' the very forces of nature ; a sportive, fantastic bewitchery of Nature ; like a caprice
' of the spirit of Nature itself, through whose teasing play there gleams the pathos of
' the comic ; **an in**dication that what in the human world is apparent chance, is divine
' foresight and providence, which the roguish Puck presents to us as a piece of jug-
' glery. There is but one genuine comedy of the *Intrighi d'amore*, of love's caprices ;
' —the *Midsummer Night's Dream*. Lavinia [in the Italian comedy] is introduced
' as a byeplay to vindicate the theme of love-tangles. Lavinia loves the silly fop Gia-
' laise, a Neapolitan, who, in turn, is silly for Lavinia's maid, Pasquina ; who raves for
' Flavio, the son of Manilio. Flavio, disguised as a Moor, escapes from his father
' and hires out to a Neapolitan in order to be near Lavinia to whom he has lost his
' heart. Manilio recovers his son, the Moor, like a black meal bug in a meal bag,
' wherein he was about to be conveyed to Lavinia's presence. Finally, Lavinia's and
' Flavio's souls coalesce in marriage. Thus portrayed, the whims of love and the
' caprices of the heart are barren imbecilities, the mental abortions of a lunatic.
' Think for a minute of Puck and his " Love-in-idleness " * squeezed on the slumber-
' ing eyelids of the lovers !
    ' Must we not believe that the mighty British poet was born, serenely and smil-
' ingly to accomplish, with regard to the stage, that purpose, to which, in regard to its
' prototype, his own Hamlet succumbed ?—namely, to put right the stage world which
' in the Italian comedy was out of joint ?'

HALLIWELL (*Memoranda*, pp. 9–12, 1879) has given many allusions to various
scenes and phrases in the *Midsummer Night's Dream* to be found in the literature
of the seventeenth century, but as they are all subsequent to 1600 they belong to
Dramatic History, and illustrate no Shakespearian question other than the popularity
of the play.

The following extracts from the *THE FAERIE QUEENE* are the passages to
which, it is to be presumed, Dr JOHNSON referred when he said : ' Fairies in [Shake-
speare's] time were much in fashion ; common tradition had made them familiar,
' and Spenser's poem had made them great.'
    In the Second Book, Tenth Canto we are told (line 631) :—

' —how first *Prometheus* did create
A man, of many partes from beasts deriued,
And then stole fire from heauen, to animate
His worke, for which he was by *Ioue* deprived
Of life him selfe, and hart-strings of an Ægle riued.

' That man so made, he called *Elfe*, to weet
Quick, the first authour of all Elfin kind :
Who wandring through the world with wearie feet,
Did in the gardins of *Adonis* find

---

* ' This flower, the emblem of capricious phantasy, is the key of the whole play.
Neither Schöll nor Ulrici has adequately appreciated this.'

A goodly creature, whom he deemd in mind
To be no earthly wight, but either Spright,
Or Angell, th' authour of all woman kind;
Therefore a *Fay* he her according hight,
Of whom all *Faeryes* spring, and fetch their lignage right.

' Of these a mightie people shortly grew,
And puissaunt kings, which all the world warrayd,
And to them selues all Nations did subdew:
The first and eldest, which that scepter swayd,
Was *Elfin*; him all *India* obayd,
And all that now *America* men call:
Next him was noble *Elfinan*, who layd
Cleopolis foundation first of all:
But *Elfiline* enclosd it with a golden wall.

' His sonne was *Elfinell*, who ouercame
The wicked *Gobbelines* in bloudy field:
But *Elfant* was of most renowmed fame,
Who all of Christall did *Panthea* build:
Then *Elfar*, who two brethren gyants kild,
The one of which had two heads, th' other three:
Then *Elfinor*, who was in Magick skild;
He built by art vpon the glassy See
A bridge of bras, whose sound heauens thunder seem'd to bee.

' He left three sonnes, the which in order raynd,
And all their Ofspring, in their dew descents,
Euen seuen hundred Princes, which maintaynd
With mightie deedes their sundry gouernments;
That were too long their infinite contents
Here to record, ne much materiall:
Yet should they be most famous moniments,
And braue ensample, both of martiall,
And ciuill rule to kings and states imperiall.

' After all these *Elficleos* did rayne,
The wise *Elficleos* in great Maiestie,
Who mightily that scepter did sustayne,
And with rich spoiles and famous victorie,
Did high aduaunce the crowne of *Faery*:
He left two sonnes, of which faire *Elferon*
The eldest brother did vntimely dy;
Whose emptie place the mightie *Oberon*
Doubly supplide, in spousall, and dominion.

' Great was his power and glorie ouer all,
Which him before, that sacred seat did fill,
That yet remaines his wide memoriall:
He dying left the fairest *Tanaquill*.

Him to succeede therein, by his last will :
Fairer and nobler liueth none this howre,
Ne like in grace, ne like in learned skill;
Therefore they *Glorian* call that glorious flowre :
Long mayst thou *Glorian* liue, in glory and great powre.'

ROBIN GOODFELLOW

KEIGHTLEY (*Fairy Myth.* 1833, ii, 127) : 'Shakespeare seems to have attempted
' a blending of the Elves of the village with the Fays of romance. His Fairies agree
' with the former in their diminutive stature,—diminished, indeed, to dimensions inap-
' preciable by village gossips,—in their fondness for dancing, their love of cleanliness,
' and in their child-abstracting propensities. Like the Fays, they form a community,
' ruled over by the princely Oberon and the fair Titania. There is a court and chiv-
' alry; Oberon, . . . like earthly monarchs, has his jester, " the shrewd and knavish
' " sprite, called Robin Good-fellow." '

'The name of Robin Goodfellow,' says HALLIWELL (*Introd.* p. 37, 1841), 'had,
it appears, been familiar to the English as early as the thirteenth century, being men-
' tioned in a tale preserved in a manuscript of that date in the Bodleian Library at
' Oxford.'

W. A. WRIGHT (*Preface*, p. xvii): 'Tyndale, in his *Obedience of a Christian*
' *Man* (Parker, Soc. ed. p. 321), says, " The pope is kin to Robin Goodfellow, which
' " sweepeth the house, washeth the dishes, and purgeth all, by night; but when day
' " cometh, there is nothing found clean." And again, in his *Exposition of the 1st*
' *Epistle of St. John* (Parker Soc. ed. p. 139), " By reason whereof the scripture . . .
' " is become a maze unto them, in which they wander as in a mist, or (as we say) led
' " by Robin Goodfellow, that they cannot come to the right way, no, though they turn
' " their caps." '

In Reginald Scot's *The discouerie of witchcraft*, &c., 1584, Robin Goodfellow is
many times mentioned by name. 'I hope you understand,' says Scot, speaking of the
birth of Merlin (4 Booke, chap. 10, p. 67, ed. Nicholson), 'that they affirme and saie,
' that *Incubus* is a spirit; and I trust you know that a spirit hath no flesh nor bones,
' &c: and that he neither dooth eate nor drinke. In deede your grandams maides
' were woont to set a boll of milke before him and his cousine Robin good-fellow, for
' grinding of malt or mustard, and sweeping the house at midnight; and you haue
' also heard that he would chafe exceedingly, if the maid or good-wife of the house,
' hauing compassion of his nakedness, laid anie clothes for him, beesides his messe of
' white bread and milke, which was his standing fee. For in that case he saith;
' What have we here ? Hemton hamten, here will I neuer more tread nor stampen.'

Again, in a passage quoted in this edition to illustrate *urchins*, in *The Tempest*, I,
ii, 385, Scot says (7 Booke, chap. xv, p. 122, ed. Nicholson) : 'It is a common saieng;
' A lion feareth no bugs. But in our childhood our mothers maids haue so terrified
' vs with an ouglie divell having hornes on his head, fier in his mouth . . . eies like a
' bason, fanges like a dog, clawes like a beare, a skin like a Niger, and a voice roring
' like a lion, whereby we start and are afraid when we heare one crie Bough: and they
' have so fraied us with bull beggers, spirits, witches, urchens, elves, hags, fairies, satyrs,
' pans, faunes, sylens, kit with the cansticke, tritons, centaurs, dwarfes, giants, imps,

19

'calcars, conjurors, nymphes, changlings, *Incubus*, Robin good-fellowe, the spoorne,
'the mare, the man in the oke, the hell waine, the fierdrake, the puckle, Tom thombe,
'hobgoblin, Tom tumbler, boneles, and such other bugs, that we are afraid of our
'owne shadowes; in so much as some never feare the divell, but in a darke night;
'and then a polled sheepe is a perillous beast, and manie times is taken for our fathers
'soule, speciallie in a churchyard, where a right hardie man heretofore scant durst
'passe by night, but his haire would stand upright.'

Again, in a noteworthy passage (7 Booke, chap. 2, p. 105, ed. Nicholson) : 'And
'know you this by the waie, that heretofore Robin goodfellow, and Hob gobblin were
'as terrible, and also as credible to the people, as hags and witches be now : and in
'time to come, a witch will be as much derided and contemned, and as plainlie per-
'ceived, as the illusion and knaverie of Robin goodfellow.   And in truth, they that
'mainteine walking spirits, with their transformation, &c : have no reason to denie
'Robin goodfellow, upon whom there hath gone as manie and as credible tales, as
'upon witches; saving that it hath not pleased the translators of the Bible, to call
'spirits by the name of Robin goodfellow, as they have termed divinors, soothsaiers,
'poisoners, and couseners by the name of witches.'

HALLIWELL (*Mem.* p. 27, 1879) notes that Tarlton, in his *'Newes out of Purga-*
'*torie,* 1589, says of Robin Goodfellow that he was "famozed in everie old wives
'" chronicle, for his mad merrye prankes." '   And again (p. 27), 'Nash, in his *Ter-*
'*rors of the Night,* 1594, observes that the Robin Goodfellowes, elfes, fairies, hob-
'goblins of our latter age, did most of their merry pranks in the night : then ground
'they malt, and had hempen shirts for their labours, daunst in greene meadows,
'pincht maids in their sleep that swept not their houses cleane, and led poor travel-
'lers out of their way notoriously.'

W. A. WRIGHT (*Preface,* p. xix) quotes from Harsnet's *Declaration of Popish
Imposture* (p. 134), a passage to the same effect as the former quotation from Scot, in
regard to the necessity of 'duly setting out the bowle of curds and creame for Robin
'Goodfellow.'   But although it has been assumed that Shakespeare was familiar with
Harsnet's book when he wrote *King Lear,* its date, 1603, is too late for this present
play.   The same is true also of Burton's *Anatomy of Melancholy,* 1621, albeit a passage cited by W. A. WRIGHT from Part I, Sec. ii, Mem. I, Subs. ii, contains one
noteworthy sentence; speaking of hobgoblins and Robin Goodfellows, and the
Ambulones' that mislead travellers, Burton says : 'These have several names in
several places; we commonly call them *pucks.'*

COLLIER edited for the Percy Society, 1841, a rare tract, called *Robin Goodfellow,
his Mad Pranks and Merry Jests,* dated 1628.   Of it, in his edition, he says, 'there
'is little doubt that it originally came out at least forty years earlier,' and added that
'a ballad inserted in the Introduction to that Reprint, shows how Shakespeare availed
'himself of popular superstitions.'   HALLIWELL (*Fairy Myth.* p. 120, 1845, ed.
Shak. Soc.) agrees with Collier in the probability that this tract is of a much earlier
production than 1628, and, 'although we have no proof of the fact, [it] had most
'likely been seen by Shakespeare in some form or other.'

R. G. WHITE, among editors and critics, has given the most attention to this claim
of precedence, and has, I think, quite demolished it.   The task seems scarcely worth

the pains. The Robin Goodfellow of the 'Mad Prankes,' like the Oberon of romance, has nothing in common, but the name, with Shakespeare's Puck. He is merely a low, lying buffoon, whose coarse jokes are calculated to evoke the horse laughter of boors. Nevertheless, as COLLIER afterwards asserted in a note to *The Devil and the Scold*, in his Roxburghe Ballads, that the '*Mad Prankes* had been published before '1588,' R. G. WHITE's settlement of the question deserves a place here. He says (*Introd.* p. 9) : ' Collier's reasons for this decision, which has not been questioned ' hitherto, are to be found only in the following passage in his Introduction to the edi-' tion of the *Mad Prankes*, published by the Percy Society : " There is no doubt that ' " *Robin Goodfellow, his Mad Prankes and Merry Jests* was printed before 1588. ' " Tarlton, the celebrated comic actor, died late in that year, and just after his decease ' " (as is abundantly established by internal evidence, though the work has no date) ' " came out in [*sic*] a tract called *Tarlton's Newes out of Purgatorie, &c., Published* ' " *by an old companion of his Robin Goodfellow ;* and on sign. A 3 we find it asserted ' " that Robin Goodfellow was ' famozed in every old wives chronicle for his mad ' " ' merrye prankes,' as if at that time the incidents detailed in the succeeding pages ' " were all known, and had been frequently related. Four years earlier Robin Good ' " fellow had been mentioned by Anthony Munday in his comedy of *Two Italian* ' " *Gentlemen,* printed in 1584, and there his other familiar name of Hobgoblin is ' " also assigned to him."

' . . . The assertion in the *Newes out of Purgatorie*, that Robin Goodfellow and ' his tricks were told of in every old wife's chronicle, certainly does show that the ' incidents related in the *Merry Pranks* were, at least in a measure, " known, and ' " had been frequently related " previous to the appearance of the former publica-' tion; but it neither establishes any sort of connection between the two works, nor ' has the slightest bearing upon the question of the order in which they were written; ' . . . to suppose that the old wives derived their stories of Robin from the author of ' *Mad Pranks*, is just to reverse that order of events which results from the very nature ' of things ; it is the author who records and puts into shape the old wives' stories. . . . ' There is, then, no reason for believing that the *Merry Pranks* is an older composi-' tion than the *Newes out of Purgatorie*, but there are reasons which lead to the con-' clusion that it was written after *A Midsummer Night's Dream*. . . . The style of the ' *Merry Pranks* is not that of a time previous to [1594, the date White assigns to cer-' tain passages in *A Midsummer Night's Dream*]. Its simplicity and directness, and ' its comparative freedom from the multitude of compound prepositions and adverbs ' which deform the sentences and obscure the thoughts of earlier writers, point to a ' period not antecedent to that of the translation of our Bible for its production. . . . ' To this evidence, afforded by the style of the narrative, the songs embodied in ' the book add some of another kind, and perhaps more generally appreciable. One, for instance, beginning, " When Virtue was a country maide," contains these ' lines :—

> " She whift her pipe, she drunke her can,
> The pot wos nere out of her span,
> She married a tobacco man,
> A stranger, a stranger."

' But tobacco had never been seen in England until 1586, only two years before the ' publication of the *Newes out of Purgatorie ;* and Aubrey, writing at least after ' 1650, says in his Ashmolean MSS. that " within a period of thirty-five years it was ' " sold for its weight in silver." But it is not necessary to go to the gossiping anti-

' quary for evidence that before 1594 or 1598 a " country maide " could not command
' the luxury of a pipe, or that rapidly as the noxious weed came into use, she could
' not then marry " a tobacco man."

   ' In the narrative we are told that Robin sung another of the songs " to the tune
' " of *What care I how faire she be ?*"   But the writer of the song to which this is a
' burthen, George Wither, was not born until 1588, the very year in which the *Newes*
' *out of Purgatorie* was published; and this song, although written a short time (we
' know not how long) before, was first published in 1619 in Wither's *Fidelia.* . . . As
' bearing upon the question of date, the following lines, in one of the songs, are also
' important :—

> " O give the poore some bread, cheese, or butter
>   Bacon hempe or *flaxe.*
>   Some pudding bring, or other thing :
>   My need doth make me *aske.*"

' Here the last word should plainly be, and originally was, *axe* (the early form
' of ' ask '), which is demanded by the rhyme, and which would have been given had
' the edition of 1628 been printed from one much earlier; for *axe* was in common use
' in the first years of the seventeenth century. The song, which is clearly many years
' older than the volume in which it appears, was written out for the press by some one
' who used the new orthography even at the cost of the old rhyme.'   [WHITE overlooks the possibility that this change in orthography might apply to all the rest of the
volume. The spelling of the ed. of 1628 might have been changed throughout from
one forty years older, to make it more saleable. I am entirely of White's way of
thinking, only this last argument, I am afraid, does not help him.—ED.]

   ' But, perhaps, the most important passage in the *Mad Pranks*, with regard to its
' relation to *A Midsummer Night's Dream*, is the last sentence of the First Part :—
' " The second part shall shew many incredible things done by Robin Goodfellow, or
' " otherwise called Hob-goblin, and his companions, by turning himself into diverse
' " sundry shapes."   For the evidence that Robin Goodfellow was not called Hob-
' goblin until Shakespeare gave him that name, which before had pertained to another
' spirit, even if not to one of another sort, is both clear and cogent. Scot says [*vide*
' *supra*] " Robin Goodfellow *and* Hobgoblin *were* as terrible," &c., and [he enume-
rates them in another passage, also given above, as two separate ' bugs ']. This was
' in 1584, only four years before the publication of the *Newes out of Purgatorie*,
' which Collier would have refer to the *Mad Pranks* in which Robin Goodfellow and
' Hobgoblin are made one. Again, in the passage from Nashe's *Terrors of the Night*,
' published in 1594, the very year in which a part, at least, of the fairy poetry of this
' play was written, Robin Goodfellows, elves, fairies, hobgoblins are enumerated as
' distinct classes of spirits; and Spenser, just before, had distinguished the Puck from
' the Hobgoblin in his *Epithalamion.* . . . Shakespeare was the first to make Robin a
' Puck and a Hobgoblin, when he wrote : " Those that Hobgoblin call you, and sweet
' " Puck, You do their work, and they shall have good luck," and since that the
' merry knave has borne the alias.

   ' We are thus led to the conclusion not only that this interesting tract, the *Mad*
Prankes, was written after the publication of the *Newes out of Purgatorie* in 1588,
and after the performance of *A Midsummer Night's Dream*, but that it was in a
measure founded upon this very play. . . . It seems that the writer . . . was incited
to his task by the popularity of this comedy, . . . and that he did his best to gather
' all the old wives' tales about Robin Goodfellow into a clumsily-designed story,

which he interspersed, . . . with such songs, old or new, as were in vogue at the time. . . .

'It seems, then, that [Shakespeare] was indebted only to popular tradition for the more important part of the rude material which he worked into a structure of such 'fanciful and surpassing beauty. . . . The plot of *A Midsummer Night's Dream* has 'no prototype in ancient or modern story.'

HALLIWELL (*Introd.* p. 28, 1841): 'Mr. Collier has in his possession an unique 'black-letter ballad, entitled *The Merry Puck, or Robin Goodfellow*, which, from 'several passages, may be fairly concluded to have been before the public previously 'to the appearance of the *Midsummer Night's Dream*.' This ballad Halliwell reprints. W. A. WRIGHT (*Preface*, p. xix) gives, without comment the following stanza (p. 36) :—

> 'Sometimes he'd counterfeit a voyce,
>   and travellers call astray,
> Sometimes a walking fire he'd be
>   and lead them from their way.'

HALLIWELL again reprinted it in his *Fairy Mythology*, p. 155, 1845, but omitted all allusion to it in his folio edition 1856, and in his *Memoranda of the Midsummer Night's Dream*, 1879.

PERCY (*Reliques of Ant. Eng. Poet.* 1765, iii, 202): 'ROBIN GOODFELLOW, alias 'PUCKE, alias HOBGOBLIN, in the creed of ancient superstition, was a kind of merry 'sprite, whose character and achievements are recorded in this ballad, and in those 'well-known lines of Milton's *L'Allegro*, which the antiquarian Peck supposes to be 'owing to it :—

> "Tells how the drudging GOBLIN swet
> To earn his cream-bowle duly set ;
> When in one night, ere glimpse of morne.
> His shadowy flail hath thresh'd the corn
> That ten day-labourers could not end ;
> Then lies him down the lubbar fiend,
> And stretch'd out all the chimney's length,
> Basks at the fire his hairy strength,
> And crop-full out of doors he flings,
> Ere the first cock his matin rings."

'The reader will observe that our simple ancestors had reduced all these whimsies 'to a kind of system, as regular, and perhaps more consistent, than many parts of 'classic mythology; a proof of the extensive influence and vast antiquity of these 'superstitions. Mankind, and especially the common people, could not everywhere 'have been so unanimously agreed concerning these arbitrary notions, if they had not 'prevailed among them for many ages. Indeed, a learned friend in Wales assures the editor that the existence of Fairies and Goblins is alluded to in the most ancient British Bards, who mention them under various names, one of the most common of which signifies "The spirits of the mountains."

'This song (which Peck attributes to Ben Jonson, tho' it is not found among his works) is given from an ancient black-letter copy in the British Musæum. It seems to have been originally intended for some Masque.'

> From Oberon, in farye land,
>   The king of ghosts and shadowes there,

Mad Robin I, at his command,
 Am sent to viewe the night-sports here.
  What revell rout
  Is kept about,
 In every corner where I go,
  I will o'ersee,
  And merry bee,
 And make good sport, with ho, ho, ho!

More swift than lightening I can flye
 About this aery welkin soone,
And, in a minute's space, descrye
 Each thing that's done belowe the moone.
  There's not a hag
  Or ghost shall wag,
 Cry, ware Goblins! where I go;
  But Robin I
  Their feates will spy,
 And send them home, with ho, ho, ho!

Whene'er such wanderers I meete,
 As from their night-sports they trudge home;
With counterfeiting voice I greete
 And call them on, with me to roame
  Thro' woods, thro' lakes,
  Thro' bogs, thro' brakes;
 Or else, unseene, with them I go,
  All in the nicke,
  To play some tricke,
 And frolicke it, with ho, ho, ho!

Sometimes I meete them like a man;
 Sometimes an ox; sometimes a hound;
And to a horse I turn me can;
 To trip and trot about them round.
  But if, to ride,
  My backe they stride,
 More swift than wind away I go,
  Ore hedge and lands, [*qu.* launds?—En ?
  Thro' pools and ponds
 I whirry, laughing, ho, ho, ho!

When lads and lasses merry be,
 With possets and with juncates fine;
Unseene of all the company,
 I eat their cakes and sip their wine;
  And, to make sport,
  I [sneeze] and snort
 And out the candles I do blow.
  The maids I kiss;
  They shrieke—Who's this?
 I answer nought, but ho, ho, ho!

Yet now and then, the maids to please,
  At midnight I card up their wooll;
And while they sleepe, and take their ease,
  With wheel to threads their flax I pull.
      I grind at mill
      Their malt up still;
  I dress their hemp, I spin their tow,
      If any 'wake,
      And would me take,
  I wend me, laughing, ho, ho, ho!

When house or harth doth sluttish lye,
  I pinch the maiden black and blue;
The bed-clothes from the bed pull I,
  And lay them naked all to view.
      'Twixt sleep and wake,
      I do them take,
  And on the key-cold floor them throw.
      If out they cry,
      Then forth I fly,
  And loudly laugh out, ho, ho, ho!

When any need to borrowe ought,
  We lend them what they do require;
And for the use demand we nought;
  Our owne is all we do desire.
      If to repay,
      They do delay,
  Abroad amongst them then I go,
      And night by night,
      I them affright
  With pinchings, dreames, and ho, ho, ho!

When lazie queans have nought to do,
  But study how to cog and lye;
To make debate and mischief too,
  'Twixt one another secretlye:
      I marke their gloze,
      And it disclose,
  To them whom they have wronged so;
      When I have done,
      I get me gone.
  And leave them scolding, ho, ho, ho!

When men do traps and engins set
  In loop-holes, where the vermine creepe,
Who from their folds and houses, get
  Their ducks, and geese, and lambes asleep:

I spy the gin,
And enter in,
And seeme a vermine taken so.
But when they there
Approach me neare,
I leap out laughing, ho, ho, ho !

By wells and rills, in meadowes greene,
We nightly dance our hey-day guise;
And to our fairye king, and queene,
We chant our moon-light harmonies.
When larks 'gin sing,
Away we fling;
And babes new-borne steal as we go,
An elfe in bed
We leave instead,
And wend us laughing, ho, ho, ho !

From hag-bred Merlin's time, have I
Thus nightly revell'd to and fro;
And for my pranks men call me by
The name of Robin Good-fellow.
Fiends, ghosts, and sprites,
Who haunt the nightes,
The hags and goblins do me know;
And beldames old
My feates have told,
So *Vale, Vale ;* ho, ho, ho !

[The foregoing song, clearly *post*-Shakespearian, would not have been reprinted here had it not been repeatedly referred to by editors and commentators.

COLLIER owned a version in a MS of the time ; which was 'the more curious,' says Collier (p. 185), ' because it has the initials B. J. at the end. It contains some ' variations and an additional stanza.'

In HALLIWELL's *Fairy Mythology* (*Shakespeare Society.* 1841) many extracts from poems and dramas may be found, but as they also are all of a later date than the present play, a reference to them is sufficient.]

DURATION OF THE ACTION

HALLIWELL (*Introduction*, &c., 1841, p. 3): The period of the action is four days, concluding with the night of the new moon. But Hermia and Lysander receive the edict of Theseus four days before the new moon; they fly from Athens 'tomorrow 'night'; they become the sport of the fairies, along with Helena and Demetrius, *during one night only*, for Oberon accomplishes all in one night, before ' the first cock ' crows,' and the lovers are discovered by Theseus the morning before that which would have rendered this portion of the plot chronologically consistent.

W. A. WRIGHT (*Preface*, p. xxii): In the play itself the time is about May-day, but Shakespeare, from haste or inadvertence, has fallen into some confusion in regard to it. Theseus' opening words point to April 27, four days before the new moon which was to behold the night of his marriage with Hippolyta. . . . The next night, which would be April 28, Lysander appoints for Hermia to escape with him from Athens. . . . The night of the second day is occupied with the adventures in the wood, and in the morning the lovers are discovered by Theseus and his huntsmen, and it is supposed that they have risen early to observe the rite of May. So that the morning of the third day is the 1st of May, and the last two days of April are lost altogether. Titania's reference to the ' middle-summer's spring ' must therefore be to the summer of the preceding year. It is a curious fact, on which, however, I would not lay too much stress, that in 1592 there was a new moon on the 1st of May; so that if *A Midsummer Night's Dream* was written so as to be acted on a May day, when the actual age of the moon corresponded with its age in the play, it must have been written for May day, 1592.

P. A. DANIEL (*Trans. New Shakspere Soc.* 1877-9, Part ii, p. 147): **Day 1.**— Act I, Sc. i. Athens. In the first two speeches the proposed duration of the action seems pretty clearly set forth. By [them] I understand that four clear days are to intervene between the time of this scene and the day of the wedding. The night of this day No. 1 would, however, suppose five *nights* to come between.

Day 2.—Act II, Act III, and part of Sc. i, Act IV, are on the morrow night in the wood, and are occupied with the adventures of the lovers; with Oberon, Titania, and Puck; the Clowns. Daybreak being at hand, the fairies trip after the nights' shade and leave the lovers and Bottom asleep.

Day 3.—Act IV, Sc. i, continued. Morning. May-day. Theseus, Hippolyta, &c. enter and awake the lovers with their hunting horns.

In Act I. it will be remembered that four days were to elapse before Theseus's nuptials and Hermia's resolve; but here we see the plot is altered, for we are now only in the second day from the opening scene, and only one clear day has intervened between day No. 1 and this, the wedding-day.

Act IV, Sc. ii. Athens. Later in the day.

Act V. In the Palace. Evening.

According to the opening speeches of Theseus and Hippolyta in Act I, we should have expected the dramatic action to have comprised five days exclusive of that Act as it is we have only three days inclusive of it.

Day 1.—Act I.

" 2.—Acts II, III, and part of Sc. i, Act IV

" 3.—Part of Sc. i, Act IV, Sc. ii, Act IV, and Act V.

FURNIVALL (*Introd. Leopold Shakspere*, 1877, p. xxvii): Note in this *Dream* the first of those inconsistencies as to the time of the action of the play that became so markt a feature in later plays, like *The Merchant of Venice*, where three months and more are crowded into 39 hours. Here Theseus and Hippolyta say that ' four happy ' days' and ' four nights' are to pass before ' the night of our solemnities;' but, in the hurry of the action of the play, Shakspere forgets this, and makes only two nights so pass. Theseus speaks to Hippolyta, and gives judgement on Hermia's case, on April 29. 'Tomorrow night,' April 30, the lovers meet, and sleep in the forest, and are found there on May-day morning by Theseus. They and he all go to Athens and get married that day, and go to bed at midnight, the fairies stopping with them till the break of the fourth day, May 2.

FLEAY (*Robinson's Epit. of Lit.* 1 Apr. 1879): All editors and commentators, as far as I know, agree that the ' four days' of I, i cannot be reconciled with the action of the play. I demur. The marriage of Theseus is on the 1st of May; the play opens on the 27th of April, but at line 137 I take it a new scene must begin [see note *ad loc.*]; and there is no reason why it should not be on the 28th or 29th of April. I would place it on the 28th. On the 29th the lovers go to the wood, and, in IV, i, 114, when the fairies leave, it is the morning of the 30th. But at this point Titania's music has struck 'more dead than common sleep' on the lovers. Yet in a few minutes enter Theseus, the horns sound, and they awake. Why this dead sleep if it has to last but a few minutes? Surely Act III ends with the fairies' exit, and the lovers sleep through the 30th of April and wake on May morning. . . . At the end of Act III there is in the Folio a curious stage-direction, which would come in well after *Sleepers lie still*, at the division I propose : *They sleep all the Act*, i. e. while the music is playing. But if this reasoning seems insufficient, let the reader turn to IV, i, 99, where Oberon says he will be at Theseus's wedding *tomorrow* midnight. This must be said on the 30th of April. . . . There must therefore be an interval of 24 hours somewhere, and this is only possible during the dead sleep of the lovers. If any one would ask why make them sleep during this time, I would answer that the 30th of April, 1592, was a Sunday.

HENRY A. CLAPP (*Atlantic Monthly*, March, 1885): *A Midsummer Night's Dream* is the only one of Shakespeare's plays in which I have discovered an inexplicable variance between the different parts of his scheme of time. . . . It is this same 'tomorrow night' which teems with wonders for all the chief persons of the piece; the whole of Acts II. and III. is included within it, and in Scene i. of Act IV. day breaks upon the following morn. . . . It is a single night, as is said over and over again by the text in diverse ways. . . . Parts of three successive days have therefore been occupied in the action, and a whole day has somehow dropped out. . . . On the whole, I think we must believe that the explanation lies in the nature of the play, whose characters, even when clothed with human flesh and blood, have little solidity or reality. I fancy that Shakespeare would smilingly plead guilty as an accessory after the fact to the blunder, and charge the principal fault upon Puck and his crew, who would doubtless rejoice in the annihilation of a mortal's day.

ENGLISH CRITICISMS

SAMUEL PEPYS, 1662, September 29:—To the King's Theatre, where we saw Midsummer Night's Dream,' which I had never seen before, nor shall ever again, for it is the most insipid, ridiculous play, that ever I saw in my life—(Vol. ii, p. 51, ed. Bright, ap. Ingleby).

HAZLITT (*Characters*, &c., 1817, p. 128) : Puck is the leader of the fairy band. He is the Ariel of the *Midsummer Night's Dream ;* and yet as unlike as can be to the Ariel in *The Tempest.* No other poet could have made two such different characters out of the same fanciful materials and situations. Ariel is a minister of retribution, who is touched with a sense of pity at the woes he inflicts. Puck is a mad-cap sprite, full of wantonness and mischief, who laughs at those whom he misleads—' Lord, what fools these mortals be !' Ariel cleaves the air, and executes his mission with the zeal of a winged messenger; Puck is borne along on his fairy errand like the light and glittering gossamer before the breeze. He is, indeed, a most Epicurean little gentleman, dealing in quaint devices, and faring in dainty delights. Prospero and his world of spirits are a set of moralists, but with Oberon and his fairies we are launched at once into the empire of the butterflies. How beautifully is this race of beings contrasted with the men and women actors in the scene, by a single epithet which Titania gives to the latter, ' the human mortals '! It is astonishing that Shakespeare should be considered, not only by foreigners but by many of our own critics, as a gloomy and heavy writer, who painted nothing but ' gorgons and ' hydras and chimeras dire.' His subtlety exceeds that of all other dramatic writers, insomuch that a celebrated person of the present day said he regarded him rather as a metaphysician than a poet. His delicacy and sportive gaiety are infinite. In the *Midsummer Night's Dream* alone we should imagine there is more sweetness and beauty of description than in the whole range of French poetry put together. What we mean is this, that we will produce out of that single play ten passages, to which we do not think any ten passages in the works of the French poets can be opposed, displaying equal fancy and imagery. Shall we mention the remonstrance of Helena to Hermia, or Titania's description of her fairy train, or her disputes with Oberon about the Indian boy, or Puck's account of himself and his employments, or the Fairy Queen's exhortation to the elves to pay due attendance upon her favorite, Bottom ; or Hippolyta's description of a chase, or Theseus's answer ? The two last are as heroical and spirited as the others are full of luscious tenderness. The reading of this play is like wandering in a grove by moonlight; the descriptions breathe a sweetness like odours thrown upon beds of flowers. . . . It has been suggested to us that this play would do admirably to get up as a Christmas after-piece. . . . Alas, the experiment has been tried and has failed, . . . from the nature of things. The *Midsummer Night's Dream,* when acted, is converted from a delightful fiction into a dull pantomime. All that is finest in the play is lost in the representation. The spectacle was grand; but the spirit was evaporated, the genius was fled.—Poetry and the stage do not agree well together. The attempt to reconcile them in this instance fails not only of effect, but of decorum. The *ideal* can have no place upon the stage, which is a picture without perspective ; everything there is in the foreground. That which was merely an airy shape, a dream, a passing thought, immediately becomes an unmanageable reality. Where all is left to the imagination (as is the case in reading) every circumstance, near or remote, has an equal chance of being

kept in mind, and tells according to the mixed impression of all that has been suggested. But the imagination cannot sufficiently qualify the actual impressions of the senses. Any offence given to the eye is not to be got rid of by explanation. Thus, Bottom's head in the play is a fantastic illusion, produced by magic spells; on the stage it is an ass's head, and nothing more; certainly a very strange costume for a gentleman to appear in. Fancy cannot be embodied any more than a simile can be painted; and it is as idle to attempt it as to personate *Wall* or *Moonshine*. Fairies are not incredible, but fairies six feet high are so. Monsters are not shocking if they are seen at a proper distance. When ghosts appear at midday, when apparitions stalk along Cheapside, then may the *Midsummer Night's Dream* be represented without injury at Covent Garden or at Drury Lane. The boards of a theatre and the regions of fancy are not the same thing.

AUGUSTINE SKOTTOWE (*Life of Shakespeare*, &c., 1824, i, 255): Few plays consist of such incongruous materials as *A Midsummer Night's Dream*. It comprises no less than four histories: that of Theseus and Hippolyta; of the four Athenian lovers; the actors; and the fairies. It is not, indeed, absolutely necessary to separate Theseus and Hippolyta from the lovers, nor the actors from the fairies, but the link of connection is extremely slender. Nothing can be more irregularly wild than to bring into contact the Fairy mythology of modern Europe and the early events of Grecian history, or to introduce Snug, Bottom, Flute, Snout, and Starveling, 'hard-handed ' men which never laboured in their minds till now,' as amateur actors in the classic city of Athens.

Of the characters constituting the serious action of this play Theseus and Hippolyta are entirely devoid of interest. Lysander and Demetrius, and Hermia and Helena, scarcely merit notice, except on account of the frequent combination of elegance, delicacy, and vigour, in their complaints, lamentations, and pleadings, and the ingenuity displayed in the management of their cross-purposed love through three several changes. . . . Bottom and his companions are probably highly-drawn caricatures of some of the monarchs of the scene whom Shakespeare found in favour and popularity when he first appeared in London, and in the bickerings, jealousies, and contemptible conceits which he has represented we are furnished with a picture of the green-room politics of the Globe.

[P. 263.] Of all spirits it was peculiar to fairies to be actuated by the feelings and passions of mankind. The loves, jealousies, quarrels, and caprices of the dramatic king give a striking exemplification of this infirmity. Oberon is by no means backward in the assertion of supremacy over his royal consort, who, to do her justice, is as little disposed as any earthly beauty tacitly to acquiesce in the pretensions of her redoubted lord. But knowledge, we have been gravely told, is power, and the animating truth is exemplified by the issue of the contest between Oberon and Titania: his majesty's acquaintance with the secret virtues of herbs and flowers compels the wayward queen to yield what neither love nor duty could force from her. . . .

[P. 274.] An air of peculiar lightness distinguishes the poet's treatment of this extremely fanciful subject from his subsequent and bolder flights into the regions of the spiritual world. He rejected from the drama on which he engrafted it, everything calculated to detract from its playfulness or to encumber it with seriousness, and, giving the rein to the brilliancy of youthful imagination, he scattered from his superabundant wealth, the choicest flowers of fancy over the fairies' paths; his fairies move amidst the fragrance of enameled meads, graceful, lovely, and enchanting. It is

equally to Shakespeare's praise that *A Midsummer Night's Dream* is not more highly distinguished by the richness and variety, than for the propriety and harmony which characterises the arrangement of the materials out of which he constructed this vivid and animated picture of fairy mythology.

THOMAS CAMPBELL (*Introductory Notice*, 1838): Addison says, ' When I look at ' the tombs of departed greatness every emotion of envy dies within me.' I have never been so sacrilegious as to envy Shakespeare, in the bad sense of the word, but if there can be such an emotion as *sinless envy*, I feel it towards him ; and if I thought that the sight of his tombstone would kill so pleasant a feeling, I should keep out of the way of it. Of all his works, the *Midsummer Night's Dream* leaves the strongest impression on my mind that this miserable world must have, for once at least, contained a happy man. This play is so purely delicious, so little intermixed with the painful passions from which Poetry distils her sterner sweets, so fragrant with hilarity, so bland and yet so bold, that I cannot imagine Shakespeare's mind to have been in any other frame than that of healthful ecstasy when the sparks of inspiration thrilled through his brain in composing it. I have heard, however, an old cold critic object that Shakespeare might have foreseen it would never be a good acting play, for where could you get actors tiny enough to couch in flower blossoms ? Well ! I believe no manager was ever so fortunate as to get recruits from Fairy-land, and yet I am told that *A Midsummer Night's Dream* was some twenty years ago revived at Covent Garden, though altered, of course, not much for the better, by Reynolds, and that it had a run of eighteen nights ; a tolerably good reception. But supposing that it never could have been acted, I should only thank Shakespeare the more that he wrote here as a poet and not as a playwright. And as a birth of his imagination, whether it was to suit the stage or not, can we suppose the poet himself to have been insensible of its worth ? Is a mother blind to the beauty of her own child ? No ! nor could Shakespeare be unconscious that posterity would dote on this, one of his loveliest children. How he must have chuckled and laughed in the act of placing the ass's head on Bottom's shoulders ! He must have foretasted the mirth of generations unborn at Titania's doating on the metamorphosed weaver, and on his calling for a repast of sweet peas. His animal spirits must have bounded with the hunter's joy whilst he wrote Theseus's description of his well-tuned dogs and of the glory of the chase. He must have been as happy as Puck himself whilst he was describing the merry Fairy, and all this time he must have been self-assured that his genius ' *was to cast a girdle* ' *round the earth*,' and that souls, not yet in being, were to enjoy the revelry of his fancy.

But nothing can be more irregular, says a modern critic, Augustine Skottowe, than to bring into contact the fairy mythology of modern Europe and the early events of Grecian history. Now, in the plural number, Shakespeare is not amenable to this charge, for he alludes to only one event in that history, namely, to the marriage of Theseus and Hippolyta ; and as to the introduction of fairies, I am not aware that he makes any of the Athenian personages believe in their existence, though they are subject to their influence. Let us be candid on the subject. If there were fairies in modern Europe, which no rational believer in fairy tales will deny, why should those fine creatures not have existed previously in Greece, although the poor blind heathen Greeks, on whom the gospel of Gothic mythology had not yet dawned, had no conception of them ? If Theseus and Hippolyta had talked believingly about the dapper elves, there would have been some room for critical complaint ; but otherwise the

fairies have as good a right to be in Greece in the days of Theseus, as to play their pranks anywhere else or at any other time.

There are few plays, says the same critic, which consist of such incongruous materials as *A Midsummer Night's Dream*. It comprises four histories—that of Theseus and Hippolyta, that of the four Athenian Lovers, that of the Actors, and that of the Fairies, and the link of connection between them is exceedingly *slender*. In answer to this, I say that the plot contains nothing about any of the four parties concerned approaching to the pretension of a history. Of Theseus and Hippolyta my critic says that they are uninteresting, but when he wrote that judgement he must have fallen asleep after the hunting scene. Their felicity is seemingly secure, and it throws a tranquil assurance that all will end well. But the bond of sympathy between Theseus and his four loving subjects is anything but slender. It is, on the contrary, most natural and probable for a newly-married pair to have patronised their amorous lieges during their honeymoon. Then comes the question, What *natural* connection can a party of fairies have with human beings? This is indeed a posing interrogation, and I can only reply that fairies are an odd sort of beings, whose connection with mortals can never be set down but as supernatural.

Very soon Mr Augustine Skottowe blames Shakespeare for introducing common mechanics as amateur actors during the reign of Theseus in classic Athens. I dare say Shakespeare troubled himself little about Greek antiquities, but here the poet happens to be right and his critic to be wrong. Athens was not a classical city in the days of Theseus; and, about seven hundred years later than his reign, the players of Attica roved about in carts, besmearing their faces with the lees of wine. I have little doubt that, long after the time of Theseus, there were many prototypes of Bottom the weaver and Snug the joiner in the itinerant acting companies of Attica.

C. A. BROWN (*Shakespeare's Autobiographical Poems*, 1838, p. 268): How must Spenser have been enchanted with this poetry! [i. e. the present play]. But can we believe that the multitude were enchanted? or, if they were, could poetry compensate, in their eyes, for its inapplicability for the stage? Before the invention of machinery, an audience must indeed have carried to the theatre more imagination than is requisite at the present day; yet, still I cannot but think that these ideal beings, in representation, claimed too much of so rare a quality, and that it failed at the first, as when it was last attempted in London. Hazlitt has dwelt on the unmanageable nature of this 'dream' for the stage; and was it not equally unmanageable at all times? . . .

Regarding it as certain that Shakespeare was, at one period, unsuccessful as a dramatic poet, we have the more reason to love his nature, which never led him, throughout his works, especially in the *Poems to his Friend*, where he speaks much of himself, into querulousness at the bad taste of the town, and angry invectives against actors and audiences, so common to the disappointed playwrights of his time.

COLLIER: There is every reason to believe that [this play] was popular; in 1622, the year before it was reprinted in the first folio, it is thus mentioned by Taylor, the Water-poet, in his *Sir Gregory Nonsense* :—'I say, as it is applausfully written, and commended to posterity, in the Midsummer Night's Dream :—if we offend, it is with our good will; we came with no intent but to offend, and show our simple skill.'

HALLAM (*Lit. of Europe*, 1839, ii, 387): The beautiful play of *Midsummer Night's Dream* . . . evidently belongs to the earlier period of Shakespeare's genius;

poetical as we account it, more than dramatic, yet rather so, because the indescribable profusion of imaginative poetry in this play overpowers our senses till we can hardly observe anything else, than from any deficiency of dramatic excellence. For in reality the structure of the fable, consisting as it does of three if not four actions, very distinct in their subjects and personages, yet wrought into each other without effort or confusion, displays the skill, or rather instinctive felicity, of Shakespeare, as much as in any play he has written. No preceding dramatist had attempted to fabricate a complex plot; for low comic scenes, interspersed with a serious action upon which they have no influence, do not merit notice. The *Menæchmi* of Plautus had been imitated by others, as well as by Shakespeare; but we speak here of original invention.

The *Midsummer Night's Dream* is, I believe, altogether original in one of the most beautiful conceptions that ever visited the mind of a poet, the fairy machinery. A few before him had dealt in a vulgar and clumsy manner with popular superstition; but the sportive, beneficent, invisible population of the air and earth, long since established in the creed of childhood and of those simple as children, had never for a moment been blended with 'human mortals' among the personages of the drama. . . . The language of *Midsummer Night's Dream* is equally novel with the machinery. It sparkles in perpetual brightness with all the hues of the rainbow; yet there is nothing overcharged or affectedly ornamented. Perhaps no play of Shakespeare has fewer blemishes, or is from beginning to end in so perfect keeping; none in which so few lines could be erased, or so few expressions blamed. His own peculiar idiom, the dress of his mind, which began to be discernible in *The Two Gentlemen of Verona*, is more frequently manifested in the present play. The expression is seldom obscure, but it is never in poetry, and hardly in prose, the expression of other dramatists, and far less of the people. And here, without reviving the debated question of Shakespeare's learning, I must venture to think that he possessed rather more acquaintance with the Latin language than many believe. The phrases, unintelligible and improper, except in the sense of their primitive roots, which occur so copiously in his plays, seem to be unaccountable on the supposition of absolute ignorance. In the *Midsummer Night's Dream* these are much less frequent than in his later dramas. But here we find several instances. Thus, 'things base and vile, 'holding no *quantity*,' for value; rivers, that 'have overborne their *continents*,' the *continente ripa* of Horace; '*compact* of imagination;' 'something of great *constancy*,' for consistency; 'sweet Pyramus *translated* there;' 'the law of Athens, which by no 'means we may *extenuate*.' I have considerable doubts whether any of these expressions would be found in the contemporary prose of Elizabeth's reign, which was less overrun by pedantry than that of her successor; but, could authority be produced for Latinisms so forced, it is still not very likely that one, who did not understand their proper meaning, would have introduced them into poetry. It would be a weak answer that we do not detect in Shakespeare any imitations of the Latin poets. His knowledge of the language may have been chiefly derived, like that of schoolboys, from the Dictionary, and insufficient for the thorough appreciation of their beauties. But, if we should believe him well acquainted with Virgil or Ovid, it would be by no means surprising that his learning does not display itself in imitation. Shakespeare seems, now and then, to have a tinge on his imagination from former passages; but he never distinctly imitates, though, as we have seen, he has sometimes adopted. The streams of invention flowed too fast from his own mind to leave him time to accommodate the words of a foreign language to our own. He knew that to create would be easier, and pleasanter, and better.

CHARLES KNIGHT (*Supplementary Notice*, 1840, p. 382): We can conceive that with scarcely what can be called a model before him, Shakespeare's early dramatic attempts must have been a series of experiments to establish a standard by which he could regulate what he addressed to a mixed audience. The plays of his middle and mature life, with scarcely an exception, are acting plays; and they are so, not from the absence of the higher poetry, but from the predominance of character and passion in association with it. But even in those plays which call for a considerable exercise of the unassisted imaginative faculty in an audience, such as *The Tempest* and *A Midsummer Night's Dream*, where the passions are not powerfully roused and the senses are not held enchained by the interests of the plot, he is still essentially dramatic. What has been called of late years the dramatic poem—that something between the epic and the dramatic, which is held to form an apology for whatever is episodical or incongruous the author may choose to introduce—was unattempted by him. *The Faithful Shepherdess* of Fletcher—a poet who knew how to accommodate himself to the taste of a mixed audience more readily than Shakespeare—was condemned on the first night of its appearance. Seward, one of his editors, calls this the scandal of our nation. And yet it is extremely difficult to understand how the event could have been otherwise; for *The Faithful Shepherdess* is essentially undramatic. Its exquisite poetry was, therefore, thrown away upon an impatient audience—its occasional indelicacy could not propitiate them. Milton's *Comus* is, in the same way, essentially undramatic; and none but such a refined audience as that at Ludlow Castle could have endured its representation. But the *Midsummer Night's Dream* is composed altogether upon a different principle. It exhibits all that congruity of parts—that natural progression of scenes—that subordination of action and character to one leading design—that ultimate harmony evolved out of seeming confusion—which constitute the dramatic spirit. With 'audience fit, though few,'—with a stage not encumbered with decorations—with actors approaching (if it were so possible) to the idea of grace and archness which belong to the fairy troop—the subtle and evanescent beauties of this drama might not be wholly lost in the representation. But under the most favourable circumstances much would be sacrificed. It is in the closet that we must not only suffer our senses to be overpowered by its 'indescribable profusion of 'imaginative poetry,' but trace the instinctive felicity of Shakespeare in the 'structure 'of the fable.' If the *Midsummer Night's Dream could* be acted, there can be no doubt how well it would act. Our imagination must amend what is wanting. . . .

To offer an analysis of this subtle and ethereal drama would, we believe, be as unsatisfactory as the attempts to associate it with the realities of the stage. With scarcely an exception, the proper understanding of the other plays of Shakespeare may be assisted by connecting the apparently separate parts of the action, and by developing and reconciling what seems obscure and anomalous in the features of the characters. But to follow out the caprices and illusions of the loves of Demetrius and Lysander, of Helena and Hermia; to reduce to prosaic description the consequence of the jealousies of Oberon and Titania; to trace the Fairy Queen under the most fantastic of deceptions, . . . and, finally, to go along with the scene till the illusions disappear, . . . such an attempt as this would be worse than unreverential criticism. No,—the *Midsummer Night's Dream* must be left to its own influences.

THE EDINBURGH REVIEW (April, 1848, p. 422): The play consists of several groups, which at first sight appear to belong not so much to the same landscape as to different compartments of the same canvas. Between them, however, a coherence

and connection are soon discovered, of which we have rather hints and glimpses and a general impression than full assurance. We do not say that this connection is not cheerfully admitted on all hands, but it is noticed as a kind of paradox, as though it were not the result of obedience to any discernible law. [See Knight, *supra*. —Ed.] . . .

[P. 425.] Practically, we come to the old division of the characters into three parties, the Heroes (the Lovers being included), the Fairies, and the Artizans. But of these three equivalent, incoherent elements, which is the principal? Whose action is the main action? We look for a key to the composition; on which set of figures are we to fix the eye? It is worthy of remark that ever since Shakespeare's own day some difficulty seems to have been felt, perhaps unconsciously, as to the dominant action of the *Midsummer Night's Dream*. [From the appearance of the piece called *The Merry Conceited Humours of Bottom the Weaver* and from the incident connected with the performance of *A Midsummer Night's Dream* in 1631 (see 'John Spencer,' *post*) the Reviewer says that] we must come to the strange conclusion that at this time the Artizans were thought to constitute the main action. . . .

[P. 426.] Let us examine the two groups, first presented to our notice. The first of these consists of the Heroes,—Theseus and his very unhistorical court. These are themselves fanciful and unsubstantial; not, indeed, creatures of the elements, yet scarcely the men and women of flesh and blood with whom Shakespeare has elsewhere peopled his living stage. We cannot but suspect there is a meaning in their mythological origin. Shakespeare has neither drawn them from history, his resource when he wished to paint the broader realities of life, nor from the lights and shadows, the gay gallantry and devoted love, of the Italian novel. They are apparently selected purely for their want of association. Their humanity is of the most delicately refined order; their perplexities the turbulence of still life. Moreover, the components of the group, the pairs of Athenian lovers, seem only to be so distributed in order to be confused. There are no distinctive features in their members. Lysander differs in nothing from Demetrius, Helena in nothing but height from Hermia. Finally, they speak a great deal of poetry, and poetry more exquisite never dropped from human pen; but it is purely objective, and not in the slightest degree modified by the character of the particular speaker. Turn we now to the second group. If the first were as far as possible removed from every-day experience, these are types of a class ever ready to our hand. They are of the earth, earthy. Bottom sat at a Stratford loom, Starveling on a Stratford tailoring-board; between them they perhaps made the doublet which captivated the eyes of Richard Hathaway's daughter, or the hose that were torn in the park of the Lucys. If the former personages were all of one string, the characters of the latter are stamped with curious marks of difference. The πολυπραγμοσύνη of Bottom,—he would now-a-days be a Chartist celebrity,—the discretion of Snug, the fickleness of Starveling are (as Hazlitt has shown) minutely and fancifully discriminated. And most strongly too is the homely idiomatic prose of their dialogue contrasted with the blinding brilliancy of those rhymed verses which speak the eternal language of love by the mouths of the Athenian ladies and their lovers. In short, they are the very counterpart of the former group; and it is this that we wish to establish, an intentional antagonism between the two. They seem to us, in their respective delicacy and coarseness, to mark the two extreme phases of life, the highest and the lowest, as presented to the imaginative faculty; the lowest, as it may be seen by experience,—the highest, as it may be conceived of in dreams.

We must ask our readers to notice particularly that the first act is nearly equally

20

divided between these two actions; one occupying the first half, the other the second. The two parties, without in the smallest degree intermingling, arrange themselves so as to admit of certain complications, the dominant feeling in the one case being refined sentiment; in the other a ridiculous ambition.

In Act II we are presented for the first time with a new creation, that of the Fairies. Henceforward, the first two actions, so remarkably separated in Act I, are gradually interwoven with the third, though nowhere with each other. In the beings of whom this third group is composed, nothing is so characteristic as the humanity of their motives and passions—humanity modified by the peculiarities of the fairy race— such as might be expected in a duodecimo edition of mankind. We find working in them splenetic jealousy, love, hatred, revenge, all the passions of men,—the littlenesses of soul brought out by each, being, as we think, designedly exaggerated. Their movements too are eminently significant of a vigorous dramatic action, the story being almost epical in form,—the tale of the μῆνις ’Ωβερῶνος; of which, as it gradually and uniformly advances, we are enabled to trace in the play the origin, developement, and consequences. The hypothesis, then, which we wish to put forward is, that the *fairies* are the primary conception of the piece, and their action the main action; that Shakespeare wished to represent this fanciful creation in contact with two strongly-marked extremes of human nature; the instruments by which they influence them being, aptly enough, in one case the ass's head, in the other the ' little western flower.'

It is necessary to this idea, that the two actions of the Heroes and the Artizans should be considered completely subordinate, and their separate relations among themselves as not having been created relatively to the whole piece, but principally to the intended action of the Fairies upon them. We shall then have the singular arrangement of the first Act purposely designed to exhibit successively the characteristics of the two groups in marked opposition, before exposing them to the influence of the Fairies. Finally, the interlude of Pyramus and Thisbe is the ingenious machinery by which, after the stage has ceased to be occupied by the fairy action, these two otherwise independent groups are wrought together and amalgamated.

Some difficulty may yet present itself as to the form of the piece, furnished as it were with a preface and supplement; but we think this can be satisfactorily accounted for. We are not aware whether the *time* employed in the *Midsummer Night's Dream* has been generally noticed. The *Midsummer Night's Dream* is a dream on the night of Midsummer Day; a night sanctified to the operations of fairies, as Hallowe'en was to those of witches. The play is distributed into three distinguishable portions, those included in Act I—in Acts II, III and the first scene of Act IV—and in the last scene of Act IV together with Act V. The second, and by far the most important division, comprehends all the transactions of the Midsummer Night; its action is carefully restricted to the duration of these twelve witching hours (Oberon having, as he says, to perform all before ' the first cock crow '), while those of the first and third portions take place at distances of two days and one day respectively. Here then we have a stringent reason for Shakespeare's arrangement. He could not introduce us to the two subordinate groups, show us their isolated relations, and in the end interweave them by a consistent process, without separating them, when operating *per se*, from the main action. He could, for instance, neither account for the appearance of the lovers in the wood without a previous exposition of their difficulties, and of the agreement to fly on the ' morrow deep midnight,' nor for that of the stage-struck artizans, without some intimation of the intention to act a play, which made a rehearsal necessary. He could not follow his usual practice of developing

together the relations and position of all his characters, because the limitation to twelve hours would not admit it—and out of these twelve hours he could not remove the fairy action. So that the first and last sections of the drama, in which the main action does not proceed and only the subordinate groups appear, have nothing to do with the Midsummer Night's Dream, but are merely exegetical of it.

There are some minor indications of the truth of our theory. The very title, for instance, solely applicable as it is to that part of the drama in which the fairies appear, seems not a little significant. . . . Nor is the distribution of blank and rhymed verse unobservable. . . . We have occasionally fancied that, where the objectively poetical element prevails, the dialogue is mostly written in rhyme ; where the dramatic, in the ordinary blank verse of Shakespeare. Both Heroes and Fairies speak in blank and rhymed verse, but not indifferently. The relations of the subordinate group are generally, though not invariably, conveyed through the imaginative rhymed lines, while the Fairies, the dramatic personages, rarely quit the vigorous versification we are so well accustomed to.

We are desirous that the Fairies should assume in this play a position commensurate with the influence they must always exercise over English literature. Great as is the direct importance of combined purity and beauty in a national mythology, the indirect value is even greater. We have escaped much, as well as gained much, if our imagination has conversed with a more delicate creation than the sensuous divinities of Greece, or the vulgar spectres of the Walpurgis-Nacht. But whether the *entente cordiale* between England and Fairy-land be for good or for evil, we must at any rate acknowledge that the connection virtually began on that very Midsummer Night which witnessed the quarrel between Oberon and Titania.

HARTLEY COLERIDGE (*Essays*, &c., 1851, ii, 138): I know not any play of Shakespeare's in which the language is so uniformly unexceptionable as this. It is all poetry, and sweeter poetry was never written. One defect there may be. Perhaps the distress of Hermia and Helena, arising from Puck's blundering application of Love-in-Idleness, is too serious, too real for so fantastic a source. Yet their altercation is so very, very beautiful, so girlish, so loveable that one cannot wish it away. The characters might be arranged by a chromatic scale, gradually shading from the thick-skinned Bottom and the rude mechanicals, the absolute old father, the proud and princely Theseus and his warrior bride, to the lusty, high-hearted wooers, and so to the sylph-like maidens, till the line melts away in Titania and her fairy train, who seem as they were made of the moonshine wherein they gambol.

CHARLES COWDEN-CLARKE (*Shakespeare Characters*, 1863, p. 97): What a rich set of fellows those 'mechanicals' are ! and how individual are their several characteristics ! Bully Bottom, the epitome of all the conceited donkeys that ever strutted or straddled on this stage of the world. In his own imagination equal to the performance of anything separately, and of all things collectively ; the meddler, the director, the dictator. He is for dictating every movement, and directing everybody, —when he is not helping himself. He is a choice arabesque impersonation of that colouring of conceit, which by the half-malice of the world has been said to tinge the disposition of actors as invariably as the rouge does their cheeks. . . .

The character of Bottom is well worthy of a close analysis, to notice in how extraordinary a manner Shakespeare has carried out all the concurring qualities to compound a thoroughly conceited man. Conceited people, moreover, being upon such

amiable terms with themselves, are ordinarily good-natured, if not good-tempered. And so with Bottom; whether he carry an amendment or not, with his companions he is always placable; and if foiled, away he starts for some other point,—nothing disturbs his equanimity. His temper and self-possession never desert him. . . . Combined with his amusing and harmless quality of conceit, the worthy Bottom displays no inconsiderable store of imagination in his intercourse with the little people of the fairy world. How pleasantly he falls in with their several natures and qualities; dismissing them one by one with a gracious speech, like a prince at his levee. . . .

Then there is Snug, the joiner, who can board and lodge only one idea at a time, and that tardily. . . . To him succeeds Starveling, the tailor, a melancholy man, and who questions the feasibility and the propriety of everything proposed.

If, as some writers have asserted, Shakespeare was a profound practical metaphysician, it is scarcely too much to conclude that all this dovetailing of contingencies, requisite to perfectionate these several characters, was all foreseen and provided in his mind, and not the result of mere accident. By an intuitive power, that always confounds us when we examine its effects, I believe that whenever Shakespeare adopted any distinctive class of character, his 'mind's eye' took in at a glance all the concomitant minutiæ of features requisite to complete its characteristic identity. 'As from a watch-tower' he comprehended the whole course of human action,—its springs, its motives, its consequences; and he has laid down for us a trigonometrical chart of it. I believe that he did nothing without anxious premeditation; and that they who really study,—not simply read him,—must come to the same conclusion. Not only was he not satisfied with preserving the integrity of his characters while they were in speech and action before the audience; but we constantly find them carrying on their peculiarities,—*out* of the scene,—by hints of action, and casual remarks from others. Was there no design in all this? no contrivance? no foregone conclusion? nay, does it not manifest consummate intellectual power, with a sleepless assiduity? . . .

As Ariel is the etherialised impersonation of swift obedience, with an attachment perfectly feminine in its character—Puck, Robin Goodfellow, is an abstraction of all the 'quips, and cranks, and wanton wiles,' of all the tricks and practical jokes in vogue among 'human mortals.' Puck is the patron saint of 'skylarking.' . . . The echo of his laugh has reverberated from age to age, striking the promontories and headlands of eternal poetry; and to those whose spirits are finely touched, it is still heard through the mist of temporal cares and toils,—dimly heard, and at fitful intervals; for the old faith is that fairy presence has ceased for ever, and exists only in the record of those other elegant fancies that were the offspring of the young world of imagination.

General E. A. HITCHCOCK (*Remarks on the Sonnets of Shakespeare. Showing that they belong to the Hermetic Class of Writings, &c*, New York, 1866, p. 95): Here are three, the spirit in man, the dull substance of the flesh, and the over-soul, 'and these three are conceived as one,' but with a disturbing sense of the body interposed, as it were, between the two spirits, where it stands like a *wall* of separation, the wall being now conceived of as the man, and then as the vestment of the universe itself—which, we read, is to be rolled up like a scroll, etc., when God shall be all in all. This consummation does not appear in the *Sonnets* themselves, though, as a doctrine, it is everywhere implied by the Poet's deep sense of the unity. It is mystically shown, however, in the ancient fable of Pyramus and Thisbe, as the reader is

expected to see by the manner in which the poet uses that fable in the Interlude introduced in the closing Act of *Midsummer Night's Dream.* It may not be amiss to remind the reader of the dramas that it was usual with our poet to express the most profound truths through dramatic characters, and yet partially screen them from common inspection by the circumstances, or the sort of character made the vehicle of them,—such as Jaques and others. The reader need not be surprised, therefore, to find the *dramatis personæ* of the ' merry and tragical ' Interlude to be boorish and idiotic, while it is worth remarking that even the *wall*, as also the other parts, are all represented by men, unconscious of their calling. We now turn to the drama, and remark, that it was designed by the poet that a secret meaning should be inferred by the reader. This appears from several decisive passages, besides the general inference to be drawn from the fact, that the Interlude, more than all the rest of the play, if taken literally, is what Hippolyta says of it—the silliest stuff that was ever seen. No reasonable man can imagine that the author of so many beauties as are seen in this drama could have introduced the absurd nonsense of the Interlude without having in his mind a secret purpose, which is to be divined by the aid of the reader's imagination—according to the answer of Theseus to the remark of Hippolyta, just recited. But the imagination must be here understood as a poetic creative gift or endowment, and not limited to mere ' fancy's images;' for Hippolyta herself, though here speaking of the play, gives us a clue to something deeper than what appears on the surface. She, in allusion to all the marvels the bridal party had just heard, observes, ' But all the story of the *night* told over, And all their minds transfigured ' so together, *More witnesseth than fancy's images, And grows to something of great* ' *constancy.*' This is plainly a hint that these ' fables and fairy toys,' as Theseus calls them, may be the vehicle of some *constant* truth or principle. Again :—' Gentles, ' perchance you wonder at this show; But wonder on, till truth makes all things ' plain.' That is, when the truth, signified in the ' show ' becomes manifest, all wonder will cease, for the object of its introduction will be understood. . . . We consider now, that we have no need to dwell upon the points in detail suggested by the closing Act of the drama, which contains the doctrine we have set out as mystically contained in the Sonnets. The curious reader, who desires to exercise his own thought, while following that of the poet, expressed through the imprisoning forms of language, will see, with the indications we have given, the purpose of the ' mirthful tragedy' of Pyramus and Thisbe. He will see the signification of the two characters or principles, figured in Pyramus and Thisbe, with the *wall*, ' the vile wall which did the ' lovers sunder.' Through this *wall* (the dull substance of the flesh), the lovers may indeed communicate, but only by a ' whisper, very secretly;' because the intercourse of spirit with spirit is a secret act of the soul in a sense of its unity with the spirit. The student will readily catch the meaning of the ' moon-shine,' or *nature*-light, in this representation, the moon being always taken as nature in all mystic writings. He will see the symbolism of the ' dog '—the *watch*-dog, of course,—representing the moral guard in a nature-life; as also the bush of *thorns*, ever ready to illustrate the doctrine that the way of the transgressor is hard. The student will notice the hint that the lovers meet by moonlight and at a tomb—a symbolic indication of the greatest mystery in life (to be found in death); and he will understand the office of the lion, which tears, not Thisbe herself, but only her ' mantle,' or what the poet calls the ' extern' of life; and finally will observe that the two principles both disappear; for the unity cannot become mystically visible, until the two principles are mystically lost sight of. It should not escape notice that the two principles are co-equal; that

a mote will turn the balance, which Pyramus, which Thisbe, is the better'—simply figured as man and woman. The student of *Midsummer Night's Dream* may observe two very marked features in the play: one where the 'juice,' which induces so many absurdities, cross-purposes, and monstrosities, is described as the juice of (a certain flower called love-in-) *idleness :* the other where we see that all of the irregularities resulting from idleness are cured by the simple anointment of the eyes by what is called 'Dian's bud,'—which has such 'force and blessed power' as to bring all of the faculties back to nature and truth,—of which Dian is one of the accepted figures in all mystic writings. The readers of this play, who look upon these indications as purely arbitrary and without distinct meaning, may, indeed, perceive some of the scattered beauties of this *fairy* drama, but must certainly miss its true import.

A. C. SWINBURNE ('The Three Stages of Shakespeare,' *The Fortnightly Rev.*, Jan. 1876): But in the final poem which concludes and crowns the first epoch of Shakespeare's work, the special graces and peculiar glories of each that went before are gathered together as in one garland 'of every hue and every scent.' The young genius of the master of all poets finds its consummation in the *Midsummer Night's Dream.* The blank verse is as full, sweet, and strong as the best of Biron's or Romeo's; the rhymed verse as clear, pure, and true as the simplest and truest melody of *Venus and Adonis* or the *Comedy of Errors.* But here each kind of excellence is equal throughout; there are here no purple patches on a gown of serge, but one seamless and imperial robe of a single dye. Of the lyric and prosaic part, the counterchange of loves and laughters, of fancy fine as air and imagination high as heaven, what need can there be for any one to shame himself by the helpless attempt to say some word not utterly unworthy? Let it suffice to accept this poem as a land-mark of our first stage, and pause to look back from it on what lies behind us of partial or of perfect work.

F. J. FURNIVALL (*Introd. to Leopold Shakespeare*, 1877, p. xxvi): Here at length we have Shakspere's genius in the full glow of fancy and delightful fun. The play is an enormous advance on what has gone before. But it is a poem, a dream, rather than a play; its freakish fancy of fairy-land fitting it for the choicest chamber of the student's brain, while its second part, the broadest farce, is just the thing for the public stage. E. A. Poe writes: 'When I am asked for a definition of poetry, I think of 'Titania and Oberon of the *Midsummer Night's Dream*.' And certainly anything must be possible to the man who could in one work range from the height of Titania to the depth of Bottom. The links with the *Errors* are, that all the wood scenes are a comedy of errors, with three sets of people, as in the *Errors* (and four in *Love's Labour's Lost*). Then we have the vixen Hermia to match the shrewish Adriana, the quarrel with husband and wife, and Titania's 'these are the forgeries of jealousy' to compare with Adriana's jealousy in the *Errors.* Adriana offers herself to Antipholus of Syracuse, but he refuses her for her sister Luciana, as Helena offers herself to Demetrius, and he refuses her for her friend Hermia. Hermia bids Demetrius love Helena, as Luciana bids Antipholus of Syracuse love his supposed wife Adriana. In the background of the *Errors* we have the father Ægeon with the sentence of death or fine pronounced by Duke Solinus. In the *Dream* we have in the background the father Egeus with the sentence of death or celibacy on Hermia pronounced by Duke Theseus. In both plays the scene is Eastern; in the *Errors*, Ephe-

sus; and in the *Dream*, Athens. We have an interesting connection with Chaucer, in that the Theseus and Hippolyta are taken from his *Knight's Tale*, and used again in *The Two Noble Kinsmen;* also the May-day and St. Valentine, and the wood birds here may be from Chaucer's *Parlement of Foules*. The fairies, too, are in Chaucer's *Wife of Bath's Tale*. As links with *Love's Labour's Lost* we notice the comedy of errors in the earlier play, the forest scene, and the rough country sub-play, while as opposed to the *Love's Labour's Lost*'s 'Jack hath not Gill,' the fairies tell us here 'Jack shall have Gill.' The fairies are the centre of the drama; the human characters are just the sport of their whims and fancies, a fact which is much altered when we come to Shakspere's use of fairy-land again in his *Tempest*, where the aërial beings are but ministers of the wise man's rule for the highest purposes. The finest character here is undoubtedly Theseus. In his noble words about the countrymen's play, the true gentleman is shown. His wife's character is but poor beside his. Though the story is Greek, yet the play is full of English life. It is Stratford which has given Shakspere the picture of the sweet country school-girls working at one flower, warbling one song, growing together like a double cherry, seeming parted, but yet a union in partition. It is Stratford that has given him the picture of the hounds with 'Ears that sweep away the morning dew.' It is Stratford that has given him his out-door woodland life, his clowns' play, and the clowns themselves, Bottom, with his inimitable conceit, and his fellows, Snug and Quince, &c. It is Stratford that has given him all Puck's fairy-lore, the cowslips tall, the red-hipt bumble bee, Oberon's bank, the pansy love-in-idleness, and all the lovely imagery of the play. But wonderful as the mixture of delicate and aërial fancy with the coarsest and broadest comedy is, clearly as it evidences the coming of a new being on this earth to whom anything is possible, it is yet clear that the play is quite young. The undignified quarreling of the ladies, Hermia with her 'painted May-pole,' her threat to scratch Helena's eyes,—Helena with her retorts 'She was a vixen when she went to school,' &c., the comical comparison of the moon tumbling through the earth (III, ii, 52) incongruously put into an accusation of murder, the descent to bathos in Shakspere's passage about his own art, from 'the poet's eye in a fine frenzy rolling' to 'how easy 'is a bush supposed a bear,' would have been impossible to Shakspere in his later developement. Those who contend for the later date of the play, from the beauty of most of the fancy, and the allusion to the effects of the rains and the floods, which they make those of 1594, must allow, I think, that the framework of the play is considerably before the date of *King John* and *The Merchant of Venice*. Possibly two dates may be allowed for the play, tho' I don't think them needful. . . .

With the *Dream* I propose to close the first Group of Shakspere's Comedies, those in which the Errors arising from mistaken identity make so much of the fun. And the name of the group may well be 'the Comedy of Errors or Mistaken-Identity Group.'

HUDSON (*Introduction*, 1880, p. 7): The whole play is indeed a sort of ideal dream; and it is from the fairy personages that its character as such mainly proceeds. All the materials of the piece are ordered and assimilated to that central and governing idea. This it is that explains and justifies the distinctive features of the work, such as the constant preponderance of the lyrical over the dramatic, and the free playing of the action unchecked by the conditions of outward fact and reality. Accordingly a sort of lawlessness is, as it ought to be, the very law of the performance. . . . In keeping with this central dream-idea, the actual order of things everywhere gives place to the spontaneous issues and capricious turnings of the dreaming

mind; the lofty and the low, the beautiful and the grotesque, the world of fancy and of fact, all the strange diversities that enter into 'such stuff as dreams are made of,' running and frisking together, and interchanging their functions and properties; so that the whole seems confused, flitting, shadowy, and indistinct, as fading away in the remoteness and fascination of moonlight. The very scene is laid in a veritable dream-land, called Athens indeed, but only because Athens was the greatest bee-hive of beautiful visions then known; or rather it is laid in an ideal forest near an ideal Athens,—a forest peopled with sportive elves and sprites and fairies feeding on moon-light and music and fragrance; a place where Nature herself is preternatural; where everything is idealised even to the sunbeams and the soil; where the vegetation pro ceeds by enchantment, and there is magic in the germination of the seed and secre-tion of the sap. . . .

[Page 9.] In further explication of this peculiar people [the Fairies], it is to be noted that there is nothing of reflection or conscience or even of a spiritualised intel-ligence in their proper life; they have all the attributes of the merely natural and sensitive soul, but no attributes of the properly rational and moral soul. They wor-ship the clean, the neat, the pretty, the pleasant, whatever goes to make up the idea of purely sensuous beauty; this is a sort of religion with them; whatever of con-science they have adheres to this; so that herein they not unfitly represent the whole-some old notion which places cleanliness next to godliness. Everything that is trim, dainty, elegant, graceful, agreeable, and sweet to the senses, they delight in; flowers, fragrances, dewdrops, and moonbeams, honey-bees, butterflies, and nightingales, dancing, play, and song,—these are their joy; out of these they weave their highest delectation; amid these they 'fleet the time carelessly,' without memory or forecast and with no thought or aim beyond the passing pleasure of the moment. On the other hand, they have an instinctive repugnance to whatever is foul, ugly, sluttish, awkward, ungainly, or misshapen; they wage unrelenting war against bats, spiders, hedgehogs, spotted snakes, blindworms, long-legg'd spinners, beetles, and all such disagreeable creatures; to 'kill cankers in the musk-rosebuds' and to 'keep back the clamorous owl,' are regular parts of their business. . . . Thus these beings embody the ideal of the mere natural soul, or rather the purely sensuous fancy which shapes and governs the pleasing or the vexing delusions of sleep. They lead a merry, luxurious life, given up entirely to the pleasures of happy sensation,—a happiness that has no moral element, nothing of reason or conscience in it. They are indeed a sort of per-sonified dreams; and so the Poet places them in a kindly or at least harmless rela-tion to mortals as the bringers of dreams. Their very kingdom is located in the aro matic, flower-scented Indies, a land where mortals are supposed to live in a half-dreamy state. From thence they come, 'following darkness,' just as dreams naturally do; or, as Oberon words it, 'tripping after the night's shade, swifter than the wander-'ing Moon.' It is their nature to shun the daylight, though they do not fear it, and to prefer the dark, as this is their appropriate worktime; but most of all they love the dusk and twilight, because this is the best dreaming-time, whether the dreamer be asleep or awake. And all the shifting phantom-jugglery of dreams, all the sweet soothing witcheries, and all the teasing and tantalising imagery of dream-land, rightly belong to their province.

[P. 15.] Any very firm or strong delineation of character, any deep passion, earnest purpose, or working of powerful motives, would clearly go at odds with the spirit of such a performance as [the present play]. It has room but for love and beauty and delight, for whatever is most poetical in nature and fancy, and for such

tranquil stirrings of thought and feeling as may flow out in musical expression. Any such tuggings of mind or heart as would ruffle and discompose the smoothness of lyrical division would be quite out of keeping in a course of dream-life. The characters here, accordingly, are drawn with light, delicate, vanishing touches; some of them being dreamy and sentimental, some gay and frolicsome, and others replete with amusing absurdities, while all are alike dipped in fancy or sprinkled with humour. And for the same reason the tender distresses of unrequited or forsaken love here touch not our moral sense at all, but only at the most our human sympathies; love itself being represented as but the effect of some visual enchantment, which the King of Fairydom can inspire, suspend, or reverse at pleasure. Even the heroic personages are fitly shown in an unheroic aspect; we see them but in their unbendings, when they have daffed their martial robes aside, to lead the train of day-dreamers, and have a nuptial jubilee. In their case, great care and art were required to make the play what it has been blamed for being, that is, to keep the dramatic sufficiently under, and lest the law of a part should override the law of the whole.

So, likewise, in the transformation of Bottom and the dotage of Titania, all the resources of fancy were needed to prevent the unpoetical from getting the upper hand, and thus swamping the genius of the piece. As it is, what words can fitly express the effect with which the extremes of the grotesque and the beautiful are here brought together? What an inward quiet laughter springs up and lubricates the fancy at Bottom's droll confusion of his two natures, when he talks now as an ass, now as a man, and anon as a mixture of both; his thoughts running at the same time on honey-bags and thistles, the charms of music and of good dry oats! Who but Shakespeare or Nature could have so interfused the lyrical spirit, not only with, but into and through, a series or cluster of the most irregular and fantastic drolleries? But, indeed, this embracing and kissing of the most ludicrous and the most poetical, the enchantment under which they meet, and the airy, dream-like grace that hovers over their union, are altogether inimitable and indescribable. In this singular wedlock the very diversity of the elements seems to link them the closer, while this linking in turn heightens that diversity; Titania being thereby drawn on to finer issues of soul, and Bottom to larger expressions of stomach. The union is so very improbable as to seem quite natural; we cannot conceive how anything but a dream could possibly have married things so contrary; and that they could not have come together save in a dream, is a sort of proof that they *were* dreamed together.

And so throughout, the execution is in strict accordance with the plan. The play from beginning to end is a perfect festival of whatever dainties and delicacies poetry may command,—a continued revelry and jollification of soul, where the understanding is lulled asleep, that the fancy may run riot in unrestrained enjoyment. The bringing together of four parts so dissimilar as those of the Duke and his warrior Bride, of the Athenian ladies and their lovers, of the amateur players and their woodland rehearsal, and of the fairy bickerings and overreaching; and the carrying of them severally to a point where they all meet and blend in lyrical respondence; all this is done in the same freedom from the laws that govern the drama of character and life. Each group of persons is made to parody itself into concert with the others; while the frequent intershootings of fairy influence lift the whole into the softest regions of fancy. At last the Interlude comes in as an amusing burlesque on all that has gone before; as in our troubled dreams we sometimes end with a dream that we have been dreaming, and our perturbations sink to rest in the sweet assurance that they were but the phantoms and unrealities of a busy sleep. . . .

[Page 21.] Partly for reasons already stated, and partly for others that I scarce know how to state, *A Midsummer Night's Dream* is a most effectual poser to criticism. Besides that its very essence is irregularity, so that it cannot be fairly brought to the test of rules, the play forms properly a class by itself; literature has nothing else really like it; nothing therefore with which it may be compared, and its merits adjusted. For so the Poet has here exercised powers apparently differing even in kind, not only from those of any other writer, but from those displayed in any other of his own writings. Elsewhere, if his characters are penetrated with the ideal, their whereabout lies in the actual, and the work may in some measure be judged by that life which it claims to represent; here the whereabout is as ideal as the characters; all is in the land of dreams,—a place for dreamers, not for critics. For who can tell what a dream ought or ought not to be, or when the natural conditions of dream-life are or are not rightly observed? How can the laws of time and space, as involved in the transpiration of human character,—how can these be applied in a place where the mind is thus absolved from their proper jurisdiction? Besides, the whole thing swarms with enchantment; all the sweet witchery of Shakespeare's sweet genius is concentrated in it, yet disposed with so subtle and cunning a hand, that we can as little grasp it as get away from it; its charms, like those of a summer evening, are such as we may see and feel, but cannot locate or define; cannot say they are here or they are there; the moment we yield ourselves up to them, they seem to be everywhere; the moment we go to master them, they seem to be nowhere.

WILLIAM WINTER (Augustin Daly's *Arrangement for Representation*, 1888; *Preface*, p. 12): The student of [this play] as often as he thinks upon this lofty and lovely expression of a most luxuriant and happy poetic fancy, must necessarily find himself impressed with its exquisite purity of spirit, its affluence of invention, its extraordinary wealth of contrasted characters, its absolute symmetry of form, and its great beauty of poetic diction. The essential, wholesome cleanliness and sweetness of Shakespeare's mind, unaffected by the gross animalism of his times, appear conspicuously in this play. No single trait of the piece impresses the reader more agreeably than its frank display of the spontaneous, natural, and entirely delightful exultation of Theseus and Hippolyta in their approaching nuptials. They are grand creatures both, and they rejoice in each other and in their perfectly accordant love. Nowhere in Shakespeare is there a more imperial man than Theseus; nor, despite her feminine impatience of dulness, a woman more beautiful and more essentially woman-like than Hippolyta. It is thought that the immediate impulse of this comedy, in Shakespeare's mind, was the marriage of his friend and benefactor, the Earl of Southampton, with Elizabeth Vernon. . . . In old English literature it is seen that such a theme often proved suggestive of ribaldry; but Shakespeare could preserve the sanctity, even while he revelled in the passionate ardor, of love, and *A Midsummer Night's Dream*, while it possesses all the rosy glow, the physical thrill, and the melting tenderness of such pieces as Herrick's *Nuptial Song*, is likewise fraught with all the moral elevation and unaffected chastity of such pieces as Milton's *Comus*. The atmosphere is free and bracing; the tone honest; the note true. Then, likewise, the fertility and felicity of the poet's invention,—intertwining the loves of earthly sovereigns and of their subjects with the dissensions of fairy monarchs, the pranks of mischievous elves, the protective care of attendant sprites, and the comic but kind-hearted and well-meant fealty of boorish peasants,—arouse lively interest and keep it steadily alert. In no other of his works has Shakespeare more brilliantly shown that

complete dominance of theme which is manifested in the perfect preservation of pro-portion. The strands of action are braided with astonishing grace. The fourfold story is never allowed to lapse into dulness or obscurity. There is caprice, but no distortion. The supernatural machinery is never wrested toward the production of startling or monstrous effects, but it deftly impels each mortal personage in the natu-ral line of human development. The dream-spirit is maintained throughout, and perhaps it is for this reason,—that the poet was living and thinking and writing in the free, untrammelled world of his own spacious and airy imagination, and not in any definite sphere of this earth,—that *A Midsummer Night's Dream* is so radically superior to the other comedies written by him at about this period.

[P. 14.] With reference to the question of suitable method in the acting of [this play], it may be observed that too much stress can scarcely be laid upon the fact that this comedy was conceived and written absolutely in the spirit of a dream. It ought not, therefore, to be treated as a rational manifestation of orderly design. It pos-sesses, indeed, a coherent and symmetrical plot and a definite purpose; but, while it moves toward a final result of absolute order, it presupposes intermediary progress through a realm of motley shapes and fantastic vision. Its persons are creatures of fancy, and all effort to make them solidly actual, to set them firmly upon the earth, and to accept them as realities of common life, is labour ill-bestowed. . . .

To body forth the forms of things is, in this case, manifestly, a difficult task; and yet the true course is obvious. Actors who yield themselves to the spirit of whim, and drift along with it, using a delicate method and avoiding insistence upon prosy realism, will succeed with this piece,—provided, also, that their audience can be fan-ciful, and can accept the performance, not as a comedy of ordinary life, but as a vision seen in a dream. The play is full of intimations that this was Shakespeare's mood.

[In *Noctes Shaksperianæ*, a collection of *Papers* by the *Winchester College Shak-spere Society* (London, 1887), is to be found, on p. 208, a paper by O. T. PERKINS, 'Ghostland and Fairyland.' It is too long for insertion here, and extracts would but mangle it. It is to be commended to all to whom the charm of Shakespeare's fairies is ever fresh, and to whom, with the author, there comes no doubt that 'as Shake-'speare wrote he felt the breath of the Warwickshire lanes, and heard the babble of 'its clear streams, and remembered the country he had known as a boy.'—ED.]

BOTTOM

HAZLITT (*Characters of Shakespeare's Plays*, 1817, p. 126): Bottom the Weaver is a character that has not had justice done him. He is the most romantic of me-chanics. . . . It has been observed that Shakespeare's characters are constructed upon deep physiological principles; and there is something in this play which looks very like it. Bottom follows a sedentary trade, and he is accordingly represented as con-ceited, serious, and fantastical. He is ready to undertake anything and everything, as if it was as much a matter of course as the motion of his loom and shuttle. He is for playing the tyrant, the lover, the lady, the lion. Snug the Joiner is the moral man of the piece, who proceeds by measurement and discretion in all things. You see him with his rule and compasses in his hand. Starveling the Tailor keeps the peace, and objects to the lion and the drawn sword. Starveling does not start the objections himself, but seconds them when made by others, as if he had not spirit to

express his fears without encouragement. It is too much to suppose all this intentional; but it very luckily falls out so. Nature includes all that is implied in the most subtle analytical distinctions; and the same distinctions will be found in Shakespeare. Bottom, who is not only chief actor, but stage-manager, for the occasion, has a device to obviate the danger of frightening the ladies, . . . and seems to have understood the subject of dramatic illusion at least as well as any modern essayist. If our holiday mechanic rules the roast among his fellows, he is no less at home in his new character of an ass. He instinctively acquires a most learned taste and grows fastidious in the choice of dried peas and bottled hay.

MAGINN (*Shakespeare Papers*, 1860, p. 121): One part of Bottom's character is easily understood, and is often well acted. Among his own companions he is the cock of the walk. His genius is admitted without hesitation. When he is lost in the wood, Quince gives up the play as marred. . . . Flute declares that he has the best wit of any handicraftman in the city. . . . It is no wonder that this perpetual flattery fills him with a most inordinate opinion of his own powers. There is not a part in the play which he cannot perform. . . . The wit of the courtiers, or the presence of the Duke, has no effect upon his nerves. He alone speaks to the audience in his own character, not for a moment sinking the personal consequence of Bottom in the assumed part of Pyramus. He sets Theseus right on a point of the play with cool importance; and replies to a jest of Demetrius (which he does not understand) with the self-command of ignorant indifference. We may be sure that he was abundantly contented with his appearance, and retired to drink in, with ear well deserving of the promotion it had attained under the patronage of Robin Goodfellow, the applause of his companions. It is true that Oberon designates him as a ' hateful fool '; that Puck stigmatises him as the greatest blockhead of the set; that the audience of wits and courtiers before whom he has performed vote him to be an ass; but what matter is that? He mixes not with them; he hears not their sarcasms; he could not understand their criticisms; and, in the congenial company of the crew of patches and base mechanicals who admire him, lives happy in the fame of being *the* Nicholas Bottom, who, by consent, to him universal and world-encompassing, is voted to be *the* Pyramus,—*the* prop of the stage,—*the* sole support of the drama.

Self-conceit, as great and undisguised as that of poor Bottom, is to be found in all classes and in all circles, and is especially pardonable in what it is considered genteel or learned to call ' the histrionic profession.' The triumphs of the player are evanescent. In no other department of intellect, real or simulated, does the applause bestowed upon the living artist bear so melancholy a disproportion to the repute awaiting him after the generation passes which has witnessed his exertions. According to the poet himself, the poor player ' Struts and frets his hour upon the stage, And ' then is heard no more.' Shakespeare's own rank as a performer was not high, and his reflections on the business of an actor are in general splenetic and discontented. He might have said,—though indeed it would not have fitted with the mood of mind of the despairing tyrant into whose mouth the reflection is put,—that the well-graced actor, who leaves the scene not merely after strutting and fretting, but after exhibiting power and genius to the utmost degree at which his art can aim, amid the thundering applause,—or, what is a deeper tribute, the breathless silence of excited and agitated thousands,—is destined ere long to an oblivion as undisturbed as that of his humbler fellow-artist, whose prattle is voted, without contradiction, to be tedious. Kemble is fading fast from our view. The gossip connected with everything about

Johnson keeps Garrick before us, but the interest concerning him daily becomes less and less. Of Betterton, Booth, Quin, we remember little more than the names. The Lowins and Burbadges of the days of Shakespeare are known only to the dramatic antiquary, or the poring commentator, anxious to preserve every scrap of information that may bear upon the elucidation of a text, or aid towards the history of the author. With the sense of this transitory fame before them, it is only natural that players should grasp at as much as comes within their reach while they have the power of doing so. . . . Pardon therefore the wearers of the sock and buskin for being obnoxious to such criticism as that lavished by Quince on Bottom. . . . It would take a long essay on the mixture of legends derived from all ages and countries to account for the production of such a personage as the ' Duke ycleped Theseus ' and his following; and the fairy mythology of the most authentic superstitions would be ransacked in vain to discover exact authorities for the Shakespearian Oberon and Titania. But no matter whence derived, the author knew well that in his hands the chivalrous and classical, the airy and the imaginative, were safe. It was necessary for his drama to introduce among his fairy party a creature of earth's mould, and he has so done it as in the midst of his mirth to convey a picturesque satire on the fortune which governs the world, and upon those passions which elsewhere he had with agitating pathos to depict As Romeo, the gentleman, is *the* unlucky man of Shakespeare, so here does he exhibit Bottom, the blockhead, as *the* lucky man, as him on whom Fortune showers her favours beyond measure. This is the part of the character which cannot be performed. It is here that the greatest talent of the actor must fail in answering the demand made by the author upon our imagination. . . . The mermaid chanting on the back of her dolphin; the fair vestal throned in the west; the bank blowing with wild thyme, and decked with oxlip and nodding violet; the roundelay of the fairies singing their queen to sleep; and a hundred images beside of aërial grace and mythic beauty, are showered upon us; and in the midst of these splendours is tumbled in Bottom the weaver, blockhead by original formation, and rendered doubly ridiculous by his partial change into a literal jackass. He, the most unfitted for the scene of all conceivable personages, makes his appearance, not as one to be expelled with loathing and derision, but to be instantly accepted as the chosen lover of the Queen of the Fairies. The gallant train of Theseus traverse the forest, but they are not the objects of such fortune. The lady, under the oppression of the glamour cast upon her eyes by the juice of love-in-idleness, reserves her rapture for an absurd clown. Such are the tricks of Fortune. . . . Abstracting the poetry, we see the same thing every day in the plain prose of the world. Many is the Titania driven by some unintelligible magic so to waste her love. Some juice, potent as that of Puck,—the true Cupid of such errant passions, often converts in the eyes of woman the grossest defects into resistless charms. The lady of youth and beauty will pass by attractions best calculated to captivate the opposite sex, to fling herself at the feet of age or ugliness. Another, decked with graces, accomplishments, and the gifts of genius, and full of all the sensibilities of refinement, will squander her affections on some good-for-nothing *roué*, whose degraded habits and pursuits banish him far away from the polished scenes which she adorns. The lady of sixteen quarters will languish for him who has no arms but those which nature has bestowed; from the midst of the gilded *salon* a soft sigh may be directed towards the thin-clad tenant of a garret; and the heiress of millions may wish them sunken in the sea if they form a barrier between her and the penniless lad toiling for his livelihood, ' Lord of his presence, and no ⁴ land beside.' . . . Ill-mated loves are generally of short duration on the side of the

nobler party, and she awakes to lament her folly. The fate of those who suffer like Titania is the hardest. . . . Woe to the unhappy lady who is obliged to confess, when the enchantment has passed by, that she was 'enamoured of an *ass!*' She must indeed 'loathe his visage,' and the memory of all connected with him is destined ever to be attended by a strong sensation of disgust.

But the ass himself of whom she was enamoured has not been the less a favourite of Fortune, less happy and self-complacent, because of her late repentance. He proceeds onward as luckily as ever. Bottom, during the time that he attracts the attentions of Titania, never for a moment thinks there is anything extraordinary in the matter. He takes the love of the Queen of the Fairies as a thing of course, orders about her tiny attendants as if they were so many apprentices at his loom, and dwells in Fairy Land, unobservant of its wonders, as quietly as if he were still in his work-shop. Great is the courage and self-possession of an ass-head. Theseus would have bent in reverent awe before Titania. Bottom treats her as carelessly as if she were the wench of the next-door tapster. Even Christopher Sly, when he finds himself transmuted into a lord, shows some signs of astonishment. He does not accommo-date himself to surrounding circumstances. . . . In the *Arabian Nights' Entertain-ments* a similar trick is played by the Caliph Haroun Alraschid upon Abou Hassan, and he submits, with much reluctance, to believe himself the Commander of the Faithful. But having in vain sought how to explain the enigma, he yields to the belief, and then performs all the parts assigned to him, whether of business or pleas-ure, of counsel or gallantry, with the easy self-possession of a practised gentleman. Bottom has none of the scruples of the tinker of Burton-Heath, or the *bon vivant* of Bagdad. He sits down among the fairies as one of themselves without any astonish-ment; but so far from assuming, like Abou Hassan, the manners of the court where he has been so strangely intruded, he brings the language and bearing of the booth into the glittering circle of Queen Titania. He would have behaved in the same manner on the throne of the caliph, or in the bedizened chamber of the lord; and the ass-head would have victoriously carried him through. . . .

Adieu, then, Bottom the weaver! and long may you go onward prospering in your course! But the prayer is needless, for you carry about you the infallible talisman of the ass-head. You will be always sure of finding a Queen of the Fairies to heap her favours upon you, while to brighter eyes and nobler natures she remains invisible or averse. Be you ever the chosen representative of the romantic and the tender before dukes and princesses; and if the judicious laugh at your efforts, despise them in return, setting down their criticism to envy. This you have a right to do. Have they, with all their wisdom and wit, captivated the heart of a Titania as you have done? Not they—nor will they ever. Prosper, therefore, with undoubting heart, despising the babble of the wise. Go on your path rejoicing; assert loudly your claim to fill every character in life; and may you be quite sure that as long as the noble race of the Bottoms continues to exist, the chances of extraordinary good luck will fall to their lot, while in the ordinary course of life they will never be unattended by the plausive criticism of a Peter Quince.

J. A. HERAUD (*Shakespeare, His Inner Life*, p. 178, 1865): Here we have Bottom in the part of theatrical reader and manager. He has been pondering the drama, until he conjures up fears for its success, takes exceptions to incidentals, and suggests rem-edies. Bottom is not only critical, he is inventive. With a little practice and encour-agement we shall see him writing a play himself. Indeed, with a trifling exaggera-

tion, the scene is only a caricature of what frequently happened in the Green-rooms of theatres in the poet's own day, and has happened since in that of every other. Here is instinct rashly mistaken for aptitude, and aptitude for knowledge, by the uninstructed artisan, who has to substitute shrewdness for experience. And thus it is with the neophyte actor and the ignorant manager, whose sole aim is to thrust aside the author, and reign independent of his control; altering and supplementing, according to their limited lights, what he has conceived in the fullness of the poetic faculty. . . . Soon, however, the poor players discover that their manager wears the ass's head, though he never suspects it himself; and even the poor faery queen, the temporarily-demented drama, is fain to place herself under his guardianship. She cannot help it under the circumstances; and, therefore, she gives him all the pretty pickings, the profits, and the perquisites of the theatre, leaving the author scarcely the gleaning. The fairies have charge of the presumptuous ignoramus, with the fairy queen's direction.

In a far different fashion Shakespeare conducted matters at his own theatre. There the poet presided, and the world has witnessed the result. The argument needs no other elucidation.

D. WILSON (*Caliban, the Missing Link*, 1873, p. 262): What inimitable power and humorous depth of irony are there in the Athenian weaver and prince of clownish players! Vain, conceited, consequential; he is nevertheless no mere empty lout, but rather the impersonation of characteristics which have abounded in every age, and find ample scope for their display in every social rank. Bottom is the work of the same master hand which wrought for us the Caliban and Miranda, the Puck and Ariel, of such diverse worlds. He is the very embodiment and idealisation of that self-esteem which is a human virtue by no means to be dispensed with, though it needs some strong counterpoise in the well-balanced mind. In the weak, vain man, who fancies everybody is thinking of him and looking at him, it takes the name of shyness, and claims nearest kin to modesty. With robust, intensitive vulgarity it assumes an air of universal philanthropy and good-fellowship. In the man of genius it reveals itself in very varying phases; gives to Pope his waspish irritability as a satirist, and crops out anew in the transparent mysteries of publication of his laboured-impromptu private letters; betrays itself in the self-laudatory exclusiveness which carried Wordsworth through long years of detraction and neglect to his final triumph; in the morbid introversions of Byron, and his assumed defiance of 'the world's dread 'laugh'; in the sturdy self-assertion of Burns, the honest faith of the peasant bard that 'The rank is but the guinea stamp, The man's the gowd for a' that!' In Ben Jonson it gave character to the whole man. Goldsmith and Chatterton Hogg and Hugh Miller, only differed from their fellows in betraying the self-esteem which more cunning adepts learn to disguise under many a mask, even from themselves. It shines in modest prefaces, writes autobiographies and diaries by the score, and publishes poems by the hundred,—'Obliged by hunger and request of friends.' Nick Bottom is thus a representative man, 'not one, but all mankind's epitome.' He is a natural genius. If he claims the lead, it is not without a recognised fitness to fulfill the duties he assumes. He is one whom nothing can put out. 'I have a device to make all well,' is his prompt reply to every difficulty, and the device, such as it is, is immediately forthcoming. . . . Bottom is as completely conceived, in all perfectness of consistency, as any character Shakespeare has drawn; ready-witted, unbounded in his self-confidence, and with a conceit nursed into the absolute proportions which we wir

ness by the admiring deference of his brother clowns. Yet this is no more than the recognition of true merit. Their admiration of his parts is rendered ungrudgingly, as it is received by him simply as his due. Peter Quince appears as responsible manager of the theatricals, and indeed is doubtless the author of 'the most lament- 'able comedy.' For Nick Bottom, though equal to all else, makes no pretension to the poetic art. . . .

But fully to appreciate the ability and self-possession of Nick Bottom in the most unwonted circumstances, we must follow the translated mechanical to Titania's bower, where the enamoured queen lavishes her favours on her strange lover. His cool prosaic commonplaces fit in with her rhythmical fancies as naturally as the dull grey of the dawn meets and embraces the sunrise. . . .

We cannot but note the quaint blending of the ass with the rude Athenian 'thick- 'skin'; as though the creator of Caliban had his own theory of evolution; and has here an eye to the more fitting progenitor of man. Titania would know what her sweet love desires to eat. 'Truly a peck of provender; I could munch your good 'dry oats.' The puzzled fairy queen would fain devise some fitter dainty for her lover. But no! Bottom has not achieved the dignity of that sleek smooth head, and those fair large ears, which Titania has been caressing and decorating with musk- roses, to miss their befitting provender. 'I had rather have a handful or two of dry 'peas.' It comes so naturally to him to be an ass! . . .

There are Bottoms everywhere. Nor are they without their uses. Vanity becomes admirable when carried out with such sublime unconsciousness; and here it is a van- ity resting on some solid foundation, and finding expression in the assumption of a leadership which his fellows recognise as his own by right. If he will play the lion's part, 'let him roar again!' Look where we will, we may chance to come on 'sweet 'bully Bottom.' In truth, there is so much of genuine human nature in this hero of *A Midsummer Night's Dream,* that it may not always be safe to peep into the look- ing-glass, lest evolution reassert itself for our special behoof, and his familiar counte- nance greet us, 'Hail, fellow, well met, give me your neif!'

J. WEISS (*Wit, Humor, and Shakespeare,* Boston, 1876, p. 110): It is also a sug- gestion of the subtlest humor when Titania summons her fairies to wait upon Bottom; for the fact is that the soul's airy and nimble fancies are constantly detailed to serve the donkeyism of this world. 'Be kind and courteous to this gentleman.' Divine gifts stick musk-roses in his sleek, smooth head. The world is a peg that keeps all spiritual being tethered. John Watt agonises to teach this *vis inertiae* to drag itself by the car-load; Palissy starves for twenty years to enamel its platter; Franklin charms its house against thunder; Raphael contributes halos to glorify its ignorance of divinity; all the poets gather for its beguilement, hop in its walk, and gambol before it, scratch its head, bring honey-bags, and light its farthing dip at glow-worms' eyes. Bottom's want of insight is circled round by fulness of insight, his clumsiness by dexterity. In matter of eating, he really prefers provender; 'good hay, sweet 'hay, hath no fellow.' But how shrewdly Bottom manages this holding of genius to his service! He knows how to send it to be oriental with the blossoms and the sweets, giving it the characteristic counsel not to fret itself too much in the action.

You see there is nothing sour and cynical about Bottom. His daily peck of oats, with plenty of munching-time, travels to the black cell where the drop of gall gets secreted into the ink of starving thinkers, and sings content to it on oaten straw. Bottom, full-ballasted, haltered to a brown-stone-fronted crib, with digestion always

waiting upon appetite, tosses a tester to Shakespeare, who might, if the tradition be true, have held his horse in the purlieus of the Curtain or Rose Theatre; perhaps he sub-let the holding while he slipped in to show Bottom how he is a deadly earnest fool; and the boxes crow and clap their unconsciousness of being put into the poet's celestial stocks. All this time Shakespeare is divinely restrained from bitterness by the serenity which overlooks a scene. If, like the ostrich, he had been only the largest of the birds which do not fly, he might have wrangled for his rations of ten-penny nails and leather, established perennial indigestion in literature, and furnished plumes to jackdaws. But he flew closest to the sun, and competed with the dawn for a first taste of its sweet and fresh impartiality.

Professor J. MACMILLAN BROWN ('An Early Rival of Shakespeare,' *New Zealand Maga.*, April, 1877, p. 102): Shakespeare, with all his tolerance, was unable to refrain from retaliation, but it is with no venomous pen he retaliates. . . . In the *Midsummer Night's Dream* he takes this early school of amateur player-poets, and pillories them in Bottom, Quince, Snug, Flute, Snout, and Starveling; and with the elfin machinery he borrows from Greene, turns his caricature, Bottom, into everlasting ridicule.

[Prof. Brown exaggerates, I think, the loan of elfin machinery from Greene, even granting that *James IV* preceded the present play, which is doubtful. GROSART (*Introd.* to Greene's *Works*, p. xxxix) says it is 'unknown which was earlier;' see the extracts from *James IV supra* in 'Source of the Plot.' In the conjecture that Greene was portrayed in Bottom, BROWN anticipates FLEAY, who observes (*Life and Work*, p. 18), 'Bottom and his scratch company have long been recognised as a personal 'satire, and the following marks would seem to indicate that Greene and the Sussex' 'company were the butts at which it was aimed. Bottom is a Johannes Factotum 'who expects a pension for his playing; his comrades are unlettered rustics who once 'obtain an audience at Theseus' court. The Earl of Sussex' men were so inferior a 'company that they acted at Court but once, viz. in January, 1591–2, and the only 'new play which can be traced to them at this date is *George a Greene*, in which 'Greene acted the part of the Pinner himself. This only shows that the circumstances 'of the fictitious and real events are not discrepant; but when we find Bottom saying 'that he will get a ballad written on his adventure, and "it shall be called Bottom's '"Dream, because it hath no bottom," and that peradventure he shall "sing it at her '"(?) death," we surely may infer an allusion to Greene's *Maiden's Dream* (*Sta-*'*tioners' Registers*, 6th Dec. 1591), apparently so called because it hath no maiden 'in it, and sung at the death of Sir Christopher Hatton.'—ED.]

HUDSON (*Introduction*, 1880, p. 20): But Bottom's metamorphosis is the most potent drawer out of his genius. The sense of his new head-dress stirs up all the manhood within him, and lifts his character into ludicrous greatness at once. Hitherto the seeming to be a man has made him content to be little better than an ass; but no sooner is he conscious of seeming an ass than he tries his best to be a man; while all his efforts that way only go to approve the fitness of his present seeming to his former being.

Schlegel happily remarks, that 'the droll wonder of Bottom's metamorphosis is 'merely the translation of a metaphor in its literal sense.' The turning of a figure of speech thus into visible form is a thing only to be thought of or imagined; so that no attempt to paint or represent it to the senses can ever succeed. We can bear—at

21

least we often have to bear—that a man should seem an ass to the mind's eye; but that he should seem such to the eye of the body is rather too much, save as it is done in those fable-pictures which have long been among the playthings of the nursery. So a child, for instance, takes great pleasure in fancying the stick he is riding to be a horse, when he would be frightened out of his wits were the stick to quicken and expand into an actual horse. In like manner we often delight in indulging fancies and giving names, when we should be shocked were our fancies to harden into facts; we enjoy visions in our sleep that would only disgust or terrify us, should we awake and find them solidified into things. The effect of Bottom's transformation can hardly be much otherwise, if set forth in visible, animated shape. Delightful to think of, it is scarcely tolerable to look upon; exquisitely true in idea, it has no truth, or even veri-similitude, when reduced to fact; so that, however gladly imagination receives it, sense and understanding revolt at it.

F. A. MARSHALL (*Irving Shakespeare*, 1888, *Introd.* ii, 325): As far as the human characters of this play are concerned, with the exception of 'sweet-faced' Nick Bottom and his amusing companions, very little can be said in their praise. Theseus and Hippolyta, Lysander and Hermia, Demetrius and Helena, are all alike essentially uninteresting. Neither in the study, nor on the stage, do they attract much of our sympathy. Their loves do not move us; not even so much as those of Biron and Rosaline, Proteus and Julia, Valentine and Silvia. If we read the play at home, we hurry over the tedious quarrels of the lovers, anxious to assist at the rehearsal of the tragi-comedy of 'Pyramus and Thisbe.' The mighty dispute that rages between Oberon and Titania about the changeling boy does not move us in the the least degree. We are much more anxious to know how Nick Bottom will acquit himself in the tragical scene between Pyramus and Thisbe. It is in the comic portion of this play that Shakespeare manifests his dramatic genius; here it is that his power of charac-terisation, his close observation of human nature, his subtle humour, make themselves felt.

GERMAN CRITICISMS

SCHLEGEL (*Lectures*, &c., trans. by J. BLACK, 1815, ii, 176): The *Midsummer Night's Dream* and *The Tempest* may be in so far compared together, that in both the influence of a wonderful world of spirits is interwoven with the turmoil of human passions and with the farcical adventures of folly. The *Midsummer Night's Dream* is certainly an earlier production; but *The Tempest*, according to all appearance, was written in Shakespeare's later days; hence most critics, on the supposition that the poet must have continued to improve with increasing maturity of mind, have given the last piece a great preference over the former. I cannot, however, altogether agree with them in this; the internal worth of these two works, in my opinion, are pretty equally balanced, and a predilection for the one or the other can only be governed by personal taste. The superiority of *The Tempest* in regard to profound and original characterisation is obvious; as a whole we must always admire the masterly skill which Shakespeare has here displayed in the economy of his means, and the dexter-ity with which he has disguised his preparations, the scaffolding for the wonderful aërial structure In the *Midsummer Night's Dream* again there flows a luxuriant

vein of the boldest and most fantastical invention; the most extraordinary combination of the most dissimilar ingredients seems to have arisen without effort by some ingenious and lucky accident, and the colours are of such clear transparency that we think the whole of the variegated fabric may be blown away with a breath. The fairy world here described resembles those elegant pieces of Arabesque where little Genii, with butterfly wings, rise, half embodied, above the flower cups. Twilight, moonlight, dew, and spring-perfumes are the elements of these tender spirits; they assist Nature in embroidering her carpet with green leaves, many-coloured flowers, and dazzling insects; in the human world they merely sport in a childish and wayward manner with their beneficent or noxious influences. Their most violent rage dissolves in good-natured raillery; their passions, stripped of all earthly matter, are merely an ideal dream. To correspond with this, the loves of mortals are painted as a poetical enchantment, which, by a contrary enchantment, may be immediately suspended and then renewed again. The different parts of the plot, the wedding of Theseus, the disagreement of Oberon and Titania, the flight of the two pair of lovers, and the theatrical operations of the mechanics, are so lightly and happily interwoven that they seem necessary to each other for the formation of a whole. . . . The droll wonder of the transmutation of Bottom is merely the translation of a metaphor in its literal sense; but in his behavior during the tender homage of the Fairy Queen we have a most amusing proof how much the consciousness of such a head-dress heightens the effect of his usual folly. Theseus and Hippolyta are, as it were, a splendid frame for the picture; they take no part in the action, but appear with a stately pomp. The discourse of the hero and his Amazon, as they course through the forest with their noisy hunting-train, works upon the imagination, like the fresh breath of morning, before which the shades of night disappear. *Pyramus and Thisbe* is not unmeaningly chosen as the grotesque play within the play; it is exactly like the pathetic part of the piece, a secret meeting of two lovers in the forest, and their separation by an unfortunate accident, and closes the whole with the most amusing parody.

GERVINUS (*Shakespeare*, Leipzig, 1849, i, 246): Shakespeare depicts [his fairies] as creatures devoid of refined feelings and of morality; just as we too in dreams meet with no check to our tender emotions and are freed from moral impulse and responsibility. Careless and unprincipled themselves, they tempt mortals to be unfaithful. The effects of the confusion which they have set on foot make no impression on them; with the mental torture of the lovers they have no jot of sympathy; but over their blunders they rejoice, and at their fondness they wonder. Furthermore, the poet depicts his fairies as creatures devoid of high intellectuality. If their speeches are attentively read, it will be noted that nowhere is there a thoughtful reflection ascribed to them. On one solitary occasion Puck makes a sententious observation on the infidelity of man, and whoever has penetrated the nature of these beings will instantly feel that the observation is out of harmony. . . . Titania has no inner, spiritual relations to her friend, the mother of the little Indian boy, but merely pleasure in her shape, her grace, and gifts of mimicry.

[Page 252.] In the old Romances of Chivalry, in Chaucer, in Spenser, the Fairies are wholly different creatures, without definite character or purpose; they harmonise with the whole world of chivalry in an unvarying monotony and lack of consistency. Whereas, in the Saxon Elfin-lore, Shakespeare found that which would enable him to cast aside the romantic art of the pastoral poets, and pass over to the rude popular taste of his country-folk. From Spenser's *Faerie Queene* he could learn the melody

of speech, the art of description, the brilliancy of romantic pictures, and the charm
of visionary scenes; but all the haughty, pretentious, romantic devices of this Elfin-
world he cast aside and grasped the little pranks of Robin Goodfellow, wherein the
simple faith of the common people had been preserved in pure and unpretentious
form.   Thus, also, with us, in Germany, at the time of the Reformation, when the
Home-life of the people was restored, the chivalric and romantic conceptions of the
spiritual world of nature, were cast aside and men returned to popular beliefs, and
we can read nothing which reminds us of Shakespeare's Fairy realm so strongly as
the Theory of Elemental Spirits by our own Paracelsus.   [This extraordinary statement should be seen in the original to vindicate the accuracy of the translation : ' man
' kann nichts lesen, was an Shakespeare's Elfenreich so sehr erinnert, wie unseres
' Paracelsus Theorie der Elementargeister.'—ED.]   Indeed, it may be said that from
the time when Shakespeare took to himself the dim ideas of these myths and their
simple expression in prose and verse, the Saxon taste of the common people dominated in him more and more.   In *Romeo and Juliet* and in *The Merchant of Venice*
his sympathies with the one side and with the other are counterbalanced, almost of
necessity, inasmuch as the poet is working exclusively with Italian materials.   But it
was the contemporaneous working on the Historical Plays which first fully and absolutely made the poet native to his home, and the scenes among the common folk in
Henry the Fourth and *Fifth* reveal how comfortably he felt there.

ULRICI (*Shakespeare's Dramatic Art*, vol. ii, p. 72.   Trans. by L. DORA SCHMITZ,
London, 1876, Bohn's ed.) : In the first place, it is self-evident that the play is based
upon the comic view of life, that is to say, upon Shakespeare's idea of comedy.   This
is here expressed without reserve and in the clearest manner possible, in so far as it
is not only in particular cases that the maddest freaks of accident come into conflict
with human capriciousness, folly, and perversity, thus thwarting one another in turn,
but that the principal spheres of life are made mutually to parody one another in
mirthful irony.   This last feature distinguishes *A Midsummer Night's Dream* from
other comedies.   Theseus and Hippolyta appear obviously to represent the grand,
heroic, historical side of human nature.   In place, however, of maintaining their
greatness, power, and dignity, it is exhibited rather as spent in the common every-
day occurrence of a marriage, which can claim no greater significance than it pos
sesses for ordinary mortals; their heroic greatness parodies itself, inasmuch as it
appears to exist for no other purpose than to be married in a suitable fashion.

[P. 74]   Hence A. SCHÖLL (*Blätter für lit. Unterhaltung*, 184) very justly remarks that, ' When Demetrius and Lysander make fun of the candour with which
' these true-hearted *dilettanti* cast aside their masks during their performance, we can-
' not avoid recalling to mind that they themselves had shortly before, in the wood, no
' less quickly fallen out of their own parts.   [See Schlegel, above.—ED.]   When these
' gentlemen consider Pyramus a bad lover, they forget that they had previously been
' no better themselves; they had then declaimed about love as unreasonably as here
' Pyramus and Thisbe.   Like the latter, they were separated from their happiness by
' a wall which was no wall but a delusion, they drew daggers which were as harmless as those of Pyramus, and were, in spite of all their efforts, no better than the
' mechanics, that is to say, they were the means of making others laugh, the elves
' and ourselves.   Nay, Puck makes the maddest game of these good citizens, for
Bottom is more comfortable in the enchanted wood than they.   The merry Puck
' has, indeed, by a mad prank had his laugh over the awkward mechanical and the

lovely fairy queen, but in deceiving the foolish mortals has at the same time deceived himself. For although he, the elf, has driven Lysander and Demetrius and the ter-' rified mechanics about the wood, the elves have, in turn, been unceremoniously sent ' hither and thither to do the errands of Bottom, the ruling favourite of Titania; ' Bottom had wit enough to chaff the small Masters Cobweb, Peaseblossom, and ' Mustard-seed, as much as Puck had chaffed him and his fellows. Thus no party ' can accuse the other of anything, and in the end we do not know whether the mor-' tals have been dreaming of elves, the elves of mortals, or we ourselves of both.' In fact, the whole play is a bantering game, in which all parties are quizzed in turn, and which, at the same time, makes game of the audience as well.

[P. 76.] The marriage festival of Theseus and Hippolyta forms, so to say, a splendid golden frame to the whole picture, with which all the several scenes stand in some sort of connection. Within it we have the gambols of the elves among one another, which, like a gay ribbon, are woven into the plans of the loving couples and into the doings of the mechanics; hence they represent a kind of relation between these two groups, while the blessings, which at the beginning they intended to bestow, and in the end actually do bestow, upon the house and lineage of Theseus make them partakers of the marriage feast, and give them a well-founded place in the drama. The play within the play, lastly, occupies the same position as a part of the wedding festivities. . . .

Human life appears conceived as a fantastic midsummer night's dream. As in a dream, the airy picture flits past our minds with the quickness of wit; the remotest regions, the strangest and most motley figures mix with one another, and, in form and composition, make an exceedingly curious medley; as in a dream they thwart, embarrass, and disembarrass one another in turn, and,—owing to their con-stant change of character and wavering feelings and passions,—vanish, like the figures of a dream, into an uncertain chiaroscuro; as in a dream, the play within the play holds up its puzzling concave mirror to the whole; and as, doubtless, in real dreams the shadow of reason comments upon the individual images in a state of half doubt, half belief,—at one time denying them their apparent reality, at another again, allowing itself to be carried away by them,—so this piece, in its tendency to parody, while flitting past our sight is, at the same time, always criticising itself.

Dr. H. WOELFFEL (*Album des literarischen Vereins in Nürnberg für 1852*, p. 126): If we gather, as it were, into one focus all the separate, distinguishing traits of these two characters [Lysander and Demetrius], if we seek to read the secret of their nature in their eyes, we shall unquestionably find it to be this, viz. in Lysander the poet wished to represent a noble magnanimous nature sensitive to the charms of the loveliness of soul and of spiritual beauty, but in Demetrius he has given us a nature fundamentally less noble; in its final analysis, even unlovely, and sensitive only to the impression of physical beauty. If there could be any doubt that these two characters are the opposites of each other, the poet has in a noteworthy way decided the question. The effect of the same magic juice on the two men is that Demetrius is rendered faithful, Lysander unfaithful—an incontrovertible sign that their natures, like their affections, are diametrically opposite.

This conclusion will be fully confirmed if we consider the two female characters, and from their traits and bearing, their features and demeanour, decipher their natures. Nay, in good sooth, the very names Hermia and Helena seem to corrobo-' rate our view. For, just as Hermes, the messenger of the gods, harmonises heaven

and earth, and, as Horace sings, first brought gentler customs and spiritual beauty to rude primitive man,—so the name Hermia hints of a charm which, born in Heaven, outshines physical beauty, and is as unattainable to common perception as is the sky to him who bends his eyes upon the earth. But since the days of Homer and of Troy, Helen has been the symbol of the charm of earthly beauty. And it is to Lysander that the poet gives Hermia, and to the earthborn Demetrius, Helena.

KREISSIG (*Vorlesungen*, &c., iii, 103, 1862) : When foreigners question the musical euphony of the English language, Englishmen are wont to point to *A Midsummer Night's Dream*, just as we Germans in turn point to the *First Part of Faust*. Such questions do not really admit of discussion. But the most pronounced contemner, however, of the scrunching, lisping, and hissing sounds of English words must be here fairly astonished at the abundance of those genuine beauties, which any good translation can convey, those similes scattered in such original and dazzling wealth, those profound thoughts, those vigorous and lovely expressions, genuine jewels as they are, with which Titania and Oberon seem to have overspread the tinted glittering garment of this delicious story. Note, for instance, the compliment to the ' fair vestal ' throned by the West,' the picture of Titania's bower, the bank whereon the wild thyme blows, the grand daybreak after the night of wild dreams, and, above all, the glorification of the poet by Theseus.

K. ELZE (*Essays*, &c., trans. by L. DORA SCHMITZ, p. 32, 1874) : It is, of course, out of the question to suppose that Jonson's Masques influenced *A Midsummer Night's Dream ;* it could more readily be conceived that the latter exercised an influence upon Jonson. At least in the present play, the two portions, masque and anti-masque, are divided in an almost Jonsonian manner. The love-stories of Theseus and of the Athenian youths,—to use Schlegel's words,—' form, as it were, a ' splendid frame to the picture.' Into this frame, which corresponds to the actual masque, the anti-masque is inserted, and the latter again is divided into the semi-choruses of the fairies (for they too belong to the anti-masque) and the clowns. Shakespeare has, of course, treated the whole with the most perfect artistic freedom. The two parts do not, as is frequently the case in masques, proceed internally unconnected by the side of each other, but are most skilfully interwoven. The anti-masque, in the scenes between Oberon and Titania, rises to the full poetic height of the masque, while the latter, in the dispute between Hermia and Helena, does not indeed enter the domain of the comic, but still diminishes in dignity, and Theseus in the Fifth Act actually descends to the jokes of the clowns. The Bergomask dance performed by the clowns forcibly reminds us of the outlandish nothings of the anti-masque, as pointed out by Jonson. Moreover, we feel throughout the play that like the masques it was originally intended for a private entertainment. The resemblance to the masques is still heightened by the completely lyrical, not to say operatic stamp, of the *Midsummer Night's Dream*. There is no action which develops of internal necessity, and the poet has here, as Gervinus says, ' completely laid aside his great ' art of finding a motive for every action.' . . . In a word, exactly as in the masques, everything is an occurrence and a living picture rather than a plot, and the delineation of the characters is accordingly given only with slight touches. . . . Yet, however imperceptible may be [the transition from masque to anti-masque] Shakespeare's play stands far above all masques, those of Jonson not excepted, and differs from them in

essential poir ts. Above all, it is obvious that Shakespeare has transferred the subject from the domain of learned poetry into the popular one, and has thus given it an imperishable and universally attractive substance. Just as he transformed the vulgar chronicle-histories into truly dramatic plays, so in the *Midsummer Night's Dream* he raised the masque to the highest form of art, as, in fact, his greatness in general consists in having carried all the existing dramatic species to the highest point of perfection. The difference between learned and popular poetry can nowhere appear more distinct than in comparing the present play with Jonson's masques. Jonson also made Oberon the principal character of a masque,—but what a contrast! Almost all the figures, all the images and allusions, are the exclusive property of the scholar, and can be neither understood by the people nor touch a sympathetic chord in their hearts. In the very first lines two Virgilian satyrs, Chromis and Mnasil, are introduced, who, even to Shakespeare's best audience, must have been unknown and unintelligible, and deserved to be hissed off the stage by the groundlings. Hence Jonson found it necessary to furnish his masques with copious notes, which would do honour to a German philosopher; Shakespeare never penned a note. Shakespeare in *A Midsummer Night's Dream* by no means effaced the mythological background, and the fabulous world of spirits peculiar to the masque, but has taken care to treat it all in an intelligible and charming manner. . . . Most genuinely national, Shakespeare shows himself in the anti-masque; whose clowns are no sylvans, fauns, or cyclops, but English tradesmen such as the poet may have become acquainted with in Stratford and London,—such as performed in the ' Coventry Plays.'

W. Oechelhäuser (*Einführungen in Shakespeare's Bühnen-Dramen*, 2te Aufl. 1885, ii, 277) [After quoting with approval Ulrici's theory, given above, that this play is a succession of parodies, the author, who is widely known as the advocate of a correct representation of Shakespeare's plays on the stage, continues:] In the word *parody* is the key to the only true comprehension and representation of the *Summernight's Dream;* but observe, there must be no attempt at a mere comic representation of love, least of all at a representation of true, genuine love, but at a *parody of love.* Above all, there is *nothing* in the whole play which is to be taken seriously; *every action and situation in it is a parody, and all persons, without exception, heroes as well as lovers, fairies as well as clowns, are exponents of this parody.*

In the midst of fairies and clowns there is no place for a serious main action. But if this be granted, then (and this it is which I now urge) let the true coloring be given to the main action when put upon the stage, and let it not, as has been hitherto the case, vaguely fluctuate between jest and earnest.

[P. 279.] There is, perhaps, no other piece which affords to managers and to actors alike, better opportunities for manifold comic effects and for a display of versatility than this very *Summernight's Dream.* It need scarcely be said that my interpretation of this tendency of the piece to parody does not contemplate a descent to low comicality, to a parody *à la* Offenbach.

If, accordingly, in the light of this interpretation, we consider more closely the presentation of the different characters, we shall find that the rôle of Duke Theseus does not in the main demand any especial exaggeration. The dignified and benevolent words which the poet, especially in the Fifth Act, puts in his mouth must be in harmony with the exterior representation of the rôle. The enlivening effect will be perceived readily enough without any aid from Theseus, as a reflex of the whole situation wherein he is placed. The old, legendary, Greek hero bears himself like an honour-

able, courteous, and, in spite of his scoffings at lovers, very respectably enamoured *bonhomme ;* of the Greek or of the Hero, nothing but the name.

An exaggeration, somewhat more pronounced than that of Theseus is required for the Amazonian queen Hippolyta. Here the contrast between classicality and an appearance in Comedy is more striking; moreover there are various indications in the play which lead directly to the conclusion that the poet intended to give this rôle a palpably comic tone. The jealous Titania speaks of her derisively as 'the bouncing 'Amazon, Your buskin'd mistress and your warrior love' [or, as it is given, very inadequately, in Schlegel's translation : 'Die Amazone, Die strotzende, hochaufge 'schürzte Dame, Dein Heldenliebchen.' It is needless to note that there is no trace here of 'buskin'd,' and that in the word substituted for it there is a vulgarity which no jealous fit could ever extort from Titania's refined fairy mouth. *Strotzende* does duty well enough for 'bouncing,' albeit Oechelhäuser would substitute for it, *fett, quatschelig*, 'fat, dumpy,' in which there is only a trace of 'bouncing.'—ED.] . . The rôles of Theseus and Hippolyta acquire the genuine and befitting shade of comicality, when they are represented as a stout middle-aged pair of lovers, past their maturity, for such was unquestionably the design of the poet, and was in harmony with their active past life. The words of Titania, just quoted, refer to that corporeal superabundance which is wont to accompany mature years. But Theseus always speaks with the sedateness of ripe age. The mutual jealous recriminations of Oberon and Titania acquire herein the comic coloring which was clearly intended; thus too the amorous impatience of the elderly lovers which runs through the whole piece.

Utterly different from this must the tendency to parody be expressed in the acts and words of the pairs of youthful lovers. First of all, every actor must rid himself of any preconceived notion that he is here dealing with ideal characters, or with ordinary, lofty personages of deep and warm feelings. Here there is nought but the jesting parody of love's passion. . . . One of Hermia's characteristics is lack of respect for her father, who complains of her 'stubborn harshness'; as also her pert questions and answers to the Duke, whose threats of death or enduring spinsterhood she treats with open levity, and behind the Duke's back snaps her fingers at both of them. . . .

[P. 283] Actresses, therefore, need not fruitlessly try to make two fondly and devotedly loving characters out of Hermia or Helena, or hope to cloak Helena's chase after Demetrius in the guise of true womanliness; it is impossible and will only prove tedious. . . .

[P. 285] There is a rich opportunity in Hermia's blustering father, Egeus. Here the colours should be well laid on. It is plain that Theseus is merely making merry with him when he says to Hermia : 'To you your father should be as a god,' &c.; and to Egeus's appeals Theseus responds merely jocosely, as Wehl observed. [See Wehl's description of the first performance of this play in Berlin, *post.*—ED.]

[P. 287] As regards the Interlude, the colours may be laid heavily on the Artisans, but nothing vulgar in acting or movement, especially in the dance at the close, must be tolerated. Their most prominent trait is naïveté; not the smallest suspicion have they of their boorishness; the more seriously they perform, the more laughable are they. . . . The spectators on the stage of the Interlude must fall into the plan and accompany the clowns' play with their encouragement and applause. For the public at large there lies in this clowns' comedy the chief attraction of the piece.

NOTABLE PERFORMANCES

FEODOR WEHL (*Didaskalien*, Leipsic, 1867, p. 2): When Tieck, in the hey-day of his life, was in Dresden, he pleaded enthusiastically for a performance of the 'Summernight's Dream.' But actors, managers, and theatre-goers shook their heads. 'The thing is impossible,' said the knowing ones. 'The idea is a chimera,—a dream 'of Queen Mab,—it can never be realised.'

Tieck flung himself angrily back in his chair, and held his peace.

Years passed by.

At last Tieck was summoned to Berlin, to the Court of Friedrich Wilhelm the Fourth, and among the pieces of poetry which he there read to attentive ears was Shakespeare's 'Summernight's Dream.' At the conclusion of the reading, which had given the keenest delight to the illustrious audience, the King asked: 'Is it really 'a fact that this piece cannot be performed on the stage?

Tieck, as he himself often afterwards humourously related, was thunderstruck. He felt his heart beat to the very tip of his tongue, and for a minute language failed him. For more than twenty years, almost a lifetime, his cherished idea had been repelled with cold opposition, prosaic arguments, or sympathetic shrugs. And now a monarch, intellectual and powerful, had asked if the play could not be performed! Tieck's head swam; before his eyes floated the vision of a fulfillment, at the close of his life, of one of the dearest wishes of his heart. 'Your majesty!' he cried at last, 'Your majesty! If I only had permission and the means, it would make the most 'enchanting performance on earth!'

'Good then, set to work, Master Ludovico,' replied Friedrich Wilhelm, in his pleasant, jesting way. 'I give you full power, and will order Kuestner (the Superintendent at that time of the Royal Theatre) to place the theatre and all his soupes (actors) at your disposal.'

It was the happiest day of Ludwig Tieck's life! The aged poet, crippled with rheumatism, reached his home, intoxicated with joy. The whole night he was thinking, pondering, ruminating, scene-shifting. The next day he arranged the Comedy, read it to the actors who were to take part in it, and consulted with FELIX MENDELSSOHN BARTHOLDY about the needful music.

The aged Master Ludwig was rejuvenated; vanished were his years, his feebleness, his valetudinarianism. Day after day he wrote, he spoke, he drove hither and thither,—his whole soul was in the work which he was now to make alive.

At last the day came which was to reveal it to the doubting and astonished eyes of the public. And what a public! All that Berlin could show of celebrities in Science, in Art, in intellect, in acknowledged or in struggling Authorship, in talent, in genius, in beauty, and grace,—all were invited to the royal palace at Potsdam, where the first representation was to take place.

The present writer was so fortunate as to be one of the invited guests, and never can he forget the impression then made on him.

The stage was set as far as possible in the Old English style, only, as was natural, it was furnished in the most beautiful and tasteful way. In the Orchestra stood Mendelssohn, beaming with joy, behind him sat Tieck, with kindling looks, handsome, and transfigured like a god. Around was gathered the glittering court, and in the rear the rising rows of invited guests.

What an assemblage! There sat the great Humboldt, the learned Boekh, Bach-

mann, the historians Raumer and Ranke, all the Professors of the University, the poets Kopisch, Kugler, Bettina von Arnim, Paalzow, Theodor Mundt, Willibad Alexis, Rellstab, Crelinger, Varnhagen von Ense, and the numberless host of the other guests

It was a time when all the world was enthusiastic over Friedrich Wilhelm the Fourth. His gift as a public speaker, his wit, his love and knowledge of Art had charmed all classes, and filled them with hope. All hearts went out to meet him as he entered, gay, joyous, smiling, and took his place among the guests.

Verily, we seemed transported to the age of Versailles in the days of the Louises. It was a gala-day for the realm, fairer and more brilliant than any hitherto in its history.

What pleasure shone in all faces, what anticipation, what suspense! An eventful moment was it when the King took his seat, and the beaming Tieck nodded to his joyous friend in the Orchestra, and the music began, that charming, original, bewitching music which clung so closely to the innermost meaning of the poetry and to the suggestions of Tieck. The Wedding March has become a popular, an immortal composition; but how lovely, how delicious, how exquisite, and here and there so full of frolic, is all the rest of it! With a master's power, which cannot be too much admired, Mendelssohn has given expression in one continuous harmony to the soft whisperings of elves, to the rustlings and flutterings of a moonlit night, to all the enchantment of love, to the clumsy nonsense of the rude mechanicals, and to the whizzings and buzzings of the mad Puck.

How it then caught the fancy of that select audience! They listened, they marvelled, they were in a dream!

And when at last the play fairly began, how like a holy benediction it fell upon all, no one stirred, no one moved, as though spellbound all sat to the very last, and then an indescribable enthusiasm burst forth, every one, from the King down to the smallest authorkin, applauded and clapped, and clapped again.

Take it for all in all, it was a day never to be forgotten, it was a day when before the eyes of an art-loving monarch, a poet revealed the miracle of a representation, and superbly proved that it was no impossibility to those who were devoted to art. In this 'Summernight's Dream' the elfin world seemed again to live; elves sprang up from the ground, from the air, from the trees, from the flowers! they fluttered in the beams of the moon! Light, shade, sound, echo, leaves and blooms, sighings and singings, and shoutings for joy! everything helped to make the wonder true and living!

Not for a second time can the like be seen.

It was the highest pinnacle of the reign of Friedrich Wilhelm the Fourth. Who could have dreamt that behind this glittering play of poetic fancy there stood dark and bloody Revolution, and fateful Death? Yet it was even so!

[After sundry suggestions as to the modulation of the voice when Mendelssohn's music accompanies the performance on the stage, Wehl gives the following extraordinary interpretation, p. 15: 'The actor who personates Theseus must have a joyous, gracious bearing. When he threatens Hermia with death or separation from the society of man, in case of her disobedience to her father, he must speak in a roguish, humorous style, and not in the sober earnestness with which the words are usually spoken.' The inference is fair that Wehl is reporting the style of Theseus's address as it was given at this celebrated performance under Tieck's direction. OECHEL-HÄUSER, as we have seen above, approves of this interpretation.—ED.]

TH. FONTANE (*Aus England*, Stuttgart, 1860, p. 49) gives an elaborate description, scene by scene, of the revival of this play by CHARLES KEAN. The most noteworthy item is, perhaps, his account of Puck who ' grows out of the ground on a
' toadstool.' ' Puck was acted by a child, a blond, roguish girl, about ten years old.
' This was well devised and accords with the traditional ideas of Robin Goodfellow.
' The Costume was well chosen: dark brownish-red garment, trimmed with blood-
' red moss and lichens; a similar crown was on the blond somewhat dishevelled hair.
' Arms thin and bare and as long as though she belonged to the Clan Campbell,
' whose arms reach to the knees. In theory I am thoroughly agreed with this way
' of representing Puck, but in practice there will be always great difficulties. This
' ten year old Miss Ellen Terry was a downright intolerable, precocious, genuine
' English ill-bred, unchildlike child. Nevertheless the impression of her mere
' appearance is so deep that I cannot now imagine a grown up Puck, with a full neck
' and round arms. Let me record the way in which, on two occasions when he has to
' hasten, Puck disappeared. The first time he seemed to stand upon a board which
' with one sudden pull, jerked him behind the coulisse; the second time he actually
' flew like an arrow through the air. Both times by machinery.' [No one can bear
an allusion to her salad days, her extremely salad days, with better grace than she
who has been ever since those days so hung upon with admiration and applause.
ED.]

In the *Introduction* to the edition of this play illustrated by J. MOYR SMITH (London, 1892, p. xii), there are full accounts of the setting on the stage at the representations by Mr. PHELPS, at Sadler's Wells, by Mr. CHARLES CALVERT, at Manchester,
and by Mr. BENSON at the Globe Theatre in London. From the account of the first
of these we learn that with Mr. Phelps was associated Mr. FREDERIC FENTON as scenic
artist. The latter says: ' In those days ' [the date is nowhere given], ' lighting was
' a serious difficulty. Very few theatres were enabled to have gas. When Phelps
' and Greenwood took the management into their hands, the lighting of Sadler's
' Wells was merely upright side-lights, about six lamps to each entrance, which were
' placed on angular frames, and revolved to darken the stage; no lights above.
' When set pieces were used, a tray of oil lamps was placed behind them, with
' coloured glasses for moonlight. For the footlights (or floats) there was a large pipe,
' with two vases at each end, with a supply of oil to charge the argand burners on
' the pipes; it was lowered out between the acts, to be trimmed as necessity required.
' . . . I obtained permission for the gas to be supplied as a permanent lighting for the
' theatre, and it was used for the first time in *A Midsummer Night's Dream*. With
' its introduction the smell of oil and sawdust, which was the prevailing odour of all
' theatres, was finally removed. . . . The effect of movement was given by a diorama
' —that is, two sets of scenes moving simultaneously. . . . For the first time used, to
' give a kind of mist, I sent to Glasgow expressly for a piece of blue net, the same
' size as the act-drop, without a seam. This after the first act, was kept down for the
' whole performance of the Dream, light being on the stage sufficient to illuminate
' the actors behind it.' In addition to this diaphanous blue net, other thicknesses of
gauze, partly painted, were used occasionally to deepen the misty effect, and to give
the illusion necessary when Oberon tells Puck to ' overcast the night.'

WILLIAM WINTER (*Old Shrines and Ivy*, 1892, p. 173): The attentive observer
of the stage version made by AUGUSTIN DALY.—and conspicuously used by him

when he revived [this play] at his Theatre on January 31, 1888,—would observe that much new and effective stage business was introduced. The disposition of the groups at the start was fresh, and so was the treatment of the quarrel between Oberon and Titania, with the disappearance of the Indian child. The moonlight effects, in the transition from Act II. to Act III. and the gradual assembly of goblins and fairies in shadowy mists through which the fire-flies glimmered, at the close of Act III., were novel and beautiful. Cuts and transpositions were made at the end of Act IV. in order to close it with the voyage of the barge of Theseus, through a summer land-scape, on the silver stream that rippled down to Athens. The Third Act was judi-ciously compressed, so that the spectator might not see too much of the perplexed and wrangling lovers. But little of the original text was omitted. The music for the choruses was selected from various English composers,—that of Mendelssohn being prescribed only for the orchestra.

COSTUME

KNIGHT (*Introductory Notice*, p. 333): For the costume of the Greeks in the heroical ages we must look to the frieze of the Parthenon. It has been justly remarked (*Elgin Marbles*, p. 165) that we are not to consider the figures of the Par-thenon frieze as affording us ‘a close representation of the national costume,’ har-mony of composition having been the principal object of the sculptors. But, never-theless, although not one figure in all the groups may be represented as fully attired according to the custom of the country, nearly all the component parts of the ancient Greek dress are to be found in the frieze. Horsemen are certainly represented with no garment but the chlamys, according to the practice of the sculptors of that age; but the tunic which was worn beneath it is seen upon others, as well as the cothurnus, or buskin, and the petasus, or Thessalian hat, which all together completed the male attire of that period. On other figures may be observed the Greek crested helmet and cuirass; the closer skull-cap, made of leather, and the large circular shield, &c. The Greeks of the heroic ages wore the sword under the left arm-pit, so that the pommel touched the nipple of the breast. It hung almost horizontally in a belt which passed over the right shoulder. It was straight, intended for cutting and thrusting, with a leaf-shaped blade, and not above twenty inches long. It had no guard, but a cross bar, which, with the scabbard, was beautifully ornamented. The hilts of the Greek swords were sometimes of ivory and gold. The Greek bow was made of two long goat's horns fastened into a handle. The original bowstrings were thongs of leather, but afterwards horse-hair was substituted. The knocks were gen-erally of gold, whilst metal and silver also ornamented the bows on other parts. The arrow-heads were sometimes pyramidal, and the shafts were furnished with feathers. They were carried in quivers, which, with the bow, were slung behind the shoulders. Some of these were square, others round, with covers to protect the arrows from dust and rain. Several which appear on fictile vases seem to have been lined with skins. The spear was generally of ash, with a leaf-shaped head of metal, and furnished with a pointed ferrule at the butt, with which it was stuck in the ground,—a method used, according to Homer, when the troops rested on their arms, or slept upon their shields. The hunting-spear (in Xenophon and Pollux) had two salient parts, sometimes three crescents, to prevent the advance of the wounded animal. On the coins of Ætolia is an undoubted hunting-spear.

The female dress consisted of the long sleeveless tunic (*stola* or *calasiris*), or a tunic with shoulder-flaps almost to the elbow, and fastened by one or more buttons down the arm (*axillaris*). Both descriptions hung in folds to the feet, which were protected by a very simple sandal (*solea* or *crepida*). Over the tunic was worn the *peplum*, a square cloth or veil fastened to the shoulders, and hanging over the bosom as low as the zone (*tænia* or *strophium*), which confined the tunic just beneath the bust. Athenian women of high rank wore hair-pins (one ornamented with a cicada, or grasshopper, is engraved in Hope's *Costume of the Ancients*, plate 138), ribands or fillets, wreaths of flowers, &c. The hair of both sexes was worn in long, formal ringlets, either of a flat and zigzagged, or of a round and corkscrew shape.

The lower orders of Greeks were clad in a short tunic of coarse materials, over which slaves wore a sort of leathern jacket, called *dipthera ;* slaves were also dis tinguished from freemen by their hair being closely shorn.

The Amazons are generally represented on the Etruscan vases in short embroidered tunics with sleeves to the wrist (the peculiar distinction of Asiatic or barbarian nations), pantaloons, ornamented with stars and flowers to correspond with the tunic, the *chlamys*, or short military cloak, and the Phrygian cap or bonnet. Hippolyta is seen so attired on horseback contending with Theseus. Vide Hope's *Costumes.*

E. W. GODWIN, F. S. A. (*The Architect*, 8 May, 1875): In affixing an approximate date for the action, I see no reason why [this play] should not be considered as wholly belonging to its author's time. The proper names . . . are no doubt eminently Greek, but the woods where Hermia and Helena 'upon faint primrose beds were ' wont to lie' are as English as the Clowns and the Fairies, than which nothing can be more English. The fact that Theseus refers to his battle with the Amazons, . . . although strictly in accordance with the classic legend, is hardly sufficient to weigh down the host of improbabilities that crowd the stage when this play is produced with costume, &c., in imitation of Greek fashions. Again, when Theseus talks of the livery of a nun, shady cloisters, and the like, he is of course distinctly referring to the votaries of Diana; and when the ladies and gentlemen swear they swear by pagan deities, although the names they give are Roman. But Puck and Bottom,— nay, even tall Helena and proud Titania,—each is quite enough to overweigh the Greek element in the play. Still, if it must be produced with classic accessories, we should do well to be true to the little there is of classic reference. Thus, although Theseus, in the heroic character we have of him, may be a myth, still the connection of his name with that of fair Helen of Troy brings the man within the range of archæology. And thus we should be led to place his union with Hippolyta only a few years before the siege of Troy If then the play of *A Midsummer Night's Dream* must needs be acted, and if it must needs be classically clothed,—and there are many reasons against both *ifs*,—the architecture, costume, and accessories may very well be the same as those in *Troilus and Cressida.* One thing is, or ought to be, quite clear, and that is that the Acropolis of Athens, as we know it, with its Parthenon, Erectheium, and Propylea, has just about as much relation to the Greeks of the time of Ulysses or Theseus as the Reform Club has to King John. We have, indeed, to travel back, not merely beyond the time of the Parthenon (438–420 B. C.), or beyond that of its predecessor (650 B. C.), but beyond the days of Hesiod and Homer (900 B. C.), past the Dorian conquest of the Achaians in Peloponnêsos, and so higher up the stream of time until we reach the early period of the Pelasgic civilisation. . . . I would accept the period 1184–900 B. C. in preference to any later or earlier

time as that wherein to seek the architecture and costume of the two plays above mentioned.

A Room in the Palace of Theseus is the only architectural scene in *A Midsummer Night's Dream,* and for the character of this interior we must turn to Assyria and Persepolis, to the descriptions of Solomon's Temple and house of the Forest of Lebanon (1005 B. C.), and the fragments of Mycenæ and other Pelasgic towns. . . .

[15 May, 1875.] The costume of Greeks and Trojans in that wide-margined period of time that I selected for the action of *Troilus and Cressida, i. e.* 1184–900 B. C., is by no means ready to our hands. . . . Although the earliest figure-painted vessels in the First Vase-Room of the [British] Museum may not take us further back than 500 B. C., and the sculptures of the Temple at Ægina may lead us certainly to no earlier period, yet by taking these as our *point de départ,* and so going up the stream of time until we reach the North-west palace at Nimroud, c. 900 B. C., we may, by the collateral assistance of Homer and Hesiod, together with such evidence as may be derived from Keltic remains, be enabled to arrive at something like a possible, if not probable, conclusion as to the costume of Achaians and Trojans in the Heroic days. . . . As to the several articles of dress, the *Iliad* supplies us with minute particulars, and from these we learn that the full armour, which was mostly made of brass, consisted of:—1, the helmet; 2, the thorax or cuirass over a linen vest; 3, the cuissots or thigh-pieces, and 4, the greaves; no mention is anywhere made of the leather, felt, or metal straps which we find depending from the lower edge of the cuirass in the armed figures on vases of a much later period. Of belts we have three kinds, the zone or waist belt, the sword belt, and the shield belt. Besides the sword and shield we have the spear, the bow, and the iron-studded mace, which last is very suggestive of the *morning-star* or *holy-water-sprinkler* of mediæval armouries. The men wore the hair long, and their skin was brown. The costume of the other sex seems to have depended for its effect not so much on quantity as on quality, and more than anything else on the proportion, articulation, and undulation of the splendour of human form. The chiton or tunic, the broad zone, the diplax, pallium or mantle sweeping the ground, the peplos or veil, the sandals, and the head-dress formed a complete toilette. Among their personal ornaments were ear-rings, diadems, or frontals, chains, brooches, and necklaces.

And now turn to the actors in this drama. Taking the Greeks first, we have Achilles presented to us as golden-haired; his sceptre is starred with gold studs; his greaves are of ductile tin; his cuissots are of silver; his cuirass of gold; his four-fold helm of sculptured (*repoussé*) brass with a golden crest of horsehair *gilded;* his shield of gold, silver, brass, and tin divided by concentric rings, each divided into four compartments; his sword is of bronze, starred with gems; and his baldrick is embroidered in various colours. Agamemnon wears, when unarmed, a fine linen vest, a purple mantle, embroidered sandals, and a lion's skin at night over his shoulders. When armed he wears a four-fold helm with horsehair plume; greaves with silver buckles; a wonderful cuirass composed of ten rows of azure steel, twenty of tin, and twelve of gold, with three dragons rising to the neck; a baldrick radiant with embroidery; a sword with gold hilt, silver sheath, and gold hangers; a broad belt with silver plates; and a shield of ten concentric bands or zones of brass, with twenty bosses and a Gorgon in the midst. Menelaus wore a leopard's skin at night. Old Nestor's mantle is of soft, warm wool, doubly lined; his shield is of gold, and he wears a scarf of divers colours. . . . Ajax is clothed in steel and carries a terrific mace, crowned with studs of iron, whilst Patroclus wears brass, silver buckled a

flaming cuirass of a thousand dyes, a sword studded with gold, and a sword-belt like a starry zone. On the Trojan side, we see Hektor with a shield *reaching from neck to ankle;* a plume or crest of white and black horsehair; a brass cuirass and spears about sixteen feet long. Paris, in curling golden tresses, comes before us in gilded armour, buckled with silver buckles; his thigh-pieces are wrought with flowers; his helmet is fastened by a strap of tough bull-hide; a leopard's skin he wears as a cloak, and his bow hangs across his shoulders. Of the fair Helen Homer says but little. . . . We see her pass out of the palace, attended by her two hand-maidens, her face and arms covered by a thin white peplos, her soft white chiton tucked up through the gold zone beneath her swelling bosom, and her embroidered diplax fastened with clasps of gold, whilst both peplos and diplax fall in multitud-inous folds until they lose themselves in a train of rippling waves. . . .

Such then is the evidence we gather from Homer as to the costume of *Troilus and Cressida* [as Godwin before remarked, it is the same for *A Midsummer Night's Dream*]; Hesiod, in so far as he refers to costume, confirms it. . . .

For the women's armlets, bracelets, necklaces, and earrings; for the woven pat-terns, and the embroidered borders of the square mantle and the chiton, we cannot be far wrong if we seek in the sculptures of the reign of Assur-nazir-pal (c. 880 B. C.). Necklaces of beads and of numerous small pendants might be used, if preferred, instead of the bolder medallion necklace. The twisted snake-like form as well as the single medallion may be used for bracelets. The hair was rolled and confined within a caul or net, made of coloured or gold thread, and a fillet not unusually of thin fine gold bound the base of the net. This fillet, in the cases of very important ladies, might expand into a frontal or diadem of thin gold, bent round the forehead from ear to ear and decorated with very delicate *repoussé* work.

PETER SQUENTZ

HALLIWELL (*Introd.*, Folio ed. 1856, p. 12): Bottom appears to have been then considered the most prominent character in the play; and 'the merry conceited ' humours of Bottom the Weaver,' with a portion of the fairy scenes, were extracted from the *Midsummer Night's Dream*, and made into a farce or droll (*The Merry conceited Humours of Bottom the Weaver, as it hath been often publikely acted by some of his Majesties Comedians, and lately privately presented by several apprentices for their harmless recreation, with great applause*, 4to, Lond. 1661), which was very frequently played ' on the sly,' after the suppression of the theatres. ' When the pub-' lique theatres were shut up,' observes Kirkman, ' and the actors forbidden to present ' us with any of their tragedies, because we had enough of that in earnest; and com-' edies, because the vices of the age were too lively and smartly represented; then ' all that we could divert ourselves with were these humours and pieces of plays, ' which passing under the name of a merry conceited fellow called Bottom the ' Weaver, Simpleton the Smith, John Swabber, or some such title, were only allowed ' us, and that but by stealth too, and under pretence of rope-dancing and the like.'— *The Wits*, 1673, an abridgement of Kirkman's *Wits, or Sport upon Sport*, 1672. Both these contain *The Humours of Bottom the Weaver*, in which Puck is transformed by name into Pugg. [In the *Dramatis Personæ* are instances of the ' doubling ' of characters, e. g. '*Oberon*, King of the Fairies, who likewise may present the Duke. ' *Titania* his Queen, the Dutchess. *Pugg.* A Spirit, a Lord. *Pyramus, Thisbe, Wall.* Who likewise may present three Fairies.'—ED.]

TIECK (*Deutsches Theater*, Berlin, 1817, ii, xvi) suggests that the foregoing Droll had, by some means, found its way to Germany, and was there translated for the stage, and brought out at Altdorf, by DANIEL SCHWENTER; 'Titania was omitted, ' Bottom changed into Pickleherring, and much added to the fun, and many phrases ' literally retained from Shakespeare, with whose play he was not acquainted.'

VOSS (*Trans.*, 1818, i, 506) thinks that Schwenter might have adopted some old legend of Folk-lore. But the literalness with which Shakespeare's words are translated renders this impossible, unless Shakespeare went to the same source.

ALBERT COHN (*Shakespeare in Germany*, 1865, p. cxxx) denies that Schwenter could have translated *The Merry Conceited Humours of Bottom*, which was not printed till 1660; Schwenter died in 1636. 'Nothing can be more probable,' says Cohn, 'than that Shakespeare's piece was brought to Germany by the English Come-' dians. Such a farce must have been especially suitable to their object. That the ' whole of the *Midsummer Night's Dream* belonged to the acting stock of the Come-' dians is very unlikely. On the contrary, they probably took from it only the comedy ' of the clowns, as may also have been done occasionally in England.'

Argument on this point is, however, somewhat superfluous, seeing that no copy of Schwenter's work has survived. Indeed all we know of it is derived from Gryphius, one of Germany's earliest dramatists, who in 1663 issued, ABSURDA COMICA, *Or Herr Peter Squentz. A Pasquinade by Andreas Gryphius*, and from the 'Address to ' the Reader,' we might be permitted to doubt (if the whole question were of any moment) whether any fragment even of Schwenter's work has survived in Gryphius's *Absurda Comica*. There need be no clashing of dates between *The Merry Conceited Humours* in 1660 and the *Absurda Comica* in 1663, and there can be no question that the latter is taken from the former. The only writer, as far as I know, who denies that Shakespeare was copied, is Dr W. Bell, who promises (*Shakespeare's Puck*, &c, 1864, iii, 181) that he will 'bring historical proof of a German origin of a ' very early date,' but I can nowhere find his promise explicitly fulfilled.

Tieck reprinted Gryphius's pasquinade in his *Deutsches Theater* (ii, 235). The address 'to the Most gracious and Highly honoured Reader' is as follows :—' Herr ' Peter Squentz, a man no longer unknown in Germany, and greatly celebrated in his ' own estimation, is herewith presented to you. Whither or not his sallies are as ' pointed, as he himself thinks, they have been hitherto in various theatres received ' and laughed at, with especial merriment by the audience, and, in consequence here ' and there, wits have been found who, without shame or scruple, have not hesitated ' to claim his parentage. Wherefore, in order that he may be no longer indebted to ' strangers, be it known that Daniel Schwenter, a man of high desert throughout Ger-' many, and skilled in all kinds of languages and in the mathematics, first introduced ' him on the stage at Altdorff, whence he travelled further and further until at last he ' encountered my dearest friend, who had him better equipped, enlarged by more ' characters, and subjected him, alongside of one of his own tragedies, to the eyes and judgement of all. But inasmuch as this friend, engrossed by weightier matters, subsequently quite forgot him, I have ventured to summon Herr Peter Squentz from the shelves of my aforesaid friend's library, and to send him in type to thee my most gracious and highly honoured reader; if thou wilt accept him with favour thou mayest forthwith expect the incomparable *Horribilicribrifax*, depicted by the same

' brush to which we owe the latest strokes on the perfected portrait of Peter Squentz,
' and herewith I remain thy ever devoted

' PHILIP-GREGORIO RIESENTOD.'

As we are here concerned only in detecting the traces of Shakespeare, it suffices to say that in the *Absurda Comica* there is nothing of the plot of *A Midsummer Night's Dream*, and that an Interlude of *Pyramus and Thisbe* is acted before King Theodore, and Cassandra, his wife, Serenus, the Prince, Violandra, the Princess, and Eubulus, the Chamberlain. The meaningless name, *Peter Squentz*, is clearly Shakespeare's *Peter Quince*, adopted apparently in ignorance that ' Quince ' is the name of the fruit, which in German is *Quitte*. The *Dramatis Personæ*, other than those just mentioned, are :—

Herr Peter Squentz, *Writer and Schoolmaster in Rumpels-Kirchen,*

	Prologus *and* Epilogus.
Pickleherring, *the King's merry counsellor,*	Piramus.
Meister Krix-over-and-over-again, *Smith,*	the Moon.
Meister Bulla Butain, *Bellowsmaker,*	Wall.
Meister Klipperling, *Joiner,*	Lion.
Meister Lüllüger, *Weaver and Head Chorister,*	Fountain.
Meister Klotz-George, *Bobbin-maker,*	Thisbe.

In this list ' Bulla Butain ' is of itself quite sufficient to stamp the play as an adaptation from Shakespeare.

In the first scene Peter Squentz unfolds the story of Pyramus and Thisbe ' as told ' by that pious father of the church, Ovidius, in his *Memorium phosis*,' and while he is distributing the characters Pickleherring asks : ' Does the lion have much to speak ?'

Peter Squentz. No, the lion has only to roar.

Pickleherring. Aha, then I will be the lion, for I am not fond of learning things by heart.

Peter Squentz. No, no ! Mons. Pickleherring has to act the chief part.

Pickleherring. Am I clever enough to be a chief person ?

Peter Squentz. Of course. But as there must be a noble, commanding, dignified man for the *Prologus* and *Epilogus*, I will take that part. . . .

Klip. Who must act the lion, then ? I think it would suit me best, because he hasn't much to say.

Kricks. Marry, I think it would sound too frightful if a fierce lion should come bounding in, and not say a word. That would frighten the ladies too horribly.

Klotz. There I agree with you. On account of the ladies you ought to say right off that you are no real lion at all, but only Klipperling, the joiner.

Pickleherring. And let your leather apron dangle out through the lion's skin. . . .

Klipperling. Never you mind, never you mind, I will roar so exquisitely that the King and Queen will say, ' dear little lionkin, roar again !'

Peter Squentz. In the meanwhile let your nails grow nice and long, and don't shave your beard, and then you will look all the more like a lion,—so that *difficultet* is over. But there's another thing; the water of my understanding will not drive the mill wheels of my brain :—the father of the church, Ovidius, writes that the moon shone, and we do not know whether the moon shines or not when we play our play.

Pickleherring. That's a hard thing.

Kricks. That's easily settled; look in the Calendar and see if the moon shines on that day.

22

Klotz. Yes, if we only had one.

Lollinger. Here, I have one. . . . Hi there, Squire Pickleherring, you understand Calendars, just look and see if the moon will shine.

Pickleherring. All right, all right, gentlemen, the moon will shine when we play. . . .

Kricks. Hark ye, what has just occurred to me. I'll tie some faggots round my waist, and carry a light in a lanthorn, and represent moon. . . .

Peter Squentz. What shall we do for a wall? . . . Piramus and Thisbe must speak through a hole in the wall.

Klipperling. I think we had better daub a fellow all over with mud and loam, and have him say that he is Wall. . . .

Peter Squentz. Squire Pickleherring you must be Pyramus.

Pickleherring. Perry must [*Birnen Most*]? what sort of a chap is that?

Peter Squentz. He is the most gentlemanlike person in the whole play—a *chevalieur*, soldier, and lover. . . .

Peter Squentz. Where shall we find a Thisbe?

Lollinger. Klotz-George can act her the best. . . .

Peter Squentz. No that won't do at all. He has a big beard. . . .

Bullabutain. You must speak small, small, small.

Klotz. Thissen [*Also?*]?

Peter Squentz. Smaller yet.

Klotz. Well, well, I'll do it right. I'll speak so small and lovely that the King and Queen will just dote on me. . . .

Peter Squentz. Gentlemen, con your parts diligently, I will finish the Comedy to-morrow, and you will get your parts, therefore, day after tomorrow.

The foregoing affords ample evidence of the source whence came *Peter Squentz.* Throughout the rest of the play there are sundry whiffs of Shakespeare, but it would be time wasted either to point them out or to read them.

JOHN SPENCER

COLLIER (*Annals of the Stage*, i, 459, 2d ed. 1879): In the autumn of 1631 a very singular circumstance occurred, connected with the history of the stage. Unless the whole story were a malicious invention by some of the many enemies of John Williams, then Bishop of Lincoln (who, previous to his disgrace, had filled the office of Lord Keeper), he had a play represented in his house in London, on Sunday, September 27th. The piece chosen, for this occasion, at least did credit to his taste, for it appears to have been Shakespeare's *Midsummer Night's Dream*,* and it was got up as a private amusement. The animosity of Laud to Williams is well known, and in the Library at Lambeth Palace is a mass of documents referring to different charges against him, thus indorsed in the handwriting of Laud himself: 'These papers concerning the Bp. of Lincoln wear delivered to me bye his Majesty's command.' One of them is an admonitory letter from a person of the name of John Spencer (who seems to have been a puritanical preacher), which purports to have been addressed

---

* One of the actors exhibited himself in an Ass's head, no doubt in the part of Bottom, and in the margin of the document relating to this event we read the words, The playe, *M. Nights Dr.*'

to some lady, not named, who was present on the occasion of the performance of the play. [To this letter is appended what] purports to be a copy of an order, or decree, made by a self-constituted Court among the Puritans, for the censure and punishment of offences of the kind :

'A COPIE OF THE ORDER, OR DECREE (*ex officio Comisarii generalis*) JOHN SPENCER.

' Forasmuch as this Courte hath beene informed, by Mr. Comisary general, of a
' greate misdemeanor committed in the house of the right honorable Lo. Bishopp of
' Lincolne, by entertaining into his house divers Knights and Ladyes, with many
' other householders servants, uppon the 27th Septembris, being the Saboth day, to
' see a playe or tragidie there acted; which began aboute tenn of the clocke at night,
' and ended about two or three of the clocke in the morning :
' Wee doe therefore order, and decree, that the Rt. honorable John, Lord Bishopp
' of Lincolne, shall, for his offence, erect a free schoole in Eaton, or else at Greate
Staughton, and endowe the same with 20*l.* per ann. for the maintenance of the
' schoolmaster for ever. . . .
' Likewise we doe order, that Mr. Wilson, because hee was a speciall plotter and
' contriver of this businees, and did in such a brutishe manner acte the same with an
' Asses head; and therefore hee shall uppon Tuisday next, from 6 of the clocke in
' the morning till six of the clocke at night, sitt in the Porters Lodge at my Lords
' Bishopps House, with his feete in the stocks and attyred with his asse head, and a
' bottle of hay sett before him, and this subscription on his breast :
' Good people I have played the beast,
And brought ill things to passe.
I was a man, but thus have made
My selfe a silly Asse.'

Regarding this remarkable incident we are without further information from any
quarter.

[As much of the above order as refers to ' Mr. Wilson' is given by INGLEBY in his
Centurie of Prayse, p. 182, ed. ii. Miss TOULMIN-SMITH, who edited the second
edition of Ingleby's volume, remarks : ' I give this doubtful " allusion," because sev-
' eral, following Collier's *Annals*, have taken for granted that it refers to the *Mid-
' summer Night's Dream*. Beyond these notices, however, there is nothing to tell
' with certainty what the play was. Near the bottom of page 3 in the margin have
' been written the words " the play M Night Dr," but these are evidently the work
' of a later hand and have been written over an erasure; they are not in the hand of
' either Laud, Lincoln, or Spencer, or of the endorser of the paper, but look like a
bad imitation of old writing. No reliance can therefore be placed on them.
' Elsewhere, Spencer speaks of the play as a *comedy;* if Wilson were not the
' author, at least he had a large share in the arrangement of it. In a *Discourse of
' Divers Petitions*, 1641, p. 19, speaking of Bp. Lincoln and this presentment, Spen-
' cer says, " one Mr. Wilson a cunning Musition having contrived a curious Comodie,
" and plotted it so, that he must needs have it acted upon the Sunday night, for he
" was to go the next day toward the Court; the Bishop put it off till nine of the
" clock at night." ']

THE FAIRY QUEEN

In 1692 *A Midsummer Night's Dream* furnished the framework of an Opera called *The Fairy Queen*, whereof 'the instrumental and vocal parts were composed 'by Mr. Purcell,' so says Downes in his *Roscius Anglicanus*, and 'the dances by Mr. 'Priest.' As this work is quite rare, and is the nearest approach that we have to a 'Players Quarto' of this play, a brief account of it may not be unacceptable. Its date is only seven years later than F_4 and fifteen years earlier than Rowe.

The Preface is a plea for the establishment of opera in England, and incidentally gives us a hint of the intoning of blank verse, which we have reason to believe was the practice of the stage. 'That Sir William Davenant's *Siege of Rhodes* was the 'first *Opera* we ever had in England,' it says, 'no man can deny; and is indeed a 'perfect *Opera* : there being this difference only between an *Opera* and a Tragedy ; 'that the one is a Story sung with a proper Action, the other spoken. And he must 'be a very ignorant Player who knows not there is a Musical Cadence in speaking ; 'and that a man may as well speak out of Tune, as sing out of Tune.'

The Opera opens with what is the Second Scene of the Comedy's First Act, where the Clowns have assembled to arrange for the Play ; Shakespeare's text is closely followed ; there are omissions, it is true, but there is no attempt at 'improvement,' and only in two instances is there what might be termed an emendation : first, where Bottom says 'To the rest,' this phrase is interpreted as a stage-direction and enclosed in brackets ; and secondly, where Bottom says 'a lover is more condoling,' the Opera has 'a lover's is,' &c., in both instances anticipating modern conjectures. At the close of this scene, in which is interwoven the subsequent arrangements for the Clowns' Interlude at the beginning of Act III, Titania enters 'leading the Indian 'boy,' for whose entertainment she commands her 'Fairy Coire' to describe, in song, 'that Happiness, that peace of mind, Which lovers only in retirement find,' and they proceed to do it in the following lively style :—

> 'Come, come, come, let us leave the Town,
>   And in some lonely place,
> Where Crouds and Noise were never known,
>   Resolve to end our days.

> 'In pleasant Shades upon the Grass
>   At Night our selves we'll lay ;
> Our Days in harmless Sport shall pass,
>   Thus Time shall slide away.'

Enter Fairies *leading in three Drunken Poets, one of them Blinded.*
  Blind Poet. Fill up the Bowl, then, *&c.*
  Fairy. Trip it, trip it in a Ring ;
Around this Mortal Dance, and Sing.
  Poet. Enough, enough,
We must play at Blind Man's Buff.
Turn me round, and stand away,
I'll catch whom I may.
  2 Fairy. About him go, so, so, so,
Pinch the Wretch from Top to Toe ;

Pinch him forty, forty times,
Pinch till he confess his Crimes.

 Poet. Hold, you damn'd tormenting Punk,
I confess—

 Both Fairies. What, what, &c.

 Poet. I'm Drunk, as I live Boys, Drunk.

 Both Fairies. What art thou, speak?

 Poet. If you will know it,
I am a scurvy Poet.

 Fairies. Pinch him, pinch him, for his Crimes,
His Nonsense, and his Dogrel Rhymes.

 Poet. Oh! oh! oh!

 1 Fairy. Confess more, more.

 Poet. I confess I'm very poor.
Nay, prithee do not pinch me so,
Good dear Devil let me go;
And as I hope to wear the Bays,
I'll write a Sonnet in thy Praise.

 Chorus. Drive 'em hence, away, away,
Let 'em sleep till break of Day.

A Fairy announces to Titania that Oberon is in sharp pursuit of the little Indian boy, whereupon Titania bids the earth open, the little boy disappears, and the act closes.

The Second Act of the Opera follows the original Second Act, in the entrances of the characters, and their speeches are mainly the same, throughout the quarrel of Oberon and Titania; the similarity continues through the description of the little Western flower, except that the compliment to Queen Elizabeth is diverted by Oberon's saying that he 'saw young Cupid in the mid-way hanging, At a fair vestal 'virgin taking aim.' At Titania's command the second Scene *changes to a Prospect of Grotto's, Arbors, and delightful Walks: The Arbors are Adorn'd with all variety of Flowers, the Grotto's supported by Terms, these lead to two Arbors on either side of the scene*, &c. &c. Then through two pages we have, pretty much like a child's fingers playing on two notes alternately on the piano, such stanzas as these:—

Come all ye Songsters of the sky,
Wake, and Assemble in this Wood;
But no ill-boding Bird be nigh,
None but the Harmless and the Good.
 May the God of Wit inspire,
  The Sacred Nine to bear a part;
 And the Blessed Heavenly Quire,
  Shew the utmost of their Art.
While Eccho shall in sounds remote,
  Repeat each Note,
   Each Note, each Note.

 Chorus. May the God, &c.

In the Third Act we have *Pyramus and Thisbe* as it is played before the Duke; at its close Robin Goodfellow drives off the clowns and puts the Ass-head on Bottom. Then ensues the scene between Titania and Bottom, for whose delectation a Fairy

Mask is brought on, and the Scene changes to '*a great Wood; a long row of large*
Trees on each side; a River in the middle; Two rows of lesser Trees of a different
'*kind just on the side of the River, which meet in the middle, and make so many*
'*Arches; Two great Dragons make a Bridge over the River; their Bodies form two*
'*Arches, through which two Swans are seen in the River at a distance.*' A troop of
Fawn, Dryades and Naiades sing as follows :—

'If Love's a Sweet Passion, why does it torment ?
If a Bitter, oh tell me whence comes my content ?
Since I suffer with pleasure, why should I complain,
Or grieve at my Fate, when I know 'tis in vain ?
    Yet so pleasing the Pain is, so soft is the Dart,
    That at once it both wounds me, and tickles my Heart.
I press her hand gently, look Languishing down,
And by Passionate Silence I make my Love known,
But oh ! how I'm blest, when so kind she does prove,
By some willing mistake to discover her Love.
    When in striving to hide, she reveals all her Flame,
    And our Eyes tell each other, what neither dares Name.'

While a Symphony's Playing, the two Swans come swimming in through the
Arches to the Bank of the River, as if they would Land; there turn them-
selves into Fairies and Dance; at the same time the Bridge vanishes, and the
Trees that were arch'd, raise themselves upright.

Four Savages Enter, fright the Fairies away, and dance an Entry.

Enter Coridon *and* Mopsa.

Co. Now the Maids and the Men are making of Hay,
We have left the dull Fools, and are stol'n away.
    Then Mopsa no more
    Be Coy as before,
But let us merrily, merrily Play,
And kiss, and kiss, the sweet time away.
    Mo. Why how now, Sir *Clown*, how came you so bold ?
I'd have you to know I'm not made of that mold.
    I tell you again,
    Maids must kiss no Men.
No, no ; no, no ; no kissing at all ;
I'le not kiss, till I kiss you for good and all.
    Co. No, no.
    Mo. No, no,
    Co. Not kiss you at all.
    Mo. Not kiss, till you kiss me for good and all.
Not kiss, *&c.*

And so this struggle continues, to be relished by an audience who witnessed a
conflict to which in daily life they were probably not accustomed.

The rest of Shakespeare's play is incorporated; the mistakes of Puck with the
love-juice, and the mischances that befall the lovers in consequence, their slumber on
the ground and their awakening by the horns of the hunters, all follow in due course.
Although we have no record whatsoever that the Opera was intended to celebrate
any nuptials, yet its appropriateness to such a celebration is as marked as in *A Mid-*
summer Night's Dream, if not even more emphatically marked—a fact which I

humbly commend to the consideration of those who contend for this interpretation
of Shakespeare's play.

The Play of *Pyramus and Thisbe* having been already given in the Second Act,
its place in the Fifth Act is supplied by an elaborate Mask, during which a ' Chinese
' enters and sings,' and to him responds a ' Chinese-woman,' and both join in a
chorus to the effect that ' We never cloy, But renew our Joy, And one Bliss another
' invites.' Then ' Six Monkeys come from between the trees and dance,' which apparently imparts so much exhilaration to ' Two Women ' that they burst into song and
demand the presence of Hymen :—

 ' Sure, the dull god of marriage does not hear;
 ' We'll rouse him with a charm. Hymen, appear !
  ' *Chorus.* Appear ! Hymen, appear !'

Hymen obeys, but complains that

 ' My torch has long been out. I hate
 ' On loose dissembled Vows to wait.
 ' Where hardly Love out-lives the Wedding-Night,
 ' False Flames, Love's Meteors, yield my Torch no light.'

There is a grand dance of twenty-four persons, then Hymen and the Two Women
sing together :—

 ' They shall be as happy as they're fair;
 ' Love shall fill all the Places of Care:
 ' And every time the Sun shall display
  ' His rising Light,
 ' It shall be to them a new Wedding-Day ;
 ' And when he sets, a new Nuptial-Night.'

This starts the Chinese man and woman dancing, which in turn starts ' The Grand
' Chorus,' in which all the dancers join, and the Mask ends.

Oberon then resumes :—
' At dead of Night we'll to the Bride-bed come,
' And sprinkle hallow'd Dew-drops round the Room.
 ' *Titania.* We'll drive the Fume about, about,
' To keep all noxious Spirits out,
' That the issue they create
' May be ever fortunate,' &c.

The Fairy King and Queen then bring the Opera to a close, pretty much in the
style of all plays in those days, by alternately threatening and cajoling the audience
until the last words are :—

 ' *Ob.* Those Beau's, who were at Nurse, chang'd by my elves.
 ' *Tit.* Shall dream of nothing, but their pretty selves.
 ' *Ob.* We'll try a Thousand charming Ways to win ye.
 ' *Tit.* If all this will not do, the Devil's in ye.'

DOWNES, in his *Roscius Anglicanus* (p. 57), says that this Opera in ornaments
' was superior to ' *King Arthur* by Dryden or *The Prophetess* by Beaumont and
Fletcher, ' especially in cloaths for all the Singers and Dancers; Scenes, Machines,
' and Decorations; all most profusely set off, and excellently performed.' ' The
' Court and Town,' he concludes, ' were wonderfully satisfy'd with it; but the
' expences in setting it out being so great, the Company got very little by it.'

PLAN OF THE WORK, &c.

IN this Edition the attempt is made to give, in the shape of TEXTUAL NOTES, on the same page with the Text, all the Various Readings of *A Midsummer Night's Dream*, from the First Quarto to the latest critical Edition of the play; then, as COMMENTARY, follow the Notes which the Editor has thought worthy of insertion, not only for the purpose of elucidating the text, but at times as illustrations of the history of Shakespearian criticism. In the APPENDIX will be found discussions of subjects, which on the score of length could not be conveniently included in the Commentary.

EDITIONS COLLATED IN THE TEXTUAL NOTES.

FISHER'S QUARTO (Ashbee's Facsimile)	..	[Q$_1$]	1600
ROBERTS'S QUARTO (Ashbee's Facsimile)	..	[Q$_2$]	1600
THE SECOND FOLIO	..	[F$_2$]	1632
THE THIRD FOLIO	..	[F$_3$]	1664
THE FOURTH FOLIO	..	[F$_4$]	1685
ROWE (First Edition)	..	[Rowe i]	1709
ROWE (Second Edition)	..	[Rowe ii]	1714
POPE (First Edition)	..	[Pope i]	1723
POPE (Second Edition)	..	[Pope ii]	1728
THEOBALD (First Edition)	..	[Theob. i]	1733
THEOBALD (Second Edition)	..	[Theob. ii]	1740
HANMER	..	[Han.]	1744
WARBURTON	..	[Warb.]	1747
JOHNSON	..	[Johns.]	1765
CAPELL	..	[Cap.] .. .. (?)	1765
JOHNSON and STEEVENS	..	[Var. '73]	1773
JOHNSON and STEEVENS	..	[Var. '78]	1778
JOHNSON and STEEVENS	..	[Var. '85]	1785
RANN	..	[Rann]	1787
MALONE	..	[Mal.]	1790
STEEVENS	..	[Steev.]	1793
REED'S STEEVENS	..	[Var. '03]	1803
REED'S STEEVENS	..	[Var. '13]	1813
BOSWELL'S MALONE	..	[Var.]	1821
KNIGHT	..	[Knt.] .. .. (?)	1840
COLLIER (First Edition)	..	[Coll. i]	1842
HALLIWELL (Folio Edition)	..	[Hal.]	1856
SINGER (Second Edition)	..	[Sing. ii]	1856
DYCE (First Edition)	..	[Dyce i]	1857
STAUNTON	..	[Sta.]	1857
COLLIER (Second Edition)	..	[Coll. ii]	1858
RICHARD GRANT WHITE (First Edition)	..	[Wh. i]	1858

CLARK and WRIGHT (*The Cambridge Edition*)	[Cam.]	..	..	.. 1863
CLARK and WRIGHT (*The Globe Edition*) ..	[Glo.]	..	..	.. 1864
KEIGHTLEY	[Ktly]	..	..	.. 1864
CHARLES and MARY COWDEN-CLARKE ..	[Cla.]	..	..	(?) 1864
DYCE (Second Edition)	[Dyce ii]	..	..	.. 1866
DYCE (Third Edition)	[Dyce iii]	..	..	.. 1875
COLLIER (Third Edition)	[Coll. iii]	..	..	.. 1877
WILLIAM ALDIS WRIGHT (*Clarendon Press Series*)	[Wrt]	..	..	.. 1877
HUDSON	[Huds.]	..	..	.. 1880
RICHARD GRANT WHITE (Second Edition) ..	[Wh. ii]	..	..	.. 1883
CAMBRIDGE (Second Edition, W. A. WRIGHT)	[Cam. ii]	..	..	.. 1891

W. HARNESS	.. 1830
W. J. ROLFE	.. 1877
W. WAGNER	.. 1881
F. A. MARSHALL (*Henry Irving Edition*)	.. 1888
K. DEIGHTON	.. 1893
A. W. VERITY (*Pitt Press Edition*)	.. 1894

The last six editions I have not collated beyond referring to them in disputed passages. The text of Shakespeare has become, within the last twenty-five years, so settled that to collate, word for word, editions which have appeared within these years, is a work of supererogation. The case is different where an editor revises his text and notes in a second or a third edition; it is then interesting to mark the effect of maturer judgement.

The TEXT is that of the FIRST FOLIO of 1623. Every word, I might say almost every letter, has been collated with the original.

In the TEXTUAL NOTES the symbol Ff indicates the agreement of the *Second*, *Third*, and *Fourth Folios*.

The omission of the apostrophe in the *Second Folio*, a peculiarity of that edition, is not generally noted.

I have not called attention to every little misprint in the Folio. The Textual Notes will show, if need be, that they are misprints by the agreement of all the Editors in their correction.

Nor is notice taken of the first Editor who adopted the modern spelling, or who substituted commas for parentheses, or changed ? to !

The sign + indicates the agreement of ROWE, POPE, THEOBALD, HANMER, WARBURTON, and JOHNSON.

When WARBURTON precedes HANMER in the Textual Notes, it indicates that HANMER has followed a suggestion of WARBURTON'S.

The words *et cet.* after any reading indicate that it is the reading of *all other* editions.

The words *et seq.* indicate the agreement of all subsequent editions.

The abbreviation (*subs.*) indicates that the reading is *substantially* given, and that immaterial variations in spelling, punctuation, or stage-directions are disregarded.

An Emendation or Conjecture which is given in the Commentary is not repeated

in the Textual Notes unless it has been adopted by an editor in his Text; nor is *conj.* added in the Textual Notes to the name of the proposer of the conjecture unless the conjecture happens to be that of an editor, in which case its omission would lead to the inference that such was the reading of his text.

COLL. (MS) refers to COLLIER'S annotated Second Folio.

QUINCY (MS) refers to an annotated Fourth Folio in the possession of MR J. P. QUINCY.

In citations from plays, other than *A Midsummer Night's Dream*, the Acts, Scenes, and Lines of *The Globe Edition* are followed.

LIST OF BOOKS FROM WHICH CITATIONS HAVE BEEN MADE.

To economise space in the Commentary I have frequently cited, with the name of an author, an abbreviated title of his work, and sometimes not even as much as that. In the following LIST, arranged alphabetically, enough of the full title is given to serve as a reference.

Be it understood that this List gives only those books wherefrom Notes have been taken at first hand; it does not include books which have been consulted or have been used in verifying quotations made by the contributors to the earlier *Variorums*, or by other critics. Were these included the List would be many times as long. Nor does it include the large number in German which I have examined, but from which, to my regret, lack of space has obliged me to forego making any extract.

E. A. ABBOTT: *Shakespearian Grammar* (3d ed.)	1870
E. ARBER: *English Garner* (vol. iii)	1880
S. BAILEY: *The Received Text of Shakespeare*	1862
C. BATTEN: 'The Academy,' 1 June	1876
T. S. BAYNES: *New Shakespearian Interpretations* (Edinburgh Review, October)	1872
T. S. BAYNES: *What Shakespeare learnt at School* (Fraser's Magazine, January)	1880
T. S. BAYNES: *Shakespeare Studies*	1894
I. S. BEISLY: *Shakspere's Garden*	1864
W. BELL: *Shakespeare's Puck, and his Folk-Lore*	1859
J. BOADEN: *On the Sonnets of Shakespeare*	1837
J. BRAND: *Popular Antiquities, &c.* (Bohn's ed.)	1873
C. A. BROWN: *Shakespeare's Autobiographical Poems*	1839
J. M. BROWN: 'New Zealand Magazine,' April	1877
J. BULLOCH: *Studies of the Text of Shakespeare*	1878
T. CAMPBELL: *Dramatic Works of Shakespeare*	1838
E. CAPELL: *Notes, &c.*	1779
R. CARTWRIGHT: *New Readings, &c.*	1866
MRS CENTLIVRE: *The Platonick Lady*	1707
G. CHALMERS: *Apology for the Believers in the Shakespeare Papers, &c.*	1797
G. CHALMERS: *Supplemental Apology, &c.*	1799
R. CHAMBERS: *Book of Days*	1863

W. CHAPPELL: *Popular Music of the Olden Time* .. .. n. d.

F. J. CHILD: *English and Scottish Popular Ballads* 1882

H. A. CLAPP: 'Atlantic Monthly,' March 1885

A. COHN: *Shakespeare in Germany* 1865

HARTLEY COLERIDGE: *Essays and Marginalia* 1851

S. T. COLERIDGE: *Notes and Lectures* 1874

J. P. COLLIER: *History of English Dramatic Poetry* (ed. ii, 1879) 1831

J. P. COLLIER: *Notes and Emendations, &c.* 1852

J. P. COLLIER: *Seven Lectures of Coleridge, &c.* 1856

J. P. COLLIER: *Bibliographical and Critical Account of the Rarest Books in English* 1865

C. COWDEN-CLARKE: *Shakespeare Characters, &c.* 1863

THE COWDEN-CLARKES: *The Shakespeare Key* 1879

RANDLE COTGRAVE: *Dictionarie of the French and English Tongues* 1632

J. CROSBY: 'The Literary World,' June 1878

P. A. DANIEL: *Notes and Emendations* 1870

P. A. DANIEL: *Trans. New Shakspere Society* 1877–9

W. B. DEVEREUX: *Lives and Letters of the Devereux Earls of Essex* .. 1853

F. DOUCE: *Illustrations of Shakespeare, &c.* 1807

E. DOWDEN: *Shakspere: His Mind and Art* 1875

N. DRAKE: *Shakespeare and His Times* 1817

DRAYTON: *Works* 1740

A. DYCE: *Remarks on Collier's and Knight's editions* 1844

A. DYCE: *Few Notes, &c.* 1853

A. DYCE: *Strictures, &c.* 1859

T. F. T. DYER: *Folk-lore of Shakespeare* 1884

J. W. EBSWORTH: *Introductions to Griggs's Photolithographic Facsimiles of the Quartos* 1880

THE EDINBURGH REVIEW, April 1848

T. EDWARDS: *Canons of Criticism* 1765

H. N. ELLACOMBE: *Plant Lore and Garden Craft of Shakespeare* 1878

A. J. ELLIS: *Early English Pronunciation* 1869

K. ELZE: *Essays* (trans. by L. Dora Schmitz) 1874

K. ELZE: *Notes on Elizabethan Dramatists* 1889

The Fairy Queen 1692

R. FARMER: *Essay on the Learning of Shakespeare* 1767

B. FIELD: 'Shakespeare Society's Papers' 1845

J. H. FENNELL: *Shakespeare Repository* 1853

F. G. FLEAY: *Shakespeare Manual* 1876

F. G. FLEAY: *Life and Work of Shakespeare* 1886

F. G. FLEAY: *History of the Stage, 1559–1642* 1890

F. G. FLEAY: 'Robinson's Epitome of Literature,' 1 April 1879

F. G. FLEAY: *Biographical Chronicle of the English Drama* 1891

JOHN FLORIO: *A Worlde of Wordes, &c.* 1598

T. FONTANE: *Aus England* 1860

F. J. FURNIVALL: *Introduction to the Leopold Shakespeare* 1877

G. G. GERVINUS: *Shakespeare* 1849

HENRY GILES: *Human Life in Shakespeare* 1868

E. W. GODWIN: 'The Architect,' May 8, 15 1875

ARTHUR GOLDING: *The XV Booke of P. Ouidius Naso, entytuled Metamorphosis translated oute of Latin into English meeter, A worke very pleasaunt and delectable. With skill, heede, and iudgement, this worke must be read, For else to the Reader it standes in small stead* 1567

G. GOULD: *Corrigenda, &c.* 1884

H. GREEN: *Shakespeare and the Emblem Writers* 1870

GREENE: *Scottish Historie of James IV.* (eds. Dyce and Grosart) 1598

Z. GREY: *Critical, Historical, and Explanatory Notes* 1754

A. B. GROSART: *Spenser's Works* 1882

E. GUEST: *History of English Rhythms* 1838

J. W. HALES: *Notes and Essays* 1884

FITZEDWARD HALL: *Modern English* 1873

FITZEDWARD HALL: ' The Nation,' 4 August 1892

H. HALLAM: *Literature of Europe* 1839

J. O. HALLIWELL: *Introduction to Midsummer Night's Dream* 1841

J. O. HALLIWELL: *Memoranda on the Midsummer Night's Dream* .. 1879

N. J. HALPIN: *Oberon's Vision* (Shakespeare Society) 1843

W. HARNESS: *Shakespeare's Dramatic Works* 1830

GEORG HART: *Die Pyramus und Thisbe-Sage* 1889–91

J. E. HARTING: *Ornithology of Shakespeare* 1871

W. HAZLITT: *Characters of Shakespeare's Plays* 1817

B. HEATH: *Revisal of Shakespeare's Text* 1765

C. C. HENSE: *Shakespeare's Sommernachtstraum erläutert* 1851

J. A. HERAUD: *Shakespeare, his Inner Life* 1865

J. G. HERR: *Scattered Notes on Shakespeare* 1879

J. HEUSER: ' Shakespeare Jahrbuch ' (vol. xxviii) 1893

E. A. HITCHCOCK: *Remarks on the Sonnets* 1866

P. HOLLAND: *Plinie's Natural History* 1635

JOSEPH HUNTER: *New Illustrations of the Life, Studies, and Writings* .. 1845

C. M. INGLEBY: *A Centurie of Prayse* 1879

The IRVING *Shakespeare* 1890

H. JOHNSON: *A Midsummer Night's Dreame*, Facsimile Reprint of the First Folio. Variant Edition 1888

T. KEIGHTLEY: *Fairy Mythology* 1833

T. KEIGHTLEY: *The Shakespeare Expositor* 1867

W. KENRICK: *Review of Johnson's Shakespeare* 1765

B. G. KINNEAR: *Cruces Shakespearianæ* 1883

J. L. KLEIN: *Geschichte des Dramas* (vol. iv) 1866

F. KREYSSIG: *Vorlesungen ueber Shakespeare* 1862

H. KURZ: ' Shakespeare Jahrbuch ' (vol. iv) 1869

G. LANGBAINE: *English Dramatic Poets* 1691

F. A. LEO: *Shakespeare-Notes* 1885

W. N. LETTSOM: *New Readings, &c.* (Blackwood's Magazine, August) .. 1853

H. LYTE: *A Nievve Herball* 1578

W. MAGINN: *Shakespeare Papers* 1860

G. P. MARSH: *Lectures on the English Language* 1860

J. MONCK MASON: *Comments, &c.* 1785

J. MONCK MASON: *Comments on Beaumont and Fletcher* 1798

GERALD MASSEY: *The Secret Drama of Shakespeare's Sonnets* 1888

R. Nares: *Glossary* (eds. Halliwell and Wright) 1867
J. Nichols: *Literary Illustrations* (vol. ii) 1817
Noctes Shaksperianæ 1887
J. B. Noyes: *Poet-Lore*, October 1892
W. Oechelhäuser: *Einführungen in Shakespeare's Bühnen-Dramen,*
    2te Aufl. 1885
J. G. Orger: *Critical Notes on Shakespeare's Comedies* .. .. .. n. d.
R. Patterson: *Insects mentioned in Shakespeare* 1838
F. Peck: *New Memoirs of Milton* 1740
Pepys's *Diary* .. .. —
T. Percy: *Reliques of Ancient English Poetry* 1765
Sir P Perring: *Hard Knots in Shakespeare* (ed. ii) 1886
J. O. Halliwell-Phillipps: *Outlines of the Life of Shakespeare* 1885
J. Plumptre: *Appendix to Observations on Hamlet* 1797
H. J. Pye: *Comments on the Commentators* 1807
J. P. Quincy: *MS Corrections in a Copy of the Fourth Folio* 1854
J. Ritson: *Cursory Criticism* 1792
J. Ritson: *Remarks, Critical and Illustrative, on the Text and Notes of the*
    last edition of Shakespeare 1783
Clement Robinson: *A Handefull of Pleasant Delites* (Arber's Reprint) .. 1584
A. Roffe: *Handbook of Shakespeare Music* 1878
E. Roffe: *The Ghost Belief of Shakespeare* 1851
W. B. Rye: *England as seen by Foreigners, &c.* 1865
A. W. Schlegel: *Lectures* (trans. by Black) 1815
A. Schmidt: *Programm der Realschule zu Koenigsberg in Pr.* 1881
A. Schmidt: *Shakespeare-Lexicon* (2d ed.) 1886
Reginald Scot: *The Discoverie of Witchcraft, &c.* (ed. Nicholson) .. 1584
Sir Philip Sidney: *The Countess of Pembroke's Arcadia* 1598
R. Simpson: *The School of Shakspere* 1878
Karl Simrock: *Die Quellen des Shakespeare, &c.* (2d ed.) 1870
W. W. Skeat: *Shakespeare's Plutarch* 1875
W. W. Skeat: *Etymological Dictionary* 1882
A. Skottowe: *Life of Shakespeare* 1824
J. Moyr Smith: *A Midsummer Night's Dream* 1892
H. Staunton: 'The Athenæum.' 27 June 1874
H. P. Stokes: *Chronological Order of Shakespeare's Plays* 1878
A. C. Swinburne: 'Fortnightly Review,' January 1876
L. Theobald: *Shakespeare Restored; or a Specimen of the Many Errors, as*
    well Committed, as unamended by Mr Pope 1726
W. J. Thoms: *Three Notelets on Shakespeare* 1865
L Tieck: *Deutsches Theater* 1817
L. Tieck: *Anmerkungen zur Uebersetzung von Schlegel* 1830
Ed. Tiessen: *Archiv f. n. Sprachen* (vol. lviii) 1877
Edvvard Topsell: *Historie of Foure-Footed Beastes* 1608
T. Tyrwhitt: *Observations and Conjectures upon Some Passages of Shake-*
    speare 1766
H. Ulrici: *Shakespeare's dramatische Kunst* 1847
J. Upton: *Critical Observations on Shakespeare* 1746
G. C. Verplanck: *The Plays of Shakespeare* 1847

W. S. WALKER: *Shakespeare's Versification* 1854
W. S. WALKER: *Critical Examination of the Text of Shakespeare* 1859
A. W. WARD: *History of English Dramatic Poetry* 1875
T. WARTON: *Observations on Spenser* 1754
T. WARTON: *History of English Poetry* 1775
A. WAY: *Promptorium Parvulorum* 1865
JULIA WEDGWOOD: 'Contemporary Review,' April 1890
F. WEHL: *Didaskalien* 1867
J. WEISS: *Wit, Humor, and Shakespeare* 1876
P. WHALLEY: *Enquiry into the Learning of Shakespeare* 1748
R. G. WHITE: *Shakespeare's Scholar* 1854
R. G. WHITE: 'Putnam's Magazine,' October 1853
T. WHITE: *More Notes on Shakespeare* (Fennell's Shakespeare Repository, 1853) 1793
W. WHITER: *Specimen of a Commentary on Shakespeare* 1794
D. WILSON: *Caliban: The Missing Link* 1873
W. WINTER: *Daly's Arrangement for Representation* 1888
W. WINTER: *Old Shrines and Ivy* 1892
H. WOELFFEL: *Album d. lit. Vereins in Nurnberg* 1852
BARTHOLOMEW YONG: *Diana of George of Montemayor* 1598
J. ZUPITZA: 'Englische Studien,' vol. viii. 1885

INDEX

23

17031